List of the Elements with Their Symbols and Atomic Masses*

Element	Symbol	Atomic Number	Atomic Mass†	Element	Symbol	Atomic Number	Atomic Mass†
Actinium	Ac	89	(227)	Mendelevium	Md	101	(256)
Aluminum	Al	13	26.98	Mercury	Hg	80	200.6
Americium	Am	95	(243)	Molybdenum	Mo	42	95.94
Antimony	Sb	51	121.8	Neodymium	Nd	60	144.2
Argon	Ar	18	39.95	Neon	Ne	10	20.18
Arsenic	As	33	74.92	Neptunium	Np	93	(237)
Astatine	At	85	(210)	Nickel	Ni	28	58.69
Barium	Ba	56	137.3	Niobium	Nb	41	92.91
Berkelium	Bk	97	(247)	Nitrogen	N	7	14.01
Beryllium	Be	4	9.012	Nobelium	No	102	(253)
Bismuth	Bi	83	209.0	Osmium	Os	76	190.2
Bohrium	Bh	107	(262)	Oxygen	O	8	16.00
Boron	B	5	10.81	Palladium	Pd	46	106.4
Bromine	Br	35	79.90	Phosphorus	P	15	30.97
Cadmium	Cd	48	112.4	Platinum	Pt	78	195.1
Calcium	Ca	20	40.08	Plutonium	Pu	94	(242)
Californium	Cf	98	(249)	Polonium	Po	84	(210)
Carbon	C	6	12.01	Potassium	K	19	39.10
Cerium	Ce	58	140.1	Praseodymium	Pr	59	140.9
Cesium	Cs	55	132.9	Promethium	Pm	61	(147)
Chlorine	Cl	17	35.45	Protactinium	Pa	91	(231)
Chromium	Cr	24	52.00	Radium	Ra	88	(226)
Cobalt	Co	27	58.93	Radon	Rn	86	(222)
Copper	Cu	29	63.55	Rhenium	Re	75	186.2
Curium	Cm	96	(247)	Rhodium	Rh	45	102.9
Dubnium	Db	105	(260)	Rubidium	Rb	37	85.47
Dysprosium	Dy	66	162.5	Ruthenium	Ru	44	101.1
Einsteinium	Es	99	(254)	Rutherfordium	Rf	104	(257)
Erbium	Er	68	167.3	Samarium	Sm	62	150.4
Europium	Eu	63	152.0	Scandium	Sc	21	44.96
Fermium	Fm	100	(253)	Seaborgium	Sg	106	(263)
Fluorine	F	9	19.00	Selenium	Se	34	78.96
Francium	Fr	87	(223)	Silicon	Si	14	28.09
Gadolinium	Gd	64	157.3	Silver	Ag	47	107.9
Gallium	Ga	31	69.72	Sodium	Na	11	22.99
Germanium	Ge	32	72.59	Strontium	Sr	38	87.62
Gold	Au	79	197.0	Sulfur	S	16	32.07
Hafnium	Hf	72	178.5	Tantalum	Ta	73	180.9
Hassium	Hs	108	(265)	Technetium	Tc	43	(99)
Helium	He	2	4.003	Tellurium	Te	52	127.6
Holmium	Ho	67	164.9	Terbium	Tb	65	158.9
Hydrogen	H	1	1.008	Thallium	Tl	81	204.4
Indium	In	49	114.8	Thorium	Th	90	232.0
Iodine	I	53	126.9	Thulium	Tm	69	168.9
Iridium	Ir	77	192.2	Tin	Sn	50	118.7
Iron	Fe	26	55.85	Titanium	Ti	22	47.88
Krypton	Kr	36	83.80	Tungsten	W	74	183.9
Lanthanum	La	57	138.9	Uranium	U	92	238.0
Lawrencium	Lr	103	(257)	Vanadium	V	23	50.94
Lead	Pb	82	207.2	Xenon	Xe	54	131.3
Lithium	Li	3	6.941	Ytterbium	Yb	70	173.0
Lutetium	Lu	71	175.0	Yttrium	Y	39	88.91
Magnesium	Mg	12	24.31	Zinc	Zn	30	65.39
Manganese	Mn	25	54.94	Zirconium	Zr	40	91.22
Meitnerium	Mt	109	(266)				

*All atomic masses have four significant figures. These values are recommended by the Committee on Teaching of Chemistry, International Union of Pure and Applied Chemistry.

†Approximate values of atomic masses for radioactive elements are given in parentheses.

IMPORTANT:

HERE IS YOUR REGISTRATION CODE TO ACCESS
YOUR McGRAW-HILL ONLINE RESOURCES.

THIS CODE gives you access to the Web resources for your course. Once the code is entered, you will be able to use the Web resources for the length of your course.

If your instructor is using **WebCT** or **Blackboard**, you will also use this code to access the McGraw-Hill content within your instructor's online course.

Access is provided if you have purchased a new book. If the registration code is missing from this book, the registration screen on our Website, and within your WebCT or Blackboard course, will tell you how to obtain a new code.

Registering for McGraw-Hill Online Resources

TO gain access to your McGraw-Hill web resources simply follow the steps below:

(1) USE YOUR WEB BROWSER TO GO TO: **http://www.mhhe.com/chang3/**

(2) CLICK ON **FIRST TIME USER**.

(3) ENTER THE REGISTRATION CODE* PRINTED ON THE TEAR-OFF BOOKMARK ON THE RIGHT.

(4) AFTER YOU HAVE ENTERED YOUR REGISTRATION CODE, CLICK **REGISTER**.

(5) FOLLOW THE INSTRUCTIONS TO SET-UP YOUR PERSONAL UserID AND PASSWORD.

(6) WRITE YOUR UserID AND PASSWORD DOWN FOR FUTURE REFERENCE.
KEEP IT IN A SAFE PLACE.

TO GAIN ACCESS to the McGraw-Hill content in your instructor's **WebCT** or **Blackboard** course simply log in to the course with the UserID and Password provided by your instructor. Enter the registration code exactly as it appears in the box to the right when prompted by the system. You will only need to use the code the first time you click on McGraw-Hill content.

Thank you, and welcome to your McGraw-Hill online Resources!

* YOUR REGISTRATION CODE CAN BE USED ONLY ONCE TO ESTABLISH ACCESS. IT IS NOT TRANSFERABLE.

0-07-241067-1 CHANG: GENERAL CHEMISTRY, 3E

How's Your Math?

Do you have the math skills you need to succeed?

Why risk not succeeding because you struggle with your math skills?

Get access to a web-based, personal math tutor:

- Available 24/7, unlimited use
- Driven by artificial intelligence
- Self-paced
- An entire month's subscription **for much less** than the cost of one hour with a human tutor

ALEKS is an inexpensive, private, infinitely patient math tutor that's accessible any time, anywhere you log on.

ALEKS®

Log On for a
FREE 48-hour Trial

www.highedstudent.aleks.com

General Chemistry

General Chemistry **Third Edition**

The Essential Concepts

Raymond Chang

Williams College

Boston   Burr Ridge, IL   Dubuque, IA   Madison, WI   New York   San Francisco   St. Louis
Bangkok   Bogotá   Caracas   Kuala Lumpur   Lisbon   London   Madrid   Mexico City
Milan   Montreal   New Delhi   Santiago   Seoul   Singapore   Sydney   Taipei   Toronto

McGraw-Hill Higher Education ✕

A Division of The **McGraw-Hill** *Companies*

GENERAL CHEMISTRY: THE ESSENTIAL CONCEPTS, THIRD EDITION

Published by McGraw-Hill, a business unit of The McGraw-Hill Companies, Inc., 1221 Avenue of the Americas, New York, NY 10020. Copyright © 2003, 2000, 1996 by The McGraw-Hill Companies, Inc. All rights reserved. No part of this publication may be reproduced or distributed in any form or by any means, or stored in a database or retrieval system, without the prior written consent of The McGraw-Hill Companies, Inc., including, but not limited to, in any network or other electronic storage or transmission, or broadcast for distance learning.

Some ancillaries, including electronic and print components, may not be available to customers outside the United States.

This book is printed on acid-free paper.

International	1 2 3 4 5 6 7 8 9 0 VNH/VNH 0 9 8 7 6 5 4 3 2
Domestic	1 2 3 4 5 6 7 8 9 0 VNH/VNH 0 9 8 7 6 5 4 3 2

ISBN 0–07–241067–1
ISBN 0–07–115115–X (ISE)

Publisher: *Kent A. Peterson*
Developmental editor: *Shirley R. Oberbroeckling*
Marketing manager: *Thomas D. Timp*
Senior project manager: *Gloria G. Schiesl*
Lead production supervisor: *Sandra Hahn*
Design manager: *Stuart D. Paterson*
Cover/interior designer: *Ellen Pettengell*
Cover image: *Gettyimages*
Senior photo research coordinator: *John C. Leland*
Photo research: *Toni Michaels/PhotoFind LLC*
Lead media project manager: *Audrey A. Reiter*
Media technology lead producer: *Steve Metz*
Compositor: *GTS Graphics, Inc.*
Typeface: *10/12 Times Roman*
Printer: *Von Hoffmann Press, Inc.*

The credits section for this book begins on page 749 and is considered an extension of the copyright page.

Library of Congress Cataloging-in-Publication Data

Chang, Raymond.
    General chemistry : the essential concepts / Raymond Chang. — 3rd ed.
      p. cm.
    Rev. ed. of: Essential chemistry. 2nd ed. ©2000.
    Includes index.
    ISBN 0–07–241067–1 (acid-free paper) — ISBN 0–07–115115–X (ISE : acid-free paper)
    1. Chemistry. I. Chang, Raymond. Essential chemistry. II. Title.

QD33.2 .C48   2003
540—dc21
                                         2002016623
                                       CIP

INTERNATIONAL EDITION ISBN 0–07–115115–X
Copyright © 2003. Exclusive rights by The McGraw-Hill Companies, Inc., for manufacture and export. This book cannot be re-exported from the country to which it is sold by McGraw-Hill. The International Edition is not available in North America.

www.mhhe.com

About the Author

Raymond Chang was born in Hong Kong and grew up in Shanghai and Hong Kong, China. He received his B.Sc. degree in chemistry from London University, England, and his Ph.D. in chemistry from Yale University. After doing postdoctoral research at Washington University and teaching for a year at Hunter College of the City University of New York, he joined the chemistry department at Williams College, where he has taught since 1968.

Professor Chang has served on the American Chemical Society Examination Committee, the National Chemistry Olympiad Examination Committee, and the Graduate Record Examination (GRE) Committee. He is an editor of *The Chemical Educator*. Professor Chang has written books on physical chemistry, industrial chemistry, and physical science. He has also coauthored books on the Chinese language, children's picture books, and a novel for juvenile readers.

For relaxation, Professor Chang maintains a forest garden; plays tennis, Ping-Pong, and the harmonica; and practices the violin.

Brief Contents

Contents

Chapter 3   Stoichiometry   57

Chapter 4   Reactions in Aqueous Solutions   91

List of Animations

Here is a list of animations correlated to *General Chemistry* with the section to which the animation relates given in parentheses. You can find the animations on the Online Learning Center for *General Chemistry* at www.mhhe.com/physsci/chemistry/chang. Just click on the *General Chemistry* book cover.

Chang Animations

Acid-base titrations (17.3)
Acid ionizations (16.5)
Activation energy (14.4)
Alpha-particle scattering (2.2)
Atomic and ionic radius (8.3)
Base ionizations (16.6)
Buffer solutions (17.2)
Chemical equilibrium (15.1)
Dissolution of a covalent compound (13.2)
Emission spectra (7.3)
Equilibrium vapor pressure (12.6)
Galvanic cells (19.2)
The gas laws (5.3)
Heat flow (6.4)
Hybridization (10.4)
Hydration (4.1)
Ionic versus covalent bonding (9.2)
Le Châtelier's principle (15.4)
Limiting reagent (3.9)
Making a solution (4.5)
Orientation of collision (14.4)

Packing spheres (12.4)
Polarity of molecules (10.2)Precipitation reactions (4.2)
Radioactive decay (21.3)
Sigma and pi bonds (10.5)
VSEPR (10.1)

McGraw-Hill Animations

Atomic line spectra (7.3)
Charles's law (5.3)
Cubic unit cells and their origins (12.4)
Dissociation of strong and weak acids (16.5)
Dissolving table salt (4.1)
Electronegativity (9.3)
Equilibrium (15.1)
Exothermic and endothermic reactions (6.2)
Formation of an ionic compound (9.2)
Formation of the covalent bond in H_2 (10.3)
Half-life (14.3)
Influence of shape on polarity (10.2)
Law of conservation of mass (2.1)
Molecular shape and orbital hybridization (10.4)
Nuclear medicine (21.7)
Operation of voltaic cell (19.2)
Oxidation-reduction reaction (4.4 and 19.1)
Phase diagrams and the states of matter (12.7)
Reaction rate and the nature of collisions (14.4)
Three states of matter (1.3)
Using a buffer (17.2)
VSEPR theory and the shapes of molecules (10.1)

Preface

The third edition of *General Chemistry: The Essential Concepts* continues the tradition of presenting only the material that is essential for a one-year general chemistry course. As before, I have included all the core topics that are necessary for a solid foundation in general chemistry without sacrificing depth, clarity, or rigor.

General Chemistry covers these core topics in the same depth and at the same level as 1100-page texts. Therefore, this book is not a condensed version of a big text. I have written it so that an instructor can cover 95 percent of the content, instead of the two-thirds or three-quarters that in my experience is typical for the big books. My hope is that this concise-but-thorough approach will appeal to efficiency-minded instructors and will please value-conscious students. Encouraging responses from users convince me that I am on the right track.

New and Improved Changes for the Third Edition

The main goal of this edition is to further improve areas that will facilitate the instructor and aid students in important areas such as organization, art program, pedagogy, readability, and media. Summarized here are the highlights of this edition.

Organization

• **New** title reflecting the content of this text is for a one-year introductory chemistry course.

• The chapter on organic chemistry (Chapter 11) has been extensively revised. Its placement enables the instructor to use organic molecules to illustrate the concepts of chemical bonding and mechanisms of reactions. For schools whose curricula involve general chemistry in the first semester and organic chemistry in the second semester, this chapter should prepare the student well for the transition.

• The chapter on coordination chemistry has been moved to near the end of the book.

• There is no general agreement on the merit of teaching molecular orbital theory in an introductory chemistry course. A recent survey by McGraw-Hill of over 200 instructors shows about a 50-50 split on this matter. For this reason, I have added a section on molecular orbital theory in Chapter 10 and have shown its application to aromatic compounds in Chapter 11. For instructors who prefer not to cover this topic, it can be omitted without disruption or loss of continuity.

Art Program

• **New** molecular art images were created using Spartan molecular modeling program. These drawings enable students to gain a better understanding of the three-dimensionality of molecules, electron distribution, and details of chemical reactions. Many other figures are new or have been revised. For example, see Figures 5.5, 9.3, 10.5, 11.12, 14.21, and 22.11.

Pedagogy

• Many "Worked Examples" have been revised to more clearly show steps in problem solving.

• A **New** periodic table icon illustrates the properties of elements according to their positions in the periodic table.

• A number of **New** end-of-chapter problems and Special Problems have been added.

• A **New** Animation Icon (⬚) points to material that is further illustrated by an animation. The list of animations has been increased in this edition.

Readability

• Material is presented in a clear-cut and concise manner.

• Main points are fully developed.

Media

• **New** Essential Study Partner (ESP) interactive student tutorial.

• Improved Online Learning Center website for instructors and students.

Instructor Resources

Test Bank

Written by Gary Wolf (Spokane Falls Community College) and edited by Marcia Gillette (Indiana University–Kokomo). This manual contains over 2000 multiple-choice and short-answer questions. The questions, which are graded in difficulty, are comparable to the problems in the text and include multistep problems that require conceptual analysis.

Computerized Test Bank

Written by Gary Wolf (Spokane Falls Community College) and edited by Marcia Gillette (Indiana University–Kokomo). This test bank contains all of the questions in the print *Test Bank* along with algorithms and over 200 algorithm-based questions that instructors can edit to create their own test templates. The Test Bank is formatted for easy integration into the following course management systems: PageOut, WebCT, and Blackboard.

Essential Study Partner

By David Harwell (University of Hawaii at Manoa), Laura Muller (Wheaton College), Norbert Pienta (University of Iowa), Kathleen Robbins (University of Las Vegas–Nevada),

and Brandon Cruickshank (Northern Arizona University). This free online study partner engages, investigates, and reinforces what the student is learning from the textbook. You will find the **Essential Study Partner** for *General Chemistry* to be a complete, interactive student study tool packed with animations and learning activities. From quizzes to interactive diagrams, you will find that this is a highly effective study partner to ensure the mastery of core concepts.

Online Learning Center

This comprehensive, book-specific website (www.mhhe.com/physsci/chemistry/chang) offers excellent tools for both the instructor and the student. Instructors can create an interactive course with the integration of this site, and a secured Instructor Center stores your essential course materials to save you prep time before class. This center offers PowerPoint images, a PowerPoint lecture outline, narratives for Chang animations, and more.

Instructor's Resource Manual with Solutions

By Brandon J. Cruickshank (Northern Arizona University) and Raymond Chang. This complete manual for teaching a general chemistry course is based on *General Chemistry*. This unique guide includes demonstrations that can be done in any classroom or assigned for homework, accompanied by discussion questions and tips to ensure success; information on relevant applications; chapter overviews and outlines; and annotated cross-references to other elements of the text package. In addition, this manual provides complete solutions to all end-of-chapter problems in the text.

Overhead Transparencies

Approximately 200 full-color text illustrations are reproduced on acetate for overhead projection.

Visual Resource Library

More than 300 images from the text can be used for stand-alone classroom presentation or can be included in a PowerPoint presentation.

Chemistry Animations Visual Resource Library

Organized by Eric Johnson (Ball State University). This instructor's CD-ROM enables you to use animations in your classroom in the way that works best for you. This multi-CD set includes over 300 animations that can be played directly from the CD or can be imported easily into your own lecture presentation. The animation library is fully searchable, and many animations are included at full-screen size.

Course-Specific PageOut

Designed specifically to help you with your individual course needs, PageOut will assist you in integrating your syllabus with *General Chemistry*, and state-of-the-art new media tools. At the heart of PageOut you will find integrated multimedia and a full-scale Online Learning Center. You can upload your original test questions and create your own custom designs. More than 60,000 professors have chosen PageOut to create customized course websites.

Primis LabBase

By Joseph Lagowski (University of Texas at Austin). More than 40 general chemistry experiments are available in this database collection of general lab experiments from the *Journal of Chemical Education* and experiments used by Professor Lagowski at the University of Texas at Austin, enabling instructors to customize their lab manuals.

Cooperative Chemistry Laboratory Manual

By Melanie Cooper (Clemson University). This innovative guide features open-ended problems designed to simulate experience in a research lab. Working in groups, students investigate one problem over a period of several weeks, so that they might complete three or four projects during the semester, rather than one preprogrammed experiment per class. The emphasis is on experimental design, analysis, problem solving, and communication.

Student Resources

Problem-Solving Workbook with Solutions

By Brandon J. Cruickshank (Northern Arizona University) and Raymond Chang. This is a success guide written for use with *General Chemistry*. It aims to help students hone their analytical and problem-solving skills by presenting detailed approaches to solving chemical problems. Solutions for all of the text's even-numbered problems are included.

OLC (Online Learning Center)

A comprehensive, exclusive website that provides a wealth of electronic resources for instructors and students alike. For students, the OLC features interactive quizzes for each chapter of the text; e-learning sessions; key-word flash-cards; NetTutor; interactive glossary with audio; ChemQuest, with Internet search exercises; visual chemistry, with Internet exercises that require students to find and manipulate molecules that are discussed in the text; links to chemical databases; and listings of professional opportunities in *General Chemistry*. You can also access the Essential Student Partner from the OLC. Log on with your passcode card at www.mhhe.com/physsci/chemistry/. The passcode card is available *free* with the purchase of a *new* textbook or you can purchase a card separately.

Essential Study Partner

By David Harwell (University of Hawaii at Manoa), Laura Muller (Wheaton College), Norbert Pienta (University of Iowa), Kathleen Robbins (University of Las Vegas–Nevada), and Brandon Cruickshank (Northern Arizona University). This online study partner engages, investigates, and reinforces what you are learning from your textbook. You will find the **Essential Study Partner** for *General Chemistry* to be a complete, interactive student study tool packed with animations and learning activities. From quizzes to interactive diagrams, you will find that this is a highly effective study partner to ensure the mastery of core concepts.

ChemSkill Builder

New! Our **online version** of ChemSkill Builder generates questions for students for every topic in the general chemistry course. The questions are presented in a randomized fashion with a constant mix of variables so that no two students will receive the same questions. The application provides feedback for students when incorrect answers are entered, and the answers can be submitted online to an instructor for grading.

CyberChem Multimedia CD-ROM

By Maha Ashour-Abdalla (University of California, Los Angeles) and Raymond Chang. This innovative CD-ROM provides a highly interactive study and tutorial package for the one-year general chemistry course. It contains 60 animations of concepts, 25 interactive lab simulations, and 25 movies focusing on real-world applications, along with interactive step-by-step problem-solving and quizzing modules. In addition, no matter where students are in the program, they will have access to a unique hyperlinked Periodic Table, with photos, information about the elements, a search function, and cross-references to these elements where they are discussed elsewhere in the program.

Schaum's Outline of College Chemistry

By Jerome Rosenberg (Michigan State University) and Lawrence Epstein (University of Pittsburgh). This helpful study aid provides students with hundreds of solved and supplementary problems for the general chemistry course.

Acknowledgments

I would like to thank the following individuals, whose comments were of great help to me in preparing this revision:

William E. Acree *University of North Texas*

Adedoyin Adeyiga *Bennett College*

Gul Afshan *Milwaukee School of Engineering*

Javier D. Agudelo *Broward Community College–North*

Edward Alexander *San Diego Mesa College*

Carol P. Anderson *University of Connecticut–Groton*

Christopher Anukwuem *Prairie View A&M University*

Marsi Archer *Missouri Southern College*

Vera Averyhart-Fullard *Olive Harvey College*

Satinder Bains *Arkansas State University–Beebe*

N. S. Bakaitis *Salem Teikyo University*

David Baker *Delta College*

Felicia Corsaro Barbieri *Gwynedd-Mercy College*

Robert Craig *Community College of Allegheny County*

Mufeed M. Basti *North Carolina A&T State University*

Robert Beer *Fordham University–Bronx*

Steven Bennett *Shawnee State University*

Mark A. Benvenuto *University of Detroit Mercy*

Vicki Bragin *Pasadena City College*

Francis W. Breivogel *Chipola Junior College*

June Bronfenbrenner *Anne Arundel Community College–Arnold*

Pamela Brown *New York City Technical College*

Joseph Bularzik *University of Notre Dame*

Cindy A. Burkhardt *Radford University*

Robert M. Burns *Florida Institute of Technology–Melbourne*

Stuart Burris *Belmont University*

Sylvester Burton *Southern University–Baton Rouge*

Patricia J. Bush *Delgado Community College*

Sheila D. Cancella *Raritan Valley Community College*

Suzanne Carpenter *Armstrong Atlantic State University*

Douglas J. Chesnut *Alfred University*

Joseph F. Chiang *SUNY–Oneonta*

Lisa E. Chirlian *Bryn Mawr College*

Ngee Sing Chong *Middle Tennessee State University*

Derald Chriss *Southern University–Baton Rouge*

Allen Clabo *Francis Marion University*

Jason Clyburne *Acadia University*

Douglas S. Cody *Nassau County Community College*

W. Lin Coker *Campbell University*

James E. Collins *Jacksonville State University*

Wheeler Conover *Southeast Community College*

Scott Davis *Mercer University–Macon*

Gary DeBoer *LeTourneau University*

Phillip Delassus *Valparaiso University*

James Digiacomo *New Jersey City University*

John P. Divincenzo *Middle Tennessee State University*

David Eaton *Essex County College*

Dennis E. Eddy *York College*

Amina K. El-Ashmawy *Collin County Community College–Plano*

Johnny Evans *Lee University*

Thomas Everton *Ventura College*

Michael F. Farona *University of North Carolina–Greensboro*

Theodore Fickel *University of Judaism*

Debbie Finocchio *University of San Diego*

Paul A. Flowers *University of North Carolina–Pembroke*

Robert Forsythe *Western Kentucky University*

Judith C. Foster *Bowdoin College*

Michelle M. Francl *Bryn Mawr College*

Leslie Fung *Loyola University–Chicago*

Herb Fynewever *California State Polytechnic University–Pomona*

Ana Gaillat *Greenfield Community College*

Roy Garvey *North Dakota State University–Fargo*

Victor L. Garza *San Antonio College*

Noel A. George *Ryerson Polytechnic University*

Marcia L. Gillette *Indiana University–Kokomo*

Jack G. Goldsmith *University of South Carolina–Aiken*

James Goll *Glenville State College*

John A. Goodwin *Coastal Carolina University*

Andrea Gorczyca *Brookhaven College*

Arthur S. Gow *University of New Haven*

John P. Graham *Arkansas Technical University*

Gregory J. Grant *University of Tennessee–Chattanooga*

Larry D. Groth *Lake Region State College*

Leo Hall *Eastern Oklahoma State College*

Victor Hall *Rose State College*

Lee D. Hansen *Brigham Young University–Provo*

Eric L. Hardegree *Abilene Christian University*

Ryan M. Harden *Central Lakes College*

C. Alton Hassell *Baylor University*

Noralynn Hassold *Kettering University*

David R. Hawkes *Lambuth University*

Yahdah Y. Hawkins *Cisco Junior College*

Daniel T. Haworth *Marquette University*

Robyn Hayes *Northern New Mexico Community College–Espanola*

Kevin D. Helm *Lee College*

Gary L. Hickernell *Keuka College*

Todd J. Hizer *Armstrong Atlantic State University*

Paul Hooker *Westminster College*

Paul A. Horton *Indian River Community College*

Narayan S. Hosmane *Northern Illinois University*

Roger K. House *Moraine Valley Community College*

Shizuka Hsieh *Oberlin College*

R. Amee Indralingam *Stetson University*

Leonard M. Itzkowitz *Nassau County Community College*

Peter Iyere *Tennessee State University*

David W. Johnson *University of Dayton*

Kevin E. Johnson *Pacific University*

Taylor B. Jones *Masters College*

Shane Kelley *Trevecca Nazarene University*

Igor V. Khudyakov *Catawba Valley Community College*

Myung Hoon Kim *Georgia Perimeter College–Dunwoody*

Pamela S. Kimbrough *Chaffey College*

Angela King *Wake Forest University*

Beata Knoedler *Springfield College*

Christa Knotts *Garrett Community College*

Joy Kobayashi *Ventura College*

Brian P. Koehler *Georgia Southern University*

Robert Kolodny *Armstrong Atlantic State University*

Lynn Koplitz *Loyola University*

Shankar Krishnamoorthy *Oklahoma City Community College*

Keith T. Kuwata *Macalester College*

Richard Langley *Stephen F. Austin State University*

Robert E. Leard *Alcorn State University*

Terrence A. Lee *Middle Tennessee State University*

Mihaela D. Leonida *Fairleigh Dickinson University–Teaneck*

Alan Levine *University of Louisiana at Lafayette*

Zhaohui Li *University of Wisconsin–Kenosha*

Clarence W. Linsey *Mid America Nazarene University*

David Lippmann *Southwest Texas State University*

Glenn V. Lo *Nicholls State University*

Bryan H. Long *Columbia State Community College*

John T. Long *Henderson State University*

Scott Luaders *Quincy University*

Yinfa Ma *Truman State University*

Catherine Macgowan *Armstrong Atlantic State University*

Madhumati Mahalingam *University of Massachusetts–Dartmouth*

Charles H. Mahler *Lycoming College*

Kristy Mardis *Pacific Lutheran University*

Vahe Marganian *Bridgewater State College*

Paul Marino *Delaware Valley College*

Jaime Martell *SUNY–Potsdam*

Albert H. Martin *Moravian College*

Ruhullah Massoudi *South Carolina State University*

Mark B. Masthay *Murray State University*

Armin Mayr *El Paso Community College–Rio Grande Campus*

Keith A. McCleary *Adrian College*

Michael W. McClure *Hopkinsville Community College*

Marcy McDonald *University of Alabama–Tuscaloosa*

Scott E. McKay *Central Missouri State University*

Christina Mewhinney *Eastfield College*

Therese Michels *Dana College*

David Miller *California State University–Northridge*

Donald M. Mitchell *Delaware Technical Community College*

Ellen M. Mitchell *Bridgewater College*

J. C. Moore *Friends University*

Linda E. Moss *Black River Technical College*

Michael R. Mueller *Rose-Hulman Institute of Technology*

Francis Mulcahy *University of Pittsburgh*

Ruth Ann Murphy *University of Mary Hardin Baylor*

Jennifer Muzyka *Centre College of Kentucky*

Rashmi Nanda *Benedictine University*

Kurt A. Neidigh *Ferrum College*

Terry Newirth *Haverford College*

Allan Nishimura *Westmont College*

Dale L. Orth *Wisconsin Lutheran College*

Jeffrey A. Orvis *Georgia Southern University*

Robert H. Paine *Rochester Institute of Technology*

James L. Pazun *Pfeiffer University*

Laura E. Pence *University of Hartford*

Thomas C. Pentecost *Aims Community College–Main Campus*

Martin Perry *Ouachita Baptist University*

Keith Peterson *Wesleyan College*

Holly Phaneuf *Salt Lake Community College*

Chester A. Pinkham *Tri-State University*

Guy Plourde *University of Northern British Columbia*

Wendy Pogozelski *State University College–Geneseo*

Suzette Polson *Kentucky State University*

Paul Popieniek *Sullivan County Community College*

Walda J. Powell *Meredith College*

James E. Privett *University of South Carolina–Sumter*

Jimmie M. Purser *Millsaps College*

Mohammad Nabi Rohni *Pace University*

Jill Rawlings *Auburn University–Montgomery*

Lisa Reece *Ozarks Technical Community College*

Preston Reeves *Texas Lutheran College*

Paul Reinbold *Southern Nazarene University*

Wanda S. Reiter *Iowa Wesleyan College*

Stacia Rink *Pacific Lutheran University*

Jill K. Robinson *University of Wyoming–Laramie*

Rene Rodriguez *Idaho State University*

Kathryn Rowberg *Purdue University–Calumet-Hammond*

Michael Russell *Mt. Hood Community College*

Ruth Russo *Whitman College*

Matthew Ryan *Purdue University–Calumet-Hammond*

Preet-Pal S. Saluja *William Rainey Harper College*

Reva Savkar *Northern Virginia Community College*

Susan M. Schelble *University of Colorado–Denver*

Udo Schnupf *Troy State University–Troy*

Jeff R. Schoonover *St. Mary's University*

Diane R. Schulman *Erie Community College South Campus–Orchard Park*

Ronald See *Indiana University of Pennsylvania–Indiana*

Neeta Sharma *Elizabethtown Community College*

Anil K. Sharma *Mississippi Valley State University*

Anne T. Sherren *North Central College*

Paul H. Short *East Mississippi Junior College–Mayhew*

Alka Shukla *Houston Community College–Southeast*

Shyam S. Shukla *Lamar University–Beaumont*

Jerry Skelton *St. Johns River Community College–Orange Park*

Charles Smith *Our Lady of the Lake University*

Joseph Sneddon *McNeese State University*

Rhodora Snow *Texas College*

Robert J. Snyder *Southern Connecticut State University*

Steven M. Socol *McHenry County College*

Mary Sohn *Florida Institute of Technology–Melbourne*

David Speckhard *Loras College*

Marjorie E. Squires *Felician College*

Jimmy H. Stanton *Cameron University*

Max Stappler *Queensborough Community College*

Roland Stout *University of Toledo–Toledo*

Douglas Strout *Alabama State University*

Xiaoping Sun *West Virginia University, Institute of Technology*

Paris Svoronos *Queensborough Community College*

Soraya Ghayourmanesh-Svoronos *Nassau County Community College*

Kathryn B. Sympson *Florida Keys Community College*

David C. Taylor *Slippery Rock University of Pennsylvania*

David L. Taylor *Northern Virginia Community College–Woodbridge*

Michael Tessmer *Southwestern College*

Suresh Tewani *LIU–Brooklyn*

Albert N. Thomson *Spelman College*

John Todd *Bowling Green State University–Bowling Green*

Marie Underdown *Montgomery College*

Ana Vande Linde *University of Wisconsin–Stout*

Margaret H. Venable *Georgia Perimeter College–Clarkston*

Rashmi Venkateswaran *University of Ottawa*

John B. Vincent *University of Alabama–Tuscaloosa*

R. Harmony Voorhies *Mesa State College*

Bill Walkenhorst *Loyola University*

Jerry Walsh *University of North Carolina–Greensboro*

Esther Waltz *Community College Allegheny County–West Mifflin*

Randall Wanke *Augustana College*

Susan A. Weiner *West Valley College*

Neil H. Weinstein *Santa Fe Community College*

Christine Whitlock *Georgia Southern University*

Will Wickun *Montana State University–Billings*

Larry Wiginton *Clarendon College*

Daniel J. Williams *Kennesaw State University*

Doyle Wilson *Mohave Community College*

Robb Wilson *Louisiana State University–Shreveport*

Joseph W. Wilson *University of Kentucky–Lexington*

Troy D. Wood *SUNY–Buffalo*

Charles M. Wynn *Eastern Connecticut State University*

Qianhui Zhang *Miami Dade Community College–North*

W. C. Zipperer *Armstrong Atlantic State University*

Lois Anne Zook *Delta State University*

I would like to thank Brandon Cruickshank for his effective animations, which will greatly help students visualize abstract concepts. As always, I have benefited much from discussions with my colleagues at Williams College and correspondence with many instructors here and abroad.

It is a pleasure to acknowledge the support given to me by the following members of McGraw-Hill's College Division staff: Jennifer Bensink, Kevin Kane, Jenni Lang, Michael Lange, Craig Marty, Sandra Hahn, James Smith, and Thomas Timp. In particular, I would like to mention Gloria Schiesl for supervising the production, Stuart Paterson for the book design, and John Leland for photo research. Kent Peterson, the publisher, supervised the overall project. Finally, my special thanks go to Shirley Oberbroeckling, the development editor, for her enthusiasm, thoroughness, and professionalism. It was a pleasure to work with her on this edition.

Raymond Chang

Guided Tour: Text

Art Program

Art throughout the text is enhanced with molecular art images using the Spartan molecular modeling program. These images enable students to gain a better understanding of the three-dimensionality of molecules and the details of chemical reactions.

Figure 10.5 *Top to bottom: As two H atoms approach each other, their 1s orbitals begin to interact and each electron begins to feel the attraction of the other proton. Gradually, the electron density builds up in the region between the two nuclei (red color). Eventually, a stable H₂ molecule is formed when the internuclear distance is 74 pm.*

The orbital diagram of the F atom is shown on p. 217.

Thus valence bond theory gives a clearer picture of chemical bond formation than the Lewis theory does. Valence bond theory states that a stable molecule forms from reacting atoms when the potential energy of the system has decreased to a minimum; the Lewis theory ignores energy changes in chemical bond formation.

The concept of overlapping atomic orbitals applies equally well to diatomic molecules other than H_2. Thus a stable F_2 molecule forms when the $2p$ orbitals (containing the unpaired electrons) in the two F atoms overlap to form a covalent bond. Similarly, the formation of the HF molecule can be explained by the overlap of the $1s$ orbital in H with the $2p$ orbital in F. In each case, VB theory accounts for the changes in potential energy as the distance between the reacting atoms changes. Because the orbitals involved are not the same kind in all cases, we can see why the bond energies and bond lengths in H_2, F_2, and HF might be different. As we stated earlier, Lewis theory treats *all* covalent bonds the same way and offers no explanation for the differences among covalent bonds.

10.4 Hybridization of Atomic Orbitals

The concept of atomic orbital overlap should apply also to polyatomic molecules. However, a satisfactory bonding scheme must account for molecular geometry. We will discuss three examples of VB treatment of bonding in polyatomic molecules.

9.7 Exceptions to the Octet Rule   279

A resonance structure with a double bond between B and F can be drawn that satisfies the octet rule for B. However, the properties of BF_3 are more consistent with a Lewis structure in which there are single bonds between B and each F, as shown here.

Although boron trifluoride is stable, it has a tendency to pick up an unshared electron pair from an atom in another compound, as shown by its reaction with ammonia:

$$:\!\overset{..}{\underset{..}{F}}\!-\!B\!-\!\overset{..}{\underset{..}{F}}\!: \;+\; :N\!-\!H \longrightarrow \;:\!\overset{..}{\underset{..}{F}}\!-\!B^-\!-\!N^+\!-\!H$$

This structure satisfies the octet rule for the B, N, and F atoms.

The B—N bond in the compound just shown is different from the covalent bonds discussed so far in the sense that both electrons are contributed by the N atom. *A covalent bond in which one of the atoms donates both electrons* is called a ***coordinate covalent bond***. Although the properties of a coordinate covalent bond do not differ from those of a normal covalent bond (because all electrons are alike no matter what their source), the distinction is useful for keeping track of valence electrons and assigning formal charges.

Odd-Electron Molecules

Some molecules contain an *odd* number of electrons. Among them are nitric oxide (NO) and nitrogen dioxide (NO_2):

$$\overset{..}{N}\!=\!\overset{..}{O} \qquad \overset{..}{O}\!=\!\overset{..}{N}^+\!-\!\overset{..}{\underset{..}{O}}:^-$$

Because we need an even number of electrons for complete pairing (to reach eight), the octet rule clearly cannot be satisfied for all the atoms in any molecule that has an odd number of electrons.

The Expanded Octet

In a number of compounds there are more than eight valence electrons around an atom. These *expanded octets* are needed only for atoms of elements in and beyond the third period of the periodic table. In addition to the $3s$ and $3p$ orbitals, elements in the third period also have $3d$ orbitals that can be used in bonding. One compound in which there is an expanded octet is sulfur hexafluoride, a very stable compound. The electron configuration of sulfur is $[Ne]3s^23p^4$. In SF_6, each of sulfur's six valence electrons forms a covalent bond with a fluorine atom, so there are 12 electrons around the central sulfur atom:

Yellow: second-period elements cannot have an expanded octet. Blue: third-period elements and beyond can have an expanded octet. Green: the noble gases usually only have an expanded octet.

In Chapter 10 we will see that these 12 electrons, or six bonding pairs, are accommodated in six orbitals that originate from the one $3s$, the three $3p$, and two of the five $3d$ orbitals. However, sulfur also forms many compounds in which it does not violate the octet rule. In sulfur dichloride, S is surrounded by only eight electrons and therefore obeys the octet rule:

$$:\!\overset{..}{\underset{..}{Cl}}\!-\!\overset{..}{\underset{..}{S}}\!-\!\overset{..}{\underset{..}{Cl}}\!:$$

Sulfur dichloride is a toxic, foul-smelling cherry-red liquid (boiling point: 59°C).

Icons

- Animation icon: This icon points to material that is further illustrated by an animation. Students can use the animations to review challenging concepts in motion. Animations are referenced and can be found on the Online Learning Center at **www.mhhe.com/chang**.
- Periodic Table icon: This icon illustrates the properties of elements according to their positions in the periodic table.

Worked Examples and Practice Exercises

Careful study of the solved examples in the body of each chapter will help to develop students' problem-solving skills. The Practice Exercise that follows each worked example allows students to check their ability to solve the type of problem illustrated in the Worked Example. Answers to the Practice Exercises can be found at the end of the chapter, following the Questions and Problems. The number of a Similar Problem is shown in the margin next to the Worked Example to encourage additional practice. Many Worked Examples have been revised to more clearly show steps in problem solving.

An air bag can protect the driver in an automobile collision.

Similar problems: 5.51, 5.52.

Example 5.6   Sodium azide (NaN_3) is used in some automobile air bags. The impact of a collision triggers the decomposition of NaN_3 as follows:

$$2NaN_3(s) \longrightarrow 2Na(s) + 3N_2(g)$$

The nitrogen gas produced quickly inflates the bag between the driver and the windshield. Calculate the volume of N_2 generated at 80°C by 823 mmHg by the decomposition of 60.0 g of NaN_3.

Reasoning and Solution   The problem requires two steps. First, we calculate the number of moles of N_2 generated from the decomposition of 60.0 g of NaN_3. Next we calculate the volume of the N_2 gas at the given temperature and pressure.

$$\text{moles of } N_2 = 60.0 \text{ g } NaN_3 \times \frac{1 \text{ mol } NaN_3}{65.02 \text{ g } NaN_3} \times \frac{3 \text{ mol } N_2}{2 \text{ mol } NaN_3}$$

$$= 1.38 \text{ mol } N_2$$

The volume of 1.38 moles of N_2 can be obtained by using the ideal gas equation:

$$V = \frac{nRT}{P} = \frac{(1.38 \text{ mol})(0.0821 \text{ L} \cdot \text{atm} \cdot \text{mol}^{-1} \cdot \text{K}^{-1})(80 + 273) \text{ K}}{(823/760) \text{ atm}}$$

$$= 36.9 \text{ L}$$

Practice Exercise   The equation for the metabolic breakdown of glucose ($C_6H_{12}O_6$) is the same as the equation for the combustion of glucose in air:

$$C_6H_{12}O_6(s) + 6O_2(g) \longrightarrow 6CO_2(g) + 6H_2O(l)$$

Calculate the volume of CO_2 produced at 37°C and 1.00 atm when 5.60 g of glucose are used up in the reaction.

5.5 Dalton's Law of Partial Pressures

Thus far we have concentrated on the behavior of pure gaseous substances, but experimental studies very often involve mixtures of gases. For example, for a study of air pollution, we may be interested in the pressure-volume-temperature relationship of a sample of air, which contains several gases. In this case, and all cases involving mixtures of gases, the total gas pressure is related to *partial pressures*, that is, *the pressures of individual gas components in the mixture*. In 1801 Dalton formulated a law, now known as *Dalton's law of partial pressures*, which states that *the total pressure of a mixture of gases is just the sum of the pressures that each gas would exert if it were present alone*. Figure 5.12 illustrates Dalton's law.

Consider a case in which two gases, A and B, are in a container of volume V. The pressure exerted by gas A, according to the ideal gas equation, is

$$P_A = \frac{n_A RT}{V}$$

As mentioned earlier, gas pressure results from the impact of gas molecules against the walls of the container.

where n_A is the number of moles of A present. Similarly, the pressure exerted by gas B is

$$P_B = \frac{n_B RT}{V}$$

Table 10.5 | Properties of Homonuclear Diatomic Molecules of the Second-Period Elements*

	Li_2	B_2	C_2	N_2	O_2	F_2	
$\sigma^*_{2p_x}$	☐	☐	☐	☐	☐	☐	$\sigma^*_{2p_x}$
$\pi^*_{2p_y}, \pi^*_{2p_z}$	☐☐	☐☐	☐☐	☐☐	↑ ↑	↑↓ ↑↓	$\pi^*_{2p_y}, \pi^*_{2p_z}$
σ_{2p_x}	☐	☐	☐	↑↓	↑↓	↑↓	σ_{2p_x}
π_{2p_y}, π_{2p_z}	☐☐	↑ ↑	↑↓ ↑↓	↑↓ ↑↓	↑↓ ↑↓	↑↓ ↑↓	π_{2p_y}, π_{2p_z}
σ^*_{2s}	☐	↑↓	↑↓	↑↓	↑↓	↑↓	σ^*_{2s}
σ_{2s}	↑↓	↑↓	↑↓	↑↓	↑↓	↑↓	σ_{2s}
Bond order	1	1	2	3	2	1	
Bond length (pm)	267	159	131	110	121	142	
Bond energy (kJ/mol)	104.6	288.7	627.6	941.4	498.7	156.9	
Magnetic properties	Diamagnetic	Paramagnetic	Diamagnetic	Diamagnetic	Paramagnetic	Diamagnetic	

*For simplicity the σ_{1s} and σ^*_{1s} orbitals are omitted. These two orbitals hold a total of four electrons. Remember that for O_2 and F_2, σ_{2p_x} is lower in energy than π_{2p_y} and π_{2p_z}.

The ground-state electron configuration of O is $1s^2 2s^2 2p^4$; thus there are 16 electrons in O_2. Using the order of increasing energies of the molecular orbitals from Table 10.5, we write the ground-state electron configuration of O_2 as

$$(\sigma_{1s})^2(\sigma^*_{1s})^2(\sigma_{2s})^2(\sigma^*_{2s})^2(\sigma_{2p_x})^2(\pi_{2p_y})^2(\pi_{2p_z})^2(\pi^*_{2p_y})^1(\pi^*_{2p_z})^1$$

According to Hund's rule, the last two electrons enter the $\pi^*_{2p_y}$ and $\pi^*_{2p_z}$ orbitals with parallel spins. Ignoring the σ_{1s} and σ^*_{1s} orbitals (because their net effects on bonding are zero), we calculate the bond order of O_2 using Equation (10.2):

$$\text{bond order} = \tfrac{1}{2}(6 - 2) = 2$$

Therefore, the O_2 molecule has a bond order of 2 and is paramagnetic, a prediction that corresponds to experimental observations.

Example 10.6   The N_2^+ ion can be prepared by bombarding the N_2 molecule with fast-moving electrons. Predict these properties of N_2^+: (a) electron configuration, (b) bond order, (c) magnetic character, and (d) bond length relative to the bond length of N_2 (is it longer or shorter?).

Reasoning and Solution   From Table 10.5 we can deduce the properties of ions generated from the homonuclear diatomic molecules.

(a) Because N_2^+ has one fewer electron than N_2, its electron configuration is

$$(\sigma_{1s})^2(\sigma^*_{1s})^2(\sigma_{2s})^2(\sigma^*_{2s})^2(\pi_{2p_y})^2(\pi_{2p_z})^2(\sigma_{2p_x})^1$$

Molecular Orbital Theory

The addition of molecular orbital theory to this edition provides the students with basic skills for future courses in organic chemistry. This material is further expanded in the aromatic compounds in the organic chemistry chapter (Chapter 11). For instructors who prefer not to cover this topic, it can be omitted without disruption or loss of continuity.

Guided Tour: Media

Instructor Media

- Online Learning Center (OLC) is a secure, book-specific website. The OLC is the doorway to a library of resources for instructors.

- PowerPoint Presentation—is organized by chapters and is ready for the classroom, or instructors can customize the lecture to reflect their teaching style.

- Course Management Systems—PageOut, WebCT, and Blackboard. All of the following tools are available on the Online Learning Center or in a cartridge for your course delivery system:
  1. Computerized Test Bank
  2. Solutions Manual
  3. Images from the text
  4. Tables from the text
  5. Narratives for the animations
  6. Essential Study Partner link to home page

- Chemistry Animation Visual Resource Library—CD-ROM set includes over 300 animations that can be used directly from the CD or imported into your own lecture presentation.

- Digital Content Manager—CD-ROM set includes electronic files of all full-color images in the text. Import the images into your own presentation, or use the PowerPoint presentation provided for each chapter.

- GradeSummit—is an Internet-based self-assessment service that provides students and faculty with diagnostic information about subject strengths and weaknesses. This detailed feedback and direction enables learners and teachers to focus study time on areas where they will be most effective.

Online Learning Center

Course Management Systems

PowerPoint

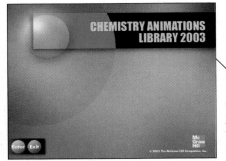

Chemistry Animation Visual Resource Library

Digital Content Manager

GradeSummit

Student Media

- Online Learning Center—is the doorway to access almost all of the media for *General Chemistry, Third Edition*. The OLC includes self-assessment quizzes, resources, key-term flashcards with audio pronunciation, and more.

- Essential Study Partner—is the essential tool for review, interaction, and self-assessment. Students can work through a basic review of core concepts in the ten units.

- NetTutor—is a tool for students when they need extra help with an end-of-chapter problem. Live sessions are scheduled or you can e-mail a tutor at any time.

- Online ChemSkill Builder—provides questions for all general chemistry topics. Students can practice problem types until they master the skill and concept. Feedback is provided for each question.

Online Learning Center

Essential Study Partner

NetTutor

Online ChemSkill Builder

A Note to the Student

General chemistry is commonly perceived to be more difficult than most other subjects. There is some justification for this perception. For one thing, chemistry has a very specialized vocabulary. At first, studying chemistry is like learning a new language. Furthermore, some of the concepts are abstract. Nevertheless, with diligence you can complete this course successfully, and you might even enjoy it. Here are some suggestions to help you form good study habits and master the material in this text.

- Attend classes regularly and take careful notes.

- If possible, always review the topics discussed in class the same day they are covered in class. Use this book to supplement your notes.

- Think critically. Ask yourself if you really understand the meaning of a term or the use of an equation. A good way to test your understanding is to explain a concept to a classmate or some other person.

- Do not hesitate to ask your instructor or your teaching assistant for help.

The third edition tools for *General Chemistry* are designed to enable you to do well in your general chemistry course. The following guide explains how to take full advantage of the text, technology, and other tools.

- Before delving into the chapter, read the chapter *outline* and the chapter *introduction* to get a sense of the important topics. Use the outline to organize your note-taking in class.

- Use the *Animation Icon* (⬚) as a guide to review challenging concepts in motion. Most of the animations are interactive.

- At the end of each chapter you will find a summary of facts and concepts and a list of key words, both of which will help you review for exams. Definitions of the key words can be studied in context on the pages cited in the end-of-chapter list or in the glossary at the back of the book.

- Use the *Key-Word flashcards* on the website (Online Learning Center, or OLC) for more practice. The flashcards include audio, so you can test your pronunciation, too. The OLC houses an extraordinary amount of resources. Go to www.mhhe.com/physsci/chemistry/chang and click on the appropriate cover to explore chapter quizzes, the e-learning sessions, opportunities, the Essential Study Partner, and more.

- Careful study of the numbered examples in the body of each chapter will improve your ability to analyze problems and correctly carry out the calculations needed to solve them. Also take the time to work through the practice exercise that follows each example to be sure you understand how to solve the type of problem illustrated in the example. The answers to the practice exercises appear at the end of the chapter, following the homework problems. For additional practice, you can turn to similar homework problems referred to in the margin next to the example.

- The questions and problems at the end of the chapter are organized by section. Generally the review questions do not require calculation. Their purpose is to help you check your understanding of new concepts introduced in the chapter. The problems enable you to test your analytical and computational skills. Even-numbered problems are similar in nature to the odd-numbered problems that precede them, except in the section titled "Additional Problems." The additional problems require that you decide how to approach the solution, and a number of them involve concepts from more than one section of the chapter. Although your instructor will most likely not assign all the problems for homework, it is to your advantage to work as many as necessary to assure yourself that you have mastered the chapter. Refer to the back of the book for answers to all even-numbered problems as well as problem-solving tutorials that are available in a separate *Student Solutions Manual*.

• For even more practice problems, use *Online ChemSkill Builder*. ChemSkill Builder is a problem-solving tutorial with hundreds of problems that include feedback.

• One of your best resources is the *Essential Study Partner*, or ESP. The ESP is a tutorial that will help you review and test your understanding of each chapter. The ESP includes animation, interactives, topic reviews with quizzes, and unit exams. Go to www.mhhe.com/physsci/chemistry/chang and click on the appropriate cover to explore the ESP.

• A live online tutorial from NetTutor is available at the *General Chemistry* website. Go to www.mhhe.com/physsci/chemistry/chang and click on the appropriate cover to discover the link to this service. Just register and you can look at frequently asked questions, join a live chat (times are predesignated), or look through the archives of past sessions.

If you follow these suggestions and stay up-to-date with your assignments, you should find that chemistry is challenging, but less difficult and much more interesting than you expected.

Raymond Chang

General Chemistry

rchimedes was a Greek mathematician and philosopher who lived from 287 to 212 B.C. According to legend, he was asked by the emperor to determine whether a crown was made of pure gold or a mixture of gold and less valuable metals, such as silver and copper. After receiving this assignment, Archimedes was taking a bath one day when he noticed that the deeper his body sank into the water, the more water overflowed the bath, and his body seemed to weigh less the more it was submerged. He concluded that the apparent loss of weight of an object submerged in a liquid is equal to the weight of the liquid it displaces. This principle, now named after him, allowed Archimedes to design the following experiment. Having previously observed that the volume of an object determines the volume of the liquid it displaces, he would weigh the crown and the same weight of gold both immersed in water in separate containers. Because every object has a different ratio of weight to volume, or *density,* the crown (if it was not made entirely of gold) and the pure gold that weighed the same amount in air would have different weights in water because they would displace different volumes of water. Supposedly, Archimedes was so excited by his brilliance that he ran from the bath shouting, "Eureka"—the Greek word for "I have found it!"

Chemistry is largely an experimental science. Like Archimedes' discovery, most advances in chemistry are the results of observations and experiments, although the setting is usually a laboratory.

Introduction

Chapter 1

ESSENTIAL CONCEPTS

 The Study of Chemistry Chemistry is the study of matter and the changes it undergoes. Elements and compounds are substances that take part in chemical transformation.

 Physical and Chemical Properties To characterize a substance, we need to know its physical properties, which can be observed without changing its identity, and chemical properties, which can be demonstrated only by chemical changes.

 Measurements and Units Chemistry is a quantitative science and requires measurements. The measured quantities (for example, mass, volume, density, and temperature) usually have units associated with them. The units used in chemistry are based on the international system (SI) of units.

 Handling Numbers Scientific notation is used to express large and small numbers, and each number in a measurement must indicate the meaningful digits, called significant figures.

 Doing Chemical Calculations A simple and effective way to perform chemical calculations is the factor-label method, which is based on dimensional analysis. In this procedure, an equation is set up in such a way that all the units cancel except the ones for the final answer.

1.1 The Study of Chemistry

Whether or not this is your first course in chemistry, you undoubtedly have some preconceived ideas about the nature of this science and about what chemists do. Most likely, you think chemistry is practiced in a laboratory by someone in a white coat who studies things in test tubes. This description is fine, up to a point. Chemistry is largely an experimental science, and a great deal of knowledge comes from laboratory research. In addition, however, today's chemist may use a computer to study the microscopic structure and chemical properties of substances or employ sophisticated electronic equipment to analyze pollutants from auto emissions or toxic substances in the soil. Many frontiers in biology and medicine are currently being explored at the level of atoms and molecules—the structural units on which the study of chemistry is based. Chemists participate in the development of new drugs and in agricultural research. What's more, they are seeking solutions to the problem of environmental pollution, along with replacements for energy sources. And most industries, whatever their products, have a basis in chemistry. For example, chemists developed the polymers (very large molecules) that manufacturers use to make a wide variety of goods, including clothing, cooking utensils, artificial organs, and toys. Indeed, because of its diverse applications, chemistry is often called the "central science."

How to Study Chemistry

Compared with other subjects, chemistry is commonly perceived to be more difficult, at least at the introductory level. There is some justification for this perception. For one thing, chemistry has a very specialized vocabulary. At first, studying chemistry is like learning a new language. Furthermore, some of the concepts are abstract. Nevertheless, with diligence you can complete this course successfully—and perhaps even pleasurably. Listed here are some suggestions to help you form good study habits and master the material:

- Attend classes regularly and take careful notes.
- If possible, always review the topics you learned in class the *same* day the topics are covered in class. Use this book to supplement your notes.
- Think critically. Ask yourself if you really understand the meaning of a term or the use of an equation. A good way to test your understanding is for you to explain a concept to a classmate or some other person.
- Do not hesitate to ask your instructor or your teaching assistant for help.

You will find that chemistry is much more than numbers, formulas, and abstract theories. It is a logical discipline brimming with interesting ideas and applications.

1.2 The Scientific Method

All sciences, including the social sciences, employ variations of what is called the *scientific method—a systematic approach to research.* For example, a psychologist who wants to know how noise affects people's ability to learn chemistry and a chemist interested in measuring the heat given off when hydrogen gas burns in air follow roughly the same procedure in carrying out their investigations. The first step is to carefully define the problem. The next step includes performing experiments, making careful observations, and recording information, or *data,* about the system—the part of the universe that is under investigation. (In these examples, the systems are the group of people the psychologist will study and a mixture of hydrogen and air.)

The data obtained in a research study may be both ***qualitative,*** *consisting of general observations about the system,* and ***quantitative,*** *comprising numbers obtained by various measurements of the system.* Chemists generally use standardized symbols and equations in recording their measurements and observations. This form of representation not only simplifies the process of keeping records, but also provides a common basis for communications with other chemists. Figure 1.1 summarizes the main steps of the research process.

When the experiments have been completed and the data have been recorded, the next step in the scientific method is interpretation, meaning that the scientist attempts to explain the observed phenomenon. Based on the data that were gathered, the researcher formulated a ***hypothesis,*** or *tentative explanation for a set of observations.* Further experiments are devised to test the validity of the hypothesis in as many ways as possible, and the process begins anew.

After a large amount of data has been collected, it is often desirable to summarize the information in a concise way, as a law. In science, a ***law*** is *a concise verbal or mathematical statement of a relationship between phenomena that is always the same under the same conditions.* For example, Sir Isaac Newton's second law of motion, which you may remember from high school science, says that force equals mass times acceleration ($F = ma$). What this law means is that an increase in the mass or in the acceleration of an object always increases the object's force proportionally and a decrease in mass or acceleration always decreases the force.

Hypotheses that survive many experimental tests of their validity may evolve into theories. A ***theory*** is *a unifying principle that explains a body of facts and/or those laws that are based on them.* Theories, too, are constantly being tested. If a theory is disproved by experiment, then it must be discarded or modified so that it becomes consistent with experimental observations. Proving or disproving a theory can take years, even centuries, in part because the necessary technology is not available. Atomic theory, which we will study in Chapter 2, is a case in point. It took more than 2000 years to work out this fundamental principle of chemistry proposed by Democritus, an ancient Greek philosopher.

Scientific progress is seldom, if ever, made in a rigid, step-by-step fashion. Sometimes a law precedes a theory; sometimes it is the other way around. Two scientists may start working on a project with exactly the same objective, but will take drastically different approaches. They may be led in vastly different directions. Scientists are, after all, human beings, and their modes of thinking and working are very much influenced by their backgrounds, training, and personalities.

The development of science has been irregular and sometimes illogical. Great discoveries are usually the result of the cumulative contributions and experience of many workers, even though the credit for formulating a theory or a law is usually given to only one individual. There is, of course, an element of luck involved in scientific discoveries, but it has been said that "chance favors the prepared mind." It takes an alert and well-trained person to recognize the significance of an accidental discovery and to take full advantage of it. More often than not, the public learns only of spectacular scientific breakthroughs. For every success story, however, there are hundreds of cases in which scientists spent years working on projects that ultimately led to a dead end. Many positive achievements came only after many wrong turns and at such a

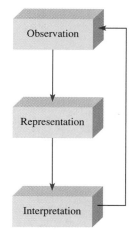

Figure 1.1 *The three levels of studying chemistry and their relationships. Observation deals with events in the macroscopic world; atoms and molecules constitute the microscopic world. Representation is a scientific shorthand for describing an experiment in symbols and chemical equations. Chemists use their knowledge of atoms and molecules to explain an observed phenomenon.*

slow pace that they went unheralded. Yet even the dead ends contribute something to the continually growing body of knowledge about the physical universe. It is the love of the search that keeps many scientists in the laboratory.

1.3 Classifications of Matter

The Chinese characters for chemistry mean "The study of change."

Matter is *anything that occupies space and has mass,* and ***chemistry*** is *the study of matter and the changes it undergoes.* All matter, at least in principle, can exist in three states: solid, liquid, and gas. Solids are rigid objects with definite shapes. Liquids are less rigid than solids and are fluid—they are able to flow and assume the shape of their containers. Like liquids, gases are fluid, but unlike liquids, they can expand indefinitely.

The three states of matter can be interconverted—they can be changed from one state into another. Upon heating, a solid will melt and become a liquid. Further heating will convert the liquid into a gas. On the other hand, cooling a gas will condense it into a liquid. Cooled further, the liquid will solidify. Figure 1.2 shows the three states of water.

Scientists also distinguish among several subcategories of matter based on composition and properties. The classifications of matter include substances, mixtures, elements, and compounds, as well as the fundamental units that make up elements and compounds—atoms and molecules—which we will consider in Chapter 2.

Substances and Mixtures

A ***substance*** is *matter that has a definite or constant composition and distinct properties.* Examples are water, silver, ethanol, table salt (sodium chloride), and carbon

Figure 1.2 *The three states of matter. A hot poker changes ice into water and steam.*

Figure 1.3 *Separating iron filings from a heterogeneous mixture. The same technique is used on a larger scale to separate iron and steel from nonmagnetic objects such as aluminum, glass, and plastics.*

(a)   (b)

dioxide. Substances differ from one another in composition and can be identified by their appearance, smell, taste, and other properties. At present, over 13 million substances are known, and the list is growing rapidly.

A **mixture** is *a combination of two or more substances in which the substances retain their distinct identities.* Some examples are air, soft drinks, milk, and cement. Mixtures do not have constant composition. Therefore, samples of air collected in different cities would probably differ in composition because of differences in altitude, pollution, and so on.

Mixtures are either homogeneous or heterogeneous. When a spoonful of sugar dissolves in water, *the composition of the mixture,* after sufficient stirring, *is the same throughout the solution.* This solution is a **homogeneous mixture.** If sand is mixed with iron filings, however, the sand grains and the iron filings remain visible and separate (Figure 1.3). This type of mixture, in which *the composition is not uniform,* is called a **heterogeneous mixture.** Adding oil to water creates another heterogeneous mixture because the liquid does not have a constant composition.

Any mixture, whether homogeneous or heterogeneous, can be created and then separated by physical means into pure components without changing the identities of the components. Thus, sugar can be recovered from a water solution by heating the solution and evaporating it to dryness. Condensing the water vapor will give us back the water component. To separate the iron-sand mixture, we can use a magnet to remove the iron filings from the sand, because sand is not attracted to the magnet (see Figure 1.3b). After separation, the components of the mixture will have the same composition and properties as they did to start with.

Elements and Compounds

A substance can be either an element or a compound. An **element** is *a substance that cannot be separated into simpler substances by chemical means.* At present, 113 elements have been positively identified. (See the list inside the front cover of this book.) Eighty-three of them occur naturally on Earth. The others have been created by scientists.

Chemists use alphabetical symbols to represent the names of the elements. The first letter of the symbol for an element is *always* capitalized, but the second letter is

Table 1.1 | Some Common Elements and Their Symbols

Name	Symbol	Name	Symbol	Name	Symbol
Aluminum	Al	Fluorine	F	Oxygen	O
Arsenic	As	Gold	Au	Phosphorus	P
Barium	Ba	Hydrogen	H	Platinum	Pt
Bromine	Br	Iodine	I	Potassium	K
Calcium	Ca	Iron	Fe	Silicon	Si
Carbon	C	Lead	Pb	Silver	Ag
Chlorine	Cl	Magnesium	Mg	Sodium	Na
Chromium	Cr	Mercury	Hg	Sulfur	S
Cobalt	Co	Nickel	Ni	Tin	Sn
Copper	Cu	Nitrogen	N	Zinc	Zn

never capitalized. For example, Co is the symbol for the element cobalt, whereas CO is the formula for carbon monoxide, which is made up of the elements carbon and oxygen. Table 1.1 shows some of the more common elements. The symbols for some elements are derived from their Latin names—for example, Au from *aurum* (gold), Fe from *ferrum* (iron), and Na from *natrium* (sodium)—although most of them are abbreviated forms of their English names.

Figure 1.4 shows the most abundant elements in Earth's crust and in the human body. As you can see, only five elements (oxygen, silicon, aluminum, iron, and calcium) comprise over 90 percent of Earth's crust. Of these five elements, only oxygen is among the most abundant elements in living systems.

Most elements can interact with one or more other elements to form compounds. We define a **compound** as *a substance composed of atoms of two or more elements chemically united in fixed proportions.* Hydrogen gas, for example, burns in oxygen gas to form water, a compound whose properties are distinctly different from those of the starting materials. Water is made up of two parts of hydrogen and one part of oxygen. This composition does not change, regardless of whether the water comes from a faucet in the United States, the Yangtze River in China, or the ice caps on Mars. Unlike mixtures, compounds can be separated only by chemical means into their pure components.

The relationships among elements, compounds, and other categories of matter are summarized in Figure 1.5.

Figure 1.4 *(a) Natural abundance of elements in Earth's crust in percent by mass. For example, oxygen's abundance is 45.5 percent. This means that in a 100-g sample of Earth's crust there are, on the average, 45.5 g of the element oxygen. (b) Abundance of elements in the human body in percent by mass.*

(a)   (b)

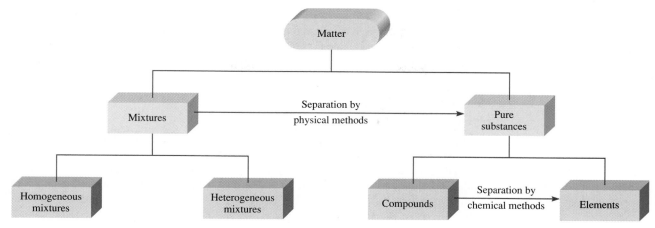

Figure 1.5 *Classification of matter.*

1.4 Physical and Chemical Properties of Matter

Substances are identified by their properties as well as by their composition. Color, melting point, boiling point, and density are physical properties. A ***physical property*** *can be measured and observed without changing the composition or identity of a substance.* For example, we can measure the melting point of ice by heating a block of ice and recording the temperature at which the ice is converted to water. Water differs from ice only in appearance and not in composition, so this is a physical change; we can freeze the water to recover the original ice. Therefore, the melting point of a substance is a physical property. Similarly, when we say that helium gas is lighter than air, we are referring to a physical property.

On the other hand, the statement "Hydrogen gas burns in oxygen gas to form water" describes a ***chemical property*** of hydrogen because *to observe this property we must carry out a chemical change,* in this case burning. After the change, the original substances, hydrogen and oxygen gas, will have vanished and a chemically different substance—water—will have taken their place. We *cannot* recover hydrogen and oxygen from water by a physical change such as boiling or freezing.

Every time we hard-boil an egg, we bring about a chemical change. When subjected to a temperature of about 100°C, the yolk and the egg white undergo reactions that alter not only their physical appearance but their chemical makeup as well. When eaten, the egg is changed again, by substances in the body called *enzymes.* This digestive action is another example of a chemical change. What happens during such a process depends on the chemical properties of the specific enzymes and of the food involved.

All measurable properties of matter fall into two categories: extensive properties and intensive properties. The measured value of an ***extensive property*** *depends on how much matter is being considered.* Mass, length, and volume are extensive properties. More matter means more mass. Values of the same extensive property can be added together. For example, two copper pennies have a combined mass that is the sum of the masses of each penny, and the volume occupied by the water in two beakers is the sum of the volumes of the water in each of the beakers.

The measured value of an ***intensive property*** *does not depend on the amount of matter being considered.* Temperature is an intensive property. Suppose that we have two

Hydrogen burning in air to form water.

beakers of water at the same temperature. If we combine them to make a single quantity of water in a larger beaker, the temperature of the larger amount of water will be the same as it was in two separate beakers. Unlike mass and volume, temperature and other intensive properties such as melting point, boiling point, and density are not additive.

1.5 Measurement

The study of chemistry depends heavily on measurement. For instance, chemists use measurements to compare the properties of different substances and to assess changes resulting from an experiment. A number of common devices enable us to make simple measurements of a substance's properties: The meterstick measures length; the buret, the pipet, the graduated cylinder, and the volumetric flask measure volume (Figure 1.6); the balance measures mass; the thermometer measures temperature. These instruments provide measurements of ***macroscopic properties,*** *which can be determined directly.* ***Microscopic properties,*** *on the atomic or molecular scale, must be determined by an indirect method,* as we will see in Chapter 2.

A measured quantity is usually written as a number with an appropriate unit. To say that the distance between New York and San Francisco by car along a certain route is 5166 is meaningless. We must specify that the distance is 5166 kilometers. In science, units are essential to stating measurements correctly.

SI Units

For many years scientists recorded measurements in *metric units,* which are related decimally, that is, by powers of 10. In 1960, however, the General Conference of

Figure 1.6 *Some common measuring devices found in a chemistry laboratory. These devices are not drawn to scale relative to one another. We will discuss the uses of these measuring devices in Chapter 4.*

Buret   Pipet   Graduated cylinder   Volumetric flask

Table 1.2 | SI Base Units

Base Quantity	Name of Unit	Symbol
Length	meter	m
Mass	kilogram	kg
Time	second	s
Electrical current	ampere	A
Temperature	kelvin	K
Amount of substance	mole	mol
Luminous intensity	candela	cd

Weights and Measures, the international authority on units, proposed a revised metric system called the ***International System of Units*** (abbreviated *SI,* from the French *System International d'Unites*). Table 1.2 shows the seven SI base units. All other SI units of measurement can be derived from these base units. Like metric units, SI units are modified in decimal fashion by a series of prefixes, as shown in Table 1.3. We use both metric and SI units in this book.

Measurements that we will utilize frequently in our study of chemistry include time, mass, volume, density, and temperature.

Mass and Weight

Mass is *a measure of the quantity of matter in an object*. The terms "mass" and "weight" are often used interchangeably, although, strictly speaking, they refer to different quantities. In scientific terms, ***weight*** is *the force that gravity exerts on an object*. An apple that falls from a tree is pulled downward by Earth's gravity. The mass of the apple is constant and does not depend on its location, but its weight does. For example, on the surface of the moon the apple would weigh only one-sixth what it does on Earth, because of the smaller mass of the moon. This is why astronauts

An astronaut on the surface of the moon.

Table 1.3 | Prefixes Used with SI Units

Prefix	Symbol	Meaning	Example
Tera-	T	1,000,000,000,000, or 10^{12}	1 terameter (Tm) = 1×10^{12} m
Giga-	G	1,000,000,000, or 10^{9}	1 gigameter (Gm) = 1×10^{9} m
Mega-	M	1,000,000, or 10^{6}	1 megameter (Mm) = 1×10^{6} m
Kilo-	k	1,000, or 10^{3}	1 kilometer (km) = 1×10^{3} m
Deci-	d	1/10, or 10^{-1}	1 decimeter (dm) = 0.1 m
Centi-	c	1/100, or 10^{-2}	1 centimeter (cm) = 0.01 m
Milli-	m	1/1,000, or 10^{-3}	1 millimeter (mm) = 0.001 m
Micro-	μ	1/1,000,000, or 10^{-6}	1 micrometer (μm) = 1×10^{-6} m
Nano-	n	1/1,000,000,000, or 10^{-9}	1 nanometer (nm) = 1×10^{-9} m
Pico-	p	1/1,000,000,000,000, or 10^{-12}	1 picometer (pm) = 1×10^{-12} m

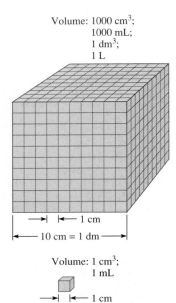

Volume: 1000 cm^3;
1000 mL;
1 dm^3;
1 L

|→| |←— 1 cm

|←——— 10 cm = 1 dm ———→|

Volume: 1 cm^3;
1 mL

|→| |←— 1 cm

Figure 1.7 *Comparison of two volumes, 1 mL and 1000 mL.*

were able to jump about rather freely on the moon's surface despite their bulky suits and equipment. The mass of an object can be determined readily with a balance, and this process, oddly, is called weighing.

The SI base unit of mass is the *kilogram* (kg), but in chemistry the smaller gram (g) is more convenient:

$$1 \text{ kg} = 1000 \text{ g} = 1 \times 10^3 \text{ g}$$

Volume

Volume is *length (m) cubed,* so its SI-derived unit is the cubic meter (m^3). Generally, however, chemists work with much smaller volumes, such as the cubic centimeter (cm^3) and the cubic decimeter (dm^3):

$$1 \text{ cm}^3 = (1 \times 10^{-2} \text{ m})^3 = 1 \times 10^{-6} \text{ m}^3$$
$$1 \text{ dm}^3 = (1 \times 10^{-1} \text{ m})^3 = 1 \times 10^{-3} \text{ m}^3$$

Another common, non-SI unit of volume is the liter (L). A **liter** is *the volume occupied by one cubic decimeter.* Chemists generally use L and mL for liquid volume. One liter is equal to 1000 milliliters (mL) or 1000 cubic centimeters:

$$1 \text{ L} = 1000 \text{ mL}$$
$$= 1000 \text{ cm}^3$$
$$= 1 \text{ dm}^3$$

and one milliliter is equal to one cubic centimeter:

$$1 \text{ mL} = 1 \text{ cm}^3$$

Figure 1.7 compares the relative sizes of two volumes.

Density

Density is *the mass of an object divided by its volume:*

$$\text{density} = \frac{\text{mass}}{\text{volume}}$$

or

$$d = \frac{m}{V} \tag{1.1}$$

where d, m, and V denote density, mass, and volume, respectively. Note that density is an intensive property that does not depend on the quantity of mass present. The reason is that V increases as m does, so the ratio of the two quantities always remains the same for a given material.

The SI-derived unit for density is the kilogram per cubic meter (kg/m^3). This unit is awkwardly large for most chemical applications. Therefore, grams per cubic centimeter (g/cm^3) and its equivalent, grams per milliliter (g/mL), are more commonly used for solid and liquid densities. Because gas densities are often very low, we express them in units of grams per liter (g/L):

$$1 \text{ g/cm}^3 = 1 \text{ g/mL} = 1000 \text{ kg/m}^3$$
$$1 \text{ g/L} = 0.001 \text{ g/mL}$$

Example 1.1 Gold is a precious metal that is chemically unreactive. It is used mainly in jewelry, dentistry, and electronic devices. A piece of gold ingot with a mass of 301 g has a volume of 15.6 cm^3. Calculate the density of gold.

Reasoning and Solution We are given the mass and volume and asked to calculate the density. Therefore, from Equation (1.1), we write

$$d = \frac{m}{V}$$

$$= \frac{301 \text{ g}}{15.6 \text{ cm}^3}$$

$$= 19.3 \text{ g/cm}^3$$

Practice Exercise* A piece of platinum metal with a density of 21.5 g/cm^3 has a volume of 4.49 cm^3. What is its mass?

Gold bars.

Similar problems: 1.17, 1.18.

Temperature

Three temperature scales are currently in use. Their units are K (kelvin), °C (degree Celsius), and °F (degree Fahrenheit). The Fahrenheit scale, which is the most commonly used scale in the United States outside the laboratory, defines the normal freezing and boiling points of water to be exactly 32°F and 212°F, respectively. The Celsius scale divides the range between the freezing point (0°C) and boiling point (100°C) of water into 100 degrees (Figure 1.8). The degree Celsius is not an SI unit, but it may be used with SI units, and we will use it often.

The size of a degree on the Fahrenheit scale is only 100/180, or 5/9, times the degree Celsius. To convert degrees Fahrenheit to degrees Celsius, we write

$$?°C = (°F - 32°F) \times \frac{5°C}{9°F} \qquad (1.2)$$

To convert degrees Celsius to degrees Fahrenheit, we write

$$?°F = \frac{9°F}{5°C} \times (°C) + 32°F \qquad (1.3)$$

The SI unit of temperature is K (kelvin)—never "degrees kelvin." The Kelvin temperature scale is discussed in Chapter 5.

100°	← Boiling point → of water	212°
37°	← Body temperature →	98.6°
25°	← Room temperature →	77°
0°	← Freezing point → of water	32°

Celsius   Fahrenheit

Figure 1.8 *Comparison of the Celsius and Fahrenheit temperature scales.*

*Answers to Practice Exercises appear at the end of the chapter.

Solder (wire on the left) is used extensively in the construction of electronic circuits.

Similar problems: 1.19, 1.20.

Example 1.2   (a) Solder is an alloy made of tin and lead that is used in electronic circuits. A certain solder has a melting point of 224°C. What is its melting point in degrees Fahrenheit? (b) Helium has the lowest boiling point of all the elements at −452°F. Convert this temperature to degrees Celsius.

Reasoning and Solution   These two parts require that we carry out temperature conversions, so we need Equations (1.2) and (1.3).

(a) This conversion is carried out by writing

$$\frac{9°F}{5°C} \times (224°C) + 32°F = 435°F$$

(b) Here we have

$$(-452°F - 32°F) \times \frac{5°C}{9°F} = -269°C$$

Practice Exercise   Convert (a) 327.5°C (the melting point of lead) to degrees Fahrenheit and (b) 172.9°F (the boiling point of ethanol) to degrees Celsius.

1.6 Handling Numbers

Having surveyed some of the units used in chemistry, we now turn to conventions for handling numbers associated with measurements: scientific notation and significant figures.

Scientific Notation

In chemistry, we often deal with numbers that are either extremely large or extremely small. For example, in 1 g of the element hydrogen there are roughly

$$602,200,000,000,000,000,000,000$$

hydrogen atoms. Each hydrogen atom has a mass of only

$$0.00000000000000000000000166 \text{ g}$$

These numbers are cumbersome to handle, and it is easy to make mistakes when using them in arithmetic computations. Consider this multiplication problem:

$$0.0000000056 \times 0.00000000048 = 0.000000000000000002688$$

It would be easy to miss one zero or add one more zero after the decimal point. To handle very large and very small numbers, we use a system called *scientific notation*. Regardless of their magnitude, all numbers can be expressed in the form

$$N \times 10^n$$

where N is a number between 1 and 10 and n is an *exponent* that can be a positive or negative *integer* (whole number). Any number expressed in this way is said to be written in scientific notation.

Suppose that we are given a certain number and asked to express it in scientific notation. Basically, this amounts to finding n. We count the number of places that the decimal point must be moved to give the number N (which is between 1 and 10). If the decimal point has to be moved to the left, then n is a positive integer; if it has to be moved to the right, n is a negative integer. These examples illustrate the use of scientific notation:

(a) Express 568.762 in scientific notation:

$$568.762 = 5.68762 \times 10^2$$

Note that the decimal point is moved to the left by two places and $n = 2$.

(b) Express 0.00000772 in scientific notation:

$$0.00000772 = 7.72 \times 10^{-6}$$

Note that the decimal point is moved to the right by six places and $n = -6$.

Keep in mind these two points. First, $n = 0$ is used for numbers that are not expressed in scientific notation. For example, 74.6×10^0 $(n = 0)$ is equivalent to 74.6. Second, the usual practice is to omit the superscript when $n = 1$. Thus the scientific notation for 74.6 is 7.46×10 and not $7.46 = 10^1$.

Next, we consider how scientific notation is handled in arithmetic operations.

> Any number raised to the power zero is equal to one.

Addition and Subtraction

To add or subtract using scientific notation, we first write each quantity—say N_1 and N_2—with the same exponent n. Then we combine N_1 and N_2; the exponents remain the same. Consider these examples:

$$(7.4 \times 10^3) + (2.1 \times 10^3) = 9.5 \times 10^3$$
$$(4.31 \times 10^4) + (3.9 \times 10^3) = (4.31 \times 10^4) + (0.39 \times 10^4)$$
$$= 4.70 \times 10^4$$
$$(2.22 \times 10^{-2}) - (4.10 \times 10^{-3}) = (2.22 \times 10^{-2}) - (0.41 \times 10^{-2})$$
$$= 1.81 \times 10^{-2}$$

Multiplication and Division

To multiply numbers expressed in scientific notation, we multiply N_1 and N_2 in the usual way, but *add* the exponents together. To divide using scientific notation, we divide N_1 and N_2 as usual and subtract the exponents. These examples show how these operations are performed:

$$(8.0 \times 10^4) \times (5.0 \times 10^2) = (8.0 \times 5.0)(10^{4+2})$$
$$= 40 \times 10^6$$
$$= 4.0 \times 10^7$$
$$(4.0 \times 10^{-5}) \times (7.0 \times 10^3) = (4.0 \times 7.0)(10^{-5+3})$$
$$= 28 \times 10^{-2}$$
$$= 2.8 \times 10^{-1}$$
$$\frac{6.9 \times 10^7}{3.0 \times 10^{-5}} = \frac{6.9}{3.0} \times 10^{7-(-5)}$$
$$= 2.3 \times 10^{12}$$

$$\frac{8.5 \times 10^4}{5.0 \times 10^9} = \frac{8.5}{5.0} \times 10^{4-9}$$
$$= 1.7 \times 10^{-5}$$

Significant Figures

Except when all the numbers involved are integers (for example, in counting the number of students in a class), it is often impossible to obtain the exact value of the quantity under investigation. For this reason, it is important to indicate the margin of error in a measurement by clearly indicating the number of *significant figures,* which are *the meaningful digits in a measured or calculated quantity.* When significant figures are used, the last digit is understood to be uncertain. For example, we might measure the volume of a given amount of liquid using a graduated cylinder with a scale that gives an uncertainty of 1 mL in the measurement. If the volume is found to be 6 mL, then the actual volume is in the range of 5 mL to 7 mL. We represent the volume of the liquid as (6 ± 1) mL. In this case, there is only one significant figure (the digit 6) that is uncertain by either plus or minus 1 mL. For greater accuracy, we might use a graduated cylinder that has finer divisions, so that the volume we measure is now uncertain by only 0.1 mL. If the volume of the liquid is now found to be 6.0 mL, we may express the quantity as (6.0 ± 0.1) mL, and the actual value is somewhere between 5.9 mL and 6.1 mL. We can further improve the measuring device and obtain more significant figures, but in every case, the last digit is always uncertain; the amount of this uncertainty depends on the particular measuring device we use.

Figure 1.9 shows a modern balance. Balances such as this one are available in many general chemistry laboratories; they readily measure the mass of objects to four decimal places. Therefore the measured mass typically will have four significant figures (for example, 0.8642 g) or more (for example, 3.9745 g). Keeping track of the number of significant figures in a measurement such as mass ensures that calculations involving the data will reflect the precision of the measurement.

Figure 1.9 *A single-pan balance.*

Guidelines for Using Significant Figures

We must always be careful in scientific work to write the proper number of significant figures. In general, it is fairly easy to determine how many significant figures a number has by following these rules:

- Any digit that is not zero is significant. Thus 845 cm has three significant figures, 1.234 kg has four significant figures, and so on.
- Zeros between nonzero digits are significant. Thus 606 m contains three significant figures, 40,501 kg contains five significant figures, and so on.
- Zeros to the left of the first nonzero digit are not significant. Their purpose is to indicate the placement of the decimal point. For example, 0.08 L contains one significant figure, 0.0000349 g contains three significant figures, and so on.
- If a number is greater than 1, then all the zeros written to the right of the decimal point count as significant figures. Thus 2.0 mg has two significant figures, 40.062 mL has five significant figures, and 3.040 dm has four significant figures. If a number is less than 1, then only the zeros that are at the end of the number and the zeros that are between nonzero digits are significant. This means that 0.090 kg has two significant figures, 0.3005 L has four significant figures, 0.00420 min has three significant figures, and so on.
- For numbers that do not contain decimal points, the trailing zeros (that is, zeros after the last nonzero digit) may or may not be significant. Thus 400 cm may have one significant figure (the digit 4), two significant figures (40), or three significant figures (400). We cannot know which is correct without more information. By using scientific notation, however, we avoid this ambiguity. In this particular case, we can express the number 400 as 4×10^2 for one significant figure, 4.0×10^2 for two significant figures, or 4.00×10^2 for three significant figures.

Example 1.3 Determine the number of significant figures in these measurements: (a) 478 cm, (b) 6.01 g, (c) 0.825 m, (d) 0.043 kg, (e) 1.310×10^{22} atoms, (f) 7000 mL.

Reasoning and Solution We follow the rules for determining significant figures.
(a) Three, (b) Three, (c) Three, (d) Two, (e) Four, (f) This is an ambiguous case. The number of significant figures may be four (7.000×10^3), three (7.00×10^3), two (7.0×10^3), or one (7×10^3).

Practice Exercise Determine the number of significant figures in each of the following measurements: (a) 24 mL, (b) 3001 g, (c) 0.0320 m^3, (d) 6.4×10^4 molecules, (e) 560 kg.

Similar problems: 1.27, 1.28.

A second set of rules specifies how to handle significant figures in calculations:

- In addition and subtraction, the answer cannot have more digits to the right of the decimal point than either of the original numbers. Consider these examples:

$$
\begin{array}{r}
89.332 \\
+\ 1.1 \\
\hline
90.432 \\
\end{array}
\quad
\begin{array}{l}
\longleftarrow \text{ one digit after the decimal point} \\
\\
\longleftarrow \text{ round off to 90.4}
\end{array}
$$

$$
\begin{array}{r}
2.097 \\
-0.12 \\
\hline
1.977 \\
\end{array}
\quad
\begin{array}{l}
\longleftarrow \text{ two digits after the decimal point} \\
\\
\longleftarrow \text{ round off to 1.98}
\end{array}
$$

The rounding-off procedure is as follows. To round off a number at a certain point we simply drop the digits that follow if the first of them is less than 5. Thus 8.724 rounds off to 8.72 if we want only two digits after the decimal point. If the first digit following the point of rounding off is equal to or greater than 5, we add 1 to the preceding digit. Thus 8.727 rounds off to 8.73, and 0.425 rounds off to 0.43.

- In multiplication and division, the number of significant figures in the final product or quotient is determined by the original number that has the *smallest* number of significant figures. These examples illustrate this rule:

$$2.8 \times 4.5039 = 12.61092 \longleftarrow \text{ round off to } 13$$

$$\frac{6.85}{112.04} = 0.0611388789 \longleftarrow \text{ round off to } 0.0611$$

- Keep in mind that *exact numbers* obtained from definitions or by counting numbers of objects can be considered to have an infinite number of significant figures. If an object has a mass of 0.2786 g, then the mass of eight such objects is

$$0.2786 \text{ g} \times 8 = 2.229 \text{ g}$$

We do *not* round off this product to one significant figure, because the number 8 is 8.00000 . . . , by definition. Similarly, to take the average of the two measured lengths 6.64 cm and 6.68 cm, we write

$$\frac{6.64 \text{ cm} + 6.68 \text{ cm}}{2} = 6.66 \text{ cm}$$

because the number 2 is 2.00000 . . . , by definition.

Example 1.4   Carry out these arithmetic operations to the correct number of significant figures: (a) 11,254.1 g + 0.1983 g, (b) 66.59 L − 3.113 L, (c) 8.16 m × 5.1355, (d) 0.0154 kg ÷ 88.3 mL, (e) 2.64×10^3 cm + 3.27×10^2 cm.

Reasoning and Solution   In addition and subtraction, the number of decimal places in the answer is determined by the number having the lowest number of decimal places. In multiplication and division, the significant figure of the answer is determined by the number having the smallest number of significant figures.

(a)   11,254.1 g
       +   0.1983 g
       ‾‾‾‾‾‾‾‾‾‾‾‾‾‾‾‾
       11,254.2983 g $\longleftarrow$ round off to 11,254.3 g

(b)   66.59 L
       − 3.113 L
       ‾‾‾‾‾‾‾‾‾‾‾
       63.477 L $\longleftarrow$ round off to 63.48 L

(c) 8.16 m × 5.1355 = 41.90568 m $\longleftarrow$ round off to 41.9 m

(d) $\dfrac{0.0154 \text{ kg}}{88.3 \text{ mL}}$ = 0.000174405436 kg/mL $\longleftarrow$ round off to 0.000174 kg/mL
                                                        or 1.74×10^{-4} kg/mL

(e) First we change 3.27×10^2 cm to 0.327×10^3 cm and then carry out the addition (2.64 cm + 0.327 cm) × 10^3. Following the procedure in (a), we find the answer is 2.97×10^3 cm.

Similar problems: 1.29, 1.30.

Practice Exercise Carry out these arithmetic operations and round off the answers to the appropriate number of significant figures: (a) 26.5862 L + 0.17 L, (b) 9.1 g − 4.682 g, (c) 7.1×10^4 dm $\times 2.2654 \times 10^2$, (d) 6.54 g ÷ 86.5542 mL, (e) $(7.55 \times 10^4$ m$) - (8.62 \times 10^3$ m$)$.

This rounding-off procedure applies to one-step calculations. In *chain calculations,* that is, calculations involving more than one step, we use a modified procedure. Consider this two-step calculation:

$$\text{First step:} \quad A \times B = C$$

$$\text{Second step:} \quad C \times D = E$$

Let us suppose that A = 3.66, B = 8.45, and D = 2.11. Depending on whether we round off C to three or four significant figures, we obtain a different number for E:

Method 1	Method 2
$3.66 \times 8.45 = 30.9$	$3.66 \times 8.45 = 30.93$
$30.9 \times 2.11 = 65.2$	$30.93 \times 2.11 = 65.3$

However, if we had carried out the calculation as $3.66 \times 8.45 \times 2.11$ on a calculator without rounding off the intermediate result, we would have obtained 65.3 as the answer for E. In general, we will show the correct number of significant figures in each step of the calculation. However, in some worked examples, only the final answer is rounded off to the correct number of significant figures. The answers for all intermediate calculations will be carried to one extra figure.

Accuracy and Precision

In discussing measurements and significant figures it is useful to distinguish between accuracy and precision. *Accuracy tells us how close a measurement is to the true value of the quantity that was measured.* **Precision** *refers to how closely two or more measurements of the same quantity agree with one another* (Figure 1.10). Suppose that three students are asked to determine the mass of a piece of copper wire that their instructor knows has a mass of 2.000 g. The results of two successive weighings by each student are

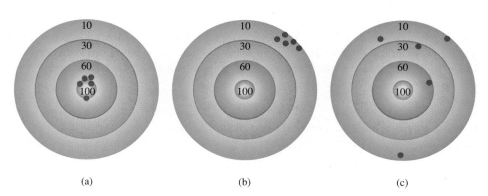

(a)   (b)   (c)

Figure 1.10 *The distribution of darts on a dart board shows the difference between precise and accurate. (a) Good accuracy and good precision. (b) Poor accuracy and good precision. (c) Poor accuracy and poor precision.*

	Student A	Student B	Student C
	1.964 g	1.972 g	2.000 g
	1.978 g	1.968 g	2.002 g
Average Value	1.971 g	1.970 g	2.001 g

Student B's results are more *precise* than those of Student A (1.972 g and 1.968 g deviate less from 1.970 g than 1.964 g and 1.978 g from 1.971 g). Neither set of results is very *accurate,* however. Student C's results are not only *precise* but also the most *accurate,* because the average value is closest to the true value. Highly accurate measurements are usually precise too. On the other hand, highly precise measurements do not necessarily guarantee accurate results. For example, an improperly calibrated meterstick or a faulty balance may give precise readings that are in error.

1.7 The Factor-Label Method of Solving Problems

Careful measurements and the proper use of significant figures, along with correct calculations, will yield accurate numerical results. But to be meaningful, the answers also must be expressed in the desired units. The procedure we will use to convert between units in solving chemistry problems is called the *factor-label method* (also called *dimensional analysis*). A simple technique requiring little memorization, the factor-label method is based on the relationship between different units that express the same physical quantity.

We know, for example, that the unit "dollar" for money is different from the unit "penny." However, we say that 1 dollar is *equivalent* to 100 pennies because they both represent the same amount of money. This equivalence allows us to write

$$1 \text{ dollar} = 100 \text{ pennies}$$

Because 1 dollar is equal to 100 pennies, it follows that their ratio has a value of 1; that is,

$$\frac{1 \text{ dollar}}{100 \text{ pennies}} = 1$$

This ratio can be read as 1 dollar per 100 pennies. This fraction is called a *unit factor* (equal to 1) because the numerator and denominator describe the same amount of money. We could also have written the ratio as

$$\frac{100 \text{ pennies}}{1 \text{ dollar}} = 1$$

This ratio reads as 100 pennies per dollar. The fraction 100 pennies/1 dollar is also a unit factor. We see that the reciprocal of any unit factor is also a unit factor. The usefulness of unit factors is that they allow us to carry out conversions between different units that measure the same quantity. Suppose that we want to convert 2.46 dollars into pennies. This problem may be expressed as

$$? \text{ pennies} = 2.46 \text{ dollars}$$

Because this is a dollar-to-penny conversion, we choose the unit factor that has the unit "dollar" in the denominator (to cancel the "dollars" in 2.46 dollars) and write

$$2.46 \text{ dollars} \times \frac{100 \text{ pennies}}{1 \text{ dollar}} = 246 \text{ pennies}$$

Note that the unit factor 100 pennies/1 dollar contains exact numbers, so it does not affect the number of significant figures in the final answer.

Next let us consider the conversion of 57.8 meters to centimeters. This problem may be expressed as

$$? \text{ cm} = 57.8 \text{ m}$$

By definition,

$$1 \text{ cm} = 1 \times 10^{-2} \text{ m}$$

Because we are converting "m" to "cm," we choose the unit factor that has meters in the denominator,

$$\frac{1 \text{ cm}}{1 \times 10^{-2} \text{ m}} = 1$$

and write the conversion as

$$? \text{ cm} = 57.8 \text{ m} \times \frac{1 \text{ cm}}{1 \times 10^{-2} \text{ m}}$$

$$= 5780 \text{ cm}$$

$$= 5.78 \times 10^{3} \text{ cm}$$

Note that scientific notation is used to indicate that the answer has three significant figures. The unit factor 1 cm/1×10^{-2} m contains exact numbers; therefore, it does not affect the number of significant figures.

In the factor-label method the units are carried through the entire sequence of calculations. Therefore, if the equation is set up correctly, then all the units will cancel except the desired one. If this is not the case, then an error must have been made somewhere, and it can usually be spotted by reviewing the solution.

Conversion factors for some of the English system units commonly used in the United States for nonscientific measurements (for example, pounds and inches) are provided inside the back cover of this book.

Example 1.5 A person's average daily intake of sucrose (a form of sugar) is 0.0833 pound (lb). What is this mass in milligrams (mg)? (1 lb = 453.6 g).

Reasoning and Solution This problem requires two conversion steps—from pounds to grams and from grams to milligrams. First, we write the unit factor for each conversion and then calculate the answer in one step:

$$? \text{ mg} = 0.0833 \text{ lb}$$

So the unit factor is

$$\frac{453.6 \text{ g}}{1 \text{ lb}} = 1$$

and

$$1 \text{ mg} = 1 \times 10^{-3} \text{ g}$$

so we must also include the unit factor

$$\frac{1 \text{ mg}}{1 \times 10^{-3} \text{ g}} = 1$$

Thus

$$? \text{ mg } = 0.0833 \text{ lb} \times \frac{453.6 \text{ g}}{1 \text{ lb}} \times \frac{1 \text{ mg}}{1 \times 10^{-3} \text{ g}} = 3.78 \times 10^4 \text{ mg}$$

Comment   Because a pound is a much larger unit for mass than milligrams, the answer is reasonable.

Similar problem: 1.37.

Practice Exercise   A roll of aluminum foil has a mass of 1.07 kg. What is its mass in pounds?

Note that unit factors may be squared or cubed, because $1^2 = 1^3 = 1$. The use of such factors is illustrated in Example 1.6.

A silver coin.

Example 1.6   The density of silver is 10.5 g/cm^3. Convert the density to units of kg/m^3.

Reasoning and Solution   This problem requires us to carry out two separate conversions involving mass and volume. We write

$$? \text{ kg/m}^3 = 10.5 \text{ g/cm}^3$$

We need two unit factors—one to convert g to kg and the other to convert cm^3 to m^3. We know that

$$1 \text{ kg } = 1000 \text{ g}$$

so

$$\frac{1 \text{ kg}}{1000 \text{ g}} = 1$$

Second, because $1 \text{ cm} = 1 \times 10^{-2}$ m, these unit factors can be generated:

$$\frac{1 \text{ cm}}{1 \times 10^{-2} \text{ m}} = 1 \quad \text{and} \quad \left(\frac{1 \text{ cm}}{1 \times 10^{-2} \text{ m}}\right)^3 = 1$$

Finally we can calculate the density in the desired units as follows:

$$? \text{ kg/m}^3 = \frac{10.5 \text{ g}}{1 \text{ cm}^3} \times \frac{1 \text{ kg}}{1000 \text{ g}} \times \left(\frac{1 \text{ cm}}{1 \times 10^{-2} \text{ m}}\right)^3 = 10{,}500 \text{ kg/m}^3$$

$$= 1.05 \times 10^4 \text{ kg/m}^3$$

Comment   Because $1 \text{ m}^3 = 1000 \text{ cm}^3$, we would expect much more mass in 1 m^3 than in 1 cm^3. Therefore, a mass of 1.05×10^4 kg compared to 10.5 g is reasonable.

Similar problems: 1.39, 1.40.

Practice Exercise   The density of the lightest metal, lithium (Li), is 5.34×10^2 kg/m^3. Convert the density to g/cm^3.

Summary

The scientific method is a systematic approach to research that begins with the gathering of information through observation and measurements. In the process, hypotheses, laws, and theories are devised and tested.

Chemists study matter and the substances of which it is composed. All substances, in principle, can exist in three states: solid, liquid, and gas. The interconversion between these states can be effected by a change in temperature.

The simplest substances in chemistry are elements. Compounds are formed by the combination of atoms of different elements. Substances have both unique physical properties that can be observed without changing the identity of the substances and unique chemical properties that, when they are demonstrated, do change the identity of the substances.

SI units are used to express physical quantities in all sciences, including chemistry. Numbers expressed in scientific notation have the form $N \times 10^n$, where N is between 1 and 10 and n is a positive or negative integer. Scientific notation helps us handle very large and very small quantities. Most measured quantities are inexact to some extent. The number of significant figures indicates the exactness of the measurement.

In the factor-label method of solving problems the units are multiplied together, divided into each other, or canceled like algebraic quantities. Obtaining the correct units for the final answer ensures that the calculation has been carried out properly.

Key Words

Accuracy p. 19
Chemical property p. 9
Chemistry p. 6
Compound p. 8
Density p. 12
Element p. 7
Extensive property p. 9
Heterogeneous mixture p. 7

Homogeneous mixture p. 7
Hypothesis p. 5
Intensive property p. 9
International System of Units p. 11
Law p. 5
Liter p. 12

Macroscopic property p. 10
Mass p. 11
Matter p. 6
Microscopic property p. 10
Mixture p. 7
Physical property p. 9
Precision p. 19

Qualitative p. 5
Quantitative p. 5
Scientific method p. 4
Significant figures p. 16
Substance p. 6
Theory p. 5
Volume p. 12
Weight p. 11

Questions and Problems

Basic Definitions

Review Questions

1.1   Define these terms: (a) matter, (b) mass, (c) weight, (d) substance, (e) mixture.

1.2   Which of these statements is scientifically correct?
"The mass of the student is 56 kg."
"The weight of the student is 56 kg."

1.3   Give an example of a homogeneous mixture and an example of a heterogeneous mixture.

1.4   What is the difference between a physical property and a chemical property?

1.5   Give an example of an intensive property and an example of an extensive property.

1.6   Define these terms: (a) element, (b) compound.

Problems

1.7   Do these statements describe chemical or physical properties? (a) Oxygen gas supports combustion. (b) Fertilizers help to increase agricultural production. (c) Water boils below 100°C on top of a mountain. (d) Lead is denser than aluminum. (e) Uranium is a radioactive element.

1.8   Does each of these describe a physical change or a chemical change? (a) The helium gas inside a balloon tends to leak out after a few hours. (b) A flashlight beam slowly gets dimmer and finally goes out. (c) Frozen orange juice is reconstituted by adding water to it. (d) The growth of plants depends on the sun's energy in a process called photosynthesis. (e) A spoonful of table salt dissolves in a bowl of soup.

1.9 Which of these properties are intensive and which are extensive? (a) length, (b) volume, (c) temperature, (d) mass.

1.10 Which of these properties are intensive and which are extensive? (a) area, (b) color, (c) density.

1.11 Classify each of these substances as an element or a compound: (a) hydrogen, (b) water, (c) gold, (d) sugar.

1.12 Classify each of these as an element or a compound: (a) sodium chloride (table salt), (b) helium, (c) alcohol, (d) platinum.

Units

Review Questions

1.13 Give the SI units for expressing these: (a) length, (b) area, (c) volume, (d) mass, (e) time, (f) force, (g) energy, (h) temperature.

1.14 Write the numbers for these prefixes: (a) mega-, (b) kilo-, (c) deci-, (d) centi-, (e) milli-, (f) micro-, (g) nano-, (h) pico-.

1.15 Define density. What units do chemists normally use for density? Is density an intensive or extensive property?

1.16 Write the equations for converting degrees Celsius to degrees Fahrenheit and degrees Fahrenheit to degrees Celsius.

Problems

1.17 A lead sphere has a mass of 1.20×10^4 g, and its volume is 1.05×10^3 cm^3. Calculate the density of lead.

1.18 Mercury is the only metal that is a liquid at room temperature. Its density is 13.6 g/mL. How many grams of mercury will occupy a volume of 95.8 mL?

1.19 Calculate the temperature in degrees Celsius of: (a) a hot summer day of 95°F, (b) a cold winter day of 12°F, (c) a fever of 102°F, (d) a furnace operating at 1852°F.

1.20 (a) Normally the human body can endure a temperature of 105°F for only short periods without permanent damage to the brain and other vital organs. What is the temperature in degrees Celsius? (b) Ethylene glycol is a liquid organic compound that is used as an antifreeze in car radiators. It freezes at −11.5°C. Calculate its freezing temperature in degrees Fahrenheit. (c) The temperature on the surface of our sun is about 6.3×10^{3}°C. What is this temperature in degrees Fahrenheit?

Scientific Notation

Problems

1.21 Express these numbers in scientific notation: (a) 0.000000027, (b) 356, (c) 0.096.

1.22 Express these numbers in scientific notation: (a) 0.749, (b) 802.6, (c) 0.000000621.

1.23 Convert these to nonscientific notation: (a) 1.52×10^4, (b) 7.78×10^{-8}.

1.24 Convert these to nonscientific notation: (a) 3.256×10^{-5}, (b) 6.03×10^6.

1.25 Express the answers to these in scientific notation:

(a) $145.75 + (2.3 \times 10^{-1})$
(b) $79,500 \div (2.5 \times 10^2)$
(c) $(7.0 \times 10^{-3}) - (8.0 \times 10^{-4})$
(d) $(1.0 \times 10^4) \times (9.9 \times 10^6)$

1.26 Express the answers to these in scientific notation:

(a) $0.0095 + (8.5 \times 10^{-3})$
(b) $653 \div (5.75 \times 10^{-8})$
(c) $850,000 - (9.0 \times 10^5)$
(d) $(3.6 \times 10^{-4}) \times (3.6 \times 10^6)$

Significant Figures

Problems

1.27 What is the number of significant figures in each of these measured quantities? (a) 4867 miles, (b) 56 mL, (c) 60,104 tons, (d) 2900 g.

1.28 What is the number of significant figures in each of these measured quantities? (a) 40.2 g/cm^3, (b) 0.0000003 cm, (c) 70 min, (d) 4.6×10^{19} atoms.

1.29 Carry out these operations as if they were calculations of experimental results, and express each answer in the correct units and with the correct number of significant figures:

(a) 5.6792 m + 0.6 m + 4.33 m
(b) 3.70 g − 2.9133 g
(c) 4.51 cm $\times$ 3.6666 cm

1.30 Carry out these operations as if they were calculations of experimental results, and express each answer in the correct units and with the correct number of significant figures:

(a) 7.310 km $\div$ 5.70 km
(b) $(3.26 \times 10^{-3}$ mg$) - (7.88 \times 10^{-5}$ mg$)$
(c) $(4.02 \times 10^6$ dm$) + (7.74 \times 10^7$ dm$)$

The Factor-Label Method

Problems

1.31 Carry out these conversions: (a) 22.6 m to decimeters, (b) 25.4 mg to kilograms.

1.32 Carry out these conversions: (a) 242 lb to milligrams, (b) 68.3 cm^3 to cubic meters.

1.33 The price of gold on a certain day in 2001 was $282 per troy ounce. How much did 1.00 g of gold cost that day? (1 troy ounce = 31.03 g.)

1.34 How many seconds are in a solar year (365.24 days)?

1.35 How many minutes does it take light from the sun to reach Earth? (The distance from the sun to Earth is 93 million mi; the speed of light = 3.00×10^8 m/s.)

1.36 A slow jogger runs a mile in 13 min. Calculate the speed in (a) in/s, (b) m/min, (c) km/h. (1 mi = 1609 m; 1 in = 2.54 cm.)

1.37 Carry out these conversions: (a) A 6.0-ft person weighs 168 lb. Express this person's height in meters and weight in kilograms. (1 lb = 453.6 g; 1 m = 3.28 ft.) (b) The current speed limit in some states in the United States is 55 miles per hour. What is the speed limit in kilometers per hour? (c) The speed of light is 3.0×10^{10} cm/s. How many miles does light travel in 1 hour? (d) Lead is a toxic substance. The "normal" lead content in human blood is about 0.40 part per million (that is, 0.40 g of lead per million grams of blood). A value of 0.80 part per million (ppm) is considered to be dangerous. How many grams of lead are contained in 6.0×10^3 g of blood (the amount in an average adult) if the lead content is 0.62 ppm?

1.38 Carry out these conversions: (a) 1.42 light-years to miles (a light-year is an astronomical measure of distance—the distance traveled by light in a year, or 365 days), (b) 32.4 yd to centimeters, (c) 3.0×10^{10} cm/s to ft/s, (d) 47.4°F to degrees Celsius, (e) −273.15°C (the lowest temperature) to degrees Fahrenheit, (f) 71.2 cm^3 to m^3, (g) 7.2 m^3 to liters.

1.39 Aluminum is a lightweight metal (density = 2.70 g/cm^3) used in aircraft construction, high-voltage transmission lines, and foils. What is its density in kg/m^3?

1.40 The density of ammonia gas under certain conditions is 0.625 g/L. Calculate its density in g/cm^3.

Additional Problems

1.41 Which of these describe physical and which describe chemical properties? (a) Iron has a tendency to rust. (b) Rainwater in industrialized regions tends to be acidic. (c) Hemoglobin molecules have a red color. (d) When a glass of water is left out in the sun, the water gradually disappears. (e) Carbon dioxide in air is converted to more complex molecules by plants during photosynthesis.

1.42 In 2000 86.6 billion pounds of sulfuric acid were produced in the United States. Convert this quantity to tons.

1.43 Suppose that a new temperature scale has been devised on which the melting point of ethanol (−117.3°C) and the boiling point of ethanol (78.3°C) are taken as 0°S and 100°S, respectively, where S is the symbol for the new temperature scale. Derive an equation relating a reading on this scale to a reading on the Celsius scale. What would this thermometer read at 25°C?

1.44 In the determination of the density of a rectangular metal bar, a student made the following measurements: length, 8.53 cm; width, 2.4 cm; height, 1.0 cm; mass, 52.7064 g. Calculate the density of the metal to the correct number of significant figures.

1.45 Calculate the mass of each of these: (a) a sphere of gold of radius 10.0 cm [the volume of a sphere of radius r is $V = (\frac{4}{3})\pi r^3$; the density of gold = 19.3 g/cm^3], (b) a cube of platinum of edge length 0.040 mm (the density of platinum = 21.4 g/cm^3), (c) 50.0 mL of ethanol (the density of ethanol = 0.798 g/mL).

1.46 A cylindrical glass tube 12.7 cm in length is filled with mercury. The mass of mercury needed to fill the tube is found to be 105.5 g. Calculate the inner diameter of the tube. (The density of mercury = 13.6 g/mL.)

1.47 This procedure was carried out to determine the volume of a flask. The flask was weighed dry and then filled with water. If the masses of the empty flask and filled flask were 56.12 g and 87.39 g, respectively, and the density of water is 0.9976 g/cm^3, calculate the volume of the flask in cubic centimeters.

1.48 A silver (Ag) object weighing 194.3 g is placed in a graduated cylinder containing 242.0 mL of water. The volume of water now reads 260.5 mL. From these data calculate the density of silver.

1.49 The experiment described in Problem 1.48 is a crude but convenient way to determine the density of some solids. Describe a similar experiment that would allow you to measure the density of ice. Specifically, what would be the requirements for the liquid used in your experiment?

1.50 The speed of sound in air at room temperature is about 343 m/s. Calculate this speed in miles per hour (mph).

1.51 The medicinal thermometer commonly used in homes can be read to ±0.1°F, whereas those in the doctor's office may be accurate to ±0.1°C. In degrees Celsius, express the percent error expected from each of these thermometers in measuring a person's body temperature of 38.9°C.

1.52 A thermometer gives a reading of 24.2°C ± 0.1°C. Calculate the temperature in degrees Fahrenheit. What is the uncertainty?

1.53 Vanillin (used to flavor vanilla ice cream and other foods) is the substance whose aroma the human nose detects in the smallest amount. The threshold limit is 2.0×10^{-11} g per liter of air. If the current price of 50 g of vanillin is $112, determine the cost to supply enough vanillin so that the aroma could be detectable in a large aircraft hangar of volume 5.0×10^7 ft^3.

1.54 A resting adult requires about 240 mL of pure oxygen/min and breathes about 12 times every minute. If inhaled air contains 20 percent oxygen by volume and exhaled air 16 percent, what is the volume of air per breath? (Assume that the volume of inhaled air is equal to that of exhaled air.)

1.55 The total volume of seawater is 1.5×10^{21} L. Assume that seawater contains 3.1 percent sodium chloride by mass and that its density is 1.03 g/mL. Calculate the total mass of sodium chloride in kilograms and in tons. (1 ton = 2000 lb; 1 lb = 453.6 g.)

1.56 Magnesium (Mg) is a valuable metal used in alloys, in batteries, and in chemical synthesis. It is obtained mostly from seawater, which contains about 1.3 g of Mg for every kilogram of seawater. Calculate the volume of seawater (in liters) needed to extract 8.0×10^4 tons of Mg, which is roughly the annual production in the United States. (Density of seawater = 1.03 g/mL.)

1.57 A student is given a crucible and asked to prove whether it is made of pure platinum. She first weighs the crucible in air and then weighs it suspended in water (density = 0.9986 g/cm^3). The readings are 860.2 g and 820.2 g, respectively. Given that the density of platinum is 21.45 g/cm^3, what should her conclusion be based on these measurements? (*Hint:* An object suspended in a fluid is buoyed up by the mass of the fluid displaced by the object. Neglect the buoyancy of air.)

1.58 At what temperature does the numerical reading on a Celsius thermometer equal that on a Fahrenheit thermometer?

1.59 The surface area and average depth of the Pacific Ocean are 1.8×10^8 km^2 and 3.9×10^3 m, respectively. Calculate the volume of water in the ocean in liters.

1.60 Percent error is often expressed as the absolute value of the difference between the true value and the experimental value, divided by the true value:

$$\text{Percent error} = \frac{|\text{true value} - \text{experimental value}|}{|\text{true value}|} \times 100\%$$

where the vertical lines indicate absolute value. Calculate the percent error for these measurements: (a) The density of alcohol (ethanol) is found to be 0.802 g/mL. (True value: 0.798 g/mL.) (b) The mass of gold in an earring is analyzed to be 0.837 g. (True value: 0.864 g.)

1.61 Osmium (Os) is the densest element known (density = 22.57 g/cm^3). Calculate the mass in pounds and kilograms of an Os sphere 15 cm in diameter (about the size of a grapefruit). See Problem 1.45 for volume of a sphere.

1.62 A 1.0-mL volume of seawater contains about 4.0×10^{-12} g of gold. The total volume of ocean water is 1.5×10^{21} L. Calculate the total amount of gold in grams that is present in seawater and its worth in dollars, assuming that the price of gold is $350 an ounce. With so much gold out there, why hasn't someone become rich by mining gold from the ocean?

1.63 The thin outer layer of Earth, called the crust, contains only 0.50 percent of Earth's total mass and yet is the source of almost all the elements (the atmosphere provides elements such as oxygen, nitrogen, and a few other gases). Silicon (Si) is the second most abundant element in Earth's crust (27.2 percent by mass). Calculate the mass of silicon in kilograms in Earth's crust. (The mass of Earth is 5.9×10^{21} tons. 1 ton = 2000 lb; 1 lb = 453.6 g.)

1.64 The diameter of a copper (Cu) atom is roughly 1.3×10^{-10} m. How many times can you divide evenly a piece of 10-cm copper wire until it is reduced to two separate copper atoms? (Assume there are appropriate tools for this procedure and that copper atoms are lined up in a straight line, in contact with each other.) Round off your answer to an integer.

1.65 One gallon of gasoline burned in an automobile's engine produces on the average 9.5 kg of carbon dioxide, which is a greenhouse gas, that is, it promotes the warming of Earth's atmosphere. Calculate the annual production of carbon dioxide in kilograms if there are 40 million cars in the United States, and each car covers a distance of 5000 miles at a consumption rate of 20 miles per gallon.

1.66 A sheet of aluminum (Al) foil has a total area of 1.000 ft^2 and a mass of 3.636 g. What is the thickness of the foil in millimeters? (Density of Al = 2.699 g/cm^3.)

1.67 Chlorine is used to disinfect swimming pools. The accepted concentration for this purpose is 1 ppm chlorine or 1 g of chlorine per million g of water. Calculate the volume of a chlorine solution (in milliliters) a homeowner should add to her swimming pool if the solution contains 6.0 percent chlorine by mass and there are 2×10^4 gallons of water in the pool. (1 gallon = 3.79 L; density of liquids = 1.0 g/mL.)

1.68 Fluoridation is the process of adding fluorine compounds to drinking water to help fight tooth decay. A concentration of 1 ppm of fluorine is sufficient for the purpose. (1 ppm means 1 g of fluorine per 1 million g of water.) The compound normally chosen for fluoridation is sodium fluoride, which is also added to some toothpastes. Calculate the quantity of sodium fluoride in kilograms needed per year for a city of 50,000 people if the daily consumption of water per person is 150 gallons. What percent of the sodium fluoride is "wasted" if each person uses only 6.0 L of water a day for drinking and cooking? (Sodium fluoride is 45.0 percent fluorine by mass. 1 gallon = 3.79 L; 1 year = 365 days; 1 ton = 2000 lb; 1 lb = 456.3 g; density of water = 1.0 g/mL.)

1.69 In water conservation, chemists spread a thin film of certain inert material over the surface of water to cut down the rate of evaporation of water in reservoirs. This technique was pioneered by Benjamin Franklin three centuries ago. Franklin found that 0.10 mL of oil could spread over the surface of water of about 40 m^2 in area. Assuming that the oil forms a *monolayer*, that is, a layer that is only one molecule thick, estimate the length of each oil molecule in nanometers. (1 nm = 1×10^{-9} m.)

1.70 Pheromones are compounds secreted by females of many insect species to attract mates. Typically, 1.0×10^{-8} g of a pheromone is sufficient to reach all targeted males within a radius of 0.50 mi. Calculate the density of the pheromone (in grams per liter) in a cylindrical air space having a radius of 0.50 mi and a height of 40 ft.

1.71 A gas company in Massachusetts charges $1.30 for 15.0 ft^3 of natural gas. (a) Convert this rate to dollars per liter of gas. (b) If it takes 0.304 ft^3 of gas to boil a liter of water, starting at room temperature (25°C), how much would it cost to boil a 2.1-L kettle of water?

Special Problems

1.72 Dinosaurs dominated life on Earth for millions of years and then disappeared very suddenly. In the experimentation and data-collecting stage, paleontologists studied fossils and skeletons found in rocks in various layers of Earth's crust. Their findings enabled them to map out which species existed on Earth during specific geologic periods. They also revealed no dinosaur skeletons in rocks formed immediately after the Cretaceous period, which dates back some 65 million years. It is therefore assumed that the dinosaurs became extinct about 65 million years ago.

Among the many hypotheses put forward to account for their disappearance were disruptions of the food chain and a dramatic change in climate caused by violent volcanic eruptions. However, there was no convincing evidence for any one hypothesis until 1977. It was then that a group of paleontologists working in Italy obtained some very puzzling data at a site near Gubbio. The chemical analysis of a layer of clay deposited above sediments formed during the Cretaceous period (and therefore a layer that records events occurring *after* the Cretaceous period) showed a surprisingly high content of the element iridium. Iridium is very rare in Earth's crust but is comparatively abundant in asteroids.

This investigation led to the hypothesis that the extinction of dinosaurs occurred as follows. To account for the quantity of iridium found, scientists suggested that a large asteroid several miles in diameter hit Earth about the time the dinosaurs disappeared. The impact of the asteroid on Earth's surface must have been so tremendous that it literally vaporized a large quantity of surrounding rocks, soils, and other objects. The resulting dust and debris floated through the air and blocked the sunlight for months or perhaps years. Without ample sunlight most plants could not grow, and the fossil record confirms that many types of plants did indeed die out at this time. Consequently, of course, many plant-eating animals gradually perished, and then, in turn, meat-eating animals began to starve. Limitation of food sources obviously affects large animals needing great amounts of food more quickly and more severely than small animals. Therefore the huge dinosaurs vanished because of lack of food.

(a) How does the study of dinosaur extinction illustrate the scientific method?
(b) Suggest two ways to test the hypothesis.
(c) In your opinion, is it justifiable to refer to the asteroid explanation as the theory of dinosaur extinction?

(d) Available evidence suggests that about 20 percent of the asteroid's mass turned to dust and spread uniformly over Earth after eventually settling out of the upper atmosphere. This dust amounted to about 0.02 g/cm^2 of Earth's surface. The asteroid very likely had a density of about 2 g/cm^3. Calculate the mass (in kilograms and tons) of the asteroid and its radius in meters, assuming that it was a sphere. (The area of Earth is 5.1×10^{14} m^2; 1 lb = 453.6 g.) (Source: *Consider a Spherical Cow—A Course in Environmental Problem Solving* by J. Harte, University Science Books, Mill Valley, CA 1988. Used with permission.)

1.73 The Mars Climate Orbiter, intended as the red planet's first weather satellite, was destroyed by heat when it entered Mars' atmosphere in 1999. The loss of the spacecraft was caused by the failure to convert English measurement units into SI units in the navigation software. Engineers who built the spacecraft specified its thrust in pounds, which is an English unit. Scientists at NASA had assumed that the thrust data they received were expressed in newtons (N). Expressed as a unit of force, 1 lb is the force due to gravitational attraction on an object of that mass and 1 N = 1 kg m/s^2. What is the relationship between these two units of force, that is, pound and newton? Why did the spacecraft enter a much lower orbit above Mars than planned? [1 lb = 0.4536 kg; acceleration due to gravity (g) = 9.81 m/s^2. Newton's second law of motion: force = mass $\times$ acceleration.]

The Martian Climate Orbiter.

Answers to Practice Exercises

1.1 96.5 g. **1.2** (a) 621.5°F, (b) 78.3°C. **1.3** (a) Two, (b) four, (c) three, (d) two, (e) three or two. **1.4** (a) 26.76 L, (b) 4.4 g, (c) 1.6×10^7 dm, (d) 0.0756 g/mL, (e) 6.69×10^4 m. **1.5** 2.36 lb. **1.6** 0.534 g/cm^3.

More than a century ago scientists became interested in the radioactivity phenomenon. A young graduate student named Marie Sklodowska Curie, with her husband Pierre Curie as her coworker, chose as the subject of her doctoral research radioactive uranium ore. In 1898 the Curies isolated a new element, which they named polonium after Marie's native country of Poland. Four months later, they discovered radium, another radioactive element.

Marie Curie in her laboratory.

In their makeshift laboratory in a shed, Marie and Pierre processed tons of pitchblende (a uranium oxide ore) to obtain just 0.1 g of radium bromide. They kept a sample of the fluorescent blue substance at their bedside, and they loved to visit the shed at night to look at the luminous test tubes glowing like fairy lights. The unusual properties of radium made it the most important element to be discovered since oxygen. Marie Curie's thesis was hailed by many scientists as "the greatest contribution to science ever made by a doctoral student."

Along with Becquerel (the discoverer of radioactivity), the Curies were awarded the Nobel Prize in Physics in 1903—the same year in which Marie received her doctoral degree. By then, Marie Curie was a household word, but the world's most famous scientist nevertheless encountered hostility and discrimination in the scientific community: She was denied admission to the French Academy of Sciences by one vote because she was a woman!

Marie Curie was the first woman to win a Nobel Prize, the first person to win two Nobel Prizes (the second in chemistry in 1911), and the first Nobel Prize winner to have a child go on to win a Nobel Prize. Today the element with the atomic number 96, curium (Cm), is named in Marie Curie's honor.

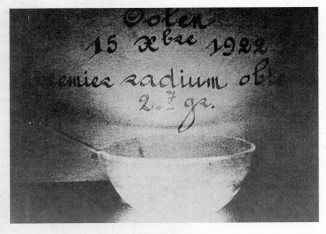

Photograph of 2.7 g radium bromide ($RaBr_2$), taken by its own light in 1922.

Atoms, Molecules, and Ions

Chapter 2

ESSENTIAL CONCEPTS

 Development of the Atomic Theory The search for the fundamental units of matter began in ancient times. The modern version of atomic theory was laid out by John Dalton, who postulated that elements are composed of extremely small particles, called atoms, and that all atoms of a given element are identical, but they are different from atoms of all other elements.

 The Structure of the Atom Through experimentation in the nineteenth and early twentieth centuries, scientists have learned that an atom is composed of three elementary particles: proton, electron, and neutron. The proton has a positive charge, the electron has a negative charge, and the neutron has no charge. Protons and neutrons are located in a small region at the center of the atom, called the nucleus, and electrons are spread out about the nucleus at some distance from it.

 Ways to Identify Atoms Atomic number is the number of protons in a nucleus; atoms of different elements have different atomic numbers. Isotopes are atoms of the same element having different numbers of neutrons. Mass number is the sum of the number of protons and neutrons in an atom. Because an atom is electrically neutral, the number of its protons is equal to the number of its electrons.

 The Periodic Table Elements can be grouped together according to their chemical and physical properties in a chart called the periodic table. The periodic table allows us to classify elements (as metals, metalloids, and nonmetals) and correlate their properties in a systematic way. It is the most useful source of chemical information.

 From Atoms to Molecules and Ions Atoms of most elements interact to form compounds, which are classified as molecules or ionic compounds made of positive (cations) and negative (anions) ions. Chemical formulas tell us the type and number of atoms present in a molecule or compound.

2.1 The Atomic Theory

In the fifth century B.C. the Greek philosopher Democritus expressed the belief that all matter consists of very small, indivisible particles, which he named *atomos* (meaning uncuttable or indivisible). Although Democritus' idea was not accepted by many of his contemporaries (notably Plato and Aristotle), somehow it endured. Experimental evidence from early scientific investigations provided support for the notion of "atomism" and gradually gave rise to the modern definitions of elements and compounds. It was in 1808 that an English scientist and schoolteacher, John Dalton, formulated a precise definition of the indivisible building blocks of matter that we call atoms.

Dalton's work marked the beginning of the modern era of chemistry. The hypotheses about the nature of matter on which Dalton's atomic theory is based can be summarized as:

1. Elements are composed of extremely small particles, called atoms. All atoms of a given element are identical, having the same size, mass, and chemical properties. The atoms of one element are different from the atoms of all other elements.

2. Compounds are composed of atoms of more than one element. In any compound, the ratio of the numbers of atoms of any two of the elements present is either an integer or a simple fraction.

3. A chemical reaction involves only the separation, combination, or rearrangement of atoms; it does not result in their creation or destruction.

Figure 2.1 is a schematic representation of the first two hypotheses.

Dalton's concept of an atom was far more detailed and specific than Democritus'. The first hypothesis states that atoms of one element are different from atoms of all other elements. Dalton made no attempt to describe the structure or composition of atoms—he had no idea what an atom is really like. But he did realize that the different properties shown by elements such as hydrogen and oxygen can be explained by assuming that hydrogen atoms are not the same as oxygen atoms.

The second hypothesis suggests that, to form a certain compound, we need not only atoms of the right kinds of elements, but the specific numbers of these atoms as well. This idea is an extension of a law published in 1799 by Joseph Proust, a French chemist. Proust's **law of definite proportions** states that *different samples of the same compound always contain its constituent elements in the same proportion by mass.* Thus, if we were

Figure 2.1 *(a) According to Dalton's atomic theory, atoms of the same element are identical, but atoms of one element are different from atoms of other elements. (b) Compound formed from atoms of elements X and Y. In this case, the ratio of the atoms of element X to the atoms of element Y is 2:1.*

Atoms of element X   Atoms of element Y

(a)

Compound of elements X and Y

(b)

to analyze samples of carbon dioxide gas obtained from different sources, we would find in each sample the same ratio by mass of carbon to oxygen. It stands to reason, then, that if the ratio of the masses of different elements in a given compound is fixed, the ratio of the atoms of these elements in the compound also must be constant.

Dalton's second hypothesis also supports another important law, the **law of multiple proportions.** According to this law, *if two elements can combine to form more than one compound, the masses of one element that combine with a fixed mass of the other element are in ratios of small whole numbers.* Dalton's theory explains the law of multiple proportions quite simply: The compounds differ in the number of atoms of each kind that combine. For example, carbon forms two stable compounds with oxygen, namely, carbon monoxide and carbon dioxide. Modern measurement techniques indicate that one atom of carbon combines with one atom of oxygen in carbon monoxide and that one atom of carbon combines with two oxygen atoms in carbon dioxide. Thus, the ratio of oxygen in carbon monoxide to oxygen in carbon dioxide is 1:2. This result is consistent with the law of multiple proportions (Figure 2.2).

Dalton's third hypothesis is another way of stating the **law of conservation of mass,** which is that *matter can be neither created nor destroyed.* Because matter is made of atoms that are unchanged in a chemical reaction, it follows that mass must be conserved as well. Dalton's brilliant insight into the nature of matter was the main stimulus for the rapid progress of chemistry during the nineteenth century.

Carbon monoxide

$$\frac{O}{C} = \frac{1}{1}$$

Carbon dioxide

$$\frac{O}{C} = \frac{2}{1}$$

Ratio of oxygen in carbon monoxide to oxygen in carbon dioxide: 1:2

Figure 2.2 *An illustration of the law of multiple proportions.*

2.2 The Structure of the Atom

On the basis of Dalton's atomic theory, we can define an **atom** as *the basic unit of an element that can enter into chemical combination.* Dalton imagined an atom that was both extremely small and indivisible. However, a series of investigations that began in the 1850s and extended into the twentieth century clearly demonstrated that atoms actually possess internal structure; that is, they are made up of even smaller particles, which are called *subatomic particles.* This research led to the discovery of three such particles—electrons, protons, and neutrons.

The Electron

In the 1890s many scientists became caught up in the study of **radiation,** *the emission and transmission of energy through space in the form of waves.* Information gained from this research contributed greatly to our understanding of atomic structure. One device used to investigate this phenomenon was a cathode ray tube, the forerunner of the television tube (Figure 2.3). It is a glass tube from which most of the air has been evacuated. When the two metal plates are connected to a high-voltage source, the negatively charged plate, called the *cathode,* emits an invisible ray. The cathode ray is drawn to the positively charged plate, called the *anode,* where it passes through a hole and continues traveling to the other end of the tube. When the ray strikes the specially coated surface, the cathode ray produces a strong fluorescence, or bright light.

In some experiments two electrically charged plates and a magnet were added to the *outside* of the cathode ray tube (see Figure 2.3). When the magnetic field is on and the electric field is off, the cathode ray strikes point A. When only the electric field is on, the ray strikes point C. When both the magnetic and the electric fields are off or when they are both on but balanced so that they cancel each other's influence, the ray strikes point B. According to electromagnetic theory, a moving charged body behaves

Figure 2.3 *A cathode ray tube with an electric field perpendicular to the direction of the cathode rays and an external magnetic field. The symbols N and S denote the north and south poles of the magnet. The cathode rays will strike the end of the tube at A in the presence of a magnetic field, at C in the presence of an electric field, and at B when there are no external fields present or when the effects of the electric field and magnetic field cancel each other.*

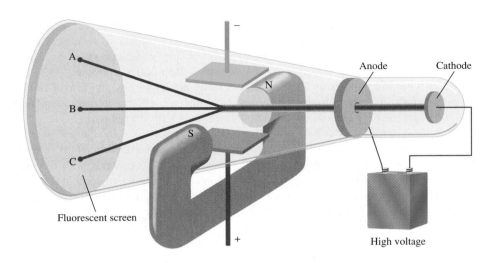

like a magnet and can interact with electric and magnetic fields through which it passes. Because the cathode ray is attracted by the plate bearing positive charges and repelled by the plate bearing negative charges, it must consist of negatively charged particles. We know these *negatively charged particles* as ***electrons.*** Figure 2.4 shows the effect of a bar magnet on the cathode ray.

Electrons are normally associated with atoms. However, they can also be studied individually.

An English physicist, J. J. Thomson, used a cathode ray tube and his knowledge of electromagnetic theory to determine the ratio of electric charge to the mass of an individual electron. The number he came up with is -1.76×10^8 C/g, where C stands for *coulomb,* which is the unit of electric charge. Thereafter, in a series of experiments carried out between 1908 and 1917, R. A. Millikan, an American physicist, found the charge of an electron to be -1.6022×10^{-19} C. From these data he calculated the mass of an electron:

$$\text{mass of an electron} = \frac{\text{charge}}{\text{charge/mass}}$$

$$= \frac{-1.6022 \times 10^{-19}\ \text{C}}{-1.76 \times 10^8\ \text{C/g}}$$

$$= 9.10 \times 10^{-28}\ \text{g}$$

which is an exceedingly small mass.

Figure 2.4 *(a) A cathode ray produced in a discharge tube. The ray itself is invisible, but the fluorescence of a zinc sulfide coating on the glass causes it to appear green. (b) The cathode ray is bent in the presence of a magnet.*

(a)

(b)

Figure 2.5 *Three types of rays emitted by radioactive elements. β rays consist of negatively charged particles (electrons) and are therefore attracted by the positively charged plate. The opposite holds true for α rays—they are positively charged and are drawn to the negatively charged plate. Because γ rays have no charge, their path is unaffected by an external electric field.*

Lead block

Radioactive substance

Radioactivity

In 1895, the German physicist Wilhelm Röntgen noticed that cathode rays caused glass and metals to emit very unusual rays. This highly energetic radiation penetrated matter, darkened covered photographic plates, and caused a variety of substances to fluoresce. Because these rays could not be deflected by a magnet, they could not contain charged particles as cathode rays do. Röntgen called them X rays.

Not long after Röntgen's discovery, Antoine Becquerel, a professor of physics in Paris, began to study fluorescent properties of substances. Purely by accident, he found that exposing thickly wrapped photographic plates to a certain uranium compound caused them to darken, even without the stimulation of cathode rays. Like X rays, the rays from the uranium compound were highly energetic and could not be deflected by a magnet, but they differed from X rays because they were generated spontaneously. One of Becquerel's students, Marie Curie, suggested the name *radioactivity* to describe this *spontaneous emission of particles and/or radiation.* Consequently, any element that spontaneously emits radiation is said to be *radioactive.*

Further investigation revealed that three types of rays are produced by the *decay,* or breakdown, of radioactive substances such as uranium. Two of the three kinds are deflected by oppositely charged metal plates (Figure 2.5). ***Alpha (α) rays*** consist of *positively charged particles,* called ***α particles,*** and therefore are deflected by the positively charged plate. ***Beta (β) rays,*** or ***β particles,*** are *electrons* and are deflected by the negatively charged plate. The third type of radioactive radiation consists of *high-energy rays* called ***gamma (γ) rays.*** Like X rays, γ rays have no charge and are not affected by an external electric or magnetic field.

The Proton and the Nucleus

By the early 1900s, two features of atoms had become clear: They contain electrons, and they are electrically neutral. To maintain electrical neutrality, an atom must contain an equal number of positive and negative charges. On the basis of this information, Thomson proposed that an atom could be thought of as a uniform, positive

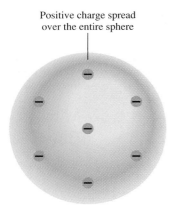

Positive charge spread
over the entire sphere

Figure 2.6 *Thomson's model of the atom, sometimes described as the "plum-pudding" model, after a traditional English dessert containing raisins. The electrons are embedded in a uniform, positively charged sphere.*

sphere of matter in which electrons are embedded (Figure 2.6). Thomson's so-called plum-pudding model was the accepted theory for a number of years.

In 1910 the New Zealand physicist Ernest Rutherford, who had earlier studied with Thomson at Cambridge University, decided to use α particles to probe the structure of atoms. Together with his associate Hans Geiger and an undergraduate named Ernest Marsden, Rutherford carried out a series of experiments using very thin foils of gold and other metals as targets for α particles from a radioactive source (Figure 2.7). They observed that the majority of particles penetrated the foil either undeflected or with only a slight deflection. They also noticed that every now and then an α particle was scattered (or deflected) at a large angle. In some instances, an α particle actually bounced back in the direction from which it had come! This was a most surprising finding, for in Thomson's model the positive charge of the atom was so diffuse (spread out) that the positive α particles were expected to pass through with very little deflection. To quote Rutherford's initial reaction when told of this discovery: "It was as incredible as if you had fired a 15-inch shell at a piece of tissue paper and it came back and hit you."

To explain the results of the α-scattering experiment, Rutherford devised a new model of atomic structure, suggesting that most of the atom must be empty space. This structure would allow most of the α particles to pass through the gold foil with little or no deflection. The atom's positive charges, Rutherford proposed, are all concentrated in the **nucleus,** *a dense central core within the atom.* Whenever an α particle came close to a nucleus in the scattering experiment, it experienced a large repulsive force and therefore a large deflection. Moreover, an α particle traveling directly toward a nucleus would experience an enormous repulsion that could completely reverse the direction of the moving particle.

The positively charged particles in the nucleus are called **protons.** In separate experiments, it was found that the charge of each proton has the same *magnitude* as that of an electron and that the mass of the proton is 1.67262×10^{-24} g—about 1840 times the mass of the oppositely charged electron.

At this stage of investigation, scientists perceived the atom as follows. The mass of a nucleus constitutes most of the mass of the entire atom, but the nucleus occupies only about $1/10^{13}$ of the volume of the atom. We express atomic (and molecular) dimensions in terms of the SI unit called the *picometer (pm),* and

$$1 \text{ pm} = 1 \times 10^{-12} \text{ m}$$

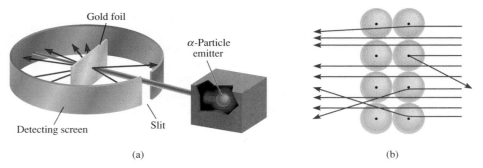

Gold foil

α-Particle
emitter

Detecting screen

Slit

(a)

(b)

Figure 2.7 *(a) Rutherford's experimental design for measuring the scattering of α particles by a piece of gold foil. Most of the α particles passed through the gold foil with little or no deflection. A few were deflected at wide angles. Occasionally an α particle was turned back. (b) Magnified view of α particles passing through and being deflected by nuclei.*

A typical atomic radius is about 100 pm, whereas the radius of an atomic nucleus is only about 5×10^{-3} pm. You can appreciate the relative sizes of an atom and its nucleus by imagining that if an atom were the size of the Houston Astrodome, the volume of its nucleus would be comparable to that of a small marble. Although the protons are confined to the nucleus of the atom, the electrons are conceived of as being spread out about the nucleus at some distance from it.

A common non-SI unit for atomic length is the angstrom (Å; 1 Å = 100 pm).

The Neutron

Rutherford's model of atomic structure left one major problem unsolved. It was known that hydrogen, the simplest atom, contains only one proton and that the helium atom contains two protons. Therefore, the ratio of the mass of a helium atom to that of a hydrogen atom should be 2:1. (Because electrons are much lighter than protons, their contribution can be ignored.) In reality, however, the ratio is 4:1.

Rutherford and others postulated that there must be another type of subatomic particle in the atomic nucleus; the proof was provided by another English physicist, James Chadwick, in 1932. When Chadwick bombarded a thin sheet of beryllium with α particles, a very high energy radiation similar to γ rays was emitted by the metal. Later experiments showed that the rays actually consisted of *electrically neutral particles having a mass slightly greater than that of protons.* Chadwick named these particles **neutrons.**

The mystery of the mass ratio could now be explained. In the helium nucleus there are two protons and two neutrons, but in the hydrogen nucleus there is only one proton and no neutrons; therefore, the ratio is 4:1.

Figure 2.8 shows the location of the elementary particles (protons, neutrons, and electrons) in an atom. There are other subatomic particles, but the electron, the proton, and the neutron are the three fundamental components of the atom that are important

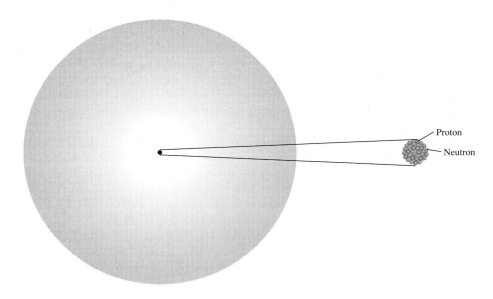

Figure 2.8 *The protons and neutrons of an atom are packed in an extremely small nucleus. Electrons are shown as "clouds" around the nucleus.*

Proton

Neutron

Table 2.1 | Mass and Charge of Subatomic Particles

Particle	Mass (g)	Charge	
		Coulomb	**Charge Unit**
Electron*	9.10939×10^{-28}	-1.6022×10^{-19}	-1
Proton	1.67262×10^{-24}	$+1.6022 \times 10^{-19}$	$+1$
Neutron	1.67493×10^{-24}	0	0

*More refined experiments have given us a more accurate value of an electron's mass than Millikan's.

in chemistry. Table 2.1 shows the masses and charges of these three elementary particles.

2.3 Atomic Number, Mass Number, and Isotopes

All atoms can be identified by the number of protons and neutrons they contain. *The number of protons in the nucleus of each atom of an element* is called the **atomic number (Z).** In a neutral atom the number of protons is equal to the number of electrons, so the atomic number also indicates the number of electrons present in the atom. The chemical identity of an atom can be determined solely by its atomic number. For example, the atomic number of nitrogen is 7; this means that each neutral nitrogen atom has 7 protons and 7 electrons. Or viewed another way, every atom in the universe that contains 7 protons is correctly named "nitrogen."

The **mass number (A)** is *the total number of neutrons and protons present in the nucleus of an atom of an element.* Except for the most common form of hydrogen, which has one proton and no neutrons, all atomic nuclei contain both protons and neutrons. In general, the mass number is given by

$$\text{mass number} = \text{number of protons} + \text{number of neutrons}$$

$$= \text{atomic number} + \text{number of neutrons}$$

The number of neutrons in an atom is equal to the difference between the mass number and the atomic number, or $(A - Z)$. For example, the mass number of fluorine is 19, and the atomic number is 9 (indicating 9 protons in the nucleus). Thus the number of neutrons in an atom of fluorine is $19 - 9 = 10$. Note that all three quantities (atomic number, number of neutrons, and mass number) must be positive integers, or whole numbers.

In most cases atoms of a given element do not all have the same mass. *Atoms that have the same atomic number but different mass numbers* are called **isotopes.** For example, there are three isotopes of hydrogen. One, simply known as hydrogen, has one proton and no neutrons. The deuterium isotope has one proton and one neutron, and tritium has one proton and two neutrons. The accepted way to denote the atomic number and mass number of an atom of element X is as follows:

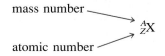

Thus, for the isotopes of hydrogen, we write

$$^1_1\text{H} \qquad ^2_1\text{H} \qquad ^3_1\text{H}$$

hydrogen   deuterium   tritium

As another example, consider two common isotopes of uranium with mass numbers of 235 and 238, respectively:

$$^{235}_{92}\text{U} \qquad ^{238}_{92}\text{U}$$

The first isotope is used in nuclear reactors and atomic bombs, whereas the second isotope lacks the properties necessary for these applications. With the exception of hydrogen, isotopes of elements are identified by their mass numbers. Thus these two isotopes are called uranium-235 (pronounced "uranium two thirty-five") and uranium-238 (pronounced "uranium two thirty-eight").

The chemical properties of an element are determined primarily by the protons and electrons in its atoms; neutrons do not take part in chemical changes under normal conditions. Therefore, isotopes of the same element have similar chemistries, forming the same types of compounds and displaying similar reactivities.

Example 2.1 Give the number of protons, neutrons, and electrons in each of these species: (a) $^{17}_8\text{O}$, (b) $^{199}_{80}\text{Hg}$, and (c) $^{200}_{80}\text{Hg}$.

Reasoning and Solution Recall that the superscript denotes mass number and the subscript denotes atomic number. Mass number is always greater than atomic number. (The only exception is ^1_1H, where the mass number is equal to the atomic number.)

(a) The atomic number is 8, so there are 8 protons. The mass number is 17, so the number of neutrons is $17 - 8 = 9$. The number of electrons is the same as the number of protons, that is, 8.

(b) The atomic number is 80, so there are 80 protons. The mass number is 199, so the number of neutrons is $199 - 80 = 119$. The number of electrons is 80.

(c) Here the number of protons is the same as in (b), or 80. The number of neutrons is $200 - 80 = 120$. The number of electrons is also the same as in (b), 80. The species in (b) and (c) are chemically similar isotopes of mercury.

Similar problems: 2.13, 2.14.

Practice Exercise How many protons, neutrons, and electrons are in this isotope of copper: $^{63}_{29}\text{Cu}$?

2.4 The Periodic Table

More than half of the elements known today were discovered between 1800 and 1900. During this period, chemists noted that many elements show very strong similarities to one another. Recognition of periodic regularities in physical and chemical behavior and the need to organize the large volume of available information about the structure and properties of elemental substances led to the development of the ***periodic table***—*a chart in which elements having similar chemical and physical properties are grouped together.* Figure 2.9 shows the modern periodic table, in which the elements are arranged by atomic number (shown above the

Figure 2.9 *The modern periodic table. The elements are arranged according to the atomic numbers above their symbols. With the exception of hydrogen (H), nonmetals appear at the far right of the table. The two rows of metals beneath the main body of the table are conventionally set apart to keep the table from being too wide. Actually, cerium (Ce) should follow lanthanum (La), and thorium (Th) should come right after actinium (Ac). The 1–18 group designation has been recommended by the International Union of Pure and Applied Chemistry (IUPAC) but is not yet in wide use. In this text we use the standard U.S. notation for group numbers (1A–8A and 1B–8B). Elements 113 and 115–118 have not yet been synthesized.*

element symbol) in *horizontal rows* called ***periods*** and in *vertical columns* known as ***groups*** or ***families,*** according to similarities in their chemical properties. Note that elements 110–112 and 114 have recently been synthesized, although they have not yet been named.

The elements can be divided into three categories—metals, nonmetals, and metalloids. A ***metal*** is *a good conductor of heat and electricity,* whereas a ***nonmetal*** is usually *a poor conductor of heat and electricity.* A ***metalloid*** *has properties that are intermediate between those of metals and nonmetals.* Figure 2.9 shows that the majority of known elements are metals; only seventeen elements are nonmetals, and eight elements are metalloids. From left to right across any period, the physical and chemical properties of the elements change gradually from metallic to nonmetallic. The periodic table is a handy tool that correlates the properties of the elements in a systematic way and helps us to make predictions about chemical behavior. We will take a closer look at this keystone of chemistry in Chapter 8.

Elements are often referred to collectively by their periodic table group number (Group 1A, Group 2A, and so on). However, for convenience, some element groups have special names. *The Group 1A elements (Li, Na, K, Rb, Cs, and Fr) are called* **alkali metals,** and *the Group 2A elements (Be, Mg, Ca, Sr, Ba, and Ra) are called* **alkaline earth metals.** *Elements in Group 7A (F, Cl, Br, I, and At) are known as* **halogens,** and *those in Group 8A (He, Ne, Ar, Kr, Xe, and Rn) are called* **noble gases** (or **rare gases**). The names of other groups or families will be introduced later.

2.5 Molecules and Ions

Of all the elements, only the six noble gases in Group 8A of the periodic table (He, Ne, Ar, Kr, Xe, and Rn) exist in nature as single atoms. For this reason, they are called *monatomic* (meaning a single atom) gases. Most matter is composed of molecules or ions formed by atoms.

Molecules

A **molecule** is an *aggregate of at least two atoms in a definite arrangement held together by chemical forces* (also called *chemical bonds*). A molecule may contain atoms of the same element or atoms of two or more elements joined in a fixed ratio, in accordance with the law of definite proportions stated in Section 2.1. Thus, a molecule is not necessarily a compound, which, by definition, is made up of two or more elements. Hydrogen gas, for example, is a pure element, but it consists of molecules made up of two H atoms each. Water, on the other hand, is a molecular compound that contains hydrogen and oxygen in a ratio of two H atoms and one O atom. Like atoms, molecules are electrically neutral.

We will discuss the nature of chemical bonds in Chapters 9 and 10.

The hydrogen molecule, symbolized as H_2, is called a **diatomic molecule** because it *contains only two atoms.* Other elements that normally exist as diatomic molecules are nitrogen (N_2) and oxygen (O_2), as well as the Group 7A elements—fluorine (F_2), chlorine (Cl_2), bromine (Br_2), and iodine (I_2). Of course, a diatomic molecule can contain atoms of different elements. Examples are hydrogen chloride (HCl) and carbon monoxide (CO).

The vast majority of molecules contain more than two atoms. They can be atoms of the same element, as in ozone (O_3), which is made up of three atoms of oxygen, or they can be combinations of two or more different elements. *Molecules containing more than two atoms* are called **polyatomic molecules.** Like ozone, water (H_2O) and ammonia (NH_3) are polyatomic molecules.

Elements that exist as diatomic molecules.

Ions

An **ion** is an *atom or a group of atoms that has a net positive or negative charge.* The number of positively charged protons in the nucleus of an atom remains the same during ordinary chemical changes (called chemical reactions), but negatively charged electrons may be lost or gained. The loss of one or more electrons from a neutral atom results in a **cation,** *an ion with a net positive charge.* For example, a sodium atom (Na) can readily lose an electron to become a sodium cation, which is represented by Na^+:

In Chapter 8 we will see why atoms of different elements gain (or lose) a specific number of electrons.

Na Atom	Na^+ Ion
11 protons	11 protons
11 electrons	10 electrons

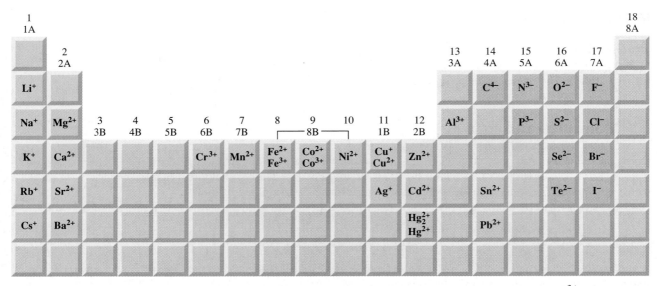

Figure 2.10 *Common monatomic ions arranged according to their positions in the periodic table. Note that the Hg_2^{2+} ion contains two atoms.*

On the other hand, an ***anion*** is *an ion whose net charge is negative* due to an increase in the number of electrons. A chlorine atom (Cl), for instance, can gain an electron to become the chloride ion Cl^-:

Cl Atom	Cl^- Ion
17 protons	17 protons
17 electrons	18 electrons

Sodium chloride (NaCl), ordinary table salt, is called an ***ionic compound*** because it is *formed from cations and anions.*

An atom can lose or gain more than one electron. Examples of ions formed by the loss or gain of more than one electron are Mg^{2+}, Fe^{3+}, S^{2-}, and N^{3-}. These ions, as well as Na^+ and Cl^-, are called ***monatomic ions*** because they *contain only one atom.* Figure 2.10 shows the charges of a number of monatomic ions. With very few exceptions, metals tend to form cations and nonmetals form anions.

In addition, two or more atoms can combine to form an ion that has a net positive or net negative charge. ***Polyatomic ions*** such as OH^- (hydroxide ion), CN^- (cyanide ion), and NH_4^+ (ammonium ion) are *ions containing more than one atom.*

2.6 Chemical Formulas

Chemists use ***chemical formulas*** to *express the composition of molecules and ionic compounds in terms of chemical symbols.* By composition we mean not only the elements present but also the ratios in which the atoms are combined. Here we are concerned with two types of formulas: molecular formulas and empirical formulas.

Molecular Formulas

A **molecular formula** *shows the exact number of atoms of each element in the smallest unit of a substance*. In our discussion of molecules, each example was given with its molecular formula in parentheses. Thus H_2 is the molecular formula for hydrogen, O_2 is oxygen, O_3 is ozone, and H_2O is water. The subscript numeral indicates the number of atoms of an element present. There is no subscript for O in H_2O because there is only one atom of oxygen in a molecule of water, and so the number "one" is omitted from the formula. Note that oxygen (O_2) and ozone (O_3) are allotropes of oxygen. An **allotrope** is *one of two or more distinct forms of an element*. Two allotropic forms of the element carbon—diamond and graphite—are dramatically different not only in properties but also in their relative cost.

Molecular Models

Molecules are too small for us to observe directly. An effective means of visualizing them is by the use of molecular models. Two standard types of molecular models are currently in use: *ball-and-stick* models and *space-filling* models (Figure 2.11). In ball-and-stick model kits, the atoms are wooden or plastic balls with holes in them. Sticks or springs are used to represent chemical bonds. The angles they form between atoms approximate the bond angles in actual molecules. With the exception of the H atom, the balls are all the same size and each type of atom is represented by a specific color. In space-filling models, atoms are represented by truncated balls held together by snap fasteners, so that the bonds are not visible. The balls are proportional in size to atoms. The first step toward building a molecular model is writing the **structural formula,** which *shows how atoms are bonded to one another in a molecule*. For example, it is

See back end paper for color code for atoms.

	Hydrogen	Water	Ammonia	Methane
Molecular formula	H_2	H_2O	NH_3	CH_4
Structural formula	H—H	H—O—H	H—N—H $\vert$ H	H—C—H (with H above and below)
Ball-and-stick model				
Space-filling model				

Figure 2.11 *Molecular and structural formulas and molecular models of four common molecules.*

known that each of the two H atoms is bonded to an O atom in the water molecule. Therefore, the structural formula of water is H—O—H. A line connecting the two atomic symbols represents a chemical bond.

Ball-and-stick models show the three-dimensional arrangement of atoms clearly, and they are fairly easy to construct. However, the balls are not proportional to the size of atoms. Furthermore, the sticks greatly exaggerate the space between atoms in a molecule. Space-filling models are more accurate because they show the variation in atomic size. Their drawbacks are that they are time-consuming to put together and they do not show the three-dimensional positions of atoms very well. We will use both models extensively in this text.

Empirical Formulas

The molecular formula of hydrogen peroxide, a substance used as an antiseptic and as a bleaching agent for textiles and hair, is H_2O_2. This formula indicates that each hydrogen peroxide molecule consists of two hydrogen atoms and two oxygen atoms. The ratio of hydrogen to oxygen atoms in this molecule is 2:2 or 1:1. The empirical formula of hydrogen peroxide is HO. Thus the ***empirical formula*** *tells us which elements are present and the simplest whole-number ratio of their atoms*, but not necessarily the actual number of atoms in a given molecule. As another example, consider the compound hydrazine (N_2H_4), which is used as a rocket fuel. The empirical formula of hydrazine is NH_2. Although the ratio of nitrogen to hydrogen is 1:2 in both the molecular formula (N_2H_4) and the empirical formula (NH_2), only the molecular formula tells us the actual number of N atoms (two) and H atoms (four) present in a hydrazine molecule.

Empirical formulas are the *simplest* chemical formulas; they are written by reducing the subscripts in molecular formulas to the smallest possible whole numbers. Molecular formulas are the *true* formulas of molecules. As we will see in Chapter 3, when chemists analyze an unknown compound, the first step is usually the determination of the compound's empirical formula.

For many molecules, the molecular formula and the empirical formula are one and the same. Some examples are water (H_2O), ammonia (NH_3), carbon dioxide (CO_2), and methane (CH_4).

H_2O_2

The word "empirical" means "derived from experiment." As we will see in Chapter 3, empirical formulas are determined experimentally.

Methanol.

Chloroform.

Similar problems: 2.33, 2.34.

Example 2.2   Write the molecular formula of methanol, an organic solvent and antifreeze, from its ball-and-stick model, shown in the margin.

Reasoning and Solution   Refer to the color code for atoms (see back end papers). There are one C atom, four H atoms, and one O atom. Therefore, the molecular formula is CH_4O. However, the standard way of writing the molecular formula for methanol is CH_3OH because it shows how the atoms are joined in the molecule.

Practice Exercise   Write the molecular formula of chloroform, which is used as a solvent and a cleansing agent. The ball-and-stick model of chloroform is shown in the margin.

Example 2.3 Write the empirical formulas for these molecules: (a) acetylene (C_2H_2), which is used in welding torches; (b) glucose ($C_6H_{12}O_6$), a substance known as blood sugar; and (c) nitrous oxide (N_2O), a gas that is used as an anesthetic gas ("laughing gas") and as an aerosol propellant for whipped creams.

Reasoning and Solution Recall that to write the empirical formula, the coefficients in the molecular formula must be converted to the smallest possible whole numbers.

(a) There are two carbon atoms and two hydrogen atoms in acetylene. Dividing the subscripts by 2, we obtain the empirical formula CH.

(b) In glucose there are 6 carbon atoms, 12 hydrogen atoms, and 6 oxygen atoms. Dividing the subscripts by 6, we obtain the empirical formula CH_2O. Note that if we had divided the subscripts by 3, we would have obtained the formula $C_2H_4O_2$. Although the ratio of carbon to hydrogen to oxygen atoms in $C_2H_4O_2$ is the same as that in $C_6H_{12}O_6$ ($1:2:1$), $C_2H_4O_2$ is not the simplest formula because its subscripts are not in the smallest whole-number ratio.

(c) Because the subscripts in N_2O are already the smallest possible whole numbers, the empirical formula for nitrous oxide is the same as its molecular formula.

Practice Exercise Write the empirical formula for caffeine ($C_8H_{10}N_4O_2$), a stimulant found in tea and coffee.

Similar problems: 2.35, 2.36.

Formula of Ionic Compounds

The formulas of ionic compounds are usually the same as their empirical formulas because ionic compounds do not consist of discrete molecular units. For example, a solid sample of sodium chloride (NaCl) consists of equal numbers of Na^+ and Cl^- ions arranged in a three-dimensional network (Figure 2.12). In such a compound there is a $1:1$ ratio of cations to anions so that the compound is electrically neutral. As you can see in Figure 2.12, no Na^+ ion in NaCl is associated

(a)   (b)   (c)

Figure 2.12 *(a) Structure of solid NaCl. (b) In reality, the cations are in contact with the anions. In both (a) and (b), the smaller spheres represent Na^+ ions and the larger spheres, Cl^- ions. (c) Crystals of NaCl.*

with just one particular Cl^- ion. In fact, each Na^+ ion is equally held by six surrounding Cl^- ions and vice versa. Thus NaCl is the empirical formula for sodium chloride. In other ionic compounds, the actual structure may be different, but the arrangement of cations and anions is such that the compounds are all electrically neutral. Note that the charges on the cation and anion are not shown in the formula for an ionic compound.

In order for ionic compounds to be electrically neutral, the sum of the charges on the cation and anion in each formula unit must be zero. If the charges on the cation and anion are numerically different, we apply the following rule to make the formula electrically neutral: *The subscript of the cation is numerically equal to the charge on the anion, and the subscript of the anion is numerically equal to the charge on the cation.* If the charges are numerically equal, then no subscripts are necessary. This rule follows from the fact that because the formulas of most ionic compounds are empirical formulas, the subscripts must always be reduced to the smallest ratios. Let us consider some examples.

- **Potassium Bromide.** The potassium cation K^+ and the bromine anion Br^- combine to form the ionic compound potassium bromide. The sum of the charges is $+1 + (-1) = 0$, so no subscripts are necessary. The formula is KBr.
- **Zinc Iodide.** The zinc cation Zn^{2+} and the iodine anion I^- combine to form zinc iodide. The sum of the charges of one Zn^{2+} ion and one I^- ion is $+2 + (-1) = +1$. To make the charges add up to zero we multiply the -1 charge of the anion by 2 and add the subscript "2" to the symbol for iodine. Therefore the formula for zinc iodide is ZnI_2.
- **Aluminum Oxide.** The cation is Al^{3+} and the oxygen anion is O^{2-}. The following diagram helps us determine the subscripts for the compound formed by the cation and the anion:

The sum of the charges is $2(+3) + 3(-2) = 0$. Thus the formula for aluminum oxide is Al_2O_3.

2.7 Naming Compounds

In addition to using formulas to show the composition of molecules and compounds, chemists have developed a system for naming substances on the basis of their composition. First we divide them into three categories: ionic compounds, molecular compounds, and acids and bases. Then we apply certain rules to derive the scientific name for a given substance.

The most reactive metals (green) and the most reactive nonmetals (blue) combine to form ionic compounds.

Ionic Compounds

In Section 2.5 we learned that ionic compounds are made up of cations (positive ions) and anions (negative ions). With the important exception of the ammonium ion, NH_4^+, all cations of interest to us are derived from metal atoms. Metal cations take their names from the elements. For example:

Element		Name of Cation	
Na	sodium	Na^+	sodium ion (or sodium cation)
K	potassium	K^+	potassium ion (or potassium cation)
Mg	magnesium	Mg^{2+}	magnesium ion (or magnesium cation)
Al	aluminum	Al^{3+}	aluminum ion (or aluminum cation)

Many ionic compounds are **binary compounds,** or *compounds formed from just two elements.* For binary ionic compounds the first element named is the metal cation, followed by the nonmetallic anion. Thus NaCl is sodium chloride. The anion is named by taking the first part of the element name (chlorine) and adding "-ide." Potassium bromide (KBr), zinc iodide (ZnI_2), and aluminum oxide (Al_2O_3) are also binary compounds. Table 2.2 shows the "-ide" nomenclature of some common monatomic anions according to their positions in the periodic table.

The "-ide" ending is also used for certain anion groups containing different elements, such as hydroxide (OH^-) and cyanide (CN^-). Thus the compounds LiOH and KCN are named lithium hydroxide and potassium cyanide. These and a number of other such ionic substances are called **ternary compounds,** meaning *compounds consisting of three elements.* Table 2.3 lists alphabetically the names of a number of common cations and anions.

Certain metals, especially the *transition metals,* can form more than one type of cation. Take iron as an example. Iron can form two cations: Fe^{2+} and Fe^{3+}. The accepted procedure for designating different cations of the *same* element is to use Roman numerals. The Roman numeral I is used for one positive charge, II for two positive charges, and so on. This is called the *Stock system.* In this system, the Fe^{2+} and Fe^{3+} ions are called iron(II) and iron(III), and the compounds $FeCl_2$ (containing the Fe^{2+} ion) and $FeCl_3$ (containing the Fe^{3+} ion) are called iron-two chloride and iron-three chloride, respectively. As another example, manganese (Mn) atoms can assume several different positive charges:

Mn^{2+}:   MnO   manganese(II) oxide
Mn^{3+}: Mn_2O_3 manganese(III) oxide
Mn^{4+}: MnO_2 manganese(IV) oxide

These compound names are pronounced manganese-two oxide, manganese-three oxide, and manganese-four oxide.

The transition metals are the elements in Groups 1B and 3B–8B (see Figure 2.9).

$FeCl_2$ (left) and $FeCl_3$ (right).

Table 2.2	The "-ide" Nomenclature of Some Common Monatomic Anions According to Their Positions in the Periodic Table			

Group 4A	Group 5A	Group 6A	Group 7A
C Carbide (C^{4-})*	N Nitride (N^{3-})	O Oxide (O^{2-})	F Fluoride (F^-)
Si Silicide (Si^{4-})	P Phosphide (P^{3-})	S Sulfide (S^{2-})	Cl Chloride (Cl^-)
		Se Selenide (Se^{2-})	Br Bromide (Br^-)
		Te Telluride (Te^{2-})	I Iodide (I^-)

*The word "carbide" is also used for the anion C_2^{2-}.

Table 2.3 | Names and Formulas of Some Common Inorganic Cations and Anions

Cation	Anion
Aluminum (Al^{3+})	Bromide (Br^-)
Ammonium (NH_4^+)	Carbonate (CO_3^{2-})
Barium (Ba^{2+})	Chlorate (ClO_3^-)
Cadmium (Cd^{2+})	Chloride (Cl^-)
Calcium (Ca^{2+})	Chromate (CrO_4^{2-})
Cesium (Cs^+)	Cyanide (CN^-)
Chromium(III) or chromic (Cr^{3+})	Dichromate ($Cr_2O_7^{2-}$)
Cobalt(II) or cobaltous (Co^{2+})	Dihydrogen phosphate ($H_2PO_4^-$)
Copper(I) or cuprous (Cu^+)	Fluoride (F^-)
Copper(II) or cupric (Cu^{2+})	Hydride (H^-)
Hydrogen (H^+)	Hydrogen carbonate or bicarbonate (HCO_3^-)
Iron(II) or ferrous (Fe^{2+})	Hydrogen phosphate (HPO_4^{2-})
Iron(III) or ferric (Fe^{3+})	Hydrogen sulfate or bisulfate (HSO_4^-)
Lead(II) or plumbous (Pb^{2+})	Hydroxide (OH^-)
Lithium (Li^+)	Iodide (I^-)
Magnesium (Mg^{2+})	Nitrate (NO_3^-)
Manganese(II) or manganous (Mn^{2+})	Nitride (N^{3-})
Mercury(I) or mercurous (Hg_2^{2+})*	Nitrite (NO_2^-)
Mercury(II) or mercuric (Hg^{2+})	Oxide (O^{2-})
Potassium (K^+)	Permanganate (MnO_4^-)
Silver (Ag^+)	Peroxide (O_2^{2-})
Sodium (Na^+)	Phosphate (PO_4^{3-})
Strontium (Sr^{2+})	Sulfate (SO_4^{2-})
Tin(II) or stannous (Sn^{2+})	Sulfide (S^{2-})
Zinc (Zn^{2+})	Sulfite (SO_3^{2-})
	Thiocyanate (SCN^-)

*Mercury(I) exists as a pair as shown.

Example 2.4 Name these ionic compounds: (a) $Cu(NO_3)_2$, (b) KH_2PO_4, and (c) NH_4ClO_3.

Reasoning and Solution Our source of reference is Table 2.3. We need to identify the cation and anion and their charges in each ionic compound.

(a) Because the nitrate ion (NO_3^-) bears one negative charge, the copper ion must have two positive charges. Therefore, the compound is copper(II) nitrate.

(b) The cation is K^+ and the anion is $H_2PO_4^-$ (dihydrogen phosphate). Because potassium only forms one type of ion (K^+), there is no need to use potassium(I) in the name. The compound is potassium dihydrogen phosphate.

(c) The cation is NH_4^+ (ammonium ion) and the anion is ClO_3^-. The compound is ammonium chlorate.

Practice Exercise Name these compounds: (a) PbO and (b) Li_2SO_3.

Similar problems: 2.43(a), (b), (e).

Example 2.5 Write chemical formulas for these compounds: (a) mercury(I) nitrite, (b) cesium sulfide, and (c) calcium phosphate.

Reasoning and Solution The Roman numerals provide useful information about the charges on cations.

(a) The mercury(I) ion is diatomic, namely, Hg_2^{2+} (see Table 2.3), and the nitrite ion is NO_2^-. Therefore, the formula is $Hg_2(NO_2)_2$.

(b) Each sulfide ion bears two negative charges, and each cesium ion bears one positive charge (cesium is in Group 1A, as is sodium). Therefore, the formula is Cs_2S.

(c) Each calcium ion (Ca^{2+}) bears two positive charges, and each phosphate ion (PO_4^{3-}) bears three negative charges. To make the sum of the charges equal zero, we must adjust the numbers of cations and anions:

$$3(+2) + 2(-3) = 0$$

Thus the formula is $Ca_3(PO_4)_2$.

Practice Exercise Write formulas for these ionic compounds: (a) rubidium sulfate and (b) barium hydride.

Similar problems: 2.45(a), (b), (h).

Molecular Compounds

Unlike ionic compounds, molecular compounds contain discrete molecular units. They are usually composed of nonmetallic elements (see Figure 2.9). Many molecular compounds are binary compounds. Naming binary molecular compounds is similar to naming binary ionic compounds. We place the name of the first element in the formula first, and the second element is named by adding "-ide" to the root of the element name. Some examples are

HCl	Hydrogen chloride	SiC	Silicon carbide
HBr	Hydrogen bromide		

It is quite common for one pair of elements to form several different compounds. In these cases, confusion in naming the compounds is avoided by the use of Greek prefixes to denote the number of atoms of each element present (Table 2.4). Consider these examples:

CO	Carbon monoxide	SO_3	Sulfur trioxide
CO_2	Carbon dioxide	NO_2	Nitrogen dioxide
SO_2	Sulfur dioxide	N_2O_4	Dinitrogen tetroxide

These guidelines are helpful when you are naming compounds with prefixes:

- The prefix "mono-" may be omitted for the first element. For example, PCl_3 is named phosphorus trichloride, not monophosphorus trichloride. Thus the absence of a prefix for the first element usually means that only one atom of that element is present in the molecule.
- For oxides, the ending "a" in the prefix is sometimes omitted. For example, N_2O_4 may be called dinitrogen tetroxide rather than dinitrogen tetraoxide.

Exceptions to the use of Greek prefixes are molecular compounds containing hydrogen. Traditionally, many of these compounds are called either by their common,

Table 2.4 Greek Prefixes Used in Naming Molecular Compounds

Prefix	Meaning
Mono-	1
Di-	2
Tri-	3
Tetra-	4
Penta-	5
Hexa-	6
Hepta-	7
Octa-	8
Nona-	9
Deca-	10

nonsystematic names or by names that do not specifically indicate the number of H atoms present:

B_2H_6	Diborane	PH_3	Phosphine
CH_4	Methane	H_2O	Water
SiH_4	Silane	H_2S	Hydrogen sulfide
NH_3	Ammonia		

Note that even the order of writing the elements in the formulas is irregular. These examples show that H is written first in water and hydrogen sulfide, whereas H is written last in the other compounds.

Writing formulas for molecular compounds is usually straightforward. Thus the name arsenic trifluoride means that there are one As atom and three F atoms in each molecule and the molecular formula is AsF_3. Note that the order of elements in the formula is the same as that in its name.

Example 2.6   Name these molecular compounds: (a) $SiCl_4$ and (b) P_4O_{10}.

Reasoning and Solution   We refer to Table 2.4 for prefixes.

(a) Because there are four chlorine atoms present, the compound is silicon tetrachloride.

(b) There are four phosphorus atoms and ten oxygen atoms present, so the compound is tetraphosphorus decoxide. Note that the "a" is omitted in "deca."

Similar problems: 2.43(c), (h), (j).

Practice Exercise   Name these molecular compounds: (a) NF_3 and (b) Cl_2O_7.

Example 2.7   Write chemical formulas for these molecular compounds: (a) carbon disulfide and (b) disilicon hexabromide.

Reasoning and Solution   We refer to Table 2.4 for prefixes.

(a) Because there are one carbon atom and two sulfur atoms present, the formula is CS_2.

(b) There are two silicon atoms and six bromine atoms present, so the formula is Si_2Br_6.

Similar problems: 2.45(f), (g).

Practice Exercise   Write chemical formulas for these molecular compounds: (a) sulfur tetrafluoride and (b) dinitrogen pentoxide.

Acids and Bases

Naming Acids

> H^+ is equivalent to one *proton,* and is sometimes referred to that way.

An ***acid*** can be described as *a substance that yields hydrogen ions (H^+) when dissolved in water.* Formulas for acids contain one or more hydrogen atoms as well as an anionic group. Anions whose names end in "-ide" have associated acids with a "hydro-" prefix and an "-ic" ending, as shown in Table 2.5. In some cases two

Table 2.5 | Some Simple Acids

Anion	Corresponding Acid
F^- (fluoride)	HF (hydrofluoric acid)
Cl^- (chloride)	HCl (hydrochloric acid)
Br^- (bromide)	HBr (hydrobromic acid)
I^- (iodide)	HI (hydroiodic acid)
CN^- (cyanide)	HCN (hydrocyanic acid)
S^{2-} (sulfide)	H_2S (hydrosulfuric acid)

different names are assigned to the same chemical formula. For instance, HCl is known as both hydrogen chloride and hydrochloric acid. The name used for this compound depends on its physical state. In the gaseous or pure liquid state, HCl is a molecular compound called hydrogen chloride. When it is dissolved in water, the molecules break up into H^+ and Cl^- ions; in this condition, the substance is called hydrochloric acid.

*Acids that contain hydrogen, oxygen, and another element (the central element) are called **oxoacids**.* The formulas of oxoacids are usually written with the H first, followed by the central element and then O, as illustrated by this series of oxoacids:

HNO₃

HNO_3	Nitric acid	H_2SO_4	Sulfuric acid
H_2CO_3	Carbonic acid	$HClO_3$	Chloric acid

Often two or more oxoacids have the same central atom but a different number of O atoms. Starting with the oxoacids whose names end with "-ic," we use these rules to name these compounds.

- Addition of one O atom to the "-ic" acid: The acid is called "per . . . -ic" acid. Thus adding an O atom to $HClO_3$ changes chloric acid to perchloric acid, $HClO_4$.
- Removal of one O atom from the "-ic" acid: The acid is called "-ous" acid. Thus nitric acid, HNO_3, becomes nitrous acid, HNO_2.
- Removal of two O atoms from the "-ic" acid: The acid is called "hypo . . . -ous" acid. Thus when $HBrO_3$ is converted to HBrO, the acid is called hypobromous acid.

H₂CO₃

The rules for naming *anions of oxoacids,* called **oxoanions,** are:

- When all the H ions are removed from the "-ic" acid, the anion's name ends with "-ate." For example, the anion CO_3^{2-} derived from H_2CO_3 is called carbonate.
- When all the H ions are removed from the "-ous" acid, the anion's name ends with "-ite." Thus the anion ClO_2^- derived from $HClO_2$ is called chlorite.
- The names of anions in which one or more but not all of the hydrogen ions have been removed must indicate the number of H ions present. For example, consider the anions derived from phosphoric acid:

H_3PO_4	Phosphoric acid	HPO_4^{2-}	Hydrogen phosphate
$H_2PO_4^-$	Dihydrogen phosphate	PO_4^{3-}	Phosphate

Table 2.6	Names of Oxoacids and Oxoanions That Contain Chlorine
Acid	**Anion**
$HClO_4$ (perchloric acid)	ClO_4^- (perchlorate)
$HClO_3$ (chloric acid)	ClO_3^- (chlorate)
$HClO_2$ (chlorous acid)	ClO_2^- (chlorite)
$HClO$ (hypochlorous acid)	ClO^- (hypochlorite)

Note that we usually omit the prefix "mono-" when there is only one H in the anion. Table 2.6 gives the names of the oxoacids and oxoanions that contain chlorine, and Figure 2.13 summarizes the nomenclature for the oxoacids and oxoanions.

Example 2.8   Name the oxoacid and oxoanion: (a) H_3PO_3 and (b) IO_4^-.

Reasoning and Solution   We refer to Figure 2.13 and Table 2.6.

(a) We start with our reference acid, phosphoric acid (H_3PO_4). Because H_3PO_3 has one fewer O atom, it is called phosphorous acid.

(b) The parent acid is HIO_4. Because the acid has one more O atom than our reference iodic acid (HIO_3), it is called periodic acid. Therefore, the anion derived from HIO_4 is called periodate.

Practice Exercise   Name the oxoacid and oxoanion: (a) HIO_3 and (b) HSO_4^-.

Similar problems: 2.44(f), 2.45(c).

Figure 2.13 *Naming oxoacids and oxoanions.*

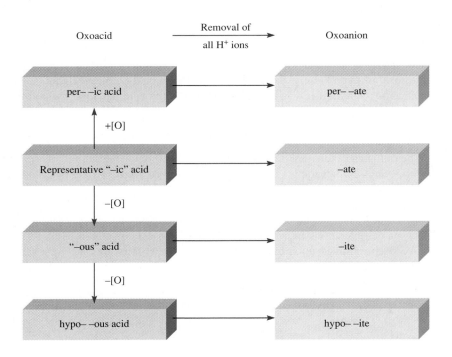

Naming Bases

A **base** can be described as *a substance that yields hydroxide ions (OH⁻) when dissolved in water*. Some examples are:

NaOH	Sodium hydroxide	$Ba(OH)_2$	Barium hydroxide
KOH	Potassium hydroxide		

Ammonia (NH_3), a molecular compound in the gaseous or pure liquid state, is also classified as a common base. At first glance this may seem to be an exception to the definition of a base. But note that as long as a substance *yields* hydroxide ions when dissolved in water, it need not contain hydroxide ions in its structure to be considered a base. In fact, when ammonia dissolves in water, NH_3 reacts partially with water to yield NH_4^+ and OH^- ions. Thus it is properly classified as a base.

Summary

Modern chemistry began with Dalton's atomic theory, which states that all matter is composed of tiny, indivisible particles called atoms; that all atoms of the same element are identical; that compounds contain atoms of different elements combined in whole-number ratios; and that atoms are neither created nor destroyed in chemical reactions (the law of conservation of mass).

Atoms of constituent elements in a particular compound are always combined in the same proportions by mass (law of definite proportions). When two elements can combine to form more than one type of compound, the masses of one element that combine with a fixed mass of the other element are in a ratio of small whole numbers (law of multiple proportions).

An atom consists of a very dense central nucleus made up of protons and neutrons, plus electrons that move about the nucleus at a relatively large distance from it. Protons are positively charged, neutrons have no charge, and electrons are negatively charged. Protons and neutrons have roughly the same mass, which is about 1840 times greater than the mass of an electron.

The atomic number of an element is the number of protons in the nucleus of an atom of the element; it determines the identity of an element. The mass number is the sum of the number of protons and the number of neutrons in the nucleus. Isotopes are atoms of the same element that have the same number of protons but different numbers of neutrons.

Chemical formulas combine the symbols for the constituent elements with whole-number subscripts to show the type and number of atoms contained in the smallest unit of a compound. The molecular formula conveys the specific number and types of atoms combined in each molecule of a compound. The empirical formula shows the simplest ratios of the atoms in a molecule.

Chemical compounds are either molecular compounds (in which the smallest units are discrete, individual molecules) or ionic compounds (in which positive and negative ions are held together by mutual attraction). Ionic compounds are made up of cations and anions, formed when atoms lose electrons and gain electrons, respectively.

The names of many inorganic compounds can be deduced from a set of simple rules. The formulas can be written from the names of the compounds.

Key Words

Questions and Problems

Structure of the Atom

Review Questions

2.1 Define these terms: (a) α particle, (b) β particle, (c) γ ray, (d) X ray.

2.2 List the types of radiation that are known to be emitted by radioactive elements.

2.3 Compare the properties of: α particles, cathode rays, protons, neutrons, and electrons. What is meant by the term "fundamental particle"?

2.4 Describe the contributions of these scientists to our knowledge of atomic structure: J. J. Thomson, R. A. Millikan, Ernest Rutherford, James Chadwick.

2.5 A sample of a radioactive element is found to be losing mass gradually. Explain what is happening to the sample.

2.6 Describe the experimental basis for believing that the nucleus occupies a very small fraction of the volume of the atom.

Problems

2.7 The diameter of a neutral helium atom is about 1×10^2 pm. Suppose that we could line up helium atoms side by side in contact with one another. Approximately how many atoms would it take to make the distance from end to end 1 cm?

2.8 Roughly speaking, the radius of an atom is about 10,000 times greater than that of its nucleus. If an atom were magnified so that the radius of its nucleus became 10 cm, what would be the radius of the atom in miles? (1 mi = 1609 m.)

Atomic Number, Mass Number, and Isotopes

Review Questions

2.9 Define these terms: (a) atomic number, (b) mass number. Why does a knowledge of atomic number enable us to deduce the number of electrons present in an atom?

2.10 Why do all atoms of an element have the same atomic number, although they may have different mass numbers? What do we call atoms of the same element with different mass numbers? Explain the meaning of each term in the symbol $_Z^A$X.

Problems

2.11 What is the mass number of an iron atom that has 28 neutrons?

2.12 Calculate the number of neutrons of ^{239}Pu.

2.13 For each of these species, determine the number of protons and the number of neutrons in the nucleus:

$$_2^3\text{He}, \; _2^4\text{He}, \; _{12}^{24}\text{Mg}, \; _{12}^{25}\text{Mg}, \; _{22}^{48}\text{Ti}, \; _{35}^{79}\text{Br}, \; _{78}^{195}\text{Pt}$$

2.14 Indicate the number of protons, neutrons, and electrons in each of these species:

$$_7^{15}\text{N}, \; _{16}^{33}\text{S}, \; _{29}^{63}\text{Cu}, \; _{38}^{84}\text{Sr}, \; _{56}^{130}\text{Ba}, \; _{74}^{186}\text{W}, \; _{80}^{202}\text{Hg}$$

2.15 Write the appropriate symbol for each of these isotopes: (a) $Z = 11, A = 23$; (b) $Z = 28, A = 64$.

2.16 Write the appropriate symbol for each of these isotopes: (a) $Z = 74, A = 186$; (b) $Z = 80; A = 201$.

The Periodic Table

Review Questions

2.17 What is the periodic table, and what is its significance in the study of chemistry? What are groups and periods in the periodic table?

2.18 Give two differences between a metal and a nonmetal.

2.19 Write the names and symbols for four elements in each of these categories: (a) nonmetal, (b) metal, (c) metalloid.

2.20 Define, with two examples, these terms: (a) alkali metals, (b) alkaline earth metals, (c) halogens, (d) noble gases.

2.21 Elements whose names end with "-ium" are usually metals; sodium is one example. Identify a nonmetal whose name also ends with "-ium".

2.22 Describe the changes in properties (from metals to nonmetals or from nonmetals to metals) as we move (a) down a periodic group and (b) across the periodic table.

2.23 Consult a handbook of chemical and physical data (ask your instructor where you can locate a copy of the handbook) to find (a) two metals less dense than water, (b) two metals more dense than mercury, (c) the densest known solid metallic element, (d) the densest known solid nonmetallic element.

2.24 Group these elements in pairs that you would expect to show similar chemical properties: K, F, P, Na, Cl, and N.

Molecules and Chemical Formulas

Review Questions

2.25 What is the difference between an atom and a molecule?

2.26 What are allotropes?

2.27 What does a chemical formula represent? What is the ratio of the atoms in these molecular formulas? (a) NO, (b) NCl_3, (c) N_2O_4, (d) P_4O_6.

2.28 Define molecular formula and empirical formula. What are the similarities and differences between the empirical formula and molecular formula of a compound?

2.29 Give an example of a case in which two molecules have different molecular formulas but the same empirical formula.

2.30 What does P_4 signify? How does it differ from 4P?

2.31 What is the difference between a molecule and a compound? Write a formula for a molecule that is also a compound and a formula for a molecule that is not a compound.

2.32 Give two examples of each of these: (a) a diatomic molecule containing atoms of the same element, (b) a diatomic molecule containing atoms of different elements, (c) a polyatomic molecule containing atoms of the same element, (d) a polyatomic molecule containing atoms of different elements.

Problems

2.33 Write the molecular formula of glycine, an amino acid present in proteins. The color codes are: black (carbon), blue (nitrogen), red (oxygen), and gray (hydrogen).

2.34 Write the molecular formula of ethanol. The color codes are: black (carbon), red (oxygen), and gray (hydrogen).

2.35 What are the empirical formulas of these compounds? (a) C_2N_2, (b) C_6H_6, (c) C_9H_{20}, (d) P_4O_{10}, (e) B_2H_6.

2.36 What are the empirical formulas of these compounds? (a) Al_2Br_6, (b) $Na_2S_2O_4$, (c) N_2O_5, (d) $K_2Cr_2O_7$.

Ionic Compounds

Review Questions

2.37 What is an ionic compound? How is electrical neutrality maintained in an ionic compound? Explain why the chemical formulas of ionic compounds are usually the same as their empirical formulas.

2.38 Compare and contrast the properties of molecular compounds and ionic compounds.

Problems

2.39 Give the number of protons and electrons in each of these ions: Na^+, Ca^{2+}, Al^{3+}, Fe^{2+}, I^-, F^-, S^{2-}, O^{2-}, N^{3-}.

2.40 Give the number of protons and electrons in each of these ions: K^+, Mg^{2+}, Fe^{3+}, Br^-, Mn^{2+}, C^{4-}, Cu^{2+}.

2.41 Which of these compounds are likely to be ionic? Which are likely to be molecular? $SiCl_4$, LiF, $BaCl_2$, B_2H_6, KCl, C_2H_4.

2.42 Which of these compounds are likely to be ionic? Which are likely to be molecular? CH_4, NaBr, BaF_2, CCl_4, ICl, CsCl, NF_3.

Naming Inorganic Compounds

Problems

2.43 Name these compounds: (a) Na_2CrO_4, (b) K_2HPO_4, (c) HBr (gas), (d) HBr (in water), (e) Li_2CO_3, (f) $K_2Cr_2O_7$,

(g) NH_4NO_2, (h) PF_3, (i) PF_5, (j) P_4O_6, (k) CdI_2, (l) $SrSO_4$, (m) $Al(OH)_3$.

2.44 Name these compounds: (a) $KClO$, (b) Ag_2CO_3, (c) $FeCl_2$, (d) $KMnO_4$, (e) $CsClO_3$, (f) HIO, (g) FeO, (h) Fe_2O_3, (i) $TiCl_4$, (j) NaH, (k) Li_3N, (l) Na_2O, (m) Na_2O_2.

2.45 Write the formulas for these compounds: (a) rubidium nitrite, (b) potassium sulfide, (c) perbromic acid, (d) magnesium phosphate, (e) calcium hydrogen phosphate, (f) boron trichloride, (g) iodine heptafluoride, (h) ammonium sulfate, (i) silver perchlorate, (j) iron(III) chromate.

2.46 Write the formulas for these compounds: (a) copper(I) cyanide, (b) strontium chlorite, (c) perchloric acid, (d) hydroiodic acid, (e) disodium ammonium phosphate, (f) lead(II) carbonate, (g) tin(II) fluoride, (h) tetraphosphorus decasulfide, (i) mercury(II) oxide, (j) mercury(I) iodide.

Additional Problems

2.47 One isotope of a metallic element has a mass number of 65 and has 35 neutrons in the nucleus. The cation derived from the isotope has 28 electrons. Write the symbol for this cation.

2.48 In which one of these pairs do the two species resemble each other most closely in chemical properties? (a) ^1_1H and $^1_1H^+$, (b) $^{14}_7N$ and $^{14}_7N^{3-}$, (c) $^{12}_6C$ and $^{13}_6C$.

2.49 This table gives numbers of electrons, protons, and neutrons in atoms or ions of a number of elements. (a) Which of the species are neutral? (b) Which are negatively charged? (c) Which are positively charged? (d) What are the conventional symbols for all the species?

Atom or ion or element	A	B	C	D	E	F	G
Number of electrons	5	10	18	28	36	5	9
Number of protons	5	7	19	30	35	5	9
Number of neutrons	5	7	20	36	46	6	10

2.50 What is wrong or ambiguous about these descriptions? (a) 1 g of hydrogen, (b) four molecules of NaCl.

2.51 These phosphorus sulfides are known: P_4S_3, P_4S_7, and P_4S_{10}. Do these compounds obey the law of multiple proportions?

2.52 Which of these are elements, which are molecules but not compounds, which are compounds but not molecules, and which are both compounds and molecules? (a) SO_2, (b) S_8, (c) Cs, (d) N_2O_5, (e) O, (f) O_2, (g) O_3, (h) CH_4, (i) KBr, (j) S, (k) P_4, (l) LiF.

2.53 Why is magnesium chloride ($MgCl_2$) not called magnesium(II) chloride?

2.54 Some compounds are better known by their common names than by their systematic chemical names. Consult a handbook, a dictionary, or your instructor for the chemical formulas of these substances: (a) dry ice, (b) table salt, (c) laughing gas, (d) marble (chalk, limestone), (e) quicklime, (f) slaked lime, (g) baking soda, (h) milk of magnesia.

2.55 Fill in the blanks in this table:

Symbol		$^{54}_{26}Fe^{2+}$			
Protons	5			79	86
Neutrons	6		16	117	136
Electrons	5		18	79	
Net charge			−3		0

2.56 (a) Which elements are most likely to form ionic compounds? (b) Which metallic elements are most likely to form cations with different charges?

2.57 Many ionic compounds contain either aluminum (a Group 3A metal) or a metal from Group 1A or Group 2A and a nonmetal—oxygen, nitrogen, or a halogen (Group 7A). Write the chemical formulas and names of all the binary compounds that can result from such combinations.

2.58 Which of these symbols provides more information about the atom: ^{23}Na or $_{11}Na$? Explain.

2.59 Write the chemical formulas and names of acids that contain Group 7A elements. Do the same for elements in Groups 3A, 4A, 5A, and 6A.

2.60 Of the 113 elements known, only two are liquids at room temperature (25°C). What are they? (*Hint:* One element is a familiar metal and the other element is in Group 7A.)

2.61 For the noble gases (the Group 8A elements) ^4_2He, $^{20}_{10}Ne$, $^{40}_{18}Ar$, $^{84}_{36}Kr$, and $^{132}_{54}Xe$, (a) determine the number of protons and neutrons in the nucleus of each atom and (b) determine the ratio of neutrons to protons in the nucleus of each atom. Describe any general trend you discover in the way this ratio changes with increasing atomic number.

2.62 List the elements that exist as gases at room temperature. (*Hint:* All except one element can be found in Groups 5A, 6A, 7A, and 8A.)

2.63 The Group 1B metals, Cu, Ag, and Au, are called coinage metals. What chemical properties make them specially suitable for making coins and jewels?

2.64 The elements in Group 8A of the periodic table are called noble gases. Can you guess the meaning of "noble" in this context?

2.65 The formula for calcium oxide is CaO. What are the formulas for magnesium oxide and strontium oxide?

2.66 A common mineral of barium is barytes, or barium sulfate ($BaSO_4$). Because elements in the same periodic group have similar chemical properties, we might expect to find some radium sulfate ($RaSO_4$) mixed with barytes because radium is the last member of Group 2A. However, the only source of radium compounds in nature is in uranium minerals. Why?

2.67 Fluorine reacts with hydrogen (H) and with deuterium (D) to form hydrogen fluoride (HF) and deuterium fluoride (DF) [deuterium ($^2_1$H) is an isotope of hydrogen]. Would a given amount of fluorine react with different masses of the two hydrogen isotopes? Does this violate the law of definite proportions? Explain.

2.68 Predict the formula and name of a binary compound formed from these elements: (a) Na and H, (b) B and O, (c) Na and S, (d) Al and F, (e) F and O, (f) Sr and Cl.

2.69 Identify each of the following elements: (a) a halogen whose anion contains 36 electrons, (b) a radioactive noble gas with 86 protons, (c) a Group 6A element whose anion contains 36 electrons, (d) an alkali metal cation that contains 36 electrons, (e) a Group 4A cation that contains 80 electrons.

Special Problem

2.70 (a) Describe Rutherford's experiment and how it led to the structure of the atom. How was he able to estimate the number of protons in a nucleus from the scattering of the α particles? (b) Consider the ^{23}Na atom. Given that the radius and mass of the nucleus are 3.04×10^{-15} m and 3.82×10^{-23} g, respectively, calculate the density of the nucleus in g/cm^3. The radius of a ^{23}Na atom is 186 pm. Calculate the density of the space occupied by the electrons in the sodium atom. Do your results support Rutherford's model of an atom? [The volume of a sphere is $(4/3)\pi r^3$.]

Answers to Practice Exercises

2.1 29 protons, 34 neutrons, and 29 electrons. **2.2** $CHCl_3$. **2.3** $C_4H_5N_2O$. **2.4** (a) Lead(II) oxide, (b) lithium sulfite. **2.5** (a) Rb_2SO_4, (b) BaH_2. **2.6** (a) Nitrogen trifluoride, (b) dichlorine heptoxide. **2.7** (a) SF_4, (b) N_2O_5. **2.8** (a) Iodic acid, (b) hydrogen sulfate ion.

When wood, paper, and wax are burned, they appear to lose mass. The decrease in mass that results from these combustion reactions was once attributed to the release of "phlogiston" into the air. For most of the eighteenth century scientists accepted the phlogiston theory. Then, in August 1774, the English chemist and clergyman Joseph Priestley isolated oxygen—which he called "dephlogisticated air"—as a product of the decomposition of mercury(II) oxide, HgO. The French chemist Antoine Lavoisier had noticed that nonmetals like phosphorus actually *gain* mass

Lavoisier igniting a mixture of hydrogen and oxygen gases.

when they burn in air. He concluded that these nonmetals must combine with something in the air. This substance turned out to be Priestley's dephlogisticated air. Lavoisier named the new element "oxygen" (from the Greek word meaning "to form acid"), because he knew that it is also a constituent of all acids.

Born in 1743, Lavoisier is generally regarded as the father of modern chemistry. He was noted for his carefully controlled experiments and for the use of quantitative measurements. By carrying out chemical reactions, such as the decomposition of mercury(II) oxide in a closed container, he showed that the total mass of the products equals the total mass of the reactants. In other words, the quantity of matter is not changed by chemical reactions. This observation is the basis of the law of conservation of mass and the principle underlying stoichiometry.

Lavoisier determined the composition of water by igniting a mixture of hydrogen and oxygen gases with an electric spark. He also served on the commission that established the metric system on which SI is based. Unfortunately, his scientific career was cut short by the French Revolution. A member of the nobility, Lavoisier was also a tax collector. For these "crimes," he was sent to the guillotine in 1794.

Stoichiometry

Chapter 3

ESSENTIAL CONCEPTS

 Atomic Mass and Molar Mass The mass of an atom, which is extremely small, is based on the carbon-12 isotope scale. An atom of the carbon-12 isotope is assigned a mass of exactly 12 atomic mass units (amu). To work with the more convenient scale of grams, chemists use the molar mass. The molar mass of carbon-12 is exactly 12 g and contains an Avogadro's number (6.022×10^{23}) of atoms. The molar masses of other elements are also expressed in grams and contain the same number of atoms. The molar mass of a molecule is the sum of the molar masses of its constituent atoms.

 Percent Composition of a Compound The makeup of a compound is most conveniently expressed in terms of its percent composition, which is the percent by mass of each element the compound contains. A knowledge of its chemical formula allows us to calculate the percent composition. Experimental determination of percent composition and the molar mass of a compound enables us to determine its chemical formula.

 Writing Chemical Equations An effective way to represent the outcome of a chemical reaction is to write a chemical equation, which uses chemical formulas to describe what happens. A chemical equation must be balanced so that we have the same number and type of atoms for the reactants, the starting materials, and the products, the substances formed at the end of the reaction.

 Mass Relationships of a Chemical Reaction A chemical equation allows us to predict the amount of product(s) formed, called the yield, knowing how much reactant(s) was (were) used. This information is of great importance for reactions run on the laboratory or industrial scale. In practice, the actual yield is almost always less than that predicted from the equation because of various complications.

3.1 Atomic Mass

Section 3.4 describes a method for determining atomic mass.

In this chapter we will use what we have learned about chemical structure and formulas in studying the mass relationships of atoms and molecules. These relationships in turn will help us to explain the composition of compounds and the ways in which the composition changes.

The mass of an atom is related to the number of electrons, protons, and neutrons it has. Knowledge of an atom's mass is important in laboratory work. But atoms are extremely small particles—even the smallest speck of dust that our unaided eyes can detect contains as many as 1×10^{16} atoms! Clearly we cannot weigh a single atom, but it is possible to determine the mass of one atom *relative* to another experimentally. The first step is to assign a value to the mass of one atom of a given element so that it can be used as a standard.

One atomic mass unit is also called one dalton.

By international agreement, an atom of the carbon isotope (called carbon-12) that has six protons and six neutrons has a mass of exactly 12 *atomic mass units (amu)*. This carbon-12 atom serves as the standard, so one atomic mass unit is defined as *a mass exactly equal to one-twelfth the mass of one carbon-12 atom:*

$$\text{mass of one carbon-12 atom} = 12 \text{ amu}$$

$$1 \text{ amu} = \frac{\text{mass of one carbon-12 atom}}{12}$$

Experiments have shown that, on average, a hydrogen atom is only 8.400 percent as massive as the standard carbon-12 atom. If we accept the mass of one carbon-12 atom to be exactly 12 amu, then the **atomic mass** (that is, *the mass of the atom in atomic mass units*) of hydrogen must be 0.08400×12.00 amu, or 1.008 amu. (Note that atomic mass is also called atomic weight.) Similar calculations show that the atomic mass of oxygen is 16.00 amu and that of iron is 55.85 amu. Thus although we do not know just how much an average iron atom's mass is, we know that it is approximately 56 times as massive as a hydrogen atom.

Average Atomic Mass

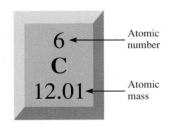

When you look up the atomic mass of carbon in a table such as the one on the inside front cover of this book, you will find that its value is not 12.00 amu but 12.01 amu. The reason for the difference is that most naturally occurring elements (including carbon) have more than one isotope. This means that when we measure the atomic mass of an element, we must generally settle for the *average* mass of the naturally occurring mixture of isotopes. For example, the natural abundances of carbon-12 and carbon-13 are 98.90 percent and 1.10 percent, respectively. The atomic mass of carbon-13 has been determined to be 13.00335 amu. Thus the average atomic mass of carbon can be calculated as follows:

average atomic mass of natural carbon
$$= (0.9890)(12.00000 \text{ amu}) + (0.0110)(13.00335 \text{ amu})$$
$$= 12.0 \text{ amu}$$

A more accurate determination gives the atomic mass of carbon as 12.01 amu. Note that in calculations involving percentages, we need to convert percentages to fractions.

For example, 98.90 percent becomes 98.90/100, or 0.9890. Because there are many more carbon-12 atoms than carbon-13 atoms in naturally occurring carbon, the average atomic mass is much closer to 12 amu than 13 amu.

It is important to understand that when we say that the atomic mass of carbon is 12.01 amu, we are referring to the *average* value. If carbon atoms could be examined individually, we would find either an atom of atomic mass 12.00000 amu or one of 13.00335 amu, but never one of 12.01 amu.

Example 3.1   Copper, a metal known since ancient times, is used in electrical cables and pennies, among other things. The atomic masses of its two stable isotopes, $^{63}_{29}Cu$ (69.09 percent) and $^{65}_{29}Cu$ (30.91 percent), are 62.93 amu and 64.9278 amu, respectively. Calculate the average atomic mass of copper. The percentages in parentheses denote the relative abundances.

Reasoning and Solution   Each isotope contributes to the atomic mass of copper according to its natural abundance. Therefore, the first step is to convert the percentages to fractions. Thus 69.09 percent becomes 0.6909 and 30.91 percent becomes 0.3091. Next we calculate the average atomic mass as follows:

$$(0.6909)(62.93 \text{ amu}) + (0.3091)(64.9278 \text{ amu}) = 63.55 \text{ amu}$$

Practice Exercise   The atomic masses of the two stable isotopes of boron, $^{10}_{5}B$ (19.78 percent) and $^{11}_{5}B$ (80.22 percent), are 10.0129 amu and 11.0093 amu, respectively. Calculate the average atomic mass of boron.

Copper.

Similar problems: 3.9, 3.10.

The atomic masses of many elements have been accurately determined to five or six significant figures. However, for our purposes we will normally use atomic masses accurate only to four significant figures (see the table of atomic masses inside the front cover).

3.2 Molar Mass of an Element and Avogadro's Number

Atomic mass units provide a relative scale for the masses of the elements. But because atoms have such small masses, no usable scale can be devised to weigh them in calibrated units of atomic mass units. In any real situation, we deal with macroscopic samples containing enormous numbers of atoms. Therefore it is convenient to have a special unit to describe a very large number of atoms. The idea of a unit to denote a particular number of objects is not new. For example, the pair (2 items), the dozen (12 items), and the gross (144 items) are all familiar units. Chemists measure atoms and molecules in moles.

In the SI system the **mole (mol)** is *the amount of a substance that contains as many elementary entities (atoms, molecules, or other particles) as there are atoms in exactly 12 g (or 0.012 kg) of the carbon-12 isotope.* The actual number of atoms in 12 g of carbon-12 is determined experimentally. This number is called **Avogadro's**

The adjective formed from the noun "mole" is "molar."

Figure 3.1 *One mole each of several common elements: copper (as pennies), iron (as nails), carbon (black charcoal powder), sulfur (yellow powder), and mercury (shiny liquid metal).*

number (N_A), in honor of the Italian scientist Amedeo Avogadro. The currently accepted value is

$$N_A = 6.0221367 \times 10^{23}$$

Generally, we round Avogadro's number to 6.022×10^{23}. Thus, just as one dozen oranges contains 12 oranges, 1 mole of hydrogen atoms contains 6.022×10^{23} H atoms. Figure 3.1 shows 1 mole each of several common elements.

We have seen that 1 mole of carbon-12 atoms has a mass of exactly 12 g and contains 6.022×10^{23} atoms. This mass of carbon-12 is its *molar mass ($\mathcal{M}$),* defined as *the mass (in grams or kilograms) of 1 mole of units* (such as atoms or molecules) of a substance. Note that the molar mass of carbon-12 (in grams) is numerically equal to its atomic mass in amu. Likewise, the atomic mass of sodium (Na) is 22.99 amu and its molar mass is 22.99 g; the atomic mass of phosphorus is 30.97 amu and its molar mass is 30.97 g; and so on. If we know the atomic mass of an element, we also know its molar mass.

Using atomic mass and molar mass, we can calculate the mass in grams of a single carbon-12 atom. From our discussion we know that 1 mole of carbon-12 atoms weighs exactly 12 grams. This enables us to write the equality

$$12.00 \text{ g carbon-12} = 1 \text{ mol carbon-12 atoms}$$

Therefore, we can write the unit factor as

$$\frac{12.00 \text{ g carbon-12}}{1 \text{ mol carbon-12 atoms}} = 1$$

(Note that we use the unit "mol" to represent "mole" in calculations.) Similarly, because there are 6.022×10^{23} atoms in 1 mole of carbon-12 atoms, we have

$$1 \text{ mol carbon-12 atoms} = 6.022 \times 10^{23} \text{ carbon-12 atoms}$$

and the unit factor is

$$\frac{1 \text{ mol carbon-12 atoms}}{6.022 \times 10^{23} \text{ carbon-12 atoms}} = 1$$

In calculations the units of molar mass are g/mol or kg/mol.

We can now calculate the mass (in grams) of 1 carbon-12 atom as:

$$1 \text{ carbon-12 atom} \times \frac{1 \text{ mol carbon-12 atoms}}{6.022 \times 10^{23} \text{ carbon-12 atoms}} \times \frac{12.00 \text{ g carbon-12}}{1 \text{ mol carbon-12 atoms}}$$

$$= 1.993 \times 10^{-23} \text{ g carbon-12}$$

We can use this result to determine the relationship between atomic mass units and grams. Because the mass of every carbon-12 atom is exactly 12 amu, the number of grams equivalent to 1 amu is

$$\frac{\text{gram}}{\text{amu}} = \frac{1.993 \times 10^{-23} \text{ g}}{1 \text{ carbon-12 atom}} \times \frac{1 \text{ carbon-12 atom}}{12 \text{ amu}}$$

$$= 1.661 \times 10^{-24} \text{ g/amu}$$

Thus

$$1 \text{ amu} = 1.661 \times 10^{-24} \text{ g}$$

and

$$1 \text{ g} = 6.022 \times 10^{23} \text{ amu}$$

This example shows that Avogadro's number can be used to convert from the atomic mass units to mass in grams and vice versa.

The notions of Avogadro's number and molar mass enable us to carry out conversions between mass and moles of atoms and between the number of atoms and mass and to calculate the mass of a single atom. We will employ these unit factors in the calculations:

$$\frac{1 \text{ mol X}}{\text{molar mass of X}} = 1 \qquad \frac{1 \text{ mol X}}{6.022 \times 10^{23} \text{ X atoms}} = 1$$

The molar masses of the elements (in g/mol) are given on the inside front cover of the book.

where X represents the symbol of an element. Figure 3.2 summarizes the relationships between the mass of an element and the number of moles of an element and between moles of an element and the number of atoms of an element. Using the proper unit factors we can convert one quantity to another.

Example 3.2 Zinc (Zn) is a silvery metal that is used in making brass (with copper) and in plating iron to prevent corrosion. How many grams of Zn are in 0.356 mole of Zn?

Reasoning and Solution To convert moles to grams, we need the unit factor

$$\frac{\text{molar mass of Zn}}{1 \text{ mol Zn}} = 1$$

The molar mass of Zn is 65.39 g, so that the mass of Zn in 0.356 mol is

$$0.356 \text{ mol Zn} \times \frac{65.39 \text{ g Zn}}{1 \text{ mol Zn}} = 23.3 \text{ g Zn}$$

Thus there are 23.3 g of Zn in 0.356 mole of Zn.

Comment The mass in 0.356 mole of Zn should be less than the molar mass of Zn.

Practice Exercise Calculate the number of grams of lead (Pb) in 12.4 moles of lead.

Zinc.

Similar problem: 3.18.

Mass of element (m) $\xrightarrow[n\mathcal{M}]{m/\mathcal{M}}$ Number of moles of element (n) $\xleftarrow[N/N_A]{nN_A}$ Number of atoms of element (N)

Figure 3.2 *The relationships between mass (in grams) of an element and number of moles of an element, and between number of moles of an element and number of atoms of an element. $\mathcal{M}$ is the molar mass (g/mol) of the element and N_A is Avogadro's number.*

Elemental sulfur (S_8) consists of eight S atoms joined in a ring.

Similar problem: 3.22.

Example 3.3   Sulfur (S) is a nonmetallic element. Its presence in coal gives rise to the acid rain phenomenon. How many atoms are in 16.3 g of S?

Reasoning and Solution   Solving this problem requires two steps. First, we need to find the number of moles of S in 16.3 g of S. Next, we need to calculate the number of S atoms from the known number of moles of S. We can combine the two steps as follows:

$$16.3 \text{ g S} \times \frac{1 \text{ mol S}}{32.07 \text{ g S}} \times \frac{6.022 \times 10^{23} \text{ S atoms}}{1 \text{ mol S}} = 3.06 \times 10^{23} \text{ S atoms}$$

Practice Exercise   Calculate the number of atoms in 0.551 g of potassium (K).

Silver.

Similar problems: 3.19, 3.20.

Example 3.4   Silver (Ag) is a precious metal used mainly in jewelry. What is the mass (in grams) of one Ag atom?

Reasoning and Solution   There are 6.022×10^{23} Ag atoms in 1 mole of Ag and the molar mass of Ag is 107.9 g. Combining these two unit factors, we calculate the mass of a single Ag atom as follows:

$$1 \text{ Ag atom} \times \frac{1 \text{ mol Ag}}{6.022 \times 10^{23} \text{ Ag atoms}} \times \frac{107.9 \text{ g}}{1 \text{ mol Ag}} = 1.792 \times 10^{-22} \text{ g}$$

Practice Exercise   What is the mass (in grams) of one iodine (I) atom?

3.3 Molecular Mass

If we know the atomic masses of the component atoms, we can calculate the mass of a molecule. The ***molecular mass*** (sometimes called *molecular weight*) is *the sum of the atomic masses (in amu) in the molecule.* For example, the molecular mass of H_2O is

$$2(\text{atomic mass of H}) + \text{atomic mass of O}$$

or

$$2(1.008 \text{ amu}) + 16.00 \text{ amu} = 18.02 \text{ amu}$$

In general, we need to multiply the atomic mass of each element by the number of atoms of that element present in the molecule and sum over all the elements.

Example 3.5 Calculate the molecular masses of these compounds: (a) sulfur dioxide (SO_2) and (b) caffeine ($C_8H_{10}N_4O_2$).

Reasoning and Solution To calculate molecular mass, we need to count the number of each type of atom in the molecule and look up its atomic mass in the periodic table (inside front cover).

(a) There are one S atom and two O atoms in sulfur dioxide, so that

$$\text{molecular mass of } SO_2 = 32.07 \text{ amu} + 2(16.00 \text{ amu})$$
$$= 64.07 \text{ amu}$$

(b) There are eight C atoms, ten H atoms, four N atoms, and two O atoms in caffeine, so that the molecular mass of $C_8H_{10}N_4O_2$ is given by

$$8(12.01 \text{ amu}) + 10(1.008 \text{ amu}) + 4(14.01 \text{ amu}) + 2(16.00 \text{ amu}) = 194.20 \text{ amu}$$

Practice Exercise What is the molecular mass of methanol (CH_4O)?

SO_2

Similar problem: 3.26.

From the molecular mass we can determine the molar mass of a molecule or compound. The molar mass of a compound (in grams) is numerically equal to its molecular mass (in amu). For example, the molecular mass of water is 18.02 amu, so its molar mass is 18.02 g. Note that 1 mole of water weighs 18.02 g and contains 6.022×10^{23} H_2O *molecules,* just as 1 mole of elemental carbon contains 6.022×10^{23} carbon *atoms.*

As Examples 3.6 and 3.7 show, a knowledge of the molar mass enables us to calculate the numbers of moles and individual atoms in a given quantity of a compound.

Example 3.6 Methane (CH_4) is the principal component of natural gas. How many moles of CH_4 are present in 6.07 g of CH_4?

Reasoning and Solution This problem is just the opposite of Example 3.2, except that we are dealing with molecules instead of atoms. Therefore, the first step is to calculate the molar mass of CH_4:

$$\text{molar mass of } CH_4 = 12.01 \text{ g} + 4(1.008 \text{ g})$$
$$= 16.04 \text{ g}$$

From the unit factor (1 mol CH_4/16.04 g CH_4) we calculate the number of moles of CH_4 as follows:

$$6.07 \text{ g } \cancel{CH_4} \times \frac{1 \text{ mol } CH_4}{16.04 \text{ g } \cancel{CH_4}} = 0.378 \text{ mol } CH_4$$

Comment Because 6.07 g is smaller than the molar mass, the answer is reasonable.

Practice Exercise Calculate the number of moles of chloroform ($CHCl_3$) in 198 g of chloroform.

Methane gas burning on a cooking range.

Similar problem: 3.28.

Urea.

Similar problem: 3.30.

Example 3.7　How many hydrogen atoms are present in 25.6 g of urea [(NH$_2$)$_2$CO], which is used as a fertilizer, in animal feed, and in the manufacture of polymers? The molar mass of urea is 60.06 g.

Reasoning and Solution　First we follow the procedure in Example 3.3 to calculate the number of molecules present in 25.6 g of urea. Next we note that there are four hydrogen atoms in every urea molecule. Combining both steps we calculate the total number of hydrogen atoms present

$$25.6 \text{ g (NH}_2)_2\text{CO} \times \frac{1 \text{ mol (NH}_2)_2\text{CO}}{60.06 \text{ g (NH}_2)_2\text{CO}} \times \frac{6.022 \times 10^{23} \text{ molecules (NH}_2)_2\text{CO}}{1 \text{ mol (NH}_2)_2\text{CO}}$$

$$\times \frac{4 \text{ H atoms}}{1 \text{ molecule (NH}_2)_2\text{CO}} = 1.03 \times 10^{24} \text{ H atoms}$$

We could calculate the number of nitrogen, carbon, and oxygen atoms by the same procedure. However, there is a shortcut. Note that the ratio of nitrogen to hydrogen atoms in urea is 2/4, or 1/2, and that of oxygen (and carbon) to hydrogen atoms is 1/4. Therefore, the number of nitrogen atoms in 25.6 g of urea is $(1/2)(1.03 \times 10^{24})$, or 5.15×10^{23} atoms. The number of oxygen atoms (and carbon atoms) is $(1/4)(1.03 \times 10^{24})$, or 2.58×10^{23} atoms.

Practice Exercise　How many H atoms are in 72.5 g of isopropanol (rubbing alcohol), C$_3$H$_8$O?

3.4 The Mass Spectrometer

The most direct and most accurate method for determining atomic and molecular masses is mass spectrometry, which is depicted in Figure 3.3. In a *mass spectrometer,* a gaseous sample is bombarded by a stream of high-energy electrons. Collisions between the electrons and the gaseous atoms (or molecules) produce positive ions by dislodging an electron from each atom or molecule. These positive ions (of mass m and charge e) are accelerated by two oppositely charged plates as they pass through the plates. The emerging ions are deflected into a circular path by a magnet. The radius of the path depends on the charge-to-mass ratio (that is, e/m). Ions of smaller e/m ratio

Figure 3.3 *Schematic diagram of one type of mass spectrometer.*

Figure 3.4 *The mass spectrum of the three isotopes of neon.*

trace a wider curve than those having a larger e/m ratio, so that ions with equal charges but different masses are separated from one another. The mass of each ion (and hence its parent atom or molecule) is determined from the magnitude of its deflection. Eventually the ions arrive at the detector, which registers a current for each type of ion. The amount of current generated is directly proportional to the number of ions, so it enables us to determine the relative abundance of isotopes.

The first mass spectrometer, developed in the 1920s by the English physicist F. W. Aston, was crude by today's standards. Nevertheless, it provided indisputable evidence of the existence of isotopes—neon-20 (atomic mass 19.9924 amu and natural abundance 90.92 percent) and neon-22 (atomic mass 21.9914 amu and natural abundance 8.82 percent). When more sophisticated and sensitive mass spectrometers became available, scientists were surprised to discover that neon has a third stable isotope with an atomic mass of 20.9940 amu and natural abundance 0.257 percent (Figure 3.4). This example illustrates how very important experimental accuracy is to a quantitative science like chemistry. Early experiments failed to detect neon-21 because its natural abundance is just 0.257 percent. In other words, only 26 in 10,000 Ne atoms are neon-21. The masses of molecules can be determined in a similar manner by the mass spectrometer.

3.5 Percent Composition of Compounds

As we have seen, the formula of a compound tells us the numbers of atoms of each element in a unit of the compound. However, suppose we needed to verify the purity of a compound for use in a laboratory experiment. From the formula we could calculate what percent of the total mass of the compound is contributed by each element. Then, by comparing the result to the percent composition obtained experimentally for our sample, we could determine the purity of the sample.

The ***percent composition*** is the *percent by mass of each element in a compound.* Percent composition is obtained by dividing the mass of each element in 1 mole of the compound by the molar mass of the compound and multiplying by 100 percent. Mathematically, the percent composition of an element in a compound is expressed as

$$\text{percent composition of an element} = \frac{n \times \text{molar mass of element}}{\text{molar mass of compound}} \times 100\% \quad (3.1)$$

where n is the number of moles of the element in 1 mole of the compound. For example, in 1 mole of hydrogen peroxide (H_2O_2) there are 2 moles of H atoms and 2 moles of O atoms. The molar masses of H_2O_2, H, and O are 34.02 g, 1.008 g, and 16.00 g, respectively. Therefore, the percent composition of H_2O_2 is calculated as follows:

$$\%H = \frac{2 \times 1.008 \text{ g}}{34.02 \text{ g}} \times 100\% = 5.926\%$$

$$\%O = \frac{2 \times 16.00 \text{ g}}{34.02 \text{ g}} \times 100\% = 94.06\%$$

The sum of the percentages is 5.926 percent + 94.06 percent = 99.99 percent. The small discrepancy from 100 percent is due to the way we rounded off the molar masses of the elements. Note that the empirical formula (HO) would give us the same results.

H_3PO_4

Similar problem: 3.42.

Example 3.8   Phosphoric acid (H_3PO_4) is a colorless, syrupy liquid used in detergents, fertilizers, and toothpastes, and in carbonated beverages for a "tangy" flavor. Calculate the percent composition by mass of H, P, and O in this compound.

Reasoning and Solution   The percent by mass of each element is given by the total mass of the atoms of the element in 1 mole of the compound divided by the molar mass of the compound, multiplied by 100 percent. The molar mass of H_3PO_4 is 97.99 g/mol. Therefore, the percent by mass of each of the elements in H_3PO_4 is

$$\%H = \frac{3(1.008 \text{ g})}{97.99 \text{ g}} \times 100\% = 3.086\%$$

$$\%P = \frac{30.97 \text{ g}}{97.99 \text{ g}} \times 100\% = 31.61\%$$

$$\%O = \frac{4(16.00 \text{ g})}{97.99 \text{ g}} \times 100\% = 65.31\%$$

The sum of the percentages is (3.086% + 31.61% + 65.31%) = 100.01%. The small discrepancy from 100 percent is due to the way we rounded off.

Practice Exercise   Calculate the percent composition by mass of each of the elements in sulfuric acid (H_2SO_4).

The procedure used in Example 3.8 can be reversed if necessary. Given the percent composition by mass of a compound, we can determine the empirical formula of the compound (Figure 3.5).

Example 3.9 Ascorbic acid (vitamin C) cures scurvy and may help prevent the common cold. It is composed of 40.92 percent carbon (C), 4.58 percent hydrogen (H), and 54.50 percent oxygen (O) by mass. Determine its empirical formula.

Reasoning and Solution To solve a problem of this type, it is convenient to assume that we started with 100 g of the compound so that each percentage can be directly converted to grams. Therefore, in this sample there will be 40.92 g of C, 4.58 g of H, and 54.50 g of O. Next, we need to calculate the number of moles of each element in the compound. Let n_C, n_H, and n_O be the number of moles of elements present. Using the molar masses of these elements, we write

$$n_C = 40.92 \text{ g C} \times \frac{1 \text{ mol C}}{12.01 \text{ g C}} = 3.407 \text{ mol C}$$

$$n_H = 4.58 \text{ g H} \times \frac{1 \text{ mol H}}{1.008 \text{ g H}} = 4.54 \text{ mol H}$$

$$n_O = 54.50 \text{ g O} \times \frac{1 \text{ mol O}}{16.00 \text{ g O}} = 3.406 \text{ mol O}$$

Thus, we arrive at the formula $C_{3.407}H_{4.54}O_{3.406}$, which gives the identity and the ratios of atoms present. However, because chemical formulas are written with whole numbers, we cannot have 3.407 C atoms, 4.54 H atoms, and 3.406 O atoms. We can make some of the subscripts whole numbers by dividing all the subscripts by the *smallest* subscript (3.406):

$$\text{C:} \frac{3.407}{3.406} = 1 \qquad \text{H:} \frac{4.54}{3.406} = 1.33 \qquad \text{O:} \frac{3.406}{3.406} = 1$$

This gives us $CH_{1.33}O$ as the formula for ascorbic acid. Next, we need to convert 1.33, the subscript for H, into an integer. This can be done by a trial-and-error procedure:

$$1.33 \times 1 = 1.33$$
$$1.33 \times 2 = 2.66$$
$$1.33 \times 3 = 3.99 \approx 4$$

Because 1.33×3 gives us an integer (4), we can multiply all the subscripts by 3 and obtain $C_3H_4O_3$ as the empirical formula for ascorbic acid.

Comment The molecular formula of ascorbic acid is $C_6H_8O_6$.

Practice Exercise Determine the empirical formula of a compound having this percent composition by mass: K: 24.75 percent; Mn: 34.77 percent; O: 40.51 percent.

Figure 3.5 *Procedure for calculating the empirical formula of a compound from its percent compositions.*

Ascorbic acid.

Similar problems: 3.51, 3.52.

Chemists often want to know the actual mass of an element in a certain mass of a compound. For example, in the mining industry, this information will tell them about the quality of the ore. Because the percent composition by mass of the element in the substance can be readily calculated, such a problem can be solved in a rather direct way.

Chalcopyrite.

Example 3.10 Chalcopyrite ($CuFeS_2$) is a principal ore of copper. Calculate the number of kilograms of Cu in 3.71×10^3 kg of chalcopyrite.

Reasoning and Solution A little thought will convince you that the product of the percent composition by mass of an element in a compound and the mass of the compound should yield the mass of the element in that compound. The molar masses of Cu and $CuFeS_2$ are 63.55 g and 183.5 g, respectively, so the percent composition by mass of Cu is

$$\%Cu = \frac{63.55 \text{ g}}{183.5 \text{ g}} \times 100\% = 34.63\%$$

To calculate the mass of Cu in a 3.71×10^3-kg sample of $CuFeS_2$, we need to convert the percentage to a fraction (that is, convert 34.63 percent to 34.63/100, or 0.3463) and write

$$\text{mass of Cu in } CuFeS_2 = 0.3463 \times 3.71 \times 10^3 \text{ kg} = 1.28 \times 10^3 \text{ kg}$$

This calculation can be simplified by combining the above two steps as:

$$\text{mass of Cu in } CuFeS_2 = 3.71 \times 10^3 \text{ kg } CuFeS_2 \times \frac{63.55 \text{ g Cu}}{183.5 \text{ g } CuFeS_2}$$

$$= 1.28 \times 10^3 \text{ kg Cu}$$

Practice Exercise Calculate the number of grams of Al in 371 g of Al_2O_3.

Similar problem: 3.50.

3.6 Experimental Determination of Empirical Formulas

The fact that we can determine the empirical formula of a compound if we know the percent composition enables us to identify compounds experimentally. The procedure is as follows. First, chemical analysis tells us the number of grams of each element present in a given amount of a compound. Then we convert the quantities in grams to number of moles of each element. Finally, using the method given in Example 3.9, we find the empirical formula of the compound.

As a specific example let us consider the compound ethanol. When ethanol is burned in an apparatus such as that shown in Figure 3.6, carbon dioxide (CO_2) and water (H_2O) are given off. Because neither carbon nor hydrogen was in the inlet gas, we can conclude that both carbon (C) and hydrogen (H) were present in

Figure 3.6 *Apparatus for determining the empirical formula of ethanol. The absorbers are substances that can retain water and carbon dioxide, respectively.*

ethanol and that oxygen (O) may also be present. (Molecular oxygen was added in the combustion process, but some of the oxygen may also have come from the original ethanol sample.)

The masses of CO_2 and of H_2O produced can be determined by measuring the increase in mass of the CO_2 and H_2O absorbers, respectively. Suppose that in one experiment the combustion of 11.5 g of ethanol produced 22.0 g of CO_2 and 13.5 g of H_2O. We can calculate the mass of carbon and hydrogen in the original 11.5-g sample of ethanol as:

$$\text{mass of C} = 22.0 \text{ g } CO_2 \times \frac{1 \text{ mol } CO_2}{44.01 \text{ g } CO_2} \times \frac{1 \text{ mol C}}{1 \text{ mol } CO_2} \times \frac{12.01 \text{ g C}}{1 \text{ mol C}}$$

$$= 6.00 \text{ g C}$$

$$\text{mass of H} = 13.5 \text{ g } H_2O \times \frac{1 \text{ mol } H_2O}{18.02 \text{ g } H_2O} \times \frac{2 \text{ mol H}}{1 \text{ mol } H_2O} \times \frac{1.008 \text{ g H}}{1 \text{ mol H}}$$

$$= 1.51 \text{ g H}$$

Thus, 11.5 g of ethanol contains 6.00 g of carbon and 1.51 g of hydrogen. The remainder must be oxygen, whose mass is

$$\text{mass of O} = \text{mass of sample} - (\text{mass of C} + \text{mass of H})$$

$$= 11.5 \text{ g} - (6.00 \text{ g} + 1.51 \text{ g})$$

$$= 4.0 \text{ g}$$

The number of moles of each element present in 11.5 g of ethanol is

$$\text{moles of C} = 6.00 \text{ g C} \times \frac{1 \text{ mol C}}{12.01 \text{ g C}} = 0.500 \text{ mol C}$$

$$\text{moles of H} = 1.51 \text{ g H} \times \frac{1 \text{ mol H}}{1.008 \text{ g H}} = 1.50 \text{ mol H}$$

$$\text{moles of O} = 4.0 \text{ g O} \times \frac{1 \text{ mol O}}{16.00 \text{ g O}} = 0.25 \text{ mol O}$$

The formula of ethanol is therefore $C_{0.50}H_{1.5}O_{0.25}$ (we round off the number of moles to two significant figures). Because the number of atoms must be an integer, we divide the subscripts by 0.25, the smallest subscript, and obtain for the empirical formula C_2H_6O.

Now we can better understand the word "empirical," which literally means "based only on observation and measurement." The empirical formula of ethanol is determined from analysis of the compound in terms of its component elements. No knowledge of how the atoms are linked together in the compound is required.

The molecular formula of ethanol is the same as its empirical formula.

Determination of Molecular Formulas

The formula calculated from percent composition by mass is always the empirical formula because the coefficients in the formula are always reduced to the smallest whole numbers. To calculate the actual, molecular formula we must know the *approximate* molar mass of the compound in addition to its empirical formula. Knowing that the molar mass of a compound must be an integral multiple of the molar mass of its empirical formula, we can use the molar mass to find the molecular formula.

Example 3.11   A sample of a compound of nitrogen (N) and oxygen (O) contains 1.52 g of N and 3.47 g of O. The molar mass of this compound is known to be between 90 g and 95 g. Determine the molecular formula and the accurate molar mass of the compound.

Reasoning and Solution   First we need to determine the empirical formula of the compound and then compare its molar mass with the true molar mass of the compound, that is, the experimentally determined molar mass. This comparison will reveal the relationship between the empirical formula and the molecular formula.

Let n_N and n_O be the number of moles of nitrogen and oxygen. Then

$$n_N = 1.52 \text{ g N} \times \frac{1 \text{ mol N}}{14.01 \text{ g N}} = 0.108 \text{ mol N}$$

$$n_O = 3.47 \text{ g O} \times \frac{1 \text{ mol O}}{16.00 \text{ g O}} = 0.217 \text{ mol O}$$

Thus the formula of the compound is $N_{0.108}O_{0.217}$. We divide the subscripts by the smaller subscript, 0.108. After rounding off, we obtain NO_2 as the empirical formula. The molecular formula will be the same as the empirical formula or some integral multiple of it (for example, two, three, four, or more times the empirical formula). The molar mass of the empirical formula NO_2 is

$$\text{empirical molar mass} = 14.01 \text{ g} + 2(16.00 \text{ g}) = 46.02 \text{ g}$$

Next we determine the number of (NO_2) units present in the molecular formula. This number is found by taking the ratio

$$\frac{\text{molar mass}}{\text{empirical molar mass}} = \frac{95 \text{ g}}{46.02 \text{ g}} = 2.1 \approx 2$$

Therefore the molar mass of the compound is twice the molar mass of the empirical formula. Consequently, there are two NO_2 units in each molecule of the compound, and the molecular formula is (NO_2)$_2$, or N_2O_4. The molar mass of the compound is 2(46.02 g), or 92.04 g, which is between 90 g and 95 g.

Comment   Note that in determining the molecular formula from the empirical formula, we need only know the approximate molar mass of the compound. The reason is that the true molar mass is an integral multiple ($1\times$, $2\times$, $3\times$, . . .) of the molar mass corresponding to the empirical formula. Thus the ratio (molar mass/empirical molar mass) will always be close to an integer.

Practice Exercise   A sample of a compound of boron (B) and hydrogen (H) contains 6.444 g of B and 1.803 g of H. The molar mass of the compound is about 30 g. What is its molecular formula?

N_2O_4

Similar problem: 3.54.

3.7 Chemical Reactions and Chemical Equations

Having discussed the masses of atoms and molecules, we turn next to what happens to atoms and molecules in a **chemical reaction,** *a process in which a substance (or substances) is changed into one or more new substances.* To communicate with one another about chemical reactions, chemists have devised a standard way to represent them using chemical equations. A **chemical equation** *uses chemical symbols to show what happens during a chemical reaction.* In this section we will learn how to write chemical equations and balance them.

Writing Chemical Equations

Consider what happens when hydrogen gas (H_2) burns in air (which contains oxygen, O_2) to form water (H_2O). This reaction can be represented by the chemical equation

$$H_2 + O_2 \longrightarrow H_2O \tag{3.2}$$

where the "plus" sign means "reacts with" and the arrow means "to yield." Thus, this symbolic expression can be read: "Molecular hydrogen reacts with molecular oxygen to yield water." The reaction is assumed to proceed from left to right as the arrow indicates.

Equation (3.2) is not complete, however, because there are twice as many oxygen atoms on the left side of the arrow (two) as on the right side (one). To conform with the law of conservation of mass, there must be the same number of each type of atom on both sides of the arrow; that is, we must have as many atoms after the reaction ends as we did before it started. We can *balance* Equation (3.2) by placing the appropriate coefficient (2 in this case) in front of H_2 and H_2O:

$$2H_2 + O_2 \longrightarrow 2H_2O \tag{3.3}$$

When the coefficient is 1, as in the case of O_2, it is not shown.

This *balanced chemical equation* shows that "two hydrogen molecules can combine or react with one oxygen molecule to form two water molecules" (Figure 3.7). Because the ratio of the number of molecules is equal to the ratio of the number of moles, the equation can also be read as "2 moles of hydrogen molecules react with 1 mole of oxygen molecules to produce 2 moles of water molecules." We know the mass of a mole of each of these substances, so we can also interpret the equation as "4.04 g of H_2 react with 32.00 g of O_2 to give 36.04 g of H_2O." These three ways of reading the equation are summarized in Table 3.1.

We refer to H_2 and O_2 in Equation (3.3) as **reactants,** which are *the starting materials in a chemical reaction.* Water is the **product,** which is *the substance formed as a result of a chemical reaction.* A chemical equation, then, is just the chemist's shorthand description of a reaction. In a chemical equation the reactants are conventionally written on the left and the products on the right of the arrow:

$$\text{reactants} \longrightarrow \text{products}$$

To provide additional information, chemists often indicate the physical states of the reactants and products by using the letters *g, l,* and *s* to denote gas, liquid, and solid, respectively. For example,

Two hydrogen molecules + One oxygen molecule $\longrightarrow$ Two water molecules

$2H_2$ + O_2 $\longrightarrow$ $2H_2O$

Figure 3.7 *Three ways of representing the combustion of hydrogen. In accordance with the law of conservation of mass, the number of each type of atom must be the same on both sides of the equation.*

Table 3.1 | Interpretation of a Chemical Equation		
$2H_2$	$+ O_2$	$\longrightarrow 2H_2O$
Two molecules	+ one molecule $\longrightarrow$ two molecules	
2 moles	+ 1 mole $\longrightarrow$ 2 moles	
$2(2.02\ g) = 4.04\ g + 32.00\ g$	$\longrightarrow 2(18.02\ g) = 36.04\ g$	
36.04 g reactants		36.04 g product

$$2CO(g) + O_2(g) \longrightarrow 2CO_2(g)$$
$$2HgO(s) \longrightarrow 2Hg(l) + O_2(g)$$

To represent what happens when sodium chloride (NaCl) is added to water, we write

$$NaCl(s) \xrightarrow{\text{H}_2\text{O}} NaCl(aq)$$

where *aq* denotes the aqueous (that is, water) environment. Writing H_2O above the arrow symbolizes the physical process of dissolving a substance in water, although it is sometimes left out for simplicity.

On heating, mercury(II) oxide (HgO) decomposes to form mercury and oxygen.

Balancing Chemical Equations

Suppose we want to write an equation to describe a chemical reaction that we have just carried out in the laboratory. How should we go about doing it? Because we know the identities of the reactants, we can write their chemical formulas. The identities of products are more difficult to establish. For simple reactions it is often possible to guess the product(s). For more complicated reactions involving three or more products, chemists may need to perform further tests to establish the presence of specific compounds.

Once we have identified all the reactants and products and have written the correct formulas for them, we assemble them in the conventional sequence—reactants on the left separated by an arrow from products on the right. The equation written at this point is likely to be *unbalanced;* that is, the number of each type of atom on one side of the arrow differs from the number on the other side. In general, we can balance a chemical equation by these steps:

- Identify all reactants and products and write their correct formulas on the left side and right side of the equation, respectively.
- Begin balancing the equation by trying different coefficients to make the number of atoms of each element the same on both sides of the equation. We can change the coefficients (the numbers preceding the formulas) but not the subscripts (the numbers within formulas). Changing the subscripts would change the identity of the substance. For example, $2NO_2$ means "two molecules of nitrogen dioxide," but if we double the subscripts, we have N_2O_4, which is the formula of dinitrogen tetroxide, a completely different compound.
- First, look for elements that appear only once on each side of the equation with the same number of atoms on each side: The formulas containing these elements must have the same coefficient. Therefore, there is no need to adjust the coefficients of these elements at this point. Next, look for elements that appear

only once on each side of the equation but in unequal numbers of atoms. Balance these elements. Finally, balance elements that appear in two or more formulas on the same side of the equation.

- Check your balanced equation to be sure that you have the same total number of each type of atoms on both sides of the equation arrow.

Let's consider a specific example. In the laboratory, small amounts of oxygen gas can be prepared by heating potassium chlorate ($KClO_3$). The products are oxygen gas (O_2) and potassium chloride (KCl). From this information, we write

$$KClO_3 \longrightarrow KCl + O_2$$

(For simplicity, we omit the physical states of reactants and products.) All three elements (K, Cl, and O) appear only once on each side of the equation, but only for K and Cl do we have equal numbers of atoms on both sides. Thus $KClO_3$ and KCl must have the same coefficient. The next step is to make the number of O atoms the same on both sides of the equation. Because there are three O atoms on the left and two O atoms on the right of the equation, we can balance the O atoms by placing a 2 in front of $KClO_3$ and a 3 in front of O_2.

$$2KClO_3 \longrightarrow KCl + 3O_2$$

Finally, we balance the K and Cl atoms by placing a 2 in front of KCl:

$$2KClO_3 \longrightarrow 2KCl + 3O_2 \tag{3.4}$$

As a final check, we can draw up a balance sheet for the reactants and products where the number in parentheses indicates the number of atoms of each element:

Reactants	Products
K (2)	K (2)
Cl (2)	Cl (2)
O (6)	O (6)

Note that this equation could also be balanced with coefficients that are multiples of 2 (for $KClO_3$), 2 (for KCl), and 3 (for O_2); for example,

$$4KClO_3 \longrightarrow 4KCl + 6O_2$$

However, it is common practice to use the *simplest* possible set of whole-number coefficients to balance the equation. Equation (3.4) conforms to this convention.

Now let us consider the combustion (that is, burning) of the natural gas component ethane (C_2H_6) in oxygen or air, which yields carbon dioxide (CO_2) and water. The unbalanced equation is

$$C_2H_6 + O_2 \longrightarrow CO_2 + H_2O$$

We see that the number of atoms is not the same on both sides of the equation for any of the elements (C, H, and O). In addition, C and H appear only once on each side of the equation; O appears in two compounds on the right side (CO_2 and H_2O). To balance the C atoms, we place a 2 in front of CO_2:

$$C_2H_6 + O_2 \longrightarrow 2CO_2 + H_2O$$

To balance the H atoms, we place a 3 in front of H_2O:

$$C_2H_6 + O_2 \longrightarrow 2CO_2 + 3H_2O$$

Heating potassium chlorate produces oxygen, which supports the combustion of wood splint.

C_2H_6

At this stage, the C and H atoms are balanced, but the O atoms are not because there are seven O atoms on the right-hand side and only two O atoms on the left-hand side of the equation. This inequality of O atoms can be eliminated by writing $\frac{7}{2}$ in front of the O_2 on the left-hand side:

$$C_2H_6 + \tfrac{7}{2}O_2 \longrightarrow 2CO_2 + 3H_2O$$

The "logic" for using $\frac{7}{2}$ as a coefficient is that there were seven oxygen atoms on the right-hand side of the equation, but only a pair of oxygen atoms (O_2) on the left. To balance them we ask how many *pairs* of oxygen atoms are needed to equal seven oxygen atoms. Just as 3.5 pairs of shoes equal seven shoes, $\frac{7}{2}O_2$ molecules equal seven O atoms. As this tally shows, the equation is now balanced:

Reactants	Products
C (2)	C (2)
H (6)	H (6)
O (7)	O (7)

However, we normally prefer to express the coefficients as whole numbers rather than as fractions. Therefore, we multiply the entire equation by 2 to convert $\frac{7}{2}$ to 7:

$$2C_2H_6 + 7O_2 \longrightarrow 4CO_2 + 6H_2O$$

The final tally is

Reactants	Products
C (4)	C (4)
H (12)	H (12)
O (14)	O (14)

Note that the coefficients used in balancing the last equation are the smallest possible set of whole numbers.

NH₃

Example 3.12   The first step in the synthesis of nitric acid (HNO_3), an important chemical used in the manufacture of fertilizers, drugs, and explosives, is the reaction between ammonia (NH_3) and oxygen to produce nitric oxide (NO) and water. Write a balanced equation for this reaction.

Reasoning and Solution   We follow the procedure described on p. 72. The unbalanced equation is

$$NH_3 + O_2 \longrightarrow NO + H_2O$$

We see that N appears only once on each side of the equation and in equal number. Therefore, NH_3 and NO must have the same coefficient. To balance H, we place a 2 in front of NH_3 and a 3 in front of H_2O:

$$2NH_3 + O_2 \longrightarrow NO + 3H_2O$$

To balance N, we place a 2 in front of NO:

$$2NH_3 + O_2 \longrightarrow 2NO + 3H_2O$$

There are two O atoms on the left and five O atoms on the right of the arrow. This inequality of O atoms can be eliminated by writing $\frac{5}{2}$ in front of O_2:

$$2NH_3 + \frac{5}{2}O_2 \longrightarrow 2NO + 3H_2O$$

As in the case of ethane described earlier, we multiply the entire equation by 2 to convert $\frac{5}{2}$ to 5

$$4NH_3 + 5O_2 \longrightarrow 4NO + 6H_2O$$

The final tally is

Reactants	Products
N (4)	N (4)
H (12)	H (12)
O (10)	O (10)

Practice Exercise Balance the equation representing the reaction between iron(III) oxide, Fe_2O_3, and carbon monoxide (CO) to yield iron (Fe) and carbon dioxide (CO_2).

Similar problems: 3.59, 3.60.

3.8 Amounts of Reactants and Products

A basic question raised in the chemical laboratory and the chemical industry is, "How much product will be formed from specific amounts of starting materials (reactants)?" Or in some cases, we might ask the reverse question: "How much starting material must be used to obtain a specific amount of product?" To interpret a reaction quantitatively, we need to apply our knowledge of molar masses and the mole concept. *Stoichiometry* is *the quantitative study of reactants and products in a chemical reaction.*

Whether the units given for reactants (or products) are moles, grams, liters (for gases), or some other units, we use moles to calculate the amount of product formed in a reaction. This approach is called the *mole method,* which means simply that *the stoichiometric coefficients in a chemical equation can be interpreted as the number of moles of each substance.* For example, the combustion of carbon monoxide in air produces carbon dioxide:

$$2CO(g) + O_2(g) \longrightarrow 2CO_2(g)$$

For stoichiometric calculations we would read this equation as "2 moles of carbon monoxide gas combine with 1 mole of oxygen gas to form 2 moles of carbon dioxide gas."

The mole method consists of these steps:

1. Write correct formulas for all reactants and products, and balance the resulting equation.
2. Convert the quantities of some or all given or known substances (usually reactants) into moles.
3. Use the coefficients in the balanced equation to calculate the number of moles of the sought or unknown quantities (usually products) in the problem.
4. Using the calculated numbers of moles and the molar masses, convert the unknown quantities to whatever units are required (typically grams).
5. Check that your answer is reasonable in physical terms.

Figure 3.8 *The mole method. First convert the quantity of reactant (in grams or other units) to number of moles. Next, use the mole ratio in the balanced equation to calculate the number of moles of product formed. Finally, convert moles of product to grams of product.*

Step 1 is a prerequisite to any stoichiometric calculation. We must know the identities of the reactants and products, and their mass relationships must not violate the law of conservation of mass (that is, we must have a balanced equation). Step 2 is the critical process of converting grams (or other units) of substances to number of moles. This conversion allows us to analyze the actual reaction in terms of moles only.

To complete step 3, we need the balanced equation furnished by step 1. The key point here is that the coefficients in a balanced equation tell us the ratio in which moles of one substance react with or form moles of another substance. Step 4 is similar to step 2, except that it deals with the quantities sought in the problem. Step 5 is often underestimated but is very important. Figure 3.8 shows the steps involved in stoichiometric calculations using the mole method.

In stoichiometry we use the symbol ⇌, which means "stoichiometrically equivalent to" or simply "equivalent to." In the balanced equation for the formation of carbon dioxide, 2 moles of CO react with 1 mole of O_2, so 2 moles of CO are equivalent to 1 mole of O_2:

$$2 \text{ mol CO} \Leftrightarrow 1 \text{ mol } O_2$$

In terms of the factor-label method, we can write the unit factor as

$$\frac{2 \text{ mol CO}}{1 \text{ mol } O_2} = 1 \qquad \text{or} \qquad \frac{1 \text{ mol } O_2}{2 \text{ mol CO}} = 1$$

Similarly, because 2 moles of CO (or 1 mole of O_2) produce 2 moles of CO_2, we can say that 2 moles of CO (or 1 mole of O_2) are equivalent to 2 moles of CO_2:

$$2 \text{ mol CO} \Leftrightarrow 2 \text{ mol } CO_2 \qquad\qquad 1 \text{ mol } O_2 \Leftrightarrow 2 \text{ mol } CO_2$$

so

$$\frac{2 \text{ mol CO}}{2 \text{ mol } CO_2} = 1 \qquad\qquad \frac{1 \text{ mol } O_2}{2 \text{ mol } CO_2} = 1$$

Lithium reacting with water to produce hydrogen gas.

Example 3.13   All alkali metals react with water to produce hydrogen gas and the corresponding alkali metal hydroxide. A typical reaction is that between lithium and water:

$$2\text{Li}(s) + 2\text{H}_2\text{O}(l) \longrightarrow 2\text{LiOH}(aq) + \text{H}_2(g)$$

(a) How many moles of H_2 will be formed by the complete reaction of 6.23 moles of Li with water? (b) How many grams of H_2 will be formed by the complete reaction of 80.57 g of Li with water?

Reasoning and Solution   We follow the steps listed on p. 75.

(a)

Step 1. The balanced equation is given in the problem.

Step 2. No conversion is needed because the amount of the starting material, Li, is given in moles.

Step 3. Because 2 moles of Li produce 1 mole of H_2, or 2 mol Li $\approx$ 1 mol H_2, we calculate moles of H_2 produced as follows:

$$\text{moles of } H_2 \text{ produced} = 6.23 \text{ mol Li} \times \frac{1 \text{ mol } H_2}{2 \text{ mol Li}}$$

$$= 3.12 \text{ mol } H_2$$

Step 4. This step is not required.

Step 5. We began with 6.23 moles of Li and produced 3.12 moles of H_2. Because 2 moles of Li produce 1 mole of H_2, 3.12 moles is a reasonable quantity.

(b)

Step 1. The reaction is the same as in (a).

Step 2. The number of moles of Li is given by

$$\text{moles of Li} = 80.57 \text{ g Li} \times \frac{1 \text{ mol Li}}{6.941 \text{ g Li}} = 11.61 \text{ mol Li}$$

Step 3. Because 2 moles of Li produce 1 mole of H_2, or 2 mol Li $\approx$ 1 mol H_2, we calculate the number of moles of H_2 produced as follows:

$$\text{moles of } H_2 \text{ produced} = 11.61 \text{ mol Li} \times \frac{1 \text{ mol } H_2}{2 \text{ mol Li}} = 5.805 \text{ mol } H_2$$

Step 4. From the molar mass of H_2 (2.016 g), we calculate the mass of H_2 produced:

$$\text{mass of } H_2 \text{ produced} = 5.805 \text{ mol } H_2 \times \frac{2.016 \text{ g } H_2}{1 \text{ mol } H_2} = 11.70 \text{ g } H_2$$

Step 5. Because the molar mass of H_2 is smaller than that of Li and 2 moles of Li are needed to produce 1 mole of H_2, we expect the answer to be smaller than 80.57 g.

Similar problem: 3.72.

Practice Exercise The reaction between nitric oxide (NO) and oxygen to form nitrogen dioxide (NO_2) is a key step in photochemical smog formation:

$$2NO(g) + O_2(g) \longrightarrow 2NO_2(g)$$

(a) How many moles of NO_2 are formed by the complete reaction of 0.254 mole of O_2?

(b) How many grams of NO_2 are formed by the complete reaction of 1.44 g of NO?

After some practice, you may find it convenient to combine steps 2, 3, and 4 in a single equation, as Example 3.14 shows.

$C_6H_{12}O_6$

Example 3.14 The food we eat is degraded, or broken down, in our bodies to provide energy for growth and function. A general overall equation for this very complex process represents the degradation of glucose ($C_6H_{12}O_6$) to carbon dioxide (CO_2) and water (H_2O):

$$C_6H_{12}O_6 + 6O_2 \longrightarrow 6CO_2 + 6H_2O$$

If 856 g of $C_6H_{12}O_6$ are consumed by a person over a certain period, what is the mass of CO_2 produced?

Reasoning and Solution We follow the steps listed on p. 75.

Step 1. The balanced equation is given.

Steps 2, 3, and 4. From the balanced equation we see that 1 mol $C_6H_{12}O_6 \backsim 6$ mol CO_2. The molar masses of $C_6H_{12}O_6$ and CO_2 are 180.2 g and 44.01 g, respectively. We combine all of these data into one equation:

$$\text{mass of } CO_2 \text{ produced} = 856 \text{ g } C_6H_{12}O_6 \times \frac{1 \text{ mol } C_6H_{12}O_6}{180.2 \text{ g } C_6H_{12}O_6} \times \frac{6 \text{ mol } CO_2}{1 \text{ mol } C_6H_{12}O_6}$$

$$\times \frac{44.01 \text{ g } CO_2}{1 \text{ mol } CO_2} = 1.25 \times 10^3 \text{ g } CO_2$$

Step 5. Because 1 mole of $C_6H_{12}O_6$ produces 6 moles of CO_2 and the molar mass of $C_6H_{12}O_6$ is four times that of CO_2, we expect the mass of CO_2 formed to be greater than 856 g. Therefore, the answer is reasonable.

Similar problem: 3.76.

Practice Exercise Methanol (CH_3OH) burns in air according to the equation

$$2CH_3OH + 3O_2 \longrightarrow 2CO_2 + 4H_2O$$

If 209 g of methanol are used up in a combustion process, what is the mass of H_2O produced?

3.9 Limiting Reagents and Yields of Reactions

When a chemist carries out a reaction, the reactants are usually not present in exact *stoichiometric amounts,* that is, *in the proportions indicated by the balanced equation.* Because the goal of a reaction is to produce the maximum quantity of a useful compound from a given quantity of starting material, frequently a large excess of one reactant is supplied to ensure that the more expensive reactant is completely converted to the desired product. Consequently some reactant will be left over at the end of the reaction. *The reactant used up first in a reaction* is called the *limiting reagent,* because the maximum amount of product formed depends on how much of this reactant was originally present. When this reactant is used up, no more product can be formed. *Excess reagents* are the *reactants present in quantities greater than necessary to react with the quantity of the limiting reagent.*

The concept of the limiting reagent is analogous to the relationship between men and women in a dance contest at a club. If there are fourteen men and only nine women, then only nine female/male pairs can compete. Five men will be left without

Limiting reagent

partners. The number of women thus *limits* the number of men that can dance in the contest, and there is an *excess* of men.

Consider the formation of nitrogen dioxide (NO_2) from nitric oxide (NO) and oxygen:

$$2NO(g) + O_2(g) \longrightarrow 2NO_2(g)$$

Suppose initially we have 8 moles of NO and 7 moles of O_2 (Figure 3.9). One way to determine which of the two reactants is the limiting reagent is to calculate the number of moles of NO_2 obtained based on the initial quantities of NO and O_2. From our definition, we see that only the limiting reagent will yield the smaller amount of the product. Starting with 8 moles of NO, we find the number of moles of NO_2 produced is

$$8 \text{ mol NO} \times \frac{2 \text{ mol NO}_2}{2 \text{ mol NO}} = 8 \text{ mol NO}_2$$

and starting with 7 moles of O_2, we have

$$7 \text{ mol O}_2 \times \frac{2 \text{ mol NO}_2}{1 \text{ mol O}_2} = 14 \text{ mol NO}_2$$

Because NO results in a smaller amount of NO_2, it must be the limiting reagent.

In stoichiometric calculations, the first step is to decide which reactant is the limiting reagent. After the limiting reagent has been identified, the rest of the problem can be solved as outlined in Section 3.8. Example 3.15 illustrates this approach. We will not include step 5 in the calculations, but you should always examine the reasonableness of *any* chemical calculation.

Example 3.15   Urea [$(NH_2)_2CO$] is prepared by reacting ammonia with carbon dioxide:

$$2NH_3(g) + CO_2(g) \longrightarrow (NH_2)_2CO(aq) + H_2O(l)$$

In one process, 637.2 g of NH_3 are allowed to react with 1142 g of CO_2. (a) Which of the two reactants is the limiting reagent? (b) Calculate the mass of $(NH_2)_2CO$ formed. (c) How much of the excess reagent (in grams) is left at the end of the reaction?

Reasoning and Solution   (a) Because we cannot tell by inspection which of the two reactants is the limiting reagent, we have to proceed by first converting their masses into numbers of moles. The molar masses of NH_3 and CO_2 are 17.03 g and 44.01 g, respectively. Thus

$$\text{moles of NH}_3 = 637.2 \text{ g NH}_3 \times \frac{1 \text{ mol NH}_3}{17.03 \text{ g NH}_3}$$

$$= 37.42 \text{ mol NH}_3$$

$$\text{moles of CO}_2 = 1142 \text{ g CO}_2 \times \frac{1 \text{ mol CO}_2}{44.01 \text{ g CO}_2}$$

$$= 25.95 \text{ mol CO}_2$$

To determine which of the reactants is the limiting reagent, we first calculate the number of moles of $(NH_2)_2CO$ produced from the given amounts of NH_3 and CO_2.

NO_2 is a dark brown gas with a choking odor whose color is sometimes visible in polluted air.

Before reaction has started

After reaction is complete

NO O_2 NO_2

Figure 3.9 *At the start of the reaction, there were eight NO molecules and seven O_2 molecules. At the end, all the NO molecules are gone and only three O_2 molecules are left. Therefore, NO is the limiting reagent and O_2 is the excess reagent. Each molecule can also be treated as one mole of the substance in this reaction.*

(NH₂)₂CO

In practice, chemists usually choose the more expensive chemical (NH₃ in this case) as the limiting reagent so that all or most of it will be consumed in the reaction.

From the balanced equation, we see that 2 mol NH_3 ⇌ 1 mol $(NH_2)_2CO$ and 1 mol CO_2 ⇌ 1 mol $(NH_2)_2CO$. Starting with 37.42 moles of NH_3, we write

$$37.42 \text{ mol NH}_3 \times \frac{1 \text{ mol } (NH_2)_2CO}{2 \text{ mol NH}_3} = 18.71 \text{ mol } (NH_2)_2CO$$

and from 25.95 moles of CO_2,

$$25.95 \text{ mol CO}_2 \times \frac{1 \text{ mol } (NH_2)_2CO}{1 \text{ mol CO}_2} = 25.95 \text{ mol } (NH_2)_2CO$$

It follows, therefore, that NH_3 must be the limiting reagent because it produces a smaller amount of $(NH_2)_2CO$.

(b) The amount of $(NH_2)_2CO$ produced is determined by the amount of limiting reagent present. Thus we write

$$\text{mass of } (NH_2)_2CO = 37.42 \text{ mol NH}_3 \times \frac{1 \text{ mol } (NH_2)_2CO}{2 \text{ mol NH}_3}$$

$$\times \frac{60.06 \text{ g } (NH_2)_2CO}{1 \text{ mol } (NH_2)_2CO} = 1124 \text{ g } (NH_2)_2CO$$

(c) The number of moles of the excess reagent (CO_2) left is

$$25.95 \text{ mol CO}_2 - \left(37.42 \text{ mol NH}_3 \times \frac{1 \text{ mol CO}_2}{2 \text{ mol NH}_3}\right) = 7.24 \text{ mol CO}_2$$

and

$$\text{mass of CO}_2 \text{ left over} = 7.24 \text{ mol CO}_2 \times \frac{44.01 \text{ g CO}_2}{1 \text{ mol CO}_2}$$

$$= 319 \text{ g CO}_2$$

Practice Exercise   The reaction between aluminum and iron(III) oxide can generate temperatures approaching 3000°C and is used in welding metals:

$$2Al + Fe_2O_3 \longrightarrow Al_2O_3 + 2Fe$$

In one process, 124 g of Al are reacted with 601 g of Fe_2O_3. (a) Calculate the mass (in grams) of Al_2O_3 formed. (b) How much of the excess reagent is left at the end of the reaction?

Similar problem: 3.82.

Yields of Reactions

The amount of limiting reagent present at the start of a reaction determines the **theoretical yield** of the reaction, that is, *the amount of product that would result if all the limiting reagent reacted*. The theoretical yield, then, is the *maximum* obtainable yield, predicted by the balanced equation. In practice, the *amount of product obtained* is almost always less than the theoretical yield. Therefore chemists define the **actual yield** as *the quantity of product that actually results from a reaction*. There are many reasons for the difference between actual and theoretical yields. For instance, many reactions are reversible, and so they do not proceed 100 percent from left to right. Even when a reaction is 100 percent complete, it may be difficult to recover all of the product from the reaction medium (say, from an aqueous solution). Some reactions are complex in the sense that the products formed may react further among themselves or with the reactants to form still other products. These additional reactions will reduce the yield of the first reaction.

To determine how efficient a given reaction is, chemists often figure the **percent yield,** which describes *the proportion of the actual yield to the theoretical yield.* It is calculated as follows:

$$\% \text{ yield} = \frac{\text{actual yield}}{\text{theoretical yield}} \times 100\% \qquad (3.5)$$

Percent yields may range from a fraction of 1 percent to 100 percent. Chemists strive to maximize the percent yield of product in a reaction. Other factors that can affect the percent yield of a reaction include temperature and pressure. We will study these effects later.

Example 3.16 Titanium is a strong, lightweight, corrosion-resistant metal that is used in rockets, aircraft, jet engines, and bicycle frames. It is prepared by the reaction of titanium(IV) chloride with molten magnesium between 950°C and 1150°C:

$$TiCl_4(g) + 2Mg(l) \longrightarrow Ti(s) + 2MgCl_2(l)$$

In a certain industrial operation 3.54×10^7 g of $TiCl_4$ are reacted with 1.13×10^7 g of Mg. (a) Calculate the theoretical yield of Ti in grams. (b) Calculate the percent yield if 7.91×10^6 g of Ti are actually obtained.

Reasoning and Solution We follow the procedure in Example 3.15 to find out which of the two reactants is the limiting agent. This knowledge will enable us to calculate the theoretical yield. The percent yield can then be obtained by applying Equation (3.5).

(a) First we calculate the number of moles of $TiCl_4$ and Mg initially present:

$$\text{moles of } TiCl_4 = 3.54 \times 10^7 \text{ g } TiCl_4 \times \frac{1 \text{ mol } TiCl_4}{189.7 \text{ g } TiCl_4} = 1.87 \times 10^5 \text{ mol } TiCl_4$$

$$\text{moles of Mg} = 1.13 \times 10^7 \text{ g Mg} \times \frac{1 \text{ mol Mg}}{24.31 \text{ g Mg}} = 4.65 \times 10^5 \text{ mol Mg}$$

Next, we must determine which of the two substances is the limiting reagent. From the balanced equation we see that 1 mol $TiCl_4 \rightleftharpoons$ 1 mol Ti and 2 mol Mg $\rightleftharpoons$ 1 mol Ti. Therefore, the amounts of Ti obtained from these two reactants are

$$1.87 \times 10^5 \text{ mol } TiCl_4 \times \frac{1 \text{ mol Ti}}{1 \text{ mol } TiCl_4} = 1.87 \times 10^5 \text{ mol Ti}$$

and

$$4.65 \times 10^5 \text{ mol Mg} \times \frac{1 \text{ mol Ti}}{2 \text{ mol Mg}} = 2.33 \times 10^5 \text{ mol Ti}$$

The frame of this bicycle is made of titanium.

Therefore, $TiCl_4$ is the limiting reagent because it produces a smaller amount of Ti. The theoretical mass of Ti formed is

$$1.87 \times 10^5 \ \text{mol Ti} \times \frac{47.88 \text{ g Ti}}{1 \text{ mol Ti}} = 8.95 \times 10^6 \text{ g Ti}$$

(b) To find the percent yield, we write

$$\% \text{ yield} = \frac{\text{actual yield}}{\text{theoretical yield}} \times 100\%$$

$$= \frac{7.91 \times 10^6 \text{ g}}{8.95 \times 10^6 \text{ g}} \times 100\%$$

$$= 88.4\%$$

Similar problems: 3.85, 3.87.

Practice Exercise   Industrially, vanadium metal, which is used in steel alloys, can be obtained by reacting vanadium(V) oxide with calcium at high temperatures:

$$5Ca + V_2O_5 \longrightarrow 5CaO + 2V$$

In one process 1.54×10^3 g of V_2O_5 react with 1.96×10^3 g of Ca. (a) Calculate the theoretical yield of V. (b) Calculate the percent yield if 803 g of V are obtained.

Summary

Atomic masses are measured in atomic mass units (amu), a relative unit based on a value of exactly 12 for the carbon-12 isotope. The atomic mass given for the atoms of a particular element is usually the average of the naturally occurring isotope distribution of that element. The molecular mass of a molecule is the sum of the atomic masses of the atoms in the molecule. Both atomic mass and molecular mass can be accurately determined with a mass spectrometer.

A mole is an Avogadro's number (6.022×10^{23}) of atoms, molecules, or other particles. The molar mass (in grams) of an element or a compound is numerically equal to the mass of the atom, molecule, or formula unit (in amu) and contains an Avogadro's number of atoms (in the case of elements), molecules, or simplest formula units (in the case of ionic compounds).

The percent composition by mass of a compound is the percent by mass of each element present. If we know the percent composition by mass of a compound, we can deduce the empirical formula of the compound and also the molecular formula of the compound if the approximate molar mass is known.

Chemical changes, called chemical reactions, are represented by chemical equations. Substances that undergo change—the reactants—are written on the left and substances formed—the products—appear on the right of the arrow. Chemical equations must be balanced, in accordance with the law of conservation of mass. The number of atoms of each type of element in the reactants and products must be equal.

Stoichiometry is the quantitative study of products and reactants in chemical reactions. Stoichiometric calculations are best done by expressing both the known and unknown quantities in terms of moles and then converting to other units if necessary. A limiting reagent is the reactant that is present in the smallest stoichiometric amount.

It limits the amount of product that can be formed. The amount of product obtained in a reaction (the actual yield) may be less than the maximum possible amount (the theoretical yield). The ratio of the two is expressed as the percent yield.

Key Words

Actual yield p. 80	Chemical equation	Mole (mol) p. 59	Product p. 71
Atomic mass p. 58	p. 71	Molecular mass p. 62	Reactant p. 71
Atomic mass unit (amu)	Chemical reaction p. 70	Mole method p. 75	Stoichiometric amount
p. 58	Excess reagent p. 78	Percent composition	p. 78
Avogadro's number (N_A)	Limiting reagent p. 78	p. 65	Stoichiometry p. 75
p. 59	Molar mass ($\mathcal{M}$) p. 60	Percent yield p. 81	Theoretical yield p. 80

Questions and Problems

Atomic Mass and Avogadro's Number

Review Questions

3.1   What is an atomic mass unit?

3.2   What is the mass (in amu) of a carbon-12 atom?

3.3   When we look up the atomic mass of carbon, we find that its value is 12.01 amu rather than 12.00 amu as defined. Why?

3.4   Define the term "mole." What is the unit for mole in calculations? What does the mole have in common with the pair, the dozen, and the gross?

3.5   What does Avogadro's number represent?

3.6   Define molar mass. What are the commonly used units for molar mass?

3.7   Calculate the charge (in coulombs) and mass (in grams) of 1 mole of electrons.

3.8   Explain clearly what is meant by the statement "The atomic mass of gold is 197.0 amu."

Problems

3.9   The atomic masses of $^{35}_{17}Cl$ (75.53%) and $^{37}_{17}Cl$ (24.47%) are 34.968 amu and 36.956 amu, respectively. Calculate the average atomic mass of chlorine. The percentages in parentheses denote the relative abundances.

3.10   The atomic masses of $^{6}_{3}Li$ and $^{7}_{3}Li$ are 6.0151 amu and 7.0160 amu, respectively. Calculate the natural abundances of these two isotopes. The average atomic mass of Li is 6.941 amu.

3.11   Earth's population is about 6.5 billion. Suppose that every person on Earth participates in a process of counting identical particles at the rate of two particles per second. How many years would it take to count 6.0×10^{23} particles? Assume that there are 365 days in a year.

3.12   The thickness of a piece of paper is 0.0036 in. Suppose a certain book has an Avogadro's number of pages; calculate the thickness of the book in light-years. (*Hint:* See Problem 1.38 for the definition of light-year.)

3.13   What is the mass in grams of 13.2 amu?

3.14   How many amu are there in 8.4 g?

3.15   How many atoms are there in 5.10 moles of sulfur (S)?

3.16   How many moles of cobalt atoms are there in 6.00×10^9 (6 billion) Co atoms?

3.17   How many moles of calcium (Ca) atoms are in 77.4 g of Ca?

3.18   How many grams of gold (Au) are there in 15.3 moles of Au?

3.19   What is the mass in grams of a single atom of each of these elements? (a) Hg, (b) Ne.

3.20   What is the mass in grams of a single atom of each of these elements? (a) As, (b) Ni.

3.21   What is the mass in grams of 1.00×10^{12} lead (Pb) atoms?

3.22   How many atoms are present in 3.14 g of copper (Cu)?

3.23   Which of these has more atoms: 1.10 g of hydrogen atoms or 14.7 g of chromium atoms?

3.24   Which of these has a greater mass: 2 atoms of lead or 5.1×10^{-23} mole of helium?

Molecular Mass

Problems

3.25   Calculate the molecular mass (in amu) of each of these substances: (a) CH_4, (b) H_2O, (c) H_2O_2, (d) C_6H_6, (e) PCl_5.

3.26   Calculate the molar mass of these substances (in grams): (a) S_8, (b) CS_2, (c) $CHCl_3$ (chloroform), (d) $C_6H_8O_6$ (ascorbic acid, or vitamin C).

3.27   Calculate the molar mass of a compound if 0.372 mole of it has a mass of 152 g.

3.28   How many moles of ethane (C_2H_6) are present in 0.334 g of C_2H_6?

3.29   Calculate the numbers of C, H, and O atoms in 1.50 g of glucose ($C_6H_{12}O_6$), a sugar.

3.30   Urea $[(NH_2)_2CO]$ is a compound used for fertilizer and many other things. Calculate the number of N, C, O, and H atoms in 1.68×10^4 g of urea.

3.31   Pheromones are a special type of compound secreted by the females of many insect species to attract the males for mating. One pheromone has the molecular formula $C_{19}H_{38}O$. Normally, the amount of this pheromone secreted by a female insect is about 1.0×10^{-12} g. How many molecules are there in this quantity?

3.32   The density of water is 1.00 g/mL at 4°C. How many water molecules are present in 2.56 mL of water at this temperature?

Mass Spectrometry

Review Questions

3.33   Describe the operation of a mass spectrometer.

3.34   Describe how you would determine the isotopic abundance of an element from its mass spectrum.

Problems

3.35   Carbon has two stable isotopes, $^{12}_{6}C$ and $^{13}_{6}C$, and fluorine has only one stable isotope, $^{19}_{9}F$. How many peaks would you observe in the mass spectrum of the positive ion of CF_4^+? Assume no decomposition of the ion into smaller fragments.

3.36   Hydrogen has two stable isotopes, ^1_1H and ^2_1H, and sulfur has four stable isotopes, $^{32}_{16}S$, $^{33}_{16}S$, $^{34}_{16}S$, $^{36}_{16}S$. How many peaks would you observe in the mass spectrum of the positive ion of hydrogen sulfide, H_2S^+? Assume no decomposition of the ion into smaller fragments.

Percent Composition and Chemical Formulas

Review Questions

3.37   Define percent composition by mass of a compound.

3.38   Describe how the knowledge of the percent composition by mass of an unknown compound of high purity can help us identify the compound.

3.39   What does the word "empirical" in empirical formula mean?

3.40   If we know the empirical formula of a compound, what additional information do we need in order to determine its molecular formula?

Problems

3.41   Tin (Sn) exists in Earth's crust as SnO_2. Calculate the percent composition by mass of Sn and O in SnO_2.

3.42   Sodium, a member of the alkali metal family (Group 1A), is very reactive and thus is never found in nature in its elemental state. It forms ionic compounds with members of the halogen family (Group 7A). Calculate the percent composition by mass of all the elements in each of these compounds: (a) NaF, (b) NaCl, (c) NaBr, (d) NaI.

3.43   For many years the organic compound chloroform ($CHCl_3$) was used as an inhalation anesthetic in spite of the fact that it is also a toxic substance that may cause severe liver, kidney, and heart damage. Calculate the percent composition by mass of this compound.

3.44   Allicin is the compound responsible for the characteristic smell of garlic. An analysis of the compound gives this percent composition by mass: C: 44.4%; H: 6.21%; S: 39.5%; O: 9.86%. Calculate its empirical formula. What is its molecular formula given that its molar mass is about 162 g?

3.45   Cinnamic alcohol is used mainly in perfumery, particularly soaps and cosmetics. Its molecular formula is $C_9H_{10}O$. (a) Calculate the percent composition by mass of C, H, and O in cinnamic alcohol. (b) How many molecules of cinnamic alcohol are contained in a sample of mass 0.469 g?

3.46   All the substances listed here are fertilizers that contribute nitrogen to the soil. Which of these is the richest source of nitrogen on a mass percentage basis?

  (a) Urea, $(NH_2)_2CO$
  (b) Ammonium nitrate, NH_4NO_3
  (c) Guanidine, $HNC(NH_2)_2$
  (d) Ammonia, NH_3

3.47   The formula for rust can be represented by Fe_2O_3. How many moles of Fe are present in 24.6 g of the compound?

3.48   How many grams of sulfur (S) are needed to combine with 246 g of mercury (Hg) to form HgS?

3.49   Calculate the mass in grams of iodine (I_2) that will react completely with 20.4 g of aluminum (Al) to form aluminum iodide (AlI_3).

3.50   Tin(II) fluoride (SnF_2) is often added to toothpaste as an ingredient to prevent tooth decay. What is the mass of F in grams in 24.6 g of the compound?

3.51   What are the empirical formulas of the compounds with these compositions? (a) 2.1% H, 65.3% O, 32.6% S; (b) 20.2% Al, 79.8% Cl; (c) 40.1% C, 6.6% H, 53.3% O; (d) 18.4% C, 21.5% N, 60.1% K.

3.52   Peroxyacylnitrate (PAN) is one of the components of smog. It is a compound of C, H, N, and O. Determine the percent composition of oxygen and the empirical formula from this percent composition by mass: 19.8% C, 2.50% H, 11.6% N.

3.53   The molar mass of caffeine is 194.19 g. Is the molecular formula of caffeine $C_4H_5N_2O$ or $C_8H_{10}N_4O_2$?

3.54 Monosodium glutamate (MSG), a food flavor enhancer, has been blamed for "Chinese restaurant syndrome," the symptoms of which are headaches and chest pains. MSG has the following composition by mass: 35.51% C, 4.77% H, 37.85% O, 8.29% N, and 13.60% Na. What is its molecular formula if its molar mass is about 169 g?

Chemical Reactions and Chemical Equations

Review Questions

3.55 Define the following terms: chemical reaction, reactant, product.

3.56 What is the difference between a chemical reaction and a chemical equation?

3.57 Why must a chemical equation be balanced? What law is obeyed by a balanced chemical equation?

3.58 Write the symbols used to represent gas, liquid, solid, and the aqueous phase in chemical equations.

Problems

3.59 Balance these equations using the method outlined in Section 3.7:

(a) $C + O_2 \rightarrow CO$
(b) $CO + O_2 \rightarrow CO_2$
(c) $H_2 + Br_2 \rightarrow HBr$
(d) $K + H_2O \rightarrow KOH + H_2$
(e) $Mg + O_2 \rightarrow MgO$
(f) $O_3 \rightarrow O_2$
(g) $H_2O_2 \rightarrow H_2O + O_2$
(h) $N_2 + H_2 \rightarrow NH_3$
(i) $Zn + AgCl \rightarrow ZnCl_2 + Ag$
(j) $S_8 + O_2 \rightarrow SO_2$

3.60 Balance these equations using the method outlined in Section 3.7:

(a) $N_2O_5 \rightarrow N_2O_4 + O_2$
(b) $KNO_3 \rightarrow KNO_2 + O_2$
(c) $NH_4NO_3 \rightarrow N_2O + H_2O$
(d) $NH_4NO_2 \rightarrow N_2 + H_2O$
(e) $NaHCO_3 \rightarrow Na_2CO_3 + H_2O + CO_2$
(f) $P_4O_{10} + H_2O \rightarrow H_3PO_4$
(g) $HCl + CaCO_3 \rightarrow CaCl_2 + H_2O + CO_2$
(h) $Al + H_2SO_4 \rightarrow Al_2(SO_4)_3 + H_2$
(i) $CO_2 + KOH \rightarrow K_2CO_3 + H_2O$
(j) $CH_4 + O_2 \rightarrow CO_2 + H_2O$

Amounts of Reactants and Products

Review Questions

3.61 Define these terms: stoichiometry, stoichiometric amounts, limiting reagent, excess reagent, theoretical yield, actual yield, percent yield.

3.62 On what law is stoichiometry based?

3.63 Describe the basic steps involved in the mole method.

3.64 Why is it essential to use balanced equations in solving stoichiometric problems?

Problems

3.65 Consider the combustion of carbon monoxide (CO) in oxygen gas:

$$2CO(g) + O_2(g) \longrightarrow 2CO_2(g)$$

Starting with 3.60 moles of CO, calculate the number of moles of CO_2 produced if there is enough oxygen gas to react with all of the CO.

3.66 Silicon tetrachloride ($SiCl_4$) can be prepared by heating Si in chlorine gas:

$$Si(s) + 2Cl_2(g) \longrightarrow SiCl_4(l)$$

In one reaction, 0.507 mole of $SiCl_4$ is produced. How many moles of molecular chlorine were used in the reaction?

3.67 The annual production of sulfur dioxide from burning coal and fossil fuels, auto exhaust, and other sources is about 26 million tons. The equation for the reaction is

$$S(s) + O_2(g) \longrightarrow SO_2(g)$$

How much sulfur, present in the original materials, would result in that quantity of SO_2?

3.68 When baking soda (sodium bicarbonate or sodium hydrogen carbonate, $NaHCO_3$) is heated, it releases carbon dioxide gas, which is responsible for the rising of cookies, doughnuts, and bread. (a) Write a balanced equation for the decomposition of the compound (one of the products is Na_2CO_3). (b) Calculate the mass of $NaHCO_3$ required to produce 20.5 g of CO_2.

3.69 When potassium cyanide (KCN) reacts with acids, a deadly poisonous gas, hydrogen cyanide (HCN), is given off. Here is the equation:

$$KCN(aq) + HCl(aq) \longrightarrow KCl(aq) + HCN(g)$$

If a sample of 0.140 g of KCN is treated with an excess of HCl, calculate the amount of HCN formed, in grams.

3.70 Fermentation is a complex chemical process of wine making in which glucose is converted into ethanol and carbon dioxide:

$$\underset{\text{glucose}}{C_6H_{12}O_6} \longrightarrow \underset{\text{ethanol}}{2C_2H_5OH} + 2CO_2$$

Starting with 500.4 g of glucose, what is the maximum amount of ethanol in grams and in liters that can be obtained by this process? (Density of ethanol = 0.789 g/mL.)

3.71   Each copper(II) sulfate unit is associated with five water molecules in crystalline copper(II) sulfate pentahydrate ($CuSO_4 \cdot 5H_2O$). When this compound is heated in air above 100°C, it loses the water molecules and also its blue color:

$$CuSO_4 \cdot 5H_2O \longrightarrow CuSO_4 + 5H_2O$$

If 9.60 g of $CuSO_4$ are left after heating 15.01 g of the blue compound, calculate the number of moles of H_2O originally present in the compound.

3.72   For many years the recovery of gold (that is, the separation of gold from other materials) involved the treatment of gold by isolation from other substances, using potassium cyanide:

$$4Au + 8KCN + O_2 + 2H_2O \longrightarrow 4KAu(CN)_2 + 4KOH$$

What is the minimum amount of KCN in moles needed to extract 29.0 g (about an ounce) of gold?

3.73   Limestone ($CaCO_3$) is decomposed by heating to quicklime (CaO) and carbon dioxide. Calculate how many grams of quicklime can be produced from 1.0 kg of limestone.

3.74   Nitrous oxide (N_2O) is also called "laughing gas." It can be prepared by the thermal decomposition of ammonium nitrate (NH_4NO_3). The other product is H_2O. (a) Write a balanced equation for this reaction. (b) How many grams of N_2O are formed if 0.46 mole of NH_4NO_3 is used in the reaction?

3.75   The fertilizer ammonium sulfate [$(NH_4)_2SO_4$] is prepared by the reaction between ammonia (NH_3) and sulfuric acid:

$$2NH_3(g) + H_2SO_4(aq) \longrightarrow (NH_4)_2SO_4(aq)$$

How many kilograms of NH_3 are needed to produce 1.00×10^5 kg of $(NH_4)_2SO_4$?

3.76   A common laboratory preparation of oxygen gas is the thermal decomposition of potassium chlorate ($KClO_3$). Assuming complete decomposition, calculate the number of grams of O_2 gas that can be obtained starting with 46.0 g of $KClO_3$. (The products are KCl and O_2.)

Limiting Reagents

Review Questions

3.77   Define limiting reagent and excess reagent. What is the significance of the limiting reagent in predicting the amount of the product obtained in a reaction?

3.78   Give an everyday example that illustrates the limiting reagent concept.

Problems

3.79   Nitric oxide (NO) reacts with oxygen gas to give nitrogen dioxide (NO_2), a dark-brown gas:

$$2NO(g) + O_2(g) \longrightarrow 2NO_2(g)$$

In one experiment 0.886 mole of NO is mixed with 0.503 mole of O_2. Calculate which of the two reactants is the limiting reagent. Calculate also the number of moles of NO_2 produced.

3.80   The depletion of ozone (O_3) in the stratosphere has been a matter of great concern among scientists in recent years. It is believed that ozone can react with nitric oxide (NO) that is discharged from the high-altitude jet plane, the SST. The reaction is

$$O_3 + NO \longrightarrow O_2 + NO_2$$

If 0.740 g of O_3 reacts with 0.670 g of NO, how many grams of NO_2 would be produced? Which compound is the limiting reagent? Calculate the number of moles of the excess reagent remaining at the end of the reaction.

3.81   Propane (C_3H_8) is a component of natural gas and is used in domestic cooking and heating. (a) Balance the following equation representing the combustion of propane in air:

$$C_3H_8 + O_2 \longrightarrow CO_2 + H_2O$$

(b) How many grams of carbon dioxide can be produced by burning 3.65 moles of propane? Assume that oxygen is the excess reagent in this reaction.

3.82   Consider the reaction

$$MnO_2 + 4HCl \longrightarrow MnCl_2 + Cl_2 + 2H_2O$$

If 0.86 mole of MnO_2 and 48.2 g of HCl react, which reagent will be used up first? How many grams of Cl_2 will be produced?

Reaction Yield

Review Questions

3.83   Why is the yield of a reaction determined only by the amount of the limiting reagent?

3.84   Why is the actual yield of a reaction almost always smaller than the theoretical yield?

Problems

3.85   Hydrogen fluoride is used in the manufacture of Freons (which destroy ozone in the stratosphere) and in the production of aluminum metal. It is prepared by the reaction

$$CaF_2 + H_2SO_4 \longrightarrow CaSO_4 + 2HF$$

In one process 6.00 kg of CaF_2 are treated with an excess of H_2SO_4 and yield 2.86 kg of HF. Calculate the percent yield of HF.

3.86 Nitroglycerin ($C_3H_5N_3O_9$) is a powerful explosive. Its decomposition may be represented by

$$4C_3H_5N_3O_9 \longrightarrow 6N_2 + 12CO_2 + 10H_2O + O_2$$

This reaction generates a large amount of heat and many gaseous products. It is the sudden formation of these gases, together with their rapid expansion, that produces the explosion. (a) What is the maximum amount of O_2 in grams that can be obtained from 2.00×10^2 g of nitroglycerin? (b) Calculate the percent yield in this reaction if the amount of O_2 generated is found to be 6.55 g.

3.87 Titanium(IV) oxide (TiO_2) is a white substance produced by the action of sulfuric acid on the mineral ilmenite ($FeTiO_3$):

$$FeTiO_3 + H_2SO_4 \longrightarrow TiO_2 + FeSO_4 + H_2O$$

Its opaque and nontoxic properties make it suitable as a pigment in plastics and paints. In one process 8.00×10^3 kg of $FeTiO_3$ yielded 3.67×10^3 kg of TiO_2. What is the percent yield of the reaction?

3.88 Ethylene (C_2H_4), an important industrial organic chemical, can be prepared by heating hexane (C_6H_{14}) at 800°C:

$$C_6H_{14} \longrightarrow C_2H_4 + \text{other products}$$

If the yield of ethylene production is 42.5 percent, what mass of hexane must be reacted to produce 481 g of ethylene?

Additional Problems

3.89 The atomic mass of element X is 33.42 amu. A 27.22-g sample of X combines with 84.10 g of another element Y to form a compound XY. Calculate the atomic mass of Y.

3.90 The aluminum sulfate hydrate $[Al_2(SO_4)_3 \cdot xH_2O]$ contains 8.20 percent of Al by mass. Calculate x, that is, the number of water molecules associated with each $Al_2(SO_4)_3$ unit.

3.91 An iron bar weighed 664 g. After the bar had been standing in moist air for a month, exactly one-eighth of the iron turned to rust (Fe_2O_3). Calculate the final mass.

3.92 A certain metal oxide has the formula MO. A 39.46-g sample of the compound is strongly heated in an atmosphere of hydrogen to remove oxygen as water molecules. At the end, 31.70 g of the metal M is left over. If O has an atomic mass of 16.00 amu, calculate the atomic mass of M and identify the element.

3.93 An impure sample of zinc (Zn) is treated with an excess of sulfuric acid (H_2SO_4) to form zinc sulfate ($ZnSO_4$) and molecular hydrogen (H_2). (a) Write a balanced equation

for the reaction. (b) If 0.0764 g of H_2 is obtained from 3.86 g of the sample, calculate the percent purity of the sample. (c) What assumptions must you make in (b)?

3.94 One of the reactions that occurs in a blast furnace, where iron ore is converted to cast iron, is

$$Fe_2O_3 + 3CO \longrightarrow 2Fe + 3CO_2$$

Suppose that 1.64×10^3 kg of Fe are obtained from a 2.62×10^3-kg sample of Fe_2O_3. Assuming that the reaction goes to completion, what is the percent purity of Fe_2O_3 in the original sample?

3.95 Carbon dioxide (CO_2) is the gas that is mainly responsible for global warming (the so-called greenhouse effect). The burning of fossil fuels is a major cause of the increased concentration of CO_2 in the atmosphere. Carbon dioxide is also the end product of metabolism (see Example 3.14). Using glucose as an example of food, calculate the annual production of CO_2 in grams from human metabolism, assuming that each person consumes 5.0×10^2 g of glucose per day. The world's population is 6.5 billion, and there are 365 days in a year.

3.96 Carbohydrates are compounds containing carbon, hydrogen, and oxygen in which the hydrogen to oxygen ratio is 2:1. A certain carbohydrate contains 40.0 percent of carbon by mass. Calculate the empirical and molecular formulas of the compound if the approximate molar mass is 178 g.

3.97 Industrially, nitric acid is produced by the Ostwald process represented by the following equations:

$$4NH_3(g) + 5O_2(g) \longrightarrow 4NO(g) + 6H_2O(l)$$
$$2NO(g) + O_2(g) \longrightarrow 2NO_2(g)$$
$$2NO_2(g) + H_2O(l) \longrightarrow HNO_3(aq) + HNO_2(aq)$$

What mass of NH_3 (in g) must be used to produce 1.00 ton of HNO_3 by this procedure, assuming an 80 percent yield in each step? (1 ton = 2000 lb; 1 lb = 453.6 g.)

3.98 Suppose you are given a cube made of magnesium (Mg) metal of edge length 1.0 cm. (a) Calculate the number of Mg atoms in the cube. (b) Atoms are spherical. Therefore, the Mg atoms in the cube cannot fill all of the available space. If only 74 percent of the space inside the cube is taken up by Mg atoms, calculate the radius in picometers of an Mg atom. (The density of Mg is 1.74 g/cm^3 and the volume of a sphere of radius r is $\frac{4}{3}\pi r^3$.)

3.99 A certain sample of coal contains 1.6 percent sulfur by mass. When the coal is burned, the sulfur is converted to sulfur dioxide. To prevent air pollution, this sulfur dioxide is treated with calcium oxide (CaO) to form calcium sulfite ($CaSO_3$). Calculate the daily mass (in kilograms) of CaO needed by a power plant that uses 6.60×10^6 kg of coal per day.

3.100 The following is a crude but effective method for estimating the *order of magnitude* of Avogadro's number using stearic acid ($C_{18}H_{36}O_2$). When stearic acid is added to water, its molecules collect at the surface and form a monolayer; that is, the layer is only one molecule thick. The cross-sectional area of each stearic acid molecule has been measured to be 0.21 nm². In one experiment it is found that 1.4×10^{-4} g of stearic acid is needed to form a monolayer over water in a dish of diameter 20 cm. Based on these measurements, what is Avogadro's number? (The area of a circle of radius r is πr^2.)

3.101 Octane (C_8H_{18}) is a component of gasoline. Complete combustion of octane yields CO_2 and H_2O. Incomplete combustion produces CO and H_2O, which not only reduces the efficiency of the engine using the fuel but is also toxic. In a certain test run, 1.000 gallon of octane is burned in an engine. The total mass of CO, CO_2, and H_2O produced is 11.53 kg. Calculate the efficiency of the process; that is, calculate the fraction of octane converted to CO_2. The density of octane is 2.650 kg/gallon.

3.102 Heating 2.40 g of the oxide of metal X (molar mass of X = 55.9 g/mol) in carbon monoxide (CO) yields the pure metal and carbon dioxide. The mass of the metal product is 1.68 g. From the data given, show that the simplest formula of the oxide is X_2O_3 and write a balanced equation for the reaction.

3.103 A mixture of $CuSO_4 \cdot 5H_2O$ and $MgSO_4 \cdot 7H_2O$ is heated until all the water is lost. If 5.020 g of the mixture gives 2.988 g of the anhydrous salts, what is the percent by mass of $CuSO_4 \cdot 5H_2O$ in the mixture?

3.104 When 0.273 g of Mg is heated strongly in a nitrogen (N_2) atmosphere, a chemical reaction occurs. The product of the reaction weighs 0.378 g. Calculate the empirical formula of the compound containing Mg and N. Name the compound.

3.105 A mixture of methane (CH_4) and ethane (C_2H_6) of mass 13.43 g is completely burned in oxygen. If the total mass of CO_2 and H_2O produced is 64.84 g, calculate the fraction of CH_4 in the mixture.

Special Problems

3.106 (a) A research chemist used a mass spectrometer to study the two isotopes of an element. Over time, she recorded a number of mass spectra of these isotopes. On analysis, she noticed that the ratio of the taller peak (the more abundant isotope) to the shorter peak (the less abundant isotope) gradually increased with time. Assuming that the mass spectrometer was functioning normally, what do you think was causing this change?

(b) Mass spectrometry can be used to identify the formulas of molecules having small molecular masses. To illustrate this point, identify the molecule which most likely accounts for the observation of a peak in a mass spectrum at: 16 amu, 17 amu, 18 amu, and 64 amu.

(c) Note that there are (among others) two likely molecules that would give rise to a peak at 44 amu, namely, C_3H_8 and CO_2. In such cases, a chemist might try to look for other peaks generated when some of the molecules break apart in the spectrometer. For example, if a chemist sees a peak at 44 amu and also one at 15 amu, which molecule is producing the 44-amu peak? Why?

(d) Using the following precise atomic masses: ¹H(1.00797 amu), ¹²C(12.00000 amu), and ¹⁶O(15.99491 amu), how precisely must the masses of C_3H_8 and CO_2 be measured to distinguish between them?

(e) Every year millions of dollars' worth of gold is stolen. In most cases the gold is melted down and shipped abroad. This way the gold retains its value while losing all means of identification. Gold is a highly unreactive metal that exists in nature in the uncombined form. During the mineralization of gold, that is, the formation of gold nuggets from microscopic gold particles, various elements such as cadmium (Cd), lead (Pb), and zinc (Zn) are incorporated into the nuggets. The amounts and types of the impurities or trace elements in gold vary according to the location where it was mined. Based on this knowledge, describe how you would identify the source of a piece of gold suspected of being stolen from Fort Knox, the federal gold depository.

3.107 Potash is any potassium mineral that is used for its potassium content. Most of the potash produced in the United States goes into fertilizer. The major sources of potash are potassium chloride (KCl) and potassium sulfate (K_2SO_4). Potash production is often reported as the potassium oxide (K_2O) equivalent or the amount of K_2O that could be made from a given mineral. (a) If KCl costs $0.055 per kg, for what price (dollar per kg) must K_2SO_4 be sold in order to supply the same amount of potassium on a per dollar basis? (b) What mass (in kg) of K_2O contains the same number of moles of K atoms as 1.00 kg of KCl?

Answers to Practice Exercises

3.1 10.81 amu. **3.2** 2.57×10^3 g. **3.3** 8.49×10^{21} K atoms. **3.4** 2.107×10^{-22} g. **3.5** 32.04 amu. **3.6** 1.66 moles. **3.7** 5.81×10^{24} H atoms. **3.8** H: 2.055%; S: 32.69%; O: 65.25%. **3.9** $KMnO_4$ (potassium permanganate). **3.10** 196 g. **3.11** B_2H_6. **3.12** $Fe_2O_3 + 3CO \longrightarrow 2Fe + 3CO_2$. **3.13** (a) 0.508 mole, (b) 2.21 g. **3.14** 235 g. **3.15** (a) 234 g, (b) 234 g. **3.16** (a) 863 g, (b) 93.0%.

After his defeat at Waterloo in 1815, Napoleon was exiled to St. Helena, a small island in the Atlantic, where he spent the last 6 years of his life. In the 1960s, samples of his hair were analyzed and found to contain a high level of arsenic, suggesting that he might have been poisoned. The prime suspects are the governor of St. Helena, with whom Napoleon did not get along, and the French royal family, who wanted to prevent his return to France.

Elemental arsenic is not particularly harmful. The commonly used poison is actually arsenic(III) oxide, As_2O_3. This white compound dissolves in water, is tasteless, and, if administered over a period of time, is hard to detect. It was once known as the "inheritance powder": adding As_2O_3 to grandfather's wine would hasten his demise so the grandson could inherit the estate!

In 1832 the English chemist James Marsh devised a procedure for detecting arsenic that now bears his name. In the Marsh test, hydrogen, formed by the reaction between zinc and sulfuric acid, is allowed to react with a sample of the suspected poison. If As_2O_3 is present, it reacts with hydrogen to form a toxic gas, arsine (AsH_3). When arsine gas is heated, it decomposes to form arsenic, which is recognized by its metallic luster. The Marsh test is an effective deterrent to murder by As_2O_3. But it was invented too late to do Napoleon any good if, in fact, he was a victim of deliberate arsenic poisoning.

A lock of Napoleon's hair.

Doubts about the conspiracy theory of Napoleon's death developed in the early 1990s, when a sample of the wallpaper from his drawing room was found to contain copper arsenite, a green pigment that was commonly used at the time Napoleon lived. It has been suggested that the damp climate on St. Helena promoted the growth of molds on the wallpaper. To rid themselves of arsenic, the molds could have converted it to trimethyl arsine [$(CH_3)_3As$], a volatile and highly poisonous compound. Prolonged exposure to this vapor could have ruined Napoleon's health and would also account for the presence of arsenic in his body, though it may not have been the primary cause of his death. This provocative theory is supported by the fact that Napoleon's regular guests suffered from gastrointestinal disturbances and other symptoms of arsenic poisoning and that their health seemed to improve whenever they spent hours working outdoors in the garden, their main hobby on the island.

We will probably never know whether Napoleon died from arsenic poisoning, intentional or accidental, but the exercise in historical sleuthing provides a fascinating example of the use of chemical analysis. Chemical analysis plays an essential part in a range of endeavors from pure research to practical applications like quality control of commercial products, disease diagnosis, and forensic science.

Reactions in Aqueous Solutions

Chapter 4

ESSENTIAL CONCEPTS

 Reactions in Aqueous Solution Many chemical and almost all biological reactions occur in the aqueous medium. Substances (solutes) that dissolve in water (solvent) can be divided into two categories: electrolytes and nonelectrolytes, depending on their ability to conduct electricity.

 Three Major Types of Reactions In a precipitation reaction, the product, an insoluble substance, separates from solution. Acid-base reactions involve the transfer of a proton (H^+) from an acid to a base. In an oxidation-reduction reaction, or redox reaction, electrons are transferred from a reducing agent to an oxidizing agent. These three types of reactions represent the majority of reactions in chemical and biological systems.

 Solution Stoichiometry Quantitative studies of reactions in solution require that we know the concentration of the solution, which is usually represented by the molarity unit. These studies include gravimetric analysis, which involves the measurement of mass, and titrations in which the unknown concentration of a solution is determined by reaction with a solution of known concentration.

4.1 General Properties of Aqueous Solutions

Many chemical reactions and virtually all biological processes take place in an aqueous environment. Therefore it is important to understand the properties of different substances in solution with water. To start with, what exactly is a solution? A *solution* is *a homogeneous mixture of two or more substances. The substance present in a smaller amount* is called the **solute,** whereas *the substance present in a larger amount* is called the **solvent.** A solution may be gaseous (such as air), solid (such as an alloy), or liquid (seawater, for example). In this section we will discuss only **aqueous solutions,** in which *the solute initially is a liquid or a solid and the solvent is water.*

Electrolytes versus Nonelectrolytes

All solutes that dissolve in water fit into one of two categories: electrolytes and nonelectrolytes. An **electrolyte** is *a substance that, when dissolved in water, results in a solution that can conduct electricity.* A **nonelectrolyte** *does not conduct electricity when dissolved in water.* Figure 4.1 shows an easy and straightforward method of distinguishing between electrolytes and nonelectrolytes. A pair of platinum electrodes is immersed in a beaker of water. To light the bulb, electric current must flow from one electrode to the other, thus completing the circuit. Pure water is a very poor conductor of electricity. However, if we add a small amount of sodium chloride (NaCl), the bulb will glow as soon as the salt dissolves in the water. Solid NaCl, an ionic compound, breaks up into Na^+ and Cl^- ions when it dissolves in water. The Na^+ ions are attracted to the negative electrode and the Cl^- ions to the positive electrode. This movement sets up an electrical current that is equivalent to the flow of electrons along

(a)   (b)   (c)

Figure 4.1 *An arrangement for distinguishing between electrolytes and nonelectrolytes. A solution's ability to conduct electricity depends on the number of ions it contains. (a) A nonelectrolyte solution does not contain ions, and the lightbulb is not lit. (b) A weak electrolyte solution contains a small number of ions, and the lightbulb is dimly lit. (c) A strong electrolyte solution contains a large number of ions, and the lightbulb is brightly lit. The molar amounts of the dissolved solutes are equal in all three cases.*

Table 4.1 | Classification of Solutes in Aqueous Solution

Strong Electrolyte	Weak Electrolyte	Nonelectrolyte
HCl	CH_3COOH	$(NH_2)_2CO$ (urea)
HNO_3	HF	CH_3OH (methanol)
$HClO_4$	HNO_2	C_2H_5OH (ethanol)
H_2SO_4*	NH_3	$C_6H_{12}O_6$ (glucose)
NaOH	$H_2O^\dagger$	$C_{12}H_{22}O_{11}$ (sucrose)
$Ba(OH)_2$		
Ionic compounds		

*H_2SO_4 has two ionizable H^+ ions.

†Pure water is an extremely weak electrolyte.

a metal wire. Because the NaCl solution conducts electricity, we say that NaCl is an electrolyte. Pure water contains very few ions, so it cannot conduct electricity.

Comparing the lightbulb's brightness for the *same molar amounts* of dissolved substances helps us distinguish between strong and weak electrolytes. A characteristic of strong electrolytes is that the solute is assumed to be 100 percent dissociated into ions in solution. (By *dissociation* we mean the breaking up of the compound into cations and anions.) Thus we can represent sodium chloride dissolving in water as

$$NaCl(s) \xrightarrow{H_2O} Na^+(aq) + Cl^-(aq)$$

What this equation says is that all the sodium chloride that enters the solution ends up as Na^+ and Cl^- ions; there are no undissociated NaCl units in solution.

Table 4.1 lists examples of strong electrolytes, weak electrolytes, and nonelectrolytes. Ionic compounds, such as sodium chloride, potassium iodide (KI), and calcium nitrate [$Ca(NO_3)_2$], are strong electrolytes. It is interesting to note that human body fluids contain many strong and weak electrolytes.

Water is a very effective solvent for ionic compounds. Although water is an electrically neutral molecule, it has a positive end (the H atoms) and a negative end (the O atom), or positive and negative "poles"; for this reason it is often referred to as a *polar* solvent. When an ionic compound such as sodium chloride dissolves in water, the three-dimensional network of the ions in the solid is destroyed, and the Na^+ and Cl^- ions are separated from each other. In solution, each Na^+ ion is surrounded by a number of water molecules orienting their negative ends toward the cation. Similarly, each Cl^- ion is surrounded by water molecules with their positive ends oriented toward the anion (Figure 4.2). *The process in which an ion is surrounded by water molecules arranged in a specific manner* is called **hydration.** Hydration helps to stabilize ions in solution and prevents cations from combining with anions.

Hydration

Acids and bases are also electrolytes. Some acids, including hydrochloric acid (HCl) and nitric acid (HNO_3), are strong electrolytes. These acids ionize completely in water; for example, when hydrogen chloride gas dissolves in water, it forms hydrated H^+ and Cl^- ions:

$$HCl(g) \xrightarrow{H_2O} H^+(aq) + Cl^-(aq)$$

In other words, *all* the dissolved HCl molecules separate into hydrated H^+ and Cl^- ions in solution. Thus when we write HCl(*aq*), it is understood that it is a solution of

Figure 4.2 *Hydration of Na⁺ and Cl⁻ ions.*

Figure 4.2 *Hydration of Na⁺ and Cl⁻ ions.*

CH₃COOH

only $H^+(aq)$ and $Cl^-(aq)$ ions and there are no hydrated HCl molecules present. On the other hand, certain acids, such as acetic acid (CH_3COOH), which is found in vinegar, ionize to a much lesser extent. We represent the ionization of acetic acid as

$$CH_3COOH(aq) \rightleftharpoons CH_3COO^-(aq) + H^+(aq)$$

in which CH_3COO^- is called the acetate ion. (In this book we will use the term *dissociation* for ionic compounds and *ionization* for acids and bases.) By writing the formula of acetic acid as CH_3COOH we indicate that the ionizable proton is in the COOH group.

The double arrow $\rightleftharpoons$ in an equation means that the reaction is **reversible;** that is, the *reaction can occur in both directions.* Initially, a number of CH_3COOH molecules break up to yield CH_3COO^- and H^+ ions. As time goes on, some of the CH_3COO^- and H^+ ions recombine to form CH_3COOH molecules. Eventually, a state is reached in which the acid molecules break up as fast as the ions recombine. Such *a chemical state, in which no net change can be observed* (although continuous activity is taking place on the molecular level), is called **chemical equilibrium.** Acetic acid, then, is a weak electrolyte because its ionization in water is incomplete. By contrast, in a hydrochloric acid solution the H^+ and Cl^- ions have no tendency to recombine to form molecular HCl. We use the single arrow to represent complete ionizations.

In the next three sections we will study three types of reactions in the aqueous medium (precipitation, acid-base, and oxidation-reduction) that are of great importance to industrial, environmental, and biological processes. They also play a role in our daily experience.

4.2 Precipitation Reactions

Precipitation reactions

One common type of reaction that occurs in aqueous solution is the **precipitation reaction,** which *results in the formation of an insoluble product, or precipitate.* A **precipitate** is *an insoluble solid that separates from the solution.* Precipitation reactions usually involve ionic compounds. For example, when an aqueous solution of lead nitrate [$Pb(NO_3)_2$] is added to an aqueous solution of sodium iodide (NaI), a yellow precipitate of lead iodide (PbI_2) is formed:

$$Pb(NO_3)_2(aq) + 2NaI(aq) \longrightarrow PbI_2(s) + 2NaNO_3(aq)$$

Sodium nitrate remains in solution. Figure 4.3 shows this reaction in progress.

Solubility

How can we predict whether a precipitate will form when a compound is added to a solution or when two solutions are mixed? It depends on the **solubility** of the solute,

Table 4.2	Solubility Characteristics of Ionic Compounds in Water at 25°C

1. All alkali metal (Group 1A) compounds are soluble.
2. All ammonium (NH_4^+) compounds are soluble.
3. All compounds containing nitrate (NO_3^-), chlorate (ClO_3^-), and perchlorate (ClO_4^-) are soluble.
4. Most hydroxides (OH^-) are insoluble. The exceptions are the alkali metal hydroxides and barium hydroxide [$Ba(OH)_2$]. Calcium hydroxide [$Ca(OH)_2$] is slightly soluble.
5. Most compounds containing chlorides (Cl^-), bromides (Br^-), or iodides (I^-) are soluble. The exceptions are those containing Ag^+, Hg_2^{2+}, and Pb^{2+}.
6. All carbonates (CO_3^{2-}), phosphates (PO_4^{3-}), and sulfides (S^{2-}) are insoluble; the exceptions are those of alkali metals and the ammonium ion.
7. Most sulfates (SO_4^{2-}) are soluble. Calcium sulfate ($CaSO_4$) and silver sulfate (Ag_2SO_4) are slightly soluble. Barium sulfate ($BaSO_4$), mercury(II) sulfate ($HgSO_4$), and lead sulfate ($PbSO_4$) are insoluble.

Figure 4.3 *Formation of yellow PbI_2 precipitate as a solution of $Pb(NO_3)_2$ is added to a solution of NaI.*

which is defined as *the maximum amount of solute that will dissolve in a given quantity of solvent at a specific temperature*. Chemists refer to substances as soluble, slightly soluble, or insoluble in a qualitative sense. A substance is said to be soluble if a fair amount of it visibly dissolves when added to water. If not, the substance is described as slightly soluble or insoluble. All ionic compounds are strong electrolytes, but they are not equally soluble.

Table 4.2 classifies a number of common ionic compounds as soluble or insoluble. Keep in mind, however, that even insoluble compounds dissolve to a certain extent. Figure 4.4 shows the colors of several precipitates.

Example 4.1   Classify these ionic compounds as soluble or insoluble: (a) silver sulfate (Ag_2SO_4), (b) calcium carbonate ($CaCO_3$), and (c) sodium phosphate (Na_3PO_4).

Reasoning and Solution   We refer to Table 4.2 to determine which compounds are insoluble.
(a) According to Table 4.2, Ag_2SO_4 is insoluble. (b) This is a carbonate and Ca is a Group 2A metal. Therefore, $CaCO_3$ is insoluble. (c) Sodium is an alkali metal (Group 1A) so Na_3PO_4 is soluble.

Practice Exercise   Classify these ionic compounds as soluble or insoluble: (a) CuS, (b) $Ca(OH)_2$, and (c) $Zn(NO_3)_2$.

Similar problems: 4.13, 4.14.

Molecular Equations and Ionic Equations

The equation describing the precipitation of lead iodide on page 94 is called a ***molecular equation*** because *the formulas of the compounds are written as though all species existed as molecules or whole units*. A molecular equation is useful

Figure 4.4 *Appearance of several precipitates. From left to right: CdS, PbS, Ni(OH)$_2$, and Al(OH)$_3$.*

because it identifies the reagents (that is, lead nitrate and sodium iodide). If we wanted to bring about this reaction in the laboratory, the molecular equation would be the one to use. However, a molecular equation does not accurately describe what actually is happening at the microscopic level.

As pointed out earlier, when ionic compounds dissolve in water, they break apart into their component cations and anions. To be more realistic, the equations should show the dissociation of dissolved ionic compounds into ions. Therefore, returning to the reaction between sodium iodide and lead nitrate, we would write

$$Pb^{2+}(aq) + 2NO_3^-(aq) + 2Na^+(aq) + 2I^-(aq) \longrightarrow PbI_2(s) + 2Na^+(aq) + 2NO_3^-(aq)$$

This equation is an example of an ***ionic equation,*** which *shows dissolved species as free ions.* An ionic equation includes ***spectator ions,*** or *ions that are not involved in the overall reaction,* in this case the Na$^+$ and NO$_3^-$ ions. Spectator ions appear on both sides of the equation and are unchanged in the chemical reaction, so they can be canceled. To focus on the change that occurs, we write the ***net ionic equation,*** which *shows only the species that actually take part in the reaction:*

$$Pb^{2+}(aq) + 2I^-(aq) \longrightarrow PbI_2(s)$$

Looking at another example, we find that when an aqueous solution of barium chloride (BaCl$_2$) is added to an aqueous solution of sodium sulfate (Na$_2$SO$_4$), a white precipitate of barium sulfate (BaSO$_4$) is formed (Figure 4.5). The molecular equation for this reaction is

$$BaCl_2(aq) + Na_2SO_4(aq) \longrightarrow BaSO_4(s) + 2NaCl(aq)$$

The ionic equation for the reaction is

$$Ba^{2+}(aq) + 2Cl^-(aq) + 2Na^+(aq) + SO_4^{2-}(aq) \longrightarrow$$
$$BaSO_4(s) + 2Na^+(aq) + 2Cl^-(aq)$$

Canceling the spectator ions (Na$^+$ and Cl$^-$) on both sides of the equation gives us the net ionic equation

$$Ba^{2+}(aq) + SO_4^{2-}(aq) \longrightarrow BaSO_4(s)$$

Figure 4.5 *Formation of BaSO$_4$ precipitate.*

These steps summarize the procedure for writing ionic and net ionic equations.

- Write a balanced molecular equation for the reaction.
- Rewrite the equation to show the dissociated ions that form in solution. We assume that all strong electrolytes, when dissolved in solution, are completely dissociated into cations and anions. This procedure gives us the ionic equation.
- Identify and cancel spectator ions on both sides of the equation to arrive at the net ionic equation.

Example 4.2   A potassium phosphate solution is mixed with a calcium nitrate solution. Predict the products and write a net ionic equation for the reaction

$$K_3PO_4(aq) + Ca(NO_3)_2(aq) \longrightarrow ?$$

Reasoning and Solution   We refer to Table 2.3 for the charges on cations and anions. When these ionic compounds dissolve in water, they produce the following ions: K^+, PO_4^{3-}, Ca^{2+}, and NO_3^-. According to Table 4.2, Ca^{2+} and PO_4^{3-} ions can form an insoluble compound, calcium phosphate $[Ca_3(PO_4)_2]$. Therefore, this is a precipitation reaction. The other product, potassium nitrate (KNO_3), is soluble and remains in solution. The molecular equation is

$$2K_3PO_4(aq) + 3Ca(NO_3)_2(aq) \longrightarrow 6KNO_3(aq) + Ca_3(PO_4)_2(s)$$

and the ionic equation is

$$6K^+(aq) + 2PO_4^{3-}(aq) + 3Ca^{2+}(aq) + 6NO_3^-(aq) \longrightarrow$$
$$6K^+(aq) + 6NO_3^-(aq) + Ca_3(PO_4)_2(s)$$

Canceling the spectator K^+ and NO_3^- ions, we obtain the net ionic equation

$$3Ca^{2+}(aq) + 2PO_4^{3-}(aq) \longrightarrow Ca_3(PO_4)_2(s)$$

Note that because we balanced the molecular equation first, the net ionic equation is balanced in terms of the number of atoms on each side and the number of positive and negative charges on the left-hand side.

Practice Exercise   Predict the precipitate produced by the following reaction, and write a net ionic equation for the reaction

$$Al(NO_3)_3(aq) + NaOH(aq) \longrightarrow ?$$

Precipitate formed by the reaction between $K_3PO_4(aq)$ and $Ca(NO_3)_2(aq)$.

Similar problems: 4.15, 4.16.

4.3 Acid-Base Reactions

Acids and bases are as familiar as aspirin and milk of magnesia although many people do not know their chemical names—acetylsalicylic acid (aspirin) and magnesium hydroxide (milk of magnesia). In addition to being the basis of many medicinal and household products, acid-base chemistry is important in industrial processes and essential in sustaining biological systems. Before we can discuss acid-base reactions, we need to know more about acids and bases themselves.

General Properties of Acids and Bases

Figure 4.6 *A piece of blackboard chalk, which is mostly $CaCO_3$, reacts with hydrochloric acid.*

In Section 2.7 we defined acids as substances that ionize in water to produce H^+ ions and bases as substances that ionize in water to produce OH^- ions. These definitions were formulated in the late nineteenth century by the Swedish chemist Svante Arrhenius to classify substances whose properties in aqueous solutions were well known.

Acids

- Acids have a sour taste; for example, vinegar owes its sourness to acetic acid, and lemons and other citrus fruits contain citric acid.
- Acids cause color changes in plant dyes; for example, they change the color of litmus from blue to red.
- Acids react with certain metals, such as zinc, magnesium, and iron, to produce hydrogen gas. A typical reaction is that between hydrochloric acid and magnesium:

$$2HCl(aq) + Mg(s) \longrightarrow MgCl_2(aq) + H_2(g)$$

- Acids react with carbonates and bicarbonates, such as Na_2CO_3, $CaCO_3$, and $NaHCO_3$, to produce carbon dioxide gas (Figure 4.6). For example,

$$2HCl(aq) + CaCO_3(s) \longrightarrow CaCl_2(aq) + H_2O(l) + CO_2(g)$$
$$HCl(aq) + NaHCO_3(s) \longrightarrow NaCl(aq) + H_2O(l) + CO_2(g)$$

- Aqueous acid solutions conduct electricity.

Bases

- Bases have a bitter taste.
- Bases feel slippery; for example, soaps, which contain bases, exhibit this property.
- Bases cause color changes in plant dyes; for example, they change the color of litmus from red to blue.
- Aqueous base solutions conduct electricity.

Brønsted Acids and Bases

Arrhenius's definitions of acids and bases are limited in that they apply only to aqueous solutions. Broader definitions were proposed by the Danish chemist Johannes Brønsted in 1932; a **Brønsted acid** is *a proton donor,* and a **Brønsted base** is *a proton acceptor*. Note that Brønsted's definitions do not require acids and bases to be in aqueous solution.

Hydrochloric acid is a Brønsted acid because it donates a proton in water:

$$HCl(aq) \longrightarrow H^+(aq) + Cl^-(aq)$$

Note that the H^+ ion is a hydrogen atom that has lost its electron; that is, it is just a bare proton. The size of a proton is about 10^{-15} m, compared to a diameter of 10^{-10} m for an average atom or ion. Such an exceedingly small charged particle cannot exist as a separate entity in aqueous solution owing to its strong attraction for the negative pole (the O atom) in H_2O. Consequently, the proton exists in the hydrated form as shown in Figure 4.7. Therefore, the ionization of hydrochloric acid should be written as

$$HCl(aq) + H_2O(l) \longrightarrow H_3O^+(aq) + Cl^-(aq)$$

Figure 4.7 *Ionization of HCl in water to form the hydronium ion and the chloride ion.*

HCl + H_2O $\longrightarrow$ H_3O^+ + Cl^-

The *hydrated proton, H_3O^+*, is called the **hydronium ion.** This equation shows a reaction in which a Brønsted acid (HCl) donates a proton to a Brønsted base (H_2O).

Experiments show that the hydronium ion is further hydrated so that the proton may have several water molecules associated with it. Because the acidic properties of the proton are unaffected by the degree of hydration, in this text we will generally use $H^+(aq)$ to represent the hydrated proton. This notation is for convenience, but H_3O^+ is closer to reality. Keep in mind that both notations represent the same species in aqueous solution.

Acids commonly used in the laboratory include hydrochloric acid (HCl), nitric acid (HNO_3), acetic acid (CH_3COOH), sulfuric acid (H_2SO_4), and phosphoric acid (H_3PO_4). The first three are **monoprotic acids;** that is, *each unit of the acid yields one hydrogen ion upon ionization:*

$$HCl(aq) \longrightarrow H^+(aq) + Cl^-(aq)$$
$$HNO_3(aq) \longrightarrow H^+(aq) + NO_3^-(aq)$$
$$CH_3COOH(aq) \rightleftharpoons CH_3COO^-(aq) + H^+(aq)$$

As mentioned earlier, because the ionization of acetic acid is incomplete (note the double arrows), it is a weak electrolyte. For this reason it is called a weak acid (see Table 4.1). On the other hand, HCl and HNO_3 are strong acids because they are strong electrolytes, so they are completely ionized in solution (note the use of single arrows).

Sulfuric acid (H_2SO_4) is a **diprotic acid** because *each unit of the acid gives up two H^+ ions,* in two separate steps:

$$H_2SO_4(aq) \longrightarrow H^+(aq) + HSO_4^-(aq)$$
$$HSO_4^-(aq) \rightleftharpoons H^+(aq) + SO_4^{2-}(aq)$$

H_2SO_4 is a strong electrolyte or strong acid (the first step of ionization is complete), but HSO_4^- is a weak acid or weak electrolyte, and we need a double arrow to represent its incomplete ionization.

Triprotic acids, which *yield three H^+ ions,* are relatively few in number. The best known triprotic acid is phosphoric acid, whose ionizations are

$$H_3PO_4(aq) \rightleftharpoons H^+(aq) + H_2PO_4^-(aq)$$
$$H_2PO_4^-(aq) \rightleftharpoons H^+(aq) + HPO_4^{2-}(aq)$$
$$HPO_4^{2-}(aq) \rightleftharpoons H^+(aq) + PO_4^{3-}(aq)$$

All three species (H_3PO_4, $H_2PO_4^-$, and HPO_4^{2-}) in this case are weak acids, and we use the double arrows to represent each ionization step. Anions such as $H_2PO_4^-$ and HPO_4^{2-} are found in aqueous solutions of phosphates such as NaH_2PO_4 and Na_2HPO_4.

Figure 4.8 *Ionization of ammonia in water to form the ammonium ion and the hydroxide ion.*

Table 4.1 shows that the bases sodium hydroxide (NaOH) and barium hydroxide [Ba(OH)$_2$] are strong electrolytes. This means that they are completely ionized in solution:

$$NaOH(s) \xrightarrow{H_2O} Na^+(aq) + OH^-(aq)$$
$$Ba(OH)_2(s) \xrightarrow{H_2O} Ba^{2+}(aq) + 2OH^-(aq)$$

The OH$^-$ ion can accept a proton as follows:

$$H^+(aq) + OH^-(aq) \longrightarrow H_2O(l)$$

Thus OH$^-$ is a Brønsted base.

Ammonia (NH$_3$) is classified as a Brønsted base because it can accept a H$^+$ ion (Figure 4.8):

$$NH_3(aq) + H_2O(l) \rightleftharpoons NH_4^+(aq) + OH^-(aq)$$

Ammonia is a weak electrolyte (and therefore a weak base) because only a small fraction of dissolved NH$_3$ molecules react with water to form NH$_4^+$ and OH$^-$ ions.

The most commonly used strong base in the laboratory is sodium hydroxide. It is cheap and soluble. (In fact, all of the alkali metal hydroxides are soluble.) The most commonly used weak base is aqueous ammonia solution, which is sometimes erroneously called ammonium hydroxide; there is no evidence that the species NH$_4$OH actually exists. All of the Group 2A elements form hydroxides of the type M(OH)$_2$, where M denotes an alkaline earth metal. Of these hydroxides, only Ba(OH)$_2$ is soluble. Magnesium and calcium hydroxides are used in medicine and industry. Hydroxides of other metals, such as Al(OH)$_3$ and Zn(OH)$_2$ are insoluble and are less commonly used.

A bottle of aqueous ammonia.

Example 4.3   Classify each of these species as a Brønsted acid or base: (a) HBr, (b) NO$_2^-$, and (c) HCO$_3^-$.

Reasoning and Solution   Remember that a Brønsted acid must contain at least one ionizable proton. If a compound or an ion does not contain any H atoms, then it cannot be a Brønsted acid.

(a) HBr, a strong acid, dissolves in water to yield H$^+$ and Br$^-$ ions:

$$HBr(aq) \longrightarrow H^+(aq) + Br^-(aq)$$

Therefore HBr is a Brønsted acid.

(b) In solution the nitrite ion can accept a proton to form nitrous acid:

$$NO_2^-(aq) + H^+(aq) \rightleftharpoons HNO_2(aq)$$

This property makes NO_2^- a Brønsted base.

(c) The bicarbonate ion is a Brønsted acid because it ionizes in solution as follows:

$$HCO_3^-(aq) \rightleftharpoons H^+(aq) + CO_3^{2-}(aq)$$

It is also a Brønsted base because it can accept a proton to form carbonic acid:

$$HCO_3^-(aq) + H^+(aq) \rightleftharpoons H_2CO_3(aq)$$

Comment   The HCO_3^- species is said to be *amphoteric* because it possesses both acidic and basic properties. The double arrows show that both reactions are reversible.

Similar problems: 4.25, 4.26.

Practice Exercise   Classify each of these species as a Brønsted acid or base: (a) SO_4^{2-}, (b) HI, and (c) HPO_4^{2-}.

Acid-Base Neutralization

A **neutralization reaction** is *a reaction between an acid and a base*. Generally, aqueous acid-base reactions produce water and a **salt**, which is *an ionic compound made up of a cation other than H^+ and an anion other than OH^- or O^{2-}*:

$$\text{acid} + \text{base} \longrightarrow \text{salt} + \text{water}$$

Acid-base reactions generally go to completion.

All salts are strong electrolytes. The substance we know as table salt, NaCl, is a familiar example. It is a product of the acid-base reaction

$$HCl(aq) + NaOH(aq) \longrightarrow NaCl(aq) + H_2O(l)$$

However, because both the acid and the base are strong electrolytes, they are completely ionized in solution. The ionic equation is

$$H^+(aq) + Cl^-(aq) + Na^+(aq) + OH^-(aq) \longrightarrow Na^+(aq) + Cl^-(aq) + H_2O(l)$$

Therefore, the reaction can be represented by the net ionic equation

$$H^+(aq) + OH^-(aq) \longrightarrow H_2O(l)$$

Both Na^+ and Cl^- are spectator ions.

If we had started this reaction with equal molar amounts of the acid and the base, at the end of the reaction we would have only a salt and no leftover acid or base. This is a characteristic of acid-base neutralization reactions.

These are also examples of acid-base neutralization reactions, represented by molecular equations:

$$HF(aq) + KOH(aq) \longrightarrow KF(aq) + H_2O(l)$$
$$H_2SO_4(aq) + 2NaOH(aq) \longrightarrow Na_2SO_4(aq) + 2H_2O(l)$$
$$HNO_3(aq) + NH_3(aq) \longrightarrow NH_4NO_3(aq)$$

The last equation looks different because it does not show water as a product. However, if we express $NH_3(aq)$ as $NH_4^+(aq)$ and $OH^-(aq)$, as discussed earlier, then the equation becomes

$$HNO_3(aq) + NH_4^+(aq) + OH^-(aq) \longrightarrow NH_4NO_3(aq) + H_2O(l)$$

4.4 Oxidation-Reduction Reactions

Whereas acid-base reactions can be characterized as proton-transfer processes, the class of reactions called **oxidation-reduction** (or **redox**) **reactions** are considered *electron-transfer reactions.* Oxidation-reduction reactions are very much a part of the world around us. They range from the combustion of fossil fuels to the action of household bleach. In addition, most metallic and nonmetallic elements are obtained from their ores by the process of oxidation or reduction.

Consider the formation of calcium oxide (CaO) from calcium and oxygen:

$$2Ca(s) + O_2(g) \longrightarrow 2CaO(s)$$

Calcium oxide (CaO) is an ionic compound made up of Ca^{2+} and O^{2-} ions. In this reaction, two Ca atoms give up or transfer four electrons to two O atoms (in O_2). For convenience, this process can be considered as two separate steps, one involving the loss of four electrons by the two Ca atoms and the other being the gain of four electrons by an O_2 molecule:

$$2Ca \longrightarrow 2Ca^{2+} + 4e^-$$
$$O_2 + 4e^- \longrightarrow 2O^{2-}$$

Each of these steps is called a **half-reaction,** which *explicitly shows the electrons involved.* The sum of the half-reactions gives the overall reaction:

$$2Ca + O_2 + 4e^- \longrightarrow 2Ca^{2+} + 2O^{2-} + 4e^-$$

or if we cancel the electrons that appear on both sides of the equation,

$$2Ca + O_2 \longrightarrow 2Ca^{2+} + 2O^{2-}$$

Finally, the Ca^{2+} and O^{2-} ions combine to form CaO:

$$2Ca^{2+} + 2O^{2-} \longrightarrow 2CaO$$

By convention, we do not show the charges in the formula of an ionic compound, so that calcium oxide is normally represented as CaO rather than $Ca^{2+}O^{2-}$.

The half-reaction that involves loss of electrons is called an **oxidation reaction.** The term "oxidation" was originally used by chemists to denote combinations of elements with oxygen. However, it now has a broader meaning that includes reactions not involving oxygen. *The half-reaction that involves gain of electrons* is called a **reduction reaction.** In the formation of calcium oxide, calcium is oxidized. It is said to act as a **reducing agent** because it *donates electrons* to oxygen and causes oxygen to be reduced. Oxygen is reduced and acts as an **oxidizing agent** because it *accepts electrons* from calcium, causing calcium to be oxidized. Note that the extent of oxidation in a redox reaction must be equal to the extent of reduction; that is, the number of electrons lost by a reducing agent must be equal to the number of electrons gained by an oxidizing agent.

A common type of redox reaction is the reaction between metals and acids, represented by

$$\text{metal} + \text{acid} \longrightarrow \text{salt} + \text{molecular hydrogen}$$

Consider the following reactions (Figure 4.9):

$$Fe(s) + 2HCl(aq) \longrightarrow FeCl_2(aq) + H_2(g)$$
$$Zn(s) + 2HCl(aq) \longrightarrow ZnCl_2(aq) + H_2(g)$$
$$Mg(s) + 2HCl(aq) \longrightarrow MgCl_2(aq) + H_2(g)$$

Oxidizing agents are always reduced, and reducing agents are always oxidized. This statement may be somewhat confusing, but it is simply a consequence of the definitions of the two processes.

(a)   (b)   (c)

Figure 4.9 *Reactions of (a) iron (Fe), (b) zinc (Zn), and (c) magnesium (Mg) with hydrochloric acid to form hydrogen gas and the metal chlorides (FeCl$_2$, ZnCl$_2$, MgCl$_2$). The reactivity of these metals is reflected in the rate of hydrogen gas evolution, which is slowest for the least reactive metal, Fe, and fastest for the most reactive metal, Mg.*

In these reactions, the metal acts as the reducing agent by transferring two electrons to the H$^+$ ions supplied by HCl. In the case of iron, the half-reactions are

$$Fe \longrightarrow Fe^{2+} + 2e^-$$
$$2H^+ + 2e^- \longrightarrow H_2$$

and we can represent the overall reaction as a net ionic equation:

$$Fe + 2H^+ \longrightarrow Fe^{2+} + H_2$$

A useful mnemonic for redox is OILRIG: Oxidation Is Loss (of electrons) and Reduction Is Gain (of electrons).

Oxidation Number

The definitions of oxidation and reduction in terms of loss and gain of electrons apply to the formation of ionic compounds such as CaO. However, these definitions do not accurately characterize the formation of hydrogen chloride (HCl) and sulfur dioxide (SO$_2$):

$$H_2(g) + Cl_2(g) \longrightarrow 2HCl(g)$$
$$S(s) + O_2(g) \longrightarrow SO_2(g)$$

Because HCl and SO$_2$ are not ionic but molecular compounds, no electrons are actually transferred in the formation of these compounds, as they are in the case of CaO. Nevertheless, chemists find it convenient to treat these reactions as redox reactions because experimental measurements show that there is a *partial* transfer of electrons (from H to Cl in HCl and from S to O in SO$_2$).

To keep track of electrons in redox reactions, it is useful to assign oxidation numbers to the reactants and products. **Oxidation number** (also called **oxidation state**) refers to *the number of charges an atom would have in a molecule (or an ionic compound) if electrons were transferred completely.* For example, we can rewrite the last two equations for the formation of HCl and SO$_2$ as follows:

$$\overset{0}{H_2}(g) + \overset{0}{Cl_2}(g) \longrightarrow \overset{+1-1}{2HCl}(g)$$
$$\overset{0}{S}(s) + \overset{0}{O_2}(g) \longrightarrow \overset{+4-2}{SO_2}(g)$$

The numbers above the element symbols are the oxidation numbers. In both of the reactions shown, there is no charge on the atoms in the reactant molecules. Thus their

Sulfur burning in air to form sulfur dioxide.

oxidation number is zero. For the product molecules, however, it is assumed that complete electron transfer has taken place and atoms have gained or lost electrons. The oxidation numbers reflect the number of electrons "transferred."

This format allows us to identify elements that are oxidized (oxidation number increase) and reduced (oxidation number decrease) at a glance. The elements that show an increase in oxidation number, hydrogen and sulfur in the examples shown on p. 103, are oxidized in the reactions. Chlorine and oxygen are reduced, so their oxidation numbers show a decrease from their initial values. Note that the sum of the oxidation numbers of H and Cl in HCl (+1 and −1) is zero. Likewise, if we add the charges on S (+4) and two atoms of O [2 × (−2)], the total is zero. The reason is that the HCl and SO_2 molecules are neutral, so the charges must cancel.

We use these rules to assign oxidation numbers:

1. In free elements (that is, in the uncombined state), each atom has an oxidation number of zero. Thus each atom in H_2, Br_2, Na, Be, K, O_2, and P_4 has the same oxidation number: zero.

2. For ions composed of only one atom, the oxidation number is equal to the charge on the ion. Thus Li^+ has an oxidation number of +1; Ba^{2+} ion, +2; Fe^{3+} ion, +3; I^- ion, −1; O^{2-} ion, −2; and so on. All alkali metals have an oxidation number of +1, and all alkaline earth metals have an oxidation number of +2 in their compounds. Aluminum has an oxidation number of +3 in all its compounds.

3. The oxidation number of oxygen in most compounds (for example, MgO and H_2O) is −2, but in hydrogen peroxide (H_2O_2) and peroxide ion (O_2^{2-}), its oxidation number is −1.

4. The oxidation number of hydrogen is +1, except when it is bonded to metals in binary compounds. In these cases (for example, LiH, NaH, and CaH_2), its oxidation number is −1.

5. Fluorine has an oxidation number of −1 in *all* its compounds. Other halogens (Cl, Br, and I) have negative oxidation numbers when they occur as halide ions in their compounds. When combined with oxygen—for example in oxoacids and oxoanions (see Section 2.7)—they have positive oxidation numbers.

6. In a neutral molecule, the sum of the oxidation numbers of all the atoms must be zero. In a polyatomic ion, the sum of oxidation numbers of all the elements in the ion must be equal to the net charge of the ion. For example, in the ammonium ion, NH_4^+, the oxidation number of N is −3 and that of H is +1. Thus the sum of the oxidation numbers is $-3 + 4(+1) = +1$, which is equal to the net charge of the ion.

Example 4.4 Assign oxidation numbers to all the elements in the compounds and ion: (a) Li_2O, (b) HNO_3, and (c) $Cr_2O_7^{2-}$.

Reasoning and Solution We refer to the rules just presented for assigning the oxidation number of the elements. Remember that the sum of the oxidation numbers of the elements in a species must be equal to the net charge of the species.

(a) By rule 2 we see that lithium has an oxidation number of +1 (Li^+) and oxygen's oxidation number is −2 (O^{2-}).

(b) This is the formula for nitric acid, which yields a H^+ ion and a NO_3^- ion in solution. From rule 4 we see that H has an oxidation number of $+1$. Thus the other group (the nitrate ion) must have a net oxidation number of -1. Because oxygen has an oxidation number of -2, the oxidation number of nitrogen (labeled x) is given by

$$-1 = x + 3(-2)$$

or $$x = +5$$

(c) From rule 6 we see that the sum of the oxidation numbers in $Cr_2O_7^{2-}$ must be -2. We know that the oxidation number of O is -2, so all that remains is to determine the oxidation number of Cr, which we call y. The sum of the oxidation numbers for the ion is

$$-2 = 2(y) + 7(-2)$$

or $$y = +6$$

Similar problems: 4.37, 4.38.

Practice Exercise Assign oxidation numbers to all the elements in the compound and ion: (a) PF_3 and (b) MnO_4^-.

Figure 4.10 shows the known oxidation numbers of the familiar elements, arranged according to their positions in the periodic table. We can summarize the content of this figure as:

- Metallic elements have only positive oxidation numbers, whereas nonmetallic elements may have either positive or negative oxidation numbers.
- The highest oxidation number an element in Groups 1A–7A can have is its group number. For example, the halogens are in Group 7A, so their highest possible oxidation number is $+7$.
- The transition metals (Groups 1B, 3B–8B) usually have several possible oxidation numbers.

Activity Series

The reactions between metals and hydrochloric acid described in the preceding section are examples of *displacement reactions,* so called because *one ion or atom in the reaction is replaced (displaced) by another ion (or atom) of another element.* In the metal-HCl reaction, the H^+ ions are displaced by a metal ion.

Displacement reactions are among the most common redox reactions. A metal in a compound can also be displaced by another metal in the uncombined state. For example, when metallic zinc is added to a solution containing copper sulfate ($CuSO_4$), it displaces Cu^{2+} ions from the solution as shown in Figure 4.11 on p. 107:

$$\overset{0}{Zn}(s) + \overset{+2}{Cu}SO_4(aq) \longrightarrow \overset{+2}{Zn}SO_4(aq) + \overset{0}{Cu}(s)$$

The net ionic equation is

$$\overset{0}{Zn}(s) + \overset{+2}{Cu^{2+}}(aq) \longrightarrow \overset{+2}{Zn^{2+}}(aq) + \overset{0}{Cu}(s)$$

Figure 4.10 *The oxidation numbers of elements in their compounds. The more common oxidation numbers are in color.*

Similarly, metallic copper displaces silver ions from a solution containing silver nitrate ($AgNO_3$) (also shown in Figure 4.11):

$$\overset{0}{Cu}(s) + 2\overset{+1}{Ag}NO_3(aq) \longrightarrow \overset{+2}{Cu}(NO_3)_2(aq) + 2\overset{0}{Ag}(s)$$

The net ionic equation is

$$\overset{0}{Cu}(s) + 2\overset{+1}{Ag}^+(aq) \longrightarrow \overset{+2}{Cu}^{2+}(aq) + 2\overset{0}{Ag}(s)$$

Reversing the roles of the metals would result in no reaction. In other words, copper metal will not displace zinc ions from zinc sulfate, and silver metal will not displace copper ions from copper nitrate.

An easy way to predict whether a metal or hydrogen displacement reaction will actually occur is to refer to an *activity series,* like the one shown in Figure 4.12. Basically, an activity series is *a convenient summary of the results of many possible displacement reactions,* similar to those already discussed. According to this series, any metal above

The Zn bar is in aqueous solution of $CuSO_4$

Cu^{2+} ions are converted to Cu atoms. Zn atoms enter the solution as Zn^{2+} ions.

When a piece of copper wire is placed in an aqueous $AgNO_3$ solution Cu atoms enter the solution as Cu^{2+} ions, and Ag^+ ions are converted to solid Ag.

Figure 4.11 *Metal displacement reactions in solution.*

hydrogen will displace it from water or from an acid, but metals below hydrogen will not react with either water or an acid. In fact, any species listed in the series will react with a compound containing any species listed below it. For example, Zn is above Cu, so zinc metal will displace copper ions from copper sulfate.

Using this series, we can predict the reaction of all alkali metals (Group 1A) and some alkaline earth metals (Group 2A) with cold water. For example, sodium (Group 1A) and calcium (Group 2A) react with water as shown in Figure 4.13:

$$\overset{0}{2Na}(s) + \overset{+1}{2H_2}\overset{+1}{O}(l) \longrightarrow \overset{+1}{2NaOH}(aq) + \overset{0}{H_2}(g)$$

$$\overset{0}{Ca}(s) + \overset{+1}{2H_2}\overset{+1}{O}(l) \longrightarrow \overset{+2}{Ca(OH)}_2(s) + \overset{0}{H_2}(g)$$

Halogen Displacement Reactions

Another activity series summarizes the halogens' behavior in halogen displacement reactions:

$$F_2 > Cl_2 > Br_2 > I_2$$

The halogens.

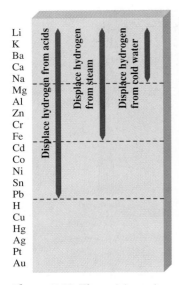

Figure 4.12 *The activity series for metals. The metals are arranged according to their ability to displace hydrogen from an acid or water. Li (lithium) is the most reactive metal, and Au (gold) is the least reactive.*

(a)   (b)

Figure 4.13 *Reactions of (a) sodium (Na) and (b) calcium (Ca) with cold water. Note that the reaction is more vigorous with Na than with Ca.*

Industrially, bromine (a fuming red liquid) is prepared by the action of chlorine on seawater, which is a rich source of Br^- ions.

The power of these elements as oxidizing agents decreases as we move down Group 7A from fluorine to iodine, so molecular fluorine can replace chloride, bromide, and iodide ions in solution. In fact, molecular fluorine is so reactive that it also attacks water; thus these reactions cannot be carried out in aqueous solutions. On the other hand, molecular chlorine can displace bromide and iodide ions in aqueous solution. The equations for the displacement reactions are

$$\overset{0}{Cl_2}(g) + 2\overset{-1}{K}\overset{}{Br}(aq) \longrightarrow 2\overset{-1}{K}\overset{}{Cl}(aq) + \overset{0}{Br_2}(l)$$

$$\overset{0}{Cl_2}(g) + 2\overset{-1}{Na}I(aq) \longrightarrow 2\overset{-1}{Na}Cl(aq) + \overset{0}{I_2}(s)$$

The ionic equations are

$$\overset{0}{Cl_2}(g) + 2\overset{-1}{Br}(aq) \longrightarrow 2\overset{-1}{Cl^-}(aq) + \overset{0}{Br_2}(l)$$

$$\overset{0}{Cl_2}(g) + 2\overset{-1}{I^-}(aq) \longrightarrow 2\overset{-1}{Cl^-}(aq) + \overset{0}{I_2}(s)$$

Molecular bromine, in turn, can displace iodide ion in solution:

$$\overset{0}{Br_2}(l) + 2\overset{-1}{I^-}(aq) \longrightarrow 2\overset{-1}{Br^-}(aq) + \overset{0}{I_2}(s)$$

Reversing the roles of the halogens produces no reaction. Thus bromine cannot displace chloride ions, and iodine cannot displace bromide and chloride ions.

4.5 Concentration of Solutions

To study solution stoichiometry we must know how much of the reactants are present in a solution and also how to control the amounts of reactants used to bring about a reaction in aqueous solution.

The **concentration of a solution** is *the amount of solute present in a given quantity of solvent or solution.* (For this discussion we will assume the solute is a liquid or a solid and the solvent is a liquid.) The concentration of a solution can be expressed in many different ways, as we will see in Chapter 13. Here we will consider one of the most commonly used units in chemistry, *molarity (M),* or *molar concentration,* which is *the number of moles of solute in 1 liter (L) of solution.* Molarity is defined by the equation

$$M = \text{molarity} = \frac{\text{moles of solute}}{\text{liters of soln}} \tag{4.1}$$

where "soln" denotes "solution." Thus, a 1.46 molar glucose ($C_6H_{12}O_6$) solution, written 1.46 M $C_6H_{12}O_6$, contains 1.46 moles of the solute ($C_6H_{12}O_6$) in 1 L of the solution; a 0.52 molar urea [$(NH_2)_2CO$] solution, written as 0.52 M $(NH_2)_2CO$, contains 0.52 mole of $(NH_2)_2CO$ (the solute) in 1 L of solution; and so on.

Of course, we do not always work with solution volumes of exactly 1 L. This is not a problem as long as we remember to convert the volume of the solution to liters. Thus, a 500-mL solution containing 0.730 mole of $C_6H_{12}O_6$ also has a concentration of 1.46 M:

$$M = \text{molarity} = \frac{0.730 \text{ mol}}{0.500 \text{ L}}$$
$$= 1.46 \text{ mol/L} = 1.46 \text{ } M$$

As you can see, the unit of molarity is moles per liter, so a 500-mL solution containing 0.730 mole of $C_6H_{12}O_6$ is equivalent to 1.46 mol/L or 1.46 M. Note that concentration, like density, is an intensive property, so its value does not depend on how much of the solution is present.

The procedure for preparing a solution of known molarity is as follows. First, the solute is accurately weighed and transferred to a volumetric flask through a funnel (Figure 4.14). Next, water is added to the flask, which is carefully swirled to dissolve the solid. After *all* the solid has dissolved, more water is added slowly to bring the level of solution exactly to the volume mark. Knowing the volume of the solution in the flask and the quantity of compound (the number of moles) dissolved, we can calculate the molarity of the solution using Equation (4.1). Note that this procedure does not require knowing the amount of water added, as long as the volume of the final solution is known.

Making a solution.

Marker showing known volume of solution

Meniscus

(a)   (b)   (c)

Figure 4.14 *Preparing a solution of known molarity. (a) A known amount of a solid solute is put into the volumetric flask; then water is added through a funnel. (b) The solid is slowly dissolved by gently swirling the flask. (c) After the solid has completely dissolved, more water is added to bring the level of solution to the mark. Knowing the volume of the solution and the amount of solute dissolved in it, we can calculate the molarity of the prepared solution.*

A $K_2Cr_2O_7$ solution.

Similar problems: 4.47, 4.52.

Example 4.5   How many grams of potassium dichromate ($K_2Cr_2O_7$) are required to prepare a 250-mL solution whose concentration is 2.16 M?

Reasoning and Solution   This problem requires the conversion from volume and concentration to number of moles and then to grams. First, we determine the number of moles of $K_2Cr_2O_7$ in 250 mL of a 2.16 M solution:

$$\text{moles of } K_2Cr_2O_7 = 250 \text{ mL soln} \times \frac{2.16 \text{ mol } K_2Cr_2O_7}{1000 \text{ mL soln}}$$

$$= 0.540 \text{ mol } K_2Cr_2O_7$$

The molar mass of $K_2Cr_2O_7$ is 294.2 g, so the mass needed is given by

$$\text{grams of } K_2Cr_2O_7 \text{ needed} = 0.540 \text{ mol } K_2Cr_2O_7 \times \frac{294.2 \text{ g } K_2Cr_2O_7}{1 \text{ mol } K_2Cr_2O_7}$$

$$= 159 \text{ g } K_2Cr_2O_7$$

Practice Exercise   What is the molarity of an 85.0-mL ethanol (C_2H_5OH) solution containing 1.77 g of ethanol?

Dilution of Solutions

Concentrated solutions are often stored in the laboratory stockroom for use as needed. Frequently we dilute these "stock" solutions before working with them. ***Dilution*** is *the procedure for preparing a less concentrated solution from a more concentrated one.*

Suppose that we want to prepare 1 L of a 0.400 M $KMnO_4$ solution from a solution of 1.00 M $KMnO_4$. For this purpose we need 0.400 mole of $KMnO_4$. Because there is 1.00 mole of $KMnO_4$ in 1 L, or 1000 mL, of a 1.00 M $KMnO_4$ solution, there is 0.400 mole of $KMnO_4$ in 0.400 $\times$ 1000 mL, or 400 mL of the same solution:

$$\frac{1.00 \text{ mol}}{1000 \text{ mL soln}} = \frac{0.400 \text{ mol}}{400 \text{ mL soln}}$$

Therefore, we must withdraw 400 mL from the 1.00 M $KMnO_4$ solution and dilute it to 1000 mL by adding water (in a 1-L volumetric flask). This method gives us 1 L of the desired solution of 0.400 M $KMnO_4$.

In carrying out a dilution process, it is useful to remember that adding more solvent to a given amount of the stock solution changes (decreases) the concentration of the solution without changing the number of moles of solute present in the solution (Figure 4.15). In other words,

moles of solute before dilution = moles of solute after dilution

Because molarity is defined as moles of solute in one liter of solution, we see that the number of moles of solute is given by

$$\underbrace{\frac{\text{moles of solute}}{\text{liters of soln}}}_{M} \times \underbrace{\text{volume of soln (in liters)}}_{V} = \text{moles of solute}$$

Two $KMnO_4$ solutions of different concentrations.

(a)   (b)

or

$$MV = \text{moles of solute}$$

Because all the solute comes from the original stock solution, we can conclude that

$$\underset{\substack{\text{moles of solute} \\ \text{before dilution}}}{M_i V_i} = \underset{\substack{\text{moles of solute} \\ \text{after dilution}}}{M_f V_f} \tag{4.2}$$

where M_i and M_f are the initial and final concentrations of the solution in molarity and V_i and V_f are the initial and final volumes of the solution, respectively. Of course, the units of V_i and V_f must be the same (mL or L) for the calculation to work. To check the reasonableness of your results, be sure that $M_i > M_f$ and $V_f > V_i$.

Example 4.6   Describe how you would prepare 5.00×10^2 mL of 1.75 M H_2SO_4 solution, starting with an 8.61 M stock solution of H_2SO_4.

Reasoning and Solution   Because the concentration of the final solution is less than that of the original one, this is a dilution process. We prepare for the calculation by tabulating our data:

$$M_i = 8.61\ M \qquad M_f = 1.75\ M$$
$$V_i = ? \qquad\qquad V_f = 5.00 \times 10^2\ \text{mL}$$

Substituting in Equation (4.2),

$$(8.61\ M)(V_i) = (1.75\ M)(5.00 \times 10^2\ \text{mL})$$

$$V_i = \frac{(1.75\ M)(5.00 \times 10^2\ \text{mL})}{8.61\ M}$$

$$= 102\ \text{mL}$$

Thus we must dilute 102 mL of the 8.61 M H_2SO_4 solution with sufficient water to give a final volume of 5.00×10^2 mL in a 500-mL volumetric flask to obtain the desired concentration.

Comment   The final volume is greater than the initial volume; therefore the answer is reasonable.

Similar problems: 4.57, 4.58.

Practice Exercise   How would you prepare 2.00×10^2 mL of a 0.866 M NaOH solution, starting with a 5.07 M stock solution?

4.6 Solution Stoichiometry

In Chapter 3 we studied stoichiometric calculations in terms of the mole method, which treats the coefficients in a balanced equation as the number of moles of reactants and products. In working with solutions of known molarity, we have to use the relationship $MV =$ moles of solute. We will examine two types of common solution stoichiometry here: gravimetric analysis and acid-base titration.

Gravimetric Analysis

Gravimetric analysis is *an analytical technique based on the measurement of mass.* One type of gravimetric analysis experiment involves the formation, isolation, and mass determination of a precipitate. Generally this procedure is applied to ionic compounds. A sample substance of unknown composition is dissolved in water and allowed to react with another substance to form a precipitate. The precipitate is filtered off, dried, and weighed. Knowing the mass and chemical formula of the precipitate formed, we can calculate the mass of a particular chemical component (that is, the anion or cation) of the original sample. From the mass of the component and the mass of the original sample, we can determine the percent composition by mass of the component in the original compound.

A reaction that is often studied in gravimetric analysis, because the reactants can be obtained in pure form, is

$$AgNO_3(aq) + NaCl(aq) \longrightarrow NaNO_3(aq) + AgCl(s)$$

The net ionic equation is

$$Ag^+(aq) + Cl^-(aq) \longrightarrow AgCl(s)$$

The purpose of this exercise is to check the purity of the NaCl sample.

The precipitate is silver chloride (see Table 4.2). As an example, let us say that we wanted to determine *experimentally* the percent by mass of Cl in NaCl. First we would accurately weigh out a sample of NaCl and dissolve it in water. Next, we would add enough $AgNO_3$ solution to the NaCl solution to cause the precipitation of all the Cl^- ions present in solution as AgCl. In this procedure NaCl is the limiting reagent and $AgNO_3$ the excess reagent. The AgCl precipitate is separated from the solution by filtration, dried, and weighed. From the measured mass of AgCl, we can calculate the mass of Cl using the percent by mass of Cl in AgCl. Because this same amount of Cl was present in the original NaCl sample, we can calculate the percent by mass of Cl in NaCl. Figure 4.16 shows how this procedure is performed.

Gravimetric analysis is a highly accurate technique, because the mass of a sample can be measured accurately. However, this procedure is applicable only to reactions that go to completion, or have nearly 100 percent yield. Thus, if AgCl were slightly soluble instead of being insoluble, it would not be possible to remove all the Cl^- ions from the NaCl solution and the subsequent calculation would be in error.

Example 4.7   A 0.5662-g sample of an ionic compound containing chloride ions and an unknown metal is dissolved in water and treated with an excess of $AgNO_3$. If 1.0882 g of AgCl precipitate forms, what is the percent by mass of Cl in the original compound?

Reasoning and Solution First we determine the mass of Cl in AgCl (this is also the mass of Cl in the original compound because all of the Cl^- ions end up in AgCl) according to the procedure in Example 3.10. We multiply the mass of AgCl by the percent by mass of Cl in AgCl:

$$\text{mass of Cl} = 1.0882 \text{ g AgCl} \times \frac{1 \text{ mol AgCl}}{143.4 \text{ g AgCl}} \times \frac{1 \text{ mol Cl}}{1 \text{ mol AgCl}} \times \frac{35.45 \text{ g Cl}}{1 \text{ mol Cl}}$$

$$= 0.2690 \text{ g Cl}$$

Next, we calculate the percent by mass of Cl in the unknown sample as

$$\%\text{Cl by mass} = \frac{\text{mass of Cl}}{\text{mass of sample}} \times 100\%$$

$$= \frac{0.2690 \text{ g}}{0.5662 \text{ g}} \times 100\%$$

$$= 47.51\%$$

Comment Note that gravimetric analysis does not establish the whole identity of the unknown. However, knowing the percent by mass of Cl greatly helps us in narrowing the possibilities. As a comparison, calculate the percent by mass of Cl in KCl.

Similar problems: 4.65, 4.66.

Practice Exercise A sample of 0.3220 g of an ionic compound containing the bromide ion (Br^-) is dissolved in water and treated with an excess of $AgNO_3$. If the mass of the AgBr precipitate that forms is 0.6964 g, what is the percent by mass of Br in the original compound?

(a)

(b)

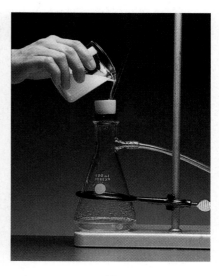
(c)

Figure 4.16 *Basic steps for gravimetric analysis. (a) A solution containing a known amount of NaCl in a beaker. (b) The precipitation of AgCl upon the addition of $AgNO_3$ solution from a measuring cylinder. In this reaction, $AgNO_3$ is the excess reagent and NaCl is the limiting reagent. (c) The solution containing the AgCl precipitate is filtered through a preweighed sintered-disk crucible, which allows the liquid (but not the precipitate) to pass through. The crucible is then removed from the apparatus, dried in an oven, and weighed again. The difference between this mass and that of the empty crucible gives the mass of the AgCl precipitate.*

Acid-Base Titrations

Quantitative studies of acid-base neutralization reactions are most conveniently carried out using a procedure known as *titration*. In a titration experiment, *a solution of accurately known concentration,* called a **standard solution,** *is added gradually to another solution of unknown concentration, until the chemical reaction between the two solutions is complete.* If we know the volumes of the standard and unknown solutions used in the titration, along with the concentration of the standard solution, we can calculate the concentration of the unknown solution.

Sodium hydroxide is one of the bases commonly used in the laboratory. However, because it is difficult to obtain solid sodium hydroxide in a pure form, a solution of sodium hydroxide must be *standardized* before it can be used in accurate analytical work. We can standardize the sodium hydroxide solution by titrating it against an acid solution of accurately known concentration. The acid often chosen for this task is a monoprotic acid called potassium hydrogen phthalate (KHP), for which the molecular formula is $KHC_8H_4O_4$. KHP is a white, soluble solid that is commercially available in highly pure form. The reaction between KHP and sodium hydroxide is

Potassium hydrogen phthalate.

$$KHC_8H_4O_4(aq) + NaOH(aq) \longrightarrow KNaC_8H_4O_4(aq) + H_2O(l)$$

The net ionic equation is

$$HC_8H_4O_4^-(aq) + OH^-(aq) \longrightarrow C_8H_4O_4^{2-}(aq) + H_2O(l)$$

Do we need to know the volume of the KHP solution?

The procedure for the titration is shown in Figure 4.17. First, a known amount of KHP is transferred to an Erlenmeyer flask and some distilled water is added to make up a solution. Next, NaOH solution is carefully added to the KHP solution from a

Figure 4.17 *(a) Apparatus for acid-base titration. A NaOH solution from the buret is added to a KHP solution in an Erlenmeyer flask. (b) A reddish-pink color is observed when the equivalence point is reached. The color has been intensified here for visual display.*

(a)   (b)

buret until we reach the **equivalence point,** that is, *the point at which the acid has completely reacted with or been neutralized by the base.* The point is usually signaled by a sharp color change of an indicator that has been added to the acid solution. In acid-base titrations, **indicators** are *substances that have distinctly different colors in acidic and basic media.* One commonly used indicator is phenolphthalein, which is colorless in acidic and neutral solutions but reddish pink in basic solutions. At the equivalence point, all the KHP has been neutralized by the NaOH and the solution is still colorless. However, if we add just one more drop of NaOH solution from the buret, the solution will immediately turn pink because the solution is now basic. Knowing the mass (and hence the number of moles) of KHP reacted, we can calculate the concentration of the NaOH solution.

Example 4.8 In a titration experiment, a student finds that 23.48 mL of a NaOH solution are needed to completely neutralize 0.5468 g of KHP. What is the concentration (in molarity) of the NaOH solution?

Reasoning and Solution From the balanced equation we see that 1 mole of KHP neutralizes 1 mole of NaOH. Therefore, the number of moles in 0.5468 g of KHP gives the number of moles of NaOH in 23.48 mL of solution. From this information we can calculate the number of moles of NaOH in 1000 mL or 1 L of solution and hence the molarity.

First we calculate the number of moles of KHP used in the titration:

$$\text{moles of KHP} = 0.5468 \text{ g KHP} \times \frac{1 \text{ mol KHP}}{204.2 \text{ g KHP}}$$

$$= 2.678 \times 10^{-3} \text{ mol KHP}$$

Because 1 mol KHP $\approx$ 1 mol NaOH, there must be 2.678×10^{-3} mole of NaOH in 23.48 mL of NaOH solution. Finally, we calculate the molarity of the NaOH as follows:

$$\text{molarity of NaOH soln} = \frac{2.678 \times 10^{-3} \text{ mol NaOH}}{23.48 \text{ mL soln}} \times \frac{1000 \text{ mL soln}}{1 \text{ L soln}}$$

$$= 0.1141 \ M$$

Similar problems: 4.71(a), 4.72(a).

Practice Exercise How many grams of KHP are needed to neutralize 18.64 mL of a 0.1004 *M* NaOH solution?

The reaction between NaOH and KHP is one of the simplest types of acid-base neutralization known. Suppose, though, that instead of KHP, a diprotic acid such as H_2SO_4 is used. The reaction is represented by

$$2NaOH(aq) + H_2SO_4(aq) \longrightarrow Na_2SO_4(aq) + 2H_2O(l)$$

Because 2 mol NaOH $\approx$ 1 mol H_2SO_4, we need twice as much NaOH to react completely with a H_2SO_4 solution of the *same* molar concentration. On the other hand, we would need twice the amount of HCl to neutralize a $Ba(OH)_2$ solution compared to a NaOH solution having the same concentration and volume because 1 mole of $Ba(OH)_2$ yields 2 moles of OH^- ions:

$$2HCl(aq) + Ba(OH)_2(aq) \longrightarrow BaCl_2(aq) + 2H_2O(l)$$

H_2SO_4 has two ionizable protons.

Example 4.9   How many milliliters of a 0.610 *M* NaOH solution are needed to neutralize 20.0 mL of a 0.245 *M* H_2SO_4 solution? The equation for the neutralization reaction was just shown.

Reasoning and Solution   This problem is similar to Example 4.8 except the stoichiometry is different. Because we are given both the volume and concentration of the H_2SO_4 solution, we can calculate the number of moles of H_2SO_4 consumed in the reaction:

$$\text{moles of } H_2SO_4 = \frac{0.245 \text{ mol } H_2SO_4}{1 \text{ L soln}} \times \frac{1 \text{ L soln}}{1000 \text{ mL soln}} \times 20.0 \text{ mL soln}$$

$$= 4.90 \times 10^{-3} \text{ mol } H_2SO_4$$

From the stoichiometry we see that 1 mol $H_2SO_4 \backsimeq$ 2 mol NaOH. Therefore, the number of moles of NaOH reacted must be $2 \times 4.90 \times 10^{-3}$ mol, or 9.80×10^{-3} mol. From the definition of molarity [see Equation (4.1)] we have

$$\text{liters of soln} = \frac{\text{moles of solute}}{\text{molarity}}$$

or

$$\text{volume of NaOH} = \frac{9.80 \times 10^{-3} \text{ mol NaOH}}{0.610 \text{ mol/L}}$$

$$= 0.0161 \text{ L or } 16.1 \text{ mL}$$

Similar problems: 4.71(b), 4.72(b).

Practice Exercise   How many milliliters of a 1.28 *M* H_2SO_4 solution are needed to neutralize 60.2 mL of a 0.427 *M* KOH solution?

Summary

Aqueous solutions are electrically conducting if the solutes are electrolytes. If the solutes are nonelectrolytes, the solutions do not conduct electricity.

From general rules about solubilities of ionic compounds, we can predict whether a precipitate will form in a reaction.

Arrhenius acids ionize in water to give H^+ ions, and Arrhenius bases ionize in water to give OH^- ions. Brønsted acids donate protons, and Brønsted bases accept protons. The reaction of an acid with a base is called neutralization.

In redox reactions, oxidation and reduction always occur simultaneously. Oxidation is characterized by the loss of electrons, reduction by the gain of electrons. Oxidation numbers help us keep track of charge distribution and are assigned to all atoms in a compound or ion according to a specific set of rules. Oxidation can be defined as an increase in oxidation number; reduction can be defined as a decrease in oxidation number. The activity series helps us predict the outcome of a displacement reaction, which is a type of redox reaction.

The concentration of a solution is the amount of solute present in a given amount of solution. Molarity expresses concentration as the number of moles of solute in 1 L of solution. Adding a solvent to a solution, a process known as dilution, decreases the concentration (molarity) of the solution without changing the total number of moles of solute present in the solution.

Gravimetric analysis is a technique for determining the identity of a compound and/or the concentration of a solution by measuring mass. Gravimetric experiments often involve precipitation reactions.

In acid-base titration, a solution of known concentration (say, a base) is added gradually to a solution of unknown concentration (say, an acid) with the goal of determining the unknown concentration. The point at which the reaction in the titration is complete is called the equivalence point.

Key Words

Questions and Problems

Aqueous Solutions

Review Questions

4.1 Define the following terms: solute, solvent, solution, electrolyte, nonelectrolyte, hydration, reversible reaction, chemical equilibrium.

4.2 What is the difference between the following symbols used in chemical equations: $\longrightarrow$ and $\rightleftharpoons$?

4.3 Water, as we know, is an extremely weak electrolyte and therefore cannot conduct electricity. Yet we are often cautioned not to operate electrical appliances when our hands are wet. Why?

4.4 LiF is a strong electrolyte. When we write LiF(*aq*), what species are actually present in the solution?

Problems

4.5 Identify each of these substances as a strong electrolyte, a weak electrolyte, or a nonelectrolyte: (a) H_2O, (b) KCl, (c) HNO_3, (d) CH_3COOH, (e) $C_{12}H_{22}O_{11}$.

4.6 Identify each of these substances as a strong electrolyte, weak electrolyte, or a nonelectrolyte: (a) $Ba(NO_3)_2$, (b) Ne, (c) NH_3, (d) NaOH.

4.7 The passage of electricity through an electrolyte solution is caused by the movement of (a) electrons only, (b) cations only, (c) anions only, (d) both cations and anions.

4.8 Predict and explain which of these systems are electrically conducting: (a) solid NaCl, (b) molten NaCl, (c) an aqueous solution of NaCl.

4.9 You are given a water-soluble compound X. Describe how you would determine whether it is an electrolyte or a nonelectrolyte. If it is an electrolyte, how would you determine whether it is strong or weak?

4.10 Explain why a solution of HCl in benzene does not conduct electricity but HCl in water does.

Precipitation Reactions

Review Questions

4.11 What is the difference between an ionic equation and a molecular equation?

4.12 What is the advantage of writing net ionic equations for precipitation reactions?

Problems

4.13 Characterize these compounds as soluble or insoluble in water: (a) $Ca_3(PO_4)_2$, (b) $Mn(OH)_2$, (c) $AgClO_3$, (d) K_2S.

4.14 Characterize these compounds as soluble or insoluble in water: (a) $CaCO_3$, (b) $ZnSO_4$, (c) $Hg(NO_3)_2$, (d) $HgSO_4$, (e) NH_4ClO_4.

4.15 Write ionic and net ionic equations for these reactions:

(a) $2AgNO_3(aq) + Na_2SO_4(aq) \longrightarrow$
(b) $BaCl_2(aq) + ZnSO_4(aq) \longrightarrow$
(c) $(NH_4)_2CO_3(aq) + CaCl_2(aq) \longrightarrow$

4.16 Write ionic and net ionic equations for these reactions:

(a) $Na_2S(aq) + ZnCl_2(aq) \longrightarrow$
(b) $2K_3PO_4(aq) + 3Sr(NO_3)_2(aq) \longrightarrow$
(c) $Mg(NO_3)_2(aq) + 2NaOH(aq) \longrightarrow$

4.17 Which of these processes will likely result in a precipitation reaction? (a) Mixing a $NaNO_3$ solution with a $CuSO_4$ solution. (b) Mixing a $BaCl_2$ solution with a K_2SO_4 solution. Write the net ionic equation.

4.18 On the basis of the solubility rules given in this chapter, suggest one method by which you might separate (a) K^+ from Ag^+, (b) Ba^{2+} from Pb^{2+}, (c) NH_4^+ from Ca^{2+}, (d) Ba^{2+} from Cu^{2+}. All cations are assumed to be in aqueous solution, and the common anion is the nitrate ion.

Acid-Base Reactions

Review Questions

4.19 List the general properties of acids and bases.

4.20 Give Arrhenius's and Brønsted's definitions of an acid and a base.

4.21 Give an example each of a monoprotic acid, a diprotic acid, and a triprotic acid.

4.22 What are the characteristics of an acid-base neutralization reaction?

4.23 Give four examples of a salt.

4.24 Identify the following as a weak or strong acid or base: (a) NH_3, (b) H_3PO_4, (c) $LiOH$, (d) $HCOOH$ (formic acid), (e) H_2SO_4, (f) HF, (g) $Ba(OH)_2$.

Problems

4.25 Identify each of these species as a Brønsted acid or base, or both: (a) HI, (b) CH_3COO^-, (c) $H_2PO_4^-$.

4.26 Identify each of these species as a Brønsted acid or base, or both: (a) PO_4^{3-}, (b) ClO_2^-, (c) NH_4^+, (d) HCO_3^-.

4.27 Balance the following equations and write the corresponding ionic and net ionic equations (if appropriate):

(a) $HBr(aq) + NH_3(aq) \longrightarrow$ (HBr is a strong acid)
(b) $Ba(OH)_2(aq) + H_3PO_4(aq) \longrightarrow$
(c) $HClO_4(aq) + Mg(OH)_2(s) \longrightarrow$

4.28 Balance these equations and write the corresponding ionic and net ionic equations (if appropriate):

(a) $CH_3COOH(aq) + KOH(aq) \longrightarrow$
(b) $H_2CO_3(aq) + NaOH(aq) \longrightarrow$
(c) $HNO_3(aq) + Ba(OH)_2(aq) \longrightarrow$

Oxidation-Reduction Reactions

Review Questions

4.29 Define the following terms: half-reaction, oxidation reaction, reduction reaction, reducing agent, oxidizing agent, redox reaction.

4.30 Is it possible to have a reaction in which oxidation occurs and reduction does not? Explain.

Problems

4.31 For the complete redox reactions given here, (i) break down each reaction into its half-reactions; (ii) identify the oxidizing agent; (iii) identify the reducing agent.

(a) $2Sr + O_2 \longrightarrow 2SrO$
(b) $2Li + H_2 \longrightarrow 2LiH$
(c) $2Cs + Br_2 \longrightarrow 2CsBr$
(d) $3Mg + N_2 \longrightarrow Mg_3N_2$

4.32 For the complete redox reactions given here, write the half-reactions and identify the oxidizing and reducing agents:

(a) $4Fe + 3O_2 \longrightarrow 2Fe_2O_3$
(b) $Cl_2 + 2NaBr \longrightarrow 2NaCl + Br_2$
(c) $Si + 2F_2 \longrightarrow SiF_4$
(d) $H_2 + Cl_2 \longrightarrow 2HCl$

Oxidation Numbers

Review Questions

4.33 Define oxidation number. Explain why, except for ionic compounds, oxidation number does not have any physical significance.

4.34 (a) Without referring to Figure 4.10, give the oxidation numbers of the alkali and alkaline earth metals in their compounds. (b) Give the highest oxidation numbers that the Groups 3A–7A elements can have.

Problems

4.35 Arrange these species in order of increasing oxidation number of the sulfur atom: (a) H_2S, (b) S_8, (c) H_2SO_4, (d) S^{2-}, (e) HS^-, (f) SO_2, (g) SO_3.

4.36 Indicate the oxidation number of phosphorus in each of these compounds: (a) HPO_3, (b) H_3PO_2, (c) H_3PO_3, (d) H_3PO_4, (e) $H_4P_2O_7$, (f) $H_5P_3O_{10}$.

4.37 Give the oxidation number of the underlined atom in these molecules and ions: (a) $\underline{Cl}F$, (b) $\underline{I}F_7$, (c) $\underline{C}H_4$, (d) $\underline{C}_2H_2$, (e) $\underline{C}_2H_4$, (f) $K_2\underline{Cr}O_4$, (g) $K_2\underline{Cr}_2O_7$, (h) $K\underline{Mn}O_4$, (i) $Na\underline{H}CO_3$, (j) $\underline{Li}_2$, (k) $Na\underline{I}O_3$, (l) $\underline{K}O_2$, (m) $\underline{P}F_6^-$, (n) $K\underline{Au}Cl_4$.

4.38 Give the oxidation number of the underlined atom in the following molecules and ions: (a) $\underline{Cs}_2O$, (b) $Ca\underline{I}_2$, (c) $\underline{Al}_2O_3$, (d) $H_3\underline{As}O_3$, (e) $\underline{Ti}O_2$, (f) $\underline{Mo}O_4^{2-}$, (g) $\underline{Pt}Cl_4^{2-}$, (h) $\underline{Pt}Cl_6^{2-}$, (i) $\underline{Sn}F_2$, (j) $\underline{Cl}F_3$, (k) $\underline{Sb}F_6^-$.

4.39 Give the oxidation number for each of these species: H_2, Se_8, P_4, O, U, As_4, B_{12}.

4.40 Give the oxidation number of the underlined atom in these molecules and ions: (a) $Mg_3\underline{N}_2$, (b) $Cs\underline{O}_2$, (c) $Ca\underline{C}_2$, (d) $\underline{C}O_3^{2-}$, (e) $\underline{C}_2O_4^{2-}$, (f) $Zn\underline{O}_2^{2-}$, (g) $Na\underline{B}H_4$, (h) $\underline{W}O_4^{2-}$.

4.41 Nitric acid is a strong oxidizing agent. State which of these species is *least* likely to be produced when nitric acid reacts with a strong reducing agent such as zinc metal, and explain why: N_2O, NO, NO_2, N_2O_4, N_2O_5, NH_4^+.

4.42 On the basis of oxidation number considerations, one of these oxides would not react with molecular oxygen: NO, N_2O, SO_2, SO_3, P_4O_6. Which one is it? Why?

4.43 Predict the outcome of the reactions represented by these equations by using the activity series, and balance the equations:

(a) $Cu(s) + HCl(aq) \longrightarrow$
(b) $I_2(s) + NaBr(aq) \longrightarrow$
(c) $Mg(s) + CuSO_4(aq) \longrightarrow$
(d) $Cl_2(g) + KBr(aq) \longrightarrow$

4.44 Which of these metals can react with water? (a) Au, (b) Li, (c) Hg, (d) Ca, (e) Pt.

Concentration of Solutions

Review Questions

4.45 Define molarity. Why is molarity a convenient concentration unit?

4.46 Describe steps involved in preparing a solution of known molar concentration using a volumetric flask.

Problems

4.47 Calculate the mass of NaOH in grams required to prepare a 5.00×10^2-mL solution of concentration 2.80 M.

4.48 A quantity of 5.25 g of NaOH is dissolved in a sufficient amount of water to make up exactly 1 L of solution. What is the molarity of the solution?

4.49 How many moles of $MgCl_2$ are present in 60.0 mL of a 0.100 M $MgCl_2$ solution?

4.50 How many grams of KOH are present in 35.0 mL of a 5.50 M solution?

4.51 Calculate the molarity of each of these solutions: (a) 29.0 g of ethanol (C_2H_5OH) in 545 mL of solution, (b) 15.4 g of sucrose ($C_{12}H_{22}O_{11}$) in 74.0 mL of solution, (c) 9.00 g of sodium chloride (NaCl) in 86.4 mL of solution.

4.52 Calculate the molarity of each of these solutions: (a) 6.57 g of methanol (CH_3OH) in 1.50×10^2 mL of solution, (b) 10.4 g of calcium chloride ($CaCl_2$) in 2.20×10^2 mL of solution, (c) 7.82 g of naphthalene ($C_{10}H_8$) in 85.2 mL of benzene solution.

4.53 Calculate the volume in milliliters of a solution required to provide: (a) 2.14 g of sodium chloride from a 0.270 M solution, (b) 4.30 g of ethanol from a 1.50 M solution, (c) 0.85 g of acetic acid (CH_3COOH) from a 0.30 M solution.

4.54 Determine how many grams of each of these solutes would be needed to make 2.50×10^2 mL of a 0.100 M solution: (a) cesium iodide (CsI), (b) sulfuric acid (H_2SO_4), (c) sodium carbonate (Na_2CO_3), (d) potassium dichromate ($K_2Cr_2O_7$), (e) potassium permanganate ($KMnO_4$).

Dilution of Solutions

Review Questions

4.55 Describe the basic steps involved in the dilution of a solution of known concentration.

4.56 Write the equation that enables you to calculate the concentration of a diluted solution. Define all the terms.

Problems

4.57 Describe how to prepare 1.00 L of a 0.646 M HCl solution, starting with a 2.00 M HCl solution.

4.58 A quantity of 25.0 mL of a 0.866 M KNO_3 solution is poured into a 500-mL volumetric flask, and water is added until the volume of the solution is exactly 500 mL. What is the concentration of the final solution?

4.59 How would you prepare 60.0 mL of 0.200 M HNO_3 from a stock solution of 4.00 M HNO_3?

4.60 You have 505 mL of a 0.125 M HCl solution, and you want to dilute it to exactly 0.100 M. How much water should you add?

4.61 A 35.2-mL, 1.66 M $KMnO_4$ solution is mixed with a 16.7-mL, 0.892 M $KMnO_4$ solution. Calculate the concentration of the final solution.

4.62 A 46.2-mL, 0.568 M calcium nitrate [$Ca(NO_3)_2$] solution is mixed with an 80.5-mL, 1.396 M calcium nitrate solution. Calculate the concentration of the final solution.

Gravimetric Analysis

Review Questions

4.63 Define gravimetric analysis. Describe the basic steps involved in gravimetric analysis. How does such an experiment help us determine the identity of a compound or the purity of a compound if its formula is known?

4.64 Distilled water must be used in the gravimetric analysis of chlorides. Why?

Problems

4.65 The volume 30.0 mL of 0.150 M $CaCl_2$ is added to 15.0 mL of 0.100 M $AgNO_3$. What is the mass in grams of AgCl precipitate formed?

4.66 A sample of 0.6760 g of an unknown compound containing barium ions (Ba^{2+}) is dissolved in water and treated with an excess of Na_2SO_4. If the mass of the $BaSO_4$ precipitate formed is 0.4105 g, what is the percent by mass of Ba in the original unknown compound?

4.67 How many grams of NaCl are required to precipitate practically all the Ag^+ ions from 2.50×10^2 mL of 0.0113 M $AgNO_3$ solution? Write the net ionic equation for the reaction.

4.68 The concentration of Cu^{2+} ions in the water (which also contains sulfate ions) discharged from a certain industrial plant is determined by adding excess sodium sulfide (Na_2S) solution to 0.800 L of the water. The molecular equation is

$$Na_2S(aq) + CuSO_4(aq) \longrightarrow Na_2SO_4(aq) + CuS(s)$$

Write the net ionic equation and calculate the molar concentration of Cu^{2+} in the water sample if 0.0177 g of solid CuS is formed.

Acid-Base Titrations

Review Questions

4.69 Define acid-base titration, standard solution, equivalence point.

4.70 Describe the basic steps involved in acid-base titration. Why is this technique of great practical value? How does an acid-base indicator work?

Problems

4.71 Calculate the volume in milliliters of a 1.420 M NaOH solution required to titrate these solutions:

(a) 25.00 mL of a 2.430 M HCl solution
(b) 25.00 mL of a 4.500 M H_2SO_4 solution
(c) 25.00 mL of a 1.500 M H_3PO_4 solution

4.72 What volume of a 0.50 M KOH solution is needed to neutralize completely each of the following?

(a) 10.0 mL of a 0.30 M HCl solution
(b) 10.0 mL of a 0.20 M H_2SO_4 solution
(c) 15.0 mL of a 0.25 M H_3PO_4 solution

Additional Problems

4.73 Classify these reactions according to the types discussed in the chapter:

(a) $Cl_2 + 2OH^- \longrightarrow Cl^- + ClO^- + H_2O$
(b) $Ca^{2+} + CO_3^{2-} \longrightarrow CaCO_3$
(c) $NH_3 + H^+ \longrightarrow NH_4^+$
(d) $2CCl_4 + CrO_4^{2-} \longrightarrow 2COCl_2 + CrO_2Cl_2 + 2Cl^-$
(e) $Ca + F_2 \longrightarrow CaF_2$
(f) $2Li + H_2 \longrightarrow 2LiH$
(g) $Ba(NO_3)_2 + Na_2SO_4 \longrightarrow 2NaNO_3 + BaSO_4$
(h) $CuO + H_2 \longrightarrow Cu + H_2O$
(i) $Zn + 2HCl \longrightarrow ZnCl_2 + H_2$
(j) $2FeCl_2 + Cl_2 \longrightarrow 2FeCl_3$

4.74 Using the apparatus shown in Figure 4.1, a student found that the lightbulb was brightly lit when the electrodes were immersed in a sulfuric acid solution. However, after the addition of a certain amount of barium hydroxide [$Ba(OH)_2$] solution, the light began to dim even though $Ba(OH)_2$ is also a strong electrolyte. Explain.

4.75 Someone gave you a colorless liquid. Describe three chemical tests you would perform on the liquid to show that it is water.

4.76 You are given two colorless solutions, one containing NaCl and the other sucrose ($C_{12}H_{22}O_{11}$). Suggest a chemical and a physical test that would distinguish between these two solutions.

4.77 Chlorine (Cl_2) is used to purify drinking water. Too much chlorine is harmful to humans. The excess chlorine is often removed by treatment with sulfur dioxide (SO_2). Balance the following equation that represents this procedure:

$$Cl_2 + SO_2 + H_2O \longrightarrow Cl^- + SO_4^{2-} + H^+$$

4.78 Before aluminum was obtained by electrolytic reduction from its ore (Al_2O_3), the metal was produced by chemical reduction of $AlCl_3$. Which metals would you use to reduce Al^{3+} to Al?

4.79 Oxygen (O_2) and carbon dioxide (CO_2) are colorless and odorless gases. Suggest two chemical tests that would allow you to distinguish between them.

4.80 Based on oxidation number, explain why carbon monoxide (CO) is flammable but carbon dioxide (CO_2) is not.

4.81 Which of these aqueous solutions would you expect to be the best conductor of electricity at 25°C? Explain your answer.

(a) 0.20 M NaCl
(b) 0.60 M CH_3COOH
(c) 0.25 M HCl
(d) 0.20 M $Mg(NO_3)_2$

4.82 A 5.00×10^2-mL sample of 2.00 M HCl solution is treated with 4.47 g of magnesium. Calculate the concentration of the acid solution after all the metal has reacted. Assume that the volume remains unchanged.

4.83 Calculate the volume (in liters) of a 0.156 M $CuSO_4$ solution that would react with 7.89 g of zinc.

4.84 Sodium carbonate (Na_2CO_3) can be obtained in very pure form and can be used to standardize acid solutions. What is the molarity of an HCl solution if 28.3 mL of the solution is required to react with 0.256 g of Na_2CO_3?

4.85 A 3.664-g sample of a monoprotic acid was dissolved in water and required 20.27 mL of a 0.1578 M NaOH solution for neutralization. Calculate the molar mass of the acid.

4.86 Acetic acid (CH_3COOH) is an important ingredient of vinegar. A sample of 50.0 mL of a commercial vinegar is titrated against a 1.00 M NaOH solution. What is the concentration (in M) of acetic acid present in the vinegar if 5.75 mL of the base were required for the titration?

4.87 Calculate the mass of precipitate formed when 2.27 L of 0.0820 M $Ba(OH)_2$ are mixed with 3.06 L of 0.0664 M Na_2SO_4.

4.88 Milk of magnesia is an aqueous suspension of magnesium hydroxide [$Mg(OH)_2$] used to treat acid indigestion. Calculate the volume of a 0.035 M HCl solution (a typical acid concentration in an upset stomach) needed to react with two spoonfuls of milk of magnesia [approximately 10.0 mL at 0.080 g $Mg(OH)_2$/mL].

4.89 A 1.00-g sample of a metal X (that is known to form X^{2+} ions) was added to a 0.100 L of 0.500 M H_2SO_4. After all the metal had reacted, the remaining acid required 0.0334 L of 0.500 M NaOH solution for neutralization. Calculate the molar mass of the metal and identify the element.

4.90 A 60.0-mL 0.513 M glucose ($C_6H_{12}O_6$) solution is mixed with 120.0 mL of 2.33 M glucose solution. What is the concentration of the final solution? Assume the volumes are additive.

4.91 You are given a soluble compound of unknown molecular formula. (a) Describe three tests that would show that the compound is an acid. (b) Once you have established that the compound is an acid, describe how you would determine its molar mass using an NaOH solution of known concentration. (Assume the acid is monoprotic.) (c) How would you find out whether the acid is weak or strong? You are provided with a sample of NaCl and an apparatus like that shown in Figure 4.1 for comparison.

4.92 Someone spilled concentrated sulfuric acid on the floor of a chemistry laboratory. To neutralize the acid, would it be preferable to pour concentrated sodium hydroxide solution or spray solid sodium bicarbonate over the acid? Explain your choice and the chemical basis for the action.

4.93 These are common household compounds: table salt (NaCl), table sugar (sucrose), vinegar (contains acetic acid), baking soda ($NaHCO_3$), washing soda ($Na_2CO_3 \cdot 10H_2O$), boric acid (H_3BO_3, used in eyewash), epsom salt ($MgSO_4 \cdot 7H_2O$), sodium hydroxide (used in drain openers), ammonia, milk of magnesia [$Mg(OH)_2$], and calcium carbonate. Based on what you have learned in this chapter, describe test(s) that would allow you to identify each of these compounds.

4.94 A 0.8870-g sample of a mixture of NaCl and KCl is dissolved in water, and the solution is then treated with an excess of $AgNO_3$ to yield 1.913 g of AgCl. Calculate the percent by mass of each compound in the mixture.

4.95 Phosphoric acid (H_3PO_4) is an important industrial chemical used in fertilizers, detergents, and in the food industry. It is produced by two different methods. In the *electric furnace method* elemental phosphorus (P_4) is burned in air to form P_4O_{10}, which is then reacted with water to give H_3PO_4. In the *wet process* the mineral phosphate rock

$[Ca_5(PO_4)_3F]$ is reacted with sulfuric acid to give H_3PO_4 (and HF and $CaSO_4$). Write equations for these processes and classify each step as precipitation, acid-base, or redox reaction.

4.96 Give a chemical explanation for each of these: (a) When calcium metal is added to a sulfuric acid solution, hydrogen gas is generated. After a few minutes, the reaction slows down and eventually stops even though none of the reactants is used up. Explain. (b) In the activity series aluminum is above hydrogen, yet the metal appears to be unreactive toward steam and hydrochloric acid. Why? (c) Sodium and potassium lie above copper in the activity series. Explain why Cu^{2+} ions in a $CuSO_4$ solution are not converted to metallic copper upon the addition of these metals. (d) A metal M reacts slowly with steam. There is no visible change when it is placed in a pale green iron(II) sulfate solution. Where should we place M in the activity series?

4.97 A number of metals are involved in redox reactions in biological systems in which the oxidation state of the metals changes. Which of these metals are most likely to take part in such reactions: Na, K, Mg, Ca, Mn, Fe, Co, Cu, Zn? Explain.

4.98 The recommended procedure for preparing a very dilute solution is not to weigh out a very small mass or measuring a very small volume of a stock solution. Instead, it is done by a series of dilutions. A sample of 0.8214 g of $KMnO_4$ was dissolved in water and made up to the volume in a 500-mL volumetric flask. A 2.000-mL sample of this solution was transferred to a 1000-mL volumetric flask and diluted to the mark with water. Next, 10.00 mL of the diluted solution were transferred to a 250-mL flask and diluted to the mark with water. (a) Calculate the concentration (in molarity) of the final solution. (b) Calculate the mass of $KMnO_4$ needed to directly prepare the final solution.

4.99 The Marsh test for arsenic described on p. 90 involves these steps: (a) The generation of hydrogen gas when sulfuric acid is added to zinc. (b) The reaction of hydrogen with As(III) oxide to produce arsine. (c) Conversion of arsine to arsenic by heating. Write equations representing these steps and identify the type of reaction in each step.

4.100 Acetylsalicylic acid ($C_9H_8O_4$) is a monoprotic acid commonly known as "aspirin." A typical aspirin tablet, however, contains only a small amount of the acid. In an experiment to determine its composition, an aspirin tablet was crushed and dissolved in water. It took 12.25 mL of 0.1466 M NaOH to neutralize the solution. Calculate the number of grains of aspirin in the tablet. (One grain = 0.0648 g.)

4.101 This "cycle of copper" experiment is performed in some general chemistry laboratories. The series of reactions starts with copper and ends with metallic copper. The steps are: (1) A piece of copper wire of known mass is allowed to react with concentrated nitric acid [the products are copper(II) nitrate, nitrogen dioxide, and water]. (2) The copper(II) nitrate is treated with a sodium hydroxide solution to form copper(II) hydroxide precipitate. (3) On heating, copper(II) hydroxide decomposes to yield copper(II) oxide. (4) The copper(II) oxide is reacted with concentrated sulfuric acid to yield copper(II) sulfate. (5) Copper(II) sulfate is treated with an excess of zinc metal to form metallic copper. (6) The remaining zinc metal is removed by treatment with hydrochloric acid, and metallic copper is filtered, dried, and weighed. (a) Write a balanced equation for each step and classify the reactions. (b) Assuming that a student started with 65.6 g of copper, calculate the theoretical yield at each step. (c) Considering the nature of the steps, comment on why it is possible to recover most of the copper used at the start.

4.102 Ammonium nitrate (NH_4NO_3) is one of the most important nitrogen-containing fertilizers. Its purity can be analyzed by titrating a solution of NH_4NO_3 with a standard NaOH solution. In one experiment a 0.2041-g sample of industrially prepared NH_4NO_3 required 24.42 mL of 0.1023 M NaOH for neutralization. (a) Write a net ionic equation for the reaction. (b) What is the percent purity of the sample?

4.103 Hydrogen halides (HF, HCl, HBr, HI) are highly reactive compounds that have many industrial and laboratory uses. (a) In the laboratory, HF and HCl can be generated by reacting CaF_2 and NaCl with concentrated sulfuric acid. Write appropriate equations for the reactions. (*Hint:* These are not redox reactions.) (b) Why is it that HBr and HI cannot be prepared similarly, that is, by reacting NaBr and NaI with concentrated sulfuric acid? (*Hint:* H_2SO_4 is a stronger oxidizing agent than both Br_2 and I_2.) (c) HBr can be prepared by reacting phosphorus tribromide (PBr_3) with water. Write an equation for this reaction.

4.104 Referring to Figure 4.14, explain why one must first dissolve the solid completely before making up the solution to the correct volume.

Special Problem

4.105 Magnesium is a valuable, lightweight metal. It is used as a structural metal and in alloys, in batteries, and in chemical synthesis. Although magnesium is plentiful in Earth's crust, it is cheaper to "mine" the metal from seawater. Magnesium forms the second most abundant cation in the sea (after sodium); there are about 1.3 g of magnesium in 1 kg of seawater. The method of obtaining magnesium from seawater employs all three types of reactions discussed in this chapter: precipitation, acid-base, and redox reactions. In the first stage in the recovery of magnesium, limestone ($CaCO_3$) is heated at high temperatures to produce quicklime, or calcium oxide (CaO):

$$CaCO_3(s) \longrightarrow CaO(s) + CO_2(g)$$

When calcium oxide is treated with seawater, it forms calcium hydroxide [$Ca(OH)_2$], which is slightly soluble and ionizes to give Ca^{2+} and OH^- ions:

$$CaO(s) + H_2O(l) \longrightarrow Ca^{2+}(aq) + 2OH^-(aq)$$

The surplus hydroxide ions cause the much less soluble magnesium hydroxide to precipitate:

$$Mg^{2+}(aq) + 2OH^-(aq) \longrightarrow Mg(OH)_2(s)$$

The solid magnesium hydroxide is filtered and reacted with hydrochloric acid to form magnesium chloride ($MgCl_2$):

$$Mg(OH)_2(s) + 2HCl(aq) \longrightarrow MgCl_2(aq) + 2H_2O(l)$$

After the water is evaporated, the solid magnesium chloride is melted in a steel cell. The molten magnesium

Magnesium hydroxide is obtained from seawater in settling ponds at the Dow Chemical Company in Freeport, Texas.

chloride contains both Mg^{2+} and Cl^- ions. In a process called electrolysis, an electric current is passed through the cell to reduce the Mg^{2+} ions and oxidize the Cl^- ions. The half-reactions are

$$Mg^{2+} + 2e^- \longrightarrow Mg$$
$$2Cl^- \longrightarrow Cl_2 + 2e^-$$

The overall reaction is

$$MgCl_2(l) \longrightarrow Mg(s) + Cl_2(g)$$

This is how magnesium metal is produced. The chlorine gas generated can be converted to hydrochloric acid and recycled through the process.

(a) Identify the precipitation, acid-base, and redox processes.

(b) Instead of calcium oxide, why don't we simply add sodium hydroxide to precipitate magnesium hydroxide?

(c) Sometimes a mineral called dolomite (a combination of $CaCO_3$ and $MgCO_3$) is substituted for limestone ($CaCO_3$) to bring about the precipitation of magnesium hydroxide. What is the advantage of using dolomite?

(d) What are the advantages of mining magnesium from the ocean rather than from Earth's crust?

Answers to Practice Exercises

4.1 (a) Insoluble, (b) slightly soluble, (c) soluble. **4.2** $Al^{3+}(aq)$ + $3OH^-(aq) \longrightarrow Al(OH)_3(s)$. **4.3** (a) Brønsted base, (b) Brønsted acid, (c) Brønsted acid and Brønsted base. **4.4** (a) P: $+3$, F: -1; (b) Mn: $+7$, O: -2. **4.5** 0.452 M. **4.6** Dilute 34.2 mL of the stock solution to 200 mL. **4.7** 92.02%. **4.8** 0.3822 g. **4.9** 10.0 mL.

The first manned balloon flight was made by the Montgolfier brothers in June 1783 in France, using a crude version of today's hot-air balloon. Two months later, the French scientist Jacques Charles carried out another balloon trial in Paris. Exploiting his knowledge of recent discoveries in the study of gases, Charles decided to inflate the balloon with hydrogen, which is lighter than air. He made his balloon out of silk and coated it with a rubber solution to prevent the gas from escaping into the atmosphere. The hydrogen gas was prepared by reacting 500 lb of sulfuric acid (H_2SO_4) with 1000 lb of iron. It took several days to inflate the balloon to a diameter of 13 ft. The flight lasted 45 min and ended 15 mi away from the launch site. The local inhabitants were so terrified by the balloon that they tore it to shreds.

The success of the first balloon flights had far-reaching effects. Around the world the potential value of balloons for travel and in warfare was recognized quickly. The invention of the balloon also allowed scientists to reach altitudes greater than 10,000 ft for the purpose of measuring the temperature and pressure of the atmosphere and collecting air samples. Their studies provided a further experimental basis for understanding gas behavior and gave support to the work that led to the overthrow of the phlogiston theory (see p. 56). Thus the first manned balloon flights were partly responsible for the birth of modern chemistry.

The first free flight of a manned hot-air balloon took place over Paris in 1783.

Gases

Chapter 5

ESSENTIAL CONCEPTS

 Properties of Gases Gases assume the volume and shape of their containers; they are easily compressible; they mix evenly and completely; and they have much lower densities than liquids and solids.

 Gas Pressures Pressure is one of the most readily measurable properties of a gas. A barometer measures atmospheric pressure and a manometer measures the pressure of a gas in the laboratory.

 The Gas Laws Over the years, a number of laws have been developed to explain the physical behavior of gases. These laws show the relationships among the pressure, temperature, volume, and amount of a gas.

 The Ideal Gas Equation The molecules of an ideal gas possess no volume and exert no forces on one another. At low pressures and high temperatures, most gases can be assumed to behave ideally; their physical behavior is described by the ideal gas equation.

 Kinetic Molecular Theory of Gases Macroscopic properties like pressure and temperature of a gas can be related to the kinetic motion of molecules. The kinetic molecular theory of gases assumes that the molecules are ideal, the number of molecules is very large, and that their motions are totally random.

 Nonideal Behavior of Gases To account for the behavior of real gases, the ideal gas equation is modified to include the finite volume of molecules and the attractive forces among them.

5.1 Substances That Exist as Gases

We live at the bottom of an ocean of air whose composition by volume is roughly 78 percent N_2, 21 percent O_2, and 1 percent other gases, including CO_2. In the 1990s, the chemistry of this vital mixture of gases became a source of great interest because of the detrimental effects of environmental pollution. Here we will focus generally on the behavior of substances that exist as gases under normal atmospheric conditions, which are defined as 25°C and 1 atmosphere (atm) pressure (see Section 5.2).

Only 11 elements are gases under normal atmospheric conditions. Table 5.1 lists these, along with a number of gaseous compounds. Note that the elements hydrogen, nitrogen, oxygen, fluorine, and chlorine exist as gaseous diatomic molecules. Another form of oxygen, ozone (O_3), is also a gas at room temperature. All the elements in Group 8A, the noble gases, are monatomic gases: He, Ne, Ar, Kr, Xe, and Rn.

Of the gases listed in Table 5.1, only O_2 is essential for our survival. Hydrogen cyanide (HCN) is a deadly poison. Carbon monoxide (CO), hydrogen sulfide (H_2S), nitrogen dioxide (NO_2), O_3, and sulfur dioxide (SO_2) are somewhat less toxic. The gases He and Ne are chemically inert; that is, they do not react with any other substance. Most gases are colorless. Exceptions are F_2, Cl_2, and NO_2. The dark-brown color of NO_2 is sometimes visible in polluted air. All gases have the following physical characteristics:

- Gases assume the volume and shape of their containers.
- Gases are the most compressible of the states of matter.
- Gases will mix evenly and completely when confined to the same container.
- Gases have much lower densities than liquids and solids.

NO₂ gas.

A gas is a substance that is normally in the gaseous state at ordinary temperatures and pressures; a vapor is the gaseous form of any substance that is a liquid or solid at normal temperatures and pressures. Thus, at 25°C and 1 atm pressure, we speak of water vapor and oxygen gas.

Table 5.1 | Some Substances Found as Gases at 1 Atm and 25°C

Elements	Compounds
H_2 (molecular hydrogen)	HF (hydrogen fluoride)
N_2 (molecular nitrogen)	HCl (hydrogen chloride)
O_2 (molecular oxygen)	HBr (hydrogen bromide)
O_3 (ozone)	HI (hydrogen iodide)
F_2 (molecular fluorine)	CO (carbon monoxide)
Cl_2 (molecular chlorine)	CO_2 (carbon dioxide)
He (helium)	NH_3 (ammonia)
Ne (neon)	NO (nitric oxide)
Ar (argon)	NO_2 (nitrogen dioxide)
Kr (krypton)	N_2O (nitrous oxide)
Xe (xenon)	SO_2 (sulfur dioxide)
Rn (radon)	H_2S (hydrogen sulfide)
	HCN (hydrogen cyanide)*

*The boiling point of HCN is 26°C, but it is close enough to qualify as a gas at ordinary atmospheric conditions.

5.2 Pressure of a Gas

Gases exert pressure on any surface with which they come in contact, because gas molecules are constantly in motion. We humans have adapted so well physiologically to the pressure of the air around us that we are usually unaware of it, perhaps as fish are not conscious of the water's pressure on them.

It is easy to demonstrate atmospheric pressure. One everyday example is the ability to drink a liquid through a straw. Sucking air out of the straw reduces the pressure inside the straw. The greater atmospheric pressure on the liquid pushes it up into the straw to replace the air that has been sucked out.

SI Units of Pressure

Pressure is one of the most readily measurable properties of a gas. In order to understand how we measure the pressure of a gas, it is helpful to know how the units of measurement are derived. We begin with velocity and acceleration.

Velocity is defined as the change in distance with elapsed time; that is,

$$\text{velocity} = \frac{\text{distance moved}}{\text{elapsed time}}$$

The SI unit for velocity is m/s, although we also use cm/s.

Acceleration is the change in velocity with time, or

$$\text{acceleration} = \frac{\text{change in velocity}}{\text{elapsed time}}$$

Acceleration is measured in m/s^2 (or cm/s^2).

The second law of motion, formulated by Sir Isaac Newton in the late seventeenth century, defines another term, from which the units of pressure are derived, namely, *force*. According to this law,

$$\text{force} = \text{mass} \times \text{acceleration}$$

In this context, the *SI unit of force* is the **newton (N),** where

$$1 \text{ N} = 1 \text{ kg m/s}^2$$

1 N is roughly equivalent to the force exerted by Earth's gravity on an apple.

Finally, we define **pressure** as *force applied per unit area*:

$$\text{pressure} = \frac{\text{force}}{\text{area}}$$

The SI unit of pressure is the **pascal (Pa),** defined as *one newton per square meter:*

$$1 \text{ Pa} = 1 \text{ N/m}^2$$

Atmospheric Pressure

The atoms and molecules of the gases in the atmosphere, like those of all other matter, are subject to Earth's gravitational pull. As a consequence, the atmosphere is much denser near the surface of Earth than at high altitudes. (The air outside the pressurized cabin of an airplane at 9 km is too thin to breathe.) In fact, the density of air decreases very rapidly with increasing distance from Earth. Measurements show that

Column of air

Figure 5.1 *The pressure exerted by a column of air extending from Earth's surface (sea level) to the upper atmosphere is the atmospheric pressure.*

76 cm

Atmospheric pressure

Figure 5.2 *A barometer for measuring atmospheric pressure. Above the mercury in the tube is a vacuum. The column of mercury is supported by atmospheric pressure.*

Similar problem: 5.14.

about 50 percent of the atmosphere lies within 6.4 km of Earth's surface, 90 percent within 16 km, and 99 percent within 32 km. Not surprisingly, the denser the air is, the greater the pressure it exerts. The force experienced by any area exposed to Earth's atmosphere is equal to *the weight of the column of air above it.* **Atmospheric pressure** is *the pressure exerted by Earth's atmosphere* (Figure 5.1). The actual value of atmospheric pressure depends on location, temperature, and weather conditions.

How is atmospheric pressure measured? The **barometer** is probably the most familiar *instrument for measuring atmospheric pressure.* A simple barometer consists of a long glass tube, closed at one end and filled with mercury. If the tube is carefully inverted in a dish of mercury so that no air enters the tube, some mercury will flow out of the tube into the dish, creating a vacuum at the top (Figure 5.2). The weight of the mercury remaining in the tube is supported by atmospheric pressure acting on the surface of the mercury in the dish. **Standard atmospheric pressure (1 atm)** is equal to *the pressure that supports a column of mercury exactly 760 mm (or 76 cm) high at 0°C at sea level.* In other words, the standard atmosphere equals a pressure of 760 mmHg, where mmHg represents the pressure exerted by a column of mercury 1 mm high. The mmHg unit is also called the *torr,* after the Italian scientist Evangelista Torricelli, who invented the barometer. Thus

$$1 \text{ torr} = 1 \text{ mmHg}$$

and

$$1 \text{ atm} = 760 \text{ mmHg}$$
$$= 760 \text{ torr}$$

The relation between atmospheres and pascals (see Appendix 1) is

$$1 \text{ atm} = 101,325 \text{ Pa}$$
$$= 1.01325 \times 10^5 \text{ Pa}$$

and because 1000 Pa = 1 kPa (kilopascal)

$$1 \text{ atm} = 1.01325 \times 10^2 \text{ kPa}$$

Example 5.1   The pressure outside a jet plane flying at high altitude falls considerably below standard atmospheric pressure. Therefore, the air inside the cabin must be pressurized to protect the passengers. What is the pressure in atmospheres in the cabin if the barometer reading is 688 mmHg?

Reasoning and Solution   This is a conversion problem and we need the unit factor (1 atm/760 mmHg). Thus the pressure in atmospheres is

$$\text{pressure} = 668 \text{ mmHg} \times \frac{1 \text{ atm}}{760 \text{ mmHg}}$$

$$= 0.905 \text{ atm}$$

Practice Exercise   Convert 749 mmHg to atmospheres.

A **manometer** is *a device used to measure the pressure of gases other than the atmosphere.* The principle of operation of a manometer is similar to that of a barometer. There are two types of manometers, shown in Figure 5.3. The *closed-tube*

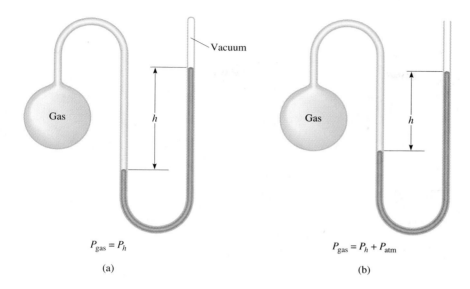

Vacuum

Gas

h

$P_{gas} = P_h$

(a)

Gas

h

$P_{gas} = P_h + P_{atm}$

(b)

Figure 5.3 *Two types of manometers used to measure gas pressures. (a) Gas pressure is less than atmospheric pressure. (b) Gas pressure is greater than atmospheric pressure.*

manometer is normally used to measure pressures below atmospheric pressure [Figure 5.3(a)], whereas the *open-tube manometer* is better suited for measuring pressures equal to or greater than atmospheric pressure [Figure 5.3(b)].

Nearly all barometers and most manometers use mercury as the working fluid, despite the fact that it is a toxic substance with a harmful vapor. The reason is that mercury has a very high density (13.6 g/mL) compared with most other liquids. Because the height of the liquid in a column is inversely proportional to the liquid's density, this property enables the construction of manageably small barometers and manometers.

5.3 The Gas Laws

The gas laws we will study in this chapter are the product of countless experiments on the physical properties of gases that were carried out over several centuries. Each of these generalizations regarding the macroscopic behavior of gaseous substances represents a milestone in the history of science. Together they have played a major role in the development of many ideas in chemistry.

Gas laws

The Pressure-Volume Relationship: Boyle's Law

In the seventeenth century, Robert Boyle studied the behavior of gases systematically and quantitatively. In one series of studies, Boyle investigated the pressure-volume relationship of a gas sample using an apparatus like that shown in Figure 5.4. In Figure 5.4(a) the pressure exerted on the gas by the mercury added to the tube is equal to atmospheric pressure. In Figure 5.4(b) an increase in pressure due to the addition of more mercury results in a decrease in the volume of the gas and in unequal levels of mercury in the tube. Boyle noticed that when temperature is held constant, the volume (V) of a given amount of a gas decreases as the total applied pressure (P)—atmospheric pressure plus the pressure due to the added mercury—is increased. This relationship between pressure and volume is evident in

The pressure applied to a gas is equal to the gas pressure.

Figure 5.4 *Apparatus for studying the relationship between pressure and volume of a gas. In (a) the pressure of the gas is equal to the atmospheric pressure. The pressure exerted on the gas increases from (a) to (d) as mercury is added, and the volume of the gas decreases, as predicted by Boyle's law. The extra pressure exerted on the gas is shown by the difference in the mercury levels (h mmHg). The temperature of the gas is kept constant.*

Gas

(a)   (b)   (c)   (d)

Figure 5.4, parts (b), (c), and (d). Conversely, if the applied pressure is decreased, the gas volume becomes larger.

The mathematical expression showing an inverse relationship between pressure and volume is

$$P \propto \frac{1}{V}$$

where the symbol $\propto$ means *proportional to*. To change $\propto$ to an equals sign, we must write

$$P = k_1 \times \frac{1}{V} \tag{5.1a}$$

where k_1 is a constant called the *proportionality constant*. Equation (5.1a) is an expression of **Boyle's law,** which states that *the pressure of a fixed amount of gas maintained at constant temperature is inversely proportional to the volume of the gas.* We can rearrange Equation (5.1a) and obtain

$$PV = k_1 \tag{5.1b}$$

This form of Boyle's law says that the product of the pressure and volume of a gas at constant temperature and amount of gas is a constant. Figure 5.5 is a schematic representation of Boyle's law. The quantity n is the number of moles of the gas and R is a constant to be defined in Section 5.4. Thus the proportionality constant k_1 in Equations (5.1) is equal to nRT.

The concept of one quantity being proportional to another and the use of a proportionality constant can be clarified through the following analogy. The daily income of a movie theater depends on both the price of the tickets (in dollars per ticket) and the number of tickets sold. Assuming that the theater charges one price for all tickets, we write

income = (dollar/ticket) × number of tickets sold

Figure 5.5 *Schematic illustrations of Boyle's law, Charles's law, and Avogadro's law.*

Figure 5.6 *Graphs showing variation of the volume of a gas sample with the pressure exerted on the gas, at constant temperature. (a) P versus V. Note that the volume of the gas doubles as the pressure is halved; (b) P versus 1/V.*

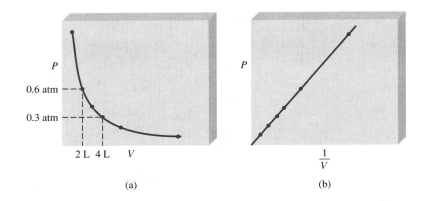

(a)   (b)

Because the number of tickets sold varies from day to day, the income on a given day is said to be proportional to the number of tickets sold:

$$\text{income} \propto \text{number of tickets sold}$$
$$= C \times \text{number of tickets sold}$$

where C, the proportionality constant, is the price per ticket.

Figure 5.6 shows two conventional ways of expressing Boyle's findings graphically. Figure 5.6(a) is a graph of the equation $PV = k_1$; Figure 5.6(b) is a graph of the equivalent equation $P = k_1 \times 1/V$. Note that the latter is a linear equation of the form $y = mx + b$, where $b = 0$.

Although the individual values of pressure and volume can vary greatly for a given sample of gas, as long as the temperature is held constant and the amount of the gas does not change, P times V is always equal to the same constant. Therefore, for a given sample of gas under two different sets of conditions at constant temperature, we have

$$P_1V_1 = k_1 = P_2V_2$$

or

$$P_1V_1 = P_2V_2 \tag{5.2}$$

where V_1 and V_2 are the volumes at pressures P_1 and P_2, respectively.

The Temperature-Volume Relationship: Charles's and Gay-Lussac's Law

Boyle's law depends on the temperature of the system remaining constant. But suppose the temperature changes: How does a change in temperature affect the volume and pressure of a gas? Let us first look at the effect of temperature on the volume of a gas. The earliest investigators of this relationship were French scientists, Jacques Charles and Joseph Gay-Lussac. Their studies showed that, at constant pressure, the volume of a gas sample expands when heated and contracts when cooled (Figure 5.7). The quantitative relations involved in changes in gas temperature and volume turn out to be remarkably consistent. For example, we observe an interesting phenomenon when we study the temperature-volume relationship at various pressures. At any given pressure, the plot of volume versus temperature yields

Low   High
temperature temperature

Figure 5.7 *Variation of the volume of a gas sample with temperature, at constant pressure. The pressure exerted on the gas is the sum of the atmospheric pressure and the pressure due to the weight of the mercury.*

Capillary tubing

Mercury

Gas

Figure 5.8 *Variation of the volume of a gas sample with changing temperature, at constant pressure. Each line represents the variation at a certain pressure. The pressures increase from P_1 to P_4. All gases ultimately condense (become liquids) if they are cooled to sufficiently low temperatures; the solid portions of the lines represent the temperature region above the condensation point. When these lines are extrapolated, or extended (the dashed portions), they all intersect at the point representing zero volume and $-273.15°C$.*

a straight line. By extending the line to zero volume, we find the intercept on the temperature axis to be $-273.15°C$. At any other pressure, we obtain a different straight line for the volume-temperature plot, but we get the *same* zero-volume temperature intercept at $-273.15°C$ (Figure 5.8). (In practice, we can measure the volume of a gas over only a limited temperature range, because all gases condense at low temperatures to form liquids.)

In 1848 the Scottish physicist Lord Kelvin realized the significance of this phenomenon. He identified $-273.15°C$ as **absolute zero,** *theoretically the lowest attainable temperature.* Then he set up an **absolute temperature scale,** now called the **Kelvin temperature scale,** with *absolute zero as the starting point.* On the Kelvin scale, one kelvin (K) is equal *in magnitude* to one degree Celsius. The only difference between the absolute temperature scale and the Celsius scale is that the zero position is shifted. Important points on the two scales match up as follows:

	Kelvin Scale	**Celsius Scale**
Absolute zero	0 K	$-273.15°C$
Freezing point of water	273.15 K	$0°C$
Boiling point of water	373.15 K	$100°C$

The conversion between °C and K is given by

$$? \, K = (°C + 273.15°C)\frac{1K}{1°C} \tag{5.3}$$

In most calculations we will use 273 instead of 273.15 as the term relating K and °C. By convention, we use T to denote absolute (kelvin) temperature and t to indicate temperature on the Celsius scale.

The dependence of the volume of a gas on temperature is given by

$$V \propto T$$
$$V = k_2 T$$

or

$$\frac{V}{T} = k_2 \tag{5.4}$$

where k_2 is the proportionality constant. Equation (5.4) is known as ***Charles's and Gay-Lussac's law,*** or simply ***Charles's law,*** which states that *the volume of a fixed amount of gas maintained at constant pressure is directly proportional to the absolute temperature of the gas.* Charles's law is also illustrated in Figure 5.5. We see that the proportionality constant k_2 in Equation (5.4) is equal to nR/P.

Another form of Charles's law shows that at constant amount of gas and volume, the pressure of a gas is proportional to temperature

$$P \propto T$$

$$P = k_3 T$$

or $$\frac{P}{T} = k_3 \qquad (5.5)$$

From Figure 5.5 we see that $k_3 = nR/V$.

Just as we did for pressure-volume relationships at constant temperature, we can compare two sets of volume-temperature conditions for a given sample of gas at constant pressure. From Equation (5.4) we can write

$$\frac{V_1}{T_1} = k_2 = \frac{V_2}{T_2}$$

or $$\frac{V_1}{T_1} = \frac{V_2}{T_2} \qquad (5.6)$$

where V_1 and V_2 are the volumes of the gases at temperatures T_1 and T_2 (both in kelvins), respectively. In all subsequent calculations we assume that the temperatures given in °C are exact, so that they do not affect the number of significant figures.

The Volume-Amount Relationship: Avogadro's Law

Avogadro's name first appeared in Section 3.2.

The work of the Italian scientist Amedeo Avogadro complemented the studies of Boyle, Charles, and Gay-Lussac. In 1811 he published a hypothesis stating that at the same temperature and pressure, equal volumes of different gases contain the same number of molecules (or atoms if the gas is monatomic). It follows that the volume of any given gas must be proportional to the number of moles of molecules present; that is,

$$V \propto n$$

$$V = k_4 n \qquad (5.7)$$

where n represents the number of moles and k_4 is the proportionality constant. Equation (5.7) is the mathematical expression of ***Avogadro's law,*** which states that *at constant pressure and temperature, the volume of a gas is directly proportional to the number of moles of the gas present.* From Figure 5.5 we see that $k_4 = RT/P$.

According to Avogadro's law we see that when two gases react with each other, their reacting volumes have a simple ratio to each other. If the product is a gas, its volume is related to the volume of the reactants by a simple ratio (a fact demonstrated earlier by Gay-Lussac). For example, consider the synthesis of ammonia from molecular hydrogen and molecular nitrogen:

$$3H_2(g) + N_2(g) \longrightarrow 2NH_3(g)$$
$$\text{3 mol} \qquad \text{1 mol} \qquad \text{2 mol}$$

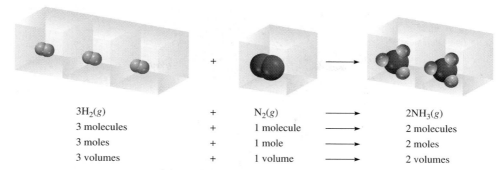

$3H_2(g)$ + $N_2(g)$ ⟶ $2NH_3(g)$
3 molecules   +   1 molecule ⟶ 2 molecules
3 moles   +   1 mole ⟶ 2 moles
3 volumes   +   1 volume ⟶ 2 volumes

Figure 5.9 *Volume relationship of gases in a chemical reaction. The ratio of the volumes of molecular hydrogen to molecular nitrogen is 3:1, and that of ammonia (the product) to molecular hydrogen and molecular nitrogen combined (the reactants) is 2:4, or 1:2.*

Because, at the same temperature and pressure, the volumes of gases are directly proportional to the number of moles of the gases present, we can now write

$$3H_2(g) + N_2(g) \longrightarrow 2NH_3(g)$$
3 volumes   1 volume   2 volumes

The volume ratio of molecular hydrogen to molecular nitrogen is 3:1, and that of ammonia (the product) to molecular hydrogen and molecular nitrogen (the reactants) is 2:4, or 1:2 (Figure 5.9).

5.4 The Ideal Gas Equation

Let us summarize the gas laws we have discussed so far:

Boyle's law: $V \propto \dfrac{1}{P}$ (at constant n and T)

Charles's law: $V \propto T$ (at constant n and P)

Avogadro's law: $V \propto n$ (at constant P and T)

We can combine all three expressions to form a single master equation for the behavior of gases:

$$V \propto \frac{nT}{P}$$

$$= R\,\frac{nT}{P}$$

or

$$PV = nRT \tag{5.8}$$

where R, *the proportionality constant,* is called the **gas constant.** Equation (5.8), which is called the **ideal gas equation,** *describes the relationship among the four variables P, V, T, and n. An* **ideal gas** *is a hypothetical gas whose pressure-volume-temperature behavior can be completely accounted for by the ideal gas equation.* The molecules of an ideal gas do not attract or repel one another, and their volume is negligible compared with the volume of the container. Although there is no such thing

Figure 5.10 *A comparison of the molar volume at STP (which is approximately 22.4 L) with a basketball.*

in nature as an ideal gas, discrepancies in the behavior of real gases over reasonable temperature and pressure ranges do not significantly affect calculations. Thus we can safely use the ideal gas equation to solve many gas problems.

Before we can apply the ideal gas equation to a real system, we must evaluate the gas constant R. At 0°C (273.15 K) and 1 atm pressure, many real gases behave like an ideal gas. Experiments show that under these conditions, 1 mole of an ideal gas occupies 22.414 L, which is somewhat greater than the volume of a basketball, as shown in Figure 5.10. *The conditions 0°C and 1 atm* are called **standard temperature and pressure,** often abbreviated **STP.** From Equation (5.8) we can write

The gas constant can be expressed in different units (see Appendix 1).

$$R = \frac{PV}{nT}$$

$$= \frac{(1 \text{ atm})(22.414 \text{ L})}{(1 \text{ mol})(273.15 \text{ K})}$$

$$= 0.082057 \frac{\text{L} \cdot \text{atm}}{\text{K} \cdot \text{mol}}$$

$$= 0.082057 \text{ L} \cdot \text{atm/K} \cdot \text{mol}$$

The dots between L and atm and between K and mol remind us that both L and atm are in the numerator and both K and mol are in the denominator. For most calculations, we will round off the value of R to three significant figures ($0.0821 \text{ L} \cdot \text{atm/K} \cdot \text{mol}$) and use 22.41 L for the molar volume of a gas at STP.

SF_6

Example 5.2   Sulfur hexafluoride (SF_6) is a colorless, odorless, very unreactive gas. Calculate the pressure (in atm) exerted by 1.82 moles of the gas in a steel vessel of volume 5.43 L at 69.5°C.

Reasoning and Solution   This problem provides information about the number of moles, volume, and temperature of a gas, but no change in any of the quantities occurs. Therefore, to calculate the pressure we can use the ideal gas equation, which can be rearranged to give

$$P = \frac{nRT}{V}$$

$$= \frac{(1.82 \text{ mol})(0.0821 \text{ L} \cdot \text{atm/K} \cdot \text{mol})(69.5 + 273) \text{ K}}{5.43 \text{ L}}$$

$$= 9.42 \text{ atm}$$

Remember to convert °C to K when solving gas-law problems.

Similar problems: 5.32, 5.33.

Practice Exercise Calculate the volume (in liters) occupied by 2.12 moles of nitric oxide (NO) at 6.54 atm and 76°C.

By using the fact that the molar volume of a gas occupies 22.41 L at STP, we can calculate the volume of a gas at STP without using the ideal gas equation.

Example 5.3 Calculate the volume (in liters) occupied by 7.40 g of CO_2 at STP.

Reasoning and Solution Recognizing that 1 mole of an ideal gas occupies 22.41 L at STP, we calculate the volume as follows:

$$V = 7.40 \text{ g } CO_2 \times \frac{1 \text{ mol } CO_2}{44.01 \text{ g } CO_2} \times \frac{22.41 \text{ L}}{1 \text{ mol } CO_2}$$

$$= 3.77 \text{ L}$$

Comment Because 7.40 g is smaller than the molar mass of CO_2 (44.01 g), we would expect the volume to be smaller than 22.41 L at STP. The problem can be solved in a different way by first converting 7.40 g of CO_2 to number of moles of CO_2 and then applying the ideal gas equation. Try it.

CO_2

Similar problems: 5.41, 5.43.

Practice Exercise What is the volume (in liters) occupied by 49.8 g of HCl at STP?

The ideal gas equation is useful for problems that do not involve changes in P, V, T, and n for a gas sample. At times, however, we need to deal with changes in pressure, volume, and temperature, or even in the amount of a gas. When conditions change, we must employ a modified form of the ideal gas equation that takes into account the initial and final conditions. We derive the modified equation as follows. From Equation (5.8),

$$R = \frac{P_1 V_1}{n_1 T_1} \quad \text{(before change)}$$

and

$$R = \frac{P_2 V_2}{n_2 T_2} \quad \text{(after change)}$$

so that

$$\frac{P_1 V_1}{n_1 T_1} = \frac{P_2 V_2}{n_2 T_2}$$

The subscripts 1 and 2 denote the initial and final states of the gas.

If $n_1 = n_2$, as is usually the case because the amount of gas normally does not change, the equation then becomes

$$\frac{P_1 V_1}{T_1} = \frac{P_2 V_2}{T_2} \qquad (5.9)$$

Example 5.4   A small bubble rises from the bottom of a lake, where the temperature and pressure are 8°C and 6.4 atm, to the water's surface, where the temperature is 25°C and pressure is 1.0 atm. Calculate the final volume (in mL) of the bubble if its initial volume was 2.1 mL.

Reasoning and Solution   Note that this problem involves the change of temperature, pressure, and volume, but not the number of moles of a gas. Thus we can use Equation (5.9) to calculate the final volume. We start by writing

Initial Conditions	**Final Conditions**
$P_1 = 6.4$ atm	$P_2 = 1.0$ atm
$V_1 = 2.1$ mL	$V_2 = ?$
$T_1 = (8 + 273)$ K $= 281$ K	$T_2 = (25 + 273)$ K $= 298$ K

The amount of the gas in the bubble remains constant, so that $n_1 = n_2$. To calculate the final volume, V_2, we rearrange Equation (5.9) as follows:

$$V_2 = V_1 \times \frac{P_1}{P_2} \times \frac{T_2}{T_1}$$

$$= 2.1 \text{ mL} \times \frac{6.4 \text{ atm}}{1.0 \text{ atm}} \times \frac{298 \text{ K}}{281 \text{ K}}$$

$$= 14 \text{ mL}$$

Thus, the bubble's volume increases from 2.1 mL to 14 mL because of the decrease in water pressure and the increase in temperature.

Practice Exercise   A gas initially at 4.0 L, 1.2 atm, and 66°C undergoes a change so that its final volume and temperature become 1.7 L and 42°C. What is its final pressure? Assume the number of moles remains unchanged.

We can use any appropriate units for volume (or pressure) as long as we use the same units on both sides of the equation.

Similar problems: 5.35, 5.39.

Density and Molar Mass of a Gaseous Substance

The ideal gas equation can be applied to determine the density or molar mass of a gaseous substance. Rearranging Equation (5.8), we write

$$\frac{n}{V} = \frac{P}{RT}$$

The number of moles of the gas, n, is given by

$$n = \frac{m}{\mathcal{M}}$$

in which m is the mass of the gas in grams and $\mathcal{M}$ is its molar mass. Therefore

$$\frac{m}{\mathcal{M}V} = \frac{P}{RT}$$

Because density, d, is mass per unit volume, we can write

$$d = \frac{m}{V} = \frac{P\mathcal{M}}{RT} \qquad (5.10)$$

Equation (5.10) allows us to calculate the density of a gas (given in units of grams per liter). More often, the density of a gas can be measured, so this equation can be rearranged for us to calculate the molar mass of a gaseous substance:

$$\mathcal{M} = \frac{dRT}{P} \qquad (5.11)$$

In a typical experiment, a bulb of known volume is filled with the gaseous substance under study. The temperature and pressure of the gas sample are recorded, and the total mass of the bulb plus gas sample is determined (Figure 5.11). The bulb is then evacuated (emptied) and weighed again. The difference in mass is the mass of the gas. The density of the gas is equal to its mass divided by the volume of the bulb. Then we can calculate the molar mass of the substance using Equation (5.11).

Figure 5.11 *An apparatus for measuring the density of a gas. A bulb of known volume is filled with the gas under study at a certain temperature and pressure. First the bulb is weighed, and then it is emptied (evacuated) and weighed again. The difference in masses gives the mass of the gas. Knowing the volume of the bulb, we can calculate the density of the gas. Under atmospheric conditions, 100 mL of air weigh about 0.12 g, an easily measured quantity.*

Example 5.5 A chemist has synthesized a greenish-yellow gaseous compound of chlorine and oxygen and finds that its density is 7.71 g/L at 36°C and 2.88 atm. Calculate the molar mass of the compound and determine its molecular formula.

Reasoning and Solution Because we are given the density of the gas and the temperature and pressure, we can use Equation (5.11) to calculate its molar mass as:

$$\mathcal{M} = \frac{dRT}{P}$$

$$= \frac{(7.71 \text{ g/L})(0.0821 \text{ L} \cdot \text{atm/K} \cdot \text{mol})(36 + 273) \text{ K}}{2.88 \text{ atm}}$$

$$= 67.9 \text{ g/mol}$$

We can determine the molecular formula of the compound by trial and error, using only the knowledge of the molar masses of chlorine (35.45 g) and oxygen (16.00 g). We know that a compound containing one Cl atom and one O atom would have a molar mass of 51.45 g, which is too low, while the molar mass of a compound made up of two Cl atoms and one O atom is 86.90 g, which is too high. Thus the compound must contain one Cl atom and two O atoms and have the formula ClO_2, which has a molar mass of 67.45 g.

Practice Exercise The density of a gaseous organic compound is 3.38 g/L at 40°C and 1.97 atm. What is its molar mass?

ClO_2

Similar problems: 5.47, 5.49.

Gas Stoichiometry

In Chapter 3 we used relationships between amounts (in moles) and masses (in grams) of reactants and products to solve stoichiometry problems. When the reactants and/or products are gases, we can also use the relationships between amounts (moles, n) and volume (V) to solve such problems.

An air bag can protect the driver in an automobile collision.

Example 5.6   Sodium azide (NaN_3) is used in some automobile air bags. The impact of a collision triggers the decomposition of NaN_3 as follows:

$$2NaN_3(s) \longrightarrow 2Na(s) + 3N_2(g)$$

The nitrogen gas produced quickly inflates the bag between the driver and the windshield. Calculate the volume of N_2 generated at 80°C and 823 mmHg by the decomposition of 60.0 g of NaN_3.

Reasoning and Solution   The problem requires two steps. First, we calculate the number of moles of N_2 generated from the decomposition of 60.0 g of NaN_3. Next we calculate the volume of the N_2 gas at the given temperature and pressure.

$$\text{moles of } N_2 = 60.0 \text{ g NaN}_3 \times \frac{1 \text{ mol NaN}_3}{65.02 \text{ g NaN}_3} \times \frac{3 \text{ mol } N_2}{2 \text{ mol NaN}_3}$$

$$= 1.38 \text{ mol } N_2$$

The volume of 1.38 moles of N_2 can be obtained by using the ideal gas equation:

$$V = \frac{nRT}{P} = \frac{(1.38 \text{ mol})(0.0821 \text{ L} \cdot \text{atm/K} \cdot \text{mol})(80 + 273) \text{ K}}{(823/760) \text{ atm}}$$

$$= 36.9 \text{ L}$$

Similar problems: 5.51, 5.52.

Practice Exercise   The equation for the metabolic breakdown of glucose ($C_6H_{12}O_6$) is the same as the equation for the combustion of glucose in air:

$$C_6H_{12}O_6(s) + 6O_2(g) \longrightarrow 6CO_2(g) + 6H_2O(l)$$

Calculate the volume of CO_2 produced at 37°C and 1.00 atm when 5.60 g of glucose are used up in the reaction.

5.5 Dalton's Law of Partial Pressures

Thus far we have concentrated on the behavior of pure gaseous substances, but experimental studies very often involve mixtures of gases. For example, for a study of air pollution, we may be interested in the pressure-volume-temperature relationship of a sample of air, which contains several gases. In this case, and all cases involving mixtures of gases, the total gas pressure is related to *partial pressures,* that is, *the pressures of individual gas components in the mixture.* In 1801 Dalton formulated a law, now known as *Dalton's law of partial pressures,* which states that *the total pressure of a mixture of gases is just the sum of the pressures that each gas would exert if it were present alone.* Figure 5.12 illustrates Dalton's law.

As mentioned earlier, gas pressure results from the impact of gas molecules against the walls of the container.

Consider a case in which two gases, A and B, are in a container of volume V. The pressure exerted by gas A, according to the ideal gas equation, is

$$P_A = \frac{n_A RT}{V}$$

where n_A is the number of moles of A present. Similarly, the pressure exerted by gas B is

$$P_B = \frac{n_B RT}{V}$$

Volume and temperature are constant

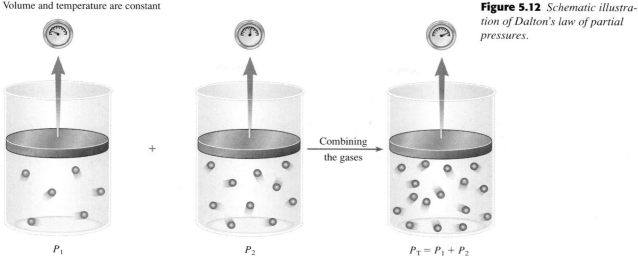

Figure 5.12 *Schematic illustration of Dalton's law of partial pressures.*

In a mixture of gases A and B, the total pressure P_T is the result of the collisions of both types of molecules, A and B, with the walls of the container. Thus, according to Dalton's law,

$$P_T = P_A + P_B$$
$$= \frac{n_A RT}{V} + \frac{n_B RT}{V}$$
$$= \frac{RT}{V}(n_A + n_B)$$
$$= \frac{nRT}{V}$$

where n, the total number of moles of gases present, is given by $n = n_A + n_B$, and P_A and P_B are the partial pressures of gases A and B, respectively. For a mixture of gases, then, P_T depends only on the total number of moles of gas present, not on the nature of the gas molecules.

In general, the total pressure of a mixture of gases is given by

$$P_T = P_1 + P_2 + P_3 + \cdots$$

where $P_1, P_2, P_3, \ldots$ are the partial pressures of components 1, 2, 3, ... To see how each partial pressure is related to the total pressure, consider again the case of a mixture of two gases A and B. Dividing P_A by P_T, we obtain

$$\frac{P_A}{P_T} = \frac{n_A RT/V}{(n_A + n_B)RT/V}$$
$$= \frac{n_A}{n_A + n_B}$$
$$= X_A$$

where X_A is called the mole fraction of gas A. The ***mole fraction*** is *a dimensionless quantity that expresses the ratio of the number of moles of one component to the* *number of moles of all components present.* It is always smaller than 1, except when A is the only component present. In that case, $n_B = 0$ and $X_A = n_A/n_A = 1$. We can express the partial pressure of A as

$$P_A = X_A P_T$$

Similarly,

$$P_B = X_B P_T$$

Note that the sum of the mole fractions for a mixture of gases must be unity. If only two components are present, then

$$X_A + X_B = \frac{n_A}{n_A + n_B} + \frac{n_B}{n_A + n_B} = 1$$

If a system contains more than two gases, then the partial pressure of the *i*th component is related to the total pressure by

$$P_i = X_i P_T \tag{5.12}$$

where X_i is the mole fraction of substance i.

How are partial pressures determined? A manometer can measure only the total pressure of a gaseous mixture. To obtain the partial pressures, we need to know the mole fractions of the components, which would involve elaborate chemical analyses. The most direct method of measuring partial pressures is using a mass spectrometer. The relative intensities of the peaks in a mass spectrum are directly proportional to the amounts, and hence to the mole fractions, of the gases present.

Example 5.7   A mixture of gases contains 4.46 moles of neon (Ne), 0.74 mole of argon (Ar), and 2.15 moles of xenon (Xe). Calculate the partial pressures of the gases if the total pressure is 2.00 atm at a certain temperature.

Reasoning and Solution   To calculate the partial pressures, we apply Dalton's law. First, we determine the mole fraction of Ne as follows

$$X_{Ne} = \frac{n_{Ne}}{n_{Ne} + n_{Ar} + n_{Xe}} = \frac{4.46 \text{ mol}}{4.46 \text{ mol} + 0.74 \text{ mol} + 2.15 \text{ mol}}$$

$$= 0.607$$

From Equation (5.12)
$$P_{Ne} = X_{Ne} P_T$$
$$= 0.607 \times 2.00 \text{ atm}$$
$$= 1.21 \text{ atm}$$

Similarly,
$$P_{Ar} = 0.10 \times 2.00 \text{ atm}$$
$$= 0.20 \text{ atm}$$

and
$$P_{Xe} = 0.293 \times 2.00 \text{ atm}$$
$$= 0.586 \text{ atm}$$

Comment   To check the correctness of the answer, make sure that the sum of the partial pressures is equal to the given total pressure; that is, $(1.21 + 0.20 + 0.586)$ atm $= 2.00$ atm.

Similar problems: 5.57, 5.58.

Practice Exercise A sample of natural gas contains 8.24 moles of methane (CH_4), 0.421 mole of ethane (C_2H_6), and 0.116 mole of propane (C_3H_8). If the total pressure of the gases is 1.37 atm, what are the partial pressures of the gases?

Dalton's law of partial pressures is useful for calculating volumes of gases collected over water. For example, when potassium chlorate ($KClO_3$) is heated, it decomposes to KCl and O_2:

$$2KClO_3(s) \longrightarrow 2KCl(s) + 3O_2(g)$$

The oxygen gas can be collected over water, as shown in Figure 5.13. Initially, the inverted bottle is completely filled with water. As oxygen gas is generated, the gas bubbles rise to the top and displace water from the bottle. This method of collecting a gas is based on the assumptions that the gas does not react with water and that it is not appreciably soluble in it. These assumptions are valid for oxygen gas, but not for gases such as NH_3, which dissolves readily in water. The oxygen gas collected in this way is not pure, however, because water vapor is also present in the bottle. The total gas pressure is equal to the sum of the pressures exerted by the oxygen gas and the water vapor:

$$P_T = P_{O_2} + P_{H_2O}$$

Consequently, we must allow for the pressure caused by the presence of water vapor when we calculate the amount of O_2 generated. Table 5.2 shows the pressure of water vapor at various temperatures.

Table 5.2 | Pressure of Water Vapor at Various Temperatures

Temperature (°C)	Water Vapor Pressure (mmHg)
0	4.58
5	6.54
10	9.21
15	12.79
20	17.54
25	23.76
30	31.82
35	42.18
40	55.32
45	71.88
50	92.51
55	118.04
60	149.38
65	187.54
70	233.7
75	289.1
80	355.1
85	433.6
90	525.76
95	633.90
100	760.00

Bottle being filled with oxygen gas

$KClO_3$ and MnO_2

Bottle filled with water ready to be placed in the plastic basin

Bottle full of oxygen gas

Figure 5.13 *An apparatus for collecting gas over water. The oxygen generated by heating potassium chlorate ($KClO_3$) in the presence of a small amount of manganese dioxide (MnO_2), which speeds up the reaction, is bubbled through water and collected in a bottle as shown. Water originally present in the bottle is pushed into the trough by the oxygen gas.*

Example 5.8   Oxygen gas generated by the decomposition of potassium chlorate is collected as shown in Figure 5.13. The volume of oxygen collected at 24°C and atmospheric pressure of 762 mmHg is 128 mL. Calculate the mass (in grams) of oxygen gas obtained. The pressure of the water vapor at 24°C is 22.4 mmHg.

Reasoning and Solution   From the total pressure and the vapor pressure of water we can calculate the partial pressure of O_2.

$$P_{O_2} = P_T - P_{H_2O}$$
$$= 762 \text{ mmHg} - 22.4 \text{ mmHg}$$
$$= 740 \text{ mmHg}$$
$$= 740 \text{ mmHg} \times \frac{1 \text{ atm}}{760 \text{ mmHg}}$$
$$= 0.974 \text{ atm}$$

From the ideal gas equation we write

$$PV = nRT = \frac{m}{\mathcal{M}} RT$$

O_2 is only slightly soluble in water.

where m and $\mathcal{M}$ are the mass of O_2 collected and the molar mass of O_2, respectively. Rearranging the equation we get

$$m = \frac{PV\mathcal{M}}{RT} = \frac{(0.974 \text{ atm}) (0.128 \text{ L}) (32.00 \text{ g/mol})}{(0.0821 \text{ L} \cdot \text{atm/K} \cdot \text{mol}) (273 + 24) \text{ K}}$$
$$= 0.164 \text{ g}$$

Similar problems: 5.61, 5.62.

Practice Exercise   Hydrogen gas generated when calcium metal reacts with water is collected as shown in Figure 5.13. The volume of gas collected at 30°C and pressure of 988 mmHg is 641 mL. What is the mass (in grams) of the hydrogen gas obtained? The pressure of water vapor at 30°C is 31.82 mmHg.

5.6 The Kinetic Molecular Theory of Gases

The gas laws help us to predict the behavior of gases, but they do not explain what happens at the molecular level to cause the changes we observe in the macroscopic world. For example, why does a gas expand on heating?

In the nineteenth century, a number of physicists, notably the Austrian physicist Ludwig Boltzmann and the Scottish physicist James Clerk Maxwell, found that the physical properties of gases can be explained in terms of the motion of individual molecules. This molecular movement is a form of *energy,* which we define as the capacity to do work or to produce change. In mechanics, *work* is defined as force times distance. Because energy can be measured as work, we can write

$$\text{energy} = \text{work done}$$
$$= \text{force} \times \text{distance}$$

The ***joule (J)*** is *the SI unit of energy*

$$1 \text{ J} = 1 \text{ kg m}^2/\text{s}^2$$
$$= 1 \text{ N m}$$

Alternatively, energy can be expressed in kilojoules (kJ):

$$1 \text{ kJ} = 1000 \text{ J}$$

As we will see in Chapter 6, there are many different kinds of energy. **Kinetic energy (KE)** is the type of energy expended by a moving object, or *energy of motion.*

The findings of Maxwell, Boltzmann, and others resulted in *a number of generalizations about gas behavior* that have since been known as the **kinetic molecular theory of gases,** or simply the *kinetic theory of gases.* Central to the kinetic theory are these assumptions:

1. A gas is composed of molecules that are separated from each other by distances far greater than their own dimensions. The molecules can be considered to be "points"; that is, they possess mass but have negligible volume.

2. Gas molecules are in constant motion in random directions, and they frequently collide with one another. Collisions among molecules are perfectly elastic. In other words, energy can be transferred from one molecule to another as a result of a collision. Nevertheless, the total energy of all the molecules in a system remains the same.

3. Gas molecules exert neither attractive nor repulsive forces on one another.

4. The average kinetic energy of the molecules is proportional to the temperature of the gas in kelvins. Any two gases at the same temperature will have the same average kinetic energy. The average kinetic energy of a molecule is given by

$$\overline{KE} = \tfrac{1}{2}m\overline{u^2}$$

where m is the mass of the molecule and u is its speed. The horizontal bar denotes an average value. The quantity $\overline{u^2}$ is called mean square speed; it is the average of the square of the speeds of all the molecules:

$$\overline{u^2} = \frac{u_1^2 + u_2^2 + \cdots + u_N^2}{N}$$

where N is the number of molecules.

Assumption 4 enables us to write

$$\overline{KE} \propto T$$
$$\tfrac{1}{2}m\overline{u^2} \propto T$$
$$\tfrac{1}{2}m\overline{u^2} = CT \tag{5.13}$$

where C is the proportionality constant and T is the absolute temperature.

According to the kinetic molecular theory, gas pressure is the result of collisions between molecules and the walls of their container. It depends on the frequency of collision per unit area and on how "hard" the molecules strike the wall. The theory also provides a molecular interpretation of temperature. According to Equation (5.13), the absolute temperature of a gas is a measure of the average kinetic energy of the molecules. In other words, the absolute temperature is an index of the random motion of the molecules—the higher the temperature, the more energetic the molecules. Because it is related to the temperature of the gas sample, random molecular motion is sometimes referred to as thermal motion.

Application to the Gas Laws

Although the kinetic theory of gases is based on a rather simple model, the mathematical details involved are very complex. However, on a qualitative basis, it is possible to use the theory to account for the general properties of substances in the gaseous state. The following examples illustrate the range of its utility.

- **Compressibility of Gases.** Because molecules in the gas phase are separated by large distances (assumption 1), gases can be compressed easily to occupy less volume.
- **Boyle's Law.** The pressure exerted by a gas results from the impact of its molecules on the walls of the container. The collision rate, or the number of molecular collisions with the walls per second, is proportional to the number density (that is, number of molecules per unit volume) of the gas. Decreasing the volume of a given amount of gas increases its number density and hence its collision rate. For this reason, the pressure of a gas is inversely proportional to the volume it occupies; as volume decreases, pressure increases and vice versa.
- **Charles's Law.** Because the average kinetic energy of gas molecules is proportional to the sample's absolute temperature (assumption 4), raising the temperature increases the average kinetic energy. Consequently, molecules will collide with the walls of the container more frequently and with greater impact if the gas is heated, and thus the pressure increases. The volume of gas will expand until the gas pressure is balanced by the constant external pressure (see Figure 5.7).
- **Avogadro's Law.** We have shown that the pressure of a gas is directly proportional to both the density and the temperature of the gas. Because the mass of the gas is directly proportional to the number of moles (n) of the gas, we can represent density by n/V. Therefore

$$P \propto \frac{n}{V} T$$

For two gases, 1 and 2, we write

$$P_1 \propto \frac{n_1 T_1}{V_1} = C \frac{n_1 T_1}{V_1}$$

$$P_2 \propto \frac{n_2 T_2}{V_2} = C \frac{n_2 T_2}{V_2}$$

where C is the proportionality constant. Thus, for two gases under the same conditions of pressure, volume, and temperature (that is, when $P_1 = P_2$, $T_1 = T_2$, and $V_1 = V_2$), it follows that $n_1 = n_2$, which is a mathematical expression of Avogadro's law.
- **Dalton's Law of Partial Pressures.** If molecules do not attract or repel one another (assumption 3), then the pressure exerted by one type of molecule is unaffected by the presence of another gas. Consequently, the total pressure is given by the sum of individual gas pressures.

> Another way of stating Avogadro's law is that at the same temperature and pressure, equal volumes of gases, whether they are the same or different gases, contain equal numbers of molecules.

Distribution of Molecular Speeds

The kinetic theory of gases allows us to investigate molecular motion in more detail. Suppose we have a large number of gas molecules, say, 1 mole, in a container. As

Figure 5.14 *(a) The distribution of speeds for nitrogen gas at three different temperatures. At the higher temperatures, more molecules are moving at faster speeds. (b) The distribution of speeds for three gases at 300 K. At a given temperature, the lighter molecules are moving faster, on the average.*

long as we hold the temperature constant, the average kinetic energy and the mean-square speed will remain unchanged as time passes. As you might expect, the motion of the molecules is totally random and unpredictable. At a given instant, how many molecules are moving at a particular speed? To answer this question Maxwell analyzed the behavior of gas molecules at different temperatures.

Figure 5.14(a) shows typical *Maxwell speed distribution curves* for nitrogen gas at three different temperatures. At a given temperature, the distribution curve tells us the number of molecules moving at a certain speed. The peak of each curve represents the *most probable speed,* that is, the speed of the largest number of molecules. Note that the most probable speed increases as temperature increases (the peak shifts toward the right). Furthermore, the curve also begins to flatten out with increasing temperature, indicating that larger numbers of molecules are moving at greater speed. Figure 5.14(b) shows the speed distributions of three gases at the *same* temperature. The difference in the curves can be explained by noting that lighter molecules move faster, on average, than heavier ones.

Root-Mean-Square Speed

How fast does a molecule move, on the average, at any temperature T? One way to estimate molecular speed is to calculate the **root-mean-square (rms) speed** **(u_{rms}),** which is *an average molecular speed.* One of the results of the kinetic theory of gases is that the total kinetic energy of a mole of any gas equals $\frac{3}{2}RT$. Earlier we saw that the average kinetic energy of one molecule is $\frac{1}{2}mu^2$ and so we can write

$$N_A(\tfrac{1}{2}\overline{mu^2}) = \tfrac{3}{2}RT$$

where N_A is Avogadro's number. Because $N_A m = \mathcal{M}$, where $\mathcal{M}$ is the molar mass, this equation can be rearranged to give

$$\overline{u^2} = \frac{3RT}{\mathcal{M}}$$

Taking the square root of both sides gives

$$\sqrt{\overline{u^2}} = u_{rms} = \sqrt{\frac{3RT}{\mathcal{M}}} \tag{5.14}$$

Equation (5.14) shows that the root-mean-square speed of a gas increases with the square root of its temperature (in kelvins). Because $\mathcal{M}$ appears in the denominator, it follows that the heavier the gas, the more slowly its molecules move. If we substitute 8.314 J/K · mol for R (see Appendix 1) and convert the molar mass to kg/mol, then u_{rms} will be calculated in meters per second (m/s).

Example 5.9   Calculate the root-mean-square speeds of helium atoms and nitrogen molecules in m/s at 25°C.

Reasoning and Solution   We need Equation (5.14) to calculate the root-mean-square speeds. Note that because R has the units J/K · mol, the molar mass must be expressed as kg/mol. Thus for He the molar mass should be 4.003×10^{-3} kg/mol, not 4.003 g/mol.

$$u_{rms} = \sqrt{\frac{3RT}{\mathcal{M}}}$$

$$= \sqrt{\frac{3(8.314 \text{ J/K} \cdot \text{mol})(298 \text{ K})}{4.003 \times 10^{-3} \text{ kg/mol}}}$$

$$= \sqrt{1.86 \times 10^6 \text{ J/kg}}$$

Using the conversion factor,

$$1 \text{ J} = 1 \text{ kg m}^2/\text{s}^2$$

we get

$$u_{rms} = \sqrt{1.86 \times 10^6 \text{ kg m}^2/\text{kg} \cdot \text{s}^2}$$

$$= \sqrt{1.86 \times 10^6 \text{ m}^2/\text{s}^2}$$

$$= 1.36 \times 10^3 \text{ m/s}$$

The procedure is the same for N_2, which has a molar mass of 28.02 g/mol, or 2.802×10^{-2} kg/mol:

$$u_{rms} = \sqrt{\frac{3(8.314 \text{ J/K} \cdot \text{mol})(298 \text{ K})}{2.802 \times 10^{-2} \text{ kg/mol}}}$$

$$= \sqrt{2.65 \times 10^5 \text{ m}^2/\text{s}^2}$$

$$= 515 \text{ m/s}$$

Comment   Because of its smaller mass, a helium atom, on the average, moves about 2.6 times faster than a nitrogen molecule at the same temperature (1360 ÷ 515 = 2.64).

Practice Exercise   Calculate the root-mean-square speed of molecular chlorine in m/s at 20°C.

Similar problems: 5.71, 5.72.

The calculation in Example 5.9 has an interesting relationship to the composition of Earth's atmosphere. Unlike Jupiter, Earth does not have appreciable amounts of hydrogen or helium in its atmosphere. Why is this the case? A smaller planet than Jupiter, Earth has a weaker gravitational attraction for these lighter molecules. A fairly straightforward calculation shows that to escape Earth's gravitational field, a molecule must possess an escape velocity equal to or greater than 1.1×10^4 m/s. Because the average speed of helium is considerably greater than that of molecular nitrogen or molecular oxygen, more helium atoms escape from Earth's atmosphere into outer space. Consequently, only a trace amount of helium is present in our atmosphere. On the other hand, Jupiter, with a mass about 320 times greater than that of Earth, retains both heavy and light gases in its atmosphere.

Jupiter. The interior of this massive planet consists mainly of liquid hydrogen.

Gas Diffusion

A direct demonstration of random motion is provided by ***diffusion,*** *the gradual mixing of molecules of one gas with molecules of another by virtue of their kinetic properties.* Diffusion always proceeds from a region of higher concentration to one of lower concentration. Despite the fact that molecular speeds are very great, the diffusion process takes a relatively long time to complete. For example, when a bottle of concentrated ammonia solution is opened at one end of a lab bench, it takes some time before a person at the other end of the bench can smell it. The reason is that a molecule experiences numerous collisions while moving from one end of the bench to the other, as shown in Figure 5.15. Thus, diffusion of gases always happens gradually, and not instantly as molecular speeds seem to suggest. Furthermore, because the root-mean-square speed of a light gas is greater than that of a heavier gas (see Example 5.9), a lighter gas will diffuse through a certain space more quickly than will a heavier gas. Figure 5.16 illustrates gaseous diffusion.

Figure 5.15 *The path traveled by a single gas molecule. Each change in direction represents a collision with another molecule.*

Figure 5.16 *A demonstration of gas diffusion. NH_3 gas (from a bottle containing aqueous ammonia) combines with HCl gas (from a bottle containing hydrochloric acid) to form solid NH_4Cl. Because NH_3 is lighter and therefore diffuses faster, solid NH_4Cl first appears nearer the HCl bottle (on the right).*

5.7 Deviation from Ideal Behavior

The gas laws and the kinetic molecular theory assume that molecules in the gaseous state do not exert any force, either attractive or repulsive, on one another. The other assumption is that the volume of the molecules is negligibly small compared with that of the container. A gas that satisfies these two conditions is said to exhibit *ideal behavior.*

Although we can assume that real gases behave like an ideal gas, we cannot expect them to do so under all conditions. For example, without intermolecular forces, gases could not condense to form liquids. The important question is: Under what conditions will gases most likely exhibit nonideal behavior?

Figure 5.17 shows PV/RT plotted against P for three real gases and an ideal gas at a given temperature. This graph provides a test of ideal gas behavior. According to the ideal gas equation (for 1 mole of gas), PV/RT equals 1, regardless of the actual gas pressure. (When $n = 1$, $PV = nRT$ becomes $PV = RT$, or $PV/RT = 1$.) For real gases, this is true only at moderately low pressures (≤ 5 atm); significant deviations occur as pressure increases. Attractive forces operate among molecules at relatively short distances. At atmospheric pressure, the molecules in a gas are far apart and the attractive forces are negligible. At high pressures, the density of the gas increases; the molecules are much closer to one another. Intermolecular forces can then be significant enough to affect the motion of the molecules, and the gas will not behave ideally.

Another way to observe the nonideal behavior of gases is to lower the temperature. Cooling a gas decreases the molecules' average kinetic energy, which in a sense deprives molecules of the drive they need to break from their mutual attraction.

To study real gases accurately, then, we need to modify the ideal gas equation, taking into account intermolecular forces and finite molecular volumes. Such an analysis was first made by the Dutch physicist J. D. van der Waals in 1873. Besides being mathematically simple, van der Waals' treatment provides us with an interpretation of real gas behavior at the molecular level.

Consider the approach of a particular molecule toward the wall of a container (Figure 5.18). The intermolecular attractions exerted by its neighbors tend to soften

Figure 5.17 *Plot of PV/RT versus P of 1 mole of a gas at 0°C. For 1 mole of an ideal gas, PV/RT is equal to 1, no matter what the pressure of the gas is. For real gases, we observe various deviations from ideality at high pressures. At very low pressures, all gases exhibit ideal behavior; that is, their PV/RT values all converge to 1 as P approaches zero.*

the impact made by this molecule against the wall. The overall effect is a lower gas pressure than we would expect for an ideal gas. Van der Waals suggested that the pressure exerted by an ideal gas, P_{ideal}, is related to the experimentally measured pressure, P_{real}, by the equation

$$P_{ideal} = \underset{\uparrow}{P_{real}} + \underset{\uparrow}{\frac{an^2}{V^2}}$$

observed   correction
pressure   term

where a is a constant and n and V are the number of moles and volume of the gas, respectively. The correction term for pressure (an^2/V^2) can be understood as follows. The intermolecular interaction that gives rise to nonideal behavior depends on how frequently any two molecules approach each other closely. The number of such "encounters" increases with the square of the number of molecules per unit volume, $(n/V)^2$, because the presence of each of the two molecules in a particular region is proportional to n/V. The quantity P_{ideal} is the pressure we would measure if there were no intermolecular attractions, and so a is just a proportionality constant.

Another correction concerns the volume occupied by the gas molecules. In the ideal gas equation, V represents the volume of the container. However, each molecule does occupy a finite, although small, intrinsic volume, so the effective volume of the gas becomes $(V - nb)$, where n is the number of moles of the gas and b is a constant. The term nb represents the volume occupied by n moles of the gas.

Having taken into account the corrections for pressure and volume, we can rewrite the ideal gas equation as follows:

$$\underbrace{\left(P + \frac{an^2}{V^2}\right)}_{\substack{\text{corrected} \\ \text{pressure}}}\underbrace{(V - nb)}_{\substack{\text{corrected} \\ \text{volume}}} = nRT \qquad (5.15)$$

Equation (5.15), *relating P, V, T, and n for a nonideal gas,* is known as the **van der Waals equation.** The van der Waals constants a and b are selected to give the best possible agreement between Equation (5.15) and observed behavior of a particular gas.

Table 5.3 lists the values of a and b for a number of gases. The value of a indicates how strongly molecules of a given type of gas attract one another. We see that helium atoms have the weakest attraction for one another, because helium has the smallest a value. There is also a rough correlation between molecular size and b. Generally, the larger the molecule (or atom), the greater b is, but the relationship between b and molecular (or atomic) size is not a simple one.

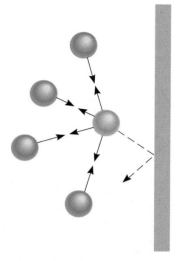

Figure 5.18 *Effect of intermolecular forces on the pressure exerted by a gas. The speed of a molecule that is moving toward the container wall (red sphere) is reduced by the attractive forces exerted by its neighbors (green spheres). Consequently, the impact this molecule makes with the wall is not as great as it would be if no intermolecular forces were present. In general, the measured gas pressure is lower than the pressure the gas would exert if it behaved ideally.*

Example 5.10   Given that 3.50 moles of NH_3 occupy 5.20 L at 47°C, calculate the pressure of the gas (in atm) using (a) the ideal gas equation and (b) the van der Waals equation.

Reasoning and Solution   Part (a) is a direct application of the ideal gas equation similar to Example 5.2. For part (b) we need Equation (5.15) and Table 5.3.

Table 5.3 | Van der Waals Constants of Some Common Gases

Gas	a $\left(\dfrac{\text{atm} \cdot \text{L}^2}{\text{mol}^2}\right)$	b $\left(\dfrac{\text{L}}{\text{mol}}\right)$
He	0.034	0.0237
Ne	0.211	0.0171
Ar	1.34	0.0322
Kr	2.32	0.0398
Xe	4.19	0.0266
H_2	0.244	0.0266
N_2	1.39	0.0391
O_2	1.36	0.0318
Cl_2	6.49	0.0562
CO_2	3.59	0.0427
CH_4	2.25	0.0428
CCl_4	20.4	0.138
NH_3	4.17	0.0371
H_2O	5.46	0.0305

Similar problems: 5.79, 5.80.

(a) We have the following data:

$$V = 5.20 \text{ L}$$
$$T = (47 + 273) \text{ K} = 320 \text{ K}$$
$$n = 3.50 \text{ mol}$$
$$R = 0.0821 \text{ L} \cdot \text{atm/K} \cdot \text{mol}$$

We substitute these values in the ideal gas equation:

$$P = \frac{nRT}{V}$$
$$= \frac{(3.50 \text{ mol})(0.0821 \text{ L} \cdot \text{atm/K} \cdot \text{mol})(320 \text{ K})}{5.20 \text{ L}}$$
$$= 17.7 \text{ atm}$$

(b) From Table 5.3, we have

$$a = 4.17 \text{ atm} \cdot \text{L}^2/\text{mol}^2$$
$$b = 0.0371 \text{ L/mol}$$

It is convenient to calculate the correction terms for Equation (5.15) first. They are

$$\frac{an^2}{V^2} = \frac{(4.17 \text{ atm} \cdot \text{L}^2/\text{mol}^2)(3.50 \text{ mol})^2}{(5.20 \text{ L})^2} = 1.89 \text{ atm}$$

$$nb = (3.50 \text{ mol})(0.0371 \text{ L/mol}) = 0.130 \text{ L}$$

Finally, substituting in the van der Waals equation, we write

$$(P + 1.89 \text{ atm})(5.20 \text{ L} - 0.130 \text{ L}) = (3.50 \text{ mol})(0.0821 \text{ L} \cdot \text{atm/K} \cdot \text{mol})(320 \text{ K})$$
$$P = 16.2 \text{ atm}$$

Comment   The actual pressure measured under these conditions is 16.0 atm. Thus, the pressure calculated by the van der Waals equation (16.2 atm) is closer to the actual value than that calculated by the ideal gas equation (17.7 atm). Note that the attraction between NH_3 molecules results in a lower gas pressure than that of an ideal gas under the same conditions.

Practice Exercise   Using the data shown in Table 5.3, calculate the pressure exerted by 4.37 moles of molecular chlorine confined in a volume of 2.45 L at 38°C. Compare the pressure with that calculated using the ideal gas equation.

Summary

Under atmospheric conditions, a number of elemental substances are gases: H_2, N_2, O_2, O_3, F_2, Cl_2, and the Group 8A elements (the noble gases).

Gases exert pressure because their molecules move freely and collide with any surface in their paths. Gas pressure units include millimeters of mercury (mmHg), torr, pascals, and atmospheres. One atmosphere equals 760 mmHg, or 760 torr.

The pressure-volume relationships of ideal gases are governed by Boyle's law: Volume is inversely proportional to pressure (at constant T and n). The temperature-volume relationships of ideal gases are described by Charles's and Gay-Lussac's law: Volume is directly proportional to temperature (at constant P and n). Absolute zero ($-273.15°C$) is the lowest theoretically attainable temperature. On the Kelvin temperature scale, 0 K is absolute zero. In all gas law calculations, temperature must be expressed in kelvins. The amount-volume relationships of ideal gases are described by Avogadro's law: Equal volumes of gases contain equal numbers of molecules (at the same T and P).

The ideal gas equation, $PV = nRT$, combines the laws of Boyle, Charles, and Avogadro. This equation describes the behavior of an ideal gas.

Dalton's law of partial pressures states that in a mixture of gases each gas exerts the same pressure as it would if it were alone and occupied the same volume.

The kinetic molecular theory, a mathematical way of describing the behavior of gas molecules, is based on the following assumptions: Gas molecules are separated by distances far greater than their own dimensions, they possess mass but have negligible volume, they are in constant motion, and they frequently collide with one another. The molecules neither attract nor repel one another. A Maxwell speed distribution curve shows how many gas molecules are moving at various speeds at a given temperature. As temperature increases, more molecules move at greater speeds.

The van der Waals equation is a modification of the ideal gas equation that takes into account the nonideal behavior of real gases. It corrects for two facts: Real gas molecules do exert forces on each other and they do have volume. The van der Waals constants are determined experimentally for each gas.

Key Words

Absolute temperature scale p. 133
Absolute zero p. 133
Atmospheric pressure p. 128
Avogadro's law p. 134
Barometer p. 128
Boyle's law p. 130
Charles's and Gay-Lussac's law p. 134

Charles's law p. 134
Dalton's law of partial pressures p. 140
Diffusion p. 149
Gas constant p. 135
Ideal gas p. 135
Ideal gas equation p. 135
Joule (J) p. 144
Kelvin temperature scale p. 133

Kinetic energy (KE) p. 145
Kinetic molecular theory of gases p. 145
Manometer p. 128
Mole fraction p. 141
Newton (N) p. 127
Partial pressure p. 140
Pascal (Pa) p. 127
Pressure p. 127

Root-mean-square (rms) speed (u_{rms}) p. 147
Standard atmospheric pressure (1 atm) p. 128
Standard temperature and pressure (STP) p. 136
van der Waals equation p. 151

Questions and Problems

Substances That Exist as Gases

Review Questions

5.1 Name five elements and five compounds that exist as gases at room temperature.

5.2 Describe the physical characteristics of gases.

Pressure

Review Questions

5.3 Define pressure and give the common units for pressure.

5.4 Describe how a barometer and a manometer are used to measure pressure.

5.5 Why is mercury a more suitable substance to use in a barometer than water?

5.6 Explain why the height of mercury in a barometer is independent of the cross-sectional area of the tube.

5.7 Would it be easier to drink water with a straw on the top or at the foot of Mt. Everest? Explain.

5.8 Is the atmospheric pressure in a mine that is 500 m below sea level greater or less than 1 atm?

5.9 What is the difference between a gas and a vapor? At 25°C, which of the following substances in the gas phase should be properly called a gas and which should be called a vapor: molecular nitrogen (N_2) or mercury?

5.10 If the maximum distance that water may be brought up a well by using a suction pump is 34 ft (10.3 m), explain how it is possible to obtain water and oil from hundreds of feet below the surface of Earth.

5.11 Why is it that if the barometer reading falls in one part of the world, it must rise somewhere else?

5.12 Why do astronauts have to wear protective suits when they are on the surface of the moon?

Problems

5.13 Convert 562 mmHg to kPa and 2.0 kPa to mmHg.

5.14 The atmospheric pressure at the summit of Mt. McKinley is 606 mmHg on a certain day. What is the pressure in atm?

The Gas Laws

Review Questions

5.15 State these gas laws in words and also in the form of an equation: Boyle's law, Charles's law, Avogadro's law. In each case, indicate the conditions under which the law is applicable and also give the units for each quantity in the equation.

5.16 Explain why a helium weather balloon expands as it rises in the air. Assume that the temperature remains constant.

Problems

5.17 A gas occupying a volume of 725 mL at a pressure of 0.970 atm is allowed to expand at constant temperature until its pressure becomes 0.541 atm. What is its final volume?

5.18 At 46°C a sample of ammonia gas exerts a pressure of 5.3 atm. What is the pressure when the volume of the gas is reduced to one-tenth (0.10) of the original value at the same temperature?

5.19 The volume of a gas is 5.80 L, measured at 1.00 atm. What is the pressure of the gas in mmHg if the volume is changed to 9.65 L? (The temperature remains constant.)

5.20 A sample of air occupies 3.8 L when the pressure is 1.2 atm. (a) What volume does it occupy at 6.6 atm? (b) What pressure is required to compress it to 0.075 L? (The temperature is kept constant.)

5.21 Convert these temperatures to Kelvin: 0°C, 37°C, 100°C, −225°C.

5.22 Convert these temperatures to degrees Celsius: 77 K (boiling point of nitrogen), 4.2 K (boiling point of helium), 6.0×10^3 K (surface temperature of the sun).

5.23 A 36.4-L volume of methane gas is heated from 25°C to 88°C at constant pressure. What is the final volume of the gas?

5.24 Under constant-pressure conditions a sample of hydrogen gas initially at 88°C and 9.6 L is cooled until its final volume is 3.4 L. What is its final temperature?

5.25 Ammonia burns in oxygen gas to form nitric oxide (NO) and water vapor. How many volumes of NO are obtained from one volume of ammonia at the same temperature and pressure?

5.26 Molecular chlorine and molecular fluorine combine to form a gaseous product. Under the same conditions of temperature and pressure it is found that one volume of Cl_2 reacts with three volumes of F_2 to yield two volumes of the product. What is the formula of the product?

The Ideal Gas Equation

Review Questions

5.27 Describe the characteristics of an ideal gas.

5.28 Write the ideal gas equation and also state it in words. Give the units for each term in the equation.

5.29 What are standard temperature and pressure (STP)? What is the significance of STP in relation to the volume of 1 mole of an ideal gas?

5.30 Why is the density of a gas much lower than that of a liquid or solid under atmospheric conditions? What units are normally used to express the density of gases?

Problems

5.31 A sample of nitrogen gas kept in a container of volume 2.3 L and at a temperature of 32°C exerts a pressure of 4.7 atm. Calculate the number of moles of gas present.

5.32 A sample of 6.9 moles of carbon monoxide gas is present in a container of volume 30.4 L. What is the pressure of the gas (in atm) if the temperature is 62°C?

5.33 What volume would 5.6 moles of sulfur hexafluoride (SF_6) gas occupy if the temperature and pressure of the gas are 128°C and 9.4 atm?

5.34 A certain amount of gas at 25°C and at a pressure of 0.800 atm is contained in a glass vessel. Suppose that the vessel can withstand a pressure of 2.00 atm. How high can you increase the temperature of the gas without bursting the vessel?

5.35 A gas-filled balloon having a volume of 2.50 L at 1.2 atm and 25°C is allowed to rise to the stratosphere (about 30 km above the surface of Earth), where the temperature and pressure are −23°C and 3.00×10^{-3} atm, respectively. Calculate the final volume of the balloon.

5.36 The temperature of 2.5 L of a gas initially at STP is increased to 250°C at constant volume. Calculate the final pressure of the gas in atm.

5.37 The pressure of 6.0 L of an ideal gas in a flexible container is decreased to one-third of its original pressure, and its absolute temperature is decreased by one-half. What is the final volume of the gas?

5.38 A gas evolved during the fermentation of glucose (wine making) has a volume of 0.78 L when measured at 20.1°C and 1.00 atm. What was the volume of this gas at the fermentation temperature of 36.5°C and 1.00 atm pressure?

5.39 An ideal gas originally at 0.85 atm and 66°C was allowed to expand until its final volume, pressure, and temperature were 94 mL, 0.60 atm, and 45°C, respectively. What was its initial volume?

5.40 The volume of a gas at STP is 488 mL. Calculate its volume at 22.5 atm and 150°C.

5.41 A gas at 772 mmHg and 35.0°C occupies a volume of 6.85 L. Calculate its volume at STP.

5.42 Dry ice is solid carbon dioxide. A 0.050-g sample of dry ice is placed in an evacuated 4.6-L vessel at 30°C. Calculate the pressure inside the vessel after all the dry ice has been converted to CO_2 gas.

5.43 A volume of 0.280 L of a gas at STP weighs 0.400 g. Calculate the molar mass of the gas.

5.44 A quantity of gas weighing 7.10 g at 741 torr and 44°C occupies a volume of 5.40 L. What is its molar mass?

5.45 The ozone molecules present in the stratosphere absorb much of the harmful radiation from the sun. Typically, the temperature and pressure of ozone in the stratosphere are 250 K and 1.0×10^{-3} atm, respectively. How many ozone molecules are present in 1.0 L of air under these conditions?

5.46 Assuming that air contains 78 percent N_2, 21 percent O_2, and 1 percent Ar, all by volume, how many molecules of each type of gas are present in 1.0 L of air at STP?

5.47 A 2.10-L vessel contains 4.65 g of gas at 1.00 atm and 27.0°C. (a) Calculate the density of the gas in grams per liter. (b) What is the molar mass of the gas?

5.48 Calculate the density of hydrogen bromide (HBr) gas in grams per liter at 733 mmHg and 46°C.

5.49 A certain anesthetic contains 64.9 percent C, 13.5 percent H, and 21.6 percent O by mass. At 120°C and 750 mmHg, 1.00 L of the gaseous compound weighs 2.30 g. What is the molecular formula of the compound?

5.50 A compound has the empirical formula SF_4. At 20°C, 0.100 g of the gaseous compound occupies a volume of 22.1 mL and exerts a pressure of 1.02 atm. What is its molecular formula?

5.51 Dissolving 3.00 g of an impure sample of calcium carbonate in hydrochloric acid produced 0.656 L of carbon dioxide (measured at 20.0°C and 792 mmHg). Calculate the percent by mass of calcium carbonate in the sample. State any assumptions.

5.52 Calculate the mass in grams of hydrogen chloride produced when 5.6 L of molecular hydrogen measured at STP react with an excess of molecular chlorine gas.

5.53 A quantity of 0.225 g of a metal M (molar mass = 27.0 g/mol) liberated 0.303 L of molecular hydrogen (measured at 17°C and 741 mmHg) from an excess of hydrochloric acid. Deduce from these data the corresponding equation and write formulas for the oxide and sulfate of M.

5.54 A compound of P and F was analyzed as follows: Heating 0.2324 g of the compound in a 378-cm^3 container turned all of it to gas, which had a pressure of 97.3 mmHg at 77°C. Then the gas was mixed with calcium chloride solution, which turned all of the F to 0.2631 g of CaF_2. Determine the molecular formula of the compound.

Dalton's Law of Partial Pressures

Review Questions

5.55 Define Dalton's law of partial pressures and mole fraction. Does mole fraction have units?

5.56 A sample of air contains only nitrogen and oxygen gases whose partial pressures are 0.80 atm and 0.20 atm, respectively. Calculate the total pressure and the mole fractions of the gases.

Problems

5.57 A mixture of gases contains CH_4, C_2H_6, and C_3H_8. If the total pressure is 1.50 atm and the numbers of moles of the gases present are 0.31 mole for CH_4, 0.25 mole for C_2H_6, and 0.29 mole for C_3H_8, calculate the partial pressures of the gases.

5.58 A 2.5-L flask at 15°C contains a mixture of three gases, N_2, He, and Ne, at partial pressures of 0.32 atm for N_2, 0.15 atm for He, and 0.42 atm for Ne. (a) Calculate the total pressure of the mixture. (b) Calculate the volume in liters at STP occupied by He and Ne if the N_2 is removed selectively.

5.59 Dry air near sea level has the following composition by volume: N_2, 78.08 percent; O_2, 20.94 percent; Ar, 0.93 percent; CO_2, 0.05 percent. The atmospheric pressure is 1.00 atm. Calculate (a) the partial pressure of each gas in atm and (b) the concentration of each gas in moles per liter at 0°C. (*Hint:* Because volume is proportional to the number of moles present, mole fractions of gases can be expressed as ratios of volumes at the same temperature and pressure.)

5.60 A mixture of helium and neon gases is collected over water at 28.0°C and 745 mmHg. If the partial pressure of helium is 368 mmHg, what is the partial pressure of neon? (Vapor pressure of water at 28°C = 28.3 mmHg.)

5.61 A piece of sodium metal undergoes complete reaction with water as follows:

$$2Na(s) + 2H_2O(l) \longrightarrow 2NaOH(aq) + H_2(g)$$

The hydrogen gas generated is collected over water at 25.0°C. The volume of the gas is 246 mL measured at 1.00 atm. Calculate the number of grams of sodium used in the reaction. (Vapor pressure of water at 25°C = 0.0313 atm.)

5.62 A sample of zinc metal is allowed to react completely with an excess of hydrochloric acid:

$$Zn(s) + 2HCl(aq) \longrightarrow ZnCl_2(aq) + H_2(g)$$

The hydrogen gas produced is collected over water at 25.0°C using an arrangement similar to that shown in Figure 5.13. The volume of the gas is 7.80 L, and the pressure is 0.980 atm. Calculate the amount of zinc metal in grams consumed in the reaction. (Vapor pressure of water at 25°C = 23.8 mmHg.)

5.63 Helium is mixed with oxygen gas for deep sea divers. Calculate the percent by volume of oxygen gas in the mixture if the diver has to submerge to a depth where the total pressure is 4.2 atm. The partial pressure of oxygen is maintained at 0.20 atm at this depth.

5.64 A sample of ammonia (NH_3) gas is completely decomposed to nitrogen and hydrogen gases over heated iron wool. If the total pressure is 866 mmHg, calculate the partial pressures of N_2 and H_2.

Kinetic Molecular Theory of Gases

Review Questions

5.65 What are the basic assumptions of the kinetic molecular theory of gases?

5.66 What is thermal motion?

5.67 What does the Maxwell speed distribution curve tell us? Does Maxwell's theory work for a sample of 200 molecules? Explain.

5.68 Write the expression for the root-mean-square speed for a gas at temperature T. Define each term in the equation and show the units that are used in the calculations.

5.69 Which of the following two statements is correct? (a) Heat is produced by the collision of gas molecules against one another. (b) When a gas is heated, the molecules collide with one another more often.

5.70 As we know, UF_6 is a much heavier gas than helium. Yet at a given temperature, the average kinetic energies of the samples of the two gases are the same. Explain.

Problems

5.71 Compare the root-mean-square speeds of O_2 and UF_6 at 65°C.

5.72 The temperature in the stratosphere is −23°C. Calculate the root-mean-square speeds of N_2, O_2, and O_3 molecules in this region.

5.73 The average distance traveled by a molecule between successive collisions is called *mean free path*. For a given amount of a gas, how does the mean free path of a gas depend on (a) density, (b) temperature at constant volume,

(c) pressure at constant temperature, (d) volume at constant temperature, and (e) size of the atoms?

5.74 At a certain temperature the speeds of six gaseous molecules in a container are 2.0 m/s, 2.2 m/s, 2.6 m/s, 2.7 m/s, 3.3 m/s, and 3.5 m/s. Calculate the root-mean-square speed and the average speed of the molecules. These two average values are close to each other, but the root-mean-square value is always the larger of the two. Why?

Nonideal Gas Behavior

Review Questions

5.75 Give two pieces of evidence to show that gases do not behave ideally under all conditions.

5.76 Under what set of conditions would a gas be expected to behave most ideally? (a) High temperature and low pressure, (b) high temperature and high pressure, (c) low temperature and high pressure, (d) low temperature and low pressure.

5.77 Write the van der Waals equation for a real gas. Explain clearly the meaning of the corrective terms for pressure and volume.

5.78 The temperature of a real gas that is allowed to expand into a vacuum usually drops. Explain.

Problems

5.79 Using the data shown in Table 5.3, calculate the pressure exerted by 2.50 moles of CO_2 confined in a volume of 5.00 L at 450 K. Compare the pressure with that calculated using the ideal gas equation.

5.80 At 27°C 10.0 moles of a gas in a 1.50-L container exert a pressure of 130 atm. Is this an ideal gas?

5.81 Under the same conditions of temperature and pressure, which of these gases would behave most ideally: Ne, N_2, or CH_4? Explain.

5.82 From these data collected at 0°C, comment on whether carbon dioxide behaves as an ideal gas.

P (atm)	0.0500	0.100	0.151	0.202	0.252
V (L)	448.2	223.8	148.8	110.8	89.0

Additional Problems

5.83 Discuss the following phenomena in terms of the gas laws: (a) the pressure in an automobile tire increasing on a hot day, (b) the "popping" of a paper bag, (c) the expansion of a weather balloon as it rises in the air, (d) the loud noise heard when a lightbulb shatters.

5.84 Nitroglycerin, an explosive, decomposes according to the equation

$$4C_3H_5(NO_3)_3(s) \longrightarrow$$
$$12CO_2(g) + 10H_2O(g) + 6N_2(g) + O_2(g)$$

Calculate the total volume of gases produced when collected at 1.2 atm and 25°C from 2.6×10^2 g of nitroglycerin. What are the partial pressures of the gases under these conditions?

5.85 The empirical formula of a compound is CH. At 200°C, 0.145 g of this compound occupies 97.2 mL at a pressure of 0.74 atm. What is the molecular formula of the compound?

5.86 When ammonium nitrite (NH_4NO_2) is heated, it decomposes to give nitrogen gas. This property is used to inflate some tennis balls. (a) Write a balanced equation for the reaction. (b) Calculate the quantity (in grams) of NH_4NO_2 needed to inflate a tennis ball to a volume of 86.2 mL at 1.20 atm and 22°C.

5.87 The percent by mass of bicarbonate (HCO_3^-) in a certain Alka-Seltzer product is 32.5 percent. Calculate the volume of CO_2 generated (in milliliters) at 37°C and 1.00 atm when a person ingests a 3.29-g tablet. (*Hint:* The reaction is between HCO_3^- and HCl acid in the stomach.)

5.88 The boiling point of liquid nitrogen is −196°C. On the basis of this information alone, do you think nitrogen is an ideal gas?

5.89 In the metallurgical process of refining nickel, the metal is first combined with carbon monoxide to form tetracarbonylnickel, which is a gas at 43°C:

$$Ni(s) + 4CO(g) \longrightarrow Ni(CO)_4(g)$$

This reaction separates nickel from other solid impurities. (a) Starting with 86.4 g of Ni, calculate the pressure of $Ni(CO)_4$ in a container of volume 4.00 L. (Assume the above reaction goes to completion.) (b) On further heating the sample above 43°C, it is observed that the pressure of the gas increases much more rapidly than predicted based on the ideal gas equation. Explain.

5.90 The partial pressure of carbon dioxide varies with seasons. Would you expect the partial pressure in the Northern Hemisphere to be higher in the summer or winter? Explain.

5.91 A healthy adult exhales about 5.0×10^2 mL of a gaseous mixture with each breath. Calculate the number of molecules present in this volume at 37°C and 1.1 atm. List the major components of this gaseous mixture.

5.92 Sodium bicarbonate ($NaHCO_3$) is called baking soda because when heated, it releases carbon dioxide gas, which is responsible for the rising of cookies, doughnuts, and bread. (a) Calculate the volume (in liters) of CO_2 produced by heating 5.0 g of $NaHCO_3$ at 180°C and 1.3 atm. (b) Ammonium bicarbonate (NH_4HCO_3) has also been used for the same purpose. Suggest one advantage and one disadvantage of using NH_4HCO_3 instead of $NaHCO_3$ for baking.

5.93 A barometer having a cross-sectional area of 1.00 cm^2 at sea level measures a pressure of 76.0 cm of mercury. The pressure exerted by this column of mercury is equal to the pressure exerted by all the air on 1 cm^2 of Earth's surface. Given that the density of mercury is 13.6 g/mL, and the average radius of Earth is 6371 km, calculate the total mass of Earth's atmosphere in kilograms. (*Hint:* The surface area of a sphere is $4\pi r^2$, in which r is the radius of the sphere.)

5.94 Some commercial drain cleaners contain two components: sodium hydroxide and aluminum powder. When the mixture is poured down a clogged drain, the following reaction occurs:

$$2NaOH(aq) + 2Al(s) + 6H_2O(l) \longrightarrow$$
$$2NaAl(OH)_4(aq) + 3H_2(g)$$

The heat generated in this reaction helps melt away obstructions such as grease, and the hydrogen gas released stirs up the solids clogging the drain. Calculate the volume of H_2 formed at STP if 3.12 g of Al is treated with an excess of NaOH.

5.95 The volume of a sample of pure HCl gas was 189 mL at 25°C and 108 mmHg. It was completely dissolved in about 60 mL of water and titrated with an NaOH solution; 15.7 mL of the NaOH solution were required to neutralize the HCl. Calculate the molarity of the NaOH solution.

5.96 Propane (C_3H_8) burns in oxygen to produce carbon dioxide gas and water vapor. (a) Write a balanced equation for this reaction. (b) Calculate the number of liters of carbon dioxide measured at STP that could be produced from 7.45 g of propane.

5.97 Consider the apparatus shown here. When a small amount of water is introduced into the flask by squeezing the bulb of the medicine dropper, water is squirted upward out of the long glass tubing. Explain this observation. (*Hint:* Hydrogen chloride gas is soluble in water.)

5.98 Nitric oxide (NO) reacts with molecular oxygen as follows:

$$2NO(g) + O_2(g) \longrightarrow 2NO_2(g)$$

Initially NO and O_2 are separated as shown in the figure. When the valve is opened, the reaction quickly goes to completion. Determine what gases remain at the end and calculate their partial pressures. Assume that the temperature remains constant at 25°C.

4.00 L at
0.500 atm

2.00 L at
1.00 atm

5.99 The apparatus shown in the diagram can be used to measure atomic and molecular speed. Suppose that a beam of metal atoms is directed at a rotating cylinder in a vacuum. A small opening in the cylinder allows the atoms to strike a target area. Because the cylinder is rotating, atoms traveling at different speeds will strike the target at different positions. In time, a layer of the metal will deposit on the target area, and the variation in its thickness is found to correspond to Maxwell's speed distribution. In one experiment it is found that at 850°C some bismuth (Bi) atoms struck the target at a point 2.80 cm from the spot directly opposite the slit. The diameter of the cylinder is 15.0 cm and it is rotating at 130 revolutions per second. (a) Calculate the speed (m/s) at which the target is moving. (*Hint:* The circumference of a circle is given by $2\pi r$, in which r is the radius.) (b) Calculate the time (in seconds) it takes for the target to travel 2.80 cm. (c) Determine the speed of the Bi atoms. Compare your result in (c) with the u_{rms} of Bi at 850°C. Comment on the difference.

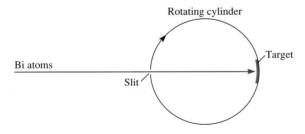

Rotating cylinder

Bi atoms

Slit

Target

5.100 Acidic oxides such as carbon dioxide react with basic oxides like calcium oxide (CaO) and barium oxide (BaO) to form salts (metal carbonates). (a) Write equations repre-

senting these two reactions. (b) A student placed a mixture of BaO and CaO of combined mass 4.88 g in a 1.46-L flask containing carbon dioxide gas at 35°C and 746 mmHg. After the reactions were complete, she found that the CO_2 pressure had dropped to 252 mmHg. Calculate the percent composition of the mixture.

5.101 The running engine of an automobile produces carbon monoxide (CO), a toxic gas, at the rate of about 188 g CO per hour. A car is left idling at 20°C in a poorly ventilated garage that is 6.0 m long, 4.0 m wide, and 2.2 m high. (a) Calculate the rate of CO production in moles per minute. (b) How long would it take to build up a lethal concentration of CO of 1000 ppmv (parts per million by volume)?

5.102 Air entering the lungs ends up in tiny sacs called alveoli. It is from the alveoli that oxygen diffuses into the blood. The average radius of the alveoli is 0.0050 cm and the air inside contains 14 percent oxygen. Assuming that the pressure in the alveoli is 1.0 atm and the temperature is 37°C, calculate the number of oxygen molecules in one of the alveoli. (*Hint:* The volume of a sphere of radius r is $\frac{4}{3}\pi r^3$.)

5.103 It is said that every breath we take, on average, contains molecules that were once exhaled by Wolfgang Amadeus Mozart (1756–1791). These calculations demonstrate the validity of this statement. (a) Calculate the total number of molecules in the atmosphere. (*Hint:* Use the result in Problem 5.93 and 29.0 g/mol as the molar mass of air.) (b) Assuming the volume of every breath (inhale or exhale) is 500 mL, calculate the number of molecules exhaled in each breath at 37°C, which is human body temperature. (c) If Mozart's lifespan was exactly 35 years, what is the number of molecules exhaled in that period? (Given that an average person breathes 12 times per minute.) (d) Calculate the fraction of molecules in the atmosphere that were breathed out by Mozart. How many of Mozart's molecules do we breathe in with every inhalation of air? Round off your answer to one significant figure. (e) List three important assumptions in these calculations.

5.104 Estimate the distance (in nanometers) between molecules of water vapor at 100°C and 1.0 atm. Assume ideal behavior. Repeat the calculation for liquid water at 100°C, given that the density of water is 0.96 g/cm^3 at that temperature. Comment on your results. (Assume water molecule to be a sphere with a diameter of 0.3 nm.) (*Hint:* First calculate the number density of water molecules. Next, convert the number density to linear density, that is, number of molecules in one direction.)

Special Problems

5.105 Apply your knowledge of the kinetic theory of gases to these situations.

(a) Does a single molecule have a temperature?

(b) Two flasks of volumes V_1 and V_2 ($V_2 > V_1$) contain the same number of helium atoms at the same temperature. (i) Compare the root-mean-square (rms) speeds and average kinetic energies of the helium (He) atoms in the flasks. (ii) Compare the frequency and the force with which the He atoms collide with the walls of their containers.

(c) Equal numbers of He atoms are placed in two flasks of the same volume at temperature T_1 and T_2 ($T_2 > T_1$) (i) Compare the rms speeds of the atoms in the two flasks. (ii) Compare the frequency and the force with which the He atoms collide with the walls of their containers.

(d) Equal numbers of He and neon (Ne) atoms are placed in two flasks of the same volume and the temperature of both gases is 74°C. Comment on the validity of these statements: (i) The rms speed of He is equal to that of Ne. (ii) The average kinetic energies of the two gases are equal. (iii) The rms speed of each He atom is 1.47×10^3 m/s.

5.106 Referring to the plot in Figure 5.17, (a) why do the plots of the gases dip before they rise? (b) Why do they all converge to 1 at very low P? (c) What is the meaning of the intercept on the ideal gas line? Does the intercept mean that the gas has become an ideal gas?

5.107 Graham's law of gas diffusion states that at the same temperature and pressure, the rates of diffusion of gases are inversely proportional to the square roots of their molar masses. Thus, for gases 1 and 2, we have

$$\frac{r_1}{r_2} = \sqrt{\frac{\mathcal{M}_2}{\mathcal{M}_1}}$$

where r_1 and r_2 are the rates of diffusion of the gases and $\mathcal{M}_1$ and $\mathcal{M}_2$ are their molar masses, respectively. Based on your knowledge of the kinetic theory of gases, derive Graham's law.

Answers to Practice Exercises

5.1 0.986 atm. **5.2** 9.29 L. **5.3** 30.6 L. **5.4** 2.6 atm. **5.5** 44.1 g/mol. **5.6** 4.75 L. **5.7** CH_4: 1.29 atm; C_2H_6: 0.0657 atm; C_3H_8: 0.0181 atm. **5.8** 0.0653 g. **5.9** 321 m/s. **5.10** 30.0 atm; 45.5 atm using the ideal gas equation.

As the giant airship hovered near the landing site, the ground crew scurried about, waiting for the thunderstorms to abate so the dirigible could dock. After two and a half days' journey, the *Hindenburg,* the pride of Germany's trans-Atlantic air fleet, had arrived at Lakehurst, New Jersey, on May 6, 1937. At 7 P.M., as the trial lines were being dropped for descent, the *Hindenburg* suddenly burst into flames. Within minutes, the 800-ft-long craft (about the length of three football fields) was reduced to a tangled mass of charred wreckage. Of the 97 on board, 35 died and many more were injured.

The *Hindenburg* was designed to be used with helium gas, which is chemically unreactive, to minimize the threat of an explosion. However, on the eve of World War II, the United States, which has most of the world's supply of helium, would not sell the gas to Germany. And so the *Hindenburg* was filled with hydrogen gas.

What caused the explosion? It is believed that leaking hydrogen, probably from a vent valve, ignited when the airship was struck by lightning. Quickly, the 7 million cubic feet of hydrogen inside the vessel combined with the surrounding air and burned. This chemical reaction violently released the energy stored in molecules of hydrogen. All airships nowadays, such as those you see at football games, are filled with helium, which is nearly as light as hydrogen.

Today hydrogen is used in a different type of air travel—to outer space. Together with oxygen, it makes an excellent rocket propellant. Even though a gallon of gasoline has three times the energy content of a gallon of liquid hydrogen, hydrogen is the fuel of choice for rockets because it weighs ten times less. In space flights, weight consideration is crucial. The large external tank of the space shuttle *Discovery* supplies about 380,000 gallons of liquid hydrogen and 140,000 gallons of liquid oxygen to the three engines that provide the thrust for liftoff.

The Hindenburg *was destroyed in a spectacular fire at Lakehurst, New Jersey, in 1937.*

Liquid hydrogen is the fuel used for space shuttles.

Energy Relationships in Chemical Reactions

Chapter 6

ESSENTIAL CONCEPTS

 Energy The many different forms of energy are, at least in principle, interconvertible.

 First Law of Thermodynamics The first law of thermodynamics, which is based on the law of conservation of energy, relates the internal energy change of a system to the heat change and the work done. It can also be expressed to show the relationship between the internal energy change and enthalpy change of a process.

 Thermochemistry Most chemical reactions involve the absorption or release of heat. At constant pressure, the heat change is equal to the enthalpy change. The heat change is measured by a calorimeter. Constant-pressure and constant-volume calorimeters are devices for measuring heat changes under the stated conditions.

 Standard Enthalpy of Reaction Standard enthalpy of reaction is the enthalpy change when the reaction is carried out at 1 atm pressure. It can be calculated from the standard enthalpies of formation of reactants and products. Hess's law enables us to measure the standard enthalpy of formation of a compound in an indirect way.

6.1 The Nature of Energy and Types of Energy

"Energy" is a much-used term that represents a rather abstract concept. For instance, when we feel tired, we might say we haven't any *energy;* and we read about the need to find alternatives to nonrenewable *energy* sources. Unlike matter, energy is known and recognized by its effects. It cannot be seen, touched, smelled, or weighed.

Energy is usually defined as *the capacity to do work.* In Chapter 5 we defined work as "force × distance," but we will soon see that there are other kinds of work. All forms of energy are capable of doing work (that is, of exerting a force over a distance), but not all of them are equally relevant to chemistry. The energy contained in tidal waves, for example, can be harnessed to perform useful work, but the relationship between tidal waves and chemistry is minimal. Chemists define **work** as *directed energy change resulting from a process.* Kinetic energy—the energy produced by a moving object—is one form of energy that is of particular interest to chemists. Others include radiant energy, thermal energy, chemical energy, and potential energy.

Radiant energy, or *solar energy, comes from the sun* and is Earth's primary energy source. Solar energy heats the atmosphere and Earth's surface, stimulates the growth of vegetation through the process known as photosynthesis, and influences global climate patterns.

Thermal energy is *the energy associated with the random motion of atoms and molecules.* In general, thermal energy can be calculated from temperature measurements. The more vigorous the motion of the atoms and molecules in a sample of matter, the hotter the sample is and the greater its thermal energy. However, we need to distinguish carefully between thermal energy and temperature. A cup of coffee at 70°C has a higher temperature than a bathtub filled with warm water at 40°C, but much more thermal energy is stored in the bathtub water because it has a much larger volume and greater mass than the coffee and therefore more water molecules and more molecular motion.

Chemical energy is *stored within the structural units of chemical substances;* its quantity is determined by the type and arrangement of constituent atoms. When substances participate in chemical reactions, chemical energy is released, stored, or converted to other forms of energy.

Potential energy is *energy available by virtue of an object's position.* For instance, because of its altitude, a rock at the top of a cliff has more potential energy and will make a bigger splash if it falls into the water below than a similar rock located partway down the cliff. Chemical energy can be considered a form of potential energy because it is associated with the relative positions and arrangements of atoms within a given substance.

All forms of energy can be converted (at least in principle) from one form to another. We feel warm when we stand in sunlight because radiant energy is converted to thermal energy on our skin. When we exercise, chemical energy stored in our bodies is used to produce kinetic energy. When a ball starts to roll downhill, its potential energy is converted to kinetic energy. You can undoubtedly think of many other examples. Although energy can assume many different forms that are interconvertible, scientists have concluded that energy can be neither destroyed nor created. When one form of energy disappears, some other form of energy (of equal magnitude) must appear, and vice versa. The principle is summarized by the *law of conservation of energy:* *the total quantity of energy in the universe is assumed constant.*

Kinetic energy was introduced in Section 5.6.

As the water falls over the dam, its potential energy is converted to kinetic energy. Use of this energy to generate electricity is called hydroelectric power.

6.2 Energy Changes in Chemical Reactions

Often the energy changes that take place during chemical reactions are of as much practical interest as the mass relationships we discussed in Chapter 3. For example, combustion reactions involving fuels such as natural gas and oil are carried out in daily life more for the thermal energy they release than for the particular quantities of the products, which are water and carbon dioxide.

Almost all chemical reactions absorb or produce (release) energy, generally in the form of heat. It is important to understand the distinction between thermal energy and heat. **Heat** is *the transfer of thermal energy between two bodies that are at different temperatures.* Thus we often speak of the "heat flow" from a hot object to a cold one. Although "heat" itself implies the transfer of energy, we customarily talk of "heat absorbed" or "heat released" when describing the energy changes that occur during a process. *The study of heat changes in chemical reactions* is called **thermochemistry.**

To analyze energy changes associated with chemical reactions we must first define the **system,** or *specific part of the universe that is of interest to us.* For chemists, systems usually include substances involved in chemical and physical changes. For example, in an acid-base neutralization experiment, the system may be a beaker containing 50 mL of HCl to which 50 mL of NaOH is added. *The rest of the universe outside the system* is called the **surroundings.**

There are three types of systems. An **open system** *can exchange mass and energy (usually in the form of heat) with its surroundings.* For example, an open system may consist of a quantity of water in an open container, as shown in Figure 6.1a. If we close the flask, as in Figure 6.1b, so that no water vapor can escape from or condense into the container, we create a **closed system,** which *allows the transfer of energy (heat) but not mass.* By placing the water in a totally insulated container, we construct an **isolated system,** which *does not allow the transfer of either mass or energy,* as shown in Figure 6.1c.

The combustion of acetylene (C_2H_2) gas in oxygen is one of many familiar chemical reactions that release considerable quantities of energy (Figure 6.2):

$$2C_2H_2(g) + 5O_2(g) \longrightarrow 4CO_2(g) + 2H_2O(l) + \text{energy}$$

In this case we label the reacting mixture (acetylene, oxygen, carbon dioxide, and water) the system and the rest of the universe the surroundings. Because energy cannot

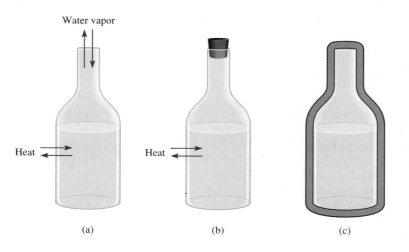

Water vapor

Heat   Heat

(a)   (b)   (c)

Figure 6.1 *Three systems represented by water in a flask: (a) an open system, which allows the exchange of both energy and mass with surroundings; (b) a closed system, which allows the exchange of energy but not mass; and (c) an isolated system, which allows neither energy nor mass to be exchanged (here the flask is enclosed by a vacuum jacket).*

Figure 6.2 *An oxyacetylene torch has a high flame temperature (3000°C) and is used to weld metals.*

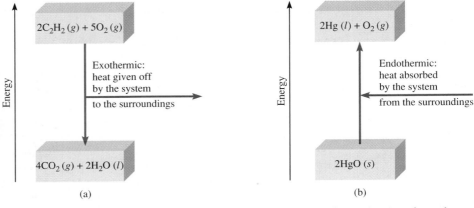

Figure 6.3 *(a) Energy-level diagram for an exothermic reaction: the combustion of acetylene gas. (b) Energy-level diagram for an endothermic reaction: the decomposition of mercury(II) oxide. The scales in (a) and (b) are not the same. Therefore, the heat released in the combustion of acetylene is not equal to the heat absorbed in the decomposition of mercury(II) oxide.*

On heating, HgO decomposes to give Hg and O_2.

be created or destroyed, any energy lost by the system must be gained by the surroundings. Thus the heat generated by the combustion process is transferred from the system to its surroundings. *Any process that gives off heat* (that is, *transfers thermal energy to the surroundings*) is called an **exothermic process** (*exo-* is a prefix meaning "out of"). Figure 6.3a shows the energy change for the combustion of acetylene gas.

Now consider another reaction, the decomposition of mercury(II) oxide (HgO) at high temperatures:

$$\text{energy} + 2\text{HgO}(s) \longrightarrow 2\text{Hg}(l) + \text{O}_2(g)$$

This is an example of an **endothermic process** (*endo-* is a prefix meaning "into"), *in which heat has to be supplied to the system* (that is, HgO) *by the surroundings* (Figure 6.3b).

From Figure 6.3 you can see that in exothermic reactions the total energy of the products is less than the total energy of the reactants. The difference in the energies is the heat supplied by the system to the surroundings. Just the opposite happens in endothermic reactions. Here, the difference in the energies of the products and reactants is equal to the heat supplied to the system by the surroundings.

6.3 Introduction to Thermodynamics

Thermochemistry is part of a broader subject called **thermodynamics,** which is *the scientific study of the interconversion of heat and other kinds of energy.* The laws of thermodynamics provide useful guidelines for understanding the energetics and directions of processes. In this section we will concentrate on the first law of thermodynamics, which is particularly relevant to the study of thermochemistry. We will continue our discussion of thermodynamics in Chapter 18.

In thermodynamics, we study changes in the **state of a system,** which is defined by *the values of all relevant macroscopic properties, for example, composition, energy, temperature, pressure, and volume.* Energy, pressure, volume, and temperature are said

Figure 6.4 *The gain in gravitational potential energy that occurs when a peson climbs from the base to the top of a mountain is independent of the path taken.*

to be *state functions*—*properties that are determined by the state of the system, regardless of how that condition was achieved.* In other words, when the state of a system changes, the magnitude of change in any state function depends only on the initial and final states of the system and not on how the change is accomplished.

The state of a given amount of a gas is specified by its volume, pressure, and temperature. Consider a gas at 2 atm, 300 K, and 1 L (the initial state). Suppose a process is carried out at constant temperature such that the gas pressure decreases to 1 atm. According to Boyle's law, its volume must increase to 2 L. The final state then corresponds to 1 atm, 300 K, and 2 L. The change in volume (ΔV) is

$$\begin{aligned} \Delta V &= V_f - V_i \\ &= 2\,\text{L} - 1\,\text{L} \\ &= 1\,\text{L} \end{aligned}$$

> The Greek letter *delta*, Δ, symbolizes change. We use Δ in this text always to mean final − initial.

where V_i and V_f denote the initial and final volume, respectively. No matter how we arrive at the final state (for example, the pressure of the gas can be increased first and then decreased to 1 atm), the change in volume is always 1 L. Thus the volume of a gas is a state function. In a similar manner we could show that pressure and temperature are also state functions.

Energy is another state function. Using potential energy as an example, we find that the net increase in gravitational potential energy when we go from the same starting point to the top of a mountain is always the same, regardless of how we get there (Figure 6.4).

> Recall that an object possesses potential energy by virtue of its position or chemical composition.

The First Law of Thermodynamics

The *first law of thermodynamics,* which is based on the law of conservation of energy, states that *energy can be converted from one form to another, but cannot be created or destroyed.* How do we know this is so? It would be impossible to prove the validity of the first law of thermodynamics if we had to determine the total energy content of the universe. Even determining the total energy content of 1 g of iron, say, would be extremely difficult. Fortunately, we can test the validity of the first law by measuring only the *change* in the internal energy of a system between its *initial state* and its *final state* in a process. The change in internal energy ΔE is given by

$$\Delta E = E_f - E_i$$

where E_i and E_f are the internal energies of the system in the initial and final states, respectively.

The internal energy of a system has two components: kinetic energy and potential energy. The kinetic energy component consists of various types of molecular motion and the movement of electrons within molecules. Potential energy is determined by the attractive interactions between electrons and nuclei and by repulsive interactions between electrons and between nuclei in individual molecules, as well as by interaction between molecules. It is impossible to measure all these contributions accurately, so we cannot calculate the total energy of a system with any certainty. Changes in energy, on the other hand, can be determined experimentally.

Consider the reaction between 1 mole of sulfur and 1 mole of oxygen gas to produce 1 mole of sulfur dioxide:

$$S(s) + O_2(g) \longrightarrow SO_2(g)$$

In this case our system is composed of the reactant molecules S and O_2 and the product molecules SO_2. We do not know the internal energy content of either the reactant molecules or the product molecules, but we can accurately measure the *change* in energy content, ΔE, given by

$\Delta E = E(\text{product}) - E(\text{reactants})$

$\quad = $ energy content of 1 mol $SO_2(g) - $ [energy content of 1 mol $S(s) + 1$ mol $O_2(g)$]

We find that this reaction gives off heat. Therefore, the energy of the product is less than that of the reactants, and ΔE is negative.

Interpreting the release of heat in this reaction to mean that some of the chemical energy contained in the reactants has been converted to thermal energy, we conclude that the transfer of energy from the reactants to the surroundings does not change the total energy of the universe. That is, the sum of the energy changes must be zero:

$$\Delta E_{\text{sys}} + \Delta E_{\text{surr}} = 0$$

or

$$\Delta E_{\text{sys}} = -\Delta E_{\text{surr}}$$

where the subscripts "sys" and "surr" denote system and surroundings, respectively. Thus, if one system undergoes an energy change ΔE_{sys}, the rest of the universe, or the surroundings, must undergo a change in energy that is equal in magnitude but opposite in sign ($-\Delta E_{\text{surr}}$); energy gained in one place must have been lost somewhere else. Furthermore, because energy can be changed from one form to another, the energy lost by one system can be gained by another system in a different form. For example, the energy lost by burning oil in a power plant may ultimately turn up in our homes as electrical energy, heat, light, and so on.

In chemistry, we are normally interested in the energy changes associated with the system (which may be a flask containing reactants and products), not with its surroundings. Therefore, a more useful form of the first law is

$$\Delta E = q + w \tag{6.1}$$

(We drop the subscript "sys" for simplicity.) Equation (6.1) says that the change in the internal energy ΔE of a system is the sum of the heat exchange q between the system and the surroundings and the work done w on (or by) the system. The sign conventions for q and w are as follows: q is positive for an endothermic process and negative for an exothermic process and w is positive for work done on the system by the surroundings and negative for work done by the system on the surroundings. Equation (6.1) may seem abstract, but it is actually quite logical. If a system loses heat to the surroundings or does work on the surroundings,

Sulfur burns in air to form SO_2.

For convenience, we sometimes omit the word "internal" when discussing the energy of a system.

Table 6.1	Sign Conventions for Work and Heat	
Process		**Sign**
Work done by the system on the surroundings		−
Work done on the system by the surroundings		+
Heat absorbed by the system from the surroundings (endothermic process)		+
Heat absorbed by the surroundings from the system (exothermic process)		−

we would expect its internal energy to decrease because they are energy-depleting processes. For this reason, both q and w are negative. Conversely, if heat is added to the system or if work is done on the system, then the internal energy of the system would increase. In this case, both q and w are positive. Table 6.1 summarizes the sign conventions for q and w.

Work and Heat

We have seen that work can be defined as force F multiplied by distance d:

$$w = Fd \qquad (6.2)$$

In thermodynamics, work has a broader meaning that includes mechanical work (for example, a crane lifting a steel beam), electrical work (a battery supplying electrons to light the bulb of a flashlight), and so on. In this section we will concentrate on mechanical work; in Chapter 19 we will discuss the nature of electrical work.

A useful example of mechanical work is the expansion of a gas (Figure 6.5). Suppose that a gas is in a cylinder fitted with a weightless, frictionless movable piston, at a certain temperature, pressure, and volume. As it expands, the gas pushes the piston upward against a constant opposing external atmospheric pressure P. The work done by the gas on the surroundings is

$$w = -P\Delta V \qquad (6.3)$$

where ΔV, the change in volume, is given by $V_f - V_i$. The minus sign in Equation (6.3) takes care of the convention for w. For gas expansion, $\Delta V > 0$, so $-P\Delta V$ is a negative quantity. For gas compression (work done on the system), $\Delta V < 0$, and $-P\Delta V$ is a positive quantity.

Figure 6.5 *The expansion of a gas against a constant external pressure (such as atmospheric pressure). The gas is in a cylinder fitted with a weightless movable piston. The work done is given by* $-P\Delta V$.

Equation (6.3) derived from the fact that pressure × volume can be expressed as (force/area) × volume; that is,

$$P \times V = \underbrace{\frac{F}{d^2}}_{\text{pressure}} \times \underbrace{d^3}_{\text{volume}} = Fd = w$$

where F is the opposing force and d has the dimension of length, d^2 has the dimensions of area, and d^3 has the dimensions of volume. Thus the product of pressure and volume is equal to force times distance, or work. You can see that for a given increase in volume (that is, for a certain value of ΔV), the work done depends on the magnitude of the external, opposing pressure P. If P is zero (that is, if the gas is expanding against a vacuum), the work done must also be zero. If P is some positive, nonzero value, then the work done is given by $-P\Delta V$.

According to Equation (6.3), the units for work done by or on a gas are liters atmospheres. To express the work done in the more familiar unit of joules, we use the conversion factor (see Appendix 1).

$$1 \text{ L} \cdot \text{atm} = 101.3 \text{ J}$$

Example 6.1   A certain gas expands in volume from 2.0 L to 6.0 L at constant temperature. Calculate the work done by the gas if it expands (a) against a vacuum and (b) against a constant pressure of 1.2 atm.

Reasoning and Solution   Recall that the work done in gas expansion is equal to the product of the external, opposing pressure and the change in volume.

(a) Because the external pressure is zero, the work w done is

$$w = -P\Delta V$$
$$= -0(6.0 - 2.0) \text{ L}$$
$$= 0$$

(b) The external, opposing pressure is 1.2 atm, so

$$w = -P\Delta V$$
$$= -(1.2 \text{ atm})(6.0 - 2.0) \text{ L}$$
$$= -4.8 \text{ atm} \cdot \text{L} = -4.8 \text{ L} \cdot \text{atm}$$

To convert the answer to joules, we write

$$w = -4.8 \text{ L} \cdot \text{atm} \times \frac{101.3 \text{ J}}{1 \text{ L} \cdot \text{atm}}$$
$$= -4.9 \times 10^2 \text{ J}$$

Similar problem: 6.76.

Practice Exercise   A gas expands from 264 mL to 971 mL at constant temperature. Calculate the work done (in joules) by the gas if it expands (a) against a vacuum and (b) against a constant pressure of 4.00 atm.

Because the temperature is kept constant, you can use Boyle's law to show that the final pressure is the same in (a) and (b).

Example 6.1 shows that work is not a state function. Although the initial and final states are the same in (a) and (b), the amount of work done is different because the external, opposing pressures are different. We *cannot* write $\Delta w = w_f - w_i$ for a change. Work done depends not only on the initial state and final state, but also on how the process is carried out.

The other component of internal energy is q, heat. Like work, heat is not a state function. Suppose that the change can be brought about in a system in two different ways. In one case the work done is zero, so that we have

$$\Delta E = q_1 + w_1$$
$$= q_1$$

In the second case work is done and heat is transferred, so that

$$\Delta E = q_2 + w_2$$

Because ΔE is the same for both cases (note that the initial and final states are the same in both cases), it follows that $q_1 \neq q_2$. This simple example shows that the heat associated with a given process, like work, depends on how the process is carried out; that is, we *cannot* write $\Delta q = q_f - q_i$. It is important to note that although neither heat nor work is a state function, their sum $(q + w)$ is equal to ΔE and, as we saw earlier, E is a state function.

In summary, heat and work are not state functions because they are not properties of a system. They manifest themselves only during a process (during a change). Thus their values depend on the path of the process and vary accordingly.

We use lowercase letters (such as w and q) to represent thermodynamic quantities that are not state functions.

Example 6.2   The work done when a gas is compressed in a cylinder like that shown in Figure 6.5 is 462 J. During this process, there is a heat transfer of 128 J from the gas to the surroundings. Calculate the energy change for this process.

Reasoning and Solution   This is an application of the first law of thermodynamics [Equation (6.1)]. Compression is work done on the gas, so the sign for w is positive; $w = 462$ J. From the direction of heat transfer (system to surroundings), we know that q is negative:

$$\Delta E = q + w$$
$$= -128\,\text{J} + 462\,\text{J}$$
$$= 334\,\text{J}$$

As a result of the compression and heat transfer, the energy of the gas increases by 334 J.

Practice Exercise   A gas expands and does *P-V* work on the surroundings equal to 279 J. At the same time, it absorbs 216 J of heat from the surroundings. What is the change in energy of the system?

Similar problems: 6.15, 6.16.

6.4 Enthalpy of Chemical Reactions

Our next step is to see how the first law of thermodynamics can be applied to processes carried out under different conditions. Specifically, we will consider two situations in which the volume of the system is kept constant and in which the pressure applied on the system is kept constant because these are the commonly encountered cases.

If a chemical reaction is run at constant volume, then $\Delta V = 0$ and no P-V work will result from this change. From Equation (6.1) it follows that

$$\Delta E = q - P\Delta V$$
$$= q_v \tag{6.4}$$

We add the subscript "v" to remind us that this is a constant-volume process. This equality may seem strange at first, for we showed earlier that q is not a state function. The process is carried out under constant-volume conditions, however, so that the heat change can have only a specific value, which is equal to ΔE.

Enthalpy

Constant-volume conditions are often inconvenient and sometimes impossible to achieve. Most reactions occur under conditions of constant pressure (usually atmospheric pressure). If such a reaction results in a net increase in the number of moles of a gas, then the system does work on the surroundings (expansion). This follows from the fact that for the gas formed to enter the atmosphere, it must push the surrounding gas back. Conversely, if more gas molecules are consumed than are produced, work is done on the system by the surroundings (compression). Finally, no work is done if there is no net change in the number of moles of gases from reactants to products.

In general, for a constant-pressure process we write

$$\Delta E = q + w$$
$$= q_p - P\Delta V$$
or
$$q_p = \Delta E + P\Delta V \tag{6.5}$$

where the subscript "p" denotes constant-pressure condition.

We now introduce a new thermodynamic function of a system called *enthalpy (H),* which is defined by the equation

$$H = E + PV \tag{6.6}$$

where E is the internal energy of the system and P and V are the pressure and volume of the system, respectively. Because E and PV have energy units, enthalpy also has energy units. Furthermore, E, P, and V are all state functions, that is, the changes in $(E + PV)$ depend only on the initial and final states. It follows, therefore, that the change in H, or ΔH, also depends only on the initial and final states. Thus H is a state function.

For any process, the change in enthalpy according to Equation (6.6) is given by

$$\Delta H = \Delta E + \Delta(PV) \tag{6.7}$$

If the pressure is held constant, then

$$\Delta H = \Delta E + P\Delta V \tag{6.8}$$

Comparing Equation (6.8) with Equation (6.5), we see that for a constant-pressure process, $q_p = \Delta H$. Again, although q is not a state function, the heat change at constant pressure is equal to ΔH because the "path" is defined and therefore it can have only a specific value.

We now have two quantities—ΔE and ΔH—that can be associated with a reaction. If the reaction occurs under constant-volume conditions, then the heat change, q_v, is equal to ΔE. On the other hand, when the reaction is carried out at constant pressure, the heat change, q_p, is equal to ΔH.

In Section 6.5 we will discuss ways to measure heat changes at constant volume and constant pressure.

Enthalpy of Reactions

Because most reactions are constant-pressure processes, we can equate the heat change in these cases to the change in enthalpy. For any reaction of the type

$$\text{reactants} \longrightarrow \text{products}$$

we define the change in enthalpy, called the **enthalpy of reaction, ΔH,** as *the difference between the enthalpies of the products and the enthalpies of the reactants:*

$$\Delta H = H(\text{products}) - H(\text{reactants}) \tag{6.9}$$

The enthalpy of reaction can be positive or negative, depending on the process. For an endothermic process (heat absorbed by the system from the surroundings), ΔH is positive (that is, $\Delta H > 0$). For an exothermic process (heat released by the system to the surroundings), ΔH is negative (that is, $\Delta H < 0$).

An analogy for enthalpy change is a change in the balance in your bank account. Suppose your initial balance is $100. After a transaction (deposit or withdrawal), the change in your bank balance, ΔX, is given by

$$\Delta X = X_{\text{final}} - X_{\text{initial}}$$

where X represents the bank balance. If you deposit $80 into your account, then $\Delta X = \$180 - \$100 = \$80$. This corresponds to an endothermic reaction. (The balance increases and so does the enthalpy of the system.) On the other hand, a withdrawal of $60 means $\Delta X = \$40 - \$100 = -\$60$. The negative sign of ΔX means your balance has decreased. Similarly, a negative value of ΔH reflects a decrease in enthalpy of the system as a result of an exothermic process. The difference between this analogy and Equation (6.9) is that while you always know your exact bank balance, there is no way to know the enthalpies of individual products and reactants. In practice, we can only measure the *difference* in their values.

Now let us apply the idea of enthalpy changes to two common processes, the first involving a physical change, the second a chemical change.

This analogy assumes that you will not overdraw your bank account. The enthalpy of a substance *cannot* be negative.

Thermochemical Equations

At 0°C and a pressure of 1 atm, ice melts to form liquid water. Measurements show that for every mole of ice converted to liquid water under these conditions, 6.01 kilojoules (kJ) of energy are absorbed by the system (ice). Because ΔH is a positive value, this is an endothermic process, as expected for an energy-absorbing change of melting ice (Figure 6.6a). The equation for this physical change is

$$H_2O(s) \longrightarrow H_2O(l) \qquad \Delta H = 6.01 \text{ kJ}$$

As another example, consider the combustion of methane (CH_4), the principal component of natural gas:

$$CH_4(g) + 2O_2(g) \longrightarrow CO_2(g) + 2H_2O(l) \qquad \Delta H = -890.4 \text{ kJ}$$

From experience we know that burning natural gas releases heat to its surroundings, so it is an exothermic process and ΔH must have a negative value. This enthalpy change is represented in Figure 6.6b.

The equations for the melting of ice and the combustion of methane are examples of **thermochemical equations,** which *show the enthalpy changes as well as the mass relationships.* These guidelines are helpful in writing and interpreting thermochemical equations:

Heat flow

Methane gas burning from a Bunsen burner.

Figure 6.6 *(a) Melting 1 mole of ice at 0°C (an endothermic process) results in an enthalpy increase in the system of 6.01 kJ. (b) Burning 1 mole of methane in oxygen gas (an exothermic process) results in an enthalpy decrease in the system of 890.4 kJ.*

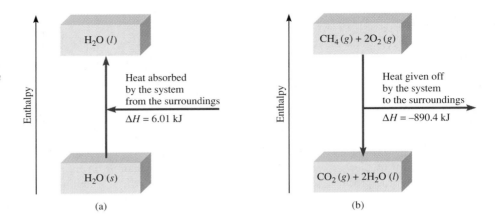

- The stoichiometric coefficients always refer to the number of moles of a substance. Thus, the equation for the melting of ice may be "read" as: When 1 mole of liquid water is formed from 1 mole of ice at 0°C, the enthalpy change is 6.01 kJ. For the combustion of methane, we interpret the equation this way: When 1 mole of gaseous methane reacts with 2 moles of gaseous oxygen to form 1 mole of gaseous carbon dioxide and 2 moles of liquid water, the enthalpy change is −890.4 kJ.

- When we reverse an equation, we change the roles of reactants and products. Consequently, the magnitude of ΔH for the equation remains the same, but its sign changes. For example, if a reaction consumes thermal energy from its surroundings (that is, if it is endothermic), then the reverse reaction must release thermal energy back to its surroundings (that is, it must be exothermic) and the enthalpy change expression must also change its sign. Thus, reversing the melting of ice and the combustion of methane, the thermochemical equations are

$$H_2O(l) \longrightarrow H_2O(s) \qquad\qquad \Delta H = -6.01 \text{ kJ}$$

$$CO_2(g) + 2H_2O(l) \longrightarrow CH_4(g) + 2O_2(g) \qquad\qquad \Delta H = 890.4 \text{ kJ}$$

and what was an endothermic process becomes exothermic, and vice versa.

- If we multiply both sides of a thermochemical equation by a factor n, then ΔH must also change by the same factor. Thus, for the melting of ice, if $n = 2$, then

$$2H_2O(s) \longrightarrow 2H_2O(l) \qquad\qquad \Delta H = 2(6.01 \text{ kJ}) = 12.0 \text{ kJ}$$

- When writing thermochemical equations, we must always specify the physical states of all reactants and products, because they help determine the actual enthalpy changes. For example, in the equation for the combustion of methane, if we show water vapor rather than liquid water as a product,

$$CH_4(g) + 2O_2(g) \longrightarrow CO_2(g) + 2H_2O(g) \qquad\qquad \Delta H = -802.4 \text{ kJ}$$

the enthalpy change is −802.4 kJ rather than −890.4 kJ because 88.0 kJ are needed to convert 2 moles of liquid water to water vapor; that is,

$$2H_2O(l) \longrightarrow 2H_2O(g) \qquad\qquad \Delta H = 88.0 \text{ kJ}$$

Example 6.3 Given the thermochemical equation

$$SO_2(g) + \tfrac{1}{2}O_2(g) \longrightarrow SO_3(g) \qquad \Delta H = -99.1 \text{ kJ}$$

calculate the heat evolved when 74.6 g of SO_2 (molar mass = 64.07 g/mol) is converted to SO_3.

Reasoning and Solution The thermochemical equation shows that for every mole of SO_2 burned, 99.1 kJ of heat are given off (note the negative sign for ΔH). We need to convert 74.6 g of SO_2 to number of moles and calculate the heat produced as follows

$$74.6 \text{ g } SO_2 \times \frac{1 \text{ mol } SO_2}{64.07 \text{ g } SO_2} \times \frac{-99.1 \text{ kJ}}{1 \text{ mol } SO_2} = -115 \text{ kJ}$$

Similar problems: 6.23, 6.24.

Practice Exercise Calculate the heat evolved when 266 g of white phosphorus (P_4) burn in air according to the equation

$$P_4(s) + 5O_2(g) \longrightarrow P_4O_{10}(s) \qquad \Delta H = -3013 \text{ kJ}$$

A Comparison of ΔH and ΔE

Finally it is instructive to examine the relationship between ΔH and ΔE for a process. To do so let us consider a specific reaction, that between sodium metal and water

$$2Na(s) + 2H_2O(l) \longrightarrow 2NaOH(aq) + H_2(g) \qquad \Delta H = -367.5 \text{ kJ}$$

This thermochemical equation says that when 2 moles of sodium react with an excess of water, 367.5 kJ of heat are given off. Note that one of the products is hydrogen gas, which must push back air to enter the atmosphere. Consequently, some of the energy produced by the reaction is used to do work of pushing back a volume of air (ΔV) against atmospheric pressure (P) (Figure 6.7). To calculate the change in internal energy, we rearrange Equation (6.8) as:

$$\Delta E = \Delta H - P\Delta V$$

If we assume the temperature to be 25°C and ignore the small change in the volume of the solution, we can show that the volume of 1 mole of H_2 gas at 1.0 atm is 24.5 L, so that $-P\Delta V = -24.5$ L · atm or -2.5 kJ. Finally,

$$\Delta E = -367.5 \text{ kJ} - 2.5 \text{ kJ}$$
$$= -370.0 \text{ kJ}$$

Sodium reacting with water.

Recall 1 L · atm = 101.3 J.

(a)   (b)

Figure 6.7 *(a) A beaker of water inside a cylinder fitted with a movable piston. The pressure inside is equal to the atmospheric pressure. (b) After the sodium metal has reacted with water, hydrogen gas pushes the piston upward (doing work on the surroundings) until the pressure inside is again equal to that outside.*

For reactions that do not result in a change in the number of moles of gases from reactants to products, $\Delta E = \Delta H$.

This calculation shows that ΔE and ΔH are approximately the same. The reason ΔH is smaller than ΔE is that some of the internal energy released is used to do gas expansion work, so less heat is evolved. For reactions that do not involve gases, ΔV is usually very small and so ΔE is practically the same as ΔH.

Another way to calculate the internal energy change of a gaseous reaction is to assume ideal gas behavior and constant temperature. In this case,

$$\Delta E = \Delta H - \Delta(PV)$$
$$= \Delta H - \Delta(nRT)$$
$$= \Delta H - RT\Delta n \qquad (6.10)$$

where Δn is defined as

Δn = number of moles of product gases − number of moles of reactant gases

Carbon monoxide burns in air to form carbon dioxide.

Example 6.4   Calculate the change in internal energy when 2 moles of CO are converted to 2 moles of CO_2 at 1 atm and 25°C:

$$2CO(g) + O_2(g) \longrightarrow 2CO_2(g) \qquad \Delta H = -566.0 \text{ kJ}$$

Reasoning and Solution   We are given the enthalpy change and asked to calculate ΔE, which requires Equation (6.10). First we calculate Δn, which is the change in the number of moles of gases

$$\Delta n = 2 \text{ mol} - 3 \text{ mol}$$
$$= -1 \text{ mol}$$

From Equation (6.10) we write

$$\Delta E = \Delta H - RT\Delta n$$
$$= -566.0 \text{ kJ} - (8.314 \text{ J/K} \cdot \text{mol})(298 \text{ K})(-1 \text{ mol})$$
$$= -563.5 \text{ kJ}$$

Comment   Note that we used 8.314 J/K · mol for R (see Appendix 1) because the units of ΔE and ΔH are in joules. The fact that ΔH is greater than ΔE (in magnitude) means that more heat is generated than the decrease in the internal energy of the reacting system. The reason is that there is a net decrease in the gas volume from reactants to product, which means that work is done by the surroundings on the system. This work of compression contributes to the extra heat produced.

Practice Exercise   What is ΔE for the formation of 1 mole of CO carried out at 1 atm and 25°C?

$$C(\text{graphite}) + \tfrac{1}{2}O_2(g) \longrightarrow CO(g) \qquad \Delta H = -110.5 \text{ kJ}$$

Similar problems: 6.25, 6.26.

6.5 Calorimetry

In the laboratory, heat changes in physical and chemical processes are measured with a *calorimeter*, a closed container designed specifically for this purpose. Our discussion of **calorimetry,** *the measurement of heat changes,* will depend on an understanding of specific heat and heat capacity, so let us consider them first.

Specific Heat and Heat Capacity

The *specific heat (s)* of a substance is *the amount of heat required to raise the temperature of one gram of the substance by one degree Celsius.* The *heat capacity (C)* of a substance is *the amount of heat required to raise the temperature of a given quantity of the substance by one degree Celsius.* Specific heat is an intensive property, whereas heat capacity is an extensive property. The relationship between the heat capacity and specific heat of a substance is

$$C = ms \qquad (6.11)$$

where m is the mass of the substance in grams. For example, the specific heat of water is 4.184 J/g · °C, and the heat capacity of 60.0 g of water is

$$(60.0 \text{ g})(4.184 \text{ J/g} \cdot °\text{C}) = 251 \text{ J/°C}$$

Note that specific heat has the units J/g · °C and heat capacity has the units J/°C. Table 6.2 shows the specific heat of some common substances.

If we know the specific heat and the amount of a substance, then the change in the sample's temperature (Δt) will tell us the amount of heat (q) that has been absorbed or released in a particular process. The equation for calculating the heat change is given by

$$q = ms\Delta t \qquad (6.12)$$

or

$$q = C\Delta t \qquad (6.13)$$

where m is the mass of the sample in grams and Δt is the temperature change:

$$\Delta t = t_{\text{final}} - t_{\text{initial}}$$

The sign convention for q is the same as that for enthalpy change; q is positive for endothermic processes and negative for exothermic processes.

Table 6.2 The Specific Heats of Some Common Substances	
Substance	**Specific Heat (J/g · °C)**
Al	0.900
Au	0.129
C (graphite)	0.720
C (diamond)	0.502
Cu	0.385
Fe	0.444
Hg	0.139
H_2O	4.184
C_2H_5OH (ethanol)	2.46

Example 6.5 A 466-g sample of water is heated from 8.50°C to 74.60°C. Calculate the amount of heat absorbed by the water.

Reasoning and Solution Using the mass and specific heat of water and the temperature change in Equation (6.12), we calculate the heat change as follows:

$$q = ms\Delta t$$
$$= (466 \text{ g})(4.184 \text{ J/g} \cdot °\text{C})(74.60°\text{C} - 8.50°\text{C})$$
$$= 1.29 \times 10^5 \text{ J}$$
$$= 129 \text{ kJ}$$

Practice Exercise An iron bar of mass 869 g cools from 94°C to 5°C. Calculate the heat released (in kilojoules) by the metal.

Similar problem: 6.32.

Constant-Volume Calorimetry

Heat of combustion is usually measured by placing a known mass of a compound in a steel container called a *constant-volume bomb calorimeter,* which is filled with oxygen at about 30 atm of pressure. The closed bomb is immersed in a known amount of water, as shown in Figure 6.8. The sample is ignited electrically, and

"Constant volume" refers to the volume of the container, which does not change during the reaction. Note that the container remains intact after the measurement. The term "bomb calorimeter" connotes the explosive nature of the reaction (on a small scale) in the presence of excess oxygen gas.

Figure 6.8 *A constant-volume bomb calorimeter. The calorimeter is filled with oxygen gas before it is placed in the bucket. The sample is ignited electrically, and the heat produced by the reaction can be accurately determined by measuring the temperature increase in the known amount of surrounding water.*

the heat produced by the combustion reaction can be calculated accurately by recording the rise in temperature of the water. The heat given off by the sample is absorbed by the water and the bomb. The special design of the calorimeter allows us to assume that no heat (or mass) is lost to the surroundings during the time it takes to make measurements. Therefore we can call the bomb and the water in which it is submerged an isolated system. Because no heat enters or leaves the system throughout the process, the heat change of the system (q_{system}) must be zero and we can write

$$q_{system} = q_{cal} + q_{rxn}$$
$$= 0 \qquad\qquad (6.14)$$

where q_{cal} and q_{rxn} are the heat changes for the calorimeter and the reaction, respectively. Thus

$$q_{rxn} = -q_{cal} \qquad\qquad (6.15)$$

Note that C_{cal} comprises both the bomb and the surrounding water.

To calculate q_{cal}, we need to know the heat capacity of the calorimeter (C_{cal}) and the temperature rise, that is

$$q_{cal} = C_{cal}\Delta t$$

The quantity C_{cal} is calibrated by burning a substance with an accurately known heat of combustion. For example, it is known that the combustion of 1 g of benzoic acid (C_6H_5COOH) releases 26.42 kJ of heat. If the temperature rise is 4.673°C, then the heat capacity of the calorimeter is given by

$$C_{cal} = \frac{q_{cal}}{\Delta t}$$

$$= \frac{26.42 \text{ kJ}}{4.673°C} = 5.654 \text{ kJ/°C}$$

Once C_{cal} has been determined, the calorimeter can be used to measure the heat of combustion of other substances.

Note that because reactions in a bomb calorimeter occur under constant-volume rather than constant-pressure conditions, the heat changes *do not* correspond to the enthalpy change ΔH (see Section 6.4). It is possible to correct the measured heat changes so that they correspond to ΔH values, but the corrections usually are quite small so we will not concern ourselves with the details here. Finally, it is interesting to note that the energy contents of food and fuel (usually expressed in calories where 1 cal = 4.184 J) are measured with constant-volume calorimeters.

Example 6.6   A quantity of 1.435 g of naphthalene ($C_{10}H_8$), a pungent-smelling substance used in moth repellents, was burned in a constant-volume bomb calorimeter. Consequently, the temperature of the water rose from 20.17°C to 25.84°C. If the heat capacity of the calorimeter is 10.17 kJ/°C, calculate the heat of combustion of naphthalene on a molar basis; that is, find the molar heat of combustion in kJ/mol.

Reasoning and Solution   First we calculate the heat absorbed by the calorimeter.

$$q_{cal} = C_{cal}\Delta t$$

$$= (10.17 \text{ kJ/°C})(25.84°C - 20.17°C)$$

$$= 57.66 \text{ kJ}$$

Because the calorimeter is an isolated system, this must also be the heat released by the reaction [see Equation (6.15)]. Therefore

$$q_{rxn} = -57.66 \text{ kJ}$$

The molar mass of naphthalene is 128.2 g, so the heat of combustion of 1 mole of naphthalene or its molar heat of combustion is

$$\text{molar heat of combustion} = \frac{-57.66 \text{ kJ}}{1.435 \text{ g } C_{10}H_8} \times \frac{128.2 \text{ g } C_{10}H_8}{1 \text{ mol } C_{10}H_8}$$

$$= -5.151 \times 10^3 \text{ kJ/mol}$$

Practice Exercise   A quantity of 1.922 g of methanol (CH_3OH) was burned in a constant-volume bomb calorimeter. Consequently, the temperature of the water rose by 4.20°C. If the heat capacity of the calorimeter was 10.4 kJ/°C, calculate the molar heat of combustion of methanol.

$C_{10}H_8$

Similar problem: 6.35.

Constant-Pressure Calorimetry

A simpler device than the constant-volume calorimeter is the constant-pressure calorimeter, which is used to determine the heat changes for noncombustion reactions. A crude constant-pressure calorimeter can be constructed from two

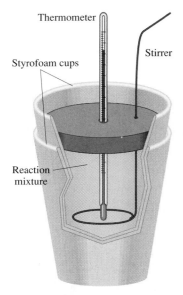

Thermometer

Stirrer

Styrofoam cups

Reaction
mixture

Figure 6.9 *A constant-pressure
calorimeter made of two Styrofoam coffee cups. The outer cup
helps to insulate the reacting
mixture from the surroundings.
Two solutions of known volume
containing the reactants at the
same temperature are carefully
mixed in the calorimeter. The
heat produced or absorbed by
the reaction can be determined
by measuring the temperature
change.*

Table 6.3 | Heats of Some Typical Reactions Measured at Constant Pressure

Type of Reaction	Example	ΔH (kJ)
Heat of neutralization	$HCl(aq) + NaOH(aq) \longrightarrow NaCl(aq) + H_2O(l)$	-56.2
Heat of ionization	$H_2O(l) \longrightarrow H^+(aq) + OH^-(aq)$	56.2
Heat of fusion	$H_2O(s) \longrightarrow H_2O(l)$	6.01
Heat of vaporization	$H_2O(l) \longrightarrow H_2O(g)$	$44.0*$
Heat of reaction	$MgCl_2(s) + 2Na(l) \longrightarrow 2NaCl(s) + Mg(s)$	-180.2

*Measured at 25°C. At 100°C, the value is 40.79 kJ.

Styrofoam coffee cups, as shown in Figure 6.9. This device measures the heat
effects of a variety of reactions, such as acid-base neutralization, as well as the
heat of solution and heat of dilution. Because the pressure is constant, the heat
change for the process (q_{rxn}) is equal to the enthalpy change (ΔH). As in the case
of a constant-volume calorimeter, we treat the calorimeter as an isolated system.
Furthermore, we neglect the small heat capacity of the coffee cups in our calculations. Table 6.3 lists some reactions that have been studied with the constant-
pressure calorimeter.

Example 6.7   A quantity of 1.00×10^2 mL of 0.500 M HCl is mixed with 1.00×10^2 mL of 0.500 M NaOH in a constant-pressure calorimeter. The initial temperature
of the HCl and NaOH solutions is the same, 22.50°C, and the final temperature
of the mixed solution is 25.86°C. Calculate the heat change for the neutralization
reaction

$$NaOH(aq) + HCl(aq) \longrightarrow NaCl(aq) + H_2O(l)$$

Assume that the densities and specific heats of the solutions are the same as those for
water (1.00 g/mL and 4.184 J/g · °C, respectively).

Reasoning and Solution   Assuming that no heat is lost to the surroundings and ignoring the heat capacity of the calorimeter, we write

$$q_{system} = q_{soln} + q_{rxn}$$
$$= 0$$

or

$$q_{rxn} = -q_{soln}$$

where q_{soln} is the heat absorbed by the combined solution. Because the density of the
solution is 1.00 g/mL, the mass of a 100-mL solution is 100 g. Thus

$$q_{soln} = (1.00 \times 10^2 \text{ g} + 1.00 \times 10^2 \text{ g})(4.184 \text{ J/g} \cdot °C)(25.86°C - 22.50°C)$$
$$= 2.81 \times 10^3 \text{ J}$$
$$= 2.81 \text{ kJ}$$

and $q_{rxn} = -2.81$ kJ

From the molarities given, we know there is 0.0500 mole of HCl in 1.00×10^2 mL of the HCl solution and 0.0500 mole of NaOH in 1.00×10^2 mL of the NaOH solution. Therefore the heat of neutralization when 1.00 mole of HCl reacts with 1.00 mole of NaOH is

$$\text{heat of neutralization} = \frac{-2.81 \text{ kJ}}{0.0500 \text{ mol}} = -56.2 \text{ kJ/mol}$$

Comment Because the reaction takes place at constant pressure, the heat given off is equal to the enthalpy change.

Practice Exercise A quantity of 4.00×10^2 mL of 0.600 M HNO_3 is mixed with 4.00×10^2 mL of 0.300 M $Ba(OH)_2$ in a constant-pressure calorimeter of negligible heat capacity. The initial temperature of both solutions is the same at 18.46°C. What is the final temperature of the solution? (Use the result in Example 6.7 for your calculation.)

Similar problem: 6.36.

6.6 Standard Enthalpy of Formation and Reaction

So far we have learned that we can determine the enthalpy change that accompanies a reaction by measuring the heat absorbed or released (at constant pressure). From Equation (6.9) we see that ΔH can also be calculated if we know the actual enthalpies of all reactants and products. However, as mentioned earlier, there is no way to measure the *absolute* value of the enthalpy of a substance. Only values *relative* to an arbitrary reference can be determined. This problem is similar to the one geographers face in expressing the elevations of specific mountains or valleys. Rather than trying to devise some type of "absolute" elevation scale (perhaps based on distance from the center of Earth?), by common agreement all geographic heights and depths are expressed relative to sea level, an arbitrary reference with a defined elevation of "zero" meters or feet. Similarly, chemists have agreed on an arbitrary reference point for enthalpy.

The "sea level" reference point for all enthalpy expressions is called the **standard enthalpy of formation (ΔH_f°),** which is defined as *the heat change that results when 1 mole of a compound is formed from its elements at a pressure of 1 atm*. Elements are said to be in the **standard state** at *1 atm,* hence the term "standard enthalpy." The superscript "o" represents standard-state conditions (1 atm), and the subscript "f" stands for formation. Although the standard state does not specify a temperature, we will always use ΔH_f° values measured at 25°C.

Table 6.4 lists the standard enthalpies of formation for a number of elements and compounds. (For a more complete list of ΔH_f° values, see Appendix 2.) By convention, *the standard enthalpy of formation of any element in its most stable form is zero.* Take the element oxygen as an example. Molecular oxygen (O_2) is more stable than the other allotropic form of oxygen, ozone (O_3), at 1 atm and 25°C. Thus we can write $\Delta H_f^\circ(O_2) = 0$, but $\Delta H_f^\circ(O_3) \neq 0$. Similarly, graphite is a more stable allotropic form of carbon than diamond at 1 atm and 25°C, so $\Delta H_f^\circ(C, \text{graphite}) = 0$ and $\Delta H_f^\circ(C, \text{diamond}) \neq 0$.

The importance of the standard enthalpies of formation is that once we know their values, we can readily calculate the **standard enthalpy of reaction, ΔH_{rxn}°,** defined

Graphite.

Diamond.

Table 6.4	Standard Enthalpies of Formation of Some Inorganic Substances at 25°C		
Substance	ΔH°_{f} **(kJ/mol)**	**Substance**	ΔH°_{f} **(kJ/mol)**
Ag(s)	0	$H_2O_2(l)$	−187.6
AgCl(s)	−127.04	Hg(l)	0
Al(s)	0	$I_2(s)$	0
$Al_2O_3(s)$	−1669.8	HI(g)	25.94
$Br_2(l)$	0	Mg(s)	0
HBr(g)	−36.2	MgO(s)	−601.8
C(graphite)	0	$MgCO_3(s)$	−1112.9
C(diamond)	1.90	$N_2(g)$	0
CO(g)	−110.5	$NH_3(g)$	−46.3
$CO_2(g)$	−393.5	NO(g)	90.4
Ca(s)	0	$NO_2(g)$	33.85
CaO(s)	−635.6	$N_2O_4(g)$	9.66
$CaCO_3(s)$	−1206.9	$N_2O(g)$	81.56
$Cl_2(g)$	0	O(g)	249.4
HCl(g)	−92.3	$O_2(g)$	0
Cu(s)	0	$O_3(g)$	142.2
CuO(s)	−155.2	S(rhombic)	0
$F_2(g)$	0	S(monoclinic)	0.30
HF(g)	−268.61	$SO_2(g)$	−296.1
H(g)	218.2	$SO_3(g)$	−395.2
$H_2(g)$	0	$H_2S(g)$	−20.15
$H_2O(g)$	−241.8	ZnO(s)	−347.98
$H_2O(l)$	−285.8		

as *the enthalpy of a reaction carried out at 1 atm.* For example, consider the hypothetical reaction

$$a\text{A} + b\text{B} \longrightarrow c\text{C} + d\text{D}$$

where a, b, c, and d are stoichiometric coefficients. For this reaction ΔH°_{rxn} is given by

$$\Delta H^{\circ}_{rxn} = [c\Delta H^{\circ}_{f}(\text{C}) + d\Delta H^{\circ}_{f}(\text{D})] - [a\Delta H^{\circ}_{f}(\text{A}) + b\Delta H^{\circ}_{f}(\text{B})] \qquad (6.16)$$

where a, b, c, and d are in moles. We can generalize Equation (6.16) as:

$$\Delta H^{\circ}_{rxn} = \Sigma n\Delta H^{\circ}_{f}(\text{products}) - \Sigma m\Delta H^{\circ}_{f}(\text{reactants}) \qquad (6.17)$$

where m and n denote the stoichiometric coefficients (in moles) for the reactants and products, and Σ (sigma) means "the sum of."

In order to use Equation (6.16) to calculate ΔH°_{rxn} we must know the ΔH°_{f} values of the compounds that take part in the reaction. To determine these values we can apply the direct method or the indirect method.

The Direct Method

This method of measuring ΔH°_{f} works for compounds that can be readily synthesized from their elements. Suppose we want to know the enthalpy of formation of carbon dioxide. We must measure the enthalpy of the reaction when carbon (graphite) and molecular oxygen in their standard states are converted to carbon dioxide in its standard state:

$$C(graphite) + O_2(g) \longrightarrow CO_2(g) \qquad \Delta H^\circ_{rxn} = -393.5 \text{ kJ}$$

As we know from experience, this combustion easily goes to completion. From Equation (6.16) we can write

$$\Delta H^\circ_{rxn} = (1 \text{ mol}) \Delta H^\circ_f(CO_2, g) - [(1 \text{ mol}) \Delta H^\circ_f(C, graphite) + (1 \text{ mol}) \Delta H^\circ_f(O_2, g)]$$

$$= -393.5 \text{ kJ}$$

Because both graphite and O_2 are stable allotropic forms of the elements, it follows that $\Delta H^\circ_f(C, graphite)$ and $\Delta H^\circ_f(O_2, g)$ are zero. Therefore

$$\Delta H^\circ_{rxn} = (1 \text{ mol}) \Delta H^\circ_f(CO_2, g) = -393.5 \text{ kJ}$$

or

$$\Delta H^\circ_f(CO_2, g) = -393.5 \text{ kJ/mol}$$

Note that arbitrarily assigning zero ΔH°_f for each element in its most stable form at the standard state does not affect our calculations in any way. Remember, in thermochemistry we are interested only in enthalpy *changes,* because they can be determined experimentally whereas the absolute enthalpy values cannot. The choice of a zero "reference level" for enthalpy makes calculations easier to handle. Again referring to the terrestrial altitude analogy, we find that Mt. Everest is 8708 ft higher than Mt. McKinley. This difference in altitude is unaffected by the decision to set sea level at 0 ft or at 1000 ft.

Other compounds that can be studied by the direct method are SF_6, P_4O_{10}, and CS_2. The equations representing their syntheses are

$$S(rhombic) + 3F_2(g) \longrightarrow SF_6(g)$$
$$P_4(white) + 5O_2(g) \longrightarrow P_4O_{10}(s)$$
$$C(graphite) + 2S(rhombic) \longrightarrow CS_2(l)$$

Note that S(rhombic) and P(white) are the most stable allotropes of sulfur and phosphorus, respectively, at 1 atm and 25°C, so their ΔH°_f values are zero.

P_4

The Indirect Method

Many compounds cannot be directly synthesized from their elements. In some cases, the reaction proceeds too slowly, or side reactions produce substances other than the desired compound. In these cases ΔH°_f can be determined by an indirect approach, which is based on the law of heat summation (or simply Hess's law). **Hess's law** can be stated as follows: *When reactants are converted to products, the change in enthalpy is the same whether the reaction takes place in one step or in a series of steps.* In other words, if we can break down the reaction of interest into a series of reactions for which ΔH°_{rxn} can be measured, we can calculate ΔH°_{rxn} for the overall reaction. Hess's law is based on the fact that because H is a state function, ΔH depends only on the initial and final state (that is, only on the nature of reactants and products). The enthalpy change would be the same whether the overall reaction takes one step or many steps.

A useful analogy of Hess's law is as follows. Suppose you go from the first floor to the sixth floor of a building by elevator. The gain in your gravitational potential energy (which corresponds to the enthalpy change for the overall process) is the same

White phosphorus burns in air to form P_4O_{10}.

whether you go directly there or stop at each floor on your way up (breaking the reaction into a series of steps).

Let's say we are interested in the standard enthalpy of formation of methane (CH_4). We might represent the synthesis of CH_4 from its elements as

$$C(graphite) + 2H_2(g) \longrightarrow CH_4(g)$$

However, this reaction does not take place as written, so we cannot measure the enthalpy change directly. We must employ an indirect route, based on Hess's law. To begin with, the following reactions involving C(graphite), H_2, and CH_4 with O_2 have all been studied and the ΔH°_{rxn} values are accurately known

(a) $\qquad C(graphite) + O_2(g) \longrightarrow CO_2(g) \qquad\qquad \Delta H^\circ_{rxn} = -393.5 \text{ kJ}$

(b) $\qquad 2H_2(g) + O_2(g) \longrightarrow 2H_2O(l) \qquad\qquad \Delta H^\circ_{rxn} = -571.6 \text{ kJ}$

(c) $\qquad CH_4(g) + 2O_2(g) \longrightarrow CO_2(g) + 2H_2O(l) \qquad \Delta H^\circ_{rxn} = -890.4 \text{ kJ}$

Because we want to obtain one equation containing only C and H_2 as reactants and CH_4 as product, we must reverse (c) to get

(d) $\qquad CO_2(g) + 2H_2O(l) \longrightarrow CH_4(g) + 2O_2(g) \qquad \Delta H^\circ_{rxn} = 890.4 \text{ kJ}$

The next step is to add Equations (a), (b), and (d):

(a) $\qquad C(graphite) + \cancel{O_2(g)} \longrightarrow \cancel{CO_2(g)} \qquad\qquad \Delta H^\circ_{rxn} = -393.5 \text{ kJ}$

(b) $\qquad 2H_2(g) + \cancel{O_2(g)} \longrightarrow \cancel{2H_2O(l)} \qquad\qquad \Delta H^\circ_{rxn} = -571.6 \text{ kJ}$

Remember to reverse the sign of ΔH when you reverse an equation.

(d) $\qquad \cancel{CO_2(g)} + \cancel{2H_2O(l)} \longrightarrow CH_4(g) + \cancel{2O_2(g)} \qquad \Delta H^\circ_{rxn} = 890.4 \text{ kJ}$

(e) $\qquad C(graphite) + 2H_2(g) \longrightarrow CH_4(g) \qquad\qquad \Delta H^\circ_{rxn} = -74.7 \text{ kJ}$

As you can see, all the nonessential species (O_2, CO_2, and H_2O) cancel in this operation. Because this equation represents the synthesis of 1 mole of CH_4 from its elements, we have $\Delta H^\circ_f(CH_4) = -74.7 \text{ kJ/mol}$.

The general rule in applying Hess's law is that we should arrange a series of chemical equations (corresponding to a series of steps) in such a way that, when added together, all species will cancel except for the reactants and products that appear in the overall reaction. This means that we want the elements on the left and the compound of interest on the right of the arrow. To achieve this, we often need to multiply some or all of the equations representing the individual steps by the appropriate coefficients.

C_2H_2

Example 6.8   Calculate the standard enthalpy of formation of acetylene (C_2H_2) from its elements:

$$2C(graphite) + H_2(g) \longrightarrow C_2H_2(g)$$

The equations for each step and the corresponding enthalpy changes are:

(a) $\qquad C(graphite) + O_2(g) \longrightarrow CO_2(g) \qquad\qquad \Delta H^\circ_{rxn} = -393.5 \text{ kJ}$

(b) $\qquad H_2(g) + \frac{1}{2}O_2(g) \longrightarrow H_2O(l) \qquad\qquad \Delta H^\circ_{rxn} = -285.8 \text{ kJ}$

(c) $\qquad 2C_2H_2(g) + 5O_2(g) \longrightarrow 4CO_2(g) + 2H_2O(l) \qquad \Delta H^\circ_{rxn} = -2598.8 \text{ kJ}$

Reasoning and Solution Because we want to obtain one equation containing the elements C and H_2 as reactants and C_2H_2 as product, which represents the synthesis of acetylene from its elements, we need to eliminate O_2, CO_2, and H_2O from Equations (a), (b), and (c). We note that (c) contains 5 moles of O_2, 4 moles of CO_2, and 2 moles of H_2O. First we reverse (c) to get C_2H_2 on the product side:

(d) $\qquad\qquad 4CO_2(g) + 2H_2O(l) \longrightarrow 2C_2H_2(g) + 5O_2(g) \qquad \Delta H^\circ_{rxn} = +2598.8 \text{ kJ}$

Next, we multiply (a) by 4 and (b) by 2 and carry out the addition of 4(a) +2(b) + (d):

$$
\begin{array}{lll}
& 4C(\text{graphite}) + \cancel{4O_2(g)} \longrightarrow \cancel{4CO_2(g)} & \Delta H^\circ_{rxn} = -1574.0 \text{ kJ} \\
+ & 2H_2(g) + \cancel{O_2(g)} \longrightarrow \cancel{2H_2O(l)} & \Delta H^\circ_{rxn} = -571.6 \text{ kJ} \\
+ & \cancel{4CO_2(g)} + \cancel{2H_2O(l)} \longrightarrow 2C_2H_2(g) + \cancel{5O_2(g)} & \Delta H^\circ_{rxn} = +2598.8 \text{ kJ} \\ \hline
& 4C(\text{graphite}) + 2H_2(g) \longrightarrow 2C_2H_2(g) & \Delta H^\circ_{rxn} = +453.2 \text{ kJ}
\end{array}
$$

or $\qquad\qquad 2C(\text{graphite}) + H_2(g) \longrightarrow C_2H_2(g) \qquad\qquad \Delta H^\circ_{rxn} = +226.6 \text{ kJ}$

Because this equation represents the synthesis of C_2H_2 from its elements, we arrive at the result $\Delta H^\circ_f(C_2H_2) = \Delta H^\circ_{rxn}/\text{mol} = +226.6 \text{ kJ/mol}$. (When the equation is divided by 2, the value of ΔH°_{rxn} is halved.)

Similar problems: 6.53, 6.54.

Practice Exercise Calculate the standard enthalpy of formation of carbon disulfide (CS_2) from its elements, given that

$$
\begin{array}{lll}
C(\text{graphite}) + O_2(g) \longrightarrow CO_2(g) & \Delta H^\circ_{rxn} = -393.5 \text{ kJ} \\
S(\text{rhombic}) + O_2(g) \longrightarrow SO_2(g) & \Delta H^\circ_{rxn} = -296.4 \text{ kJ} \\
CS_2(l) + 3O_2(g) \longrightarrow CO_2(g) + 2SO_2(g) & \Delta H^\circ_{rxn} = -1073.6 \text{ kJ}
\end{array}
$$

We can calculate the enthalpy of reactions from values of ΔH°_f as shown in Example 6.9.

Example 6.9 Pentaborane-9, B_5H_9, is a colorless, highly reactive liquid that will burst into flame or even explode when exposed to oxygen. The reaction is

$$2B_5H_9(l) + 12O_2(g) \longrightarrow 5B_2O_3(s) + 9H_2O(l)$$

Pentaborane-9 was considered as a potential rocket fuel in the 1950s because it produces a large amount of heat per gram. However, the solid B_2O_3 formed by the combustion of B_5H_9 is an abrasive that would quickly destroy the nozzle of the rocket, and so the idea was abandoned. Calculate the kilojoules of heat released per gram of the compound reacted with oxygen. The standard enthalpy of formation of B_5H_9 is 73.2 kJ/mol.

Reasoning and Solution To calculate the standard enthalpy of a reaction, we look up the ΔH°_f values of the reactants and products in Appendix 2 and use the coefficients in the balanced equation in Equation (6.16) as follows

$$\Delta H^\circ_{rxn} = [(5 \text{ mol}) \Delta H^\circ_f(B_2O_3) + (9 \text{ mol}) \Delta H^\circ_f(H_2O)]$$
$$- [(2 \text{ mol}) \Delta H^\circ_f(B_5H_9) + (12 \text{ mol}) \Delta H^\circ_f(O_2)]$$

$$\Delta H^{\circ}_{rxn} = [(5 \text{ mol})(-1263.6 \text{kJ/mol} + (9 \text{ mol})(-285.8 \text{kJ/mol})]$$
$$- [(2 \text{ mol})(73.2 \text{ kJ/mol}) + (12 \text{ mol})(0)]$$

$$= -9036.6 \text{ kJ}$$

This is the amount of heat released for every 2 moles of B_5H_9 reacted. The heat released per gram of B_5H_9 reacted is

$$\text{heat released per gram of } B_5H_9 = \frac{1 \text{ mol } B_5H_9}{63.12 \text{ g } B_5H_9} \times \frac{-9036.6 \text{ kJ}}{2 \text{ mol } B_5H_9}$$

$$= -71.58 \text{ kJ/g } B_5H_9$$

Practice Exercise   Benzene (C_6H_6) burns in air to produce carbon dioxide and liquid water. Calculate the heat released (in kilojoules) per gram of the compound reacted with oxygen. The standard enthalpy of formation of benzene is 49.04 kJ/mol.

Similar problem: 6.52.

B_5H_9 is called a high-energy compound because of the large amount of energy it releases upon combustion. In general, compounds that have positive ΔH°_f values tend to release more heat as a result of combustion and to be less stable than those with negative ΔH°_f values.

Summary

Energy is the capacity to do work. There are many forms of energy and they are all interconvertible. The law of conservation of energy states that the total amount of energy in the universe always stays the same.

Any process that gives off heat to the surroundings is called an exothermic process; any process that absorbs heat from the surroundings is an endothermic process.

The state of a system is defined by variables such as composition, volume, temperature, and pressure. The change in a state function for a system depends only on the initial and final states of the system, and not on the path by which the change is accomplished. Energy is a state function; work and heat are not.

Energy can be converted from one form to another, but it cannot be created or destroyed (first law of thermodynamics). In chemistry we are concerned mainly with thermal energy, electrical energy, and mechanical energy, which is usually associated with pressure-volume work.

The change in enthalpy (ΔH, usually given in kilojoules) is a measure of the heat of a reaction (or any other process) at constant pressure. Enthalpy is a state function. A change in enthalpy ΔH is equal to $\Delta E + P\Delta V$ for a constant-pressure process. For chemical reactions at constant temperature, ΔH is given by $\Delta E + RT\Delta n$, where Δn is the difference between moles of gaseous products and moles of gaseous reactants.

Constant-volume and constant-pressure calorimeters are used to measure heat changes of physical and chemical processes.

Hess's law states that the overall enthalpy change in a reaction is equal to the sum of enthalpy changes for the individual steps that make up the overall reaction. The standard enthalpy of a reaction can be calculated from the standard enthalpies of formation of reactants and products.

Key Words

Calorimetry p. 174
Chemical energy p. 162
Closed system p. 163
Endothermic process
    p. 164
Energy p. 162
Enthalpy (H) p. 170
Enthalpy of reaction (ΔH)
    p. 171
Exothermic process
    p. 164

First law of
    thermodynamics
    p. 165
Heat p. 163
Heat capacity (C) p. 175
Hess's law p. 181
Isolated system p. 163
Law of conservation of
    energy p. 162
Open system p. 163
Potential energy p. 162

Radiant energy p. 162
Specific heat (s) p. 175
Standard enthalpy of
    formation (ΔH_f°) p. 179
Standard enthalpy of
    reaction (ΔH_{rxn}°)
    p. 179
Standard state p. 179
State function p. 165
State of a system p. 164
Surroundings p. 163

System p. 163
Thermal energy p. 162
Thermochemical equation
    p. 171
Thermochemistry p. 163
Thermodynamics p. 164
Work p. 162

Questions and Problems

Definitions

Review Questions

6.1 Define these terms: system, surroundings, open system, closed system, isolated system, thermal energy, chemical energy, potential energy, kinetic energy, law of conservation of energy.

6.2 What is heat? How does heat differ from thermal energy? Under what condition is heat transferred from one system to another?

6.3 What are the units for energy commonly employed in chemistry?

6.4 A truck initially traveling at 60 km per hour is brought to a complete stop at a traffic light. Does this change violate the law of conservation of energy? Explain.

6.5 These are various forms of energy: chemical, heat, light, mechanical, and electrical. Suggest ways of interconverting these forms of energy.

6.6 Describe the interconversions of forms of energy occurring in these processes: (a) You throw a softball up into the air and catch it. (b) You switch on a flashlight. (c) You ride the ski lift to the top of the hill and then ski down. (d) You strike a match and let it burn down.

Energy Changes in Chemical Reactions

Review Questions

6.7 Define these terms: thermochemistry, exothermic process, endothermic process.

6.8 Stoichiometry is based on the law of conservation of mass. On what law is thermochemistry based?

6.9 Describe two exothermic processes and two endothermic processes.

6.10 Decomposition reactions are usually endothermic, whereas combination reactions are usually exothermic. Give a qualitative explanation for these trends.

First Law of Thermodynamics

Review Questions

6.11 On what law is the first law of thermodynamics based? Explain the sign conventions in the equation $\Delta E = q + w$.

6.12 Explain what is meant by a state function. Give two examples of quantities that are state functions and two that are not.

6.13 The internal energy of an ideal gas depends only on its temperature. Do a first-law analysis of this process. A sample of an ideal gas is allowed to expand at constant temperature against atmospheric pressure. (a) Does the gas do work on its surroundings? (b) Is there heat exchange between the system and the surroundings? If so, in which direction? (c) What is ΔE for the gas for this process?

6.14 Consider these changes.

(a) $Hg(l) \longrightarrow Hg(g)$
(b) $3O_2(g) \longrightarrow 2O_3(g)$
(c) $CuSO_4 \cdot 5H_2O(s) \longrightarrow CuSO_4(s) + 5H_2O(g)$
(d) $H_2(g) + F_2(g) \longrightarrow 2HF(g)$

At constant pressure, in which of the reactions is work done by the system on the surroundings? By the surroundings on the system? In which of them is no work done?

Problems

6.15 A gas expands and does P-V work on the surroundings equal to 325 J. At the same time, it absorbs 127 J of heat from the surroundings. Calculate the change in energy of the gas.

6.16 The work done to compress a gas is 74 J. As a result, 26 J of heat is given off to the surroundings. Calculate the change in energy of the gas.

6.17 Calculate the work done when 50.0 g of tin dissolves in excess acid at 1.00 atm and 25°C:

$$Sn(s) + 2H^+(aq) \longrightarrow Sn^{2+}(aq) + H_2(g)$$

Assume ideal gas behavior.

6.18 Calculate the work done in joules when 1.0 mole of water vaporizes at 1.0 atm and 100°C. Assume that the volume of liquid water is negligible compared with that of steam at 100°C, and ideal gas behavior.

Enthalpy of Chemical Reactions

Review Questions

6.19 Define these terms: enthalpy, enthalpy of reaction. Under what condition is the heat of a reaction equal to the enthalpy change of the same reaction?

6.20 In writing thermochemical equations, why is it important to indicate the physical state (that is, gaseous, liquid, solid, or aqueous) of each substance?

6.21 Explain the meaning of this thermochemical equation:

$$4NH_3(g) + 5O_2(g) \longrightarrow 4NO(g) + 6H_2O(g)$$
$$\Delta H = -904 \text{ kJ}$$

6.22 Consider this reaction:

$$2CH_3OH(l) + 3O_2(g) \longrightarrow 4H_2O(l) + 2CO_2(g)$$
$$\Delta H = -1452.8 \text{ kJ}$$

What is the value of ΔH if (a) the equation is multiplied throughout by 2, (b) the direction of the reaction is reversed so that the products become the reactants and vice versa, (c) water vapor instead of liquid water is formed as the product?

Problems

6.23 The first step in the industrial recovery of zinc from the zinc sulfide ore is roasting, that is, the conversion of ZnS to ZnO by heating:

$$2ZnS(s) + 3O_2(g) \longrightarrow 2ZnO(s) + 2SO_2(g)$$
$$\Delta H = -879 \text{ kJ}$$

Calculate the heat evolved (in kJ) per gram of ZnS roasted.

6.24 Determine the amount of heat (in kJ) given off when 1.26×10^4 g of ammonia are produced according to the equation

$$N_2(g) + 3H_2(g) \longrightarrow 2NH_3(g) \quad \Delta H = -92.6 \text{ kJ}$$

6.25 Consider the reaction

$$N_2(g) + 3H_2(g) \longrightarrow 2NH_3(g) \quad \Delta H = -92.6 \text{ kJ}$$

If 2.0 moles of N_2 react with 6.0 moles of H_2 to form NH_3, calculate the work done (in joules) against a pressure of 1.0 atm at 25°C. What is ΔE for this reaction? Assume the reaction goes to completion.

6.26 Consider the reaction

$$H_2(g) + Cl_2(g) \longrightarrow 2HCl(g) \qquad \Delta H = -184.6 \text{ kJ}$$

If 3 moles of H_2 react with 3 moles of Cl_2 to form HCl, calculate the work done (in joules) against a pressure of 1.0 atm at 25°C. What is ΔE for this reaction? Assume the reaction goes to completion.

Calorimetry

Review Questions

6.27 What is the difference between specific heat and heat capacity? What are the units for these two quantities? Which is the intensive property and which is the extensive property?

6.28 Consider two metals A and B, each having a mass of 100 g and an initial temperature of 20°C. The specific heat of A is larger than that of B. Under the same heating conditions, which metal would take longer to reach a temperature of 21°C?

6.29 Define calorimetry and describe two commonly used calorimeters.

6.30 In a calorimetric measurement, why is it important that we know the heat capacity of the calorimeter? How is this value determined?

Problems

6.31 A piece of silver of mass 362 g has a heat capacity of 85.7 J/°C. What is the specific heat of silver?

6.32 A 6.22-kg piece of copper metal is heated from 20.5°C to 324.3°C. Calculate the heat absorbed (in kJ) by the metal.

6.33 Calculate the amount of heat liberated (in kJ) from 366 g of mercury when it cools from 77.0°C to 12.0°C.

6.34 A sheet of gold weighing 10.0 g and at a temperature of 18.0°C is placed flat on a sheet of iron weighing 20.0 g and at a temperature of 55.6°C. What is the final temperature of the combined metals? Assume that no heat is lost to the surroundings. (*Hint:* The heat gained by the gold must be equal to the heat lost by the iron. The specific heats of the metals are given in Table 6.2.)

6.35 A 0.1375-g sample of solid magnesium is burned in a constant-volume bomb calorimeter that has a heat capacity of 3024 J/°C. The temperature increases by 1.126°C. Calculate the heat given off by the burning Mg, in kJ/g and in kJ/mol.

6.36 A quantity of 2.00×10^2 mL of 0.862 *M* HCl is mixed with 2.00×10^2 mL of 0.431 *M* $Ba(OH)_2$ in a constant-pressure calorimeter of negligible heat capacity. The initial temperature of the HCl and $Ba(OH)_2$ solutions is the same at 20.48°C. For the process

$$H^+(aq) + OH^-(aq) \longrightarrow H_2O(l)$$

the heat of neutralization is -56.2 kJ. What is the final temperature of the mixed solution?

Standard Enthalpy of Formation and Reaction

Review Questions

6.37 What is meant by the standard-state condition?

6.38 How are the standard enthalpies of an element and of a compound determined?

6.39 What is meant by the standard enthalpy of a reaction?

6.40 Write the equation for calculating the enthalpy of a reaction. Define all the terms.

6.41 State Hess's law. Explain, with one example, the usefulness of Hess's law in thermochemistry.

6.42 Describe how chemists use Hess's law to determine the ΔH_f° of a compound by measuring its heat (enthalpy) of combustion.

Problems

6.43 Which of these standard enthalpy of formation values is not zero at 25°C? $Na(s)$, $Ne(g)$, $CH_4(g)$, $S_8(s)$, $Hg(l)$, $H(g)$.

6.44 The ΔH_f° values of the two allotropes of oxygen, O_2 and O_3, are zero and 142.2 kJ/mol, respectively, at 25°C. Which is the more stable form at this temperature?

6.45 Which is the more negative quantity at 25°C: ΔH_f° for $H_2O(l)$ or ΔH_f° for $H_2O(g)$?

6.46 Predict the value of ΔH_f° (greater than, less than, or equal to zero) for these elements at 25°C: (a) $Br_2(g)$; $Br_2(l)$, (b) $I_2(g)$; $I_2(s)$.

6.47 Calculate the heats of combustion for these reactions from the standard enthalpies of formation listed in Appendix 2:

(a) $2H_2(g) + O_2(g) \longrightarrow 2H_2O(l)$
(b) $2C_2H_2(g) + 5O_2(g) \longrightarrow 4CO_2(g) + 2H_2O(l)$

6.48 Calculate the heats of combustion for these reactions from the standard enthalpies of formation listed in Appendix 2:

(a) $C_2H_4(g) + 3O_2(g) \longrightarrow 2CO_2(g) + 2H_2O(l)$
(b) $2H_2S(g) + 3O_2(g) \longrightarrow 2H_2O(l) + 2SO_2(g)$

6.49 Methanol, ethanol, and *n*-propanol are three common alcohols. When 1.00 g of each of these alcohols is burned in air, heat is liberated as: (a) methanol (CH_3OH), -22.6 kJ; (b) ethanol (C_2H_5OH), -29.7 kJ; (c) *n*-propanol (C_3H_7OH), -33.4 kJ. Calculate the heats of combustion of these alcohols in kJ/mol.

6.50 The standard enthalpy change for this reaction is 436.4 kJ:

$$H_2(g) \longrightarrow H(g) + H(g)$$

Calculate the standard enthalpy of formation of atomic hydrogen (H).

6.51 From the standard enthalpies of formation, calculate ΔH_{rxn}° for the reaction

$$C_6H_{12}(l) + 9O_2(g) \longrightarrow 6CO_2(g) + 6H_2O(l)$$

For $C_6H_{12}(l)$, $\Delta H_f^\circ = -151.9$ kJ/mol.

6.52 At 850°C, $CaCO_3$ undergoes substantial decomposition to yield CaO and CO_2. Assuming that the ΔH_f° values of the reactant and products are the same at 850°C as they are at 25°C, calculate the enthalpy change (in kJ) if 66.8 g of CO_2 are produced in one reaction.

6.53 From these data,

$$S(\text{rhombic}) + O_2(g) \longrightarrow SO_2(g) \qquad \Delta H_{rxn}^\circ = -296.06 \text{ kJ}$$
$$S(\text{monoclinic}) + O_2(g) \longrightarrow SO_2(g) \qquad \Delta H_{rxn}^\circ = -296.36 \text{ kJ}$$

calculate the enthalpy change for the transformation

$$S(\text{rhombic}) \longrightarrow S(\text{monoclinic})$$

(Monoclinic and rhombic are different allotropic forms of elemental sulfur.)

6.54 From these data,

$$C(\text{graphite}) + O_2(g) \longrightarrow CO_2(g) \qquad \Delta H_{rxn}^\circ = -393.5 \text{ kJ}$$
$$H_2(g) + \tfrac{1}{2}O_2(g) \longrightarrow H_2O(l) \qquad \Delta H_{rxn}^\circ = -285.8 \text{ kJ}$$
$$2C_2H_6(g) + 7O_2(g) \longrightarrow 4CO_2(g) + 6H_2O(l)$$
$$\Delta H_{rxn}^\circ = -3119.6 \text{ kJ}$$

calculate the enthalpy change for the reaction

$$2C(\text{graphite}) + 3H_2(g) \longrightarrow C_2H_6(g)$$

6.55 From these heats of combustion,

$$CH_3OH(l) + \tfrac{3}{2}O_2(g) \longrightarrow CO_2(g) + 2H_2O(l)$$
$$\Delta H_{rxn}^\circ = -726.4 \text{ kJ}$$
$$C(\text{graphite}) + O_2(g) \longrightarrow CO_2(g) \qquad \Delta H_{rxn}^\circ = -393.5 \text{ kJ}$$
$$H_2(g) + \tfrac{1}{2}O_2(g) \longrightarrow H_2O(l) \qquad \Delta H_{rxn}^\circ = -285.8 \text{ kJ}$$

calculate the enthalpy of formation of methanol (CH_3OH) from its elements:

$$C(\text{graphite}) + 2H_2(g) + \tfrac{1}{2}O_2(g) \longrightarrow CH_3OH(l)$$

6.56 Calculate the standard enthalpy change for the reaction

$$2Al(s) + Fe_2O_3(s) \longrightarrow 2Fe(s) + Al_2O_3(s)$$

given that

$$2Al(s) + \tfrac{3}{2}O_2(g) \longrightarrow Al_2O_3(s) \qquad \Delta H_{rxn}^\circ = -1601 \text{ kJ}$$
$$2Fe(s) + \tfrac{3}{2}O_2(g) \longrightarrow Fe_2O_3(s) \qquad \Delta H_{rxn}^\circ = -821 \text{ kJ}$$

Additional Problems

6.57 The convention of arbitrarily assigning a zero enthalpy value for the most stable form of each element in the standard state at 25°C is a convenient way of dealing with enthalpies of reactions. Explain why this convention cannot be applied to nuclear reactions.

6.58 Consider these two reactions:

$$A \longrightarrow 2B \qquad \Delta H^{\circ}_{rxn} = \Delta H_1$$
$$A \longrightarrow C \qquad \Delta H^{\circ}_{rxn} = \Delta H_2$$

Determine the enthalpy change for the process

$$2B \longrightarrow C$$

6.59 The standard enthalpy change ΔH° for the thermal decomposition of silver nitrate according to this equation is +78.67 kJ:

$$AgNO_3(s) \longrightarrow AgNO_2(s) + \tfrac{1}{2}O_2(g)$$

The standard enthalpy of formation of $AgNO_3(s)$ is −123.02 kJ/mol. Calculate the standard enthalpy of formation of $AgNO_2(s)$.

6.60 Hydrazine, N_2H_4, decomposes according to this reaction:

$$3N_2H_4(l) \longrightarrow 4NH_3(g) + N_2(g)$$

(a) Given that the standard enthalpy of formation of hydrazine is 50.42 kJ/mol, calculate ΔH° for its decomposition. (b) Both hydrazine and ammonia burn in oxygen to produce $H_2O(l)$ and $N_2(g)$. Write balanced equations for each of these processes and calculate ΔH° for each of them. On a mass basis (per kg), would hydrazine or ammonia be the better fuel?

6.61 Photosynthesis produces glucose, $C_6H_{12}O_6$, and oxygen from carbon dioxide and water:

$$6CO_2 + 6H_2O \longrightarrow C_6H_{12}O_6 + 6O_2$$

(a) How would you determine experimentally the ΔH°_{rxn} value for this reaction? (b) Solar radiation produces about 7.0×10^{14} kg glucose a year on Earth. What is the corresponding ΔH° change?

6.62 2.10-mole sample of crystalline acetic acid, initially at 17.0°C, is allowed to melt at 17.0°C and is then heated to 118.1°C (its normal boiling point) at 1.00 atm. The sample is allowed to vaporize at 118.1°C and is then rapidly quenched to 17.0°C, so that it recrystallizes. Calculate ΔH° for the total process as described.

6.63 Calculate the work done in joules by the reaction

$$2Na(s) + 2H_2O(l) \longrightarrow 2NaOH(aq) + H_2(g)$$

when 0.34 g of Na reacts with water to form hydrogen gas at 0°C and 1.0 atm.

6.64 You are given these data:

$$H_2(g) \longrightarrow 2H(g) \qquad \Delta H^{\circ} = 436.4 \text{ kJ}$$
$$Br_2(g) \longrightarrow 2Br(g) \qquad \Delta H^{\circ} = 192.5 \text{ kJ}$$
$$H_2(g) + Br_2(g) \longrightarrow 2HBr(g) \qquad \Delta H^{\circ} = -103.3 \text{ kJ}$$

Calculate ΔH° for the reaction

$$H(g) + Br(g) \longrightarrow HBr(g)$$

6.65 Methanol (CH_3OH) is an organic solvent and is also used as a fuel in some automobile engines. From these data, calculate the standard enthalpy of formation of methanol:

$$2CH_3OH(l) + 3O_2(g) \longrightarrow 2CO_2(g) + 4H_2O(l)$$
$$\Delta H^{\circ}_{rxn} = -1452.8 \text{ kJ}$$

6.66 A 44.0-g sample of an unknown metal at 99.0°C was placed in a constant-pressure calorimeter containing 80.0 g of water at 24.0°C. The final temperature of the system was found to be 28.4°C. Calculate the specific heat of the metal. (The heat capacity of the calorimeter is 12.4 J/°C.)

6.67 A 1.00-mole sample of ammonia at 14.0 atm and 25°C in a cylinder fitted with a movable piston expands against a constant external pressure of 1.00 atm. At equilibrium, the pressure and volume of the gas are 1.00 atm and 23.5 L, respectively. (a) Calculate the final temperature of the sample. (b) Calculate q, w, and ΔE for the process. The specific heat of ammonia is 0.0258 J/g · °C.

6.68 Producer gas (carbon monoxide) is prepared by passing air over red-hot coke:

$$C(s) + \tfrac{1}{2}O_2(g) \longrightarrow CO(g)$$

Water gas (mixture of carbon monoxide and hydrogen) is prepared by passing steam over red-hot coke:

$$C(s) + H_2O(g) \longrightarrow CO(g) + H_2(g)$$

For many years, both producer gas and water gas were used as fuels in industry and for domestic cooking. The large-scale preparation of these gases was carried out alternately, that is, first producer gas, then water gas, and so on. Using thermochemical reasoning, explain why this procedure was chosen.

6.69 Compare the heat produced by the complete combustion of 1 mole of methane (CH_4) with a mole of water gas (0.50 mole H_2 and 0.50 mole CO) under the same conditions. On the basis of your answer, would you prefer methane over water gas as a fuel? Can you suggest two other reasons why methane is preferable to water gas as a fuel?

6.70 The so-called hydrogen economy is based on hydrogen produced from water using solar energy. The gas is then burned as a fuel:

$$2H_2(g) + O_2(g) \longrightarrow 2H_2O(l)$$

A primary advantage of hydrogen as a fuel is that it is nonpolluting. A major disadvantage is that it is a gas and therefore is harder to store than liquids or solids. Calculate the volume of hydrogen gas at 25°C and 1.00 atm required to produce an amount of energy equivalent to that produced by the combustion of a gallon of octane (C_8H_{18}). The density of octane is 2.66 kg/gal, and its standard enthalpy of formation is −249.9 kJ/mol.

6.71 If energy is conserved, how can there be an energy crisis?

6.72 The combustion of what volume of ethane (C_2H_6), measured at 23.0°C and 752 mmHg, would be required to heat 855 g of water from 25.0°C to 98.0°C?

6.73 The heat of vaporization of a liquid (ΔH_{vap}) is the energy required to vaporize 1.00 g of the liquid at its boiling point. In one experiment, 60.0 g of liquid nitrogen (boiling point −196°C) are poured into a Styrofoam cup containing 2.00×10^2 g of water at 55.3°C. Calculate the molar heat of vaporization of liquid nitrogen if the final temperature of the water is 41.0°C.

6.74 For which of these reactions does $\Delta H^\circ_{rxn} = \Delta H^\circ_f$?

    (a) $H_2(g) + S(rhombic) \longrightarrow H_2S(g)$
    (b) $C(diamond) + O_2(g) \longrightarrow CO_2(g)$
    (c) $H_2(g) + CuO(s) \longrightarrow H_2O(l) + Cu(s)$
    (d) $O(g) + O_2(g) \longrightarrow O_3(g)$

6.75 Calculate the work done (in joules) when 1.0 mole of water is frozen at 0°C and 1.0 atm. The volumes of 1 mole of water and ice at 0°C are 0.0180 L and 0.0196 L, respectively.

6.76 A quantity of 0.020 mole of a gas initially at 0.050 L and 20°C undergoes a constant-temperature expansion until its volume is 0.50 L. Calculate the work done (in joules) by the gas if it expands (a) against a vacuum and (b) against a constant pressure of 0.20 atm. (c) If the gas in (b) is allowed to expand unchecked until its pressure is equal to the external pressure, what would its final volume be before it stopped expanding, and what would be the work done?

6.77 Calculate the standard enthalpy of formation for diamond, given that

$$C(graphite) + O_2(g) \longrightarrow CO_2(g) \qquad \Delta H^\circ = -393.5 \text{ kJ}$$
$$C(diamond) + O_2(g) \longrightarrow CO_2(g) \qquad \Delta H^\circ = -395.4 \text{ kJ}$$

6.78 (a) For most efficient use, refrigerator freezer compartments should be fully packed with food. What is the thermochemical basis for this recommendation? (b) Starting at the same temperature, tea and coffee remain hot longer in a thermal flask than chicken noodle soup. Explain.

6.79 Calculate the standard enthalpy change for the fermentation process. (See Problem 3.70.)

6.80 Portable hot packs are available for skiers and people engaged in other outdoor activities in a cold climate. The air-permeable paper packet contains a mixture of powdered iron, sodium chloride, and other components, all moistened by a little water. The exothermic reaction that produces the heat is a very common one—the rusting of iron:

$$4Fe(s) + 3O_2(g) \longrightarrow 2Fe_2O_3(s)$$

When the outside plastic envelope is removed, O_2 molecules penetrate the paper, causing the reaction to begin. A typical packet contains 250 g of iron to warm your hands or feet for up to 4 hours. How much heat (in kJ) is produced by this reaction? (*Hint:* See Appendix 2 for ΔH°_f values.)

6.81 A person ate 0.50 pound of cheese (an energy intake of 4000 kJ). Suppose that none of the energy was stored in his body. What mass (in grams) of water would he need to perspire in order to maintain his original temperature? (It takes 44.0 kJ to vaporize 1 mole of water.)

6.82 The total volume of the Pacific Ocean is estimated to be 7.2×10^8 km^3. A medium-sized atomic bomb produces 1.0×10^{15} J of energy upon explosion. Calculate the number of atomic bombs needed to release enough energy to raise the temperature of the water in the Pacific Ocean by 1°C.

6.83 A 19.2-g quantity of dry ice (solid carbon dioxide) is allowed to sublime (evaporate) in an apparatus like the one shown in Figure 6.5. Calculate the expansion work done against a constant external pressure of 0.995 atm and at a constant temperature of 22°C. Assume that the initial volume of dry ice is negligible and that CO_2 behaves like an ideal gas.

6.84 The enthalpy of combustion of benzoic acid (C_6H_5COOH) is commonly used as the standard for calibrating constant-volume bomb calorimeters; its value has been accurately determined to be −3226.7 kJ/mol. When 1.9862 g of benzoic acid are burned, the temperature rises from 21.84°C to 25.67°C. What is the heat capacity of the calorimeter?

6.85 Lime is a term that includes calcium oxide (CaO, also called quicklime) and calcium hydroxide [Ca(OH)$_2$, also called slaked lime]. It is used in the steel industry to remove acidic impurities, in air-pollution control to remove acidic oxides such as SO_2, and in water treatment. Quicklime is made industrially by heating limestone ($CaCO_3$) above 2000°C:

$$CaCO_3(s) \longrightarrow CaO(s) + CO_2(g) \qquad \Delta H^\circ = 177.8 \text{ kJ}$$

Slaked lime is produced by treating quicklime with water:

$$CaO(s) + H_2O(l) \longrightarrow Ca(OH)_2(s) \qquad \Delta H^\circ = -65.2 \text{ kJ}$$

The exothermic reaction of quicklime with water and the rather small specific heats of both quicklime (0.946 J/g · °C) and slaked lime (1.20 J/g · °C) make it hazardous to store and transport lime in vessels made of wood. Wooden sailing ships carrying lime would occasionally catch fire when water leaked into the hold. (a) If a 500-g sample of water reacts with an equimolar amount of CaO (both at an initial temperature of 25°C), what is the final temperature of the product, Ca(OH)$_2$? Assume that the product absorbs all of the heat released in the reaction. (b) Given that the standard enthalpies of formation of CaO and H_2O are −635.6 kJ/mol and −285.8 kJ/mol, respectively, calculate the standard enthalpy of formation of Ca(OH)$_2$.

6.86 Calcium oxide (CaO) is used to remove sulfur dioxide generated by coal-burning power stations:

$$2CaO(s) + 2SO_2(g) + O_2(g) \longrightarrow 2CaSO_4(s)$$

Calculate the enthalpy change for this process if 6.6×10^5 g of SO_2 are removed by this process every day.

6.87 A balloon 16 m in diameter is inflated with helium. Calculate the work done (in joules) during the inflation process if the atmospheric pressure is 98.7 kPa.

6.88 (a) A person drinks four glasses of cold water (3.0°C) every day. The volume of each glass is 2.5×10^2 mL. How much heat (in kJ) does the body have to supply to raise the temperature of the water to 37°C, the body temperature? (b) How much heat would your body lose if you were to ingest 8.0×10^2 g of snow at 0°C to quench thirst? (The amount of heat necessary to melt snow is 6.01 kJ/mol.)

6.89 Why are cold, damp air and hot, humid air more uncomfortable than dry air at the same temperatures? (The specific heats of water vapor and air are approximately 1.9 J/g · °C and 1.0 J/g · °C, respectively.)

6.90 From the enthalpy of formation for CO_2 and the information given here, calculate the standard enthalpy of formation for carbon monoxide (CO).

$$CO(g) + \tfrac{1}{2}O_2(g) \longrightarrow CO_2(g) \qquad \Delta H° = -283.0 \text{ kJ}$$

Why can't we obtain it directly by measuring the enthalpy of the following reaction?

$$C(graphite) + \tfrac{1}{2}O_2(g) \longrightarrow CO(g)$$

6.91 In the nineteenth century two scientists named Dulong and Petit noticed that for a solid element, the product of its mo-

lar mass and its specific heat is approximately 25 J/°C. This observation, now called Dulong and Petit's law, was used to estimate the specific heat of metals. Verify the law for the metals listed in Table 6.2. The law does not apply to one of the metals. Which one is it? Why?

6.92 Determine the standard enthalpy of formation of ethanol (C_2H_5OH) from its standard enthalpy of combustion (-1367.4 kJ/mol).

6.93 A gas company in Massachusetts charges \$1.30 for 15 ft³ of natural gas (CH_4) measured at 20°C and 1.0 atm. Calculate the cost of heating 200 mL of water (enough to make a cup of coffee or tea) from 20°C to 100°C. Assume that only 50 percent of the heat generated by the combustion is used to heat the water; the rest of the heat is lost to the surroundings.

6.94 Calculate the internal energy of a Goodyear blimp filled with helium gas at 1.2×10^5 Pa. The volume of the blimp is 5.5×10^3 m³. If all the energy were used to heat 10.0 tons of copper at 21°C, calculate the final temperature of the metal. (*Hint:* See Section 5.7 for help in calculating the internal energy of a gas. 1 ton $= 9.072 \times 10^5$ g.)

6.95 The standard enthalpy change for the reaction is 941.4 kJ:

$$N_2(g) \longrightarrow N(g) + N(g)$$

Calculate the standard enthalpy of formation of atomic nitrogen.

6.96 Acetylene (C_2H_2) can be made by reacting calcium carbide (CaC_2) with water. (a) Write an equation for the reaction. (b) What is the maximum amount of heat (in joules) that can be obtained from the combustion of acetylene, starting with 74.6 g of CaC_2?

Special Problems

6.97 (a) A snowmaking machine contains a mixture of compressed air and water vapor at about 20 atm. When the mixture is sprayed into the atmosphere it expands so rapidly that, as a good approximation, no heat exchange occurs between the system (air and water) and its surroundings. (In thermodynamics, such a process is called an adiabatic process.) Do a first law of thermodynamics analysis to show how snow is formed under these conditions.

(b) If you have ever pumped air into a bicycle tire, you probably noticed a warming effect at the valve stem. The action of the pump compresses the air inside the

pump and the tire. The process is rapid enough to be treated as an adiabatic process. Apply the first law of thermodynamics to account for the warming effect.

(c) A driver's manual states that the stopping distance quadruples as the speed doubles; that is, if it takes 30 ft to stop a car traveling at 25 mph, then it would take 120 ft to stop a car moving at 50 mph. Justify this statement by using the first law of thermodynamics. Assume that when a car is stopped, its kinetic energy ($\tfrac{1}{2}mu^2$) is totally converted to heat.

6.98 When 1 mole of naphthalene ($C_{10}H_8$) is burned in a constant-volume bomb calorimeter at 298 K, 5150 kJ of heat is evolved. Calculate ΔE and ΔH for the reaction.

Answers to Practice Exercises

6.1 (a) 0, (b) -286 J. **6.2** -63 J. **6.3** -6.47×10^3 kJ. **6.4** -111.7 kJ. **6.5** -34.3 kJ. **6.6** -728 kJ/mol. **6.7** $22.49°C$. **6.8** 87.3 kJ/mol. **6.9** -41.83 kJ/g.

Fraunhofer's original drawing, showing the dark absorption lines in the sun's emission spectrum.

We often hear that a particular star, such as our sun, is known to contain certain elements. How is this information obtained?

In the early nineteenth century, Joseph Fraunhofer, a German physicist, studied the spectrum of colors produced by the sun and observed a number of dark lines. Each color in the continuous spectrum corresponds to light waves of a specific length, so the dark lines must represent wavelengths that are absorbed before sunlight reaches Earth. Such lines are called *absorption lines*.

We now know that the "Fraunhofer lines" represent sunlight that has been absorbed by atoms of gases through which solar radiation passes. Because of structural differences between atoms of different elements, each element has a distinctive spectrum in which light is emitted or absorbed at specific wavelengths. Much like fingerprints, which can be used to identify people because each individual's fingerprints are unique, spectral lines have enabled scientists to determine the composition of the sun and other celestial phenomena, including the gases responsible for the Fraunhofer lines. In fact, one element was identified in the solar spectrum before it was found on Earth. In 1868, the French physicist Pierre Janssen detected a dark line in the solar spectrum that could not be matched with any known element. The new element, named *helium* (from the Greek *helios,* meaning "sun"), was discovered on Earth 27 years later.

The Rosette Nebula has a reddish color because of its bright emission lines from hydrogen gas.

The Electronic Structure of Atoms

Chapter 7

ESSENTIAL CONCEPTS

 Planck's Quantum Theory To explain the dependence of radiation emitted by objects on wavelength, Planck proposed that atoms and molecules could emit (or absorb) energy only in discrete quantities called quanta. Planck's theory revolutionized physics.

 The Advent of Quantum Mechanics Planck's work led to the explanation of the photoelectric effect by Einstein, who postulated that light consists of particles called photons, and the emission spectrum of the hydrogen atom by Bohr. Further advancements to quantum theory were made by de Broglie, who demonstrated that an electron possesses both particle and wave properties, and Heisenberg, who derived an inherent limitation to measuring submicroscopic systems. These developments culminated in the Schrödinger equation, which describes the behavior and energy of electrons, atoms, and molecules.

 The Hydrogen Atom The solution to the Schrödinger equation for the hydrogen atom shows quantized energies for the electron and a set of wave functions called atomic orbitals. The atomic orbitals are labeled with specific quantum numbers; the orbitals tell us the regions in which an electron can be located. The results obtained for hydrogen, with minor modifications, can be applied to more complex atoms.

 The Building-Up Principle The periodic table can be constructed by increasing atomic number and adding electrons in a stepwise fashion. Specific guidelines (the Pauli exclusion principle and Hund's rule) help us write ground-state electron configurations of the elements, which tell us how electrons are distributed among the atomic orbitals.

7.1 From Classical Physics to Quantum Theory

Early attempts to understand atoms and molecules met with only limited success. By assuming that molecules behave like rebounding balls, physicists were able to predict and explain some macroscopic phenomena, such as the pressure exerted by a gas. However, their model did not account for the stability of molecules; that is, it could not explain the forces that hold atoms together. It took a long time to realize—and an even longer time to accept—that the properties of atoms and molecules are *not* governed by the same laws that work so well for larger objects.

The new era in physics started in 1900 with a young German physicist named Max Planck. While analyzing the data on the radiation emitted by solids heated to various temperatures, Planck discovered that atoms and molecules emit energy only in certain discrete quantities, or *quanta*. Physicists had always assumed that energy is continuous, which meant that any amount of energy could be released in a radiation process, so Planck's *quantum theory* turned physics upside down. Indeed, the flurry of research that ensued altered our concept of nature forever.

To understand quantum theory, we must know something about the nature of waves. A *wave* can be thought of as *a vibrating disturbance by which energy is transmitted*. The speed of a wave depends on the type of wave and the nature of the medium through which the wave is traveling (for example, air, water, or a vacuum). *The distance between identical points on successive waves* is called the *wavelength* λ (lambda). The *frequency* ν (nu) of the wave is *the number of waves that pass through a particular point in one second*. The *amplitude* is *the vertical distance from the midline of a wave to the peak or trough* (Figure 7.1a). Figure 7.1b shows two waves that have the same amplitude but different wavelengths and frequencies.

An important property of a wave traveling through space is its speed (u), which is given by the product of its wavelength and its frequency:

$$u = \lambda\nu \tag{7.1}$$

(a)

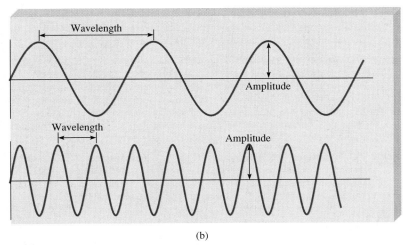

(b)

Figure 7.1 *(a) Wavelength and amplitude. (b) Two waves having different wavelengths and frequencies. The wavelength of the top wave is three times that of the lower wave, but its frequency is only one-third that of the lower wave. Both waves have the same amplitude.*

The inherent "sensibility" of Equation (7.1) can be seen by analyzing the physical dimensions involved in the three terms. The wavelength (λ) expresses the length of a wave, or distance/wave. The frequency (ν) indicates the number of these waves that pass any reference point per unit of time, or waves/time. Thus the product of these terms results in dimensions of distance/time, which is speed:

$$\frac{\text{distance}}{\text{wave}} \times \frac{\text{waves}}{\text{time}} = \frac{\text{distance}}{\text{time}}$$

$$\lambda \quad \times \quad \nu \quad = \quad u$$

Wavelength is usually expressed in units of meters, centimeters, or nanometers, and frequency is measured in hertz (Hz), in which

$$1 \text{ Hz} = 1 \text{ cycle/s}$$

The word "cycle" may be left out and the frequency expressed as, for example, 25/s (read as "25 per second").

Electromagnetic Radiation

All types of *radiant energy,* also called **electromagnetic radiation,** move through a vacuum at a speed of about 3.00×10^8 m/s, which is called the *speed of light (c).* The speed changes somewhat depending on the medium, but the variations are insignificant for our calculations. The waves associated with electromagnetic radiation are called electromagnetic waves.

Sound waves and water waves are not electromagnetic waves, but X rays and radio waves are.

Figure 7.2 shows various types of electromagnetic radiation, which differ from one another in wavelength and frequency. The long radio waves are emitted by large antennas, such as those used by broadcasting stations. The shorter, visible light waves are produced by the motions of electrons within atoms and molecules. The shortest waves, which also have the highest frequency, are associated with γ (gamma) rays (see Chapter 2), which result from changes within the nucleus of the atom. As we will see shortly, the higher the frequency, the more energetic the radiation. Thus, ultraviolet radiation, X rays, and γ rays are high-energy radiation.

Example 7.1   The wavelength of the green light from a traffic signal is centered at 522 nm. What is the frequency of this radiation?

Reasoning and Solution   Rearranging Equation (7.1) and replacing u with c because we are dealing with electromagnetic waves, we get

$$\nu = \frac{c}{\lambda}$$

Recalling that $1 \text{ nm} = 1 \times 10^{-9}$ m (see Table 1.3) and that $c = 3.00 \times 10^8$ m/s, we write

$$\nu = \frac{3.00 \times 10^8 \text{ m/s}}{522 \text{ nm} \times \dfrac{1 \times 10^{-9} \text{ m}}{1 \text{ nm}}}$$

$$= 5.75 \times 10^{14}\text{/s}, \quad \text{or} \quad 5.75 \times 10^{14} \text{ Hz}$$

Similar problems: 7.7, 7.8.

Comment   Thus, 5.75×10^{14} waves pass a fixed point every second. This very high frequency is in accordance with the very high speed of light.

Practice Exercise   What is the wavelength (in meters) of an electromagnetic wave whose frequency is 3.64×10^7 Hz?

Planck's Quantum Theory

When solids are heated, they emit radiation over a wide range of wavelengths. The dull red glow of an electric heater and the bright white light of a tungsten lightbulb are examples of radiation from solids heated to different temperatures.

Measurements taken in the latter part of the nineteenth century showed that the amount of radiant energy emitted depends on the wavelength. Attempts to account for this dependence in terms of established wave theory and thermodynamic laws were only partially successful. One theory explained short-wavelength dependence but failed to account for the longer wavelengths. Another theory accounted for the longer wavelengths but failed for short wavelengths. It seemed that something fundamental was missing from the laws of classical physics.

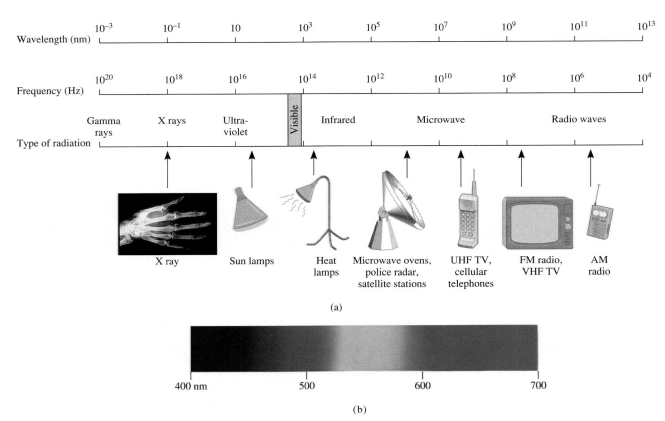

Figure 7.2 *(a) Types of electromagnetic radiation. Gamma rays have the shortest wavelength and highest frequency; radio waves have the longest wavelength and the lowest frequency. Each type of radiation is spread over a specific range of wavelengths (and frequencies). (b) Visible light ranges from a wavelength of 400 nm (violet) to 700 nm (red).*

Planck solved the problem with an assumption that departed drastically from accepted concepts. Classical physics assumed that atoms and molecules could emit (or absorb) any arbitrary amount of radiant energy. Planck said that atoms and molecules could emit (or absorb) energy only in discrete quantities, like small packages or bundles. Planck gave the name **quantum** to *the smallest quantity of energy that can be emitted (or absorbed) in the form of electromagnetic radiation.* The energy E of a single quantum of energy is given by

$$E = h\nu \tag{7.2}$$

in which h is called *Planck's constant* and ν is the frequency of radiation. The value of Planck's constant is 6.63×10^{-34} J s.

According to Planck's quantum theory, energy is always emitted in multiples of $h\nu$; for example, $h\nu$, $2\,h\nu$, $3\,h\nu$, . . . , but never, for example, $1.67\,h\nu$ or $4.98\,h\nu$. At the time Planck presented his theory, he could not explain why energies should be fixed or quantized in this manner. Starting with this hypothesis, however, he had no trouble correlating the experimental data for emission by solids over the *entire* range of wavelengths; they all supported the quantum theory.

The idea that energy should be quantized, or "bundled," may seem strange at first, but the concept of quantization has many analogies. For example, an electric charge is also quantized; there can be only whole-number multiples of e, the charge of one electron. Matter itself is quantized, for the numbers of electrons, protons, and neutrons and the numbers of atoms in a sample of matter must also be integers. Our money system is based on a "quantum" of value called a penny. Even processes in living systems involve quantized phenomena. The eggs laid by hens are quantized, and pregnant cats give birth to an integral number of kittens, not to one-half or three-quarters of a kitten.

7.2 The Photoelectric Effect

In 1905, only 5 years after Planck presented his quantum theory, Albert Einstein used the theory to solve another mystery in physics, the *photoelectric effect*. Experiments had already demonstrated that electrons were ejected from the surface of certain metals exposed to light of at least a certain minimum frequency, called the threshold frequency (Figure 7.3). The number of electrons ejected was proportional to the intensity (or brightness) of the light, but the energies of the ejected electrons were not. Below the threshold frequency no electrons were ejected, no matter how intense the light.

The observed phenomenon could not be explained by the wave theory of light. Einstein, however, made an extraordinary assumption. He suggested that a beam of light is a stream of particles. These *particles of light* are called **photons.** Using Planck's quantum theory of radiation as a starting point, Einstein deduced that each photon must possess energy E, given by the equation

$$E = h\nu$$

in which ν is the frequency of light. [This equation has the same form as Equation (7.2) because, as we will see shortly, electromagnetic radiation is emitted as well as absorbed in the form of photons.]

Electrons are held in a metal by attractive forces, and so to remove them from the metal we must employ light of a sufficiently high frequency (which corresponds to sufficiently high energy) to break them free. Shining a beam of light onto a metal surface can be thought of as shooting a beam of particles—photons—at the metal atoms.

Incident light

Metal

Voltage source

Meter

Figure 7.3 *An apparatus for studying the photoelectric effect. Light of a certain frequency falls on a clean metal surface. Ejected electrons are attracted toward the positive electrode. The flow of electrons is registered by a detecting meter.*

If the frequency of the photons is such that $h\nu$ is exactly equal to the binding energy of the electrons in the metal, then the light will have just enough energy to knock the electrons loose. If we use light of a higher frequency, then not only will the electrons be knocked loose, but they will also acquire some kinetic energy. This situation is summarized by the equation

$$h\nu = \text{KE} + \text{BE} \qquad (7.3)$$

in which KE is the kinetic energy of the ejected electron and BE is the binding energy of the electron in the metal. Rewriting Equation (7.3) as

$$\text{KE} = h\nu - \text{BE}$$

shows that the more energetic the photon is (that is, the higher its frequency), the greater the kinetic energy of the ejected electron.

Now consider two beams of light having the same frequency (which is greater than the threshold frequency) but different intensities. The more intense beam of light consists of a larger number of photons; consequently, the number of electrons ejected from the metal's surface is greater than the number of electrons produced by the weaker beam of light. Thus the more intense the light, the greater the number of electrons emitted by the target metal; the higher the frequency of the light, the greater the kinetic energy of the emitted electrons.

Example 7.2   Calculate the energy (in joules) of (a) a photon with a wavelength of 5.00×10^4 nm (infrared region) and (b) a photon with a wavelength of 5.00×10^{-2} nm (X-ray region).

Reasoning and Solution   (a) The solution to this problem takes two steps. To calculate the energy of a photon, we need Equation (7.2). Because we are given the wavelength of the photon, we need to first convert it to frequency using Equation (7.1). Combining these two steps we get

$$E = h\nu$$

$$= h\frac{c}{\lambda}$$

$$= \frac{(6.63 \times 10^{-34}\,\text{J}\cdot\text{s})\,(3.00 \times 10^8\,\text{m/s})}{(5.00 \times 10^4\,\text{nm})\,\dfrac{1 \times 10^{-9}\,\text{m}}{1\,\text{nm}}}$$

$$= 3.98 \times 10^{-21}\,\text{J}$$

This is the energy of a single photon with a 5.00×10^4 nm wavelength.

(b) Following the same procedure as in (a), we can show that the energy of the photon that has a wavelength of 5.00×10^{-2} nm is 3.98×10^{-15} J. Thus an "X-ray" photon is 1×10^6, or a million times, more energetic than an "infrared" photon in our case.

Practice Exercise   The energy of a photon is 5.87×10^{-20} J. What is its wavelength (in nanometers)?

Similar problem: 7.15.

Einstein's theory of light posed a dilemma for scientists. On the one hand, it explains the photoelectric effect satisfactorily. On the other hand, the particle theory of light is not consistent with the known wave behavior of light. The only way to resolve the dilemma is to accept the idea that light possesses *both* particlelike and wavelike properties. Depending on the experiment, light behaves either as a wave or as a stream of particles. This concept was totally alien to the way physicists had thought about matter and radiation, and it took a long time for them to accept it. It turns out that the property of dual nature (particles and waves) is not unique to light but is characteristic of all matter, including submicroscopic particles like electrons, as we will see in Section 7.4.

7.3 Bohr's Theory of the Hydrogen Atom

Einstein's work paved the way for the solution of yet another nineteenth-century "mystery" in physics: the emission spectra of atoms.

Emission Spectra

Ever since the seventeenth century, when Newton showed that sunlight is composed of various color components that can be recombined to produce white light, scientists have studied the characteristics of **emission spectra,** that is, *either continuous or line spectra of radiation emitted by substances.* The emission spectrum of a substance can be seen by energizing a sample of material either with thermal energy or with some other form of energy (such as a high-voltage electrical discharge if the substance is gaseous). A "red-hot" or "white-hot" iron bar freshly removed from a high-temperature source glows in a characteristic way. This visible glow is the portion of its emission spectrum that is sensed by eye. The warmth of the same iron bar represents another portion of its emission spectrum—in the infrared region. A feature common to the emission spectra of the sun and of a heated solid is that their spectra are continuous; that is, all wavelengths of light are represented in the spectra (see the visible region in Figure 7.2).

Emission spectra

The emission spectra of atoms in the gas phase, on the other hand, do not show a continuous spread of wavelengths from red to violet; rather, the atoms *emit light only at specific wavelengths.* Such spectra are called **line spectra** because the radiation is identified by the appearance of bright lines in the spectra. Figure 7.4 is a schematic diagram of a discharge tube that is used to study emission spectra and Figure 7.5 shows the color emitted by hydrogen atoms in a discharge tube.

Every element has a unique emission spectrum. The characteristic lines in atomic spectra can be used in chemical analysis to identify unknown atoms, much as fingerprints are used to identify people. When the lines of the emission spectrum of a known element exactly match the lines of the emission spectrum of an unknown sample, the identity of the sample is quickly established. Although the utility of this procedure was recognized some time ago, the origin of these lines was unknown until the early 1900s.

Emission Spectrum of the Hydrogen Atom

In 1913, not too long after Planck's and Einstein's discoveries, the Danish physicist Niels Bohr offered a theoretical explanation for the emission spectrum of the hydrogen atom. Bohr's treatment is very complex and is no longer considered to be correct in all its details. Thus we will concentrate only on his important assumptions and final results, which do account for the spectral lines.

When Bohr first tackled this problem, physicists already knew that the atom contains electrons and protons. They thought of an atom as an entity in which electrons whirled

Figure 7.4 *Top: An experimental arrangement for studying the emission spectra of atoms and molecules. The gas under study is in a discharge tube containing two electrodes. As electrons flow from the negative electrode to the positive electrode, they collide with the gas. This collision process eventually leads to the emission of light by the atoms (or molecules). The emitted light is separated into its components by a prism. Each component color is focused at a definite position, according to its wavelength, and forms a colored image of the slit on the photographic plate. The colored images are called spectral lines. Bottom: The line emission spectrum of hydrogen atoms in the visible region.*

Figure 7.5 *Color emitted by hydrogen atoms in a discharge tube. The color observed results from the combination of the colors emitted in the visible spectrum.*

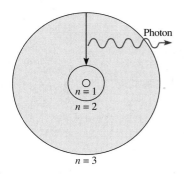

Figure 7.6 *The emission process in an excited hydrogen atom, according to Bohr's theory. An electron originally in a higher-energy orbit (n = 3) falls back to a lower-energy orbit (n = 2). As a result, a photon with energy hν is given off. The value of hν is equal to the difference in energies of the two orbits occupied by the electron in the emission process. For simplicity, only three orbits are shown.*

around the nucleus in circular orbits at high velocities. This was an appealing model because it resembled the well-understood motions of the planets around the sun. In the hydrogen atom, it was believed that the electrostatic attraction between the positive "solar" proton and the negative "planetary" electron pulls the electron inward and that this force is balanced exactly by the acceleration caused by the circular motion of the electron.

Bohr's model for the atom included the idea of electrons moving in circular orbits, but he imposed a rather severe restriction: The single electron in the hydrogen atom could be located only in certain orbits. Because each orbit has a particular energy associated with it, the energies associated with electron motion in the permitted orbits must be fixed in value, or quantized.

The emission of radiation by an energized hydrogen atom, then, can be attributed to the electron dropping from a higher-energy orbit to a lower one and giving up a quantum of energy (a photon) in the form of light (Figure 7.6). Using arguments based on electrostatic interaction and Newton's laws of motion, Bohr showed that the energies that the electron in the hydrogen atom can possess are given by

$$E_n = -R_H\left(\frac{1}{n^2}\right) \qquad (7.4)$$

in which R_H, the Rydberg constant, has the value 2.18×10^{-18} J. The number n is an integer called the *principal quantum number;* it has the values $n = 1, 2, 3, \ldots$.

The negative sign in Equation (7.4) may seem strange, for it implies that all the allowable energies of the electron are negative. Actually, this sign is nothing more than an arbitrary convention; it signifies that the energy of the electron in the atom is *lower* than the energy of a *free electron,* or an electron that is infinitely far from the nucleus. The energy of a free electron is arbitrarily assigned a value of zero. Mathematically, this corresponds to setting n equal to infinity (∞) in Equation (7.4), so that $E_\infty = 0$. As the electron gets closer to the nucleus (as n decreases), E_n becomes larger in absolute value, but also more negative. The most negative value, then, is reached when $n = 1$, which corresponds to the most stable orbit. We call this the **ground state** or the **ground level,** which refers to *the lowest energy state of a system* (which is an atom in our discussion). The stability of the electron diminishes for $n = 2, 3, \ldots$, and each of these is called an **excited state** or **excited level,** which is *higher in energy than the ground state.* An electron in a hydrogen atom that occupies an orbit with n greater than 1 is said to be in an excited state. The radius of each circular orbit depends on n^2. Thus, as n increases from 1 to 2 to 3, the orbit radius increases in size very rapidly. The higher the excited state, the farther away the electron is from the nucleus (and the less tightly it is held by the nucleus).

Figure 7.7 *A mechanical analogy for the emission processes. The ball can rest on any step but not between steps.*

Bohr's theory enables us to explain the line spectra of the hydrogen atom. Radiant energy absorbed by the atom causes the electron to move from a lower-energy orbit (characterized by a smaller n value) to a higher-energy orbit (characterized by a larger n value). Conversely, radiant energy (in the form of a photon) is emitted when the electron moves from a higher-energy orbit to a lower-energy orbit. The quantized movement of the electron from one orbit to another is analogous to the movement of a tennis ball either up or down a set of stairs (Figure 7.7). The ball can be on any of a variety of different stairs but never between stairs. The journey from a lower stair to a higher one is an energy-requiring process, whereas that from a higher stair to a lower stair is an energy-releasing process. The quantity of energy involved in either type of change is determined by the distance between the beginning and ending stairs. Similarly, the amount of energy needed to move an electron in the Bohr atom depends on the difference in energy levels of the initial and final states.

Let us now apply Equation (7.4) to the emission process in a hydrogen atom. Suppose that the electron is initially in an excited state characterized by the principal quantum number n_i. During emission, the electron drops to a lower-energy state characterized by the principal quantum number n_f (the subscripts i and f denote the initial and final states, respectively). This lower-energy state may be either another excited state or the ground state. The difference between the energies of the initial and final states is

$$\Delta E = E_f - E_i$$

From Equation (7.4),

$$E_f = -R_H\left(\frac{1}{n_f^2}\right)$$

and

$$E_i = -R_H\left(\frac{1}{n_i^2}\right)$$

Therefore

$$\Delta E = \left(\frac{-R_H}{n_f^2}\right) - \left(\frac{-R_H}{n_i^2}\right)$$

$$= R_H\left(\frac{1}{n_i^2} - \frac{1}{n_f^2}\right)$$

Table 7.1	The Various Series in Atomic Hydrogen Emission Spectrum		
Series	n_f	n_i	**Spectrum Region**
Lyman	1	2, 3, 4, . . .	Ultraviolet
Balmer	2	3, 4, 5, . . .	Visible and ultraviolet
Paschen	3	4, 5, 6, . . .	Infrared
Brackett	4	5, 6, 7, . . .	Infrared

Because this transition results in the emission of a photon of frequency ν and energy $h\nu$ (see Figure 7.6), we can write

$$\Delta E = h\nu = R_H\left(\frac{1}{n_i^2} - \frac{1}{n_f^2}\right) \tag{7.5}$$

When a photon is emitted, $n_i > n_f$. Consequently the term in parentheses is negative and ΔE is negative (energy is lost to the surroundings). When energy is absorbed, $n_i < n_f$ and the term in parentheses is positive, so ΔE is positive. Each spectral line in the emission spectrum corresponds to a particular transition in a hydrogen atom. When we study a large number of hydrogen atoms, we observe all possible transitions and hence the corresponding spectral lines. The brightness of a spectral line depends on how many photons of the same wavelength are emitted.

The emission spectrum of hydrogen covers a wide range of wavelengths from the infrared to the ultraviolet. Table 7.1 lists the series of transitions in the hydrogen spectrum; they are named after their discoverers. The Balmer series is particularly easy to study because a number of its lines fall in the visible range.

Figure 7.8 *The energy levels in the hydrogen atom and the various emission series. Each energy level corresponds to the energy associated with an allowed energy state for an orbit, as postulated by Bohr and shown in Figure 7.4 for the Balmer series. The emission lines are labeled according to the scheme in Table 7.1.*

Figure 7.6 shows a single transition. However, it is more informative to express transitions as shown in Figure 7.8. Each horizontal line represents an *energy level* associated with the particular orbit. The orbits are labeled with their principal quantum numbers.

Example 7.3   What is the wavelength (in nanometers) of a photon emitted during a transition from the $n_i = 5$ state to the $n_f = 2$ state in the hydrogen atom?

Reasoning and Solution   Because $n_f < n_i$, this is an emission process and the fact that $n_f = 2$ means that the transition will give rise to a spectral line in the Balmer series (see Figure 7.8). We can calculate the energy change and hence the energy of the emitted photon using Equation (7.5)

$$\Delta E = R_H \left(\frac{1}{n_i^2} - \frac{1}{n_f^2} \right)$$

$$= 2.18 \times 10^{-18} \text{ J} \left(\frac{1}{5^2} - \frac{1}{2^2} \right)$$

$$= -4.58 \times 10^{-19} \text{ J}$$

The negative sign indicates that this is energy associated with an emission process. To calculate the wavelength we will omit the minus sign for ΔE because the wavelength of the photon must be positive. Because $\Delta E = h\nu$ or $\nu = \Delta E/h$, we can calculate the wavelength of the photon by writing

$$\lambda = \frac{c}{\nu}$$

$$= \frac{ch}{\Delta E}$$

$$= \frac{(3.00 \times 10^8 \text{ m/s})(6.63 \times 10^{-34} \text{ J s})}{4.58 \times 10^{-19} \text{ J}}$$

$$= 4.34 \times 10^{-7} \text{ m}$$

$$= 4.34 \times 10^{-7} \text{ m} \times \left(\frac{1 \times 10^9 \text{ nm}}{1 \text{ m}} \right) = 434 \text{ nm}$$

which is in the visible region of the electromagnetic region (see Figure 7.2).

Similar problems: 7.31, 7.32.

Practice Exercise   What is the wavelength (in namometers) of a photon emitted during a transition from $n_i = 6$ to $n_f = 4$ state in the H atom?

7.4 The Dual Nature of the Electron

Physicists were mystified but intrigued by Bohr's theory. They questioned why the energies of the hydrogen electron are quantized. Or, phrasing the question in a more concrete way, why is the electron in a Bohr atom restricted to orbiting the nucleus at certain fixed distances? For a decade no one, not even Bohr himself, had a logical explanation. In 1924 the French physicist Louis de Broglie provided a solution to this puzzle. De Broglie reasoned as follows: If light waves can behave like a stream of

Figure 7.9 *The standing waves generated by plucking a guitar string. Each dot represents a node. The length of the string (ℓ) must be equal to a whole number times one-half the wavelength ($\lambda/2$).*

(a)

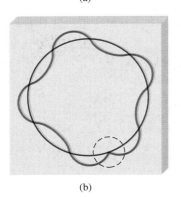

(b)

Figure 7.10 *(a) The circumference of the orbit is equal to an integral number of wavelengths. This is an allowed orbit. (b) The circumference of the orbit is not equal to an integral number of wavelengths. As a result, the electron wave does not close on itself evenly. This is a nonallowed orbit.*

particles (photons), then perhaps particles such as electrons can possess wave properties. According to de Broglie, an electron bound to the nucleus behaves like a *standing wave*. Standing waves can be generated by plucking, say, a guitar string (Figure 7.9). The waves are described as standing, or stationary, because they do not travel along the string. Some points on the string, called **nodes**, do not move at all; that is, *the amplitude of the wave at these points is zero*. There is a node at each end, and there may be nodes between the ends. The greater the frequency of vibration, the shorter the wavelength of the standing wave and the greater the number of nodes. As Figure 7.9 shows, only certain wavelengths can be in any of the allowed motions of the string.

De Broglie argued that if an electron does behave like a standing wave in the hydrogen atom, the length of the wave must fit the circumference of the orbit exactly (Figure 7.10). Otherwise the wave would partially cancel itself on each successive orbit; eventually the amplitude of the wave would be reduced to zero, and the wave would not exist.

The relation between the circumference of an allowed orbit ($2\pi r$) and the wavelength (λ) of the electron is given by

$$2\pi r = n\lambda \tag{7.6}$$

in which r is the radius of the orbit, λ is the wavelength of the electron wave, and $n = 1, 2, 3, \ldots$. Because n is an integer, it follows that r can have only certain values as n increases from 1 to 2 to 3 and so on. And because the energy of the electron depends on the size of the orbit (or the value of r), its value must be quantized.

De Broglie's reasoning led to the conclusion that waves can behave like particles and particles can exhibit wave properties. De Broglie deduced that particle and wave properties are related by the expression

$$\lambda = \frac{h}{mu} \tag{7.7}$$

in which λ, m, and u are the wavelengths associated with a moving particle, its mass, and its velocity, respectively. Equation (7.7) implies that a particle in motion can be treated as a wave and a wave can exhibit the properties of a particle.

Example 7.4 Calculate the wavelength of the "particle" in the following two cases:
(a) The fastest serve in tennis is about 140 miles per hour, or 62 m/s. Calculate the wavelength associated with a 6.0×10^{-2}-kg tennis ball traveling at this velocity.
(b) Calculate the wavelength associated with an electron moving at 62 m/s.

Reasoning and Solution (a) This is a direct application of the de Broglie relation [Equation (7.7)]. We start with

$$\lambda = \frac{h}{mu}$$

$$= \frac{6.63 \times 10^{-34} \text{ J} \cdot \text{s}}{6.0 \times 10^{-2} \text{ kg} \times 62 \text{ m/s}}$$

The conversion factor is $1 \text{ J} = 1 \text{ kg} \cdot \text{m}^2/\text{s}^2$ (see Section 5.6). Therefore

$$\lambda = 1.8 \times 10^{-34} \text{ m}$$

This is an exceedingly small wavelength, because the size of an atom itself is on the order of 1×10^{-10} m. For this reason, the wave properties of such a tennis ball cannot be detected by any existing measuring device.

(b) In this case

$$\lambda = \frac{h}{mu}$$

$$= \frac{6.63 \times 10^{-34} \text{ J} \cdot \text{s}}{9.1095 \times 10^{-31} \text{ kg} \times 62 \text{ m/s}}$$

where 9.1095×10^{-31} kg is the mass of an electron. Proceeding as in (a) we obtain $\lambda = 1.2 \times 10^{-5}$ m, or 1.2×10^4 nm, which is in the infrared region.

Practice Exercise Calculate the wavelength (in nanometers) of a H atom (mass = 1.674×10^{-27} kg) moving at 7.00×10^2 cm/s.

The operation of an electron microscope is based on the wave properties of the electron.

Similar problems: 7.40, 7.42.

Example 7.4 shows that although de Broglie's equation can be applied to diverse systems, the wave properties become observable only for submicroscopic objects. This distinction is brought about by the smallness of Planck's constant, h, which appears in the numerator in Equation (7.7).

Shortly after de Broglie advanced his equation, Clinton Davisson and Lester Germer in the United States and G. P. Thomson in England demonstrated that electrons do indeed possess wavelike properties. By passing a beam of electrons through a thin piece of gold foil, Thomson obtained a set of concentric rings on a screen, similar to the pattern observed when X rays (which are waves) were used. Figure 7.11 on p. 206 shows such patterns for aluminum.

7.5 Quantum Mechanics

The spectacular success of Bohr's theory was followed by a series of disappointments. Bohr's approach could not account for the emission spectra of atoms containing more than one electron, such as atoms of helium and lithium. Nor did it explain why extra lines appear in the hydrogen emission spectrum when a magnetic field is applied. Another problem arose with the discovery that electrons are wavelike: How can the "position" of a wave be specified? We cannot define the precise location of a wave because a wave extends in space.

In reality, Bohr's theory accounted for the observed emission spectra of He$^+$ and Li^{2+} ions, as well as that of hydrogen. However, all three systems have one feature in common—each contains a single electron. Thus the Bohr model worked successfully only for the hydrogen atom and for "hydrogenlike ions."

Figure 7.11 *Left: X-ray diffraction pattern of aluminum foil. Right: Electron diffraction of aluminum foil. The similarity of these two patterns shows that electrons can behave like X rays and display wave properties.*

The dual nature of electrons was particularly troublesome because of the electron's exceedingly small mass. To describe the problem of trying to locate a subatomic particle that behaves like a wave, the German physicist Werner Heisenberg formulated what is now known as the **Heisenberg uncertainty principle:** *It is impossible to know simultaneously both the momentum (mass times velocity) and the position of a particle with certainty.* In other words, to get a precise measurement of the momentum of a particle we must settle for less precise knowledge of the particle's position, and vice versa. Applying the Heisenberg uncertainty principle to the hydrogen atom, we see that it is inherently impossible to know simultaneously both the precise location and precise momentum of the electron. Thus it is not appropriate to imagine the electron circling the nucleus in well-defined orbits.

To be sure, Bohr made a significant contribution to our understanding of atoms, and his suggestion that the energy of an electron in an atom is quantized remains unchallenged. But his theory did not provide a complete description of electronic behavior in atoms. In 1926 the Austrian physicist Erwin Schrödinger, using a complicated mathematical technique, formulated an equation that describes the behavior and energies of submicroscopic particles in general, an equation analogous to Newton's laws of motion for macroscopic objects. The *Schrödinger equation* requires advanced calculus to solve, and we will not discuss it here. It is important to know, however, that the equation incorporates both particle behavior, in terms of mass m, and wave behavior, in terms of a *wave function* ψ (psi), which depends on the location in space of the system (such as an electron in an atom).

The wave function itself has no direct physical meaning. However, the probability of finding the electron in a certain region in space is proportional to the square of the wave function, ψ^2. The idea of relating ψ^2 to probability stemmed from a wave theory analogy. According to wave theory, the intensity of light is proportional to the square of the amplitude of the wave, or ψ^2. The most likely place to find a photon is where the intensity is greatest, that is, where the value of ψ^2 is greatest. A similar argument associates ψ^2 with the likelihood of finding an electron in regions surrounding the nucleus.

Schrödinger's equation began a new era in physics and chemistry, for it launched a new field, *quantum mechanics* (also called *wave mechanics*). We now refer to the developments in quantum theory from 1913—the time Bohr presented his analysis for the hydrogen atom—to 1926 as "old quantum theory."

Quantum Mechanical Treatment of the Hydrogen Atom

The Schrödinger equation, when solved for the hydrogen atom, specifies the possible energy states the electron can occupy and identifies the corresponding wave functions. These energy states and wave functions are characterized by a set of quantum numbers (to be discussed shortly). With these values we can construct a comprehensive model of the hydrogen atom.

Although quantum mechanics tells us that we cannot pinpoint an electron in an atom, it does define the region where the electron might be at a given time. The concept of *electron density* *gives the probability that an electron will be found in a particular region of an atom.* The square of the wave function defines the distribution of electron density in space around the nucleus. Regions of high electron density represent a high probability of locating the electron, whereas the opposite holds for regions of low electron density (Figure 7.12).

To distinguish the quantum mechanical description of an atom from Bohr's model, we speak of an *atomic orbital,* rather than an orbit. An atomic orbital can be thought of as *the wave function of an electron in an atom.* When we say that an electron is in a certain orbital, we mean that the distribution of the electron density or the probability of locating the electron in space is described by the square of the wave function associated with that orbital. An atomic orbital, therefore, has a characteristic energy, as well as a characteristic distribution of electron density.

The Schrödinger equation works nicely for the simple hydrogen atom with its one proton and one electron, but it turns out that it cannot be solved exactly for any atom containing more than one electron! Fortunately, chemists and physicists have learned to get around this kind of difficulty by approximation. For example, although the behavior of electrons in *many-electron atoms* (that is, *atoms containing two or more electrons*) is not the same as in the hydrogen atom, we assume that the difference is probably not too great. Thus we can use the energies and wave functions obtained from the hydrogen atom as good approximations of the behavior of electrons in more complex atoms. In fact, this approach provides fairly good descriptions of electronic behavior in many-electron atoms.

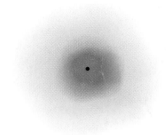

Figure 7.12 *A representation of the electron density distribution surrounding the nucleus in the hydrogen atom. It shows a high probability of finding the electron closer to the nucleus.*

Although the helium atom has only two electrons, in quantum mechanics it is regarded as a many-electron atom.

7.6 Quantum Numbers

In quantum mechanics, three *quantum numbers* are required to *describe the distribution of electrons in hydrogen and other atoms.* These numbers are derived from the mathematical solution of the Schrödinger equation for the hydrogen atom. They are called the *principal quantum number,* the *angular momentum quantum number,* and the *magnetic quantum number.* These quantum numbers will be used to describe atomic orbitals and to label electrons that reside in them. A fourth quantum number—the *spin quantum number*—describes the behavior of a specific electron and completes the description of electrons in atoms.

The Principal Quantum Number (*n*)

The principal quantum number (*n*) can have integral values 1, 2, 3, and so forth; it corresponds to the quantum number in Equation (7.4). In a hydrogen atom, the value

Equation (7.4) holds only for the hydrogen atom.

of n determines the energy of an orbital. As we will see shortly, this is not the case for a many-electron atom. The principal quantum number also relates to the average distance of the electron from the nucleus in a particular orbital. The larger n is, the greater the average distance of an electron in the orbital from the nucleus and therefore the larger the orbital.

The Angular Momentum Quantum Number (ℓ)

The angular momentum quantum number (ℓ) tells us the "shape" of the orbitals (see Section 7.7). The values of ℓ depend on the value of the principal quantum number, n. For a given value of n, ℓ has possible integral values from 0 to $(n - 1)$. If $n = 1$, there is only one possible value of ℓ; that is, $\ell = n - 1 = 1 - 1 = 0$. If $n = 2$, there are two values of ℓ, given by 0 and 1. If $n = 3$, there are three values of ℓ, given by 0, 1, and 2. The value of ℓ is generally designated by the letters $s, p, d, \ldots$ as follows:

ℓ	0	1	2	3	4	5
Name of orbital	s	p	d	f	g	h

Thus if $\ell = 0$, we have an s orbital; if $\ell = 1$, we have a p orbital; and so on.

A collection of orbitals with the same value of n is frequently called a shell. One or more orbitals with the same n and ℓ values are referred to as a subshell. For example, the shell with $n = 2$ is composed of two subshells, $\ell = 0$ and 1 (the allowed values for $n = 2$). These subshells are called the $2s$ and $2p$ subshells where 2 denotes the value of n, and s and p denote the values of ℓ.

Remember that the "2" in 2s refers to the value of n and the "s" symbolizes the value of ℓ.

The Magnetic Quantum Number (m_ℓ)

The magnetic quantum number (m_ℓ) describes the orientation of the orbital in space (to be discussed in Section 7.7). Within a subshell, the value of m_ℓ depends on the value of the angular momentum quantum number, ℓ. For a certain value of ℓ, there are $(2\ell + 1)$ integral values of m_ℓ as follows:

$$-\ell, (-\ell + 1), \ldots 0, \ldots (+\ell - 1), +\ell$$

If $\ell = 0$, then $m_\ell = 0$. If $\ell = 1$, then there are $[(2 \times 1) + 1]$, or three values of m_ℓ, namely, -1, 0, and 1. If $\ell = 2$, there are $[(2 \times 2) + 1]$, or five values of m_ℓ, namely, $-2, -1, 0, 1$, and 2. The number of m_ℓ values indicates the number of orbitals in a subshell with a particular ℓ value.

To conclude our discussion of these three quantum numbers, let us consider a situation in which $n = 2$ and $\ell = 1$. The values of n and ℓ indicate that we have a $2p$ subshell, and in this subshell we have *three* $2p$ orbitals (because there are three values of m_ℓ, given by -1, 0, and 1).

(a)   (b)

Figure 7.13 *The (a) clockwise and (b) counterclockwise spins of an electron. The magnetic fields generated by these two spinning motions are analogous to those from the two magnets. The upward and downward arrows are used to denote the direction of spin.*

The Electron Spin Quantum Number (m_s)

Experiments on the emission spectra of hydrogen and sodium atoms indicated that lines in the emission spectra could be split by the application of an external magnetic field. The only way physicists could explain these results was to assume that electrons act like tiny magnets. If electrons are thought of as spinning on their own axes, as Earth does, their magnetic properties can be accounted for. According to electromagnetic theory, a spinning charge generates a magnetic field, and it is this motion that causes an electron to behave like a magnet. Figure 7.13 shows the two possible spinning motions of an electron, one clockwise and the other counterclockwise. To take the electron spin into account, it is necessary to introduce a fourth quantum number, called the electron spin quantum number (m_s), which has a value of $+\frac{1}{2}$ or $-\frac{1}{2}$.

7.7 Atomic Orbitals

Table 7.2 shows the relation between quantum numbers and atomic orbitals. We see that when $\ell = 0$, $(2\ell + 1) = 1$, and there is only one value of m_ℓ, thus we have an s orbital. When $\ell = 1$, $(2\ell + 1) = 3$, so there are three values of m_ℓ or three p orbitals, labeled p_x, p_y, and p_z. When $\ell = 2$, $(2\ell + 1) = 5$ and there are five values of m_ℓ, and the corresponding five d orbitals are labeled with more elaborate subscripts. In the following sections we will consider the s, p, and d orbitals separately.

That the wave function for an orbital theoretically has no outer limit as one moves outward from the nucleus raises interesting philosophical questions regarding the sizes of atoms. Chemists have agreed on an operational definition of atomic size, as we will see in later chapters.

s Orbitals

One of the important questions we ask when studying the properties of atomic orbitals is, What are the shapes of the orbitals? Strictly speaking, an orbital does not have a well-defined shape because the wave function characterizing the orbital extends from the nucleus to infinity. In that sense, it is difficult to say what an orbital looks like. On the other hand, it is certainly convenient to think of orbitals as having specific shapes, particularly in discussing the formation of chemical bonds between atoms, as we will do in Chapters 9 and 10.

Although in principle an electron can be found anywhere, we know that most of the time it is quite close to the nucleus. Figure 7.14(a) shows the electron density falls off rapidly as the distance from the nucleus increases. Roughly speaking, there is about a 90 percent probability of finding the electron within a sphere of radius 100 pm (1 pm = 1 $\times 10^{-12}$ m) surrounding the nucleus. Thus, we can represent the $1s$ orbital by drawing a **boundary surface diagram** that *encloses about 90 percent of the total electron density in an orbital,* as shown in Figure 7.14(b). A $1s$ orbital represented in this manner is merely a sphere. A more realistic way of viewing electron density distribution is shown in Figure 7.14(c). The $1s$ orbital is divided into successive spherical thin shells. A plot of the probability of finding the electron in each shell, called *radial probability,* as a

n	ℓ	m_ℓ	Number of Orbitals	Atomic Orbital Designations
1	0	0	1	$1s$
2	0	0	1	$2s$
	1	$-1, 0, 1$	3	$2p_x$, $2p_y$, $2p_z$
3	0	0	1	$3s$
	1	$-1, 0, 1$	3	$3p_x$, $3p_y$, $3p_z$
	2	$-2, -1, 0, 1, 2$	5	$3d_{xy}$, $3d_{yz}$, $3d_{xz}$, $3d_{x^2-y^2}$, $3d_{z^2}$
⋮	⋮	⋮	⋮	⋮

Table 7.2 | Relation Between Quantum Numbers and Atomic Orbitals

An s subshell has one orbital, a p subshell has three orbitals, and a d subshell has five orbitals.

Figure 7.14 *(a) The hydrogen 1s orbital. The electron density falls off rapidly as the distance from the nucleus increases. (b) Boundary surface diagram of the hydrogen 1s orbital. (c) The 1s orbital is divided into successive spherical thin shells. A plot of the probability of finding the electron in each shell, called radial probability, as a function of distance shows a maximum at 52.9 pm from the nucleus.*

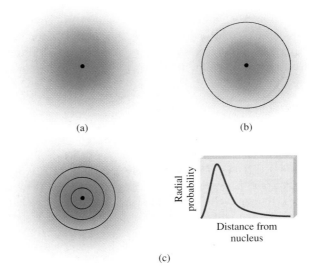

(a)   (b)

(c)

function of distance shows a maximum at 52.9 pm from the nucleus. Interestingly, this is equal to the radius of the innermost ($n = 1$) orbit in the Bohr model.

Figure 7.15 shows boundary surface diagrams for the 1s, 2s, and 3s hydrogen atomic orbitals. All s orbitals are spherical in shape but differ in size, which increases as the principal quantum number increases. Although the details of electron density variation within each boundary surface are lost, there is no serious disadvantage. For us the most important features of atomic orbitals are their shapes and *relative* sizes, which are adequately represented by boundary surface diagrams.

p Orbitals

It should be clear that the *p* orbitals start with the principal quantum number $n = 2$. If $n = 1$, then the angular momentum quantum number ℓ can assume only the value of zero; therefore, there is only a 1s orbital. As we saw earlier, when $\ell = 1$, the magnetic quantum number m_ℓ can have values of -1, 0, 1. Starting with $n = 2$ and $\ell = 1$, we therefore have three 2p orbitals: $2p_x$, $2p_y$, and $2p_z$ (Figure 7.16). The letter subscripts indicate the axes along which the orbitals are oriented. These three *p* orbitals are identical in size, shape, and energy; they differ from one another only in orientation. Note, however, that there is no simple relation between the values of m_ℓ and the *x*, *y*, and *z* directions. For our purpose, you need only remember that because there are three possible values of m_ℓ, there are three *p* orbitals with different orientations.

The boundary surface diagrams of *p* orbitals in Figure 7.16 show that each *p* orbital can be thought of as two lobes on opposite sides of the nucleus. Like s orbitals, *p* orbitals increase in size from 2p to 3p to 4p orbital and so on.

d Orbitals and Other Higher-Energy Orbitals

When $\ell = 2$, there are five values of m_ℓ, which correspond to five *d* orbitals. The lowest value of *n* for a *d* orbital is 3. Because ℓ can never be greater than $n - 1$, when $n = 3$ and $\ell = 2$, we have five 3d orbitals ($3d_{xy}$, $3d_{yz}$, $3d_{xz}$, $3d_{x^2-y^2}$, and $3d_{z^2}$), shown in Figure 7.17. As in the case of the *p* orbitals, the different orientations of the

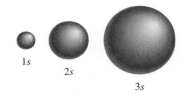

1s
2s
3s

Figure 7.15 *Boundary surface diagrams of the hydrogen 1s, 2s, and 3s orbitals. Each sphere contains about 90 percent of the total electron density. All s orbitals are spherical. Roughly speaking, the size of an orbital is proportional to n^2, where n is the principal quantum number.*

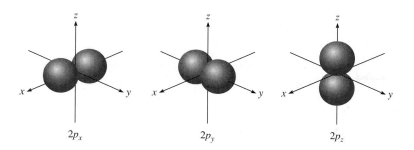

2p_x   2p_y   2p_z

Figure 7.16 *The boundary sur-face diagrams of the three 2p orbitals. These orbitals are iden-tical in shape and energy, but their orientations are different. The p orbitals of higher principal quantum numbers have a similar shape.*

d orbitals correspond to the different values of m_ℓ, but again there is no direct correspondence between a given orientation and a particular m_ℓ value. All the $3d$ orbitals in an atom are identical in energy. The d orbitals for which n is greater than 3 ($4d$, $5d$, . . .) have similar shapes.

Orbitals having higher energy than d orbitals are labeled f, g, . . . , and so on. The f orbitals are important in accounting for the behavior of elements with atomic numbers greater than 57, but their shapes are difficult to represent. In general chemistry we are not concerned with orbitals having ℓ values greater than 3 (the g orbitals and beyond).

Example 7.5 List the values of n, ℓ, and m_ℓ for orbitals in the $4d$ subshell.

Reasoning and Solution As we saw earlier, the number given in the designation of the subshell is the principal quantum number, so in this case $n = 4$. Because we are dealing with d orbitals, $\ell = 2$. The values of m_ℓ can vary from $-\ell$ to ℓ. Therefore, m_ℓ can be -2, -1, 0, 1, 2 (which correspond to the five d orbitals).

Practice Exercise Give the values of the quantum numbers associated with the orbitals in the $3p$ subshell.

Similar problem: 7.55.

Example 7.6 What is the total number of orbitals associated with the principal quan-tum number $n = 3$?

Reasoning and Solution For $n = 3$, the possible values of ℓ are 0, 1, and 2. Thus there is one $3s$ orbital ($n = 3$, $\ell = 0$, and $m_\ell = 0$); there are three $3p$ orbitals ($n = 3$,

---

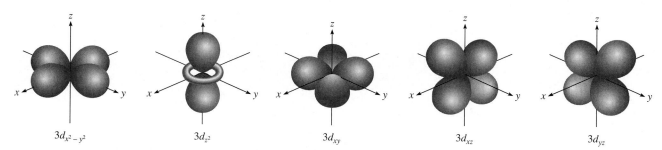

3d_{x^2-y^2}   3d_{z^2}   3d_{xy}   3d_{xz}   3d_{yz}

Figure 7.17 *Boundary surface diagrams of the five 3d orbitals. Although the $3d_{z^2}$ orbital looks different, it is equivalent to the other four orbitals in all other respects. The d orbitals of higher principal quantum numbers have similar shapes.*

Similar problem: 7.60.

$\ell = 1$, and $m_\ell = -1, 0, 1$); there are five $3d$ orbitals ($n = 3$, $\ell = 2$, and $m_\ell = -2$, $-1, 0, 1, 2$). Therefore the total number of orbitals is $1 + 3 + 5 = 9$.

Comment   Can you verify the general relationship that the total number of orbitals in a shell labeled n is n^2?

Practice Exercise   What is the total number of orbitals associated with the principal quantum number $n = 4$?

The Energies of Orbitals

Now that we have some understanding of the shapes and sizes of atomic orbitals, we are ready to inquire into their relative energies and look at how energy levels affect the actual arrangement of electrons in atoms.

According to Equation (7.4), the energy of an electron in a hydrogen atom is determined solely by its principal quantum number. Thus the energies of hydrogen orbitals increase as follows (Figure 7.18):

$$1s < 2s = 2p < 3s = 3p = 3d < 4s = 4p = 4d = 4f < \cdots$$

Although the electron density distributions are different in the $2s$ and $2p$ orbitals, hydrogen's electron has the same energy whether it is in the $2s$ orbital or a $2p$ orbital. The $1s$ orbital in a hydrogen atom corresponds to the most stable condition, the ground state. An electron residing in this orbital is most strongly held by the nucleus because it is closest to the nucleus. An electron in the $2s$, $2p$, or higher orbitals in a hydrogen atom is in an excited state.

The energy picture is more complex for many-electron atoms than for hydrogen. The energy of an electron in such an atom depends on its angular momentum quantum number as well as on its principal quantum number (Figure 7.19). For many-electron atoms, the $3d$ energy level is very close to the $4s$ energy level. The total energy of an atom, however, depends not only on the sum of the orbital energies but also on the energy of repulsion between the electrons in these orbitals (each orbital can accommodate up to

Figure 7.18 *Orbital energy levels in the hydrogen atom. Each short horizontal line represents one orbital. Orbitals with the same principal quantum number (n) all have the same energy.*

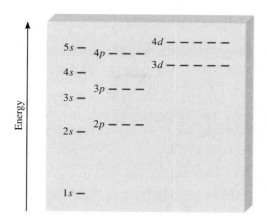

Figure 7.19 *Orbital energy levels in the many-electron atom. Note that the energy level depends on both n and ℓ values.*

two electrons, as we will see in Section 7.8). It turns out that the total energy of an atom is lower when the $4s$ subshell is filled before a $3d$ subshell. Figure 7.20 depicts the order in which atomic orbitals are filled in a many-electron atom. We will consider specific examples in Section 7.8.

7.8 Electron Configuration

The four quantum numbers n, ℓ, m_ℓ, and m_s enable us to label completely an electron in any orbital in an atom. In a sense, we can regard the set of four quantum numbers as the "address" of an electron in the atom, somewhat in the same way that a street address, city, state, and postal ZIP code specify the address of an individual. For example, the four quantum numbers for a $2s$ orbital electron are $n = 2$, $\ell = 0$, $m_\ell = 0$, and $m_s = +\frac{1}{2}$ or $-\frac{1}{2}$. It is inconvenient to write out all the individual quantum numbers, and so we use the simplified notation (n, ℓ, m_ℓ, m_s). For the example here, the quantum numbers are either $(2, 0, 0, +\frac{1}{2})$ or $(2, 0, 0, -\frac{1}{2})$. The value of m_s has no effect on the energy, size, shape, or orientation of an orbital, but it determines how electrons are arranged in an orbital.

Figure 7.20 *The order in which atomic subshells are filled in a many-electron atom. Start with the 1s orbital and move downward, following the direction of the arrows. Thus the order goes as follows: 1s < 2s < 2p < 3s < 3p < 4s < 3d < . . . .*

Example 7.7   Write the four quantum numbers for an electron in a $3p$ orbital.

Reasoning and Solution   To start with, we know that the principal quantum number n is 3 and the angular momentum quantum number ℓ must be 1 (because we are dealing with a p orbital). For $\ell = 1$, there are three values of m_ℓ given by -1, 0, 1. Because the electron spin quantum number m_s can be either $+\frac{1}{2}$ or $-\frac{1}{2}$, we conclude that there are six possible ways to designate the electron:

$$(3, 1, -1, +\tfrac{1}{2}) \qquad (3, 1, -1, -\tfrac{1}{2})$$
$$(3, 1, 0, +\tfrac{1}{2}) \qquad (3, 1, 0, -\tfrac{1}{2})$$
$$(3, 1, 1, +\tfrac{1}{2}) \qquad (3, 1, 1, -\tfrac{1}{2})$$

Practice Exercise   Write the four quantum numbers for an electron in a $5p$ orbital.

Similar problems: 7.55, 7.56.

The hydrogen atom is a particularly simple system because it contains only one electron. The electron may reside in the $1s$ orbital (the ground state), or it may be found in some higher-energy orbital (an excited state). For many-electron atoms, however, we must know the **electron configuration** of the atom, that is, *how the electrons are distributed among the various atomic orbitals,* in order to understand electronic behavior. We will use the first 10 elements (hydrogen to neon) to illustrate the rules for writing electron configurations for atoms in the *ground state.* (Section 7.9 will describe how these rules can be applied to the remainder of the elements in the periodic table.) For this discussion, recall that the number of electrons in an atom is equal to its atomic number Z.

Figure 7.18 indicates that the electron in a ground-state hydrogen atom must be in the $1s$ orbital, so its electron configuration is $1s^1$.

The electron configuration can also be represented by an *orbital diagram* that shows the spin of the electron (see Figure 7.13):

$$\text{H} \quad \boxed{\uparrow}$$
$$1s^1$$

Remember that the direction of electron spin has no effect on the energy of the electron.

The upward arrow denotes one of the two possible spinning motions of the electron. (Alternatively, we could have represented the electron with a downward arrow.) The box represents an atomic orbital.

The Pauli Exclusion Principle

For many-electron atoms we use the **Pauli exclusion principle** (after the Austrian physicist Wolfgang Pauli) to determine electron configurations. This principle states that *no two electrons in an atom can have the same four quantum numbers.* If two electrons in an atom have the same n, ℓ, and m_ℓ values (that is, these two electrons are in the *same* atomic orbital), then they must have different values of m_s. In other words, only two electrons may occupy the same atomic orbital, and these electrons must have opposite spins. Consider the helium atom, which has two electrons. The three possible ways of placing two electrons in the $1s$ orbital are as follows:

$$\text{He} \quad \boxed{\uparrow\uparrow} \quad \boxed{\downarrow\downarrow} \quad \boxed{\uparrow\downarrow}$$
$$\quad\quad 1s^2 \quad\; 1s^2 \quad\; 1s^2$$
$$\quad\quad (a) \quad\;\; (b) \quad\;\; (c)$$

Diagrams (a) and (b) are ruled out by the Pauli exclusion principle. In (a), both electrons have the same upward spin and would have the quantum numbers $(1, 0, 0, +\frac{1}{2})$; in (b), both electrons have downward spins and would have the quantum numbers $(1, 0, 0, -\frac{1}{2})$. Only the configuration in (c) is physically acceptable, because one electron has the quantum numbers $(1, 0, 0, +\frac{1}{2})$ and the other has $(1, 0, 0, -\frac{1}{2})$. Thus the helium atom has the following configuration:

$$\text{He} \quad \boxed{\uparrow\downarrow}$$
$$1s^2$$

Note that $1s^2$ is read "one s two," not "one s squared."

Diamagnetism and Paramagnetism

The Pauli exclusion principle is one of the fundamental principles of quantum mechanics. It can be tested by a simple observation. If the two electrons in the $1s$ orbital of a helium atom had the same, or parallel, spins ($\uparrow\uparrow$ or $\downarrow\downarrow$), their net magnetic fields would reinforce each other. Such an arrangement would make the helium atom *para-magnetic* [Figure 7.21(a)]. ***Paramagnetic*** substances are those that *are attracted by a magnet*. On the other hand, if the electron spins are paired, or antiparallel to each other ($\uparrow\downarrow$ or $\downarrow\uparrow$), the magnetic effects cancel out and the atom is ***diamagnetic*** [Figure 7.21(b)]. Diamagnetic substances are slightly *repelled by a magnet*.

Measurements of magnetic properties provide the most direct evidence for specific electron configurations of elements. Advances in instrument design during the last 20 years or so enable us to determine not only whether an atom is paramagnetic but also how many unpaired electrons are present. By experiment we find that the helium atom is diamagnetic in the ground state, in accord with the Pauli exclusion principle. A useful general rule to keep in mind is that any atom with an *odd* number of electrons *must be* paramagnetic, because we need an even number of electrons for complete pairing. On the other hand, atoms containing an even number of electrons may be either diamagnetic or paramagnetic. We will see the reason for this behavior shortly.

As another example, consider the lithium atom ($Z = 3$), which has three electrons. The third electron cannot go into the $1s$ orbital because it would inevitably have the same four quantum numbers as one of the first two electrons. Therefore, this electron "enters" the next (energetically) higher orbital, which is the $2s$ orbital (see Figure 7.19). The electron configuration of lithium is $1s^2 2s^1$, and its orbital diagram is

$$\text{Li} \quad \boxed{\uparrow\downarrow} \quad \boxed{\uparrow}$$
$$1s^2 \qquad 2s^1$$

The lithium atom contains one unpaired electron and is therefore paramagnetic.

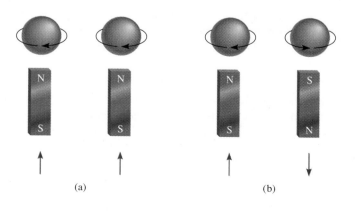

Figure 7.21 *The (a) parallel and (b) antiparallel spins of two electrons. In (a) the two magnetic fields reinforce each other, and the atom is paramagnetic. In (b) the two magnetic fields cancel each other, and the atom is diamagnetic.*

(a)   (b)

The Shielding Effect in Many-Electron Atoms

Experimentally we find that the $2s$ orbital lies at a lower energy level than the $2p$ orbital in a many-electron atom. Why? In comparing the electron configurations of $1s^2 2s^1$ and $1s^2 2p^1$, we note that, in both cases, the $1s$ orbital is filled with two electrons. Because the $2s$ and $2p$ orbitals are larger than the $1s$ orbital, an electron in either of these orbitals will spend more time away from the nucleus (on the average) than an electron in the $1s$ orbital. Thus, we can speak of a $2s$ or $2p$ electron being partly "shielded" from the attractive force of the nucleus by the $1s$ electrons. The important consequence of the shielding effect is that it *reduces* the electrostatic attraction between protons in the nucleus and the electron in the $2s$ or $2p$ orbital.

The manner in which the electron density varies as we move from the nucleus outward depends on the type of orbital. The density near the nucleus is greater for the $2s$ electron than for the $2p$ electron. In other words, a $2s$ electron spends more time near the nucleus than a $2p$ electron does (on the average). For this reason, the $2s$ orbital is said to be more "penetrating" than the $2p$ orbital, and it is less shielded by the $1s$ electrons. In fact, for the same principal quantum number n, the penetrating power decreases as the angular momentum quantum number ℓ increases, or

$$s > p > d > f > \cdots$$

Because the stability of an electron is determined by the strength of its attraction to the nucleus, it follows that a $2s$ electron will be lower in energy than a $2p$ electron. To put it another way, less energy is required to remove a $2p$ electron than a $2s$ electron because a $2p$ electron is not held quite as strongly by the nucleus. The hydrogen atom has only one electron and, therefore, is without such a shielding effect.

Continuing our discussion of atoms of the first 10 elements, we go next to beryllium ($Z = 4$). The ground-state electron configuration of beryllium is $1s^2 2s^2$, or

Beryllium atoms are diamagnetic, as we would expect.

The electron configuration of boron ($Z = 5$) is $1s^2 2s^2 2p^1$, or

Note that the unpaired electron can be in the $2p_x$, $2p_y$, or $2p_z$ orbital. The choice is completely arbitrary because the three p orbitals are equivalent in energy. As the diagram shows, boron atoms are paramagnetic.

Hund's Rule

The electron configuration of carbon ($Z = 6$) is $1s^2 2s^2 2p^2$. The following are different ways of distributing two electrons among three p orbitals:

$2p_x 2p_y 2p_z$	$2p_x 2p_y 2p_z$	$2p_x 2p_y 2p_z$
(a)	(b)	(c)

None of the three arrangements violates the Pauli exclusion principle, so we must determine which one will give the greatest stability. The answer is provided by **Hund's rule,** which states that *the most stable arrangement of electrons in subshells is the one with the greatest number of parallel spins*. The arrangement shown in (c) satisfies this condition. In both (a) and (b) the two spins cancel each other. Thus the orbital diagram for carbon is

$$C \quad \boxed{\uparrow\downarrow} \quad \boxed{\uparrow\downarrow} \quad \boxed{\uparrow\,|\,\uparrow\,|\;}$$
$$1s^2 \qquad 2s^2 \qquad 2p_x 2p_y 2p_z$$

Qualitatively, we can understand why (c) is preferred to (a). In (a), the two electrons are in the same $2p_x$ orbital, and their proximity results in a greater mutual repulsion than when they occupy two separate orbitals, say $2p_x$ and $2p_y$. The choice of (c) over (b) is more subtle but can be justified on theoretical grounds. The fact that carbon atoms are paramagnetic, each containing two unpaired electrons, is in accord with Hund's rule.

The electron configuration of nitrogen ($Z = 7$) is $1s^2 2s^2 2p^3$:

$$N \quad \boxed{\uparrow\downarrow} \quad \boxed{\uparrow\downarrow} \quad \boxed{\uparrow\,|\,\uparrow\,|\,\uparrow}$$
$$1s^2 \qquad 2s^2 \qquad 2p^3$$

Again, Hund's rule dictates that all three $2p$ electrons have spins parallel to one another; the nitrogen atom is therefore paramagnetic, containing three unpaired electrons.

The electron configuration of oxygen ($Z = 8$) is $1s^2 2s^2 2p^4$. An oxygen atom is paramagnetic because it has two unpaired electrons:

$$O \quad \boxed{\uparrow\downarrow} \quad \boxed{\uparrow\downarrow} \quad \boxed{\uparrow\downarrow\,|\,\uparrow\,|\,\uparrow}$$
$$1s^2 \qquad 2s^2 \qquad 2p^4$$

The electron configuration of fluorine ($Z = 9$) is $1s^2 2s^2 2p^5$. The nine electrons are arranged as:

$$F \quad \boxed{\uparrow\downarrow} \quad \boxed{\uparrow\downarrow} \quad \boxed{\uparrow\downarrow\,|\,\uparrow\downarrow\,|\,\uparrow}$$
$$1s^2 \qquad 2s^2 \qquad 2p^5$$

The fluorine atom is thus paramagnetic, having one unpaired electron.

In neon ($Z = 10$), the $2p$ subshell is completely filled. The electron configuration of neon is $1s^2 2s^2 2p^6$, and *all* the electrons are paired:

$$Ne \quad \boxed{\uparrow\downarrow} \quad \boxed{\uparrow\downarrow} \quad \boxed{\uparrow\downarrow\,|\,\uparrow\downarrow\,|\,\uparrow\downarrow}$$
$$1s^2 \qquad 2s^2 \qquad 2p^6$$

The neon atom thus should be diamagnetic, and experimental observation bears out this prediction.

General Rules for Assigning Electrons to Atomic Orbitals

Based on the preceding examples we can formulate some general rules for determining the maximum number of electrons that can be assigned to the various subshells and orbitals for a given value of n:

- Each shell or principal level of quantum number n contains n subshells. For example, if $n = 2$, then there are two subshells (two values of ℓ) of angular momentum quantum numbers 0 and 1.

- Each subshell of quantum number ℓ contains $2\ell + 1$ orbitals. For example, if $\ell = 1$, then there are three p orbitals.
- No more than two electrons can be placed in each orbital. Therefore, the maximum number of electrons is simply twice the number of orbitals that are employed.
- A quick way to determine the maximum number of electrons that an atom can have in a principal level n is to use the formula $2n^2$.

Example 7.8   What is the maximum number of electrons that can be present in the principal level for which $n = 3$?

Reasoning and Solution   When $n = 3$, then $\ell = 0, 1,$ and 2. The number of orbitals for each value of ℓ is given by

Value of ℓ	Number of Orbitals $(2\ell + 1)$
0	1
1	3
2	5

The total number of orbitals is nine. Because each orbital can accommodate two electrons, the maximum number of electrons that can reside in the orbitals is 2×9, or 18. Or, using the formula $2n^2$, we have $n = 3$, so $2(3^2) = 18$.

Similar problems: 7.61, 7.62.

Practice Exercise   Calculate the total number of electrons that can be present in the principal level for which $n = 4$.

Example 7.9   An oxygen atom has a total of eight electrons. Write the four quantum numbers for each of the eight electrons in the ground state.

Reasoning and Solution   We start with $n = 1$, so $\ell = 0$, a subshell corresponding to the $1s$ orbital. This orbital can accommodate a total of two electrons. Next, $n = 2$, and ℓ may be either 0 or 1. The $\ell = 0$ subshell contains one $2s$ orbital, which can accommodate two electrons. The remaining four electrons are placed in the $\ell = 1$ subshell, which contains three $2p$ orbitals. The orbital diagram is

$$\text{O} \quad \boxed{\uparrow\downarrow} \quad \boxed{\uparrow\downarrow} \quad \boxed{\uparrow\downarrow\,|\,\uparrow\,|\,\uparrow}$$
$$\quad\quad\; 1s^2 \quad\; 2s^2 \quad\quad 2p^4$$

The results are summarized in the following table:

Electron	n	ℓ	m_ℓ	m_s	Orbital
1	1	0	0	$+\frac{1}{2}$	$\Big\}\,1s$
2	1	0	0	$-\frac{1}{2}$	
3	2	0	0	$+\frac{1}{2}$	$\Big\}\,2s$
4	2	0	0	$-\frac{1}{2}$	

Electron	n	ℓ	m_ℓ	m_s	Orbital
5	2	1	-1	$+\frac{1}{2}$	
6	2	1	0	$+\frac{1}{2}$	$2p_x, 2p_y, 2p_z$
7	2	1	1	$+\frac{1}{2}$	
8	2	1	1	$-\frac{1}{2}$	

Of course, the placement of the eighth electron in the orbital labeled $m_\ell = 1$ is completely arbitrary. It would be equally correct to assign it to $m_\ell = 0$ or $m_\ell = -1$.

Practice Exercise   Write a complete set of quantum numbers for each of the electrons in boron (B).

Similar problem: 7.83.

At this point let's summarize what our examination has revealed about ground-state electron configurations and the properties of electrons in atoms:

- No two electrons in the same atom can have the same four quantum numbers. This is the Pauli exclusion principle.
- Each orbital can be occupied by a maximum of two electrons. They must have opposite spins, or different electron spin quantum numbers.
- The most stable arrangement of electrons in a subshell is the one that has the greatest number of parallel spins. This is Hund's rule.
- Atoms in which one or more electrons are unpaired are paramagnetic. Atoms in which all the electron spins are paired are diamagnetic.
- In a hydrogen atom, the energy of the electron depends only on its principal quantum number n. In a many-electron atom, the energy of an electron depends on both n and its angular momentum quantum number ℓ.
- In a many-electron atom the subshells are filled in the order shown in Figure 7.20.
- For electrons of the same principal quantum number, their penetrating power, or proximity to the nucleus, decreases in the order $s > p > d > f$. This means that, for example, more energy is required to separate a $3s$ electron from a many-electron atom than is required to remove a $3p$ electron.

7.9 The Building-Up Principle

Here we will extend the rules used in writing electron configurations for the first 10 elements to the rest of the elements. This process is based on the Aufbau principle. (The German word "Aufbau" means "building up.") The *Aufbau principle* dictates that *as protons are added one by one to the nucleus to build up the elements, electrons are similarly added to the atomic orbitals.* Through this process we gain a detailed knowledge of the ground-state electron configurations of the elements. As we will see later, knowledge of electron configurations helps us to understand and predict the properties of the elements; it also explains why the periodic table works so well.

Table 7.3 gives the ground-state electron configurations of elements from H ($Z = 1$) through Mt ($Z = 109$). The electron configurations of all elements except hydrogen and helium are represented by a **noble gas core,** which *shows in brackets the noble gas element that most nearly precedes the element being considered,* followed by the symbol for the highest filled subshells in the outermost shells. Notice that the

The noble gases.

Table 7.3 | **The Ground-State Electron Configurations of the Elements***

Atomic Number	Symbol	Electron Configuration	Atomic Number	Symbol	Electron Configuration	Atomic Number	Symbol	Electron Configuration
1	H	$1s^1$	37	Rb	$[Kr]5s^1$	73	Ta	$[Xe]6s^24f^{14}5d^3$
2	He	$1s^2$	38	Sr	$[Kr]5s^2$	74	W	$[Xe]6s^24f^{14}5d^4$
3	Li	$[He]2s^1$	39	Y	$[Kr]5s^24d^1$	75	Re	$[Xe]6s^24f^{14}5d^5$
4	Be	$[He]2s^2$	40	Zr	$[Kr]5s^24d^2$	76	Os	$[Xe]6s^24f^{14}5d^6$
5	B	$[He]2s^22p^1$	41	Nb	$[Kr]5s^14d^4$	77	Ir	$[Xe]6s^24f^{14}5d^7$
6	C	$[He]2s^22p^2$	42	Mo	$[Kr]5s^14d^5$	78	Pt	$[Xe]6s^14f^{14}5d^9$
7	N	$[He]2s^22p^3$	43	Tc	$[Kr]5s^24d^5$	79	Au	$[Xe]6s^14f^{14}5d^{10}$
8	O	$[He]2s^22p^4$	44	Ru	$[Kr]5s^14d^7$	80	Hg	$[Xe]6s^24f^{14}5d^{10}$
9	F	$[He]2s^22p^5$	45	Rh	$[Kr]5s^14d^8$	81	Tl	$[Xe]6s^24f^{14}5d^{10}6p^1$
10	Ne	$[He]2s^22p^6$	46	Pd	$[Kr]4d^{10}$	82	Pb	$[Xe]6s^24f^{14}5d^{10}6p^2$
11	Na	$[Ne]3s^1$	47	Ag	$[Kr]5s^14d^{10}$	83	Bi	$[Xe]6s^24f^{14}5d^{10}6p^3$
12	Mg	$[Ne]3s^2$	48	Cd	$[Kr]5s^24d^{10}$	84	Po	$[Xe]6s^24f^{14}5d^{10}6p^4$
13	Al	$[Ne]3s^23p^1$	49	In	$[Kr]5s^24d^{10}5p^1$	85	At	$[Xe]6s^24f^{14}5d^{10}6p^5$
14	Si	$[Ne]3s^23p^2$	50	Sn	$[Kr]5s^24d^{10}5p^2$	86	Rn	$[Xe]6s^24f^{14}5d^{10}6p^6$
15	P	$[Ne]3s^23p^3$	51	Sb	$[Kr]5s^24d^{10}5p^3$	87	Fr	$[Rn]7s^1$
16	S	$[Ne]3s^23p^4$	52	Te	$[Kr]5s^24d^{10}5p^4$	88	Ra	$[Rn]7s^2$
17	Cl	$[Ne]3s^23p^5$	53	I	$[Kr]5s^24d^{10}5p^5$	89	Ac	$[Rn]7s^26d^1$
18	Ar	$[Ne]3s^23p^6$	54	Xe	$[Kr]5s^24d^{10}5p^6$	90	Th	$[Rn]7s^26d^2$
19	K	$[Ar]4s^1$	55	Cs	$[Xe]6s^1$	91	Pa	$[Rn]7s^25f^26d^1$
20	Ca	$[Ar]4s^2$	56	Ba	$[Xe]6s^2$	92	U	$[Rn]7s^25f^36d^1$
21	Sc	$[Ar]4s^23d^1$	57	La	$[Xe]6s^25d^1$	93	Np	$[Rn]7s^25f^46d^1$
22	Ti	$[Ar]4s^23d^2$	58	Ce	$[Xe]6s^24f^15d^1$	94	Pu	$[Rn]7s^25f^6$
23	V	$[Ar]4s^23d^3$	59	Pr	$[Xe]6s^24f^3$	95	Am	$[Rn]7s^25f^7$
24	Cr	$[Ar]4s^13d^5$	60	Nd	$[Xe]6s^24f^4$	96	Cm	$[Rn]7s^25f^76d^1$
25	Mn	$[Ar]4s^23d^5$	61	Pm	$[Xe]6s^24f^5$	97	Bk	$[Rn]7s^25f^9$
26	Fe	$[Ar]4s^23d^6$	62	Sm	$[Xe]6s^24f^6$	98	Cf	$[Rn]7s^25f^{10}$
27	Co	$[Ar]4s^23d^7$	63	Eu	$[Xe]6s^24f^7$	99	Es	$[Rn]7s^25f^{11}$
28	Ni	$[Ar]4s^23d^8$	64	Gd	$[Xe]6s^24f^75d^1$	100	Fm	$[Rn]7s^25f^{12}$
29	Cu	$[Ar]4s^13d^{10}$	65	Tb	$[Xe]6s^24f^9$	101	Md	$[Rn]7s^25f^{13}$
30	Zn	$[Ar]4s^23d^{10}$	66	Dy	$[Xe]6s^24f^{10}$	102	No	$[Rn]7s^25f^{14}$
31	Ga	$[Ar]4s^23d^{10}4p^1$	67	Ho	$[Xe]6s^24f^{11}$	103	Lr	$[Rn]7s^25f^{14}6d^1$
32	Ge	$[Ar]4s^23d^{10}4p^2$	68	Er	$[Xe]6s^24f^{12}$	104	Rf	$[Rn]7s^25f^{14}6d^2$
33	As	$[Ar]4s^23d^{10}4p^3$	69	Tm	$[Xe]6s^24f^{13}$	105	Db	$[Rn]7s^25f^{14}6d^3$
34	Se	$[Ar]4s^23d^{10}4p^4$	70	Yb	$[Xe]6s^24f^{14}$	106	Sg	$[Rn]7s^25f^{14}6d^4$
35	Br	$[Ar]4s^23d^{10}4p^5$	71	Lu	$[Xe]6s^24f^{14}5d^1$	107	Bh	$[Rn]7s^25f^{14}6d^5$
36	Kr	$[Ar]4s^23d^{10}4p^6$	72	Hf	$[Xe]6s^24f^{14}5d^2$	108	Hs	$[Rn]7s^25f^{14}6d^6$
						109	Mt	$[Rn]7s^25f^{14}6d^7$

*The symbol [He] is called the helium core and represents $1s^2$. [Ne] is called the neon core and represents $1s^22s^22p^6$. [Ar] is called the argon core and represents $[Ne]3s^23p^6$. [Kr] is called the krypton core and represents $[Ar]4s^23d^{10}4p^6$. [Xe] is called the xenon core and represents $[Kr]5s^24d^{10}5p^6$. [Rn] is called the radon core and represents $[Xe]6s^24f^{14}5d^{10}6p^6$.

electron configurations of the highest filled subshells in the outermost shells for the elements sodium ($Z = 11$) through argon ($Z = 18$) follow a pattern similar to those of lithium ($Z = 3$) through neon ($Z = 10$).

As mentioned in Section 7.7, the $4s$ subshell is filled before the $3d$ subshell in a many-electron atom (see Figure 7.20). Thus the electron configuration of potassium ($Z = 19$) is $1s^2 2s^2 2p^6 3s^2 3p^6 4s^1$. Because $1s^2 2s^2 2p^6 3s^2 3p^6$ is the electron configuration of argon, we can simplify the electron configuration of potassium by writing [Ar]$4s^1$, where [Ar] denotes the "argon core." Similarly, we can write the electron configuration of calcium ($Z = 20$) as [Ar]$4s^2$. The placement of the outermost electron in the $4s$ orbital (rather than in the $3d$ orbital) of potassium is strongly supported by experimental evidence. The following comparison also suggests that this is the correct configuration. The chemistry of potassium is very similar to that of lithium and sodium, the first two alkali metals. The outermost electron of both lithium and sodium is in an s orbital (there is no ambiguity in assigning their electron configurations); therefore, we expect the last electron in potassium to occupy the $4s$ rather than the $3d$ orbital.

The elements from scandium ($Z = 21$) to copper ($Z = 29$) are transition metals. **Transition metals** either *have incompletely filled d subshells or readily give rise to cations that have incompletely filled d subshells.* Consider the first transition metal series, from scandium through copper. In this series additional electrons are placed in the $3d$ orbitals, according to Hund's rule. However, there are two irregularities. The electron configuration of chromium ($Z = 24$) is [Ar]$4s^1 3d^5$ and not [Ar]$4s^2 3d^4$, as we might expect. A similar break in the pattern is observed for copper, whose electron configuration is [Ar]$4s^1 3d^{10}$ rather than [Ar]$4s^2 3d^9$. The reason for these irregularities is that slightly greater stability is associated with the half-filled ($3d^5$) and completely filled ($3d^{10}$) subshells. Electrons in the same subshell (in this case, the d orbitals) have equal energy but different spatial distributions. Consequently, their shielding of one another is relatively small, and the electrons are more strongly attracted by the nucleus when they have the $3d^5$ configuration. According to Hund's rule, the orbital diagram for Cr is

The transition metals.

Cr [Ar] $\boxed{\uparrow}$ $\boxed{\uparrow\,\uparrow\,\uparrow\,\uparrow\,\uparrow}$

$\quad\quad\quad\quad 4s^1 \quad\quad\quad 3d^5$

Thus, Cr has a total of six unpaired electrons. The orbital diagram for copper is

Cu [Ar] $\boxed{\uparrow}$ $\boxed{\uparrow\downarrow\,\uparrow\downarrow\,\uparrow\downarrow\,\uparrow\downarrow\,\uparrow\downarrow}$

$\quad\quad\quad\quad 4s^1 \quad\quad\quad 3d^{10}$

Again, extra stability is gained in this case by having the $3d$ subshell completely filled.

For elements Zn ($Z = 30$) through Kr ($Z = 36$), the $4s$ and $4p$ subshells fill in a straightforward manner. With rubidium ($Z = 37$), electrons begin to enter the $n = 5$ energy level.

The electron configurations in the second transition metal series [yttrium ($Z = 39$) to silver ($Z = 47$)] are also irregular, but we will not be concerned with the details here.

The sixth period of the periodic table begins with cesium ($Z = 55$) and barium ($Z = 56$), whose electron configurations are [Xe]$6s^1$ and [Xe]$6s^2$, respectively. Next we come to lanthanum ($Z = 57$). From Figure 7.20 we would expect that after filling the $6s$ orbital we would place the additional electrons in $4f$ orbitals. In reality, the energies of the $5d$ and $4f$ orbitals are very close; in fact, for lanthanum $4f$ is slightly higher in energy than $5d$. Thus lanthanum's electron configuration is [Xe]$6s^2 5d^1$ and not [Xe]$6s^2 4f^1$.

Following lanthanum are the 14 elements known as the **lanthanides,** or **rare earth series** [cerium ($Z = 58$) to lutetium ($Z = 71$)]. The rare earth metals *have*

Figure 7.22 *Classification of groups of elements in the periodic table according to the type of outermost subshell being filled with electrons.*

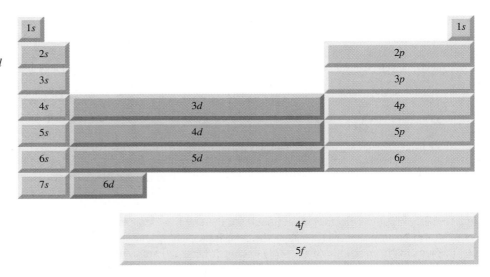

incompletely filled 4f subshells or readily give rise to cations that have incompletely filled 4f subshells. In this series, the added electrons are placed in 4f orbitals. After the 4f subshell is completely filled, the next electron enters the 5d subshell of lutetium. Note that the electron configuration of gadolinium ($Z = 64$) is $[Xe]6s^2 4f^7 5d^1$ rather than $[Xe]6s^2 4f^8$. Like chromium, gadolinium gains extra stability by having a half-filled subshell ($4f^7$).

The third transition metal series, including lanthanum and hafnium ($Z = 72$) and extending through gold ($Z = 79$), is characterized by the filling of the 5d subshell. The 6s and 6p subshells are filled next, which takes us to radon ($Z = 86$).

The *last row of elements* is the ***actinide series,*** which starts at thorium ($Z = 90$). *Most of these elements are not found in nature but have been synthesized.*

With few exceptions, you should be able to write the electron configuration of any element, using Figure 7.20 as a guide. Elements that require particular care are the transition metals, the lanthanides, and the actinides. As we noted earlier, at larger values of the principal quantum number *n,* the order of subshell filling may reverse from one element to the next. Figure 7.22 groups the elements according to the type of subshell in which the outermost electrons are placed.

Example 7.10   Write the ground-state electron configurations for sulfur and for palladium, which is diamagnetic.

Reasoning and Solution   Sulfur ($Z = 16$)

1. Sulfur has 16 electrons.

2. It takes 10 electrons to complete the first and second periods ($1s^2 2s^2 2p^6$). This leaves us 6 electrons to fill the 3s orbital and partially fill the 3p orbitals. Thus the electron configuration of S is

$$1s^2 2s^2 2p^6 3s^2 3p^4$$

or $[Ne]3s^2 3p^4$.

Palladium ($Z = 46$)

1. Palladium has 46 electrons.

2. We need 36 electrons to complete the fourth period, leaving us with 10 more electrons to distribute among the $5s$ and $4d$ orbitals. The three choices are (a) $4d^{10}$, (b) $4d^9 5s^1$, and (c) $4d^8 5s^2$. Because palladium is diamagnetic, its electron configuration must be

$$1s^2 2s^2 2p^6 3s^2 3p^6 4s^2 3d^{10} 4p^6 4d^{10}$$

or simply [Kr]$4d^{10}$. The configurations in (b) and (c) both give paramagnetic Pd atoms.

Practice Exercise Write the ground-state electron configuration for phosphorus (P).

Similar problems: 7.81, 7.82.

Summary

Several major discoveries by physicists in the early twentieth century made a profound impact on our understanding of atoms and molecules and revolutionized physics. Planck's quantum theory explained the way heated solids emit radiant energy. Radiation was already known to have the properties of waves, described in terms of wavelength, frequency, and amplitude. Quantum theory suggested that radiant energy is emitted by atoms and molecules in small discrete amounts (quanta), rather than over a continuous range. This behavior is governed by the relationship $E = h\nu$. Energy is always emitted in whole-number multiples of $h\nu$ ($1\ h\nu$, $2\ h\nu$, $3\ h\nu$, . . .).

Einstein extended the quantum theory to explain the photoelectric effect by proposing that light can behave like a stream of particles (photons).

Bohr developed a model of the hydrogen atom in which the energy of the single electron orbiting the nucleus is quantized—limited to certain energy values determined by an integer, the principal quantum number. An electron in its most stable energy state is said to be in the ground state, and an electron having more energy than its most stable state is said to be in an excited state. In the Bohr model, an electron emits a photon when it drops from a higher-energy orbit (an excited state) to a lower-energy orbit (a less excited state). This model accounts for the specific energies represented by the lines in the hydrogen emission spectrum.

The idea that light has the characteristics of both particles and waves was extended by de Broglie to all matter in motion by the equation $\lambda = h/mu$.

The Schrödinger equation describes the motions and energies of submicroscopic particles in terms of quantum theory. It tells us the possible energy states of the electron in a hydrogen atom and the probability of its location in a particular region surrounding the nucleus. These results can be applied with reasonable accuracy to many-electron atoms.

An atomic orbital is a wave function and the square of the wave function defines the distribution of electron density in space. Orbitals are represented by boundary surface diagrams. Four quantum numbers are needed to characterize completely each electron in an atom: the principal quantum number n, which determines the main energy level, or shell, of the orbital; the angular momentum quantum number ℓ, which determines the shape of the orbital; the magnetic quantum number m_ℓ, which determines the orientation of the orbital in space; and the electron spin quantum number m_s, which relates only to the electron's spin on its own axis.

The single s orbital in each energy level is spherical and centered on the nucleus. The three p orbitals each have two lobes, and the pairs of lobes are tangential to each other. There are five d orbitals, which have more complex shapes and orientations in space.

The energy of the electron in a hydrogen atom is determined solely by its principal quantum number. In many-electron atoms, both the principal quantum number and the angular momentum quantum number determine the energy of an electron. No two electrons in the same atom can have the same four quantum numbers (the Pauli exclusion principle). The most stable arrangement of electrons in a subshell is the one that has the greatest number of parallel spins (Hund's rule). Atoms with one or more unpaired spins are paramagnetic. Atoms with all electron spins paired are diamagnetic.

Using the Aufbau (building-up) principle we can determine the ground-state electron configurations of all elements based on atomic number and location in the periodic table.

Key Words

Actinide series p. 222	Electron configuration	Heisenberg uncertainty	Pauli exclusion principle
Amplitude p. 194	p. 214	principle p. 206	p. 214
Atomic orbital p. 207	Electron density p. 207	Hund's rule p. 217	Photon p. 197
Aufbau principle	Emission spectrum	Lanthanide series p. 221	Quantum p. 197
p. 219	p. 199	Line spectrum p. 199	Quantum numbers p. 207
Boundary surface diagram	Excited level (or state)	Many-electron atom	Rare earth series p. 221
p. 209	p. 201	p. 207	Transition metals p. 221
Diamagnetic p. 215	Frequency (ν) p. 194	Noble gas core p. 219	Wave p. 194
Electromagnetic radiation	Ground level (or state)	Node p. 204	Wavelength (λ) p. 194
p. 195	p. 201	Paramagnetic p. 215	

Questions and Problems

Quantum Theory and Electromagnetic Radiation

Review Questions

7.1 What is a wave? Explain these terms that are associated with waves: wavelength, frequency, amplitude.

7.2 What are the units for wavelength and frequency of electromagnetic waves? What is the speed of light in meters per second and miles per hour?

7.3 List the types of electromagnetic radiation, starting with the radiation having the longest wavelength and ending with the radiation having the shortest wavelength.

7.4 Give the high and low wavelength values that define the visible region of the electromagnetic spectrum.

7.5 Briefly explain Planck's quantum theory and explain what a quantum is. What are the units for Planck's constant?

7.6 Give two everyday examples that illustrate the concept of quantization.

Problems

7.7 Convert 8.6×10^{13} Hz to nanometers, and 566 nm to hertz.

7.8 (a) What is the frequency of light having a wavelength of 456 nm? (b) What is the wavelength (in nanometers) of radiation having a frequency of 2.45×10^9 Hz? (This is the type of radiation used in microwave ovens.)

7.9 The average distance between Mars and Earth is about 1.3×10^8 miles. How long would it take TV pictures transmitted from the *Viking* space vehicle on Mars' surface to reach Earth? (1 mile = 1.61 km.)

7.10 How many minutes would it take a radio wave to travel from the planet Venus to Earth? (Average distance from Venus to Earth = 28 million miles.)

7.11 The SI unit of time is the second, which is defined as 9,192,631,770 cycles of radiation associated with a certain emission process in the cesium atom. Calculate the wavelength of this radiation (to three significant figures). In which region of the electromagnetic spectrum is this wavelength found?

7.12 The SI unit of length is the meter, which is defined as the length equal to 1,650,763.73 wavelengths of the light emitted by a particular energy transition in krypton atoms. Calculate the frequency of the light to three significant figures.

The Photoelectric Effect

Review Questions

7.13 Explain what is meant by the photoelectric effect.

7.14 What are photons? What role did Einstein's explanation of the photoelectric effect play in the development of the particle-wave interpretation of the nature of electromagnetic radiation?

Problems

7.15 A photon has a wavelength of 624 nm. Calculate the energy of the photon in joules.

7.16 The blue color of the sky results from the scattering of sunlight by air molecules. The blue light has a frequency of about 7.5×10^{14} Hz. (a) Calculate the wavelength, in nm, associated with this radiation, and (b) calculate the energy, in joules, of a single photon associated with this frequency.

7.17 A photon has a frequency of 6.0×10^4 Hz. (a) Convert this frequency into wavelength (nm). Does this frequency fall in the visible region? (b) Calculate the energy (in joules) of this photon. (c) Calculate the energy (in joules) of 1 mole of photons all with this frequency.

7.18 What is the wavelength, in nm, of radiation that has an energy content of 1.0×10^3 kJ/mol? In which region of the electromagnetic spectrum is this radiation found?

7.19 When copper is bombarded with high-energy electrons, X rays are emitted. Calculate the energy (in joules) associated with the photons if the wavelength of the X rays is 0.154 nm.

7.20 A particular electromagnetic radiation has a frequency of 8.11×10^{14} Hz. (a) What is its wavelength in nanometers? in meters? (b) To what region of the electromagnetic spectrum would you assign it? (c) What is the energy (in joules) of one quantum of this radiation?

Bohr's Theory of the Hydrogen Atom

Review Questions

7.21 What are emission spectra? What are line spectra?

7.22 What is an energy level? Explain the difference between a ground state and an excited state.

7.23 Briefly describe Bohr's theory of the hydrogen atom and how it explains the appearance of an emission spectrum. How does Bohr's theory differ from concepts of classical physics?

7.24 Explain the meaning of the negative sign in Equation (7.4).

Problems

7.25 Explain why elements produce their own characteristic colors when they emit photons.

7.26 When some compounds containing copper are heated in a flame, green light is emitted. How would you determine whether the light is of one wavelength or a mixture of two or more wavelengths?

7.27 Would it be possible for a fluorescent material to emit radiation in the ultraviolet region after absorbing visible light? Explain your answer.

7.28 Explain how astronomers are able to tell which elements are present in distant stars by analyzing the electromagnetic radiation emitted by the stars.

7.29 Consider the following energy levels of a hypothetical atom:

E_4 _____ -1.0×10^{-19} J

E_3 _____ -5.0×10^{-19} J

E_2 _____ -10×10^{-19} J

E_1 _____ -15×10^{-19} J

(a) What is the wavelength of the photon needed to excite an electron from E_1 to E_4? (b) What is the value (in joules) of the quantum of energy of a photon needed to excite an electron from E_2 to E_3? (c) When an electron drops from the E_3 level to the E_1 level, the atom is said to undergo emission. Calculate the wavelength of the photon emitted in this process.

7.30 The first line of the Balmer series occurs at a wavelength of 656.3 nm. What is the energy difference between the two energy levels involved in the emission that results in this spectral line?

7.31 Calculate the wavelength of a photon emitted by a hydrogen atom when its electron drops from the $n = 5$ state to the $n = 3$ state.

7.32 Calculate the frequency and wavelength of the emitted photon when an electron drops from the $n = 4$ to the $n = 2$ level in a hydrogen atom.

7.33 Careful spectral analysis shows that the familiar yellow light of sodium lamps (such as street lamps) is made up of photons of two wavelengths, 589.0 nm and 589.6 nm. What is the difference in energy (in joules) between photons possessing these wavelengths?

7.34 An electron in an orbit of principal quantum number n_i in a hydrogen atom makes a transition to an orbit of principal quantum number 2. The photon emitted has a wavelength of 434 nm. Calculate n_i.

Particle-Wave Duality

Review Questions

7.35 Explain what is meant by the statement that matter and radiation have a "dual nature."

7.36 How does de Broglie's hypothesis account for the fact that the energies of the electron in a hydrogen atom are quantized?

7.37 Why is Equation (7.7) meaningful only when applied to submicroscopic particles such as electrons and atoms and not to macroscopic objects?

7.38 Does a baseball in flight possess wave properties? If so, why can we not determine its wave properties?

Problems

7.39 Thermal neutrons are neutrons that move at speeds comparable to those of air molecules at room temperature. These neutrons are most effective in initiating a nuclear chain reaction among ^{235}U isotopes. Calculate the wavelength (in nanometers) associated with a beam of neutrons moving at 7.00×10^2 m/s. (Mass of a neutron $= 1.675 \times 10^{-27}$ kg.)

7.40 Protons in certain particle accelerators can be accelerated to speeds near that of light. Estimate the wavelength (in nanometers) of such a proton moving at 2.90×10^8 m/s. (Mass of a proton $= 1.673 \times 10^{-27}$ kg.)

7.41 What is the de Broglie wavelength in centimeters of a 12.4-g hummingbird flying at 1.20×10^2 mph? (1 mi $=$ 1.61 km.)

7.42 What is the de Broglie wavelength associated with a 2.5-g Ping-Pong ball traveling at 35 mph?

Quantum Mechanics and the Hydrogen Atom

Review Questions

7.43 What are the inadequacies of Bohr's theory?

7.44 What is the Heisenberg uncertainty principle? What is the Schrödinger equation?

7.45 What is the physical significance of the wave function?

7.46 How is the concept of electron density used to describe the position of an electron in the quantum mechanical treatment of an atom?

7.47 Define atomic orbital. How does an atomic orbital differ from an orbit?

7.48 Describe the characteristics of an s orbital, a p orbital, and a d orbital.

7.49 Why is a boundary surface diagram useful in representing an atomic orbital?

7.50 Describe the four quantum numbers used to characterize an electron in an atom.

7.51 Which quantum number defines a shell? Which quantum numbers define a subshell?

7.52 Which of the four quantum numbers (n, ℓ, m_ℓ, m_s) determine the energy of an electron in (a) a hydrogen atom and (b) a many-electron atom?

Problems

7.53 An electron in a certain atom is in the $n = 2$ quantum level. List the possible values of ℓ and m_ℓ that the electron can have.

7.54 An electron in an atom is in the $n = 3$ quantum level. List the possible values of ℓ and m_ℓ that the electron can have.

7.55 Give the values of the quantum numbers associated with these orbitals: (a) $2p$, (b) $3s$, (c) $5d$.

7.56 For these subshells give the values of the quantum numbers (n, ℓ, and m_ℓ) and the number of orbitals in each subshell: (a) $4p$, (b) $3d$, (c) $3s$, (d) $5f$.

7.57 Discuss the similarities and differences between a $1s$ and a $2s$ orbital.

7.58 What is the difference between a $2p_x$ and a $2p_y$ orbital?

7.59 List the possible subshells and orbitals associated with the principal quantum number n, if $n = 5$.

7.60 List the possible subshells and orbitals associated with the principal quantum number n, if $n = 6$.

7.61 Calculate the total number of electrons that can occupy (a) one s orbital, (b) three p orbitals, (c) five d orbitals, (d) seven f orbitals.

7.62 What is the total number of electrons that can be held in all orbitals having the same principal quantum number n?

7.63 What is the maximum number of electrons that can be found in each of the following subshells: $3s$, $3d$, $4p$, $4f$, $5f$?

7.64 Indicate the total number of (a) p electrons in N ($Z = 7$); (b) s electrons in Si ($Z = 14$); and (c) $3d$ electrons in S ($Z = 16$).

7.65 Make a chart of all allowable orbitals in the first four principal energy levels of the hydrogen atom. Designate each by type (for example, s, p) and indicate how many orbitals of each type there are.

7.66 Why do the $3s$, $3p$, and $3d$ orbitals have the same energy in a hydrogen atom but different energies in a many-electron atom?

7.67 For each of these pairs of hydrogen orbitals, indicate which is higher in energy: (a) $1s$, $2s$; (b) $2p$, $3p$; (c) $3d_{xy}$, $3d_{yz}$; (d) $3s$, $3d$; (e) $5s$, $4f$.

7.68 Which orbital in each of these pairs is lower in energy in a many-electron atom? (a) $2s$, $2p$; (b) $3p$, $3d$; (c) $3s$, $4s$; (d) $4d$, $5f$.

Electron Configuration and the Periodic Table

Review Questions

7.69 Define these terms: electron configuration, the Pauli exclusion principle, Hund's rule.

7.70 Explain the meaning of diamagnetic and paramagnetic. Give an example of an atom that is diamagnetic and one that is paramagnetic.

7.71 Explain the meaning of the symbol $4d^6$.

7.72 What is meant by the term "shielding of electrons" in an atom? Using the Li atom as an example, describe the effect of shielding on the energy of electrons in an atom.

7.73 Define and give an example for each of these terms: transition metals, lanthanides, actinides.

7.74 Explain why the ground-state electron configurations for Cr and Cu are different from what we might expect.

7.75 Explain what is meant by a noble gas core. Write the electron configuration of a "xenon core."

7.76 Comment on the correctness of this statement: The probability of finding two electrons with the same four quantum numbers in an atom is zero.

Problems

7.77 Which of the following sets of quantum numbers are unacceptable? Explain your answers. (a) $(1, 0, \frac{1}{2}, -\frac{1}{2})$, (b) $(3, 0, 0, +\frac{1}{2})$, (c) $(2, 2, 1, +\frac{1}{2})$, (d) $(4, 3, -2, +\frac{1}{2})$, (e) $(3, 2, 1, 1)$.

7.78 The ground-state electron configurations listed here are incorrect. Explain what mistakes have been made in each and write the correct electron configurations.

Al: $1s^2 2s^2 2p^4 3s^2 3p^3$
B: $1s^2 2s^2 2p^6$
F: $1s^2 2s^2 2p^6$

7.79 The atomic number of an element is 73. Are the atoms of this element diamagnetic or paramagnetic?

7.80 Indicate the number of unpaired electrons present in each of these atoms: B, Ne, P, Sc, Mn, Se, Zr, Ru, Cd, I, W, Pb, Ce, Ho.

7.81 Write the ground-state electron configurations for these elements: B, V, Ni, As, I, Au.

7.82 Write the ground-state electron configurations for these elements: Ge, Fe, Zn, Ru, W, Tl.

7.83 The electron configuration of a neutral atom is $1s^2 2s^2 2p^6 3s^2$. Write a complete set of quantum numbers for each of the electrons. Name the element.

7.84 Which of these species has the most unpaired electrons: S^+, S, or S^-? Explain how you arrived at your answer.

Additional Problems

7.85 When a compound containing cesium ion is heated in a Bunsen burner flame, photons with an energy of 4.30×10^{-19} J are emitted. What color is the cesium flame?

7.86 Discuss the current view of the correctness of the following statements. (a) The electron in the hydrogen atom is in an orbit that never brings it closer than 100 pm to the nucleus. (b) Atomic absorption spectra result from transitions of electrons from lower to higher energy levels. (c) A many-electron atom behaves somewhat like a solar system that has a number of planets.

7.87 Distinguish carefully between these terms: (a) wavelength and frequency, (b) wave properties and particle properties, (c) quantization of energy and continuous variation in energy.

7.88 What is the maximum number of electrons in an atom that can have the following quantum numbers? Label each electron with a particular orbital. (a) $n = 2$, $m_s = +\frac{1}{2}$; (b)

$n = 4$, $m_\ell = +1$; (c) $n = 3$, $\ell = 2$; (d) $n = 2$, $\ell = 0$, $m_s = -\frac{1}{2}$; (e) $n = 4$, $\ell = 3$, $m_\ell = -2$.

7.89 Identify the following individuals and their contributions to the development of quantum theory: de Broglie, Einstein, Bohr, Planck, Heisenberg, Schrödinger.

7.90 What properties of electrons are used in the operation of an electron microscope?

7.91 In a photoelectric experiment a student uses a light source whose frequency is greater than that needed to eject electrons from a certain metal. However, after continuously shining the light on the same area of the metal for a long period the student notices that the maximum kinetic energy of ejected electrons begins to decrease, even though the frequency of the light is held constant. How would you account for this behavior?

7.92 A certain pitcher's fastballs have been clocked at about 100 mph. (a) Calculate the wavelength of a baseball (in nanometers) of mass 0.141 kg at this speed. (b) What is the wavelength of a hydrogen atom at the same speed? (1 mi = 1.61 km.)

7.93 Considering only the ground-state electron configuration, are there more elements that have diamagnetic or paramagnetic atoms? Explain.

7.94 Photodissociation of water

$$H_2O + h\nu \longrightarrow H_2(g) + \tfrac{1}{2}O_2(g)$$

has been suggested as a source of hydrogen. The ΔH°_{rxn} for the reaction, calculated from thermochemical data, is 285.8 kJ/mol of water decomposed. Calculate the maximum wavelength (in nm) that would provide the necessary energy. In principle, is it feasible to use sunlight as a source of energy for this process?

7.95 Spectral lines of the Lyman and Balmer series do not overlap. Verify this statement by calculating the longest wavelength associated with the Lyman series and the shortest wavelength associated with the Balmer series (in nm).

7.96 The He^+ ion contains only one electron and is therefore a hydrogen-like ion. Calculate the wavelengths, in increasing order, of the first four transitions in the Balmer series of the He^+ ion. Compare these wavelengths with the same transitions in an H atom. Comment on the differences. (The Rydberg constant for He^+ is 8.72×10^{-18} J.)

7.97 Ozone (O_3) in the stratosphere absorbs the harmful radiation from the sun by undergoing decomposition: $O_3 \rightarrow O + O_2$. (a) Referring to Table 6.4, calculate the ΔH° for this process. (b) Calculate the maximum wavelength of photons (in nm) that possess enough energy to cause the decomposition of ozone photochemically.

7.98 An electron in an excited state in a hydrogen atom can return to the ground state in two different ways: (a) via a direct transition in which a photon of wavelength λ_1 is emitted and (b) via an intermediate excited state reached by the emission of a photon of wavelength λ_2. This intermediate excited state then decays to the ground state by emitting another photon of wavelength λ_3. Derive an equation that relates λ_1 to λ_2 and λ_3.

7.99 The ionization energy of a certain element is 412 kJ/mol. However, when the atoms of this element are in the first excited state ($n = 2$), the ionization energy is only 126 kJ/mol. Based on this information, calculate the wavelength of light emitted in a transition from $n = 2$ to $n = 1$. (*Hint:* Ionization energy is the minimum energy required to remove a ground-state electron from an atom.)

7.100 Alveoli are the tiny sacs of air in the lungs (see Problem 5.96) whose average diameter is 5.0×10^{-5} m. Consider an oxygen molecule (mass $= 5.3 \times 10^{-26}$ kg) trapped within a sac. Calculate the uncertainty in the velocity of the oxygen molecule. [*Hint:* The maximum uncertainty in the position of the molecule is given by the diameter of the sac. The mathematical form of the Heisenberg uncertainty principle is $\Delta x \Delta p = h/4\pi$, where Δx is the uncertainty in position, Δp is the uncertainty in momentum (mass $\times$ velocity), and h is the Planck constant.]

7.101 How many photons at 660 nm must be absorbed to melt 5.0×10^2 g of ice? On average, how many H_2O molecules does one photon convert from ice to water? (*Hint:* It takes 334 J to melt 1 g of ice at 0°C.)

7.102 The sun is surrounded by a white circle of gaseous material called the corona, which becomes visible during a total eclipse of the sun. The temperature of the corona is in the millions of degrees Celsius, which is high enough to break up molecules and remove some or all of the electrons from atoms. One way astronomers have been able to estimate the temperature of the corona is by studying the emission lines of ions of certain elements. For example, the emission spectrum of Fe^{14+} ions has been recorded and analyzed. Knowing that it takes 3.5×10^4 kJ/mol to convert Fe^{13+} to Fe^{14+}, estimate the temperature of the sun's corona. (*Hint:* The average kinetic energy of 1 mole of a gas is $\frac{3}{2}RT$.)

7.103 The UV light that is responsible for tanning the skin falls in the 320- to 400-nm region. Calculate the total energy (in joules) absorbed by a person exposed to this radiation for 2.0 hours, given that there are 2.0×10^{16} photons hitting Earth's surface per square centimeter per second over an 80-nm (320 nm to 400 nm) range and that the exposed body area is 0.45 m^2. Assume that only half of the radiation is absorbed and the other half is reflected by the body. (*Hint:* Use an average wavelength of 360 nm in calculating the energy of a photon.)

7.104 A photoelectric experiment was performed by separately shining a laser at 450 nm (blue light) and a laser at 560 nm (yellow light) on a clean metal surface and measuring the number and kinetic energy of the ejected electrons. Which light would generate more electrons? Which light would eject electrons with greater kinetic energy? Assume that the same amount of energy is delivered to the metal surface by each laser and that the frequencies of the laser lights exceed the threshold frequency.

7.105 Calculate the de Broglie wavelength (in nm) of a N_2 molecule at 300 K.

Special Problems

7.106 Describe in your own words each of the following: (a) Planck's quantum theory, (b) the photoelectric effect, (c) Bohr's model for the hydrogen atom, (d) de Broglie's particle-wave relation, (e) the Heisenberg uncertainty principle. How does each of them contribute to the quantum mechanical description of atoms?

7.107 An electron in a hydrogen atom is excited from the ground state to the $n = 4$ state. Comment on the correctness of the following statements (true or false).

(a) $n = 4$ is the first excited state.
(b) It takes more energy to ionize (remove) the electron from $n = 4$ than from the ground state.
(c) The electron is farther from the nucleus (on average) in $n = 4$ than from the ground state.

(d) The wavelength of light emitted when the electron drops from $n = 4$ to $n = 1$ is longer than that from $n = 4$ to $n = 2$.
(e) The wavelength the atom absorbs in going from $n = 1$ to $n = 4$ is the same as that emitted as it goes from $n = 4$ to $n = 1$.

7.108 When an electron makes a transition between energy levels of a hydrogen atom, there are no restrictions on the initial and final values of the principal quantum number n. However, there is a quantum mechanical rule that restricts the initial and final values of the orbital angular momentum ℓ. This is the *selection rule*, which states that $\Delta \ell = \pm 1$, that is, in a transition, the value of ℓ can only increase or decrease by one. According to this rule, which of the following transitions are allowed: (a) $1s \longrightarrow 2s$, (b) $2p \longrightarrow 1s$, (c) $1s \longrightarrow 3d$, (d) $3d \longrightarrow 4f$, (e) $4d \longrightarrow 3s$?

Answers to Practice Exercises

7.1 8.24 m. **7.2** 3.39 $\times$ 10^3 nm. **7.3** 2.63 $\times$ 10^3 nm. **7.4** 56.6 nm. **7.5** $n = 3$, $\ell = 1$, $m_\ell = -1, 0, 1$. **7.6** 16. **7.7** $(5, 1, -1, +\frac{1}{2})$, $(5, 1, 0, +\frac{1}{2})$, $(5, 1, 1, +\frac{1}{2})$, $(5, 1, -1, -\frac{1}{2})$, $(5, 1, 0, -\frac{1}{2})$, $(5, 1, 1, -\frac{1}{2})$. **7.8** 32. **7.9** $(1, 0, 0, +\frac{1}{2})$, $(1, 0, 0, -\frac{1}{2})$, $(2, 0, 0, +\frac{1}{2})$, $(2, 0, 0, -\frac{1}{2})$, $(2, 1, -1, +\frac{1}{2})$. There are five other acceptable ways to write the quantum numbers for the last electron. **7.10** $[Ne]3s^2 3p^3$.

In 1848, 14-year-old Dmitri Ivanovich Mendeleev took a journey that changed his life. Mendeleev was born in Siberia, the last of 14 children. The family was destitute, but his mother decided that Dmitri Ivanovich should go to Moscow to complete his education. After hitchhiking 14,000 miles across Russia, Mendeleev was refused admission to the University of Moscow because he was a Siberian. Mother and son went on to St. Petersburg, where Mendeleev obtained a grant to train as a secondary school teacher. Eventually he became a professor of inorganic chemistry at the University of St. Petersburg.

Mendeleev was noted for cutting his hair only once a year.

Over time Mendeleev noticed a relationship between the molecular masses and physical properties of similar compounds. He wondered whether an analogous relationship existed between atomic masses and the properties of elements. In 1869 Mendeleev drew up a table of elements arranged according to increasing atomic mass. The table revealed a periodic relationship among the elements' properties. Now known as the periodic table of elements, Mendeleev's chart proved to be the key for unlocking the mysteries of atomic structure and chemical bonding.

Mendeleev had discovered a fundamental law of nature. He also realized that there were many more elements to be discovered and predicted their properties using the periodic table. In 1875, when the French chemist Lecoq discovered gallium and gave its density as 4.7 g/cm^3, Mendeleev pointed out that the density should be 5.9 g/cm^3, which is accurate to within 1 percent of the true value. The scientific community was astounded that a theorist knew the properties of a new element better than the chemist who had discovered it!

In addition to formulating the periodic law, Mendeleev wrote textbooks in chemistry and philosophy, helped to develop the oil industry in Russia, and systematized weights and measurements. In his lifetime he was the most famous scientist in Russia. When he died in 1907, students bore huge periodic tables above the funeral procession.

Today Mendeleev's work on the periodic table is regarded as the most important contribution to chemistry in the nineteenth century. Element 101 is named mendelevium [Md] in his honor.

The Periodic Table

Chapter 8

ESSENTIAL CONCEPTS

Development of the Periodic Table In the nineteenth century chemists noticed a regular, periodic recurrence of chemical and physical properties of elements. In particular, the periodic table drawn up by Mendeleev grouped the elements accurately and was able to predict the properties of several elements that had not been discovered.

Periodic Classification of the Elements Elements are grouped according to their outer-shell electron configurations, which account for their similar chemical behavior. Special names are assigned to these various groups.

Periodic Variation in Properties Overall, physical properties such as atomic and ionic radii of the elements vary in a regular and periodic fashion. Similar variation is also noted in their chemical properties. Chemical properties of special importance are ionization energy, which measures the tendency of an atom of an element to lose an electron, and electron affinity, which measures the tendency of an atom to accept an electron. Ionization energy and electron affinity form the basis for understanding chemical bond formation.

8.1 Development of the Periodic Table

In the nineteenth century, when chemists had only a vague idea of atoms and molecules and did not know of the existence of electrons and protons, they devised the periodic table using their knowledge of atomic masses. Accurate measurements of the atomic masses of many elements had already been made. Arranging elements according to their atomic masses in a periodic table seemed logical to chemists, who felt that chemical behavior should somehow be related to atomic mass.

In 1864 the English chemist John Newlands noticed that when the known elements were arranged in order of atomic mass, every eighth element had similar properties. Newlands referred to this peculiar relationship as the *law of octaves*. However, this "law" turned out to be inadequate for elements beyond calcium, and Newlands's work was not accepted by the scientific community.

Five years later the Russian chemist Dmitri Mendeleev and the German chemist Lothar Meyer independently proposed a much more extensive tabulation of the elements, based on the regular, periodic recurrence of properties. Mendeleev's classification was a great improvement over Newlands's for two reasons. First, it grouped the elements together more accurately, according to their properties. Equally important, it made possible the prediction of the properties of several elements that had not yet been discovered. For example, Mendeleev proposed the existence of an unknown element that he called eka-aluminum. (*Eka* is a Sanskrit word meaning "first"; thus eka-aluminum would be the first element under aluminum in the same group.) When gallium was discovered four years later, its properties closely matched the predicted properties of eka-aluminum as shown here.

	Eka-Aluminum (Ea)	Gallium (Ga)
Atomic mass	68 amu	69.9 amu
Melting point	Low	29.78°C
Density	5.9 g/cm^3	5.94 g/cm^3
Formula of oxide	Ea_2O_3	Ga_2O_3

Nevertheless, the early versions of the periodic table had some glaring inconsistencies. For example, the atomic mass of argon (39.95 amu) is greater than that of potassium (39.10 amu). If elements were arranged solely according to increasing atomic mass, argon would appear in the position occupied by potassium in our modern periodic table (see the inside front cover). But no chemist would place argon, an inert gas, in the same group as lithium and sodium, two very reactive metals. This and other discrepancies suggested that some fundamental property other than atomic mass is the basis of the observed periodicity. This property turned out to be associated with atomic number.

Using data from α-scattering experiments (see Section 2.2), Rutherford was able to estimate the number of positive charges in the nucleus of a few elements, but until 1913 there was no general procedure for determining atomic numbers. In that year a young English physicist, Henry Moseley, discovered a correlation between atomic number and the frequency of X rays generated by the bombardment of the element under study with high-energy electrons. With a few exceptions, Moseley found that the order of increasing atomic number is the same as the order of increasing atomic mass. For example, calcium is the twentieth element in increasing order of atomic mass, and it has an atomic number of 20. The discrepancies that bothered scientists now made sense. The atomic number of argon is 18 and that of potassium is 19, so potassium should follow argon in the periodic table.

A modern periodic table usually shows the atomic number along with the element symbol. As you already know, the atomic number also indicates the number of electrons in the atoms of an element. Electron configurations of elements help explain the recurrence of physical and chemical properties. The importance and usefulness of the periodic table lie in the fact that we can use our understanding of the general properties and trends within a group or a period to predict with considerable accuracy the properties of any element, even though that element may be unfamiliar to us.

Appendix 4 explains the names and symbols of the elements.

8.2 Periodic Classification of the Elements

Figure 8.1 shows the periodic table together with the outermost ground-state electron configurations of the elements. (The electron configurations of the elements are also given in Table 7.3.) Starting with hydrogen, we see that the subshells are filled in the order shown in Figure 7.20. According to the type of subshell being filled, the elements can be divided into categories—the representative elements, the noble gases, the transition elements (or transition metals), the lanthanides, and the actinides. Referring to Figure 8.1, the **representative elements** (also called *main group elements*) are *the elements in Groups 1A through 7A, all of which have incompletely filled s or p subshells of the highest principal quantum number.* With the exception of helium, the **noble gases** *(the Group 8A elements) all have a completely filled p subshell.* (The electron configurations are $1s^2$ for helium and ns^2np^6 for the other noble gases, in which n is the principal quantum number

Figure 8.1 *The ground-state electron configurations of the elements. For simplicity, only the configurations of the outer shells are shown.*

Table 8.1 | Electron Configurations of Group 1A and Group 2A Elements

Group 1A	Group 2A
Li $[He]2s^1$	Be $[He]2s^2$
Na $[Ne]3s^1$	Mg $[Ne]3s^2$
K $[Ar]4s^1$	Ca $[Ar]4s^2$
Rb $[Kr]5s^1$	Sr $[Kr]5s^2$
Cs $[Xe]6s^1$	Ba $[Xe]6s^2$
Fr $[Rn]7s^1$	Ra $[Rn]7s^2$

For the representative elements, the valence electrons are simply those electrons at the highest principal energy level *n*.

for the outermost shell.) The transition metals are the elements in Groups 1B and 3B through 8B, which have incompletely filled *d* subshells or readily produce cations with incompletely filled *d* subshells. (These metals are sometimes referred to as the *d*-block transition elements.) The Group 2B elements are Zn, Cd, and Hg, which are neither representative elements nor transition metals. The lanthanides and actinides are sometimes called *f*-block transition elements because they have incompletely filled *f* subshells.

A clear pattern emerges when we examine the electron configurations of the elements in a particular group. The electron configurations for Groups 1A and 2A elements are shown in Table 8.1. We see that all members of the Group 1A alkali metals have similar outer electron configurations; each has a noble gas core and an ns^1 configuration of the outer electron. Similarly, the Group 2A alkaline earth metals have a noble gas core and an ns^2 configuration of the outer electrons. *The outer electrons of an atom, which are those involved in chemical bonding,* are often called the **valence electrons.** Having the same number of valence electrons accounts for similarities in chemical behavior among the elements within each of these groups. This observation holds true also for the halogens (the Group 7A elements), which have outer electron configurations of ns^2np^5 and exhibit very similar properties. We must be careful, however, in predicting properties for Groups 3A through 6A. For example, the elements in Group 4A all have the same outer electron configuration, ns^2np^4, but there is much variation in chemical properties among these elements: Carbon is a nonmetal, silicon and germanium are metalloids, and tin and lead are metals.

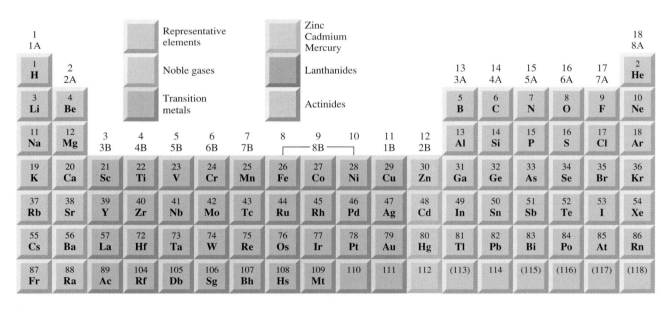

Figure 8.2 *Classification of the elements. Note that the Group 2B elements are often classified as transition metals even though they do not exhibit the characteristics of the transition metals.*

As a group, the noble gases behave very similarly. With the exception of krypton and xenon, these elements are totally inert chemically. The reason is that these elements all have completely filled outer ns^2np^6 subshells, a condition that represents great stability. Although the outer electron configuration of the transition metals is not always the same within a group and there is no regular pattern in the change of the electron configuration from one metal to the next in the same period, all transition metals share many characteristics that set them apart from other elements. The reason is that these metals all have an incompletely filled d subshell. Likewise, the lanthanide (and the actinide) elements resemble one another within the series because they have incompletely filled f subshells. Figure 8.2 on p. 234 distinguishes the groups of elements discussed here.

Example 8.1 A neutral atom of a certain element has 15 electrons. Without consulting a periodic table, answer the following questions: (a) What is the ground-state electron configuration of the element? (b) How should the element be classified? (c) Are the atoms of this element diamagnetic or paramagnetic?

Reasoning and Solution (a) Using the building-up principle and knowing the maximum capacity of s and p subshells, we can write the ground-state electron configuration of the element as $1s^22s^22p^63s^23p^3$.

(b) Because the $3p$ subshell is not completely filled, this is a representative element. Based on the information given, we cannot say whether it is a metal, a nonmetal, or a metalloid.

(c) According to Hund's rule, the three electrons in the $3p$ orbitals have parallel spins. Therefore, the atoms of this element are paramagnetic, with three unpaired spins. (Remember, we saw in Chapter 7 that any atom that contains an odd number of electrons must be paramagnetic.)

Similar problem: 8.13.

Practice Exercise A neutral atom of a certain element has 20 electrons. (a) Write the ground-state configuration of the element, (b) classify the element, and (c) determine whether the atoms of this element are diamagnetic or paramagnetic.

Electron Configurations of Cations and Anions

Because many ionic compounds are made up of monatomic anions and/or cations, it is helpful to know how to write the electron configurations of these ionic species. The procedure for writing the electron configurations of ions requires only a slight extension of the method used for neutral atoms. We will group the ions in two categories for discussion.

Ions Derived from Representative Elements

In the formation of a cation from the neutral atom of a representative element, one or more electrons are removed from the highest occupied n shell. Here are the electron configurations of some neutral atoms and their corresponding cations:

$$\text{Na: } [\text{Ne}]3s^1 \qquad \text{Na}^+\text{: } [\text{Ne}]$$
$$\text{Ca: } [\text{Ar}]4s^2 \qquad \text{Ca}^{2+}\text{: } [\text{Ar}]$$
$$\text{Al: } [\text{Ne}]3s^23p^1 \qquad \text{Al}^{3+}\text{: } [\text{Ne}]$$

Note that each ion has a stable noble gas configuration.

In the formation of an anion, one or more electrons are added to the highest partially filled n shell. Consider these examples:

H: $1s^1$	H$^-$: $1s^2$ or [He]
F: $1s^2 2s^2 2p^5$	F$^-$: $1s^2 2s^2 2p^6$ or [Ne]
O: $1s^2 2s^2 2p^4$	O^{2-}: $1s^2 2s^2 2p^6$ or [Ne]
N: $1s^2 2s^2 2p^3$	N^{3-}: $1s^2 2s^2 2p^6$ or [Ne]

Again, all the anions have stable noble gas configurations. Thus a characteristic of most representative elements is that ions derived from their neutral atoms have the noble gas outer electron configuration ns^2np^6. *Ions, or atoms and ions, that have the same number of electrons, and hence the same ground-state electron configuration, are said to be **isoelectronic**.* Thus H$^-$ and He are isoelectronic; F$^-$, Na$^+$, and Ne are isoelectronic; and so on.

Cations Derived from Transition Metals

In Section 7.9 we saw that in the first-row transition metals (Sc to Cu), the $4s$ orbital is always filled before the $3d$ orbitals. Consider manganese, whose electron configuration is [Ar]$4s^2 3d^5$. When the Mn^{2+} ion is formed, we might expect the two electrons to be removed from the $3d$ orbitals to yield [Ar]$4s^2 3d^3$. In fact, the electron configuration of Mn^{2+} is [Ar]$3d^5$! The reason is that the electron-electron and electron-nucleus interactions in a neutral atom can be quite different from those in its ion. Thus, whereas the $4s$ orbital is always filled before the $3d$ orbital in Mn, electrons are removed from the $4s$ orbital in forming Mn^{2+} because the $3d$ orbital is more stable than the $4s$ orbital in transition metal ions. Therefore, when a cation is formed from an atom of a transition metal, electrons are always removed first from the ns orbital and then from the $(n-1)d$ orbitals.

Keep in mind that most transition metals can form more than one cation and that frequently the cations are not isoelectronic with the preceding noble gases.

Bear in mind that the order of electron filling does not determine or predict the order of electron removal for transition metals.

8.3 Periodic Variation in Physical Properties

As we have seen, the electron configurations of the elements show a periodic variation with increasing atomic number. Consequently, the elements also display periodic variations in their physical and chemical behavior. In this section and Sections 8.4 and 8.5, we will examine some physical properties of elements in a group and across a period and properties that influence the chemical behavior of elements. Before we do so, however, let's look at the concept of effective nuclear charge, which will help us to understand these topics better.

Effective Nuclear Charge

In Chapter 7, we discussed the shielding effect that electrons close to the nucleus have on outer-shell electrons in many-electron atoms. The presence of shielding electrons reduces the electrostatic attraction between the positively charged protons in the nucleus and the outer electrons. Moreover, the repulsive forces between electrons in a many-electron atom further offset the attractive force exerted by the nucleus. The concept of effective nuclear charge enables us to account for the effects of shielding on periodic properties.

Consider, for example, the helium atom, which has the electron configuration $1s^2$. Helium's two protons give the nucleus a charge of $+2$, but the full attractive force of this charge on the two $1s$ electrons is partially offset by electron-electron repulsion. Consequently, we say that each $1s$ electron is shielded from the nucleus by the other $1s$ electron. The *effective nuclear charge*, Z_{eff}, is given by

$$Z_{eff} = Z - \sigma$$

in which Z is the actual nuclear charge (that is, it is the atomic number) and σ (sigma) is called the *shielding constant*. σ is greater than zero but smaller than Z.

One example of the effects of electron shielding is the energy required to remove an electron from a many-electron atom. Measurements show that it takes 2373 kJ of energy to remove the first electron from 1 mole of He atoms and 5251 kJ of energy to remove the remaining electron from 1 mole of He^+ ions. The reason it takes so much more energy to remove the second electron is that with only one electron present, there is no shielding and the electron feels the full effect of the $+2$ nuclear charge.

For atoms with three or more electrons, the electrons in a given shell are shielded by electrons in inner shells (that is, shells closer to the nucleus) but not by electrons in outer shells. Thus, for lithium, whose electron configuration is $1s^22s^1$, the $2s$ electron is shielded by the two $1s$ electrons, but the $2s$ electron does not have a shielding effect on the $1s$ electrons. In addition, filled inner shells shield outer electrons more effectively than electrons in the same subshell shield one another. As we will see, electron shielding has important consequences for atomic size and the formation of ions and molecules.

Atomic Radius

A number of physical properties, including density, melting point, and boiling point, are related to the sizes of atoms, but atomic size is difficult to define. As we saw in Chapter 7, the electron density in an atom extends far beyond the nucleus. In practice, we normally think of atomic size as the volume containing about 90 percent of the total electron density around the nucleus.

Several techniques allow us to estimate the size of an atom. First consider the metallic elements. The structure of metals is quite varied, but all metals share one structural characteristic: Their atoms are linked to one another in an extensive three-dimensional network. Thus the **atomic radius** of a metal is *one-half the distance between the two nuclei in two adjacent atoms* (Figure 8.3a). *For elements that exist as diatomic molecules, the atomic radius is one-half the distance between the nuclei of the two atoms in a particular molecule* (Figure 8.3b).

Figure 8.4 shows the atomic radii of many elements according to their positions in the periodic table, and Figure 8.5 plots the atomic radii of these elements against their atomic numbers. Periodic trends are clearly evident. In studying the trends, bear in mind that the atomic radius is determined to a large extent by how strongly the outer-shell electrons are held by the nucleus. The larger the effective nuclear charge, the more strongly held these electrons are and the smaller the atomic radius is. Consider the second-period elements from Li to F. Moving from left to right, we find that the number

The increase in effective nuclear charge from left to right across a period and from bottom to top in a group for representative elements.

(b)

Figure 8.3 (a) In metals such as beryllium, the atomic radius is defined as one-half the distance between the centers of two adjacent atoms. (b) For elements that exist as diatomic molecules, such as iodine, the radius of the atom is defined as one-half the distance between the centers of the atoms in the molecule.

Atomic and ionic radius

Figure 8.4 *Atomic radii (in picometers) of representative elements according to their positions in the periodic table. Note that there is no general agreement on the size of atomic radii. We focus only on the trends in atomic radii, not on their precise values.*

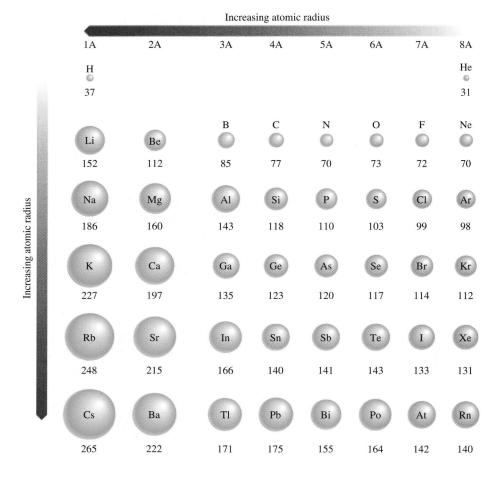

of electrons in the inner shell ($1s^2$) remains constant while the nuclear charge increases. The electrons that are added to counterbalance the increasing nuclear charge are ineffective in shielding one another. Consequently, the effective nuclear charge increases steadily whereas the principal quantum number remains constant ($n = 2$). For example, the outer $2s$ electron on lithium is shielded from the nucleus (which has 3 protons) by the two $1s$ electrons. As an approximation, we assume that the shielding effect of the two $1s$ electrons is to cancel two positive charges in the nucleus. Thus the $2s$ electron only feels the attraction caused by one proton in the nucleus, or the effective nuclear charge is $+1$. In beryllium ($1s^2 2s^2$) each of the $2s$ electrons is shielded by the inner two $1s$ electrons, which cancel two of the four positive charges in the nucleus. Because the $2s$ electrons do not shield each other as effectively, the net result is that the effective nuclear charge of each $2s$ electron is greater than $+1$. Thus as the effective nuclear charge increases, the atomic radius decreases steadily from lithium to fluorine.

As we go down a group—say, Group 1A—we find that atomic radius increases with increasing atomic number. For alkali metals the outermost electron resides in the ns orbital. Because the orbital size increases with the increasing principal quantum number n, the size of the metal atoms increases from Li to Cs. The same reasoning can be applied to elements in other groups.

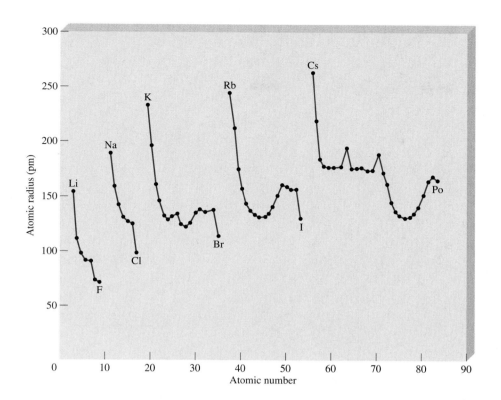

Figure 8.5 *Plot of atomic radii (in picometers) of elements against their atomic numbers.*

Example 8.2   Referring to a periodic table, arrange the following atoms in order of increasing radius: P, Si, N.

Reasoning and Solution   Note that N and P are in the same group (Group 5A) and that N is above P. Therefore, the radius of N is smaller than that of P (atomic radius increases as we go down a group). Both Si and P are in the third period, and Si is to the left of P. Therefore, the radius of P is smaller than that of Si (atomic radius decreases as we move from left to right across a period). Thus the order of increasing radius is N < P < Si.

Practice Exercise   Arrange the following atoms in order of decreasing radius: C, Li, Be.

Similar problems: 8.37, 8.38.

Ionic Radius

Ionic radius is *the radius of a cation or an anion.* Ionic radius affects the physical and chemical properties of an ionic compound. For example, the three-dimensional structure of an ionic compound depends on the relative sizes of its cations and anions.

When a neutral atom is converted to an ion, we expect a change in size. If the atom forms an anion, its size (or radius) increases, because the nuclear charge remains the same but the repulsion resulting from the additional electron(s) enlarges the domain of the electron cloud. On the other hand, a cation is smaller than the neutral atom, because removing one or more electrons reduces electron-electron repulsion but

Figure 8.6 *Comparison of atomic radii with ionic radii. (a) Alkali metals and alkali metal cations. (b) Halogens and halide ions.*

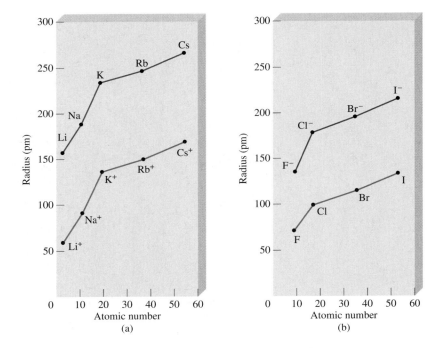

(a)

(b)

the nuclear charge remains the same, so the electron cloud shrinks. Figure 8.6 shows the changes in size when alkali metals are converted to cations and halogens are converted to anions; Figure 8.7 shows the changes in size when a lithium atom reacts with a fluorine atom to form a LiF unit.

Figure 8.8 shows the radii of ions derived from the more familiar elements, arranged according to their positions in the periodic table. We see that in some places there are parallel trends between atomic radii and ionic radii. For example, from the top to the bottom of the periodic table both the atomic radius and the ionic radius increase. For ions derived from elements in different groups, the comparison in size is meaningful only if the ions are isoelectronic. If we examine isoelectronic ions, we find that cations are smaller than anions. For example, Na^+ is smaller than F^-. Both ions have the same number of electrons, but Na ($Z = 11$) has more protons than F ($Z = 9$). The larger effective nuclear charge of Na^+ results in a smaller radius.

Focusing on isoelectronic cations, we see that the radii of *tripositive ions* (that is, ions that bear three positive charges) are smaller than those of *dipositive ions* (that is, ions that bear two positive charges), which in turn are smaller than *unipositive ions* (that is, ions that bear one positive charge). This trend is nicely illustrated by the sizes of three isoelectronic ions in the third period: Al^{3+}, Mg^{2+}, and Na^+ (see Figure 8.8). The Al^{3+} ion has the same number of electrons as Mg^{2+}, but it has one more proton. Thus the electron cloud in Al^{3+} is pulled inward more than that in Mg^{2+}. The

Figure 8.7 *Changes in the sizes of Li and F when they react to form LiF.*

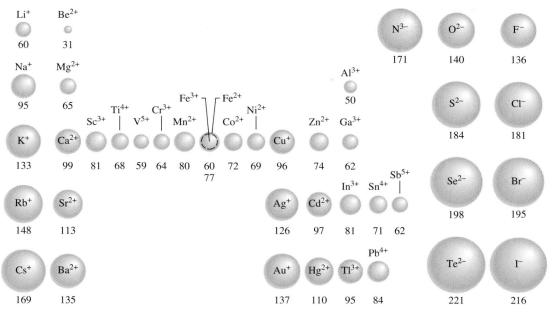

Figure 8.8 *The radii (in picometers) of ions of familiar elements arranged according to the elements' positions in the periodic table.*

smaller radius of Mg^{2+} compared with that of Na^+ can be similarly explained. Turning to isoelectronic anions, we find that the radius increases as we go from ions with uninegative charge (that is, $-$) to those with dinegative charge (that is, $2-$), and so on. Thus the oxide ion is larger than the fluoride ion because oxygen has one fewer proton than fluorine; the electron cloud is spread out to a greater extent in O^{2-}.

Example 8.3 For each of the following pairs, indicate which one of the two species is larger: (a) N^{3-} or F^-; (b) Mg^{2+} or Ca^{2+}; (c) Fe^{2+} or Fe^{3+}.

Reasoning and Solution In comparing ionic radii, it is useful to classify the ions into three categories: (1) isoelectronic ions, (2) ions carrying the same charges and are generated from atoms of the same periodic group, and (3) ions carrying different charges but are generated from the same atom. In case (1), ions carrying a greater negative charge are always larger; in case (2), ions from atoms having a greater atomic number are always larger; in case (3), ions having a greater positive charge are always smaller.

(a) N^{3-} and F^- are isoelectronic anions. Because N^{3-} has only seven protons and F^- has nine, N^{3-} is larger.

(b) Both Mg and Ca belong to Group 2A (the alkaline earth metals). The Ca^{2+} ion is larger than Mg^{2+} because Ca's valence electrons are in a larger shell ($n = 4$) than are Mg's ($n = 3$).

(c) Both ions have the same nuclear charge, but Fe^{2+} has one more electron and hence greater electron-electron repulsion. The radius of Fe^{2+} is larger.

Similar problems: 8.43, 8.44.

Practice Exercise Select the smaller ion in each of the following pairs: (a) K^+, Li^+; (b) Au^+, Au^{3+}; (c) P^{3-}, N^{3-}.

8.4 Ionization Energy

As we will see throughout this book, the chemical properties of any atom are determined by the configuration of the atom's valence electrons. The stability of these outermost electrons is reflected directly in the atom's ionization energies. *Ionization energy* is *the minimum energy required to remove an electron from a gaseous atom in its ground state.* The magnitude of ionization energy is a measure of the effort required to force an atom to give up an electron, or of how "tightly" the electron is held in the atom. The higher the ionization energy, the more difficult it is to remove the electron.

For a many-electron atom, the amount of energy required to remove the first electron from the atom in its ground state,

$$\text{energy} + \text{X}(g) \longrightarrow \text{X}^+(g) + e^-$$

is called the *first ionization energy* (I_1). In this equation, X represents a gaseous atom of any element and e^- is an electron. Unlike an atom in the condensed liquid and solid phases, an atom in the gaseous phase is virtually uninfluenced by its neighbors. The *second ionization energy* (I_2) and the *third ionization energy* (I_3) are shown in these equations:

$$\text{energy} + \text{X}^+(g) \longrightarrow \text{X}^{2+}(g) + e^- \qquad \text{second ionization}$$
$$\text{energy} + \text{X}^{2+}(g) \longrightarrow \text{X}^{3+}(g) + e^- \qquad \text{third ionization}$$

The pattern continues for the removal of subsequent electrons.

When an electron is removed from a neutral atom, the repulsion among the remaining electrons decreases. Because the nuclear charge remains constant, more energy is

Table 8.2 | The Ionization Energies (kJ/mol) of the First 20 Elements

Z	Element	First	Second	Third	Fourth	Fifth	Sixth
1	H	1,312					
2	He	2,373	5,251				
3	Li	520	7,300	11,815			
4	Be	899	1,757	14,850	21,005		
5	B	801	2,430	3,660	25,000	32,820	
6	C	1,086	2,350	4,620	6,220	38,000	47,261
7	N	1,400	2,860	4,580	7,500	9,400	53,000
8	O	1,314	3,390	5,300	7,470	11,000	13,000
9	F	1,680	3,370	6,050	8,400	11,000	15,200
10	Ne	2,080	3,950	6,120	9,370	12,200	15,000
11	Na	495.9	4,560	6,900	9,540	13,400	16,600
12	Mg	738.1	1,450	7,730	10,500	13,600	18,000
13	Al	577.9	1,820	2,750	11,600	14,800	18,400
14	Si	786.3	1,580	3,230	4,360	16,000	20,000
15	P	1,012	1,904	2,910	4,960	6,240	21,000
16	S	999.5	2,250	3,360	4,660	6,990	8,500
17	Cl	1,251	2,297	3,820	5,160	6,540	9,300
18	Ar	1,521	2,666	3,900	5,770	7,240	8,800
19	K	418.7	3,052	4,410	5,900	8,000	9,600
20	Ca	589.5	1,145	4,900	6,500	8,100	11,000

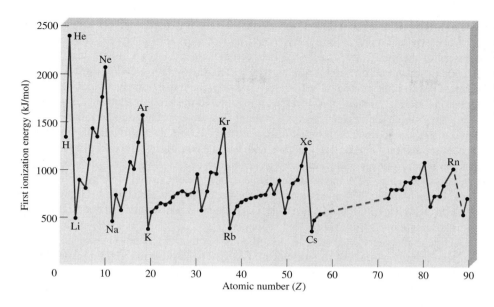

Figure 8.9 *Variation of the first ionization energy with atomic number. Note that the noble gases have high ionization energies, whereas the alkali metals and alkaline earth metals have low ionization energies.*

needed to remove another electron from the positively charged ion. Thus, for the same element, ionization energies always increase in this order:

$$I_1 < I_2 < I_3 < \cdots$$

Table 8.2 lists the ionization energies of the first 20 elements in kilojoules per mole (kJ/mol), that is, the amount of energy in kilojoules needed to remove 1 mole of electrons from 1 mole of gaseous atoms (or ions). By convention, energy absorbed by atoms (or ions) in the ionization process has a positive value. Thus ionization energies are all positive quantities. Figure 8.9 shows the variation of the first ionization energy with atomic number. The plot clearly exhibits the periodicity in the stability of the most loosely held electron. Note that, apart from small irregularities, the ionization energies of elements in a period increase with increasing atomic number. We can explain this trend by referring to the increase in effective nuclear charge from left to right (as in the case of atomic radii variation). A larger effective nuclear charge means a more tightly held outer electron, and hence a higher first ionization energy. A notable feature of Figure 8.9 is the peaks, which correspond to the noble gases. Such high ionization energies are consistent with the fact that most noble gases are chemically unreactive. In fact, helium has the highest first ionization energy of all the elements.

The Group 1A elements (the alkali metals) at the bottom of the graph in Figure 8.9 have the lowest ionization energies. Each of these metals has one valence electron (the outermost electron configuration is ns^1) that is effectively shielded by the completely filled inner shells. Consequently, it is energetically easy to remove an electron from the atom of an alkali metal to form a unipositive ion (Li^+, Na^+, K^+, . . .).

The Group 2A elements (the alkaline earth metals) have higher first ionization energies than the alkali metals do. The alkaline earth metals have two valence electrons (the outermost electron configuration is ns^2). Because these two s electrons do not shield each other well, the effective nuclear charge for an alkaline earth metal atom is larger than that for the preceding alkali metal atom. Most alkaline earth compounds contain dipositive ions (Mg^{2+}, Ca^{2+}, Sr^{2+}, Ba^{2+}), which are isoelectronic with the unipositive alkali metal ions to their left in the same period.

The increase in first ionization energy from left to right across a period and from bottom to top in a group for representative elements.

From the preceding discussion, we see that the metals have relatively low ionization energies, whereas nonmetals possess much higher ionization energies. The ionization energies of the metalloids usually fall between those of metals and nonmetals. The difference in ionization energies suggests why metals always form cations and nonmetals form anions in ionic compounds. For a given group, the ionization energy decreases with increasing atomic number (that is, as we move down the group). Elements in the same group have similar outer electron configurations. However, as the principal quantum number n increases, so does the average distance of a valence electron from the nucleus. A greater separation between the electron and the nucleus means a weaker attraction, so that the electron becomes increasingly easier to remove as we go from element to element down a group. Thus the metallic character of the elements within a group increases from top to bottom. This trend is particularly noticeable for elements in Groups 3A to 7A. For example, in Group 4A we note that carbon is a nonmetal, silicon and germanium are metalloids, and tin and lead are metals.

Although the general trend in the periodic table is that of ionization energies increasing from left to right, some irregularities do exist. The first occurs in going from Group 2A to 3A (for example, from Be to B and from Mg to Al). The Group 3A elements all have a single electron in the outermost p subshell (ns^2np^1), which is well shielded by the inner electrons and the ns^2 electrons. Therefore, less energy is needed to remove a single p electron than to remove a paired s electron from the same principal energy level. This explains the *lower* ionization energies in Group 3A elements compared with those in Group 2A in the same period. The second irregularity occurs between Groups 5A and 6A (for example, from N to O and from P to S). In the Group 5A elements (ns^2np^3) the p electrons are in three separate orbitals according to Hund's rule. In Group 6A (ns^2np^4) the additional electron must be paired with one of the three p electrons. The proximity of two electrons in the same orbital results in greater electrostatic repulsion, which makes it easier to ionize an atom of the Group 6A element, even though the nuclear charge has increased by one unit. Thus the ionization energies in Group 6A are *lower* than those in Group 5A in the same period.

Example 8.4   (a) Which atom in Group 6A should have a smaller first ionization energy: oxygen or sulfur? (b) Which atom in the second period should have a higher second ionization energy: lithium or beryllium?

Reasoning and Solution   (a) Oxygen and sulfur are members of Group 6A. They have the same valence electron configuration (ns^2np^4), but the $3p$ electron in sulfur is farther from the nucleus and experiences less nuclear attraction than the $2p$ electron in oxygen. Thus, following the general rule that the ionization energy of elements decreases as we move down a periodic group, we predict that sulfur should have a smaller first ionization energy. Table 8.2 confirms our reasoning.

(b) The electron configurations of Li and Be are $1s^2 2s^1$ and $1s^2 2s^2$, respectively. For the second ionization process we write

$$Li^+(g) \longrightarrow Li^{2+}(g) + e^-$$
$$1s^2 \qquad\qquad 1s^1$$

$$Be^+(g) \longrightarrow Be^{2+}(g) + e^-$$
$$1s^2 2s^1 \qquad\qquad 1s^2$$

Because $1s$ electrons shield $2s$ electrons much more effectively than they shield each other, we predict that it should be much easier to remove a $2s$ electron from Be^+ than to remove a $1s$ electron from Li^+. This is consistent with the data in Table 8.2.

Similar problem: 8.53.

Practice Exercise   (a) Which of the following atoms should have a larger first ionization energy: N or P? (b) Which of the following atoms should have a smaller second ionization energy: Na or Mg?

8.5 Electron Affinity

Another property that greatly influences the chemical behavior of atoms is their ability to accept one or more electrons. This ability is called *electron affinity,* which is *the negative of the energy change that occurs when an electron is accepted by an atom of an element in the gaseous state.*

$$X(g) + e^- \longrightarrow X^-(g)$$

As we saw in Section 8.4, a positive ionization energy means that energy must be supplied to remove an electron. A positive electron affinity, on the other hand, signifies that energy is liberated when an electron is added to an atom. To clarify this apparent paradox, let us consider the process in which a fluorine atom accepts an electron in the gaseous state:

$$F(g) + e^- \longrightarrow F^-(g) \qquad \Delta H = -328 \text{ kJ}$$

The sign of ΔH indicates this is an exothermic process, but the electron affinity of fluorine is assigned a value of $+328$ kJ/mol. Thus we can think of electron affinity as the energy that must be supplied to remove an electron from a negative ion. For the removal of an electron from a fluoride ion, we have

$$F^-(g) \longrightarrow F(g) + e^- \qquad \Delta H = +328 \text{ kJ}$$

Keep in mind that a large positive electron affinity means that the negative ion is very stable (that is, the atom has a great tendency to accept an electron), just as a high ionization energy of an atom means that the atom is very stable.

Experimentally, electron affinity is determined by removing the additional electron from an anion. In contrast to ionization energies, however, electron affinities are difficult to measure because the anions of many elements are unstable. Table 8.3 shows the electron affinities of some representative elements and the noble gases, and Figure 8.10 plots the electron affinities of the first 56 elements versus atomic number. The overall trend is an increase in the tendency to accept electrons (that is, electron affinity values become more positive) from left to right across a period. The electron affinities of metals are generally lower than those of nonmetals. The values vary little within a given group. The halogens (Group 7A) have the highest electron affinity values. This is not surprising when we realize that by accepting an electron, each halogen atom assumes the stable electron configuration of the noble gas immediately to its right. For example, the electron configuration of F^- is $1s^2 2s^2 2p^6$ or [Ne]; for Cl^- it is [Ne]$3s^2 3p^6$ or [Ar]; and so on. Calculations show that the noble gases all have electron affinities of less than zero. Thus the anions of these gases, if formed, would be inherently unstable.

Table 8.3	Electron Affinities (kJ/mol) of Some Representative Elements and the Noble Gases*						
1A	2A	3A	4A	5A	6A	7A	8A
H 73							He <0
Li 60	Be ≤0	B 27	C 122	N 0	O 141	F 328	Ne <0
Na 53	Mg ≤0	Al 44	Si 134	P 72	S 200	Cl 349	Ar <0
K 48	Ca 2.4	Ga 29	Ge 118	As 77	Se 195	Br 325	Kr <0
Rb 47	Sr 4.7	In 29	Sn 121	Sb 101	Te 190	I 295	Xe <0
Cs 45	Ba 14	Tl 30	Pb 110	Bi 110	Po ?	At ?	Rn <0

*The electron affinities of the noble gases, Be, and Mg have not been determined experimentally, but are believed to be close to zero or negative.

Electron affinity is positive if the reaction is exothermic and negative if the reaction is endothermic.

The electron affinity of oxygen has a positive value (141 kJ/mol), which means that the process

$$O(g) + e^- \longrightarrow O^-(g) \qquad \Delta H = -141 \text{ kJ}$$

is favorable (exothermic). On the other hand, the electron affinity of the O^- ion is highly negative (-780 kJ/mol), which means the process

$$O^-(g) + e^- \longrightarrow O^{2-}(g) \qquad \Delta H = +780 \text{ kJ}$$

is endothermic even though the O^{2-} ion is isoelectronic with the noble gas Ne. This process is unfavorable in the gas phase because the resulting increase in electron-electron repulsion outweighs the stability gained by achieving a noble gas configuration. However, note that O^{2-} is common in ionic compounds (for example, Li_2O and MgO); in solids, the O^{2-} ion is stabilized by the neighboring cations.

Example 8.5 Why are the electron affinities of the alkaline earth metals, shown in Table 8.3, either negative or small positive values?

Reasoning and Solution The valence configuration of the alkaline earth metals is ns^2. For the process

$$M(g) + e^- \longrightarrow M^-(g)$$

where M denotes a member of the Group 2A family, the extra electron must enter the np subshell, which is effectively shielded by the two ns electrons (the np electrons are farther away from the nucleus than the ns electrons) and the inner electrons. Consequently, alkaline earth metals have little tendency to pick up an extra electron.

Similar problem: 8.61.

Practice Exercise Is it likely that Ar will form the anion Ar^-?

Figure 8.10 *A plot of electron affinity against atomic number for the first 56 elements.*

8.6 Variation in Chemical Properties of the Representative Elements

Ionization energy and electron affinity help chemists understand the types of reactions that elements undergo and the nature of the elements' compounds. Utilizing these concepts, we can survey the chemical behavior of the elements systematically, paying particular attention to the relationship between chemical properties and electron configuration.

General Trends in Chemical Properties

We have said that elements in the same group resemble one another in chemical behavior because they have similar outer electron configurations. This statement, although correct in the general sense, must be applied with caution. Chemists have long known that the first member of each group (that is, the elements in the second period from lithium to fluorine) differs from the rest of the members of the same group. For example, lithium, although exhibiting many of the properties characteristic of the alkali metals, is the only metal in Group 1A that does not form more than one compound with oxygen. Generally, the reason for the differences is the unusually small size of the first member of each group compared with the other members in the same group.

Another trend in chemical behavior of the representative elements is the ***diagonal relationship***. Diagonal relationship refers to *similarities that exist between pairs of elements in different groups and periods of the periodic table.* Specifically, the first three members of the second period (Li, Be, and B) exhibit many similarities to the elements located diagonally below them in the periodic table (Figure 8.11). The chemistry of lithium resembles that of magnesium in some ways; the same holds for beryllium and aluminum and for boron and silicon. We will see some examples illustrating this relationship later.

Figure 8.11 *Diagonal relationships in the periodic table.*

In comparing the properties of elements in the same group, bear in mind that the comparison is most valid if we are dealing with elements of the same type. This guideline applies to the elements in Groups 1A and 2A, which are all metals, and to the elements in Group 7A, which are all nonmetals. But Groups 3A through 6A include nonmetals, metalloids, and metals, so it is natural to expect some variations in chemical properties even though the members of the same group have similar outer electron configurations.

Now let us take a closer look at the chemical properties of hydrogen and the representative elements.

Hydrogen ($1s^1$)

There is no totally suitable position for hydrogen in the periodic table. Traditionally hydrogen is shown in Group 1A of the periodic table, but you should *not* think of it as a member of that group. Like the alkali metals, it has a single *s* valence electron and forms a unipositive ion (H^+), which is hydrated in solution. On the other hand, hydrogen also forms the hydride ion (H^-) in ionic compounds such as NaH and CaH_2. In this respect, hydrogen resembles the halogens, all of which form uninegative ions (F^-, Cl^-, Br^-, and I^-). Ionic hydrides react with water to produce hydrogen gas and the corresponding metal hydroxides:

$$2NaH(s) + 2H_2O(l) \longrightarrow 2NaOH(aq) + 2H_2(g)$$
$$CaH_2(s) + 2H_2O(l) \longrightarrow Ca(OH)_2(s) + 2H_2(g)$$

The most important compound of hydrogen is, of course, water, which is formed when hydrogen burns in air:

$$2H_2(g) + O_2(g) \longrightarrow 2H_2O(l)$$

Group 1A Elements (ns^1, $n \geq 2$)

Figure 8.12 shows the Group 1A elements, the alkali metals. All these elements have low ionization energies and therefore a great tendency to lose the single valence electron. In fact, in the vast majority of their compounds they are unipositive ions. These metals are so reactive that they are never found in the free, uncombined state in nature. They react with water to produce hydrogen gas and the corresponding metal hydroxide:

$$2M(s) + 2H_2O(l) \longrightarrow 2MOH(aq) + H_2(g)$$

in which M denotes an alkali metal. When exposed to air, they gradually lose their shiny appearance as they combine with oxygen gas to form oxides. Lithium forms lithium oxide (containing the O^{2-} ion):

$$4Li(s) + O_2(g) \longrightarrow 2Li_2O(s)$$

The other alkali metals all form *peroxides* (containing the O_2^{2-} ion) in addition to oxides. For example,

$$2Na(s) + O_2(g) \longrightarrow Na_2O_2(s)$$

Potassium, rubidium, and cesium also form *superoxides* (containing the O_2^- ion):

$$K(s) + O_2(g) \longrightarrow KO_2(s)$$

The reason that different types of oxides are formed when alkali metals react with oxygen has to do with the stability of the oxides in the solid state. Because these

Lithium (Li)   *Sodium (Na)*

Potassium (K)   *Rubidium (Rb)*   *Cesium (Cs)*

Figure 8.12 *The Group 1A elements: the alkali metals. Francium (not shown) is radioactive.*

oxides are all ionic compounds, their stability depends on how strongly the cations and anions attract one another. Lithium tends to form predominantly lithium oxide because this compound is more stable than lithium peroxide. The formation of other alkali metal oxides can be similarly explained.

Group 2A Elements (ns^2, $n \geq 2$)

Figure 8.13 shows the Group 2A elements. As a group, the alkaline earth metals are somewhat less reactive than the alkali metals. Both the first and the second ionization energies decrease from beryllium to barium. Thus the tendency is to form M^{2+} ions (in which M denotes an alkaline metal atom), and hence the metallic character increases as we go down the group. Most beryllium compounds (BeH_2 and beryllium halides, such as $BeCl_2$) and some magnesium compounds (MgH_2, for example) are molecular rather than ionic in nature. The reactivities of alkaline earth metals with water vary quite markedly. Beryllium does not react with water; magnesium reacts slowly with steam; and calcium, strontium, and barium are reactive enough to attack cold water:

$$Ba(s) + 2H_2O(l) \longrightarrow Ba(OH)_2(aq) + H_2(g)$$

The reactivities of the alkaline earth metals toward oxygen also increase from Be to Ba. Beryllium and magnesium form oxides (BeO and MgO) only at elevated temperatures, whereas CaO, SrO, and BaO are formed at room temperature.

Magnesium reacts with acids to liberate hydrogen gas:

Beryllium (Be)   *Magnesium (Mg)*   *Calcium (Ca)*

Strontium (Sr)   *Barium (Ba)*   *Radium (Ra)*

Figure 8.13 *The Group 2A elements: the alkaline earth metals.*

$$Mg(s) + 2H^+(aq) \longrightarrow Mg^{2+}(aq) + H_2(g)$$

Calcium, strontium, and barium also react with acids to produce hydrogen gas. However, because these metals also attack water, two different reactions will occur simultaneously.

The chemical properties of calcium and strontium provide an interesting example of periodic group similarity. Strontium-90, a radioactive isotope, is a major product of an atomic bomb explosion. If an atomic bomb is exploded in the atmosphere, the strontium-90 formed will eventually settle on land and water alike, and reach our bodies via a relatively short food chain. For instance, cows eat contaminated grass and drink contaminated water, and then they pass along strontium-90 in their milk. Because calcium and strontium are chemically similar, Sr^{2+} ions can replace Ca^{2+} ions in our bodies—for example, in bones. Constant exposure of the body to high-energy radiation emitted by the strontium-90 isotopes can lead to anemia, leukemia, and other chronic illnesses.

Group 3A Elements (ns^2np^1, $n \geq 2$)

The first member of Group 3A, boron, is a metalloid; the rest are metals (Figure 8.14). Boron does not form binary ionic compounds and is unreactive toward oxygen gas and water. The next element, aluminum, readily forms aluminum oxide when exposed to air:

$$4Al(s) + 3O_2(g) \longrightarrow 2Al_2O_3(s)$$

Aluminum that has a protective layer of aluminum oxide is less reactive than elemental aluminum. Aluminum forms only tripositive ions. It reacts with hydrochloric acid as follows:

$$2Al(s) + 6H^+(aq) \longrightarrow 2Al^{3+}(aq) + 3H_2(g)$$

Boron (B)

Aluminum (Al)

Gallium (Ga)

Indium (In)

Figure 8.14 *The Group 3A elements. The low melting point of gallium (29.8°C) causes it to melt when held in hand.*

The other Group 3A metallic elements form both unipositive and tripositive ions. Moving down the group, we find that the unipositive ion becomes more stable than the tripositive ion.

The metallic elements in Group 3A also form many molecular compounds. For example, aluminum reacts with hydrogen to form AlH_3, which resembles BeH_2 in properties. (Here is an example of the diagonal relationship.) Thus, as we move from left to right across the periodic table, we begin to see a gradual shift from metallic to nonmetallic character in the representative elements.

Group 4A Elements (ns²np², n ≥ 2)

The first member of Group 4A, carbon, is a nonmetal, and the next two members, silicon and germanium, are metalloids (Figure 8.15). The metallic elements of this group, tin and lead, do not react with water but react with acids (hydrochloric acid, for example) to liberate hydrogen gas:

$$Sn(s) + 2H^+(aq) \longrightarrow Sn^{2+}(aq) + H_2(g)$$
$$Pb(s) + 2H^+(aq) \longrightarrow Pb^{2+}(aq) + H_2(g)$$

The Group 4A elements form compounds in both the +2 and +4 oxidation states. For carbon and silicon, the +4 oxidation state is the more stable one. For example, CO_2 is more stable than CO, and SiO_2 is a stable compound, but SiO does not exist under normal conditions. As we move down the group, however, the trend in stability is reversed. In tin compounds the +4 oxidation state is only slightly more stable than the +2 oxidation state. In lead compounds the +2 oxidation state is unquestionably the more stable one. The outer electron configuration of lead is $6s^26p^2$, and lead tends to lose only the $6p$ electrons (to form Pb^{2+}) rather than both the $6p$ and $6s$ electrons (to form Pb^{4+}).

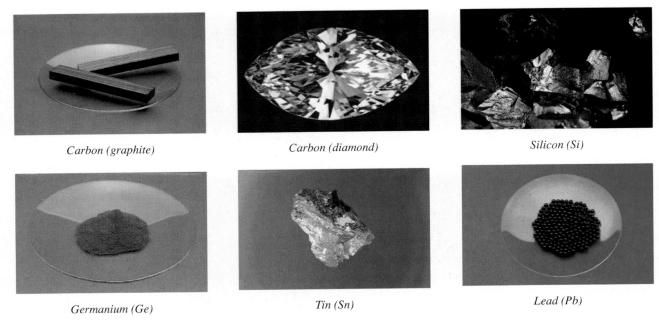

Carbon (graphite)   Carbon (diamond)   Silicon (Si)

Germanium (Ge)   Tin (Sn)   Lead (Pb)

Figure 8.15 *The Group 4A elements.*

Group 5A Elements (ns^2np^3, $n \geq 2$)

In Group 5A, nitrogen and phosphorus are nonmetals, arsenic and antimony are metalloids, and bismuth is a metal (Figure 8.16). Thus we expect a larger variation in properties as we move down the group. Elemental nitrogen is a diatomic gas (N_2). It forms a number of oxides (NO, N_2O, NO_2, N_2O_4, and N_2O_5), of which only N_2O_5 is a solid; the others are gases. Nitrogen has a tendency to accept three electrons to form the nitride ion, N^{3-} (thus achieving the electron configuration $1s^2 2s^2 2p^6$, which is isoelectronic with neon). Most metallic nitrides (Li_3N and Mg_3N_2, for example) are ionic compounds. Phosphorus exists as P_4 molecules. It forms two solid oxides with the formulas P_4O_6 and P_4O_{10}. The important oxoacids HNO_3 and H_3PO_4 are formed when the following oxides react with water:

$$N_2O_5(s) + H_2O(l) \longrightarrow 2HNO_3(aq)$$
$$P_4O_{10}(s) + 6H_2O(l) \longrightarrow 4H_3PO_4(aq)$$

Arsenic, antimony, and bismuth have extensive three-dimensional structures. Bismuth is a far less reactive metal than those in the preceding groups.

Group 6A Elements (ns^2np^4, $n \geq 2$)

The first three members of Group 6A (oxygen, sulfur, and selenium) are nonmetals, and the last two (tellurium and polonium) are metalloids (Figure 8.17). Oxygen is a diatomic gas; elemental sulfur and selenium have the molecular formulas S_8 and Se_8, respectively; tellurium and polonium have more extensive three-dimensional structures. (Polonium is a radioactive element that is difficult to study in the laboratory.) Oxygen has a tendency to accept two electrons to form the oxide ion (O^{2-}) in many ionic compounds. (The electron configuration of O^{2-} is $1s^2 2s^2 2p^6$, which is isoelectronic with Ne.) Sulfur, selenium, and tellurium also form dinegative anions: S^{2-}, Se^{2-}, and Te^{2-}. The elements in this group (especially oxygen)

Nitrogen (N_2)

White and red phosphorus (P)

Arsenic (As)

Antimony (Sb)

Bismuth (Bi)

Figure 8.16 *The Group 5A elements. Molecular nitrogen is a colorless, odorless gas.*

form a large number of molecular compounds with nonmetals. The important compounds of sulfur are SO_2, SO_3, and H_2S. Sulfuric acid is formed when sulfur trioxide dissolves in water:

$$SO_3(g) + H_2O(l) \longrightarrow H_2SO_4(aq)$$

Group 7A Elements (ns^2np^5, $n \geq 2$)

All the halogens are nonmetals with the general formula X_2, in which X denotes a halogen element (Figure 8.18). Because of their great reactivity, the halogens are never found in the elemental form in nature. (The last member of Group 7A is astatine, a radioactive element. Little is known about its properties.) Fluorine is so reactive that it attacks water to generate oxygen:

$$2F_2(g) + 2H_2O(l) \longrightarrow 4HF(aq) + O_2(g)$$

Sulfur (S$_8$)   *Selenium (Se$_8$)*   *Tellurium (Te)*

Figure 8.17 *The Group 6A elements sulfur, selenium, and tellurium. Molecular oxygen is a colorless, odorless gas. Polonium (not shown) is radioactive.*

Actually the reaction between molecular fluorine and water is quite complex; the products formed depend on reaction conditions. The reaction shown here is one of several possible changes. The halogens have high ionization energies and large positive electron affinities. These facts suggest that they would preferentially form anions of the type X^-. Anions derived from the halogens (F^-, Cl^-, Br^-, and I^-) are called *halides*. They are isoelectronic with the noble gases. For example, F^- is isoelectronic with Ne, Cl^- with Ar, and so on. The vast majority of the alkali metal and alkaline earth metal halides are ionic compounds. The halogens also form many molecular compounds among themselves (such as ICl and BrF$_3$) and with nonmetallic elements in other groups (such as NF$_3$, PCl$_5$, and SF$_6$). The halogens react with hydrogen to form hydrogen halides:

$$H_2(g) + X_2(g) \longrightarrow 2HX(g)$$

When this reaction involves fluorine, it is explosive, but it becomes less and less violent as we substitute chlorine and iodine. The hydrogen halides dissolve in water to form hydrohalic acids. Of this series, hydrofluoric acid (HF) is a weak acid (that is, it is a weak electrolyte), but the other acids (HCl, HBr, and HI) are all strong acids (or strong electrolytes).

Figure 8.18 *The Group 7A elements chlorine, bromine, and iodine. Fluorine is a greenish-yellow gas that attacks ordinary glassware. Astatine is radioactive.*

Figure 8.19 *All noble gases are colorless and odorless. These pictures show the colors emitted by the gases from a discharge tube.*

Helium (He)   *Neon (Ne)*   *Argon (Ar)*

Krypton (Kr)   *Xenon (Xe)*

Group 8A Elements (ns²np⁶, n ≥ 2)

Group 8A Elements (ns^2np^6, $n \geq 2$)

All noble gases exist as monatomic species (Figure 8.19). The electron configurations of the noble gases show that their atoms have completely filled outer *ns* and *np* subshells, indicating great stability. (Helium is $1s^2$.) The Group 8A ionization energies are among the highest of all elements, and these gases have no tendency to accept extra electrons. For a number of years these elements were called inert gases, and rightly so. Until 1963 no one had been able to prepare a compound containing any of these elements. But an experiment carried out by the British chemist Neil Bartlett shattered chemists' long-held views of these elements. When he exposed xenon to platinum hexafluoride, a strong oxidizing agent, the following reaction took place (Figure 8.20):

$$Xe(g) + PtF_6(g) \longrightarrow XePtF_6(s)$$

(a)   (b)

Figure 8.21 *Crystals of xenon tetrafluoride (XeF₄).*

Since then, a number of xenon compounds (XeF_4, XeO_3, XeO_4, $XeOF_4$) and a few krypton compounds (KrF_2, for example) have been prepared (Figure 8.21). Despite the immense interest in the chemistry of the noble gases, however, their compounds do not have any commercial applications, and they are not involved in natural biological processes. In 2000, chemists prepared a compound containing argon (HArF), which is stable only at very low temperatures. No compounds of helium or neon are known.

Properties of Oxides across a Period

One way to compare the properties of the representative elements across a period is to examine the properties of a series of similar compounds. Because oxygen combines with almost all elements, we will compare the properties of oxides of the third-period elements to see how metals differ from metalloids and nonmetals. Table 8.4 lists a few general characteristics of the oxides.

We observed earlier that oxygen has a tendency to form the oxide ion. This tendency is greatly favored in reactions with metals that have low ionization energies, namely, those in Groups 1A and 2A and aluminum. Thus Na_2O, MgO, and Al_2O_3 are ionic compounds, as indicated by their high melting points and boiling points. They have extensive three-dimensional structures in which each cation is surrounded by a specific number of anions, and vice versa. As the ionization energies of the elements increase from left to right, so does the molecular nature of the oxides that are formed.

Table 8.4 | Some Properties of Oxides of the Third-Period Elements

	Na₂O	MgO	Al₂O₃	SiO₂	P₄O₁₀	SO₃	Cl₂O₇
Type of compound	←	Ionic	→	←	Molecular		→
Structure	←— Extensive three-dimensional —→			←	Discrete molecular units		→
Melting point (°C)	1275	2800	2045	1610	580	16.8	−91.5
Boiling point (°C)	?	3600	2980	2230	?	44.8	82
Acid-base nature	Basic	Basic	Amphoteric	←	Acidic		→

Silicon is a metalloid; its oxide (SiO_2) also has a huge three-dimensional network, although no ions are present. The oxides of phosphorus, sulfur, and chlorine are molecular compounds composed of small discrete units. The weak attractions among these molecules result in relatively low melting points and boiling points.

Most oxides can be classified as acidic or basic, depending on whether they produce acids or bases when dissolved in water or react as acids or bases in certain processes. Some oxides are **amphoteric oxides,** which means that they *display both acidic and basic properties.* The first two oxides of the third period, Na_2O and MgO, are basic oxides. For example, Na_2O reacts with water to form sodium hydroxide (which is a base):

$$Na_2O(s) + H_2O(l) \longrightarrow 2NaOH(aq)$$

Magnesium oxide is quite insoluble; it does not react with water to any appreciable extent. However, it does react with acids in a manner that resembles an acid-base reaction:

$$MgO(s) + 2HCl(aq) \longrightarrow MgCl_2(aq) + H_2O(l)$$

Note that the products of this reaction are a salt ($MgCl_2$) and water, the usual products of an acid-base neutralization.

Aluminum oxide is even less soluble than magnesium oxide; it too does not react with water. However, it shows basic properties by reacting with acids:

$$Al_2O_3(s) + 6HCl(aq) \longrightarrow 2AlCl_3(aq) + 3H_2O(l)$$

It also exhibits acidic properties by reacting with bases:

$$Al_2O_3(s) + 2NaOH(aq) + 3H_2O(l) \longrightarrow 2NaAl(OH)_4(aq)$$

Note that this acid-base neutralization produces a salt but no water.

Thus Al_2O_3 is classified as an amphoteric oxide because it has properties of both acids and bases. Other amphoteric oxides are ZnO, BeO, and Bi_2O_3.

Silicon dioxide is insoluble and does not react with water. It has acidic properties, however, because it reacts with very concentrated bases:

$$SiO_2(s) + 2NaOH(aq) \longrightarrow Na_2SiO_3(aq) + H_2O(l)$$

For this reason, concentrated bases such as NaOH should not be stored in Pyrex glassware, which is made of SiO_2.

The remaining third-period oxides are acidic, as indicated by their reactions with water to form phosphoric acid (H_3PO_4), sulfuric acid (H_2SO_4), and perchloric acid ($HClO_4$):

$$P_4O_{10}(s) + 6H_2O(l) \longrightarrow 4H_3PO_4(aq)$$
$$SO_3(g) + H_2O(l) \longrightarrow H_2SO_4(aq)$$
$$Cl_2O_7(s) + H_2O(l) \longrightarrow 2HClO_4(aq)$$

This brief examination of oxides of the third-period elements shows that as the metallic character of the elements decreases from left to right across the period, their oxides change from basic to amphoteric to acidic. Normal metallic oxides are usually basic, and most oxides of nonmetals are acidic. The intermediate properties of the oxides (as shown by the amphoteric oxides) are exhibited by elements whose positions are intermediate within the period. Note also that because the metallic character of the elements increases as we move down a particular group of the representative elements, we would expect oxides of elements with higher atomic numbers to be more basic than the lighter elements. This is indeed the case.

Summary

Newlands, Mendeleev, and Meyer devised the periodic table by arranging elements in order of increasing mass. Some discrepancies in the early version of the periodic table were resolved when Moseley showed that elements should be arranged according to increasing atomic number.

Electron configuration directly influences the properties of the elements. The modern periodic table classifies the elements according to their atomic numbers, and thus also by their electron configurations. The configuration of the outermost electrons (called valence electrons) directly affects the properties of the atoms of the representative elements.

Periodic variations in the physical properties of the elements reflect differences in structure. The metallic character of elements decreases across a period from metals through the metalloids to nonmetals and increases from top to bottom within a particular group of representative elements. Atomic size defined in terms of atomic radius also varies periodically, decreasing from left to right and increasing from top to bottom within a group.

Ionization energy is a measure of the tendency of an atom to resist the loss of an electron. The higher the ionization energy, the stronger the nucleus's hold on the electron. Electron affinity is a measure of the tendency of an atom to gain an electron. The more positive the electron affinity, the greater the tendency for the atom to gain an electron. In general, metals have low ionization energies and nonmetals have high electron affinities.

Noble gases are very stable because their outer ns and np subshells are completely filled. The metals of the representative elements (in Groups 1A, 2A, and 3A) tend to lose electrons until their cations become isoelectronic with the noble gases preceding them in the periodic table. The nonmetals of the representative elements (in Groups 5A, 6A, and 7A) tend to accept electrons until their anions become isoelectronic with the noble gases following them in the periodic table.

Key Words

Amphoteric oxide p. 257	Electron affinity p. 245	Isoelectronic p. 236	Valence electrons p. 234
Atomic radius p. 237	Ionic radius p. 239	Noble gases p. 233	
Diagonal relationship p. 247	Ionization energy p. 242	Representative elements p. 233	

Questions and Problems

Development of the Periodic Table

Review Questions

8.1   Briefly describe the significance of Mendeleev's periodic table.

8.2   What is Moseley's contribution to the modern periodic table?

8.3   Describe the general layout of a modern periodic table.

8.4   What is the most important relationship among elements in the same group in the periodic table?

Periodic Classification of the Elements

Review Questions

8.5   Which of these elements are metals, nonmetals, and metalloids: As, Xe, Fe, Li, B, Cl, Ba, P, I, Si?

8.6   Compare the physical and chemical properties of metals and nonmetals.

8.7   Draw a rough sketch of a periodic table (no details are required). Indicate where metals, nonmetals, and metalloids are located.

8.8 What is a representative element? Give names and symbols of four representative elements.

8.9 Without referring to a periodic table, write the name and symbol for an element in each of these groups: 1A, 2A, 3A, 4A, 5A, 6A, 7A, 8A, transition metals.

8.10 Indicate whether these elements exist as atomic species, molecular species, or extensive three-dimensional structures in their most stable state at 25°C and 1 atm, and write the molecular or empirical formula for the elements: phosphorus, iodine, magnesium, neon, arsenic, sulfur, carbon, selenium, and oxygen.

8.11 You are given a dark shiny solid and asked to determine whether it is iodine or a metallic element. Suggest a nondestructive test that would allow you to arrive at the correct answer.

8.12 Define valence electrons. For representative elements, the number of valence electrons of an element is equal to its group number. Show that this is true for the following elements: Al, Sr, K, Br, P, S, C.

8.13 A neutral atom of a certain element has 17 electrons. Without consulting a periodic table, (a) write the ground-state electron configuration of the element, (b) classify the element, (c) determine whether the atoms of this element are diamagnetic or paramagnetic.

8.14 In the periodic table, the element hydrogen is sometimes grouped with the alkali metals (as in this book) and sometimes with the halogens. Explain why hydrogen can resemble the Group 1A and the Group 7A elements.

8.15 Write the outer electron configurations for (a) the alkali metals, (b) the alkaline earth metals, (c) the halogens, (d) the noble gases.

8.16 Use the first-row transition metals (Sc to Cu) as an example to illustrate the characteristics of the electron configurations of transition metals.

Problems

8.17 Group these electron configurations in pairs that would represent similar chemical properties of their atoms:

(a) $1s^2 2s^2 2p^6 3s^2$
(b) $1s^2 2s^2 2p^3$
(c) $1s^2 2s^2 2p^6 3s^2 3p^6 4s^2 3d^{10} 4p^6$
(d) $1s^2 2s^2$
(e) $1s^2 2s^2 2p^6$
(f) $1s^2 2s^2 2p^6 3s^2 3p^3$

8.18 Group these electron configurations in pairs that would represent similar chemical properties of their atoms:

(a) $1s^2 2s^2 2p^5$
(b) $1s^2 2s^1$
(c) $1s^2 2s^2 2p^6$
(d) $1s^2 2s^2 2p^6 3s^2 3p^5$
(e) $1s^2 2s^2 2p^6 3s^2 3p^6 4s^1$
(f) $1s^2 2s^2 2p^6 3s^2 3p^6 4s^2 3d^{10} 4p^6$

8.19 Without referring to a periodic table, write the electron configurations of elements with these atomic numbers: (a) 9, (b) 20, (c) 26, (d) 33. Classify the elements.

8.20 Specify in what group of the periodic table each of these elements is found: (a) $[Ne]3s^1$, (b) $[Ne]3s^2 3p^3$, (c) $[Ne]3s^2 3p^6$, (d) $[Ar]4s^2 3d^8$.

8.21 An ion M^{2+} derived from a metal in the first transition metal series has four and only four electrons in the $3d$ subshell. What element might M be?

8.22 A metal ion with a net $+3$ charge has five electrons in the $3d$ subshell. Identify the metal.

Electron Configurations of Ions

Review Questions

8.23 What is the characteristic of the electron configuration of stable ions derived from representative elements?

8.24 What do we mean when we say that two ions or an atom and an ion are isoelectronic?

8.25 What is wrong with the statement "The atoms of element X are isoelectronic with the atoms of element Y"?

8.26 Give three examples of first-row transition metal (Sc to Cu) ions whose electron configurations are represented by the argon core.

Problems

8.27 Write ground-state electron configurations for these ions: (a) Li^+, (b) H^-, (c) N^{3-}, (d) F^-, (e) S^{2-}, (f) Al^{3+}, (g) Se^{2-}, (h) Br^-, (i) Rb^+, (j) Sr^{2+}, (k) Sn^{2+}.

8.28 Write ground-state electron configurations for these ions, which play important roles in biochemical processes in our bodies: (a) Na^+, (b) Mg^{2+}, (c) Cl^-, (d) K^+, (e) Ca^{2+}, (f) Fe^{2+}, (g) Cu^{2+}, (h) Zn^{2+}.

8.29 Write ground-state electron configurations for these transition metal ions: (a) Sc^{3+}, (b) Ti^{4+}, (c) V^{5+}, (d) Cr^{3+}, (e) Mn^{2+}, (f) Fe^{2+}, (g) Fe^{3+}, (h) Co^{2+}, (i) Ni^{2+}, (j) Cu^+, (k) Cu^{2+}, (l) Ag^+, (m) Au^+, (n) Au^{3+}, (o) Pt^{2+}.

8.30 Name the ions with $+3$ charges that have these electron configurations: (a) $[Ar]3d^3$, (b) $[Ar]$, (c) $[Kr]4d^6$, (d) $[Xe]4f^{14}5d^6$.

8.31 Which of these species are isoelectronic with each other: C, Cl^-, Mn^{2+}, B^-, Ar, Zn, Fe^{3+}, Ge^{2+}?

8.32 Group the species that are isoelectronic: Be^{2+}, F^-, Fe^{2+}, N^{3-}, He, S^{2-}, Co^{3+}, Ar.

Periodic Variation in Physical Properties

Review Questions

8.33 Define atomic radius. Does the size of an atom have a precise meaning?

8.34 How does atomic radius change as we move (a) from left to right across the period and (b) from top to bottom in a group?

8.35 Define ionic radius. How does the size change when an atom is converted to (a) an anion and (b) a cation?

8.36 Explain why, for isoelectronic ions, the anions are larger than the cations.

Problems

8.37 On the basis of their positions in the periodic table, select the atom with the larger atomic radius in each of these pairs: (a) Na, Cs; (b) Be, Ba; (c) N, Sb; (d) F, Br; (e) Ne, Xe.

8.38 Arrange the following atoms in order of decreasing atomic radius: Na, Al, P, Cl, Mg.

8.39 Which is the largest atom in Group 4A?

8.40 Which is the smallest atom in Group 7A?

8.41 Why is the radius of the lithium atom considerably larger than the radius of the hydrogen atom?

8.42 Use the second period of the periodic table as an example to show that the sizes of atoms decrease as we move from left to right. Explain the trend.

8.43 In each of the following pairs, indicate which one of the two species is smaller: (a) Cl or Cl^-, (b) Na or Na^+, (c) O^{2-} or S^{2-}, (d) Mg^{2+} or Al^{3+}, (e) Au^+ or Au^{3+}.

8.44 List these ions in order of increasing ionic radius: N^{3-}, Na^+, F^-, Mg^{2+}, O^{2-}.

8.45 Explain which of these ions is larger, and why: Cu^+ or Cu^{2+}.

8.46 Explain which of these anions is larger, and why: Se^{2-} or Te^{2-}.

8.47 Give the physical states (gas, liquid, or solid) of the representative elements in the fourth period at 1 atm and 25°C: K, Ca, Ga, Ge, As, Se, Br.

8.48 The boiling points of neon and krypton are −245.9°C and −152.9°C, respectively. Using these data, estimate the boiling point of argon. (*Hint:* The properties of argon are intermediate between those of neon and krypton.)

Ionization Energy

Review Questions

8.49 Define ionization energy. Ionization energy is usually measured in the gaseous state. Why? Why is the second ionization energy always greater than the first ionization energy for any element?

8.50 Sketch an outline of the periodic table and show group and period trends in the first ionization energy of the elements. What types of elements have the highest ionization energies and what types the lowest ionization energies?

Problems

8.51 Use the third period of the periodic table as an example to illustrate the change in first ionization energies of the elements as we move from left to right. Explain the trend.

8.52 Ionization energy usually increases from left to right across a given period. Aluminum, however, has a lower ionization energy than magnesium. Explain.

8.53 The first and second ionization energies of K are 419 kJ/mol and 3052 kJ/mol, and those of Ca are 590 kJ/mol and 1145 kJ/mol, respectively. Compare their values and comment on the differences.

8.54 Two atoms have the electron configurations $1s^2 2s^2 2p^6$ and $1s^2 2s^2 2p^6 3s^1$. The first ionization energy of one is 2080 kJ/mol, and that of the other is 496 kJ/mol. Pair each ionization energy with one of the given electron configurations. Justify your choice.

8.55 A hydrogen-like ion is an ion containing only one electron. The energies of the electron in a hydrogen-like ion are given by

$$E_n = -(2.18 \times 10^{-18}\ \text{J})Z^2\left(\frac{1}{n^2}\right)$$

in which n is the principal quantum number and Z is the atomic number of the element. Calculate the ionization energy (in kilojoules per mole) of the He^+ ion.

8.56 Plasma is a state of matter in which a gaseous system consists of positive ions and electrons. In the plasma state, a mercury atom would be stripped of its 80 electrons and exist as Hg^{80+}. Use the equation in Problem 8.55 to calculate the energy required for the last step of ionization; that is,

$$Hg^{79+}(g) \longrightarrow Hg^{80+}(g) + e^-$$

Electron Affinity

Review Questions

8.57 (a) Define electron affinity. Electron affinity is usually measured with atoms in the gaseous state. Why? (b) Ionization energy is always a positive quantity, whereas electron affinity may be either positive or negative. Explain.

8.58 Explain the trends in electron affinity from aluminum to chlorine (see Table 8.3).

Problems

8.59 Arrange the elements in each of these groups in order of increasing electron affinity: (a) Li, Na, K; (b) F, Cl, Br, I.

8.60 Which of these elements would you expect to have the greatest electron affinity? He, K, Co, S, Cl.

8.61 From the electron-affinity values for the alkali metals, do you think it is possible for these metals to form an anion like M^-, where M represents an alkali metal?

8.62 Explain why alkali metals have a greater affinity for electrons than alkaline earth metals do.

Variation in Chemical Properties

Review Questions

8.63 Explain what is meant by the diagonal relationship. List two pairs of elements that show this relationship.

8.64 Which elements are more likely to form acidic oxides? basic oxides? amphoteric oxides?

Problems

8.65 Use the alkali metals and alkaline earth metals as examples to show how we can predict the chemical properties of elements simply from their electron configurations.

8.66 Based on your knowledge of the chemistry of the alkali metals, predict some of the chemical properties of francium, the last member of the group.

8.67 As a group, the noble gases are very stable chemically (only Kr and Xe are known to form some compounds). Why?

8.68 Why are the Group 1B elements more stable than the Group 1A elements even though they seem to have the same outer electron configuration ns^1 in which n is the principal quantum number of the outermost shell?

8.69 How do the chemical properties of oxides change as we move across a period from left to right? as we move down a particular group?

8.70 Predict (and give balanced equations for) the reactions between each of these oxides and water: (a) Li_2O, (b) CaO, (c) CO_2.

8.71 Write formulas and give names for the binary hydrogen compounds of the second-period elements (Li to F). Describe the changes in physical and chemical properties of these compounds as we move across the period from left to right.

8.72 Which oxide is more basic, MgO or BaO? Why?

Additional Problems

8.73 State whether each of these properties of the representative elements generally increases or decreases (a) from left to right across a period and (b) from top to bottom in a group: metallic character, atomic size, ionization energy, acidity of oxides.

8.74 With reference to the periodic table, name (a) a halogen element in the fourth period, (b) an element similar to phosphorus in chemical properties, (c) the most reactive metal in the fifth period, (d) an element that has an atomic number smaller than 20 and is similar to strontium.

8.75 Why do elements that have high ionization energies usually have more positive electron affinities?

8.76 Arrange the following isoelectronic species in order of (a) increasing ionic radius and (b) increasing ionization energy: O^{2-}, F^-, Na^+, Mg^{2+}.

8.77 Write the empirical (or molecular) formulas of compounds that the elements in the third period (sodium to chlorine) are expected to form with (a) molecular oxygen and (b) molecular chlorine. In each case indicate whether you expect the compound to be ionic or molecular.

8.78 Element M is a shiny and highly reactive metal (melting point 63°C), and element X is a highly reactive nonmetal (melting point −7.2°C). They react to form a compound with the empirical formula MX, a colorless, brittle solid that melts at 734°C. When dissolved in water or when molten, the substance conducts electricity. When chlorine gas is bubbled through an aqueous solution containing MX, a reddish-brown liquid appears and Cl^- ions are formed. From these observations, identify M and X. (You may need to consult a handbook of chemistry for the melting-point values.)

8.79 Match each of the elements on the right with its description on the left:

(a) A dark-red liquid   Calcium (Ca)
(b) A colorless gas that burns   Gold (Au)
     in oxygen gas   Hydrogen (H_2)
(c) A reactive metal that   Neon (Ne)
     attacks water   Bromine (Br_2)
(d) A shiny metal that is used
     in jewelry
(e) A totally inert gas

8.80 Arrange these species in isoelectronic pairs: O^+, Ar, S^{2-}, Ne, Zn, Cs^+, N^{3-}, As^{3+}, N, Xe.

8.81 In which of these are the species written in decreasing radius? (a) Be, Mg, Ba, (b) N^{3-}, O^{2-}, F^-, (c) Tl^{3+}, Tl^{2+}, Tl^+.

8.82 Which of these properties show a clear periodic variation? (a) first ionization energy, (b) molar mass of the elements, (c) number of isotopes of an element, (d) atomic radius.

8.83 When carbon dioxide is bubbled through a clear calcium hydroxide solution, the solution becomes milky. Write an equation for the reaction and explain how this reaction illustrates that CO_2 is an acidic oxide.

8.84 You are given four substances: a fuming red liquid, a dark metallic-looking solid, a pale-yellow gas, and a yellow-green gas that attacks glass. You are told that these substances are the first four members of Group 7A, the halogens. Name each one.

8.85 For each pair of elements listed below, give three properties that show their chemical similarity: (a) sodium and potassium and (b) chlorine and bromine.

8.86 Name the element that forms compounds, under appropriate conditions, with every other element in the periodic table except He and Ne.

8.87 Explain why the first electron affinity of sulfur is 200 kJ/mol but the second electron affinity is −649 kJ/mol.

8.88 The H^- ion and the He atom have two $1s$ electrons each. Which of the two species is larger? Explain.

8.89 Acidic oxides are those that react with water to produce acid solutions, whereas the reactions of basic oxides with water produce basic solutions. Nonmetallic oxides are usually acidic, whereas those of metals are basic. Predict the products of the reactions of these oxides with water:

Na_2O, BaO, CO_2, N_2O_5, P_4O_{10}, SO_3. Write an equation for each of the reactions.

8.90 Write the formulas and names of the oxides of the second-period elements (Li to N). Identify the oxides as acidic, basic, or amphoteric.

8.91 State whether each of the elements listed here is a gas, a liquid, or a solid under atmospheric conditions. Also state whether it exists in the elemental form as atoms, as molecules, or as a three-dimensional network: Mg, Cl, Si, Kr, O, I, Hg, Br.

8.92 What factors account for the unique nature of hydrogen?

8.93 The formula for calculating the energies of an electron in a hydrogen-like ion is

$$E_n = -(2.18 \times 10^{-18}\ \text{J})Z^2\left(\frac{1}{n^2}\right)$$

This equation cannot be applied to many-electron atoms. One way to modify it for the more complex atoms is to replace Z with $(Z - \sigma)$, in which Z is the atomic number and σ is a positive dimensionless quantity called the shielding constant. Consider the helium atom as an example. The physical significance of σ is that it represents the extent of shielding that the two $1s$ electrons exert on each other. Thus the quantity $(Z - \sigma)$ is appropriately called the "effective nuclear charge." Calculate the value of σ if the first ionization energy of helium is 3.94×10^{-18} J per atom. (Ignore the minus sign in the given equation in your calculation.)

8.94 A technique called photoelectron spectroscopy is used to measure the ionization energy of atoms. A sample is irradiated with ultraviolet (UV) light, and electrons are ejected from the valence shell. The kinetic energies of the ejected electrons are measured. Because the energy of the UV photon and the kinetic energy of the ejected electron are known, we can write

$$h\nu = \text{IE} + \tfrac{1}{2}mu^2$$

in which ν is the frequency of the UV light, and m and u are the mass and velocity of the electron, respectively. In one experiment the kinetic energy of the ejected electron from potassium is found to be 5.34×10^{-19} J using a UV source of wavelength 162 nm. Calculate the ionization energy of potassium. How can you be sure that this ionization energy corresponds to the electron in the valence shell (that is, the most loosely held electron)?

8.95 A student is given samples of three elements, X, Y, and Z, which could be an alkali metal, a member of Group 4A, and a member of Group 5A. She makes the following observations: Element X has a metallic luster and conducts electricity. It reacts slowly with hydrochloric acid to produce hydrogen gas. Element Y is a light-yellow solid that does not conduct electricity. Element Z has a metallic luster and conducts electricity. When exposed to air, it

slowly forms a white powder. A solution of the white powder in water is basic. What can you conclude about the elements from these observations?

8.96 Using these melting-point data, estimate the melting point of francium, which is a radioactive element:

Metal	Li	Na	K	Rb	Cs
melting point (°C)	180.5	97.8	63.3	38.9	28.4

(*Hint:* Plot melting point versus atomic number.)

8.97 Experimentally, the electron affinity of an element can be determined by using a laser light to ionize the anion of the element in the gas phase:

$$X^-(g) + h\nu \longrightarrow X(g) + e^-$$

Referring to Table 8.3, calculate the photon wavelength (in nanometers) corresponding to the electron affinity for chlorine. In what region of the electromagnetic spectrum does this wavelength fall?

8.98 Name an element in Group 1A or Group 2A that is an important constituent of each of these substances: (a) remedy for acid indigestion, (b) coolant in nuclear reactors, (c) Epsom salt, (d) baking powder, (e) gunpowder, (f) a light alloy, (g) fertilizer that also neutralizes acid rain, (h) cement, and (i) grit for icy roads. You may need to ask your instructor about some of the items.

8.99 Explain why the electron affinity of nitrogen is approximately zero, although the elements on either side, carbon and oxygen, have substantial positive electron affinities.

8.100 Little is known of the chemistry of astatine, the last member of Group 7A. Describe the physical characteristics that you would expect this halogen to have. Predict the products of the reaction between sodium astatide (NaAt) and sulfuric acid. (*Hint:* Sulfuric acid is an oxidizing agent.)

8.101 The ionization energies of sodium (in kJ/mol), starting with the first and ending with the eleventh, are 495.9, 4560, 6900, 9540, 13,400, 16,600, 20,120, 25,490, 28,930, 141,360, 170,000. Plot the log of ionization energy (*y* axis) versus the number of ionization (*x* axis); for example, log 495.9 is plotted versus 1 (labeled I_1, the first ionization energy), log 4560 is plotted versus 2 (labeled I_2, the second ionization energy), and so on. (a) Label I_1 through I_{11} with the electrons in orbitals such as $1s$, $2s$, $2p$, and $3s$. (b) What can you deduce about electron shells from the breaks in the curve?

8.102 Calculate the maximum wavelength of light (in nanometers) required to ionize a single sodium atom.

8.103 The first four ionization energies of an element are approximately 738 kJ/mol, 1450 kJ/mol, 7.7×10^3 kJ/mol, and 1.1×10^4 kJ/mol. To which periodic group does this element belong? Why?

8.104 Match each of the elements on the right with its description on the left:

(a) A greenish-yellow gas
    that reacts with water
(b) A soft metal that reacts
    with water to produce
    hydrogen
(c) A metalloid that is hard
    and has a high melting point
(d) A colorless, odorless gas
(e) A more reactive metal than
    iron, which does not corrode
    in air

Nitrogen (N_2)
Boron (B)
Aluminum (Al)
Fluorine (F_2)
Sodium (Na)

Special Problems

8.105 In the late 1800s the British physicist Lord Rayleigh accurately determined the atomic masses of a number of elements, but he obtained a puzzling result with nitrogen. One of his methods of preparing nitrogen was by the thermal decomposition of ammonia:

$$2NH_3(g) \longrightarrow N_2(g) + 3H_2(g)$$

Another method was to start with air and remove oxygen, carbon dioxide, and water vapor from it. Invariably, the nitrogen from air was a little denser (by about 0.5 percent) than the nitrogen from ammonia.

Later the English chemist Sir William Ramsay carried out an experiment in which he passed nitrogen, which he had obtained from air by Raleigh's procedure, over red-hot magnesium to convert it to magnesium nitride:

$$3Mg(s) + N_2(g) \longrightarrow Mg_3N_2(s)$$

After all of the nitrogen had reacted with magnesium, Ramsay was left with an unknown gas that would not combine with anything. The atomic mass of this gas was determined to be 39.95 amu. Ramsay called the gas *argon,* which means "the lazy one" in Greek.

(a) Later Rayleigh and Ramsay, with the help of Sir William Crookes, the inventor of the discharge tube, showed that argon was a new element. Describe the type of experiment performed that led them to the conclusion.
(b) Why did it take so long to discover argon?
(c) Once argon had been discovered, why did it take relatively little time to discover the rest of the noble gases?
(d) Why was helium the last noble gas to be discovered on Earth?
(e) The only confirmed compound of radon is radon fluoride, RnF. Give two reasons why there are so few known radon compounds.

8.106 On the same graph, plot the effective nuclear charge (in parentheses) and atomic radius (see Figure 8.4) versus atomic number for the second-period elements: Li(1.30), Be(1.95), B(2.60), C(3.25), N(3.90), O(4.55), F(5.20), Ne(5.85). Comment on the trends.

Answers to Practice Exercises

8.1 (a) [Ar]$4s^2$; (b) representative element; (c) diamagnetic. **8.2** Li > Be > C. **8.3** (a) Li$^+$, (b) Au^{3+}, (c) N^{3-}. **8.4** (a) N, (b) Mg. **8.5** No.

While lecturing to an introductory chemistry class—or so the story goes—Gilbert N. Lewis had an inspiration: Suppose atoms form molecules by sharing electrons. This chemical bonding scheme might account for the fact that some compounds are more stable than their component atoms alone. Based on this insight, Lewis developed the so-called octet rule, which proved to be very useful in explaining the properties and reactions of compounds.

Born in 1875, Lewis grew up in Nebraska. After studying at Harvard University and in Europe, he taught at the Massachusetts Institute of Technology before joining the faculty at the University of California, Berkeley, in 1912. There he molded the chemistry department into one of the world's leading teaching and research facilities.

Although Lewis is best known for his work in chemical bonding, he also made outstanding contributions in acid-base chemistry, thermodynamics, and spectroscopy (the study of interaction between light and matter).

Lewis first sketched his idea about the octet rule on the back of an envelope.

Professor Lewis in his laboratory.

In addition, he was the first person to prepare heavy water (D_2O) and to investigate its properties. In spite of his distinguished achievements, Lewis never won a Nobel Prize. He died in 1946.

264

Chemical Bonding I: The Covalent Bond

Chapter 9

ESSENTIAL CONCEPTS

 The Covalent Bond Lewis postulated the formation of a covalent bond in which atoms share one or more pairs of electrons. The octet rule was formulated to predict the correctness of Lewis structures. This rule says that an atom other than hydrogen tends to form bonds until it is surrounded by eight valence electrons.

 Characteristics of Lewis Structures In addition to covalent bonds, a Lewis structure also shows lone pairs, which are pairs of electrons not involved in bonding, on atoms and formal charges, which are the result of bookkeeping of electrons used in bonding. A resonance structure is one of two or more Lewis structures for a single molecule that cannot be described fully with only one Lewis structure.

 Exceptions to the Octet Rule The octet rule applies mainly to the second-period elements. The three categories of exceptions to the octet rule are the incomplete octet, in which an atom in a molecule has fewer than eight valence electrons, the odd-electron molecules, which have an odd number of valence electrons, and the expanded octet, in which an atom has more than eight valence electrons. These exceptions can be explained by more refined theories of chemical bonding.

 Thermochemistry Based on Bond Energy From a knowledge of the strength of covalent bonds or bond energies, it is possible to estimate the enthalpy change of a reaction.

9.1 Lewis Dot Symbols

The development of the periodic table and concept of electron configuration gave chemists a rationale for molecule and compound formation. This explanation, formulated by Gilbert Lewis, is that atoms combine in order to achieve a more stable electron configuration. Maximum stability results when an atom is isoelectronic with a noble gas.

When atoms interact to form a chemical bond, only their outer regions are in contact. For this reason, when we study chemical bonding, we are concerned primarily with the valence electrons of the atoms. To keep track of valence electrons in a chemical reaction, and to make sure that the total number of electrons does not change, chemists use a system of dots devised by Lewis, called the Lewis dot symbols. A *Lewis dot symbol consists of the symbol of an element and one dot for each valence electron in an atom of the element.* Figure 9.1 shows the Lewis dot symbols for the representative elements and the noble gases. Note that, except for helium, the number of valence electrons each atom has is the same as the group number of the element. For example, Li is a Group 1A element and has one dot for one valence electron; Be, a Group 2A element, has two valence electrons (two dots); and so on. Elements in the same group have similar outer electron configurations and hence similar Lewis dot symbols. The transition metals, lanthanides, and actinides all have incompletely filled inner shells, and in general, we cannot write simple Lewis dot symbols for them.

In this chapter we will learn to use electron configurations and the periodic table to predict the type of bond that atoms will form, as well as the number of bonds an atom of a particular element can form and the stability of the product.

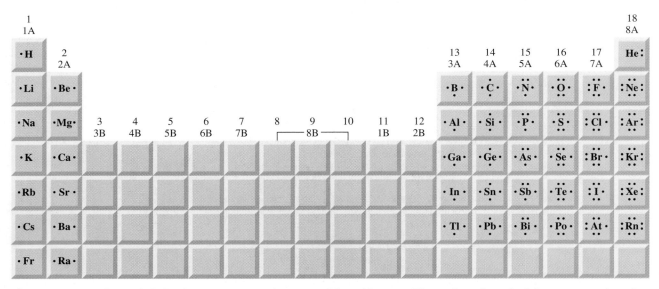

Figure 9.1 *Lewis dot symbols for the representative elements and the noble gases. The number of unpaired dots corresponds to the number of bonds an atom of the element can form in a compound.*

9.2 The Covalent Bond

Although the concept of molecules goes back to the seventeenth century, it was not until early in the twentieth century that chemists began to understand how and why molecules form. The first major breakthrough was Gilbert Lewis's suggestion that a chemical bond involves electron sharing by atoms. He depicted the formation of a chemical bond in H_2 as

$$H\cdot + \cdot H \longrightarrow H:H$$

This type of electron pairing is an example of a ***covalent bond,*** *a bond in which two electrons are shared by two atoms.* ***Covalent compounds*** *are compounds that contain only covalent bonds.* For the sake of simplicity, the shared pair of electrons is often represented by a single line. Thus the covalent bond in the hydrogen molecule can be written as H—H. In a covalent bond, each electron in a shared pair is attracted to the nuclei of both atoms. This attraction holds the two atoms in H_2 together and is responsible for the formation of covalent bonds in other molecules.

Covalent bonding between many-electron atoms involves only the valence electrons. Consider the fluorine molecule, F_2. The electron configuration of F is $1s^2 2s^2 2p^5$. The $1s$ electrons are low in energy and stay near the nucleus most of the time. For this reason they do not participate in bond formation. Thus each F atom has seven valence electrons (the $2s$ and $2p$ electrons). According to Figure 9.1, there is only one unpaired electron on F, so the formation of the F_2 molecule can be represented as follows:

$$:\!\overset{..}{\underset{..}{F}}\!\cdot + \cdot\overset{..}{\underset{..}{F}}\!: \longrightarrow :\!\overset{..}{\underset{..}{F}}\!:\!\overset{..}{\underset{..}{F}}\!: \quad \text{or} \quad :\!\overset{..}{\underset{..}{F}}\!\!-\!\!\overset{..}{\underset{..}{F}}\!:$$

Note that only two valence electrons participate in the formation of F_2. The other, nonbonding electrons, are called ***lone pairs***—*pairs of valence electrons that are not involved in covalent bond formation.* Thus each F in F_2 has three lone pairs of electrons:

$$\text{lone pairs} \longrightarrow :\!\overset{..}{\underset{..}{F}}\!\!-\!\!\overset{..}{\underset{..}{F}}\!: \longleftarrow \text{lone pairs}$$

The structures we use to represent covalent compounds, such as H_2 and F_2, are called Lewis structures. A ***Lewis structure*** is *a representation of covalent bonding in which shared electron pairs are shown either as lines or as pairs of dots between two atoms, and lone pairs are shown as pairs of dots on individual atoms.* Only valence electrons are shown in a Lewis structure.

Let us consider the Lewis structure of the water molecule. Figure 9.1 shows the Lewis dot symbol for oxygen with two unpaired dots or two unpaired electrons, as we expect that O might form two covalent bonds. Because hydrogen has only one electron, it can form only one covalent bond. Thus the Lewis structure for water is

$$H:\overset{..}{\underset{..}{O}}:H \quad \text{or} \quad H\!\!-\!\!\overset{..}{\underset{..}{O}}\!\!-\!\!H$$

In this case, the O atom has two lone pairs. The hydrogen atom has no lone pairs because its only electron is used to form a covalent bond.

In the F_2 and H_2O molecules, the F and O atoms achieve the stable noble gas configuration by sharing electrons:

Ionic vs. covalent bonding

This discussion applies only to representative elements. Remember that for these elements, the number of valence electrons is equal to the group number (Groups 1A–7A).

74 pm   161 pm

H_2 HI

Figure 9.2 *Bond length in H_2 and HI.*

The formation of these molecules illustrates the **octet rule,** formulated by Lewis: *An atom other than hydrogen tends to form bonds until it is surrounded by eight valence electrons.* In other words, a covalent bond forms when there are not enough electrons for each individual atom to have a complete octet. By sharing electrons in a covalent bond, the individual atoms can complete their octets. The requirement for hydrogen is that it attain the electron configuration of helium, or a total of two electrons.

The octet rule works mainly for elements in the second period of the periodic table. These elements have only $2s$ and $2p$ subshells, which can hold a total of eight electrons. When an atom of one of these elements forms a covalent compound, it can attain the noble gas electron configuration [Ne] by sharing electrons with other atoms in the same compound. Later, we will discuss a number of important exceptions to the octet rule that give us further insight into the nature of chemical bonding.

Atoms can form different types of covalent bonds. In a **single bond,** *two atoms are held together by one electron pair.* Many compounds are held together by **multiple bonds,** that is, bonds formed when *two atoms share two or more pairs of electrons.* If *two atoms share two pairs of electrons,* the covalent bond is called a **double bond.** Double bonds are found in molecules of carbon dioxide (CO_2) and ethylene (C_2H_4):

A **triple bond** arises when *two atoms share three pairs of electrons,* as in the nitrogen molecule (N_2):

The acetylene molecule (C_2H_2) also contains a triple bond, in this case between two carbon atoms:

Note that in ethylene and acetylene all the valence electrons are used in bonding; there are no lone pairs on the carbon atoms. In fact, with the exception of carbon monoxide, stable molecules containing carbon do not have lone pairs on the carbon atoms.

Multiple bonds are shorter than single covalent bonds. **Bond length** is defined as the *distance between the nuclei of two covalently bonded atoms in a molecule* (Figure 9.2). Table 9.1 shows some experimentally determined bond lengths. For a given pair of atoms, such as carbon and nitrogen, triple bonds are shorter than double bonds, which, in turn, are shorter than single bonds. The shorter multiple bonds are also more stable than single bonds, as we will see later.

Shortly you will be introduced to the rules for writing proper Lewis structures. Here we simply want to become familiar with the language associated with them.

Table 9.1 | Average Bond Lengths of Some Common Single, Double, and Triple Bonds

Bond Type	Bond Length (pm)
C—H	107
C—O	143
C=O	121
C—C	154
C=C	133
C≡C	120
C—N	143
C=N	138
C≡N	116
N—O	136
N=O	122
O—H	96

9.3 Electronegativity

A covalent bond, as we have said, is the sharing of an electron pair by two atoms. In a molecule like H_2, in which the atoms are identical, we expect the electrons to be equally shared—that is, the electrons spend the same amount of time in the vicinity of each atom. However, in the covalently bonded HF molecule, the H and F atoms do not share the bonding electrons equally because H and F are different atoms:

$$H—\ddot{\underset{\cdot\cdot}{F}}:$$

The bond in HF is called a **polar covalent bond,** or simply a *polar bond,* because *the electrons spend more time in the vicinity of one atom than the other.* Experimental evidence indicates that in the HF molecule the electrons spend more time near the F atom. We can think of this unequal sharing of electrons as a partial electron transfer or a shift in electron density, as it is more commonly described, from H to F (Figure 9.3). This "unequal sharing" of the bonding electron pair results in a relatively greater electron density near the fluorine atom and a correspondingly lower electron density near hydrogen. The HF bond and other polar bonds can be thought of as being intermediate between a (nonpolar) covalent bond, in which the sharing of electrons is exactly equal, and an **ionic bond,** in which the *transfer of the electron(s) is nearly complete.*

A property that helps us distinguish a nonpolar covalent bond from a polar covalent bond is **electronegativity,** *the ability of an atom to attract toward itself the electrons in a chemical bond.* Elements with high electronegativity have a greater tendency to attract electrons than do elements with low electronegativity. As we might expect, electronegativity is related to electron affinity and ionization energy. Thus an atom such as fluorine, which has a high electron affinity (tends to pick up electrons easily) and a high ionization energy (does not lose electrons easily), has a high electronegativity. On the other hand, sodium has a low electron affinity, a low ionization energy, and a low electronegativity.

Electronegativity is a relative concept, meaning that an element's electronegativity can be measured only in relation to the electronegativity of other elements. The American chemist Linus Pauling devised a method for calculating *relative* electronegativities of most elements. These values are shown in Figure 9.4. A careful examination of this chart reveals trends and relationships among electronegativity values of different elements. In general, electronegativity increases from left to right across a period in the periodic table, as the metallic character of the elements decreases. Within each group, electronegativity decreases with increasing atomic number, and increasing metallic character. Note that the transition metals do not follow these trends. The most electronegative elements—the halogens, oxygen, nitrogen, and sulfur—are found in the upper right-hand corner of the periodic table, and the least electronegative elements (the alkali and alkaline earth metals) are clustered near the lower left-hand corner. These trends are readily apparent on a graph, as shown in Figure 9.5.

Atoms of elements with widely different electronegativities tend to form ionic bonds (such as those that exist in NaCl and CaO compounds) with each other because the atom of the less electronegative element gives up its electron(s) to the atom of the more electronegative element. An ionic bond generally joins an atom of a metallic element and an atom of a nonmetallic element. Atoms of elements with comparable electronegativities tend to form polar covalent bonds with each other because the shift in electron density is usually small. Most covalent bonds involve atoms of nonmetallic elements. Only atoms of the same element, which have

Figure 9.3 *Electron density distribution in the HF molecule. The distribution varies according to the colors of the rainbow. The most electron-rich region is red; the most electron-poor region is blue.*

Electronegativity values have no units.

Increasing electronegativity

1A																	8A
H 2.1	2A											3A	4A	5A	6A	7A	
Li 1.0	**Be** 1.5											**B** 2.0	**C** 2.5	**N** 3.0	**O** 3.5	**F** 4.0	
Na 0.9	**Mg** 1.2	3B	4B	5B	6B	7B	⌐—— 8B ——⌐			1B	2B	**Al** 1.5	**Si** 1.8	**P** 2.1	**S** 2.5	**Cl** 3.0	
K 0.8	**Ca** 1.0	**Sc** 1.3	**Ti** 1.5	**V** 1.6	**Cr** 1.6	**Mn** 1.5	**Fe** 1.8	**Co** 1.9	**Ni** 1.9	**Cu** 1.9	**Zn** 1.6	**Ga** 1.6	**Ge** 1.8	**As** 2.0	**Se** 2.4	**Br** 2.8	
Rb 0.8	**Sr** 1.0	**Y** 1.2	**Zr** 1.4	**Nb** 1.6	**Mo** 1.8	**Tc** 1.9	**Ru** 2.2	**Rh** 2.2	**Pd** 2.2	**Ag** 1.9	**Cd** 1.7	**In** 1.7	**Sn** 1.8	**Sb** 1.9	**Te** 2.1	**I** 2.5	
Cs 0.7	**Ba** 0.9	**La-Lu** 1.0-1.2	**Hf** 1.3	**Ta** 1.5	**W** 1.7	**Re** 1.9	**Os** 2.2	**Ir** 2.2	**Pt** 2.2	**Au** 2.4	**Hg** 1.9	**Tl** 1.8	**Pb** 1.9	**Bi** 1.9	**Po** 2.0	**At** 2.2	
Fr 0.7	**Ra** 0.9																

Increasing electronegativity (vertical, left axis)

Figure 9.4 *The electronegativities of common elements.*

> An ionic bond generally involves an atom of a metallic element and an atom of a nonmetallic element, whereas a covalent bond generally involves two atoms of nonmetallic elements.

the same electronegativity, can be joined by a pure covalent bond. These trends and characteristics are what we would expect, given our knowledge of ionization energies and electron affinities.

There is no sharp distinction between a polar bond and an ionic bond, but the following rule is helpful in distinguishing between them. An ionic bond forms when the electronegativity difference between the two bonding atoms is 2.0 or more. This rule applies to most but not all ionic compounds. Sometimes chemists use the quantity *percent ionic character* to describe the nature of a bond. A purely ionic bond would have 100 percent ionic character, although no such bond is known, whereas a nonpolar or purely covalent bond has 0 percent ionic character.

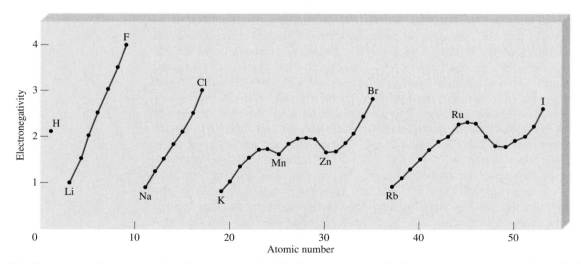

Figure 9.5 *Variation of electronegativity with atomic number. The halogens have the highest electronegativities, and the alkali metals the lowest.*

Electronegativity and electron affinity are related but different concepts. Both indicate the tendency of an atom to attract electrons. However, electron affinity refers to an isolated atom's attraction for an additional electron, whereas electronegativity signifies the ability of an atom in a chemical bond (with another atom) to attract the shared electrons. Furthermore, electron affinity is an experimentally measurable quantity, whereas electronegativity is an estimated number that cannot be measured.

Example 9.1   Classify these bonds as ionic, polar covalent, or covalent: (a) the bond in HCl, (b) the bond in KF, and (c) the CC bond in H_3CCH_3.

Reasoning and Solution   (a) In Figure 9.4 we see that the electronegativity difference between H and Cl is 0.9, which is appreciable but not large enough (by the 2.0 rule) to qualify HCl as an ionic compound. Therefore, the bond between H and Cl is polar covalent.

(b) The electronegativity difference between K and F is 3.2, well above the 2.0 mark; therefore, the bond between K and F is ionic.

(c) The two C atoms are identical in every respect—they are bonded to each other and each is bonded to three H atoms. Therefore, the bond between them is purely covalent.

Practice Exercise   Which of these bonds is covalent, which is polar covalent, and which is ionic? (a) the bond in CsCl, (b) the bond in H_2S, (c) the NN bond in H_2NNH_2.

The most electronegative elements are the nonmetals (red) and the least electronegative elements are the alkali and alkaline earth metals (green). Beryllium, the first member of Group 2A, forms mostly covalent compounds.

Similar problems: 9.13, 9.14.

Electronegativity and Oxidation Number

In Chapter 4 we introduced the rules for assigning oxidation numbers of elements in their compounds. The concept of electronegativity is the basis for these rules. In essence, oxidation number refers to the number of charges an atom would have if electrons were transferred completely to the more electronegative of the bonded atoms in a molecule.

Consider the NH_3 molecule, in which the N atom forms three single bonds with the H atoms. Because N is more electronegative than H, electron density will be shifted from H to N. If the transfer were complete, each H would donate an electron to N, which would have a total charge of 3− while each H would have a charge of 1+. Thus we assign an oxidation number of −3 to N and an oxidation number of +1 to H in NH_3.

The same procedure is applied to molecules containing multiple bonds. Thus for the CO_2 molecule, whose Lewis structure is shown on p. 268, each O atom receives two electrons from the C atom because of the C=O double bond. The C atom would have a charge of 4+ (an oxidation number of +4) and each O atom would have a charge of 2− (an oxidation number of −2).

Oxygen usually has an oxidation number of −2 in its compounds, except in hydrogen peroxide (H_2O_2), whose Lewis structure is

$$H\!-\!\overset{\cdot\cdot}{\underset{\cdot\cdot}{O}}\!-\!\overset{\cdot\cdot}{\underset{\cdot\cdot}{O}}\!-\!H$$

A bond between identical atoms makes no contribution to the oxidation number of those atoms because the electron pair of that bond is *equally* shared. Because H has an oxidation number of +1, each O atom has an oxidation number of −1.

Can you see now why fluorine always has an oxidation number of -1? It is the most electronegative element known, and it *always* forms a single bond in its compounds. Therefore, it would bear a $1-$ charge if electron transfer were complete.

9.4 Writing Lewis Structures

Although the octet rule and Lewis structures do not present a complete picture of covalent bonding, they do help to explain the bonding scheme in many compounds and account for the properties and reactions of molecules. For this reason, you should practice writing Lewis structures of compounds. The basic steps are as follows:

1. Write the skeletal structure of the compound, using chemical symbols and placing bonded atoms next to one another. For simple compounds, this task is fairly easy. For more complex compounds, we must either be given the information or make an intelligent guess about it. In general, the least electronegative atom occupies the central position. Hydrogen and fluorine usually occupy the terminal (end) positions in the Lewis structure.

2. Count the total number of valence electrons present, referring, if necessary, to Figure 9.1. For polyatomic anions, add the number of negative charges to that total. (For example, for the CO_3^{2-} ion we add two electrons because the $2-$ charge indicates that there are two more electrons than are provided by the neutral atoms.) For polyatomic cations, we subtract the number of positive charges from this total. (Thus, for NH_4^+ we subtract one electron because the $1+$ charge indicates a loss of one electron from the group of neutral atoms.)

3. Draw a single covalent bond between the central atom and each of the surrounding atoms. Complete the octets of the atoms bonded to the central atom. (Remember that the valence shell of a hydrogen atom is complete with only two electrons.) Electrons belonging to the central or surrounding atoms must be shown as lone pairs if they are not involved in bonding. The total number of electrons to be used is that determined in step 2.

4. If the octet rule is not satisfied for the central atom, try adding double or triple bonds between the surrounding atoms and the central atom, using the lone pairs from the surrounding atoms.

Examples 9.2 and 9.3 illustrate the four-step procedure for writing Lewis structures of compounds and an ion.

NF$_3$ is a colorless, odorless, unreactive gas.

Example 9.2   Write the Lewis structure for nitrogen trifluoride (NF_3) in which all three F atoms are bonded to the N atom.

Reasoning and Solution   We follow the listed procedure for writing Lewis structure.

Step 1. The N atom is less electronegative than F, so the skeletal structure of NF_3 is

$$F \quad N \quad F$$
$$F$$

Step 2. The outer-shell electron configurations of N and F are $2s^2 2p^3$ and $2s^2 2p^5$, respectively. Thus there are $5 + (3 \times 7)$, or 26, valence electrons to account for in NF_3.

Step 3. We draw a single covalent bond between N and each F, and complete the octets for the F atoms. We place the remaining two electrons on N:

$$:\ddot{F}-\overset{..}{\underset{\underset{:\ddot{F}:}{|}}{N}}-\ddot{F}:$$

Because this structure satisfies the octet rule for all the atoms, step 4 is not required. To check, we count the valence electrons in NF_3 (in chemical bonds and in lone pairs). The result is 26, the same as the number of valence electrons on three F atoms and one N atom.

Similar problem: 9.21.

Practice Exercise Write the Lewis structure for carbon disulfide (CS_2).

Example 9.3 Write the Lewis structure for the carbonate ion (CO_3^{2-}).

Reasoning and Solution We follow the listed procedure for writing Lewis structures and note that this is an anion with two negative charges.

Step 1. We can deduce the skeletal structure of the carbonate ion by recognizing that C is less electronegative than O. Therefore, it is most likely to occupy a central position as follows:

O

O   C   O

Step 2. The outer-shell electron configurations of C and O are $2s^2 2p^2$ and $2s^2 2p^4$, respectively, and the ion itself has two negative charges. Thus the total number of electrons is $4 + (3 \times 6) + 2$, or 24.

Step 3. We draw a single covalent bond between C and each O and comply with the octet rule for the O atoms:

$$:\overset{..}{\underset{..}{O}}:$$

This structure shows all 24 electrons.

Step 4. Although the octet rule is satisfied for the O atoms, it is not for the C atom. Therefore we move a lone pair from one of the O atoms to form another bond with C. Now the octet rule is also satisfied for the C atom:

$$\left[\begin{array}{c} :\ddot{O}: \\ \| \\ :\ddot{O}-C-\ddot{O}: \end{array} \right]^{2-}$$

We use the brackets to indicate that the 2− charge is on the whole structure.

As a final check, we verify that there are 24 valence electrons in the Lewis structure of the carbonate ion.

Similar problem: 9.22.

Practice Exercise Write the Lewis structure for the nitrite ion (NO_2^-).

CO_3^{2-}

9.5 Formal Charge and Lewis Structure

By comparing the number of electrons in an isolated atom with the number of electrons that are associated with the same atom in a Lewis structure, we can determine the distribution of electrons in the molecule and draw the most plausible Lewis structure. The bookkeeping procedure is as follows: In an isolated atom, the number of electrons associated with the atom is simply the number of valence electrons. (As usual, we need not be concerned with the inner electrons.) In a molecule, electrons associated with the atom are the nonbonding electrons plus the electrons in the bonding pair(s) between the atom and other atom(s). However, because electrons are shared in a bond, we must divide the electrons in a bonding pair equally between the atoms forming the bond. An atom's **_formal charge_** is _the electrical charge difference between the valence electrons in an isolated atom and the number of electrons assigned to that atom in a Lewis structure._

To assign the number of electrons on an atom in a Lewis structure, we proceed as:

- All the atom's nonbonding electrons are assigned to the atom.
- We break the bond(s) between the atom and other atom(s) and assign half of the bonding electrons to the atom.

Let us illustrate the concept of formal charge using the ozone molecule (O_3). Proceeding by steps, as we did in Examples 9.2 and 9.3, we arrive at the Lewis structure

$$\ddot{O}{=}\ddot{O}{-}\ddot{O}{:}$$

The formal charge on each atom in O_3 can now be calculated according to the following scheme:

Liquid ozone below its boiling point ($-111.3°C$). Ozone is a toxic, light-blue gas with a pungent odor.

$$\ddot{O}{\approx}\ddot{O}{\approx}\ddot{O}{:}$$

Valence e^-	6	6	6
e^- assigned to atom	6	5	7
Difference (formal charge)	0	1+	1−

By the same token, the breaking of a triple bond transfers three electrons to each of the bonding atoms.

where the wiggly red lines denote the breaking of the bonds. Note that the breaking of a single bond results in the transfer of an electron, the breaking of a double bond results in a transfer of two electrons to each of the bonding atoms, and so on. Thus the formal charges of the atoms in O_3 are

$$\ddot{O}{=}\overset{+}{\ddot{O}}{-}\ddot{O}{:}^{-}$$

For single positive and negative charges, we normally omit the numeral 1.

When you write formal charges, these rules are helpful:

- For molecules, the sum of the formal charges must add up to zero because they are electrically neutral species. (This rule applies, for example, to the O_3 molecule.)
- For cations, the sum of the formal charges must equal the positive charge.
- For anions, the sum of the formal charges must equal the negative charge.

Keep in mind that formal charges do not represent actual charge separation within the molecule. In the O_3 molecule, for example, there is no evidence that the central atom bears a net 1+ charge or that one of the end atoms bears a 1− charge. Writing these charges on the atoms in the Lewis structure merely helps us keep track of the valence electrons in the molecule.

Example 9.4   Write formal charges for the carbonate ion.

Reasoning and Solution   The Lewis structure for the carbonate ion was developed in Example 9.3:

$$\left[\begin{array}{c} :\ddot{O}: \\ \\ :\ddot{O} \overset{\displaystyle \sim}{} C \overset{\displaystyle \sim}{} \ddot{O}: \end{array}\right]^{2-}$$

The formal charges on the atoms can be calculated using the given procedure.

The C atom   The C atom has four valence electrons and there are no nonbonding electrons on the atom in the Lewis structure. The breaking of the double bond and two single bonds results in the transfer of four electrons to the C atom. Therefore, the formal charge is $4 - 4 = 0$.

The O atom in C=O   The O atom has six valence electrons and there are four nonbonding electrons on the atom. The breaking of the double bond results in the transfer of two electrons to the O atom. Here the formal charge is $6 - 4 - 2 = 0$.

The O atom in C—O   This atom has six nonbonding electrons and the breaking of the single bond transfers another electron to it. Therefore, the formal charge is $6 - 6 - 1 = 1-$.

Thus the Lewis structure for CO_3^{2-} with formal charges is

$$\begin{array}{c} :\ddot{O}: \\ \| \\ {}^{-}:\ddot{O} - C - \ddot{O}:{}^{-} \end{array}$$

Note that the sum of the formal charges is $2-$, the same as the charge on the carbonate ion.

Practice Exercise   Write formal charges for the nitrite ion (NO_2^-).

Similar problem: 9.22.

Sometimes there is more than one acceptable Lewis structure for a given species. In such cases, we can often select the most plausible Lewis structure by using formal charges and the following guidelines:

- For neutral molecules, a Lewis structure in which there are no formal charges is preferable to one in which formal charges are present.
- Lewis structures with large formal charges ($2+$, $3+$, and/or $2-$, $3-$, and so on) are less plausible than those with small formal charges.
- Among Lewis structures having similar distributions of formal charges, the most plausible structure is the one in which negative formal charges are placed on the more electronegative atoms.

Example 9.5   Formaldehyde (CH_2O), a liquid with a disagreeable odor, traditionally has been used to preserve laboratory species. Draw the most likely Lewis structure for the compound.

CH_2O

Similar problem: 9.23.

Reasoning and Solution The two possible skeletal structures are

$$H \quad C \quad O \quad H \qquad\qquad \begin{array}{c} H \\ C \quad O \\ H \end{array}$$

$$\text{(a)} \qquad\qquad\qquad \text{(b)}$$

We can draw a Lewis structure for each of these possibilities:

$$H-\overset{\cdot\cdot}{\underset{}{C}}=\overset{+}{\underset{}{O}}-H \qquad\qquad \begin{array}{c} H \\ \diagdown \\ C=\overset{\cdot\cdot}{\underset{\cdot\cdot}{O}} \\ \diagup \\ H \end{array}$$

$$\text{(a)} \qquad\qquad\qquad \text{(b)}$$

Although both structures satisfy the octet rule, (b) is the more likely structure because it carries no formal charges.

Comment Can you suggest two other reasons why (a) is less plausible?

Practice Exercise Draw the most reasonable Lewis structure of a molecule that contains a N atom, a C atom, and a H atom.

9.6 The Concept of Resonance

Our drawing of the Lewis structure for ozone (O_3) satisfied the octet rule for the central atom because we placed a double bond between it and one of the two end O atoms. In fact, we can put the double bond at either end of the molecule, as shown by these two equivalent Lewis structures:

$$\overset{\cdot\cdot}{\underset{\cdot\cdot}{O}}=\overset{\cdot\cdot}{\overset{+}{O}}-\overset{\cdot\cdot}{\underset{\cdot\cdot}{O}}:^{-} \qquad\qquad ^{-}:\overset{\cdot\cdot}{\underset{\cdot\cdot}{O}}-\overset{\cdot\cdot}{\overset{+}{O}}=\overset{\cdot\cdot}{O}$$

However, neither one of these two Lewis structures accounts for the known bond lengths in O_3.

We would expect the O—O bond in O_3 to be longer than the O=O bond because double bonds are known to be shorter than single bonds. Yet experimental evidence shows that both oxygen-to-oxygen bonds are equal in length (128 pm). We resolve this discrepancy by using *both* Lewis structures to represent the ozone molecule:

$$\overset{\cdot\cdot}{\underset{\cdot\cdot}{O}}=\overset{\cdot\cdot}{\overset{+}{O}}-\overset{\cdot\cdot}{\underset{\cdot\cdot}{O}}:^{-} \longleftrightarrow ^{-}:\overset{\cdot\cdot}{\underset{\cdot\cdot}{O}}-\overset{\cdot\cdot}{\overset{+}{O}}=\overset{\cdot\cdot}{O}$$

Each of these structures is called a resonance structure. A *resonance structure,* then, is *one of two or more Lewis structures for a single molecule that cannot be represented accurately by only one Lewis structure.* The double-headed arrow indicates that the structures shown are resonance structures.

The term *resonance* itself means *the use of two or more Lewis structures to represent a particular molecule.* Like the medieval European traveler to Africa who described a rhinoceros as a cross between a griffin and a unicorn, two familiar but imaginary animals, we describe ozone, a real molecule, in terms of two familiar but nonexistent structures.

A common misconception about resonance is the notion that a molecule such as ozone somehow shifts quickly back and forth from one resonance structure to the other. Keep in mind that *neither* resonance structure adequately represents the actual molecule, which has its own unique, stable structure. "Resonance" is a human invention, designed to address the limitations in these simple bonding models. To extend the animal analogy, a rhinoceros is a distinct creature, not some oscillation between mythical griffin and unicorn!

The carbonate ion provides another example of resonance:

$$: \ddot{O} : \qquad \qquad : \ddot{O} :^- \qquad \qquad : \ddot{O} :^-$$
$$^- : \ddot{O} - C - \ddot{O} :^- \longleftrightarrow \ddot{O} = C - \ddot{O} :^- \longleftrightarrow ^- : \ddot{O} - C = \ddot{O}$$

According to experimental evidence, all carbon-to-oxygen bonds in CO_3^{2-} are equivalent. Therefore, the properties of the carbonate ion are best explained by considering its resonance structures together.

The concept of resonance applies equally well to organic systems. A good example is the benzene molecule (C_6H_6):

[benzene Lewis structures with H and C atoms shown, two resonance forms connected by ↔]

The hexagonal structure of benzene was first proposed by the German chemist August Kekulé (1829–1896).

If one of these resonance structures corresponded to the actual structure of benzene, there would be two different bond lengths between adjacent C atoms, one characteristic of the single bond and the other of the double bond. In fact, the distance between all adjacent C atoms in benzene is 140 pm, which is shorter than a C—C bond (154 pm) and longer than a C=C bond (133 pm).

A simpler way of drawing the structure of the benzene molecule and other compounds containing the "benzene ring" is to show only the skeleton and not the carbon and hydrogen atoms. By this convention the resonance structures are represented by

[hexagon resonance structures ↔]

An electron micrograph of benzene molecule, which shows clearly the ring structure.

Note that the C atoms at the corners of the hexagon and the H atoms are all omitted, although they are understood to exist. Only the bonds between the C atoms are shown.

Remember this important rule for drawing resonance structures: The positions of electrons, but not those of atoms, can be rearranged in different resonance structures. In other words, the same atoms must be bonded to one another in all the resonance structures for a given species.

NO_3^-

Similar problems: 9.29, 9.34.

Example 9.6   Draw resonance structures (including formal charges) for the nitrate ion (NO_3^-), which has this skeletal arrangement:

$$O$$

$$O \quad N \quad O$$

Reasoning and Solution   Because nitrogen has five valence electrons and each oxygen has six valence electrons and there is a net negative charge, the total number of valence electrons is $5 + (3 \times 6) + 1 = 24$. These resonance structures are all equivalent:

$$
\overset{..}{:}\overset{..}{O}\overset{..}{:}{}^{-} \qquad\qquad \overset{..}{:}\overset{}{O}\overset{..}{:} \qquad\qquad \overset{..}{:}\overset{..}{O}\overset{..}{:}{}^{-}
$$

$$
\overset{..}{O}{=}\overset{+}{N}{-}\overset{..}{\underset{..}{O}}{:}{}^{-} \longleftrightarrow {}^{-}\overset{..}{:}\overset{..}{\underset{..}{O}}{-}\overset{+}{N}{-}\overset{..}{\underset{..}{O}}{:}{}^{-} \longleftrightarrow {}^{-}\overset{..}{:}\overset{..}{\underset{..}{O}}{-}\overset{+}{N}{=}\overset{..}{O}
$$

Practice Exercise   Draw resonance structures for the nitrite ion (NO_2^-).

Finally, note that although it is more accurate to show all the resonance structures of an ion or a compound, we often use only one Lewis structure for simplicity.

9.7 Exceptions to the Octet Rule

As mentioned earlier, the octet rule applies mainly to the second-period elements. Exceptions to the octet rule fall into three categories, characterized by an incomplete octet, an odd number of electrons, and more than eight valence electrons around the central atom.

The Incomplete Octet

Beryllium, unlike the other Group 2A elements, forms mostly covalent compounds of which BeH₂ is an example.

In some compounds the number of electrons surrounding the central atom in a stable molecule is fewer than eight. Consider, for example, beryllium, which is a Group 2A (and a second-period) element. The electron configuration of beryllium is $1s^2 2s^2$; it has two valence electrons in the $2s$ orbital. In the gas phase, beryllium hydride (BeH_2) exists as discrete molecules. The Lewis structure of BeH_2 is

$$H—Be—H$$

As you can see, only four electrons surround the Be atom, and there is no way to satisfy the octet rule for beryllium in this molecule.

Elements in Group 3A, particularly boron and aluminum, also tend to form compounds in which they are surrounded by fewer than eight electrons. Take boron as an example. Because its electron configuration is $1s^2 2s^2 2p^1$, it has a total of three valence electrons. Boron forms with the halogens a class of compounds of the general formula BX_3, in which X is a halogen atom. Thus, in boron trifluoride there are only six electrons around the boron atom:

Boron trifluoride is a colorless gas with a pungent odor.

$$
\begin{array}{c}
\overset{..}{:}\overset{}{F}\overset{..}{:} \\
| \\
\overset{..}{:}\overset{}{F}{-}B \\
\overset{..}{} \quad | \\
\overset{..}{:}\overset{}{F}\overset{..}{:} \\
\end{array}
$$

A resonance structure with a double bond between B and F can be drawn that satisfies the octet rule for B. However, the properties of BF_3 are more consistent with a Lewis structure in which there are single bonds between B and each F, as shown here.

Although boron trifluoride is stable, it has a tendency to pick up an unshared electron pair from an atom in another compound, as shown by its reaction with ammonia:

$$
:\overset{..}{\underset{..}{F}}: \quad H
$$

$$
:\overset{..}{\underset{..}{F}}\!-\!B \;\; + \;\; :\overset{}{N}\!-\!H \;\longrightarrow\; :\overset{..}{\underset{..}{F}}\!-\!\overset{-}{B}\!-\!\overset{+}{N}\!-\!H
$$

This structure satisfies the octet rule for the B, N, and F atoms.

The B—N bond in the compound just shown is different from the covalent bonds discussed so far in the sense that both electrons are contributed by the N atom. *A covalent bond in which one of the atoms donates both electrons* is called a **coordinate covalent bond.** Although the properties of a coordinate covalent bond do not differ from those of a normal covalent bond (because all electrons are alike no matter what their source), the distinction is useful for keeping track of valence electrons and assigning formal charges.

Odd-Electron Molecules

Some molecules contain an *odd* number of electrons. Among them are nitric oxide (NO) and nitrogen dioxide (NO_2):

$$
\overset{..}{\underset{.}{N}}\!=\!\overset{..}{\underset{..}{O}} \qquad \overset{..}{\underset{..}{O}}\!=\!\overset{+}{N}\!-\!\overset{..}{\underset{..}{O}}: {}^{-}
$$

Because we need an even number of electrons for complete pairing (to reach eight), the octet rule clearly cannot be satisfied for all the atoms in any molecule that has an odd number of electrons.

The Expanded Octet

In a number of compounds there are more than eight valence electrons around an atom. These *expanded octets* are needed only for atoms of elements in and beyond the third period of the periodic table. In addition to the 3s and 3p orbitals, elements in the third period also have 3d orbitals that can be used in bonding. One compound in which there is an expanded octet is sulfur hexafluoride, a very stable compound. The electron configuration of sulfur is $[Ne]3s^2 3p^4$. In SF_6, each of sulfur's six valence electrons forms a covalent bond with a fluorine atom, so there are 12 electrons around the central sulfur atom:

Yellow: second-period elements cannot have an expanded octet. Blue: third-period elements and beyond can have an expanded octet. Green: the noble gases usually only have an expanded octet.

$$
\begin{array}{ccc}
 & :\overset{..}{\underset{}{F}}: & \\
:\overset{..}{F} & \diagdown\;|\;\diagup & \overset{..}{F}: \\
 & S & \\
:\overset{..}{F} & \diagup\;|\;\diagdown & \overset{..}{F}: \\
 & :\overset{}{\underset{..}{F}}: &
\end{array}
$$

In Chapter 10 we will see that these 12 electrons, or six bonding pairs, are accommodated in six orbitals that originate from the one 3s, the three 3p, and two of the five 3d orbitals. However, sulfur also forms many compounds in which it does not violate the octet rule. In sulfur dichloride, S is surrounded by only eight electrons and therefore obeys the octet rule:

Sulfur dichloride is a toxic, foul-smelling cherry-red liquid (boiling point: 59°C).

$$
:\overset{..}{\underset{..}{Cl}}\!-\!\overset{..}{\underset{..}{S}}\!-\!\overset{..}{\underset{..}{Cl}}:
$$

AlI₃ has a tendency to dimerize or form two units as Al₂I₆.

Similar problem: 9.40.

Example 9.7   Draw the Lewis structure for aluminum triiodide (AlI_3).

Reasoning and Solution   The outer-shell electron configuration of Al is $3s^2 3p^1$. The Al atom forms three covalent bonds with the I atoms as follows:

$$\ddot{\textrm{:}}\,\ddot{\textrm{I}}\,\ddot{\textrm{:}}$$
$$|$$
$$\textrm{:}\ddot{\textrm{I}}\textrm{—Al}$$
$$|$$
$$\textrm{:}\ddot{\textrm{I}}\textrm{:}$$

Comment   Although the octet rule is satisfied for the I atoms, there are only six valence electrons around the Al atom. This molecule is an example of the incomplete octet.

Practice Exercise   Draw the Lewis structure for BeF_2.

PF₅ is a reactive gaseous compound.

Similar problem: 9.42.

Example 9.8   Draw the Lewis structure for phosphorus pentafluoride (PF_5), in which all five F atoms are bonded directly to the P atom.

Reasoning and Solution   The outer-shell electron configurations for P and F are $3s^2 3p^3$ and $2s^2 2p^5$, respectively, and so the total number of valence electrons is $5 + (5 \times 7)$, or 40. Phosphorus, like sulfur, is a third-period element, and therefore it can have an expanded octet. The Lewis structure of PF_5 is

$$
\begin{array}{c}
\textrm{:}\ddot{\textrm{F}}\textrm{:}\quad \ddot{\textrm{F}}\textrm{:} \\
| \;\diagup \\
\textrm{:}\ddot{\textrm{F}}\textrm{—P} \\
| \;\diagdown \\
\textrm{:}\ddot{\textrm{F}}\textrm{:}\quad \ddot{\textrm{F}}\textrm{:}
\end{array}
$$

Comment   Although the octet rule is satisfied for the F atoms, there are 10 valence electrons around the P atom, giving it an expanded octet.

Practice Exercise   Draw the Lewis structure for arsenic pentafluoride (AsF_5).

9.8 Bond Energy

A measure of the stability of a molecule is its ***bond energy,*** which is *the enthalpy change required to break a particular bond in 1 mole of gaseous molecules.* (Bond energies in solids and liquids are affected by neighboring molecules.) The experimentally determined bond energy of the diatomic hydrogen molecule, for example, is

$$H_2(g) \longrightarrow H(g) + H(g) \qquad \Delta H° = 436.4 \text{ kJ}$$

This equation tells us that breaking the covalent bonds in 1 mole of gaseous H_2 molecules requires 436.4 kJ of energy. For the less stable chlorine molecule,

$$Cl_2(g) \longrightarrow Cl(g) + Cl(g) \qquad \Delta H° = 242.7 \text{ kJ}$$

Bond energies can also be directly measured for diatomic molecules containing unlike elements, such as HCl,

$$HCl(g) \longrightarrow H(g) + Cl(g) \qquad \Delta H° = 431.9 \text{ kJ}$$

as well as for molecules containing double and triple bonds:

$$O_2(g) \longrightarrow O(g) + O(g) \qquad \Delta H° = 498.7 \text{ kJ}$$
$$N_2(g) \longrightarrow N(g) + N(g) \qquad \Delta H° = 941.4 \text{ kJ}$$

The Lewis structure of O_2 is $\ddot{O}=\ddot{O}$.

Measuring the strength of covalent bonds in polyatomic molecules is more complicated. For example, measurements show that the energy needed to break the first O—H bond in H_2O is different from that needed to break the second O—H bond:

$$H_2O(g) \longrightarrow H(g) + OH(g) \qquad \Delta H° = 502 \text{ kJ}$$
$$OH(g) \longrightarrow H(g) + O(g) \qquad \Delta H° = 427 \text{ kJ}$$

In each case, an O—H bond is broken, but the first step is more endothermic than the second. The difference between the two $\Delta H°$ values suggests that the second O—H bond itself undergoes change, because of the changes in chemical environment.

Now we can understand why the bond energy of the same O—H bond in two different molecules such as methanol (CH_3OH) and water (H_2O) will not be the same: their environments are different. Thus for polyatomic molecules we speak of the *average* bond energy of a particular bond. For example, we can measure the energy of the O—H bond in 10 different polyatomic molecules and obtain the average O—H bond energy by dividing the sum of the bond energies by 10. Table 9.2 lists the average bond energies of a number of diatomic and polyatomic molecules. As stated earlier, triple bonds are stronger than double bonds, which, in turn, are stronger than single bonds.

Use of Bond Energies in Thermochemistry

A comparison of the thermochemical changes that take place during a number of reactions (Chapter 6) reveals a strikingly wide variation in the enthalpies of different reactions. For example, the combustion of hydrogen gas in oxygen gas is fairly exothermic:

$$H_2(g) + \tfrac{1}{2}O_2(g) \longrightarrow H_2O(l) \qquad \Delta H° = -285.8 \text{ kJ}$$

On the other hand, the formation of glucose ($C_6H_{12}O_6$) from water and carbon dioxide, best achieved by photosynthesis, is highly endothermic:

$$6CO_2(g) + 6H_2O(l) \longrightarrow C_6H_{12}O_6(s) + 6O_2(g) \quad \Delta H° = 2801 \text{ kJ}$$

We can account for such variations by looking at the stability of individual reactant and product molecules. After all, most chemical reactions involve the making and breaking of bonds. Therefore, knowing the bond energies and hence the stability of molecules tells us something about the thermochemical nature of reactions that molecules undergo.

Table 9.2	Some Bond Energies of Diatomic Molecules* and Average Bond Energies for Bonds in Polyatomic Molecules		
Bond	**Bond Energy (kJ/mol)**	**Bond**	**Bond Energy (kJ/mol)**
H—H	436.4	C—S	255
H—N	393	C=S	477
H—O	460	N—N	193
H—S	368	N=N	418
H—P	326	N≡N	941.4
H—F	568.2	N—O	176
H—Cl	431.9	N—P	209
H—Br	366.1	O—O	142
H—I	298.3	O=O	498.7
C—H	414	O—P	502
C—C	347	O=S	469
C=C	620	P—P	197
C≡C	812	P=P	489
C—N	276	S—S	268
C=N	615	S=S	352
C≡N	891	F—F	156.9
C—O	351	Cl—Cl	242.7
C=O†	745	Br—Br	192.5
C—P	263	I—I	151.0

*Bond energies for diatomic molecules (in color) have more significant figures than bond energies for bonds in polyatomic molecules because the bond energies of diatomic molecules are directly measurable quantities and not averaged over many compounds.

†The C=O bond energy in CO_2 is 799 kJ/mol.

In many cases, it is possible to predict the approximate enthalpy of reaction by using the average bond energies. Because energy is always required to break chemical bonds and chemical bond formation is always accompanied by a release of energy, we can estimate the enthalpy of a reaction by counting the total number of bonds broken and formed in the reaction and recording all the corresponding energy changes. The enthalpy of reaction in the *gas phase* is given by

$$\Delta H^\circ = \Sigma BE(\text{reactants}) - \Sigma BE(\text{products})$$

$$= \text{total energy input} - \text{total energy released} \tag{9.1}$$

where BE stands for average bond energy and Σ is the summation sign. As written, Equation (9.1) takes care of the sign convention for ΔH°. Thus, if the total energy input is greater than the total energy released, ΔH° is positive and the reaction is endothermic. On the other hand, if more energy is released than absorbed, ΔH° is negative and the reaction is exothermic (Figure 9.6). If reactants and products are all diatomic molecules, then Equation (9.1) will yield accurate results because the bond energies of diatomic molecules are accurately known. If some or all of the reactants and products are polyatomic molecules, Equation (9.1) will yield only approximate results because the bond energies used will be averages.

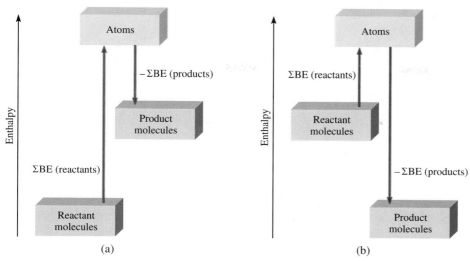

Figure 9.6 *Bond energy changes in (a) an endothermic reaction and (b) an exothermic reaction.*

Example 9.9 Estimate the enthalpy change for the combustion of hydrogen gas:

$$2H_2(g) + O_2(g) \longrightarrow 2H_2O(g)$$

Reasoning and Solution Keep in mind that bond breaking is an energy absorbing (endothermic) process and bond making is an energy releasing (exothermic) process. Therefore, the overall energy change is the difference between these two opposing processes, as described by Equation (9.1). We start by counting the number of bonds broken and the number of bonds formed and referring to Table 9.2. This is best done by creating a table:

Type of Bonds Broken	Number of Bonds Broken	Bond Energy (kJ/mol)	Energy Change (kJ)
H—H (H_2)	2	436.4	872.8
O=O (O_2)	1	498.7	498.7

Type of Bonds Formed	Number of Bonds Formed	Bond Energy (kJ/mol)	Energy Change (kJ)
O—H (H_2O)	4	460	1840

Next, we obtain the total energy input and total energy released:

$$\text{total energy input} = 872.8 \text{ kJ} + 498.7 \text{ kJ} = 1372 \text{ kJ}$$

$$\text{total energy released} = 1840 \text{ kJ}$$

Using Equation (9.1) we write

$$\Delta H° = 1372 \text{ kJ} - 1840 \text{ kJ} = -468 \text{ kJ}$$

This result is only an estimate because the bond energy of O—H is an average quantity. Alternatively, we can use Equation (6.16) and the data in Appendix 2 to calculate the enthalpy of reaction:

$$\Delta H^\circ = 2\Delta H^\circ_f\,(H_2O) - [2\Delta H^\circ_f(H_2) + \Delta H^\circ_f(O_2)]$$
$$= (2\ mol)(-241.8\ kJ/mol) - 0 - 0$$
$$= -483.6\ kJ$$

Comment   The estimated value based on average bond energies is quite close to the true value calculated using ΔH°_f data. In general, Equation (9.1) works best for reactions that are either quite endothermic or quite exothermic, that is, for $\Delta H^\circ > 100$ kJ or for $\Delta H^\circ < -100$ kJ.

Similar problem: 9.50.

Practice Exercise   For the reaction

$$H_2(g) + C_2H_4(g) \longrightarrow C_2H_6(g)$$

(a) Estimate the enthalpy of reaction, using the bond energy values in Table 9.2.
(b) Calculate the enthalpy of reaction, using standard enthalpies of formation. (ΔH°_f for H_2, C_2H_4, and C_2H_6 are 0, 52.3 kJ/mol, and -84.7 kJ/mol, respectively.)

Summary

A Lewis dot symbol shows the number of valence electrons possessed by an atom of a given element. Lewis dot symbols are useful mainly for the representative elements.

In a covalent bond, two electrons (one pair) are shared by two atoms. In multiple covalent bonds, two or three electron pairs are shared by two atoms. Some bonded atoms possess lone pairs, that is, pairs of valence electrons not involved in bonding. The arrangement of bonding electrons and lone pairs around each atom in a molecule is represented by the Lewis structure.

The octet rule predicts that atoms form enough covalent bonds to surround themselves with eight electrons each. When one atom in a covalently bonded pair donates two electrons to the bond, the Lewis structure can include the formal charge on each atom as a means of keeping track of the valence electrons. There are exceptions to the octet rule, particularly for covalent beryllium compounds, elements in Group 3A, and elements in the third period and beyond in the periodic table. Electronegativity is a measure of the ability of an atom to attract electrons in a chemical bond.

For some molecules or polyatomic ions, two or more Lewis structures based on the same skeletal structure satisfy the octet rule and appear chemically reasonable. Such resonance structures taken together represent the molecule or ion.

The strength of a covalent bond is measured in terms of its bond energy. Bond energies can be used to estimate the enthalpy of reactions.

Key Words

Bond energy p. 280
Bond length p. 268
Coordinate covalent bond
  p. 279
Covalent bond p. 267
Covalent compounds
  p. 267

Double bond p. 268
Electronegativity p. 269
Formal charge p. 274
Ionic bond p. 269
Lewis dot symbol p. 266
Lewis structure p. 267
Lone pair p. 267

Multiple bond p. 268
Octet rule p. 268
Polar covalent bond
  p. 269
Resonance p. 276

Resonance structure
  p. 276
Single bond p. 268
Triple bond p. 268

Questions and Problems

Lewis Dot Symbols

Review Questions

9.1 What is a Lewis dot symbol? To what elements does the symbol mainly apply?

9.2 Use the second member of each group from Group 1A to Group 7A to show that the number of valence electrons on an atom of the element is the same as its group number.

9.3 Without referring to Figure 9.1, write Lewis dot symbols for atoms of these elements: (a) Be, (b) K, (c) Ca, (d) Ga, (e) O, (f) Br, (g) N, (h) I, (i) As, (j) F.

9.4 Write Lewis dot symbols for these ions: (a) Li^+, (b) Cl^-, (c) S^{2-}, (d) Ba^{2+}, (e) N^{3-}.

9.5 Write Lewis dot symbols for these atoms and ions: (a) I, (b) I^-, (c) S, (d) S^{2-}, (e) P, (f) P^{3-}, (g) Na, (h) Na^+, (i) Mg, (j) Mg^{2+}, (k) Al, (l) Al^{3+}, (m) Pb, (n) Pb^{2+}.

The Covalent Bond

Review Questions

9.6 What is Lewis's contribution to our understanding of the covalent bond?

9.7 Define these terms: lone pairs, Lewis structure, the octet rule, bond length.

9.8 What is the difference between a Lewis dot symbol and a Lewis structure?

9.9 How many lone pairs are on the underlined atoms in these compounds? H<u>Br</u>, H<u>2</u>S, <u>C</u>H<u>4</u>.

9.10 Distinguish among single, double, and triple bonds in a molecule, and give an example of each.

Electronegativity and Bond Type

Review Questions

9.11 Define electronegativity, and explain the difference between electronegativity and electron affinity. Describe in general how the electronegativities of the elements change according to position in the periodic table.

9.12 What is a polar covalent bond? Name two compounds that contain one or more polar covalent bonds.

Problems

9.13 List these bonds in order of increasing ionic character: the lithium-to-fluorine bond in LiF, the potassium-to-oxygen bond in K_2O, the nitrogen-to-nitrogen bond in N_2, the sulfur-to-oxygen bond in SO_2, the chlorine-to-fluorine bond in ClF_3.

9.14 Arrange these bonds in order of increasing ionic character: carbon to hydrogen, fluorine to hydrogen, bromine to hydrogen, sodium to chlorine, potassium to fluorine, lithium to chlorine.

9.15 Four atoms are arbitrarily labeled D, E, F, and G. Their electronegativities are: D = 3.8, E = 3.3, F = 2.8, and G = 1.3. If the atoms of these elements form the molecules DE, DG, EG, and DF, how would you arrange these molecules in order of increasing covalent bond character?

9.16 List these bonds in order of increasing ionic character: cesium to fluorine, chlorine to chlorine, bromine to chlorine, silicon to carbon.

9.17 Classify these bonds as ionic, polar covalent, or covalent, and give your reasons: (a) the CC bond in H_3CCH_3, (b) the KI bond in KI, (c) the NB bond in H_3NBCl_3, (d) the ClO bond in ClO_2.

9.18 Classify these bonds as ionic, polar covalent, or covalent, and give your reasons: (a) the SiSi bond in $Cl_3SiSiCl_3$, (b) the SiCl bond in $Cl_3SiSiCl_3$, (c) the CaF bond in CaF_2, (d) the NH bond in NH_3.

Lewis Structure and the Octet Rule

Review Questions

9.19 Summarize the essential features of the Lewis octet rule. The octet rule applies mainly to the second-period elements. Explain.

9.20 Explain the concept of formal charge. Do formal charges on a molecule represent actual separation of charges?

Problems

9.21 Write Lewis structures for these molecules: (a) ICl, (b) PH_3, (c) P_4 (each P is bonded to three other P atoms), (d) H_2S, (e) N_2H_4, (f) $HClO_3$, (g) $COBr_2$ (C is bonded to O and Br atoms).

9.22 Write Lewis structures for these ions: (a) O_2^{2-}, (b) C_2^{2-}, (c) NO^+, (d) NH_4^+. Show formal charges.

9.23 These Lewis structures are incorrect. Explain what is wrong with each one and give a correct Lewis structure for the molecule. (Relative positions of atoms are shown correctly.)

(a) $H—\overset{..}{C}=\overset{..}{N}$

(b) $H=C=C=H$

(c) $\overset{..}{O}—Sn—\overset{..}{O}$

(d) $:\overset{..}{F}\diagdown_{B}\diagup\overset{..}{F}:$
    $\quad\quad\overset{|}{:\overset{..}{F}:}$

(e) $H—\overset{..}{O}=\overset{..}{F}:$

(f) $H\diagdown_{\overset{..}{C}—\overset{..}{F}:}$
    $\quad\overset{..}{O}$

(g) $:\overset{..}{F}\diagdown_{N}\diagup\overset{..}{F}:$
    $\quad\quad:\overset{..}{F}:$

9.24 The skeletal structure of acetic acid in this structure is correct, but some of the bonds are wrong. (a) Identify the incorrect bonds and explain what is wrong with them. (b) Write the correct Lewis structure for acetic acid.

$$
\begin{array}{ccc}
\text{H} & :\text{O}: & \\
| & | & \\
\text{H}=\text{C}-\text{C}-\text{O}-\text{H} \\
| & & \\
\text{H} &
\end{array}
$$

Resonance

Review Questions

9.25 Define bond length, resonance, and resonance structure.

9.26 Is it possible to "trap" a resonance structure of a compound for study? Explain.

9.27 The resonance concept is sometimes described by analogy to a mule, which is a cross between a horse and a donkey. Compare this analogy with that used in this chapter, that is, the description of a rhinoceros as a cross between a griffin and a unicorn. Which description is more appropriate? Why?

9.28 Describe the general rules for drawing plausible resonance structures. What are the other two reasons for choosing (b) in Example 9.5?

Problems

9.29 Write Lewis structures for these species, including all resonance forms, and show formal charges: (a) HCO_2^-, (b) $CH_2NO_2^-$. Relative positions of the atoms are as follows:

$$
\begin{array}{cccc}
\text{O} & & \text{H} & \text{O} \\
\text{H} \quad \text{C} & & & \text{C} \quad \text{N} \\
\text{O} & & \text{H} & \text{O}
\end{array}
$$

9.30 Draw three resonance structures for the chlorate ion, ClO_3^-. Show formal charges.

9.31 Write three resonance structures for hydrazoic acid, HN_3. The atomic arrangement is HNNN. Show formal charges.

9.32 Draw two resonance structures for diazomethane, CH_2N_2. Show formal charges. The skeletal structure of the molecule is

$$
\begin{array}{c}
\text{H} \\
\quad\quad \text{C} \quad \text{N} \quad \text{N} \\
\text{H}
\end{array}
$$

9.33 Draw three reasonable resonance structures for the OCN^- ion. Show formal charges.

9.34 Draw three resonance structures for the molecule N_2O in which the atoms are arranged in the order NNO. Indicate formal charges.

Exceptions to the Octet Rule

Review Questions

9.35 Why does the octet rule not hold for many compounds containing elements in the third period of the periodic table and beyond?

9.36 Give three examples of compounds that do not satisfy the octet rule. Write a Lewis structure for each.

9.37 Because fluorine has seven valence electrons ($2s^2 2p^5$), seven covalent bonds in principle could form around the atom. Such a compound might be FH_7 or FCl_7. These compounds have never been prepared. Why?

9.38 What is a coordinate covalent bond? Is it different from a normal covalent bond?

Problems

9.39 The BF_3 molecule has an incomplete octet around B. Draw three resonance structures of the molecule in which the octet rule is satisfied for both the B and the F atoms. Show formal charges.

9.40 In the vapor phase, beryllium chloride consists of discrete molecular units $BeCl_2$. Is the octet rule satisfied for Be in this compound? If not, can you form an octet around Be by drawing another resonance structure? How plausible is this structure?

9.41 Of the noble gases, only Kr, Xe, and Rn are known to form a few compounds with O and/or F. Write Lewis structures for these molecules: (a) XeF_2, (b) XeF_4, (c) XeF_6, (d) $XeOF_4$, (e) XeO_2F_2. In each case Xe is the central atom.

9.42 Write a Lewis structure for $SbCl_5$. Is the octet rule obeyed in this molecule?

9.43 Write Lewis structures for SeF_4 and SeF_6. Is the octet rule satisfied for Se?

9.44 Write Lewis structures for the reaction

$$AlCl_3 + Cl^- \longrightarrow AlCl_4^-$$

What kind of bond is between Al and Cl in the product?

Bond Energies

Review Questions

9.45 Define bond energy. Bond energies of polyatomic molecules are average values. Why?

9.46 Explain why the bond energy of a molecule is usually defined in terms of a gas-phase reaction. Why are bond-breaking processes always endothermic and bond-forming processes always exothermic?

Problems

9.47 From these data, calculate the average bond energy for the N—H bond:

$$NH_3(g) \longrightarrow NH_2(g) + H(g) \qquad \Delta H° = 435 \text{ kJ}$$
$$NH_2(g) \longrightarrow NH(g) + H(g) \qquad \Delta H° = 381 \text{ kJ}$$
$$NH(g) \longrightarrow N(g) + H(g) \qquad \Delta H° = 360 \text{ kJ}$$

9.48 For the reaction

$$O(g) + O_2(g) \longrightarrow O_3(g) \qquad \Delta H° = -107.2 \text{ kJ}$$

calculate the average bond energy in O_3.

9.49 The bond energy of $F_2(g)$ is 156.9 kJ/mol. Calculate $\Delta H_f°$ for $F(g)$.

9.50 (a) For the reaction

$$2C_2H_6(g) + 7O_2(g) \longrightarrow 4CO_2(g) + 6H_2O(g)$$

predict the enthalpy of reaction from the average bond energies in Table 9.2. (b) Calculate the enthalpy of reaction from the standard enthalpies of formation (see Appendix 2) of the reactant and product molecules, and compare the result with your answer for part (a).

Additional Problems

9.51 Match each of these energy changes with one of the processes given: ionization energy, electron affinity, bond energy, standard enthalpy of formation.

(a) $F(g) + e^- \longrightarrow F^-(g)$
(b) $F_2(g) \longrightarrow 2F(g)$
(c) $Na(g) \longrightarrow Na^+(g) + e^-$
(d) $Na(s) + \frac{1}{2}F_2(g) \longrightarrow NaF(s)$

9.52 The formulas for the fluorides of the third-period elements are NaF, MgF_2, AlF_3, SiF_4, PF_5, SF_6, and ClF_3. Classify these compounds as covalent or ionic.

9.53 Use the ionization energy (see Table 8.3) and electron affinity (see Table 8.4) values to calculate the energy change (in kilojoules) for these reactions:

(a) $Li(g) + I(g) \longrightarrow Li^+(g) + I^-(g)$
(b) $Na(g) + F(g) \longrightarrow Na^+(g) + F^-(g)$
(c) $K(g) + Cl(g) \longrightarrow K^+(g) + Cl^-(g)$

9.54 Describe some characteristics of an ionic compound such as KF that would distinguish it from a covalent compound such as CO_2.

9.55 Write Lewis structures for BrF_3, ClF_5, and IF_7. Identify those in which the octet rule is not obeyed.

9.56 Write three reasonable resonance structures of the azide ion N_3^- in which the atoms are arranged as NNN. Show formal charges.

9.57 The amide group plays an important role in determining the structure of proteins:

$$\begin{array}{c} \ddot{\,}\,O\,\ddot{\,} \\ \parallel \\ -\ddot{N}-C- \\ \mid \\ H \end{array}$$

Draw another resonance structure of this group. Show formal charges.

9.58 Give an example of an ion or molecule containing Al that (a) obeys the octet rule, (b) has an expanded octet, and (c) has an incomplete octet.

9.59 Draw four reasonable resonance structures for the PO_3F^{2-} ion. The central P atom is bonded to the three O atoms and to the F atom. Show formal charges.

9.60 Attempts to prepare these as stable species under atmospheric conditions have failed. Suggest reasons for the failure.

$$CF_2 \qquad CH_5 \qquad FH_2^- \qquad PI_5$$

9.61 Draw reasonable resonance structures for these sulfur-containing ions: (a) HSO_4^-, (b) SO_4^{2-}, (c) HSO_3^-, (d) SO_3^{2-}.

9.62 True or false: (a) Formal charges represent actual separation of charges; (b) $\Delta H_{rxn}°$ can be estimated from bond energies of reactants and products; (c) all second-period elements obey the octet rule in their compounds; (d) the resonance structures of a molecule can be separated from one another.

9.63 A rule for drawing plausible Lewis structures is that the central atom is invariably less electronegative than the surrounding atoms. Explain why this is so.

9.64 Using this information:

$$C(s) \longrightarrow C(g) \qquad \Delta H_{rxn}° = 716 \text{ kJ}$$
$$2H_2(g) \longrightarrow 4H(g) \qquad \Delta H_{rxn}° = 872.8 \text{ kJ}$$

and the fact that the average C—H bond energy is 414 kJ/mol, estimate the standard enthalpy of formation of methane (CH_4).

9.65 Based on energy considerations, which of these two reactions will occur more readily?

(a) $Cl(g) + CH_4(g) \longrightarrow CH_3Cl(g) + H(g)$
(b) $Cl(g) + CH_4(g) \longrightarrow CH_3(g) + HCl(g)$

(*Hint:* Refer to Table 9.2, and assume that the average bond energy of the C—Cl bond is 338 kJ/mol.)

9.66 Which of these molecules has the shortest nitrogen-to-nitrogen bond? Explain.

$$N_2H_4 \qquad N_2O \qquad N_2 \qquad N_2O_4$$

9.67 Most organic acids can be represented as RCOOH, in which COOH is the carboxyl group and R is the rest of the molecule. (For example, R is CH_3 in acetic acid, CH_3COOH.)

(a) Draw a Lewis structure of the carboxyl group. (b) Upon ionization, the carboxyl group is converted to the carboxylate group, COO^-. Draw resonance structures of the carboxylate group.

9.68 Which of these molecules or ions are isoelectronic: NH_4^+, C_6H_6, CO, CH_4, N_2, $B_3N_3H_6$?

9.69 These species have been detected in interstellar space: (a) CH, (b) OH, (c) C_2, (d) HNC, (e) HCO. Draw Lewis structures of these species and indicate whether they are diamagnetic or paramagnetic.

9.70 The amide ion, NH_2^-, is a Brønsted base. Represent the reaction between the amide ion and water in terms of Lewis structures.

9.71 Draw Lewis structures of these organic molecules: (a) tetrafluoroethylene (C_2F_4), (b) propane (C_3H_8), (c) butadiene ($CH_2CHCHCH_2$), (d) propyne (CH_3CCH), (e) benzoic acid (C_6H_5COOH). (*Hint:* To draw C_6H_5COOH, replace an H atom in benzene with a COOH group.)

9.72 The triiodide ion (I_3^-) in which the I atoms are arranged as III is stable, but the corresponding F_3^- ion does not exist. Explain.

9.73 Compare the bond energy of F_2 with the energy change for this process:

$$F_2(g) \longrightarrow F^+(g) + F^-(g)$$

Which is the preferred dissociation for F_2, energetically speaking?

9.74 Methyl isocyanate, CH_3NCO, is used to make certain pesticides. In December 1984, water leaked into a tank containing this substance at a chemical plant to produce a toxic cloud that killed thousands of people in Bhopal, India. Draw Lewis structures for this compound, showing formal charges.

9.75 The chlorine nitrate molecule ($ClONO_2$) is believed to be involved in the destruction of ozone in the Antarctic stratosphere. Draw a plausible Lewis structure for the molecule.

9.76 Several resonance structures of the molecule CO_2 are given here. Explain why some of them are likely to be of little importance in describing the bonding in this molecule.

(a) $\overset{..}{\underset{..}{O}}=C=\overset{..}{\underset{..}{O}}$

(b) $\overset{+}{:}O\equiv C-\overset{..}{\underset{..}{O}}:^-$

(c) $:O\equiv\overset{+}{C}\quad\overset{..}{\underset{..}{O}}:^-$

(d) $^-:\overset{..}{\underset{..}{O}}-\overset{2+}{C}-\overset{..}{\underset{..}{O}}:^-$

9.77 Draw a Lewis structure for each of these organic molecules in which the carbon atoms are bonded to each other by single bonds: C_2H_6, C_4H_{10}, C_5H_{12}.

9.78 Draw Lewis structures for these chlorofluorocarbons (CFCs), which are partly responsible for the depletion of ozone in the stratosphere: $CFCl_3$, CF_2Cl_2, CHF_2Cl, CF_3CHF_2.

9.79 Draw Lewis structures for these organic molecules, in each of which there is one C=C bond and the rest of the carbon atoms are joined by C—C bonds: C_2H_3F, C_3H_6, C_4H_8.

9.80 Calculate $\Delta H°$ for the reaction

$$H_2(g) + I_2(g) \longrightarrow 2HI(g)$$

using (a) Equation (9.1) and (b) Equation (6.16), given that $\Delta H_f°$ for $I_2(g)$ is 61.0 kJ/mol.

9.81 Draw Lewis structures of these organic molecules: (a) methanol (CH_3OH); (b) ethanol (CH_3CH_2OH); (c) tetraethyllead [$Pb(CH_2CH_3)_4$], which is used in "leaded" gasoline; (d) methylamine (CH_3NH_2); (e) mustard gas ($ClCH_2CH_2SCH_2CH_2Cl$), a poisonous gas used in World War I; (f) urea [$(NH_2)_2CO$], a fertilizer; (g) glycine (NH_2CH_2COOH), an amino acid.

9.82 Write Lewis structures for these four isoelectronic species: (a) CO, (b) NO^+, (c) CN^-, (d) N_2. Show formal charges.

9.83 Oxygen forms three types of ionic compounds in which the anions are oxide (O^{2-}), peroxide (O_2^{2-}), and superoxide (O_2^-). Draw Lewis structures of these ions.

9.84 Comment on the correctness of this statement: All compounds containing a noble gas atom violate the octet rule.

9.85 (a) From these data:

$F_2(g) \longrightarrow 2F(g)$	$\Delta H_{rxn}° = 156.9$ kJ
$F^-(g) \longrightarrow F(g) + e^-$	$\Delta H_{rxn}° = 333$ kJ
$F_2^-(g) \longrightarrow F_2(g) + e^-$	$\Delta H_{rxn}° = 290$ kJ

calculate the bond energy of the F_2^- ion. (b) Explain the difference between the bond energies of F_2 and F_2^-.

9.86 Write three resonance structures for the isocyanate ion (CNO^-). Rank them in importance.

9.87 The only known argon-containing compound is HArF, which was prepared in 2000. Draw a Lewis structure of the compound.

Special Problems

9.88 The neutral hydroxyl radical (OH) plays an important role in atmospheric chemistry. It is highly reactive and has a tendency to combine with an H atom from other compounds, causing them to break up. Thus it is sometimes called a "detergent" radical because it helps to clean up the atmosphere.

(a) Write the Lewis structure for the radical.
(b) Refer to Table 9.2 and explain why the radical has a high affinity for H atoms.
(c) Estimate the enthalpy change for the following reaction:

$$OH(g) + CH_4(g) \longrightarrow CH_3(g) + H_2O(g)$$

(d) The radical is generated when sunlight hits water vapor. Calculate the maximum wavelength (in nanometers) required to break up an O—H bond in H_2O.

9.89 Ethylene dichloride ($C_2H_4Cl_2$) is used to make vinyl chloride (C_2H_3Cl), which in turn is used to manufacture the plastic poly(vinyl chloride) (PVC), found in piping, siding, floor tiles, and toys.

(a) Write the Lewis structures of ethylene dichloride and vinyl chloride. Classify the bonds as covalent or polar.
(b) Poly(vinyl chloride) is a polymer; that is, it is a molecule with very high molar mass (on the order of thousands to millions of grams). It is formed by joining many vinyl chloride molecules together. The repeating unit in poly(vinyl chloride) is —CH_2—$CHCl$—. Draw a portion of the molecule showing three such repeating units.
(c) Calculate the enthalpy change when 1.0×10^3 kg of vinyl chloride react to form poly(vinyl chloride). Comment on your answer in relation to industrial design for such a process.

Answers to Practice Exercises

9.1 (a) Ionic, (b) polar covalent, (c) covalent.

9.2 S=C=S .   **9.3** [O=N—O]⁻ .   **9.4** O=N—O:⁻ .

9.5 H—C≡N: .   **9.6** O=N—O:⁻ ⟷ ⁻:O—N=O .

9.7 :F—Be—F: .   **9.8** :F—As(F)(F):F:

9.9 (a) −119 kJ, (b) −137.0 kJ.

In an effort to create unusual molecules believed to exist in interstellar space, chemists at Rice University used a high-powered laser to vaporize graphite. Among the products formed during this 1985 experiment was a stable species whose molar mass corresponds to the formula C_{60}. Such a molecule would have to have an exotic shape because of its size and the fact that only one type of atom is present. Working with paper, scissors, and Scotch tape, the researchers assembled a closed sphere from 20 paper hexagons and 12 paper pentagons. Significantly, this sphere contains 60 vertices, which correspond to 60 carbon atoms.

The geometry of a buckyball C_{60} (left) resembles a soccer ball (right). Scientists arrived at this structure by fitting together paper cutouts of enough hexagons and pentagons to accommodate 60 carbon atoms at the points where they intersect.

Because of its shape, the chemists named the new molecule "buckminsterfullerene," in honor of the late R. Buckminster Fuller, whose engineering and architectural designs often incorporated geodesic domes. But buckminsterfullerene is something of a mouthful, and the new molecule, which also resembles a soccer ball, quickly became known as "buckyball." Subsequent spectroscopic and X-ray measurements have confirmed the predicted structure.

Buckyball has generated tremendous interest in the scientific community for a number of reasons. First, it is a stable allotrope of carbon, as are diamond and graphite. It turns out that buckyball can be prepared under less extreme conditions, and it is a natural component of soot. In addition, buckyball is the most symmetrical molecule known. Furthermore, since buckyball was discovered, chemists have created similar "fullerenes" with more than 60 carbon atoms, like C_{70} and C_{76}.

Buckyball and the other fullerenes represent a whole new concept in molecular architecture. Studies have already shown that these molecules (and compounds derived from them) can act as high-temperature superconductors and lubricants, and they may well find uses as catalysts and antiviral drugs.

One fascinating discovery, made in 1991 by Japanese scientists, was the identification of structural relatives of buckyball. These molecules can be hundreds of nanometers long with a tubular shape and are dubbed "buckytubes" or "nanotubes" (because of their size). Nanotubes are several times stronger than steel wires of similar dimensions and one day they might be used to make ultralight bicycles, rocket motor castings, and tennis racquets. They could also serve as casting molds for very thin metal wires to be used in microscopic integrated circuits or as "bottles" for storing molecules.

The structure of a buckytube that consists of a single layer of carbon atoms. Note that the truncated buckyball "cap," which has been separated from the rest of the buckytube in this view, has a different structure from the cylindrical portion of the tube. Chemists have devised ways to open the cap in order to place other molecules inside the tube.

Chemical Bonding II: Molecular Geometry and Hybridization of Atomic Orbitals

Chapter 10

ESSENTIAL CONCEPTS

 Molecular Geometry Molecular geometry refers to the three-dimensional arrangement of atoms in a molecule. For relatively small molecules, in which the central atom contains two to six bonds, geometries can be reliably predicted by the valence-shell electron-pair repulsion (VSEPR) model. This model is based on the assumption that chemical bonds and lone pairs tend to remain as far apart as possible to minimize repulsion.

 Dipole Moments In a diatomic molecule the difference in the electronegativities of bonding atoms results in a polar bond and a dipole moment. The dipole moment of a molecule made up of three or more atoms depends on both the polarity of the bonds and molecular geometry. Dipole moment measurements can help us distinguish between different possible geometries of a molecule.

 Hybridization of Atomic Orbitals Hybridization is the quantum mechanical description of chemical bonding. Atomic orbitals are hybridized, or mixed, to form hybrid orbitals. These orbitals then interact with other atomic orbitals to form chemical bonds. Various molecular geometries can be generated by different hybridizations. The hybridization concept accounts for the exception to the octet rule and also explains the formation of double and triple bonds.

 Molecular Orbital Theory Molecular orbital theory describes bonding in terms of the combination of atomic orbitals to form orbitals that are associated with the molecule as a whole. Molecules are stable if the number of electrons in bonding molecular orbitals is greater than that in antibonding molecular orbitals. We write electron configurations for molecular orbitals as we do for atomic orbitals, using the Pauli exclusion principle and Hund's rule.

10.1 Molecular Geometry

Molecular geometry is the three-dimensional arrangement of atoms in a molecule. A molecule's geometry affects its physical and chemical properties, such as melting point, boiling point, density, and the types of reactions it undergoes. In general, bond lengths and bond angles must be determined by experiment. However, there is a simple procedure that enables us to predict with considerable success the overall geometry of a molecule or ion if we know the number of electrons surrounding a central atom in its Lewis structure. The basis of this approach is the assumption that electron pairs in the valence shell of an atom repel one another. The **valence shell** is *the outermost electron-occupied shell of an atom; it holds the electrons that are usually involved in bonding.* In a covalent bond, a pair of electrons, often called the *bonding pair,* is responsible for holding two atoms together. However, in a polyatomic molecule, where there are two or more bonds between the central atom and the surrounding atoms, the repulsion between electrons in different bonding pairs causes them to remain as far apart as possible. The geometry that the molecule ultimately assumes (as defined by the positions of all the atoms) minimizes the repulsion. This approach to the study of molecular geometry is called the **valence-shell electron-pair repulsion (VSEPR) model,** because *it accounts for the geometric arrangements of electron pairs around a central atom in terms of the electrostatic repulsion between electron pairs.*

VSEPR

Two general rules govern the use of the VSEPR model:

- As far as electron-pair repulsion is concerned, double bonds and triple bonds can be treated like single bonds. This approximation is good for qualitative purposes. However, you should realize that in reality multiple bonds are "larger" than single bonds; that is, because there are two or three bonds between two atoms, the electron density occupies more space.
- If a molecule has two or more resonance structures, we can apply the VSEPR model to any one of them. Formal charges are usually not shown.

With this model in mind, we can predict the geometry of molecules (and ions) in a systematic way. For this purpose, it is convenient to divide molecules into two categories, according to whether or not the central atom has lone pairs.

Molecules in Which the Central Atom Has No Lone Pairs

For simplicity we will consider molecules that contain atoms of only two elements, A and B, of which A is the central atom. These molecules have the general formula AB_x, where x is an integer 2, 3, ... (If $x = 1$, we have the diatomic molecule AB, which is linear by definition.) In the vast majority of cases, x is between 2 and 6.

Table 10.1 shows five possible arrangements of electron pairs around the central atom A. As a result of mutual repulsion, the electron pairs stay as far from one another as possible. Note that the table shows arrangements of the electron pairs but not the positions of the atoms that surround the central atom. Molecules in which the central atom has no lone pairs have one of these five arrangements of bonding pairs. Using Table 10.1 as a reference, let us take a close look at the geometry of molecules with the formulas AB_2, AB_3, AB_4, AB_5, and AB_6.

Table 10.1 | Arrangement of Electron Pairs about a Central Atom (A) in a Molecule and Geometry of Some Simple Molecules and Ions in Which the Central Atom Has No Lone Pairs

Number of Electron Pairs	Arrangement of Electron Pairs*	Molecular Geometry*	Examples
2	Linear (180°)	Linear (B—A—B)	$BeCl_2$, $HgCl_2$
3	Trigonal planar (120°)	Trigonal planar	BF_3
4	Tetrahedral (109.5°)	Tetrahedral	CH_4, NH_4^+
5	Trigonal bipyramidal (90°, 120°)	Trigonal bipyramidal	PCl_5
6	Octahedral (90°, 90°)	Octahedral	SF_6

*The colored lines are used only to show the overall shapes; they do not represent bonds.

AB₂: Beryllium Chloride (BeCl₂)

The Lewis structure of beryllium chloride in the gaseous state is

Because the bonding pairs repel each other, they must be at opposite ends of a straight line in order for them to be as far apart as possible. Thus, the ClBeCl angle is predicted to be 180°, and the molecule is linear (see Table 10.1). The "ball-and-stick" model of $BeCl_2$ is

> **The blue and yellow spheres are for atoms in general.**

AB₃: Boron Trifluoride (BF₃)

Boron trifluoride contains three covalent bonds, or bonding pairs. In the most stable arrangement, the three BF bonds point to the corners of an equilateral triangle with B in the center of the triangle:

According to Table 10.1, the geometry of BF_3 is *trigonal planar* because the three end atoms are at the corners of an equilateral triangle which is planar:

Planar

Thus, each of the three FBF angles is 120°, and all four atoms lie in the same plane.

AB₄: Methane (CH₄)

The Lewis structure of methane is

$$
\begin{array}{c}
\text{H} \\
| \\
\text{H}-\text{C}-\text{H} \\
| \\
\text{H}
\end{array}
$$

Because there are four bonding pairs, the geometry of CH_4 is tetrahedral (see Table 10.1). A *tetrahedron* has four sides (the prefix *tetra* means "four"), or faces, all of which are equilateral triangles. In a tetrahedral molecule, the central atom (C in this case) is located at the center of the tetrahedron and the other four atoms are at the corners. The bond angles are all 109.5°.

Tetrahedral

AB₅: Phosphorus Pentachloride (PCl₅)

The Lewis structure of phosphorus pentachloride (in the gas phase) is

$$\text{Cl}_5\text{P}$$

The only way to minimize the repulsive forces among the five bonding pairs is to arrange the PCl bonds in the form of a trigonal bipyramid (see Table 10.1). A trigonal bipyramid can be generated by joining two tetrahedrons along a common triangular base:

Trigonal
bipyramidal

The central atom (P in this case) is at the center of the common triangle with the surrounding atoms positioned at the five corners of the trigonal bipyramid. The atoms that are above and below the triangular plane are said to occupy *axial* positions, and those that are in the triangular plane are said to occupy *equatorial* positions. The angle between any two equatorial bonds is 120°; that between an axial bond and an equatorial bond is 90°, and that between the two axial bonds is 180°.

AB₆: Sulfur Hexafluoride (SF₆)

The Lewis structure of sulfur hexafluoride is

$$\text{F}_6\text{S}$$

The most stable arrangement of the six SF bonding pairs is in the shape of an octahedron, shown in Table 10.1. An octahedron has eight sides (the prefix *octa* means "eight"). It can be generated by joining two square pyramids on a common base. The central atom (S in this case) is at the center of the square base and the surrounding atoms are at the six corners. All bond angles are 90° except the one made by the bonds between the central atom and the pairs of atoms that are diametrically opposite each other. That angle is 180°. Because the six bonds are equivalent in an octahedral molecule, we cannot use the terms "axial" and "equatorial" as in a trigonal bipyramidal molecule.

Octahedral

Molecules in Which the Central Atom Has One or More Lone Pairs

Determining the geometry of a molecule is more complicated if the central atom has both lone pairs and bonding pairs. In such molecules there are three types of repulsive forces—those between bonding pairs, those between lone pairs, and those between a bonding pair and a lone pair. In general, according to the VSEPR model, the repulsive forces decrease in the following order:

lone-pair vs. lone-pair > lone-pair vs. bonding- > bonding-pair vs. bonding-
repulsion   pair repulsion   pair repulsion

Electrons in a bond are held by the attractive forces exerted by the nuclei of the two bonded atoms. These electrons have less "spatial distribution" than lone pairs; that is, they take up less space than lone-pair electrons, which are associated with only one particular atom. Because lone-pair electrons in a molecule occupy more space, they experience greater repulsion from neighboring lone pairs and bonding pairs. To keep track of the total number of bonding pairs and lone pairs, we designate molecules with lone pairs as AB_xE_y, where A is the central atom, B is a surrounding atom, and E is a lone pair on A. Both x and y are integers; $x = 2, 3, \ldots$, and $y = 1, 2, \ldots$. Thus the values of x and y indicate the number of surrounding atoms and number of lone pairs on the central atom, respectively. The simplest such molecule would be a triatomic molecule with one lone pair on the central atom and the formula is AB_2E.

As the next examples show, in most cases the presence of lone pairs on the central atom makes it difficult to predict the bond angles accurately.

AB₂E: Sulfur Dioxide (SO₂)

The Lewis structure of sulfur dioxide is

SO_2

Because VSEPR treats double bonds as though they were single, the SO_2 molecule can be viewed as consisting of three electron pairs on the central S atom. Of these, two are bonding pairs and one is a lone pair. In Table 10.1 we see that the overall arrangement of three electron pairs is trigonal planar. But because one of the electron pairs is a lone pair, the SO_2 molecule has a "bent" shape.

S
O   O

Because the lone-pair versus bonding-pair repulsion is greater than the bonding-pair versus bonding-pair repulsion, the two sulfur-to-oxygen bonds are pushed together slightly and the OSO angle is less than 120°.

AB₃E: Ammonia (NH₃)

The ammonia molecule contains three bonding pairs and one lone pair:

H—N—H
|
H

As Table 10.1 shows, the overall arrangement of four electron pairs is tetrahedral. But in NH_3 one of the electron pairs is a lone pair, so the geometry of NH_3 is trigonal pyramidal (so called because it looks like a pyramid, with the N atom at the apex).

Because the lone pair repels the bonding pairs more strongly, the three NH bonding pairs are pushed closer together:

Thus the HNH angle in ammonia is smaller than the ideal tetrahedral angle of 109.5° (Figure 10.1).

AB_2E_2: Water (H_2O)

A water molecule contains two bonding pairs and two lone pairs:

$$H—\overset{..}{\underset{..}{O}}—H$$

The overall arrangement of the four electron pairs in water is tetrahedral, the same as in ammonia. However, unlike ammonia, water has two lone pairs on the central O atom. These lone pairs tend to be as far from each other as possible. Consequently, the two OH bonding pairs are pushed toward each other, and we predict an even greater deviation from the tetrahedral angle than in NH_3. As Figure 10.1 shows, the HOH angle is 104.5°. The geometry of H_2O is bent:

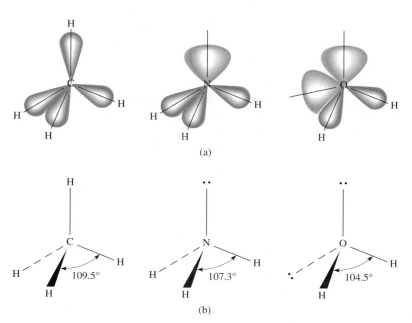

Figure 10.1 *(a) The relative sizes of bonding pairs and lone pairs in CH_4, NH_3, and H_2O. (b) The bond angles in CH_4, NH_3, and H_2O. Note that dashed lines represent bond axes behind the plane of the paper, wedged lines represent bond axes in front of the plane of the paper, and thin solid lines represent bonds in the plane of the paper.*

SF_4

AB_4E: Sulfur Tetrafluoride (SF_4)

The Lewis structure of SF_4 is

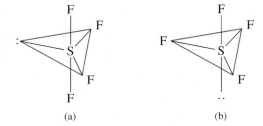

The central sulfur atom has five electron pairs whose arrangement, according to Table 10.1, is trigonal bipyramidal. In the SF_4 molecule, however, one of the electron pairs is a lone pair, so the molecule must have one of the following geometries:

(a)   (b)

In (a) the lone pair occupies an equatorial position, and in (b) it occupies an axial position. The axial position has three neighboring pairs at 90° and one at 180°, while the equatorial position has two neighboring pairs at 90° and two more at 120°. The repulsion is smaller for (a), and indeed (a) is the structure observed experimentally. This shape is sometimes described as a distorted tetrahedron (or seesaw if you turn the structure 90° to the right to view it). The angle between the axial F atoms and S is 186°, and that between the equatorial F atoms and S is 116°.

Table 10.2 shows the geometries of simple molecules in which the central atom has one or more lone pairs, including some that we have not discussed.

Geometry of Molecules with More Than One Central Atom

So far we have discussed the geometry of molecules having only one central atom. The overall geometry of molecules with more than one central atom is difficult to define in most cases. Often we can only describe the shape around each of the central atoms. For example, consider methanol, CH_3OH, whose Lewis structure is shown below:

$$H-\overset{\overset{\displaystyle H}{|}}{\underset{\underset{\displaystyle H}{|}}{C}}-\overset{..}{\underset{..}{O}}-H$$

The two central (nonterminal) atoms in methanol are C and O. We can say that the three CH and the CO bonding pairs are tetrahedrally arranged about the C atom. The HCH and OCH bond angles are approximately 109°. The O atom here is like the one in water in that it has two lone pairs and two bonding pairs. Therefore, the HOC portion of the molecule is bent, and the angle HOC is approximately equal to 105° (Figure 10.2).

Guidelines for Applying the VSEPR Model

Having studied the geometries of molecules in two categories (central atoms with and without lone pairs), let us consider some rules for applying the VSEPR model to all types of molecules:

Figure 10.2 *The geometry of* CH_3OH.

Table 10.2 | Geometry of Simple Molecules and Ions in Which the Central Atom Has One or More Lone Pairs

Class of Molecule	Total Number of Electron Pairs	Number of Bonding Pairs	Number of Lone Pairs	Arrangement of Electron Pairs*	Geometry	Examples
AB_2E	3	2	1	Trigonal planar	Bent	SO_2
AB_3E	4	3	1	Tetrahedral	Trigonal pyramidal	NH_3
AB_2E_2	4	2	2	Tetrahedral	Bent	H_2O
AB_4E	5	4	1	Trigonal bipyramidal	Distorted tetrahedron (or seesaw)	SF_4
AB_3E_2	5	3	2	Trigonal bipyramidal	T-shaped	ClF_3
AB_2E_3	5	2	3	Trigonal bipyramidal	Linear	I_3^-
AB_5E	6	5	1	Octahedral	Square pyramidal	BrF_5
AB_4E_2	6	4	2	Octahedral	Square planar	XeF_4

*The colored lines are used to show the overall shape, not bonds.

- Write the Lewis structure of the molecule, considering only the electron pairs around the central atom (that is, the atom that is bonded to more than one other atom).
- Count the number of electron pairs around the central atom (bonding pairs and lone pairs). Treat double and triple bonds as though they were single bonds. Refer to Table 10.1 to predict the overall arrangement of the electron pairs.
- Use Tables 10.1 and 10.2 to predict the geometry of the molecule.
- In predicting bond angles, note that a lone pair repels another lone pair or a bonding pair more strongly than a bonding pair repels another bonding pair. Remember that in general there is no easy way to predict bond angles accurately when the central atom possesses one or more lone pairs.

The VSEPR model generates reliable predictions of the geometries of a variety of molecular structures. Chemists use the VSEPR approach because of its simplicity. Although there are some theoretical concerns about whether "electron-pair repulsion" actually determines molecular shapes, the assumption that it does leads to useful (and generally reliable) predictions. We need not ask more of any model at this stage in the study of chemistry.

Example 10.1   Use the VSEPR model to predict the geometry of these molecules and ions: (a) AsH_3, (b) OF_2, (c) $AlCl_4^-$, (d) I_3^-, (e) C_2H_4.

Reasoning and Solution   To predict the geometry using the VSEPR model, we need to draw the Lewis structure of the compound and focus only on the lone pairs on the central atom.

(a) The Lewis structure of AsH_3 is

$$H\!-\!\overset{..}{As}\!-\!H$$
$$\underset{H}{|}$$

AsH_3

This molecule has three bonding pairs and one lone pair, a combination similar to that in ammonia. Therefore, the geometry of AsH_3 is trigonal pyramidal, like NH_3. We cannot predict the HAsH angle accurately, but we know that it must be less than 109.5° because the repulsion of the bonding electron pairs by the lone pair on As is greater than the repulsion between bonding pairs.

(b) The Lewis structure of OF_2 is

$$:\!\overset{..}{\underset{..}{F}}\!-\!\overset{..}{\underset{..}{O}}\!-\!\overset{..}{\underset{..}{F}}\!:$$

Because there are two lone pairs on the O atom, the OF_2 molecule must have a bent shape, like that of H_2O. Again, all we can say about angle size is that the FOF angle should be less than 109.5° due to the stronger repulsive force between the lone pairs and the bonding pairs.

(c) The Lewis structure of $AlCl_4^-$ is

$$\left[\begin{array}{c} :\!\overset{..}{Cl}\!: \\ | \\ :\!\overset{..}{\underset{..}{Cl}}\!-\!Al\!-\!\overset{..}{\underset{..}{Cl}}\!: \\ | \\ :\!\overset{..}{\underset{..}{Cl}}\!: \end{array}\right]^{-}$$

$AlCl_4^-$

Because the central Al atom has no lone pairs and all four Al—Cl bonds are equivalent, the $AlCl_4^-$ ion should be tetrahedral, and the ClAlCl angles should all be 109.5°.

(d) The Lewis structure of I_3^- is

$$\left[\,:\overset{\cdot\cdot}{\underset{\cdot\cdot}{I}}-\overset{\cdot\cdot}{\underset{\cdot\cdot}{I}}-\overset{\cdot\cdot}{\underset{\cdot\cdot}{I}}:\,\right]^-$$

The central I atom has two bonding pairs and three lone pairs. From Table 10.2 we see that the three lone pairs all lie in the triangular plane, and the I_3^- ion should be linear.

(e) The Lewis structure of C_2H_4 is

$$\overset{H}{\underset{H}{\diagdown}}C=C\overset{H}{\underset{H}{\diagup}}$$

The C=C bond is treated as though it were a single bond. Because there are no lone pairs present, the arrangement around each C atom has a trigonal planar shape like BF_3, discussed earlier. Thus the predicted bond angles in C_2H_4 are all 120°.

$$\overset{H}{\underset{H}{\diagdown}}\overset{120°}{C}=\overset{H}{\underset{H}{C}}\,\,120°$$

Comment (1) The I_3^- ion is one of the few structures for which the bond angle (180°) can be predicted accurately even though the central atom contains lone pairs. (2) In C_2H_4, all six atoms lie in the same plane. The overall planar geometry is not predicted by the VSEPR model, but we will see why the molecule prefers to be planar later. In reality, the angles are close, but not equal, to 120° because the bonds are not all equivalent.

Practice Exercise Use the VSEPR model to predict the geometry of (a) $SiBr_4$, (b) CS_2, and (c) NO_3^-.

C_2H_4

Similar problems: 10.7, 10.8, 10.9.

10.2 Dipole Moments

In Section 9.2 we learned that hydrogen fluoride is a covalent compound with a polar bond. There is a shift of electron density from H to F because the F atom is more electronegative than the H atom (see Figure 9.4). The shift of electron density is symbolized by placing a crossed arrow ($\longmapsto$) above the Lewis structure to indicate the direction of the shift. For example,

$$\overset{\longmapsto}{H-\overset{\cdot\cdot}{\underset{\cdot\cdot}{F}}:}$$

The consequent charge separation can be represented as

$$\overset{\delta+\,\,\,\delta-}{H-\overset{\cdot\cdot}{\underset{\cdot\cdot}{F}}:}$$

where δ (delta) denotes a partial charge. This separation of charges can be confirmed in an electric field (Figure 10.3). When the field is turned on, HF molecules orient

Figure 10.3 *Behavior of polar molecules (a) in the absence of an external electric field and (b) when the electric field is turned on. Nonpolar molecules are not affected by an electric field.*

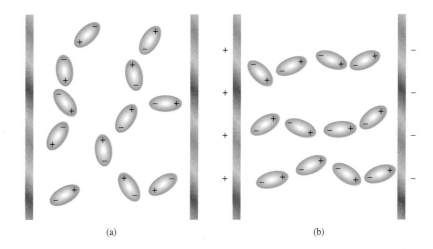

(a)   (b)

their negative ends toward the positive plate and their positive ends toward the negative plate. This alignment of molecules can be detected experimentally.

A quantitative measure of the polarity of a bond is its ***dipole moment (μ),*** which is *the product of the charge Q and the distance r between the charges:*

$$\mu = Q \times r \tag{10.1}$$

In a diatomic molecule like HF, the charge Q is equal to $\delta+$ and $\delta-$.

To maintain electrical neutrality, the charges on both ends of an electrically neutral diatomic molecule must be equal in magnitude and opposite in sign. However, in Equation (10.1), Q refers only to the magnitude of the charge and not to its sign, so μ is always positive. Dipole moments are usually expressed in debye units (D), named for the Dutch-American chemist and physicist Peter Debye. The conversion factor is

$$1 \text{ D} = 3.336 \times 10^{-30} \text{ C m}$$

where C is coulomb and m is meter.

Polarity of molecules

Diatomic molecules containing atoms of *different* elements (for example, HCl, CO, and NO) *have dipole moments* and are called ***polar molecules.*** Diatomic molecules containing atoms of the *same* element (for example, H_2, O_2, and F_2) are examples of ***nonpolar molecules*** because they *do not have dipole moments*. For a molecule made up of three or more atoms both the polarity of the bonds and the molecular geometry determine whether there is a dipole moment. Even if polar bonds are present, the molecule will not necessarily have a dipole moment. Carbon dioxide (CO_2), for example, is a triatomic molecule, so its geometry is either linear or bent:

Each carbon-to-oxygen bond is polar, with the electron density shifted toward the more electronegative oxygen atom. However, the linear geometry of the molecule results in the cancellation of the two bond moments.

$$\overset{\longleftarrow \;+\; \longrightarrow}{O=C=O}$$

linear molecule
(no dipole moment)

$$\underset{O}{\overset{\times C \times}{\diagdown}} \underset{O}{\diagup}$$

resultant
dipole moment

bent molecule
(has a dipole moment)

The arrows show the shift of electron density from the less electronegative carbon atom to the more electronegative oxygen atom. In each case, the dipole moment of the entire molecule is made up of two *bond moments,* that is, individual dipole moments in the polar C=O bonds. The bond moment is a *vector quantity,* which means that it has both magnitude and direction. The measured dipole moment is equal to the vector sum of the

Molecule	Geometry	Dipole Moment (D)
HF	Linear	1.92
HCl	Linear	1.08
HBr	Linear	0.78
HI	Linear	0.38
H_2O	Bent	1.87
H_2S	Bent	1.10
NH_3	Trigonal pyramidal	1.46
SO_2	Bent	1.60

Table 10.3 | Dipole Moments of Some Polar Molecules

H_2O

bond moments. The two bond moments in CO_2 are equal in magnitude. Because they point in opposite directions in a linear CO_2 molecule, the sum or resultant dipole moment would be zero. On the other hand, if the CO_2 molecule were bent, the two bond moments would partially reinforce each other, so that the molecule would have a dipole moment. Experimentally it is found that carbon dioxide has no dipole moment. Therefore, we conclude that the carbon dioxide molecule is linear. The linear nature of carbon dioxide has been confirmed through other experimental measurements.

Contrary to CO_2, sulfur dioxide (SO_2) has a bent geometry. Therefore, it is a polar molecule because the two bond moments do not cancel each other:

The VSEPR model predicts that CO_2 is a linear molecule.

resultant
dipole moment

Indeed, the dipole moment of SO_2 is 1.6 D. Table 10.3 lists the dipole moments of several polar molecules.

Because SO_2 has a bent geometry, the bond moments do not cancel and the molecule has a dipole moment.

Example 10.2   Predict whether each of these molecules has a dipole moment: (a) IBr, (b) BF_3 (trigonal planar), (c) CH_2Cl_2 (tetrahedral).

Reasoning and Solution   Keep in mind that the dipole moment of a molecule depends both on the difference in electronegativities of the elements present and its geometry.

(a) Because IBr (iodine bromide) is diatomic, it has a linear geometry. Bromine is more electronegative than iodine (see Figure 9.4), so IBr is polar with bromine at the negative end.

I—Br

Thus the molecule does have a dipole moment.

(b) Because fluorine is more electronegative than boron, each B—F bond in BF_3 (boron trifluoride) is polar and the three bond moments are equal. However, the symmetry of a trigonal planar shape means that the three bond moments exactly cancel one another:

The electron density is symmetrically distributed in the BF_3 molecule.

An analogy is an object that is pulled in the directions shown by the three bond moments. If the forces are equal, the object will not move. Consequently BF_3 has no dipole moment; it is a nonpolar molecule.

(c) The Lewis structure of CH_2Cl_2 (methylene chloride) is

This molecule is similar to CH_4 in that it has an overall tetrahedral shape. However, because not all the bonds are identical, there are three different bond angles: HCH, HCCl, and ClCCl. These bond angles are close to, but not equal to, 109.5°. Because chlorine is more electronegative than carbon, which is more electronegative than hydrogen, the bond moments do not cancel and the molecule possesses a dipole moment:

Thus CH_2Cl_2 is a polar molecule.

Practice Exercise   Does the $AlCl_3$ molecule have a dipole moment?

The electron density in CH_2Cl_2 is shifted toward the electronegative Cl atoms.

Similar problems: 10.19, 10.21, 10.22.

10.3 Valence Bond Theory

The VSEPR model, based largely on Lewis structures, provides a relatively simple and straightforward method for predicting the geometry of molecules. But as we noted earlier, the Lewis theory of chemical bonding does not clearly explain why chemical bonds exist. Relating the formation of a covalent bond to the pairing of electrons was a step in the right direction, but it did not go far enough. For example, the Lewis theory describes the single bond between the H atoms in H_2 and that between the F atoms in F_2 in essentially the same way—as the pairing of two electrons. Yet these two molecules have quite different bond energies and bond lengths (436.4 kJ/mol and 74 pm for H_2 and 150.6 kJ/mol and 142 pm for F_2). These and many other facts cannot be explained by the Lewis theory. For a more complete explanation of chemical bond formation we look to quantum mechanics. In fact, the quantum mechanical study of chemical bonding also provides a means for understanding molecular geometry.

At present, two quantum mechanical theories are used to describe covalent bond formation and the electronic structure of molecules. *Valence bond (VB) theory* assumes that the electrons in a molecule occupy atomic orbitals of the individual atoms. It enables us to retain a picture of individual atoms taking part in the bond formation. The second theory, called *molecular orbital (MO) theory,* assumes the formation of molecular orbitals from the atomic orbitals. Neither theory perfectly explains all aspects of bonding, but each has contributed something to our understanding of many observed molecular properties.

Let us start our discussion of valence bond theory by considering the formation of a H_2 molecule from two H atoms. The Lewis theory describes the H—H bond in terms of the pairing of the two electrons on the H atoms. In the framework of valence bond theory, the covalent H—H bond is formed by the *overlap* of the two $1s$ orbitals in the H atoms. By overlap, we mean that the two orbitals share a common region in space.

What happens to two H atoms as they move toward each other and form a bond? Initially, when the two atoms are far apart, there is no interaction. We say that the potential energy of this system (that is, the two H atoms) is zero. As the atoms approach each other, each electron is attracted by the nucleus of the other atom; at the same time, the electrons repel each other, as do the nuclei. While the atoms are still separated, attraction is stronger than repulsion, so that the potential energy of the system *decreases* (that is, it becomes negative) as the atoms approach each other (Figure 10.4). This trend continues until the potential energy reaches a minimum value. At this point, when the system has the lowest potential energy, it is most stable. This condition corresponds to substantial overlap of the $1s$ orbitals and the formation of a stable H_2 molecule. If the distance between nuclei were to decrease further, the potential energy would rise steeply and finally become positive as a result of the increased electron-electron and nuclear-nuclear repulsions. In accord with the law of conservation of energy, the decrease in potential energy as a result of H_2 formation must be accompanied by a release of energy. Experiments show that as a H_2 molecule is formed from two H atoms, heat is given off. The converse is also true. To break a H—H bond, energy must be supplied to the molecule. Figure 10.5 is another way of viewing the formation of an H_2 molecule.

Recall that an object has potential energy by virtue of its position.

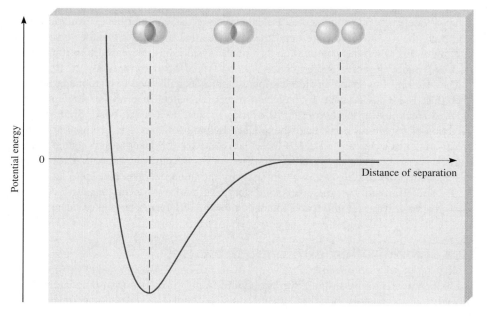

Figure 10.4 *Change in potential energy of two H atoms with changes in their distance of separation. At the point of minimum potential energy, the H_2 molecule is in its most stable state, and the bond length is 74 pm. The spheres represent the 1s orbitals.*

Potential energy

0

Distance of separation

Figure 10.5 *Top to bottom: As two H atoms approach each other, their 1s orbitals begin to interact and each electron begins to feel the attraction of the other proton. Gradually, the electron density builds up in the region between the two nuclei (red color). Eventually, a stable H_2 molecule is formed when the internuclear distance is 74 pm.*

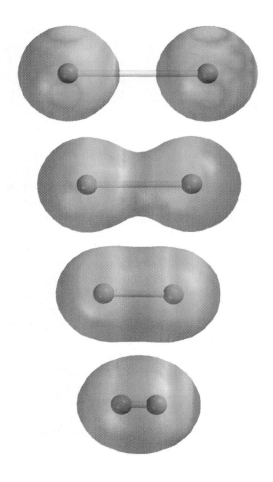

The orbital diagram of the F atom is shown on p. 217.

Thus valence bond theory gives a clearer picture of chemical bond formation than the Lewis theory does. Valence bond theory states that a stable molecule forms from reacting atoms when the potential energy of the system has decreased to a minimum; the Lewis theory ignores energy changes in chemical bond formation.

The concept of overlapping atomic orbitals applies equally well to diatomic molecules other than H_2. Thus a stable F_2 molecule forms when the $2p$ orbitals (containing the unpaired electrons) in the two F atoms overlap to form a covalent bond. Similarly, the formation of the HF molecule can be explained by the overlap of the $1s$ orbital in H with the $2p$ orbital in F. In each case, VB theory accounts for the changes in potential energy as the distance between the reacting atoms changes. Because the orbitals involved are not the same kind in all cases, we can see why the bond energies and bond lengths in H_2, F_2, and HF might be different. As we stated earlier, Lewis theory treats *all* covalent bonds the same way and offers no explanation for the differences among covalent bonds.

10.4 Hybridization of Atomic Orbitals

The concept of atomic orbital overlap should apply also to polyatomic molecules. However, a satisfactory bonding scheme must account for molecular geometry. We will discuss three examples of VB treatment of bonding in polyatomic molecules.

sp³ **Hybridization**

Consider the CH_4 molecule. Focusing only on the valence electrons, we can represent the orbital diagram of C as

Hybridization

$$\boxed{\uparrow\downarrow}\quad \boxed{\uparrow\,|\,\uparrow\,|\,}$$
$$2s\qquad\quad 2p$$

Because the carbon atom has two unpaired electrons (one in each of the two $2p$ orbitals), it can form only two bonds with hydrogen in its ground state. Although the species CH_2 is known, it is very unstable. To account for the four C—H bonds in methane, we can try to promote (that is, energetically excite) an electron from the $2s$ orbital to the $2p$ orbital:

$$\boxed{\uparrow}\quad \boxed{\uparrow\,|\,\uparrow\,|\,\uparrow}$$
$$2s\qquad\quad 2p$$

Now there are four unpaired electrons on C that could form four C—H bonds. However, the geometry is wrong, because three of the HCH bond angles would have to be 90° (remember that the three $2p$ orbitals on carbon are mutually perpendicular), and yet *all* HCH angles are 109.5°.

To explain the bonding in methane, VB theory uses hypothetical ***hybrid orbitals,*** which are *atomic orbitals obtained when two or more nonequivalent orbitals of the same atom combine in preparation for covalent bond formation.* **Hybridization** is the term applied to *the mixing of atomic orbitals in an atom (usually a central atom) to generate a set of hybrid orbitals.* We can generate four equivalent hybrid orbitals for carbon by mixing the $2s$ orbital and the three $2p$ orbitals:

$$\boxed{\uparrow\,|\,\uparrow\,|\,\uparrow\,|\,\uparrow}$$
$$\underbrace{\qquad\qquad\qquad}_{sp^3\ \text{orbitals}}$$

Because the new orbitals are formed from one s and three p orbitals, they are called sp^3 hybrid orbitals. Figure 10.6 shows the shape and orientations of the sp^3 orbitals.   *sp³* **is pronounced "s-p three."**

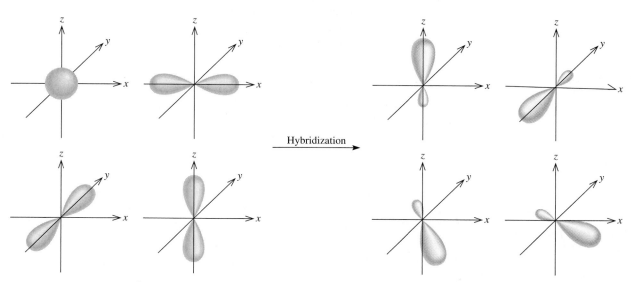

Hybridization

Figure 10.6 *Formation of sp^3 hybrid orbitals.*

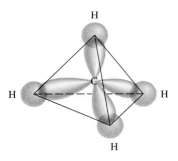

Figure 10.7 *Formation of four bonds between the carbon sp^3 hybrid orbitals and the hydrogen 1s orbitals in CH$_4$.*

These four hybrid orbitals are directed toward the four corners of a regular tetrahedron. Figure 10.7 shows the formation of four covalent bonds between the carbon sp^3 hybrid orbitals and the hydrogen 1s orbitals in CH$_4$. Thus CH$_4$ has a tetrahedral shape, and all the HCH angles are 109.5°. Note that although energy is required to bring about hybridization, this input is more than compensated for by the energy released upon the formation of C—H bonds. (Recall that bond formation is an exothermic process.)

This analogy is useful for understanding hybridization. Suppose that we have a beaker of a red solution and three beakers of blue solution and that the volume of each is 50 mL. The red solution corresponds to one 2s orbital, the blue solutions represent three 2p orbitals, and the four equal volumes symbolize four separate orbitals. By mixing the solutions we obtain 200 mL of a purple solution, which can be divided into four 50-mL portions (that is, the hybridization process generates four sp^3 orbitals). Just as the purple color is made up of the red and blue components of the original solutions, the sp^3 hybrid orbitals possess both s and p orbital characteristics.

Another example of sp^3 hybridization is ammonia (NH$_3$). Table 10.1 shows that the arrangement of four electron pairs is tetrahedral, so that the bonding in NH$_3$ can be explained by assuming that N, like C in CH$_4$, is sp^3-hybridized. The ground-state electron configuration of N is $1s^2 2s^2 2p^3$, so that the orbital diagram for the sp^3-hybridized N atom is

sp^3 orbitals

Three of the four hybrid orbitals form covalent N—H bonds, and the fourth hybrid orbital accommodates the lone pair on nitrogen (Figure 10.8). Repulsion between the lone-pair electrons and electrons in the bonding orbitals decreases the HNH bond angles from 109.5° to 107.3°.

It is important to understand the relationship between hybridization and the VSEPR model. We use hybridization to describe the bonding scheme only when the arrangement of electron pairs has been predicted using VSEPR. If the VSEPR model predicts a tetrahedral arrangement of electron pairs, then we assume that one s and three p orbitals are hybridized to form four sp^3 hybrid orbitals. The following are examples of other types of hybridization.

sp Hybridization

The beryllium chloride (BeCl$_2$) molecule is predicted to be linear by VSEPR. The orbital diagram for the valence electrons in Be is

2s   2p

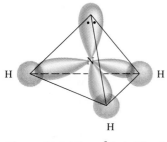

Figure 10.8 *The sp^3-hybridized N atom in NH$_3$. Three sp^3 hybrid orbitals form bonds with the H atoms. The fourth is occupied by nitrogen's lone pair.*

We know that in its ground state Be does not form covalent bonds with Cl because its electrons are paired in the 2s orbital. So we turn to hybridization for an explanation of Be's bonding behavior. First we promote a 2s electron to a 2p orbital, resulting in

2s   2p

Figure 10.9 *Formation of sp hybrid orbitals.*

Now there are two Be orbitals available for bonding, the 2s and 2p. However, if two Cl atoms were to combine with Be in this excited state, one Cl atom would share a 2s electron and the other Cl would share a 2p electron, making two nonequivalent BeCl bonds. This scheme contradicts experimental evidence. In the actual $BeCl_2$ molecule, the two BeCl bonds are identical in every respect. Thus the 2s and 2p orbitals must be mixed, or hybridized, to form two equivalent *sp* hybrid orbitals:

sp orbitals   empty 2p orbitals

Figure 10.9 shows the shape and orientation of the *sp* orbitals. These two hybrid orbitals lie on the same line, the *x*-axis, so that the angle between them is 180°. Each of the BeCl bonds is then formed by the overlap of a Be *sp* hybrid orbital and a Cl 3p orbital, and the resulting $BeCl_2$ molecule has a linear geometry (Figure 10.10).

sp^2 Hybridization

Next we will look at the BF_3 (boron trifluoride) molecule, known to have planar geometry based on VSEPR. Considering only the valence electrons, the orbital diagram of B is

2s   2p

First, we promote a 2s electron to an empty 2p orbital:

2s   2p

Mixing the 2s orbital with the two 2p orbitals generates three sp^2 hybrid orbitals:

sp^2 orbitals   empty 2p orbital

These three sp^2 orbitals lie in the same plane, and the angle between any two of them is 120° (Figure 10.11). Each of the BF bonds is formed by the overlap of a boron sp^2 hybrid orbital and a fluorine 2p orbital (Figure 10.12). The BF_3 molecule is planar with all the FBF angles equal to 120°. This result conforms to experimental findings and also to VSEPR predictions.

sp^2 is pronounced "s-p two."

Figure 10.10 *The linear geometry of $BeCl_2$ can be explained by assuming that Be is sp-hybridized. The two sp hybrid orbitals overlap with the two chlorine 3p orbitals to form two covalent bonds.*

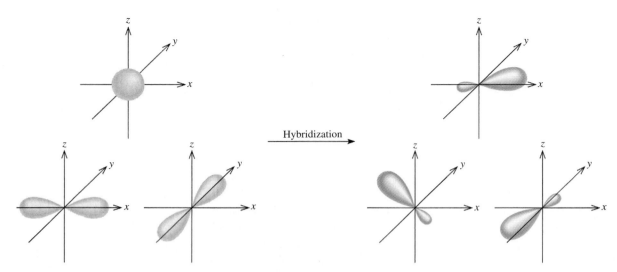

Figure 10.11 *Formation of sp² hybrid orbitals.*

You may have noticed an interesting connection between hybridization and the octet rule. Regardless of the type of hybridization, an atom starting with one *s* and three *p* orbitals would still possess four orbitals, enough to accommodate a total of eight electrons in a compound. For elements in the second period of the periodic table, eight is the maximum number of electrons that an atom of any of these elements can accommodate in the valence shell. It is for this reason that the octet rule is usually obeyed by the second-period elements.

The situation is different for an atom of a third-period element. If we use only the 3*s* and 3*p* orbitals of the atom to form hybrid orbitals in a molecule, then the octet rule applies. However, in some molecules the same atom may use one or more 3*d* orbitals, in addition to the 3*s* and 3*p* orbitals, to form hybrid orbitals. In these cases the octet rule does not hold. We will see specific examples of the participation of the 3*d* orbital in hybridization shortly.

To summarize our discussion of hybridization, we note that

- The concept of hybridization is not applied to isolated atoms. It is a theoretical model used only to explain covalent bonding.
- Hybridization is the mixing of at least two nonequivalent atomic orbitals, for example, *s* and *p* orbitals. Therefore, a hybrid orbital is not a pure atomic orbital. Hybrid orbitals and pure atomic orbitals have very different shapes.
- The number of hybrid orbitals generated is equal to the number of pure atomic orbitals that participate in the hybridization process.
- Hybridization requires an input of energy; however, the system more than recovers this energy during bond formation.
- Covalent bonds in polyatomic molecules and ions are formed by the overlap of hybrid orbitals, or of hybrid orbitals with unhybridized ones. Therefore, the hybridization bonding scheme is still within the framework of valence bond theory; electrons in a molecule are assumed to occupy hybrid orbitals of the individual atoms.

Figure 10.12 *The sp² hybrid orbitals of boron overlap with the 2p orbitals of fluorine. The BF₃ molecule is planar, and all the FBF angles are 120°.*

Table 10.4 summarizes *sp*, *sp²*, and *sp³* hybridization (as well as other types that we will discuss shortly).

Table 10.4 | Important Hybrid Orbitals and Their Shapes

Pure Atomic Orbitals of the Central Atom	Hybridization of the Central Atom	Number of Hybrid Orbitals	Shape of Hybrid Orbitals	Examples
s, p	sp	2	Linear	$BeCl_2$
s, p, p	sp^2	3	Planar	BF_3
s, p, p, p	sp^3	4	Tetrahedral	CH_4, NH_4^+
s, p, p, p, d	sp^3d	5	Trigonal bipyramidal	PCl_5
s, p, p, p, d, d	sp^3d^2	6	Octahedral	SF_6

Procedure for Hybridizing Atomic Orbitals

Before going on to discuss the hybridization of *d* orbitals, let us specify what we need to know in order to apply hybridization to bonding in polyatomic molecules in general. In essence, hybridization simply extends Lewis theory and the VSEPR model. To assign a suitable state of hybridization to the central atom in a molecule, we must have some idea about the geometry of the molecule. The steps are:

1. Draw the Lewis structure of the molecule.
2. Predict the overall arrangement of the electron pairs (both bonding pairs and lone pairs) using the VSEPR model (see Table 10.1).
3. Deduce the hybridization of the central atom by matching the arrangement of the electron pairs with those of the hybrid orbitals shown in Table 10.4.

Example 10.3   Determine the hybridization state of the central (underlined) atom in each of these molecules: (a) $\underline{Hg}Cl_2$, (b) $\underline{Al}I_3$, and (c) $\underline{P}F_3$. Describe the hybridization process and determine the molecular geometry in each case.

Reasoning and Solution   We follow the given outline for predicting the hybridization state of the central atom.

(a) The ground-state electron configuration of Hg is $[Xe]6s^2 4f^{14} 5d^{10}$; therefore, the Hg atom has two valence electrons (the $6s$ electrons). The Lewis structure of $HgCl_2$ is

$$: \overset{..}{\underset{..}{Cl}} \!\!-\!\! Hg \!\!-\!\! \overset{..}{\underset{..}{Cl}} :$$

There are no lone pairs on the Hg atom, so the arrangement of the two electron pairs is linear (see Table 10.1). From Table 10.4 we conclude that Hg is *sp*-hybridized because it has the geometry of the two *sp* hybrid orbitals. The hybridization process can be imagined to take place as follows. First we draw the orbital diagram for the ground state of Hg:

$$\quad\quad 6s \quad\quad\quad\quad 6p$$

By promoting a $6s$ electron to the $6p$ orbital, we get the excited state:

$$\quad\quad 6s \quad\quad\quad\quad 6p$$

The $6s$ and $6p$ orbitals then mix to form two *sp* hybrid orbitals:

$$\quad sp \text{ orbitals} \quad\quad\quad \text{empty } 6p$$
$$\text{orbitals}$$

The two Hg—Cl bonds are formed by the overlap of the Hg *sp* hybrid orbitals with the $3p$ orbitals of the Cl atoms. Thus $HgCl_2$ is a linear molecule.

(b) The ground-state electron configuration of Al is $[Ne]3s^2 3p^1$. Therefore, Al has three valence electrons. The Lewis structure of AlI_3 is

---

Table 7.3 gives the electron configurations of all the elements.

$HgCl_2$

There are three bonding pairs and no lone pair on the Al atom. From Table 10.1 we see that the shape of three electron pairs is trigonal planar, and from Table 10.4 we conclude that Al must be sp^2-hybridized in AlI_3. The orbital diagram of the ground-state Al atom is

By promoting a $3s$ electron into the $3p$ orbital we obtain the following excited state:

The $3s$ and two $3p$ orbitals then mix to form three sp^2 hybrid orbitals:

The sp^2 hybrid orbitals overlap with the $5p$ orbitals of I to form three covalent Al—I bonds. We predict that the AlI_3 molecule is trigonal planar and all the IAlI angles are 120°.

(c) The ground-state electron configuration of P is [Ne]$3s^2 3p^3$. Therefore, the P atom has five valence electrons. The Lewis structure of PF_3 is

$$F - \overset{..}{\underset{\underset{F}{|}}{P}} - F$$

There are three bonding pairs and one lone pair on the P atom. In Table 10.1 we see that the overall arrangement of four electron pairs is tetrahedral, and from Table 10.4 we conclude that P must be sp^3-hybridized. The orbital diagram of the ground-state P atom is

By mixing the $3s$ and $3p$ orbitals, we obtain four sp^3 hybrid orbitals.

AlI_3

PF_3

Similar problems: 10.31, 10.32.

As in the case of NH_3, one of the sp^3 hybrid orbitals is used to accommodate the lone pair on P. The other three sp^3 hybrid orbitals form covalent P—F bonds with the $2p$ orbitals of F. We predict the geometry of the molecule to be trigonal pyramidal; the FPF angle should be somewhat less than $109.5°$.

Practice Exercise   Determine the hybridization state of the underlined atoms in these compounds: (a) $\underline{Si}Br_4$ and (b) $\underline{B}Cl_3$.

Hybridization of *s, p,* and *d* Orbitals

We have seen that hybridization neatly explains bonding that involves *s* and *p* orbitals. For elements in the third period and beyond, however, we cannot always account for molecular geometry by assuming that only *s* and *p* orbitals hybridize. To understand the formation of molecules with trigonal bipyramidal and octahedral geometries, for instance, we must include *d* orbitals in the hybridization concept.

Consider the SF_6 molecule as an example. In Section 10.1 we saw that this molecule has octahedral geometry, which is also the arrangement of the six electron pairs. Table 10.4 shows that the S atom is sp^3d^2-hybridized in SF_6. The ground-state electron configuration of S is $[Ne]3s^2 3p^4$:

$$3s \qquad 3p \qquad\qquad\qquad 3d$$

Because the $3d$ level is quite close in energy to the $3s$ and $3p$ levels, we can promote $3s$ and $3p$ electrons to two of the $3d$ orbitals:

$$3s \qquad 3p \qquad\qquad 3d$$

SF_6

sp^3d^2 is pronounced "s-p three d two."

Mixing the $3s$, three $3p$, and two $3d$ orbitals generates six sp^3d^2 hybrid orbitals:

$$sp^3d^2 \text{ orbitals} \qquad\qquad \text{empty } 3d \text{ orbitals}$$

The six S—F bonds are formed by the overlap of the hybrid orbitals of the S atom and the $2p$ orbitals of the F atoms. Because there are 12 electrons around the S atom, the octet rule is violated. The use of *d* orbitals in addition to *s* and *p* orbitals to form an expanded octet (see Section 9.7) is an example of *valence-shell expansion*. Second-period elements, unlike third-period elements, do not have $2d$ energy levels, so they can never expand their valence shells. (Recall that when $n = 2$, $\ell = 0$ and 1. Thus we can only have $2s$ and $2p$ orbitals.) Hence atoms of second-period elements can never be surrounded by more than eight electrons in any of their compounds.

Example 10.4 Describe the hybridization state of phosphorus in phosphorus pentabromide (PBr₅).

Reasoning and Solution We follow the outline listed on p. 312 for predicting the hybridization state of the central atom. The Lewis structure of PBr₅ is

PBr₅

$$\begin{array}{c} :\!\overset{\displaystyle\cdots}{Br}: \\[2pt] :\!\overset{\displaystyle\cdots}{Br}\diagdown \;\; \Big| \\[2pt] \overset{\displaystyle}{P}\!\!-\!\!\overset{\displaystyle\cdots}{Br}: \\[2pt] :\!\overset{\displaystyle\cdots}{Br}\diagup \;\; \Big| \\[2pt] :\!\overset{\displaystyle\cdots}{Br}: \end{array}$$

Table 10.1 indicates that the arrangement of five electron pairs is trigonal bipyramidal. Referring to Table 10.4, we find that this is the shape of five sp^3d hybrid orbitals. Thus P must be sp^3d-hybridized in PBr₅. The ground-state electron configuration of P is [Ne]$3s^2 3p^3$. To describe the hybridization process, we start with the orbital diagram for the ground state of P:

Promoting a 3s electron into a 3d orbital results in the following excited state:

Mixing the one 3s, three 3p, and one 3d orbitals generates five sp^3d hybrid orbitals:

These hybrid orbitals overlap the 4p orbitals of Br to form five covalent P—Br bonds. Because there are no lone pairs on the P atom, the geometry of PBr₅ is trigonal bipyramidal.

Practice Exercise Describe the hybridization state of Se in SeF₆.

Similar problem: 10.40.

10.5 Hybridization in Molecules Containing Double and Triple Bonds

The concept of hybridization is useful also for molecules with double and triple bonds. Consider the ethylene molecule, C_2H_4, as an example. In Example 10.1 we saw that C_2H_4 contains a carbon-carbon double bond and has planar geometry. Both the geometry and the bonding can be understood if we assume that each carbon atom is sp^2-hybridized. Figure 10.13 shows orbital diagrams of this hybridization process. We assume that only the $2p_x$ and $2p_y$ orbitals combine with the 2s orbital, and that the $2p_z$ orbital remains unchanged. Figure 10.14 shows that the $2p_z$ orbital is perpendicular to the plane of the hybrid orbitals. Now how do we account for the bonding of the C atoms? As Figure 10.15(a) shows, each carbon atom uses the three sp^2 hybrid orbitals to form two

Sigma and pi bonds

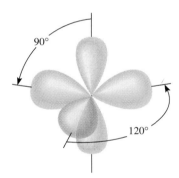

Figure 10.13 *The sp^2 hybridization of a carbon atom. The 2s orbital is mixed with only two 2p orbitals to form three equivalent sp^2 hybrid orbitals. This process leaves an electron in the unhybridized orbital, the $2p_z$ orbital.*

Figure 10.14 *Each carbon atom in the C_2H_4 molecule has three sp^2 hybrid orbitals (green) and one unhybridized $2p_z$ orbital (gray), which is perpendicular to the plane of the hybrid orbitals.*

bonds with the two hydrogen 1s orbitals and one bond with the sp^2 hybrid orbital of the adjacent C atom. In addition, the two unhybridized $2p_z$ orbitals of the C atoms form another bond by overlapping sideways [Figure 10.15(b)].

A distinction is made between the two types of covalent bonds in C_2H_4. The three bonds formed by each C atom in Figure 10.15(a) are all **sigma bonds (σ bonds),** *covalent bonds formed by orbitals overlapping end-to-end, with the electron density concentrated between the nuclei of the bonding atoms.* The second type is called a **pi bond (π bond),** which is defined as *a covalent bond formed by sideways overlapping orbitals with electron density concentrated above and below the plane of the nuclei of the bonding atoms.* The two C atoms form a pi bond as shown in Figure 10.15(b). It is this pi bond formation that gives ethylene its planar geometry. Figure 10.15(c) shows the orientation of the sigma and pi bonds. Figure 10.16 is yet another way of looking at the planar C_2H_4 molecule and the formation of the pi bond. Although we normally represent the carbon-carbon double bond as C=C (as in a Lewis structure), it is important to keep in mind that the two bonds are different types: One is a sigma bond and the other is a pi bond.

The acetylene molecule (C_2H_2) contains a carbon-carbon triple bond. Because the molecule is linear, we can explain its geometry and bonding by assuming that each C atom is *sp*-hybridized by mixing the 2s with the $2p_x$ orbital (Figure 10.17). As Figure 10.18 shows, the two *sp* hybrid orbitals of each C atom form one sigma bond with a hydrogen 1s orbital and another sigma bond with the other C atom. In addition, two pi bonds are formed by the sideways overlap of the unhybridized $2p_y$ and $2p_z$ orbitals. Thus the C≡C bond is made up of one sigma bond and two pi bonds.

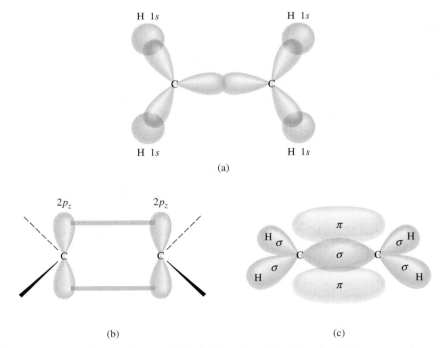

Figure 10.15 *Bonding in ethylene, C_2H_4. (a) Top view of the sigma bonds between carbon atoms and between carbon and hydrogen atoms. All the atoms lie in the same plane. (b) Side view showing how the two $2p_z$ orbitals on the two carbon atoms overlap, leading to the formation of a pi bond. (c) The interactions in (a) and (b) lead to the formation of the sigma bonds and the pi bond in ethylene. Note that the pi bond lies above and below the plane of the molecule.*

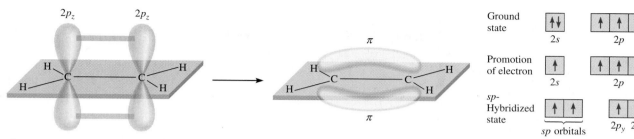

Figure 10.16 *Another view of pi bond formation in the C_2H_4 molecule. Note that all six atoms are in the same plane. It is the overlap of the $2p_z$ orbitals that causes the molecule to assume a planar structure.*

Figure 10.17 *The sp hybridization of a carbon atom. The 2s orbital is mixed with only one 2p orbital to form two sp hybrid orbitals. This process leaves an electron in each of the two unhybridized 2p orbitals, namely, the $2p_y$ and $2p_z$ orbitals.*

The following rule helps us predict hybridization in molecules containing multiple bonds: If the central atom forms a double bond, it is sp^2-hybridized; if it forms two double bonds or a triple bond, it is sp-hybridized. Note that this rule applies only to atoms of the second-period elements. Atoms of third-period elements and beyond that form multiple bonds present a more complicated picture and will not be dealt with here.

Example 10.5 Describe the bonding in the formaldehyde molecule whose Lewis structure is

$$\begin{array}{c} H \\ \diagdown \\ C = \ddot{O} \\ \diagup \\ H \end{array}$$

Assume that the O atom is sp^2-hybridized.

Reasoning and Solution We note that the C atom (a second-period element) has a double bond; therefore, it is sp^2-hybridized. The orbital diagram of the O atom is

$$\underbrace{\boxed{\uparrow}\;\boxed{\uparrow\downarrow}\;\boxed{\uparrow\downarrow}}_{sp^2} \qquad \underbrace{\boxed{\uparrow}}_{2p_z}$$

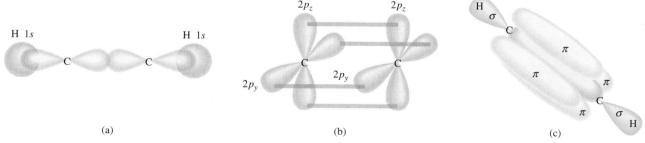

(a)   (b)   (c)

Figure 10.18 *Bonding in acetylene, C_2H_2. (a) Top view showing the overlap of the sp orbitals between the C atoms and the overlap of the sp orbital with the 1s orbital between the C and H atoms. All the atoms lie along a straight line; therefore acetylene is a linear molecule. (b) Side view showing the overlap of the two $2p_y$ orbitals and of the two $2p_z$ orbitals of the two carbon atoms, which leads to the formation of two pi bonds. (c) Formation of the sigma and pi bonds as a result of the interactions in (a) and (b).*

Similar problems: 10.36, 10.78.

Figure 10.19 *Bonding in the formaldehyde molecule. A sigma bond is formed by the overlap of the sp² hybrid orbital of carbon and the sp² hybrid orbital of oxygen; a pi bond is formed by the overlap of the 2pz orbitals of the carbon and oxygen atoms. The two lone pairs on oxygen are placed in the other two sp² orbitals of oxygen.*

Two of the sp^2 orbitals of the C atom form two sigma bonds with the H atoms, and the third sp^2 orbital forms a sigma bond with a sp^2 orbital of the O atom. The $2p_z$ orbital of the C atom overlaps with the $2p_z$ orbital of the O atom to form a pi bond. The two lone pairs on the O atom are placed in its two remaining sp^2 orbitals (Figure 10.19).

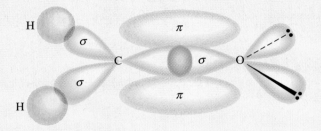

Practice Exercise   Describe the bonding in the hydrogen cyanide molecule, HCN. Assume that N is sp-hybridized.

10.6 Molecular Orbital Theory

Valence bond theory is one of the two quantum mechanical approaches that explain bonding in molecules. It accounts, at least qualitatively, for the stability of the covalent bond in terms of overlapping atomic orbitals. Using the concept of hybridization, valence bond theory can explain molecular geometries predicted by the VSEPR model. However, the assumption that electrons in a molecule occupy atomic orbitals of the individual atoms can only be an approximation, because each bonding electron in a molecule must be in an orbital that is characteristic of the molecule as a whole.

In some cases, valence bond theory cannot satisfactorily account for observed properties of molecules. Consider the oxygen molecule, whose Lewis structure is

$$\overset{\cdot\,\cdot}{\underset{\cdot\,\cdot}{O}}=\overset{\cdot\,\cdot}{\underset{\cdot\,\cdot}{O}}$$

According to this description, all the electrons in O_2 are paired and the molecule should therefore be diamagnetic. But experiments have shown that the oxygen molecule is paramagnetic, with two unpaired electrons (Figure 10.20). This finding suggests a fundamental deficiency in valence bond theory, one that justifies searching for an alternative bonding approach that accounts for the properties of O_2 and other molecules that do not match the predictions of valence bond theory.

Magnetic and other properties of molecules are sometimes better explained by another quantum mechanical approach called *molecular orbital (MO) theory.* Molecular orbital theory describes covalent bonds in terms of **molecular orbitals,** which *result from interaction of the atomic orbitals of the bonding atoms and are associated with the entire molecule.* The difference between a molecular orbital and an atomic orbital is that an atomic orbital is associated with only one atom.

Bonding and Antibonding Molecular Orbitals

According to MO theory, the overlap of the 1s orbitals of two hydrogen atoms leads to the formation of two molecular orbitals: one bonding molecular orbital and one

Figure 10.20 *Liquid oxygen caught between the poles of a magnet, because the O_2 molecules are paramagnetic.*

antibonding molecular orbital. A **bonding molecular orbital** has *lower energy and greater stability than the atomic orbitals from which it was formed.* An **antibonding molecular orbital** has *higher energy and lower stability than the atomic orbitals from which it was formed.* As the names "bonding" and "antibonding" suggest, placing electrons in a bonding molecular orbital yields a stable covalent bond, whereas placing electrons in an antibonding molecular orbital results in an unstable bond.

In the bonding molecular orbital the electron density is greatest between the nuclei of the bonding atoms. In the antibonding molecular orbital, on the other hand, the electron density decreases to zero between the nuclei. We can understand this distinction if we recall that electrons in orbitals have wave characteristics. A property unique to waves allows waves of the same type to interact in such a way that the resultant wave has either an enhanced amplitude or a diminished amplitude. In the former case, we call the interaction *constructive interference;* in the latter case, it is *destructive interference* (Figure 10.21).

The formation of bonding molecular orbitals corresponds to constructive interference (the increase in amplitude is analogous to the buildup of electron density between the two nuclei). The formation of antibonding molecular orbitals corresponds to destructive interference (the decrease in amplitude is analogous to the decrease in electron density between the two nuclei). The constructive and destructive interactions between the two 1s orbitals in the H_2 molecule, then, lead to the formation of a sigma bonding molecular orbital (σ_{1s}) and a sigma antibonding molecular orbital $\sigma_{1s}^{\star}$:

a sigma bonding molecular orbital   a sigma antibonding molecular orbital

σ_{1s} $\sigma_{1s}^{\star}$

formed from 1s orbitals   formed from 1s orbitals

where the star denotes an antibonding molecular orbital.

In a **sigma molecular orbital** (bonding or antibonding) *the electron density is concentrated symmetrically around a line between the two nuclei of the bonding atoms.* Two electrons in a sigma molecular orbital form a sigma bond (see Section 10.5). Remember that a single covalent bond (such as H—H or F—F) is almost always a sigma bond.

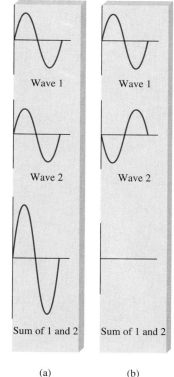

Wave 1   Wave 1

Wave 2   Wave 2

Sum of 1 and 2   Sum of 1 and 2

(a)   (b)

Figure 10.21 *Constructive interference (a) and destructive interference (b) of two waves of the same wavelength and amplitude.*

The two electrons in the sigma molecular orbital are paired. The Pauli exclusion principle applies to molecules as well as to atoms.

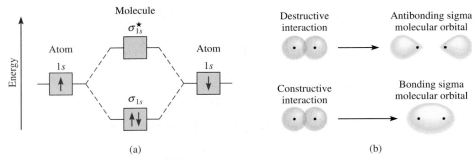

Figure 10.22 *(a) Energy levels of bonding and antibonding molecular orbitals in the H_2 molecule. Note that the two electrons in the σ_{1s} orbital must have opposite spins in accord with the Pauli exclusion principle. Keep in mind that the higher the energy of the molecular orbital, the less stable the electrons in that molecular orbital. (b) Constructive and destructive interactions between the two hydrogen 1s orbitals lead to the formation of a bonding and an antibonding molecular orbital. In the bonding molecular orbital, there is a buildup between the nuclei of electron density, which acts as a negatively charged "glue" to hold the positively charged nuclei together.*

Figure 10.22 shows the *molecular orbital energy level diagram*—that is, the relative energy levels of the orbitals produced in the formation of the H_2 molecule—and the constructive and destructive interactions between the two 1s orbitals. Notice that in the antibonding molecular orbital there is a *node* between the nuclei that signifies zero electron density. The nuclei are repelled by each other's positive charges, rather than held together. Electrons in the antibonding molecular orbital have higher energy (and less stability) than they would have in the isolated atoms. On the other hand, electrons in the bonding molecular orbital have less energy (and hence greater stability) than they would have in the isolated atoms.

Although we have used the hydrogen molecule to illustrate molecular orbital formation, the concept is equally applicable to other molecules. In the H_2 molecule we consider only the interaction between 1s orbitals; with more complex molecules we need to consider additional atomic orbitals as well. Nevertheless, for all s orbitals, the process is the same as for 1s orbitals. Thus, the interaction between two 2s or 3s orbitals can be understood in terms of the molecular orbital energy level diagram and the formation of bonding and antibonding molecular orbitals similar to those shown in Figure 10.22.

For p orbitals, the process is more complex because they can interact with each other in two different ways. For example, two 2p orbitals can approach each other end-to-end to produce a sigma bonding and a sigma antibonding molecular orbital, as shown in Figure 10.23(a). Alternatively, the two p orbitals can overlap sideways to generate a bonding and an antibonding pi molecular orbital [Figure 10.23(b)].

a pi bonding
molecular orbital

a pi antibonding
molecular orbital

π_{2p}

$\pi_{2p}^{\star}$

formed from
2p orbitals

formed from
2p orbitals

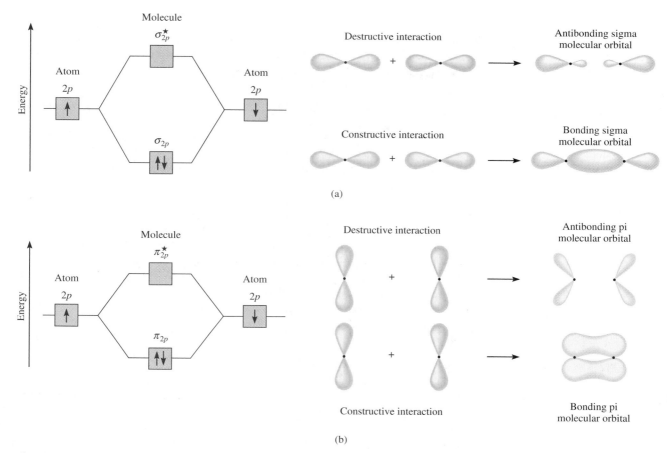

Figure 10.23 *Two possible interactions between two equivalent p orbitals and the corresponding molecular orbitals. (a) When the p orbitals overlap end-to-end, a sigma bonding and a sigma antibonding molecular orbital form. (b) When the p orbitals overlap side-to-side, a pi bonding and a pi antibonding molecular orbital form. Normally, a sigma bonding molecular orbital is more stable than a pi bonding molecular orbital, since side-to-side interaction leads to a smaller overlap of the p orbitals than does end-to-end interaction. We assume that the $2p_x$ orbitals take part in the sigma molecular orbital formation. The $2p_y$ and $2p_z$ orbitals can interact to form only π molecular orbitals. The behavior shown in (b) represents the interaction between the $2p_y$ orbitals or the $2p_z$ orbitals.*

In a *pi molecular orbital* (bonding or antibonding), *the electron density is concentrated above and below an imaginery line joining the two nuclei of the bonding atoms.* Two electrons in a pi molecular orbital form a pi bond (see Section 10.5). A double bond is almost always composed of a sigma bond and a pi bond; a triple bond is always a sigma bond plus two pi bonds.

Molecular Orbital Configurations

To understand properties of molecules, we must know how electrons are distributed among molecular orbitals. The procedure for determining the electron configuration of a molecule is analogous to the one we use to determine the electron configurations of atoms (see Section 7.8).

Rules Governing Molecular Electron Configuration and Stability

In order to write the electron configuration of a molecule, we must first arrange the molecular orbitals in order of increasing energy. Then we can use these guidelines to fill the molecular orbitals with electrons. The rules also help us understand the stabilities of the molecular orbitals.

- The number of molecular orbitals formed is always equal to the number of atomic orbitals combined.
- The more stable the bonding molecular orbital, the less stable the corresponding antibonding molecular orbital.
- The filling of molecular orbitals proceeds from low to high energies. In a stable molecule, the number of electrons in bonding molecular orbitals is always greater than that in antibonding molecular orbitals because we place electrons first in the lower-energy bonding molecular orbitals.
- Like an atomic orbital, each molecular orbital can accommodate up to two electrons with opposite spins in accordance with the Pauli exclusion principle.
- When electrons are added to molecular orbitals of the same energy, the most stable arrangement is predicted by Hund's rule; that is, electrons enter these molecular orbitals with parallel spins.
- The number of electrons in the molecular orbitals is equal to the sum of all the electrons on the bonding atoms.

Hydrogen and Helium Molecules

Later in this section we will study molecules formed by atoms of the second-period elements. Before we do, it will be instructive to predict the relative stabilities of the simple species H_2^+, H_2, He_2^+, and He_2, using the energy-level diagrams shown in Figure 10.24. The σ_{1s} and $\sigma_{1s}^\star$ orbitals can accommodate a maximum of four electrons. The total number of electrons increases from one for H_2^+ to four for He_2. The Pauli exclusion principle stipulates that each molecular orbital can accommodate a maximum of two electrons with opposite spins. We are concerned only with the ground-state electron configurations in these cases.

To evaluate the stabilities of these species we determine their **bond order,** defined as

$$\text{bond order} = \frac{1}{2}\left(\begin{array}{c}\text{number of electrons} \\ \text{in bonding MOs}\end{array} - \begin{array}{c}\text{number of electrons} \\ \text{in antibonding MOs}\end{array}\right) \quad (10.2)$$

The quantitative measure of the strength of a bond is bond energy (Section 9.8).

The bond order indicates the strength of a bond. For example, if there are two electrons in the bonding molecular orbital and none in the antibonding molecular orbital, the bond order is one, which means that there is one covalent bond and that the molecule is stable.

Figure 10.24 *Energy levels of the bonding and antibonding molecular orbitals in H_2^+, H_2, He_2^+, and He_2. In all these species, the molecular orbitals are formed by the interaction of two 1s orbitals.*

Note that the bond order can be a fraction, but a bond order of zero (or a negative value) means the bond has no stability and the molecule cannot exist. Bond order can be used only qualitatively for purposes of comparison. For example, a bonding sigma molecular orbital with two electrons and a bonding pi molecular orbital with two electrons would each have a bond order of one. Yet, these two bonds must differ in bond strength (and bond length) because of the differences in the extent of atomic orbital overlap.

We are ready now to make predictions about the stability of H_2^+, H_2, He_2^+, and He_2 (see Figure 10.24). The H_2^+ molecular ion has only one electron, in the σ_{1s} orbital. Because a covalent bond consists of two electrons in a bonding molecular orbital, H_2^+ has only half of one bond, or a bond order of $\frac{1}{2}$. Thus, we predict that the H_2^+ molecule may be a stable species. The electron configuration of H_2^+ is written as $(\sigma_{1s})^1$.

The H_2 molecule has two electrons, both of which are in the σ_{1s} orbital. According to our scheme, two electrons equal one full bond; therefore, the H_2 molecule has a bond order of one, or one full covalent bond. The electron configuration of H_2 is $(\sigma_{1s})^2$.

As for the He_2^+ molecular ion, we place the first two electrons in the σ_{1s} orbital and the third electron in the $\sigma_{1s}^\star$ orbital. Because the antibonding molecular orbital is destabilizing, we expect He_2^+ to be less stable than H_2. Roughly speaking, the instability resulting from the electron in the $\sigma_{1s}^\star$ orbital is balanced by one of the σ_{1s} electrons. The bond order is $\frac{1}{2}(2-1) = \frac{1}{2}$ and the overall stability of He_2^+ is similar to that of the H_2^+ molecule. The electron configuration of He_2^+ is $(\sigma_{1s})^2(\sigma_{1s}^\star)^1$.

In He_2 there would be two electrons in the σ_{1s} orbital and two electrons in the $\sigma_{1s}^\star$ orbital, so the molecule would have a bond order of zero and no net stability. The electron configuration of He_2 would be $(\sigma_{1s})^2(\sigma_{1s}^\star)^2$.

To summarize, we can arrange our examples in order of decreasing stability:

$$H_2 > H_2^+, \; He_2^+ > He_2$$

We know that the hydrogen molecule is a stable species. Our simple molecular orbital method predicts that H_2^+ and He_2^+ also possess some stability, because both have bond orders of $\frac{1}{2}$. Indeed, their existence has been confirmed by experiment. In turns out that H_2^+ is somewhat more stable than He_2^+, because there is only one electron in the hydrogen molecular ion and therefore it has no electron-electron repulsion. Furthermore, H_2^+ also has less nuclear repulsion than He_2^+. Our prediction about He_2 is that it would have no stability, but in 1993 He_2 gas was found to exist. The "molecule" is extremely unstable and has only a transient existence under specially created conditions.

The superscript in $(\sigma_{1s})^1$ indicates that there is one electron in the sigma bonding molecular orbital.

Homonuclear Diatomic Molecules of Second-Period Elements

We are now ready to study the ground-state electron configuration of molecules containing second-period elements. We will consider only the simplest case, that of **homonuclear diatomic molecules,** or *diatomic molecules containing atoms of the same elements.*

Figure 10.25 shows the molecular orbital energy level diagram for the first member of the second period, Li_2. These molecular orbitals are formed by the overlap of $1s$ and $2s$ orbitals. We will use this diagram to build up all the diatomic molecules, as we will see shortly.

The situation is more complex when the bonding also involves p orbitals. Two p orbitals can form either a sigma bond or a pi bond. Because there are three p orbitals for each atom of a second-period element, we know that one sigma and two pi molecular orbitals will result from the constructive interaction. The sigma molecular orbital is formed by the overlap of the $2p_x$ orbitals along the internuclear axis, that is, the x-axis. The $2p_y$ and $2p_z$ orbitals are perpendicular to the x-axis, and they will overlap

Figure 10.25 *Molecular orbital energy level diagram for the Li_2 molecule. The six electrons in Li_2 (Li's electron configuration is $1s^2 2s^1$) are in the σ_{1s}, $\sigma_{1s}^{\star}$, and σ_{2s} orbitals. Because there are two electrons each in σ_{1s} and $\sigma_{1s}^{\star}$ (just as in He_2), there is no net bonding or antibonding effect. Therefore, the single covalent bond in Li_2 is formed by the two electrons in the bonding molecular orbital σ_{2s}.*

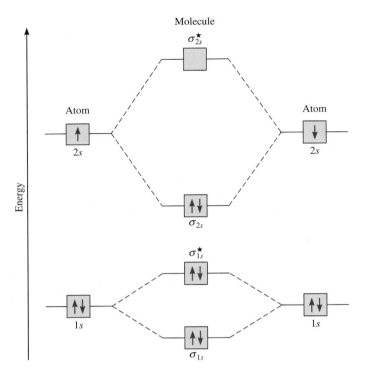

sideways to give two pi molecular orbitals. The molecular orbitals are called σ_{2p_x}, π_{2p_y}, and π_{2p_z} orbitals, where the subscripts indicate which atomic orbitals take part in forming the molecular orbitals. As shown in Figure 10.23, overlap of the two p orbitals is normally greater in a σ molecular orbital than in a π molecular orbital, so we would expect the former to be lower in energy. However, the energies of molecular orbitals actually increase as:

$$\sigma_{1s} < \sigma_{1s}^{\star} < \sigma_{2s} < \sigma_{2s}^{\star} < \pi_{2p_y} = \pi_{2p_z} < \sigma_{2p_x} < \pi_{2p_y}^{\star} = \pi_{2p_z}^{\star} < \sigma_{2p_x}^{\star}$$

The inversion of the σ_{2p_x} orbital and the π_{2p_y} and π_{2p_z} is due to the complex interaction between the 2s and 2p orbitals. As a result, the σ_{2p_x} orbital is *higher* in energy than the π_{2p_y} and π_{2p_z} orbitals for the lighter molecules B_2, C_2, and N_2, but *lower* in energy than the π_{2p_y} and π_{2p_z} orbitals for O_2 and F_2.

Table 10.5 summarizes the general properties of the stable homonuclear diatomic molecules of the second period.

The Oxygen Molecule (O_2)

As we saw earlier, valence bond theory does not account for the magnetic properties of the oxygen molecule. To show the two unpaired electrons on O_2, we need to draw an alternative to the resonance structure presented on p. 318:

$$\cdot \ddot{O} - \ddot{O} \cdot$$

This structure is unsatisfactory on at least two counts. First, it implies, the presence of a single covalent bond, but experimental evidence strongly suggests that there is a double bond in this molecule. Second, it places seven valence electrons around each oxygen atom, a violation of the octet rule.

Table 10.5 | Properties of Homonuclear Diatomic Molecules of the Second-Period Elements*

Orbital	Li₂	B₂	C₂	N₂	O₂	F₂	Orbital
$\sigma^\star_{2p_x}$	□	□	□	□	□	□	$\sigma^\star_{2p_x}$
$\pi^\star_{2p_y},\ \pi^\star_{2p_z}$	□ □	□ □	□ □	□ □	↑ | ↑	↑↓ | ↑↓	$\pi^\star_{2p_y},\ \pi^\star_{2p_z}$
σ_{2p_x} / $\pi_{2p_y},\ \pi_{2p_z}$	□	□	□	↑↓	↑↓ | ↑↓	↑↓ | ↑↓	$\pi_{2p_y},\ \pi_{2p_z}$
$\pi_{2p_y},\ \pi_{2p_z}$ / σ_{2p_x}	□ □	↑ | ↑	↑↓ | ↑↓	↑↓ | ↑↓	↑↓	↑↓	σ_{2p_x}
$\sigma^\star_{2s}$	□	↑↓	↑↓	↑↓	↑↓	↑↓	$\sigma^\star_{2s}$
σ_{2s}	↑↓	↑↓	↑↓	↑↓	↑↓	↑↓	σ_{2s}
Bond order	1	1	2	3	2	1	
Bond length (pm)	267	159	131	110	121	142	
Bond energy (kJ/mol)	104.6	288.7	627.6	941.4	498.7	156.9	
Magnetic properties	Diamagnetic	Paramagnetic	Diamagnetic	Diamagnetic	Paramagnetic	Diamagnetic	

*For simplicity the σ_{1s} and $\sigma^\star_{1s}$ orbitals are omitted. These two orbitals hold a total of four electrons. Remember that for O_2 and F_2, σ_{2p_x} is lower in energy than π_{2p_y} and π_{2p_z}.

The ground-state electron configuration of O is $1s^2 2s^2 2p^4$; thus there are 16 electrons in O_2. Using the order of increasing energies of the molecular orbitals from Table 10.5, we write the ground-state electron configuration of O_2 as

$$(\sigma_{1s})^2(\sigma^\star_{1s})^2(\sigma_{2s})^2(\sigma^\star_{2s})^2(\sigma_{2p_x})^2(\pi_{2p_y})^2(\pi_{2p_z})^2(\pi^\star_{2p_y})^1(\pi^\star_{2p_z})^1$$

According to Hund's rule, the last two electrons enter the $\pi^\star_{2p_y}$ and $\pi^\star_{2p_z}$ orbitals with parallel spins. Ignoring the σ_{1s} and σ_{2s} orbitals (because their net effects on bonding are zero), we calculate the bond order of O_2 using Equation (10.2):

$$\text{bond order} = \tfrac{1}{2}(6 - 2) = 2$$

Therefore, the O_2 molecule has a bond order of 2 and is paramagnetic, a prediction that corresponds to experimental observations.

Example 10.6 The N_2^+ ion can be prepared by bombarding the N_2 molecule with fast-moving electrons. Predict these properties of N_2^+: (a) electron configuration, (b) bond order, (c) magnetic character, and (d) bond length relative to the bond length of N_2 (is it longer or shorter?).

Reasoning and Solution From Table 10.5 we can deduce the properties of ions generated from the homonuclear diatomic molecules.

(a) Because N_2^+ has one fewer electron than N_2, its electron configuration is

$$(\sigma_{1s})^2(\sigma^\star_{1s})^2(\sigma_{2s})^2(\sigma^\star_{2s})^2(\pi_{2p_y})^2(\pi_{2p_z})^2(\sigma_{2p_x})^1$$

(b) The bond order of N_2^+ is found by using Equation (10.2):

$$\text{bond order} = \tfrac{1}{2}(9 - 4) = 2.5$$

(c) N_2^+ has one unpaired electron, so it is paramagnetic.

(d) Because the electrons in the bonding molecular orbitals are responsible for holding the atoms together, N_2^+ should have a weaker and, therefore, longer bond than N_2. (In fact, the bond length of N_2^+ is 112 pm, compared with 110 pm for N_2.)

Practice Exercise Which of these species has a longer bond length: F_2 or F_2^-?

Similar problems: 10.55, 10.56.

Summary

The VSEPR model for predicting molecular geometry is based on the assumption that valence-shell electron pairs repel one another and tend to stay as far apart as possible. According to the VSEPR model, molecular geometry can be predicted from the number of bonding electron pairs and lone pairs. Lone pairs repel other pairs more strongly than bonding pairs do and thus distort bond angles from those of the ideal geometry.

The dipole moment is a measure of the charge separation in molecules containing atoms of different electronegativities. The dipole moment of a molecule is the resultant of whatever bond moments are present in a molecule. Information about molecular geometry can be obtained from dipole moment measurements.

In valence bond theory, hybridized atomic orbitals are formed by the combination and rearrangement of orbitals of the same atom. The hybridized orbitals are all of equal energy and electron density, and the number of hybridized orbitals is equal to the number of pure atomic orbitals that combine. Valence-shell expansion can be explained by assuming hybridization of s, p, and d orbitals.

In sp hybridization, the two hybrid orbitals lie in a straight line; in sp^2 hybridization, the three hybrid orbitals are directed toward the corners of a triangle; in sp^3 hybridization, the four hybrid orbitals are directed toward the corners of a tetrahedron; in sp^3d hybridization, the five hybrid orbitals are directed toward the corners of a trigonal bipyramid; in sp^3d^2 hybridization, the six hybrid orbitals are directed toward the corners of an octahedron.

In an sp^2-hybridized atom (for example, carbon), the one unhybridized p orbital can form a pi bond with another p orbital. A carbon-carbon double bond consists of a sigma bond and a pi bond. In an sp-hybridized carbon atom, the two unhybridized p orbitals can form two pi bonds with two p orbitals on another atom (or atoms). A carbon-carbon triple bond consists of one sigma bond and two pi bonds.

Molecular orbital theory describes bonding in terms of the combination and rearrangement of atomic orbitals to form orbitals that are associated with the molecule as a whole. Bonding molecular orbitals increase electron density between the nuclei and are lower in energy than individual atomic orbitals. Antibonding molecular orbitals have a region of zero electron density between the nuclei, and an energy level higher than that of the individual atomic orbitals. Molecules are stable if the number of electrons in bonding molecular orbitals is greater than that in antibonding molecular orbitals.

We write electron configurations for molecular orbitals as we do for atomic orbitals, referring to the Pauli exclusion principle and Hund's rule.

Key Words

Antibonding molecular orbital p. 319	Homonuclear diatomic molecule p. 323	Pi bond (π bond) p. 316	Sigma molecular orbital p. 319
Bond order p. 322	Hybrid orbital p. 307	Pi molecular orbital p. 321	Valence shell p. 292
Bonding molecular orbital p. 319	Hybridization p. 307	Polar molecule p. 302	Valence-shell electron-pair repulsion (VSEPR) model p. 292
Dipole moment (μ) p. 302	Molecular orbital p. 318 Nonpolar molecule p. 302	Sigma bond (σ bond) p. 316	

Questions and Problems

Molecular Geometry

Review Questions

10.1 How is the geometry of a molecule defined and why is the study of molecular geometry important?

10.2 Sketch the shape of a linear triatomic molecule, a trigonal planar molecule containing four atoms, a tetrahedral molecule, a trigonal bipyramidal molecule, and an octahedral molecule. Give the bond angles in each case.

10.3 How many atoms are directly bonded to the central atom in a tetrahedral molecule, a trigonal bipyramidal molecule, and an octahedral molecule?

10.4 Discuss the basic features of the VSEPR model. Explain why the magnitude of repulsion decreases in this order: lone pair-lone pair > lone pair-bonding pair > bonding pair-bonding pair.

10.5 In the trigonal bipyramidal arrangement, why does a lone pair occupy an equatorial position rather than an axial position?

10.6 The geometry of CH_4 could be square planar, with the four H atoms at the corners of a square and the C atom at the center of the square. Sketch this geometry and compare its stability with that of a tetrahedral CH_4 molecule.

Problems

10.7 Predict the geometries of these species using the VSEPR method: (a) PCl_3, (b) $CHCl_3$, (c) SiH_4, (d) $TeCl_4$.

10.8 Predict the geometries of these species: (a) $AlCl_3$, (b) $ZnCl_2$, (c) $ZnCl_4^{2-}$.

10.9 Predict the geometry of these molecules and ion using the VSEPR method: (a) $HgBr_2$, (b) N_2O (arrangement of atoms is NNO), (c) SCN^- (arrangement of atoms is SCN).

10.10 Predict the geometries of these ions: (a) NH_4^+, (b) NH_2^-, (c) CO_3^{2-}, (d) ICl_2^-, (e) ICl_4^-, (f) AlH_4^-, (g) $SnCl_5^-$, (h) H_3O^+, (i) BeF_4^{2-}.

10.11 Describe the geometry around each of the three central atoms in the CH_3COOH molecule.

10.12 Which of these species are tetrahedral? $SiCl_4$, SeF_4, XeF_4, CI_4, $CdCl_4^{2-}$

Dipole Moments

Review Questions

10.13 Define dipole moment. What are the units and symbol for dipole moment?

10.14 What is the relationship between the dipole moment and bond moment? How is it possible for a molecule to have bond moments and yet be nonpolar?

10.15 Explain why an atom cannot have a permanent dipole moment.

10.16 The bonds in beryllium hydride (BeH_2) molecules are polar, and yet the dipole moment of the molecule is zero. Explain.

Problems

10.17 Referring to Table 10.3, arrange the following molecules in order of increasing dipole moment: H_2O, H_2S, H_2Te, H_2Se.

10.18 The dipole moments of the hydrogen halides decrease from HF to HI (see Table 10.3). Explain this trend.

10.19 List these molecules in order of increasing dipole moment: H_2O, CBr_4, H_2S, HF, NH_3, CO_2.

10.20 Does the molecule OCS have a higher or lower dipole moment than CS_2?

10.21 Which of these molecules has a higher dipole moment?

(a)   (b)

10.22 Arrange these compounds in order of increasing dipole moment:

(a)   (b)   (c)   (d)

Valence Bond Theory

Review Questions

10.23 What is valence bond theory? How does it differ from the Lewis concept of chemical bonding?

10.24 Use valence bond theory to explain the bonding in Cl_2 and HCl. Show how the atomic orbitals overlap when a bond is formed.

10.25 Draw a potential energy curve for the bond formation in F_2.

Hybridization

Review Questions

10.26 What is the hybridization of atomic orbitals? Why is it impossible for an isolated atom to exist in the hybridized state?

10.27 How does a hybrid orbital differ from a pure atomic orbital? Can two $2p$ orbitals of an atom hybridize to give two hybridized orbitals?

10.28 What is the angle between these two hybrid orbitals on the same atom? (a) sp and sp hybrid orbitals, (b) sp^2 and sp^2 hybrid orbitals, (c) sp^3 and sp^3 hybrid orbitals

10.29 How would you distinguish between a sigma bond and a pi bond?

10.30 Which of these pairs of atomic orbitals of adjacent nuclei can overlap to form a sigma bond? Which overlap to form a pi bond? Which cannot overlap (no bond)? Consider the x-axis to be the internuclear axis, that is, the line joining the nuclei of the two atoms. (a) $1s$ and $1s$, (b) $1s$ and $2p_x$, (c) $2p_x$ and $2p_y$, (d) $3p_y$ and $3p_y$, (e) $2p_x$ and $2p_x$, (f) $1s$ and $2s$

Problems

10.31 Describe the bonding scheme of the AsH_3 molecule in terms of hybridization.

10.32 What is the hybridization state of Si in SiH_4 and in H_3Si—SiH_3?

10.33 Describe the change in hybridization (if any) of the Al atom in this reaction:

$$AlCl_3 + Cl^- \longrightarrow AlCl_4^-$$

10.34 Consider the reaction

$$BF_3 + NH_3 \longrightarrow F_3B-NH_3$$

Describe the changes in hybridization (if any) of the B and N atoms as a result of this reaction.

10.35 What hybrid orbitals are used by nitrogen atoms in these species? (a) NH_3, (b) H_2N—NH_2, (c) NO_3^-

10.36 What are the hybrid orbitals of the carbon atoms in these molecules?

(a) H_3C—CH_3
(b) H_3C—CH=CH_2

(c) CH_3—C≡C—CH_2OH
(d) CH_3CH=O
(e) CH_3COOH

10.37 Specify which hybrid orbitals are used by carbon atoms in these species: (a) CO, (b) CO_2, (c) CN^-.

10.38 What is the hybridization state of the central N atom in the azide ion, N_3^-? (Arrangement of atoms: NNN.)

10.39 The allene molecule H_2C=C=CH_2 is linear (the three C atoms lie on a straight line). What are the hybridization states of the carbon atoms? Draw diagrams to show the formation of sigma bonds and pi bonds in allene.

10.40 Describe the hybridization of phosphorus in PF_5.

10.41 How many sigma bonds and pi bonds are there in each of these molecules?

(a)   (b)   (c)

10.42 How many pi bonds and sigma bonds are there in the tetracyanoethylene molecule?

Molecular Orbital Theory

Review Questions

10.43 What is molecular orbital theory? How does it differ from valence bond theory?

10.44 Define these terms: bonding molecular orbital, antibonding molecular orbital, pi molecular orbital, sigma molecular orbital.

10.45 Sketch the shapes of these molecular orbitals: σ_{1s}, $\sigma_{1s}^\star$, π_{2p}, and $\pi_{2p}^\star$. How do their energies compare?

10.46 Explain the significance of bond order. Can bond order be used for quantitative comparisons of the strengths of chemical bonds?

Problems

10.47 Explain in molecular orbital terms the changes in H—H internuclear distance that occur as the molecular H_2 is ionized first to H_2^+ and then to H_2^{2+}.

10.48 The formation of H_2 from two H atoms is an energetically favorable process. Yet statistically there is less than a 100 percent chance that any two H atoms will undergo the reaction. Apart from energy considerations, how would you account for this observation based on the electron spins in the two H atoms?

10.49 Draw a molecular orbital energy level diagram for each of these species: He_2, HHe, He_2^+. Compare their relative stabilities in terms of bond orders. (Treat HHe as a diatomic molecule with three electrons.)

10.50 Arrange these species in order of increasing stability: Li_2, Li_2^+, Li_2^-. Justify your choice with a molecular orbital energy level diagram.

10.51 Use molecular orbital theory to explain why the Be_2 molecule does not exist.

10.52 Which of these species has a longer bond, B_2 or B_2^+? Explain in terms of molecular orbital theory.

10.53 Acetylene (C_2H_2) has a tendency to lose two protons (H^+) and form the carbide ion (C_2^{2-}), which is present in a number of ionic compounds, such as CaC_2 and MgC_2. Describe the bonding scheme in the C_2^{2-} ion in terms of molecular orbital theory. Compare the bond order in C_2^{2-} with that in C_2.

10.54 Compare the Lewis and molecular orbital treatments of the oxygen molecule.

10.55 Explain why the bond order of N_2 is greater than that of N_2^+, but the bond order of O_2 is less than that of O_2^+.

10.56 Compare the relative stability of these species and indicate their magnetic properties (that is, diamagnetic or paramagnetic): O_2, O_2^+, O_2^- (superoxide ion), O_2^{2-} (peroxide ion).

10.57 Use molecular orbital theory to compare the relative stabilities of F_2 and F_2^+.

10.58 A single bond is almost always a sigma bond, and a double bond is almost always made up of a sigma bond and a pi bond. There are very few exceptions to this rule. Show that the B_2 and C_2 molecules are examples of the exceptions.

Additional Problems

10.59 Which of these species is not likely to have a tetrahedral shape? (a) $SiBr_4$, (b) NF_4^+, (c) SF_4, (d) $BeCl_4^{2-}$, (e) BF_4^-, (f) $AlCl_4^-$

10.60 Draw the Lewis structure of mercury(II) bromide. Is this molecule linear or bent? How would you establish its geometry?

10.61 Sketch the bond moments and resultant dipole moments for these molecules: H_2O, PCl_3, XeF_4, PCl_5, SF_6.

10.62 Although both carbon and silicon are in Group 4A, very few $Si=Si$ bonds are known. Account for the instability of silicon-to-silicon double bonds in general. (*Hint:* Compare the atomic radii of C and Si in Figure 8.3. What effect would the larger size have on pi bond formation?)

10.63 Predict the geometry of sulfur dichloride (SCl_2) and the hybridization of the sulfur atom.

10.64 Antimony pentafluoride, SbF_5, reacts with XeF_4 and XeF_6 to form ionic compounds, $XeF_3^+SbF_6^-$ and $XeF_5^+SbF_6^-$.

Describe the geometries of the cations and anion in these two compounds.

10.65 Draw Lewis structures and give the other information requested for the following molecules: (a) BF_3. Shape: planar or nonplanar? (b) ClO_3^-. Shape: planar or nonplanar? (c) H_2O. Show the direction of the resultant dipole moment. (d) OF_2. Polar or nonpolar molecule? (e) SeO_2. Estimate the OSeO bond angle.

10.66 Predict the bond angles for these molecules: (a) $BeCl_2$, (b) BCl_3, (c) CCl_4, (d) CH_3Cl, (e) Hg_2Cl_2 (arrangement of atoms: ClHgHgCl), (f) $SnCl_2$, (g) H_2O_2, (h) SnH_4.

10.67 Briefly compare the VSEPR and hybridization approaches to the study of molecular geometry.

10.68 Describe the hybridization state of arsenic in arsenic pentafluoride (AsF_5).

10.69 Draw Lewis structures and give the other information requested for these: (a) SO_3. Polar or nonpolar molecule? (b) PF_3. Polar or nonpolar molecule? (c) F_3SiH. Show the direction of the resultant dipole moment. (d) SiH_3^-. Planar or pyramidal shape? (e) Br_2CH_2. Polar or nonpolar molecule?

10.70 Which of these molecules are linear? ICl_2^-, IF_2^+, OF_2, SnI_2, $CdBr_2$

10.71 Draw the Lewis structure for the $BeCl_4^{2-}$ ion. Predict its geometry and describe the hybridization state of the Be atom.

10.72 The N_2F_2 molecule can exist in either of these two forms:

(a) What is the hybridization of N in the molecule?
(b) Which structure has a dipole moment?

10.73 Cyclopropane (C_3H_6) has the shape of a triangle in which a C atom is bonded to two H atoms and two other C atoms at each corner. Cubane (C_8H_8) has the shape of a cube in which a C atom is bonded to one H atom and three other C atoms at each corner. (a) Draw Lewis structures of these molecules. (b) Compare the CCC angles in these molecules with those predicted for an sp^3-hybridized C atom. (c) Would you expect these molecules to be easy to make?

10.74 The compound 1,2-dichloroethane ($C_2H_4Cl_2$) is nonpolar, while *cis*-dichloroethylene ($C_2H_2Cl_2$) has a dipole moment:

1,2-dichloroethane   *cis*-dichloroethylene

The reason for the difference is that groups connected by a single bond can rotate with respect to each other, but no rotation occurs when a double bond connects the groups. On the basis of bonding considerations, explain why rotation occurs in 1,2-dichloroethane but not in *cis*-dichloroethylene.

10.75 Does the following molecule have a dipole moment?

$$Cl\diagdown \atop H \diagup C=C=C \diagup H \atop \diagdown Cl$$

(*Hint:* See the answer to Problem 10.39.)

10.76 The compounds carbon tetrachloride (CCl_4) and silicon tetrachloride ($SiCl_4$) are similar both in geometry and hybridization. However, CCl_4 does not react with water but $SiCl_4$ does. Explain the difference in their chemical reactivities. (*Hint:* The first step of the reaction is believed to be the addition of a water molecule to the Si atom in $SiCl_4$.)

10.77 Write the ground-state electron configuration for B_2. Is the molecule diamagnetic or paramagnetic?

10.78 What are the hybridization states of the C and N atoms in this molecule?

$$\begin{array}{c} NH_2 \\ | \\ C \\ N \diagup \diagdown C - H \\ \| \qquad \| \\ O = C \diagdown \diagup C - H \\ N \\ | \\ H \end{array}$$

10.79 Use molecular orbital theory to explain the difference between the bond energies of F_2 and F_2^- (see Problem 9.85).

10.80 The ionic character of the bond in a diatomic molecule can be estimated by the formula

$$\frac{\mu}{ed} \times 100\%$$

where μ is the experimentally measured dipole moment (in C m), e is the electronic charge, and d is the bond length in meters. (The quantity ed is the hypothetical dipole moment for the case in which the transfer of an electron from the less electronegative to the more electronegative atom is complete.) Given that the dipole moment and bond length of HF are 1.92 D and 91.7 pm, respectively, calculate the percent ionic character of the molecule.

10.81 The geometries discussed in this chapter all lend themselves to fairly straightforward elucidation of bond angles. The exception is the tetrahedron, because its bond angles are hard to visualize. Consider the CCl_4 molecule, which has a tetrahedral geometry and is nonpolar. By equating the bond moment of a particular C—Cl bond to the resultant bond moments of the other three C—Cl bonds in opposite directions, show that the bond angles are all equal to 109.5°.

Special Problems

10.82 Progesterone is a hormone responsible for female sex characteristics. In the usual shorthand structure, each point where lines meet represents a C atom, and most H atoms are not shown. Draw the complete structure of the molecule, showing all C and H atoms. Indicate which C atoms are sp^2- and sp^3-hybridized.

$$\begin{array}{c} CH_3 \\ CH_3 \quad C=O \\ CH_3 \\ \\ O \end{array}$$

10.83 Greenhouse gases absorb (and trap) outgoing infared radiation (heat) from Earth and contribute to global warming. The molecule of a greenhouse gas either possesses a permanent dipole moment or has a changing dipole moment during its vibrational motions. Consider three of the vibrational modes of carbon dioxide

$$\overset{\leftarrow}{O}=C=\vec{O} \qquad \vec{O}=\overset{\leftarrow}{C}=\vec{O} \qquad \overset{\uparrow}{O}=\underset{\downarrow}{C}=\overset{\uparrow}{O}$$

where the arrows indicate the movement of the atoms. (During a complete cycle of vibration, the atoms move toward one extreme position and then reverse their direction to the other extreme position.) Which of the above vibrations are responsible for CO_2 to behave as a greenhouse gas? Which of the following molecules can act as a greenhouse gas: N_2, O_2, CO, NO_2, and N_2O?

Answers to Practice Exercises

10.1 (a) Tetrahedral, (b) linear, (c) trigonal planar. **10.2** No. **10.3** (a) sp^3, (b) sp^2. **10.4** sp^3d^2. **10.5** The C atom is sp-hybridized. It forms a sigma bond with the H atom and another sigma bond with the N atom. The two unhybridized p orbitals on the C atom are used to form two pi bonds with the N atom. The lone pair on the N atom is placed in the sp orbital. **10.6** F_2^-.

Can you imagine that there is a kind of ice that burns in air? Yes, there is such a thing. It is called *methane hydrate,* and there is enough of it to meet America's energy needs for years. But scientists have yet to figure out how to mine it without causing an environmental disaster.

Bacteria in the sediments beneath the ocean floor consume organic material and generate methane gas. Under high pressure and low temperature conditions, methane forms methane hydrate, which consists of single molecules of the natural gas trapped within

Methane hydrate. The methane molecule is trapped in a cage of frozen water molecules (blue spheres) held together by hydrogen bonds.

crystalline cages formed by frozen water molecules. A lump of methane hydrate looks like a gray ice cube, but if a lighted match is put to it, it will burn.

Oil companies have known about methane hydrate since the 1930s, when they began using high-pressure pipelines to transport natural gas in cold climates. Unless water is carefully removed before the gas enters the pipeline, chunks of methane hydrate will impede the flow of gas.

The total reserve of the methane hydrate in the world's oceans is estimated to be 10^{13} tons of carbon content, about twice the amount of carbon in all the coal, oil, and natural gas on land. However, harvesting the energy stored in methane hydrate presents a tremendous engineering challenge. It is believed that methane hydrate acts as a kind of cement to keep the ocean floor sediments together. Tampering with the hydrate deposits could cause underwater landslides, leading to the discharge of methane into the atmosphere. This event could have serious consequences for the environment, because methane is a potent greenhouse gas; that is, it has the ability to warm Earth's atmosphere. In fact, scientists have speculated that the abrupt release of methane hydrates may have hastened the end of the last Ice Age about 10,000 years ago. As the great blanket of continental ice melted, global sea levels swelled by more than 90 m, submerging Arctic regions rich in hydrate deposits. The relatively warm ocean water would have melted the hydrates, unleashing tremendous amounts of methane, which led to global warming.

Methane hydrate burning in air.

Introduction to Organic Chemistry   **Chapter 11**

ESSENTIAL CONCEPTS

 Organic Compounds Organic compounds contain primarily carbon and hydrogen atoms, plus nitrogen, oxygen, sulfur, and atoms of other elements. The parent compounds of all organic compounds are the hydrocarbons—the alkanes (containing only single bonds), the alkenes (containing carbon-carbon double bonds), the alkynes (containing carbon-carbon triple bonds), and the aromatic hydrocarbons (containing the benzene ring).

 Functional Groups The reactivity of organic compounds can be reliably predicted by the presence of functional groups, which are groups of atoms that are largely responsible for the chemical behavior of the compounds.

 Chirality Certain organic compounds can exist as nonsuperimposable mirror-image twins. These compounds are said to be chiral or optically active. The pure optical isomer of a compound can rotate plane-polarized light. Optical isomers have identical physical properties but exhibit different chemical properties toward another chiral substance.

11.1 Classes of Organic Compounds

Common elements in organic compounds.

Carbon can form more compounds than any other element because carbon atoms are able not only to form single, double, and triple carbon-carbon bonds, but also to link up with each other in chains and ring structures. *The branch of chemistry that deals with carbon compounds is* **organic chemistry.**

Classes of organic compounds can be distinguished according to functional groups they contain. A **functional group** is *a group of atoms that is largely responsible for the chemical behavior of the parent molecule.* Different molecules containing the same kind of functional group or groups undergo similar reactions. Thus, by learning the characteristic properties of a few functional groups, we can study and understand the properties of many organic compounds. In the second half of this chapter we will discuss the functional groups known as alcohols, ethers, aldehydes and ketones, carboxylic acids, and amines.

All organic compounds are derived from a group of compounds known as **hydrocarbons** because they are *made up of only hydrogen and carbon.* On the basis of structure, hydrocarbons are divided into two main classes—aliphatic and aromatic. **Aliphatic hydrocarbons** *do not contain the benzene group, or the benzene ring,* whereas **aromatic hydrocarbons** *contain one or more benzene rings.*

11.2 Aliphatic Hydrocarbons

Aliphatic hydrocarbons are divided into alkanes, alkenes, and alkynes, discussed in this section (Figure 11.1).

Alkanes

Alkanes *have the general formula* C_nH_{2n+2}, *where* $n = 1, 2, \ldots$. The essential characteristic of alkane hydrocarbon molecules is that *only single covalent bonds are present.* The alkanes are known as **saturated hydrocarbons** because they *contain the maximum number of hydrogen atoms that can bond with the number of carbon atoms present.*

> For a given number of carbon atoms, the saturated hydrocarbon contains the largest number of hydrogen atoms.

The simplest alkane (that is, with $n = 1$) is methane CH_4, which is a natural product of the anaerobic bacterial decomposition of vegetable matter under water. Because

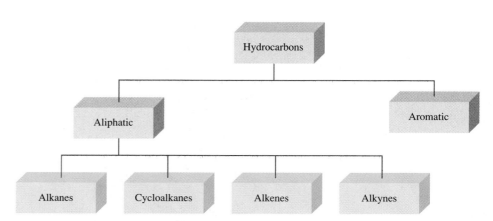

Figure 11.1 *Classification of hydrocarbons.*

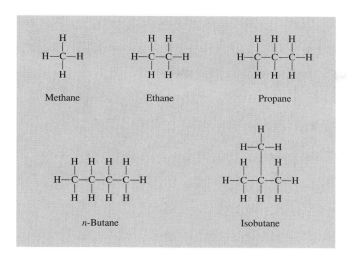

Figure 11.2 *Structures of the first four alkanes. Note that butane can exist in two structurally different forms, called structural isomers.*

Methane　　Ethane　　Propane

n-Butane　　　Isobutane

it was first collected in marshes, methane became known as "marsh gas." A rather improbable but proven source of methane is termites. When these voracious insects consume wood, the microorganisms that inhabit their digestive system break down cellulose (the major component of wood) into methane, carbon dioxide, and other compounds. An estimated 170 million tons of methane are produced annually by termites! It is also produced in some sewage treatment processes. Commercially, methane is obtained from natural gas.

Termites are a natural source of methane.

Figure 11.2 shows the structures of the first four alkanes ($n = 1$ to $n = 4$). Natural gas is a mixture of methane, ethane, and a small amount of propane. We discussed the bonding scheme of methane in Chapter 10. Indeed the carbon atoms in all the alkanes can be assumed to be sp^3-hybridized. The structures of ethane and propane are straightforward, for there is only one way to join the carbon atoms in these molecules. Butane, however, has two possible bonding schemes resulting in the **structural isomers** *n*-butane (*n* stands for normal) and isobutane, *molecules that have the same molecular formula, but different structures.* Alkanes such as the structural isomers of butane are described as having the straight chain or branched chain structures. *n*-Butane is a straight-chain alkane because the carbon atoms are joined along one line. In a branched-chain alkane like isobutane, one or more carbon atoms are bonded to at least three other carbon atoms.

In the alkane series, as the number of carbon atoms increases, the number of structural isomers increases rapidly. For example, butane, C_4H_{10}, has two isomers; decane, $C_{10}H_{22}$, has 75 isomers; and the alkane $C_{30}H_{62}$ has over 400 million, or 4×10^8, possible isomers! Obviously, most of these isomers do not exist in nature nor have they been synthesized. Nevertheless, the numbers help to explain why carbon is found in so many more compounds than any other element.

Example 11.1 How many structural isomers can be identified for pentane, C_5H_{12}?

Reasoning and Solution For small alkanes (eight or fewer carbon atoms), it is relatively easy to determine the number of structural isomers by trial and error. The first step is to write the straight-chain structure:

n-pentane

2-methylbutane

2,2-dimethylpropane

Similar problems: 11.11, 11.12.

$$\begin{array}{ccccc}
& H & H & H & H & H \\
& | & | & | & | & | \\
H- & C- & C- & C- & C- & C-H \\
& | & | & | & | & | \\
& H & H & H & H & H
\end{array}$$

n-pentane
(b.p. 36.1°C)

The second structure, by necessity, must be a branched chain:

2-methylbutane
(b.p. 27.9°C)

Yet another branched-chain structure is possible:

2,2-dimethylpropane
(b.p. 9.5°C)

We can draw no other structure for an alkane having the molecular formula C_5H_{12}. Thus pentane has three structural isomers, in which the numbers of carbon and hydrogen atoms remain unchanged despite the differences in structure.

Practice Exercise   How many structural isomers are there in the alkane C_6H_{14}?

Table 11.1 shows the melting and boiling points of the straight-chain isomers of the first 10 alkanes. The first four are gases at room temperature; and pentane through decane are liquids. As molecular size increases, so does the boiling point.

Drawing Chemical Structures

Shortly we will discuss the nomenclature of alkanes.

Before proceeding further, it is useful to learn different ways of drawing the structure of organic compounds. Consider the alkane 2-methylbutane (C_5H_{12}). To see how atoms are connected in this molecule, we need to first write the "expanded" molecular formula, $CH_3CH(CH_3)CH_2CH_3$, and then draw its structural formula, shown in Figure 11.3(a). While informative, this structure is time-consuming to draw. Therefore, chemists have devised ways to simplify the representation. Figure 11.3(b) is an abbreviated version and the structure shown in Figure 11.3(c) is called the *skeletal structure* in which all the C and H letters are omitted. A carbon atom is assumed to be at each intersection of two lines (bonds) and at the end of each line. Because every C atom forms four bonds, we can always deduce the number of H atoms bonded to any C atom. One of the two end CH_3 groups is represented by a vertical line. What is lacking in these structures, however, is the three dimensionality of the molecule, which is shown by the molecular model in Figure 11.3(d). Depending on the purpose of discussion, any of these representations can be used to describe the properties of the molecule.

Skeletal structure is the simplest structure. Atoms other than C and H must be shown explicitly in a skeletal structure.

Table 11.1 | The First 10 Straight-Chain Alkanes

Name of Hydrocarbon	Molecular Formula	Number of Carbon Atoms	Melting Point (°C)	Boiling Point (°C)
Methane	CH_4	1	−182.5	−161.6
Ethane	CH_3—CH_3	2	−183.3	−88.6
Propane	CH_3—CH_2—CH_3	3	−189.7	−42.1
Butane	CH_3—$(CH_2)_2$—CH_3	4	−138.3	−0.5
Pentane	CH_3—$(CH_2)_3$—CH_3	5	−129.8	36.1
Hexane	CH_3—$(CH_2)_4$—CH_3	6	−95.3	68.7
Heptane	CH_3—$(CH_2)_5$—CH_3	7	−90.6	98.4
Octane	CH_3—$(CH_2)_6$—CH_3	8	−56.8	125.7
Nonane	CH_3—$(CH_2)_7$—CH_3	9	−53.5	150.8
Decane	CH_3—$(CH_2)_8$—CH_3	10	−29.7	174.0

Crude oil is a source of hydro-carbons.

Conformation of Ethane

Molecular geometry gives the spatial arrangement of atoms in a molecule. However, atoms are not held rigidly in position because of internal molecular motions. For this reason, even a simple molecule like ethane may be structurally more complicated than we think.

The two C atoms in ethane are sp^3-hybridized and they are joined by a sigma bond (see Section 10.4). Sigma bonds have cylindrical symmetry, that is, the overlap of the sp^3 orbitals is the same regardless of the rotation of the C—C bond. Yet this bond rotation is not totally free because of the interactions between the H atoms on different C atoms. Figure 11.4 shows the two extreme *conformations* of ethane. **Conformations** are *different spatial arrangements of a molecule that are generated by rotation about single bonds.* In the staggered conformation, the three H atoms on one C atom are pointing away from the H atoms on the other C atom, whereas in the eclipsed conformation the two groups of H atoms are aligned parallel to one another.

Figure 11.3 *Different representations of 2-methylbutane. (a) Structural formula. (b) Abbreviated structure. (c) Skeletal structure. (d) Molecular model.*

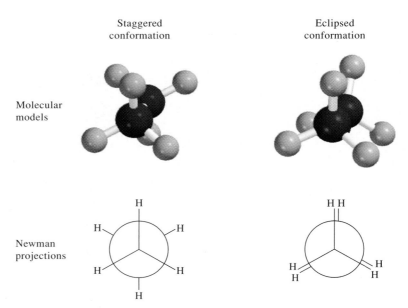

Staggered conformation

Eclipsed conformation

Molecular models

Newman projections

Figure 11.4 *Molecular models and Newman projections of the staggered and eclipsed conformations of ethane. The dihedral angle in the staggered form is 60° and that in the eclipsed form is 0°. The C—C bond is rotated slightly in the Newman projection of the eclipsed form in order to see the H atoms attached to the back C atom. The proximity of the H atoms on the two C atoms in the eclipsed form results in a greater repulsion, and hence its instability relative to the staggered form. However, recent theoretical analysis shows that the situation is more complicated than just the repulsion. Nevertheless, in general we can always assume that the staggered conformation is more stable than the eclipsed conformation.*

A simpler and effective way of viewing these two conformations is by using the Newman projection, also shown in Figure 11.4. Look at the C—C bond end-on. The two C atoms are represented by a circle. The C—H bonds attached to the front C atom are the lines going to the center of the circle, and the C—H bonds attached to the rear C atom appear as lines going to the edge of the circle. The eclipsed form of ethane is less stable than the staggered form. Figure 11.5 shows the variation in the potential energy of ethane as a function of rotation. The rotation of one CH_3 group relative to the other is described in terms of the angle between the C—H bonds on front and back carbons, called the *dihedral* angle. The dihedral angle for the first eclipsed conformation is zero. A clockwise rotation of 60° about the C—C bond generates a staggered conformation, which is converted to another eclipsed conformation by a similar rotation and so on.

Conformational analysis of molecules is of great importance in understanding the details of reactions ranging from simple hydrocarbons to proteins and DNAs.

Alkane Nomenclature

The names of the straight-chain alkanes are fairly easy to remember. As Table 11.1 shows, except for the first four members, the number of carbon atoms in each alkane is reflected in the Greek prefix (see Table 2.4). When one or more hydrogen atoms are replaced by other groups, the name of the compound must indicate the locations of carbon atoms where replacements are made. Let us first consider groups that are themselves derived from alkanes. For example, when a hydrogen atom is removed from

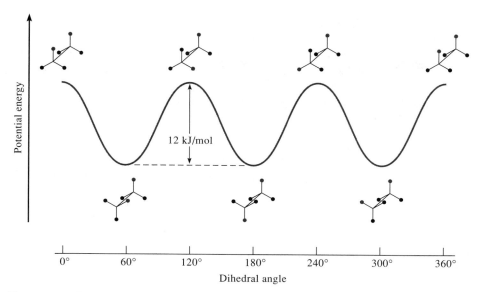

Figure 11.5 *Potential energy diagram for the internal rotation in ethane. Here the dihedral angle is defined by the angle between the two C—H bonds (with the red spheres representing the H atoms). Dihedral angles of 0°, 120°, 240°, and 360° represent the eclipsed conformation, while those of 60°, 180°, and 300° represent the staggered conformation. Thus a rotation of 60° changes the eclipsed conformation to the staggered one and vice versa. The staggered conformation is more stable than the eclipsed conformation by 12 kJ/mol. However, these two forms interconvert rapidly and cannot be separated from each other.*

methane, we are left with the CH_3 fragment, which is called a *methyl* group. Similarly, removing a hydrogen atom from the ethane molecule gives an *ethyl* group, or C_2H_5. Because they are derived from alkanes, they are also called *alkyl groups*. Table 11.2 lists the names of several common alkyl groups.

With the aid of Tables 11.1 and 11.2 and the rule that the locations of other groups are denoted by numbering the atoms along the longest carbon chain, we are now ready to name some substituted alkanes. Note that we always start numbering from the end of the longest chain that is closer to the carbon atom bearing the substituted group. Consider the following example:

$$\overset{\displaystyle CH_3}{\underset{\displaystyle H}{H_3\overset{1}{C}-\overset{2}{C}-\overset{3}{C}H_2-\overset{4}{C}H_3}}$$

Because the longest carbon chain contains four C atoms, the parent name of the compound is butane. We note that a methyl group is attached to carbon number 2; therefore, the name of the compound is 2-methylbutane. Here is another example:

$$\overset{\displaystyle CH_3 \qquad CH_3}{\underset{\displaystyle CH_3}{H_3\overset{1}{C}-\overset{2}{C}-\overset{3}{C}H_2-\overset{4}{C}H-\overset{5}{C}H_2-\overset{6}{C}H_3}}$$

Table 11.2 | Common Alkyl Groups

Name	Formula
Methyl	$-CH_3$
Ethyl	$-CH_2-CH_3$
n-Propyl	$-(CH_2)_2-CH_3$
n-Butyl	$-(CH_2)_3-CH_3$
Isopropyl	$\underset{\displaystyle CH_3}{\overset{\displaystyle CH_3}{-C-H}}$
t-Butyl*	$\underset{\displaystyle CH_3}{\overset{\displaystyle CH_3}{-C-CH_3}}$

*The letter t stands for tertiary.

Table 11.3 | Names of Common Substituent Groups

Functional Group	Name
—NH$_2$	Amino
—F	Fluoro
—Cl	Chloro
—Br	Bromo
—I	Iodo
—NO$_2$	Nitro
—CH=CH$_2$	Vinyl

The parent name of this compound is hexane. Because two methyl groups are attached to carbon number 2 and one methyl group is attached to carbon number 4, we call the compound 2,2,4-trimethylhexane. Note that the total number of methyl groups is indicated with the prefix "tri."

Of course, we can have many different types of substituents on alkanes. Table 11.3 lists the names of some common functional groups. Thus the compound

$$\underset{1}{H_3C}\overset{Br}{\underset{2}{-CH}}\overset{NO_2}{\underset{3}{-CH}}\underset{4}{-CH_3}$$

is called 2-bromo-3-nitrobutane. If the Br is attached to the first carbon atom

$$\underset{1}{Br-CH_2}\underset{2}{-CH_2}\overset{NO_2}{\underset{3}{-CH}}\underset{4}{-CH_3}$$

the compound becomes 1-bromo-3-nitrobutane.

Reactions of Alkanes

Alkanes are generally not considered to be very reactive substances. However, under suitable conditions they do react. For example, natural gas, gasoline, and fuel oil are alkanes that undergo highly exothermic combustion reactions:

$$CH_4(g) + 2O_2(g) \longrightarrow CO_2(g) + 2H_2O(l) \qquad \Delta H° = -890.4 \text{ kJ}$$
$$2C_2H_6(g) + 7O_2(g) \longrightarrow 4CO_2(g) + 6H_2O(l) \qquad \Delta H° = -3119 \text{ kJ}$$

These, and similar combustion reactions, have long been utilized in industrial processes and in domestic heating and cooking.

Besides combustion, alkanes undergo substitution reactions in which one or more H atoms are replaced by other atoms, usually the halogens. For example, when a mixture of methane and chlorine is heated above 100°C or irradiated with light of a suitable wavelength, methyl chloride is produced:

$$CH_4(g) + Cl_2(g) \longrightarrow \underset{\text{methyl chloride}}{CH_3Cl(g)} + HCl(g)$$

If an excess of chlorine gas is present, the reaction can proceed further:

$$CH_3(g) + Cl_2(g) \longrightarrow \underset{\text{methylene chloride}}{CH_2Cl_2(l)} + HCl(g)$$

$$CH_2Cl_2(g) + Cl_2(g) \longrightarrow \underset{\text{chloroform}}{CHCl_3(l)} + HCl(g)$$

$$CHCl_3(l) + Cl_2(g) \longrightarrow \underset{\text{carbon tetrachloride}}{CCl_4(l)} + HCl(g)$$

A great deal of experimental evidence suggests that the initial step of the first halogenation reaction occurs as follows:

$$Cl_2 + \text{energy} \longrightarrow Cl\cdot + Cl\cdot$$

The systematic names of methyl chloride, methylene chloride, and chloroform are monochloromethane, dichloromethane, and trichloromethane, respectively.

Thus the covalent bond in Cl_2 breaks and two chlorine atoms form. We know it is the Cl—Cl bond that breaks when the mixture is heated or irradiated because the bond energy of Cl_2 is 242.7 kJ/mol, whereas about 414 kJ/mol are needed to break C—H bonds in CH_4.

A chlorine atom is a *radical,* which *contains an unpaired electron* (shown by a single dot). Chlorine atoms are highly reactive and attack methane molecules according to the equation

$$CH_4 + Cl\cdot \longrightarrow \cdot CH_3 + HCl$$

This reaction produces hydrogen chloride and the methyl radical $\cdot CH_3$. The methyl radical is another reactive species; it combines with molecular chlorine to give methyl chloride and a chlorine atom:

$$\cdot CH_3 + Cl_2 \longrightarrow CH_3Cl + Cl\cdot$$

The production of methylene chloride from methyl chloride and any further reactions can be explained in the same way. The actual mechanism is more complex than the scheme we have shown because "side reactions" that do not lead to the desired products often take place, such as

$$Cl\cdot + Cl\cdot \longrightarrow Cl_2$$
$$\cdot CH_3 + \cdot CH_3 \longrightarrow C_2H_6$$

Alkanes in which one or more hydrogen atoms have been replaced by a halogen atom are called *alkyl halides.* Among the large number of alkyl halides, the best known are chloroform ($CHCl_3$), carbon tetrachloride (CCl_4), methylene chloride (CH_2Cl_2), and the chlorofluorohydrocarbons such as CF_2Cl_2.

Cycloalkanes

Alkanes whose carbon atoms are joined in rings are known as *cycloalkanes.* They have the general formula C_nH_{2n}, where $n = 3, 4, \ldots$. The simplest cycloalkane is cyclopropane, C_3H_6 (Figure 11.6). Many biologically significant substances such as antibiotics, sugars, cholesterol, and hormones contain one or more such ring systems. Theoretical analysis shows that cyclohexane can assume two different conformations called the chair and boat that are relatively free of strain (Figure 11.7). By "strain" we mean that bonds are compressed, stretched, or twisted out of normal geometric shapes predicted by sp^3 hybridization.

In addition to C, atoms such as N, O, and S may also occupy the ring positions in these compounds.

Cyclopropane   Cyclobutane   Cyclopentane   Cyclohexane

Figure 11.6 *Structures of the first four cycloalkanes and their skeletal forms. With the exception of cyclopropane, all cycloalkanes have nonplanar ring structures.*

Figure 11.7 *The cyclohexane molecule can exist in various conformations. The most stable conformation is the chair form; the boat form is less stable. Two types of H atoms are labeled axial and equatorial, respectively.*

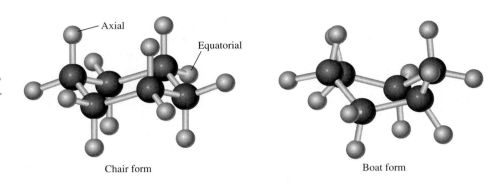

Chair form   Boat form

Alkenes

The **alkenes** (also called *olefins*) *contain at least one carbon-carbon double bond. Alkenes have the general formula* C_nH_{2n}, *where* $n = 2, 3, \ldots$. The simplest alkene is C_2H_4, ethylene, in which both carbon atoms are sp^2-hybridized and the double bond is made up of a sigma bond and a pi bond (see Section 10.5).

Geometric Isomers of Alkenes

In a compound such as ethane, C_2H_6, the rotation of the two methyl groups about the carbon-carbon single bond (which is a sigma bond) is quite free. The situation is different for molecules that contain carbon-carbon double bonds, such as ethylene, C_2H_4. In addition to the sigma bond, there is a pi bond between the two carbon atoms. Rotation about the carbon-carbon linkage does not affect the sigma bond, but it does move the two $2p_z$ orbitals out of alignment for overlap and, hence, partially or totally destroys the pi bond (see Figure 10.15). This process requires an input of energy on the order of 270 kJ/mol. For this reason, the rotation of a carbon-carbon double bond is considerably restricted, but not impossible. Consequently, molecules containing carbon-carbon double bonds (that is, the alkenes) may have **geometric isomers,** *which have the same type and number of atoms and the same chemical bonds but different spatial arrangements. Such isomers cannot be interconverted without breaking a chemical bond.*

The molecule dichloroethylene, $ClHC{=}CHCl$, can exist as one of the two geometric isomers called *cis*-dichloroethylene and *trans*-dichloroethylene:

In cis-dichloroethylene (top), the bond moments reinforce one another and the molecule is polar. The opposite holds for trans-dichloroethylene and the molecule is nonpolar.

resultant dipole moment

cis-dichloroethylene
$\mu = 1.89$ D
b.p. 60.3°C

trans-dichloroethylene
$\mu = 0$
b.p. 47.5°C

where the term *cis* means that two particular atoms (or groups of atoms) are adjacent to each other, and *trans* means that the two atoms (or groups of atoms) are across from each other. Generally, *cis* and *trans* isomers have distinctly different physical and chemical properties. Heat or irradiation with light is commonly used to bring about the conversion of one geometric isomer to another, a process called *cis-trans isomerization,* or geometric isomerization (Figure 11.8).

Figure 11.8 *Breaking and remaking the pi bond. When a compound containing a C*=*C bond is heated or excited by light, the weaker pi bond is broken. This allows the free rotation of the single carbon-to-carbon sigma bond. A rotation of 180° converts a cis isomer to a trans isomer or the other way around. Note that a dashed line represents a bond axis behind the plane of the paper, the wedged line represents a bond axis in front of the paper, and the thin solid line represents bonds in the plane of the paper. The letters A and B represent atoms (other than H) or groups of atoms. Here we have a cis to trans isomerization.*

Cis-trans Isomerization in the Vision Process

The molecules in the retina that respond to light are rhodopsin, which has two components called 11-*cis* retinal and opsin (Figure 11.9). Retinal is the light-sensitive component and opsin is a protein molecule. Upon receiving a photon in the visible region the 11-*cis* retinal isomerizes to the all-*trans* retinal by breaking a carbon-carbon pi bond. With the pi bond broken, the remaining carbon-carbon sigma bond is free to rotate and transforms into the all-*trans* retinal. At this point an electrical impulse is generated and transmitted to the brain, which forms a visual image. The all-*trans* retinal does not fit into the binding site on opsin and eventually separates from the protein. In time, the trans isomer is converted back to 11-*cis* retinal by an enzyme (in the absence of light) and rhodopsin is regenerated by the binding of the *cis* isomer to opsin and the visual cycle can begin again.

Alkene Nomenclature

In naming alkenes we indicate the positions of the carbon-carbon double bonds. The names of compounds containing C=C bonds end with -*ene*. As with the alkanes, the name of the parent compound is determined by the number of carbon atoms in the longest chain (see Table 11.1), as shown here:

$$CH_2{=}CH{-}CH_2{-}CH_3 \qquad H_3C{-}CH{=}CH{-}CH_3$$

1-butene   2-butene

The numbers in the names of alkenes refer to the lowest numbered carbon atom in the chain that is part of the C=C bond of the alkene. The name "butene" means that there

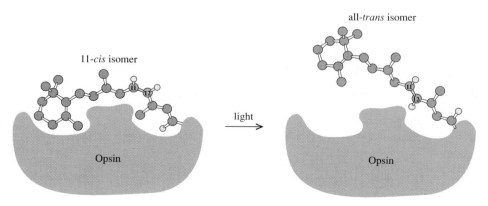

Figure 11.9 *The primary event in the vision process is the conversion of 11-cis retinal to the all-trans isomer on rhodopsin. The double bond at which the isomerization occurs is shown in green. For simplicity, most of the H atoms are omitted. In the absence of light, this transformation takes place about once in a thousand years!*

are four carbon atoms in the longest chain. Alkene nomenclature must specify whether a given molecule is *cis* or *trans* if it is a geometric isomer, such as

Note that geometric isomers are always structural isomers but the reverse is not true.

4-methyl-*cis*-2-hexene

4-methyl-*trans*-2-hexene

Properties and Reactions of Alkenes

Ethylene is an extremely important substance because it is used in large quantities in manufacturing organic polymers (very large molecules) and in preparing many other organic chemicals. Ethylene and other alkenes are prepared industrially by the *cracking* process, that is, the thermal decomposition of a large hydrocarbon into smaller molecules. When ethane is heated to about 800°C in the presence of platinum, it undergoes the following reaction:

$$C_2H_6(g) \xrightarrow[\text{catalyst}]{\text{Pt}} CH_2{=}CH_2(g) + H_2(g)$$

(a)

(b)

Figure 11.10 *Oils and fats are molecules with long hydrocarbon chains. (a) Oils contain one or more C=C bonds. The cis form of the hydrocarbon chains prevents close packing of the molecules. Therefore, oils are liquids. (b) Upon hydrogenation, the saturated hydrocarbon chains stack well together. As a result, fats have higher density than oils and are solids at room temperature.*

Figure 11.11 *When ethylene gas is bubbled through an aqueous bromine solution, the reddish-brown color gradually disappears due to the formation of 1,2-dibromoethane, which is colorless.*

The platinum acts as a *catalyst,* a substance that speeds up a reaction without being used up in the process and therefore does not appear on either side of the equation. Other alkenes can be prepared by cracking the higher members of the alkane family.

Alkenes are classified as **unsaturated hydrocarbons,** *compounds with double or triple carbon-carbon bonds.* Unsaturated hydrocarbons commonly undergo **addition reactions** in which *one molecule adds to another to form a single product.* An example of an addition reaction is **hydrogenation,** which is *the addition of molecular hydrogen to compounds containing* C=C *and* C≡C *bonds*

$$H_2 + \quad \overset{H}{\underset{H}{}}C=C\overset{H}{\underset{H}{}} \longrightarrow H-\overset{\overset{H}{|}}{\underset{\underset{H}{|}}{C}}-\overset{\overset{H}{|}}{\underset{\underset{H}{|}}{C}}-H$$

Hydrogenation is an important process in the food industry. Vegetable oils have considerable nutritional value, but many oils must be hydrogenated to eliminate some of the C=C bonds before they can be used to prepare food. Upon exposure to air, *polyunsaturated* molecules—molecules with many C=C bonds—undergo oxidation to yield unpleasant-tasting products (vegetable oil that has oxidized is said to be rancid). In the hydrogenation process, a small amount of nickel catalyst is added to the oil and the mixture is exposed to hydrogen gas at high temperature and pressure. Afterward, the nickel is removed by filtration. Hydrogenation reduces the number of double bonds in the molecule but does not completely eliminate them. If all the double bonds are eliminated, the oil becomes hard and brittle (Figure 11.10). Under controlled conditions, suitable cooking oils and margarine may be prepared by hydrogenation from vegetable oils extracted from cottonseed, corn, and soybeans.

Other addition reactions to the C=C bond involve the hydrogen halides and halogens (Figure 11.11):

$$H_2C=CH_2 + X_2 \longrightarrow CH_2X-CH_2X$$
$$H_2C=CH_2 + HX \longrightarrow CH_3-CH_2X$$

in which X represents a halogen atom. Figure 11.12 shows the electron density maps of HCl and ethylene. When the two molecules react, the interaction is between the

Figure 11.12 *The addition reaction between HCl and ethylene. The initial interaction is between the positive end of HCl (blue) and the electron-rich region of ethylene (red), which is associated with the pi electrons of the* C=C *bond.*

electron-rich region (pi electrons of the double bond) and the electron-poor region of HCl, which is the H atom. The steps are:

$$H_2C=CH_2 + HCl \longrightarrow H_2\overset{\overset{\displaystyle H}{\displaystyle |}}{C}-\overset{+}{C}H_2 + Cl^- \longrightarrow CH_3-CH_2Cl$$

The addition of a hydrogen halide to an unsymmetrical alkene such as propylene is more complicated because two products are possible:

The electron density is higher on the carbon atom in the CH_2 group in propylene.

In reality, however, only 2-bromopropane is formed. This phenomenon was observed in all reactions between unsymmetrical reagents and alkenes. In 1871 the Russian chemist Vladimir Markovnikov postulated a generalization that enables us to predict the outcome of such an addition reaction. This generalization, now known as *Markovnikov's rule,* states that in the addition of unsymmetrical (that is, polar) reagents to alkenes, the positive portion of the reagent (usually hydrogen) adds to the carbon atom that already has the most hydrogen atoms. As the marginal figure shows, the C atom to which the two H atoms are attached has a higher electron density. Therefore, this is the site for the H^+ ion (from HBr) to form a C—H bond, followed by the formation of the C—Br bond on the other C atom.

Finally we note that ethylene undergoes a different type of addition reaction that leads to the formation of a polymer. In this process, first an *initiator* molecule (R_2) is heated to produce two radicals:

$$R_2 \longrightarrow 2R\cdot$$

It is the pi bond that is broken in the polymerization of ethylene.

The reactive radical attacks an ethylene molecule to generate a new radical:

$$R\cdot + CH_2=CH_2 \longrightarrow R-CH_2-CH_2\cdot$$

which further reacts with another ethylene molecule, and so on:

$$R-CH_2-CH_2\cdot + CH_2=CH_2 \longrightarrow R-CH_2-CH_2-CH_2-CH_2\cdot$$

Very quickly a long chain of CH_2 groups is built. Eventually, this process is terminated by the combination of two long-chain radicals to give the polymer called polyethylene (Figure 11.13):

$$R(CH_2-CH_2)_nCH_2CH_2\cdot + R(CH_2-CH_2)_nCH_2CH_2\cdot \longrightarrow$$
$$R(CH_2-CH_2)_nCH_2CH_2-CH_2CH_2(CH_2-CH_2)_nR$$

where $(CH_2-CH_2)_n$ is a convenient shorthand convention for representing the repeating unit in the polymer. The value of n is understood to be very large, on the order of hundreds.

Figure 11.13 *Structure of polyethylene. Each carbon atom is sp^3-hybridized.*

Under different conditions, it is possible to prepare polyethylene with branched chains. Today, many different forms of polyethylene with widely different physical properties are known. Polyethylene is mainly used in films in frozen food packaging and other product wrappings. A specially treated type of polyethylene, called Tyvek, is used in home insulation and mailer envelopes.

A mailing envelope made of Tyvek.

Alkynes

Alkynes *contain at least one carbon-carbon triple bond.* They have the *general formula* C_nH_{2n-2}, *where* $n = 2, 3, \ldots$.

Alkyne Nomenclature

Names of compounds containing C≡C bonds end with *-yne*. Again the name of the parent compound is determined by the number of carbon atoms in the longest chain (see Table 11.1 for names of alkane counterparts). As in the case of alkenes, the names of alkynes indicate the position of the carbon-carbon triple bond, as, for example, in

$$HC\equiv C-CH_2-CH_3 \qquad H_3C-C\equiv C-CH_3$$
$$\text{1-butyne} \qquad\qquad\qquad \text{2-butyne}$$

Properties and Reactions of Alkynes

The simplest alkyne is ethyne, better known as acetylene (C_2H_2). The structure and bonding of C_2H_2 were discussed in Section 10.5. Acetylene is a colorless gas (b.p. $-84°C$) prepared by the reaction between calcium carbide and water:

$$CaC_2(s) + 2H_2O(l) \longrightarrow C_2H_2(g) + Ca(OH)_2(aq)$$

Acetylene has many important uses in industry. Because of its high heat of combustion

$$2C_2H_2(g) + 5O_2(g) \longrightarrow 4CO_2(g) + 2H_2O(l) \quad \Delta H° = -2599.2 \text{ kJ}$$

acetylene burned in an "oxyacetylene torch" gives an extremely hot flame (about 3000°C). Thus, oxyacetylene torches are used to weld metals (see p. 164).

Acetylene is unstable and has a tendency to decompose:

$$C_2H_2(g) \longrightarrow 2C(s) + H_2(g)$$

In the presence of a suitable catalyst or when the gas is kept under pressure, this reaction can occur with explosive violence. To be transported safely, it must be dissolved in an inert organic solvent such as acetone at moderate pressure. In the liquid state, acetylene is very sensitive to shock and is highly explosive.

Being an unsaturated hydrocarbon, acetylene can be hydrogenated to yield ethylene:

$$C_2H_2(g) + H_2(g) \longrightarrow C_2H_4(g)$$

It undergoes these addition reactions with hydrogen halides and halogens:

$$C_2H_2(g) + HX(g) \longrightarrow CH_2=CHX(g)$$
$$C_2H_2(g) + X_2(g) \longrightarrow CHX=CHX(g)$$
$$C_2H_2(g) + 2X_2(g) \longrightarrow CHX_2-CHX_2(g)$$

The reaction of calcium carbide with water produces acetylene, a flammable gas.

Propyne. Can you account for Markovnikov's rule in this molecule?

Methylacetylene (propyne), CH_3—$C\equiv C$—H, is the next member in the alkyne family. It undergoes reactions similar to those of acetylene. The addition reactions of propyne also obey Markovnikov's rule:

$$CH_3-C\equiv C-H + HBr \longrightarrow$$

propyne   2-bromopropene

11.3 Aromatic Hydrocarbons

Benzene (C_6H_6) is the parent compound of this large family of organic substances. As we saw in Section 9.6, the properties of benzene are best represented by both of the following resonance structures (p. 277):

Benzene is a planar hexagonal molecule with carbon atoms situated at the six corners. All carbon-carbon bonds are equal in length and strength, as are all carbon-hydrogen bonds, and the CCC and HCC angles are all 120°. Therefore, each carbon atom is sp^2-hybridized; it forms three sigma bonds with two adjacent carbon atoms and a hydrogen atom (Figure 11.14). This arrangement leaves an unhybridized $2p_z$ orbital on each carbon atom, perpendicular to the plane of the benzene molecule, or *benzene ring,* as it is often called. So far the description resembles the configuration of ethylene (C_2H_4), discussed in Section 10.5, except that in this case there are six unhybridized $2p_z$ orbitals in a cyclic arrangement.

Because of their similar shape and orientation, each $2p_z$ orbital overlaps two others, one on each adjacent carbon atom. According to the rules listed on p. 322, the interaction of six $2p_z$ orbitals leads to the formation of six pi molecular orbitals, of which three are bonding and three antibonding. A benzene molecule in the ground state therefore has six electrons in the three pi bonding molecular orbitals, two electrons with paired spins in each orbital (Figure 11.15).

In the ethylene molecule, the overlap of the two $2p_z$ orbitals gives rise to a bonding and an antibonding molecular orbital, which are localized over the two C atoms. The interaction of the $2p_z$ orbitals in benzene, however, leads to the formation of **delocalized molecular orbitals,** *which are not confined between two adjacent bonding atoms, but actually extend over three or more atoms.* Therefore, electrons residing in any of these orbitals are free to move around the benzene ring. For this reason, the structure of benzene is sometimes represented as

Pi electron density (red color) above and below the plane of the benzene molecule. For simplicity, only the framework of the molecule is shown.

in which the circle indicates that the pi bonds between carbon atoms are not confined to individual pairs of atoms; rather, the pi electron densities are evenly distributed throughout the benzene molecule. As we will see shortly, electron delocalization imparts extra stability to aromatic hydrocarbons.

We can now state that each carbon-to-carbon linkage in benzene contains a sigma bond and a "partial" pi bond. The bond order between any two adjacent carbon atoms is therefore between 1 and 2. Thus molecular orbital theory offers an alternative to the resonance approach, which is based on valence bond theory.

Nomenclature of Aromatic Compounds

The naming of monosubstituted benzenes, that is, benzenes in which one H atom has been replaced by another atom or a group of atoms, is quite straightforward, as shown below:

| CH$_2$CH$_3$ | Cl | NH$_2$ | NO$_2$ |

ethylbenzene   chlorobenzene   aminobenzene   nitrobenzene
(aniline)

If more than one substituent is present, we must indicate the location of the second group relative to the first. The systematic way to accomplish this is to number the carbon atoms as follows:

Figure 11.14 *The sigma bond framework in the benzene molecule. Each carbon atom is sp^2-hybridized and forms sigma bonds with two adjacent carbon atoms and another sigma bond with a hydrogen atom.*

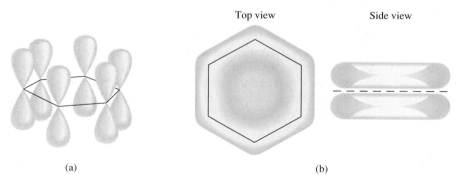

Top view   Side view

(a)   (b)

Figure 11.15 *(a) The six 2p$_z$ orbitals on the carbon atoms in benzene. (b) The delocalized molecular orbital formed by the overlap of the 2p$_z$ orbitals. The delocalized molecular orbital possesses pi symmetry and lies above and below the plane of the benzene ring. Actually, these 2p$_z$ orbitals can combine in six different ways to yield three bonding molecular orbitals and three antibonding molecular orbitals. The one shown here is the most stable.*

Three different dibromobenzenes are possible:

| 1,2-dibromobenzene | 1,3-dibromobenzene | 1,4-dibromobenzene |
| (*o*-dibromobenzene) | (*m*-dibromobenzene) | (*p*-dibromobenzene) |

The prefixes *o*- (*ortho*-), *m*- (*meta*-), and *p*- (*para*-) are also used to denote the relative positions of the two substituted groups, as shown above for the dibromobenzenes. Compounds in which the two substituted groups are different are named accordingly. Thus,

is named 3-bromonitrobenzene, or *m*-bromonitrobenzene.

Finally we note that the group containing benzene minus a hydrogen atom (C_6H_5) is called the *phenyl* group. Thus the following molecule is called 2-phenylpropane:

This compound is also called isopropyl benzene (see Table 11.2).

$$CH_3 - CH - CH_3$$

Properties and Reactions of Aromatic Compounds

Benzene is a colorless, flammable liquid obtained chiefly from petroleum and coal tar. Perhaps the most remarkable chemical property of benzene is its relative inertness. Although it has the same empirical formula as acetylene (CH) and a high degree of unsaturation, it is much less reactive than either ethylene or acetylene. The stability of benzene is the result of electron delocalization. In fact, benzene can be hydrogenated, but only with difficulty. The following reaction is carried out at significantly higher temperatures and pressures than are similar reactions for the alkenes:

A catalyst is a substance that can speed up the rate of a reaction without itself being used up. More on this topic in Chapter 14.

$$+ \; 3H_2 \;\xrightarrow[\text{catalyst}]{\text{Pt}}$$

cyclohexane

We saw earlier that alkenes react readily with halogens and hydrogen halides to form addition products, because the pi bond in C=C can be broken more easily. The most common reactions of halogens with benzene is substitution reaction. For example,

Note that if the reaction were an addition reaction, electron delocalization would be destroyed in the product

and the molecule would not have the aromatic characteristic of chemical unreactivity.

Alkyl groups can be introduced into the ring system by allowing benzene to react with an alkyl halide using $AlCl_3$ as the catalyst:

An enormously large number of compounds can be generated from substances in which benzene rings are fused together. Some of these *polycyclic* aromatic hydrocarbons are shown in Figure 11.16. The best known of these compounds is naphthalene, which is used in mothballs. These and many other similar compounds are present in coal tar. Some of the compounds with several rings are powerful carcinogens—they can cause cancer in humans and other animals.

Naphthalene   Anthracene   Phenanthrene   Naphthacene

Benz(*a*)anthracene*   Dibenz(*a,h*)anthracen*   Benzo(*a*)pyrene

Figure 11.16 *Some polycyclic aromatic hydrocarbons. Compounds denoted by * are potent carcinogens. An enormous number of such compounds exist in nature.*

11.4 Chemistry of the Functional Groups

We now examine some organic functional groups, groups that are responsible for most of the reactions of the parent compounds. In particular, we focus on oxygen-containing and nitrogen-containing compounds.

Alcohols

C_2H_5OH

All *alcohols* contain *the hydroxyl* functional *group, —OH*. Some common alcohols are shown in Figure 11.17. Ethyl alcohol, or ethanol, is by far the best known. It is produced biologically by the fermentation of sugar or starch. In the absence of oxygen the enzymes present in bacterial cultures or yeast catalyze the reaction

$$C_6H_{12}O_6(aq) \xrightarrow{\text{enzymes}} \underset{\text{ethanol}}{2CH_3CH_2OH(aq)} + 2CO_2(g)$$

This process gives off energy, which microorganisms, in turn, use for growth and other functions.

Commercially, ethanol is prepared by an addition reaction in which water is combined with ethylene at about 280°C and 300 atm:

$$CH_2{=}CH_2(g) + H_2O(g) \xrightarrow{H_2SO_4} CH_3CH_2OH(g)$$

Ethanol has countless applications as a solvent for organic chemicals and as a starting compound for the manufacture of dyes, synthetic drugs, cosmetics, and explosives. It is also a constituent of alcoholic beverages. Ethanol is the only nontoxic (more properly, the least toxic) of the straight-chain alcohols; our bodies produce an enzyme, called *alcohol dehydrogenase,* which helps metabolize ethanol by oxidizing it to acetaldehyde:

$$CH_3CH_2OH \xrightarrow{\text{alcohol dehydrogenase}} \underset{\text{acetaldehyde}}{CH_3CHO} + H_2$$

This equation is a simplified version of what actually takes place; the H atoms are taken up by other molecules, so that no H_2 gas is evolved.

Ethanol can also be oxidized by inorganic oxidizing agents, such as acidified potassium dichromate, to acetic acid:

Figure 11.17 *Common alcohols. Note that all the compounds contain the OH group. The properties of phenol are quite different from those of the aliphatic alcohols.*

$$3CH_3CH_2OH + 2K_2Cr_2O_7 + 8H_2SO_4 \longrightarrow 3CH_3COOH + 2Cr_2(SO_4)_3$$

orange-yellow green

$$+ 2K_2SO_4 + 11H_2O$$

This reaction has been employed by law enforcement agencies to test drivers suspected of being drunk. A sample of the driver's breath is drawn into a device called a breath analyzer, where it is reacted with an acidic potassium dichromate solution. From the color change (orange-yellow to green) it is possible to determine the alcohol content in the driver's blood.

Ethanol is called an aliphatic alcohol because it is derived from an alkane (ethane). The simplest aliphatic alcohol is methanol, CH_3OH. Called *wood alcohol*, it was prepared at one time by the dry distillation of wood. It is now synthesized industrially by the reaction of carbon monoxide and molecular hydrogen at high temperatures and pressures:

$$CO(g) + 2H_2(g) \xrightarrow[\text{catalyst}]{Fe_2O_3} CH_3OH(l)$$

methanol

Methanol is highly toxic. Ingestion of only a few milliliters can cause nausea and blindness. Ethanol intended for industrial use is often mixed with methanol to prevent people from drinking it. Ethanol containing methanol or other toxic substances is called *denatured alcohol.*

The alcohols are very weakly acidic; they do not react with strong bases, such as NaOH. The alkali metals react with alcohols to produce hydrogen:

$$2CH_3OH + 2Na \longrightarrow 2CH_3ONa + H_2$$

sodium methoxide

However, the reaction is much less violent than that between Na and water:

$$2H_2O + 2Na \longrightarrow 2NaOH + H_2$$

Two other familiar aliphatic alcohols are 2-propanol (or isopropanol), commonly known as rubbing alcohol, and ethylene glycol, which is used as an antifreeze. Most alcohols—especially those with low molar masses—are highly flammable.

Ethers

Ethers contain the R—O—R' linkage, where R and R' are a hydrocarbon (aliphatic or aromatic) group. They are formed by the reaction between two alcohols

$$CH_3OH + HOCH_3 \xrightarrow[\text{catalyst}]{H_2SO_4} CH_3OCH_3 + H_2O$$

dimethyl ether

This reaction is an example of a *condensation reaction,* which is characterized by the joining of two molecules and the elimination of a small molecule, usually water.

Like alcohols, ethers are extremely flammable. When left standing in air, they have a tendency to slowly form explosive peroxides:

$$C_2H_5OC_2H_5 + O_2 \longrightarrow C_2H_5O-\overset{\overset{\displaystyle CH_3}{|}}{\underset{\underset{\displaystyle H}{|}}{C}}-O-O-H$$

diethyl ether   1-ethyoxyethyl hydroperoxide

Left: A $K_2Cr_2O_7$ solution.
Right: A $Cr_2(SO_4)_3$ solution.

Alcohols react more slowly with sodium metal than water does.

CH_3OCH_3

Diethyl ether, commonly known as "ether," was used as an anesthetic for many years. It produces unconsciousness by depressing the activity of the central nervous system. The major disadvantages of diethyl ether are its irritating effects on the respiratory system and the occurrence of postanesthetic nausea and vomiting. "Neothyl," or methyl propyl ether, $CH_3OCH_2CH_2CH_3$, is currently favored as an anesthetic because it is relatively free of side effects.

Aldehydes and Ketones

Under mild oxidation conditions, it is possible to convert alcohols to aldehydes and ketones:

$$CH_3OH + \tfrac{1}{2}O_2 \longrightarrow H_2C{=}O + H_2O$$
formaldehyde

$$C_2H_5OH + \tfrac{1}{2}O_2 \longrightarrow \begin{matrix} H_3C \\ \ \\ H \end{matrix}{>}C{=}O + H_2O$$
acetaldehyde

$$CH_3{-}\overset{\overset{\displaystyle H}{|}}{\underset{\underset{\displaystyle OH}{|}}{C}}{-}CH_3 + \tfrac{1}{2}O_2 \longrightarrow \begin{matrix} H_3C \\ \ \\ H_3C \end{matrix}{>}C{=}O + H_2O$$
acetone

CH₃CHO

The functional group in these compounds is the *carbonyl group,* $>C{=}O$. In an ***aldehyde*** *at least one hydrogen atom is bonded to the carbon in the carbonyl group.* In a ***ketone,*** *the carbon atom in the carbonyl group is bonded to two hydrocarbon groups.*

The simplest aldehyde, formaldehyde ($H_2C{=}O$) has a tendency to *polymerize;* that is, the individual molecules join together to form a compound of high molar mass. This action gives off much heat and is often explosive, so formaldehyde is usually prepared and stored in aqueous solution (to reduce the concentration). This rather disagreeable-smelling liquid is used as a starting material in the polymer industry and in the laboratory as a preservative for animal specimens. Interestingly, the higher molar mass aldehydes, such as cinnamic aldehyde

Cinnamic aldehyde gives cinnamon its characteristic aroma.

$$\text{(benzene ring)}{-}CH{=}CH{-}C{\overset{\displaystyle H}{\underset{\displaystyle O}{\big<}}}$$

have a pleasant odor and are used in the manufacture of perfumes.

Ketones generally are less reactive than aldehydes. The simplest ketone is acetone, a pleasant-smelling liquid that is used mainly as a solvent for organic compounds and nail polish remover.

Carboxylic Acids

Under appropriate conditions both alcohols and aldehydes can be oxidized to ***carboxylic acids,*** *acids that contain the carboxyl group,* —COOH:

Figure 11.18 *Some common carboxylic acids. Note that they all contain the COOH group. (Glycine is one of the amino acids found in proteins.)*

Formic acid Acetic acid Butyric acid Benzoic acid

Glycine Oxalic acid Citric acid

$$CH_3CH_2OH + O_2 \longrightarrow CH_3COOH + H_2O$$
$$CH_3CHO + \tfrac{1}{2}O_2 \longrightarrow CH_3COOH$$

These reactions occur so readily, in fact, that wine must be protected from atmospheric oxygen while in storage. Otherwise, it would soon turn to vinegar due to the formation of acetic acid. Figure 11.18 shows the structure of some of the common carboxylic acids.

Carboxylic acids are widely distributed in nature; they are found in both the plant and animal kingdoms. All protein molecules are made of amino acids, a special kind of carboxylic acid containing an amino group ($-NH_2$) and a carboxyl group ($-COOH$).

Unlike the inorganic acids HCl, HNO_3, and H_2SO_4, carboxylic acids are usually weak. They react with alcohols to form pleasant-smelling esters:

CH_3COOH

$$CH_3COOH + HOCH_2CH_3 \longrightarrow CH_3-\overset{\overset{\text{O}}{\|}}{C}-O-CH_2CH_3 + H_2O$$

acetic acid   ethanol   ethyl acetate

Other common reactions of carboxylic acids are neutralization

$$CH_3COOH + NaOH \longrightarrow CH_3COONa + H_2O$$

and formation of acid halides, such as acetyl chloride

$$CH_3COOH + PCl_5 \longrightarrow CH_3COCl + HCl + POCl_3$$
acetyl chloride   phosphoryl chloride

Acid halides are reactive compounds used as intermediates in the preparation of many other organic compounds.

Esters

Esters have the general formula R′COOR, in which R′ can be H, an alkyl, or an aromatic hydrocarbon group and R is an alkyl or an aromatic hydrocarbon group. Esters are used in the manufacture of perfumes and as flavoring agents in the

The odor of fruits is mainly due to the ester compounds in them.

confectionery and soft-drink industries. Many fruits owe their characteristic smell and flavor to the presence of esters. For example, bananas contain isopentyl acetate [$CH_3COOCH_2CH_2CH(CH_3)_2$], oranges contain octyl acetate ($CH_3COOC_8H_{17}$), and apples contain methyl butyrate ($CH_3CH_2CH_2COOCH_3$).

The functional group in esters is —COOR. In the presence of an acid catalyst, such as HCl, esters undergo a reaction with water (a *hydrolysis* reaction) to regenerate a carboxylic acid and an alcohol. For example, in acid solution, ethyl acetate is converted to acetic acid:

$$\underset{\text{ethyl acetate}}{CH_3COOC_2H_5} + H_2O \rightleftharpoons \underset{\text{acetic acid}}{CH_3COOH} + \underset{\text{ethanol}}{C_2H_5OH}$$

However, this reaction does not go to completion because the reverse reaction, that is, the formation of an ester from an alcohol and an acid, also occurs to an appreciable extent. On the other hand, when the hydrolysis reaction is run in aqueous NaOH solution, ethyl acetate is converted to sodium acetate, which does not react with ethanol, so this reaction goes to completion from left to right:

$$\underset{\text{ethyl acetate}}{CH_3COOC_2H_5} + NaOH \longrightarrow \underset{\text{sodium acetate}}{CH_3COO^-Na^+} + \underset{\text{ethanol}}{C_2H_5OH}$$

The term **saponification** (meaning *soapmaking*) was originally used to describe the reaction between an ester and sodium hydroxide to yield soap molecules (sodium stearate):

$$\underset{\text{ethyl stearate}}{C_{17}H_{35}COOC_2H_5} + NaOH \longrightarrow \underset{\text{sodium stearate}}{C_{17}H_{35}COO^-Na^+} + C_2H_5OH$$

Saponification is now *a general term for alkaline hydrolysis of any type of ester.* Soap molecules are characterized by a long nonpolar hydrocarbon chain and a polar head (the —COO$^-$ group). The hydrocarbon chain is readily soluble in oily substances, while the ionic carboxylate group (—COO$^-$) remains outside the oily nonpolar surface. Figure 11.19 shows the action of soap.

(a)	(b)	(c)

Figure 11.19 *The cleansing action of soap. The soap molecule is represented by a polar head and zigzag hydrocarbon tail. An oily spot (a) can be removed by soap (b) because the nonpolar tail dissolves in the oil, and the entire system becomes soluble in water because the exterior portion is now ionic.*

Amines

Amines are *organic bases. They have the general formula R$_3$N, in which one of the R groups must be an alkyl group or an aromatic hydrocarbon group.* Like ammonia, amines are Brønsted base that react with water as follows:

$$RNH_2 + H_2O \longrightarrow RNH_3^+ + OH^-$$

Like all bases, the amines form salts when allowed to react with acids:

$$CH_3NH_2 + HCl \longrightarrow CH_3NH_3^+Cl^-$$
methylamine   methylammonium chloride

These salts are usually colorless, odorless solids that are soluble in water. Many of the aromatic amines are carcinogenic.

CH$_3$NH$_2$

Summary of Functional Groups

Table 11.4 summarizes the common functional groups, including the C=C and C≡C groups. Organic compounds commonly contain more than one functional group. Generally, the reactivity of a compound is determined by the number and types of functional groups in its makeup.

Table 11.4 | Important Functional Groups and Their Reactions

Functional Group	Name	Typical Reactions
$\diagdown$C=C$\diagup$	Carbon-carbon double bond	Addition reactions with halogens, hydrogen halides, and water; hydrogenation to yield alkanes
—C≡C—	Carbon-carbon triple bond	Addition reactions with halogens, hydrogen halides; hydrogenation to yield alkenes and alkanes
—X: (X = F, Cl, Br, I)	Halogen	Exchange reactions: $CH_3CH_2Br + KI \longrightarrow CH_3CH_2I + KBr$
—O—H	Hydroxyl	Esterification (formation of an ester) with carboxylic acids; oxidation to aldehydes, ketones, and carboxylic acids
$\diagdown$C=O	Carbonyl	Reduction to yield alcohols; oxidation of aldehydes to yield carboxylic acids
:O: ‖ —C—O—H	Carboxyl	Esterification with alcohols; reaction with phosphorus pentachloride to yield acid chlorides
:O: ‖ —C—O—R (R = hydrocarbon)	Ester	Hydrolysis to yield acids and alcohols
R —N R (R = H or hydrocarbon)	Amine	Formation of ammonium salts with acids

An artery becoming blocked by cholesterol.

Example 11.2   Cholesterol is a major component of gallstones, and it is believed that the cholesterol level in the blood is a contributing factor in certain types of heart disease. From the structure of the compound, predict its reaction with (a) Br_2, (b) H_2 (in the presence of a Pt catalyst), (c) CH_3COOH.

Reasoning and Solution   To predict the type of reactions a molecule may undergo, we must first identify the functional groups present. There are two functional groups in cholesterol: the hydroxyl group and the carbon-carbon double bond.

(a) The reaction with bromine results in the addition of bromine to the double-bonded carbons, which become single-bonded.

(b) This is a hydrogenation reaction. Again, the carbon-carbon double bond is converted to a carbon-carbon single bond.

(c) The acid reacts with the hydroxyl group to form an ester and water. Figure 11.20 shows the products of these reactions.

Figure 11.20 *The products formed by the reaction of cholesterol with (a) molecular bromine, (b) molecular hydrogen, and (c) acetic acid.*

Comment   Note that although cholesterol is much larger than ethylene, they react the same way toward Br_2 and H_2 because both contain the C=C bond. Likewise, cholesterol and methanol react similarly with CH_3COOH because they both possess the hydroxyl functional group. Remember it is the functional groups that determines the types of reactions a molecule will undergo.

Practice Exercise   Predict the products of the following reaction:

$$CH_3OH + CH_3CH_2COOH \longrightarrow ?$$

Similar problem: 11.41.

11.5 Chirality—The Handedness of Molecules

Many organic compounds can exist as mirror-image twins, in which one partner may cure disease, quell headache, or smell good, whereas its mirror-reversed counterpart may be poisonous, smell repugnant, or simply be inert. Compounds that come as mirror-image pairs are sometimes compared with left-handed and right-handed gloves and are referred to as **chiral,** or *handed,* molecules. Although every molecule can have a mirror image, the difference between chiral and *achiral* (meaning nonchiral) molecules is that only the twins of the former are nonsuperimposable.

Consider the substituted methanes CH_2ClBr and $CHFClBr$. Figure 11.21 shows perspective drawings of these two molecules and their mirror images. The two mirror images of Figure 11.21(a) are superimposable, but those of Figure 11.21(b) are not, no matter how we rotate the molecules. Thus the $CHFClBr$ molecule is chiral. Careful observation shows that most simple chiral molecules contain at least one *asymmetric* carbon atom—that is, a carbon atom bonded to four different atoms or groups of atoms.

The nonsuperimposable mirror images of a chiral compound are called **optical isomers** or **enantiomers.** Like geometric isomers, optical isomers come in pairs. However, the optical isomers of a compound have identical physical and chemical properties, such as melting point, boiling point, and chemical reactivity toward molecules that are not chiral themselves. Chiral molecules are said to be optically active because of their ability to rotate the plane of polarization of polarized light as it passes through them. Unlike ordinary light, which vibrates in all directions, plane-polarized light vibrates only in a single plane. To *study the interaction between plane-polarized light and optical isomers* we use a **polarimeter,** shown schematically in Figure 11.22. A beam of unpolarized light first passes through a Polaroid sheet, called the polarizer, and then through a sample tube containing a solution of a chiral compound. As the polarized light passes through the sample tube, its plane of polarization is rotated either to the right or to the left. This rotation can be measured directly by turning the analyzer in the appropriate direction until minimal light transmission is achieved (Figure 11.23). If the plane of polarization is rotated to the right, the isomer is said to be dextrorotatory (D); it is levorotatory (L) if the rotation is to the left. The D and L isomers of a chiral substance always rotate the

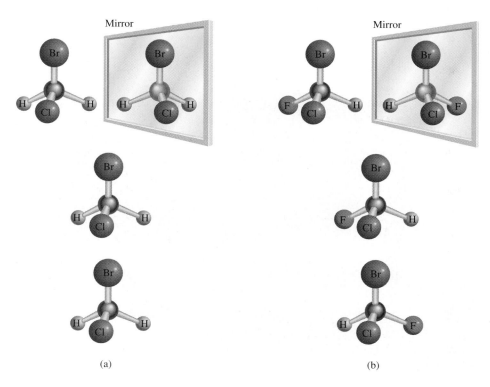

(a)

(b)

Figure 11.21 *(a) The CH_2ClBr molecule and its mirror image. Since the molecule and its mirror image are superimposable, the molecule is said to be achiral. (b) The CHFClBr molecule and its mirror image. Because the molecule and its mirror image are not superimposable, no matter how we rotate one with respect to the other, the molecule is said to be chiral.*

Figure 11.22 *Operation of a polarimeter. Initially, the tube is filled with an achiral compound. The analyzer is rotated so that its plane of polarization is perpendicular to that of the polarizer. Under this condition, no light reaches the observer. Next, a chiral compound is placed in the tube as shown. The plane of polarization of the polarized light is rotated as it travels though the tube so that some light reaches the observer. Rotating the analyzer (either to the left or to the right) until no light reaches the observer again allows the angle of optical rotation to be measured.*

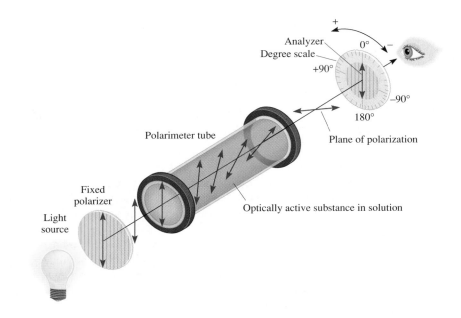

light by the same amount, but in opposite directions. Thus, in *an equimolar mixture of two enantiomers,* called a ***racemic mixture,*** the net rotation is zero.

Chirality plays an important role in biological systems. Protein molecules have many asymmetric carbon atoms and their functions are often influenced by their chirality. Because the enantiomers of a chiral compound usually behave very differently from each other in the body, chiral twins are coming under increasing scrutiny among pharmaceutical manufacturers. More than half of the most prescribed drugs in 2001 are chiral. In

Figure 11.23 *With one Polaroid sheet over a picture, light passes through. With a second sheet of Polaroid placed over the first so that the axes of polarization of the sheets are perpendicular, little or no light passes through. If the axes of polarization of the two sheets were parallel, light would pass through.*

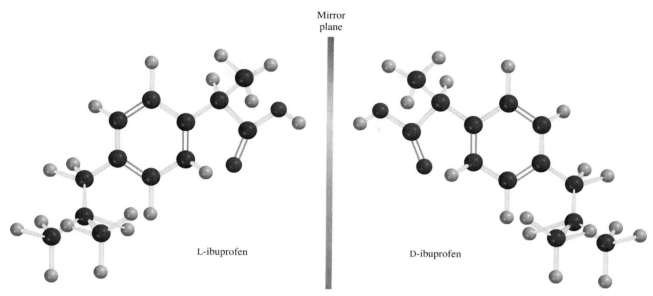

Figure 11.24 *The L and D forms of ibuprofen are mirror images of each other. There is only one asymmetric carbon atom in the molecule. Can you spot it?*

most of these cases only one enantiomer of the drug works as a medicine, whereas the other form is useless or less effective or may even cause serious side effects. The best-known case in which the use of a racemic mixture of a drug had tragic consequences occurred in Europe in the late 1950s. The drug thalidomide was prescribed for pregnant women there as an antidote to morning sickness. But by 1962, the drug had to be withdrawn from the market after thousands of deformed children had been born to mothers who had taken it. Only later did researchers discover that the sedative properties of thalidomide belong to D-thalidomide and that L-thalidomide is a potent mutagen. (A *mutagen* is a substance that causes gene mutation, usually leading to deformed offspring.)

Figure 11.24 shows the two enantiomeric forms of another drug, ibuprofen. This popular pain reliever is sold as a racemic mixture, but only the one on the left is potent. The other form is ineffective but also harmless. Organic chemists today are actively researching ways to synthesize enantiomerically pure drugs, or "chiral drugs." Chiral drugs contain only one enantiomeric form both for efficiency and for protection against possible side effects from its mirror-image twin.

Example 11.3   Is this molecule chiral?

$$\begin{array}{c} Cl \\ | \\ H-C-CH_2-CH_3 \\ | \\ CH_3 \end{array}$$

Reasoning and Solution   We note that the central carbon atom is bonded to a hydrogen atom, a chlorine atom, a —CH_3 group, and a —CH_2—CH_3 group. Therefore the central carbon atom is asymmetric and the molecule is chiral.

Similar problems: 11.45, 11.46.

Practice Exercise Is this molecule chiral?

$$
\begin{array}{c}
\text{Br} \\
| \\
\text{I}-\text{C}-\text{CH}_2-\text{CH}_3 \\
| \\
\text{Br}
\end{array}
$$

Summary

Because carbon atoms can link up with other carbon atoms in straight and branched chains, carbon can form more compounds than any other element.

Alkanes and cycloalkanes are saturated hydrocarbons. Methane, CH_4, is the simplest of the alkanes, a family of hydrocarbons with the general formula C_nH_{2n+2}. The cycloalkanes are a subfamily of alkanes whose carbon atoms are joined in a ring. Ethylene, $CH_2\!=\!CH_2$, is the simplest of alkenes, a class of hydrocarbons containing carbon-carbon double bonds and having the general formula C_nH_{2n}. Unsymmetrical alkenes can exist as *cis* and *trans* isomers. Acetylene, $CH\!\equiv\!CH$, is the simplest of the alkynes, which are compounds that have the general formula C_nH_{2n-2} and contain carbon-carbon triple bonds. Compounds that contain one or more benzene rings are called aromatic hydrocarbons. The stability of the benzene molecule is the result of electron delocalization.

Functional groups determine the chemical reactivity of molecules in which they are found. Classes of compounds characterized by their functional groups include alcohols, ethers, aldehydes and ketones, carboxylic acids and esters, and amines.

Chirality refers to molecules that have nonsuperimposable mirror images. Most chiral molecules contain one or more asymmetric carbon atoms. Chiral molecules are widespread in biological systems and are important in drug design.

Key Words

Addition reaction p. 345	Aromatic hydrocarbon	Ester p. 355	Polarimeter p. 359
Alcohol p. 352	p. 334	Ether p. 353	Racemic mixture p. 360
Aldehyde p. 354	Carboxylic acid p. 354	Functional group p. 334	Radical p. 341
Aliphatic hydrocarbon	Chiral p. 359	Geometric isomers p. 342	Saponification p. 356
p. 334	Conformations p. 337	Hydrocarbon p. 334	Saturated hydrocarbon
Alkane p. 334	Cycloalkane p. 341	Hydrogenation p. 345	p. 334
Alkene p. 342	Delocalized molecular	Ketone p. 354	Structural isomer p. 335
Alkyne p. 347	orbitals p. 348	Optical isomer p. 359	Unsaturated hydrocarbon
Amine p. 357	Enantiomer p. 359	Organic chemistry p. 334	p. 345

Questions and Problems

Aliphatic Hydrocarbons

Review Questions

11.1   Explain why carbon is able to form so many more compounds than any other element.

11.2   What is the difference between aliphatic and aromatic hydrocarbons?

11.3   What do "saturated" and "unsaturated" mean when applied to hydrocarbons? Give examples of a saturated hydrocarbon and an unsaturated hydrocarbon.

11.4 What are structural isomers?

11.5 Use ethane as an example to explain the meaning of conformations. What are Newman projections? How do the conformations of a molecule differ from structural isomers?

11.6 Draw skeletal structures of the boat and chair forms of cyclohexane.

11.7 Alkenes exhibit geometric isomerism because rotation about the C=C bond is restricted. Explain.

11.8 Why is it that alkanes and alkynes, unlike alkenes, have no geometric isomers?

11.9 What is Markovnikov's rule?

11.10 Describe reactions that are characteristic of alkanes, alkenes, and alkynes.

Problems

11.11 Draw all possible structural isomers for this alkane: C_7H_{16}.

11.12 How many distinct chloropentanes, $C_5H_{11}Cl$, could be produced in the direct chlorination of *n*-pentane, $CH_3(CH_2)_3CH_3$? Draw the structure of each molecule.

11.13 Draw all possible isomers for the molecule C_4H_8.

11.14 Draw all possible isomers for the molecule C_3H_5Br.

11.15 The structural isomers of pentane, C_5H_{12}, have quite different boiling points (see Example 11.1). Explain the observed variation in boiling point, in terms of structure.

11.16 Discuss how you can determine which of these compounds might be alkanes, cycloalkanes, alkenes, or alkynes, without drawing their formulas: (a) C_6H_{12}, (b) C_4H_6, (c) C_5H_{12}, (d) C_7H_{14}, (e) C_3H_4.

11.17 Draw Newman projections of the staggered and eclipsed conformations of propane. Rank them in stability.

11.18 Draw Newman projections of four different conformations of butane. Rank them in stability. (*Hint:* Two of the conformations represent the most stable forms and the other two least stable forms.)

11.19 Draw the structures of *cis*-2-butene and *trans*-2-butene. Which of the two compounds would have the higher heat of hydrogenation? Explain.

11.20 Would you expect cyclobutadiene to be a stable molecule? Explain.

11.21 How many different isomers can be derived from ethylene if two hydrogen atoms are replaced by a fluorine atom and a chlorine atom? Draw their structures and name them. Indicate which are structural isomers and which are geometric isomers.

11.22 Suggest two chemical tests that would help you distinguish between these two compounds:

(a) $CH_3CH_2CH_2CH_2CH_3$
(b) $CH_3CH_2CH_2CH=CH_2$

11.23 Sulfuric acid (H_2SO_4) adds to the double bond of alkenes as H^+ and $^-OSO_3H$. Predict the products when sulfuric acid reacts with (a) ethylene and (b) propylene.

11.24 Acetylene is an unstable compound. It has a tendency to form benzene as follows:

$$3C_2H_2(g) \longrightarrow C_6H_6(l)$$

Calculate the standard enthalpy change in kilojoules for this reaction at 25°C.

11.25 Predict products when HBr is added to (a) 1-butene and (b) 2-butene.

11.26 Geometric isomers are not restricted to compounds containing the C=C bond. For example, certain disubstituted cycloalkanes can exist in the cis and the trans forms. Label the following molecules as the *cis* and *trans* isomer, of the same compound:

11.27 Write the structural formulas for these organic compounds: (a) 3-methylhexane, (b) 1,3,5-trichlorocyclohexane, (c) 2,3-dimethylpentane, (d) 2-phenyl-4-bromopentane, (e) 3,4,5-trimethyloctane.

11.28 Name these compounds:

(a)
$$\overset{\overset{\displaystyle CH_3}{|}}{CH_3-CH-CH_2-CH_2-CH_3}$$

(b)
$$\overset{\overset{\displaystyle C_2H_5 \quad CH_3 \; CH_3}{| \qquad | \quad \; |}}{CH_3-CH--CH-CH-CH_3}$$

(c)
$$CH_3-CH_2-\overset{\overset{\displaystyle |}{CH}}{}-CH_2-CH_3$$
$$\overset{|}{CH_2-CH_2-CH_3}$$

(d)
$$\overset{\overset{\displaystyle CH_3}{|}}{CH_2=CH-CH-CH=CH_2}$$

(e)
$$CH_3-C\equiv C-CH_2-CH_3$$

Aromatic Hydrocarbons

Review Questions

11.29 Comment on the extra stability of benzene compared to ethylene. Why does ethylene undergo addition reactions while benzene usually undergoes substitution reactions?

11.30 Benzene and cyclohexane molecules both contain six-membered rings. Benzene is a planar molecule, and cyclohexane is nonplanar. Explain.

Problems

11.31 Write structures for these compounds: (a) 1-bromo-3-methylbenzene, (b) 1-chloro-2-propyl-benzene, (c) 1,2,4,5-tetramethylbenzene.

11.32 Name these compounds:

(a) [structure: benzene ring with Cl at top, Cl at right, CH₃ at bottom]

(b) [structure: benzene ring with NO₂ at top, CH₂CH₃ at lower right, NO₂ at bottom]

(c) [structure: benzene ring with CH₃ at top, CH₃ at right, H₃C at left, CH₃ at bottom]

(d) [structure: benzene ring with CH₃—CH—CH=CH₂]

Chemistry of the Functional Groups

Review Questions

11.33 What are functional groups? Why is it logical and useful to classify organic compounds according to their functional groups?

11.34 Draw the Lewis structure for each of these functional groups: alcohol, ether, aldehyde, ketone, carboxylic acid, ester, amine.

Problems

11.35 Draw structures for molecules with these formulas: (a) CH_4O, (b) C_2H_6O, (c) $C_3H_6O_2$, (d) C_3H_8O.

11.36 Classify each of these molecules as alcohol, aldehyde, ketone, carboxylic acid, amine, or ether:

(a) $CH_3—O—CH_2—CH_3$ (b) $H_3C—CH_2—NH_2$

(c) $CH_3—CH_2—\overset{O}{\underset{H}{\overset{\|}{C}}}$ (d) $CH_3—\underset{O}{\overset{\|}{C}}—CH_2—CH_3$

(e) $H—\overset{O}{\overset{\|}{C}}—OH$ (f) $H_3C—CH_2CH_2—OH$

(g) [benzene ring]—$CH_2—\overset{NH_2}{\underset{H}{\overset{|}{C}}}—\overset{O}{\overset{\|}{C}}—OH$

11.37 Generally aldehydes are more susceptible to oxidation in air than are ketones. Use acetaldehyde and acetone as

examples and show why ketones such as acetone are more stable than aldehydes in this respect.

11.38 Complete this equation and identify the products:

$$HCOOH + CH_3OH \longrightarrow$$

11.39 A compound has the empirical formula $C_5H_{12}O$. Upon controlled oxidation, it is converted into a compound of empirical formula $C_5H_{10}O$, which behaves as a ketone. Draw possible structures for the original compound and the final compound.

11.40 A compound having the molecular formula $C_4H_{10}O$ does not react with sodium metal. In the presence of light, the compound reacts with Cl_2 to form three compounds having the formula C_4H_9OCl. Draw a structure for the original compound that is consistent with this information.

11.41 Predict the product or products of each of these reactions:

(a) $CH_3CH_2OH + HCOOH \longrightarrow$
(b) $H—C\equiv C—CH_3 + H_2 \longrightarrow$
(c) $\underset{H}{\overset{C_2H_5}{}}C=C\underset{H}{\overset{H}{}}$ + HBr $\longrightarrow$

11.42 Identify the functional groups in each of these molecules:

(a) $CH_3CH_2COCH_2CH_2CH_3$
(b) $CH_3COOC_2H_5$
(c) $CH_3CH_2OCH_2CH_2CH_3$

Chirality

Review Questions

11.43 What factor determines whether a carbon atom in a compound is asymmetric?

11.44 Give examples of a chiral substituted alkane and an achiral substituted alkane.

Problems

11.45 Which of these amino acids are chiral:

(a) $CH_3CH(NH_2)COOH$,   (b) $CH_2(NH_2)COOH$,
(c) $CH_2(OH)CH(NH_2)COOH$?

11.46 Indicate the asymmetric carbon atoms in these compounds:

(a) $CH_3—CH_2—\overset{CH_3}{\overset{|}{C}}H—\underset{NH_2}{\overset{|}{C}}H—\overset{O}{\overset{\|}{C}}—NH_2$

(b)

Additional Problems

11.47 Draw all the possible structural isomers for the molecule having the formula C_7H_7Cl. The molecule contains one benzene ring.

11.48 Given these data

$$C_2H_4(g) + 3O_2(g) \longrightarrow 2CO_2(g) + 2H_2O(l)$$
$$\Delta H° = -1411 \text{ kJ}$$

$$2C_2H_2(g) + 5O_2(g) \longrightarrow 4CO_2(g) + 2H_2O(l)$$
$$\Delta H° = -2599 \text{ kJ}$$

$$H_2(g) + \tfrac{1}{2}O_2(g) \longrightarrow H_2O(l)$$
$$\Delta H° = -285.8 \text{ kJ}$$

calculate the heat of hydrogenation for acetylene:

$$C_2H_2(g) + H_2(g) \longrightarrow C_2H_4(g)$$

11.49 State which member of each of these pairs of compounds is the more reactive and explain why: (a) propane and cyclopropane, (b) ethylene and methane, (c) acetaldehyde and acetone.

11.50 Like ethylene, tetrafluoroethylene (C_2F_4) undergoes polymerization reaction to form polytetrafluoroethylene (Teflon). Draw a repeating unit of the polymer.

11.51 An organic compound is found to contain 37.5 percent carbon, 3.2 percent hydrogen, and 59.3 percent fluorine by mass. These pressure and volume data were obtained for 1.00 g of this substance at 90°C:

P (atm)	V (L)
2.00	0.332
1.50	0.409
1.00	0.564
0.50	1.028

The molecule is known to have no dipole moment. (a) What is the empirical formula of this substance? (b) Does this substance behave as an ideal gas? (c) What is its molecular formula? (d) Draw the Lewis structure of this molecule and describe its geometry. (e) What is the systematic name of this compound?

11.52 State at least one commercial use for each of the following compounds: (a) 2-propanol (isopropanol), (b) acetic acid, (c) naphthalene, (d) methanol, (e) ethanol, (f) ethylene glycol, (g) methane, (h) ethylene.

11.53 How many liters of air (78 percent N_2, 22 percent O_2 by volume) at 20°C and 1.00 atm are needed for the complete combustion of 1.0 L of octane, C_8H_{18}, a typical gasoline component that has a density of 0.70 g/mL?

11.54 How many carbon-carbon sigma bonds are present in each of these molecules? (a) 2-butyne, (b) anthracene (see Figure 11.7), (c) 2,3-dimethylpentane

11.55 How many carbon-carbon sigma bonds are present in each of these molecules? (a) benzene, (b) cyclobutane, (c) 2-methyl-3-ethylpentane

11.56 The combustion of 20.63 mg of compound Y, which contains only C, H, and O, with excess oxygen gave 57.94 mg of CO_2 and 11.85 mg of H_2O. (a) Calculate how many milligrams of C, H, and O were present in the original sample of Y. (b) Derive the empirical formula of Y. (c) Suggest a plausible structure for Y if the empirical formula is the same as the molecular formula.

11.57 Draw all the structural isomers of compounds with the formula $C_4H_8Cl_2$. Indicate which isomers are chiral and give them systematic names.

11.58 The combustion of 3.795 mg of liquid B, which contains only C, H, and O, with excess oxygen gave 9.708 mg of CO_2 and 3.969 mg of H_2O. In a molar mass determination, 0.205 g of B vaporized at 1.00 atm and 200.0°C and occupied a volume of 89.8 mL. Derive the empirical formula, molar mass, and molecular formula of B and draw three plausible structures.

11.59 Beginning with 3-methyl-1-butyne, show how you would prepare these compounds:

(a) $\begin{array}{cc} & \text{Br} \quad \text{CH}_3 \\ & | \quad\quad | \\ \text{CH}_2{=}\text{C}-\text{CH}-\text{CH}_3 \end{array}$

(b) $\begin{array}{c} \quad\quad\quad\quad \text{CH}_3 \\ \quad\quad\quad\quad | \\ \text{CH}_2\text{Br}-\text{CBr}_2-\text{CH}-\text{CH}_3 \end{array}$

(c) $\begin{array}{cc} & \text{Br} \quad \text{CH}_3 \\ & | \quad\quad | \\ \text{CH}_3-\text{CH}-\text{CH}-\text{CH}_3 \end{array}$

11.60 Write structural formulas for these compounds: (a) *trans*-2-pentene, (b) 2-ethyl-1-butene, (c) 4-ethyl-*trans*-2-heptene, (d) 3-phenylbutyne.

11.61 Suppose benzene contained three distinct single bonds and three distinct double bonds. How many different structural isomers would there be for dichlorobenzene ($C_6H_4Cl_2$)? Draw all your proposed structures.

11.62 Write the structural formula of an aldehyde that is a structural isomer of acetone.

11.63 Draw structures for these compounds: (a) cyclopentane, (b) *cis*-2-butene, (c) 2-hexanol, (d) 1,4-dibromobenzene, (e) 2-butyne.

11.64 Name the classes to which these compounds belong:

(a) C_4H_9OH (b) $CH_3OC_2H_5$
(c) C_2H_5CHO (d) C_6H_5COOH
(e) CH_3NH_2

11.65 Ethanol, C_2H_5OH, and dimethyl ether, CH_3OCH_3, are structural isomers. Compare their melting points, boiling points, and solubilities in water.

11.66 Amines are Brønsted bases. The unpleasant smell of fish is due to the presence of certain amines. Explain why cooks often add lemon juice to suppress the odor of fish (in addition to enhancing the flavor).

11.67 You are given two bottles, each containing a colorless liquid. You are told that one liquid is cyclohexane and the other is benzene. Suggest one chemical test that would allow you to distinguish between these two liquids.

11.68 Give the chemical names of these organic compounds and write their formulas: marsh gas, grain alcohol, wood alcohol, rubbing alcohol, antifreeze, mothballs, chief ingredient of vinegar.

11.69 The compound $CH_3—C\equiv C—CH_3$ is hydrogenated to an alkene using platinum as the catalyst. Predict whether the product is the pure *trans* isomer, the pure *cis* isomer, or a mixture of *cis* and *trans* isomers. (*Hint:* Prior to the reaction the compound is bound to the surface of the metal and the H_2 molecule first dissociates into H atoms, which then approach the $C\equiv C$ bond from the top.)

11.70 How many asymmetric carbon atoms are present in each of these compounds?

(a)

$$H—\overset{\overset{\displaystyle H}{|}}{C}—\overset{\overset{\displaystyle H}{|}}{\underset{\underset{\displaystyle Cl}{|}}{C}}—\overset{\overset{\displaystyle H}{|}}{\underset{\underset{\displaystyle H}{|}}{C}}—Cl$$

(b)

$$H_3C—\overset{\overset{\displaystyle OH}{|}}{\underset{\underset{\displaystyle H}{|}}{C}}—————\overset{\overset{\displaystyle CH_3}{|}}{\underset{\underset{\displaystyle H}{|}}{C}}—CH_2OH$$

(c)

(structure of a sugar ring with CH_2OH, C, O, OH, H, OH, H, HO, C, C, H, OH groups)

11.71 Isopropanol is prepared by reacting propylene (CH_3CHCH_2) with sulfuric acid, followed by treatment with water. (a) Show the sequence of steps leading to the product. What is the role of sulfuric acid? (b) Draw the structure of an alcohol that is an isomer of isopropanol. (c) Is isopropanol a chiral molecule? (d) What property of isopropanol makes it useful as a rubbing alcohol?

11.72 When a mixture of methane and bromine vapor is exposed to light, this reaction occurs slowly:

$$CH_4(g) + Br_2(g) \longrightarrow CH_3Br(g) + HBr(g)$$

Suggest a mechanism for this reaction. (*Hint:* Bromine vapor is deep red; methane is colorless.)

Special Problems

11.73 Octane number is assigned to gasoline to indicate the tendency of "knocking" in the automobile's engine. The higher the octane number, the more smoothly the fuel will burn without knocking. Branched-chain aliphatic hydrocarbons have higher octane numbers than straight-chain aliphatic hydrocarbons, and aromatic hydrocarbons have the highest octane numbers.

(a) Arrange these compounds in the order of decreasing octane numbers: 2,2,4-trimethylpentane, toluene (methylbenzene), *n*-heptane, and 2-methylhexane.
(b) Oil refineries carry out *catalytic reforming* in which a straight-chain hydrocarbon, in the presence of a catalyst, is converted to an aromatic molecule and a useful by-product. Write an equation for the conversion from *n*-heptane to toluene.
(c) Until recently (2000), methyl *tert*-butyl ether had been widely used as an antiknocking agent to enhance the octane number of gasoline. Write the structural formula of the compound.

11.74 Fats and oils are names for the same class of compounds, called triglycerides, which contain three ester groups

$$CH_2—O—\overset{\overset{\displaystyle O}{\|}}{C}—R$$
$$CH—O—\overset{\overset{\displaystyle O}{\|}}{C}—R'$$
$$CH_2—O—\overset{\overset{\displaystyle O}{\|}}{C}—R''$$

A fat or oil

in which R, R′, and R″ represent long hydrocarbon chains.

(a) Suggest a reaction that leads to the formation of a triglyceride molecule, starting with glycerol and carboxylic acids (see p. 378 for structure of glycerol).
(b) In the old days, soaps were made by hydrolyzing animal fat with lye (a sodium hydroxide solution). Write an equation for this reaction.
(c) The difference between fats and oils is that at room temperature, the former are solid and the latter are liquids. Fats are usually produced by animals, whereas oils are commonly found in plants. The melting points of these substances are determined by the number of

C=C bonds (or the extent of unsaturation) present—the larger the number of C=C bonds, the lower the melting point and the more likely the substance is a liquid. Explain.

(d) One way to convert liquid oil to solid fat is to hydrogenate the oil, a process by which some or all of the C=C bonds are converted to C—C bonds. This procedure prolongs shelf life of the oil by removing the more reactive C=C group and facilitates packaging. How would you carry out such a process (that is, what reagents and catalyst would you employ)?

(e) The degree of unsaturation of oil can be determined by reacting the oil with iodine, which reacts with the C=C as follows:

$$-\overset{|}{\underset{|}{C}}-\overset{|}{C}=\overset{|}{C}-\overset{|}{\underset{|}{C}}-\ +\ I_2\ \longrightarrow\ -\overset{|}{\underset{|}{C}}-\overset{\overset{I}{|}}{\underset{|}{C}}-\overset{\overset{I}{|}}{\underset{|}{C}}-\overset{|}{\underset{|}{C}}-$$

The procedure is to add a known amount of iodine to the oil and allow the reaction to go to completion. The amount of excess (unreacted) iodine is determined by titrating the remaining iodine with a standard sodium thiosulfate ($Na_2S_2O_3$) solution:

$$I_2 + 2Na_2S_2O_3 \longrightarrow Na_2S_4O_6 + 2NaI$$

The number of grams of iodine that reacts with 100 g of oil is called the *iodine number*. In one case, 43.8 g of I_2 were treated with 35.3 g of corn oil. The excess iodine required 20.6 mL of 0.142 M $Na_2S_2O_3$ for neutralization. Calculate the iodine number of the corn oil.

Answers to Practice Exercises

11.1 5. **11.2** $CH_3CH_2COOCH_3$ and H_2O. **11.3** No.

Copper and aluminum are good conductors of electricity, but they do possess some electrical resistance. In fact, up to 20 percent of electrical energy may be lost (in the form of heat) when these metals are used as electrical transmission cables. Wouldn't it be marvelous if we could produce cables that possessed no electrical resistance?

Actually, it has been known for many years that certain metals and alloys totally lose their resistance when cooled to very low temperatures (around the boiling point of liquid helium, or 4 K). However, it is not practical to use these substances, called superconductors, for transmission of electric power because of the prohibitive cost of maintaining electrical cables at such low temperatures.

In 1986 two physicists in Switzerland discovered a new class of materials that are superconducting at around 30 K. Although 30 K is still a very low temperature, the improvement over the 4 K range was so dramatic that their work generated immense interest and triggered a flurry of research activity. Within months, scientists synthesized compounds containing copper, barium, and a rare earth metal that are superconducting at about 90 K, which is well above the boiling point of liquid nitrogen (77 K). One of the distinct properties of a superconductor is the capacity to levitate magnetized objects.

The levitation of a magnet above a high-temperature superconductor immersed in liquid nitrogen.

There is good reason to be enthusiastic about the new class of superconductors. Because liquid nitrogen is quite cheap (a liter of liquid nitrogen costs less than a liter of milk), we can envision electrical power being transmitted over hundreds of miles with no loss of energy. The levitation effect can be used in the operation of trains that ride above (but not in contact with) the tracks. Such trains are fast as well as quiet and smooth-running. Superconductors can be used to build very fast computers, called supercomputers, whose speed is limited by how fast electric current flows. The enormous magnetic fields that can be created in superconductors will result in more powerful particle accelerators, efficient devices for nuclear fusion, and more accurate magnetic resonance image techniques for medical diagnosis.

Although some progress has been made since their discovery, only limited uses have been found for these high-temperature superconductors. The reason is that it has proven difficult to make durable wires out of these materials. Furthermore, the mechanism of this interesting phenomenon is still not fully understood. In another promising development, in 2001 scientists in Japan reported that the compound magnesium diboride (MgB_2) becomes superconducting at 39 K. Although it operates at a lower temperature, this compound shows considerable promise because it is cheap and it can be more easily made into wires and thin films. It seems likely that similar compounds with better performance will be discovered before too long.

Today, high-temperature superconductivity is one of the most actively researched areas in physics and chemistry.

An experimental levitation train that operates on superconducting material at liquid helium temperature.

Intermolecular Forces and Liquids and Solids

Chapter 12

ESSENTIAL CONCEPTS

Intermolecular Forces Intermolecular forces, which are responsible for the nonideal behavior of gases, also account for the existence of the condensed states of matter—liquids and solids. These forces exist between polar molecules, between ions and polar molecules, and between nonpolar molecules. A special type of intermolecular force, called the hydrogen bond, describes the interaction between the hydrogen atom in a polar bond and an electronegative atom such as O, N, or F. The melting point and boiling point of a substance are good measures of the strength of intermolecular forces.

The Liquid State Liquids tend to assume the shapes of their containers. The surface tension of a liquid is the energy required to increase its surface area. It manifests itself in capillary action, which is responsible for the rise (or depression) of a liquid in a narrow tubing. Viscosity, another characteristic of liquids, is a measure of a liquid's resistance to flow. It always decreases with increasing temperature. The structure of water is unique in that its solid state (ice) is less dense than its liquid state. This behavior is due to the ability of a water molecule to form four hydrogen bonds with its neighbors.

The Crystalline State A crystalline solid possesses rigid and long-range order. Different crystal structures can be generated by packing identical spheres in three dimensions.

Bonding in Solids Atoms, molecules, or ions are held in a solid by different types of bonding. Electrostatic forces are responsible for ionic solids, intermolecular forces are responsible for molecular solids, covalent bonds are responsible for covalent solids, and a special type of interaction, which involves electrons being delocalized over the entire crystal, accounts for the existence of metals and their abilities to conduct heat and electricity.

Phase Transitions The states of matter can be interconverted by heating or cooling. Two phases are in equilibrium at the transition temperature such as boiling or freezing. Solids can also be directly converted to vapor by sublimation. Above a certain temperature, called the critical temperature, the gas form of a substance cannot be made to liquefy. The pressure-temperature relationships of solid, liquid, and vapor phases are best represented by a phase diagram, which shows the conditions at which a substance can exist in one, two, or all three phases.

12.1 The Kinetic Molecular Theory of Liquids and Solids

In Chapter 5 we used the kinetic molecular theory to explain the behavior of gases. That explanation was based on the understanding that a gaseous system is a collection of molecules in constant, random motion. In gases, the distances between molecules are so great (compared with their diameters) that at ordinary temperatures and pressures (say, 25°C and 1 atm), there is no appreciable interaction between the molecules. This rather simple description explains several characteristic properties of gases. Because there is a great deal of empty space in a gas, that is, space not occupied by molecules, gases can be readily compressed. The lack of strong forces between molecules enables a gas to expand to the volume of its container. The large amount of empty space also explains why gases have very low densities under normal conditions.

Liquids and solids are quite a different story. The principal difference between the condensed state (liquids or solids) and the gaseous state lies in the distance between molecules. In a liquid the molecules are so close together that there is very little empty space. Thus liquids are much more difficult to compress than gases and much denser under normal conditions. Molecules in a liquid are held together by one or more types of attractive forces, which will be discussed in Section 12.2. A liquid also has a definite volume, because molecules in a liquid do not break away from the attractive forces. The molecules can, however, move past one another freely, and so a liquid can flow, can be poured, and assumes the shape of its container.

In a solid, molecules are held rigidly in position with virtually no freedom of motion. Many solids are characterized by long-range order; that is, the molecules are arranged in regular configurations in three dimensions. There is even less empty space in a solid than in a liquid. Thus solids are almost incompressible and possess definite shape and volume. With very few exceptions (water being the most important), the density of the solid is higher than that of the liquid for a given substance. Table 12.1 summarizes some of the characteristic properties of the three states of matter.

12.2 Intermolecular Forces

Attractive forces between molecules, called **intermolecular forces,** are responsible for the nonideal behavior of gases that was described in Chapter 5. They also account for the existence of the condensed states of matter—liquids and solids. As the temperature

Table 12.1	Characteristic Properties of Gases, Liquids, and Solids			
State of Matter	**Volume/Shape**	**Density**	**Compressibility**	**Motion of Molecules**
Gas	Assumes the volume and shape of its container	Low	Very compressible	Very free motion
Liquid	Has a definite volume but assumes the shape of its container	High	Only slightly compressible	Slide past one another freely
Solid	Has a definite volume and shape	High	Virtually incompressible	Vibrate about fixed positions

of a gas drops, the average kinetic energy of its molecules decreases. Eventually, at a sufficiently low temperature, the molecules no longer have enough energy to break away from one another's attraction. At this point, the molecules aggregate to form small drops of liquid. This phenomenon of going from the gaseous to the liquid state is known as *condensation*.

Figure 12.1 *Molecules that have a permanent dipole moment tend to align with opposite polarities in the solid phase for maximum attractive interaction.*

In contrast to intermolecular forces, ***intramolecular forces*** *hold atoms together in a molecule.* (Chemical bonding, discussed in Chapters 9 and 10, involves intramolecular forces.) Intramolecular forces stabilize individual molecules, whereas intermolecular forces are primarily responsible for the bulk properties of matter (for example, melting point and boiling point).

Generally, intermolecular forces are much weaker than intramolecular forces. Thus it usually requires much less energy to evaporate a liquid than to break the bonds in the molecules of the liquid. For example, it takes about 41 kJ of energy to vaporize 1 mole of water at its boiling point; it takes about 930 kJ of energy to break the two O—H bonds in 1 mole of water molecules. The boiling points of substances often reflect the strength of the intermolecular forces operating among the molecules. At the boiling point, enough energy must be supplied to overcome the attraction among molecules so that they can enter the vapor phase. If it takes more energy to separate molecules of substance A than of substance B because A molecules are held together by stronger intermolecular forces than B molecules, then the boiling point of A is higher than that of B. The same principle applies also to the melting points of the substances. In general the melting points of substances increase with the strength of the intermolecular forces.

To understand the properties of condensed matter, we must understand the different types of intermolecular forces. *Dipole-dipole, dipole-induced dipole, and dispersion forces* make up what chemists commonly refer to as ***van der Waals forces.*** Ions and dipoles are attracted to one another by electrostatic forces called *ion-dipole forces;* they are not van der Waals forces. *Hydrogen bonding* is a particularly strong type of dipole-dipole interaction. Because only a few elements can participate in hydrogen bond formation, it is treated as a separate category. Depending on the physical state of a substance (that is, gas, liquid, or solid), the nature of chemical bonds, and the types of elements present, more than one type of interaction may contribute to the total attraction between molecules, as we will soon see.

Dipole-Dipole Forces

Dipole-dipole forces are *forces that act between polar molecules,* that is, between molecules that possess dipole moments (see Section 10.2). Their origin is electrostatic. The larger the dipole moments, the greater the force. Figure 12.1 shows the orientation of polar molecules in a solid. In liquids the molecules are not held as rigidly as in a solid, but they tend to align in such a way that, on the average, the attractive interaction is at a maximum.

Ion-Dipole Forces

Electrostatic interaction also explains ***ion-dipole forces,*** which *occur between an ion (either a cation or an anion) and a polar molecule* (Figure 12.2). The strength of this interaction depends on the charge and the size of the ion and on the magnitude of the dipole moment and the size of the molecule. The charges on cations are generally more concentrated, because cations are usually smaller than anions. Therefore, for equal charges, a cation interacts more strongly with dipoles than does an anion.

Figure 12.2 *Two types of ion-dipole interaction.*

Figure 12.3 *Deflection of water stream by a charged ebonite rod.*

Hydration, discussed in Section 4.1, is one example of ion-dipole interaction. In an aqueous NaCl solution, the Na^+ and Cl^- ions are surrounded by water molecules, which have a large dipole moment (1.87 D). When an ionic compound like NaCl dissolves, the water molecules act as an electrical insulator that keeps the ions apart. On the other hand, carbon tetrachloride (CCl_4) is a nonpolar molecule, and therefore it lacks the ability to participate in ion-dipole interaction. What we find in practice is that carbon tetrachloride is a poor solvent for ionic compounds, as are most nonpolar liquids.

A convincing demonstration of ion-dipole attraction is that shown in Figure 12.3. Water in a buret is allowed to run into a beaker. If a negatively charged rod (for example, an ebonite rod rubbed with fur) is brought near the stream of water, the water is deflected *toward* the rod. When the ebonite rod is replaced by a positively charged rod (for example, a glass rod rubbed with silk), the water is again attracted to the rod. With the ebonite rod in position, the water molecules orient themselves so that the positive ends of the dipoles lie closer to the rod and they are attracted to the rod's negative charge. When the ebonite rod is replaced with the glass rod, the negative ends of the dipoles are oriented toward the rod. Similar deflection is observed for other polar liquids. No deflection occurs when nonpolar liquids such as hexane (C_6H_{14}) are used instead of water.

Dispersion Forces

So far only ionic species and polar molecules have been discussed. What kind of attractive interaction exists between nonpolar molecules? To learn the answer to this question, consider the arrangement shown in Figure 12.4. If we place an ion or a polar molecule near an atom (or a nonpolar molecule), the electron distribution of the atom (or molecule) is distorted by the force exerted by the ion or the polar molecule. The resulting dipole in the atom (or molecule) is said to be an ***induced dipole*** because *the separation of positive and negative charges in the atom (or a nonpolar molecule) is due to the proximity of an ion or a polar molecule*. The attractive interaction between an ion and the induced dipole is called *ion-induced dipole interaction,* and the attractive interaction between a polar molecule and the induced dipole is called *dipole-induced dipole interaction.*

The likelihood of a dipole moment being induced depends not only on the charge on the ion or the strength of the dipole but also on the polarizability of the atom or molecule. ***Polarizability*** is *the ease with which the electron distribution in the atom (or molecule) can be distorted.* Generally, the larger the number of electrons and the more diffuse the electron cloud in the atom or molecule, the greater its polarizability. By *diffuse cloud* we mean an electron cloud that is spread over an appreciable volume, so that the electrons are not held tightly by the nucleus.

Polarizability enables gases containing atoms or nonpolar molecules (for example, He and N_2) to condense. In a helium atom the electrons are moving at some distance from the nucleus. At any instant it is likely that the atom has a dipole moment created by the specific positions of the electrons. This dipole moment is called an *instantaneous dipole* because it lasts for just a tiny fraction of a second. In the next instant the electrons are in different locations and the atom has a new instantaneous dipole and so on. Averaged over time (that is, the time it takes to make a dipole moment measurement), however, the atom has no dipole moment because the instantaneous dipoles all cancel one another. In a collection of He atoms an instantaneous dipole of one He atom can induce a dipole in each of its nearest neighbors (Figure 12.5). At the next moment, a different instantaneous dipole can create temporary dipoles in the

(a)

Cation   Induced dipole

(b)

Dipole   Induced dipole

(c)

Figure 12.4 *(a) Spherical charge distribution in a helium atom. (b) Distortion caused by the approach of a cation. (c) Distortion caused by the approach of a dipole.*

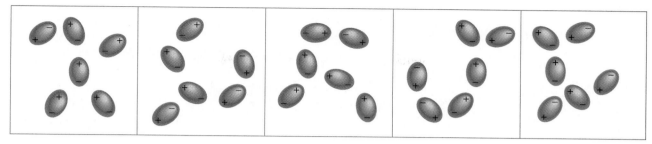

Figure 12.5 *Induced dipoles interacting with each other. Such patterns exist only momentarily; new arrangements are formed in the next instant. This type of interaction is responsible for the condensation of nonpolar gases.*

surrounding He atoms. The important point is that this kind of interaction results in an attraction between the He atoms. At very low temperatures (and reduced atomic speeds), this attraction is enough to hold the atoms together, causing the helium gas to condense. The attraction between nonpolar molecules can be explained similarly.

A quantum mechanical interpretation of temporary dipoles was provided by Fritz London, a German physicist, in 1930. London showed that the magnitude of this attractive interaction is directly proportional to the polarizability of the atom or molecule. As we might expect, *the attractive forces that arise as a result of temporary dipoles induced in the atoms or molecules,* called **dispersion forces,** may be quite weak. This is certainly true of helium, which has a boiling point of only 4.2 K, or −269°C. (Note that helium has only two electrons, which are tightly held in the 1s orbital. Therefore, the helium atom has a small polarizability.)

Dispersion forces usually increase with molar mass because molecules with larger molar mass tend to have more electrons, and dispersion forces increase in strength with the number of electrons. Furthermore, larger molar mass often means bigger atoms whose electron distributions are more easily disturbed because the outer electrons are less tightly held by the nuclei. Table 12.2 compares the melting points of some substances that consist of nonpolar molecules. As expected, the melting point increases as the number of electrons in the molecule increases. Because these are all nonpolar molecules, the only attractive intermolecular forces present are the dispersion forces.

In many cases, dispersion forces are comparable to or even greater than the dipole-dipole forces between polar molecules. For a dramatic illustration, let us compare the boiling points of CH_3F (−78.4°C) and CCl_4 (76.5°C). Although CH_3F has a dipole moment of 1.8 D, it boils at a much lower temperature than CCl_4, a nonpolar molecule. CCl_4 boils at a higher temperature simply because it contains more electrons. As a result, the dispersion forces between CCl_4 molecules are stronger than the dispersion forces plus the dipole-dipole forces between CH_3F molecules. (Keep in mind that dispersion forces exist among species of all types, whether they are neutral or bear a net charge and whether they are polar or nonpolar.)

Strictly speaking, the forces between two nonbonded atoms should be called "interatomic forces." However, for simplicity we use the term "intermolecular forces" for both atoms and molecules.

Example 12.1 What type(s) of intermolecular forces exist between these pairs? (a) HBr and H_2S, (b) Cl_2 and CBr_4, (c) I_2 and NO_3^-, and (d) NH_3 and C_6H_6?

Reasoning and Solution To identify intermolecular forces, it is useful to classify the species being considered as (1) nonpolar molecules, (2) polar molecules, and (3) ions. Keep in mind that dispersion forces exist between all species.

Table 12.2 Melting Points of Similar Nonpolar Compounds

Compound	Melting Point (°C)
CH_4	−182.5
CF_4	−150.0
CCl_4	−23.0
CBr_4	90.0
CI_4	171.0

(a) Both HBr and H_2S are polar molecules, so the forces between them are dipole-dipole forces. There are also dispersion forces between the molecules.

(b) Both Cl_2 and CBr_4 are nonpolar, so there are only dispersion forces between these molecules.

(c) I_2 is nonpolar, so the forces between it and the ion NO_3^- are ion-induced dipole forces and dispersion forces.

(d) NH_3 is polar, and C_6H_6 is nonpolar. The forces are dipole-induced dipole forces and dispersion forces.

Practice Exercise Name the type(s) of intermolecular forces that exist between molecules (or basic units) in each of these species: (a) LiF, (b) CH_4, and (c) SO_2.

Similar problem: 12.10.

The Hydrogen Bond

The **hydrogen bond** is *a special type of dipole-dipole interaction between the hydrogen atom in a polar bond, such as N—H, O—H, or F—H, and an electronegative O, N, or F atom.* This interaction is written

$$A—H\cdots B \qquad \text{or} \qquad A—H\cdots A$$

A and B represent O, N, or F; A—H is one molecule or part of a molecule and B is a part of another molecule; and the dotted line represents the hydrogen bond. The three atoms usually lie along a straight line, but the angle AHB (or AHA) can deviate as much as 30° from linearity.

The average energy of a hydrogen bond is quite large for a dipole-dipole interaction (up to 40 kJ/mol). Thus, hydrogen bonds are a powerful force in determining the structures and properties of many compounds. Figure 12.6 shows several examples of hydrogen bonding.

Early evidence of hydrogen bonding came from the study of the boiling points of compounds. Normally, the boiling points of a series of similar compounds containing elements in the same group increase with increasing molar mass. But as Figure 12.7 shows, the hydrogen compounds of Groups 5A, 6A, and 7A elements do not follow this trend. In each of these series, the lightest compound (NH_3, H_2O, HF) has the

The three most electronegative elements that take part in hydrogen bonding.

Figure 12.6 *Hydrogen bonding among water, ammonia, and hydrogen fluoride molecules. Solid lines represent covalent bonds, and dotted red lines represent hydrogen bonds.*

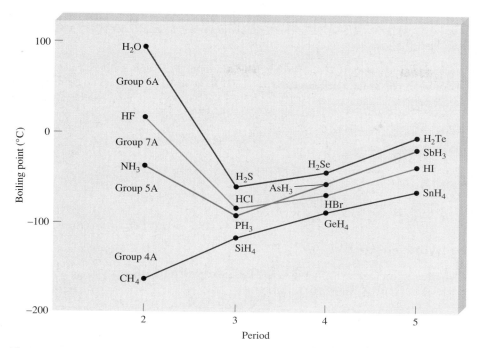

Figure 12.7 *Boiling points of the hydrogen compounds of Groups 4A, 5A, 6A, and 7A elements. Although normally we expect the boiling point to increase as we move down a group, we see that three compounds (NH$_3$, H$_2$O, and HF) behave differently. The anomaly can be explained in terms of intermolecular hydrogen bonding.*

highest boiling point, contrary to our expectations based on molar mass. The reason is that there is extensive hydrogen bonding between molecules in these compounds. In solid HF, for example, molecules do not exist as individual units; instead, they form long zigzag chains held together by hydrogen bonds:

In the liquid state the zigzag chains are broken, but the molecules are still hydrogen-bonded to one another. Molecules held together by hydrogen bonding are more difficult to break apart, so liquid HF has an unusually high boiling point.

The strength of the hydrogen bond is determined by the coulombic interaction between the lone-pair electrons of the electronegative atom and the hydrogen nucleus. It may seem odd that the boiling point of HF is lower than that of water. Fluorine is more electronegative than oxygen, and so we would expect a stronger hydrogen bond to exist in liquid HF than in H$_2$O. But H$_2$O is unique in that each molecule takes part in *four* intermolecular hydrogen bonds, and the H$_2$O molecules are therefore held together more strongly. We will return to this very important property of water in Section 12.3. Example 12.2 shows the type of species that can form hydrogen bonds with water.

HCOOH forms hydrogen bonds with two H₂O molecules.

Similar problem: 12.12.

Example 12.2   Which of these can form hydrogen bonds with water? CH_3OCH_3, CH_4, F^-, HCOOH, Na^+

Reasoning and Solution   For hydrogen bonds to form, an H atom must be bonded to the atom of one of the electronegative elements (N, O, and F). There are no such atoms in CH_4 or Na^+. Therefore, only CH_3OCH_3, F^-, and HCOOH can form hydrogen bonds with water.

Practice Exercise   Which of these species are capable of hydrogen bonding among themselves? (a) H_2S, (b) C_6H_6, (c) CH_3OH

The intermolecular forces discussed so far are all attractive in nature. Keep in mind, though, that molecules also exert repulsive forces on one another. Thus when two molecules approach each other, the repulsion between their electrons and between their nuclei comes into play. The magnitude of the repulsive force rises very steeply as the distance separating the molecules in a condensed phase decreases. This is why liquids and solids are so hard to compress. In these phases, the molecules are already in close contact with one another, and so they greatly resist being compressed further.

12.3 The Liquid State

A great many interesting and important chemical reactions occur in water and other liquid solvents, as we will discover in subsequent chapters. In this section we will look at two phenomena associated with liquids: surface tension and viscosity, both of which are attributable to intermolecular forces. We will also discuss the structure and properties of water.

Surface Tension

We have seen that liquids tend to assume the shapes of their containers. Why, then, does water bead up on a newly waxed car instead of forming a sheet over it? The answer to this question lies in intermolecular forces.

Molecules within a liquid are pulled in all directions by intermolecular forces; there is no tendency for them to be pulled in any one way. However, molecules at the surface are pulled downward and sideways by other molecules, but not upward away from the surface (Figure 12.8). These intermolecular attractions thus tend to pull the molecules into the liquid and cause the surface to tighten like an elastic film. Because there is little or no attraction between polar water molecules and the wax molecules

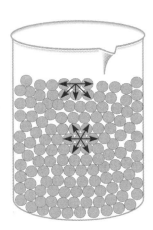

Figure 12.8 *Intermolecular forces acting on a molecule in the surface layer of a liquid and in the interior region of the liquid.*

(which are essentially nonpolar) on a freshly waxed car, a drop of water assumes the shape of a small round bead (Figure 12.9).

A measure of the elastic force in the surface of a liquid is surface tension. The **surface tension** of a liquid is *the amount of energy required to stretch or increase the surface by unit area.* Liquids in which there are strong intermolecular forces also have high surface tensions. For example, because of hydrogen bonding, water has a considerably greater surface tension than most common liquids.

Another way in which surface tension manifests itself is in *capillary action.* Figure 12.10(a) shows water rising spontaneously in a capillary tube. A thin film of water adheres to the wall of the glass tube. The surface tension of water causes this film to contract, and as it does, it pulls the water up the tube. Two types of forces bring about capillary action. One is *the intermolecular attraction between like molecules* (in this case, the water molecules), called **cohesion.** The other, which is called **adhesion,** is *an attraction between unlike molecules,* such as those in water and in the sides of a glass tube. If adhesion is stronger than cohesion, as it is in Figure 12.10(a), the contents of the tube will be pulled upward. This process continues until the adhesive force is balanced by the weight of the water in the tube. This action is by no means universal among liquids, as Figure 12.10(b) shows. In mercury, cohesion is greater than the adhesion between mercury and glass, so that a depression in the liquid level actually occurs when a capillary tube is dipped into mercury.

Figure 12.9 *Water beads on an apple, which has a waxy surface.*

Viscosity

The expression "slow as molasses in January" owes its truth to another physical property of liquids called viscosity. **Viscosity** is *a measure of a fluid's resistance to flow.* The greater the viscosity, the more slowly the liquid flows. The viscosity of a liquid usually decreases as temperature increases; thus hot molasses flows much faster than cold molasses.

(a) (b)

Figure 12.10 *(a) When adhesion is greater than cohesion, the liquid (for example, water) rises in the capillary tube. (b) When cohesion is greater than adhesion, as it is for mercury, a depression of the liquid in the capillary tube results. Note that the meniscus in the tube of water is concave, or rounded downward, whereas that in the tube of mercury is convex, or rounded upward.*

Table 12.3 | Viscosity of Some Common Liquids at 20°C

Liquid	Viscosity (N s/m^2)*
Acetone (C_3H_6O)	3.16×10^{-4}
Benzene (C_6H_6)	6.25×10^{-4}
Carbon tetrachloride (CCl_4)	9.69×10^{-4}
Ethanol (C_2H_5OH)	1.20×10^{-3}
Diethyl ether ($C_2H_5OC_2H_5$)	2.33×10^{-4}
Glycerol ($C_3H_8O_3$)	1.49
Mercury (Hg)	1.55×10^{-3}
Water (H_2O)	1.01×10^{-3}
Blood	4×10^{-3}

*The SI units of viscosity are newton-second per meter squared.

Liquids that have strong intermolecular forces have higher viscosities than those that have weak intermolecular forces (Table 12.3). Water has a higher viscosity than many other liquids because of its ability to form hydrogen bonds. Interestingly, the viscosity of glycerol is significantly higher than that of all the other liquids shown. Glycerol has the structure

$$CH_2—OH$$
$$|$$
$$CH—OH$$
$$|$$
$$CH_2—OH$$

Glycerol is a clear, odorless, syrupy liquid used to make explosives, ink, and lubricants.

Like water, glycerol can form hydrogen bonds. Each glycerol molecule has three —OH groups that can participate in hydrogen bonding with other glycerol molecules. Furthermore, because of their shape, the molecules have a great tendency to become entangled rather than to slip past one another as the molecules in less viscous liquids do. These interactions contribute to its high viscosity.

The Structure and Properties of Water

If water did not have the ability to form hydrogen bonds, it would be a gas at room temperature.

Water is so common a substance on Earth that we often overlook its unique nature. All life processes involve water. Water is an excellent solvent for many ionic compounds, and for other substances capable of forming hydrogen bonds with water.

As Table 6.2 shows, water has a high specific heat. The reason for this is that to raise the temperature of water (that is, to increase the average kinetic energy of water molecules), we must first break the many intermolecular hydrogen bonds. Thus water can absorb a substantial amount of heat while its temperature rises only slightly. The converse is also true: Water can give off much heat with only a slight decrease in its temperature. For this reason, the huge quantities of water that are present in our lakes and oceans can effectively moderate the climate of adjacent land areas by absorbing heat in the summer and giving off heat in the winter, with only small changes in the temperature of the body of water.

The most striking property of water is that its solid form is less dense than its liquid form: An ice cube floats at the surface of water in a glass. This is a virtually unique property. The density of almost all other substances is greater in the solid state than in the liquid state (Figure 12.11).

Figure 12.11 *Left: Ice cubes float on water. Right: Solid benzene sinks to the bottom of liquid benzene.*

To understand why water is different, we have to examine the electronic structure of the H_2O molecule. As we saw in Chapter 9, there are two pairs of nonbonding electrons, or two lone pairs, on the oxygen atom:

$$\overset{\displaystyle \cdot\cdot\,\cdot\cdot}{O}\diagdown\diagup$$
$$H \qquad H$$

Although many compounds are capable of forming intermolecular hydrogen bonds, the difference between H_2O and other polar molecules, such as NH_3 and HF, is that each oxygen atom can form *two* hydrogen bonds, the same as the number of lone electron pairs on the oxygen atom. Thus water molecules are joined together in an extensive three-dimensional network in which each oxygen atom is approximately tetrahedrally bonded to four hydrogen atoms, two by covalent bonds and two by hydrogen bonds. This equality in the number of hydrogen atoms and lone pairs is not characteristic of NH_3 or HF or, for that matter, of any other molecule capable of forming hydrogen bonds. Consequently, these other molecules can form rings or chains, but not three-dimensional structures.

The highly ordered three-dimensional structure of ice (Figure 12.12) prevents the molecules from getting too close to one another. But consider what happens when ice melts. At the melting point, a number of water molecules have enough kinetic energy to pull free of the intermolecular hydrogen bonds. These molecules become trapped in the cavities of the three-dimensional structure, which is broken down into smaller clusters. As a result there are more molecules per unit volume in liquid water than in ice. Thus, because density = mass/volume, the density of water is greater than that of ice. With further heating, more water molecules are released from intermolecular hydrogen bonding, so that just above the melting point water's density tends to rise with temperature. Of course, at the same time, water expands as it is being heated, with the result that its density is decreased. These two processes—the trapping of free water molecules in cavities and thermal expansion—act in opposite directions. From 0°C to 4°C, the trapping prevails and water becomes progressively denser. Beyond 4°C, however, thermal expansion predominates and the density of water decreases with increasing temperature (Figure 12.13).

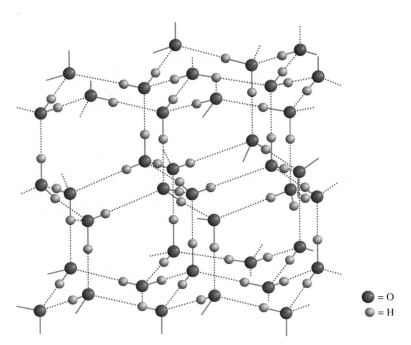

Figure 12.12 *The three-dimensional structure of ice. Each O atom is bonded to four H atoms. The covalent bonds are shown by short solid lines and the weaker hydrogen bonds by long dotted lines between O and H. The empty space in the structure accounts for the low density of ice.*

Figure 12.13 *Plot of density versus temperature for liquid water. The maximum density of water is reached at 4°C. The density of ice at 0°C is about 0.92 g/cm³.*

12.4 Crystal Structure

Solids can be divided into two categories: crystalline and amorphous. Ice is a ***crystalline solid,*** which *possesses rigid and long-range order; its atoms, molecules, or ions occupy specific positions.* The arrangement of atoms, molecules, or ions in a crystalline solid is such that the net attractive intermolecular forces are at their maximum. The forces responsible for the stability of any crystal can be ionic forces, covalent bonds, van der Waals forces, hydrogen bonds, or a combination of these forces. *Amorphous solids* such as glass lack a well-defined arrangement and long-range molecular order. In this section we will concentrate on the structure of crystalline solids.

The basic repeating structural unit of a crystalline solid is a ***unit cell.*** Figure 12.14 shows a unit cell and its extension in three dimensions. Each sphere *represents an atom, an ion, or a molecule* and is called a ***lattice point.*** In many crystals, the lattice point does not actually contain an atom, ion, or molecule. Rather, there may be several atoms, ions, or molecules identically arranged about each lattice point. For simplicity, however, we can assume that each lattice point is occupied by an atom. Every crystalline solid can be described in terms of one of the seven types of unit cells shown in Figure 12.15. The geometry of the cubic unit cell is particularly simple because all sides and all angles are equal. Any of the unit cells, when repeated in space in all three dimensions, forms the lattice structure characteristic of a crystalline solid.

Packing Spheres

We can understand the general geometric requirements for crystal formation by considering the different ways of packing a number of identical spheres (Ping-Pong balls, for example) to form an ordered three-dimensional structure. The way the spheres are arranged in layers determines what type of unit cell we have.

Packing spheres

In the simplest case, a layer of spheres can be arranged as shown in Figure 12.16(a). The three-dimensional structure can be generated by placing a layer above and below this layer in such a way that spheres in one layer are directly over the spheres in the layer below it. This procedure can be extended to generate many, many layers, as in the case of a crystal. Focusing on the sphere marked with *x,* we see that it is in contact with four spheres in its own layer, one sphere in the layer above, and one sphere in the layer

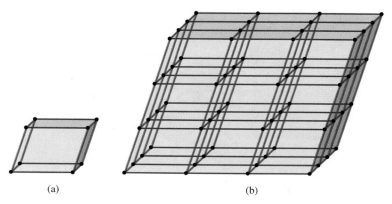

(a)   (b)

Figure 12.14 *(a) A unit cell and (b) its extension in three dimensions. The black spheres represent either atoms or molecules.*

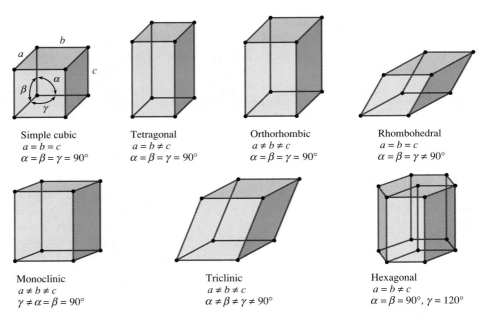

Figure 12.15 *The seven types of unit cells. Angle α is defined by edges b and c, angle β by edges a and c, and angle γ by edges a and b.*

below. Each sphere in this arrangement is said to have a coordination number of 6 because it has six immediate neighbors. The ***coordination number*** is defined as *the number of atoms (or ions) surrounding an atom (or ion) in a crystal lattice*. The basic, repeating unit in this array of spheres is called a *simple cubic cell* (scc) [Figure 12.16(b)].

The other types of cubic cells are the *body-centered cubic cell* (bcc) and the *face-centered cubic cell* (fcc) (Figure 12.17). A body-centered cubic arrangement differs from a simple cube in that the second layer of spheres fits into the depressions of the first layer and the third layer into the depressions of the second layer. The coordination number of each sphere in this structure is 8 (each sphere is in contact with four spheres in the layer above and four spheres in the layer below). In the face-centered cubic cell there are spheres at the center of each of the six faces of the cube in addition to the eight corner spheres and the coordination number of each sphere is 12.

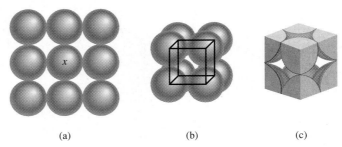

(a)   (b)   (c)

Figure 12.16 *Arrangement of identical spheres in a simple cubic cell. (a) Top view of one layer of spheres. (b) Definition of a simple cubic cell. (c) Because each sphere is shared by eight unit cells and there are eight corners in a cube, there is the equivalent of one complete sphere inside a simple cubic unit cell.*

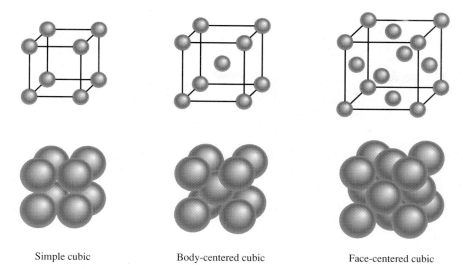

Simple cubic   Body-centered cubic   Face-centered cubic

Figure 12.17 *Three types of cubic cells. In reality, the spheres representing atoms, molecules, or ions are in contact with one another in these cubic cells.*

Because every unit cell in a crystalline solid is adjacent to other unit cells, most of a cell's atoms are shared by neighboring cells. For example, in all types of cubic cells, each corner atom belongs to eight unit cells [Figure 12.18(a)]; a face-centered atom is shared by two unit cells [Figure 12.18(b)]. Because each corner sphere is shared by eight unit cells and there are eight corners in a cube, there will be the equivalent of only one complete sphere inside a simple cubic unit cell (Figure 12.19). A body-centered cubic cell contains the equivalent of two complete spheres, one in the center and eight shared corner spheres. A face-centered cubic cell contains four complete spheres—three from the six face-centered atoms and one from the eight shared corner spheres.

Figure 12.19 also summarizes the relationship between the atomic radius r and the edge length a of a simple cubic cell, a body-centered cubic cell, and a face-centered cubic cell. This relationship can be used to determine the density of a crystal, as Example 12.3 shows.

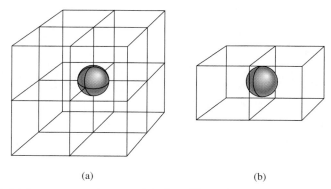

(a)   (b)

Figure 12.18 *(a) A corner atom in any cell is shared by eight unit cells. (b) A face-centered atom in a cubic cell is shared by two unit cells.*

scc

$a = 2r$

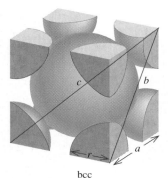

bcc

$b^2 = a^2 + a^2$
$c^2 = a^2 + b^2$
$\quad = 3a^2$
$c = \sqrt{3}a = 4r$
$a = \dfrac{4r}{\sqrt{3}}$

fcc

$b = 4r$
$b^2 = a^2 + a^2$
$16r^2 = 2a^2$
$a = \sqrt{8}r$

Figure 12.19 *The number of spheres within a simple cubic cell (scc) is one, a body-centered cell (bcc) is two, and a face-centered cell is four. Also shown are the relationships between the edge length (a) of the cell and radius (r) of the sphere.*

Example 12.3   Gold (Au) crystallizes in a face-centered cubic structure and has a density of 19.3 g/cm³. Calculate the atomic radius of gold in picometers.

Reasoning and Solution   Several steps are required to solve this problem. First we determine the volume of a unit cell from the known density. Next we calculate the edge length of the unit cell from its volume. Finally, from the edge length of the cell, we find the radius of one Au atom.

Step 1.   Each unit cell has eight corners and six faces. The total number of atoms within such a cell is, according to Figure 12.18, $8 \times \frac{1}{8} + 6 \times \frac{1}{2} = 4$. The mass of a unit cell is

$$m = \frac{4 \text{ atoms}}{1 \text{ unit cell}} \times \frac{197.0 \text{ g Au}}{1 \text{ mol Au}} \times \frac{1 \text{ mol}}{6.022 \times 10^{23} \text{ atoms}}$$
$$= 1.31 \times 10^{-21} \text{ g/unit cell}$$

From the definition of density we calculate the volume of the unit cell as

$$V = \frac{m}{d} = \frac{1.31 \times 10^{-21} \text{ g/unit cell}}{19.3 \text{ g/cm}^3} = 6.79 \times 10^{-23} \text{ cm}^3$$

$$= 6.79 \times 10^{-23} \text{ cm}^3 \times \left(\frac{1 \times 10^{-2} \text{ m}}{1 \text{ cm}}\right)^3 \left(\frac{1 \text{ pm}}{1 \times 10^{-12} \text{ m}}\right)^3$$

$$= 6.79 \times 10^7 \text{ pm}^3$$

Step 2.   We take the cubic root of the volume of the unit cell to obtain the edge length (*a*) of the cell

$$a = \sqrt[3]{V}$$
$$= \sqrt[3]{6.79 \times 10^7 \text{ pm}^3}$$
$$= 408 \text{ pm}$$

Remember that density is an intensive property, so that it is the same for one unit cell or 1 cm³ of the substance.

Step 3. From Figure 12.19 we see that the radius of an Au sphere (r) is related to the edge length by

$$a = \sqrt{8}\, r$$

Therefore,

$$r = \frac{a}{\sqrt{8}} = \frac{408\ \text{pm}}{\sqrt{8}} = 144\ \text{pm}$$

Similar problem: 12.48.

Practice Exercise When silver crystallizes, it forms face-centered cubic cells. The unit cell edge length is 408.7 pm. Calculate the density of silver.

12.5 Bonding in Solids

The structure and properties of crystalline solids, such as melting point, density, and hardness, are determined by the attractive forces that hold the particles together. We can classify crystals according to the types of forces between particles: ionic, molecular, covalent, and metallic (Table 12.4).

Ionic Crystals

Ionic crystals consist of ions held together by ionic bonds. The structure of an ionic crystal depends on the charges on the cation and anion and on their radii. We have already discussed the structure of sodium chloride, which has a face-centered cubic lattice (see Figure 2.12). Figure 12.20 shows the structures of three other ionic crystals: CsCl, ZnS, and CaF_2. Because Cs^+ is considerably larger than Na^+, CsCl has

These giant potassium dihydrogen phosphate crystals were grown in the laboratory. The largest one weighs 701 lb!

Table 12.4 | Types of Crystals and General Properties

Type of Crystal	Force(s) Holding the Units Together	General Properties	Examples
Ionic	Electrostatic attraction	Hard, brittle, high melting point, poor conductor of heat and electricity	NaCl, LiF, MgO, $CaCO_3$
Molecular*	Dispersion forces, dipole-dipole forces, hydrogen bonds	Soft, low melting point, poor conductor of heat and electricity	Ar, CO_2, I_2, H_2O, $C_{12}H_{22}O_{11}$ (sucrose)
Covalent	Covalent bond	Hard, high melting point, poor conductor of heat and electricity	C (diamond),† SiO_2 (quartz)
Metallic	Metallic bond	Soft to hard, low to high melting point, good conductor of heat and electricity	All metallic elements; for example, Na, Mg, Fe, Cu

*Included in this category are crystals made up of individual atoms.
†Diamond is a good thermal conductor.

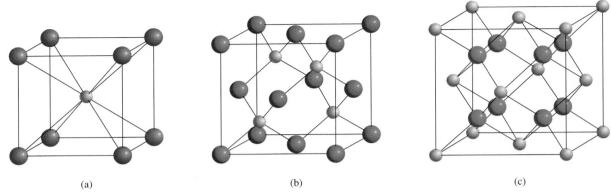

(a)   (b)   (c)

Figure 12.20 *Crystal structures of (a) CsCl, (b) ZnS, and (c) CaF₂. In each case, the cation is the smaller sphere.*

the simple cubic lattice structure. ZnS has the *zincblende* structure, which is based on the face-centered cubic lattice. If the S^{2-} ions are located at the lattice points, the Zn^{2+} ions are located one-fourth of the distance along each body diagonal. Other ionic compounds that have the zincblende structure include CuCl, BeS, CdS, and HgS. CaF_2 has the *fluorite* structure. The Ca^{2+} ions are located at the lattice points, and each F^- ion is tetrahedrally surrounded by four Ca^{2+} ions. The compounds SrF_2, BaF_2, $BaCl_2$, and PbF_2 also have the fluorite structure.

Ionic solids have high melting points, an indication of the strong cohesive force holding the ions together. These solids do not conduct electricity because the ions are fixed in position. However, in the molten state (that is, when melted) or dissolved in water, the ions are free to move and the resulting liquid is electrically conducting.

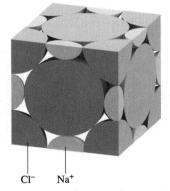

Cl⁻   Na⁺

Figure 12.21 *Portions of Na⁺ and Cl⁻ ions within a face-centered cubic unit cell.*

Similar problem: 12.47.

Example 12.4   How many Na^+ and Cl^- ions are in each NaCl unit cell?

Reasoning and Solution   NaCl has a structure based on a face-centered cubic lattice. As Figure 2.12 shows, one whole Na^+ ion is at the center of the unit cell, and there are twelve Na^+ ions at the edges. Because each edge Na^+ ion is shared by four unit cells, the total number of Na^+ ions is $1 + (12 \times \frac{1}{4}) = 4$. Similarly, there are six Cl^- ions at the face centers and eight Cl^- ions at the corners. Each face-centered ion is shared by two unit cells, and each corner ion is shared by eight unit cells (see Figure 12.18), so the total number of Cl^- ions is $(6 \times \frac{1}{2}) + (8 \times \frac{1}{8}) = 4$. Thus there are four Na^+ ions and four Cl^- ions in each NaCl unit cell. Figure 12.21 shows the portions of the Na^+ and Cl^- ions *within* a unit cell.

Practice Exercise   How many atoms are in a body-centered cube, assuming that all atoms occupy lattice points?

Molecular Crystals

Molecular crystals consist of atoms or molecules held together by van der Waals forces and/or hydrogen bonding. An example of a molecular crystal is solid sulfur dioxide (SO_2), in which the predominant attractive force is dipole-dipole interaction.

Intermolecular hydrogen bonding is mainly responsible for the three-dimensional ice lattice (see Figure 12.12). Other examples of molecular crystals are I_2, P_4, and S_8.

In general, except in ice, molecules in molecular crystals are packed together as closely as their size and shape allow. Because van der Waals forces and hydrogen bonding are generally quite weak compared with covalent and ionic bonds, molecular crystals are more easily broken apart than ionic and covalent crystals. Indeed, most molecular crystals melt below 200°C.

Sulfur.

Covalent Crystals

In covalent crystals (sometimes called covalent network crystals), atoms are held together entirely by covalent bonds in an extensive three-dimensional network. No discrete molecules are present, as in the case of molecular solids. Well-known examples are the two allotropes of carbon: diamond and graphite (see Figure 8.15). In diamond each carbon atom is tetrahedrally bonded to four other atoms (Figure 12.22). The strong covalent bonds in three dimensions contribute to diamond's unusual hardness (it is the hardest material known) and high melting point (3550°C). In graphite, carbon atoms are arranged in six-membered rings. The atoms are all sp^2-hybridized; each atom is covalently bonded to three other atoms. The unhybridized $2p$ orbital is used in pi bonding. In fact, the electrons in these $2p$ orbitals are free to move around, making graphite a good conductor of electricity in the planes of the bonded carbon atoms. The layers are held together by the weak van der Waals forces. The covalent bonds in graphite account for its hardness; however, because the layers can slide over one another, graphite is slippery to the touch and is effective as a lubricant. It is also used in pencils and in ribbons made for computer printers and typewriters.

Another type of covalent crystal is quartz (SiO_2). The arrangement of silicon atoms in quartz is similar to that of carbon in diamond, but in quartz there is an oxygen atom between each pair of Si atoms. Because Si and O have different electronegativities (see Figure 9.4), the Si—O bond is polar. Nevertheless, SiO_2 is similar to diamond in many respects, such as hardness and high melting point (1610°C).

Quartz.

Metallic Crystals

In a sense, the structure of metallic crystals is the simplest to deal with, because every lattice point in a crystal is occupied by an atom of the same metal. The bonding in

(a)

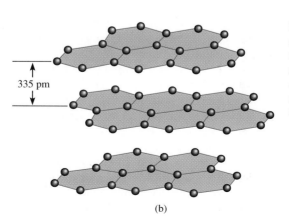

335 pm

(b)

Figure 12.22 *(a) The structure of diamond. Each carbon is tetrahedrally bonded to four other carbon atoms. (b) The structure of graphite. The distance between successive layers is 335 pm.*

Figure 12.23 *A cross section of a metallic crystal. Each circled positive charge represents the nucleus and inner electrons of a metal atom. The gray area surrounding the positive metal ions indicates the mobile sea of electrons.*

metals is quite different from that in other types of crystals. In a metal the bonding electrons are spread (or *delocalized*) over the entire crystal. In fact, metal atoms in a crystal can be imagined as an array of positive ions immersed in a sea of delocalized valence electrons (Figure 12.23). The great cohesive force resulting from delocalization is responsible for a metal's strength, which increases as the number of electrons available for bonding increases. For example, the melting point of sodium, with one valence electron, is 97.6°C, whereas that of aluminum, with three valence electrons, is 660°C. The mobility of the delocalized electrons makes metals good conductors of heat and electricity.

12.6 Phase Changes

The discussions in Chapter 5 and in this chapter have given us an overview of the properties of the three states of matter: gas, liquid, and solid. Each of these states is often referred to as a ***phase,*** which is *a homogeneous part of the system in contact with other parts of the system but separated from them by a well-defined boundary.* An ice cube floating in water makes up two phases of water—the solid phase (ice) and the liquid phase (water). ***Phase changes,*** *transformations from one phase to another,* occur when energy (usually in the form of heat) is added or removed. Phase changes are physical changes that are characterized by changes in molecular order; molecules in the solid state have the most order, and those in the gas phase have the greatest randomness. Keeping in mind the relationship between energy change and the increase or decrease in molecular order will help us understand the nature of phase changes.

Liquid-Vapor Equilibrium

Vapor Pressure

The difference between a gas and a vapor is explained on p. 126.

Equilibrium vapor pressure

Molecules in a liquid are not fixed in a rigid lattice. Although they lack the total freedom of gaseous molecules, these molecules are in constant motion. Because liquids are denser than gases, the collision rate among molecules is much higher in the liquid phase than in the gas phase. *At any given temperature, a certain number of the molecules in a liquid possess sufficient kinetic energy to escape from the surface.* This process is called ***evaporation,*** or ***vaporization.***

When a liquid evaporates, its gaseous molecules exert a vapor pressure. Consider the apparatus shown in Figure 12.24. Before the evaporation process starts, the mercury levels in the U-shaped manometer are equal. As soon as some molecules leave the liquid, a vapor phase is established. The vapor pressure is measurable only when a fair amount of vapor is present. The process of evaporation does not continue indefinitely, however. Eventually, the mercury levels stabilize and no further changes are seen.

What happens at the molecular level during evaporation? In the beginning, the traffic is only one way: Molecules are moving from the liquid to the empty space. Soon the molecules in the space above the liquid establish a vapor phase. *As the concentration of molecules in the vapor phase increases, some molecules return to the liquid phase,* a process called ***condensation.*** Condensation occurs because a molecule striking the liquid surface becomes trapped by intermolecular forces in the liquid.

The rate of evaporation is constant at any given temperature, and the rate of condensation increases with increasing concentration of molecules in the vapor phase.

Figure 12.25 *Comparison of the rates of evaporation and condensation at constant temperature.*

Figure 12.24 *Apparatus for measuring the vapor pressure of a liquid (a) before the evaporation begins and (b) at equilibrium, when no further change is evident. In (b) the number of molecules leaving the liquid is equal to the number of molecules returning to the liquid. The difference in the mercury levels (h) gives the equilibrium vapor pressure of the liquid at the specified temperature.*

A state of **dynamic equilibrium,** in which *the rate of a forward process is exactly balanced by the rate of the reverse process,* is reached when the rates of condensation and evaporation become equal (Figure 12.25). *The vapor pressure measured under dynamic equilibrium of condensation and evaporation* is called the **equilibrium vapor pressure.** We often use the simpler term "vapor pressure" when we talk about the equilibrium vapor pressure of a liquid. This practice is acceptable as long as we know the meaning of the abbreviated term.

Equilibrium vapor pressure is independent of the amount of liquid as long as there is some liquid present.

It is important to note that the equilibrium vapor pressure is the *maximum* vapor pressure a liquid exerts at a given temperature and that it is constant at constant temperature. Vapor pressure does change with temperature, however. Plots of vapor pressure versus temperature for three different liquids are shown in Figure 12.26. We know that the number of molecules with higher kinetic energies is greater at the higher temperature and therefore so is the evaporation rate. For this reason, the vapor pressure of a liquid always

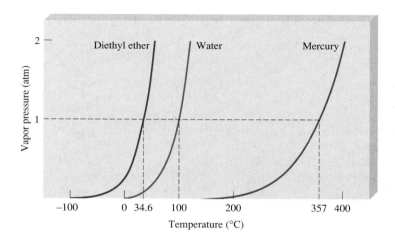

Figure 12.26 *The increase in vapor pressure with temperature for three liquids. The normal boiling points of the liquids (at 1 atm) are shown on the horizontal axis.*

increases with temperature. For example, the vapor pressure of water is 17.5 mmHg at 20°C, but it rises to 760 mmHg at 100°C.

Heat of Vaporization and Boiling Point

A measure of how strongly molecules are held in a liquid is its **molar heat of vaporization (ΔH_{vap})**, defined as *the energy* (usually in kilojoules) *required to vaporize one mole of a liquid*. The molar heat of vaporization is directly related to the strength of intermolecular forces that exist in the liquid. If the intermolecular attraction is strong, it takes a lot of energy to free the molecules from the liquid phase. Consequently, the liquid has a relatively low vapor pressure and a high molar heat of vaporization.

The quantitative relationship between the vapor pressure P of a liquid and the absolute temperature T is given by the Clausius-Clapeyron equation

$$\ln P = -\frac{\Delta H_{vap}}{RT} + C \qquad (12.1)$$

in which ln is the natural logarithm, R is the gas constant (8.314 J/K · mol), and C is a constant. The Clausius-Clapeyron equation has the form of the linear equation $y = mx + b$;

$$\ln P = \left(-\frac{\Delta H_{vap}}{R}\right)\left(\frac{1}{T}\right) + C$$

$$\begin{array}{ccccc} \updownarrow & & \updownarrow & \updownarrow & \updownarrow \\ y & = & m & x & + \ b \end{array}$$

By measuring the vapor pressure of a liquid at different temperatures and plotting ln P versus $1/T$, we determine the slope of the line described by the equation, which is equal to $-\Delta H_{vap}/R$. (ΔH_{vap} is assumed to be independent of temperature.) This is the method used to determine heats of vaporization. Figure 12.27 shows plots of ln P versus $1/T$ for water and diethyl ether ($C_2H_5OC_2H_5$). Note that the straight line for water has a steeper slope because water has a larger ΔH_{vap} (Table 12.5).

If we know the values of ΔH_{vap} and P of a liquid at one temperature, we can use the Clausius-Clapeyron equation to calculate the vapor pressure of the liquid at a

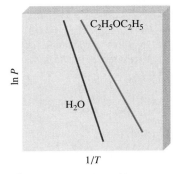

Figure 12.27 *Plots of ln P versus 1/T for water and diethyl ether.*

Table 12.5	Molar Heats of Vaporization for Selected Liquids	
Substance	**Boiling Point* (°C)**	**ΔH_{vap} (kJ/mol)**
Argon (Ar)	−186	6.3
Methane (CH_4)	−164	9.2
Diethyl ether ($C_2H_5OC_2H_5$)	34.6	26.0
Ethanol (C_2H_5OH)	78.3	39.3
Benzene (C_6H_6)	80.1	31.0
Water (H_2O)	100	40.79
Mercury (Hg)	357	59.0

*Measured at 1 atm.

different temperature. At temperatures T_1 and T_2 the vapor pressures are P_1 and P_2. From Equation (12.1) we can write

$$\ln P_1 = -\frac{\Delta H_{vap}}{RT_1} + C \qquad (12.2)$$

$$\ln P_2 = -\frac{\Delta H_{vap}}{RT_2} + C \qquad (12.3)$$

Subtracting Equation (12.3) from Equation (12.2) we obtain

$$\ln P_1 - \ln P_2 = -\frac{\Delta H_{vap}}{RT_1} - \left(-\frac{\Delta H_{vap}}{RT_2}\right)$$

$$= \frac{\Delta H_{vap}}{R}\left(\frac{1}{T_2} - \frac{1}{T_1}\right)$$

Hence

$$\ln \frac{P_1}{P_2} = \frac{\Delta H_{vap}}{R}\left(\frac{1}{T_2} - \frac{1}{T_1}\right)$$

$$= \frac{\Delta H_{vap}}{R}\left(\frac{T_1 - T_2}{T_1 T_2}\right) \qquad (12.4)$$

Example 12.5   Diethyl ether is a volatile, highly flammable organic liquid that is used mainly as a solvent. The vapor pressure of diethyl ether is 401 mmHg at 18°C. Calculate its vapor pressure at 32°C.

Reasoning and Solution   We are given the vapor pressure of diethyl ether at one temperature and asked to find the pressure at another temperature. Therefore, we need Equation (12.4). From Table 12.5, $\Delta H_{vap} = 26.0$ kJ/mol $= 26,000$ J/mol. The data are

$$P_1 = 401 \text{ mmHg} \qquad P_2 = ?$$
$$T_1 = 18°C = 291 \text{ K} \qquad T_2 = 32°C = 305 \text{ K}$$

From Equation (12.4) we have

$$\ln \frac{401}{P_2} = \frac{26,000 \text{ J/mol}}{8.314 \text{ J/K} \cdot \text{mol}}\left[\frac{291 \text{ K} - 305 \text{ K}}{(291 \text{ K})(305 \text{ K})}\right]$$

Taking the antilog of both sides (see Appendix 3), we obtain

$$\frac{401}{P_2} = e^{-0.493} = 0.611$$

Hence

$$P_2 = 656 \text{ mmHg}$$

Comment   We expect the vapor pressure to be greater at the higher temperature. Therefore, the answer is reasonable.

Practice Exercise   The vapor pressure of ethanol is 100 mmHg at 34.9°C. What is its vapor pressure at 63.5°C? (ΔH_{vap} for ethanol is 39.3 kJ/mol.)

$C_2H_5OC_2H_5$

Similar problem: 12.80.

Isopropanol (C$_3$H$_7$OH) is also called rubbing alcohol.

A practical way to demonstrate the molar heat of vaporization is by rubbing alcohol on your hands. The heat from your hands increases the kinetic energy of the alcohol molecules. The alcohol evaporates rapidly, extracting heat from your hands and cooling them. The process is similar to perspiration, which is one means by which the human body maintains a constant temperature. Because of the strong intermolecular hydrogen bonding that exists in water, a considerable amount of energy is needed to vaporize the water in perspiration from the body's surface. This energy is supplied by the heat generated in various metabolic processes.

You have already seen that the vapor pressure of a liquid increases with temperature. For every liquid there exists a temperature at which the liquid begins to boil. The **boiling point** is *the temperature at which the vapor pressure of a liquid is equal to the external pressure.* The *normal* boiling point of a liquid is the boiling point when the external pressure is 1 atm.

At the boiling point, bubbles form within the liquid. When a bubble forms, the liquid originally occupying that space is pushed aside, and the level of the liquid in the container is forced to rise. The pressure exerted *on* the bubble is largely atmospheric pressure, plus some *hydrostatic pressure* (that is, pressure caused by the presence of liquid). The pressure *inside* the bubble is due solely to the vapor pressure of the liquid. When the vapor pressure equals the external pressure, the bubble rises to the surface of the liquid and bursts. If the vapor pressure in the bubble were lower than the external pressure, the bubble would collapse before it could rise. We can thus conclude that the boiling point of a liquid depends on the external pressure. (We usually ignore the small contribution caused by the hydrostatic pressure.) For example, at 1 atm water boils at 100°C, but if the pressure is reduced to 0.5 atm, water boils at only 82°C.

Because the boiling point is defined in terms of the vapor pressure of the liquid, we expect the boiling point to be related to the molar heat of vaporization: The higher ΔH_{vap}, the higher the boiling point. The data in Table 12.5 roughly confirm our prediction. Ultimately, both the boiling point and ΔH_{vap} are determined by the strength of intermolecular forces. For example, argon (Ar) and methane (CH$_4$), which have weak dispersion forces, have low boiling points and small molar heats of vaporization. Diethyl ether (C$_2$H$_5$OC$_2$H$_5$) has a dipole moment, and the dipole-dipole forces account for its moderately high boiling point and ΔH_{vap}. Both ethanol (C$_2$H$_5$OH) and water have strong hydrogen bonding, which accounts for their high boiling points and large ΔH_{vap} values. Strong metallic bonding causes mercury to have the highest boiling point and ΔH_{vap} of this group of liquids. Interestingly, the boiling point of benzene, which is nonpolar, is comparable to that of ethanol. Benzene has a large polarizability, and the dispersion forces among benzene molecules can be as strong as or even stronger than dipole-dipole forces and/or hydrogen bonds.

Critical Temperature and Pressure

The opposite of evaporation is condensation. In principle, a gas can be made to liquefy by either one of two techniques. By cooling a sample of gas we decrease the kinetic energy of its molecules, and eventually molecules aggregate to form small drops of liquid. Alternatively, we may apply pressure to the gas. Under compression, the average distance between molecules is reduced so that they are held together by mutual attraction. Industrial liquefaction processes combine these two methods.

Every substance has a **critical temperature (T$_c$),** *above which its gas form cannot be made to liquefy, no matter how great the applied pressure.* This is also *the highest*

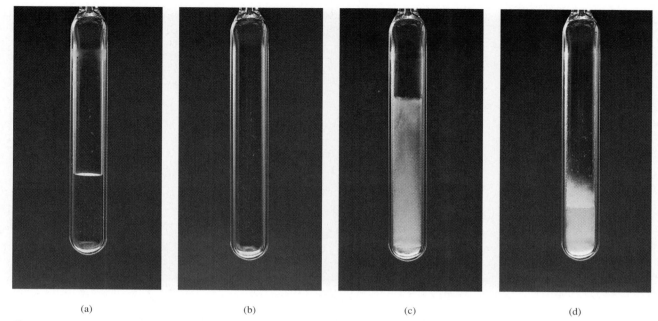

(a) (b) (c) (d)

Figure 12.28 *The critical phenomenon of sulfur hexafluoride. (a) Below the critical temperature the clear liquid phase is visible. (b) Above the critical temperature the liquid phase has disappeared. (c) The substance is cooled just below its critical temperature. The fog represents the condensation of vapor. (d) Finally, the liquid phase reappears.*

temperature at which a substance can exist as a liquid. The minimum pressure that must be applied to bring about liquefaction at the critical temperature is called the **critical pressure (P_c).** The existence of the critical temperature can be qualitatively explained as follows. The intermolecular attraction is a finite quantity for any given substance. Below T_c, this force is sufficiently strong to hold the molecules together (under some appropriate pressure) in a liquid. Above T_c, molecular motion becomes so energetic that the molecules can always break away from this attraction. Figure 12.28 shows what happens when sulfur hexafluoride is heated above its critical temperature (45.5°C) and then cooled down to below 45.5°C.

Table 12.6 lists the critical temperatures and critical pressures of a number of common substances. Benzene, ethanol, mercury, and water, which have strong intermolecular forces, also have high critical temperatures compared with the other substances listed in the table.

Intermolecular forces are independent of temperature; the kinetic energy of molecules increases with temperature.

Liquid-Solid Equilibrium

The transformation of liquid to solid is called *freezing,* and the reverse process is called *melting* or *fusion.* The **melting point** of a solid (or the freezing point of a liquid) is *the temperature at which solid and liquid phases coexist in equilibrium.* The *normal* melting point (or the *normal* freezing point) of a substance is the melting point (or freezing point) measured at 1 atm pressure. We generally omit the word "normal" in referring to the melting point of a substance at 1 atm.

The most familiar liquid-solid equilibrium is that of water and ice. At 0°C and 1 atm, the dynamic equilibrium is represented by

"Fusion" refers to the process of melting. Thus a "fuse" breaks an electrical circuit when a metallic strip melts due to the heat generated by excessively high electrical current.

$$\text{ice} \rightleftharpoons \text{water}$$

Table 12.6	Critical Temperatures and Critical Pressures of Selected Substances	
Substance	T_c(°C)	P_c(atm)
Ammonia (NH_3)	132.4	111.5
Argon (Ar)	−186	6.3
Benzene (C_6H_6)	288.9	47.9
Carbon dioxide (CO_2)	31.0	73.0
Ethanol (C_2H_5OH)	243	63.0
Diethyl ether ($C_2H_5OC_2H_5$)	192.6	35.6
Mercury (Hg)	1462	1036
Methane (CH_4)	−83.0	45.6
Molecular hydrogen (H_2)	−239.9	12.8
Molecular nitrogen (N_2)	−147.1	33.5
Molecular oxygen (O_2)	−118.8	49.7
Sulfur hexafluoride (SF_6)	45.5	37.6
Water (H_2O)	374.4	219.5

A practical illustration of this dynamic equilibrium is provided by a glass of ice water. As the ice cubes melt to form water, some of the water between the ice cubes may freeze, thus joining the cubes together. This is not a true dynamic equilibrium; because the glass is not kept at 0°C, all the ice cubes will eventually melt away.

The energy (usually in kilojoules) *required to melt 1 mole of a solid* is called the ***molar heat of fusion (ΔH_{fus})***. Table 12.7 shows the molar heats of fusion for the substances listed in Table 12.5. A comparison of the data in the two tables shows that for each substance ΔH_{fus} is smaller than ΔH_{vap}. This is consistent with the fact that molecules in a liquid are fairly closely packed together, so that some energy is needed to bring about the rearrangement from solid to liquid. On the other hand, when a liquid evaporates, its molecules are completely separated from one another and considerably more energy is required to overcome the attractive forces.

Table 12.7	Molar Heats of Fusion for Selected Substances	
Substance	Melting Point* (°C)	ΔH_{fus} (kJ/mol)
Argon (Ar)	−190	1.3
Methane (CH_4)	−183	0.84
Diethyl ether ($C_2H_5OC_2H_5$)	−116.2	6.90
Ethanol (C_2H_5OH)	−117.3	7.61
Benzene (C_6H_6)	5.5	10.9
Water (H_2O)	0	6.01
Mercury (Hg)	−39	23.4

*Measured at 1 atm.

Solid-Vapor Equilibrium

Solids, too, undergo evaporation and therefore possess a vapor pressure. Consider the following dynamic equilibrium:

$$\text{solid} \rightleftharpoons \text{vapor}$$

The process in which molecules go directly from the solid into the vapor phase is called **sublimation,** and the reverse process (that is, *from vapor directly to solid*) is called **deposition.** Naphthalene (the substance used to make mothballs) has a fairly high vapor pressure for a solid (1 mmHg at 53°C); thus its pungent vapor quickly permeates an enclosed space. Generally, because molecules are more tightly held in a solid, the vapor pressure of a solid is much less than that of the corresponding liquid. *The energy* (usually in kilojoules) *required to sublime 1 mole of a solid,* called the **molar heat of sublimation (ΔH_{sub}),** is given by the sum of the molar heats of fusion and vaporization:

$$\Delta H_{\text{sub}} = \Delta H_{\text{fus}} + \Delta H_{\text{vap}} \tag{12.5}$$

Strictly speaking, Equation (12.5), which is an illustration of Hess's law, holds if all the phase changes occur at the same temperature. The enthalpy, or heat change, for the overall process is the same whether the substance changes directly from the solid to the vapor form or goes from solid to liquid and then to vapor. Figure 12.29 summarizes the types of phase changes discussed in this section.

Solid iodine in equilibrium with its vapor.

12.7 Phase Diagrams

The overall relationships among the solid, liquid, and vapor phases are best represented in a single graph known as a phase diagram. A *phase diagram* summarizes *the conditions under which a substance exists as a solid, liquid, or gas.* In this section we will briefly discuss the phase diagrams of water and carbon dioxide.

Water

Figure 12.30(a) shows the phase diagram of water. The graph is divided into three regions, each of which represents a pure phase. The line separating any two regions indicates conditions under which these two phases can exist in equilibrium. For example, the curve between the liquid and vapor phases shows the variation of vapor pressure with temperature. The other two curves similarly indicate conditions for equilibrium between ice and liquid water and between ice and water vapor. (Note that the solid-liquid boundary line has a negative slope.) The point at which all three curves meet is called the **triple point.** For water, this point is at 0.01°C and 0.006 atm. This is the only *temperature and pressure at which all three phases can be in equilibrium with one another.*

Phase diagrams enable us to predict changes in the melting point and boiling point of a substance as a result of changes in the external pressure; we can also anticipate directions of phase transitions brought about by changes in temperature and pressure. The normal melting point and boiling point of water, measured at 1 atm, are 0°C and 100°C, respectively. What would happen if melting and boiling were carried out at some other pressure? Figure 12.30(b) shows clearly that increasing the pressure above

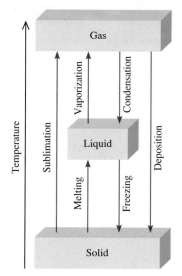

Figure 12.29 *The various phase changes that a substance can undergo.*

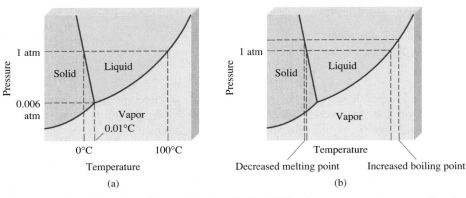

Figure 12.31 *The phase diagram of carbon dioxide. Note that the solid-liquid boundary line has a positive slope. The liquid phase is not stable below 5.2 atm, so that only the solid and vapor phases can exist under atmospheric conditions.*

Figure 12.30 *(a) The phase diagram of water. Each solid line between two phases specifies the conditions of pressure and temperature under which the two phases can exist in equilibrium. The point at which all three phases can exist in equilibrium (0.006 atm and 0.01°C) is called the triple point. (b) This phase diagram tells us that increasing the pressure on ice lowers its melting point and that increasing the pressure of liquid water raises its boiling point.*

Figure 12.32 *Under atmospheric conditions, solid carbon dioxide does not melt; it only sublimes. The cold carbon dioxide gas causes nearby water vapor to condense and form a fog.*

1 atm will raise the boiling point and lower the melting point. A decrease in pressure will lower the boiling point and raise the melting point.

Carbon Dioxide

The phase diagram of carbon dioxide (Figure 12.31) is similar to that of water, with one important exception—the slope of the curve between solid and liquid is positive. In fact, this holds true for almost all other substances. Water behaves differently because ice is *less* dense than liquid water. The triple point of carbon dioxide is at 5.2 atm and −57°C.

An interesting observation can be made about the phase diagram in Figure 12.31. As you can see, the entire liquid phase lies well above atmospheric pressure; therefore, it is impossible for solid carbon dioxide to melt at 1 atm. Instead, when solid CO_2 is heated to −78°C at 1 atm, it sublimes. In fact, solid carbon dioxide is called dry ice because it looks like ice and *does not melt* (Figure 12.32). Because of this property, dry ice is useful as a refrigerant.

Summary

All substances exist in one of three states: gas, liquid, or solid. The major difference between the condensed states and the gaseous state is the distance of separation between molecules.

Intermolecular forces act between molecules or between molecules and ions. Generally, these forces are much weaker than bonding forces. Dipole-dipole forces and ion-dipole forces attract molecules with dipole moments to other polar molecules or ions. Dispersion forces are the result of temporary dipole moments induced in ordinarily nonpolar molecules. The extent to which a dipole moment can be induced in a molecule is determined by its polarizability. The term "van der Waals forces" refers to dipole-dipole, dipole-induced dipole, and dispersion forces.

Hydrogen bonding is a relatively strong dipole-dipole force that acts between a polar bond containing a hydrogen atom and the bonded electronegative atoms, N, O, or F. Hydrogen bonds between water molecules are particularly strong.

Liquids tend to assume a geometry that ensures the minimum surface area. Surface tension is the energy needed to expand a liquid surface area; strong intermolecular forces lead to greater surface tension. Viscosity is a measure of the resistance of a liquid to flow; it decreases with increasing temperature.

Water molecules in the solid state form a three-dimensional network in which each oxygen atom is covalently bonded to two hydrogen atoms and is hydrogen-bonded to two hydrogen atoms. This unique structure accounts for the fact that ice is less dense than liquid water. Water is also ideally suited for its ecological role by its high specific heat, another property imparted by its strong hydrogen bonding. Large bodies of water are able to moderate the climate by giving off and absorbing substantial amounts of heat with only small changes in the water temperature.

All solids are either crystalline (with a regular structure of atoms, ions, or molecules) or amorphous (without a regular structure). The basic structural unit of a crystalline solid is the unit cell, which is repeated to form a three-dimensional crystal lattice.

The four types of crystals and the forces that hold their particles together are ionic crystals, held together by ionic bonding; molecular crystals, van der Waals forces and/or hydrogen bonding; covalent crystals, covalent bonding; and metallic crystals, metallic bonding.

A liquid in a closed vessel eventually establishes a dynamic equilibrium between evaporation and condensation. The vapor pressure over the liquid under these conditions is the equilibrium vapor pressure, which is often referred to simply as vapor pressure. At the boiling point, the vapor pressure of a liquid equals the external pressure. The molar heat of vaporization of a liquid is the energy required to vaporize 1 mole of the liquid. It can be determined by measuring the vapor pressure of the liquid as a function of temperature and using Equation (12.1). The molar heat of fusion of a solid is the energy required to melt 1 mole of the solid.

For every substance there is a temperature, called the critical temperature, above which its gas form cannot be made to liquefy.

The relationships among the three phases of a single substance are represented by a phase diagram, in which each region represents a pure phase and the boundaries between the regions show the temperatures and pressures at which the two phases are in equilibrium. At the triple point, all three phases are in equilibrium.

Key Words

Adhesion p. 377
Boiling point p. 392
Cohesion p. 377
Condensation p. 388
Coordination number p. 382
Critical pressure (P_c) p. 393
Critical temperature (T_c) p. 392
Crystalline solid p. 381
Deposition p. 395
Dipole-dipole forces p. 371
Dispersion forces p. 373
Dynamic equilibrium p. 389
Equilibrium vapor pressure p. 389
Evaporation p. 388
Hydrogen bond p. 374
Induced dipole p. 372
Intermolecular forces p. 370
Intramolecular forces p. 371
Ion-dipole forces p. 371
Lattice point p. 381
Melting point p. 393
Molar heat of fusion (ΔH_{fus}) p. 394
Molar heat of sublimation (ΔH_{sub}) p. 395
Molar heat of vaporization (ΔH_{vap}) p. 390
Phase p. 388
Phase changes p. 388
Phase diagram p. 395
Polarizability p. 372
Sublimation p. 395
Surface tension p. 377
Triple point p. 395
Unit cell p. 381
van der Waals forces p. 371
Vaporization p. 388
Viscosity p. 377

Questions and Problems

Intermolecular Forces

Review Questions

12.1   Define these terms and give an example for each category: (a) dipole-dipole interaction, (b) dipole-induced dipole interaction, (c) ion-dipole interaction, (d) dispersion forces, (e) van der Waals forces.

12.2   Explain the term "polarizability." What kind of molecules tend to have high polarizabilities? What is the relationship between polarizability and intermolecular forces?

12.3   Explain the difference between the temporary dipole moment induced in a molecule and the permanent dipole moment in a polar molecule.

12.4   Give some evidence that all molecules exert attractive forces on one another.

12.5   What type of physical properties would you need to consider in comparing the strength of intermolecular forces in solids and in liquids?

12.6   Which elements can take part in hydrogen bonding?

Problems

12.7   The compounds Br_2 and ICl have the same number of electrons, yet Br_2 melts at $-7.2°C$, whereas ICl melts at $27.2°C$. Explain.

12.8   If you lived in Alaska, state which of these natural gases you would keep in an outdoor storage tank in winter and explain why: methane (CH_4), propane (C_3H_8), or butane (C_4H_{10}).

12.9   The binary hydrogen compounds of the Group 4A elements are CH_4 ($-162°C$), SiH_4 ($-112°C$), GeH_4 ($-88°C$), and SnH_4 ($-52°C$). The temperatures in parentheses are the corresponding boiling points. Explain the increase in boiling points from CH_4 to SnH_4.

12.10   List the types of intermolecular forces that exist in each of these species: (a) benzene (C_6H_6), (b) CH_3Cl, (c) PF_3, (d) NaCl, (e) CS_2.

12.11   Ammonia is both a donor and an acceptor of hydrogen in hydrogen bond formation. Draw a diagram to show the hydrogen bonding of an ammonia molecule with two other ammonia molecules.

12.12   Which of these species are capable of hydrogen bonding among themselves: (a) C_2H_6, (b) HI, (c) KF, (d) BeH_2, (e) CH_3COOH?

12.13   Arrange these in order of increasing boiling point: RbF, CO_2, CH_3OH, CH_3Br. Explain your arrangement.

12.14   Diethyl ether has a boiling point of $34.5°C$, and 1-butanol has a boiling point of $117°C$.

diethyl ether

1-butanol

Both of these compounds have the same numbers and types of atoms. Explain the difference in their boiling points.

12.15   Which member of each of these pairs of substances would you expect to have a higher boiling point: (a) O_2 or N_2, (b) SO_2 or CO_2, (c) HF or HI?

12.16   State which substance in each of these pairs you would expect to have the higher boiling point and explain why: (a) Ne or Xe, (b) CO_2 or CS_2, (c) CH_4 or Cl_2, (d) F_2 or LiF, (e) NH_3 or PH_3.

12.17   Explain in terms of intermolecular forces why (a) NH_3 has a higher boiling point than CH_4 and (b) KCl has a higher melting point than I_2.

12.18   What kind of attractive forces must be overcome to (a) melt ice, (b) boil molecular bromine, (c) melt solid iodine, and (d) dissociate F_2 into F atoms?

12.19   These nonpolar molecules have the same number and type of atoms. Which one would you expect to have a higher boiling point?

$$CH_3—CH_2—CH_2—CH_3 \qquad CH_3—\overset{\overset{\displaystyle CH_3}{|}}{CH}—CH_3$$

(*Hint:* Molecules that can be stacked together more easily have greater intermolecular attraction.)

12.20   Explain the difference in the melting points of these compounds:

m.p. 45°C   m.p. 115°C

(*Hint:* Only one of the two can form intramolecular hydrogen bonds.)

The Liquid State

Review Questions

12.21 Explain why liquids, unlike gases, are virtually incompressible.

12.22 Define surface tension. What is the relationship between the intermolecular forces that exist in a liquid and its surface tension?

12.23 Despite the fact that stainless steel is much denser than water, a stainless-steel razor blade can be made to float on water. Why?

12.24 Use water and mercury as examples to explain adhesion and cohesion.

12.25 A glass can be filled slightly above the rim with water. Explain why the water does not overflow.

12.26 Draw diagrams showing the capillary action of (a) water and (b) mercury in three tubes of different radii.

12.27 What is viscosity? What is the relationship between the intermolecular forces that exist in a liquid and its viscosity?

12.28 Why does the viscosity of a liquid decrease with increasing temperature?

12.29 Why is ice less dense than water?

12.30 Outdoor water pipes have to be drained or insulated in winter in a cold climate. Why?

Problems

12.31 Predict which of these liquids has the greater surface tension: ethanol (C_2H_5OH) or dimethyl ether (CH_3OCH_3).

12.32 Predict the viscosity of ethylene glycol

$$CH_2 - OH$$
$$|$$
$$CH_2 - OH$$

relative to that of ethanol and glycerol (see Table 12.3).

Crystalline Solids

Review Questions

12.33 Define these terms: crystalline solid, lattice point, unit cell, coordination number.

12.34 Describe the geometries of these cubic cells: simple cubic cell, body-centered cubic cell, face-centered cubic cell. Which of these cells would give the highest density for the same type of atoms?

12.35 Describe, with examples, these types of crystals: (a) ionic crystals, (b) covalent crystals, (c) molecular crystals, (d) metallic crystals.

12.36 A solid is hard, brittle, and electrically nonconducting. Its melt (the liquid form of the substance) and an aqueous solution containing the substance do conduct electricity. Classify the solid.

12.37 A solid is soft and has a low melting point (below 100°C). The solid, its melt, and a solution containing the substance are all nonconductors of electricity. Classify the solid.

12.38 A solid is very hard and has a high melting point. Neither the solid nor its melt conducts electricity. Classify the solid.

12.39 Why are metals good conductors of heat and electricity? Why does the ability of a metal to conduct electricity decrease with increasing temperature?

12.40 Classify the solid states of the elements in the second period of the periodic table.

12.41 The melting points of the oxides of the third-period elements are given in parentheses: Na_2O (1275°C), MgO (2800°C), Al_2O_3 (2045°C), SiO_2 (1610°C), P_4O_{10} (580°C), SO_3 (16.8°C), Cl_2O_7 (−91.5°C). Classify these solids.

12.42 Which of these are molecular solids and which are covalent solids: Se_8, HBr, Si, CO_2, C, P_4O_6, B, SiH_4?

Problems

12.43 What is the coordination number of each sphere in (a) a simple cubic lattice, (b) a body-centered cubic lattice, and (c) a face-centered cubic lattice? Assume the spheres to be of equal size.

12.44 Calculate the number of spheres in these unit cells: simple cubic, body-centered cubic, and face-centered cubic cells. Assume that the spheres are of equal size and that they are only at the lattice points.

12.45 Metallic iron crystallizes in a cubic lattice. The unit cell edge length is 287 pm. The density of iron is 7.87 g/cm^3. How many iron atoms are there within a unit cell?

12.46 Barium metal crystallizes in a body-centered cubic lattice (the Ba atoms are at the lattice points only). The unit cell edge length is 502 pm, and the density of Ba is 3.50 g/cm^3. Using this information, calculate Avogadro's number. (*Hint:* First calculate the volume occupied by 1 mole of Ba atoms in the unit cells. Next calculate the volume occupied by one of the Ba atoms in the unit cell.)

12.47 Vanadium crystallizes in a body-centered cubic lattice (the V atoms occupy only the lattice points). How many V atoms are in a unit cell?

12.48 Europium crystallizes in a body-centered cubic lattice (the Eu atoms occupy only the lattice points). The density of Eu is 5.26 g/cm^3. Calculate the unit cell edge length in picometers.

12.49 Crystalline silicon has a cubic structure. The unit cell edge length is 543 pm. The density of the solid is 2.33 g/cm^3. Calculate the number of Si atoms in one unit cell.

12.50 A face-centered cubic cell contains 8 X atoms at the corners of the cell and 6 Y atoms at the faces. What is the empirical formula of the solid?

12.51 Classify the crystalline form of these substances as ionic crystals, covalent crystals, molecular crystals, or metallic crystals: (a) CO_2, (b) B, (c) S_8, (d) KBr, (e) Mg, (f) SiO_2, (g) LiCl, (h) Cr.

12.52 Explain why diamond is harder than graphite.

Phase Changes

Review Questions

12.53 Define phase change. Name all possible changes that can occur among the vapor, liquid, and solid states of a substance.

12.54 What is the equilibrium vapor pressure of a liquid? How does it change with temperature?

12.55 Use any one of the phase changes to explain what is meant by dynamic equilibrium.

12.56 Define these terms: (a) molar heat of vaporization, (b) molar heat of fusion, (c) molar heat of sublimation. What are their units?

12.57 How is the molar heat of sublimation related to the molar heats of vaporization and fusion? On what law is this relation based?

12.58 What can we learn about the strength of intermolecular forces in a liquid from its molar heat of vaporization?

12.59 The greater the molar heat of vaporization of a liquid, the greater its vapor pressure. True or false?

12.60 Define boiling point. How does the boiling point of a liquid depend on external pressure? Referring to Table 5.2, what is the boiling point of water when the external pressure is 187.5 mmHg?

12.61 As a liquid is heated at constant pressure, its temperature rises. This trend continues until the boiling point of the liquid is reached. No further rise in the temperature of the liquid can be induced by heating. Explain.

12.62 Define critical temperature. What is the significance of critical temperature in the liquefaction of gases?

12.63 What is the relationship between intermolecular forces in a liquid and the liquid's boiling point and critical temperature? Why is the critical temperature of water greater than that of most other substances?

12.64 How do the boiling points and melting points of water and carbon tetrachloride vary with pressure? Explain any difference in behavior of these two substances.

12.65 Why is solid carbon dioxide called dry ice?

12.66 The vapor pressure of a liquid in a closed container depends on which of these: (a) the volume above the liquid, (b) the amount of liquid present, (c) temperature?

12.67 Referring to Figure 12.26, estimate the boiling points of diethyl ether, water, and mercury at 0.5 atm.

12.68 Wet clothes dry more quickly on a hot, dry day than on a hot, humid day. Explain.

12.69 Which of the following phase transitions gives off more heat: (a) 1 mole of steam to 1 mole of water at 100°C or (b) 1 mole of water to 1 mole of ice at 0°C?

12.70 A beaker of water is heated to boiling by a Bunsen burner. Would adding another burner raise the boiling point of water? Explain.

Problems

12.71 Calculate the amount of heat (in kilojoules) required to convert 74.6 g of water to steam at 100°C.

12.72 How much heat (in kilojoules) is needed to convert 866 g of ice at −10°C to steam at 126°C? (The specific heats of ice and steam are 2.03 J/g · °C and 1.99 J/g · °C, respectively.)

12.73 How is the rate of evaporation of a liquid affected by (a) temperature, (b) the surface area of liquid exposed to air, (c) intermolecular forces?

12.74 The molar heats of fusion and sublimation of molecular iodine are 15.27 kJ/mol and 62.30 kJ/mol, respectively. Estimate the molar heat of vaporization of liquid iodine.

12.75 These compounds are liquid at −10°C; their boiling points are given: butane, −0.5°C; ethanol, 78.3°C; toluene, 110.6°C. At −10°C, which of these liquids would you expect to have the highest vapor pressure? Which the lowest?

12.76 Freeze-dried coffee is prepared by freezing a sample of brewed coffee and then removing the ice component by vacuum-pumping the sample. Describe the phase changes taking place during these processes.

12.77 A student hangs wet clothes outdoors on a winter day when the temperature is −15°C. After a few hours, the clothes are found to be fairly dry. Describe the phase changes in this drying process.

12.78 Steam at 100°C causes more serious burns than water at 100°C. Why?

12.79 Vapor pressure measurements at several different temperatures are shown here for mercury. Determine graphically the molar heat of vaporization for mercury.

t (°C)	200	250	300	320	340
P (mmHg)	17.3	74.4	246.8	376.3	557.9

12.80 The vapor pressure of benzene, C_6H_6, is 40.1 mmHg at 7.6°C. What is its vapor pressure at 60.6°C? The molar heat of vaporization of benzene is 31.0 kJ/mol.

12.81 The vapor pressure of liquid X is lower than that of liquid Y at 20°C, but higher at 60°C. What can you deduce about the relative magnitude of the molar heats of vaporization of X and Y?

Phase Diagrams

Review Questions

12.82 What is a phase diagram? What useful information can be obtained from a phase diagram?

12.83 Explain how water's phase diagram differs from those of most substances. What property of water causes the difference?

Problems

12.84 The blades of ice skates are quite thin, so the pressure exerted on ice by a skater can be substantial. Explain how this fact helps a person to skate on ice.

12.85 A length of wire is placed on top of a block of ice. The ends of the wire extend over the edges of the ice, and a heavy weight is attached to each end. It is found that the ice under the wire gradually melts, so that the wire slowly moves through the ice block. At the same time, the water above the wire refreezes. Explain the phase changes that accompany this phenomenon.

12.86 Consider the phase diagram of water shown below. Label the regions. Predict what would happen if we did the following: (a) Starting at A, we raise the temperature at constant pressure. (b) Starting at C, we lower the temperature at constant pressure. (c) Starting at B, we lower the pressure at constant temperature.

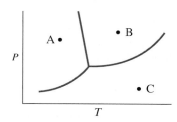

12.87 The boiling point and freezing point of sulfur dioxide are $-10°C$ and $-72.7°C$ (at 1 atm), respectively. The triple point is $-75.5°C$ and 1.65×10^{-3} atm, and its critical point is at $157°C$ and 78 atm. On the basis of this information, draw a rough sketch of the phase diagram of SO_2.

Additional Problems

12.88 Name the kinds of attractive forces that must be overcome to (a) boil liquid ammonia, (b) melt solid phosphorus (P_4), (c) dissolve CsI in liquid HF, (d) melt potassium metal.

12.89 Which of these indicates very strong intermolecular forces in a liquid: (a) a very low surface tension, (b) a very low critical temperature, (c) a very low boiling point, (d) a very low vapor pressure?

12.90 At $-35°C$, liquid HI has a higher vapor pressure than liquid HF. Explain.

12.91 From these properties of elemental boron, classify it as one of the crystalline solids discussed in Section 12.5: high melting point ($2300°C$), poor conductor of heat and electricity, insoluble in water, very hard substance.

12.92 Referring to Figure 12.31, determine the stable phase of CO_2 at (a) 4 atm and $-60°C$ and (b) 0.5 atm and $-20°C$.

12.93 A solid contains X, Y, and Z atoms in a cubic lattice with X atoms in the corners, Y atoms in the body-centered positions, and Z atoms on the faces of the cell. What is the empirical formula of the compound?

12.94 A CO_2 fire extinguisher is located on the outside of a building in Massachusetts. During the winter months, one can hear a sloshing sound when the extinguisher is gently shaken. In the summertime the sound is often absent. Explain. Assume that the extinguisher has no leaks and that it has not been used.

12.95 What is the vapor pressure of mercury at its normal boiling point ($357°C$)?

12.96 A flask containing water is connected to a powerful vacuum pump. When the pump is turned on, the water begins to boil. After a few minutes, the same water begins to freeze. Eventually, the ice disappears. Explain what happens at each step.

12.97 The liquid-vapor boundary line in the phase diagram of any substance always stops abruptly at a certain point. Why?

12.98 Given the phase diagram of carbon shown here, answer these questions: (a) How many triple points are there and what are the phases that can coexist at each triple point? (b) Which has a higher density, graphite or diamond? (c) Synthetic diamond can be made from graphite. Using the phase diagram, how would you go about making diamond?

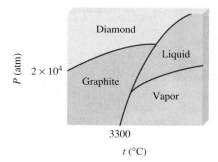

12.99 Estimate the molar heat of vaporization of a liquid whose vapor pressure doubles when the temperature is raised from $85°C$ to $95°C$.

12.100 A student is given four samples of solids W, X, Y, and Z, all of which have a metallic luster. She is told that the solids are gold, lead sulfide, mica (which is quartz, or SiO_2), and iodine. The results of her investigation are (a) W is a good electrical conductor; X, Y, and Z are poor electrical conductors; (b) when the solids are hit with a hammer, W flattens out, X shatters into many pieces, Y is smashed into a powder, and Z is not affected; (c) when the solids are heated with a Bunsen burner, Y melts with some sublimation, but X, W, and Z do not melt; (d) in treatment with 6 M HNO_3, X dissolves; there is no effect on W, Y, or Z. On the basis of her studies, identify the solids.

12.101 Which of these statements is false: (a) Dipole-dipole interactions between molecules are greatest if the molecules possess only temporary dipole moments. (b) All compounds containing hydrogen atoms can participate in hydrogen bond formation. (c) Dispersion forces exist between all atoms, molecules, and ions. (d) The extent of ion-induced dipole interaction depends only on the charge on the ion.

12.102 The south pole of Mars is covered with dry ice, which partly sublimes during the summer. The CO_2 vapor recondenses in the winter when the temperature drops to 150 K. Given that the heat of sublimation of CO_2 is 25.9 kJ/mol, calculate the atmospheric pressure on the surface of Mars. [*Hint:* Use Figure 12.31 to determine the normal sublimation temperature of dry ice and Equation (12.4), which also applies to sublimation.]

12.103 The standard enthalpy of formation of gaseous molecular bromine is 30.7 kJ/mol. Use this information to calculate the molar heat of vaporization of molecular bromine at 25°C.

12.104 Heats of hydration, that is, heat changes that occur when ions become hydrated in solution, are largely due to ion-dipole interactions. The heats of hydration for the alkali metal ions are Li^+, −520 kJ/mol; Na^+, −405 kJ/mol; K^+, −321 kJ/mol. Account for the trend in these values.

12.105 A beaker of water is placed in a closed container. Predict the effect on the vapor pressure of the water when (a) its temperature is lowered, (b) the volume of the container is doubled, (c) more water is added to the beaker.

12.106 Ozone (O_3) is a strong agent that can oxidize all the common metals except gold and platinum. A convenient test for ozone is based on its action on mercury. When exposed to ozone, mercury becomes dull looking and sticks to glass tubing (instead of flowing freely through it). Write a balanced equation for the reaction. What property of mercury is altered by its interaction with ozone?

12.107 A pressure cooker is a sealed container that allows steam to escape when it exceeds a predetermined pressure. How does this device reduce the time needed for cooking?

12.108 A 1.20-g sample of water is injected into an evacuated 5.00-L flask at 65°C. What percentage of the water will be vapor when the system reaches equilibrium? Assume ideal behavior of water vapor and that the volume of liquid water is negligible. The vapor pressure of water at 65°C is 187.5 mmHg.

12.109 Swimming coaches sometimes suggest that a drop of alcohol (ethanol) placed in an ear plugged with water "draws out the water." Explain this action from a molecular point of view.

Special Problems

12.110 A quantitative measure of how efficiently spheres pack into unit cells is called *packing efficiency,* which is the percentage of the cell space occupied by the spheres. Calculate the packing efficiencies of a simple cubic cell, a body-centered cubic cell, and a face-centered cubic cell. (*Hint:* Refer to Figure 12.19 and use the relationship that the volume of a sphere is $\frac{4}{3}\pi r^3$, in which r is the radius of the sphere.)

12.111 A chemistry instructor performed the following mystery demonstration. Just before the students arrived in class, she heated some water to boiling in an Erlenmeyer flask. She then removed the flask from the flame and closed the flask with a rubber stopper. After the class commenced, she held the flask in front of the students and announced that she could make the water boil simply by rubbing an ice cube on the outside walls of the flask. To the amazement of everyone, it worked. Can you give the explanation for this phenomenon?

12.112 Silicon used in computer chips must have an impurity level below 10^{-9} (that is, fewer than one impurity atom for every 10^9 Si atoms). Silicon is prepared by the reduction of quartz (SiO_2) with coke (a form of carbon made by the destructive distillation of coal) at about 2000°C:

$$SiO_2(s) + 2C(s) \longrightarrow Si(l) + 2CO(g)$$

Next, solid silicon is separated from other solid impurities by treatment with hydrogen chloride at 350°C to form gaseous trichlorosilane ($SiCl_3H$):

$$Si(s) + 3HCl(g) \longrightarrow SiCl_3H(g) + H_2(g)$$

Finally, ultrapure Si can be obtained by reversing the above reaction at 1000°C:

$$SiCl_3H(g) + H_2(g) \longrightarrow Si(s) + 3HCl(g)$$

(a) Trichlorosilane has a vapor pressure of 0.258 atm at −2°C. What is its normal boiling point? Is trichlorosilane's boiling point consistent with the type of intermolecular forces that exist among its molecules? (The molar heat of vaporization of trichlorosilane is 28.8 kJ/mol.) (b) What types of crystals do Si and SiO_2 form? (c) Silicon has a diamond crystal structure (see Figure 12.22). Each cubic unit cell (edge length $a = 543$ pm) contains eight Si atoms. If there are 1.0×10^{13} boron atoms per cubic centimeter in a sample of pure silicon, how many Si atoms are there for every B atom in the sample? Does this sample satisfy the 10^{-9} purity requirement for the electronic grade silicon?

Answers to Practice Exercises

12.1 (a) Ionic and dispersion forces, (b) dispersion forces, (c) dipole-dipole and dispersion forces. **12.2** CH_3OH. **12.3** 10.50 g/cm^3. **12.4** 2. **12.5** 369 mmHg.

When Svante August Arrhenius (1859–1927) presented his doctoral thesis at the University of Uppsala (Sweden) in May 1884, it was given the lowest possible rating. Consequently he was not awarded an academic appointment at the institution. Arrhenius was bitterly disappointed.

In his thesis, Arrhenius proposed an ionic theory of electrolyte solutions. Experimental work showed that the electrical conductance of certain salt solutions increased with dilution. Arrhenius suggested that these solutions contained a mixture of "active" (electrolyte) and "inactive" (nonelectrolyte) parts. On dilution, the number of active parts increased; in other words, electrolytic dissociation increased with dilution. Incidentally, his theory also explained that it is the ions that carry electric current in solution.

Arrhenius was a precocious child who taught himself to read.

Older chemists and physicists ridiculed the idea of separated electrical charges in solution, but the new generation of scientists readily accepted the notion. The concept of electrolytic dissociation easily accommodated the discovery of electrons and the incorporation of electrons into chemical bonding theories during the early years of the twentieth century. For his work, Arrhenius was awarded the Nobel Prize in chemistry in 1903.

In addition to his theory of electrolyte solutions, Arrhenius made several valuable contributions to chemistry, the most notable of which was an equation showing the effect of temperature on reaction rates (see Chapter 14).

In his later years Arrhenius began writing about science for the laity and the applications of chemistry to astronomy, biology, and geology. He advanced the idea of *panspermia,* which speculates that life did not begin on Earth but instead came from other planets.

Physical Properties of Solutions

Chapter 13

ESSENTIAL CONCEPTS

 Solutions There are many types of solutions; the most common is the liquid solution in which the solvent is a liquid and the solute is a solid or a liquid. Molecules that possess similar types of intermolecular forces readily mix with each other. Solubility is a quantitative measure of the amount of a solute dissolved in a solvent at a specific temperature.

 Concentration Units The four common concentration units for solutions are percent by mass, mole fraction, molarity, and molality. Each one has its advantages and limitations.

 Effect of Temperature and Pressure on Solubility Temperature generally has a marked influence on the solubility of a substance. Pressure can affect the solubility of a gas in a liquid but has little effect if the solute is a solid or liquid.

 Colligative Properties The presence of a solute affects the vapor pressure, boiling point, and freezing point of the solvent. In addition, when a solution is separated from the solvent by a semipermeable membrane, osmosis, the passage of solvent molecules from the solvent to the solution, occurs. Equations have been derived that relate the extent of the changes in these properties to the concentration of the solution.

13.1 Types of Solutions

Most chemical reactions take place not between pure solids, liquids, or gases, but among ions and molecules dissolved in water or other solvents. In Section 4.1 we noted that a solution is a homogeneous mixture of two or more substances. Because this definition places no restriction on the nature of the substances involved, we can distinguish six types of solutions, depending on the original states (solid, liquid, or gas) of the solution components. Table 13.1 gives examples of each of these types.

Our focus here will be on solutions involving at least one liquid component—that is, gas-liquid, liquid-liquid, and solid-liquid solutions. And, perhaps not too surprisingly, the liquid solvent in most of the solutions we will study is water.

Chemists also characterize solutions by their capacity to dissolve a solute. *A solution that contains the maximum amount of a solute in a given solvent, at a specific temperature,* is called a **saturated solution.** Before the saturation point is reached, the solution is said to be **unsaturated;** it *contains less solute than it has the capacity to dissolve.* A third type, a **supersaturated solution,** *contains more solute than is present in a saturated solution.* Supersaturated solutions are not very stable. In time, some of the solute will come out of a supersaturated solution as crystals. *The process in which dissolved solute comes out of solution and forms crystals* is called **crystallization.** Note that both precipitation and crystallization describe the separation of excess solid substance from a supersaturated solution. However, solids formed by the two processes differ in appearance. We normally think of precipitates as being made up of small particles, whereas crystals may be large and well formed (Figure 13.1).

13.2 A Molecular View of the Solution Process

In liquids and solids, molecules are held together by intermolecular attractions. These forces also play a central role in the formation of solutions. When one substance (the solute) dissolves in another (the solvent), particles of the solute disperse throughout the solvent. The solute particles occupy positions that are normally taken by solvent molecules. The ease with which a solute particle replaces a solvent molecule depends on the relative strengths of three types of interactions:

- solvent-solvent interaction
- solute-solute interaction
- solvent-solute interaction

Figure 13.1 *In a supersaturated sodium acetate solution (top), sodium acetate crystals rapidly form when a small seed crystal is added.*

Table 13.1	Types of Solutions		
Solute	**Solvent**	**State of Resulting Solution**	**Examples**
Gas	Gas	Gas	Air
Gas	Liquid	Liquid	Soda water (CO_2 in water)
Gas	Solid	Solid	H_2 gas in palladium
Liquid	Liquid	Liquid	Ethanol in water
Solid	Liquid	Liquid	NaCl in water
Solid	Solid	Solid	Brass (Cu/Zn), solder (Sn/Pb)

Solvent — Step 1 ΔH_1 — Step 2 ΔH_2 — Solute

Step 3 | ΔH_3

Solution

Figure 13.2 *A molecular view of the solution process portrayed as taking place in three steps: First the solvent and solute molecules are separated (steps 1 and 2). Then the solvent and solute molecules mix (step 3).*

For simplicity, we can imagine the solution process taking place in three distinct steps (Figure 13.2). Step 1 is the separation of solvent molecules, and step 2 entails the separation of solute molecules. These steps require energy input to break attractive intermolecular forces; therefore, they are endothermic. In step 3 the solvent and solute molecules mix. This step may be exothermic or endothermic. The heat of solution ΔH_{soln} is given by

$$\Delta H_{soln} = \Delta H_1 + \Delta H_2 + \Delta H_3 \qquad (13.1)$$

Dissolution of a covalent compound

This equation is an application of Hess's law.

If the solute-solvent attraction is stronger than the solvent-solvent attraction and solute-solute attraction, the solution process is favorable; that is, it is exothermic ($\Delta H_{soln} < 0$). If the solute-solvent interaction is weaker than the solvent-solvent and solute-solute interactions, the solution process is endothermic ($\Delta H_{soln} > 0$).

You may wonder why a solute dissolves in a solvent at all if the attraction among its own molecules is stronger than that between its molecules and the solvent molecules. The solution process, like all physical and chemical processes, is governed by two factors. One is energy, which determines whether a solution process is exothermic or endothermic. The second factor is an inherent tendency toward disorder in all natural events. In much the same way that a deck of new playing cards becomes mixed up after it has been shuffled a few times, when solute and solvent molecules mix to form a solution, there is an increase in randomness or disorder. In the pure state, the solvent and solute possess a fair degree of order, characterized by the more or less regular arrangement of atoms, molecules, or ions in three-dimensional space. Much of this order is destroyed when the solute dissolves in the solvent (see Figure 13.2). Therefore, the solution process is accompanied by an increase in disorder or randomness. It is the increase in disorder of the system that favors the solubility of any substance, even if the solution process is endothermic.

Solubility is a measure of the amount of a solute that will dissolve in a solvent at a specific temperature. The saying "like dissolves like" helps in predicting the solubility of a substance in a solvent. What this expression means is that two substances with intermolecular forces of similar type and magnitude are likely to be soluble in each other. For example, both carbon tetrachloride (CCl_4) and benzene (C_6H_6) are nonpolar liquids. The only intermolecular forces present in these substances are

CH₃OH

C₂H₅OH

CH₂(OH)CH₂(OH)

(NH₂)₂CO

Similar problem: 13.9.

dispersion forces (see Section 12.2). When these two liquids are mixed, they readily dissolve in each other, because the attraction between CCl_4 and C_6H_6 molecules is comparable in magnitude to that between CCl_4 molecules and between C_6H_6 molecules. When *two liquids are completely soluble in each other in all proportions,* as in this case, they are said to be ***miscible.*** Alcohols such as methanol, ethanol, and ethylene glycol are miscible with water because of their ability to form hydrogen bonds with water molecules:

$$
\begin{array}{ccc}
\text{H} & \text{H H} & \text{H H} \\
| & | \; | & | \; | \\
\text{H—C—O—H} & \text{H—C—C—O—H} & \text{H—O—C—C—O—H} \\
| & | \; | & | \; | \\
\text{H} & \text{H H} & \text{H H} \\
\text{methanol} & \text{ethanol} & \text{1,2-ethylene glycol}
\end{array}
$$

The rules given in Table 4.2 (p. 95) enable us to predict the solubility of a particular ionic compound in water. When sodium chloride dissolves in water, the ions are stabilized in solution by hydration, which involves ion-dipole interaction. In general, we predict that ionic compounds should be much more soluble in polar solvents, such as water, liquid ammonia, and liquid hydrogen fluoride, than in nonpolar solvents, such as benzene and carbon tetrachloride. Because the molecules of nonpolar solvents lack a dipole moment, they cannot effectively solvate the Na^+ and Cl^- ions. (***Solvation*** is *the process in which an ion or a molecule is surrounded by solvent molecules arranged in a specific manner.* When the solvent is water, the process is called *hydration.*) The predominant intermolecular interaction between ions and nonpolar compounds is ion-induced dipole interaction, which is much weaker than ion-dipole interaction. Consequently, ionic compounds usually have extremely low solubility in nonpolar solvents.

Example 13.1   Predict the relative solubilities in these cases: (a) Br_2 in benzene (C_6H_6) $(\mu = 0 \text{ D})$ and in water $(\mu = 1.87 \text{ D})$, (b) KCl in carbon tetrachloride $(\mu = 0 \text{ D})$ and in liquid ammonia $(\mu = 1.46 \text{ D})$, (c) urea $(NH_2)_2CO$ in carbon disulfide $(\mu = 0)$ and in water.

Reasoning and Solution   In predicting solubility, we generally follow the "like dissolves like" principle.

(a) Br_2 is a nonpolar molecule and therefore should be more soluble in C_6H_6, which is also nonpolar, than in water. The only intermolecular forces between Br_2 and C_6H_6 are dispersion forces.

(b) KCl is an ionic compound. For it to dissolve, the individual K^+ and Cl^- ions must be stabilized by ion-dipole interaction. Because carbon tetrachloride has no dipole moment, potassium chloride should be more soluble in liquid ammonia, a polar molecule with a large dipole moment.

(c) From the structure of urea we expect it to be a polar molecule. Because carbon disulfide is nonpolar, the forces between molecules of urea and carbon disulfide are dipole-induced dipole and dispersion. On the other hand, urea can form hydrogen bonds with water, so it should be more soluble in that solvent.

Practice Exercise   Is iodine (I_2) more soluble in water or in carbon disulfide (CS_2)?

13.3 Concentration Units

Quantitative study of a solution requires that we know its *concentration,* that is, the amount of solute present in a given amount of solution. Chemists use several different concentration units, each of which has advantages as well as limitations. Let us examine the three most common units of concentration: percent by mass, molarity, and molality.

Types of Concentration Units

Percent by Mass

The **percent by mass** (also called the *percent by weight* or the *weight percent*) is defined as

$$\text{percent by mass of solute} = \frac{\text{mass of solute}}{\text{mass of solute} + \text{mass of solvent}} \times 100\%$$

$$= \frac{\text{mass of solute}}{\text{mass of soln}} \times 100\% \qquad (13.2)$$

The percent by mass has no units because it is a ratio of two similar quantities.

Molarity (M)

The molarity unit was defined in Section 4.5 as the number of moles of solute in 1 L of solution; that is,

$$\text{molarity} = \frac{\text{moles of solute}}{\text{liters of soln}} \qquad (13.3)$$

Thus, molarity has the units of mole per liter (mol/L).

For calculations involving molarity, see Example 4.5.

Molality (m)

Molality is *the number of moles of solute dissolved in 1 kg (1000 g) of solvent—*that is,

$$\text{molality} = \frac{\text{moles of solute}}{\text{mass of solvent (kg)}} \qquad (13.4)$$

For example, to prepare a 1 *molal,* or 1 *m,* sodium sulfate (Na_2SO_4) aqueous solution, we need to dissolve 1 mole (142.0 g) of the substance in 1000 g (1 kg) of water. Depending on the nature of the solute-solvent interaction, the final volume of the solution will be either greater or less than 1000 mL. It is also possible, though very unlikely, that the final volume could be equal to 1000 mL.

Example 13.2 Calculate the molality of a sulfuric acid solution containing 24.4 g of sulfuric acid in 198 g of water. The molar mass of sulfuric acid is 98.08 g.

Reasoning and Solution From the known molar mass of sulfuric acid, we can calculate the molality in two steps. First we need to find the number of grams of sulfuric

H_2SO_4

acid dissolved in 1000 g (1 kg) of water. Next we must convert the number of grams into the number of moles. Combining these two steps we write

$$\text{molality} = \frac{\text{moles of solute}}{\text{mass of solvent (kg)}}$$

$$= \frac{24.4 \text{ g } H_2SO_4}{198 \text{ g } H_2O} \times \frac{1000 \text{ g } H_2O}{1 \text{ kg } H_2O} \times \frac{1 \text{ mol } H_2SO_4}{98.08 \text{ g } H_2SO_4}$$

$$= 1.26 \text{ mol } H_2SO_4/\text{kg } H_2O$$

$$= 1.26 \text{ } m$$

Similar problem: 13.15.

Practice Exercise What is the molality of a solution containing 7.78 g of urea [(NH₂)₂CO] in 203 g of water?

Comparison of Concentration Units

The choice of a concentration unit is based on the purpose of the measurement. The advantage of molarity is that it is generally easier to measure the volume of a solution, using precisely calibrated volumetric flasks, than to weigh the solvent, as we saw in Section 4.5. For this reason, molarity is often preferred over molality. On the other hand, molality is independent of temperature, because the concentration is expressed in number of moles of solute and mass of solvent, whereas the volume of a solution typically increases with increasing temperature. A solution that is 1.0 M at 25°C may become 0.97 M at 45°C because of the increase in volume. This concentration dependence on temperature can significantly affect the accuracy of an experiment.

Percent by mass is similar to molality in that it is independent of temperature. Furthermore, because it is defined in terms of ratio of mass of solute to mass of solution, we do not need to know the molar mass of the solute in order to calculate the percent by mass.

Glucose is soluble in water because of its ability to form hydrogen bonds with water molecules (like dissolves like).

Example 13.3 Calculate the molarity of a 0.396 m glucose ($C_6H_{12}O_6$) solution. The molar mass of glucose is 180.2 g, and the density of the solution is 1.16 g/mL.

Reasoning and Solution The mass of the solution must be converted to volume in working problems of this type. The density of the solution is used as a conversion factor. Because a 0.396 m glucose solution contains 0.396 mole of glucose in 1 kg of water, the total mass of the solution is

$$\left(0.396 \text{ mol } C_6H_{12}O_6 \times \frac{180.2 \text{ g}}{1 \text{ mol } C_6H_{12}O_6}\right) + 1000 \text{ g } H_2O \text{ soln} = 1071 \text{ g}$$

From the known density of the solution (1.16 g/mL), we can calculate the volume and hence the molarity of the solution as follows:

$$\text{molarity} = \frac{0.396 \text{ mol } C_6H_{12}O_6}{1071 \text{ g soln}} \times \frac{1.16 \text{ g soln}}{1 \text{ mL soln}} \times \frac{1000 \text{ mL soln}}{1 \text{ L soln}}$$

$$= \frac{0.429 \text{ mol } C_6H_{12}O_6}{1 \text{ L soln}}$$

$$= 0.429 \text{ } M$$

Similar problems: 13.21, 13.22.

Practice Exercise Calculate the molarity of a 1.74 *m* sucrose ($C_{12}H_{22}O_{11}$) solution whose density is 1.12 g/mL.

Example 13.4 The density of a 2.45 *M* aqueous solution of methanol (CH_3OH) is 0.976 g/mL. What is the molality of the solution? The molar mass of methanol is 32.04 g.

Reasoning and Solution To calculate the molality, we need to know the number of moles of methanol in 1 kg of solvent. This can be done in two steps. First, we calculate the mass of the solvent in 1 L of the solution. The total mass of 1 L of a 2.45 *M* solution of methanol is given by volume × density or

$$1 \text{ L soln} \times \frac{1000 \text{ mL soln}}{1 \text{ L soln}} \times \frac{0.976 \text{ g soln}}{1 \text{ mL soln}} = 976 \text{ g soln}$$

Because this solution contains 2.45 moles of methanol, the amount of water in the solution is

$$976 \text{ g soln} - \left(2.45 \text{ mol } CH_3OH \times \frac{32.04 \text{ g } CH_3OH}{1 \text{ mol } CH_3OH} \right) = 898 \text{ g } H_2O$$

Now the molality of the solution can be calculated:

$$\begin{aligned} \text{molality} &= \frac{2.45 \text{ mol } CH_3OH}{898 \text{ g } H_2O} \times \frac{1000 \text{ g } H_2O}{1 \text{ kg } H_2O} \\ &= \frac{2.73 \text{ mol } CH_3OH}{1 \text{ kg } H_2O} \\ &= 2.73 \ m \end{aligned}$$

Practice Exercise Calculate the molality of a 5.86 *M* ethanol (C_2H_5OH) solution whose density is 0.927 g/mL.

CH_3OH

Similar problem: 13.16(a).

Example 13.5 Calculate the molality of a 35.4 percent (by mass) aqueous solution of phosphoric acid (H_3PO_4). The molar mass of phosphoric acid is 98.00 g.

Reasoning and Solution In solving this type of problem it is convenient to assume that we start with 100.0 g of the solution. If the mass of phosphoric acid is 35.4 percent, or 35.4 g, the percent and mass of water must be 100.0% − 35.4% = 64.6% and 64.6 g. From the known molar mass of phosphoric acid, we can calculate the molality in two steps. First we need to find the number of grams of phosphoric acid dissolved in 1000 g (1 kg) of water. Next, we must convert the number of grams into the number of moles. Combining these two steps we write

$$\begin{aligned} \text{molality} &= \frac{35.4 \text{ g } H_3PO_4}{64.6 \text{ g } H_2O} \times \frac{1000 \text{ g } H_2O}{1 \text{ kg } H_2O} \times \frac{1 \text{ mol } H_3PO_4}{98.00 \text{ g } H_3PO_4} \\ &= 5.59 \text{ mol } H_3PO_4/\text{kg } H_2O \\ &= 5.59 \ m \end{aligned}$$

Practice Exercise Calculate the molality of a 44.6 percent (by mass) aqueous solution of sodium chloride.

H_3PO_4

Similar problem: 13.16(b).

13.4 Effect of Temperature on Solubility

Solubility always refers to a saturated solution.

Recall that solubility is defined as the maximum amount of a solute that will dissolve in a given quantity of solvent *at a specific temperature*. For most substances, temperature affects solubility. In this section we will consider the effects of temperature on the solubility of solids and gases.

Solid Solubility and Temperature

Figure 13.3 shows the temperature dependence of the solubility of some ionic compounds in water. In most but certainly not all cases, the solubility of a solid substance increases with temperature. However, there is no clear correlation between the sign of ΔH_{soln} and the variation of solubility with temperature. For example, the solution process of $CaCl_2$ is exothermic and that of NH_4NO_3 is endothermic. But the solubility of both compounds increases with increasing temperature. In general, the effect of temperature on solubility is best determined experimentally.

Gas Solubility and Temperature

The solubility of gases in water usually decreases with increasing temperature (Figure 13.4). When water is heated in a beaker, you can see bubbles of air forming on the side of the glass before the water boils. As the temperature rises, the dissolved air molecules begin to "boil out" of the solution long before the water itself boils.

The reduced solubility of molecular oxygen in hot water has a direct bearing on *thermal pollution*, that is, *the heating of the environment*—usually waterways—*to temperatures that are harmful to its living inhabitants*. It is estimated that every year in the United States some 100,000 billion gallons of water are used for industrial cooling, mostly in electric power and nuclear power production. This process heats up the

Figure 13.3 *Temperature dependence of the solubility of some ionic compounds in water.*

water, which is then returned to the rivers and lakes from which it was taken. Ecologists have become increasingly concerned about the effect of thermal pollution on aquatic life. Fish, like all other cold-blooded animals, have much more difficulty coping with rapid temperature fluctuation in the environment than humans do. An increase in water temperature accelerates their rate of metabolism, which generally doubles with each 10°C rise. The speedup of metabolism increases the fish's need for oxygen at the same time that the supply of oxygen decreases because of its lower solubility in heated water. Effective ways to cool power plants while doing only minimal damage to the biological environment are being sought.

On the lighter side, a knowledge of the variation of gas solubility with temperature can improve one's performance in a popular recreational sport—fishing. On a hot summer day, an experienced fisherman usually picks a deep spot in the river or lake to cast the bait. Because the oxygen content is greater in the deeper, cooler region, most fish will be found there.

Figure 13.4 *Temperature dependence of O_2 gas solubility in water. Note that the solubility decreases as temperature increases. The pressure of the gas over the solution is 1 atm.*

13.5 Effect of Pressure on the Solubility of Gases

For all practical purposes, external pressure has no influence on the solubilities of liquids and solids, but it does greatly affect the solubility of gases. The quantitative relationship between gas solubility and pressure is given by **Henry's law**, which states that *the solubility of a gas in a liquid is proportional to the pressure of the gas over the solution:*

$$c \propto P$$
$$c = kP \tag{13.5}$$

Here c is the molar concentration (moles per liter) of the dissolved gas; P is the pressure (in atmospheres) of the gas over the solution; and, for a given gas, k is a constant that depends only on temperature. The constant k has the units mol/L · atm. You can see that when the pressure of the gas is 1 atm, c is *numerically* equal to k.

Henry's law can be understood qualitatively in terms of the kinetic molecular theory. The amount of gas that will dissolve in a solvent depends on how frequently the molecules in the gas phase collide with the liquid surface and become trapped by the condensed phase. Suppose we have a gas in dynamic equilibrium with a solution [Figure 13.5(a)]. At every instant, the number of gas molecules entering the solution

Each gas has a different k value in a specific solvent at a given temperature.

(a)

(b)

Figure 13.5 *A molecular interpretation of Henry's law. When the partial pressure of the gas over the solution increases from (a) to (b), the concentration of the dissolved gas also increases according to Equation (13.5).*

is equal to the number of dissolved molecules moving into the gas phase. When the partial pressure is increased, more molecules dissolve in the liquid because more molecules are striking the surface of the liquid. This process continues until the concentration of the solution is again such that the number of molecules leaving the solution per second equals the number entering the solution [Figure 13.5(b)]. Because of the increased concentration of molecules in both the gas and solution phases, this number is greater in (b) than in (a), where the partial pressure is lower.

A practical demonstration of Henry's law is the effervescence of a soft drink when the cap of the bottle is removed. Before the beverage bottle is sealed, it is pressurized with a mixture of air and CO_2 saturated with water vapor. Because of the high partial pressure of CO_2 in the pressurizing gas mixture, the amount dissolved in the soft drink is many times the amount that would dissolve under normal atmospheric conditions. When the cap is removed, the pressurized gases escape, eventually the pressure in the bottle falls to atmospheric pressure, and the amount of CO_2 remaining in the beverage is determined only by the normal atmospheric partial pressure of CO_2, 0.0003 atm. The excess dissolved CO_2 comes out of solution, causing the effervescence.

The effervescence of a soft drink. The bottle was shaken before being opened to dramatize the escape of CO_2.

Example 13.6   The solubility of nitrogen gas at 25°C and 1 atm is 6.8×10^{-4} mol/L. What is the concentration of nitrogen dissolved in water under atmospheric conditions? The partial pressure of nitrogen gas in the atmosphere is 0.78 atm.

Reasoning and Solution   To calculate the concentration of dissolved nitrogen gas under atmospheric conditions, we need Henry's constant for the gas using the given equation. From Equation (13.5)

$$c = kP$$

$$6.8 \times 10^{-4} \text{ mol/L} = k \,(1 \text{ atm})$$

$$k = 6.8 \times 10^{-4} \text{ mol/L} \cdot \text{atm}$$

Therefore, the solubility of nitrogen gas in water is

$$c = (6.8 \times 10^{-4} \text{ mol/L} \cdot \text{atm})(0.78 \text{ atm})$$

$$= 5.3 \times 10^{-4} \text{ mol/L}$$

$$= 5.3 \times 10^{-4} \, M$$

Comment   The decrease in solubility is the result of lowering the pressure from 1 atm to 0.78 atm.

Practice Exercise   Calculate the molar concentration of oxygen in water at 25°C for a partial pressure of 0.22 atm. The Henry's law constant for oxygen is 1.3×10^{-3} mol/L · atm.

Similar problem: 13.35.

Most gases obey Henry's law, but there are some important exceptions. For example, if the dissolved gas *reacts* with water, higher solubilities can result. The solubility of ammonia is much higher than expected because of the reaction

$$NH_3 + H_2O \rightleftharpoons NH_4^+ + OH^-$$

Carbon dioxide also reacts with water, as:

$$CO_2 + H_2O \rightleftharpoons H_2CO_3$$

13.6 Colligative Properties

Several important properties of solutions depend on the number of solute particles in solution and not on the nature of the solute particles. These properties are called **colligative properties** (or collective properties) because they are bound together by a common origin; that is, they all depend on the number of solute particles present, whether these particles are atoms, ions, or molecules. The colligative properties are vapor-pressure lowering, boiling-point elevation, freezing-point depression, and osmotic pressure. For our discussion of colligative properties of nonelectrolyte solutions it is important to keep in mind that we are talking about relatively dilute solutions, that is, solutions whose concentrations are $\leq 0.2\ M$.

Vapor-Pressure Lowering

If a solute is **nonvolatile** (that is, it *does not have a measurable vapor pressure*), the vapor pressure of its solution is always less than that of the pure solvent. Thus the relationship between solution vapor pressure and solvent vapor pressure depends on the concentration of the solute in the solution. This relationship is given by **Raoult's law** (after the French chemist Francois Raoult), which states that *the partial pressure of a solvent over a solution, P_1, is given by the vapor pressure of the pure solvent, P_1°, times the mole fraction of the solvent in the solution, X_1*:

To review the concept of equilibrium vapor pressure as it applies to pure liquids, see Section 12.6.

$$P_1 = X_1 P_1^\circ \tag{13.6}$$

Mole fraction is defined on p. 141.

In a solution containing only one solute, $X_1 = 1 - X_2$, in which X_2 is the mole fraction of the solute (see Section 5.5). Equation (13.6) can therefore be rewritten as

$$P_1 = (1 - X_2)P_1^\circ$$
$$P_1^\circ - P_1 = \Delta P = X_2 P_1^\circ \tag{13.7}$$

We see that the *decrease* in vapor pressure, ΔP, is directly proportional to the concentration (measured in mole fraction) of the solute present.

Example 13.7 At 25°C, the vapor pressure of pure water is 23.76 mmHg and that of a urea solution is 22.98 mmHg. Estimate the molality of the solution.

Reasoning and Solution From the vapor pressure lowering we first determine the mole fraction of urea, which then enables us to estimate the molality of the solution. From Equation (13.7) we write

$$\Delta P = (23.76 - 22.98)\ \text{mmHg} = X_2(23.76\ \text{mmHg})$$
$$X_2 = 0.033$$

By definition
$$X_2 = \frac{n_2}{n_1 + n_2}$$

where n_1 and n_2 are the numbers of moles of solvent and solute, respectively. Because the mole fraction of urea in this solution is only 0.033, the solution is dilute, and we can assume that n_1 is much larger than n_2. Therefore we can write

$$X_2 = \frac{n_2}{n_1 + n_2} \simeq \frac{n_2}{n_1} \quad (n_1 \gg n_2)$$

$$n_2 = n_1 X_2$$

The number of moles of water in 1 kg of water is

$$1000 \text{ g H}_2\text{O} \times \frac{1 \text{ mol H}_2\text{O}}{18.02 \text{ g H}_2\text{O}} = 55.49 \text{ mol H}_2\text{O}$$

and the number of moles of urea present in 1 kg of water is

$$n_2 = n_1 X_2 = (55.49 \text{ mol})(0.033)$$

$$= 1.8 \text{ mol}$$

Similar problems: 13.49, 13.50.

Thus, the concentration of the urea solution is 1.8 m.

Practice Exercise   The vapor pressure of a glucose ($C_6H_{12}O_6$) solution is 17.01 mmHg at 20°C, while that of pure water is 17.25 mmHg at the same temperature. Calculate the molality of the solution.

Why is the vapor pressure of a solution less than that of its pure solvent? As was mentioned in Section 13.2, one driving force in physical and chemical processes is the increase in disorder—the greater the disorder created, the more favorable the process. Vaporization increases the disorder of a system because molecules in a vapor have less order than those in a liquid. Because a solution is more disordered than a pure solvent, the difference in disorder between a solution and a vapor is less than that between a pure solvent and a vapor. Thus solvent molecules have less of a tendency to leave a solution than to leave the pure solvent to become vapor, and the vapor pressure of a solution is less than that of the solvent.

If both components of a solution are **volatile** (that is, *have measurable vapor pressure*), the vapor pressure of the solution is the sum of the individual partial pressures. Raoult's law holds equally well in this case:

$$P_A = X_A P_A^\circ$$

$$P_B = X_B P_B^\circ$$

in which P_A and P_B are the partial pressures over the solution for components A and B; P_A° and P_B° are the vapor pressures of the pure substances; and X_A and X_B are their mole fractions. The total pressure is given by Dalton's law of partial pressure [Equation (5.12)]:

$$P_T = P_A + P_B$$

Benzene and toluene have similar structures and therefore similar intermolecular forces:

benzene   toluene

In a solution of benzene and toluene, the vapor pressure of each component obeys Raoult's law. Figure 13.6 shows the dependence of the total vapor pressure (P_T) in a benzene-toluene solution on the composition of the solution. Note that we need only express the composition of the solution in terms of the mole fraction of one component. For every value of $X_{benzene}$, the mole fraction of toluene is given by ($1 - X_{benzene}$). The benzene-toluene solution is one of the few examples of an ***ideal solution***, which is *any solution that obeys Raoult's law*. One characteristic of an ideal solution is that the heat of solution, ΔH_{soln}, is always zero.

Boiling-Point Elevation

Because the presence of a *nonvolatile* solute lowers the vapor pressure of a solution, it must also affect the boiling point of the solution. The boiling point of a solution is the temperature at which its vapor pressure equals the external atmospheric pressure (see Section 12.6). Figure 13.7 shows the phase diagram of water and the changes that occur in an aqueous solution. Because at any temperature the vapor pressure of the solution is lower than that of the pure solvent, the liquid-vapor curve for the solution lies below that for the pure solvent. Consequently, the solution curve (dotted line) intersects the horizontal line that marks $P = 1$ atm at a *higher* temperature than the normal boiling point of the pure solvent. This graphical analysis shows that the boiling point of the solution is higher than that of water. The *boiling-point elevation*, ΔT_b, is defined as

$$\Delta T_b = T_b - T_b^\circ$$

in which T_b is the boiling point of the solution and T_b° the boiling point of the pure solvent. Because ΔT_b is proportional to the vapor-pressure lowering, it is also proportional to the concentration (molality) of the solution. That is,

$$\Delta T_b \propto m$$
$$\Delta T_b = K_b m \qquad (13.8)$$

in which m is the molality of the solution and K_b is the *molal boiling-point elevation constant*. The units of K_b are $°C/m$.

Figure 13.6 *The dependence of the partial pressures of benzene and toluene on their mole fractions in a benzene-toluene solution ($X_{toluene} = 1 - X_{benzene}$) at 80°C. This solution is said to be ideal because the vapor pressures obey Raoult's law.*

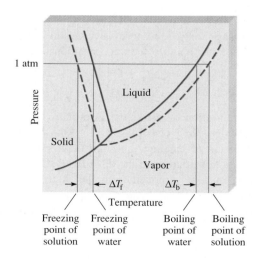

Figure 13.7 *Phase diagram illustrating the boiling-point elevation and freezing-point depression of aqueous solutions. The dashed curves pertain to the solution, and the solid curves to the pure solvent. As you can see, the boiling point of the solution is higher than that of water, and the freezing point of the solution is lower than that of water.*

Table 13.2	Molal Boiling-Point Elevation and Freezing-Point Depression Constants of Several Common Liquids				
Solvent	**Normal Freezing Point (°C)***	K_f (°C/m)	**Normal Boiling Point (°C)***	K_b (°C/m)	
Water	0	1.86	100	0.52	
Benzene	5.5	5.12	80.1	2.53	
Ethanol	−117.3	1.99	78.4	1.22	
Acetic acid	16.6	3.90	117.9	2.93	
Cyclohexane	6.6	20.0	80.7	2.79	

*Measured at 1 atm.

It is important to understand the choice of concentration unit here. We are dealing with a system (the solution) whose temperature is *not* kept constant, so we cannot express the concentration units in molarity because molarity changes with temperature.

Table 13.2 lists the value of K_b for several common solvents. Using the boiling-point elevation constant for water and Equation (13.8), you can see that if the molality of an aqueous solution is 1.00 m, the boiling point will be 100.52°C.

Freezing-Point Depression

De-icing of airplanes is based on freezing-point depression.

A nonscientist may remain forever unaware of the boiling-point elevation phenomenon, but a careful observer living in a cold climate is familiar with freezing-point depression. Ice on frozen roads and sidewalks melts when sprinkled with salts such as NaCl or $CaCl_2$. This method of thawing succeeds because it depresses the freezing point of water.

Figure 13.7 shows that lowering the vapor pressure of the solution shifts the solid-liquid curve to the left. Consequently, this line intersects the horizontal line at a temperature *lower* than the freezing point of water. The *freezing-point depression*, ΔT_f, is defined as

$$\Delta T_f = T_f^\circ - T_f$$

in which T_f° is the freezing point of the pure solvent, and T_f the freezing point of the solution. Again, ΔT_f is proportional to the concentration of the solution:

$$\Delta T_f \propto m$$
$$\Delta T_f = K_f m \qquad (13.9)$$

in which m is the concentration of the solute in molality units, and K_f is the *molal freezing-point depression constant* (see Table 13.2). Like K_b, K_f has the units °C/m.

A qualitative explanation of the freezing-point depression phenomenon is as follows. Freezing involves a transition from the disordered state to the ordered state. For this to happen, energy must be removed from the system. Because a solution has greater disorder than the solvent, more energy needs to be removed from it to create order than in the case of a pure solvent. Therefore, the solution has a lower freezing point than the solvent. Note that when a solution freezes, the solid that separates is the solvent component.

Whereas the solute must be nonvolatile in the case of boiling-point elevation, no such restriction applies to freezing-point depression. For example, methanol (CH_3OH), a fairly volatile liquid that boils at only 65°C, has sometimes been used as an antifreeze in automobile radiators.

Example 13.8 Ethylene glycol (EG), $CH_2(OH)CH_2(OH)$, is a common automobile antifreeze. It is water soluble and fairly nonvolatile (b.p. 197°C). Calculate the freezing point of a solution containing 651 g of this substance in 2505 g of water. Would you keep this substance in your car radiator during the summer? The molar mass of ethylene glycol is 62.01 g.

Reasoning and Solution This calculation requires two steps. First we determine the molality of the solution. Next we use Equation (13.9) to calculate the depression in freezing point.

The number of moles of EG in 651 g EG is

$$651 \text{ g EG} \times \frac{1 \text{ mol EG}}{62.01 \text{ g EG}} = 10.50 \text{ mol EG}$$

This is the number of moles in 2505 g or 2.505 kg of H_2O. Therefore, the number of moles of EG in 1 kg of H_2O, or the molality, is

$$\frac{10.50 \text{ mol EG}}{2.505 \text{ kg H}_2\text{O}} = 4.19 \text{ mol EG/kg H}_2\text{O} = 4.19 \, m$$

From Equation (13.9) and Table 13.2 we write

$$\Delta T_f = (1.86°C/m)(4.19 \, m)$$
$$= 7.79°C$$

Pure water freezes at 0°C, so the solution will freeze at −7.79°C. We can calculate boiling-point elevation using Equation (13.8) in the same way as follows:

$$\Delta T_b = (0.52°C/m)(4.19 \, m)$$
$$= 2.2°C$$

Because the solution will boil at (100 + 2.2)°C, 102.2°C, it would be preferable to leave the antifreeze in your car radiator in summer to prevent the solution from boiling.

Practice Exercise Calculate the boiling point and freezing point of a solution containing 478 g of ethylene glycol in 3202 g of water.

In cold climate regions, antifreeze must be used in car radiators in winter.

Similar problems: 13.56, 13.59.

Osmotic Pressure

Many chemical and biological processes depend on the selective passage of solvent molecules through a porous membrane from a dilute solution to a more concentrated one. Figure 13.8 illustrates this phenomenon. The left compartment of the apparatus contains pure solvent; the right compartment contains a solution. The two compartments are separated by a **semipermeable membrane,** which *allows solvent molecules to pass through but blocks the passage of solute molecules.* At the start, the water levels in the two tubes are equal [see Figure 13.8(a)]. After some time, the level in the right tube begins to rise; this continues until equilibrium is reached. *The net movement*

Figure 13.8 *Osmotic pressure. (a) The levels of the pure solvent (left) and of the solution (right) are equal at the start. (b) During osmosis, the level on the solution side rises as a result of the net flow of solvent from left to right. The osmotic pressure is equal to the hydrostatic pressure exerted by the column of fluid in the right tube at equilibrium. Basically the same effect occurs when the pure solvent is replaced by a more dilute solution than that on the right.*

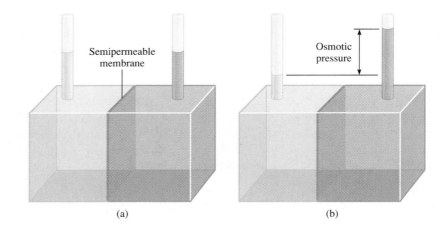

(a)   (b)

of solvent molecules through a semipermeable membrane from a pure solvent or from a dilute solution to a more concentrated solution is called **osmosis.** The **osmotic pressure (π)** of a solution is *the pressure required to stop osmosis.* As shown in Figure 13.8(b), this pressure can be measured directly from the difference in the final fluid levels.

What causes water to move spontaneously from left to right in this case? Compare the vapor pressure of pure water and that of water from a solution (Figure 13.9). Because the vapor pressure of pure water is higher, there is a net transfer of water from the left beaker to the right one. Given enough time, the transfer will continue to completion. A similar force causes water to move into the solution during osmosis.

Although osmosis is a common and well-studied phenomenon, relatively little is known about how the semipermeable membrane stops some molecules yet allows others to pass. In some cases, it is simply a matter of size. A semipermeable membrane may have pores small enough to let only the solvent molecules through. In other cases, a different mechanism may be responsible for the membrane's selectivity—for example, the solvent's greater "solubility" in the membrane.

The osmotic pressure of a solution is given by

$$\pi = MRT \tag{13.10}$$

Figure 13.9 *(a) Unequal vapor pressures inside the container lead to a net transfer of water from the left beaker (which contains pure water) to the right one (which contains a solution). (b) At equilibrium, all the water in the left beaker has been transferred to the right beaker. This driving force for solvent transfer is analogous to the osmotic phenomenon shown in Figure 13.8.*

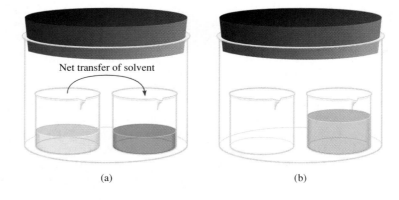

(a)   (b)

in which M is the molarity of solution, R is the gas constant (0.0821 L · atm/K · mol), and T is the absolute temperature. The osmotic pressure, π, is expressed in atmospheres. Because osmotic pressure measurements are carried out at constant temperature, we express the concentration here in terms of the more convenient units of molarity rather than molality.

Like boiling-point elevation and freezing-point depression, osmotic pressure is directly proportional to the concentration of solution. This is what we would expect, bearing in mind that all colligative properties depend only on the number of solute particles in solution. If two solutions are of equal concentration and, hence, of the same osmotic pressure, they are said to be *isotonic*. If two solutions are of unequal osmotic pressures, the more concentrated solution is said to be *hypertonic* and the more dilute solution is described as *hypotonic* (Figure 13.10).

The osmotic pressure phenomenon manifests itself in many interesting applications. To study the contents of red blood cells, which are protected from the external environment by a semipermeable membrane, biochemists use a technique called hemolysis. The red blood cells are placed in a hypotonic solution. Because the hypotonic solution is less concentrated than the interior of the cell, water moves into the cells, as shown in Figure 13.10(b). The cells swell and eventually burst, releasing hemoglobin and other molecules.

Home preserving of jam and jelly provides another example of the use of osmotic pressure. A large quantity of sugar is actually essential to the preservation process because the sugar helps to kill bacteria that may cause botulism. As Figure 13.10(c) shows, when a bacterial cell is in a hypertonic (high-concentration) sugar solution, the intracellular water tends to move out of the bacterial cell to the more concentrated solution by osmosis. This process, known as *crenation,* causes the cell to shrink and, eventually, to cease functioning. The natural acidity of fruits also inhibits bacteria growth.

Osmotic pressure also is the major mechanism for transporting water upward in plants. Because leaves constantly lose water to the air, in a process called *transpiration,* the solute concentrations in leaf fluids increase. Water is pushed up through the trunk, branches, and stems of trees by osmotic pressure. Up to 10 to 15 atm pressure is necessary to transport water to the leaves at the tops of California's redwoods, which reach about 120 m in height. (The capillary action discussed in Section 12.3 is responsible for the rise of water only up to a few centimeters.)

California redwoods.

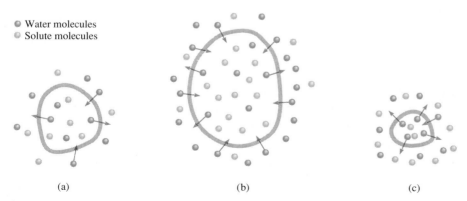

- Water molecules
- Solute molecules

(a)   (b)   (c)

Figure 13.10 *A cell in (a) an isotonic solution, (b) a hypotonic solution, and (c) a hypertonic solution. The cell remains unchanged in (a), swells in (b), and shrinks in (c).*

Using Colligative Properties to Determine Molar Mass

The colligative properties of nonelectrolyte solutions provide a means of determining the molar mass of a solute. Theoretically, any of the four colligative properties are suitable for this purpose. In practice, however, only freezing-point depression and osmotic pressure are used because they show the most pronounced changes.

Can you account for naphthalene's high solubility in benzene?

Example 13.9   A 7.85-g sample of a compound with the empirical formula C_5H_4 is dissolved in 301 g of benzene. The freezing point of the solution is 1.05°C below that of pure benzene. What are the molar mass and molecular formula of this compound?

Reasoning and Solution   Solving this problem requires three steps. First, we calculate the molality of the solution from its freezing point. Next, we determine the number of moles of the compound in 7.58 g. Finally, comparing the experimental molar mass with the molar mass of the empirical formula leads to the molecular formula.

From Equation (13.9) and Table 13.2 we can write

$$\text{molality} = \frac{\Delta T_f}{K_f} = \frac{1.05°C}{5.12°C/m} = 0.205 \ m$$

Because there is 0.205 mole of the solute in 1 kg of solvent, the number of moles of solute in 301 g, or 0.301 kg, of solvent is

$$\frac{0.205 \ \text{mol}}{1 \ \text{kg solvent}} \times 0.301 \ \text{kg solvent} = 0.0617 \ \text{mol}$$

Finally, we calculate the molar mass of the solute as:

$$\frac{7.85 \ \text{g}}{0.0617 \ \text{mol}} = 127 \ \text{g/mol}$$

Because the formula mass of C_5H_4 is 64 g and the molar mass is 127 g, the molecular formula is $C_{10}H_8$ (naphthalene).

Practice Exercise   A solution of 0.85 g of an organic compound in 100.0 g of benzene has a freezing point of 5.16°C. What are the molality of the solution and the molar mass of the solute?

Similar problem: 13.57.

Example 13.10   A solution is prepared by dissolving 35.0 g of hemoglobin (Hb) in enough water to make up 1 L in volume. If the osmotic pressure of the solution is found to be 10.0 mmHg at 25°C, calculate the molar mass of hemoglobin.

Reasoning and Solution   The information given enables us to determine the molarity of the solution. Because the volume of the solution is 1 L, we can calculate the molar mass from the number of moles and mass of Hb.

First we calculate the molarity of the solution using Equation (13.10).

$$\pi = MRT$$

$$M = \frac{\pi}{RT}$$

$$= \frac{10.0 \text{ mmHg} \times \dfrac{1 \text{ atm}}{760 \text{ mmHg}}}{(0.0821 \text{ L} \cdot \text{atm/K} \cdot \text{mol})(298 \text{ K})}$$

$$= 5.38 \times 10^{-4} \, M$$

The volume of the solution is 1 L, so it must contain 5.38×10^{-4} mole of Hb. We use this quantity to calculate the molar mass:

$$\text{moles of Hb} = \frac{\text{mass of Hb}}{\text{molar mass of Hb}}$$

$$\text{molar mass of Hb} = \frac{\text{mass of Hb}}{\text{moles of Hb}}$$

$$= \frac{35.0 \text{ g}}{5.38 \times 10^{-4} \text{ mol}}$$

$$= 6.51 \times 10^{4} \text{ g/mol}$$

Similar problems: 13.64, 13.66.

Practice Exercise A 202-mL benzene solution containing 2.47 g of an organic polymer has an osmotic pressure of 8.63 mmHg at 21°C. Calculate the molar mass of the polymer.

A pressure such as 10.0 mmHg in Example 13.10 can be measured easily and accurately. For this reason, osmotic pressure measurements can be used to determine the molar masses of large molecules, such as proteins. To see how much more practical the osmotic pressure technique is than freezing-point depression would be, let us estimate the change in freezing point of the same hemoglobin solution. If an aqueous solution is quite dilute, we can assume that molarity is roughly equal to molality. (Molarity would be equal to molality if the density of the solution were 1 g/mL.) Hence, from Equation (13.9) we write

The density of mercury is 13.6 g/mL. Therefore 10 mmHg corresponds to a column of water 13.6 cm in height.

$$\Delta T_f = (1.86°C/m)(5.38 \times 10^{-4} \, m)$$

$$= 1.00 \times 10^{-3}°C$$

The thousandth-of-a-degree freezing-point depression is too small a temperature change to measure accurately. For this reason, the freezing-point depression technique is more suitable for determining the molar mass of smaller and more soluble molecules, those having molar masses of 500 g or less, because the freezing-point depressions of their solutions are much greater.

Colligative Properties of Electrolytes

The colligative properties of electrolytes require a slightly different approach than the one used for the colligative properties of nonelectrolytes. The reason is that electrolytes

(a)

(b)

Figure 13.11 *(a) Free ions and (b) ion pairs in solution. Such an ion pair bears no net charge and therefore cannot conduct electricity in solution.*

Table 13.3	The van't Hoff Factor of 0.0500 M Electrolyte Solutions at 25°C	
Electrolyte	i (measured)	i (calculated)
Sucrose*	1.0	1.0
HCl	1.9	2.0
NaCl	1.9	2.0
$MgSO_4$	1.3	2.0
$MgCl_2$	2.7	3.0
$FeCl_3$	3.4	4.0

*Sucrose is a nonelectrolyte. It is listed here for comparison only.

dissociate into ions in solution, and so one unit of an electrolyte compound separates into two or more particles when it dissolves. (Remember, it is the number of solute particles that determines the colligative properties of a solution.) For example, each unit of NaCl dissociates into two ions—Na^+ and Cl^-. Thus the colligative properties of a 0.1 m NaCl solution should be twice as great as those of a 0.1 m solution containing a nonelectrolyte, such as sucrose. Similarly, we would expect a 0.1 m $CaCl_2$ solution to depress the freezing point by three times as much as a 0.1 m sucrose solution. To account for this effect we must modify the equations for colligative properties as follows:

$$\Delta T_b = iK_b m \qquad (13.11)$$

$$\Delta T_f = iK_f m \qquad (13.12)$$

$$\pi = iMRT \qquad (13.13)$$

The variable i is the *van't Hoff factor*, which is defined as

$$i = \frac{\text{actual number of particles in soln after dissociation}}{\text{number of formula units initially dissolved in soln}} \qquad (13.14)$$

Every unit of NaCl or KNO_3 that dissociates yields two ions ($i = 2$); every unit of Na_2SO_4 or $MgCl_2$ that dissociates produces three ions ($i = 3$).

Thus i should be 1 for all nonelectrolytes. For strong electrolytes such as NaCl and KNO_3, i should be 2, and for strong electrolytes such as Na_2SO_4 and $MgCl_2$, i should be 3.

In reality, the colligative properties of electrolyte solutions are usually smaller than anticipated because at higher concentrations, electrostatic forces come into play, drawing cations and anions together. *A cation and an anion held together by electrostatic forces* is called an **ion pair.** The formation of an ion pair reduces the number of particles in solution by one, causing a reduction in the colligative properties (Figure 13.11). Table 13.3 shows the experimentally measured values of i and those calculated assuming complete dissociation. As you can see, the agreement is close but not perfect, indicating that the extent of ion-pair formation in these solutions is appreciable.

Example 13.11   The osmotic pressure of 0.010 M potassium iodide solution (KI) at 25°C is 0.465 atm. Calculate the van't Hoff factor for KI at this concentration.

Reasoning and Solution   If KI were to dissociate completely into free ions, the osmotic pressure of the solution would be $2(0.010\ M)(0.0821\ \text{L} \cdot \text{atm/K} \cdot \text{mol})(298\ \text{K})$ or 0.489 atm. Because the observed osmotic pressure is only 0.465 atm, this must mean some extent of ion-pair formation. From Equation (13.13) we have

$$i = \frac{\pi}{MRT}$$

$$= \frac{0.465\ \text{atm}}{(0.010\ M)(0.0821\ \text{L} \cdot \text{atm/K} \cdot \text{mol})(298\ \text{K})}$$

$$= 1.90$$

Similar problem: 13.79.

Practice Exercise   The freezing-point depression of a 0.100 m $MgSO_4$ solution is 0.225°C. Calculate the van't Hoff factor of $MgSO_4$ at this concentration.

Summary

Solutions are homogeneous mixtures of two or more substances, which may be solids, liquids, or gases. The ease of dissolution of a solute in a solvent is governed by intermolecular forces. Energy and the increase in disorder that result when molecules of the solute and solvent mix to form a solution are the forces driving the solution process.

The concentration of a solution can be expressed as percent by mass, mole fraction, molarity, and molality. The circumstances dictate which units are appropriate.

A rise in temperature usually increases the solubility of solid and liquid substances and decreases the solubility of gases. According to Henry's law, the solubility of a gas in a liquid is directly proportional to the partial pressure of the gas over the solution.

Raoult's law states that the partial pressure of a substance A over a solution is related to the mole fraction (X_A) of A and to the vapor pressure (P_A°) of pure A as: $P_A = X_A P_A^\circ$. An ideal solution obeys Raoult's law over the entire range of concentration. In practice, very few solutions exhibit ideal behavior.

Vapor-pressure lowering, boiling-point elevation, freezing-point depression, and osmotic pressure are colligative properties of solutions; that is, they are properties that depend only on the number of solute particles that are present and not on their nature. In electrolyte solutions, the interaction between ions leads to the formation of ion pairs. The van't Hoff factor provides a measure of the extent of dissociation of electrolytes in solution.

Key Words

Colligative properties p. 415	Miscible p. 408	Percent by mass p. 409	Supersaturated solution p. 406
Crystallization p. 406	Molality p. 409	Raoult's law p. 415	Thermal pollution p. 412
Henry's law p. 413	Nonvolatile p. 415	Saturated solution p. 406	Unsaturated solution p. 406
Ideal solution p. 417	Osmosis p. 420	Semipermeable membrane p. 419	Volatile p. 416
Ion pair p. 424	Osmotic pressure (π) p. 420	Solvation p. 408	

Questions and Problems

The Solution Process

Review Questions

13.1 Briefly describe the solution process at the molecular level. Use the dissolution of a solid in a liquid as an example.

13.2 Basing your answer on intermolecular force considerations, explain what "like dissolves like" means.

13.3 What is solvation? What are the factors that influence the extent to which solvation occurs? Give two examples of solvation, including one that involves ion-dipole interaction and another in which dispersion forces come into play.

13.4 As you know, some solution processes are endothermic and others are exothermic. Provide a molecular interpretation for the difference.

13.5 Explain why the solution process invariably leads to an increase in disorder.

13.6 Describe the factors that affect the solubility of a solid in a liquid. What does it mean to say that two liquids are miscible?

Problems

13.7 Why is naphthalene ($C_{10}H_8$) more soluble than CsF in benzene?

13.8 Explain why ethanol (C_2H_5OH) is not soluble in cyclohexane (C_6H_{12}).

13.9 Arrange these compounds in order of increasing solubility in water: O_2, LiCl, Br_2, methanol (CH_3OH).

13.10 Explain the variations in solubility in water of the alcohols listed here:

Compound	Solubility in Water g/100 g, 20°C
CH_3OH	∞
CH_3CH_2OH	∞
$CH_3CH_2CH_2OH$	∞
$CH_3CH_2CH_2CH_2OH$	9
$CH_3CH_2CH_2CH_2CH_2OH$	2.7

Note: ∞ means the alcohol and water are completely miscible in all proportions.

Concentration Units

Review Questions

13.11 Define these concentration terms and give their units: percent by mass, molarity, molality. Compare their advantages and disadvantages.

13.12 Outline the steps required for conversion among molarity, molality, and percent by mass.

Problems

13.13 Calculate the percent by mass of the solute in each of these aqueous solutions: (a) 5.50 g of NaBr in 78.2 g of solution, (b) 31.0 g of KCl in 152 g of water, (c) 4.5 g of toluene in 29 g of benzene.

13.14 Calculate the amount of water (in grams) that must be added to (a) 5.00 g of urea [(NH_2)$_2$CO] in the preparation of a 16.2 percent by mass solution and (b) 26.2 g of $MgCl_2$ in the preparation of a 1.5 percent by mass solution.

13.15 Calculate the molality of each of these solutions: (a) 14.3 g of sucrose ($C_{12}H_{22}O_{11}$) in 676 g of water, (b) 7.20 moles of ethylene glycol ($C_2H_6O_2$) in 3546 g of water.

13.16 Calculate the molality of each of the following aqueous solutions: (a) 2.50 *M* NaCl solution (density of solution = 1.08 g/mL), (b) 48.2 percent by mass KBr solution.

13.17 Calculate the molalities of these aqueous solutions: (a) 1.22 *M* sugar ($C_{12}H_{22}O_{11}$) solution (density of solution = 1.12 g/mL), (b) 0.87 *M* NaOH solution (density of solution = 1.04 g/mL), (c) 5.24 *M* NaHCO$_3$ solution (density of solution = 1.19 g/mL).

13.18 For dilute aqueous solutions in which the density of the solution is roughly equal to that of the pure solvent, the molarity of the solution is equal to its molality. Show that this statement is correct for a 0.010 *M* urea [(NH_2)$_2$CO] solution.

13.19 The alcohol content of hard liquor is normally given in terms of the "proof," which is defined as twice the percentage by volume of ethanol (C_2H_5OH) present. Calculate the number of grams of alcohol present in 1.00 L of 75 proof gin. The density of ethanol is 0.798 g/mL.

13.20 The concentrated sulfuric acid we use in the laboratory is 98.0 percent H_2SO_4 by mass. Calculate the molality and molarity of the acid solution. The density of the solution is 1.83 g/mL.

13.21 Calculate the molarity and the molality of NH_3 for a solution of 30.0 g of NH_3 in 70.0 g of water. The density of the solution is 0.982 g/mL.

13.22 The density of an aqueous solution containing 10.0 percent ethanol (C_2H_5OH) by mass is 0.984 g/mL. (a) Calculate the molality of this solution. (b) Calculate its molarity. (c) What volume of the solution would contain 0.125 mole of ethanol?

Effect of Temperature and Pressure on Solubility

Review Questions

13.23 How do the solubilities of most ionic compounds in water change with temperature?

13.24 What is the effect of pressure on the solubility of a liquid in liquid and of a solid in liquid?

Problems

13.25 A 3.20-g sample of a salt dissolves in 9.10 g of water to give a saturated solution at 25°C. What is the solubility (in g salt/100 g of H_2O) of the salt?

13.26 The solubility of KNO_3 is 155 g per 100 g of water at 75°C and 38.0 g at 25°C. What mass (in grams) of KNO_3 will crystallize out of solution if exactly 100 g of its saturated solution at 75°C are cooled to 25°C?

Gas Solubility

Review Questions

13.27 Discuss the factors that influence the solubility of a gas in a liquid. Explain why the solubility of a gas in a liquid usually decreases with increasing temperature.

13.28 What is thermal pollution? Why is it harmful to aquatic life?

13.29 What is Henry's law? Define each term in the equation, and give its units. Explain the law in terms of the kinetic molecular theory of gases.

13.30 Give two exceptions to Henry's law.

13.31 A student is observing two beakers of water. One beaker is heated to 30°C, and the other is heated to 100°C. In each case, bubbles form in the water. Are these bubbles of the same origin? Explain.

13.32 A man bought a goldfish in a pet shop. Upon returning home, he put the goldfish in a bowl of recently boiled water that had been cooled quickly. A few minutes later the fish was dead. Explain what happened to the fish.

Problems

13.33 A beaker of water is initially saturated with dissolved air. Explain what happens when He gas at 1 atm is bubbled through the solution for a long time.

13.34 A miner working 260 m below sea level opened a carbonated soft drink during a lunch break. To his surprise, the soft drink tasted rather "flat." Shortly afterward, the miner took an elevator to the surface. During the trip up, he could not stop belching. Why?

13.35 The solubility of CO_2 in water at 25°C and 1 atm is 0.034 mol/L. What is its solubility under atmospheric conditions? (The partial pressure of CO_2 in air is 0.0003 atm.) Assume that CO_2 obeys Henry's law.

13.36 The solubility of N_2 in blood at 37°C and at a partial pressure of 0.80 atm is 5.6×10^{-4} mol/L. A deep-sea diver breathes compressed air with the partial pressure of N_2 equal to 4.0 atm. Assume that the total volume of blood in the body is 5.0 L. Calculate the amount of N_2 gas released (in liters) when the diver returns to the surface of the water, where the partial pressure of N_2 is 0.80 atm.

Colligative Properties of Nonelectrolytes

Review Questions

13.37 What are colligative properties? What is the meaning of the word "colligative" in this context?

13.38 Give two examples of a volatile liquid and two examples of a nonvolatile liquid.

13.39 Define Raoult's law. Define each term in the equation representing Raoult's law, and give its units. What is an ideal solution?

13.40 Define boiling-point elevation and freezing-point depression. Write the equations relating boiling-point elevation and freezing-point depression to the concentration of the solution. Define all the terms, and give their units.

13.41 How is the lowering in vapor pressure related to a rise in the boiling point of a solution?

13.42 Use a phase diagram to show the difference in freezing point and boiling point between an aqueous urea solution and pure water.

13.43 What is osmosis? What is a semipermeable membrane?

13.44 Write the equation relating osmotic pressure to the concentration of a solution. Define all the terms and give their units.

13.45 What does it mean when we say that the osmotic pressure of a sample of seawater is 25 atm at a certain temperature?

13.46 Explain why molality is used for boiling-point elevation and freezing-point depression calculations and molarity is used in osmotic pressure calculations.

13.47 Describe how you would use the freezing-point depression and osmotic pressure measurements to determine the molar mass of a compound. Why is the boiling-point elevation phenomenon normally not used for this purpose?

13.48 Explain why it is essential that fluids used in intravenous injections have approximately the same osmotic pressure as blood.

Problems

13.49 A solution is prepared by dissolving 396 g of sucrose ($C_{12}H_{22}O_{11}$) in 624 g of water. What is the vapor pressure of this solution at 30°C? (The vapor pressure of water is 31.8 mmHg at 30°C.)

13.50 How many grams of sucrose ($C_{12}H_{22}O_{11}$) must be added to 552 g of water to give a solution with a vapor pressure 2.0 mmHg less than that of pure water at 20°C? (The vapor pressure of water at 20°C is 17.5 mmHg.)

13.51 The vapor pressure of benzene is 100.0 mmHg at 26.1°C. Calculate the vapor pressure of a solution containing 24.6 g of camphor ($C_{10}H_{16}O$) dissolved in 98.5 g of benzene. (Camphor is a low-volatility solid.)

13.52 The vapor pressures of ethanol (C_2H_5OH) and 1-propanol (C_3H_7OH) at 35°C are 100 mmHg and 37.6 mmHg, respectively. Assume ideal behavior and calculate the partial pressures of ethanol and 1-propanol at 35°C over a solution of ethanol in 1-propanol, in which the mole fraction of ethanol is 0.300.

13.53 The vapor pressure of ethanol (C_2H_5OH) at 20°C is 44 mmHg, and the vapor pressure of methanol (CH_3OH) at the same temperature is 94 mmHg. A mixture of 30.0 g of methanol and 45.0 g of ethanol is prepared (and may be assumed to behave as an ideal solution). (a) Calculate the vapor pressure of methanol and ethanol above this solution at 20°C. (b) Calculate the mole fraction of methanol and ethanol in the vapor above this solution at 20°C.

13.54 How many grams of urea [$(NH_2)_2CO$] must be added to 450 g of water to give a solution with a vapor pressure 2.50 mmHg less than that of pure water at 30°C? (The vapor pressure of water at 30°C is 31.8 mmHg.)

13.55 What are the boiling point and freezing point of a 2.47 *m* solution of naphthalene in benzene? (The boiling point and freezing point of benzene are 80.1°C and 5.5°C, respectively.)

13.56 An aqueous solution contains the amino acid glycine (NH_2CH_2COOH). Assuming no ionization of the acid, calculate the molality of the solution if it freezes at −1.1°C.

13.57 Pheromones are compounds secreted by the females of many insect species to attract males. One of these compounds contains 80.78% C, 13.56% H, and 5.66% O. A solution of 1.00 g of this pheromone in 8.50 g of benzene freezes at 3.37°C. What are the molecular formula and molar mass of the compound? (The normal freezing point of pure benzene is 5.50°C.)

13.58 The elemental analysis of an organic solid extracted from gum arabic showed that it contained 40.0% C, 6.7% H, and 53.3% O. A solution of 0.650 g of the solid in 27.8 g of the solvent diphenyl gave a freezing-point depression of 1.56°C. Calculate the molar mass and molecular formula of the solid. (K_f for diphenyl is 8.00°C/*m*.)

13.59 How many liters of the antifreeze ethylene glycol [$CH_2(OH)CH_2(OH)$] would you add to a car radiator containing 6.50 L of water if the coldest winter temperature in your area is −20°C? Calculate the boiling point of this water-ethylene glycol mixture. The density of ethylene glycol is 1.11 g/mL.

13.60 A solution is prepared by condensing 4.00 L of a gas, measured at 27°C and 748 mmHg pressure, into 58.0 g of benzene. Calculate the freezing point of this solution.

13.61 The molar mass of benzoic acid (C_6H_5COOH) determined by measuring the freezing-point depression in benzene is twice that expected for the molecular formula, $C_7H_6O_2$. Explain this apparent anomaly.

13.62 A solution of 2.50 g of a compound of empirical formula C_6H_5P in 25.0 g of benzene is observed to freeze at 4.3°C. Calculate the molar mass of the solute and its molecular formula.

13.63 What is the osmotic pressure (in atmospheres) of a 12.36 *M* aqueous urea solution at 22.0°C?

13.64 A solution containing 0.8330 g of a protein of unknown structure in 170.0 mL of aqueous solution was found to have an osmotic pressure of 5.20 mmHg at 25°C. Determine the molar mass of the protein.

13.65 A quantity of 7.480 g of an organic compound is dissolved in water to make 300.0 mL of solution. The solution has an osmotic pressure of 1.43 atm at 27°C. The analysis of this compound shows it to contain 41.8% C, 4.7% H, 37.3% O, and 16.3% N. Calculate the molecular formula of the organic compound.

13.66 A solution of 6.85 g of a carbohydrate in 100.0 g of water has a density of 1.024 g/mL and an osmotic pressure of 4.61 atm at 20.0°C. Calculate the molar mass of the carbohydrate.

Colligative Properties of Electrolyte Solutions

Review Questions

13.67 Why is the discussion of the colligative properties of electrolyte solutions more involved than that of nonelectrolyte solutions?

13.68 Define ion pairs. What effect does ion-pair formation have on the colligative properties of a solution? How does the ease of ion-pair formation depend on (a) charges on the ions, (b) size of the ions, (c) nature of the solvent (polar versus nonpolar), (d) concentration?

13.69 In each case, indicate which of these pairs of compounds is more likely to form ion pairs in water: (a) NaCl or Na_2SO_4, (b) $MgCl_2$ or $MgSO_4$, (c) LiBr or KBr.

13.70 Define the van't Hoff factor. What information does this quantity provide?

Problems

13.71 Which of these two aqueous solutions has (a) the higher boiling point, (b) the higher freezing point, and (c) the lower vapor pressure: 0.35 *m* $CaCl_2$ or 0.90 *m* urea? State your reasons.

13.72 Consider two aqueous solutions, one of sucrose ($C_{12}H_{22}O_{11}$) and the other of nitric acid (HNO_3), both of which freeze at −1.5°C. What other properties do these solutions have in common?

13.73 Arrange these solutions in order of decreasing freezing point: (a) 0.10 m Na_3PO_4, (b) 0.35 m NaCl, (c) 0.20 m $MgCl_2$, (d) 0.15 m $C_6H_{12}O_6$, (e) 0.15 m CH_3COOH.

13.74 Arrange these aqueous solutions in order of decreasing freezing point and explain your reasons: (a) 0.50 m HCl, (b) 0.50 m glucose, (c) 0.50 m acetic acid.

13.75 What are the normal freezing points and boiling points of the following solutions: (a) 21.2 g NaCl in 135 mL of water, (b) 15.4 g of urea in 66.7 mL of water?

13.76 At 25°C the vapor pressure of pure water is 23.76 mmHg and that of seawater is 22.98 mmHg. Assuming that seawater contains only NaCl, estimate its concentration in molality units.

13.77 Both NaCl and $CaCl_2$ are used to melt ice on roads in winter. What advantages do these substances have over sucrose or urea in lowering the freezing point of water?

13.78 A 0.86 percent by mass solution of NaCl is called "physiological saline" because its osmotic pressure is equal to that of the solution in blood cells. Calculate the osmotic pressure of this solution at normal body temperature (37°C). Note that the density of the saline solution is 1.005 g/mL.

13.79 The osmotic pressure of 0.010 M solutions of $CaCl_2$ and urea at 25°C are 0.605 atm and 0.245 atm, respectively. Calculate the van't Hoff factor for the $CaCl_2$ solution.

13.80 Calculate the osmotic pressure of a 0.0500 M $MgSO_4$ solution at 22°C. (*Hint:* See Table 13.3.)

Additional Problems

13.81 Lysozyme is an enzyme that cleaves bacterial cell walls. A sample of lysozyme extracted from chicken egg white has a molar mass of 13,930 g. A quantity of 0.100 g of this enzyme is dissolved in 150 g of water at 25°C. Calculate the vapor-pressure lowering, the depression in freezing point, the elevation in boiling point, and the osmotic pressure of this solution. (The vapor pressure of water at 25°C = 23.76 mmHg.)

13.82 Solutions A and B have osmotic pressures of 2.4 atm and 4.6 atm, respectively, at a certain temperature. What is the osmotic pressure of a solution prepared by mixing equal volumes of A and B at the same temperature?

13.83 A cucumber placed in concentrated brine (salt water) shrivels into a pickle. Explain.

13.84 Two liquids A and B have vapor pressures of 76 mmHg and 132 mmHg, respectively, at 25°C. What is the total vapor pressure of the ideal solution made up of (a) 1.00 mole of A and 1.00 mole of B and (b) 2.00 moles of A and 5.00 moles of B?

13.85 Calculate the van't Hoff factor of Na_3PO_4 in a 0.40 m aqueous solution whose boiling point is 100.78°C.

13.86 A 262-mL sample of a sugar solution containing 1.22 g of the sugar has an osmotic pressure of 30.3 mmHg at 35°C. What is the molar mass of the sugar?

13.87 Consider these three mercury manometers. One of them has 1 mL of water placed on top of the mercury, another has 1 mL of a 1 m urea solution placed on top of the mercury, and the third one has 1 mL of a 1 m NaCl solution placed on top of the mercury. Identify X, Y, and Z with these solutions.

13.88 A forensic chemist is given a white powder for analysis. She dissolves 0.50 g of the substance in 8.0 g of benzene. The solution freezes at 3.9°C. Can the chemist conclude that the compound is cocaine ($C_{17}H_{21}NO_4$)? What assumptions are made in the analysis?

13.89 "Time-release" drugs have the advantage of releasing the drug to the body at a constant rate so that the drug concentration at any time is not so high as to have harmful side effects or so low as to be ineffective. A schematic diagram of a pill that works on this basis is shown here. Explain how it works.

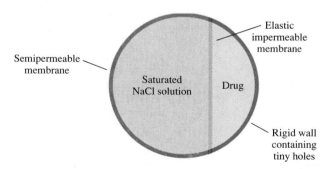

13.90 Concentrated hydrochloric acid is usually available at 37.7 percent by mass. What is its concentration in molarity? (The density of the solution is 1.19 g/mL.)

13.91 A protein has been isolated as a salt with the formula $Na_{20}P$ (this notation means that there are 20 Na^+ ions associated with a negatively charged protein P^{20-}). The osmotic pressure of a 10.0-mL solution containing 0.225 g of the protein is 0.257 atm at 25.0°C. (a) Calculate the molar mass of the protein from these data. (b) What is the actual molar mass of the protein?

13.92 A nonvolatile organic compound Z was used to make up two solutions. Solution A contains 5.00 g of Z dissolved in 100 g of water, and solution B contains 2.31 g of Z dissolved in 100 g of benzene. Solution A has a vapor pressure of 754.5 mmHg at the normal boiling point of water, and solution B has the same vapor pressure at the normal boiling point of benzene. Calculate the molar mass of Z in solutions A and B and account for the difference.

13.93 Hydrogen peroxide with a concentration of 3.0 percent (3.0 g of H_2O_2 in 100 mL of solution) is sold in drugstores for use as an antiseptic. For a 10.0-mL 3.0 percent H_2O_2 solution, calculate (a) the oxygen gas produced (in liters) at STP when the compound undergoes complete decomposition and (b) the ratio of the volume of O_2 collected to the initial volume of the H_2O_2 solution.

13.94 Before a carbonated beverage bottle is sealed, it is pressurized with a mixture of air and carbon dioxide. (a) Explain the effervescence that occurs when the cap of the bottle is removed. (b) What causes the fog to form near the mouth of the bottle right after the cap is removed?

13.95 Two beakers, one containing a 50-mL aqueous 1.0 M glucose solution and the other a 50-mL aqueous 2.0 M glucose solution, are placed under a tightly sealed bell jar as that shown in Figure 13.9 at room temperature. What are the volumes in these two beakers at equilibrium?

13.96 Explain each of these statements: (a) The boiling point of seawater is higher than that of pure water. (b) Carbon dioxide escapes from the solution when the cap is removed from a soft-drink bottle. (c) Molal concentrations and molar concentrations of dilute aqueous solutions are approximately equal. (d) In discussing the colligative properties of a solution (other than osmotic pressure), it is preferable to express the concentration in units of molality rather than in molarity. (e) Methanol (b.p. 65°C) is useful as an auto antifreeze, but it should be removed from the car radiator during the summer season.

13.97 Calculate the molarity, molality, and mole fraction of NH_3 for a solution containing 30.0 g of NH_3 in 70.0 g of water. The density of the solution is 0.982 g/mL.

13.98 A 1.32-g sample of a mixture of cyclohexane (C_6H_{12}) and naphthalene ($C_{10}H_8$) is dissolved in 18.9 g of benzene (C_6H_6). The freezing point of the solution is 2.2°C. Calculate the mass percent of the mixture.

13.99 How does each of the following affect the solubility of an ionic compound: (a) lattice energy, (b) solvent (polar versus nonpolar), (c) enthalpies of hydration of cation and anion?

13.100 A solution contains two volatile liquids A and B. Complete the following table, in which the symbol ⟷ indicates attractive intermolecular forces.

Attractive Forces	Deviation from Raoult's Law	ΔH_{soln}
A ⟷ A, B ⟷ B > A ⟷ B		
	Negative	
		Zero

A negative deviation means the vapor pressure of the solution is less than that expected from Raoult's law. The opposite holds for a positive deviation.

13.101 A mixture of ethanol and 1-propanol behaves ideally at 36°C and is in equilibrium with its vapor. If the mole fraction of ethanol in the solution is 0.62, calculate its mole fraction in the vapor phase at this temperature. (The vapor pressures of pure ethanol and 1-propanol at 36°C are 108 mmHg and 40.0 mmHg, respectively.)

13.102 For ideal solutions, the volumes are additive. This means that if 5 mL of A and 5 mL of B form an ideal solution, the volume of the solution is 10 mL. Provide a molecular interpretation for this observation. When 500 mL of ethanol (C_2H_5OH) are mixed with 500 mL of water, the final volume is less than 1000 mL. Why?

13.103 Acetic acid is a polar molecule and can form hydrogen bonds with water molecules. Therefore, it has a high solubility in water. Yet acetic acid is also soluble in benzene (C_6H_6), a nonpolar solvent that lacks the ability to form hydrogen bonds. A solution of 3.8 g of CH_3COOH in 80 g C_6H_6 has a freezing point of 3.5°C. Calculate the molar mass of the solute and explain your result.

Special Problem

13.104 Desalination is a process by which salts are removed from seawater. Three major ways to accomplish desalination are distillation, freezing, and reverse osmosis. The freezing method is based on the fact that when an aqueous solution freezes, the solid that separates from the solution is almost pure water. Reverse osmosis uses water movement from a more concentrated solution to a less concentrated one through a semipermeable membrane.

(a) With reference to Figure 13.8, draw a diagram showing how reverse osmosis can be carried out.

(b) What are the advantages and disadvantages of reverse osmosis compared to the freezing and boiling methods?

(c) What minimum pressure (in atm) must be applied to seawater at 25°C for reverse osmosis to occur? (Treat seawater as an 0.70 M NaCl solution.)

Answers to Practice Exercises

13.1 Carbon disulfide. **13.2** 0.638 m. **13.3** 1.22 M. **13.4** 8.92 m.
13.5 13.8 m. **13.6** 2.9×10^{-4} M. **13.7** 0.78 m. **13.8** 101.3°C;
−4.48°C. **13.9** 0.066 m, 1.3×10^2 g/mol. **13.10** 2.60×10^4
g/mol. **13.11** 1.21.

The ability to follow chemical reactions at the molecular level has been one of the most relentlessly pursued goals in chemistry. Accomplishing this goal means chemists will be able to understand when a certain reaction occurs and the dependence of its rate on temperature and other parameters. On the practical side, this information will help chemists control reaction rates and increase reaction yields. A complete understanding of reaction mechanism requires a detailed knowledge of the activated complex (also called the transition state), which is formed by the reactant molecule(s) prior to conversion to products. However, the transition state is a highly energetic species that could not be isolated because of its extremely short lifetime.

Reaction progress
(a)

Reaction progress
(b)

The decomposition of cyclobutane to form two ethylene molecules can take place in one of two ways. (a) The reaction proceeds via a single step, which involves the breaking of two C—C bonds simultaneously. (b) The reaction proceeds in two steps, with the formation of a short-lived intermediate in which just one bond is broken. There is only a small energy barrier for the intermediate to proceed to the final products. The correct mechanism is (b).

The situation changed in the 1980s when chemists at the California Institute of Technology began to use very short laser pulses to probe chemical reactions. Because transition states last only 10 to 1000 femtoseconds, so the laser pulses needed to probe them must be extraordinarily short. (1 femtosecond, or 1 fs, is 1×10^{-15} s. To appreciate how short this time duration is, note that there are as many femtoseconds in one second as there are seconds in about 32 million years!) One of the reactions studied was the decomposition of cyclobutane (C_4H_8) to ethylene (C_2H_4). There are two possible mechanisms. The first is a single step process in which two carbon-carbon bonds break simultaneously to form the product.

$$\begin{array}{cc} CH_2-CH_2 \\ | \quad\quad | \\ CH_2-CH_2 \end{array} \longrightarrow 2CH_2{=}CH_2$$

The second mechanism has two steps, with an intermediate

$$\begin{array}{cc} CH_2-CH_2 \\ | \quad\quad | \\ CH_2-CH_2 \end{array} \longrightarrow \begin{array}{cc} \overset{\bullet}{C}H_2 \quad \overset{\bullet}{C}H_2 \\ | \quad\quad | \\ CH_2-CH_2 \end{array} \longrightarrow 2CH_2{=}CH_2$$

where the dot represents an unpaired electron.

The Cal Tech chemists initiated the reaction with a pump laser pulse, which energized the reactant. The first probe pulse hits the molecules a few femtoseconds later and is followed by many thousands more, every 10 fs or so, for the duration of the reaction. Each probe pulse results in an absorption spectrum and changes in the spectrum reveal the motion of the molecule and the state of the chemical bonds. In this way, the chemists were effectively equipped with a camera having different shutter speeds to capture the progress of the reaction. The results show that the second mechanism is what happens. The lifetime of the intermediate is about 700 fs.

The femtosecond laser technique has been applied to unravel the mechanisms of many chemical reactions and biological processes such as photosynthesis and vision. It has created a new area in chemical kinetics that has become known as femtochemistry.

Chemical Kinetics

Chapter 14

ESSENTIAL CONCEPTS

Rate of a Reaction The rate of a reaction measures how fast a reactant is consumed or how fast a product is formed. The rate is expressed as a ratio of the change in concentration to elapsed time.

Rate Laws Experimental measurement of the rate leads to the rate law for the reaction, which expresses the rate in terms of the rate constant and the concentrations of the reactants. The dependence of rate on concentrations gives the order of a reaction. A reaction may be described as zeroth order if the rate does not depend on the concentration of the reactant or first order if it depends on the concentration of the reactant raised to the first power. Higher orders and fractional orders are also known. An important characteristic of reaction rates is the time required for the concentration of a reactant to decrease to half of its initial concentration, called the half-life. For first-order reactions, the half-life is independent of the initial concentration.

Temperature Dependence of Rate Constants To react, molecules must possess energy equal to or greater than the activation energy. The rate constant generally increases with increasing temperature. The Arrhenius equation relates the rate constant to activation energy and temperature.

Reaction Mechanism The progress of a reaction may be broken into a series of elementary steps at the molecular level, and the sequence of such steps is the mechanism of the reaction. Elementary steps may be unimolecular, involving one molecule, bimolecular, where two molecules react, or in rare cases, termolecular, involving the simultaneous encounter of three molecules. The rate of a reaction having more than one elementary step is governed by the slowest step, called the rate-determining step.

Catalysis A catalyst speeds up the rate of a reaction without itself being consumed. In heterogeneous catalysis, the reactants and the catalyst are in different phases. In homogeneous catalysis, the reactants and catalyst are dispersed in a single phase. Enzymes, which are highly efficient catalysts, play a central role in all living systems.

14.1 The Rate of a Reaction

The area of chemistry concerned with the speed, or rate, at which a chemical reaction occurs is called **chemical kinetics.** The word "kinetic" suggests movement or change; in Chapter 5 we defined kinetic energy as the energy available because of the motion of an object. Here kinetics refers to the rate of a reaction, or the **reaction rate,** which is *the change in concentration of a reactant or a product with time* (*M*/s).

We know that any reaction can be represented by the general equation

$$\text{reactants} \longrightarrow \text{products}$$

This equation tells us that, during the course of a reaction, reactant molecules are consumed while product molecules are formed. As a result, we can follow the progress of a reaction by monitoring either the decrease in concentration of the reactants or the increase in concentration of the products.

Figure 14.1 shows the progress of a simple reaction in which A molecules are converted to B molecules (for example, the conversion of *cis*-dichloroethylene to *trans*-dichloroethylene shown on p. 342):

$$A \longrightarrow B$$

The decrease in the number of A molecules and the increase in the number of B molecules with time are shown in Figure 14.2. In general, it is more convenient to express the rate in terms of change in concentration with time. Thus for the above reaction we can express the rate as

Recall that Δ denotes the difference between the final and initial states.

$$\text{rate} = -\frac{\Delta[A]}{\Delta t} \quad \text{or} \quad \text{rate} = \frac{\Delta[B]}{\Delta t}$$

in which $\Delta[A]$ and $\Delta[B]$ are the changes in concentration (in molarity) over a period Δt. Because the concentration of A *decreases* during the time interval, $\Delta[A]$ is a negative quantity. The rate of a reaction is a positive quantity, so a minus sign is needed in the rate expression to make the rate positive. On the other hand, the rate of product formation does not require a minus sign because $\Delta[B]$ is a positive quantity (the concentration of B *increases* with time).

For more complex reactions, we must be careful in writing the rate expression. Consider, for example, the reaction

$$2A \longrightarrow B$$

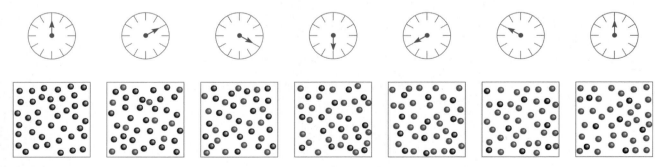

Figure 14.1 *The progress of reaction A ⟶ B at 10-s intervals over a period of 60 s. Initially, only A molecules (black spheres) are present. As time progresses, B molecules (red spheres) are formed.*

Figure 14.2 *The rate of reaction A $\longrightarrow$ B, represented as the decrease of A molecules with time and as the increase of B molecules with time.*

Two moles of A disappear for each mole of B that forms—that is, the rate of disappearance of A is twice as fast as the rate of appearance of B. We write the rate as either

$$\text{rate} = -\frac{1}{2}\frac{\Delta[A]}{\Delta t} \quad \text{or} \quad \text{rate} = \frac{\Delta[B]}{\Delta t}$$

For the reaction

$$aA + bB \longrightarrow cC + dD$$

the rate is given by

$$\text{rate} = -\frac{1}{a}\frac{\Delta[A]}{\Delta t} = -\frac{1}{b}\frac{\Delta[B]}{\Delta t} = \frac{1}{c}\frac{\Delta[C]}{\Delta t} = \frac{1}{d}\frac{\Delta[D]}{\Delta t}$$

Example 14.1 Write the rate expressions for these reactions in terms of the disappearance of the reactants and the appearance of the products:

(a) $I^-(aq) + OCl^-(aq) \longrightarrow Cl^-(aq) + OI^-(aq)$

(b) $3O_2(g) \longrightarrow 2O_3(g)$

(c) $4NH_3(g) + 5O_2(g) \longrightarrow 4NO(g) + 6H_2O(g)$

Reasoning and Solution (a) Because each of the stoichiometric coefficients equals 1,

$$\text{rate} = -\frac{\Delta[I^-]}{\Delta t} = -\frac{\Delta[OCl^-]}{\Delta t} = \frac{\Delta[Cl^-]}{\Delta t} = \frac{\Delta[OI^-]}{\Delta t}$$

(b) Here the coefficients are 3 and 2, so

$$\text{rate} = -\frac{1}{3}\frac{\Delta[O_2]}{\Delta t} = \frac{1}{2}\frac{\Delta[O_3]}{\Delta t}$$

(c) In this reaction

$$\text{rate} = -\frac{1}{4}\frac{\Delta[NH_3]}{\Delta t} = -\frac{1}{5}\frac{\Delta[O_2]}{\Delta t} = \frac{1}{4}\frac{\Delta[NO]}{\Delta t} = \frac{1}{6}\frac{\Delta[H_2O]}{\Delta t}$$

Similar problem: 14.5.

Practice Exercise Write the rate expression for the following reaction:

$$CH_4(g) + 2O_2(g) \longrightarrow CO_2(g) + 2H_2O(g)$$

Figure 14.3 *The decrease in bromine concentration as time elapses shows up as a loss of color (from left to right).*

Depending on the nature of the reaction, there are a number of ways in which to measure reaction rate. For example, in aqueous solution, molecular bromine reacts with formic acid (HCOOH) as:

$$Br_2(aq) + HCOOH(aq) \longrightarrow 2H^+(aq) + 2Br^-(aq) + CO_2(g)$$

Molecular bromine is reddish brown. All other species in the reaction are colorless. As the reaction progresses, the concentration of Br_2 steadily decreases and its color fades (Figure 14.3). Thus the change in concentration (which is evident by the intensity of the color) with time can be followed with a spectrometer. We can determine the reaction rate graphically by plotting the concentration of bromine versus time, as Figure 14.4 shows. The rate of the reaction at a particular instant is given by the slope of the tangent (which is $\Delta[Br_2]/\Delta t$) at that instant. In a certain experiment we find that the rate is 2.96×10^{-5} M/s at 100 s after the start of the reaction, 2.09×10^{-5} M/s at 200 s, and so on. Because generally the rate is proportional to the concentration of the reactant, it is not surprising that its value falls as the concentration of bromine decreases.

If one of the products or reactants of a reaction is a gas, we can use a manometer to find the reaction rate. To illustrate this method, let us consider the decomposition of hydrogen peroxide:

Figure 14.4 *The instantaneous rates of the reaction between molecular bromine and formic acid at t = 100 s, 200 s, and 300 s are given by the slopes of the tangents at these times.*

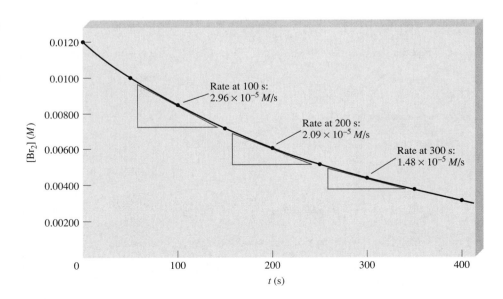

$$2H_2O_2(l) \longrightarrow 2H_2O(l) + O_2(g)$$

In this case, the rate of decomposition can be conveniently determined by measuring the rate of oxygen evolution with a manometer (Figure 14.5). The oxygen pressure can be readily converted to concentration by using the ideal gas equation [Equation (5.8)]:

$$PV = nRT$$

or

$$P = \frac{n}{V}RT = MRT$$

in which n/V gives the molarity (M) of oxygen gas. Rearranging the equation, we get

$$M = \frac{1}{RT}P$$

The reaction rate, which is given by the rate of oxygen production, can now be written as

$$\text{rate} = \frac{\Delta[O_2]}{\Delta t} = \frac{1}{RT}\frac{\Delta P}{\Delta t}$$

If a reaction either consumes or generates ions, its rate can be measured by monitoring electrical conductance. If H^+ ion is the reactant or product, we can determine the reaction rate by measuring the solution's pH as a function of time.

14.2 The Rate Laws

One way to study the effect of reactant concentration on reaction rate is to determine how the initial rate depends on the starting concentrations. In general it is preferable to measure the initial rate because as the reaction proceeds, the concentrations of the reactants decrease and it may become difficult to measure the changes accurately. Also, there may be a reverse reaction such that

$$\text{products} \longrightarrow \text{reactants}$$

which would introduce error in the rate measurement. Both of these complications are virtually absent during the early stages of the reaction.

Table 14.1 shows three rate measurements for the reaction

$$F_2(g) + 2ClO_2(g) \longrightarrow 2FClO_2(g)$$

Table 14.1	Rate Data for the Reaction Between F_2 and ClO_2	
[F_2] (M)	**[ClO_2] (M)**	**Initial Rate (M/s)**
1. 0.10	0.010	1.2×10^{-3}
2. 0.10	0.040	4.8×10^{-3}
3. 0.20	0.010	2.4×10^{-3}

Figure 14.5 *The rate of hydrogen peroxide decomposition can be measured with a manometer, which shows the increase in the oxygen gas pressure with time.*

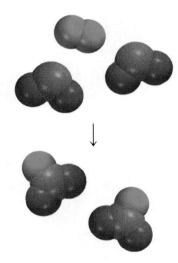

Looking at table entries 1 and 3, we see that if we double $[F_2]$ while holding $[ClO_2]$ constant, the rate doubles. Thus the rate is directly proportional to $[F_2]$. Similarly, the data in entries 1 and 2 show that when we quadruple $[ClO_2]$ at constant $[F_2]$, the rate increases by four times, so that the rate is also directly proportional to $[ClO_2]$. We can summarize these observations by writing

$$\text{rate} \propto [F_2][ClO_2]$$
$$= k[F_2][ClO_2]$$

The term k is the **rate constant,** *a constant of proportionality between the reaction rate and the concentrations of the reactants.* This equation is known as the **rate law,** *an expression relating the rate of a reaction to the rate constant and the concentrations of the reactants.* From the reactant concentrations and the initial rate, we can also calculate the rate constant. Using the first entry of data in Table 14.1, we can write

$$k = \frac{\text{rate}}{[F_2][ClO_2]}$$
$$= \frac{1.2 \times 10^{-3}\ M/s}{(0.10\ M)(0.010\ M)}$$
$$= 1.2/M \cdot s$$

For a general reaction of the type

$$a\text{A} + b\text{B} \longrightarrow c\text{C} + d\text{D}$$

the rate law takes the form

$$\text{rate} = k[\text{A}]^x[\text{B}]^y \tag{14.1}$$

If we know the values of k, x, and y, as well as the concentrations of A and B, we can use the rate law to calculate the rate of the reaction. Like k, x and y must be determined experimentally. *The sum of the powers to which all reactant concentrations appearing in the rate law are raised* is called the overall **reaction order.** In the rate law expression shown, the overall reaction order is given by $x + y$. For the reaction involving F_2 and ClO_2, the overall order is $1 + 1$, or 2. We say that the reaction is first order in F_2 and first order in ClO_2, or second order overall. Note that reaction order is *always* determined by reactant concentrations and never by product concentrations.

Reaction order enables us to appreciate better the dependence of rate on reactant concentrations. Suppose, for example, that, for a certain reaction, $x = 1$ and $y = 2$. The rate law for the reaction from Equation (14.1) is

$$\text{rate} = k[\text{A}][\text{B}]^2$$

This reaction is first order in A, second order in B, and third order overall ($1 + 2 = 3$). Let us assume that initially $[\text{A}] = 1.0\ M$ and $[\text{B}] = 1.0\ M$. The rate law tells us that if we double the concentration of A from $1.0\ M$ to $2.0\ M$ at constant $[\text{B}]$, we also double the reaction rate:

$$\text{for } [\text{A}] = 1.0\ M \qquad \text{rate}_1 = k(1.0\ M)(1.0\ M)^2$$
$$= k(1.0\ M^3)$$
$$\text{for } [\text{A}] = 2.0\ M \qquad \text{rate}_2 = k(2.0\ M)(1.0\ M)^2$$
$$= k(2.0\ M^3)$$

Hence

$$\text{rate}_2 = 2(\text{rate}_1)$$

On the other hand, if we double the concentration of B from 1.0 M to 2.0 M at constant [A], the reaction rate will increase by a factor of 4 because of the power 2 in the exponent:

$$\text{for } [B] = 1.0\ M \quad \text{rate}_1 = k(1.0\ M)(1.0\ M)^2$$
$$= k(1.0\ M^3)$$
$$\text{for } [B] = 2.0\ M \quad \text{rate}_2 = k(1.0\ M)(2.0\ M)^2$$
$$= k(4.0\ M^3)$$

Hence

$$\text{rate}_2 = 4(\text{rate}_1)$$

If, for a certain reaction, $x = 0$ and $y = 1$, then the rate law is

$$\text{rate} = k[A]^0[B]$$
$$= k[B]$$

This reaction is zero order in A, first order in B, and first order overall. Thus the rate of this reaction is *independent* of the concentration of A.

Experimental Determination of Rate Laws

If a reaction involves only one reactant, the rate law can be readily determined by measuring the initial rate of the reaction as a function of the reactant's concentration. For example, if the rate doubles when the concentration of the reactant doubles, then the reaction is first order in the reactant. If the rate quadruples when the concentration doubles, the reaction is second order in the reactant.

For a reaction involving more than one reactant, we can find the rate law by measuring the dependence of the reaction rate on the concentration of each reactant, one at a time. We fix the concentrations of all but one reactant and record the rate of the reaction as a function of the concentration of that reactant. Any changes in the rate must be due only to changes in that substance. The dependence thus observed gives us the order in that particular reactant. The same procedure is then applied to the next reactant, and so on. This approach is known as the *isolation method.*

Example 14.2   The reaction of nitric oxide with hydrogen at 1280°C is

$$2NO(g) + 2H_2(g) \longrightarrow N_2(g) + 2H_2O(g)$$

From these data collected at this temperature, determine the rate law and calculate the rate constant.

Experiment	[NO]	[H$_2$]	Initial Rate (M/s)
1	5.0×10^{-3}	2.0×10^{-3}	1.3×10^{-5}
2	10.0×10^{-3}	2.0×10^{-3}	5.0×10^{-5}
3	10.0×10^{-3}	4.0×10^{-3}	10.0×10^{-5}

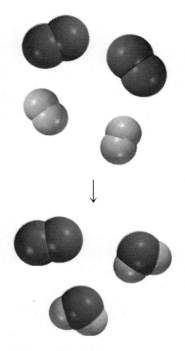

Similar problem: 14.17.

Reasoning and Solution   We assume that the rate law takes the form

$$\text{rate} = k[NO]^x[H_2]^y$$

Experiments 1 and 2 show that when we double the concentration of NO at constant concentration of H_2, the rate quadruples. Thus the reaction is second order in NO. Experiments 2 and 3 indicate that doubling $[H_2]$ at constant [NO] doubles the rate; the reaction is first order in H_2. The rate law is given by

$$\text{rate} = k[NO]^2[H_2]$$

which shows that it is a $(1 + 2)$ or third-order reaction overall.

The rate constant k can be calculated using the data from any one of the experiments. Because

$$k = \frac{\text{rate}}{[NO]^2[H_2]}$$

data from experiment 2 gives us

$$k = \frac{5.0 \times 10^{-5}\,M/s}{(10.0 \times 10^{-3}\,M)^2(2.0 \times 10^{-3}\,M)}$$

$$= 2.5 \times 10^2/M^2 \cdot s$$

Practice Exercise   The reaction of peroxydisulfate ion ($S_2O_8^{2-}$) with iodide ion (I^-) is

$$S_2O_8^{2-}(aq) + 3I^-(aq) \longrightarrow 2SO_4^{2-}(aq) + I_3^-(aq)$$

From these data collected at a certain temperature, determine the rate law and calculate the rate constant.

Experiment	$[S_2O_8^{2-}]$	$[I^-]$	Initial Rate (M/s)
1	0.080	0.034	2.2×10^{-4}
2	0.080	0.017	1.1×10^{-4}
3	0.16	0.017	2.2×10^{-4}

Example 14.2 brings out an important point about rate laws: There is no relationship between the exponents x and y and the stoichiometric coefficients in the balanced equation. We see that the stoichiometric coefficient for H_2 is 2 and yet the reaction is first order in H_2. As another example, consider the thermal decomposition of N_2O_5:

$$2N_2O_5(g) \longrightarrow 4NO_2(g) + O_2(g)$$

The rate law is

$$\text{rate} = k[N_2O_5]$$

and *not* rate $= k[N_2O_5]^2$, as we might have guessed from the balanced equation. *In general, the order of a reaction must be determined by experiment; it cannot be deduced from the overall balanced equation.*

14.3 Relation Between Reactant Concentrations and Time

Rate laws enable us to calculate the rate of a reaction from the rate constant and reactant concentrations. They can also be converted into equations that allow us to determine the concentrations of reactants at any time during the course of a reaction. We will illustrate this application by considering first one of the simplest kind of rate laws—that applying to reactions that are first order overall.

First-Order Reactions

A *first-order reaction* is *a reaction whose rate depends on the reactant concentration raised to the first power.* In a first-order reaction of the type

$$A \longrightarrow \text{product}$$

the rate is

$$\text{rate} = -\frac{\Delta[A]}{\Delta t}$$

From the rate law, we also know that

$$\text{rate} = k[A]$$

Thus

$$-\frac{\Delta[A]}{\Delta t} = k[A] \tag{14.2}$$

We can determine the units of the first-order rate constant k by transposing:

$$k = -\frac{\Delta[A]}{[A]} \frac{1}{\Delta t}$$

Because the unit for $\Delta[A]$ and $[A]$ is M and that for Δt is s, the unit for k is

$$\frac{M}{M \text{ s}} = \frac{1}{\text{s}} = \text{s}^{-1}$$

(The minus sign does not enter into the evaluation of units.) Using calculus, we can show from Equation (14.2) that

$$\ln \frac{[A]}{[A]_0} = -kt \tag{14.3}$$

in which ln is the natural logarithm, and $[A]_0$ and $[A]$ are the concentrations of A at times $t = 0$ and $t = t$, respectively. It should be understood that $t = 0$ need not correspond to the beginning of the experiment; it can be any time when we choose to monitor the change in the concentration of A.

Equation (14.3) can be rearranged as follows:

$$\ln [A] - \ln [A]_0 = -kt$$

or

$$\ln [A] = -kt + \ln [A]_0 \tag{14.4}$$

Equation (14.4) has the form of the linear equation $y = mx + b$, in which m is the slope of the line that is the graph of the equation:

For a first-order reaction, doubling the concentration of the reactant doubles the rate.

In differential form, Equation (14.2) becomes

$$-\frac{d[A]}{dt} = k[A]$$

Rearranging, we get

$$\frac{d[A]}{[A]} = -k\,dt$$

Integrating between $t = 0$ and $t = t$ gives

$$\int_{[A]_0}^{[A]} \frac{d[A]}{[A]} = -k \int_0^t dt$$

$$\ln [A] - \ln [A]_0 = -kt$$

or

$$\ln \frac{[A]}{[A]_0} = -kt$$

Figure 14.6 *First-order reaction characteristics: (a) Decrease of reactant concentration with time; (b) plot of the straight-line relationship to obtain the rate constant. The slope of the line is equal to −k.*

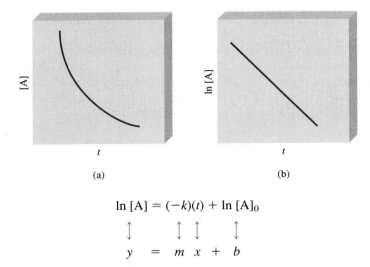

(a)   (b)

$$\ln [A] = (-k)(t) + \ln [A]_0$$

$$\updownarrow \qquad \updownarrow \ \updownarrow \qquad \updownarrow$$

$$y \quad = \quad m \ x \ + \ b$$

Thus a plot of $\ln [A]$ versus t (or y versus x) gives a straight line with a slope of $-k$ (or m). This enables us to calculate the rate constant k. Figure 14.6 shows the characteristics of a first-order reaction.

There are many known first-order reactions. All nuclear decay processes are first order (see Chapter 21). Another example is the decomposition of ethane (C_2H_6) into highly reactive methyl radicals (CH_3):

$$C_2H_6 \longrightarrow 2CH_3$$

Now let us determine graphically the order and rate constant of the decomposition of nitrogen pentoxide in carbon tetrachloride (CCl_4) solvent at 45°C:

$$2N_2O_5(CCl_4) \longrightarrow 4NO_2(g) + O_2(g)$$

This table shows the variation of N_2O_5 concentration with time, and the corresponding $\ln [N_2O_5]$ values

N_2O_5

t(s)	$[N_2O_5]$	$\ln [N_2O_5]$
0	0.91	−0.094
300	0.75	−0.29
600	0.64	−0.45
1200	0.44	−0.82
3000	0.16	−1.83

Applying Equation (14.4) we plot $\ln [N_2O_5]$ versus t, as shown in Figure 14.7. The fact that the points lie on a straight line shows that the rate law is first order. Next we determine the rate constant from the slope. We select two points far apart on the line and subtract their y and x values as:

$$\text{slope } (m) = \frac{\Delta y}{\Delta x}$$

$$= \frac{-1.50 - (-0.34)}{(2430 - 400) \text{ s}}$$

$$= -5.7 \times 10^{-4} \text{s}^{-1}$$

Because $m = -k$, we get $k = 5.7 \times 10^{-4} \text{ s}^{-1}$.

N_2O_5 *decomposes to give NO_2 (brown color).*

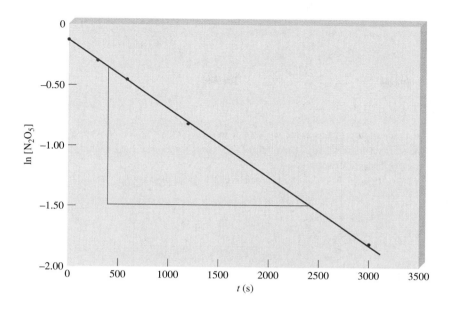

Figure 14.7 *Plot of ln [N₂O₅] versus time. The rate constant can be determined from the slope of the straight line.*

Example 14.3 The conversion of cyclopropane to propene in the gas phase is a first-order reaction with a rate constant of 6.7×10^{-4} s^{-1} at 500°C.

$$\underset{\text{cyclopropane}}{\overset{\displaystyle\underset{\displaystyle\!\diagup\;\diagdown}{CH_2}}{CH_2\!-\!CH_2}} \longrightarrow \underset{\text{propene}}{CH_2\!\!=\!\!CH\!-\!CH_3}$$

(a) If the initial concentration of cyclopropane was 0.25 *M*, what is the concentration after 8.8 min? (b) How long will it take for the concentration of cyclopropane to decrease from 0.25 *M* to 0.15 *M*? (c) How long will it take to convert 74 percent of the starting material?

Reasoning and Solution (a) We are given the initial concentration of a compound and asked to calculate its concentration at a later time, so we need Equation (14.3). Because *k* is given in units of s^{-1}, we must convert 8.8 min to seconds.

$$\ln \frac{[A]}{[A]_0} = -kt$$

$$\ln \frac{[A]}{0.25\ M} = -(6.7 \times 10^{-4}\ \text{s}^{-1})\left(8.8\ \text{min} \times \frac{60\ \text{s}}{1\ \text{min}}\right)$$

Solving the equation, we obtain

$$\ln \frac{[A]}{0.25\ M} = -0.354$$

$$\frac{[A]}{0.25\ M} = e^{-0.354}$$

$$[A] = 0.18\ M$$

(b) Again, using Equation (14.3), we have

$$\ln \frac{0.15\,M}{0.25\,M} = -(6.7 \times 10^{-4}\,\text{s}^{-1})t$$

$$t = 7.6 \times 10^2\,\text{s}$$

$$= 13\,\text{min}$$

(c) In a calculation of this type, we do not need to know the actual concentration of the starting material. If 74 percent of the starting material has reacted, then the amount left after time t is $(100\% - 74\%)$, or 26%. Thus $[A]/[A]_0 = 26\%/100\%$, or 0.26. Equation (14.3) can be rearranged to give

$$t = \frac{1}{k} \ln \frac{[A]_0}{[A]}$$

$$= \frac{1}{6.7 \times 10^{-4}\,\text{s}^{-1}} \ln \frac{1.0}{0.26}$$

$$= 2.0 \times 10^3\,\text{s}$$

$$= 33\,\text{min}$$

Similar problems: 14.24(b), 14.25(a).

Practice Exercise   The reaction $2A \longrightarrow B$ is first order in A with a rate constant of $2.8 \times 10^{-2}\,\text{s}^{-1}$ at 80°C. How long (in seconds) will it take for A to decrease from $0.88\,M$ to $0.14\,M$?

The **half-life** of a reaction, $t_{\frac{1}{2}}$, is *the time required for the concentration of a reactant to decrease to half of its initial concentration.* We can obtain an expression for $t_{\frac{1}{2}}$ for a first-order reaction as shown next. Rearranging Equation (14.3) we get

$$t = \frac{1}{k} \ln \frac{[A]_0}{[A]}$$

By the definition of half-life, when $t = t_{\frac{1}{2}}$, $[A] = [A]_0/2$, so

$$t_{\frac{1}{2}} = \frac{1}{k} \ln \frac{[A]_0}{[A]_0/2}$$

or

$$t_{\frac{1}{2}} = \frac{1}{k} \ln 2 = \frac{0.693}{k} \qquad (14.5)$$

Equation (14.5) tells us that the half-life of a first-order reaction is *independent* of the initial concentration of the reactant. Thus, it takes the same time for the concentration of the reactant to decrease from $1.0\,M$ to $0.50\,M$, say, as it does for a decrease in concentration from $0.10\,M$ to $0.050\,M$ (Figure 14.8). Measuring the half-life of a reaction is one way to determine the rate constant of a first-order reaction.

This analogy is helpful in understanding Equation (14.5). The duration of a college undergraduate's career, assuming the student does not take any time off, is 4 years. Thus the half-life of his or her stay at the college is 2 years. This half-life is not affected by how many other students are present. Similarly, the half-life of a first-order reaction is concentration independent.

The usefulness of $t_{\frac{1}{2}}$ is that it gives us an estimate of the magnitude of the rate constant—the shorter the half-life, the larger the k. Consider, for example, two radioactive isotopes that are used in nuclear medicine: ^{24}Na ($t_{\frac{1}{2}} = 14.7$ hr) and

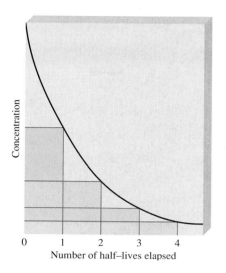

Figure 14.8 *Change in concentration of a reactant with number of half-lives for a first-order reaction.*

^{60}Co ($t_{\frac{1}{2}}$ = 5.3 yr). It is obvious that the ^{24}Na isotope decays faster because it has a shorter half-life. If we started with 1 g of each of the isotopes, most of the ^{24}Na would be gone in a week, whereas the ^{60}Co sample would be mostly intact.

Example 14.4 The decomposition of ethane (C_2H_6) to methyl radicals is a first-order reaction with a rate constant of 5.36×10^{-4} s^{-1} at 700°C:

$$C_2H_6(g) \longrightarrow 2CH_3(g)$$

Calculate the half-life of the reaction in minutes.

Reasoning and Solution For a first-order reaction, we only need the rate constant to calculate the half-life of the reaction. From Equation (14.5)

$$t_{\frac{1}{2}} = \frac{0.693}{k}$$

$$= \frac{0.693}{5.36 \times 10^{-4}\,\text{s}^{-1}}$$

$$= 1.29 \times 10^3\,\text{s}$$

$$= 21.5\,\text{min}$$

Practice Exercise Calculate the half-life of the decomposition of N_2O_5, discussed on p. 442.

Similar problem: 14.24(a).

Second-Order Reactions

A ***second-order reaction*** is a *reaction whose rate depends on the concentration of one reactant raised to the second power or on the concentrations of two different reactants, each raised to the first power.* The simpler type involves only one kind of reactant molecule:

$$A \longrightarrow \text{product}$$

for which

$$\text{rate} = -\frac{\Delta[A]}{\Delta t}$$

From the rate law

$$\text{rate} = k[A]^2$$

As before, we can determine the units of k by writing

$$k = \frac{\text{rate}}{[A]^2} = \frac{M/s}{M^2} = 1/M \cdot s$$

Another type of second-order reaction is

$$A + B \longrightarrow \text{product}$$

and the rate law is given by

$$\text{rate} = k[A][B]$$

The reaction is first order in A and first order in B, so it has an overall reaction order of 2.

Using calculus, we can obtain the following expressions for "A $\longrightarrow$ product" second-order reactions:

$$\frac{1}{[A]} = \frac{1}{[A]_0} + kt \tag{14.6}$$

(The corresponding equation for "A + B $\longrightarrow$ product" reactions is too complex for our discussion.)

We can obtain an equation for the half-life of a second-order reaction by setting $[A] = [A]_0/2$ in Equation (14.6).

$$\frac{1}{[A]_0/2} = \frac{1}{[A]_0} + kt_{\frac{1}{2}}$$

Solving for $t_{\frac{1}{2}}$ we obtain

$$t_{\frac{1}{2}} = \frac{1}{k[A]_0} \tag{14.7}$$

Note that the half-life of a second-order reaction is inversely proportional to the initial reactant concentration. This result makes sense because the half-life should be shorter in the early stage of the reaction when more reactant molecules are present to collide with each other. Measuring the half-lives at different initial concentrations is one way to distinguish between a first-order and a second-order reaction.

Example 14.5   Iodine atoms combine to form molecular iodine in the gas phase

$$I(g) + I(g) \longrightarrow I_2(g)$$

This reaction follows second-order kinetics and has the high rate constant $7.0 \times 10^9/M \cdot s$ at 23°C. (a) If the initial concentration of I was 0.086 M, calculate the concentration after 2.0 min. (b) Calculate the half-life of the reaction if the initial concentration of I is 0.60 M and if it is 0.42 M.

Reasoning and Solution (a) To calculate the concentration of a species at a later time of a second-order reaction, we need the initial concentration and the rate constant. Applying Equation (14.6)

$$\frac{1}{[A]} = \frac{1}{[A]_0} + kt$$

$$\frac{1}{[A]} = \frac{1}{0.086\ M} + (7.0 \times 10^9/M \cdot s)\left(2.0\ \text{min} \times \frac{60\ s}{1\ \text{min}}\right)$$

where [A] is the concentration at $t = 2.0$ min. Solving the equation, we get

$$[A] = 1.2 \times 10^{-12}\ M$$

This is such a low concentration that it is virtually undetectable. The very large rate constant for the reaction means that practically all the I atoms combine after only 2.0 min of reaction time.

(b) We need Equation (14.7) for this part.
For $[I]_0 = 0.60\ M$

$$t_{\frac{1}{2}} = \frac{1}{k[A]_0}$$

$$= \frac{1}{(7.0 \times 10^9/M \cdot s)(0.60\ M)}$$

$$= 2.4 \times 10^{-10}\ s$$

For $[I]_0 = 0.42\ M$

$$t_{\frac{1}{2}} = \frac{1}{(7.0 \times 10^9/M \cdot s)(0.42\ M)}$$

$$= 3.4 \times 10^{-10}\ s$$

Comment These results confirm that the half-life of a second-order reaction is not a constant but depends on the initial concentration of the reactant(s).

Similar problem: 14.26.

Practice Exercise The reaction $2A \longrightarrow B$ is second order with a rate constant of $51/M \cdot \text{min}$ at 24°C. (a) Starting with $[A] = 0.0092\ M$, how long will it take for $[A] = 3.7 \times 10^{-3}\ M$? (b) Calculate the half-life of the reaction.

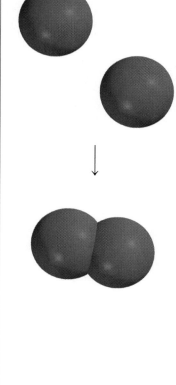

First- and second-order reactions are the most common reaction types. Reactions whose order is zero are rare. For a zero-order reaction

$$A \longrightarrow \text{product}$$

the rate law is given by

$$\text{rate} = k[A]^0$$
$$= k \qquad (14.8)$$

Thus the rate of a zero-order reaction is a *constant*, independent of reactant concentrations. Third-order and higher order reactions are quite complex; they are not presented in this book. Table 14.2 summarizes the kinetics for first-order and second-order reactions of the type $A \longrightarrow$ product.

Figure 14.9 *Dependence of rate constant on temperature. The rate constants of most reactions increase with increasing temperature.*

Table 14.2	Summary of the Kinetics of First-Order and Second-Order Reactions		
Order	Rate Law	Concentration-Time Equation	Half-Life
1	Rate = $k[A]$	$\ln\dfrac{[A]}{[A]_0} = -kt$	$\dfrac{0.693}{k}$
2	Rate = $k[A]^2$	$\dfrac{1}{[A]} = \dfrac{1}{[A]_0} + kt$	$\dfrac{1}{k[A]_0}$

14.4 Activation Energy and Temperature Dependence of Rate Constants

With very few exceptions, reaction rates increase with increasing temperature. For example, much less time is required to hard-boil an egg at 100°C (about 10 min) than at 80°C (about 30 min). Conversely, an effective way to preserve foods is to store them at subzero temperatures, thereby slowing the rate of bacterial decay. Figure 14.9 shows a typical example of the relationship between the rate constant of a reaction and temperature. To explain this behavior, we must ask how reactions get started in the first place.

The Collision Theory of Chemical Kinetics

The kinetic molecular theory of gases (p. 144) states that gas molecules frequently collide with one another. Therefore it seems logical to assume—and it is generally true—that chemical reactions occur as a result of collisions between reacting molecules. In terms of the *collision theory* of chemical kinetics, then, we expect the rate of a reaction to be directly proportional to the number of molecular collisions per second, or to the frequency of molecular collisions:

$$\text{rate} \propto \frac{\text{number of collisions}}{\text{s}}$$

This simple relationship explains the dependence of reaction rate on concentration.

Consider the reaction of A molecules with B molecules to form some product. Suppose that each product molecule is formed by the direct combination of an A molecule and a B molecule. If we doubled the concentration of A, say, then the number of A-B collisions would also double, because, in any given volume, there would be twice as many A molecules that could collide with B molecules (Figure 14.10). Consequently, the rate would increase by a factor of 2. Similarly, doubling the concentration of B molecules would increase the rate twofold. Thus we can express the rate law as

$$\text{rate} = k[A][B]$$

The reaction is first order in both A and B and obeys second-order kinetics.

The collision theory is intuitively appealing, but the relationship between rate and molecular collisions is more complicated than you might expect. The implication of the collision theory is that a reaction always occurs when an A and a B molecule collide. However, not all collisions lead to reactions. Calculations based on the kinetic molecular theory show that, at ordinary pressures (say, 1 atm) and temperatures (say,

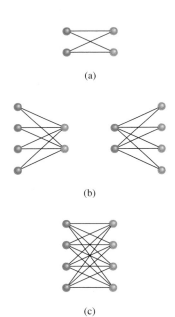

Figure 14.10 *Dependence of number of collisions on concentration. We consider here only A-B collisions, which can lead to formation of products. (a) There are four possible collisions among two A and two B molecules. (b) Doubling the number of either type of molecule (but not both) increases the number of collisions to eight. (c) Doubling both the A and B molecules increases the number of collisions to sixteen.*

298 K), there are about 1×10^{27} binary collisions (collisions between two molecules) in 1 mL of volume every second, in the gas phase. Even more collisions per second occur in liquids. If every binary collision led to a product, then most reactions would be complete almost instantaneously. In practice, we find that the rates of reactions differ greatly. This means that, in many cases, collisions alone do not guarantee that a reaction will take place.

Any molecule in motion possesses kinetic energy; the faster it moves, the greater the kinetic energy. When molecules collide, part of their kinetic energy is converted to vibrational energy. If the initial kinetic energies are large, then the colliding molecules will vibrate so strongly as to break some of the chemical bonds. This bond fracture is the first step toward product formation. If the initial kinetic energies are small, the molecules will merely bounce off each other intact. Energetically speaking, there is some minimum collision energy below which no reaction occurs.

We postulate that, to react, the colliding molecules must have a total kinetic energy equal to or greater than the **activation energy (E_a)**, which is *the minimum amount of energy required to initiate a chemical reaction.* Lacking this energy, the molecules remain intact, and no change results from the collision. *The species temporarily formed by the reactant molecules as a result of the collision before they form the product* is called the **activated complex** (also called *transition state*).

Activation energy

Figure 14.11 shows two different potential energy profiles for the reaction

$$A + B \longrightarrow C + D$$

If the products are more stable than the reactants, then the reaction will be accompanied by a release of heat; that is, the reaction is exothermic [Figure 14.11(a)]. On the other hand, if the products are less stable than the reactants, then heat will be absorbed by the reacting mixture from the surroundings and we have an endothermic reaction

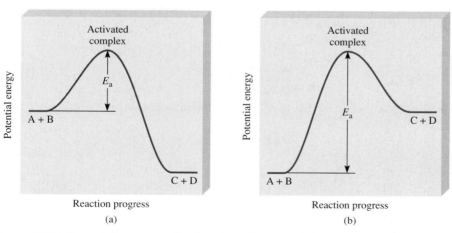

Figure 14.11 *Potential energy profiles for (a) exothermic and (b) endothermic reactions. These plots show the change in potential energy as reactants A and B are converted to products C and D. The activated complex is a highly unstable species with a high potential energy. The activation energy is defined for the forward reaction in both (a) and (b). Note that the products C and D are more stable than the reactants in (a) and less stable than the reactants in (b).*

[Figure 14.11(b)]. In both cases we plot the potential energy of the reacting system versus the progress of the reaction. Qualitatively, these plots show the potential energy changes as reactants are converted to products.

We can think of activation energy as a barrier that prevents less energetic molecules from reacting. Because the number of reactant molecules in an ordinary reaction is very large, the speeds, and hence also the kinetic energies of the molecules, vary greatly. Normally, only a small fraction of the colliding molecules—the fastest-moving ones—have enough kinetic energy to exceed the activation energy. These molecules can therefore take part in the reaction. The increase in the rate (or the rate constant) with temperature can now be explained: The speeds of the molecules obey the Maxwell distributions shown in Figure 5.14. Compare the speed distributions at two different temperatures. Because more high-energy molecules are present at the higher temperature, the rate of product formation is also greater at the higher temperature.

The Arrhenius Equation

The dependence of the rate constant of a reaction on temperature can be expressed by this equation, now known as the *Arrhenius equation:*

$$k = Ae^{-E_a/RT} \tag{14.9}$$

in which E_a is the activation energy of the reaction (in kilojoules per mole), R is the gas constant (8.314 J/K · mol), T is the absolute temperature, and e is the base of the natural logarithm scale (see Appendix 3). The quantity A represents the collision frequency and is called the *frequency factor.* It can be treated as a constant for a given reacting system over a fairly wide temperature range. Equation (14.9) shows that the rate constant is directly proportional to A and, therefore, to the collision frequency. Further, because of the minus sign associated with the exponent E_a/RT, the rate constant decreases with increasing activation energy and increases with increasing temperature. This equation can be expressed in a more useful form by taking the natural logarithm of both sides:

$$\ln k = \ln Ae^{-E_a/RT}$$

$$= \ln A - \frac{E_a}{RT} \tag{14.10}$$

Equation (14.10) can take the form of a linear equation:

$$\ln k = \left(-\frac{E_a}{R}\right)\left(\frac{1}{T}\right) + \ln A \tag{14.11}$$

$$\begin{array}{ccccc} \updownarrow & & \updownarrow & \updownarrow & \updownarrow \\ y & = & m & x & + & b \end{array}$$

Thus, a plot of $\ln k$ versus $1/T$ gives a straight line whose slope m is equal to $-E_a/R$ and whose intercept b with the ordinate (the y-axis) is $\ln A$.

Example 14.6 The rate constants for the decomposition of acetaldehyde

$$CH_3CHO(g) \longrightarrow CH_4(g) + CO(g)$$

were measured at five different temperatures. The data are shown in the table. Plot ln k versus $1/T$, and determine the activation energy (in kJ/mol) for the reaction. Note that the reaction is "$\frac{3}{2}$" order in CH_3CHO, so k has the units of $1/M^{\frac{1}{2}} \cdot s$.

$k \, (1/M^{\frac{1}{2}} \cdot s)$	$T \, (K)$
0.011	700
0.035	730
0.105	760
0.343	790
0.789	810

Reasoning and Solution According to Equation (14.11), a plot ln k versus $1/T$ yields a straight line whose slope is $-E_a/R$. Thus, we first convert the data to this table:

ln k	$1/T \, (K^{-1})$
-4.51	1.43×10^{-3}
-3.35	1.37×10^{-3}
-2.254	1.32×10^{-3}
-1.070	1.27×10^{-3}
-0.237	1.23×10^{-3}

A plot of these data yields the graph in Figure 14.12. The slope of the line is calculated from two pairs of coordinates:

$$\text{slope} = \frac{-4.00 - (-0.45)}{(1.41 - 1.24) \times 10^{-3} \, K^{-1}} = -2.09 \times 10^4 \, K$$

From the linear form of Equation (14.11)

$$\text{slope} = -\frac{E_a}{R} = -2.09 \times 10^4 \, K$$

$$E_a = (8.314 \, J/K \cdot mol)(2.09 \times 10^4 \, K)$$
$$= 1.74 \times 10^5 \, J/mol$$
$$= 1.74 \times 10^2 \, kJ/mol$$

Figure 14.12 Plot of ln k versus 1/T.

Similar problem: 14.33.

Comment It is important to note that although the rate constant itself has the units $1/M^{\frac{1}{2}} \cdot s$, the quantity $\ln k$ has no units (we cannot take the logarithm of a unit).

Practice Exercise The second-order rate constant for the decomposition of nitrous oxide (N_2O) into nitrogen molecule and oxygen atom has been measured at different temperatures:

k (1/$M \cdot s$)	$t(°C)$
1.87×10^{-3}	600
0.0113	650
0.0569	700
0.244	750

Determine graphically the activation energy for the reaction.

An equation relating the rate constants k_1 and k_2 at temperatures T_1 and T_2 can be used to calculate the activation energy or to find the rate constant at another temperature if the activation energy is known. To derive such an equation we start with Equation (14.10):

$$\ln k_1 = \ln A - \frac{E_a}{RT_1}$$

$$\ln k_2 = \ln A - \frac{E_a}{RT_2}$$

Subtracting $\ln k_2$ from $\ln k_1$ gives

$$\ln k_1 - \ln k_2 = \frac{E_a}{R}\left(\frac{1}{T_2} - \frac{1}{T_1}\right)$$

$$\ln \frac{k_1}{k_2} = \frac{E_a}{R}\left(\frac{1}{T_2} - \frac{1}{T_1}\right)$$

$$= \frac{E_a}{R}\left(\frac{T_1 - T_2}{T_1 T_2}\right) \tag{14.12}$$

Example 14.7 The rate constant of a first-order reaction is 3.46×10^{-2} s^{-1} at 298 K. What is the rate constant at 350 K if the activation energy for the reaction is 50.2 kJ/mol?

Reasoning and Solution This is an application of Equation (14.12). The data are

$$k_1 = 3.46 \times 10^{-2} \text{ s}^{-1} \qquad k_2 = ?$$
$$T_1 = 298 \text{ K} \qquad T_2 = 350 \text{ K}$$

Substituting in Equation (14.12),

$$\ln \frac{3.46 \times 10^{-2} \text{ s}^{-1}}{k_2} = \frac{50.2 \times 10^3 \text{ J/mol}}{8.314 \text{ J/K} \cdot \text{mol}}\left[\frac{298 \text{ K} - 350 \text{ K}}{(298 \text{ K})(350 \text{ K})}\right]$$

Solving the equation gives

$$\ln \frac{3.46 \times 10^{-2}\,\text{s}^{-1}}{k_2} = -3.01$$

$$\frac{3.46 \times 10^{-2}\,\text{s}^{-1}}{k_2} = e^{-3.01} = 0.0493$$

$$k_2 = 0.702\,\text{s}^{-1}$$

Comment   The rate constant is expected to be greater at a higher temperature. Therefore, the answer is reasonable.

Practice Exercise   The first-order rate constant for the reaction of methyl chloride (CH_3Cl) with water to produce methanol (CH_3OH) and hydrochloric acid (HCl) is $3.32 \times 10^{-10}\,\text{s}^{-1}$ at 25°C. Calculate the rate constant at 40°C if the activation energy is 116 kJ/mol.

Similar problem: 14.36.

For simple reactions (for example, those between atoms), we can equate the frequency factor (A) in the Arrhenius equation with the frequency of collisions between the reacting species. For more complex reactions, we must also consider the "orientation factor," that is, how reacting molecules are oriented relative to each other. The carefully studied reaction between potassium atoms (K) and methyl iodide (CH_3I) to form potassium iodide (KI) and a methyl radical (CH_3) illustrates this point:

$$K + CH_3I \longrightarrow KI + CH_3$$

This reaction is most favorable only when the K atom collides head-on with the I atom in CH_3I (Figure 14.13). Otherwise, a few or no products are formed. The nature of the orientation factor is satisfactorily dealt with in a more advanced treatment of chemical kinetics.

Orientation of collision

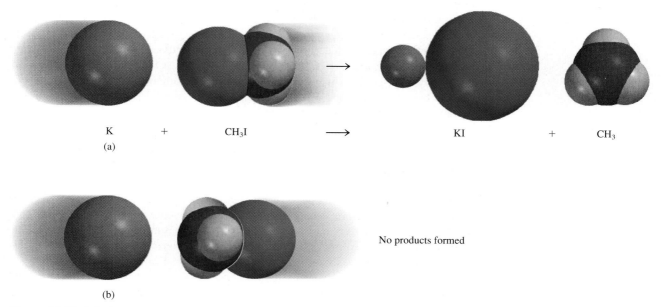

K   + CH_3I $\longrightarrow$ KI   + CH_3
(a)

(b)

No products formed

Figure 14.13 *Relative orientation of reacting molecules. Only when the K atom collides directly with the I atom will the reaction most likely occur.*

14.5 Reaction Mechanisms and Rate Laws

As we mentioned earlier, an overall balanced chemical equation does not tell us much about how a reaction actually takes place. In many cases, it merely represents the sum of *a series of simple reactions* that are often called the **elementary steps** (or *elementary reactions*) because they *represent the progress of the overall reaction at the molecular level. The sequence of elementary steps that leads to product formation* is called the **reaction mechanism.** As an example of a reaction mechanism, let us consider the reaction between nitric oxide and oxygen:

$$2NO(g) + O_2(g) \longrightarrow 2NO_2(g)$$

We know that the products are not formed directly from the collision of two NO molecules with an O_2 molecule because N_2O_2 is detected during the course of the reaction. Let us assume that the reaction actually takes place via two elementary steps as follows:

$$2NO(g) \longrightarrow N_2O_2(g)$$
$$N_2O_2(g) + O_2(g) \longrightarrow 2NO_2(g)$$

In the first elementary step, two NO molecules collide to form a N_2O_2 molecule. This event is followed by the reaction between N_2O_2 and O_2 to give two molecules of NO_2. The net chemical equation, which represents the overall change, is given by the sum of the elementary steps 1 and 2:

Step 1:	$NO + NO \longrightarrow N_2O_2$
Step 2:	$N_2O_2 + O_2 \longrightarrow 2NO_2$
Overall reaction:	$2NO + N_2O_2 + O_2 \longrightarrow N_2O_2 + 2NO_2$

The sum of the elementary steps must give the overall balanced equation.

Species such as N_2O_2 are called **intermediates** because they *appear in the mechanism of the reaction (that is, the elementary steps) but not in the overall balanced equation.* Keep in mind that an intermediate is always formed in an early elementary step and consumed in a later elementary step.

The number of molecules reacting in an elementary step determines the **molecularity of a reaction.** Each of the elementary steps just discussed is called a **bimolecular reaction,** *an elementary step that involves two molecules.* A **unimolecular reaction** is an elementary step in which only one reacting molecule participates. Consider the natural process by which ozone in the stratosphere absorbs harmful ultraviolet radiation and decomposes to form molecular oxygen:

Step 1:	$O_3(g) \longrightarrow O(g) + O_2(g)$
Step 2:	$O_3(g) + O(g) \longrightarrow 2O_2(g)$
Overall reaction:	$2O_3(g) \longrightarrow 3O_2(g)$

In this mechanism, step 1 is a unimolecular reaction and step 2 is a bimolecular reaction.

Very few **termolecular reactions,** *reactions that involve the participation of three molecules in one elementary step,* are known. The reason is that in a termolecular reaction the product forms as a result of the simultaneous encounter of three molecules, which is a far less likely event than a bimolecular collision.

Knowing the elementary steps of a reaction enables us to deduce the rate law. Suppose we have the following unimolecular elementary step:

$$A \longrightarrow products$$

Because this *is* the process that occurs at the molecular level, the larger the number of A molecules present, the faster the rate of product formation. Thus we can write the rate law directly according to the elementary step:

$$\text{rate} = k[\text{A}]$$

For a bimolecular elementary step involving A and B molecules:

$$\text{A} + \text{B} \longrightarrow \text{product}$$

the rate of product formation depends on how frequently A and B collide, which in turn depends on the concentrations of A and B. In this case we can write the rate law as

$$\text{rate} = k[\text{A}][\text{B}]$$

Similarly, for a bimolecular elementary step of the type

$$\text{A} + \text{A} \longrightarrow \text{products}$$

or

$$2\text{A} \longrightarrow \text{products}$$

the rate law becomes

$$\text{rate} = k[\text{A}]^2$$

These examples show that the reaction order for each reactant in an elementary step is equal to its stoichiometric coefficient in the chemical equation for that step. In contrast, we cannot tell by merely looking at the overall balanced equation whether the reaction occurs as shown or in a series of elementary steps. This determination is made in the laboratory.

Experimental studies of reaction mechanisms begin with the collection of data (rate measurements). Next, we analyze these data to determine the rate constant and order of the reaction, and we write the rate law. Finally, we suggest a plausible mechanism for the reaction in terms of elementary steps (Figure 14.14). The elementary steps must satisfy two requirements:

- The sum of the elementary steps must give the overall balanced equation for the reaction.
- The **rate-determining step,** which is *the slowest step in the sequence of steps leading to product formation,* should predict the same rate law as is determined experimentally.

An analogy for the rate-determining step is the flow of traffic along a narrow road. Assuming the cars cannot pass one another on the road, the rate at which the cars travel is governed by the slowest-moving car. Keep in mind that for any proposed reaction scheme, we must be able to detect the presence of any intermediate(s) formed in one or more elementary steps.

Figure 14.14 *Sequence of steps in the study of a reaction mechanism.*

Figure 14.15 *The decomposition of hydrogen peroxide is catalyzed by the iodide ion. A few drops of liquid soap have been added to the solution to dramatize the evolution of oxygen gas. (Some of the iodide ions are oxidized to molecular iodine, which then reacts with iodide ions to form the brown triiodide ion, I_3^-.)*

The decomposition of hydrogen peroxide illustrates the elucidation of reaction mechanisms by experimental studies. This reaction is facilitated by iodide ions (Figure 14.15). The overall reaction is

$$2H_2O_2(aq) \longrightarrow 2H_2O(l) + O_2(g)$$

By experiment, the rate law is found to be

$$\text{rate} = k[H_2O_2][I^-]$$

Thus the reaction is first order with respect to both H_2O_2 and I^-. You can see that decomposition does not occur in a single elementary step corresponding to the overall balanced equation. If it did, the reaction would be second order in H_2O_2 (note the coefficient 2 in the equation). What's more, the I^- ion, which is not even in the overall equation, appears in the rate law expression. How can we reconcile these facts?

We can account for the observed rate law by assuming that the reaction takes place in two separate elementary steps, each of which is bimolecular:

Step 1: $H_2O_2 + I^- \xrightarrow{k_1} H_2O + IO^-$

Step 2: $H_2O_2 + IO^- \xrightarrow{k_2} H_2O + O_2 + I^-$

If we further assume that step 1 is the rate-determining step, then the rate of the reaction can be determined from the first step alone:

$$\text{rate} = k_1[H_2O_2][I^-]$$

in which $k_1 = k$. Note that the IO^- ion is an intermediate because it does not appear in the overall balanced equation. Although the I^- ion also does not appear in the overall equation, I^- differs from IO^- in that the former is present at the start of the reaction and at its completion. The function of I^- is to speed up the reaction; that is, it is a catalyst. We will discuss catalysis in the next section. Finally, note that the sum of steps 1 and 2 gives the overall balanced equation shown earlier.

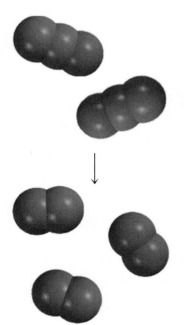

Example 14.8   The gas-phase decomposition of nitrous oxide (N_2O) is believed to occur via two elementary steps:

Step 1. $N_2O \xrightarrow{k_1} N_2 + O$

Step 2. $N_2O + O \xrightarrow{k_2} N_2 + O_2$

Experimentally the rate law is found to be rate = $k[N_2O]$: (a) Write the equation for the overall reaction. (b) Identify the intermediate. (c) What can you say about the relative rates of steps 1 and 2?

Reasoning and Solution   (a) Adding the equations for steps 1 and 2 gives the overall reaction

$$2N_2O \longrightarrow 2N_2 + O_2$$

(b) Because the O atom is produced in the first elementary step and it does not appear in the overall balanced equation, it is an intermediate.

(c) If we assume that step 1 is the rate-determining step (that is, if $k_2 \gg k_1$), then the rate of the overall reaction is given by

$$\text{rate} = k_1[\text{N}_2\text{O}]$$

and $k = k_1$.

Practice Exercise The reaction between NO_2 and CO to produce NO and CO_2 is believed to occur via two steps:

Step 1. $\qquad\qquad\qquad \text{NO}_2 + \text{NO}_2 \longrightarrow \text{NO} + \text{NO}_3$

Step 2. $\qquad\qquad\qquad \text{NO}_3 + \text{CO} \longrightarrow \text{NO}_2 + \text{CO}_2$

The experimental rate law is rate $= k[\text{NO}_2]^2$. (a) Write the equation for the overall reaction. (b) Identify the intermediate. (c) What can you say about the relative rates of steps 1 and 2?

Similar problem: 14.47.

14.6 Catalysis

We saw in studying the decomposition of hydrogen peroxide that the reaction rate depends on the concentration of iodide ions even though I^- does not appear in the overall equation. We noted there that I^- acts as a catalyst for that reaction. A **catalyst** is *a substance that increases the rate of a chemical reaction without itself being consumed.* The catalyst may react to form an intermediate, but it is regenerated in a subsequent step of the reaction.

In the laboratory preparation of molecular oxygen, a sample of potassium chlorate is heated; the reaction is (see p. 143)

$$2\text{KClO}_3(s) \longrightarrow 2\text{KCl}(s) + 3\text{O}_2(g)$$

However, this thermal decomposition is very slow in the absence of a catalyst. The rate of decomposition can be increased dramatically by adding a small amount of the catalyst manganese dioxide (MnO_2), a black powdery substance. All the MnO_2 can be recovered at the end of the reaction, just as all the I^- ions remain following H_2O_2 decomposition.

A catalyst speeds up a reaction by providing a set of elementary steps with more favorable kinetics than those that exist in its absence. From Equation (14.9) we know that the rate constant k (and hence the rate) of a reaction depends on the frequency factor A and the activation energy E_a—the larger the A or the smaller the E_a, the greater the rate. In many cases, a catalyst increases the rate by lowering the activation energy for the reaction.

Let us assume that the following reaction has a certain rate constant k and an activation energy E_a.

$$\text{A} + \text{B} \xrightarrow{\;k\;} \text{C} + \text{D}$$

In the presence of a catalyst, however, the rate constant is k_c (called the *catalytic rate constant*):

$$\text{A} + \text{B} \xrightarrow{\;k_c\;} \text{C} + \text{D}$$

A rise in temperature also increases the rate of a reaction. However, at high temperatures the products formed may undergo other reactions, thereby reducing the yield.

To extend the traffic analogy, adding a catalyst can be compared with building a tunnel through a mountain to connect two towns that were previously linked by a winding road over the mountain.

Figure 14.16 *Comparison of the activation energy barriers of (a) an uncatalyzed reaction and (b) the same reaction with a catalyst. The catalyst lowers the energy barrier but does not affect the actual energies of the reactants or products. Although the reactants and products are the same in both cases, the reaction mechanisms and rate laws are different in (a) and (b).*

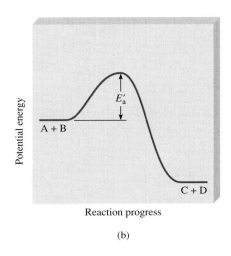

(a)   (b)

By the definition of a catalyst,

$$\text{rate}_{\text{catalyzed}} > \text{rate}_{\text{uncatalyzed}}$$

Figure 14.16 shows the potential energy profiles for both reactions. Note that the total energies of the reactants (A and B) and those of the products (C and D) are unaffected by the catalyst; the only difference between the two is a lowering of the activation energy from E_a to E'_a. Because the activation energy for the reverse reaction is also lowered, a catalyst enhances the rate of the reverse reaction to the same extent as it does the forward reaction rate.

A catalyst lowers the activation energy for both the forward and reverse reactions.

There are three general types of catalysis, depending on the nature of the rate-increasing substance: heterogeneous catalysis, homogeneous catalysis, and enzyme catalysis.

Heterogeneous Catalysis

Metals and compounds of metals that are most frequently used in heterogeneous catalysis. For color code, see Figure 8.2 on p. 234.

In *heterogeneous catalysis,* the reactants and the catalyst are in different phases. Usually the catalyst is a solid and the reactants are either gases or liquids. Heterogeneous catalysis is by far the most important type of catalysis in industrial chemistry, especially in the synthesis of many key chemicals. Here we describe two specific examples of heterogeneous catalysis.

The Manufacture of Nitric Acid

Nitric acid is one of the most important inorganic acids. It is used in the production of fertilizers, dyes, drugs, and explosives. The major industrial method of producing nitric acid is the *Ostwald process.* The starting materials, ammonia and molecular oxygen, are heated in the presence of a platinum-rhodium catalyst (Figure 14.17) to about 800°C:

$$4NH_3(g) + 5O_2(g) \longrightarrow 4NO(g) + 6H_2O(g)$$

The nitric oxide formed readily oxidizes (without catalysis) to nitrogen dioxide:

$$2NO(g) + O_2(g) \longrightarrow 2NO_2(g)$$

When dissolved in water, NO_2 forms both nitrous acid and nitric acid:

$$2NO_2(g) + H_2O(l) \longrightarrow HNO_2(aq) + HNO_3(aq)$$

Figure 14.17 *Platinum-rhodium catalyst used in the Ostwald process.*

On heating, nitrous acid is converted to nitric acid as follows:

$$3HNO_2(aq) \longrightarrow HNO_3(aq) + H_2O(l) + 2NO(g)$$

The NO generated can be recycled to produce NO_2 in the second step.

Catalytic Converters

At high temperatures inside a running car's engine, nitrogen and oxygen gases react to form nitric oxide:

$$N_2(g) + O_2(g) \rightleftharpoons 2NO(g)$$

When released into the atmosphere, NO rapidly combines with O_2 to form NO_2. Nitrogen dioxide and other gases emitted by an automobile, such as carbon monoxide (CO) and various unburned hydrocarbons, make automobile exhaust a major source of air pollution.

Most new cars are equipped with catalytic converters (Figure 14.18). An efficient catalytic converter serves two purposes: It oxidizes CO and unburned hydrocarbons to CO_2 and H_2O, and it reduces NO and NO_2 to N_2 and O_2. Hot exhaust gases into which air has been injected are passed through the first chamber of one converter to accelerate the complete burning of hydrocarbons and to decrease CO emission. (A cross section of the catalytic converter, containing Pt or Pd or a transition metal oxide

A hot platinum wire glows when held over a concentrated ammonia solution. The oxidation of ammonia to produce nitric oxide, catalyzed by platinum, is highly exothermic.

Exhaust manifold

Exhaust pipe

Tail pipe

Air compressor: source of secondary air

Catalytic converters

Figure 14.18 *A two-stage catalytic converter for an automobile.*

Figure 14.19 *A cross-sectional view of a catalytic converter. The beads contain platinum, palladium, and rhodium, which catalyze the combustion of CO and hydrocarbons.*

such as CuO or Cr_2O_3, is shown in Figure 14.19). However, because high temperatures increase NO production, a second chamber containing a different catalyst (a transition metal or a transition metal oxide) and operating at a lower temperature is required to dissociate NO into N_2 and O_2 before the exhaust is discharged through the tailpipe.

Homogeneous Catalysis

In *homogeneous catalysis* the reactants and catalyst are dispersed in a single phase, usually liquid. Acid and base catalyses are the most important type of homogeneous catalysis in liquid solution. For example, the reaction of ethyl acetate with water to form acetic acid and ethanol normally occurs too slowly to be measured.

$$\underset{\text{ethyl acetate}}{CH_3-\overset{\overset{\displaystyle O}{\|}}{C}-O-C_2H_5} + H_2O \longrightarrow \underset{\text{acetic acid}}{CH_3-\overset{\overset{\displaystyle O}{\|}}{C}-OH} + \underset{\text{ethanol}}{C_2H_5OH}$$

This reaction is zero order in water because water's concentration is very high and therefore it is unaffected by the reaction.

In the absence of the catalyst, the rate law is given by

$$\text{rate} = k[CH_3COOC_2H_5]$$

However, the reaction can be catalyzed by an acid. In the presence of hydrochloric acid, the rate is given by

$$\text{rate} = k_c[CH_3COOC_2H_5][H^+]$$

Enzyme Catalysis

Of all the intricate processes that have evolved in living systems, none is more striking or more essential than enzyme catalysis. **Enzymes** are *biological catalysts*. The amazing fact about enzymes is that not only can they increase the rate of biochemical reactions by factors ranging from 10^6 to 10^{18}, but they are also highly specific. An enzyme acts only on certain molecules, called *substrates* (that is, reactants), while leaving the rest of the system unaffected. It has been estimated that an average living cell may contain some 3000 different enzymes, each of them catalyzing a specific reaction in which a substrate is converted into the appropriate products. Enzyme catalyses are usually homogeneous with the substrate and enzyme present in the same aqueous solution.

Figure 14.20 *The lock-and-key model of an enzyme's specificity for substrate molecules.*

Substrate
+

Enzyme

Enzyme-substrate
complex

Products
+

Enzyme

An enzyme is typically a large protein molecule that contains one or more *active sites* where interactions with substrates take place. These sites are structurally compatible with specific molecules, in much the same way as a key fits a particular lock (Figure 14.20). However, an enzyme molecule (or at least its active site) has a fair amount of structural flexibility and can modify its shape to accommodate different kinds of substrates (Figure 14.21).

The mathematical treatment of enzyme kinetics is quite complex, even when we know the basic steps involved in the reaction. A simplified scheme is:

$$E + S \rightleftharpoons ES$$

$$ES \xrightarrow{k} P + E$$

in which E, S, and P represent enzyme, substrate, and product, and ES is the enzyme-substrate intermediate. Figure 14.22 shows the potential energy profile for the reaction. It is often assumed that the formation of ES and its decomposition back to enzyme and substrate molecules occur rapidly and that the rate-determining step is

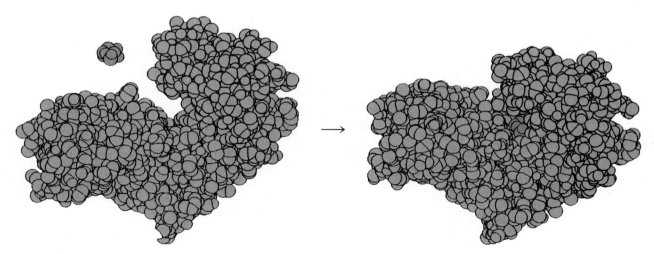

Figure 14.21 *Left to right: The binding of glucose molecule (red) to hexokinase (an enzyme in the metabolic pathway). Note how the region at the active site closes around glucose after binding. Frequently the geometries of both the substrate and the active site are altered to fit each other.*

Figure 14.22 *Comparison of (a) an uncatalyzed reaction and (b) the same reaction catalyzed by an enzyme. The plot in (b) assumes that the catalyzed reaction has a two-step mechanism, in which the second step (ES ⟶ E + P) is rate-determining.*

(a)

(b)

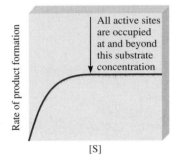

Figure 14.23 *Plot of the rate of product formation versus substrate concentration in an enzyme-catalyzed reaction.*

the formation of product. In general, the rate of such a reaction is given by the equation

$$\text{rate} = \frac{\Delta[P]}{\Delta t}$$

$$= k[ES]$$

The concentration of the ES intermediate is itself proportional to the amount of the substrate present, and a plot of the rate versus the concentration of substrate typically yields a curve such as that shown in Figure 14.23. Initially the rate rises rapidly with increasing substrate concentration. However, above a certain concentration all the active sites are occupied, and the reaction becomes zero order in the substrate. That is, the rate remains the same even though the substrate concentration increases. At and beyond this point, the rate of formation of product depends only on how fast the ES intermediate breaks down, not on the number of substrate molecules present.

Summary

The rate of a chemical reaction is shown by the change in the concentration of reactant or product molecules over time. The rate is not constant but varies continuously as concentration changes.

The rate law is an expression relating the rate of a reaction to the rate constant and the concentrations of the reactants raised to appropriate powers. The rate constant k for a given reaction changes only with temperature. Overall reaction order is the sum of the powers to which reactant concentrations are raised in the rate law. The rate law and the reaction order cannot be determined from the stoichiometry of the overall equation for a reaction; they must be determined by experiment. The half-life of a reaction (the time it takes for the concentration of a reactant to decrease by one-half) can be used to determine the rate constant of a first-order reaction.

According to collision theory, a reaction occurs when molecules collide with sufficient energy, called the activation energy, to break bonds and initiate the reaction. The rate constant and the activation energy are related by the Arrhenius equation, $k = Ae^{-E_a/RT}$.

The overall balanced equation for a reaction may be the sum of a series of simple reactions, called elementary steps. The complete series of elementary steps for a reaction is the reaction mechanism. The slowest step in a reaction mechanism is the rate-determining step.

A catalyst speeds up a reaction usually by lowering the value of E_a. A catalyst can be recovered unchanged at the end of a reaction. In heterogeneous catalysis, which is of great industrial importance, the catalyst is a solid and the reactants are gases or liquids. In homogeneous catalysis, the catalyst and the reactants are in the same phase. Enzymes are catalysts in living systems.

Key Words

Activated complex p. 449
Activation energy (E_a)
  p. 449
Bimolecular reaction
  p. 454
Catalyst p. 457
Chemical kinetics p. 434
Elementary step p. 454

Enzymes p. 460
First-order reaction
  p. 441
Half-life p. 444
Intermediate p. 454
Molecularity of a reaction
  p. 454
Rate constant p. 438

Rate-determining step
  p. 455
Rate law p. 438
Reaction mechanism
  p. 454
Reaction order p. 438
Reaction rate p. 434

Second-order reaction
  p. 445
Termolecular reaction
  p. 454
Unimolecular reaction
  p. 454

Questions and Problems

Reaction Rate

Review Questions

14.1 What is meant by the rate of a chemical reaction?

14.2 What are the units of the rate of a reaction?

14.3 What are the advantages of measuring the initial rate of a reaction?

14.4 Can you suggest two reactions that are very slow (take days or longer to complete) and two reactions that are very fast (are over in minutes or seconds)?

Problems

14.5 Write the reaction rate expressions for these reactions in terms of the disappearance of the reactants and the appearance of products:

(a) $H_2(g) + I_2(g) \longrightarrow 2HI(g)$
(b) $2H_2(g) + O_2(g) \longrightarrow 2H_2O(g)$
(c) $5Br^-(aq) + BrO_3^-(aq) + 6H^+(aq) \longrightarrow$
$$3Br_2(aq) + 3H_2O(l)$$

14.6 Consider the reaction

$$N_2(g) + 3H_2(g) \longrightarrow 2NH_3(g)$$

Suppose that at a particular moment during the reaction molecular hydrogen is reacting at the rate of 0.074 M/s. (a) At what rate is ammonia being formed? (b) At what rate is molecular nitrogen reacting?

Rate Laws

Review Questions

14.7 Explain what is meant by the rate law of a reaction.

14.8 What is meant by the order of a reaction?

14.9 What are the units for the rate constants of first-order and second-order reactions?

14.10 Write an equation relating the concentration of a reactant A at $t = 0$ to that at $t = t$ for a first-order reaction. Define all the terms and give their units.

14.11 Consider the zero-order reaction A $\longrightarrow$ product. (a) Write the rate law for the reaction. (b) What are the units for the rate constant? (c) Plot the rate of the reaction versus [A].

14.12 The rate constant of a first-order reaction is 66 s^{-1}. What is the rate constant in units of minutes?

14.13 On which of these quantities does the rate constant of a reaction depend: (a) concentrations of reactants, (b) nature of reactants, (c) temperature?

14.14 For each of these pairs of reaction conditions, indicate which has the faster rate of formation of hydrogen gas: (a) sodium or potassium with water, (b) magnesium or iron with 1.0 M HCl, (c) magnesium rod or magnesium powder with 1.0 M HCl, (d) magnesium with 0.10 M HCl or magnesium with 1.0 M HCl.

Problems

14.15 The rate law for the reaction

$$NH_4^+(aq) + NO_2^-(aq) \longrightarrow N_2(g) + 2H_2O(l)$$

is given by rate $= k[NH_4^+][NO_2^-]$. At 25°C, the rate constant is $3.0 \times 10^{-4}/M \cdot s$. Calculate the rate of the reaction at this temperature if $[NH_4^+] = 0.26\ M$ and $[NO_2^-] = 0.080\ M$.

14.16 Starting with the data in Table 14.1, (a) deduce the rate law for the reaction, (b) calculate the rate constant, and (c) calculate the rate of the reaction at the time when $[F_2] = 0.010\ M$ and $[ClO_2] = 0.020\ M$.

14.17 Consider the reaction

$$A + B \longrightarrow products$$

From these data obtained at a certain temperature, determine the order of the reaction and calculate the rate constant:

[A] (*M*)	[B] (*M*)	Rate (*M*/s)
1.50	1.50	3.20×10^{-1}
1.50	2.50	3.20×10^{-1}
3.00	1.50	6.40×10^{-1}

14.18 Consider the reaction

$$X + Y \longrightarrow Z$$

These data are obtained at 360 K:

Initial Rate of Disappearance of X (*M*/s)	[X]	[Y]
0.147	0.10	0.50
0.127	0.20	0.30
4.064	0.40	0.60
1.016	0.20	0.60
0.508	0.40	0.30

(a) Determine the order of the reaction. (b) Determine the initial rate of disappearance of X when the concentration of X is 0.30 *M* and that of Y is 0.40 *M*.

14.19 Determine the overall orders of the reactions to which these rate laws apply: (a) rate $= k[NO_2]^2$; (b) rate $= k$; (c) rate $= k[H_2][Br_2]^{\frac{1}{2}}$; (d) rate $= k[NO]^2[O_2]$.

14.20 Consider the reaction

$$A \longrightarrow B$$

The rate of the reaction is $1.6 \times 10^{-2}\ M$/s when the concentration of A is 0.35 *M*. Calculate the rate constant if the reaction is (a) first order in A, (b) second order in A.

Relationship Between Reactant Concentration and Time

Review Questions

14.21 Define the half-life of a reaction. Write the equation relating the half-life of a first-order reaction to the rate constant.

14.22 For a first-order reaction, how long will it take for the concentration of reactant to fall to one-eighth its original value? Express your answer in terms of the half-life ($t_{\frac{1}{2}}$) and in terms of the rate constant k.

Problems

14.23 What is the half-life of a compound if 75 percent of a given sample of the compound decomposes in 60 min? Assume first-order kinetics.

14.24 The thermal decomposition of phosphine (PH_3) into phosphorus and molecular hydrogen is a first-order reaction:

$$4PH_3(g) \longrightarrow P_4(g) + 6H_2(g)$$

The half-life of the reaction is 35.0 s at 680°C. Calculate (a) the first-order rate constant for the reaction and (b) the time required for 95 percent of the phosphine to decompose.

14.25 The rate constant for the second-order reaction

$$2NOBr(g) \longrightarrow 2NO(g) + Br_2(g)$$

is $0.80/M \cdot s$ at 10°C. (a) Starting with a concentration of 0.086 *M*, calculate the concentration of NOBr after 22 s. (b) Calculate the half-lives when $[NOBr]_0 = 0.072\ M$ and $[NOBr]_0 = 0.054\ M$.

14.26 The rate constant for the second-order reaction

$$2NO_2(g) \longrightarrow 2NO(g) + O_2(g)$$

is $0.54/M \cdot s$ at 300°C. (a) How long (in seconds) would it take for the concentration of NO_2 to decrease from 0.62 *M* to 0.28 *M*? (b) Calculate the half-lives at these two concentrations.

Activation Energy

Review Questions

14.27 Define activation energy. What role does activation energy play in chemical kinetics?

14.28 Write the Arrhenius equation and define all terms.

14.29 Use the Arrhenius equation to show why the rate constant of a reaction (a) decreases with increasing activation energy and (b) increases with increasing temperature.

14.30 As we know, methane burns readily in oxygen in a highly exothermic reaction. Yet a mixture of methane and oxygen gas can be kept indefinitely without any apparent change. Explain.

14.31 Sketch a potential-energy-versus-reaction-progress plot for the following reactions:

(a) $S(s) + O_2(g) \longrightarrow SO_2(g)$ $\qquad \Delta H° = -296.06$ kJ
(b) $Cl_2(g) \longrightarrow Cl(g) + Cl(g)$ $\qquad \Delta H° = 242.7$ kJ

14.32 The reaction $H + H_2 \longrightarrow H_2 + H$ has been studied for many years. Sketch a potential-energy-versus-reaction-progress diagram for this reaction.

Problems

14.33 Variation of the rate constant with temperature for the first-order reaction

$$2N_2O_5(g) \longrightarrow 2N_2O_4(g) + O_2(g)$$

is given in the following table. Determine graphically the activation energy for the reaction.

$T(K)$	$k(s^{-1})$
273	7.87×10^3
298	3.46×10^5
318	4.98×10^6
338	4.87×10^7

14.34 Given the same concentrations, the reaction

$$CO(g) + Cl_2(g) \longrightarrow COCl_2(g)$$

at 250°C is 1.50×10^3 times as fast as the same reaction at 150°C. Calculate the energy of activation for this reaction. Assume that the frequency factor is constant.

14.35 For the reaction

$$NO(g) + O_3(g) \longrightarrow NO_2(g) + O_2(g)$$

the frequency factor A is 8.7×10^{12} s^{-1} and the activation energy is 63 kJ/mol. What is the rate constant for the reaction at 75°C?

14.36 The rate constant of a first-order reaction is 4.60×10^{-4} s^{-1} at 350°C. If the activation energy is 104 kJ/mol, calculate the temperature at which its rate constant is 8.80×10^{-4} s^{-1}.

14.37 The rate constants of some reactions double with every 10-degree rise in temperature. Assume a reaction takes place at 295 K and 305 K. What must the activation energy be for the rate constant to double as described?

14.38 The rate at which tree crickets chirp is 2.0×10^2 per minute at 27°C but only 39.6 per minute at 5°C. From these data, calculate the "energy of activation" for the chirping process. (*Hint:* The ratio of rates is equal to the ratio of rate constants.)

Reaction Mechanisms

Review Questions

14.39 What do we mean by the mechanism of a reaction?

14.40 What is an elementary step?

14.41 What is the molecularity of a reaction?

14.42 Reactions can be classified as unimolecular, bimolecular, and so on. Why are there no zero-molecular reactions?

14.43 Explain why termolecular reactions are rare.

14.44 What is the rate-determining step of a reaction? Give an everyday analogy to illustrate the meaning of the term "rate determining."

14.45 The equation for the combustion of ethane (C_2H_6) is

$$2C_2H_6 + 7O_2 \longrightarrow 4CO_2 + 6H_2O$$

Explain why it is unlikely that this equation also represents the elementary step for the reaction.

14.46 Which of these species cannot be isolated in a reaction: activated complex, product, intermediate?

Problems

14.47 The rate law for the reaction

$$2NO(g) + Cl_2(g) \longrightarrow 2NOCl(g)$$

is given by rate $= k[NO][Cl_2]$. (a) What is the order of the reaction? (b) A mechanism involving these steps has been proposed for the reaction

$$NO(g) + Cl_2(g) \longrightarrow NOCl_2(g)$$
$$NOCl_2(g) + NO(g) \longrightarrow 2NOCl(g)$$

If this mechanism is correct, what does it imply about the relative rates of these two steps?

14.48 For the reaction $X_2 + Y + Z \longrightarrow XY + XZ$ it is found that doubling the concentration of X_2 doubles the reaction rate, tripling the concentration of Y triples the rate, and doubling the concentration of Z has no effect. (a) What is the rate law for this reaction? (b) Why is it that the change in the concentration of Z has no effect on the rate? (c) Suggest a mechanism for the reaction that is consistent with the rate law.

Catalysis

Review Questions

14.49 How does a catalyst increase the rate of a reaction?

14.50 What are the characteristics of a catalyst?

14.51 A certain reaction is known to proceed slowly at room temperature. Is it possible to make the reaction proceed at a faster rate without changing the temperature?

14.52 Distinguish between homogeneous catalysis and heterogeneous catalysis. Describe some important industrial processes that utilize heterogeneous catalysis.

14.53 Are enzyme-catalyzed reactions examples of homogeneous or heterogeneous catalysis?

14.54 The concentrations of enzymes in cells are usually quite small. What is the biological significance of this fact?

Problems

14.55 Most reactions, including enzyme-catalyzed reactions, proceed faster at higher temperatures. However, for a given enzyme, the rate drops off abruptly at a certain temperature. Account for this behavior.

14.56 Consider this mechanism for the enzyme-catalyzed reaction

$$E + S \underset{k_{-1}}{\overset{k_1}{\rightleftharpoons}} ES \qquad \text{(fast equilibrium)}$$

$$ES \overset{k_2}{\longrightarrow} E + P \qquad \text{(slow)}$$

Derive an expression for the rate law of the reaction in terms of the concentrations of E and S. (*Hint:* To solve for [ES], make use of the fact that, at equilibrium, the rate of the forward reaction is equal to the rate of the reverse reaction.)

Additional Problems

14.57 Suggest experimental means by which the rates of the following reactions could be followed:

(a) $CaCO_3(s) \longrightarrow CaO(s) + CO_2(g)$
(b) $Cl_2(g) + 2Br^-(aq) \longrightarrow Br_2(aq) + 2Cl^-(aq)$
(c) $C_2H_6(g) \longrightarrow C_2H_4(g) + H_2(g)$

14.58 List four factors that influence the rate of a reaction.

14.59 "The rate constant for the reaction

$$NO_2(g) + CO(g) \longrightarrow NO(g) + CO_2(g)$$

is $1.64 \times 10^{-6}/M \cdot s$." What is incomplete about this statement?

14.60 In a certain industrial process using a heterogeneous catalyst, the volume of the catalyst (in the shape of a sphere) is 10.0 cm^3. Calculate the surface area of the catalyst. If the sphere is broken down into eight spheres, each of which has a volume of 1.25 cm^3, what is the total surface area of the spheres? Which of the two geometric configurations of the catalyst is more effective? Explain. (The surface area of a sphere is $4\pi r^2$, in which r is the radius of the sphere.)

14.61 When methyl phosphate is heated in acid solution, it reacts with water:

$$CH_3OPO_3H_2 + H_2O \longrightarrow CH_3OH + H_3PO_4$$

If the reaction is carried out in water enriched with ^{18}O, the oxygen-18 isotope is found in the phosphoric acid product but not in the methanol. What does this tell us about the bond-breaking scheme in the reaction?

14.62 The rate of the reaction

$$CH_3COOC_2H_5(aq) + H_2O(l)$$
$$\longrightarrow CH_3COOH(aq) + C_2H_5OH(aq)$$

shows first-order characteristics—that is, rate = $k[CH_3COOC_2H_5]$—even though this is a second-order reaction (first order in $CH_3COOC_2H_5$ and first order in H_2O). Explain.

14.63 Explain why most metals used in catalysis are transition metals.

14.64 The bromination of acetone is acid-catalyzed:

$$CH_3COCH_3 + Br_2 \xrightarrow[\text{catalyst}]{H^+} CH_3COCH_2Br + H^+ + Br^-$$

The rate of disappearance of bromine was measured for several different concentrations of acetone, bromine, and H$^+$ ions at a certain temperature:

	[CH$_3$COCH$_3$]	[Br$_2$]	[H$^+$]	Rate of Disappearance of Br$_2$ (*M*/s)
(a)	0.30	0.050	0.050	5.7×10^{-5}
(b)	0.30	0.10	0.050	5.7×10^{-5}
(c)	0.30	0.050	0.10	1.2×10^{-4}
(d)	0.40	0.050	0.20	3.1×10^{-4}
(e)	0.40	0.050	0.050	7.6×10^{-5}

(a) What is the rate law for the reaction? (b) Determine the rate constant.

14.65 The reaction $2A + 3B \longrightarrow C$ is first order with respect to A and B. When the initial concentrations are [A] = 1.6×10^{-2} M and [B] = 2.4×10^{-3} M, the rate is 4.1×10^{-4} M/s. Calculate the rate constant of the reaction.

14.66 The decomposition of N_2O to N_2 and O_2 is a first-order reaction. At 730°C the half-life of the reaction is 3.58×10^3 min. If the initial pressure of N_2O is 2.10 atm at 730°C, calculate the total gas pressure after one half-life. Assume that the volume remains constant.

14.67 The reaction $S_2O_8^{2-} + 2I^- \longrightarrow 2SO_4^{2-} + I_2$ proceeds slowly in aqueous solution, but it can be catalyzed by the Fe^{3+} ion. Given that Fe^{3+} can oxidize I$^-$ and Fe^{2+} can reduce $S_2O_8^{2-}$, write a plausible two-step mechanism for this reaction. Explain why the uncatalyzed reaction is slow.

14.68 What are the units of the rate constant for a third-order reaction?

14.69 Consider the zero-order reaction A $\longrightarrow$ B. Sketch the following plots: (a) rate versus [A] and (b) [A] versus t.

14.70 A flask contains a mixture of compounds A and B. Both compounds decompose by first-order kinetics. The half-lives are 50.0 min for A and 18.0 min for B. If the concentrations of A and B are equal initially, how long will it take for the concentration of A to be four times that of B?

14.71 Referring to the decomposition of N_2O_5 on p. 440, explain how you would measure experimentally the partial pressure of N_2O_5 as a function of time.

14.72 The rate law for the reaction $2NO_2(g) \longrightarrow N_2O_4(g)$ is rate = $k[NO_2]^2$. Which of these changes will change the value of k? (a) The pressure of NO_2 is doubled. (b) The reaction is run in an organic solvent. (c) The volume of the container is doubled. (d) The temperature is decreased. (e) A catalyst is added to the container.

14.73 The reaction of G_2 with E_2 to form 2EG is exothermic, and the reaction of G_2 with X_2 to form 2XG is endothermic. The activation energy of the exothermic reaction is greater than that of the endothermic reaction. Sketch the potential energy profile diagrams for these two reactions on the same graph.

14.74 In the nuclear industry, workers use a rule of thumb that the radioactivity from any sample will be relatively harmless after 10 half-lives. Calculate the fraction of a radioactive sample that remains after this time. (*Hint:* Radioactive decays obey first-order kinetics.)

14.75 Briefly comment on the effect of a catalyst on each of the following: (a) activation energy, (b) reaction mechanism, (c) enthalpy of reaction, (d) rate of forward step, (e) rate of reverse step.

14.76 A quantity of 6 g of granulated Zn is added to a solution of 2 M HCl in a beaker at room temperature. Hydrogen gas is generated. For each of the following changes (at constant volume of the acid) state whether the rate of hydrogen gas evolution will be increased, decreased, or unchanged: (a) 6 g of powdered Zn is used; (b) 4 g of granulated Zn is used; (c) 2 M acetic acid is used instead of 2 M HCl; (d) temperature is raised to 40°C.

14.77 These data were collected for the reaction between hydrogen and nitric oxide at 700°C:

$$2H_2(g) + 2NO(g) \longrightarrow 2H_2O(g) + N_2(g)$$

Experiment	$[H_2]$	$[NO]$	Initial Rate (M/s)
1	0.010	0.025	2.4×10^{-6}
2	0.0050	0.025	1.2×10^{-6}
3	0.010	0.0125	0.60×10^{-6}

(a) Determine the order of the reaction. (b) Calculate the rate constant. (c) Suggest a plausible mechanism that is consistent with the rate law. (*Hint:* Assume the oxygen atom is the intermediate.)

14.78 A certain first-order reaction is 35.5 percent complete in 4.90 min at 25°C. What is its rate constant?

14.79 The decomposition of dinitrogen pentoxide has been studied in carbon tetrachloride solvent (CCl_4) at a certain temperature:

$$2N_2O_5 \longrightarrow 4NO_2 + O_2$$

$[N_2O_5]$ (M)	Initial Rate (M/s)
0.92	0.95×10^{-5}
1.23	1.20×10^{-5}
1.79	1.93×10^{-5}
2.00	2.10×10^{-5}
2.21	2.26×10^{-5}

Determine graphically the rate law for the reaction and calculate the rate constant.

14.80 The thermal decomposition of N_2O_5 obeys first-order kinetics. At 45°C, a plot of $\ln [N_2O_5]$ versus t gives a slope of -6.18×10^{-4} min^{-1}. What is the half-life of the reaction?

14.81 When a mixture of methane and bromine is exposed to light, the following reaction occurs slowly:

$$CH_4(g) + Br_2(g) \longrightarrow CH_3Br(g) + HBr(g)$$

Suggest a reasonable mechanism for this reaction. (*Hint:* Bromine vapor is deep red; methane is colorless.)

14.82 Consider this elementary step:

$$X + 2Y \longrightarrow XY_2$$

(a) Write a rate law for this reaction. (b) If the initial rate of formation of XY_2 is 3.8×10^{-3} M/s and the initial concentrations of X and Y are 0.26 M and 0.88 M, what is the rate constant of the reaction?

14.83 Consider the reaction

$$C_2H_5I(aq) + H_2O(l)$$
$$\longrightarrow C_2H_5OH(aq) + H^+(aq) + I^-(aq)$$

How could you follow the progress of the reaction by measuring the electrical conductance of the solution?

14.84 A compound X undergoes two *simultaneous* first-order reactions as follows: X $\longrightarrow$ Y with rate constant k_1 and X $\longrightarrow$ Z with rate constant k_2. The ratio of k_1/k_2 at 40°C is 8.0. What is the ratio at 300°C? Assume that the frequency factor of the two reactions is the same.

14.85 In recent years ozone in the stratosphere has been depleted at an alarmingly fast rate by chlorofluorocarbons (CFCs). A CFC molecule such as $CFCl_3$ is first decomposed by UV radiation:

$$CFCl_3 \longrightarrow CFCl_2 + Cl$$

The chlorine radical then reacts with ozone as follows:

$$Cl + O_3 \longrightarrow ClO + O_2$$
$$ClO + O \longrightarrow Cl + O_2$$

(a) Write the overall reaction for the last two steps. (b) What are the roles of Cl and ClO? (c) Why is the fluorine radical not important in this mechanism? (d) One suggestion to reduce the concentration of chlorine radicals is to add hydrocarbons such as ethane (C_2H_6) to the stratosphere. How will this work?

14.86 Consider a car fitted with a catalytic converter. The first 10 min or so after it is started are the most polluting. Why?

14.87 The reaction 2A + 3B $\longrightarrow$ C is first order with respect to A and B. When the initial concentrations are [A] = 1.6×10^{-2} M and [B] = 2.4×10^{-3} M, the rate is 4.1×10^{-4} M/s. Calculate the rate constant of the reaction.

14.88 The following mechanism has been proposed for the reaction described in Problem 14.64:

$$CH_3-\overset{\overset{\displaystyle O}{\|}}{C}-CH_3 + H_3O^+ \rightleftharpoons CH_3-\overset{\overset{\displaystyle +OH}{\|}}{C}-CH_3 + H_2O$$
$$\text{(fast equilibrium)}$$

$$CH_3-\overset{\overset{\displaystyle +OH}{\|}}{C}-CH_3 + H_2O \longrightarrow CH_3-\overset{\overset{\displaystyle OH}{|}}{C}=CH_2 + H_3O^+ \quad \text{(slow)}$$

$$CH_3-\overset{\overset{\displaystyle OH}{|}}{C}=CH_2 + Br_2 \longrightarrow CH_3-\overset{\overset{\displaystyle O}{\|}}{C}-CH_2Br + HBr \quad \text{(fast)}$$

Show that the rate law deduced from the mechanism is consistent with that shown in (a) of Problem 14.64.

14.89 The integrated rate law for the zero-order reaction A $\longrightarrow$ B is $[A] = [A]_0 - kt$. (a) Sketch the following plots: (i) rate versus [A] and (ii) [A] versus t. (b) Derive an expression for the half-life of the reaction. (c) Calculate the time in half-lives when the integrated rate law is no longer valid, that is, when $[A] = 0$.

14.90 Strictly speaking, the rate law derived for the reaction in Problem 14.77 applies only to certain concentrations of H_2. The general rate law for the reaction takes the form

$$\text{rate} = \frac{k_1[NO]^2[H_2]}{1 + k_2[H_2]}$$

in which k_1 and k_2 are constants. Derive rate law expressions under the conditions of very high and very low hydrogen concentrations. Does the result from Problem 14.77 agree with one of the rate expressions here?

14.91 (a) What can you deduce about the activation energy of a reaction if its rate constant changes significantly with a small change in temperature? (b) If a bimolecular reaction occurs every time an A and a B molecule collide, what can you say about the orientation factor and activation energy of the reaction?

14.92 The rate law for this reaction

$$CO(g) + NO_2(g) \longrightarrow CO_2(g) + NO(g)$$

is rate $= k[NO_2]^2$. Suggest a plausible mechanism for the reaction, given that the unstable species NO_3 is an intermediate.

14.93 Radioactive plutonium-239 ($t_{\frac{1}{2}} = 2.44 \times 10^5$ yr) is used in nuclear reactors and atomic bombs. If there are 5.0×10^2 g of the isotope in a small atomic bomb, how long will it take for the substance to decay to 1.0×10^2 g, too small an amount for an effective bomb? (*Hint:* Radioactive decays follow first-order kinetics.)

14.94 Many reactions involving heterogeneous catalysts are zero order; that is, rate $= k$. An example is the decomposition of phosphine (PH_3) over tungsten (W):

$$4PH_3(g) \longrightarrow P_4(g) + 6H_2(g)$$

It is found that the reaction is independent of $[PH_3]$ as long as phosphine's pressure is sufficiently high (≥ 1 atm). Explain.

14.95 Thallium(I) is oxidized by cerium(IV) as follows:

$$Tl^+ + 2Ce^{4+} \longrightarrow Tl^{3+} + 2Ce^{3+}$$

The elementary steps, in the presence of Mn(II), are as follows:

$$Ce^{4+} + Mn^{2+} \longrightarrow Ce^{3+} + Mn^{3+}$$
$$Ce^{4+} + Mn^{3+} \longrightarrow Ce^{3+} + Mn^{4+}$$
$$Tl^+ + Mn^{4+} \longrightarrow Tl^{3+} + Mn^{2+}$$

(a) Identify the catalyst, intermediates, and the rate-determining step if the rate law is rate $= k[Ce^{4+}][Mn^{2+}]$. (b) Explain why the reaction is slow without the catalyst. (c) Classify the type of catalysis (homogeneous or heterogeneous).

14.96 Consider the following elementary steps for a consecutive reaction

$$A \xrightarrow{k_1} B \xrightarrow{k_2} C$$

(a) Write an expression for the rate of change of B. (b) Derive an expression for the concentration of B under steady-state conditions; that is, when B is decomposing to C at the same rate as it is formed from A.

14.97 For gas-phase reactions, we can replace the concentration terms in Equation (14.3) with the pressures of the gaseous reactant. (a) Derive the equation

$$\ln \frac{P}{P_0} = -kt$$

where P and P_0 are the pressures at $t = t$ and $t = 0$ respectively. (b) Consider the decomposition of azomethane

$$H_3C-N=N-CH_3(g) \longrightarrow N_2(g) + C_2H_6(g)$$

The data obtained at 300°C are shown in the following table:

Time (s)	Partial Pressure of Azomethane (mmHg)
0	284
100	220
150	193
200	170
250	150
300	132

Are these values consistent with first-order kinetics? If so, determine the rate constant by plotting the data as shown in Figure 14.6(b). (c) Determine the rate constant by the half-life method.

14.98 The hydrolysis of methyl acetate

$$CH_3-\overset{\overset{\displaystyle O}{\|}}{C}-O-CH_3 + H_2O \longrightarrow CH_3-\overset{\overset{\displaystyle O}{\|}}{C}-OH + CH_3OH$$

methyl acetate   acetic acid   methanol

involves the breaking of a C—O bond. The two possibilities are

$$CH_3-\overset{\overset{\displaystyle O}{\|}}{C}+O-CH_3 \quad CH_3-\overset{\overset{\displaystyle O}{\|}}{C}-O+CH_3$$

                (a)   (b)

Suggest an experiment that would allow you to distinguish between these two possibilities.

Special Problems

14.99 Polyethylene is used in many items such as water pipes, bottles, electrical insulation, toys, and mailer envelopes. It is a *polymer,* a molecule with a very high molar mass made by joining many ethylene molecules (the basic unit is called a monomer) together (see p. 346). The initiation step is

$$R_2 \xrightarrow{k_i} 2R\cdot \qquad \text{initiation}$$

The R· species (called a radical) reacts with an ethylene molecule (M) to generate another radical

$$R\cdot + M \longrightarrow M_1\cdot$$

Reaction of $M_1\cdot$ with another monomer leads to the growth or propagation of the polymer chain:

$$M_1\cdot + M \xrightarrow{k_p} M_2\cdot \qquad \text{propagation}$$

This step can be repeated with hundreds of monomer units. The propagation terminates when two radicals combine

$$M'\cdot + M''\cdot \xrightarrow{k_t} M'—M'' \qquad \text{termination}$$

(a) The initiator used in the polymerization of ethylene is benzoyl peroxide $[(C_6H_5COO)_2]$:

$$(C_6H_5COO)_2 \longrightarrow 2C_6H_5COO\cdot$$

This is a first-order reaction. The half-life of benzoyl peroxide at 100°C is 19.8 min. (a) Calculate the rate constant (in min^{-1}) of the reaction. (b) If the half-life of benzoyl peroxide is 7.30 hr or 438 min, at 70°C, what is the activation energy (in kJ/mol) for the decomposition of benzoyl peroxide? (c) Write the rate laws for the elementary steps in the above polymerization process and identify the reactant, product, and intermediates. (d) What condition would favor the growth of long high-molar-mass polyethylenes?

14.100 Ethanol is a toxic substance that, when consumed in excess, can impair respiratory and cardiac functions by interference with the neurotransmitters of the nervous system. In the human body, ethanol is metabolized by the enzyme alcohol dehydrogenase to acetaldehyde, which causes "hangovers." (a) Based on your knowledge of enzyme kinetics, explain why binge drinking (that is, consuming too much alcohol too fast) can prove fatal. (b) Methanol is even more toxic than ethanol. It is also metabolized by alcohol dehydrogenase, and the product, formaldehyde, can cause blindness or death. An antidote to methanol poisoning is ethanol. Explain how this procedure works.

Answers to Practice Exercises

14.1 rate $= -\dfrac{\Delta[CH_4]}{\Delta t} = -\dfrac{1}{2}\dfrac{\Delta[O_2]}{\Delta t} = \dfrac{\Delta[CO_2]}{\Delta t} = \dfrac{1}{2}\dfrac{\Delta[H_2O]}{\Delta t}$.
14.2 rate $= k[S_2O_8^{2-}][I^-]$, $k = 8.1 \times 10^{-2}/M \cdot$ s. **14.3** 66 s. **14.4** 1.2×10^3 s. **14.5** (a) 3.2 min, (b) 2.1 min. **14.6** 240 kJ/mol. **14.7** 3.13×10^{-9} s^{-1}. **14.8** (a) $NO_2 + CO \longrightarrow NO + CO_2$, (b) NO_3, (c) the first step is rate-determining.

In the early 1900s there was a shortage of nitrogen compounds, which were used as agricultural fertilizers and as explosives. Chemists at that time were interested in converting atmospheric nitrogen into useable compounds (a process called nitrogen fixation), such as ammonia. In 1912 Fritz Haber, a German chemist, developed a method, which now bears his name, for synthesizing ammonia directly from nitrogen and hydrogen:

The catalytic action in the synthesis of ammonia. First the H_2 and N_2 molecules bind to the surface of the catalyst. This interaction weakens the covalent bonds and causes the molecules to dissociate. The highly reactive atoms migrate on the surface and eventually combine to form NH_3 molecules, which then leave the surface.

$$N_2(g) + 3H_2(g) \rightleftharpoons 2NH_3(g)$$

This process is sometimes called the Haber-Bosch process to give credit also to Karl Bosch, the engineer who developed the equipment for the industrial production of ammonia, which has to be carried out at 500°C and 500 atm. Haber's success was based on his knowledge of factors affecting gaseous equilibrium and the choice of proper catalysts (iron and oxides of aluminum and potassium). His effort is believed to have prolonged World War I by a few years because it allowed Germany to continue manufacturing explosives after its supply of sodium nitrate from Chile was cut off by the Allied naval blockade. In 2001 about 35 million pounds of ammonia were produced in the United States by the Haber process; most of it was used for fertilizer.

Fritz Haber was born in Prussia in 1868. Besides the synthesis of ammonia, he did important work in electrochemistry and combustion chemistry, and he was the director of the prestigious Kaiser Wilhelm Institute. Haber's notable scientific failure was his attempt to extract gold from the oceans—he had overestimated the gold concentration in seawater by a thousand times! The activity that most tarnished his name, however, was his involvement in the introduction of poison gas (chlorine) on the battlefield. The decision to award Haber the Nobel Prize in chemistry in 1918 generated considerable controversy and criticism, something that rarely happened with Nobel awards in the physical sciences.

When the Nazis came to power in 1933, Haber was expelled from Germany because he was Jewish. He died of a heart attack in Switzerland the next year.

Liquid ammonia being applied to the soil before planting.

Chemical Equilibrium   **Chapter 15**

ESSENTIAL CONCEPTS

Chemical Equilibrium Chemical equilibrium describes the state in which the rates of forward and reverse reactions are equal and the concentrations of the reactants and products remain unchanged with time. This state of dynamic equilibrium is characterized by an equilibrium constant. Depending on the nature of reacting species, the equilibrium constant can be expressed in terms of molarities (for solutions) or partial pressure (for gases). The equilibrium constant provides information about the net direction of a reversible reaction and the concentrations of the equilibrium mixture.

Factors That Affect Chemical Equilibrium Changes in concentration can affect the position of an equilibrium state; that is, the relative amounts of reactants and products. Changes in pressure and volume may have the same effect for gaseous systems at equilibrium. Only a change in temperature can alter the value of the equilibrium constant. A catalyst can establish the equilibrium state faster by speeding the forward and reverse reactions, but it can change neither the equilibrium position nor the equilibrium constant.

15.1 The Concept of Equilibrium

Few chemical reactions proceed in only one direction. Most are, at least to some extent, reversible. At the start of a reversible process, the reaction proceeds toward the formation of products. As soon as some product molecules are formed, the reverse process—that is, the formation of reactant molecules from product molecules—begins to take place. *When the rates of the forward and reverse reactions are equal and the concentrations of the reactants and products no longer change with time,* **chemical equilibrium** is reached.

Chemical equilibrium is a dynamic process. As such, it can be likened to the movement of skiers at a busy ski resort, where the number of skiers carried up the mountain on the chair lift is equal to the number coming down the slopes. Thus, although there is a constant transfer of skiers, the number of people at the top and the number at the bottom of the slope do not change.

Note that a chemical equilibrium reaction involves different substances as reactants and products. Equilibrium between two phases of the same substance is called **physical equilibrium** because *the changes that occur are physical processes.* The vaporization of water in a closed container at a given temperature is an example of physical equilibrium. In this instance, the number of H_2O molecules leaving and the number returning to the liquid phase are equal:

$$H_2O(l) \rightleftharpoons H_2O(g)$$

(Recall from Chapter 4 that the double arrow means that the reaction is reversible.) The study of physical equilibrium yields useful information, such as the equilibrium vapor pressure (see Section 12.6). However, chemists are particularly interested in chemical equilibrium processes, such as the reversible reaction involving nitrogen dioxide (NO_2) and dinitrogen tetroxide (N_2O_4). The progress of the reaction:

$$N_2O_4(g) \rightleftharpoons 2NO_2(g)$$

can be monitored easily because N_2O_4 is a colorless gas, whereas NO_2 has a dark-brown color that makes it sometimes visible in polluted air. Suppose that a known amount of N_2O_4 is injected into an evacuated flask. Some brown color appears immediately, indicating the formation of NO_2 molecules. The color intensifies as the dissociation of N_2O_4 continues until eventually equilibrium is reached. Beyond that point, no further change in color is observed. By experiment we find that we can also reach the equilibrium state by starting with pure NO_2 or with a mixture of NO_2 and N_2O_4. In each case, we observe an initial change in color, caused either by the formation of NO_2 (if the color intensifies) or by the depletion of NO_2 (if the color fades), and then the final state in which the color of NO_2 no longer changes. Depending on the temperature of the reacting system and on the initial amounts of NO_2 and N_2O_4, the concentrations of NO_2 and N_2O_4 at equilibrium differ from system to system (Figure 15.1).

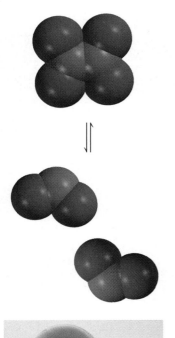

NO_2 and N_2O_4 gases at equilibrium.

The Equilibrium Constant

Table 15.1 shows some experimental data for this reaction at 25°C. The gas concentrations are expressed in molarity, which can be calculated from the number of moles of the gases present initially and at equilibrium and the volume of the flask

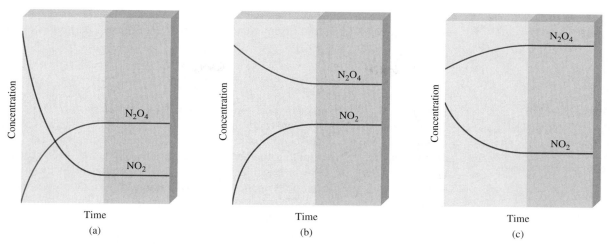

Figure 15.1 *Change in the concentrations of NO_2 and N_2O_4 with time, in three situations. (a) Initially only NO_2 is present. (b) Initially only N_2O_4 is present. (c) Initially a mixture of NO_2 and N_2O_4 is present. Note that even though equilibrium is reached in all cases, the equilibrium concentrations of NO_2 and N_2O_4 are not the same.*

in liters. Analysis of the data at equilibrium shows that although the ratio $[NO_2]/[N_2O_4]$ gives scattered values, the ratio $[NO_2]^2/[N_2O_4]$ gives a nearly constant value that averages 4.63×10^{-3}. This value is called the *equilibrium constant, K,* for the reaction at 25°C. Expressed mathematically, the equilibrium constant for the NO_2–N_2O_4 equilibrium is

$$K = \frac{[NO_2]^2}{[N_2O_4]} = 4.63 \times 10^{-3} \qquad (15.1)$$

Note that the exponent 2 for $[NO_2]$ in this expression is the same as the stoichiometric coefficient for NO_2 in the reversible reaction.

We can generalize this discussion by considering the following reversible reaction:

$$a\text{A} + b\text{B} \rightleftharpoons c\text{C} + d\text{D}$$

Table 15.1 | The NO₂–N₂O₄ System at 25°C

Initial Concentrations (*M*)		Equilibrium Concentrations (*M*)		Ratio of Concentrations at Equilibrium	
$[NO_2]$	$[N_2O_4]$	$[NO_2]$	$[N_2O_4]$	$\dfrac{[NO_2]}{[N_2O_4]}$	$\dfrac{[NO_2]^2}{[N_2O_4]}$
0.000	0.670	0.0547	0.643	0.0851	4.65×10^{-3}
0.0500	0.446	0.0457	0.448	0.102	4.66×10^{-3}
0.0300	0.500	0.0475	0.491	0.0967	4.60×10^{-3}
0.0400	0.600	0.0523	0.594	0.0880	4.60×10^{-3}
0.200	0.000	0.0204	0.0898	0.227	4.63×10^{-3}

in which a, b, c, and d are the stoichiometric coefficients for the reacting species A, B, C, and D. The equilibrium constant for the reaction at a particular temperature is

$$K = \frac{[C]^c[D]^d}{[A]^a[B]^b} \tag{15.2}$$

Equation (15.2) is the mathematical form of the *law of mass action*. It *relates the concentrations of reactants and products at equilibrium* in terms of a quantity called the **equilibrium constant.** The equilibrium constant is defined by a quotient. The numerator is obtained by multiplying together the equilibrium concentrations of the *products,* each raised to a power equal to its stoichiometric coefficient in the balanced equation. The same procedure is applied to the equilibrium concentrations of *reactants* to obtain the denominator. This formulation is based on purely empirical evidence, such as the study of reactions like NO_2–N_2O_4.

The equilibrium constant has its origin in thermodynamics, to be discussed in Chapter 18. However, we can gain some insight into K by considering the kinetics of chemical reactions. Let us suppose that this reversible reaction occurs via a mechanism of a single *elementary step* in both the forward and reverse directions:

$$A + 2B \underset{k_r}{\overset{k_f}{\rightleftharpoons}} AB_2$$

The forward rate is given by

$$\text{rate}_f = k_f[A][B]^2$$

and the reverse rate is

$$\text{rate}_r = k_r[AB_2]$$

in which k_f and k_r are the rate constants for the forward and reverse directions, respectively. At equilibrium, when no net changes occur, the two rates must be equal:

$$\text{rate}_f = \text{rate}_r$$

or

$$k_f[A][B]^2 = k_r[AB_2]$$

$$\frac{k_f}{k_r} = \frac{[AB_2]}{[A][B]^2}$$

Because both k_f and k_r are constants at a given temperature, their ratio is also a constant, which is equal to the equilibrium constant K_c.

$$\frac{k_f}{k_r} = K_c = \frac{[AB_2]}{[A][B]^2}$$

So K_c is a constant regardless of the equilibrium concentrations of the reacting species because it is always equal to k_f/k_r, the quotient of two quantities that are themselves constant at a given temperature. Because rate constants are temperature-dependent [see Equation (14.9)], it follows that the equilibrium constant must also change with temperature.

Finally, we note that if the equilibrium constant is much greater than 1 (that is, $K \gg 1$), the equilibrium will lie to the right of the reaction arrows and favor the products. Conversely, if the equilibrium constant is much smaller than 1 (that is, $K \ll 1$), the equilibrium will lie to the left and favor the reactants.

In keeping with the convention, we refer to substances on the left of the equilibrium arrows as "reactants" and those on the right as "products."

To review reaction mechanism, see Section 14.5.

The signs $\gg$ and $\ll$ mean "much greater than" and "much smaller than," respectively.

15.2 Ways of Expressing Equilibrium Constants

To use equilibrium constants, we must express them in terms of the reactant and product concentrations. Our only guidance is the law of mass action [Equation (15.2)]. However, because the concentrations of the reactants and products can be expressed in different units and because the reacting species are not always in the same phase, there may be more than one way to express the equilibrium constant for the *same* reaction. To begin with, we will consider reactions in which the reactants and products are in the same phase.

Homogeneous Equilibria

The term *homogeneous equilibrium* applies to reactions in which *all reacting species are in the same phase*. An example of homogeneous gas-phase equilibrium is the dissociation of N_2O_4. The equilibrium constant, as given in Equation (15.1), is

$$K_c = \frac{[NO_2]^2}{[N_2O_4]}$$

Note that the subscript in K_c denotes that the concentrations of the reacting species are expressed in moles per liter. The concentrations of reactants and products in gaseous reactions can also be expressed in terms of their partial pressures. From Equation (5.8) we see that at constant temperature the pressure P of a gas is directly related to the concentration in moles per liter of the gas; that is, $P = (n/V)RT$. Thus, for the equilibrium process

$$N_2O_4(g) \rightleftharpoons 2NO_2(g)$$

we can write

$$K_P = \frac{P_{NO_2}^2}{P_{N_2O_4}} \tag{15.3}$$

in which P_{NO_2} and $P_{N_2O_4}$ are the equilibrium partial pressures (in atmospheres) of NO_2 and N_2O_4, respectively. The subscript in K_P tells us that equilibrium concentrations are expressed in terms of pressure.

In general, K_c is not equal to K_P, because the partial pressures of reactants and products are not equal to their concentrations expressed in moles per liter. A simple relationship between K_P and K_c can be derived as follows. Let us consider this equilibrium in the gas phase:

$$aA(g) \rightleftharpoons bB(g)$$

in which a and b are stoichiometric coefficients. The equilibrium constant K_c is

$$K_c = \frac{[B]^b}{[A]^a}$$

and the expression for K_P is

$$K_P = \frac{P_B^b}{P_A^a}$$

in which P_A and P_B are the partial pressures of A and B. Assuming ideal gas behavior,

$$P_A V = n_A RT$$

$$P_A = \frac{n_A RT}{V}$$

in which V is the volume of the container in liters. Also

$$P_B V = n_B RT$$

$$P_B = \frac{n_B RT}{V}$$

Substituting these relations into the expression for K_P, we obtain

$$K_P = \frac{\left(\dfrac{n_B RT}{V}\right)^b}{\left(\dfrac{n_A RT}{V}\right)^a} = \frac{\left(\dfrac{n_B}{V}\right)^b}{\left(\dfrac{n_A}{V}\right)^a}(RT)^{b-a}$$

Now both n_A/V and n_B/V have the units of moles per liter and can be replaced by [A] and [B], so that

$$K_P = \frac{[B]^b}{[A]^a}(RT)^{\Delta n}$$

$$= K_c(RT)^{\Delta n} \qquad (15.4)$$

in which

$$\Delta n = b - a$$

$$= \text{moles of gaseous products} - \text{moles of gaseous reactants}$$

Because pressure is usually expressed in atmospheres, the gas constant R is given by 0.0821 L · atm/K · mol, and we can write the relationship between K_P and K_c as

$$K_P = K_c(0.0821T)^{\Delta n} \qquad (15.5)$$

To use this equation, the pressures in K_P must be in atm.

In general $K_P \neq K_c$ except in the special case when $\Delta n = 0$. In that case Equation (15.5) can be written as

$$K_P = K_c(0.0821T)^0$$

$$= K_c$$

Any number raised to the zero power is equal to 1.

As another example of homogeneous equilibrium, let us consider the ionization of acetic acid (CH_3COOH) in water:

$$CH_3COOH(aq) + H_2O(l) \rightleftharpoons CH_3COO^-(aq) + H_3O^+(aq)$$

The equilibrium constant is

$$K_c' = \frac{[CH_3COO^-][H_3O^+]}{[CH_3COOH][H_2O]}$$

(We use the prime for K_c here to distinguish it from the final form of equilibrium constant to be derived shortly.) However, in 1 L, or 1000 g, of water, there are 1000 g/(18.02 g/mol), or 55.5 mol, of water. Therefore, the "concentration" of water, or $[H_2O]$, is 55.5 mol/L, or 55.5 M. This is a large quantity compared with the concentrations of other species in solution (usually 1 M or smaller), and we can assume that it does not change appreciably during the course of a reaction. Thus we may treat $[H_2O]$ as a constant and rewrite the equilibrium constant as

$$K_c = \frac{[CH_3COO^-][H_3O^+]}{[CH_3COOH]}$$

in which

$$K_c = K_c'[H_2O]$$

Note that it is general practice not to include units for the equilibrium constant. In thermodynamics, K is defined to have no units because every concentration (molarity) or pressure (atmospheres) term is actually a ratio to a standard value, which is 1 M or 1 atm. This procedure eliminates all units but does not alter the numerical parts of the concentration or pressure. Consequently, K has no units. We will extend this practice to acid-base equilibria and solubility equilibria (to be discussed in Chapters 16 and 17).

Example 15.1 Write expressions for K_c, and K_P, if applicable, for these reversible reactions at equilibrium:

(a) $2NO(g) + O_2(g) \rightleftharpoons 2NO_2(g)$

(b) $CH_3COOH(aq) + C_2H_5OH(aq) \rightleftharpoons CH_3COOC_2H_5(aq) + H_2O(l)$

Reasoning and Solution Keep in mind these facts: (1) The K_P expression applies only to gaseous reactions and (2) the concentration of solvent (usually water) does not appear in the equilibrium constant expression.

(a) All the reacting species are gases so we can write both the K_c and K_P expressions for the reaction

$$K_c = \frac{[NO_2]^2}{[NO]^2[O_2]} \qquad K_P = \frac{P_{NO_2}^2}{P_{NO}^2 P_{O_2}}$$

(b) The equilibrium constant K_c' is given by

$$K_c' = \frac{[CH_3COOC_2H_5][H_2O]}{[CH_3COOH][C_2H_5OH]}$$

Because the water produced in the reaction is negligible compared with the water solvent, the concentration of water does not change appreciably. Thus we can write the new equilibrium constant as

$$K_c = \frac{[CH_3COOC_2H_5]}{[CH_3COOH][C_2H_5OH]}$$

where $K_c = K_c'/[H_2O]$.

Practice Exercise Write K_c and K_P for this reaction:

$$2N_2O_5(g) \rightleftharpoons 4NO_2(g) + O_2(g)$$

Reaction shown in (a).

Similar problem: 15.8.

Similar problem: 15.16.

Example 15.2   The equilibrium constant K_P for the decomposition of phosphorus pentachloride to phosphorus trichloride and molecular chlorine

$$PCl_5(g) \rightleftharpoons PCl_3(g) + Cl_2(g)$$

is found to be 1.05 at 250°C. If the equilibrium partial pressures of PCl_5 and PCl_3 are 0.875 atm and 0.463 atm, respectively, what is the equilibrium partial pressure of Cl_2 at 250°C?

Reasoning and Solution   First, we write K_P in terms of the partial pressures of the reacting species.

$$K_P = \frac{P_{PCl_3}P_{Cl_2}}{P_{PCl_5}}$$

Knowing the partial pressures, we write

$$1.05 = \frac{(0.463)(P_{Cl_2})}{(0.875)}$$

or

$$P_{Cl_2} = \frac{(1.05)(0.875)}{(0.463)} = 1.98 \text{ atm}$$

Comment   Note that we have added atm as the unit for P_{Cl_2}.

Practice Exercise   The equilibrium constant K_P for the reaction

$$2NO_2(g) \rightleftharpoons 2NO(g) + O_2(g)$$

is 158 at 1000 K. Calculate P_{O_2} if $P_{NO_2} = 0.400$ atm and $P_{NO} = 0.270$ atm.

Example 15.3   For the reaction

$$N_2(g) + 3H_2(g) \rightleftharpoons 2NH_3(g)$$

K_P is 4.3×10^{-4} at 375°C. Calculate K_c for the reaction.

Reasoning and Solution   From Equation (15.5) we write

$$K_c = \frac{K_P}{(0.0821T)^{\Delta n}}$$

Because $T = 375 + 273 = 648$ K and $\Delta n = 2 - 4 = -2$, we have

$$K_c = \frac{4.3 \times 10^{-4}}{(0.0821 \times 648)^{-2}}$$

$$= 1.2$$

Comment   Note that K_c, like K_P, is a dimensionless quantity. This example shows that we can get quite a different value for the equilibrium constant for the same reaction, depending on whether we express the concentrations in moles per liter or in atmospheres.

Similar problem: 15.13.

Practice Exercise   The equilibrium constant (K_c) for the reaction

$$N_2O_4(g) \rightleftharpoons 2NO_2(g)$$

is 4.63×10^{-3} at 25°C. What is the value of K_P at this temperature?

Heterogeneous Equilibria

A reversible reaction involving reactants and products that are in different phases leads to a **heterogeneous equilibrium.** For example, when calcium carbonate is heated in a closed vessel, this equilibrium is attained:

$$CaCO_3(s) \rightleftharpoons CaO(s) + CO_2(g)$$

The two solids and one gas constitute three separate phases. At equilibrium, we might write the equilibrium constant as

$$K'_c = \frac{[CaO][CO_2]}{[CaCO_3]} \qquad (15.6)$$

The mineral calcite is made of calcium carbonate, as are chalk and marble.

However, the "concentration" of a solid, like its density, is an intensive property and does not depend on how much of the substance is present. [Note that the units of concentration (moles per liter) can be converted to units of density (grams per cubic centimeters) and vice versa.] For this reason, the terms $[CaCO_3]$ and $[CaO]$ are themselves constants and can be combined with the equilibrium constant. We can simplify the equilibrium expression by writing

$$\frac{[CaCO_3]}{[CaO]}K'_c = K_c = [CO_2] \qquad (15.7)$$

in which K_c, the "new" equilibrium constant, is now conveniently expressed in terms of a single concentration, that of CO_2. Keep in mind that the value of K_c does not depend on how much $CaCO_3$ and CaO are present, as long as some of each is present at equilibrium (Figure 15.2).

Alternatively, we can express the equilibrium constant as

$$K_P = P_{CO_2} \qquad (15.8)$$

The equilibrium constant in this case is numerically equal to the pressure of CO_2 gas, an easily measurable quantity.

What has been said about solids also applies to liquids. Thus if a liquid is a reactant or a product, we can treat its concentration as constant and omit it from the equilibrium constant expression.

(a)   (b)

Figure 15.2 *In (a) and (b) the equilibrium pressure of CO_2 is the same at the same temperature, despite the presence of different amounts of $CaCO_3$ and CaO.*

Example 15.4 Consider this heterogeneous equilibrium:

$$CaCO_3(s) \rightleftharpoons CaO(s) + CO_2(g)$$

At 800°C, the pressure of CO_2 is 0.236 atm. Calculate (a) K_P and (b) K_c for the reaction at this temperature.

Reasoning and Solution (a) Because the concentration of solids does not appear in the equilibrium constant expression, using Equation (15.8), we write

$$K_P = P_{CO_2}$$
$$= 0.236$$

(b) From Equation (15.5), we know

$$K_P = K_c(0.0821T)^{\Delta n}$$

In this case, $T = 800 + 273 = 1073$ K and $\Delta n = 1$, so we substitute these in the equation and obtain

$$0.236 = K_c(0.0821 \times 1073)$$
$$K_c = 2.68 \times 10^{-3}$$

Similar problem: 15.18.

Practice Exercise Consider the following equilibrium at 295 K:

$$NH_4HS(s) \rightleftharpoons NH_3(g) + H_2S(g)$$

The partial pressure of each gas is 0.265 atm. Calculate K_P and K_c for the reaction.

The Form of K and the Equilibrium Equation

Before closing this section, we should note these two important rules about writing equilibrium constants:

- When the equation for a reversible reaction is written in the opposite direction, the equilibrium constant becomes the reciprocal of the original equilibrium constant. Thus if we write the NO_2–N_2O_4 equilibrium at 25°C as

$$N_2O_4(g) \rightleftharpoons 2NO_2(g)$$

The reciprocal of x is 1/x.

then

$$K_c = \frac{[NO_2]^2}{[N_2O_4]} = 4.63 \times 10^{-3}$$

However, we can represent the equilibrium equally well as

$$2NO_2(g) \rightleftharpoons N_2O_4(g)$$

and the equilibrium constant is now given by

$$K_c' = \frac{[N_2O_4]}{[NO_2]^2} = \frac{1}{K_c} = \frac{1}{4.63 \times 10^{-3}} = 216$$

You can see that $K_c = 1/K_c'$ or $K_cK_c' = 1.00$. Either K_c or K_c' is a valid equilibrium constant, but it is meaningless to say that the equilibrium constant for the NO_2–N_2O_4 system is 4.63×10^{-3}, or 216, unless we also specify how the equilibrium equation is written.

- The value of K also depends on how the equilibrium equation is balanced. Consider the following two ways of describing the same equilibrium:

$$\tfrac{1}{2}N_2O_4(g) \rightleftharpoons NO_2(g) \qquad K'_c = \frac{[NO_2]}{[N_2O_4]^{1/2}}$$

$$N_2O_4(g) \rightleftharpoons 2NO_2(g) \qquad K_c = \frac{[NO_2]^2}{[N_2O_4]}$$

Looking at the exponents we see that $K'_c = \sqrt{K_c}$. In Table 15.1 we find the average value of $K_c = 4.63 \times 10^{-3}$; therefore $K'_c = 0.0680$.

Thus if you double a chemical equation throughout, the corresponding equilibrium constant will be the square of the original value; if you triple the equation, the equilibrium constant will be the cube of the original value, and so on. The NO_2–N_2O_4 example illustrates once again the need to write the particular chemical equation when quoting the numerical value of an equilibrium constant.

Summary of the Rules for Writing Equilibrium Constant Expressions

- The concentrations of the reacting species in the condensed phase are expressed in moles per liter; in the gaseous phase, the concentrations can be expressed in moles per liter or in atmospheres. K_c is related to K_P by a simple equation [Equation (15.5)].
- The concentrations of pure solids, pure liquids (in heterogeneous equilibria), and solvents (in homogeneous equilibria) do not appear in the equilibrium constant expressions.
- The equilibrium constant (K_c or K_P) is dimensionless.
- In quoting a value for the equilibrium constant, we must specify the balanced equation and the temperature.

15.3 What Does the Equilibrium Constant Tell Us?

We have seen that the equilibrium constant for a given reaction can be calculated from known equilibrium concentrations. Once we know the value of the equilibrium constant, we can use Equation (15.2) to calculate unknown equilibrium concentrations—remembering, of course, that the equilibrium constant has a constant value only if the temperature does not change. In general, the equilibrium constant helps us to predict the direction in which a reaction mixture will proceed to achieve equilibrium and to calculate the concentrations of reactants and products once equilibrium has been reached. These uses of the equilibrium constant will be explored in this section.

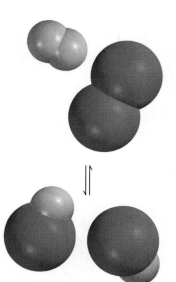

Predicting the Direction of a Reaction

The equilibrium constant K_c for the reaction

$$H_2(g) + I_2(g) \rightleftharpoons 2HI(g)$$

is 54.3 at 430°C. Suppose that in a certain experiment we place 0.243 mole of H_2, 0.146 mole of I_2, and 1.98 moles of HI all in a 1.00-L container at 430°C. Will there

Figure 15.3 *The direction of a reversible reaction to reach equilibrium depends on the relative magnitudes of Q_c and K_c.*

Reactants → Products   Equilibrium: no net change   Reactants ← Products

be a net reaction to form more H_2 and I_2 or more HI? Inserting the starting concentrations in the equilibrium constant expression, we write

$$\frac{[\text{HI}]_0^2}{[\text{H}_2]_0[\text{I}_2]_0} = \frac{(1.98)^2}{(0.243)(0.146)} = 111$$

where the subscript 0 indicates initial concentrations. Because the calculated quotient $[\text{HI}]_0^2/[\text{H}_2]_0[\text{I}_2]_0$ is greater than K_c, this system is not at equilibrium. Consequently, some of the HI will react to form more H_2 and I_2 (decreasing the value of the quotient). Thus the net reaction proceeds from right to left to reach equilibrium.

The quantity obtained by substituting the initial concentrations into the equilibrium constant expression is called the **reaction quotient** **(Q_c).** To determine in which direction the net reaction will proceed to achieve equilibrium, we compare the values of Q_c and K_c. The three possible cases are as follows (Figure 15.3):

- $Q_c < K_c$ The ratio of initial concentrations of products to reactants is too small. To reach equilibrium, reactants must be converted to products. The system proceeds from left to right (consuming reactants, forming products) to reach equilibrium.
- $Q_c = K_c$ The initial concentrations are equilibrium concentrations. The system is at equilibrium.
- $Q_c > K_c$ The ratio of initial concentrations of products to reactants is too large. To reach equilibrium, products must be converted to reactants. The system proceeds from right to left (consuming products, forming reactants) to reach equilibrium.

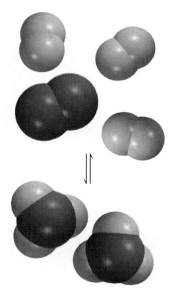

Example 15.5   At the start of a reaction, there are 0.249 mol N_2, 3.21×10^{-2} mol H_2, and 6.42×10^{-4} mol NH_3 in a 3.50-L reaction vessel at 375°C. If the equilibrium constant (K_c) for the reaction

$$N_2(g) + 3H_2(g) \rightleftharpoons 2NH_3(g)$$

is 1.2 at this temperature, decide whether the system is at equilibrium. If it is not, predict which way the net reaction will proceed.

Reasoning and Solution   The initial concentrations enable us to calculate Q_c which can be compared with K_c to determine the direction of the net reaction to reach equilibrium.

The initial concentrations of the reacting species are

$$[N_2]_0 = \frac{0.249 \text{ mol}}{3.50 \text{ L}} = 0.0711 \text{ } M$$

$$[H_2]_0 = \frac{3.21 \times 10^{-2} \text{ mol}}{3.50 \text{ L}} = 9.17 \times 10^{-3} \text{ } M$$

$$[NH_3]_0 = \frac{6.42 \times 10^{-4} \text{ mol}}{3.50 \text{ L}} = 1.83 \times 10^{-4} \text{ } M$$

Next we write

$$\frac{[NH_3]_0^2}{[N_2]_0[H_2]_0^3} = \frac{(1.83 \times 10^{-4})^2}{(0.0711)(9.17 \times 10^{-3})^3} = 0.611 = Q_c$$

Because Q_c is smaller than K_c (1.2), the system is not at equilibrium. The net result will be an increase in the concentration of NH_3 and a decrease in the concentrations of N_2 and H_2. That is, the net reaction will proceed from left to right until equilibrium is reached (see Figure 15.3).

Similar problem: 15.28.

Practice Exercise The equilibrium constant (K_c) for the formation of nitrosyl chloride, an orange-yellow compound, from nitric oxide and molecular chlorine

$$2NO(g) + Cl_2(g) \rightleftharpoons 2NOCl(g)$$

is 6.5×10^4 at 35°C. In a certain experiment, 2.0×10^{-2} mole of NO, 8.3×10^{-3} mole of Cl_2, and 6.8 moles of NOCl are mixed in a 2.0-L flask. In which direction will the system proceed to reach equilibrium?

Calculating Equilibrium Concentrations

If we know the equilibrium constant for a particular reaction, we can calculate the concentrations in the equilibrium mixture from a knowledge of the initial concentrations. Depending on the information given, the calculation may be straightforward or complex. In the most common situation, only the initial reactant concentrations are given. Let us consider this system involving a pair of geometric isomers in an organic solvent (Figure 15.4), which has an equilibrium constant (K_c) of 24.0 at 200°C:

$$\textit{cis}\text{-stilbene} \rightleftharpoons \textit{trans}\text{-stilbene}$$

Suppose that only *cis*-stilbene is initially present at a concentration of 0.850 mol/L. How do we calculate the concentrations of *cis*- and *trans*-stilbene at equilibrium? From the stoichiometry of the reaction we see that for every mole of *cis*-stilbene converted, 1 mole of *trans*-stilbene is formed. Let x be the equilibrium concentration of *trans*-stilbene in moles per liter; therefore, the equilibrium concentration of *cis*-stilbene must be $(0.850 - x)$ mol/L. It is useful to summarize the changes in concentration as:

	cis-stilbene $\rightleftharpoons$ *trans*-stilbene	
Initial (*M*):	0.850	0
Change (*M*):	$-x$	$+x$
Equilibrium (*M*):	$(0.850 - x)$	x

cis-Stilbene

trans-Stilbene

Figure 15.4 *The equilibrium between cis-stilbene and trans-stilbene. Note that both molecules have the same molecular formula ($C_{14}H_{12}$) and also the same type of bonds. However, in cis-stilbene, the benzene rings are on one side of the C=C bond and the H atoms are on the other side, whereas in trans-stilbene the benzene rings and the H atoms are across from the C=C bond. These compounds have different melting points and dipole moments.*

A positive ($+$) change represents an increase and a negative ($-$) change indicates a decrease in concentration at equilibrium. Next we set up the equilibrium constant expression

$$K_c = \frac{[\textit{trans}\text{-stilbene}]}{[\textit{cis}\text{-stilbene}]}$$

$$24.0 = \frac{x}{0.850 - x}$$

$$x = 0.816\,M$$

Having solved for x, we calculate the equilibrium concentrations of cis-stilbene and $trans$-stilbene as:

$$[\textit{cis}\text{-stilbene}] = (0.850 - 0.816)\,M = 0.034\,M$$
$$[\textit{trans}\text{-stilbene}] = 0.816\,M$$

We summarize our approach to solving equilibrium constant problems as:

1. Express the equilibrium concentrations of all species in terms of the initial concentrations and a single unknown quantity x, which represents the change in concentration.
2. Write the equilibrium constant expression in terms of the equilibrium concentrations. Knowing the value of the equilibrium constant, solve for x.
3. Having solved for x, calculate the equilibrium concentrations of all species.

Example 15.6   A mixture of 0.500 mol H_2 and 0.500 mol I_2 was placed in a 1.00-L stainless-steel flask at 430°C. The equilibrium constant K_c for the reaction $H_2(g) + I_2(g) \rightleftharpoons 2HI(g)$ is 54.3 at this temperature. Calculate concentrations of H_2, I_2, and HI at equilibrium.

Reasoning and Solution   We follow the given procedure to calculate the equilibrium concentrations.

Step 1. The stoichiometry of the reaction is 1 mol H_2 reacting with 1 mol I_2 to yield 2 mol HI. Let x be the depletion in concentration (mol/L) of H_2 and I_2 at equilibrium. It follows that the equilibrium concentration of HI must be $2x$. We summarize the changes in concentrations as:

	H_2	$+$	I_2	$\rightleftharpoons$	2HI
Initial (M):	0.500		0.500		0.000
Change (M):	$-x$		$-x$		$+2x$
Equilibrium (M):	$(0.500 - x)$		$(0.500 - x)$		$2x$

Step 2. The equilibrium constant is given by

$$K_c = \frac{[HI]^2}{[H_2][I_2]}$$

Substituting, we get

$$54.3 = \frac{(2x)^2}{(0.500 - x)(0.500 - x)}$$

Taking the square root of both sides, we get

$$7.37 = \frac{2x}{0.500 - x}$$

$$x = 0.393 \, M$$

Step 3. At equilibrium, the concentrations are

$$[H_2] = (0.500 - 0.393) \, M = 0.107 \, M$$

$$[I_2] = (0.500 - 0.393) \, M = 0.107 \, M$$

$$[HI] = 2 \times 0.393 \, M = 0.786 \, M$$

To check your answer, calculate the equilibrium constant using the equilibrium concentrations.

Practice Exercise Consider the reaction in Example 15.6. Starting with a concentration of 0.040 M for HI, calculate the concentrations of HI, H_2, and I_2 at equilibrium.

Similar problem: 15.29.

Example 15.7 For the same reaction and temperature as in Example 15.6, suppose that the initial concentrations of H_2, I_2, and HI are 0.00623 M, 0.00414 M, and 0.0224 M, respectively. Calculate the concentrations of these species at equilibrium.

Reasoning and Solution Following the procedure in Example 15.6, we proceed.

Step 1. Let x be the depletion in concentration (mol/L) for H_2 and I_2 at equilibrium. From the stoichiometry of the reaction it follows that the increase in concentration for HI must be $2x$. Next we write

	H_2	$+$	I_2	$\rightleftharpoons$	2HI
Initial (M):	0.00623		0.00414		0.0224
Change (M):	$-x$		$-x$		$+2x$
Equilibrium (M):	$(0.00623 - x)$		$(0.00414 - x)$		$(0.0224 + 2x)$

Use the initial concentrations to calculate Q_c, and then compare it with K_c to determine the direction of the net reaction toward equilibrium.

Step 2. The equilibrium constant is

$$K_c = \frac{[HI]^2}{[H_2][I_2]}$$

Substituting, we get

$$54.3 = \frac{(0.0224 + 2x)^2}{(0.00623 - x)(0.00414 - x)}$$

It is not possible to solve this equation by the square root shortcut, as the starting concentrations $[H_2]$ and $[I_2]$ are unequal. Instead, we must first carry out the multiplications

$$54.3(2.58 \times 10^{-5} - 0.0104x + x^2) = 5.02 \times 10^{-4} + 0.0896x + 4x^2$$

Collecting terms, we get

$$50.3x^2 - 0.654x + 8.98 \times 10^{-4} = 0$$

This is a quadratic equation of the form $ax^2 + bx + c = 0$. The solution for a quadratic equation (see Appendix 3) is

$$x = \frac{-b \pm \sqrt{b^2 - 4ac}}{2a}$$

Here we have $a = 50.3$, $b = -0.654$, and $c = 8.98 \times 10^{-4}$, so that

$$x = \frac{0.654 \pm \sqrt{(-0.654)^2 - 4(50.3)(8.98 \times 10^{-4})}}{2 \times 50.3}$$

$$x = 0.0114\ M \quad \text{or} \quad x = 0.00156\ M$$

The first solution is physically impossible because the amounts of H_2 and I_2 reacted would be more than those originally present. The second solution gives the correct answer. Note that in solving quadratic equations of this type, one answer is always physically impossible, so choosing a value for x is easy.

Step 3. At equilibrium, the concentrations are

$$[H_2] = (0.00623 - 0.00156)\ M = 0.00467\ M$$

$$[I_2] = (0.00414 - 0.00156)\ M = 0.00258\ M$$

$$[HI] = (0.0224 + 2 \times 0.00156)\ M = 0.0255\ M$$

Practice Exercise   At 1280°C the equilibrium constant (K_c) for the reaction

$$Br_2(g) \rightleftharpoons 2Br(g)$$

is 1.1×10^{-3}. If the initial concentrations are $[Br_2] = 6.3 \times 10^{-2}\ M$ and $[Br] = 1.2 \times 10^{-2}\ M$, calculate the concentrations of these species at equilibrium.

To check your answer, calculate the equilibrium constant using the equilibrium concentrations.

Similar problem: 15.76.

15.4 Factors That Affect Chemical Equilibrium

Chemical equilibrium represents a balance between forward and reverse reactions. In most cases, this balance is quite delicate. Changes in experimental conditions may disturb the balance and shift the equilibrium position so that more or less of the desired product is formed. When we say that an equilibrium position shifts to the right, for example, we mean that the net reaction is now from left to right. At our disposal are the following experimentally controllable variables: concentration, pressure, volume, and temperature. Here we will examine how each of these variables affects a reacting system at equilibrium. In addition, we will examine the effect of a catalyst on equilibrium.

Le Châtelier's Principle

Le Châtelier's principle

There is a general rule that helps us to predict the direction in which an equilibrium reaction will move when a change in concentration, pressure, volume, or temperature occurs. The rule, known as **Le Châtelier's principle** (after the French chemist Henri Le Châtelier), states that *if an external stress is applied to a system at equilibrium, the system adjusts in such a way that the stress is partially offset as it tries to re-establish equilibrium.* The word "stress" here means a change in concentration, pressure, volume, or temperature that removes a system from the equilibrium state. We will use Le Châtelier's principle to assess the effects of such changes.

Changes in Concentrations

Iron(III) thiocyanate [$Fe(SCN)_3$] dissolves readily in water to give a red solution. The red color is due to the presence of hydrated $FeSCN^{2+}$ ion. The equilibrium between undissociated $FeSCN^{2+}$ and the Fe^{3+} and SCN^- ions is given by

(a)   (b)   (c)   (d)

$$FeSCN^{2+}(aq) \rightleftharpoons Fe^{3+}(aq) + SCN^-(aq)$$
   red   pale yellow   colorless

Figure 15.5 *Effect of concentration change on the position of equilibrium. (a) An aqueous $Fe(SCN)_3$ solution. The color of the solution is due to both the red $FeSCN^{2+}$ and the yellow Fe^{3+} species. (b) After the addition of some NaSCN to the solution in (a), the equilibrium shifts to the left. (c) After the addition of some $Fe(NO_3)_3$ to the solution in (a), the equilibrium shifts to the left. (d) After the addition of some $H_2C_2O_4$ to the solution in (a), the equilibrium shifts to the right. The yellow color is due to the $Fe(C_2O_4)_3^{3-}$ ions.*

What happens if we add some sodium thiocyanate (NaSCN) to this solution? In this case, the stress applied to the equilibrium system is an increase in the concentration of SCN^- (from the dissociation of NaSCN). To offset this stress, some Fe^{3+} ions react with the added SCN^- ions, and the equilibrium shifts from right to left:

$$FeSCN^{2+}(aq) \longleftarrow Fe^{3+}(aq) + SCN^-(aq)$$

Consequently, the red color of the solution deepens (Figure 15.5). Similarly, if we added iron(III) nitrate [$Fe(NO_3)_3$] to the original solution, the red color would also deepen because the additional Fe^{3+} ions [from $Fe(NO_3)_3$] would shift the equilibrium from right to left. Both Na^+ and NO_3^- are colorless spectator ions.

Now suppose we add some oxalic acid ($H_2C_2O_4$) to the original solution. Oxalic acid ionizes in water to form the oxalate ion, $C_2O_4^{2-}$, which binds strongly to the Fe^{3+} ions. The formation of the stable yellow ion $Fe(C_2O_4)_3^{3-}$ removes free Fe^{3+} ions from solution. Consequently, more $FeSCN^{2+}$ units dissociate and the equilibrium shifts from left to right:

$$FeSCN^{2+}(aq) \longrightarrow Fe^{3+}(aq) + SCN^-(aq)$$

The red solution will turn yellow because of the formation of $Fe(C_2O_4)_3^{3-}$ ions.

This experiment demonstrates that at equilibrium all reactants and products are present in the reacting system. Second, increasing the concentrations of the products (Fe^{3+} or SCN^-) shifts the equilibrium to the left, and decreasing the concentration of the product Fe^{3+} shifts the equilibrium to the right. These results are just as predicted by Le Châtelier's principle.

Oxalic acid is sometimes used to remove bathtub rings that consist of rust, or Fe_2O_3.

Le Châtelier's principle simply summarizes the observed behavior of equilibrium systems; therefore, it is incorrect to say that a given equilibrium shift occurs "because of" Le Châtelier's principle.

Example 15.8 At 720°C, the equilibrium constant K_c for the reaction

$$N_2(g) + 3H_2(g) \rightleftharpoons 2NH_3(g)$$

is 2.37×10^{-3}. In a certain experiment, the equilibrium concentrations are [N_2] = 0.683 M, [H_2] = 8.80 M, and [NH_3] = 1.05 M. Suppose some of the NH_3 is added to the mixture so that its concentration is increased to 3.65 M. (a) Use Le Châtelier's principle to predict the shift in direction of the net reaction to reach a new equilibrium. (b) Confirm your prediction by calculating the reaction quotient Q_c and comparing its value with K_c.

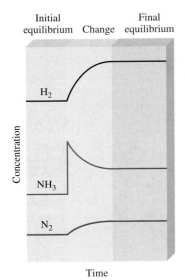

Initial
equilibrium   Change   Final
equilibrium

Figure 15.6 *Changes in concentration of H_2, N_2, and NH_3 after the addition of NH_3 to the equilibrium mixture.*

Similar problem: 15.34.

Reasoning and Solution (a) The stress applied to the system is the addition of NH_3. To offset this stress, some NH_3 reacts to produce N_2 and H_2 until a new equilibrium is established. The net reaction therefore shifts from right to left; that is,

$$N_2(g) + 3H_2(g) \longleftarrow 2NH_3(g)$$

(b) At the instant when some of the NH_3 is added, the system is no longer at equilibrium. The reaction quotient is given by

$$Q_c = \frac{[NH_3]_0^2}{[N_2]_0[H_2]_0^3}$$

$$= \frac{(3.65)^2}{(0.683)(8.80)^3}$$

$$= 2.86 \times 10^{-2}$$

Because this value is greater than 2.37×10^{-3}, the net reaction shifts from right to left until Q_c equals K_c.

Figure 15.6 shows qualitatively the changes in concentrations of the reacting species.

Practice Exercise At 430°C, the equilibrium constant (K_P) for the reaction

$$2NO(g) + O_2(g) \rightleftharpoons 2NO_2(g)$$

is 1.5×10^5. In one experiment, the initial pressures for NO, O_2, and NO_2 are 2.1×10^{-3} atm, 1.1×10^{-2} atm, and 0.14 atm, respectively. Calculate Q_P and predict the direction that the net reaction will shift to reach equilibrium.

Changes in Pressure and Volume

Changes in pressure ordinarily do not affect the concentrations of reacting species in condensed phases (say, in an aqueous solution) because liquids and solids are virtually incompressible. On the other hand, concentrations of gases are greatly affected by changes in pressure. Let us look again at Equation (5.8):

$$PV = nRT$$

$$P = \left(\frac{n}{V}\right)RT$$

Thus P and V are related to each other inversely: The greater the pressure, the smaller the volume, and vice versa. Note, too, that the term (n/V) is the concentration of the gas in moles per liter, and it varies directly with pressure.

Suppose that the equilibrium system

$$N_2O_4(g) \rightleftharpoons 2NO_2(g)$$

is in a cylinder fitted with a movable piston. What happens if we increase the pressure on the gases by pushing down on the piston at constant temperature? Because the volume decreases, the concentration (n/V) of both NO_2 and N_2O_4 increases. Because the concentration of NO_2 is squared, the increase in pressure increases the numerator more than the denominator. The system is no longer at equilibrium, so we write

$$Q_c = \frac{[NO_2]_0^2}{[N_2O_4]_0}$$

Thus $Q_c > K_c$, and the net reaction will shift to the left until $Q_c = K_c$. Conversely, a decrease in pressure (increase in volume) would result in $Q_c < K_c$; the net reaction would shift to the right until $Q_c = K_c$.

In general, an increase in pressure (decrease in volume) favors the net reaction that decreases the total number of moles of gases (the reverse reaction, in the preceding case), and a decrease in pressure (increase in volume) favors the net reaction that increases the total number of moles of gases (here, the forward reaction). For reactions in which there is no change in the number of moles of gases, a pressure (or volume) change has no effect on the position of equilibrium.

It is possible to change the pressure of a system without changing its volume. Suppose the $NO_2-N_2O_4$ system is contained in a stainless-steel vessel whose volume is constant. We can increase the total pressure in the vessel by adding an inert gas (helium, for example) to the equilibrium system. Adding helium to the equilibrium mixture at constant volume increases the total gas pressure and decreases the mole fractions of both NO_2 and N_2O_4; but the partial pressure of each gas, given by the product of its mole fraction and total pressure (see Section 5.5), does not change. Thus the presence of an inert gas in such a case does not affect the equilibrium.

The shift in equilibrium can also be predicted using Le Châtelier's principle.

Example 15.9   Consider these equilibrium systems:

(a) $2PbS(s) + 3O_2(g) \rightleftharpoons 2PbO(s) + 2SO_2(g)$

(b) $PCl_5(g) \rightleftharpoons PCl_3(g) + Cl_2(g)$

(c) $H_2(g) + CO_2(g) \rightleftharpoons H_2O(g) + CO(g)$

Predict the direction of the net reaction in each case as a result of increasing the pressure (decreasing the volume) on the system at constant temperature.

Reasoning and Solution   A change in pressure can affect only gaseous components in the reacting mixture at equilibrium.

(a) Consider only the gaseous molecules. In the balanced equation there are 3 moles of gaseous reactants and 2 moles of gaseous products. Therefore the net reaction will shift toward the products (to the right) to decrease the number of moles of gases when the pressure is increased.

(b) The number of moles of products is 2 and that of reactants is 1; therefore the net reaction will shift to the left, toward the reactant.

(c) The number of moles of products is equal to the number of moles of reactants, so a change in pressure has no effect on the equilibrium position.

Comment   In each case, the prediction is consistent with Le Châtelier's principle.

Similar problem: 15.44.

Practice Exercise   Consider the equilibrium reaction involving nitrosyl chloride, nitric oxide, and molecular chlorine

$$2NOCl(g) \rightleftharpoons 2NO(g) + Cl_2(g)$$

Predict the direction of the net reaction as a result of decreasing the pressure (increasing the volume) on the system at constant temperature.

Figure 15.7 *(a) Two bulbs containing a mixture of NO₂ and N₂O₄ gases at equilibrium. (b) When one bulb is immersed in ice water (left), its color becomes lighter, indicating the formation of colorless N₂O₄ gas. When the other bulb is immersed in hot water (right), its color darkens, indicating an increase in NO₂.*

(a)   (b)

Changes in Temperature

A change in concentration, pressure, or volume may alter the equilibrium position, but it does not change the value of the equilibrium constant. Only a change in temperature can alter the equilibrium constant.

The formation of NO_2 from N_2O_4 is an endothermic process:

$$N_2O_4(g) \longrightarrow 2NO_2(g) \qquad\qquad \Delta H^\circ = 58.0 \text{ kJ}$$

and the reverse reaction is exothermic:

$$2NO_2(g) \longrightarrow N_2O_4(g) \qquad\qquad \Delta H^\circ = -58.0 \text{ kJ}$$

At equilibrium the heat effect is zero because there is no net reaction. What happens if the following equilibrium system

$$N_2O_4(g) \rightleftharpoons 2NO_2(g)$$

is heated at constant volume? Because endothermic processes absorb heat from the surroundings, heating favors the dissociation of N_2O_4 into NO_2 molecules. Consequently, the equilibrium constant, given by

$$K_c = \frac{[NO_2]^2}{[N_2O_4]}$$

increases with temperature (Figure 15.7).

As another example, consider the equilibrium between the following ions:

$$\underset{\text{blue}}{CoCl_4^{2-}} + 6H_2O \rightleftharpoons \underset{\text{pink}}{Co(H_2O)_6^{2+}} + 4Cl^-$$

The formation of $CoCl_4^{2-}$ is endothermic. On heating, the equilibrium shifts to the left and the solution turns blue. Cooling favors the exothermic reaction [the formation of $Co(H_2O)_6^{2+}$] and the solution turns pink (Figure 15.8).

In summary, *a temperature increase favors an endothermic reaction and a temperature decrease favors an exothermic reaction.*

The Effect of a Catalyst

A catalyst increases the rate at which a reaction occurs (see Chapter 14). For a reversible reaction, a catalyst affects the rate in the forward and reverse direction to the same

Figure 15.8 *(Left) Heating favors the formation of the blue $CoCl_4^{2-}$ ion. (Right) Cooling favors the formation of the pink $Co(H_2O)_6^{2+}$ ion.*

extent. Therefore, the presence of a catalyst does not alter the equilibrium constant, nor does it shift the position of an equilibrium system. Adding a catalyst to a reaction mixture that is not at equilibrium will speed up both the forward and reverse rates to achieve an equilibrium mixture much faster. The same equilibrium mixture could be obtained without the catalyst, but we might have to wait much longer for this to happen.

Le Châtelier's principle finds many applications in industrial processes. Consider the Haber process of ammonia synthesis on page 470:

$$N_2(g) + 3H_2(g) \rightleftharpoons 2NH_3(g) \qquad \Delta H° = -92.6 \text{ kJ}$$

Because the forward reaction results in a decrease of the number of moles of gases, the reaction is carried out at very high pressures—between 500 atm and 1000 atm—to maximize the yield. The exothermic nature of ammonia formation suggests that a low-temperature operation would be desirable. However, the process is actually run at about 500°C, with catalysts, to increase the rate of the reaction even though the equilibrium constant is smaller at the high temperature. In practice, the reaction is not allowed to reach equilibrium. Instead, ammonia is constantly removed from the reacting mixture so as to shift the net reaction from left to right.

Summary of Factors That May Affect the Equilibrium Position

We have considered four ways to affect a reacting system at equilibrium. It is important to remember that, of the four, *only a change in temperature changes the value of the equilibrium constant.* Changes in concentration, pressure, and volume can alter the equilibrium concentrations of the reacting mixture, but they cannot change the equilibrium constant as long as the temperature does not change. A catalyst can help establish equilibrium faster, but it has no effect on the equilibrium constant or on the equilibrium concentrations of the reacting species.

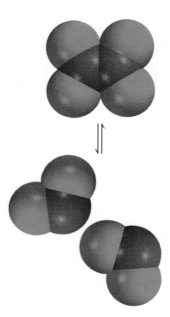

Example 15.10   Consider the following equilibrium process between dinitrogen tetrafluoride (N_2F_4) and nitrogen difluoride (NF_2):

$$N_2F_4(g) \rightleftharpoons 2NF_2(g) \qquad \qquad \Delta H° = 38.5 \text{ kJ}$$

Predict the changes in the equilibrium if (a) the reacting mixture is heated at constant volume; (b) NF_2 gas is removed from the reacting mixture at constant temperature and volume; (c) the pressure on the reacting mixture is decreased at constant temperature; and (d) an inert gas, such as helium, is added to the reacting mixture at constant volume and temperature.

Reasoning and Solution   (a) Because the forward reaction is endothermic, an increase in temperature favors the formation of NF_2. The equilibrium constant

$$K_c = \frac{[NF_2]^2}{[N_2F_4]}$$

will therefore increase with increasing temperature.

(b) The stress here is the removal of NF_2 gas. To offset it, more N_2F_4 will decompose to form NF_2. The equilibrium constant K_c remains unchanged, however.

(c) A decrease in pressure (which is accompanied by an increase in gas volume) favors the formation of more gas molecules, that is, the forward reaction. Thus, more NF_2 gas will be formed. The equilibrium constant will remain unchanged.

(d) Adding helium to the equilibrium mixture at constant volume will not shift the equilibrium position.

Similar problems: 15.45, 15.46.

Practice Exercise   Consider the equilibrium between molecular oxygen and ozone

$$3O_2(g) \rightleftharpoons 2O_3(g) \qquad\qquad \Delta H° = 284 \text{ kJ}$$

What would be the effect of (a) increasing the pressure on the system by decreasing the volume, (b) increasing pressure by adding O_2 to the system, (c) decreasing the temperature, and (d) adding a catalyst?

Summary

Chemists are concerned with dynamic equilibria between phases (physical equilibria) and between reacting substances (chemical equilibria). For the general chemical reaction

$$aA + bB \rightleftharpoons cC + dD$$

the concentrations of reactants and products at equilibrium (in moles per liter) are related by the equilibrium constant expression

$$K_c = \frac{[C]^c[D]^d}{[A]^a[B]^b}$$

The equilibrium constant may also be expressed in terms of the equilibrium partial pressures (in atmospheres) of gases, as K_P.

An equilibrium in which all reactants and products are in the same phase is a homogeneous equilibrium. If the reactants and products are not all in the same phase, the equilibrium is called heterogeneous. The concentrations of pure solids, pure liquids, and solvents are constant and do not appear in the equilibrium constant expression of a reaction. The value of K depends on how the chemical equation is balanced, and the equilibrium constant for the reverse of a particular reaction is the reciprocal of the equilibrium constant of that reaction.

The reaction quotient Q has the same form as the equilibrium constant expression, but it applies to a reaction that may not be at equilibrium. If $Q > K$, the reaction will proceed from right to left to achieve equilibrium. If $Q < K$, the reaction will proceed from left to right to achieve equilibrium.

Le Châtelier's principle states that if an external stress is applied to a system at chemical equilibrium, the system will adjust to partially offset the stress. Only a change in temperature changes the value of the equilibrium constant for a particular reaction. Changes in concentration, pressure, or volume may change the equilibrium concentrations of reactants and products. The addition of a catalyst hastens the attainment of equilibrium but does not affect the equilibrium concentrations of reactants and products.

Key Words

Chemical equilibrium p. 472	Heterogeneous equilibrium p. 479	Le Châtelier's principle p. 486	Reaction quotient (Q_c) p. 482
Equilibrium constant p. 474	Homogeneous equilibrium p. 475	Physical equilibrium p. 472	

Questions and Problems

Concept of Equilibrium

Review Questions

15.1 Define equilibrium. Give two examples of a dynamic equilibrium.

15.2 Explain the difference between physical equilibrium and chemical equilibrium. Give two examples of each.

15.3 Briefly describe the importance of equilibrium in the study of chemical reactions.

15.4 Consider the equilibrium system $3A \rightleftharpoons B$. Sketch the change in concentrations of A and B with time for these situations: (a) Initially only A is present; (b) initially only B is present; (c) initially both A and B are present (with A in higher concentration). In each case, assume that the concentration of B is higher than that of A at equilibrium.

Equilibrium Constant Expressions

Review Questions

15.5 Define homogeneous equilibrium and heterogeneous equilibrium. Give two examples of each.

15.6 What do the symbols K_c and K_P represent?

Problems

15.7 Write the expressions for the equilibrium constants K_P of these thermal decompositions:

(a) $2NaHCO_3(s) \rightleftharpoons Na_2CO_3(s) + CO_2(g) + H_2O(g)$
(b) $2CaSO_4(s) \rightleftharpoons 2CaO(s) + 2SO_2(g) + O_2(g)$

15.8 Write equilibrium constant expressions for K_c and for K_P, if applicable, for these processes:

(a) $2CO_2(g) \rightleftharpoons 2CO(g) + O_2(g)$
(b) $3O_2(g) \rightleftharpoons 2O_3(g)$

(c) $CO(g) + Cl_2(g) \rightleftharpoons COCl_2(g)$
(d) $H_2O(g) + C(s) \rightleftharpoons CO(g) + H_2(g)$
(e) $HCOOH(aq) \rightleftharpoons H^+(aq) + HCOO^-(aq)$
(f) $2HgO(s) \rightleftharpoons 2Hg(l) + O_2(g)$

15.9 Write the equilibrium constant expressions for K_c and K_P, if applicable, for these reactions:

(a) $2NO_2(g) + 7H_2(g) \rightleftharpoons 2NH_3(g) + 4H_2O(l)$
(b) $2ZnS(s) + 3O_2(g) \rightleftharpoons 2ZnO(s) + 2SO_2(g)$
(c) $C(s) + CO_2(g) \rightleftharpoons 2CO(g)$
(d) $C_6H_5COOH(aq) \rightleftharpoons C_6H_5COO^-(aq) + H^+(aq)$

Calculating Equilibrium Constants

Review Question

15.10 Write the equation relating K_c and K_P and define all the terms.

Problems

15.11 The equilibrium constant (K_c) for the reaction

$$2HCl(g) \rightleftharpoons H_2(g) + Cl_2(g)$$

is 4.17×10^{-34} at 25°C. What is the equilibrium constant for the reaction

$$H_2(g) + Cl_2(g) \rightleftharpoons 2HCl(g)$$

at the same temperature?

15.12 Consider the following equilibrium process at 700°C:

$$2H_2(g) + S_2(g) \rightleftharpoons 2H_2S(g)$$

Analysis shows that there are 2.50 moles of H_2, 1.35×10^{-5} mole of S_2, and 8.70 moles of H_2S present in a 12.0-L flask at equilibrium. Calculate the equilibrium constant K_c for the reaction.

15.13 What is the K_P at 1273°C for the reaction

$$2CO(g) + O_2(g) \rightleftharpoons 2CO_2(g)$$

if K_c is 2.24×10^{22} at the same temperature?

15.14 The equilibrium constant K_P for the reaction

$$2SO_3(g) \rightleftharpoons 2SO_2(g) + O_2(g)$$

if 5.0×10^{-4} at 302°C. What is K_c for this reaction?

15.15 Consider this reaction:

$$N_2(g) + O_2(g) \rightleftharpoons 2NO(g)$$

If the equilibrium partial pressures of N_2, O_2, and NO are 0.15 atm, 0.33 atm, and 0.050 atm, respectively, at 2200°C, what is K_P?

15.16 A reaction vessel contains NH_3, N_2, and H_2 at equilibrium at a certain temperature. The equilibrium concentrations are $[NH_3] = 0.25\ M$, $[N_2] = 0.11\ M$, and $[H_2] = 1.91\ M$. Calculate the equilibrium constant K_c for the synthesis of ammonia if the reaction is represented as

(a) $N_2(g) + 3H_2(g) \rightleftharpoons 2NH_3(g)$
(b) $\frac{1}{2}N_2(g) + \frac{3}{2}H_2(g) \rightleftharpoons NH_3(g)$

15.17 The equilibrium constant K_c for the reaction

$$I_2(g) \rightleftharpoons 2I(g)$$

is 3.8×10^{-5} at 727°C. Calculate K_c and K_P for the equilibrium

$$2I(g) \rightleftharpoons I_2(g)$$

at the same temperature.

15.18 The pressure of the reacting mixture

$$CaCO_3(s) \rightleftharpoons CaO(s) + CO_2(g)$$

at equilibrium is 0.105 atm at 350°C. Calculate K_P and K_c for this reaction.

15.19 The equilibrium constant K_P for the reaction

$$PCl_5(g) \rightleftharpoons PCl_3(g) + Cl_2(g)$$

is 1.05 at 250°C. The reaction starts with a mixture of PCl_5, PCl_3, and Cl_2 at pressures of 0.177 atm, 0.223 atm, and 0.111 atm, respectively, at 250°C. When the mixture comes to equilibrium at that temperature, which pressures will have decreased and which will have increased? Explain why.

15.20 Ammonium carbamate, $NH_4CO_2NH_2$, decomposes as:

$$NH_4CO_2NH_2(s) \rightleftharpoons 2NH_3(g) + CO_2(g)$$

Starting with only the solid, it is found that at 40°C the total gas pressure (NH_3 and CO_2) is 0.363 atm. Calculate the equilibrium constant K_P.

15.21 Consider the following reaction at 1600°C:

$$Br_2(g) \rightleftharpoons 2Br(g)$$

When 1.05 moles of Br_2 are put in a 0.980-L flask, 1.20 percent of the Br_2 undergoes dissociation. Calculate the equilibrium constant K_c for the reaction.

15.22 Pure phosgene gas ($COCl_2$), 3.00×10^{-2} mol, was placed in a 1.50-L container. It was heated at 800 K, and at equilibrium the pressure of CO was found to be 0.497 atm. Calculate the equilibrium constant K_P for the reaction

$$CO(g) + Cl_2(g) \rightleftharpoons COCl_2(g)$$

15.23 Consider the equilibrium

$$2NOBr(g) \rightleftharpoons 2NO(g) + Br_2(g)$$

If nitrosyl bromide, NOBr, is 34 percent dissociated at 25°C and the total pressure is 0.25 atm, calculate K_P and K_c for the dissociation at this temperature.

15.24 A 2.50-mol quantity of NOCl was initially placed in a 1.50-L reaction chamber at 400°C. After equilibrium was established, it was found that 28.0 percent of the NOCl had dissociated:

$$2NOCl(g) \rightleftharpoons 2NO(g) + Cl_2(g)$$

Calculate the equilibrium constant K_c for the reaction.

Calculating Equilibrium Concentrations

Review Questions

15.25 Define reaction quotient. How does it differ from equilibrium constant?

15.26 Outline the steps for calculating the concentrations of reacting species in an equilibrium reaction.

Problems

15.27 The equilibrium constant K_P for the reaction

$$2SO_2(g) + O_2(g) \rightleftharpoons 2SO_3(g)$$

is 5.60×10^4 at 350°C. SO_2 and O_2 are mixed initially at 0.350 atm and 0.762 atm, respectively, at 350°C. When the mixture equilibrates, is the total pressure less than or greater than the sum of the initial pressures, 1.112 atm?

15.28 For the synthesis of ammonia

$$N_2(g) + 3H_2(g) \rightleftharpoons 2NH_3(g)$$

the equilibrium constant K_c at 375°C is 1.2. Starting with $[H_2]_0 = 0.76\ M$, $[N_2]_0 = 0.60\ M$, and $[NH_3]_0 = 0.48\ M$, when this mixture comes to equilibrium, which gases will have increased in concentration and which will have decreased in concentration?

15.29 For the reaction

$$H_2(g) + CO_2(g) \rightleftharpoons H_2O(g) + CO(g)$$

at 700°C, $K_c = 0.534$. Calculate the number of moles of H_2 formed at equilibrium if a mixture of 0.300 mole of CO and 0.300 mole of H_2O is heated to 700°C in a 10.0-L container.

15.30 A sample of pure NO_2 gas heated to 1000 K decomposes:

$$2NO_2(g) \rightleftharpoons 2NO(g) + O_2(g)$$

The equilibrium constant K_P is 158. Analysis shows that the partial pressure of O_2 is 0.25 atm at equilibrium. Calculate the pressure of NO and NO_2 in the mixture.

15.31 The equilibrium constant K_c for the reaction

$$H_2(g) + Br_2(g) \rightleftharpoons 2HBr(g)$$

is 2.18×10^6 at 730°C. Starting with 3.20 moles HBr in a 12.0-L reaction vessel, calculate the concentrations of H_2, Br_2, and HBr at equilibrium.

15.32 The dissociation of molecular iodine into iodine atoms is represented as

$$I_2(g) \rightleftharpoons 2I(g)$$

At 1000 K, the equilibrium constant K_c for the reaction is 3.80×10^{-5}. Suppose you start with 0.0456 mole of I_2 in a 2.30-L flask at 1000 K. What are the concentrations of the gases at equilibrium?

15.33 The equilibrium constant K_c for the decomposition of phosgene, $COCl_2$, is 4.63×10^{-3} at 527°C:

$$COCl_2(g) \rightleftharpoons CO(g) + Cl_2(g)$$

Calculate the equilibrium partial pressure of all the components, starting with pure phosgene at 0.760 atm.

15.34 Consider this equilibrium process at 686°C:

$$CO_2(g) + H_2(g) \rightleftharpoons CO(g) + H_2O(g)$$

The equilibrium concentrations of the reacting species are $[CO] = 0.050\ M$, $[H_2] = 0.045\ M$, $[CO_2] = 0.086\ M$, and $[H_2O] = 0.040\ M$. (a) Calculate K_c for the reaction at 686°C. (b) If the concentration of CO_2 were raised to 0.50 mol/L by the addition of CO_2, what would be the concentrations of all the gases when equilibrium is reestablished?

15.35 Consider the heterogeneous equilibrium process:

$$C(s) + CO_2(g) \rightleftharpoons 2CO(g)$$

At 700°C, the total pressure of the system is found to be 4.50 atm. If the equilibrium constant K_P is 1.52, calculate the equilibrium partial pressures of CO_2 and CO.

15.36 The equilibrium constant K_c for the reaction

$$H_2(g) + CO_2(g) \rightleftharpoons H_2O(g) + CO(g)$$

is 4.2 at 1650°C. Initially 0.80 mol H_2 and 0.80 mol CO_2 are injected into a 5.0-L flask. Calculate the concentration of each species at equilibrium.

Le Châtelier's Principle

Review Questions

15.37 Explain Le Châtelier's principle. How can this principle help us maximize the yields of reactions?

15.38 Use Le Châtelier's principle to explain why the equilibrium vapor pressure of a liquid increases with increasing temperature.

15.39 List four factors that can shift the position of an equilibrium. Which one can alter the value of the equilibrium constant?

15.40 What is meant by "the position of an equilibrium"? Does the addition of a catalyst have any effects on the position of an equilibrium?

Problems

15.41 Consider this equilibrium system:

$$SO_2(g) + Cl_2(g) \rightleftharpoons SO_2Cl_2(g)$$

Predict how the equilibrium position would change if (a) Cl_2 gas were added to the system, (b) SO_2Cl_2 were removed from the system, (c) SO_2 were removed from the system. The temperature remains constant.

15.42 Heating solid sodium bicarbonate in a closed vessel established this equilibrium:

$$2NaHCO_3(s) \rightleftharpoons Na_2CO_3(s) + H_2O(g) + CO_2(g)$$

What would happen to the equilibrium position if (a) some of the CO_2 were removed from the system, (b) some solid Na_2CO_3 were added to the system, (c) some of the solid $NaHCO_3$ were removed from the system? The temperature remains constant.

15.43 Consider these equilibrium systems:

(a) $A \rightleftharpoons 2B$	$\Delta H° = 20.0$ kJ
(b) $A + B \rightleftharpoons C$	$\Delta H° = -5.4$ kJ
(c) $A \rightleftharpoons B$	$\Delta H° = 0.0$ kJ

Predict the change in the equilibrium constant K_c that would occur in each case if the temperature of the reacting system were raised.

15.44 What effect does an increase in pressure have on each of these systems at equilibrium?

(a) $A(s) \rightleftharpoons 2B(s)$
(b) $2A(l) \rightleftharpoons B(l)$
(c) $A(s) \rightleftharpoons B(g)$
(d) $A(g) \rightleftharpoons B(g)$
(e) $A(g) \rightleftharpoons 2B(g)$

The temperature is kept constant. In each case, the reacting mixture is in a cylinder fitted with a movable piston.

15.45 Consider the equilibrium

$$2I(g) \rightleftharpoons I_2(g)$$

What would be the effect on the position of equilibrium of (a) increasing the total pressure on the system by decreasing its volume, (b) adding I_2 to the reaction mixture, (c) decreasing the temperature?

15.46 Consider this equilibrium process:

$$PCl_5(g) \rightleftharpoons PCl_3(g) + Cl_2(g) \qquad \Delta H^\circ = 92.5 \text{ kJ}$$

Predict the direction of the shift in equilibrium when (a) the temperature is raised, (b) more chlorine gas is added to the reaction mixture, (c) some PCl_3 is removed from the mixture, (d) the pressure on the gases is increased, (e) a catalyst is added to the reaction mixture.

15.47 Consider the reaction

$$2SO_2(g) + O_2(g) \rightleftharpoons 2SO_3(g) \qquad \Delta H^\circ = -198.2 \text{ kJ}$$

Comment on the changes in the concentrations of SO_2, O_2, and SO_3 at equilibrium if we were to (a) increase the temperature, (b) increase the pressure, (c) increase SO_2, (d) add a catalyst, (e) add helium at constant volume.

15.48 In the uncatalyzed reaction

$$N_2O_4(g) \rightleftharpoons 2NO_2(g)$$

at 100°C the pressures of the gases at equilibrium are $P_{N_2O_4} = 0.377$ atm and $P_{NO_2} = 1.56$ atm. What would happen to these pressures if a catalyst were present?

15.49 Consider the gas-phase reaction

$$2CO(g) + O_2(g) \rightleftharpoons 2CO_2(g)$$

Predict the shift in the equilibrium position when helium gas is added to the equilibrium mixture (a) at constant pressure and (b) at constant volume.

15.50 Consider this reaction at equilibrium in a closed container:

$$CaCO_3(s) \rightleftharpoons CaO(s) + CO_2(g)$$

What would happen if (a) the volume is increased, (b) some CaO is added to the mixture, (c) some $CaCO_3$ is removed, (d) some CO_2 is added to the mixture, (e) a few drops of an NaOH solution are added to the mixture, (f) a few drops of an HCl solution are added to the mixture (ignore the reaction between CO_2 and water), (g) the temperature is increased?

Additional Problems

15.51 Consider the statement: The equilibrium constant of a reacting mixture of solid NH_4Cl and gaseous NH_3 and HCl is 0.316. List three important pieces of information that are missing from this statement.

15.52 Pure NOCl gas was heated at 240°C in a 1.00-L container. At equilibrium the total pressure was 1.00 atm and the NOCl pressure was 0.64 atm.

$$2NOCl(g) \rightleftharpoons 2NO(g) + Cl_2(g)$$

(a) Calculate the partial pressures of NO and Cl_2 in the system. (b) Calculate the equilibrium constant K_P.

15.53 Consider this reaction:

$$N_2(g) + O_2(g) \rightleftharpoons 2NO(g)$$

The equilibrium constant K_P for the reaction is 1.0×10^{-15} at 25°C and 0.050 at 2200°C. Is the formation of nitric oxide endothermic or exothermic? Explain your answer.

15.54 Baking soda (sodium bicarbonate) undergoes thermal decomposition as:

$$2NaHCO_3(s) \rightleftharpoons Na_2CO_3(s) + CO_2(g) + H_2O(g)$$

Would we obtain more CO_2 and H_2O by adding extra baking soda to the reaction mixture in (a) a closed vessel or (b) an open vessel?

15.55 Consider the following reaction at equilibrium:

$$A(g) \rightleftharpoons 2B(g)$$

From the following data, calculate the equilibrium constant (both K_P and K_c) at each temperature. Is the reaction endothermic or exothermic?

Temperature (°C)	[A]	[B]
200	0.0125	0.843
300	0.171	0.764
400	0.250	0.724

15.56 The equilibrium constant K_P for the reaction

$$2H_2O(g) \rightleftharpoons 2H_2(g) + O_2(g)$$

is found to be 2×10^{-42} at 25°C. (a) What is K_c for the reaction at the same temperature? (b) The very small value of K_P (and K_c) indicates that the reaction overwhelmingly favors the formation of water molecules. Explain why, despite this fact, a mixture of hydrogen and oxygen gases can be kept at room temperature without any change.

15.57 Consider the following reacting system:

$$2NO(g) + Cl_2(g) \rightleftharpoons 2NOCl(g)$$

What combination of temperature and pressure would maximize the yield of NOCl? [*Hint:* $\Delta H^\circ_f(NOCl) = 51.7$ kJ/mol. You will also need to consult Appendix 2.]

15.58 At a certain temperature and a total pressure of 1.2 atm, the partial pressures of an equilibrium mixture

$$2A(g) \rightleftharpoons B(g)$$

are $P_A = 0.60$ atm and $P_B = 0.60$ atm. (a) Calculate the K_P for the reaction at this temperature. (b) If the total pressure were increased to 1.5 atm, what would be the partial pressures of A and B at equilibrium?

15.59 The decomposition of ammonium hydrogen sulfide

$$NH_4HS(s) \rightleftharpoons NH_3(g) + H_2S(g)$$

is an endothermic process. A 6.1589-g sample of the solid is placed in an evacuated 4.000-L vessel at exactly 24°C. After equilibrium has been established, the total pressure inside is 0.709 atm. Some solid NH_4HS remains in the vessel. (a) What is the K_P for the reaction? (b) What percentage of the solid has decomposed? (c) If the volume of the vessel were doubled at constant temperature, what would happen to the amount of solid in the vessel?

15.60 Consider the reaction

$$2NO(g) + O_2(g) \rightleftharpoons 2NO_2(g)$$

At 430°C, an equilibrium mixture consists of 0.020 mole of O_2, 0.040 mole of NO, and 0.96 mole of NO_2. Calculate K_P for the reaction, given that the total pressure is 0.20 atm.

15.61 When heated, ammonium carbamate decomposes as:

$$NH_4CO_2NH_2(s) \rightleftharpoons 2NH_3(g) + CO_2(g)$$

At a certain temperature the equilibrium pressure of the system is 0.318 atm. Calculate K_P for the reaction.

15.62 A mixture of 0.47 mole of H_2 and 3.59 moles of HCl is heated to 2800°C. Calculate the equilibrium partial pressures of H_2, Cl_2, and HCl if the total pressure is 2.00 atm. The K_P for the reaction $H_2(g) + Cl_2(g) \rightleftharpoons 2HCl(g)$ is 193 at 2800°C.

15.63 Consider the reaction in a closed container:

$$N_2O_4(g) \rightleftharpoons 2NO_2(g)$$

Initially, 1 mole of N_2O_4 is present. At equilibrium, α mole of N_2O_4 has dissociated to form NO_2. (a) Derive an expression for K_P in terms of α and P, the total pressure. (b) How does the expression in (a) help you predict the shift in equilibrium caused by an increase in P? Does your prediction agree with Le Châtelier's principle?

15.64 One mole of N_2 and 3 moles of H_2 are placed in a flask at 397°C. Calculate the total pressure of the system at equilibrium if the mole fraction of NH_3 is found to be 0.21. The K_P for the reaction is 4.31×10^{-4}.

15.65 At 1130°C the equilibrium constant (K_c) for the reaction

$$2H_2S(g) \rightleftharpoons 2H_2(g) + S_2(g)$$

is 2.25×10^{-4}. If $[H_2S] = 4.84 \times 10^{-3}\ M$ and $[H_2] = 1.50 \times 10^{-3}\ M$, calculate $[S_2]$.

15.66 A quantity of 6.75 g of SO_2Cl_2 was placed in a 2.00-L flask. At 648 K, there is 0.0345 mole of SO_2 present. Calculate K_c for the reaction

$$SO_2Cl_2(g) \rightleftharpoons SO_2(g) + Cl_2(g)$$

15.67 The formation of SO_3 from SO_2 and O_2 is an intermediate step in the manufacture of sulfuric acid, and it is also responsible for the acid rain phenomenon. The equilibrium constant (K_P) for the reaction

$$2SO_2(g) + O_2(g) \rightleftharpoons 2SO_3(g)$$

is 0.13 at 830°C. In one experiment 2.00 moles of SO_2 and 2.00 moles of O_2 were initially present in a flask. What must the total pressure be at equilibrium to have an 80.0 percent yield of SO_3?

15.68 Consider the dissociation of iodine:

$$I_2(g) \rightleftharpoons 2I(g)$$

A 1.00-g sample of I_2 is heated at 1200°C in a 500-mL flask. At equilibrium the total pressure is 1.51 atm. Calculate K_P for the reaction. [*Hint:* Use the result in problem 15.63(a). The degree of dissociation α can be obtained by first calculating the ratio of observed pressure over calculated pressure, assuming no dissociation.]

15.69 Eggshells are composed mostly of calcium carbonate ($CaCO_3$) formed by the reaction

$$Ca^{2+}(aq) + CO_3^{2-}(aq) \rightleftharpoons CaCO_3(s)$$

The carbonate ions are supplied by carbon dioxide produced as a result of metabolism. Explain why eggshells are thinner in the summer, when the rate of chicken panting is greater. Suggest a remedy for this situation.

15.70 The equilibrium constant K_P for the following reaction is found to be 4.31×10^{-4} at 375°C:

$$N_2(g) + 3H_2(g) \rightleftharpoons 2NH_3(g)$$

In a certain experiment a student starts with 0.862 atm of N_2 and 0.373 atm of H_2 in a constant-volume vessel at 375°C. Calculate the partial pressures of all species when equilibrium is reached.

15.71 A quantity of 0.20 mole of carbon dioxide was heated at a certain temperature with an excess of graphite in a closed container until the following equilibrium was reached:

$$C(s) + CO_2(g) \rightleftharpoons 2CO(g)$$

Under this condition, the average molar mass of the gases was found to be 35 g/mol. (a) Calculate the mole fractions of CO and CO_2. (b) What is the K_P for the equilibrium if the total pressure was 11 atm? (*Hint:* The average molar mass is the sum of the products of the mole fraction of each gas and its molar mass.)

15.72 When dissolved in water, glucose (corn sugar) and fructose (fruit sugar) exist in equilibrium as follows:

$$\text{fructose} \rightleftharpoons \text{glucose}$$

A chemist prepared a 0.244 M fructose solution at 25°C. At equilibrium, it was found that its concentration had decreased to 0.113 M. (a) Calculate the equilibrium constant for the reaction. (b) At equilibrium, what percentage of fructose was converted to glucose?

15.73 At room temperature, solid iodine is in equilibrium with its vapor through sublimation and deposition (see Figure 8.18). Describe how you would use radioactive iodine, in either solid or vapor form, to show that there is a dynamic equilibrium between these two phases.

15.74 At 1024°C, the pressure of oxygen gas from the decomposition of copper(II) oxide (CuO) is 0.49 atm:

$$4CuO(s) \rightleftharpoons 2Cu_2O(s) + O_2(g)$$

(a) What is the K_P for the reaction? (b) Calculate the fraction of CuO decomposed if 0.16 mole of it is placed in a 2.0-L flask at 1024°C. (c) What would be the fraction if a 1.0-mole sample of CuO were used? (d) What is the smallest amount of CuO (in moles) that would establish the equilibrium?

15.75 A mixture containing 3.9 moles of NO and 0.88 mole of CO_2 was allowed to react in a flask at a certain temperature according to the equation

$$NO(g) + CO_2(g) \rightleftharpoons NO_2(g) + CO(g)$$

At equilibrium, 0.11 mole of CO_2 was present. Calculate the equilibrium constant K_c of this reaction.

15.76 The equilibrium constant K_c for the reaction

$$H_2(g) + I_2(g) \rightleftharpoons 2HI(g)$$

is 54.3 at 430°C. At the start of the reaction there are 0.714 mole of H_2, 0.984 mole of I_2, and 0.886 mole of HI in a 2.40-L reaction chamber. Calculate the concentrations of the gases at equilibrium.

15.77 On heating, a gaseous compound A dissociates as follows:

$$A(g) \rightleftharpoons B(g) + C(g)$$

In an experiment A was heated at a certain temperature until its equilibrium pressure reached $0.14P$, in which P is the total pressure. Calculate the equilibrium constant (K_P) of this reaction.

15.78 When a gas was heated at atmospheric pressure and 25°C, its color was found to deepen. Heating above 150°C caused the color to fade, and at 550°C the color was barely detectable. However, at 550°C, the color was partially restored by increasing the pressure of the system. Which of these best fits this description? Justify your choice. (a) A mixture of hydrogen and bromine, (b) pure bromine, (c) a mixture of nitrogen dioxide and dinitrogen tetroxide. (*Hint:* Bromine has a reddish color and nitrogen dioxide is a brown gas. The other gases are colorless.)

15.79 The equilibrium constant K_c for the following reaction is 0.65 at 395°C.

$$N_2(g) + 3H_2(g) \rightleftharpoons 2NH_3(g)$$

(a) What is the value of K_P for this reaction?
(b) What is the value of the equilibrium constant K_c for $2NH_3(g) \rightleftharpoons N_2(g) + 3H_2(g)$?
(c) What is K_c for $\frac{1}{2}N_2(g) + \frac{3}{2}H_2(g) \rightleftharpoons NH_3(g)$?
(d) What are the values of K_P for the reactions described in (b) and (c)?

15.80 A sealed glass bulb contains a mixture of NO_2 and N_2O_4 gases. When the bulb is heated from 20°C to 40°C, what happens to these properties of the gases: (a) color, (b) pressure, (c) average molar mass, (d) degree of dissociation (from N_2O_4 to NO_2), (e) density? Assume that volume remains constant. (*Hint:* NO_2 is a brown gas; N_2O_4 is colorless.)

15.81 At 20°C, the vapor pressure of water is 0.0231 atm. Calculate K_P and K_c for the process

$$H_2O(l) \rightleftharpoons H_2O(g)$$

15.82 Industrially, sodium metal is obtained by electrolyzing molten sodium chloride. The reaction at the cathode is $Na^+ + e^- \longrightarrow Na$. We might expect that potassium metal would also be prepared by electrolyzing molten potassium chloride. However, potassium metal is soluble in molten potassium chloride and therefore is hard to recover. Furthermore, potassium vaporizes readily at the operating temperature, creating hazardous conditions. Instead, potassium is prepared by the distillation of molten potassium chloride in the presence of sodium vapor at 892°C:

$$Na(g) + KCl(l) \rightleftharpoons NaCl(l) + K(g)$$

In view of the fact that potassium is a stronger reducing agent than sodium, explain why this approach works. (The boiling points of sodium and potassium are 892°C and 770°C, respectively.)

15.83 In the gas phase, nitrogen dioxide is actually a mixture of nitrogen dioxide (NO_2) and dinitrogen tetroxide (N_2O_4). If the density of such a mixture at 74°C and 1.3 atm is 2.9 g/L, calculate the partial pressures of the gases and K_P.

15.84 About 75 percent of hydrogen for industrial use is produced by the *steam-reforming* process. This process is carried out in two stages called primary and secondary reforming. In the primary stage, a mixture of steam and methane at about 30 atm is heated over a nickel catalyst at 800°C to give hydrogen and carbon monoxide:

$$CH_4(g) + H_2O(g) \rightleftharpoons CO(g) + 3H_2(g) \quad \Delta H° = 206 \text{ kJ}$$

The secondary stage is carried out at about 1000°C, in the presence of air, to convert the remaining methane to hydrogen:

$$CH_4(g) + \tfrac{1}{2}O_2(g) \rightleftharpoons CO(g) + 2H_2(g) \quad \Delta H° = 35.7 \text{ kJ}$$

(a) What conditions of temperature and pressure would favor the formation of products in both the primary and secondary stages? (b) The equilibrium constant K_c for the primary stage is 18 at 800°C. (i) Calculate K_P for the reaction. (ii) If the partial pressures of methane and steam were both 15 atm at the start, what are the pressures of all the gases at equilibrium?

15.85 Photosynthesis can be represented by

$$6CO_2(g) + 6H_2O(l) \rightleftharpoons C_6H_{12}O_6(s) + 6O_2(g)$$
$$\Delta H° = 2801 \text{ kJ}$$

Explain how the equilibrium would be affected by the following changes: (a) partial pressure of CO_2 is increased, (b) O_2 is removed from the mixture, (c) $C_6H_{12}O_6$ (sucrose) is removed from the mixture, (d) more water is added, (e) a catalyst is added, (f) temperature is decreased, (g) more sunlight falls on the plants.

15.86 Consider the decomposition of ammonium chloride at a certain temperature:

$$NH_4Cl(s) \rightleftharpoons NH_3(g) + HCl(g)$$

Calculate the equilibrium constant K_P if the total pressure is 2.2 atm at that temperature.

15.87 At 25°C, the equilibrium partial pressures of NO_2 and N_2O_4 are 0.15 atm and 0.20 atm, respectively. If the volume is doubled at constant temperature, calculate the partial pressures of the gases when a new equilibrium is established.

15.88 In 1899 the German chemist Ludwig Mond developed a process for purifying nickel by converting it to the volatile nickel tetracarbonyl [$Ni(CO)_4$] (b.p. = 42.2°C):

$$Ni(s) + 4CO(g) \rightleftharpoons Ni(CO)_4(g)$$

(a) Describe how you can separate nickel and its solid impurities. (b) How would you recover nickel? [ΔH_f° for $Ni(CO)_4$ is −602.9 kJ/mol.]

15.89 Consider the equilibrium reaction in Problem 15.19. A quantity of 2.50 g of PCl_5 is placed in an evacuated flask of volume 0.500 L and heated to 250°C. (a) Calculate the pressure of PCl_5 if it did not dissociate. (b) Calculate the partial pressure of PCl_5 at equilibrium. (c) What is the total pressure at equilibrium? (d) What is the degree of dissociation of PCl_5? (The degree of dissociation is given by the fraction of PCl_5 that has undergone dissociation.)

15.90 The vapor pressure of mercury is 0.0020 mmHg at 26°C. (a) Calculate K_c and K_P for the process $Hg(l) \rightleftharpoons Hg(g)$. (b) A chemist breaks a thermometer and spills mercury onto the floor of a laboratory measuring 6.1 m long, 5.3 m wide, and 3.1 m high. Calculate the mass of mercury (in grams) vaporized at equilibrium and the concentration of mercury vapor in mg/m^3. Does this concentration exceed the safety limit of 0.050 mg/m^3? (Ignore the volume of furniture and other objects in the laboratory.)

Special Problems

15.91 In this chapter we learned that a catalyst has no effect on the position of an equilibrium because it speeds up both the forward and reverse rates to the same extent. To test this statement, consider a situation in which an equilibrium of the type

$$2A(g) \rightleftharpoons B(g)$$

is established inside a cylinder fitted with a weightless piston. The piston is attached by a string to the cover of a box containing a catalyst. When the piston moves upward (expanding against atmospheric pressure), the cover is lifted and the catalyst is exposed to the gases. When the piston moves downward, the box is closed. Assume that the catalyst speeds up the forward reaction ($2A \longrightarrow B$) but does not affect the reverse process ($B \longrightarrow 2A$). Suppose the catalyst is suddenly exposed to the equilibrium system as shown on the right. Describe what would happen subsequently. How does this "thought" experiment convince you that no such catalyst can exist?

15.92 At 25°C, a mixture of NO_2 and N_2O_4 gases are in equilibrium in a cylinder fitted with a movable piston. The concentrations are: $[NO_2] = 0.0475\ M$ and $[N_2O_4] = 0.491\ M$. The volume of the gas mixture is halved by pushing down on the piston at constant temperature. Calculate the concentrations of the gases when equilibrium is re-established. Will the color become darker or lighter after the change? [*Hint:* K_c for the dissociation of N_2O_4 is 4.63×10^{-3}. $N_2O_4(g)$ is colorless and $NO_2(g)$ has a brown color.]

Answers to Practice Exercises

15.1 $K_c = \dfrac{[NO_2]^4[O_2]}{[N_2O_5]^2}$ $K_P = \dfrac{P_{NO_2}^4 P_{O_2}}{P_{N_2O_5}^2}$. **15.2** 347 atm. **15.3** 0.113. **15.4** $K_P = 0.0702$; $K_c = 1.20 \times 10^{-4}$. **15.5** From right to left. **15.6** $[HI] = 0.031\ M$, $[H_2] = 4.3 \times 10^{-3}\ M$, $[I_2] = 4.3 \times 10^{-3}\ M$. **15.7** $[Br_2] = 0.065\ M$, $[Br] = 8.4 \times 10^{-3}\ M$. **15.8** $Q_P = 4.0 \times 10^5$, the net reaction will shift from right to left. **15.9** Left to right. **15.10** The equilibrium will shift from (a) left to right, (b) left to right, and (c) right to left. (d) A catalyst has no effect on the equilibrium.

Why are British sailors called "limeys"? The nickname originated in the late eighteenth century when limes and other citrus fruits became part of British navy rations. Until that time sailors often ate only biscuits and salt-preserved meats for months at a time when they were at sea. Many developed scurvy, a disease characterized by aching joints, skin lesions, bleeding gums, and loose teeth. In 1753 it was discovered that scurvy can be cured with citrus fruit, which contains large amounts of ascorbic acid, or vitamin C ($C_6H_8O_6$).

Ascorbic acid also acts as an antioxidant, reacting with oxidizing species such as the ·OH radical, which can damage an organism's DNA. Most mammals make their own vitamin C, but humans and other primates, as well as bats and guinea pigs, must obtain it from their food. Almost any diet that includes fresh fruit and vegetables will provide an abundance of vitamin C, as long as the food is not overcooked; heat readily destroys the water-soluble, odorless white solid.

Beyond the fact that vitamin C is essential for good health, little was known about its chemistry before the twentieth century. Indeed, when the Hungarian-born biochemist Albert von Szent-Gyorgi first isolated this compound from a cabbage extract in 1928, he thought it might be a sugar, like sucrose and fructose. Consequently, Szent-Gyorgi called it "godnose," a name that was firmly rejected by the editor of a scientific journal. Later, Szent-Gyorgi and W. Norman Haworth, a British chemist who devised a method for synthesizing vitamin C, renamed it ascorbic acid. Both won Nobel Prizes in 1937 for their work on vitamin C.

The biochemistry of ascorbic acid is complex, and it is not known how much vitamin C the human body needs on a daily basis. Nevertheless, Szent-Gyorgi, together with Linus Pauling and some other scientists, advocated taking large doses of vitamin C tablets to maintain good health. They believed that a high level of vitamin C helps to prevent the common cold and protect against certain cancers. These claims remain controversial, but it is worth noting that Szent-Gyorgi and Pauling both lived to the age of 93.

Ascorbic acid. How many asymmetric C atoms are there in this molecule?

Acids and Bases

Chapter 16

ESSENTIAL CONCEPTS

 Brønsted Acids and Bases A Brønsted acid can donate a proton and a Brønsted base can accept a proton. For every Brønsted acid there exists a conjugate Brønsted base and vice versa.

 Acid-Base Properties of Water and the pH Scale Water acts both as a Brønsted acid and a Brønsted base. At 25°C, the concentrations of H^+ and OH^- ions are both at 10^{-7} *M*. The pH scale is established to express the acidity of a solution—the smaller the pH, the higher the H^+ concentration and the greater the acidity.

 Acid and Base Ionization Constants Strong acids and strong bases are assumed to ionize completely. Most weak acids and weak bases ionize to a small extent. The concentrations of the acid, conjugate base, and H^+ ion at equilibrium can be calculated from the acid ionization constant, which is the equilibrium constant for the reaction.

 Acid-Base Properties of Salts and Oxides Many salts react with water in a process called hydrolysis. From the nature of the cation and anion present in the salt it is possible to predict the pH of the resulting solution. Most oxides also react with water to produce acidic or basic solutions.

 Lewis Acids and Bases A more general definition of acids and bases characterizes an acid as a substance that can accept a pair of electrons and a base as a substance that can donate a pair of electrons. All Brønsted acids and bases are Lewis acids and bases.

16.1 Brønsted Acids and Bases

In Chapter 4 we defined a Brønsted acid as a substance capable of donating a proton, and a Brønsted base as a substance capable of accepting a proton. These definitions are generally suitable for discussion of the properties and reactions of acids and bases.

Conjugate Acid-Base Pairs

Conjugate means "joined together."

An extension of the Brønsted definition of acids and bases is the concept of the *conjugate acid-base pair,* which can be defined as *an acid and its conjugate base or a base and its conjugate acid.* The conjugate base of a Brønsted acid is the species that remains when one proton has been removed from the acid. Conversely, a conjugate acid results from the addition of a proton to a Brønsted base.

Every Brønsted acid has a conjugate base, and every Brønsted base has a conjugate acid. For example, the chloride ion (Cl^-) is the conjugate base formed from the acid HCl, and H_2O is the conjugate base of the acid H_3O^+. Similarly, the ionization of acetic acid can be represented as

$$CH_3COOH(aq) + H_2O(l) \rightleftharpoons CH_3COO^-(aq) + H_3O^+(aq)$$
$$\text{acid}_1 \qquad\qquad \text{base}_2 \qquad\qquad \text{base}_1 \qquad\qquad \text{acid}_2$$

The formula of a conjugate base always has one fewer hydrogen atom and one more negative charge (or one fewer positive charge) than the formula of the corresponding acid.

The subscripts 1 and 2 designate the two conjugate acid-base pairs. Thus the acetate ion (CH_3COO^-) is the conjugate base of CH_3COOH. Both the ionization of HCl (see Section 4.3) and the ionization of CH_3COOH are examples of Brønsted acid-base reactions.

The Brønsted definition also allows us to classify ammonia as a base because of its ability to accept a proton:

$$NH_3(aq) + H_2O(l) \rightleftharpoons NH_4^+(aq) + OH^-(aq)$$
$$\text{base}_1 \qquad\quad \text{acid}_2 \qquad\qquad \text{acid}_1 \qquad\quad \text{base}_2$$

In this case, NH_4^+ is the conjugate acid of the base NH_3, and OH^- is the conjugate base of the acid H_2O. Note that the atom in the Brønsted base that accepts a H^+ ion must have a lone pair.

A less clear-cut case is that of NaOH, which, strictly speaking, is not a Brønsted base because it cannot accept a proton. However, NaOH is a strong electrolyte that ionizes completely in solution. The hydroxide ion (OH^-) formed by that ionization *is* a Brønsted base because it can accept a proton:

$$H_3O^+(aq) + OH^-(aq) \rightleftharpoons 2H_2O(l)$$

Thus, when we call NaOH or any other metal hydroxide a base, we are actually referring to the OH^- species derived from the hydroxide.

Example 16.1 Identify the conjugate acid-base pairs in the reaction between ammonia and hydrofluoric acid in aqueous solution

$$NH_3(aq) + HF(aq) \rightleftharpoons NH_4^+(aq) + F^-(aq)$$

Reasoning and Solution NH_3 has one fewer H atom and one fewer positive charge than NH_4^+. F^- has one fewer H atom and one more negative charge than HF. Therefore, the conjugate acid-base pairs are (1) NH_4^+ and NH_3 and (2) HF and F^-.

Practice Exercise Identify the conjugate acid-base pairs for the reaction

$$CN^- + H_2O \rightleftharpoons HCN + OH^-$$

Similar problem: 16.5.

The hydronium ion (H_3O^+) represents a hydrated proton in water. In reality, the proton may be associated with more than one H_2O molecule and has more complicated formulas such as $H_5O_2^+$ or $H_9O_4^+$. However, we will always use the hydronium ion to represent the hydrated proton. In calculations, for simplicity we will use H^+ when discussing hydrogen ion concentrations. Keep in mind that both H^+ and H_3O^+ represent the *same* species in solution.

16.2 The Acid-Base Properties of Water

Water, as we know, is a unique solvent. One of its special properties is its ability to act both as an acid and as a base. Water functions as a base in reactions with acids such as HCl and CH_3COOH, and it functions as an acid in reactions with bases such as NH_3. Water is a very weak electrolyte and therefore a poor conductor of electricity, but it does undergo ionization to a small extent:

$$H_2O(l) \rightleftharpoons H^+(aq) + OH^-(aq)$$

Tap water and water from underground sources do conduct electricity because they contain many dissolved ions.

This reaction is sometimes called the *autoionization* of water. To describe the acid-base properties of water in the Brønsted framework, we express its autoionization as (Figure 16.1)

$$H\!-\!\overset{..}{\underset{H}{O}}: \ + \ H\!-\!\overset{..}{\underset{H}{O}}: \ \rightleftharpoons \ \left[H\!-\!\overset{..}{\underset{H}{O}}\!-\!H\right]^+ \ + \ H\!-\!\overset{..}{\underset{..}{O}}:^-$$

or

$$\underset{\text{acid}_1}{H_2O} + \underset{\text{base}_2}{H_2O} \rightleftharpoons \underset{\text{acid}_2}{H_3O^+} + \underset{\text{base}_1}{OH^-} \qquad (16.1)$$

The acid-base conjugate pairs are (1) H_2O (acid) and OH^- (base) and (2) H_3O^+ (acid) and H_2O (base).

Figure 16.1 *Reaction between two water molecules to form hydronium and hydroxide ions.*

The Ion Product of Water

In the study of acid-base reactions in aqueous solutions, the important quantity is the hydrogen ion concentration. Expressing the proton as H_3O^+ or H^+, we can write the equilibrium constant for the autoionization of water, Equation (16.1), as

$$K_c = \frac{[H_3O^+][OH^-]}{[H_2O]} \quad \text{or} \quad K_c = \frac{[H^+][OH^-]}{[H_2O]}$$

These two equations are the same because H_3O^+ and H^+ represent the same species in solution.

Recall that in pure water, [H_2O] = 55.5 M (see p. 476).

Because a very small fraction of water molecules are ionized, the concentration of water, that is, [H_2O], remains virtually unchanged. Therefore

$$K_c[H_2O] = K_w = [H^+][OH^-] \tag{16.2}$$

The equilibrium constant K_w is called the ***ion-product constant,*** which is *the product of the molar concentrations of H^+ and OH^- ions at a particular temperature.*

In pure water at 25°C, the concentrations of H^+ and OH^- ions are equal and found to be $[H^+] = 1.0 \times 10^{-7}$ M and $[OH^-] = 1.0 \times 10^{-7}$ M. Thus, from Equation (16.2), at 25°C

If you could randomly remove and examine 10 particles (H_2O, H^+, or OH^-) per second from a liter of water, it would take you 2 yr on average, working nonstop, to find one H^+ ion!

$$K_w = (1.0 \times 10^{-7})(1.0 \times 10^{-7}) = 1.0 \times 10^{-14}$$

Note that whether we have pure water or a solution of dissolved species, the following relation *always* holds at 25°C:

$$K_w = [H^+][OH^-]$$
$$= 1.0 \times 10^{-14}$$

Whenever $[H^+] = [OH^-]$, the aqueous solution is said to be neutral. In an acidic solution there is an excess of H^+ ions and $[H^+] > [OH^-]$. In a basic solution there is an excess of hydroxide ions, so $[H^+] < [OH^-]$. In practice we can change the concentration of either H^+ or OH^- ions in solution, but we cannot vary them independently. If we adjust the solution so that $[H^+] = 1.0 \times 10^{-6}$ M, the OH^- concentration *must* change to

$$[OH^-] = \frac{K_w}{[H^+]}$$
$$= \frac{1.0 \times 10^{-14}}{1.0 \times 10^{-6}} = 1.0 \times 10^{-8} M$$

Be aware that the calculations shown above, and indeed all the calculations involving solution concentrations discussed in Chapters 15–17, are subject to error because we have implicitly assumed ideal behavior. In reality, ion-pair formation and other types of intermolecular interactions may affect the actual concentrations of species in solution and hence also the equilibrium constant values. The situation is analogous to the relationships between ideal gas behavior and the behavior of real gases discussed in Chapter 5. Depending on temperature, volume, and amount and type of gas present, the measured gas pressure may differ from that calculated using the ideal gas equation. Similarly, the actual, or "effective," concentration of a solute may not be what we think it should be from the amount of substance originally dissolved in solution. Just as we have the van der Waals and other equations to reconcile discrepancies between the ideal gas equation and nonideal gas behavior, we can account for nonideal behavior in solution. But for our purposes, it is acceptable to ignore deviations from ideality. In many cases this approach will give us a good approximation of the chemical processes that actually take place in the solution phase.

Example 16.2 The concentration of OH^- ions in a certain household ammonia cleaning solution is 0.0025 M. Calculate the concentration of H^+ ions.

Reasoning and Solution Rearranging Equation (16.2), we write

$$[H^+] = \frac{K_w}{[OH^-]} = \frac{1.0 \times 10^{-14}}{0.0025} = 4.0 \times 10^{-12}\, M$$

Comment Because $[H^+] < [OH^-]$, the solution is basic, as we would expect from the earlier discussion of the reaction of ammonia with water.

Similar problem: 16.16(c).

Practice Exercise Calculate the concentration of OH^- ions in a HCl solution whose hydrogen ion concentration is 1.3 M.

16.3 pH—A Measure of Acidity

Because the concentrations of H^+ and OH^- ions in aqueous solutions are frequently very small numbers and therefore inconvenient to work with, the Danish biochemist Soren Sorensen in 1909 proposed a more practical measure called pH. The **pH** of a solution is defined as *the negative logarithm of the hydrogen ion concentration (in moles per liter):*

$$pH = -\log\,[H_3O^+] \quad \text{or} \quad pH = -\log\,[H^+] \tag{16.3}$$

Keep in mind that Equation (16.3) is simply a definition designed to give us convenient numbers to work with. In most cases the negative logarithm gives us a positive number for pH, which otherwise would be negative because of the small value of $[H^+]$. (See Appendix 3 for a discussion of logarithms and significant figures.) Furthermore, the term $[H^+]$ in Equation (16.3) pertains only to the *numerical part* of the expression for hydrogen ion concentration, for we cannot take the logarithm of units. Thus, like the equilibrium constant, the pH of a solution is a dimensionless quantity.

Because pH is simply a way to express hydrogen ion concentration, acidic and basic solutions at 25°C can be identified by their pH values, as:

Acidic solutions: $[H^+] > 1.0 \times 10^{-7}\, M$, pH < 7.00

Basic solutions: $[H^+] < 1.0 \times 10^{-7}\, M$, pH > 7.00

Neutral solutions: $[H^+] = 1.0 \times 10^{-7}\, M$, pH $= 7.00$

Notice that pH increases as $[H^+]$ decreases.

In the laboratory, the pH of a solution is measured with a pH meter (Figure 16.2). Table 16.1 lists the pHs for a number of common fluids.

A pOH scale analogous to the pH scale can be devised using the negative logarithm of the hydroxide ion concentration. Thus we define pOH as

$$pOH = -\log\,[OH^-] \tag{16.4}$$

Now consider again the ion-product constant for water:

$$[H^+][OH^-] = K_w = 1.0 \times 10^{-14}$$

Taking the negative logarithm of both sides, we obtain

Note that a unit pH change corresponds to a 10-fold change in [H⁺].

The pH of concentrated acid solutions can be negative. For example, the pH of a 2.0 M HCl solution is −0.30.

Table 16.1 | The pHs of Some Common Fluids

Sample	pH Value
Gastric juice in the stomach	1.0–2.0
Lemon juice	2.4
Vinegar	3.0
Grapefruit juice	3.2
Orange juice	3.5
Urine	4.8–7.5
Water exposed to air*	5.5
Saliva	6.4–6.9
Milk	6.5
Pure water	7.0
Blood	7.35–7.45
Tears	7.4
Milk of magnesia	10.6
Household ammonia	11.5

*Water exposed to air for a long period of time absorbs atmospheric CO_2 to form carbonic acid, H_2CO_3.

Figure 16.2 *A pH meter is commonly used in the laboratory to determine the pH of a solution. Although many pH meters have scales marked with values from 1 to 14, pH values can, in fact, be less than 1 and greater than 14.*

$$-(\log [H^+] + \log [OH^-]) = -\log (1.0 \times 10^{-14})$$
$$-\log [H^+] - \log [OH^-] = 14.00$$

From the definitions of pH and pOH we obtain

$$pH + pOH = 14.00 \tag{16.5}$$

Equation (16.5) provides us with another way to express the relationship between the H^+ ion concentration and the OH^- ion concentration.

Example 16.3   The concentration of H^+ ions in a bottle of table wine was 3.2×10^{-4} *M* right after the cork was removed. Only half of the wine was consumed. The other half, after it had been standing open to the air for a month, was found to have a hydrogen ion concentration equal to 1.0×10^{-3} *M*. Calculate the pH of the wine on these two occasions.

Reasoning and Solution   When the bottle was first opened, $[H^+] = 3.2 \times 10^{-4}$ *M*, which we substitute in Equation (16.3).

$$pH = -\log [H^+]$$
$$= -\log (3.2 \times 10^{-4}) = 3.49$$

On the second occasion, $[H^+] = 1.0 \times 10^{-3}$ *M*, so that

$$pH = -\log (1.0 \times 10^{-3}) = 3.00$$

Comment   The increase in hydrogen ion concentration (or decrease in pH) is largely the result of the conversion of some of the alcohol (ethanol) to acetic acid, a reaction that takes place in the presence of molecular oxygen.

Practice Exercise   Nitric acid (HNO_3) is used in the production of fertilizer, dyes, drugs, and explosives. Calculate the pH of a HNO_3 solution having a hydrogen ion concentration is 0.76 *M*.

In each case the pH has only two significant figures. The two digits to the right of the decimal in 3.49 tell us that there are two significant figures in the original number (see Appendix 3).

Similar problems: 16.17(a), (d).

Example 16.4 The pH of rainwater collected in a certain region of the northeastern United States on a particular day was 4.82. Calculate the H^+ ion concentration of the rainwater.

Reasoning and Solution From Equation (16.3)

$$4.82 = -\log [H^+]$$

Taking the antilog of both sides of this equation (see Appendix 3) gives

$$[H^+] = 10^{-4.82} \, M = 1.5 \times 10^{-5} \, M$$

Comment Because the pH is between 4 and 5, we can expect $[H^+]$ to be between $1 \times 10^{-4} \, M$ and $1 \times 10^{-5} \, M$. Therefore, the answer is reasonable.

Similar problems: 16.16(a), (b).

Practice Exercise The pH of a certain orange juice is 3.33. Calculate the H^+ ion concentration.

Example 16.5 In a NaOH solution $[OH^-]$ is $2.9 \times 10^{-4} \, M$. Calculate the pH of the solution.

Reasoning and Solution We use Equation (16.4):

$$\begin{aligned} pOH &= -\log [OH^-] \\ &= -\log (2.9 \times 10^{-4}) \\ &= 3.54 \end{aligned}$$

Now we apply Equation (16.5):

$$\begin{aligned} pH + pOH &= 14.00 \\ pH &= 14.00 - pOH \\ &= 14.00 - 3.54 = 10.46 \end{aligned}$$

Comment Alternatively, we can use Equation (16.2) to calculate $[H^+]$ and then the pH. Try it.

Similar problem: 16.17(b).

Practice Exercise The OH^- ion concentration of a blood sample is $2.5 \times 10^{-7} \, M$. What is the pH of the blood?

16.4 Strength of Acids and Bases

Strong acids are strong electrolytes, which, for most practical purposes, *are assumed to ionize completely in water* (Figure 16.3). Most of the strong acids are inorganic acids: hydrochloric acid (HCl), nitric acid (HNO_3), perchloric acid ($HClO_4$), and sulfuric acid (H_2SO_4):

In reality, no acids are known to ionize completely in water.

$$HCl(aq) + H_2O(l) \longrightarrow H_3O^+(aq) + Cl^-(aq)$$
$$HNO_3(aq) + H_2O(l) \longrightarrow H_3O^+(aq) + NO_3^-(aq)$$
$$HClO_4(aq) + H_2O(l) \longrightarrow H_3O^+(aq) + ClO_4^-(aq)$$
$$H_2SO_4(aq) + H_2O(l) \longrightarrow H_3O^+(aq) + HSO_4^-(aq)$$

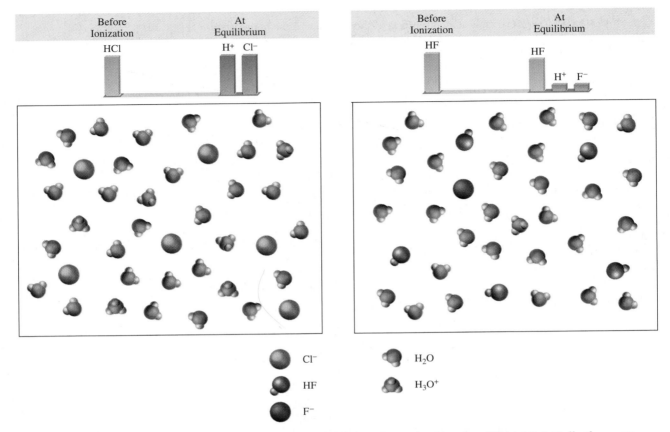

Figure 16.3 *The extent of ionization of a strong acid such as HCl (left) and a weak acid such as HF (right). Initially, there were 6 HCl and 6 HF molecules present. The strong acid is assumed to be completely ionized in solution. The proton exists in solution as the hydronium ion (H_3O^+). Compare the number of H_3O^+ ions and the conjugate bases (Cl^- and F^-) in the HCl and HF solutions.*

Note that H_2SO_4 is a diprotic acid; we show only the first stage of ionization here. At equilibrium, solutions of strong acids will not contain any nonionized acid molecules.

Most acids *ionize only to a limited extent in water*. Such acids are classified as **weak acids.** At equilibrium, aqueous solutions of weak acids contain a mixture of non-ionized acid molecules, H_3O^+ ions, and the conjugate base. Examples of weak acids are hydrofluoric acid (HF), acetic acid (CH_3COOH), and the ammonium ion (NH_4^+). Within this category, the strength of acids can vary greatly because of differences in extent of ionization. The limited ionization of weak acids is related to the equilibrium constant for ionization, which we will study in the next section.

What has been said about strong acids also applies to strong bases, which include hydroxides of alkali metals and certain alkaline earth metals, such as NaOH, KOH, and $Ba(OH)_2$. **Strong bases** *are all strong electrolytes that ionize completely in water:*

$$NaOH(s) \xrightarrow{H_2O} Na^+(aq) + OH^-(aq)$$

$$KOH(s) \xrightarrow{H_2O} K^+(aq) + OH^-(aq)$$

$$BaOH_2(s) \xrightarrow{H_2O} Ba^{2+}(aq) + 2OH^-(aq)$$

If we compare the structure of an oxoacid (see Section 2.7) such as HNO_3

$$
\begin{array}{c}
:\!\overset{\displaystyle :\!\!O\!:}{\underset{}{\|}} \\
:\!\overset{..}{\underset{..}{O}}\!-\!N\!-\!\overset{..}{\underset{..}{O}}\!-\!H
\end{array}
$$

HNO₃

with that of NaOH, we see that both compounds contain the OH group. Because sodium is much less electronegative (or much more electropositive) than oxygen, the pair of electrons shared between Na and O is transferred to oxygen, and an ionic compound involving Na^+ and OH^- ions is formed. In the case of HNO_3, N is an electronegative element and the N—O bond in N—O—H is more covalent in character, so the substance does not lose OH^-. Instead, the O—H bond becomes more polar, thereby favoring the loss of H^+.

Weak bases, like weak acids, *are weak electrolytes.* Ammonia is a weak base. It ionizes to a very limited extent in water:

$$NH_3(aq) + H_2O(l) \rightleftharpoons NH_4^+(aq) + OH^-(aq)$$

Table 16.2 lists some important conjugate acid-base pairs in order of their relative strengths. It is useful to keep the following points in mind:

- If an acid is strong, its conjugate base has no measurable strength.
- H_3O^+ is the strongest acid that can exist in aqueous solution. Acids stronger than H_3O^+ react with water to produce H_3O^+ and their conjugate bases. Thus HCl, which is a stronger acid than H_3O^+, reacts with water completely to form H_3O^+ and Cl^-:

$$HCl(aq) + H_2O(l) \longrightarrow H_3O^+(aq) + Cl^-(aq)$$

Acids weaker than H_3O^+ react with water to a much smaller extent, producing H_3O^+ and their conjugate bases. For example, the following equilibrium lies primarily to the left:

$$HF(aq) + H_2O(l) \rightleftharpoons H_3O^+(aq) + F^-(aq)$$

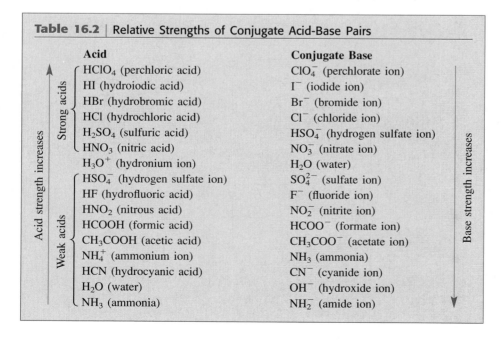

Table 16.2 | Relative Strengths of Conjugate Acid-Base Pairs

	Acid	Conjugate Base
Strong acids	$HClO_4$ (perchloric acid)	ClO_4^- (perchlorate ion)
	HI (hydroiodic acid)	I^- (iodide ion)
	HBr (hydrobromic acid)	Br^- (bromide ion)
	HCl (hydrochloric acid)	Cl^- (chloride ion)
	H_2SO_4 (sulfuric acid)	HSO_4^- (hydrogen sulfate ion)
	HNO_3 (nitric acid)	NO_3^- (nitrate ion)
	H_3O^+ (hydronium ion)	H_2O (water)
Weak acids	HSO_4^- (hydrogen sulfate ion)	SO_4^{2-} (sulfate ion)
	HF (hydrofluoric acid)	F^- (fluoride ion)
	HNO_2 (nitrous acid)	NO_2^- (nitrite ion)
	HCOOH (formic acid)	$HCOO^-$ (formate ion)
	CH_3COOH (acetic acid)	CH_3COO^- (acetate ion)
	NH_4^+ (ammonium ion)	NH_3 (ammonia)
	HCN (hydrocyanic acid)	CN^- (cyanide ion)
	H_2O (water)	OH^- (hydroxide ion)
	NH_3 (ammonia)	NH_2^- (amide ion)

Acid strength increases →

Base strength increases →

- The OH^- ion is the strongest base that can exist in aqueous solution. Bases stronger than OH^- react with water to produce OH^- and their conjugate acids. For example, the oxide ion (O^{2-}) is a stronger base than OH^-, so it reacts with water completely as follows:

$$O^{2-}(aq) + H_2O(l) \longrightarrow 2OH^-(aq)$$

For this reason the oxide ion does not exist in aqueous solutions. (Note that one of the two OH^- ions produced is identified as the conjugate acid of the O^{2-} ion.)

Example 16.6   Calculate the pH of (a) a 1.0×10^{-3} M HCl solution and (b) a 0.020 M $Ba(OH)_2$ solution.

Reasoning and Solution   (a) Because HCl is a strong acid, it is completely ionized in solution:

$$HCl(aq) \longrightarrow H^+(aq) + Cl^-(aq)$$

The concentrations of all the species (HCl, H^+, and Cl^-) before and after ionization can be represented as follows:

	$HCl(aq)$	$\longrightarrow$	$H^+(aq)$	$+$	$Cl^-(aq)$
Initial (M):	1.0×10^{-3}		0.0		0.0
Change (M):	-1.0×10^{-3}		$+1.0 \times 10^{-3}$		$+1.0 \times 10^{-3}$
Final (M):	0.0		1.0×10^{-3}		1.0×10^{-3}

A positive (+) change represents an increase and a negative (−) change indicates a decrease in concentration. Thus

$$[H^+] = 1.0 \times 10^{-3} \ M$$
$$pH = -\log(1.0 \times 10^{-3})$$
$$= 3.00$$

Ba(OH)₂ is the only strong base that contains two OH⁻ groups.

(b) $Ba(OH)_2$ is a strong base; each $Ba(OH)_2$ unit produces two OH^- ions:

$$Ba(OH)_2(aq) \longrightarrow Ba^{2+}(aq) + 2OH^-(aq)$$

The changes in the concentrations of all the species can be represented as:

	$Ba(OH)_2(aq)$	$\longrightarrow$	$Ba^{2+}(aq)$	$+$	$2OH^-(aq)$
Initial (M):	0.020		0.00		0.00
Change (M):	-0.020		$+0.020$		$+2(0.020)$
Final (M):	0.00		0.020		0.040

Thus

$$[OH^-] = 0.040 \ M$$
$$pOH = -\log 0.040 = 1.40$$

Therefore

$$pH = 14.00 - pOH$$
$$= 14.00 - 1.40$$
$$= 12.60$$

Comment Note that in both (a) and (b) we have neglected the contribution of the autoionization of water to [H$^+$] and [OH$^-$] because 1.0×10^{-7} M is so small compared with 1.0×10^{-3} M and 0.040 M.

Similar problems: 16.17(a), (c).

Practice Exercise Calculate the pH of a 1.8×10^{-2} M Ba(OH)$_2$ solution.

If we know the relative strengths of two acids, we can predict the position of equilibrium between one of the acids and the conjugate base of the other, as illustrated in Example 16.7.

Example 16.7 Predict the direction of the following reaction in aqueous solution:

$$HNO_2(aq) + CN^-(aq) \rightleftharpoons HCN(aq) + NO_2^-(aq)$$

Reasoning and Solution In Table 16.2 we see that HNO$_2$ is a stronger acid than HCN. Thus CN$^-$ is a stronger base than NO$_2^-$. The net reaction will proceed from left to right as written because HNO$_2$ is a better proton donor than HCN (and CN$^-$ is a better proton acceptor than NO$_2^-$).

Similar problems: 16.33, 16.34.

Practice Exercise Predict whether the equilibrium constant for the following reaction is greater than or smaller than 1:

$$CH_3COOH(aq) + HCOO^-(aq) \rightleftharpoons CH_3COO^-(aq) + HCOOH(aq)$$

16.5 Weak Acids and Acid Ionization Constants

As we have seen, the vast majority of acids are weak acids. Consider a weak monoprotic acid HA. Its ionization in water is represented by

$$HA(aq) + H_2O(l) \rightleftharpoons H_3O^+(aq) + A^-(aq)$$

or simply

$$HA(aq) \rightleftharpoons H^+(aq) + A^-(aq)$$

The *equilibrium constant for* this *acid ionization,* which we call the **acid ionization constant, K$_a$,** is given by

$$K_a = \frac{[H_3O^+][A^-]}{[HA]} \quad \text{or} \quad K_a = \frac{[H^+][A^-]}{[HA]} \qquad (16.6)$$

Acid ionizations

All concentrations in these equations are equilibrium concentrations.

At a given temperature, the strength of the acid HA is measured quantitatively by the magnitude of K_a. The larger K_a, the stronger the acid—that is, the greater the concentration of H$^+$ ions at equilibrium because of its ionization.

Because the ionization of weak acids is never complete, all species (the nonionized acid, the H$^+$ ions, and the A$^-$ ions) are present at equilibrium. Table 16.3 lists a number of weak acids and their K_a values in order of decreasing acid strength. Note that at equilibrium the predominant species in solution (other than the solvent)

Table 16.3 | Ionization Constants of Some Weak Acids and Their Conjugate Bases at 25°C

Name of Acid	Formula	Structure	K_a	Conjugate Base	K_b
Hydrofluoric acid	HF	H—F	7.1×10^{-4}	F^-	1.4×10^{-11}
Nitrous acid	HNO_2	O=N—O—H	4.5×10^{-4}	NO_2^-	2.2×10^{-11}
Acetylsalicylic acid (aspirin)	$C_9H_8O_4$	(structure)	3.0×10^{-4}	$C_9H_7O_4^-$	3.3×10^{-11}
Formic acid	HCOOH	(structure)	1.7×10^{-4}	$HCOO^-$	5.9×10^{-11}
Ascorbic acid*	$C_6H_8O_6$	(structure)	8.0×10^{-5}	$C_6H_7O_6^-$	1.3×10^{-10}
Benzoic acid	C_6H_5COOH	(structure)	6.5×10^{-5}	$C_6H_5COO^-$	1.5×10^{-10}
Acetic acid	CH_3COOH	(structure)	1.8×10^{-5}	CH_3COO^-	5.6×10^{-10}
Hydrocyanic acid	HCN	H—C≡N	4.9×10^{-10}	CN^-	2.0×10^{-5}
Phenol	C_6H_5OH	(structure)	1.3×10^{-10}	$C_6H_5O^-$	7.7×10^{-5}

*For ascorbic acid it is the upper left hydroxyl group that is associated with this ionization constant.

is the nonionized acid. Although the compounds listed in Table 16.3 are all weak acids, within the group there is great variation in acid strength. For example, K_a for HF (7.1×10^{-4}) is about 1.5 million times that for HCN (4.9×10^{-10}).

We can calculate K_a from the initial concentration of the acid and the pH of the solution, and we can use K_a and the initial concentration of the acid to calculate equilibrium concentrations of all the species and pH of solution. In calculating the equilibrium concentrations of a weak acid, we follow essentially the same procedure outlined in Section 15.3; the systems may be different, but the calculations are based on the same principle, the law of mass action [Equation (15.2)]. In this case, the three basic steps are:

1. Express the equilibrium concentrations of all species in terms of the initial concentrations and a single unknown x, which represents the change in concentration.

2. Write the acid ionization constant in terms of the equilibrium concentrations. Knowing the value of K_a, we can solve for x.

3. Having solved for x, calculate the equilibrium concentrations of all species and/or the pH of the solution.

Unless otherwise stated, we will assume that the temperature is 25°C for all such calculations.

Example 16.8   Calculate the concentrations of the nonionized acid and of the ions of a 0.100 M formic acid (HCOOH) solution at equilibrium.

Reasoning and Solution

Step 1. Because it is monoprotic, one HCOOH molecule ionizes to give one H^+ ion and one $HCOO^-$ ion. Let x be the equilibrium concentration of H^+ and $HCOO^-$ ions in moles per liter. Then the equilibrium concentration of HCOOH must be $(0.100 - x)$ mol/L. We can now summarize the changes in concentrations as:

$$HCOOH(aq) \rightleftharpoons H^+(aq) + HCOO^-(aq)$$

	HCOOH(aq)	H⁺(aq)	HCOO⁻(aq)
Initial (M):	0.100	0.000	0.000
Change (M):	$-x$	$+x$	$+x$
Equilibrium (M):	$(0.100 - x)$	x	x

Step 2. According to Table 16.3

$$K_a = \frac{[H^+][HCOO^-]}{[HCOOH]} = 1.7 \times 10^{-4}$$

$$\frac{x^2}{0.100 - x} = 1.7 \times 10^{-4}$$

This equation can be rewritten as

$$x^2 + 1.7 \times 10^{-4}x - 1.7 \times 10^{-5} = 0$$

which fits the quadratic equation $ax^2 + bx + c = 0$.

We can often apply a simplifying approximation to this type of problem. Because HCOOH is a weak acid, the extent of its ionization must be small. Therefore x is small compared with 0.100. As a general rule, if the quantity x that is subtracted from the original concentration of the acid (0.100 M in this case) is equal to or less than 5 percent of the original concentration, we can assume $0.100 - x \approx 0.100$. This removes x from the denominator of the equilibrium constant expression. If x is more than 5 percent of the original value, then we must solve the quadratic equation. In case of doubt, we can always solve for x by the approximate method and then check the validity of our approximation. Assuming that $0.100 - x \approx 0.100$, then

$$\frac{x^2}{0.100 - x} \approx \frac{x^2}{0.100} = 1.7 \times 10^{-4}$$

$$x^2 = 1.7 \times 10^{-5}$$

Taking the square root of both sides, we obtain

$$x = 4.1 \times 10^{-3} \ M$$

HCOOH

The validity of the 5 percent approximation is based on the fact that the K_a values are generally known to an accuracy of only ±5 percent.

The sign $\approx$ means "approximately equal to." An analogy of the approximation here is a truck loaded with coal. Losing a few lumps of coal on a delivery trip will not change the overall mass of the load.

Step 3. At equilibrium, therefore,

$$[H^+] = 4.1 \times 10^{-3} \ M$$
$$[HCOO^-] = 4.1 \times 10^{-3} \ M$$
$$[HCOOH] = (0.100 - 0.0041) \ M$$
$$= 0.096 \ M$$

To check the validity of our approximation

$$\frac{0.0041 \ M}{0.100 \ M} \times 100\% = 4.1\%$$

This shows that the quantity x is less than 5 percent of the original concentration of the acid. Thus our approximation was justified.

Comment   Note that we omitted the contribution to $[H^+]$ by water. Except in very dilute acid solutions, this omission is always valid, because the hydrogen ion concentration caused by water is negligibly small compared with that caused by the acid.

Similar problem: 16.41.

Practice Exercise   Calculate the concentration of H^+, A^-, and nonionized HA in a 0.20 M HA solution. The K_a of HA is 2.7×10^{-4}.

HNO₂

Example 16.9   Calculate the pH of a 0.052 M nitrous acid (HNO_2) solution.

Reasoning and Solution

Step 1. From Table 16.3 we see that HNO_2 is a weak acid. Letting x be the equilibrium concentration of H^+ and NO_2^- ions in moles per liter, we summarize:

	$HNO_2(aq) \rightleftharpoons$	$H^+(aq) +$	$NO_2^-(aq)$
Initial (M):	0.052	0.00	0.00
Change (M):	$-x$	$+x$	$+x$
Equilibrium (M):	$(0.052 - x)$	x	x

Step 2. From Table 16.3,

$$K_a = \frac{[H^+][NO_2^-]}{[HNO_2]} = 4.5 \times 10^{-4}$$

$$\frac{x^2}{0.052 - x} = 4.5 \times 10^{-4}$$

Applying the approximation $0.052 - x \approx 0.052$, we obtain

$$\frac{x^2}{0.052 - x} \approx \frac{x^2}{0.052} = 4.5 \times 10^{-4}$$

$$x^2 = 2.3 \times 10^{-5}$$

Taking the square root of both sides gives

$$x = 4.8 \times 10^{-3} \ M$$

To test the approximation

$$\frac{0.0048\ M}{0.052\ M} \times 100\% = 9.2\%$$

This shows that x is more than 5 percent of the original concentration, so the approximation is not valid. There are two ways to get the correct answer.

1. *The quadratic equation.* We can solve the following quadratic equation:

$$x^2 + 4.5 \times 10^{-4}x - 2.3 \times 10^{-5} = 0$$

Using the quadratic formula,

$$x = \frac{-b \pm \sqrt{b^2 - 4ac}}{2a}$$

$$= \frac{-4.5 \times 10^{-4} \pm \sqrt{(4.5 \times 10^{-4})^2 - 4(1)(-2.3 \times 10^{-5})}}{2(1)}$$

Thus

$$x = 4.6 \times 10^{-3}\ M \qquad \text{or} \qquad x = -5.0 \times 10^{-3}\ M$$

The second solution is physically impossible, because the concentrations of ions produced as a result of ionization cannot be negative. Therefore, the solution is given by the positive root—that is,

$$x = 4.6 \times 10^{-3}\ M$$

2. *The method of successive approximation.* In this method, we first solve for x by assuming $0.052 - x \approx 0.052$ as before. Next, we use the approximate value of x ($4.8 \times 10^{-3}\ M$) to find a more exact value for the concentration of HNO_2:

$$[HNO_2] = 0.052\ M - 4.8 \times 10^{-3}\ M = 0.047\ M$$

Substituting the value of $[HNO_2]$ in the expression for K_a, we write

$$\frac{x^2}{0.047} = 4.5 \times 10^{-4}$$

$$x^2 = 2.1 \times 10^{-5}$$

$$x = 4.6 \times 10^{-3}\ M$$

Using $4.6 \times 10^{-3}\ M$ for x, we can recalculate $[HNO_2]$ and solve for x again. This time we find that the answer is still $4.6 \times 10^{-3}\ M$. In general, we apply the method of successive approximation until the value of x obtained for the last step does not differ from the previous step. (In the majority of cases you will need to apply this method only twice to get the correct answer.)

Step 3. At equilibrium

$$[H^+] = 4.6 \times 10^{-3}\ M$$

and

$$pH = -\log (4.6 \times 10^{-3})$$

$$= 2.34$$

Similar problem: 16.45.

Practice Exercise What is the pH of a 0.122 M monoprotic acid whose K_a is 5.7×10^{-4}?

C_6H_5COOH

Similar problem: 16.43.

Example 16.10   The pH of a 0.100 M benzoic acid (C_6H_5COOH) solution is 2.60. What is the K_a of the acid?

Reasoning and Solution   In this case we are given the pH, from which we can obtain the equilibrium concentrations, and we are asked to calculate the acid ionization constant. We can follow the same basic steps.

Step 1. First we need to calculate the hydrogen ion concentration from the pH.

$$pH = -\log[H^+]$$
$$2.60 = -\log[H^+]$$

Taking the antilog of both sides, we get

$$[H^+] = 10^{-2.60}\,M = 2.5 \times 10^{-3}\,M$$

Next we summarize the changes:

	$C_6H_5COOH(aq)$	$\rightleftharpoons$	$H^+(aq)$	+	$C_6H_5COO^-(aq)$
Initial (M):	0.100		0.000		0.000
Change (M):	-0.0025		$+0.0025$		$+0.0025$
Equilibrium (M):	$(0.100 - 0.0025)$		0.0025		0.0025

Step 2. The acid ionization constant is given by

$$K_a = \frac{[H^+][C_6H_5COO^-]}{[C_6H_5COOH]} = \frac{(0.0025)(0.0025)}{(0.100 - 0.0025)}$$

$$= 6.4 \times 10^{-5}$$

Practice Exercise   The pH of a 0.060 M weak monoprotic acid is 3.44. Calculate the K_a of the acid.

Percent Ionization

We have seen that the magnitude of K_a indicates the strength of an acid. Another way to study the strength of an acid is to measure its ***percent ionization,*** which is

$$\text{percent ionization} = \frac{\text{ionized acid concentration at equilibrium}}{\text{initial concentration of acid}} \times 100\% \quad (16.7)$$

The stronger the acid, the greater the percent ionization. For a monoprotic acid HA, the concentration of the acid that undergoes ionization is equal to the concentration of the H^+ ions or the concentration of the A^- ions at equilibrium. Therefore, we can write the percent ionization as

$$\text{percent ionization} = \frac{[H^+]}{[HA]_0} \times 100\%$$

in which $[H^+]$ is the concentration at equilibrium and $[HA]_0$ is the initial concentration. Note that this expression is the same as that used to check the approximation in Example 16.8. Thus the percent ionization of a 0.100 M formic acid is 4.1 percent.

The extent to which a weak acid ionizes depends on the initial concentration of the acid. The more dilute the solution, the greater the percent ionization (Figure 16.4).

Figure 16.4 *Dependence of percent ionization on initial concentration of acid. Note that at very low concentrations, all acids (weak and strong) are almost completely ionized.*

In qualitative terms, when an acid is diluted, initially the number of particles (nonionized acid molecules plus ions) per unit volume is reduced. According to Le Châtelier's principle (see Section 15.4), to counteract this "stress" (that is, the dilution), the equilibrium shifts from nonionized acid to H^+ and its conjugate base to produce more particles (ions).

Diprotic and Polyprotic Acids

Diprotic and polyprotic acids may yield more than one hydrogen ion per molecule. These acids ionize in a stepwise manner, that is, they lose one proton at a time. An ionization constant expression can be written for each ionization stage. Consequently, two or more equilibrium constant expressions must often be used to calculate the concentrations of species in the acid solution. For example, for H_2CO_3 we write

$$H_2CO_3(aq) \rightleftharpoons H^+(aq) + HCO_3^-(aq) \quad K_{a_1} = \frac{[H^+][HCO_3^-]}{[H_2CO_3]}$$

$$HCO_3^-(aq) \rightleftharpoons H^+(aq) + CO_3^{2-}(aq) \quad K_{a_2} = \frac{[H^+][CO_3^{2-}]}{[HCO_3^-]}$$

Note that the conjugate base in the first ionization stage becomes the acid in the second ionization stage.

Table 16.4 shows the ionization constants of several diprotic acids and a polyprotic acid. For a given acid, the first ionization constant is much larger than the second ionization constant, and so on. This trend is reasonable because it is easier to remove an H^+ ion from a neutral molecule than to remove another H^+ from a negatively charged ion derived from the molecule.

Top to bottom: H_2CO_3, HCO_3^-, and CO_3^{2-}.

Example 16.11 Oxalic acid ($C_2H_2O_4$) is a poisonous substance used chiefly as a bleaching and cleansing agent (for example, to remove bathtub rings). Calculate the concentrations of all the species present at equilibrium in a 0.10 *M* solution.

Reasoning and Solution Oxalic acid is a diprotic acid (see Table 16.4), which has two stages of ionization. We begin with the first stage of ionization.

Step 1.

$$C_2H_2O_4(aq) \rightleftharpoons H^+(aq) + C_2HO_4^-(aq)$$

	$C_2H_2O_4$	H^+	$C_2HO_4^-$
Initial (*M*):	0.10	0.00	0.00
Change (*M*):	$-x$	$+x$	$+x$
Equilibrium (*M*):	$(0.10 - x)$	x	x

Step 2. With the ionization constant from Table 16.4, we write

$$K_{a_1} = \frac{[H^+][C_2HO_4^-]}{[C_2H_2O_4]} = 6.5 \times 10^{-2}$$

$$\frac{x^2}{0.10 - x} = 6.5 \times 10^{-2}$$

$C_2H_2O_4$

Always check the validity of the
approximation.

Because the ionization constant is fairly large, we cannot apply the approximation $0.10 - x \approx 0.10$. Instead, we must solve the quadratic equation

$$x^2 + 6.5 \times 10^{-2}x - 6.5 \times 10^{-3} = 0$$

The solution gives $x = 0.054 \ M$.

Step 3. When the equilibrium for the first stage of ionization is reached, the concentrations are

$$[H^+] = 0.054 \ M$$
$$[C_2HO_4^-] = 0.054 \ M$$
$$[C_2H_2O_4] = (0.10 - 0.054) \ M = 0.046 \ M$$

Next we consider the second stage of ionization.

Step 1. Let y be the equilibrium concentration of $C_2O_4^{2-}$ in moles per liter. Thus the equilibrium concentration of $C_2HO_4^-$ must be $(0.054 - y) \ M$. We have

$$C_2HO_4^-(aq) \rightleftharpoons H^+(aq) + C_2O_4^{2-}(aq)$$

Initial (M):	0.054	0.054	0.00
Change (M):	$-y$	$+y$	$+y$
Equilibrium (M):	$(0.054 - y)$	$(0.054 + y)$	y

Step 2. Using the ionization constant in Table 16.4, we write

$$K_{a_2} = \frac{[H^+][C_2O_4^{2-}]}{[C_2HO_4^-]} = 6.1 \times 10^{-5}$$

$$\frac{(0.054 + y)y}{(0.054 - y)} = 6.1 \times 10^{-5}$$

Because the ionization constant is small, we can make these approximations:

$$0.054 + y \approx 0.054$$
$$0.054 - y \approx 0.054$$

Next we substitute them in the equation, which reduces to

$$y = 6.1 \times 10^{-5} \ M$$

Step 3. At equilibrium, therefore

$$[C_2H_2O_4] = 0.046 \ M$$
$$[C_2HO_4^-] = (0.054 - 6.1 \times 10^{-5}) \ M \approx 0.054 \ M$$
$$[H^+] = (0.054 + 6.1 \times 10^{-5}) \ M \approx 0.054 \ M$$
$$[C_2O_4^{2-}] = 6.1 \times 10^{-5} \ M$$
$$[OH^-] = 1.0 \times 10^{-14}/0.054 = 1.9 \times 10^{-13} \ M$$

Similar problem: 16.52.

Practice Exercise Calculate the concentrations of $C_2H_2O_4$, $C_2HO_4^-$, $C_2O_4^{2-}$, and H^+ in a 0.20 M oxalic acid solution.

Table 16.4 | Ionization Constants of Some Diprotic Acids and a Polyprotic Acid and Their Conjugate Bases at 25°C

Name of Acid	Formula	Structure	K_a	Conjugate Base	K_b
Sulfuric acid	H_2SO_4	$$\begin{array}{c} O \\ \parallel \\ H-O-S-O-H \\ \parallel \\ O \end{array}$$	Very large	HSO_4^-	Very small
Hydrogen sulfate ion	HSO_4^-	$$\begin{array}{c} O \\ \parallel \\ H-O-S-O^- \\ \parallel \\ O \end{array}$$	1.3×10^{-2}	SO_4^{2-}	7.7×10^{-13}
Oxalic acid	$C_2H_2O_4$	$$\begin{array}{c} O \quad O \\ \parallel \quad \parallel \\ H-O-C-C-O-H \end{array}$$	6.5×10^{-2}	$C_2HO_4^-$	1.5×10^{-13}
Hydrogen oxalate ion	$C_2HO_4^-$	$$\begin{array}{c} O \quad O \\ \parallel \quad \parallel \\ H-O-C-C-O^- \end{array}$$	6.1×10^{-5}	$C_2O_4^{2-}$	1.6×10^{-10}
Sulfurous acid*	H_2SO_3	$$\begin{array}{c} O \\ \parallel \\ H-O-S-O-H \end{array}$$	1.3×10^{-2}	HSO_3^-	7.7×10^{-13}
Hydrogen sulfite ion	HSO_3^-	$$\begin{array}{c} O \\ \parallel \\ H-O-S-O^- \end{array}$$	6.3×10^{-8}	SO_3^{2-}	1.6×10^{-7}
Carbonic acid	H_2CO_3	$$\begin{array}{c} O \\ \parallel \\ H-O-C-O-H \end{array}$$	4.2×10^{-7}	HCO_3^-	2.4×10^{-8}
Hydrogen carbonate ion	HCO_3^-	$$\begin{array}{c} O \\ \parallel \\ H-O-C-O^- \end{array}$$	4.8×10^{-11}	CO_3^{2-}	2.1×10^{-4}
Hydrosulfuric acid	H_2S	$H-S-H$	9.5×10^{-8}	HS^-	1.1×10^{-7}
Hydrogen sulfide ion[†]	HS^-	$H-S^-$	1×10^{-19}	S^{2-}	1×10^5
Phosphoric acid	H_3PO_4	$$\begin{array}{c} O \\ \parallel \\ H-O-P-O-H \\ \mid \\ O \\ \mid \\ H \end{array}$$	7.5×10^{-3}	$H_2PO_4^-$	1.3×10^{-12}
Dihydrogen phosphate ion	$H_2PO_4^-$	$$\begin{array}{c} O \\ \parallel \\ H-O-P-O^- \\ \mid \\ O \\ \mid \\ H \end{array}$$	6.2×10^{-8}	HPO_4^{2-}	1.6×10^{-7}
Hydrogen phosphate ion	HPO_4^{2-}	$$\begin{array}{c} O \\ \parallel \\ H-O-P-O^- \\ \mid \\ O^- \end{array}$$	4.8×10^{-13}	PO_4^{3-}	2.1×10^{-2}

*H_2SO_3 has never been isolated and exists in only minute concentration in aqueous solution of SO_2. The K_a value here refers to the process $SO_2(g) + H_2O(l) \rightleftharpoons H^+(aq) + HSO_3^-(aq)$.

[†]The ionization constant of HS^- is very low and difficult to measure. The value listed here is only an estimate.

Example 16.11 shows that for diprotic acids if $K_{a_1} \gg K_{a_2}$, then the concentration of the H^+ ions at equilibrium may be assumed to result only from the first stage of ionization. Furthermore, the concentration of the conjugate base for the second-stage ionization is numerically equal to K_{a_2}.

Phosphoric acid (H_3PO_4) is an important polyprotic acid with three ionizable hydrogen atoms:

$$H_3PO_4(aq) \rightleftharpoons H^+(aq) + H_2PO_4^-(aq) \qquad K_{a_1} = \frac{[H^+][H_2PO_4^-]}{[H_3PO_4]} = 7.5 \times 10^{-3}$$

$$H_2PO_4^-(aq) \rightleftharpoons H^+(aq) + HPO_4^{2-}(aq) \qquad K_{a_2} = \frac{[H^+][HPO_4^{2-}]}{[H_2PO_4^-]} = 6.2 \times 10^{-8}$$

$$HPO_4^{2-}(aq) \rightleftharpoons H^+(aq) + PO_4^{3-}(aq) \qquad K_{a_3} = \frac{[H^+][PO_4^{3-}]}{[HPO_4^{2-}]} = 4.8 \times 10^{-13}$$

H_3PO_4

We see that phosphoric acid is a weak polyprotic acid and that its ionization constants decrease markedly for the second and third stages. Thus we can predict that, in a solution containing phosphoric acid, the concentration of the nonionized acid is the highest, and the only other species present in significant concentrations are H^+ and $H_2PO_4^-$ ions.

16.6 Weak Bases and Base Ionization Constants

Base ionizations

Weak bases are treated like weak acids. When ammonia dissolves in water, it undergoes the reaction

$$NH_3(aq) + H_2O(l) \rightleftharpoons NH_4^+(aq) + OH^-(aq)$$

The production of hydroxide ions in this *base ionization* reaction means that, in this solution at 25°C, $[OH^-] > [H^+]$, and therefore pH > 7.

Compared with the total concentration of water, very few water molecules are consumed by the reaction, so we can treat $[H_2O]$ as a constant. Thus we can write the equilibrium constant as

$$K_b = \frac{[NH_4^+][OH^-]}{[NH_3]}$$

$$= 1.8 \times 10^{-5}$$

in which **K_b,** *the equilibrium constant for base ionization,* is called the ***base ionization constant.*** Table 16.5 lists a number of common weak bases and their ionization constants. Note that the basicity of all these compounds is attributable to the lone pair of electrons on the nitrogen atom.

In solving problems involving bases, we follow the same procedure we used for weak acids. The main difference is that we calculate $[OH^-]$ first, rather than $[H^+]$.

The lone pair (red color) on the N atom accounts for ammonia's basicity.

Example 16.12   What is the pH of a 0.400 *M* ammonia solution?

Reasoning and Solution   The procedure is essentially the same as the one we use for weak acids.

Table 16.5 | Ionization Constants of Some Weak Bases and Their Conjugate Acids at 25°C

Name of Base	Formula	Structure	K_b*	Conjugate Acid	K_a
Ethylamine	$C_2H_5NH_2$		5.6×10^{-4}	$C_2H_5\overset{+}{N}H_3$	1.8×10^{-11}
Methylamine	CH_3NH_2		4.4×10^{-4}	$CH_3\overset{+}{N}H_3$	2.3×10^{-11}
Caffeine	$C_8H_{10}N_4O_2$		4.1×10^{-4}	$C_8H_{11}\overset{+}{N}_4O_2$	2.4×10^{-11}
Ammonia	NH_3		1.8×10^{-5}	NH_4^+	5.6×10^{-10}
Pyridine	C_5H_5N		1.7×10^{-9}	$C_5H_5\overset{+}{N}H$	5.9×10^{-6}
Aniline	$C_6H_5NH_2$		3.8×10^{-10}	$C_6H_5\overset{+}{N}H_3$	2.6×10^{-5}
Urea	N_2H_4CO		1.5×10^{-14}	$H_2NCO\overset{+}{N}H_3$	0.67

*The nitrogen atom with the lone pair accounts for each compound's basicity. In the case of urea, K_b can be associated with either nitrogen atom.

Step 1. Let x be the concentration in moles per liter of NH_4^+ and OH^- ions at equilibrium. Next we summarize:

$$NH_3(aq) + H_2O(l) \rightleftharpoons NH_4^+(aq) + OH^-(aq)$$

Initial (M):	0.400	0.000	0.000
Change (M):	$-x$	$+x$	$+x$
Equilibrium (M):	$(0.400 - x)$	x	x

Step 2. Using the base ionization constant listed in Table 16.5, we write

$$K_b = \frac{[NH_4^+][OH^-]}{[NH_3]} = 1.8 \times 10^{-5}$$

$$\frac{x^2}{0.400 - x} = 1.8 \times 10^{-5}$$

Always check the validity of the approximation.

Applying the approximation $0.400 - x \approx 0.400$ gives

$$\frac{x^2}{0.400 - x} \approx \frac{x^2}{0.400} = 1.8 \times 10^{-5}$$

$$x^2 = 7.2 \times 10^{-6}$$

$$x = 2.7 \times 10^{-3} \, M$$

Step 3. At equilibrium, $[OH^-] = 2.7 \times 10^{-3} \, M$. Thus

$$pOH = -\log (2.7 \times 10^{-3})$$

$$= 2.57$$

$$pH = 14.00 - 2.57$$

$$= 11.43$$

Note that we omitted the contribution to $[OH^-]$ from water. This condition is similar to that for ignoring the contribution to $[H^+]$ from water (see p. 514).

Similar problem: 16.55.

Practice Exercise   Calculate the pH of a 0.26 M methylamine solution (see Table 16.5).

16.7 The Relationship Between Conjugate Acid-Base Ionization Constants

An important relationship between the acid ionization constant and the ionization constant of its conjugate base can be derived as follows, using acetic acid as an example:

$$CH_3COOH(aq) \rightleftharpoons H^+(aq) + CH_3COO^-(aq)$$

$$K_a = \frac{[H^+][CH_3COO^-]}{[CH_3COOH]}$$

The conjugate base, CH_3COO^-, reacts with water according to the equation

$$CH_3COO^-(aq) + H_2O(l) \rightleftharpoons CH_3COOH(aq) + OH^-(aq)$$

and we can write the base ionization constant as

$$K_b = \frac{[CH_3COOH][OH^-]}{[CH_3COO^-]}$$

The product of these two ionization constants is given by

$$K_a K_b = \frac{[H^+][CH_3COO^-]}{[CH_3COOH]} \times \frac{[CH_3COOH][OH^-]}{[CH_3COO^-]}$$

$$= [H^+][OH^-]$$

$$= K_w$$

This result may seem strange at first, but it can be understood by realizing that the sum of reactions (1) and (2) here is simply the autoionization of water.

$$
\begin{array}{llr}
(1) & CH_3COOH(aq) \rightleftharpoons H^+(aq) + CH_3COO^-(aq) & K_a \\
(2) & \underline{CH_3COO^-(aq) + H_2O(l) \rightleftharpoons CH_3COOH(aq) + OH^-(aq)} & K_b \\
(3) & H_2O(l) \rightleftharpoons H^+(aq) + OH^-(aq) & K_w
\end{array}
$$

This example illustrates one of the rules for chemical equilibria: When two reactions are added to give a third reaction, the equilibrium constant for the third reaction is the product of the equilibrium constants for the two added reactions. Thus for any conjugate acid-base pair it is always true that

$$K_a K_b = K_w \tag{16.8}$$

Expressing Equation (16.8) as

$$K_a = \frac{K_w}{K_b} \qquad K_b = \frac{K_w}{K_a}$$

enables us to draw an important conclusion: The stronger the acid (the larger K_a), the weaker its conjugate base (the smaller K_b), and vice versa (see Tables 16.3, 16.4, and 16.5).

We can use Equation (16.8) to calculate the K_b of the conjugate base (CH_3COO^-) of CH_3COOH as follows. We find the K_a value of CH_3COOH in Table 16.3 and write

$$
\begin{aligned}
K_b &= \frac{K_w}{K_a} \\
&= \frac{1.0 \times 10^{-14}}{1.8 \times 10^{-5}} \\
&= 5.6 \times 10^{-10}
\end{aligned}
$$

16.8 Acid-Base Properties of Salts

As defined in Section 4.3, a salt is an ionic compound formed by the reaction between an acid and a base. Salts are strong electrolytes that completely dissociate in water and in some cases react with water. The term *salt hydrolysis* describes *the reaction of an anion or a cation of a salt, or both, with water.* Salt hydrolysis usually affects the pH of a solution.

The word "hydrolysis" is derived from the Greek words *hydro,* meaning "water," and *lysis,* meaning "to split apart."

Salts That Produce Neutral Solutions

It is generally true that salts containing an alkali metal ion or alkaline earth metal ion (except Be^{2+}) and the conjugate base of a strong acid (for example, Cl^-, Br^-, and NO_3^-) do not undergo hydrolysis to an appreciable extent, and their solutions are assumed to be neutral. For instance, when $NaNO_3$, a salt formed by the reaction of $NaOH$ with HNO_3, dissolves in water, it dissociates completely as:

In reality, all positive ions give acid solutions in water.

$$NaNO_3(s) \xrightarrow{\ H_2O\ } Na^+(aq) + NO_3^-(aq)$$

The hydrated Na^+ ion neither donates nor accepts H^+ ions. The NO_3^- ion is the conjugate base of the strong acid HNO_3, and it has no affinity for H^+ ions. Consequently, a solution containing Na^+ and NO_3^- ions is neutral, with a pH of 7.

Salts That Produce Basic Solutions

The dissociation of sodium acetate (CH_3COONa) in water is given by

$$CH_3COONa(s) \xrightarrow{\ H_2O\ } Na^+(aq) + CH_3COO^-(aq)$$

The hydrated Na^+ ion has no acidic or basic properties. The acetate ion CH_3COO^-, however, is the conjugate base of the weak acid CH_3COOH and therefore has an affinity for H^+ ions. The hydrolysis reaction is given by

$$CH_3COO^-(aq) + H_2O(l) \rightleftharpoons CH_3COOH(aq) + OH^-(aq)$$

Because this reaction produces OH^- ions, the sodium acetate solution will be basic. The equilibrium constant for this hydrolysis reaction is the base ionization constant expression for CH_3COO^-, so we write

$$K_b = \frac{[CH_3COOH][OH^-]}{[CH_3COO^-]} = 5.6 \times 10^{-10}$$

Salts That Produce Acidic Solutions

When a salt derived from a strong acid and a weak base dissolves in water, the solution becomes acidic. For example, consider the process

$$NH_4Cl(s) \xrightarrow{H_2O} NH_4^+(aq) + Cl^-(aq)$$

The Cl^- ion has no affinity for H^+ ions. The ammonium ion NH_4^+ is the weak conjugate acid of the weak base NH_3 and ionizes as:

$$NH_4^+(aq) + H_2O(l) \rightleftharpoons NH_3(aq) + H_3O^+(aq)$$

or simply

$$NH_4^+(aq) \rightleftharpoons NH_3(aq) + H^+(aq)$$

Because this reaction produces H^+ ions, the pH of the solution decreases. As you can see, the hydrolysis of the NH_4^+ ion is the same as the ionization of the NH_4^+ acid. The equilibrium constant (or ionization constant) for this process is given by

By coincidence, the K_a of the ammonium ion is the same as the K_b of the acetate ion.

$$K_a = \frac{[NH_3][H^+]}{[NH_4^+]} = \frac{K_w}{K_b} = \frac{1.0 \times 10^{-14}}{1.8 \times 10^{-5}} = 5.6 \times 10^{-10}$$

In solving salt hydrolysis problems, we follow the same procedure we used for weak acids and weak bases.

Example 16.13   Calculate the pH of a 0.15 M solution of sodium acetate (CH_3COONa). What is the percent hydrolysis?

Reasoning and Solution   Note that the salt first dissociates into cation and anion in solution. We then examine separately the possible reaction of each ion with water.

Step 1. The major species in a sodium acetate solution are Na^+ and CH_3COO^- ions. Because we started with a 0.15 M sodium acetate solution, the initial concentrations of the ions are also equal to 0.15 M:

$$CH_3COONa(s) \xrightarrow{H_2O} Na^+(aq) + CH_3COO^-(aq)$$
$$\phantom{CH_3COONa(s) \xrightarrow{H_2O}} 0.15\ M \qquad 0.15\ M$$

Of these ions, only the acetate ion will hydrolyze. We ignore the contribution to H^+ and OH^- ions by water.

Step 2. Let x be the equilibrium concentration of CH_3COOH and OH^- ions in mol/L, we summarize:

$$CH_3COO^-(aq) + H_2O(l) \rightleftharpoons CH_3COOH(aq) + OH^-(aq)$$

Initial (M):	0.15	0.00	0.00
Change (M):	$-x$	$+x$	$+x$
Equilibrium (M):	$0.15 - x$	x	x

Step 3. From the earlier discussion we write the equilibrium constant of hydrolysis, or the base ionization constant, as

$$K_b = \frac{[CH_3COOH][OH^-]}{[CH_3COO^-]} = 5.6 \times 10^{-10}$$

$$\frac{x^2}{0.15 - x} = 5.6 \times 10^{-10}$$

Because K_b is very small and the initial concentration of the base is large, we can apply the approximation $0.15 - x \approx 0.15$:

$$\frac{x^2}{0.15 - x} \approx \frac{x^2}{0.15} = 5.6 \times 10^{-10}$$

$$x = 9.2 \times 10^{-6} \, M$$

Step 4. At equilibrium:

$$[OH^-] = 9.2 \times 10^{-6} \, M$$

$$pOH = -\log(9.2 \times 10^{-6})$$

$$= 5.04$$

$$pH = 14.00 - 5.04$$

$$= 8.96$$

Thus the solution is basic, as we would expect. The percent hydrolysis is given by

$$\% \text{ hydrolysis} = \frac{9.2 \times 10^{-6} \, M}{0.15 \, M} \times 100\%$$

$$= 0.0061\%$$

Comment The result shows that only a very small amount of the anion undergoes hydrolysis. Note that the calculation of percent hydrolysis takes the same form as the test for the approximation, which is valid in this case.

Similar problem: 16.67.

Practice Exercise Calculate the pH of a 0.24 M sodium formate solution (HCOONa).

Metal Ion Hydrolysis

Salts that contain small, highly charged metal cations (for example, Al^{3+}, Cr^{3+}, Fe^{3+}, Bi^{3+}, and Be^{2+}) and the conjugate bases of strong acids also produce an acidic solution. For example, when aluminum chloride ($AlCl_3$) dissolves in water, the Al^{3+} ions take the hydrated form $Al(H_2O)_6^{3+}$ (Figure 16.5). Let us consider one bond between the metal ion and the oxygen atom of one of the six water molecules in $Al(H_2O)_6^{3+}$:

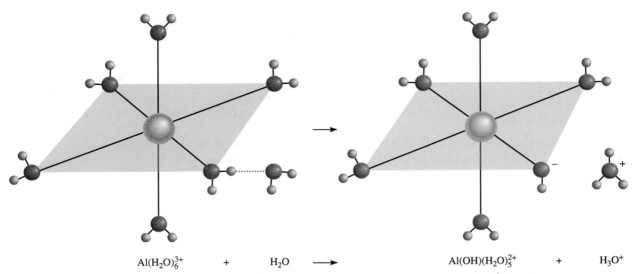

$$Al(H_2O)_6^{3+} \quad + \quad H_2O \quad \longrightarrow \quad Al(OH)(H_2O)_5^{2+} \quad + \quad H_3O^+$$

Figure 16.5 *The six H_2O molecules surround the Al^{3+} ion octahedrally. The attraction of the small Al^{3+} ion for the lone pairs on the oxygen atoms is so great that the O—H bonds in a H_2O molecule attached to the metal cation are weakened, allowing the loss of a proton (H^+) to an incoming H_2O molecule.*

The positively charged Al^{3+} ion draws electron density toward itself, making the O—H bond more polar. Consequently, the H atoms have a greater tendency to ionize than those in water molecules not involved in hydration. The resulting ionization process can be written as

> The hydrated Al^{3+} qualifies as a proton donor and thus a Brønsted acid.

$$Al(H_2O)_6^{3+}(aq) + H_2O(l) \rightleftharpoons Al(OH)(H_2O)_5^{2+}(aq) + H_3O^+(aq)$$

or simply

$$Al(H_2O)_6^{3+}(aq) \rightleftharpoons Al(OH)(H_2O)_5^{2+}(aq) + H^+(aq)$$

The equilibrium constant for the metal cation hydrolysis is given by

> Note that $Al(H_2O)_6^{3+}$ is roughly as strong an acid as CH_3COOH.

$$K_a = \frac{[Al(OH)(H_2O)_5^{2+}][H^+]}{[Al(H_2O)_6^{3+}]} = 1.3 \times 10^{-5}$$

Note that the $Al(OH)(H_2O)_5^{2+}$ species can undergo further ionization as

$$Al(OH)(H_2O)_5^{2+}(aq) \rightleftharpoons Al(OH)_2(H_2O)_4^+(aq) + H^+(aq)$$

and so on. However, generally it is sufficient to consider only the first step of hydrolysis.

The extent of hydrolysis is greatest for the smallest and most highly charged ions because a "compact" highly charged ion is more effective in polarizing the O—H bond and facilitating ionization. This is the reason that relatively large ions of low charge such as Na^+ and K^+ do not undergo hydrolysis to any appreciable extent.

Table 16.6 | Acid-Base Properties of Salts

Type of Salt	Examples	Ions That Undergo Hydrolysis	pH of Solution
Cation from strong base; anion from strong acid	NaCl, KI, KNO$_3$, RbBr, BaCl$_2$	None	≈ 7
Cation from strong base; anion from weak acid	CH$_3$COONa, KNO$_2$	Anion	> 7
Cation from weak base; anion from strong acid	NH$_4$Cl, NH$_4$NO$_3$	Cation	< 7
Cation from weak base; anion from weak acid	NH$_4$NO$_2$, CH$_3$COONH$_4$, NH$_4$CN	Anion and cation	< 7 if $K_b < K_a$
			≈ 7 if $K_b \approx K_a$
			> 7 if $K_b > K_a$
Small, highly charged cation; anion from strong acid	AlCl$_3$, Fe(NO$_3$)$_3$	Hydrated cation	< 7

Salts in Which Both the Cation and the Anion Hydrolyze

So far we have considered salts in which only one ion undergoes hydrolysis. For salts derived from a weak acid and a weak base, both the cation and the anion hydrolyze. However, whether a solution containing such a salt is acidic, basic, or neutral depends on the relative strengths of the weak acid and the weak base. Because the mathematics associated with this type of system is rather involved, we limit ourselves to making qualitative predictions about these solutions. We consider three situations:

- $K_b > K_a$. If K_b for the anion is greater than K_a for the cation, then the solution must be basic because the anion will hydrolyze to a greater extent than the cation. At equilibrium, there will be more OH$^-$ ions than H$^+$ ions.
- $K_b < K_a$. Conversely, if K_b of the anion is smaller than K_a of the cation, the solution will be acidic because cation hydrolysis will be more extensive than anion hydrolysis.
- $K_a \approx K_b$. If K_a is approximately equal to K_b, the solution will be nearly neutral.

Table 16.6 summarizes the behavior in aqueous solution of the salts discussed in this section.

Example 16.14 Predict whether these solutions will be acidic, basic, or nearly neutral: (a) NH$_4$I, (b) CaCl$_2$, (c) KCN, (d) Fe(NO$_3$)$_3$.

Reasoning and Solution We first break up the salt into its cation and anion components and then examine the possible reaction of each ion with water.

(a) The cation is NH$_4^+$, which will hydrolyze to produce NH$_3$ and H$^+$. The I$^-$ anion is the conjugate base of the strong acid HI. Therefore, I$^-$ will not hydrolyze and so the solution is acidic.

(b) The Ca^{2+} cation does not hydrolyze. The Cl$^-$ anion also does not hydrolyze. The solution will be almost neutral.

(c) The K$^+$ cation does not hydrolyze. CN$^-$ is the conjugate base of the weak acid HCN and will hydrolyze to give OH$^-$ and HCN. The solution will be basic.

(d) Fe^{3+} is a small metal ion with a high charge and hydrolyzes to produce H$^+$ ions. The NO$_3^-$ ion does not hydrolyze. Consequently, the solution will be acidic.

Similar problem: 16.63.

> **Practice Exercise**   Predict the pH of these salt solutions: (a) $LiClO_4$, (b) Na_3PO_4,
> (c) $Bi(NO_3)_3$, (d) NH_4CN.

Finally we note that some anions can act either as an acid or as a base. For example, the bicarbonate ion (HCO_3^-) can ionize or undergo hydrolysis as (see Table 16.4):

$$HCO_3^-(aq) + H_2O(l) \rightleftharpoons H_3O^+(aq) + CO_3^{2-}(aq) \quad K_a = 4.8 \times 10^{-11}$$
$$HCO_3^-(aq) + H_2O(l) \rightleftharpoons H_2CO_3(aq) + OH^-(aq) \quad K_b = 2.4 \times 10^{-8}$$

Because $K_b > K_a$, we predict that the hydrolysis reaction will outweigh the ionization process. Thus, a solution of sodium bicarbonate ($NaHCO_3$) will be basic.

16.9 Acidic, Basic, and Amphoteric Oxides

As we saw in Chapter 8, oxides can be classified as acidic, basic, or amphoteric. Our discussion of acid-base reactions would be incomplete without examining the properties of these compounds.

Figure 16.6 shows the formulas of a number of oxides of the representative elements in their highest oxidation states. Note that all alkali metal oxides and all alkaline earth metal oxides except BeO are basic. Beryllium oxide and several metallic oxides in Groups 3A and 4A are amphoteric. Nonmetallic oxides in which the oxidation number of the representative element is high are acidic (for example, N_2O_5, SO_3, and Cl_2O_7). Those in which the oxidation number of the representative element is low (for example, CO and NO) show no measurable acidic properties. No nonmetallic oxides are known to have basic properties.

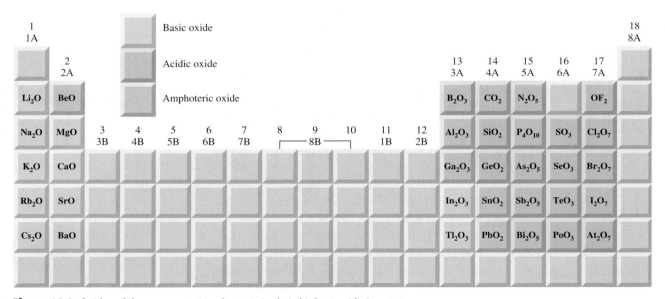

Figure 16.6 *Oxides of the representative elements in their highest oxidation states.*

The basic metallic oxides react with water to form metal hydroxides:

$$Na_2O(s) + H_2O(l) \longrightarrow 2NaOH(aq)$$
$$BaO(s) + H_2O(l) \longrightarrow Ba(OH)_2(aq)$$

The reactions between acidic oxides and water are:

$$CO_2(g) + H_2O(l) \rightleftharpoons H_2CO_3(aq)$$
$$SO_3(g) + H_2O(l) \rightleftharpoons H_2SO_4(aq)$$
$$N_2O_5(g) + H_2O(l) \rightleftharpoons 2HNO_3(aq)$$
$$P_4O_{10}(s) + 6H_2O(l) \rightleftharpoons 4H_3PO_4(aq)$$
$$Cl_2O_7(g) + H_2O(l) \rightleftharpoons 2HClO_4(aq)$$

The reaction between CO_2 and H_2O explains why when pure water is exposed to air (which contains CO_2) it gradually reaches a pH of about 5.5 (Figure 16.7). The reaction between SO_3 and H_2O is largely responsible for acid rain.

Reactions between acidic oxides and bases and those between basic oxides and acids resemble normal acid-base reactions in that the products are a salt and water:

$$\underset{\text{acidic oxide}}{CO_2(g)} + \underset{\text{base}}{2NaOH(aq)} \longrightarrow \underset{\text{salt}}{Na_2CO_3(aq)} + \underset{\text{water}}{H_2O(l)}$$

$$\underset{\text{basic oxide}}{BaO(s)} + \underset{\text{acid}}{2HNO_3(aq)} \longrightarrow \underset{\text{salt}}{Ba(NO_3)_2(aq)} + \underset{\text{water}}{H_2O(l)}$$

As Figure 16.6 shows, aluminum oxide (Al_2O_3) is amphoteric. Depending on the reaction conditions, it can behave either as an acidic oxide or as a basic oxide. For example, Al_2O_3 acts as a base with hydrochloric acid to produce a salt ($AlCl_3$) and water:

$$Al_2O_3(s) + 6HCl(aq) \longrightarrow 2AlCl_3(aq) + 3H_2O(l)$$

A forest damaged by acid rain.

Figure 16.7 *(Left) A beaker of water to which a few drops of bromothymol blue indicator have been added. (Right) As dry ice is added to the water, the CO_2 reacts to form carbonic acid, which turns the solution acidic and changes the color from blue to yellow.*

and acts as an acid with sodium hydroxide:

$$Al_2O_3(s) + 2NaOH(aq) + 3H_2O(l) \longrightarrow 2NaAl(OH)_4(aq)$$

Note that only a salt, $NaAl(OH)_4$ [containing the Na^+ and $Al(OH)_4^-$ ions], is formed in the latter reaction; no water is produced. Nevertheless, this reaction can still be classified as an acid-base reaction because Al_2O_3 neutralizes NaOH.

Some transition metal oxides in which the metal has a high oxidation number act as acidic oxides. Consider manganese(VII) oxide (Mn_2O_7) and chromium(VI) oxide (CrO_3), both of which react with water to produce acids:

$$Mn_2O_7(l) + H_2O(l) \longrightarrow 2HMnO_4(aq)$$
permanganic acid

$$CrO_3(s) + H_2O(l) \longrightarrow H_2CrO_4(aq)$$
chromic acid

The higher the oxidation state of the metal, the more covalent the compound; the lower the oxidation state, the more ionic the compound.

16.10 Lewis Acids and Bases

Acid-base properties so far have been discussed in terms of the Brønsted theory. To behave as a Brønsted base, for example, a substance must be able to accept protons. By this definition both the hydroxide ion and ammonia are bases:

$$H^+ + {}^-\!:\!\overset{..}{\underset{..}{O}}\!-\!H \longrightarrow H\!-\!\overset{..}{\underset{..}{O}}\!-\!H$$

$$H^+ + :\!\overset{\underset{|}{H}}{\underset{\underset{H}{|}}{N}}\!-\!H \longrightarrow \left[H\!-\!\overset{\underset{|}{H}}{\underset{\underset{H}{|}}{N}}\!-\!H \right]^+$$

In each case, the atom to which the proton becomes attached possesses at least one unshared pair of electrons. This characteristic property of the OH^- ion, of NH_3, and of other Brønsted bases suggests a more general definition of acids and bases.

The American chemist G. N. Lewis formulated such a definition. According to Lewis's definition, a base is *a substance that can donate a pair of electrons,* and an acid is *a substance that can accept a pair of electrons.* For example, in the protonation of ammonia, NH_3 acts as a **Lewis base** because it donates a pair of electrons to the proton H^+, which acts as a **Lewis acid** by accepting the pair of electrons. A Lewis acid-base reaction, therefore, is one that involves the donation of a pair of electrons from one species to another. Such a reaction does not produce a salt and water.

The significance of the Lewis concept is that it is much more general than other definitions; it includes as acid-base reactions many reactions that do not involve Brønsted acids. Consider, for example, the reaction between boron trifluoride (BF_3) and ammonia (Figure 16.8):

$$\underset{\text{acid}}{F\!-\!\overset{\underset{|}{F}}{\underset{\underset{F}{|}}{B}}} + \underset{\text{base}}{:\!\overset{\underset{|}{H}}{\underset{\underset{H}{|}}{N}}\!-\!H} \longrightarrow F\!-\!\overset{\underset{|}{F}}{\underset{\underset{F}{|}}{B}}\!-\!\overset{\underset{|}{H}}{\underset{\underset{H}{|}}{N}}\!-\!H$$

Figure 16.8 *A Lewis acid-base reaction involving BF_3 and NH_3. Here the ammonia and boron trifluoride gases react to form a solid product.*

In Section 10.4 we saw that the B atom in BF_3 is sp^2-hybridized. The vacant, unhybridized $2p$ orbital accepts the pair of electrons from NH_3. So BF_3 functions as an acid according to the Lewis definition, even though it does not contain an ionizable proton. Note that a coordinate covalent bond is formed between the B and N atoms (see p. 279).

Another Lewis acid containing boron is boric acid (H_3BO_3). Boric acid (a weak acid used in eyewash) is an oxoacid with the following structure:

$$\begin{array}{c} H \\ | \\ :O: \\ | \\ H-\ddot{O}-B-\ddot{O}-H \end{array}$$

H_3BO_3

The K_a of boric acid is 7.3×10^{-10}.

Note that boric acid does not ionize in water to produce an H^+ ion. Instead, its reaction with water is

$$B(OH)_3(aq) + H_2O(l) \rightleftharpoons B(OH)_4^-(aq) + H^+(aq)$$

In this Lewis acid-base reaction, boric acid accepts a pair of electrons from the hydroxide ion that is derived from the H_2O molecule.

The hydration of carbon dioxide to produce carbonic acid

$$CO_2(g) + H_2O(l) \rightleftharpoons H_2CO_3(aq)$$

can be understood in the Lewis framework as follows: The first step involves donation of a lone pair on the oxygen atom in H_2O to the carbon atom in CO_2. An orbital is vacated on the C atom to accommodate the lone pair by removal of the electron pair in the C—O pi bond. These shifts of electrons are indicated by the curved arrows.

Therefore, H_2O is a Lewis base and CO_2 is a Lewis acid. Next, a proton is transferred onto the O atom bearing a negative charge to form H_2CO_3.

Summary

Brønsted acids donate protons and Brønsted bases accept protons. Every Brønsted acid has a conjugate Brønsted base and vice versa.

The acidity of an aqueous solution is expressed as its pH, which is defined as the negative logarithm of the hydrogen ion concentration (in moles per liter). At 25°C, an acidic solution has pH < 7, a basic solution has pH > 7, and a neutral solution has pH = 7.

In aqueous solution, the following are classified as strong acids because they ionize completely: $HClO_4$, HI, HBr, HCl, H_2SO_4 (first stage of ionization), and HNO_3. Most acids are weak acids and, therefore, weak electrolytes. These are classified as strong bases because they ionize completely in aqueous solution: hydroxides of alkali metals and of alkaline earth metals (except beryllium).

The acid ionization constant K_a increases with acid strength. K_b similarly expresses the strengths of bases. Percent ionization is another measure of the strength of acids. The more dilute a solution of a weak acid, the greater the percent ionization of the acid. Diprotic and polyprotic acids yield more than one proton on ionization per molecule of the acid. The product of the ionization constant of an acid and the ionization constant of its conjugate base is equal to the ion-product constant of water; that is, $K_a K_b = K_w$.

Most salts are strong electrolytes and dissociate completely into ions in solution. The reaction of these ions with water, called salt hydrolysis, can lead to acidic or basic solutions. In salt hydrolysis, the conjugate bases of weak acids yield basic solutions, and the conjugate acids of weak bases yield acidic solutions. Small, highly charged metal ions such as Al^{3+} and Fe^{3+} hydrolyze to yield acidic solutions. Most oxides can be classified as acidic, basic, or amphoteric.

Lewis acids accept pairs of electrons and Lewis bases donate pairs of electrons. The term "Lewis acid" is generally reserved for substances that can accept electron pairs but do not contain ionizable hydrogen atoms.

Key Words

Acid ionization constant (K_a) p. 511	Conjugate acid-base pair p. 502	Lewis base p. 530	Strong base p. 508
Base ionization constant (K_b) p. 520	Ion-product constant p. 504	Percent ionization p. 516	Weak acid p. 508
		pH p. 505	Weak base p. 509
	Lewis acid p. 530	Salt hydrolysis p. 523	
		Strong acid p. 507	

Questions and Problems

Brønsted Acids and Bases

Review Questions

16.1 Define Brønsted acids and bases. How do the Brønsted definitions differ from Arrhenius's definitions of acids and bases?

16.2 For a species to act as a Brønsted base, an atom in the species must possess a lone pair of electrons. Explain why this is so.

Problems

16.3 Classify each of these species as a Brønsted acid or base, or both: (a) H_2O, (b) OH^-, (c) H_3O^+, (d) NH_3, (e) NH_4^+, (f) NH_2^-, (g) NO_3^-, (h) CO_3^{2-}, (i) HBr, (j) HCN.

16.4 What are the names and formulas of the conjugate bases of these acids: (a) HNO_2, (b) H_2SO_4, (c) H_2S, (d) HCN, (e) HCOOH (formic acid)?

16.5 Identify the acid-base conjugate pairs in each of these reactions:

(a) $CH_3COO^- + HCN \rightleftharpoons CH_3COOH + CN^-$

(b) $HCO_3^- + HCO_3^- \rightleftharpoons H_2CO_3 + CO_3^{2-}$
(c) $H_2PO_4^- + NH_3 \rightleftharpoons HPO_4^{2-} + NH_4^+$
(d) $HClO + CH_3NH_2 \rightleftharpoons CH_3NH_3^+ + ClO^-$
(e) $CO_3^{2-} + H_2O \rightleftharpoons HCO_3^- + OH^-$
(f) $CH_3COO^- + H_2O \rightleftharpoons CH_3COOH + OH^-$

16.6 Give the conjugate acid of each of these bases: (a) HS^-, (b) HCO_3^-, (c) CO_3^{2-}, (d) $H_2PO_4^-$, (e) HPO_4^{2-}, (f) PO_4^{3-}, (g) HSO_4^-, (h) SO_4^{2-}, (i) NO_2^-, (j) SO_3^{2-}.

16.7 Give the conjugate base of each of these acids: (a) $CH_2ClCOOH$, (b) HIO_4, (c) H_3PO_4, (d) $H_2PO_4^-$, (e) HPO_4^{2-}, (f) H_2SO_4, (g) HSO_4^-, (h) HCOOH, (i) HSO_3^-, (j) NH_4^+, (k) H_2S, (l) HS^-, (m) HClO.

16.8 Oxalic acid ($C_2H_2O_4$) has the following structure:

$$O=C-OH$$
$$|$$
$$O=C-OH$$

An oxalic acid solution contains these species in varying concentrations: $C_2H_2O_4$, $C_2HO_4^-$, $C_2O_4^{2-}$, and H^+. (a) Draw Lewis structures of $C_2HO_4^-$ and $C_2O_4^{2-}$. (b) Which of the four species listed here can act only as acids, which can act only as bases, and which can act as both acids and bases?

pH and pOH Calculations

Review Questions

16.9 What is the ion-product constant for water?

16.10 Write an equation relating $[H^+]$ and $[OH^-]$ in solution at 25°C.

16.11 The ion-product constant for water is 1.0×10^{-14} at 25°C and 3.8×10^{-14} at 40°C. Is the process

$$H_2O(l) \rightleftharpoons H^+(aq) + OH^-(aq)$$

endothermic or exothermic?

16.12 Define pH. Why do chemists normally choose to discuss the acidity of a solution in terms of pH rather than hydrogen ion concentration, $[H^+]$?

16.13 The pH of a solution is 6.7. From this statement alone, can you conclude that the solution is acidic? If not, what additional information would you need? Can the pH of a solution be zero or negative? If so, give examples to illustrate these values.

16.14 Define pOH. Write an equation relating pH and pOH.

Problems

16.15 Calculate the hydrogen ion concentration for solutions with these pH values: (a) 2.42, (b) 11.21, (c) 6.96, (d) 15.00.

16.16 Calculate the hydrogen ion concentration in moles per liter for each of these solutions: (a) a solution whose pH is 5.20, (b) a solution whose pH is 16.00; (c) a solution whose hydroxide concentration is $3.7 \times 10^{-9}\,M$.

16.17 Calculate the pH of each of these solutions: (a) 0.0010 M HCl, (b) 0.76 M KOH, (c) $2.8 \times 10^{-4}\,M$ Ba(OH)$_2$, (d) $5.2 \times 10^{-4}\,M$ HNO$_3$.

16.18 Calculate the pH of water at 40°C, given that K_w is 3.8×10^{-14} at this temperature.

16.19 Complete this table for a solution:

pH	$[H^+]$	Solution is
< 7		
	$< 1.0 \times 10^{-7}\,M$	
		Neutral

16.20 Fill in the word "acidic," "basic," or "neutral" for these solutions:

(a) pOH > 7; solution is _____.
(b) pOH = 7; solution is _____.
(c) pOH < 7; solution is _____.

16.21 The pOH of a solution is 9.40. Calculate the hydrogen ion concentration of the solution.

16.22 Calculate the number of moles of KOH in 5.50 mL of a 0.360 M KOH solution. What is the pOH of the solution?

16.23 A solution is made by dissolving 18.4 g of HCl in 662 mL of water. Calculate the pH of the solution. (Assume that the volume of the solution is also 662 mL.)

16.24 How much NaOH (in grams) is needed to prepare 546 mL of solution with a pH of 10.00?

Strengths of Acids and Bases

Review Questions

16.25 Explain what is meant by the strength of an acid.

16.26 Without referring to the text, write the formulas of four strong acids and four weak acids.

16.27 What are the strongest acid and strongest base that can exist in water?

16.28 H$_2$SO$_4$ is a strong acid, but HSO$_4^-$ is a weak acid. Account for the difference in strength of these two related species.

Problems

16.29 Classify each of these species as a weak or strong acid: (a) HNO$_3$, (b) HF, (c) H$_2$SO$_4$, (d) HSO$_4^-$, (e) H$_2$CO$_3$, (f) HCO$_3^-$, (g) HCl, (h) HCN, (i) HNO$_2$.

16.30 Classify each of these species as a weak or strong base: (a) LiOH, (b) CN$^-$, (c) H$_2$O, (d) ClO$_4^-$, (e) NH$_2^-$.

16.31 Which of these statements is/are true regarding a 0.10 M solution of a weak acid HA?

(a) The pH is 1.00.
(b) $[H^+] \gg [A^-]$
(c) $[H^+] = [A^-]$
(d) The pH is less than 1.

16.32 Which of these statements is/are true regarding a 1.0 M solution of a strong acid HA?

(a) $[A^-] > [H^+]$
(b) The pH is 0.00.
(c) $[H^+] = 1.0\,M$
(d) $[HA] = 1.0\,M$

16.33 Predict the direction that predominates in this reaction:

$$F^-(aq) + H_2O(l) \rightleftharpoons HF(aq) + OH^-(aq)$$

16.34 Predict whether this reaction will proceed from left to right to any measurable extent:

$$CH_3COOH(aq) + Cl^-(aq) \longrightarrow$$

16.35 Which of the following acids is the stronger: CH$_3$COOH or CH$_2$ClCOOH? Explain your choice. (*Hint:* Chlorine is more electronegative than hydrogen. How would its presence affect the ionization of the carboxyl group?)

16.36   Consider these compounds:

phenol   methanol

Experimentally, phenol is found to be a stronger acid than methanol. Explain this difference in terms of the structures of the conjugate bases. (*Hint:* A more stable conjugate base favors ionization. Only one of the conjugate bases can be stabilized by resonance.)

Weak Acid Ionization Constants

Review Questions

16.37   What does the ionization constant tell us about the strength of an acid?

16.38   List the factors on which the K_a of a weak acid depends.

16.39   Why do we normally not quote K_a values for strong acids such as HCl and HNO_3? Why is it necessary to specify temperature when giving K_a values?

16.40   Which of the following solutions has the highest pH? (a) 0.40 M HCOOH, (b) 0.40 M $HClO_4$, (c) 0.40 M CH_3COOH.

Problems

16.41   Calculate the concentrations of all the species (HCN, H^+, CN^-, and OH^-) in a 0.15 M HCN solution.

16.42   A 0.0560-g quantity of acetic acid is dissolved in enough water to make 50.0 mL of solution. Calculate the concentrations of H^+, CH_3COO^-, and CH_3COOH at equilibrium. (K_a for acetic acid = 1.8×10^{-5}.)

16.43   The pH of an HF solution is 6.20. Calculate the ratio [conjugate base]/[acid] for HF at this pH.

16.44   What is the original molarity of a solution of formic acid (HCOOH) whose pH is 3.26 at equilibrium?

16.45   Calculate the pH of a 0.050M HF solution.

16.46   Calculate the percent ionization of hydrofluoric acid at these concentrations: (a) 0.60 M, (b) 0.080 M, (c) 0.0046 M, (d) 0.00028 M. Comment on the trends.

16.47   A 0.040 M solution of a monoprotic acid is 14 percent ionized. Calculate the ionization constant of the acid.

16.48   (a) Calculate the percent ionization of a 0.20 M solution of the monoprotic acetylsalicylic acid (aspirin). (K_a = 3.0×10^{-4}.) (b) The pH of gastric juice in the stomach of a certain individual is 1.00. After a few aspirin tablets have been swallowed, the concentration of acetylsalicylic acid in the stomach is 0.20 M. Calculate the percent ionization of the acid under these conditions.

Diprotic and Polyprotic Acids

Review Questions

16.49   Malonic acid [$CH_2(COOH)_2$] is a diprotic acid. Explain what that means.

16.50   Write all the species (except water) that are present in a phosphoric acid solution. Indicate which species can act as a Brønsted acid, which as a Brønsted base, and which as both a Brønsted acid and a Brønsted base.

Problems

16.51   What are the concentrations of HSO_4^-, SO_4^{2-}, and H^+ in a 0.20 M $KHSO_4$ solution? (*Hint:* H_2SO_4 is a strong acid: K_a for $HSO_4^- = 1.3 \times 10^{-2}$.)

16.52   Calculate the concentrations of H^+, HCO_3^-, and CO_3^{2-} in a 0.025 M H_2CO_3 solution.

Weak Base Ionization Constants; K_a–K_b Relationship

Review Questions

16.53   Use NH_3 to illustrate what we mean by the strength of a base.

16.54   Write the equation relating K_a for a weak acid and K_b for its conjugate base. Use NH_3 and its conjugate acid NH_4^+ to derive the relationship between K_a and K_b.

Problems

16.55   Calculate the pH for each of these solutions: (a) 0.10 M NH_3, (b) 0.050 M pyridine.

16.56   The pH of a 0.30 M solution of a weak base is 10.66. What is the K_b of the base?

16.57   What is the original molarity of a solution of ammonia whose pH is 11.22?

16.58   In a 0.080 M NH_3 solution, what percent of the NH_3 is present as NH_4^+?

Acid-Base Properties of Salt Solutions

Review Questions

16.59   Define salt hydrolysis. Categorize salts according to how they affect the pH of a solution.

16.60   Explain why small, highly charged metal ions are able to undergo hydrolysis.

16.61 Al^{3+} is not a Brønsted acid, but $Al(H_2O)_6^{3+}$ is. Explain.

16.62   Specify which of these salts will undergo hydrolysis: KF, $NaNO_3$, NH_4NO_2, $MgSO_4$, KCN, C_6H_5COONa, RbI, Na_2CO_3, $CaCl_2$, HCOOK.

16.63   Predict the pH (>7, <7, or ≈ 7) of the aqueous solutions containing the following salts: (a) KBr, (b) $Al(NO_3)_3$, (c) $BaCl_2$, (d) $Bi(NO_3)_3$.

16.64   Which ion of the alkaline earth metals is most likely to undergo hydrolysis?

Problems

16.65 A certain salt, MX (containing the M^+ and X^- ions), is dissolved in water, and the pH of the resulting solution is 7.0. Can you say anything about the strengths of the acid and the base from which the salt is derived?

16.66 In a certain experiment a student finds that the pHs of 0.10 M solutions of three potassium salts KX, KY, and KZ are 7.0, 9.0, and 11.0, respectively. Arrange the acids HX, HY, and HZ in order of increasing acid strength.

16.67 Calculate the pH of a 0.36 M CH_3COONa solution.

16.68 Calculate the pH of a 0.42 M NH_4Cl solution.

16.69 Predict whether a solution containing the salt K_2HPO_4 will be acidic, neutral, or basic. (*Hint:* You need to consider both the ionization and hydrolysis of HPO_4^{2-}.)

16.70 Predict the pH (>7, <7, ≈ 7) of a $NaHCO_3$ solution.

Lewis Acids and Bases

Review Questions

16.71 What are the Lewis definitions of an acid and a base? In what way are they more general than the Brønsted definitions?

16.72 In terms of orbitals and electron arrangements, what must be present for a molecule or an ion to act as a Lewis acid (use H^+ and BF_3 as examples)? What must be present for a molecule or ion to act as a Lewis base (use OH^- and NH_3 as examples)?

Problems

16.73 Classify each of these following species as a Lewis acid or a Lewis base: (a) CO_2, (b) H_2O, (c) I^-, (d) SO_2, (e) NH_3, (f) OH^-, (g) H^+, (h) BCl_3.

16.74 Describe this reaction according to the Lewis theory of acids and bases:

$$AlCl_3(s) + Cl^-(aq) \longrightarrow AlCl_4^-(aq)$$

16.75 Which would be considered a stronger Lewis acid: (a) BF_3 or BCl_3, (b) Fe^{2+} or Fe^{3+}? Explain.

16.76 All Brønsted acids are Lewis acids, but the reverse is not true. Give two examples of Lewis acids that are not Brønsted acids.

Additional Problems

16.77 Classify these following oxides as acidic, basic, amphoteric, or neutral: (a) CO_2, (b) K_2O, (c) CaO, (d) N_2O_5, (e) CO, (f) NO, (g) SnO_2, (h) SO_3, (i) Al_2O_3, (j) BaO.

16.78 A typical reaction between an antacid and the hydrochloric acid in gastric juice is

$$NaHCO_3(aq) + HCl(aq) \longrightarrow$$
$$NaCl(aq) + H_2O(l) + CO_2(g)$$

Calculate the volume (in liters) of CO_2 generated from 0.350 g of $NaHCO_3$ and excess gastric juice at 1.00 atm and 37.0°C.

16.79 Addition to which of these of an equal volume of 0.60 M NaOH would lead to a solution having a lower pH: (a) water, (b) 0.30 M HCl, (c) 0.70 M KOH, (d) 0.40 M $NaNO_3$?

16.80 The pH of a 0.0642 M solution of a monoprotic acid is 3.86. Is this a strong acid?

16.81 Like water, ammonia undergoes autoionization in liquid ammonia:

$$NH_3 + NH_3 \Longrightarrow NH_4^+ + NH_2^-$$

(a) Identify the Brønsted acids and Brønsted bases in this reaction. (b) What species correspond to H^+ and OH^-, and what is the condition for a neutral solution?

16.82 HA and HB are both weak acids although HB is the stronger of the two. Will it take more volume of a 0.10 M NaOH solution to neutralize 50.0 mL of 0.10 M HB than 50.0 mL of 0.10 M HA?

16.83 A 1.87-g sample of Mg reacts with 80.0 mL of a HCl solution whose pH is -0.544. What is the pH of the solution after all the Mg has reacted? Assume volume of solution is constant.

16.84 The three common chromium oxides are CrO, Cr_2O_3, and CrO_3. If Cr_2O_3 is amphoteric, what can you say about the acid-base properties of CrO and CrO_3?

16.85 Most of the hydrides of Group 1A and Group 2A metals are ionic (the exceptions are BeH_2 and MgH_2, which are covalent compounds). (a) Describe the reaction between the hydride ion (H^-) and water in terms of a Brønsted acid-base reaction. (b) The same reaction can also be classified as a redox reaction. Identify the oxidizing and reducing agents.

16.86 Use the data in Table 16.3 to calculate the equilibrium constant for the following reaction:

$$CH_3COOH(aq) + NO_2^-(aq) \Longrightarrow$$
$$CH_3COO^-(aq) + HNO_2(aq)$$

16.87 Calculate the pH of a 0.20 M ammonium acetate (CH_3COONH_4) solution.

16.88 Novocaine, used as a local anesthetic by dentists, is a weak base ($K_b = 8.91 \times 10^{-6}$). What is the ratio of the concentration of the base to that of its acid in the blood plasma (pH = 7.40) of a patient?

16.89 In the vapor phase, acetic acid molecules associate to a certain extent to form dimers:

$$2CH_3COOH(g) \Longrightarrow (CH_3COOH)_2(g)$$

At 51°C the pressure of a certain acetic acid vapor system is 0.0342 atm in a 360-mL flask. The vapor is condensed and neutralized with 13.8 mL of 0.0568 M NaOH. (a) Calculate the degree of dissociation (α) of the dimer under these conditions:

$$(CH_3COOH)_2 \Longrightarrow 2CH_3COOH$$

(*Hint:* See Problem 15.63 for general procedure.) (b) Calculate the equilibrium constant K_P for the reaction in (a).

16.90 Calculate the concentrations of all the species in a 0.100 M Na_2CO_3 solution.

16.91 Henry's law constant for CO_2 at 38°C is 2.28×10^{-3} mol/L · atm. Calculate the pH of a solution of CO_2 at 38°C in equilibrium with the gas at a partial pressure of 3.20 atm.

16.92 Hydrocyanic acid (HCN) is a weak acid and a deadly poisonous compound that, in the gaseous form (hydrogen cyanide), is used in gas chambers. Why is it dangerous to treat sodium cyanide with acids (such as HCl) without proper ventilation?

16.93 A solution of formic acid (HCOOH) has a pH of 2.53. How many grams of formic acid are there in 100.0 mL of the solution?

16.94 Calculate the pH of a 1-L solution containing 0.150 mole of CH_3COOH and 0.100 mole of HCl.

16.95 You are given two beakers containing separately an aqueous solution of strong acid (HA) and an aqueous solution of weak acid (HB) of the same concentration. Describe how you would compare the strengths of these two acids by (a) measuring the pH, (b) measuring electrical conductance, (c) studying the rate of hydrogen gas evolution when these solutions are reacted with an active metal such as Mg or Zn.

16.96 Use Le Châtelier's principle to predict the effect of the following changes on the extent of hydrolysis of sodium nitrite ($NaNO_2$) solution: (a) HCl is added; (b) NaOH is added; (c) NaCl is added; (d) the solution is diluted.

16.97 The disagreeable odor of fish is mainly due to organic compounds (RNH_2) containing an amino group, —NH_2, in which R is the rest of the molecule. Amines are bases just like ammonia. Explain why putting some lemon juice on fish can greatly reduce the odor.

16.98 A 0.400 M formic acid (HCOOH) solution freezes at −0.758°C. Calculate the K_a of the acid at that temperature. (*Hint:* Assume that molarity is equal to molality. Carry your calculations to three significant figures and round off to two for K_a.)

16.99 Both the amide ion (NH_2^-) and the nitride ion (N^{3-}) are stronger bases than the hydroxide ion and hence do not exist in aqueous solutions. (a) Write the equations showing the reactions of these ions with water, and identify the Brønsted acid and base in each case. (b) Which of the two is the stronger base?

16.100 The atmospheric sulfur dioxide (SO_2) concentration over a certain region is 0.12 ppm by volume. Calculate the pH of the rainwater as a result of this pollutant. Assume that the dissolution of SO_2 does not affect its pressure and that the pH of rainwater is solely due to this compound.

16.101 Explain the action of smelling salt, which is ammonium carbonate [$(NH_4)_2CO_3$]. (*Hint:* The thin film of aqueous solution that lines the nasal passage is slightly basic.)

Special Problems

16.102 About half of the hydrochloric acid produced annually in the United States (3.0 billion pounds) is used for metal pickling. This process involves the removal of metal oxide layers from metal surfaces to prepare them for coating. (a) Write the overall and net ionic equations for the reaction between iron(III) oxide, which represents the rust layer over iron, and HCl. Identify the Brønsted acid and base. (b) Hydrochloric acid is also used to remove scale (which is mostly $CaCO_3$) from water pipes. Hydrochloric acid reacts with calcium carbonate in two stages; the first stage forms the bicarbonate ion, which then reacts further to form carbon dioxide. Write equations for these two stages and for the overall reaction. (c) Hydrochloric acid is used to recover oil from the ground. It dissolves rocks (often $CaCO_3$) so that the oil can flow more easily. In one process, a 15 percent (by mass) HCl solution is injected into an oil well to dissolve the rocks. If the density of the acid solution is 1.073 g/mL, what is the pH of the solution?

16.103 Hemoglobin (Hb) is a blood protein that is responsible for transporting oxygen. It can exist in the protonated form of HbH^+. The binding of oxygen can be represented by the simplified equation

$$HbH^+ + O_2 \rightleftharpoons HbO_2 + H^+$$

(a) What form of hemoglobin is favored in the lungs where oxygen concentration is highest? (b) In body tissues, where carbon dioxide is released as a result of metabolism, the medium is more acidic because of the formation of carbonic acid. What form of hemoglobin is favored under this condition? (c) When a person hyperventilates, the concentration of CO_2 in his or her blood decreases. How does this action affect the above equilibrium? Frequently a person who is hyperventilating is advised to breathe into a paper bag. Why does this action help the individual?

Answers to Practice Exercises

16.1 (1) H_2O (acid) and OH^- (base); (2) HCN (acid) and CN^- (base). **16.2** $7.7 \times 10^{-15}\,M$. **16.3** 0.12. **16.4** $4.7 \times 10^{-4}\,M$. **16.5** 7.40. **16.6** 12.56. **16.7** Smaller than 1. **16.8** $[H^+] = 7.3 \times 10^{-3}\,M$, $[A^-] = 7.3 \times 10^{-3}\,M$, $[HA] = 0.19\,M$. **16.9** 2.09. **16.10** 2.2×10^{-6}. **16.11** $[C_2H_2O_4] = 0.11\,M$, $[C_2HO_4^-] = 0.086\,M$, $[C_2O_4^{2-}] = 6.1 \times 10^{-5}\,M$, $[H^+] = 0.086\,M$. **16.12** 12.03. **16.13** 8.58. **16.14** (a) ≈ 7, (b) > 7, (c) < 7, (d) > 7.

How is the shell of a hen's egg formed? The eggshell is largely composed of calcite, which is a crystalline form of calcium carbonate ($CaCO_3$).

An average eggshell weighs about 5 g and is 40 percent calcium. Most of the calcium in an eggshell is laid down within a 16-hour period. This means that it is deposited at a rate of about 125 mg per hour. No hen can consume calcium fast enough to meet this demand. Instead, it is supplied by special bony masses in the hen's long bones, which accumulate large reserves of calcium for eggshell formation. [The inorganic calcium component of the bone is calcium phosphate, $Ca_3(PO_4)_2$, another insoluble compound.] If a hen is fed a low-calcium diet, her eggshells become progressively thinner; she may have to mobilize 10 percent of the total amount of calcium in her bones just to lay one egg! When the food supply is constantly low in calcium, egg production eventually stops.

Normally, the raw materials, the Ca^{2+} and CO_3^{2-} ions, are supplied by the blood to the shell gland. The calcification process is a precipitation reaction:

$$Ca^{2+}(aq) + CO_3^{2-}(aq) \rightleftharpoons CaCO_3(s)$$

Chicken eggs.

X-ray micrograph of an eggshell, showing columns of calcite.

In the blood, free Ca^{2+} ions are in equilibrium with calcium ions bound to proteins. As the free ions are taken up by the shell gland, more are provided by the dissociation of the protein-bound calcium.

The carbonate ions necessary for eggshell formation are a metabolic by-product. Carbon dioxide produced during metabolism is converted to carbonic acid (H_2CO_3) by the enzyme carbonic anhydrase (CA):

$$CO_2(g) + H_2O(l) \xrightarrow{CA} H_2CO_3(aq)$$

Carbonic acid ionizes stepwise to produce carbonate ions:

$$H_2CO_3(aq) \rightleftharpoons H^+(aq) + HCO_3^-(aq)$$
$$HCO_3^-(aq) \rightleftharpoons H^+(aq) + CO_3^{2-}(aq)$$

Chickens do not perspire and so must pant to cool themselves. Panting expels more CO_2 from the chicken's body than normal respiration does. According to Le Châtelier's principle, panting will shift the CO_2–H_2CO_3 equilibrium shown from right to left, thereby lowering the concentration of the CO_3^{2-} ions in solution and resulting in thin eggshells. One remedy to this problem is to give chickens carbonated water to drink in hot weather. The CO_2 dissolved in the water adds CO_2 to the chicken's body fluids and shifts the CO_2–H_2CO_3 equilibrium to the right.

Acid-Base Equilibria and Solubility Equilibria

Chapter 17

ESSENTIAL CONCEPTS

Buffer Solutions A buffer solution contains a weak acid and a salt derived from the acid. To maintain a relatively constant pH, the acid and base components of the buffer solution react with added acid or base. Buffer solutions play an important role in many chemical and biological processes.

Acid-Base Titrations The characteristics of an acid-base titration depend on the strength of the acid and base involved. Different indicators are used to determine the end point of a titration.

Solubility Equilibria Another application of the equilibrium concept is the solubility equilibria of sparingly soluble salts, which are expressed as the solubility product. The solubility of such a substance can be affected by the presence of a common cation or anion, or the pH. Complex-ion formation, an example of the Lewis acid-base type reaction, increases the solubility of an insoluble salt.

17.1 Homogeneous Versus Heterogeneous Solution Equilibria

In Chapter 16 we saw that weak acids and weak bases never ionize completely in water. Thus at equilibrium a weak acid solution, for example, contains nonionized acid as well as H^+ ions and the conjugate base. Nevertheless, all of these species are dissolved, so that the system is an example of homogeneous equilibrium (see Chapter 15).

Another important type of equilibrium, which we will study in the second half of the chapter, involves the dissolution and precipitation of slightly soluble substances. These processes are examples of heterogeneous equilibria; that is, they pertain to reactions in which the components are in more than one phase. But first we will conclude our discussion of acid-base equilibria by considering buffer solutions and taking a closer look at acid-base titrations.

17.2 Buffer Solutions

Fluids for intravenous injection must include buffer systems to maintain the proper blood pH.

Buffer solutions

A **buffer solution** is *a solution of (1) a weak acid or a weak base and (2) its salt; both components must be present. The solution has the ability to resist changes in pH upon the addition of small amounts of either acid or base.* Buffers are very important to chemical and biological systems. The pH in the human body varies greatly from one fluid to another; for example, the pH of blood is about 7.4, whereas the gastric juice in our stomachs has a pH of about 1.5. These pH values, which are crucial for the proper functioning of enzymes and the balance of osmotic pressure, are maintained by buffers in most cases.

A buffer solution must contain a relatively large concentration of acid to react with any OH^- ions that may be added to it and must contain a similar concentration of base to react with any added H^+ ions. Furthermore, the acid and the base components of the buffer must not consume each other in a neutralization reaction. These requirements are satisfied by an acid-base conjugate pair (a weak acid and its conjugate base or a weak base and its conjugate acid).

A simple buffer solution can be prepared by adding comparable amounts of acetic acid (CH_3COOH) and sodium acetate (CH_3COONa) to water. The equilibrium concentrations of both the acid and the conjugate base (from CH_3COONa) are assumed to be the same as the starting concentrations. This is so because (1) CH_3COOH is a weak acid and the extent of hydrolysis of the CH_3COO^- ion is very small and (2) the presence of CH_3COO^- ions suppresses the ionization of CH_3COOH, and the presence of CH_3COOH suppresses the hydrolysis of the CH_3COO^- ions.

A solution containing these two substances has the ability to neutralize either added acid or added base. Sodium acetate, a strong electrolyte, dissociates completely in water:

$$CH_3COONa(s) \xrightarrow{H_2O} CH_3COO^-(aq) + Na^+(aq)$$

If an acid is added, the H^+ ions will be consumed by the conjugate base in the buffer, CH_3COO^-, according to the equation

$$CH_3COO^-(aq) + H^+(aq) \longrightarrow CH_3COOH(aq)$$

If a base is added to the buffer system, the OH^- ions will be neutralized by the acid in the buffer:

$$CH_3COOH(aq) + OH^-(aq) \longrightarrow CH_3COO^-(aq) + H_2O(l)$$

(a)   (b)   (c)   (d)

Figure 17.1 *The acid-base indicator bromophenol blue (added to all solutions shown) is used to illustrate buffer action. The indicator's color is blue-purple above pH 4.6 and yellow below pH 3.0. (a) A buffer solution made up of 50 mL of 0.1 M CH₃COOH and 50 mL of 0.1 M CH₃COONa. The solution has a pH of 4.7 and turns the indicator blue-purple. (b) After the addition of 40 mL of 0.1 M HCl solution to the solution in (a), the color remains blue-purple. (c) A 100-mL CH₃COOH solution whose pH is 4.7. (d) After the addition of 6 drops (about 0.3 mL) of 0.1 M HCl solution, the color turns yellow. Without buffer action, the pH of the solution decreases rapidly to less than 3.0 upon the addition of 0.1 M HCl.*

The *buffering capacity,* that is, the effectiveness of the buffer solution, depends on the amounts of acid and conjugate base from which the buffer is made. The larger the amount, the greater the buffering capacity.

In general, a buffer system can be represented as salt/acid or conjugate base/acid. Thus the sodium acetate–acetic acid buffer system can be written as CH_3COONa/CH_3COOH or CH_3COO^-/CH_3COOH. Figure 17.1 shows this buffer system in action.

Example 17.1   Which of these solutions are buffer systems? (a) KH_2PO_4/H_3PO_4, (b) $NaClO_4/HClO_4$, (c) C_5H_5N/C_5H_5NHCl (C_5H_5N is pyridine; its K_b is given in Table 16.5). Explain your answer.

Reasoning and Solution   The criteria for a buffer system is that we must have a weak acid and its salt (containing the weak conjugate base) or a weak base and its salt (containing the weak conjugate acid).

(a) H_3PO_4 is a weak acid, and its conjugate base, $H_2PO_4^-$, is a weak base (see Table 16.4). Therefore, this is a buffer system.

(b) Because $HClO_4$ is a strong acid, its conjugate base, ClO_4^-, is an extremely weak base. This means that the ClO_4^- ion will not combine with a H^+ ion in solution to form $HClO_4$. Thus the system cannot act as a buffer system.

(c) As Table 16.5 shows, C_5H_5N is a weak base and its conjugate acid, $C_5H_5\overset{+}{N}H$ (the cation of the salt C_5H_5NHCl), is a weak acid. Therefore, this is a buffer system.

Similar problems: 17.5, 17.6.

Practice Exercise   Which of these are buffer systems? (a) KF/HF, (b) KBr/HBr, (c) $Na_2CO_3/NaHCO_3$.

Example 17.2   (a) Calculate the pH of a buffer system containing 1.0 M CH_3COOH and 1.0 M CH_3COONa. (b) What is the pH of the buffer system after the addition of 0.10 mole of gaseous HCl to 1 L of the solution? Assume that the volume of the solution does not change when the HCl is added.

Reasoning and Solution   (a) The pH of the buffer system before the addition of HCl can be calculated from the concentrations of acetic acid and its salt. Assuming that ionization of the acetic acid and hydrolysis of the acetate ions are negligible, at equilibrium we have

$$[CH_3COOH] = 1.0\,M \qquad \text{and} \qquad [CH_3COO^-] = 1.0\,M$$

$$K_a = \frac{[H^+][CH_3COO^-]}{[CH_3COOH]} = 1.8 \times 10^{-5}$$

$$[H^+] = \frac{K_a[CH_3COOH]}{[CH_3COO^-]}$$

$$= \frac{(1.8 \times 10^{-5})(1.0)}{(1.0)}$$

$$= 1.8 \times 10^{-5}\,M$$

$$pH = -\log{(1.8 \times 10^{-5})}$$

$$= 4.74$$

Thus when the concentrations of the acid and the conjugate base are the same, the hydrogen ion concentration is numerically equal to the K_a of the acid.

(b) After the addition of HCl, complete ionization of HCl acid occurs:

$$\begin{array}{cccc} HCl(aq) & \longrightarrow & H^+(aq) & + & Cl^-(aq) \\ 0.10\text{ mol} & & 0.10\text{ mol} & & 0.10\text{ mol} \end{array}$$

Originally, 1.0 mol CH_3COOH and 1.0 mol CH_3COO^- were present in 1 L of the solution. After neutralization of the HCl acid by CH_3COO^-, which we write as

$$\begin{array}{cccc} CH_3COO^-(aq) & + & H^+(aq) & \longrightarrow & CH_3COOH(aq) \\ 0.10\text{ mol} & & 0.10\text{ mol} & & 0.10\text{ mol} \end{array}$$

the number of moles of acetic acid and the number of moles of acetate ions present are

$$CH_3COOH: \quad (1.0 + 0.1)\text{ mol} = 1.1\text{ mol}$$

$$CH_3COO^-: \quad (1.0 - 0.1)\text{ mol} = 0.90\text{ mol}$$

Next we calculate the hydrogen ion concentration:

$$[H^+] = \frac{K_a[CH_3COOH]}{[CH_3COO^-]}$$

$$= \frac{(1.8 \times 10^{-5})(1.1)}{0.90}$$

$$= 2.2 \times 10^{-5}\,M$$

The pH of the solution becomes

The change in pH is 4.74 − 4.66 or 0.08 unit.

$$pH = -\log{(2.2 \times 10^{-5})}$$

$$= 4.66$$

Comment Because the volume of the solution is the same for both species, we replaced the ratio of their molar concentrations with the ratio of the number of moles present; that is, (1.1 mol/L)/(0.90 mol/L) = (1.1 mol/0.90 mol). You can show that without the buffer, the pH of the solution after the addition of HCl would be 1.00!

Similar problem: 17.14.

Practice Exercise Calculate the pH of the 0.30 M NH_3/0.36 M NH_4Cl buffer system. What is the pH after the addition of 20.0 mL of 0.050 M NaOH to 80.0 mL of the buffer solution?

Preparing a Buffer Solution with a Specific pH

Now suppose we want to prepare a buffer solution with a specific pH. How do we go about it? Referring to the acetic acid–sodium acetate buffer system, we can write the equilibrium constant as

$$K_a = \frac{[CH_3COO^-][H^+]}{[CH_3COOH]}$$

Note that this expression holds whether we have only acetic acid or a mixture of acetic acid and sodium acetate in solution. Rearranging the equation gives

$$[H^+] = \frac{K_a[CH_3COOH]}{[CH_3COO^-]}$$

Taking the negative logarithm of both sides, we obtain

$$-\log [H^+] = -\log K_a - \log \frac{[CH_3COOH]}{[CH_3COO^-]}$$

or

$$-\log [H^+] = -\log K_a + \log \frac{[CH_3COO^-]}{[CH_3COOH]}$$

So

$$pH = pK_a + \log \frac{[CH_3COO^-]}{[CH_3COOH]} \qquad (17.1)$$

in which

$$pK_a = -\log K_a \qquad (17.2)$$

Equation (17.1) is called the *Henderson-Hasselbalch equation.* In a more general form it can be expressed as

$$pH = pK_a + \log \frac{[\text{conjugate base}]}{[\text{acid}]} \qquad (17.3)$$

If the molar concentrations of the acid and its conjugate base are approximately equal, that is, [acid] ≈ [conjugate base], then

$$\log \frac{[\text{conjugate base}]}{[\text{acid}]} \approx 0$$

or

$$pH \approx pK_a$$

Thus, to prepare a buffer solution, we choose a weak acid whose pK_a is close to the desired pH. This choice not only gives the correct pH value of the buffer system, but also ensures that we have *comparable* amounts of the acid and its conjugate base present; both are prerequisites for the buffer system to function effectively.

pK_a is related to K_a as pH is related to [H$^+$]. Remember that the stronger the acid (that is, the larger the K_a), the smaller the pK_a.

Keep in mind that pK_a is a constant, but the ratio of the two concentration terms in Equation (17.3) depends on a particular solution.

Example 17.3 Describe how you would prepare a "phosphate buffer" with a pH of about 7.40.

Reasoning and Solution   The condition for a buffer solution to act effectively is that the pK_a of the acid component must be close to the desired pH value. Because phosphoric acid is a triprotic acid, we write the three stages of ionization as follows. The K_a values are obtained from Table 16.4 and the pK_a values are found by applying Equation (17.2).

$$H_3PO_4(aq) \rightleftharpoons H^+(aq) + H_2PO_4^-(aq) \qquad K_{a_1} = 7.5 \times 10^{-3}; pK_{a_1} = 2.12$$
$$H_2PO_4^-(aq) \rightleftharpoons H^+(aq) + HPO_4^{2-}(aq) \qquad K_{a_2} = 6.2 \times 10^{-8}; pK_{a_2} = 7.21$$
$$HPO_4^{2-}(aq) \rightleftharpoons H^+(aq) + PO_4^{3-}(aq) \qquad K_{a_3} = 4.8 \times 10^{-13}; pK_{a_3} = 12.32$$

The most suitable of the three buffer systems is $HPO_4^{2-}/H_2PO_4^-$, because the pK_a of the acid $H_2PO_4^-$ is closest to the desired pH. From the Henderson-Hasselbalch equation we write

$$pH = pK_a + \log \frac{[\text{conjugate base}]}{[\text{acid}]}$$

$$7.40 = 7.21 + \log \frac{[HPO_4^{2-}]}{[H_2PO_4^-]}$$

$$\log \frac{[HPO_4^{2-}]}{[H_2PO_4^-]} = 0.19$$

Taking the antilog, we obtain

$$\frac{[HPO_4^{2-}]}{[H_2PO_4^-]} = 10^{0.19} = 1.5$$

Thus, one way to prepare a phosphate buffer with a pH of 7.40 is to dissolve disodium hydrogen phosphate (Na_2HPO_4) and sodium dihydrogen phosphate (NaH_2PO_4) in a mole ratio of 1.5:1.0 in water. For example, we could dissolve 1.5 moles of Na_2HPO_4 and 1.0 mole of NaH_2PO_4 in enough water to make up a 1-L solution.

Similar problems: 17.15, 17.16.

Practice Exercise   How would you prepare a liter of "carbonate buffer" at a pH of 10.10? You are provided with carbonic acid (H_2CO_3), sodium hydrogen carbonate ($NaHCO_3$), and sodium carbonate (Na_2CO_3).

17.3 A Closer Look at Acid-Base Titrations

Acid-base titration

Having discussed buffer solutions, we can now look in more detail at the quantitative aspects of acid-base titrations (see Section 4.6). We will consider three types of reactions: (1) titrations involving a strong acid and a strong base, (2) titrations involving a weak acid and a strong base, and (3) titrations involving a strong acid and a weak base. Titrations involving a weak acid and a weak base are complicated by the hydrolysis of both the cation and the anion of the salt formed. These titrations will not be dealt with here. Figure 17.2 shows the arrangement for monitoring the pH during the course of a titration.

Strong Acid–Strong Base Titrations

The reaction between a strong acid (say, HCl) and a strong base (say, NaOH) can be represented by

Figure 17.2 *A pH meter is used to monitor an acid-base titration.*

$$NaOH(aq) + HCl(aq) \longrightarrow NaCl(aq) + H_2O(l)$$

or in terms of the net ionic equation

$$H^+(aq) + OH^-(aq) \longrightarrow H_2O(l)$$

Consider the addition of a 0.100 *M* NaOH solution (from a buret) to an Erlenmeyer flask containing 25.0 mL of 0.100 *M* HCl. For convenience, we will use only three significant figures for volume and concentration and two significant figures for pH. Figure 17.3 shows the pH profile of the titration (also known as the titration curve). Before the addition of

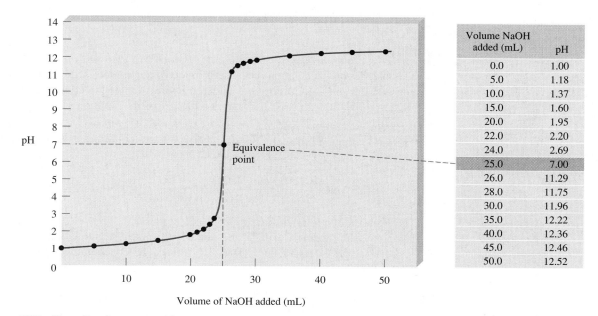

Volume NaOH added (mL)	pH
0.0	1.00
5.0	1.18
10.0	1.37
15.0	1.60
20.0	1.95
22.0	2.20
24.0	2.69
25.0	7.00
26.0	11.29
28.0	11.75
30.0	11.96
35.0	12.22
40.0	12.36
45.0	12.46
50.0	12.52

Volume of NaOH added (mL)

Figure 17.3 *pH profile of a strong acid–strong base titration. A 0.100 M NaOH solution is added from a buret to 25.0 mL of a 0.100 M HCl solution in an Erlenmeyer flask (see Figure 4.17). This curve is sometimes referred to as a titration curve.*

NaOH, the pH of the acid is given by $-\log(0.100)$, or 1.00. When NaOH is added, the pH of the solution increases slowly at first. Near the equivalence point the pH begins to rise steeply, and at the equivalence point (that is, the point at which equimolar amounts of acid and base have reacted) the curve rises almost vertically. In a strong acid–strong base titration both the hydrogen ion and hydroxide ion concentrations are very small at the equivalence point (approximately 1×10^{-7} M); consequently, the addition of a single drop of the base can cause a large increase in $[OH^-]$ and in the pH of the solution. Beyond the equivalence point, the pH again increases slowly with the addition of NaOH.

It is possible to calculate the pH of the solution at every stage of titration. Here are three sample calculations:

1. *After the addition of 10.0 mL of 0.100 M NaOH to 25.0 mL of 0.100 M HCl.* The total volume of the solution is 35.0 mL. The number of moles of NaOH in 10.0 mL is

$$10.0 \text{ mL} \times \frac{0.100 \text{ mol NaOH}}{1 \text{ L NaOH}} \times \frac{1 \text{ L}}{1000 \text{ mL}} = 1.00 \times 10^{-3} \text{ mol}$$

The number of moles of HCl originally present in 25.0 mL of solution is

$$25.0 \text{ mL} \times \frac{0.100 \text{ mol HCl}}{1 \text{ L HCl}} \times \frac{1 \text{ L}}{1000 \text{ mL}} = 2.50 \times 10^{-3} \text{ mol}$$

Keep in mind that 1 mol NaOH ⇔ 1 mol HCl.

Thus the amount of HCl left after partial neutralization is (2.50×10^{-3}) − (1.00×10^{-3}), or 1.50×10^{-3} mol. Next, the concentration of H^+ ions in 35.0 mL of solution is found as follows:

$$\frac{1.50 \times 10^{-3} \text{ mol HCl}}{35.0 \text{ mL}} \times \frac{1000 \text{ mL}}{1 \text{ L}} = 0.0429 \text{ mol HCl/L}$$

$$= 0.0429 \text{ } M \text{ HCl}$$

Thus $[H^+] = 0.0429$ M, and the pH of the solution is

$$\text{pH} = -\log 0.0429 = 1.37$$

2. *After the addition of 25.0 mL of 0.100 M NaOH to 25.0 mL of 0.100 M HCl.* This is a simple calculation, because it involves a complete neutralization reaction and the salt (NaCl) does not undergo hydrolysis. At the equivalence point, $[H^+] = [OH^-] = 1.00 \times 10^{-7}$ M and the pH of the solution is 7.00.

3. *After the addition of 35.0 mL of 0.100 M NaOH to 25.0 mL of 0.100 M HCl.* The total volume of the solution is now 60.0 mL. The number of moles of NaOH added is

$$35.0 \text{ mL} \times \frac{0.100 \text{ mol NaOH}}{1 \text{ L NaOH}} \times \frac{1 \text{ L}}{1000 \text{ mL}} = 3.50 \times 10^{-3} \text{ mol}$$

The number of moles of HCl in 25.0 mL solution is 2.50×10^{-3} mol. After complete neutralization of HCl, the number of moles of NaOH left is (3.50×10^{-3}) − (2.50×10^{-3}), or 1.00×10^{-3} mol. The concentration of NaOH in 60.0 mL of solution is

$$\frac{1.00 \times 10^{-3} \text{ mol NaOH}}{60.0 \text{ mL}} \times \frac{1000 \text{ mL}}{1 \text{ L}} = 0.0167 \text{ mol NaOH/L}$$

$$= 0.0167 \text{ } M \text{ NaOH}$$

Thus $[OH^-] = 0.0167\ M$ and $pOH = -\log 0.0167 = 1.78$. Hence, the pH of the solution is

$$pH = 14.00 - pOH$$
$$= 14.00 - 1.78$$
$$= 12.22$$

Weak Acid–Strong Base Titrations

Consider the neutralization reaction between acetic acid (a weak acid) and sodium hydroxide (a strong base):

$$CH_3COOH(aq) + NaOH(aq) \longrightarrow CH_3COONa(aq) + H_2O(l)$$

This equation can be simplified to

$$CH_3COOH(aq) + OH^-(aq) \longrightarrow CH_3COO^-(aq) + H_2O(l)$$

The acetate ion undergoes hydrolysis as follows:

$$CH_3COO^-(aq) + H_2O(l) \rightleftharpoons CH_3COOH(aq) + OH^-(aq)$$

Therefore, at the equivalence point, when we only have sodium acetate present, the pH will be *greater than* 7 as a result of the excess OH^- ions formed (Figure 17.4). Note that this situation is analogous to the hydrolysis of sodium acetate (CH_3COONa) (see p. 524).

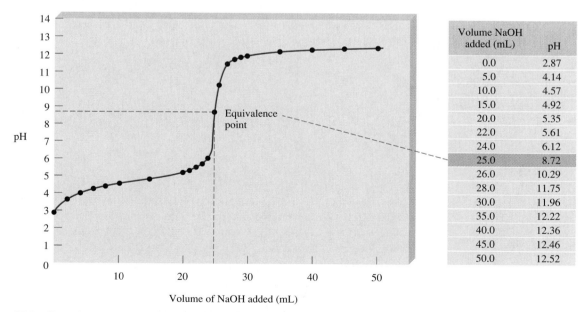

Volume NaOH added (mL)	pH
0.0	2.87
5.0	4.14
10.0	4.57
15.0	4.92
20.0	5.35
22.0	5.61
24.0	6.12
25.0	8.72
26.0	10.29
28.0	11.75
30.0	11.96
35.0	12.22
40.0	12.36
45.0	12.46
50.0	12.52

Figure 17.4 *pH profile of a weak acid–strong base titration. A 0.100 M NaOH solution is added from a buret to 25.0 mL of a 0.100 M CH₃COOH solution in an Erlenmeyer flask. Due to the hydrolysis of the salt formed, the pH at the equivalence point is greater than 7.*

Example 17.4 Calculate the pH in the titration of 25.0 mL of 0.100 M acetic acid by sodium hydroxide after the addition to the acid solution of (a) 10.0 mL of 0.10 M NaOH, (b) 25.0 mL of 0.100 M NaOH, (c) 35.0 mL of 0.100 M NaOH.

Reasoning and Solution The neutralization reaction is

$$CH_3COOH(aq) + NaOH(aq) \longrightarrow CH_3COONa(aq) + H_2O(l)$$

For each of the three stages of the titration we first calculate the number of moles of NaOH added to the acetic acid solution. Next, we calculate the number of moles of the acid (or the base) left over after neutralization. Then we determine the pH of the solution.

(a) The number of moles of NaOH in 10.0 mL is

$$10.0 \text{ mL} \times \frac{0.100 \text{ mol NaOH}}{1 \text{ L NaOH soln}} \times \frac{1 \text{ L}}{1000 \text{ mL}} = 1.00 \times 10^{-3} \text{ mol}$$

The number of moles of CH_3COOH originally present in 25.0 mL of solution is

$$25.0 \text{ mL} \times \frac{0.100 \text{ mol CH}_3\text{COOH}}{1 \text{ L CH}_3\text{COOH soln}} \times \frac{1 \text{ L}}{1000 \text{ mL}} = 2.50 \times 10^{-3} \text{ mol}$$

Thus, the amount of CH_3COOH left after all added base has been neutralized is

$$(2.50 \times 10^{-3} - 1.00 \times 10^{-3}) \text{ mol} = 1.50 \times 10^{-3} \text{ mol}$$

The amount of CH_3COONa formed is 1.00×10^{-3} mole:

$$\begin{array}{ccccc} CH_3COOH(aq) & + & NaOH(aq) & \longrightarrow & CH_3COONa(aq) + H_2O(l) \\ 1.00 \times 10^{-3} \text{ mol} & & 1.00 \times 10^{-3} \text{ mol} & & 1.00 \times 10^{-3} \text{ mol} \end{array}$$

At this stage we have a buffer system made up of CH_3COONa and CH_3COOH. To calculate the pH of this solution we write

$$[H^+] = \frac{[CH_3COOH]K_a}{[CH_3COO^-]}$$

$$= \frac{(1.50 \times 10^{-3})(1.8 \times 10^{-5})}{1.0 \times 10^{-3}}$$

$$= 2.7 \times 10^{-5} M$$

$$pH = -\log(2.7 \times 10^{-5})$$

$$= 4.57$$

Because the volume of the solution is the same for CH_3COOH and CH_3COO^-, the ratio of the number of moles present is equal to the ratio of their molar concentrations.

(b) These quantities (that is, 25.0 mL of 0.100 M NaOH reacting with 25.0 mL of 0.100 M CH$_3$COOH) correspond to the equivalence point. The number of moles of both NaOH and CH_3COOH in 25.0 mL is 2.50×10^{-3} mol, so the number of moles of the salt formed is

$$\begin{array}{ccccc} CH_3COOH(aq) & + & NaOH(aq) & \longrightarrow & CH_3COONa(aq) + H_2O(l) \\ 2.50 \times 10^{-3} \text{ mol} & & 2.50 \times 10^{-3} \text{ mol} & & 2.50 \times 10^{-3} \text{ mol} \end{array}$$

The total volume is 50.0 mL, so the concentration of the salt is

$$[CH_3COONa] = \frac{2.50 \times 10^{-3} \text{ mol}}{50.0 \text{ mL}} \times \frac{1000 \text{ mL}}{1 \text{ L}}$$

$$= 0.0500 \text{ mol/L} = 0.0500 M$$

The next step is to calculate the pH of the solution that results from the hydrolysis of the CH_3COO^- ions. Following the procedure described in Example 16.13, we find that the pH of the solution at the equivalence point is 8.72.

(c) After the addition of 35.0 mL of NaOH, the solution is well past the equivalence point. At this stage we have two species that are responsible for making the solution basic: CH_3COO^- and OH^-. However, because OH^- is a much stronger base than CH_3COO^-, we can safely neglect the CH_3COO^- ions and calculate the pH of the solution using only the concentration of the OH^- ions. Only 25.0 mL of NaOH are needed for complete neutralization, so the number of moles of NaOH left after neutralization is

$$(35.0 - 25.0) \text{ mL} \times \frac{0.100 \text{ mol NaOH}}{1 \text{ L NaOH soln}} \times \frac{1 \text{ L}}{1000 \text{ mL}} = 1.00 \times 10^{-3} \text{ mol}$$

The total volume of the combined solutions is now 60.0 mL, so we calculate OH^- concentration as follows:

$$[OH^-] = \frac{1.00 \times 10^{-3} \text{ mol}}{60.0 \text{ mL}} \times \frac{1000 \text{ mL}}{1 \text{ L}}$$

$$= 0.0167 \text{ mol/L} = 0.0167 \text{ } M$$

$$pOH = -\log 0.0167$$

$$= 1.78$$

$$pH = 14.00 - pOH$$

$$= 14.00 - 1.78$$

$$= 12.22$$

Similar problem: 17.19(b).

Practice Exercise   Exactly 100 mL of 0.10 M nitrous acid are titrated with a 0.10 M NaOH solution. Calculate the pH for (a) the initial solution, (b) the point at which 80 mL of the base has been added, (c) the equivalence point, (d) the point at which 105 mL of the base has been added.

Strong Acid–Weak Base Titrations

Consider the titration of HCl, a strong acid, with NH_3, a weak base:

$$HCl(aq) + NH_3(aq) \longrightarrow NH_4Cl(aq)$$

or simply

$$H^+(aq) + NH_3(aq) \longrightarrow NH_4^+(aq)$$

The pH at the equivalence point is *less than* 7 due to the hydrolysis of the NH_4^+ ion:

$$NH_4^+(aq) + H_2O(l) \rightleftharpoons NH_3(aq) + H_3O^+(aq)$$

or simply

$$NH_4^+(aq) \rightleftharpoons NH_3(aq) + H^+(aq)$$

Because of the volatility of an aqueous ammonia solution, it is more convenient to add hydrochloric acid from a buret to the ammonia solution. Figure 17.5 shows the titration curve for this experiment.

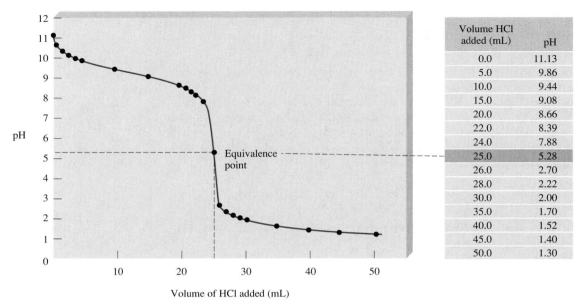

Volume HCl added (mL)	pH
0.0	11.13
5.0	9.86
10.0	9.44
15.0	9.08
20.0	8.66
22.0	8.39
24.0	7.88
25.0	5.28
26.0	2.70
28.0	2.22
30.0	2.00
35.0	1.70
40.0	1.52
45.0	1.40
50.0	1.30

Figure 17.5 *pH profile of a strong acid–weak base titration. A 0.100 M HCl solution is added from a buret to 25.0 mL of a 0.100 M NH$_3$ solution in an Erlenmeyer flask. As a result of salt hydrolysis, the pH at the equivalence point is lower than 7.*

Example 17.5 Calculate the pH at the equivalence point when 25.0 mL of 0.100 M NH$_3$ is titrated by a 0.100 M HCl solution.

Reasoning and Solution The equation for the neutralization reaction is given above. At the equivalence point, the number of moles of NH$_3$ reacted is equal to that of HCl. The number of moles of NH$_3$ in 25.0 mL of 0.100 M solution is

$$25.0 \text{ mL} \times \frac{0.100 \text{ mol NH}_3}{1 \text{ L NH}_3} \times \frac{1 \text{ L}}{1000 \text{ mL}} = 2.50 \times 10^{-3} \text{ mol}$$

Because 1 mol NH$_3$ $\backsimeq$ 1 mol HCl, at the equivalence point the number of moles of HCl reacted is also 2.50×10^{-3} mol. The number of moles of the salt (NH$_4$Cl) formed is

$$\begin{array}{ccccc} \text{HCl}(aq) & + & \text{NH}_3(aq) & \longrightarrow & \text{NH}_4\text{Cl}(aq) \\ 2.50 \times 10^{-3} \text{ mol} & & 2.50 \times 10^{-3} \text{ mol} & & 2.50 \times 10^{-3} \text{ mol} \end{array}$$

The total volume is 50.0 mL, so the concentration of NH$_4$Cl is

$$[\text{NH}_4\text{Cl}] = \frac{2.50 \times 10^{-3} \text{ mol}}{50.0 \text{ mL}} \times \frac{1000 \text{ mL}}{1 \text{ L}}$$

$$= 0.0500 \text{ mol/L} = 0.0500 \ M$$

The pH of the solution at the equivalence point is determined by the hydrolysis (or ionization) of NH$_4^+$ ions.

Step 1. We represent the hydrolysis of the cation NH$_4^+$, and let x be the equilibrium concentration of NH$_3$ and H$^+$ ions in mol/L:

$$NH_4^+(aq) \rightleftharpoons NH_3(aq) + H^+(aq)$$

	$NH_4^+(aq)$	$NH_3(aq)$	$H^+(aq)$
Initial (M):	0.0500	0.000	0.000
Change (M):	$-x$	$+x$	$+x$
Equilibrium (M):	$(0.0500 - x)$	x	x

Step 2. From Table 16.5 we obtain the K_a for NH_4^+:

$$K_a = \frac{[NH_3][H^+]}{[NH_4^+]} = 5.6 \times 10^{-10}$$

$$\frac{x^2}{0.0500 - x} = 5.6 \times 10^{-10}$$

Applying the approximation $0.0500 - x \approx 0.0500$, we get

$$\frac{x^2}{0.0500 - x} \approx \frac{x^2}{0.0500} = 5.6 \times 10^{-10}$$

$$x = 5.3 \times 10^{-6} M$$

Thus the pH is given by

$$pH = -\log(5.3 \times 10^{-6})$$
$$= 5.28$$

Practice Exercise Calculate the pH at the equivalence point in the titration of 50 mL of 0.10 M methylamine (see Table 16.5) with a 0.20 M HCl solution.

Always check the validity of the approximation.

Similar problem: 17.19(a).

17.4 Acid-Base Indicators

The equivalence point, as we have seen, is the point at which the number of moles of OH^- ions added to a solution is equal to the number of moles of H^+ ions originally present. To determine the equivalence point in a titration, then, we must know exactly how much volume of a base to add from a buret to an acid in a flask. One way to achieve this goal is to add a few drops of an acid-base indicator to the acid solution at the start of the titration. You will recall from Chapter 4 that an indicator is usually a weak organic acid or base that has distinctly different colors in its non-ionized and ionized forms. These two forms are related to the pH of the solution in which the indicator is dissolved. The **end point** of a titration *occurs when the indicator changes color.* However, not all indicators change color at the same pH, so the choice of indicator for a particular titration depends on the nature of the acid and base used in the titration (that is, whether they are strong or weak). By choosing the proper indicator for a titration, we can use the end point to determine the equivalence point, as we will see below.

Let us consider a weak monoprotic acid that we will call HIn. To be an effective indicator, HIn and its conjugate base, In^-, must have distinctly different colors. In solution, the acid ionizes to a small extent:

$$HIn(aq) \rightleftharpoons H^+(aq) + In^-(aq)$$

If the indicator is in a sufficiently acidic medium, the equilibrium, according to Le Châtelier's principle, shifts to the left and the predominant color of the indicator is that of the nonionized form (HIn). On the other hand, in a basic medium the equilibrium shifts to the right and the color of the solution will be due mainly to that of the conjugate base (In⁻). Roughly speaking, we can use the following concentration ratios to predict the perceived color of the indicator:

$$\frac{[\text{HIn}]}{[\text{In}^-]} \geq 10 \qquad \text{color of acid (HIn) predominates}$$

$$\frac{[\text{HIn}]}{[\text{In}^-]} \leq 0.1 \qquad \text{color of conjugate base (In}^-\text{) predominates}$$

If [HIn] $\approx$ [In⁻], then the indicator color is a combination of the colors of HIn and In⁻.

The end point of an indicator does not occur at a specific pH; rather, there is a range of pH within which the end point will occur. In practice, we choose an indicator whose end point range lies on the steep part of the titration curve. Because the equivalence point also lies on the steep part of the curve, this choice ensures that the pH at the equivalence point will fall within the range over which the indicator changes color. In Section 4.6 we mentioned that phenolphthalein is a suitable indicator for the titration of NaOH and HCl. Phenolphthalein is colorless in acidic and neutral solutions, but reddish pink in basic solutions. Measurements show that at pH < 8.3 the indicator is colorless but that it begins to turn reddish pink when the pH exceeds 8.3. As shown in Figure 17.3, the steepness of the pH curve near the equivalence point means that the addition of a very small quantity of NaOH (say, 0.05 mL, which is about the volume of a drop from the buret) brings about a large rise in the pH of the solution. What is important, however, is the fact that the steep portion of the pH profile includes the range over which phenolphthalein changes from colorless to reddish pink. Whenever such a correspondence occurs, the indicator can be used to locate the equivalence point of the titration (Figure 17.6).

Figure 17.6 *The titration curve of a strong acid with a strong base. Because the regions over which the indicators methyl red and phenolphthalein change color lie along the steep portion of the curve, they can be used to monitor the equivalence point of the titration. Thymol blue, on the other hand, cannot be used for the same purpose (see Table 17.1).*

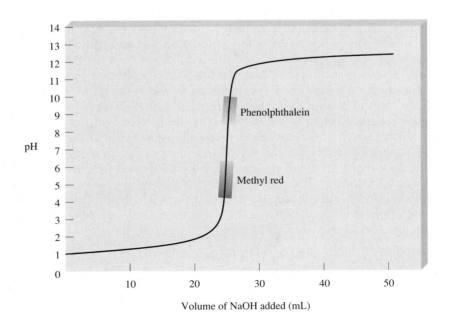

Volume of NaOH added (mL)

Table 17.1 | Some Common Acid-Base Indicators

| Indicator | Color | | pH Range* |
	In Acid	In Base	
Thymol blue	Red	Yellow	1.2–2.8
Bromophenol blue	Yellow	Bluish purple	3.0–4.6
Methyl orange	Orange	Yellow	3.1–4.4
Methyl red	Red	Yellow	4.2–6.3
Chlorophenol blue	Yellow	Red	4.8–6.4
Bromothymol blue	Yellow	Blue	6.0–7.6
Cresol red	Yellow	Red	7.2–8.8
Phenolphthalein	Colorless	Reddish pink	8.3–10.0

*The pH range is defined as the range over which the indicator changes from the acid color to the base color.

Many acid-base indicators are plant pigments. For example, by boiling chopped red cabbage in water we can extract pigments that exhibit many different colors at various pHs (Figure 17.7). Table 17.1 lists a number of indicators commonly used in acid-base titrations. The choice of a particular indicator depends on the strength of the acid and base to be titrated.

Example 17.6 Which indicator or indicators listed in Table 17.1 would you use for the acid-base titrations shown in (a) Figure 17.3, (b) Figure 17.4, and (c) Figure 17.5.

Reasoning and Solution An indicator can be used effectively only if its pH range matches the steep portion of the titration curve.

Because indicators are acids, only a small amount (a drop or two) should be used in a titration experiment.

(a) Near the equivalence point, the pH of the solution changes abruptly from 4 to 10. Therefore all the indicators except thymol blue, bromophenol blue, and methyl orange are suitable for use in the titration.

(b) Here the steep portion covers the pH range between 7 and 10; therefore, the suitable indicators are cresol red and phenolphthalein.

(c) Here the steep portion of the pH curve covers the pH range between 3 and 7; therefore, the suitable indicators are bromophenol blue, methyl orange, methyl red, and chlorophenol blue.

Practice Exercise   Referring to Table 17.1, specify which indicator or indicators you would use for these titrations: (a) HBr versus CH_3NH_2, (b) HNO_3 versus NaOH, (c) HNO_2 versus KOH.

Similar problem: 17.25.

17.5 Solubility Equilibria

Precipitation reactions are important in industry, medicine, and everyday life. For example, the preparation of many essential industrial chemicals such as sodium carbonate (Na_2CO_3) makes use of precipitation reactions. The dissolving of tooth enamel, which is mainly made of hydroxyapatite [$Ca_5(PO_4)_3OH$], in an acidic medium leads to tooth decay. Barium sulfate ($BaSO_4$), an insoluble compound that is opaque to X rays, is used to diagnose ailments of the digestive tract. Stalactites and stalagmites, which consist of calcium carbonate ($CaCO_3$), are produced by a precipitation reaction, and so are many foods, such as fudge.

The general rules for predicting the solubility of ionic compounds in water were introduced in Section 4.2. Although useful, these solubility rules do not allow us to make quantitative predictions about how much of a given ionic compound will dissolve in water. To develop a quantitative approach, we start with what we already know about chemical equilibrium.

$BaSO_4$ imaging of human large intestine.

Solubility Product

Consider a saturated solution of silver chloride that is in contact with solid silver chloride. The solubility equilibrium can be represented as

$$AgCl(s) \rightleftharpoons Ag^+(aq) + Cl^-(aq)$$

Because salts such as AgCl are treated as strong electrolytes, all the AgCl that dissolves in water is assumed to dissociate completely into Ag^+ and Cl^- ions. We know from Chapter 15 that for heterogeneous reactions the concentration of the solid is a constant. Thus we can write the equilibrium constant for the dissolution of AgCl as

$$K_{sp} = [Ag^+][Cl^-]$$

Downward-growing stalactites and upward-growing stalagmites.

in which K_{sp} is called the solubility product constant or simply the solubility product. In general, the **solubility product** of a compound is *the product of the molar concentrations of the constituent ions, each raised to the power of its stoichiometric coefficient in the equilibrium equation.*

Because each AgCl unit contains only one Ag^+ ion and one Cl^- ion, its solubility product expression is particularly simple to write. The following cases are more complex.

- MgF_2

$$MgF_2(s) \rightleftharpoons Mg^{2+}(aq) + 2F^-(aq) \qquad K_{sp} = [Mg^{2+}][F^-]^2$$

We ignore both ion-pair formation and salt hydrolysis (see p. 504).

- Ag_2CO_3

$$Ag_2CO_3(s) \rightleftharpoons 2Ag^+(aq) + CO_3^{2-}(aq) \qquad K_{sp} = [Ag^+]^2[CO_3^{2-}]$$

- $Ca_3(PO_4)_2$

$$Ca_3(PO_4)_2(s) \rightleftharpoons 3Ca^{2+}(aq) + 2PO_4^{3-}(aq) \qquad K_{sp} = [Ca^{2+}]^3[PO_4^{3-}]^2$$

Table 17.2 lists the K_{sp} values for a number of salts of low solubility. Soluble salts such as $NaCl$ and KNO_3, which have very large K_{sp} values, are not listed in the table.

For the dissolution of an ionic solid in aqueous solution, any one of the following conditions may exist: (1) The solution is unsaturated, (2) the solution is saturated, or (3) the solution is supersaturated. Following the procedure in Section 15.3, we use Q, called the *ion product,* to represent the product of the molar concentrations of the ions raised to the power of their stoichiometric coefficients. Thus for an aqueous solution containing Ag^+ and Cl^- ions at 25°C

$$Q = [Ag^+]_0[Cl^-]_0$$

Table 17.2 | Solubility Products of Some Slightly Soluble Ionic Compounds at 25°C

Compound	K_{sp}	Compound	K_{sp}
Aluminum hydroxide [$Al(OH)_3$]	1.8×10^{-33}	Lead(II) chromate ($PbCrO_4$)	2.0×10^{-14}
Barium carbonate ($BaCO_3$)	8.1×10^{-9}	Lead(II) fluoride (PbF_2)	4.1×10^{-8}
Barium fluoride (BaF_2)	1.7×10^{-6}	Lead(II) iodide (PbI_2)	1.4×10^{-8}
Barium sulfate ($BaSO_4$)	1.1×10^{-10}	Lead(II) sulfide (PbS)	3.4×10^{-28}
Bismuth sulfide (Bi_2S_3)	1.6×10^{-72}	Magnesium carbonate ($MgCO_3$)	4.0×10^{-5}
Cadmium sulfide (CdS)	8.0×10^{-28}	Magnesium hydroxide [$Mg(OH)_2$]	1.2×10^{-11}
Calcium carbonate ($CaCO_3$)	8.7×10^{-9}	Manganese(II) sulfide (MnS)	3.0×10^{-14}
Calcium fluoride (CaF_2)	4.0×10^{-11}	Mercury(I) chloride (Hg_2Cl_2)	3.5×10^{-18}
Calcium hydroxide [$Ca(OH)_2$]	8.0×10^{-6}	Mercury(II) sulfide (HgS)	4.0×10^{-54}
Calcium phosphate [$Ca_3(PO_4)_2$]	1.2×10^{-26}	Nickel(II) sulfide (NiS)	1.4×10^{-24}
Chromium(III) hydroxide [$Cr(OH)_3$]	3.0×10^{-29}	Silver bromide ($AgBr$)	7.7×10^{-13}
Cobalt(II) sulfide (CoS)	4.0×10^{-21}	Silver carbonate (Ag_2CO_3)	8.1×10^{-12}
Copper(I) bromide ($CuBr$)	4.2×10^{-8}	Silver chloride ($AgCl$)	1.6×10^{-10}
Copper(I) iodide (CuI)	5.1×10^{-12}	Silver iodide (AgI)	8.3×10^{-17}
Copper(II) hydroxide [$Cu(OH)_2$]	2.2×10^{-20}	Silver sulfate (Ag_2SO_4)	1.4×10^{-5}
Copper(II) sulfide (CuS)	6.0×10^{-37}	Silver sulfide (Ag_2S)	6.0×10^{-51}
Iron(II) hydroxide [$Fe(OH)_2$]	1.6×10^{-14}	Strontium carbonate ($SrCO_3$)	1.6×10^{-9}
Iron(III) hydroxide [$Fe(OH)_3$]	1.1×10^{-36}	Strontium sulfate ($SrSO_4$)	3.8×10^{-7}
Iron(II) sulfide (FeS)	6.0×10^{-19}	Tin(II) sulfide (SnS)	1.0×10^{-26}
Lead(II) carbonate ($PbCO_3$)	3.3×10^{-14}	Zinc hydroxide [$Zn(OH)_2$]	1.8×10^{-14}
Lead(II) chloride ($PbCl_2$)	2.4×10^{-4}	Zinc sulfide (ZnS)	3.0×10^{-23}

Figure 17.8 *Sequence of steps (a) for calculating K_{sp} from solubility data and (b) for calculating solubility from K_{sp} data.*

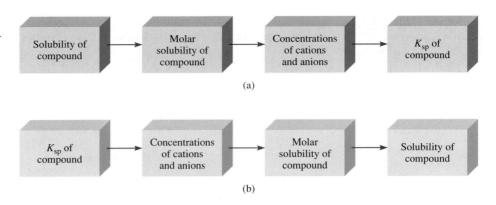

(a)

(b)

The subscript 0 reminds us that these are initial concentrations and do not necessarily correspond to those at equilibrium. The possible relationships between Q and K_{sp} are

$$Q < K_{sp}$$
$$[Ag^+]_0[Cl^-]_0 < 1.6 \times 10^{-10}$$

Unsaturated solution

$$Q = K_{sp}$$
$$[Ag^+][Cl^-] = 1.6 \times 10^{-10}$$

Saturated solution

$$Q > K_{sp}$$
$$[Ag^+]_0[Cl^-]_0 > 1.6 \times 10^{-10}$$

Supersaturated solution; AgCl will precipitate out until the product of the ionic concentrations is equal to 1.6×10^{-10}

Molar Solubility and Solubility

The value of K_{sp} indicates the solubility of an ionic compound—the smaller the value, the less soluble the compound in water. However, in using K_{sp} values to compare solubilities, you should choose compounds that have similar formulas, such as AgCl and ZnS, or CaF_2 and $Fe(OH)_2$. There are two other quantities that express a substance's solubility: **molar solubility,** which is *the number of moles of solute in 1 L of a saturated solution (moles per liter),* and **solubility,** which is *the number of grams of solute in 1 L of a saturated solution (grams per liter).* Note that all these expressions refer to the concentration of saturated solutions at some given temperature (usually 25°C). Figure 17.8 shows the relationships among solubility, molar solubility, and K_{sp}.

Both molar solubility and solubility are convenient to use in the laboratory. We can use them to determine K_{sp} by following the steps outlined in Figure 17.8(a).

Calcium sulfate is used as a drying agent and in the manufacture of paints, ceramics, and paper. A hydrated form of calcium sulfate, called plaster of Paris, is used to make casts for broken bones.

Example 17.7   The solubility of calcium sulfate is found experimentally to be 0.67 g/L. Calculate the value of K_{sp} for calcium sulfate.

Reasoning and Solution   To convert solubility to K_{sp}, we need to convert g/L to mol/L. First, we calculate the number of moles of $CaSO_4$ dissolved in 1 L of solution:

$$\frac{0.67 \text{ g CaSO}_4}{1 \text{ L soln}} \times \frac{1 \text{ mol CaSO}_4}{136.2 \text{ g CaSO}_4} = 4.9 \times 10^{-3} \text{ mol/L}$$

The solubility equilibrium

$$CaSO_4(s) \rightleftharpoons Ca^{2+}(aq) + SO_4^{2-}(aq)$$

shows that for every mole of $CaSO_4$ that dissolves, 1 mole of Ca^{2+} and 1 mole of SO_4^{2-} are produced. Thus, at equilibrium

$$[Ca^{2+}] = 4.9 \times 10^{-3} \, M \quad \text{and} \quad [SO_4^{2-}] = 4.9 \times 10^{-3} \, M$$

Now we can calculate K_{sp}:

$$\begin{aligned} K_{sp} &= [Ca^{2+}][SO_4^{2-}] \\ &= (4.9 \times 10^{-3})(4.9 \times 10^{-3}) \\ &= 2.4 \times 10^{-5} \end{aligned}$$

Similar problem: 17.37.

Practice Exercise The solubility of lead chromate ($PbCrO_4$) is 4.5×10^{-5} g/L. Calculate the solubility product of this compound.

Sometimes we are given the value of K_{sp} for a compound and asked to calculate the compound's molar solubility. For example, the K_{sp} of silver bromide (AgBr) is 7.7×10^{-13}. We can calculate its molar solubility by the same procedure as outlined for acid ionization constants. First we identify the species present at equilibrium. Here we have Ag^+ and Br^- ions. Let s be the molar solubility (in mol/L) of AgBr. Because one unit of AgBr yields one Ag^+ and one Br^- ion, at equilibrium both $[Ag^+]$ and $[Br^-]$ are equal to s. We summarize the changes in concentrations as follows:

	AgBr(s) $\rightleftharpoons$	Ag$^+$(aq) +	Br$^-$(aq)
Initial (M):		0.00	0.00
Change (M):		+s	+s
Equilibrium (M):		s	s

Silver bromide is used in photographic emulsions.

From Table 17.2 we write

$$\begin{aligned} K_{sp} &= [Ag^+][Br^-] \\ 7.7 \times 10^{-3} &= (s)(s) \\ s &= \sqrt{7.7 \times 10^{-13}} = 8.8 \times 10^{-7} \, M \end{aligned}$$

Therefore, at equilibrium

$$[Ag^+] = 8.8 \times 10^{-7} \, M$$
$$[Br^-] = 8.8 \times 10^{-7} \, M$$

Thus the molar solubility of AgBr also is $8.8 \times 10^{-7} \, M$. Knowing the molar solubility will enable us to calculate the solubility in g/L, as shown in Example 17.8.

Example 17.8 Using the data in Table 17.2, calculate the solubility of copper(II) hydroxide, $Cu(OH)_2$, in g/L.

Reasoning and Solution To convert K_{sp} to solubility, we need to first calculate the number of moles of the dissolved compound (that is, molar solubility) and then calculate the mass in grams using the molar mass of the compound.

Copper(II) hydroxide is used as a pesticide and to treat seeds.

Step 1. When $Cu(OH)_2$ dissociates, the species in solution are Cu^{2+} and OH^- ions.

Step 2. Let s be the molar solubility of $Cu(OH)_2$. Because one unit of $Cu(OH)_2$ yields one Cu^{2+} ion and two OH^- ions, at equilibrium $[Cu^{2+}]$ is s and $[OH^-]$ is $2s$. We summarize the changes in concentrations as follows:

	$Cu(OH)_2(s) \longrightarrow$	$Cu^{2+}(aq)$	$+ 2OH^-(aq)$
Initial (M):		0.00	0.00
Change (M):		$+s$	$+2s$
Equilibrium (M):		s	$2s$

Step 3.
$$K_{sp} = [Cu^{2+}][OH^-]^2$$
$$2.2 \times 10^{-20} = (s)(2s)^2$$
$$s^3 = \frac{2.2 \times 10^{-20}}{4} = 5.5 \times 10^{-21}$$

Solving for s, we get $s = 1.8 \times 10^{-7} \, M$

Knowing that the molar mass of $Cu(OH)_2$ is 97.57 g and knowing its molar solubility, we can calculate the solubility in g/L as:

$$\text{solubility of } Cu(OH)_2 = \frac{1.8 \times 10^{-7} \text{ mol } Cu(OH)_2}{1 \text{ L soln}} \times \frac{97.57 \text{ g } Cu(OH)_2}{1 \text{ mol } Cu(OH)_2}$$
$$= 1.8 \times 10^{-5} \text{ g/L}$$

Practice Exercise Calculate the solubility of silver chloride (AgCl) in g/L.

Similar problem: 17.38.

As Examples 17.7 and 17.8 show, solubility and solubility product are related. If we know one, we can calculate the other, but each quantity provides different information. Table 17.3 shows the relationship between molar solubility and solubility product for a number of ionic compounds.

When carrying out solubility and/or solubility product calculations, keep in mind these important points:

- The solubility is the quantity of a substance that dissolves in a certain quantity of water. In solubility equilibria calculations, it is usually expressed as *grams* of solute per liter of solution. Molar solubility is the number of *moles* of solute per liter of solution.
- The solubility product is an equilibrium constant.
- Molar solubility, solubility, and solubility product all refer to a *saturated solution.*

Predicting Precipitation Reactions

From a knowledge of the solubility rules (see Section 4.2) and the solubility products listed in Table 17.2, we can predict whether a precipitate will form when we mix two solutions or add a soluble compound to a solution. This ability often has practical value. In industrial and laboratory preparations, we can adjust the concentrations of ions until the ion product exceeds K_{sp} in order to obtain a given compound (in the form of a precipitate). The ability to predict precipitation reactions is also useful in medicine. For example, kidney stones, which can be extremely painful, consist largely of calcium oxalate, CaC_2O_4 ($K_{sp} = 2.3 \times 10^{-9}$). The normal physiological concentration of

A kidney stone.

Table 17.3 | Relationship Between K_{sp} and Molar Solubility (s)

Compound	K_{sp} Expression	Cation	Anion	Relation Between K_{sp} and s
AgCl	$[Ag^+][Cl^-]$	s	s	$K_{sp} = s^2; s = (K_{sp})^{\frac{1}{2}}$
BaSO$_4$	$[Ba^{2+}][SO_4^{2-}]$	s	s	$K_{sp} = s^2; s = (K_{sp})^{\frac{1}{2}}$
Ag$_2$CO$_3$	$[Ag^+]^2[CO_3^{2-}]$	$2s$	s	$K_{sp} = 4s^3; s = \left(\dfrac{K_{sp}}{4}\right)^{\frac{1}{3}}$
PbF$_2$	$[Pb^{2+}][F^-]^2$	s	$2s$	$K_{sp} = 4s^3; s = \left(\dfrac{K_{sp}}{4}\right)^{\frac{1}{3}}$
Al(OH)$_3$	$[Al^{3+}][OH^-]^3$	s	$3s$	$K_{sp} = 27s^4; s = \left(\dfrac{K_{sp}}{27}\right)^{\frac{1}{4}}$
Ca$_3$(PO$_4$)$_2$	$[Ca^{2+}]^3[PO_4^{3-}]^2$	$3s$	$2s$	$K_{sp} = 108s^5; s = \left(\dfrac{K_{sp}}{108}\right)^{\frac{1}{5}}$

calcium ions in blood plasma is about 5 mM (1 mM = 1 × 10^{-3} M). Oxalate ions ($C_2O_4^{2-}$), derived from oxalic acid present in many vegetables such as rhubarb and spinach, react with the calcium ions to form insoluble calcium oxalate, which can gradually build up in the kidneys. Proper adjustment of a patient's diet can help to reduce precipitate formation.

Example 17.9 Exactly 200 mL of 0.0040 M BaCl$_2$ are added to exactly 600 mL of 0.0080 M K$_2$SO$_4$. Will a precipitate form?

Reasoning and Solution The ions present in solution are Ba^{2+}, Cl^-, K^+, and SO_4^{2-}. According to the solubility rules on p. 95, the only precipitate that might form is BaSO$_4$:

$$Ba^{2+}(aq) + SO_4^{2-}(aq) \longrightarrow BaSO_4(s)$$

The number of moles of Ba^{2+} present in the original 200 mL of solution is

$$200 \text{ mL} \times \frac{0.0040 \text{ mol Ba}^{2+}}{1 \text{ L soln}} \times \frac{1 \text{ L}}{1000 \text{ mL}} = 8.0 \times 10^{-4} \text{ mol Ba}^{2+}$$

The total volume after combining the two solutions is 800 mL. The concentration of Ba^{2+} in the 800 mL volume is

We assume that the volumes are additive.

$$[Ba^{2+}] = \frac{8.0 \times 10^{-4} \text{ mol}}{800 \text{ mL}} \times \frac{1000 \text{ mL}}{1 \text{ L soln}}$$

$$= 1.0 \times 10^{-3} M$$

The number of moles of SO_4^{2-} in the original 600 mL solution is

$$600 \text{ mL} \times \frac{0.0080 \text{ mol SO}_4^{2-}}{1 \text{ L soln}} \times \frac{1 \text{ L}}{1000 \text{ mL}} = 4.8 \times 10^{-3} \text{ mol SO}_4^{2-}$$

The concentration of SO_4^{2-} in the 800 mL of the combined solution is

$$[SO_4^{2-}] = \frac{4.8 \times 10^{-3} \text{ mol}}{800 \text{ mL}} \times \frac{1000 \text{ mL}}{1 \text{ L soln}}$$

$$= 6.0 \times 10^{-3} M$$

Now we must compare Q with K_{sp}. From Table 17.2, the K_{sp} for $BaSO_4$ is 1.1×10^{-10}. As for Q,

$$Q = [Ba^{2+}]_0[SO_4^{2-}]_0 = (1.0 \times 10^{-3})(6.0 \times 10^{-3})$$
$$= 6.0 \times 10^{-6}$$

Therefore,

$$Q > K_{sp}$$

The solution is supersaturated because the value of Q indicates that the concentrations of the ions are too large. Thus some of the $BaSO_4$ will precipitate out of solution until

$$[Ba^{2+}][SO_4^{2-}] = 1.1 \times 10^{-10}$$

Similar problem: 17.41.

Practice Exercise   If 2.00 mL of 0.200 M NaOH are added to 1.00 L of 0.100 M $CaCl_2$, will precipitation occur?

17.6 The Common Ion Effect and Solubility

As we have noted, the solubility product is an equilibrium constant; precipitation of an ionic compound from solution occurs whenever the ion product exceeds K_{sp} for that substance. In a saturated solution of AgCl, for example, the ion product $[Ag^+][Cl^-]$ is, of course, equal to K_{sp}. Furthermore, simple stoichiometry tells us that $[Ag^+] = [Cl^-]$. But this equality does not hold in all situations.

Suppose we study a solution containing two dissolved substances that share a common ion, say, AgCl and $AgNO_3$. In addition to the dissociation of AgCl, the following process also contributes to the total concentration of the common silver ions in solution:

$$AgNO_3(s) \xrightarrow{H_2O} Ag^+(aq) + NO_3^-(aq)$$

If $AgNO_3$ is added to a saturated AgCl solution, the increase in $[Ag^+]$ will make the ion product greater than the solubility product:

$$Q = [Ag^+]_0[Cl^-]_0 > K_{sp}$$

To reestablish equilibrium, some AgCl will precipitate out of the solution, as Le Châtelier's principle would predict, until the ion product is once again equal to K_{sp}. The effect of adding a common ion, then, is a *decrease* in the solubility of the salt (AgCl) in solution. Note that in this case $[Ag^+]$ is no longer equal to $[Cl^-]$ at equilibrium; rather, $[Ag^+] > [Cl^-]$.

At a given temperature, only the solubility of a compound is altered (decreased) by the common ion effect. Its solubility product, which is an equilibrium constant, remains the same whether or not other substances are present in the solution.

Example 17.10   Calculate the solubility of silver chloride (in g/L) in a 6.5×10^{-3} M silver nitrate solution.

Reasoning and Solution   This is a common ion effect problem. Note that the presence of the common ion (Ag^+) does not affect the value of K_{sp} of AgCl. We proceed as follows.

Step 1.   The relevant species in solution are Ag^+ ions (from both AgCl and $AgNO_3$) and Cl^- ions. The NO_3^- ions are spectator ions.

Step 2. Because $AgNO_3$ is a soluble strong electrolyte, it dissociates completely:

$$AgNO_3(s) \xrightarrow{H_2O} Ag^+(aq) \quad + \quad NO_3^-(aq)$$
$$6.5 \times 10^{-3} \, M \quad 6.5 \times 10^{-3} \, M$$

Let s be the molar solubility of AgCl in $AgNO_3$ solution. We summarize the changes in concentrations as follows:

$$AgCl(s) \rightleftharpoons Ag^+(aq) + Cl^-(aq)$$

Initial (M):	6.5×10^{-3}	0.00
Change (M):	$+s$	$+s$
Equilibrium (M):	$(6.5 \times 10^{-3} + s)$	s

Step 3.
$$K_{sp} = [Ag^+][Cl^-]$$
$$1.6 \times 10^{-10} = (6.5 \times 10^{-3} + s)(s)$$

Because AgCl is quite insoluble and the presence of Ag^+ ions from $AgNO_3$ further lowers the solubility of AgCl, s must be very small compared with 6.5×10^{-3}. Therefore, applying the approximation $6.5 \times 10^{-3} + s \approx 6.5 \times 10^{-3}$, we obtain

$$1.6 \times 10^{-10} = 6.5 \times 10^{-3} \, s$$
$$s = 2.5 \times 10^{-8} \, M$$

Step 4. At equilibrium

$$[Ag^+] = (6.5 \times 10^{-3} + 2.5 \times 10^{-8}) \, M \approx 6.5 \times 10^{-3} \, M$$
$$[Cl^-] = 2.5 \times 10^{-8} \, M$$

and so our approximation was justified in step 3. Because all the Cl^- ions must come from AgCl, the amount of AgCl dissolved in $AgNO_3$ solution also is $2.5 \times 10^{-8} \, M$. Then, knowing the molar mass of AgCl (143.4 g), we can calculate the solubility of AgCl as:

$$\text{solubility of AgCl in } AgNO_3 \text{ solution} = \frac{2.5 \times 10^{-8} \text{ mol AgCl}}{1 \text{ L soln}} \times \frac{143.4 \text{ g AgCl}}{1 \text{ mol AgCl}}$$
$$= 3.6 \times 10^{-6} \text{ g/L}$$

Comment The solubility of AgCl in pure water is 1.9×10^{-3} g/L (see the Practice Exercise in Example 17.8). Therefore, the lower solubility (3.6×10^{-6} g/L) in the presence of $AgNO_3$ is reasonable.

Similar problem: 17.46.

Practice Exercise Calculate the solubility in g/L of AgBr in (a) pure water and (b) 0.0010 M NaBr.

17.7 Complex Ion Equilibria and Solubility

Lewis acid-base reactions in which a metal cation (electron-pair acceptor) combines with a Lewis base (electron-pair donor) result in the formation of complex ions:

$$\underset{\text{acid}}{Ag^+(aq)} + \underset{\text{base}}{2NH_3(aq)} \rightleftharpoons Ag(NH_3)_2^+(aq)$$

Lewis acids and bases are discussed in Section 16.10.

Figure 17.9 *(Left) An aqueous cobalt(II) chloride solution. The pink color is due to the presence of $Co(H_2O)_6^{2+}$ ions. (Right) After the addition of HCl solution, the solution turns blue because of the formation of the complex $CoCl_4^{2-}$ ions.*

According to our definition, $Co(H_2O)_6^{2+}$ itself is a complex ion. When we write $Co(H_2O)_6^{2+}$, we mean the hydrated Co^{2+} ion.

Thus, we can define a **complex ion** as *an ion containing a central metal cation bonded to one or more molecules or ions.* Complex ions are crucial to many chemical and biological processes. Here we will consider the effect of complex ion formation on solubility. In Chapter 20 we will discuss the chemistry of complex ions in more detail.

Transition metals have a particular tendency to form complex ions. For example, a solution of cobalt(II) chloride is pink because of the presence of the $Co(H_2O)_6^{2+}$ ions (Figure 17.9). When HCl is added, the solution turns blue as a result of the formation of the complex ion $CoCl_4^{2-}$:

$$Co^{2+}(aq) + 4Cl^-(aq) \rightleftharpoons CoCl_4^{2-}(aq)$$

Copper(II) sulfate ($CuSO_4$) dissolves in water to produce a blue solution. The hydrated copper(II) ions are responsible for this color; many other sulfates (Na_2SO_4, for example) are colorless. Adding a *few drops* of concentrated ammonia solution to a $CuSO_4$ solution causes a light-blue precipitate, copper(II) hydroxide, to form (Figure 17.10):

$$Cu^{2+}(aq) + 2OH^-(aq) \longrightarrow Cu(OH)_2(s)$$

Figure 17.10 *(Left) An aqueous solution of copper(II) sulfate. (Center) After the addition of a few drops of a concentrated aqueous ammonia solution, a light-blue precipitate of $Cu(OH)_2$ is formed. (Right) When more concentrated aqueous ammonia solution is added, the $Cu(OH)_2$ precipitate dissolves to form the dark-blue complex ion $Cu(NH_3)_4^{2+}$.*

Table 17.4 | Formation Constants of Selected Complex Ions in Water at 25°C

Complex Ion	Equilibrium Expression	Formation Constant (K_f)
$Ag(NH_3)_2^+$	$Ag^+ + 2NH_3 \rightleftharpoons Ag(NH_3)_2^+$	1.5×10^7
$Ag(CN)_2^-$	$Ag^+ + 2CN^- \rightleftharpoons Ag(CN)_2^-$	1.0×10^{21}
$Cu(CN)_4^{2-}$	$Cu^{2+} + 4CN^- \rightleftharpoons Cu(CN)_4^{2-}$	1.0×10^{25}
$Cu(NH_3)_4^{2+}$	$Cu^{2+} + 4NH_3 \rightleftharpoons Cu(NH_3)_4^{2+}$	5.0×10^{13}
$Cd(CN)_4^{2-}$	$Cd^{2+} + 4CN^- \rightleftharpoons Cd(CN)_4^{2-}$	7.1×10^{16}
CdI_4^{2-}	$Cd^{2+} + 4I^- \rightleftharpoons CdI_4^{2-}$	2.0×10^6
$HgCl_4^{2-}$	$Hg^{2+} + 4Cl^- \rightleftharpoons HgCl_4^{2-}$	1.7×10^{16}
HgI_4^{2-}	$Hg^{2+} + 4I^- \rightleftharpoons HgI_4^{2-}$	2.0×10^{30}
$Hg(CN)_4^{2-}$	$Hg^{2+} + 4CN^- \rightleftharpoons Hg(CN)_4^{2-}$	2.5×10^{41}
$Co(NH_3)_6^{3+}$	$Co^{3+} + 6NH_3 \rightleftharpoons Co(NH_3)_6^{3+}$	5.0×10^{31}
$Zn(NH_3)_4^{2+}$	$Zn^{2+} + 4NH_3 \rightleftharpoons Zn(NH_3)_4^{2+}$	2.9×10^9

in which the OH^- ions are supplied by the ammonia solution. If an *excess* of NH_3 is then added, the blue precipitate redissolves to produce a beautiful dark-blue solution, this time as a result of the formation of the complex ion $Cu(NH_3)_4^{2+}$, also shown in Figure 17.10:

$$Cu(OH)_2(s) + 4NH_3(aq) \rightleftharpoons Cu(NH_3)_4^{2+}(aq) + 2OH^-(aq)$$

Thus the formation of the complex ion $Cu(NH_3)_4^{2+}$ increases the solubility of $Cu(OH)_2$.

A measure of the tendency of a metal ion to form a particular complex ion is given by the **formation constant K_f** (also called the *stability constant*), which is *the equilibrium constant for complex ion formation*. The larger K_f is, the more stable the complex ion is. Table 17.4 lists the formation constants of a number of complex ions.

The formation of the $Cu(NH_3)_4^{2+}$ ion can be expressed as

$$Cu^{2+}(aq) + 4NH_3(aq) \rightleftharpoons Cu(NH_3)_4^{2+}(aq)$$

for which the formation constant is

$$K_f = \frac{[Cu(NH_3)_4^{2+}]}{[Cu^{2+}][NH_3]^4}$$
$$= 5.0 \times 10^{13}$$

The very large value of K_f in this case indicates the great stability of the complex ion in solution and accounts for the very low concentration of copper(II) ions at equilibrium.

Example 17.11 A 0.20-mole quantity of $CuSO_4$ is added to a liter of 1.20 M NH_3 solution. What is the concentration of Cu^{2+} ions at equilibrium?

Reasoning and Solution The addition of $CuSO_4$ to the NH_3 solution results in complex ion formation

$$Cu^{2+}(aq) + 4NH_3(aq) \rightleftharpoons Cu(NH_3)_4^{2+}(aq)$$

Because K_f is very large (5.0×10^{13}), the reaction lies mostly to the right. As a good approximation, we can assume that essentially all the dissolved Cu^{2+} ions end up as $Cu(NH_3)_4^{2+}$ ions. Thus the amount of NH_3 consumed in forming the complex ions is 4×0.20 mol, or 0.80 mol. (Note that 0.20 mol Cu^{2+} is initially present in solution and four NH_3 molecules are needed to form a complex with one Cu^{2+} ion.) The concentration of NH_3 at equilibrium therefore is $(1.20 - 0.80)$ M, or 0.40 M, and that of $Cu(NH_3)_4^{2+}$ is 0.20 M, the same as the initial concentration of Cu^{2+}. Because $Cu(NH_3)_4^{2+}$ does dissociate to a slight extent, we call the concentration of Cu^{2+} ions at equilibrium x and write

$$K_f = \frac{[Cu(NH_3)_4^{2+}]}{[Cu^{2+}][NH_3]^4} = 5.0 \times 10^{13}$$

$$\frac{0.20}{x(0.40)^4} = 5.0 \times 10^{13}$$

Solving for x and keeping in mind that the volume of solution is 1 L, we obtain

$$x = 1.6 \times 10^{-13} \ M = [Cu^{2+}]$$

Comment This calculation (and indeed for calculations involving complex ions in general) is more complicated than shown. This is because we have ignored other stages of dissociations in solution that lead to the formation of additional complex ions such as $Cu(NH_3)_3^{2+}$, $Cu(NH_3)_2^{2+}$, and so on. Thus the result can only be treated as a rough, ballpark value.

Practice Exercise If 2.50 g of $CuSO_4$ are dissolved in 9.0×10^2 mL of 0.30 M NH_3, what are the concentrations of Cu^{2+}, $Cu(NH_3)_4^{2+}$, and NH_3 at equilibrium?

Similar problem: 17.53.

All amphoteric hydroxides are insoluble compounds.

Finally, we note that there is a class of hydroxides, called *amphoteric hydroxides,* which can react with both acids and bases. Examples are $Al(OH)_3$, $Pb(OH)_2$, $Cr(OH)_3$, $Zn(OH)_2$, and $Cd(OH)_2$. For example, aluminum hydroxide reacts with acids and bases as:

$$Al(OH)_3(s) + 3H^+(aq) \longrightarrow Al^{3+}(aq) + 3H_2O(l)$$

$$Al(OH)_3(s) + OH^-(aq) \rightleftharpoons Al(OH)_4^-(aq)$$

The increase in solubility of $Al(OH)_3$ in a basic medium is the result of the formation of the complex ion $[Al(OH)_4^-]$ in which $Al(OH)_3$ acts as the Lewis acid and OH^-

Figure 17.11 *(Left to right) Formation of $Al(OH)_3$ precipitate when NaOH solution is added to an $Al(NO_3)_3$ solution. With further addition of NaOH solution, the precipitate dissolves due to the formation of the complex ion $Al(OH)_4^-$.*

acts as the Lewis base (Figure 17.11). Other amphoteric hydroxides behave in a similar manner.

17.8 Application of the Solubility Product Principle to Qualitative Analysis

In Section 4.6, we discussed the principle of gravimetric analysis, by which we measure the amount of an ion in an unknown sample. Here we will briefly discuss *qualitative analysis, the determination of the types of ions present in a solution.* We will focus on the cations.

Twenty common cations can be analyzed readily in aqueous solution. These cations can be divided into five groups according to the solubility products of their insoluble salts (Table 17.5). Because an unknown solution may contain any one or up to all 20 ions, analysis must be carried out systematically from group 1 through group 5. Let us consider the general procedure for separating these ions by adding precipitating reagents to an unknown solution.

Do not confuse the groups in Table 17.5, which are based on solubility products, with those in the periodic table, which are based on the electron configurations of the elements.

- *Group 1 cations.* When dilute HCl is added to the unknown solution, only the Ag^+, Hg_2^{2+}, and Pb^{2+} ions precipitate as insoluble chlorides. The other ions, whose chlorides are soluble, remain in solution.
- *Group 2 cations.* After the chloride precipitates have been removed by filtration, hydrogen sulfide is reacted with the unknown acidic solution. Under this

Table 17.5 | Separation of Cations into Groups According to Their Precipitation Reactions with Various Reagents

Group	Cation	Precipitating Reagents	Insoluble Compound	K_{sp}
1	Ag^+	HCl	AgCl	1.6×10^{-10}
	Hg_2^{2+}	↓	Hg_2Cl_2	3.5×10^{-18}
	Pb^{2+}		$PbCl_2$	2.4×10^{-4}
2	Bi^{3+}	H_2S	Bi_2S_3	1.6×10^{-72}
	Cd^{2+}	in acidic	CdS	8.0×10^{-28}
	Cu^{2+}	solutions	CuS	6.0×10^{-37}
	Sn^{2+}	↓	SnS	1.0×10^{-26}
3	Al^{3+}	H_2S	$Al(OH)_3$	1.8×10^{-33}
	Co^{2+}	in basic	CoS	4.0×10^{-21}
	Cr^{3+}	solutions	$Cr(OH)_3$	3.0×10^{-29}
	Fe^{2+}		FeS	6.0×10^{-19}
	Mn^{2+}		MnS	3.0×10^{-14}
	Ni^{2+}		NiS	1.4×10^{-24}
	Zn^{2+}	↓	ZnS	3.0×10^{-23}
4	Ba^{2+}	Na_2CO_3	$BaCO_3$	8.1×10^{-9}
	Ca^{2+}	↓	$CaCO_3$	8.7×10^{-9}
	Sr^{2+}		$SrCO_3$	1.6×10^{-9}
5	K^+	No precipitating	None	
	Na^+	reagent	None	
	NH_4^+		None	

condition, the concentration of the S^{2-} ion in solution is negligible. Therefore the precipitation of metal sulfides is best represented as

$$M^{2+}(aq) + H_2S(aq) \rightleftharpoons MS(s) + 2H^+(aq)$$

Adding acid to the solution shifts this equilibrium to the left so that only the least soluble metal sulfides, that is, those with the smallest K_{sp} values, will precipitate out of solution. These are Bi_2S_3, CdS, CuS, and SnS.

- *Group 3 cations.* At this stage, sodium hydroxide is added to the solution to make it basic. In a basic solution, the above equilibrium shifts to the right. Therefore, the more soluble sulfides (CoS, FeS, MnS, NiS, ZnS) now precipitate out of solution. Note that the Al^{3+} and Cr^{3+} ions actually precipitate as the hydroxides $Al(OH)_3$ and $Cr(OH)_3$, rather than as the sulfides, because the hydroxides are less soluble. The solution is then filtered to remove the insoluble sulfides and hydroxides.
- *Group 4 cations.* After all the group 1, 2, and 3 cations have been removed from solution, sodium carbonate is added to the basic solution to precipitate Ba^{2+}, Ca^{2+}, and Sr^{2+} ions as $BaCO_3$, $CaCO_3$, and $SrCO_3$. These precipitates too are removed from solution by filtration.
- *Group 5 cations.* At this stage, the only cations possibly remaining in solution are Na^+, K^+, and NH_4^+. The presence of NH_4^+ can be determined by adding sodium hydroxide:

$$NaOH(aq) + NH_4^+(aq) \longrightarrow Na^+(aq) + H_2O(l) + NH_3(g)$$

The ammonia gas is detected either by noting its characteristic odor or by observing a piece of wet red litmus paper turning blue when placed above (not in contact with) the solution. To confirm the presence of Na^+ and K^+ ions, we usually use a flame test, as follows: A piece of platinum wire (chosen because platinum is inert) is moistened with the solution and is then held over a Bunsen burner flame. Each type of metal ion gives a characteristic color when heated in this manner. For example, the color emitted by Na^+ ions is yellow, that of K^+ ions is violet, and that of Cu^{2+} ions is green (Figure 17.12).

Because NaOH is added in group 3 and Na_2CO_3 is added in group 4, the flame test for Na^+ ions is carried out using the original solution.

Figure 17.12 *(Left to right) Flame colors of lithium, sodium, potassium, and copper.*

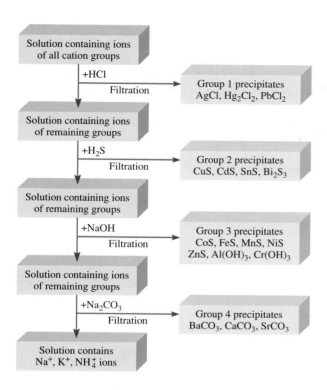

Figure 17.13 *A flowchart for the separation of cations in qualitative analysis.*

Figure 17.13 summarizes this scheme for separating metal ions.

Two points regarding qualitative analysis must be mentioned. First, the separation of the cations into groups is made as selective as possible; that is, the anions that are added as reagents must be such that they will precipitate the fewest types of cations. For example, all the cations in group 1 also form insoluble sulfides. Thus, if H_2S were reacted with the solution at the start, as many as seven different sulfides might precipitate out of solution (group 1 *and* group 2 sulfides), an undesirable outcome. Second, the removal of cations at each step must be carried out as completely as possible. For example, if we do not add enough HCl to the unknown solution to remove all the group 1 cations, they will precipitate with the group 2 cations as insoluble sulfides; this too would interfere with further chemical analysis and lead us to draw erroneous conclusions.

Summary

Equilibria involving weak acids or weak bases in aqueous solution are homogeneous. Solubility equilibria are examples of heterogeneous equilibria.

A buffer solution is a combination of a weak acid and its weak conjugate base; the solution reacts with small amounts of added acid or base in such a way that the pH of the solution remains nearly constant. Buffer systems play a vital role in maintaining the pH of body fluids.

The pH at the equivalence point of an acid-base titration depends on the hydrolysis of the salt formed in the neutralization reaction. For strong acid–strong base titrations, the pH at the equivalence point is 7; for weak acid–strong base titrations, the pH at the equivalence point is greater than 7; for strong acid–weak base titrations, the pH at the equivalence point is less than 7. Acid-base indicators are weak organic acids or bases that change color at the end point in an acid-base neutralization reaction.

The solubility product K_{sp} expresses the equilibrium between a solid and its ions in solution. Solubility can be found from K_{sp} and vice versa. The presence of a common ion decreases the solubility of a salt.

Complex ions are formed in solution by the combination of a metal cation with a Lewis base. The formation constant K_f measures the tendency toward the formation of a specific complex ion. Complex ion formation can increase the solubility of an insoluble substance.

Qualitative analysis is the identification of cations and anions in solution. It is based largely on the principles of solubility equilibria.

Key Words

Buffer solution p. 540
Complex ion p. 562
End point p. 551

Formation constant (K_f)
   p. 563
Molar solubility p. 556

Qualitative analysis
   p. 565
Solubility p. 556

Solubility product (K_{sp})
   p. 554

Questions and Problems

Buffer Solutions

Review Questions

17.1   Define buffer solution.

17.2   Define pK_a for a weak acid and explain the relationship between the value of the pK_a and the strength of the acid. Do the same for pK_b and a weak base.

17.3   The pK_as of two monoprotic acids HA and HB are 5.9 and 8.1, respectively. Which of the two is the stronger acid?

17.4   The pK_bs for the bases X^-, Y^-, and Z^- are 2.72, 8.66, and 4.57, respectively. Arrange the following acids in order of increasing strength: HX, HY, HZ.

Problems

17.5   Specify which of these systems can be classified as a buffer system: (a) KCl/HCl, (b) NH_3/NH_4NO_3, (c) Na_2HPO_4/NaH_2PO_4.

17.6   Specify which of these systems can be classified as a buffer system: (a) KNO_2/HNO_2, (b) $KHSO_4/H_2SO_4$, (c) HCOOK/HCOOH.

17.7   The pH of a bicarbonate–carbonic acid buffer is 8.00. Calculate the ratio of the concentration of carbonic acid to that of the bicarbonate ion.

17.8   Calculate the pH of these two buffer solutions: (a) 2.0 M CH_3COONa/2.0 M CH_3COOH, (b) 0.20 M CH_3COONa/0.20 M CH_3COOH. Which is the more effective buffer? Why?

17.9   Calculate the pH of the buffer system 0.15 M NH_3/0.35 M NH_4Cl.

17.10   What is the pH of the buffer 0.10 M Na_2HPO_4/0.15 M KH_2PO_4?

17.11   The pH of a sodium acetate–acetic acid buffer is 4.50. Calculate the ratio $[CH_3COO^-]/[CH_3COOH]$.

17.12   The pH of blood plasma is 7.40. Assuming the principal buffer system is HCO_3^-/H_2CO_3, calculate the ratio $[HCO_3^-]/[H_2CO_3]$. Is this buffer more effective against an added acid or an added base?

17.13   Calculate the pH of 1.00 L of the buffer 0.80 M CH_3NH_2/1.00 M CH_3NH_3Cl before and after the addition of (a) 0.070 mol NaOH and (b) 0.11 mol HCl. (See Table 16.5 for K_a value.)

17.14   Calculate the pH of 1.00 L of the buffer 1.00 M CH_3COONa/1.00 M CH_3COOH before and after the addition of (a) 0.080 mol NaOH and (b) 0.12 mol HCl. (Assume that there is no change in volume.)

17.15   A diprotic acid, H_2A, has the following ionization constants: $K_{a_1} = 1.1 \times 10^{-3}$ and $K_{a_2} = 2.5 \times 10^{-6}$. To make up a buffer solution of pH 5.80, which combination would you choose: $NaHA/H_2A$ or $Na_2A/NaHA$?

17.16   A student wishes to prepare a buffer solution at pH = 8.60. Which of these weak acids should she choose and why: HA ($K_a = 2.7 \times 10^{-3}$), HB ($K_a = 4.4 \times 10^{-6}$), or HC ($K_a = 2.6 \times 10^{-9}$)?

Acid-Base Titrations

Problems

17.17 A 0.2688-g sample of a monoprotic acid neutralizes 16.4 mL of 0.08133 M KOH solution. Calculate the molar mass of the acid.

17.18 A 5.00-g quantity of a diprotic acid is dissolved in water and made up to exactly 250 mL. Calculate the molar mass of the acid if 25.0 mL of this solution required 11.1 mL of 1.00 M KOH for neutralization. Assume that both protons of the acid are titrated.

17.19 Calculate the pH at the equivalence point for these titrations: (a) 0.10 M HCl versus 0.10 M NH$_3$, (b) 0.10 M CH$_3$COOH versus 0.10 M NaOH.

17.20 A sample of 0.1276 g of an unknown monoprotic acid was dissolved in 25.0 mL of water and titrated with 0.0633 M NaOH solution. The volume of base required to reach the equivalence point was 18.4 mL. (a) Calculate the molar mass of the acid. (b) After 10.0 mL of base had been added to the titration, the pH was determined to be 5.87. What is the K_a of the unknown acid?

Acid-Base Indicators

Review Questions

17.21 Explain how an acid-base indicator works in a titration.

17.22 What are the criteria for choosing an indicator for a particular acid-base titration?

17.23 The amount of indicator used in an acid-base titration must be small. Why?

17.24 A student carried out an acid-base titration by adding NaOH solution from a buret to an Erlenmeyer flask containing HCl solution and using phenolphthalein as indicator. At the equivalence point, he observed a faint reddish-pink color. However, after a few minutes, the solution gradually turned colorless. What do you suppose happened?

Problems

17.25 Referring to Table 17.1, specify which indicator or indicators you would use for the following titrations: (a) HCOOH versus NaOH, (b) HCl versus KOH, (c) HNO$_3$ versus NH$_3$.

17.26 The ionization constant K_a of an indicator HIn is 1.0×10^{-6}. The color of the nonionized form is red and that of the ionized form is yellow. What is the color of this indicator in a solution whose pH is 4.00? (*Hint:* The color of an indicator can be estimated by considering the ratio [HIn]/[In$^-$]. If the ratio is equal to or greater than 10, the color will be that of the nonionized form. If the ratio is equal to or smaller than 0.1, the color will be that of the ionized form.)

Solubility and Solubility Product

Review Questions

17.27 Define solubility, molar solubility, and solubility product. Explain the difference between solubility and the solubility product of a slightly soluble substance such as BaSO$_4$.

17.28 Why do we usually not quote the K_{sp} values for soluble ionic compounds?

17.29 Write balanced equations and solubility product expressions for the solubility equilibria of these compounds: (a) CuBr, (b) ZnC$_2$O$_4$, (c) Ag$_2$CrO$_4$, (d) Hg$_2$Cl$_2$, (e) AuCl$_3$, (f) Mn$_3$(PO$_4$)$_2$.

17.30 Write the solubility product expression for the ionic compound A$_x$B$_y$.

17.31 How can we predict whether a precipitate will form when two solutions are mixed?

17.32 Silver chloride has a larger K_{sp} than silver carbonate (see Table 17.2). Does this mean that the former also has a larger molar solubility than the latter?

Problems

17.33 Calculate the concentration of ions in these saturated solutions:

(a) [I$^-$] in AgI solution with [Ag$^+$] = 9.1×10^{-9} M
(b) [Al^{3+}] in Al(OH)$_3$ with [OH$^-$] = 2.9×10^{-9} M

17.34 From the solubility data given, calculate the solubility products for these compounds:

(a) SrF$_2$, 7.3×10^{-2} g/L
(b) Ag$_3$PO$_4$, 6.7×10^{-3} g/L

17.35 The molar solubility of MnCO$_3$ is 4.2×10^{-6} M. What is K_{sp} for this compound?

17.36 Using data from Table 17.2, calculate the molar solubility of calcium phosphate.

17.37 The solubility of an ionic compound M$_2$X$_3$ (molar mass = 288 g) is 3.6×10^{-17} g/L. What is K_{sp} for the compound?

17.38 Using data from Table 17.2, calculate the solubility of CaF$_2$ in g/L.

17.39 What is the pH of a saturated zinc hydroxide solution?

17.40 The pH of a saturated solution of a metal hydroxide MOH is 9.68. Calculate the K_{sp} for the compound.

17.41 A sample of 20.0 mL of 0.10 M Ba(NO$_3$)$_2$ is added to 50.0 mL of 0.10 M Na$_2$CO$_3$. Will BaCO$_3$ precipitate?

17.42 A volume of 75 mL of 0.060 M NaF is mixed with 25 mL of 0.15 M Sr(NO$_3$)$_2$. Calculate the concentrations in the final solution of NO$_3^-$, Na$^+$, Sr^{2+}, and F$^-$. (K_{sp} for SrF$_2$ = 2.0×10^{-10}.)

The Common Ion Effect

Review Questions

17.43 How does a common ion affect solubility? Use Le Châtelier's principle to explain the decrease in solubility of $CaCO_3$ in a Na_2CO_3 solution.

17.44 The molar solubility of AgCl in 6.5×10^{-3} M AgNO₃ is 2.5×10^{-8} M. In deriving K_{sp} from these data, which of these assumptions are reasonable?

  (a) K_{sp} is the same as solubility.
  (b) K_{sp} of AgCl is the same in 6.5×10^{-3} M AgNO₃ as in pure water.
  (c) Solubility of AgCl is independent of the concentration of AgNO₃.
  (d) $[Ag^+]$ in solution does not change significantly on the addition of AgCl to 6.5×10^{-3} M AgNO₃.
  (e) $[Ag^+]$ in solution after the addition of AgCl to 6.5×10^{-3} M AgNO₃ is the same as it would be in pure water.

Problems

17.45 How many grams of $CaCO_3$ will dissolve in 3.0×10^2 mL of 0.050 M Ca(NO₃)₂?

17.46 The solubility product of PbBr₂ is 8.9×10^{-6}. Determine the molar solubility (a) in pure water, (b) in 0.20 M KBr solution, (c) in 0.20 M Pb(NO₃)₂ solution.

17.47 Calculate the molar solubility of AgCl in a solution made by dissolving 10.0 g of CaCl₂ in 1.00 L of solution.

17.48 Calculate the molar solubility of BaSO₄ (a) in water and (b) in a solution containing 1.0 M SO_4^{2-} ions.

Complex Ions

Review Questions

17.49 Explain the formation of complexes in Table 17.4 in terms of Lewis acid-base theory.

17.50 Give an example to illustrate the general effect of complex ion formation on solubility.

Problems

17.51 Write the formation constant expressions for these complex ions: (a) $Zn(OH)_4^{2-}$, (b) $Co(NH_3)_6^{3+}$, (c) HgI_4^{2-}.

17.52 Explain, with balanced ionic equations, why (a) CuI₂ dissolves in ammonia solution, (b) AgBr dissolves in NaCN solution, (c) Hg₂Cl₂ dissolves in KCl solution.

17.53 If 2.50 g of CuSO₄ are dissolved in 9.0×10^2 mL of 0.30 M NH₃, what are the concentrations of Cu^{2+}, $Cu(NH_3)_4^{2+}$, and NH₃ at equilibrium?

17.54 Calculate the concentrations of Cd^{2+}, $Cd(CN)_4^{2-}$, and CN^- at equilibrium when 0.50 g of Cd(NO₃)₂ dissolves in 5.0×10^2 mL of 0.50 M NaCN.

17.55 If NaOH is added to 0.010 M Al^{3+}, which will be the predominant species at equilibrium: $Al(OH)_3$ or $Al(OH)_4^-$? The pH of the solution is 14.00. [K_f for $Al(OH)_4^- = 2.0 \times 10^{33}$.]

17.56 Calculate the molar solubility of AgI in a 1.0 M NH₃ solution. (*Hint:* You need to consider two different types of equilibria.)

Qualitative Analysis

Review Questions

17.57 Outline the general principle of qualitative analysis.

17.58 Give two examples of metal ions in each group (1 through 5) in the qualitative analysis scheme.

Problems

17.59 In a group 1 analysis, a student obtained a precipitate containing both AgCl and PbCl₂. Suggest one reagent that would allow her to separate AgCl(s) from PbCl₂(s).

17.60 In a group 1 analysis, a student adds hydrochloric acid to the unknown solution to make $[Cl^-] = 0.15$ M. Some PbCl₂ precipitates. Calculate the concentration of Pb^{2+} remaining in solution.

17.61 Both KCl and NH₄Cl are white solids. Suggest one reagent that would allow you to distinguish between these two compounds.

17.62 Describe a simple test that would enable you to distinguish between AgNO₃(s) and Cu(NO₃)₂(s).

Additional Problems

17.63 A quantity of 0.560 g of KOH is added to 25.0 mL of 1.00 M HCl. Excess Na₂CO₃ is then added to the solution. What mass (in grams) of CO₂ is formed?

17.64 A volume of 25.0 mL of 0.100 M HCl is titrated against a 0.100 M NH₃ solution added to it from a buret. Calculate the pH values of the solution (a) after 10.0 mL of NH₃ solution have been added, (b) after 25.0 mL of NH₃ solution have been added, (c) after 35.0 mL of NH₃ solution have been added.

17.65 The buffer range is defined by the equation $pH = pK_a \pm 1$. Calculate the range of the ratio [conjugate base]/[acid] that corresponds to this equation.

17.66 The pK_a of the indicator methyl orange is 3.46. Over what pH range does this indicator change from 90% HIn to 90% In⁻?

17.67 Sketch the titration curve of a weak acid versus a strong base such that shown in Figure 17.4. On your graph indicate the volume of base used at the equivalence point and also at the half-equivalence point, that is, the point at which half of the base has been added. Show how you can measure the pH of the solution at the half-equivalence point. Using Equation (17.3), explain how you can determine the pK_a of the acid by this procedure.

17.68 A 200-mL volume of NaOH solution was added to 400 mL of a 2.00 M HNO₂ solution. The pH of the mixed solution was 1.50 units greater than that of the original acid solution. Calculate the molarity of the NaOH solution.

17.69 The pK_a of butyric acid (HBut) is 4.7. Calculate K_b for the butyrate ion (But⁻).

17.70 A solution is made by mixing exactly 500 mL of 0.167 M NaOH with exactly 500 mL 0.100 M CH_3COOH. Calculate the equilibrium concentrations of H^+, CH_3COOH, CH_3COO^-, OH^-, and Na^+.

17.71 $Cd(OH)_2$ is an insoluble compound. It dissolves in a NaOH solution. Write a balanced ionic equation for this reaction. What type of reaction is this?

17.72 Calculate the pH of the 0.20 M NH_3/0.20 M NH_4Cl buffer. What is the pH of the buffer after the addition of 10.0 mL of 0.10 M HCl to 65.0 mL of the buffer?

17.73 For which of these reactions is the equilibrium constant called a solubility product?

(a) $Zn(OH)_2(s) + 2OH^-(aq) \rightleftharpoons Zn(OH)_4^{2-}(aq)$
(b) $3Ca^{2+}(aq) + 2PO_4^{3-}(aq) \rightleftharpoons Ca_3(PO_4)_2(s)$
(c) $CaCO_3(s) + 2H^+(aq) \rightleftharpoons$
  $\qquad\qquad Ca^{2+}(aq) + H_2O(l) + CO_2(g)$
(d) $PbI_2(s) \rightleftharpoons Pb^{2+}(aq) + 2I^-(aq)$

17.74 A student mixes 50.0 mL of 1.00 M $Ba(OH)_2$ with 86.4 mL of 0.494 M H_2SO_4. Calculate the mass of $BaSO_4$ formed and the pH of the mixed solution.

17.75 A 2.0-L kettle contains 116 g of calcium carbonate as boiler scale. How many times would the kettle have to be completely filled with distilled water to remove all of the deposit at 25°C?

17.76 Equal volumes of 0.12 M $AgNO_3$ and 0.14 M $ZnCl_2$ solution are mixed. Calculate the equilibrium concentrations of Ag^+, Cl^-, Zn^{2+}, and NO_3^-.

17.77 Calculate the solubility (in grams per liter) of Ag_2CO_3.

17.78 Find the approximate pH range suitable for the separation of Fe^{3+} and Zn^{2+} by precipitation of $Fe(OH)_3$ from a solution that is initially 0.010 M in Fe^{3+} and Zn^{2+}.

17.79 Which of these ionic compounds will be more soluble in acid solution than in water: (a) $BaSO_4$, (b) $PbCl_2$, (c) $Fe(OH)_3$, (d) $CaCO_3$? Explain.

17.80 Which of these substances will be more soluble in acid solution than in pure water: (a) CuI, (b) Ag_2SO_4, (c) $Zn(OH)_2$, (d) BaC_2O_4, (e) $Ca_3(PO_4)_2$?

17.81 What is the pH of a saturated solution of aluminum hydroxide?

17.82 The molar solubility of $Pb(IO_3)_2$ in a 0.10 M $NaIO_3$ solution is 2.4×10^{-11} mol/L. What is K_{sp} for $Pb(IO_3)_2$?

17.83 The solubility product of $Mg(OH)_2$ is 1.2×10^{-11}. What minimum OH^- concentration must be attained (for example, by adding NaOH) to make the Mg^{2+} concentration in a solution of $Mg(NO_3)_2$ less than 1.0×10^{-10} M?

17.84 Calculate whether a precipitate will form if 2.00 mL of 0.60 M NH_3 are added to 1.0 L of 1.0×10^{-3} M $FeSO_4$.

17.85 Both Ag^+ and Zn^{2+} form complex ions with NH_3. Write balanced equations for the reactions. However, $Zn(OH)_2$ is soluble in 6 M NaOH, and AgOH is not. Explain.

17.86 When a KI solution was added to a solution of mercury(II) chloride, a precipitate [mercury(II) iodide] was formed. A student plotted the mass of the precipitate formed versus the volume of the KI solution added and obtained the graph shown here. Explain the appearance of the graph.

17.87 Barium is a toxic substance that can cause serious deterioration of the heart's function. In a barium enema procedure, a patient drinks an aqueous suspension of 20 g $BaSO_4$. If this substance were to equilibrate with the 5.0 L of the blood in the patient's body, how many grams of $BaSO_4$ will dissolve in the blood? For a good estimate, we may assume that the temperature is at 25°C. Why is $Ba(NO_3)_2$ not chosen for this procedure?

17.88 The pK_a of phenolphthalein is 9.10. Over what pH range does this indicator change from 95% HIn to 95% In⁻?

17.89 Look up the K_{sp} values for $BaSO_4$ and $SrSO_4$ in Table 17.2. Calculate the values of $[Ba^{2+}]$, $[Sr^{2+}]$, and $[SO_4^{2-}]$ in a solution that is saturated with both compounds.

17.90 Solid NaI is slowly added to a solution that is 0.010 M in Cu^+ and 0.010 M in Ag^+. (a) Which compound will begin to precipitate first? (b) Calculate $[Ag^+]$ when CuI just begins to precipitate. (c) What percentage of Ag^+ remains in solution at this point?

17.91 Radiochemical techniques are useful in estimating the solubility product of many compounds. In one experiment, 50.0 mL of a 0.010 M $AgNO_3$ solution containing a silver isotope with a radioactivity of 74,025 counts per min per mL were mixed with 100 mL of a 0.030 M $NaIO_3$ solution. The mixed solution was diluted to 500 mL and filtered to remove all of the $AgIO_3$ precipitate. The remaining solution was found to have a radioactivity of 44.4 counts per min per mL. What is the K_{sp} of $AgIO_3$?

17.92 The molar mass of a certain metal carbonate, MCO_3, can be determined by adding an excess of HCl acid to react with the carbonate and then "back-titrating" the remaining acid with NaOH. (a) Write an equation for these reactions. (b) In a certain experiment, 20.00 mL of 0.0800 M HCl were added to a 0.1022-g sample of MCO_3. The excess HCl required 5.64 mL of 0.1000 M NaOH for neutralization. Calculate the molar mass of the carbonate and identify M.

17.93 Acid-base reactions usually go to completion. Confirm this statement by calculating the equilibrium constant for each of the following cases: (a) a strong acid reacting with a strong base, (b) a strong acid reacting with a weak base (NH_3), (c) a weak acid (CH_3COOH) reacting with a strong base, (d) a weak acid (CH_3COOH) reacting with a weak base (NH_3). (*Hint:* Strong acids exist as H^+ ions and strong bases exist as OH^- ions in solution. You need to look up K_a, K_b, and K_w.)

17.94 Calculate x, the number of molecules of water in oxalic acid hydrate, $H_2C_2O_4 \cdot xH_2O$, from the following data: 5.00 g of the compound is made up to exactly 250 mL solution and 25.0 mL of this solution requires 15.9 mL of 0.500 M NaOH solution for neutralization.

17.95 Describe how you would prepare 1-L of the buffer 0.20 M CH_3COONa/0.20 M CH_3COOH by (a) mixing a solution of CH_3COOH with a solution of CH_3COONa, (b) reacting a solution of CH_3COOH with a solution of NaOH, and (c) reacting a solution of CH_3COONa with a solution of HCl.

17.96 What reagents would you employ to separate these pairs of ions in solution: (a) Na^+ and Ba^{2+}, (b) K^+ and Pb^{2+}, (c) Zn^{2+} and Hg^{2+}?

17.97 $CaSO_4$ ($K_{sp} = 2.4 \times 10^{-5}$) has a larger K_{sp} value than that of Ag_2SO_4 ($K_{sp} = 1.4 \times 10^{-5}$). Does it follow that $CaSO_4$ also has greater solubility (g/L)?

17.98 How many milliliters of 1.0 M NaOH must be added to 200 mL of 0.10 M NaH_2PO_4 to make a buffer solution with a pH of 7.50?

17.99 The maximum allowable concentration of Pb^{2+} ions in drinking water is 0.05 ppm (that is, 0.05 g of Pb^{2+} in 1 million g of water). Is this guideline exceeded if an underground water supply is at equilibrium with the mineral anglesite, $PbSO_4$ ($K_{sp} = 1.6 \times 10^{-8}$)?

17.100 Which of these solutions has the highest $[H^+]$: (a) 0.10 M HF, (b) 0.10 M HF in 0.10 M NaF, or (c) 0.10 M HF in 0.10 M SbF_5? (*Hint:* SbF_5 reacts with F^- to form the complex ion SbF_6^-.)

17.101 Distribution curves show how the fractions of nonionized acid and its conjugate base vary as a function of pH of the medium. Plot distribution curves for CH_3COOH and its conjugate base CH_3COO^- in solution. Your graph should show fraction as the y axis and pH as the x axis. What are the fractions and pH at the point these two curves intersect?

17.102 Water containing Ca^{2+} and Mg^{2+} ions is called *hard water* and is unsuitable for some household and industrial use because these ions react with soap to form insoluble salts, or curds. One way to remove the Ca^{2+} ions from hard water is by adding washing soda ($Na_2CO_3 \cdot 10H_2O$). (a) The molar solubility of $CaCO_3$ is 9.3×10^{-5} M. What is its molar solubility in a 0.050 M Na_2CO_3 solution? (b) Why are Mg^{2+} ions not removed by this procedure? (c) The Mg^{2+} ions are removed as $Mg(OH)_2$ by adding slaked lime [$Ca(OH)_2$] to the water to produce a saturated solution. Calculate the pH of a saturated $Ca(OH)_2$ solution. (d) What is the concentration of Mg^{2+} ions at this pH? (e) In general, which ion (Ca^{2+} or Mg^{2+}) would you remove first? Why?

Special Problems

17.103 One of the most commonly used antibiotics is penicillin G (benzylpenicillinic acid), which has the following structure:

It is a weak monoprotic acid:

$$HP \rightleftharpoons H^+ + P^- \qquad K_a = 1.64 \times 10^{-3}$$

in which HP denotes the parent acid and P^- the conjugate base. Penicillin G is produced by growing molds in fermentation tanks at 25°C and a pH range of 4.5 to 5.0.

The crude form of this antibiotic is obtained by extracting the fermentation broth with an organic solvent in which the acid is soluble. (a) Identify the acidic hydrogen atom. (b) In one stage of purification, the organic extract of the crude penicillin G is treated with a buffer solution at pH = 6.50. What is the ratio of the conjugate base of penicillin G to the acid at this pH? Would you expect the conjugate base to be more soluble in water than the acid? (c) Penicillin G is not suitable for oral administration, but the sodium salt (NaP) is because it is soluble. Calculate the pH of a 0.12 M NaP solution formed when a tablet containing the salt is dissolved in a glass of water.

17.104 Amino acids are the building blocks of proteins. These compounds contain at least one amino group and one carboxyl group. Consider glycine, whose structure is shown in Figure 11.18. Depending on the pH of the solution, glycine can exist in one of three possible forms:

Fully protonated: $\overset{+}{N}H_3—CH_2—COOH$

Dipolar ion: $\overset{+}{N}H_3—CH_2—COO^-$

Fully ionized: $NH_2—CH_2—COO^-$

Predict the predominant form of glycine at pH 1.0, 7.0, and 12.0. The pK_a of the carboxyl group is 2.3 and that of the ammonium group is 9.6. [*Hint:* Use Equation (17.3).]

Answers to Practice Exercises

17.1 (a) and (c). **17.2** 9.17, 9.20. **17.3** Weigh out Na_2CO_3 and $NaHCO_3$ in a mole ratio of 0.60 to 1.0. Dissolve in enough water to make up a 1-L solution. **17.4** (a) 2.19, (b) 3.95, (c) 8.02, (d) 11.39. **17.5** 5.92. **17.6** (a) Bromophenol blue, methyl orange, methyl red, and chlorophenol blue; (b) all except thymol blue, bromophenol blue, and methyl orange; (c) cresol red and phenolphthalein. **17.7** 1.9×10^{-14}. **17.8** 1.9×10^{-3} g/L. **17.9** No. **17.10** (a) 1.7×10^{-4} g/L, (b) 1.4×10^{-7} g/L. **17.11** $[Cu^{2+}] = 1.2 \times 10^{-13}$ M, $[Cu(NH_3)_4^{2+}] = 0.017$ M, $[NH_3] = 0.23$ M.

The second law of thermodynamics was developed by the French engineer Sadi Carnot around 1820. Carnot's formulation has to do with the efficiency of heat engines, machines that convert heat to mechanical work. It helps us to answer the question, for a given amount of heat input, how much useful work can we get out of the engine?

Consider the operation of a steam turbine, a typical heat engine, at an electric power station. Superheated steam at about 560°C (or 833 K) is used to drive the turbine to run the generator. Eventually, the steam is discharged to the surroundings at about 38°C (or 311 K). Carnot showed that the maximum efficiency of this and all other heat engines is given by $(T_2 - T_1)/T_2$, where T_2 and T_1 are the upper and lower temperatures (in kelvin). Because T_1 cannot be zero and T_2 cannot be infinite, the efficiency can never approach 100 percent. (For the described steam turbine, the maximum efficiency would be 0.63, or 63 percent.) Thus it is impossible to build a heat engine that would convert heat totally into mechanical work; some heat must inevitably be discharged to the surroundings. This is the essence of the second law.

Carnot's work is significant but not directly applicable to chemistry. The foundation of "chemical thermodynamics" was laid by an American scientist, Josiah Willard Gibbs (1839–1903). In 1876 Gibbs published a 323-page paper entitled "On the Equilibrium of Heterogeneous Substances," in which he presented the basic principles of chemical equilibrium and phase equilibrium, introduced a new concept called free energy, and explained the relations governing energy changes in electrochemical cells.

The son of a Yale professor and the first person to be awarded a Ph.D. in science from an American university, Gibbs spent all of his professional life as professor of mathematical physics at Yale. Although he is arguably the most brilliant native-born American scientist, Gibbs was a modest and private individual, and he never gained the eminence as his contemporary and admirer James Maxwell did in Europe. Even today, very few people outside of chemistry and physics have ever heard of Gibbs.

The efficiency of converting heat to work, as in the case of a steam locomotive, is governed by the second law of thermodynamics.

Thermodynamics

Chapter 18

ESSENTIAL CONCEPTS

Laws of Thermodynamics The laws of thermodynamics have been successfully applied to the study of chemical and physical processes. The first law of thermodynamics is based on the law of conservation of energy. The second law of thermodynamics deals with natural or spontaneous processes. The function that predicts the spontaneity of a reaction is entropy, which is a measure of the disorder of a system. The second law states that for a spontaneous process, the change in the entropy of the universe must be positive. The third law enables us to determine absolute entropy values.

Gibbs Free Energy Gibbs free energy helps us to determine the spontaneity of a reaction by focusing only on the system. The change in Gibbs free energy for a process is made up of two terms: a change in enthalpy and a change in entropy times temperature. At constant temperature and pressure, a decrease in Gibbs free energy signals a spontaneous reaction. The change in the standard Gibbs free energy can be related to the equilibrium constant of a reaction.

Thermodynamics in Living Systems Many reactions of biological importance are nonspontaneous. By coupling such reactions to those that have a negative Gibbs free energy change with the aid of enzymes, the net reaction can be made to proceed to yield the desired products.

18.1 The Three Laws of Thermodynamics

In Chapter 6 we encountered the first of three laws of thermodynamics, which says that energy can be converted from one form to another, but it cannot be created or destroyed. One measure of these changes is the amount of heat given off or absorbed by a system during a constant-pressure process, which chemists define as a change in enthalpy (ΔH).

The second law of thermodynamics explains why chemical processes tend to favor one direction. The third law is an extension of the second law and will be examined briefly in Section 18.3.

18.2 Spontaneous Processes and Entropy

One of the main objectives in studying thermodynamics, as far as chemists are concerned, is to be able to predict whether or not a reaction will occur when reactants are brought together under a special set of conditions (for example, at a certain temperature, pressure, and concentration). This knowledge is important whether one is synthesizing compounds in a research laboratory, manufacturing chemicals on an industrial scale, or trying to understand the intricate biological processes in a cell. A reaction that *does* occur under the given set of conditions is called a *spontaneous reaction*. If a reaction does not occur under specified conditions, it is said to be non-spontaneous. We observe spontaneous physical and chemical processes every day, including many of these examples:

- A waterfall runs downhill, but never up, spontaneously.
- A lump of sugar spontaneously dissolves in a cup of coffee, but dissolved sugar does not spontaneously reappear in its original form.
- Water freezes spontaneously below 0°C, and ice melts spontaneously above 0°C (at 1 atm).
- Heat flows from a hotter object to a colder one, but the reverse never happens spontaneously.
- The expansion of a gas in an evacuated bulb is a spontaneous process [Figure 18.1(a)]. The reverse process, that is, the gathering of all the molecules into one bulb, is not spontaneous [Figure 18.1(b)].
- A piece of sodium metal reacts violently with water to form sodium hydroxide and hydrogen gas. However, hydrogen gas does not react with sodium hydroxide to form water and sodium.
- Iron exposed to water and oxygen forms rust, but rust does not spontaneously change back to iron.

These examples show that processes that occur spontaneously in one direction cannot, under the same conditions, also take place spontaneously in the opposite direction.

If we assume that spontaneous processes occur so as to decrease the energy of a system, we can explain why a ball rolls downhill and why springs in a clock unwind. Similarly, a large number of exothermic reactions are spontaneous. An example is the combustion of methane:

$$CH_4(g) + 2O_2(g) \longrightarrow CO_2(g) + 2H_2O(l) \qquad \Delta H° = -890.4 \text{ kJ}$$

Another example is the acid-base neutralization reaction:

A spontaneous reaction does not necessarily mean an instantaneous reaction.

A spontaneous and a nonspontaneous process.

(a)

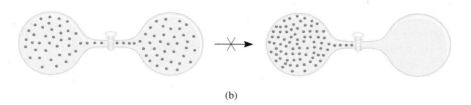

(b)

$$H^+(aq) + OH^-(aq) \longrightarrow H_2O(l) \qquad \Delta H° = -56.2 \text{ kJ}$$

But consider a solid-to-liquid phase transition such as

$$H_2O(s) \longrightarrow H_2O(l) \qquad \Delta H° = 6.01 \text{ kJ}$$

In this case, the assumption that spontaneous processes always decrease a system's energy fails. Experience tells us that ice melts spontaneously above 0°C even though the process is endothermic. Another example that contradicts our assumption is the dissolution of ammonium nitrate in water:

$$NH_4NO_3(s) \xrightarrow{\text{H}_2\text{O}} NH_4^+(aq) + NO_3^-(aq) \qquad \Delta H° = 25 \text{ kJ}$$

This process is spontaneous, and yet it is also endothermic. The decomposition of mercury(II) oxide is an endothermic reaction that is nonspontaneous at room temperature, but it becomes spontaneous when the temperature is raised:

$$2HgO(s) \longrightarrow 2Hg(l) + O_2(g) \qquad \Delta H° = 90.7 \text{ kJ}$$

From a study of the examples mentioned and many more cases, we come to the following conclusion: Exothermicity favors the spontaneity of a reaction but does not guarantee it. Just as it is possible for an endothermic reaction to be spontaneous, it is possible for an exothermic reaction to be nonspontaneous. In other words, we cannot decide whether or not a chemical reaction will occur spontaneously solely on the basis of energy changes in the system. To make this kind of prediction we need another thermodynamic quantity, which turns out to be *entropy*.

When heated, HgO decomposes to give Hg and O_2.

Entropy

In order to predict the spontaneity of a process, we need to know two things about the system. One is the change in enthalpy, which is nearly equivalent to ΔE for most processes. The other is **entropy (S)**, which is *a measure of the randomness or disorder of a system*. The greater the disorder of a system, the greater its entropy. Conversely, the more ordered a system, the smaller its entropy. One way to illustrate order and disorder is with a deck of playing cards. A new deck of cards is arranged in an ordered fashion (the cards are from ace to king and the suits are in the order of spades to hearts to diamonds to clubs). Once the deck has been shuffled, the cards are no longer in sequence by number or by suit. It is possible, but extremely unlikely, that by reshuffling the cards we can restore the original order. There are many ways for the cards to be out of sequence but only one way for them to be ordered according to our definition.

For any substance, the particles in the solid state are more ordered than those in the liquid state, which in turn are more ordered than those in the gaseous state. So for the *same* molar amount of a substance, we can write

$$S_{\text{solid}} < S_{\text{liquid}} \ll S_{\text{gas}}$$

In other words, entropy describes the extent to which atoms, molecules, or ions are distributed in a disorderly fashion in a given region in space.

One way to conceptualize order and disorder is in terms of probability. A probable event is one that can happen in many ways, and an improbable event is one that can happen in only one or a few ways. For example, referring to Figure 18.1(a), suppose that there is only one molecule in the left flask initially. Because the two flasks have equal volume, the probability of finding the molecule in either flask after the stopcock is opened is $\frac{1}{2}$. If two molecules are present to start with, the probability of finding both molecules in the *same* flask after we open the stopcock becomes $\frac{1}{2} \times \frac{1}{2}$, or $\frac{1}{4}$. Now $\frac{1}{4}$ is a high ratio, and so it would not be surprising to find both molecules in the same flask after the stopcock has been opened. However, it is not difficult to see that as the number of molecules increases, the probability (P) of finding all the molecules in the *same* flask becomes progressively smaller:

$$P = \left(\frac{1}{2}\right) \times \left(\frac{1}{2}\right) \times \left(\frac{1}{2}\right) \times \cdots$$
$$= \left(\frac{1}{2}\right)^N$$

where N is the total number of molecules present. If $N = 100$, we have

$$P = \left(\frac{1}{2}\right)^{100} = 8 \times 10^{-31}$$

If N is of the order of 6×10^{23}, the number of molecules in 1 mole of a gas, the probability becomes $\left(\frac{1}{2}\right)^{6 \times 10^{23}}$, which is such a small number that, for all practical purposes, it can be regarded as zero. On the basis of probability considerations, we would expect the gas to fill both flasks spontaneously and evenly. By the same token, we now understand that the situation depicted in Figure 18.1(b) is not spontaneous because it represents a highly improbable event. To summarize our discussion: An ordered state has a low probability of occurring and a small entropy, while a disordered state has a high probability of occurring and a large entropy.

As we will see, it is possible to determine the *absolute* entropy of a substance, something we cannot do for energy or enthalpy. *Standard entropy* is the absolute entropy of a substance at 1 atm and 25°C. It is this value that is generally used in calculations. (Recall that the standard state refers only to 1 atm. The reason for specifying 25°C is that many processes are carried out at room temperature.) Table 18.1 lists standard entropies of a few elements and compounds; Appendix 2 provides a more extensive listing. The units of entropy are J/K or J/K · mol for 1 mole of the substance. We use joules rather than kilojoules because entropy values are typically quite small. Entropies of elements and compounds are all positive (that is, $S° > 0$). By contrast, the standard enthalpy of formation ($\Delta H_f°$) for elements in their stable form is zero, and for compounds it may be positive or negative.

Table 18.1 | Standard Entropy Values ($S°$) for Some Substances at 25°C

Substance	$S°$ (J/K · mol)
$H_2O(l)$	69.9
$H_2O(g)$	188.7
$Br_2(l)$	152.3
$Br_2(g)$	245.3
$I_2(s)$	116.7
$I_2(g)$	260.6
C(diamond)	2.44
C(graphite)	5.69
He(g)	126.1
Ne(g)	146.2

Referring to Table 18.1, we can make these observations. The entropy of water vapor is greater than the entropy of liquid water, because a mole of water has a much smaller volume in the liquid state than it does in the gaseous state. In other words, water molecules are more ordered in the liquid state because there is less space for them to occupy. Similarly, bromine vapor has a higher entropy value than liquid bromine, and iodine vapor has a greater entropy value than solid iodine. By the same token, the entropy of a mole of a solid substance is smaller than that of its liquid form because molecules are more orderly arranged in the solid state. For different substances in the same phase, the molecular complexity and molar mass determine which ones have higher entropy values. Both diamond and graphite are solids, for example, but diamond has the more ordered structure (see Figure 12.22). Therefore, diamond has a smaller entropy value than graphite. Both neon and helium are monatomic gases, but neon has a greater entropy value than helium because its molar mass is greater.

Like energy and enthalpy, entropy is a state function. Consider a certain process in which a system changes from some initial state to some final state. The entropy change for the process, ΔS, is

$$\Delta S = S_f - S_i$$

where S_f and S_i are the entropies of the system in the final and initial states, respectively. If the change results in an increase in randomness, or disorder, then $S_f > S_i$ or $\Delta S > 0$.

Figure 18.2 shows several processes that lead to an increase in entropy. In each case, you can see that the system changes from a more ordered state to a less ordered one. It is clear that both melting and vaporization processes have $\Delta S > 0$. The solution process usually leads to an increase in entropy. When a sugar crystal dissolves in water,

We assume temperature is the same for the following comparisons.

State functions are properties that are determined by the state of the system (see Section 6.3).

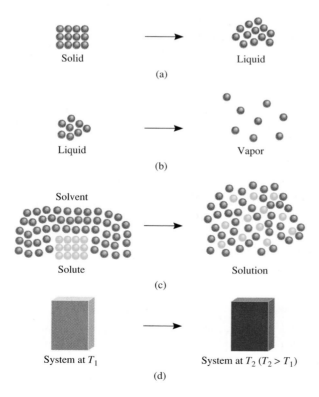

Solid → Liquid
(a)

Liquid → Vapor
(b)

Solvent
Solute → Solution
(c)

System at T_1 → System at T_2 $(T_2 > T_1)$
(d)

Figure 18.2 *Processes that lead to an increase in entropy of the system: (a) melting: $S_{liquid} > S_{solid}$; (b) vaporization: $S_{vapor} > S_{liquid}$; (c) dissolving; (d) heating: $S_{T_2} > S_{T_1}$.*

Figure 18.3 (a) A diatomic molecule can rotate about the y- and z-axes (the x-axis is along the bond). (b) Vibrational motion of a diatomic molecule. Chemical bonds can be stretched and compressed like a spring.

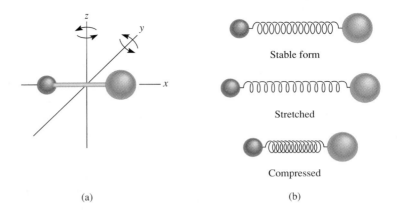

(a)

Stable form

Stretched

Compressed

(b)

the highly ordered structure of the solid and part of the ordered structure of water break down. Consequently, the solution is more disordered than the pure solute and pure solvent. When an ionic solid dissolves in water, there are two contributions to entropy increase; the solution process and the dissociation of the compound into ions:

$$NaCl(s) \xrightarrow{H_2O} Na^+(aq) + Cl^-(aq)$$

However, we must also consider hydration, which causes the water molecules to become more ordered around the ions. This process *decreases* entropy. For small, highly charged ions such as Al^{3+} and Fe^{3+}, the decrease in entropy due to hydration can outweigh the increase in entropy due to mixing and dissociation so that the entropy change for the overall process can actually be negative. Heating also increases the entropy of a system. In addition to translational motion (that is, motion from one point to another), molecules can also execute rotational motion and vibrational motion (Figure 18.3). The higher the temperature, the greater the molecular motion. This means an increase in randomness at the molecular level, which corresponds to an increase in entropy.

Bromine is a fuming liquid at room temperature.

Similar problem: 18.5.

Example 18.1   Predict whether the entropy change is greater than or less than zero for each of these processes: (a) freezing ethanol, (b) evaporating a beaker of liquid bromine at room temperature, (c) dissolving sucrose in water, (d) cooling nitrogen gas from 80°C to 20°C.

Reasoning and Solution   (a) This is a liquid-to-solid phase transition. The system becomes more ordered, so that $\Delta S < 0$.

(b) This is a liquid-to-vapor phase transition. The system becomes more disordered, and $\Delta S > 0$.

(c) A solution is usually more disordered than its components (the solute and solvent). Therefore, $\Delta S > 0$.

(d) Cooling decreases molecular motion; therefore, $\Delta S < 0$.

Comment   Note that all ΔS values apply to the system.

Practice Exercise   How does the entropy of a system change for each of these processes? (a) condensing water vapor, (b) forming sucrose crystals from a supersaturated solution, (c) heating hydrogen gas from 60°C to 80°C, and (d) subliming dry ice.

18.3 The Second Law of Thermodynamics

The connection between entropy and the spontaneity of a reaction is expressed by the **second law of thermodynamics:** *The entropy of the universe increases in a spontaneous process and remains unchanged in an equilibrium process.* Because the universe is made up of the system and the surroundings, the entropy change in the universe (ΔS_{univ}) for any process is the *sum* of the entropy changes in the system (ΔS_{sys}) and in the surroundings (ΔS_{surr}). Mathematically, we can express the second law of thermodynamics as follows:

Just talking about entropy increases its value in the universe.

For a spontaneous process: $\quad \Delta S_{univ} = \Delta S_{sys} + \Delta S_{surr} > 0 \qquad (18.1)$

For an equilibrium process: $\quad \Delta S_{univ} = \Delta S_{sys} + \Delta S_{surr} = 0 \qquad (18.2)$

For a spontaneous process, the second law says that ΔS_{univ} must be greater than zero, but it does not place a restriction on either ΔS_{sys} or ΔS_{surr}. Thus it is possible for either ΔS_{sys} or ΔS_{surr} to be negative, as long as the sum of these two quantities is greater than zero. For an equilibrium process, ΔS_{univ} is zero. In this case, ΔS_{sys} and ΔS_{surr} must be equal in magnitude, but opposite in sign. What if for some process we find that ΔS_{univ} is negative? What this means is that the process is not spontaneous in the direction described. Rather, it is spontaneous in the *opposite* direction.

Entropy Changes in the System

To calculate ΔS_{univ}, we need to know both ΔS_{sys} and ΔS_{surr}. Let us focus first on ΔS_{sys}. Suppose that the system is represented by the following reaction:

$$a\text{A} + b\text{B} \longrightarrow c\text{C} + d\text{D}$$

As is the case for the enthalpy of a reaction [see Equation (6.16)], the **standard entropy of reaction** ΔS_{rxn}° is given by

$$\Delta S_{rxn}^{\circ} = [cS^{\circ}(\text{C}) + dS^{\circ}(\text{D})] - [aS^{\circ}(\text{A}) + bS^{\circ}(\text{B})] \qquad (18.3)$$

or, in general, using Σ to represent summation and m and n for the stoichiometric coefficients in the reaction,

$$\Delta S_{rxn}^{\circ} = \Sigma n S^{\circ}(\text{products}) - \Sigma m S^{\circ}(\text{reactants}) \qquad (18.4)$$

The standard entropy values of a large number of compounds have been measured in J/K · mol. To calculate ΔS_{rxn}° (which is ΔS_{sys}), we look up their values in Appendix 2.

Example 18.2   From the standard entropy values in Appendix 2, calculate the standard entropy changes for these reactions at 25°C.

(a) $CaCO_3(s) \longrightarrow CaO(s) + CO_2(g)$

(b) $N_2(g) + 3H_2(g) \longrightarrow 2NH_3(g)$

(c) $H_2(g) + Cl_2(g) \longrightarrow 2HCl(g)$

Reasoning and Solution   We can calculate ΔS° by using Equation (18.4).

(a) $\Delta S^{\circ}_{rxn} = [S^{\circ}(CaO) + S^{\circ}(CO_2)] - [S^{\circ}(CaCO_3)]$

$= [(1\ mol)(39.8\ J/K \cdot mol) + (1\ mol)(213.6\ J/K \cdot mol)]$
$- (1\ mol)(92.9\ J/K \cdot mol)$

$= 160.5\ J/K$

Thus, when 1 mole of $CaCO_3$ decomposes to form 1 mole of CaO and 1 mole of gaseous CO_2, there is an increase in entropy equal to 160.5 J/K.

(b) $\Delta S^{\circ}_{rxn} = [2S^{\circ}(NH_3)] - [S^{\circ}(N_2) + 3S^{\circ}(H_2)]$

$= (2\ mol)(193\ J/K \cdot mol)$
$- [(1\ mol)(192\ J/K \cdot mol) + (3\ mol)(131\ J/K \cdot mol)]$

$= -199\ J/K$

This result shows that when 1 mole of gaseous nitrogen reacts with 3 moles of gaseous hydrogen to form 2 moles of gaseous ammonia, there is a decrease in entropy equal to −199 J/K.

(c) $\Delta S^{\circ}_{rxn} = [2S^{\circ}(HCl)] - [S^{\circ}(H_2) + S^{\circ}(Cl_2)]$

$= (2\ mol)(187\ J/K \cdot mol)$
$- [(1\ mol)(131\ J/K \cdot mol) + (1\ mol)(223\ J/K \cdot mol)]$

$= 20\ J/K$

Thus the formation of 2 moles of gaseous HCl from 1 mole of gaseous H_2 and 1 mole of gaseous Cl_2 results in a small increase in entropy equal to 20 J/K.

Comment   The ΔS°_{rxn} values all apply to the system.

Practice Exercise   Calculate the standard entropy change for these reactions at 25°C:

(a) $2CO(g) + O_2(g) \longrightarrow 2CO_2(g)$

(b) $3O_2(g) \longrightarrow 2O_3(g)$

(c) $2NaHCO_3(s) \longrightarrow Na_2CO_3(s) + H_2O(l) + CO_2(g)$

Similar problems: 18.11 and 18.12.

The results of Example 18.2 are consistent with those observed for many other reactions. Taken together, they support these general rules:

We omit the subscript rxn for simplicity.

- If a reaction produces more gas molecules than it consumes [Example 18.2(a)], ΔS° is positive.
- If the total number of gas molecules diminishes [Example 18.2(b)], ΔS° is negative.
- If there is no net change in the total number of gas molecules [Example 18.2(c)], then ΔS° may be positive or negative, but will be relatively small numerically.

These conclusions make sense, given that gases invariably have greater entropy than liquids and solids. For reactions involving only liquids and solids, predicting the sign of ΔS° is more difficult, but in many such cases an increase in the total number of molecules and/or ions is accompanied by an increase in entropy.

Example 18.3   Predict whether the entropy change of the system in each of these reactions is positive or negative.

(a) $2H_2(g) + O_2(g) \longrightarrow 2H_2O(l)$

(b) $NH_4Cl(s) \longrightarrow NH_3(g) + HCl(g)$

(c) $H_2(g) + Br_2(g) \longrightarrow 2HBr(g)$

Reasoning and Solution   We follow the preceding rules for determining the changes in entropy.

(a) Two gases combine to form a liquid. Therefore, ΔS is negative.

(b) Because the solid is converted to two gaseous products, ΔS is positive.

(c) We see that the same number of moles of gas is involved in the reactants as in the product. Therefore we cannot predict the sign of ΔS, but we know the change must be quite small.

Comment   All the entropy changes apply to the system.

Practice Exercise   Discuss qualitatively the sign of the entropy change expected for each of the following processes:

(a) $I_2(s) \longrightarrow 2I(g)$

(b) $2Zn(s) + O_2(g) \longrightarrow 2ZnO(s)$

(c) $N_2(g) + O_2(g) \longrightarrow 2NO(g)$

Similar problems: 18.13 and 18.14.

Entropy Changes in the Surroundings

Next we see how ΔS_{surr} is calculated. When an exothermic process takes place in the system, the heat transferred to the surroundings enhances motion of the molecules in the surroundings. Consequently, there is an increase in disorder at the molecular level, and the entropy of the surroundings increases. Conversely, an endothermic process in the system absorbs heat from the surroundings and so decreases the entropy of the surroundings because molecular motion decreases (Figure 18.4). For constant-pressure processes the heat change is equal to the enthalpy change of the system, ΔH_{sys}. Therefore, the change in entropy of the surroundings, ΔS_{surr}, is proportional to ΔH_{sys}.

$$\Delta S_{surr} \propto -\Delta H_{sys}$$

The minus sign is used because if the process is exothermic, ΔH_{sys} is negative and ΔS_{surr} is a positive quantity, indicating an increase in entropy. On the other hand, for

(a)

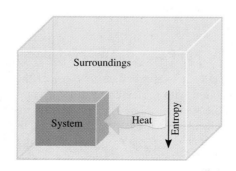

(b)

Figure 18.4 *(a) An exothermic process transfers heat from the system to the surroundings and results in an increase in the entropy of the surroundings. (b) An endothermic process absorbs heat from the surroundings and thereby decreases the entropy of the surroundings.*

an endothermic process, ΔH_{sys} is positive and the negative sign ensures that the entropy of the surroundings decreases.

The change in entropy for a given amount of heat also depends on the temperature. If the temperature of the surroundings is high, the molecules are already quite energetic. Therefore, the absorption of heat from an exothermic process in the system will have relatively little impact on molecular motion and the resulting increase in entropy will be small. However, if the temperature of the surroundings is low, then the addition of the same amount of heat will cause a more drastic increase in molecular motion and hence a larger increase in entropy. By analogy, someone coughing in a crowded restaurant will not disturb too many people, but someone coughing in a library definitely will. From the inverse relationship between ΔS_{surr} and temperature T (in kelvins)—that is, the higher the temperature, the smaller the ΔS_{surr} and vice versa—we can rewrite the previous relationship as

This equation, which can be derived from the laws of thermodynamics, assumes that both the system and the surroundings are at temperature T.

$$\Delta S_{surr} = \frac{-\Delta H_{sys}}{T} \qquad (18.5)$$

Let us now apply the procedure for calculating ΔS_{sys} and ΔS_{surr} to the synthesis of ammonia and ask whether the reaction is spontaneous at 25°C:

$$N_2(g) + 3H_2(g) \longrightarrow 2NH_3(g) \qquad \Delta H° = -92.6 \text{ kJ}$$

From Example 18.2(b) we have $\Delta S_{sys} = -199$ J/K, and substituting ΔH_{sys} (−92.6 kJ) in Equation (18.5), we obtain

$$\Delta S_{surr} = \frac{-(-92.6 \times 1000) \text{ J}}{298 \text{ K}} = 311 \text{ J/K}$$

The change in entropy of the universe is

$$\Delta S_{univ} = \Delta S_{sys} + \Delta S_{surr}$$
$$= -199 \text{ J/K} + 311 \text{ J/K}$$
$$= 112 \text{ J/K}$$

Because ΔS_{univ} is positive, we predict that the reaction is spontaneous at 25°C. It is important to keep in mind that just because a reaction is spontaneous does not mean that it will occur at an observable rate. The synthesis of ammonia is, in fact, extremely slow at room temperature. Thermodynamics can tell us whether a reaction will occur spontaneously under specific conditions, but it does not say how fast it will occur. Reaction rates are the subject of chemical kinetics (see Chapter 14).

The Third Law of Thermodynamics and Absolute Entropy

Finally it is appropriate to consider the *third law of thermodynamics* briefly in connection with the determination of entropy values. So far we have related entropy to molecular disorder—the greater the disorder or freedom of motion of the atoms or molecules in a system, the greater the entropy of the system. The most ordered arrangement of any substance with the least freedom of atomic or molecular motion is a perfect crystalline substance at absolute zero (0 K). It follows, therefore, that the lowest entropy any substance can attain is that of a perfect crystal at absolute zero.

According to the ***third law of thermodynamics,*** *the entropy of a perfect crystalline substance is zero at the absolute zero of temperature.* As the temperature increases, the freedom of motion also increases. Thus the entropy of any substance at a temperature above 0 K is greater than zero. Note also that if the crystal is impure or if it has defects, then its entropy is greater than zero even at 0 K because it would not be perfectly ordered.

The important point about the third law of thermodynamics is that it allows us to determine the *absolute* entropies of substances. Starting with the knowledge that the entropy of a pure crystalline substance is zero at 0 K, we can measure the increase in entropy of the substance when it is heated to, say, 298 K. The change in entropy, ΔS, is given by

$$\Delta S = S_f - S_i$$
$$= S_f$$

because S_i is zero. The entropy of the substance at 298 K, then, is given by ΔS or S_f, which is called the absolute entropy because this is the *true* value and not a value derived using some arbitrary reference. Thus the entropy values quoted so far are all absolute entropies. In contrast, we cannot have the absolute energy or enthalpy of a substance because the zero of energy or enthalpy is undefined. Figure 18.5 shows the change (increase) in entropy of a substance with temperature. At absolute zero, it has a zero entropy value (assuming that it is a perfect crystalline substance). As it is heated, its entropy increases gradually because of greater molecular motion. At the melting point, there is a sizable increase in entropy as the more random liquid state is formed. Further heating increases the entropy of the liquid again due to enhanced molecular motion. At the boiling point there is a large increase in entropy as a result of the liquid to gas transition. Beyond that temperature, the entropy of the gas continues to rise with increasing temperature.

The entropy increase can be calculated from the temperature change and heat capacity of the substance.

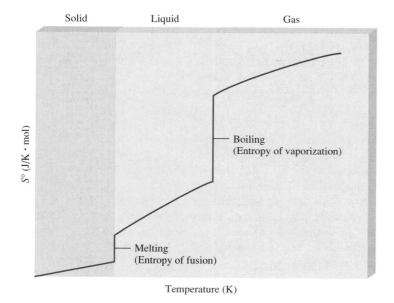

Figure 18.5 *Entropy increase of a substance as the temperature rises from absolute zero.*

18.4 Gibbs Free Energy

The second law of thermodynamics tells us that a spontaneous reaction increases the entropy of the universe; that is, $\Delta S_{\text{univ}} > 0$. In order to determine the sign of ΔS_{univ} for a reaction, however, we would need to calculate both ΔS_{sys} and ΔS_{surr}. However, we are usually concerned only with what happens in a particular system, and the calculation of ΔS_{surr} can be quite difficult. Therefore, we generally rely on another thermodynamic function to help us determine whether a reaction will occur spontaneously if we consider only the system itself.

From Equation (18.1), we know that for a spontaneous process, we have

$$\Delta S_{\text{univ}} = \Delta S_{\text{sys}} + \Delta S_{\text{surr}} > 0$$

We assume both the system and the surroundings are at temperature T.

Substituting $-\Delta H_{\text{sys}}/T$ for ΔS_{surr}, we write

$$\Delta S_{\text{univ}} = \Delta S_{\text{sys}} - \frac{\Delta H_{\text{sys}}}{T} > 0$$

Multiplying both sides of the equation by T gives

$$T\Delta S_{\text{univ}} = -\Delta H_{\text{sys}} + T\Delta S_{\text{sys}} > 0$$

Now we have a criterion for a spontaneous reaction that is expressed only in terms of the properties of the system (ΔH_{sys} and ΔS_{sys}) and we can ignore the surroundings. For convenience, we can change the above equation, multiplying it throughout by -1 and replacing the $>$ sign with $<$:

The change in unequal sign when we multiply the equation by −1 follows from the fact that 1 > 0 and −1 < 0.

$$-T\Delta S_{\text{univ}} = \Delta H_{\text{sys}} - T\Delta S_{\text{sys}} < 0$$

This equation says that for a process carried out at consant pressure and temperature T, if the changes in enthalpy and entropy of the system are such that $\Delta H_{\text{sys}} - T\Delta S_{\text{sys}}$ is less than zero, the process must be spontaneous.

In order to express the spontaneity of a reaction more directly, we can use another thermodynamic function called **Gibbs free energy (G),** or simply **free energy** (after the American physicist Josiah Willard Gibbs):

We omit the subscript sys for simplicity.

$$G = H - TS \tag{18.6}$$

All quantities in Equation (18.6) pertain to the system, and T is the temperature of the system. You can see that G has units of energy (both H and TS are in energy units). Like H and S, G is a state function.

The change in free energy (ΔG) of a system for a *constant-temperature* process is

$$\Delta G = \Delta H - T\Delta S \tag{18.7}$$

The word "free" in the term "free energy" does not mean without cost.

In this context free energy is *the energy available to do work.* Thus, if a particular reaction is accompanied by a release of usable energy (that is, if ΔG is negative), this fact alone guarantees that it is spontaneous, and there is no need to worry about what happens to the rest of the universe.

Note that we have merely organized the expression for the entropy change of the universe, eliminating ΔS_{univ} and equating the free-energy change of the system (ΔG) with $-T\Delta S_{\text{univ}}$, so that we can focus on changes in the system. We can now summarize the conditions for spontaneity and equilibrium at constant temperature and pressure in terms of ΔG:

$\Delta G < 0$ The reaction is spontaneous in the forward direction.

$\Delta G > 0$ The reaction is nonspontaneous. The reaction is spontaneous in the opposite direction.

$\Delta G = 0$ The system is at equilibrium. There is no net change.

Standard Free-Energy Changes

The ***standard free-energy of reaction*** (ΔG°_{rxn}) is *the free-energy change for a reaction when it occurs under standard-state conditions, that is, when reactants in their standard states are converted to products in their standard states.* Table 18.2 summarizes the conventions used by chemists to define the standard states of pure substances as well as solutions. To calculate ΔG°_{rxn} we start with the equation

$$a\text{A} + b\text{B} \longrightarrow c\text{C} + d\text{D}$$

The standard free-energy change for this reaction is given by

$$\Delta G^{\circ}_{rxn} = [c\Delta G^{\circ}_{f}(\text{C}) + d\Delta G^{\circ}_{f}(\text{D})] - [a\Delta G^{\circ}_{f}(\text{A}) + b\Delta G^{\circ}_{f}(\text{B})]$$

or, in general,

$$\Delta G^{\circ}_{rxn} = \Sigma n\Delta G^{\circ}_{f}(\text{products}) - \Sigma m\Delta G^{\circ}_{f}(\text{reactants}) \qquad (18.8)$$

where m and n are stoichiometric coefficients. The term $\boldsymbol{\Delta G^{\circ}_{f}}$ is the ***standard free energy of formation*** of a compound, that is, *the free-energy change that occurs when 1 mole of the compound is synthesized from its elements in their standard states.* For the combustion of graphite:

$$\text{C(graphite)} + \text{O}_2(g) \longrightarrow \text{CO}_2(g)$$

the standard free-energy change [from Equation (18.8)] is

$$\Delta G^{\circ}_{rxn} = \Delta G^{\circ}_{f}(\text{CO}_2) - [\Delta G^{\circ}_{f}(\text{C, graphite}) + \Delta G^{\circ}_{f}(\text{O}_2)]$$

As in the case of the standard enthalpy of formation (p. 179), we define the standard free energy of formation of any element in its stable form as zero. Thus

$$\Delta G^{\circ}_{f}(\text{C, graphite}) = 0 \qquad \text{and} \qquad \Delta G^{\circ}_{f}(\text{O}_2) = 0$$

Therefore the standard free-energy change for the reaction in this case is numerically equal to the standard free energy of formation of CO_2:

$$\Delta G^{\circ}_{rxn} = \Delta G^{\circ}_{f}(\text{CO}_2)$$

Note that ΔG°_{rxn} is in kJ, but ΔG°_{f} is in kJ/mol. The equation holds because the coefficient in front of ΔG°_{f} (1 in this case) has the unit "mol." Appendix 2 lists the values of ΔG°_{f} for a number of compounds.

Table 18.2 | Conventions for Standard States

State of Matter	Standard State
Gas	1 atm pressure
Liquid	Pure liquid
Solid	Pure solid
Elements*	$\Delta G^{\circ}_{f} = 0$
Solution	1 molar concentration

*The most stable allotropic form at 25°C and 1 atm.

Example 18.4   Calculate the standard free-energy changes for these reactions at 25°C.

(a) $\text{CH}_4(g) + 2\text{O}_2(g) \longrightarrow \text{CO}_2(g) + 2\text{H}_2\text{O}(l)$

(b) $2\text{MgO}(s) \longrightarrow 2\text{Mg}(s) + \text{O}_2(g)$

Reasoning and Solution (a) According to Equation (18.8), we write

$$\Delta G_{rxn}^{\circ} = [\Delta G_f^{\circ}(CO_2) + 2\Delta G_f^{\circ}(H_2O)] - [\Delta G_f^{\circ}(CH_4) + 2\Delta G_f^{\circ}(O_2)]$$

We insert the appropriate values from Appendix 2:

$$\Delta G_{rxn}^{\circ} = [(1 \text{ mol})(-394.4 \text{ kJ/mol}) + (2 \text{ mol})(-237.2 \text{ kJ/mol})]$$
$$- [(1 \text{ mol})(-50.8 \text{ kJ/mol}) + (2 \text{ mol})(0 \text{ kJ/mol})]$$

$$= -818.0 \text{ kJ}$$

(b) The equation is

$$\Delta G_{rxn}^{\circ} = [2\Delta G_f^{\circ}(Mg) + \Delta G_f^{\circ}(O_2)] - [2\Delta G_f^{\circ}(MgO)]$$

From data in Appendix 2 we write

$$\Delta G_{rxn}^{\circ} = [(2 \text{ mol})(0 \text{ kJ/mol}) + (1 \text{ mol})(0 \text{ kJ/mol})] - [(2 \text{ mol})(-569.6 \text{ kJ/mol})]$$
$$= 1139 \text{ kJ}$$

Similar problems: 18.17 and 18.18.

Practice Exercise Calculate the standard free-energy changes for these reactions at 25°C:

(a) $H_2(g) + Br_2(l) \longrightarrow 2HBr(g)$

(b) $2C_2H_6(g) + 7O_2(g) \longrightarrow 4CO_2(g) + 6H_2O(l)$

In Example 18.4 the large negative value of ΔG_{rxn}° for the combustion of methane in (a) means that the reaction is a spontaneous process under standard-state conditions, whereas the decomposition of MgO in (b) is nonspontaneous because ΔG_{rxn}° is a large, positive quantity. Remember, however, that a large, negative ΔG_{rxn}° does not tell us anything about the actual *rate* of the spontaneous process; a mixture of CH_4 and O_2 at 25°C could sit unchanged for quite some time in the absence of a spark or flame.

Applications of Equation (18.7)

In order to predict the sign of ΔG, according to Equation (18.7) we need to know both ΔH and ΔS. A negative ΔH (an exothermic reaction) and a positive ΔS (a reaction that results in an increase in disorder of the system) tend to make ΔG negative, although temperature may also influence the *direction* of a spontaneous reaction. The four possible outcomes of this relationship are:

- If both ΔH and ΔS are positive, then ΔG will be negative only when the $T\Delta S$ term is greater in magnitude than ΔH. This condition is met when T is large.
- If ΔH is positive and ΔS is negative, ΔG will always be positive, regardless of temperature.
- If ΔH is negative and ΔS is positive, then ΔG will always be negative regardless of temperature.
- If ΔH is negative and ΔS is negative, then ΔG will be negative only when $T\Delta S$ is smaller in magnitude than ΔH. This condition is met when T is small.

The temperatures that will cause ΔG to be negative for the first and last cases depend on the actual values of ΔH and ΔS of the system. Table 18.3 summarizes the effects of the possibilities just described.

We will now consider two specific applications of Equation (18.7).

Table 18.3	Factors Affecting the Sign of ΔG in the Relationship $\Delta G = \Delta H - T\Delta S$		
ΔH	ΔS	ΔG	Example
+	+	Reaction proceeds spontaneously at high temperatures. At low temperatures, reaction is spontaneous in the reverse direction.	$H_2(g) + I_2(g) \longrightarrow 2HI(g)$
+	−	ΔG is always positive. Reaction is spontaneous in the reverse direction at all temperatures.	$3O_2(g) \longrightarrow 2O_3(g)$
−	+	ΔG is always negative. Reaction proceeds spontaneously at all temperatures.	$2H_2O_2(l) \longrightarrow 2H_2O(l) + O_2(g)$
−	−	Reaction proceeds spontaneously at low temperatures. At high temperatures, the reverse reaction becomes spontaneous.	$NH_3(g) + HCl(g) \longrightarrow NH_4Cl(s)$

Temperature and Chemical Reactions

Calcium oxide (CaO), also called quicklime, is an extremely valuable inorganic substance used in steelmaking, production of calcium metal, the paper industry, water treatment, and pollution control. It is prepared by decomposing limestone ($CaCO_3$) in a kiln at a high temperature (Figure 18.6):

$$CaCO_3(s) \rightleftharpoons CaO(s) + CO_2(g)$$

The reaction is reversible, and CaO readily combines with CO_2 to form $CaCO_3$. The pressure of CO_2 in equilibrium with $CaCO_3$ and CaO increases with temperature. In the industrial preparation of quicklime, the system is never maintained at equilibrium; rather, CO_2 is constantly removed from the kiln to shift the equilibrium from left to right, promoting the formation of calcium oxide.

Le Châtelier's principle predicts that the forward, endothermic reaction is favored by heating.

The important information for the practical chemist is the temperature at which the decomposition of $CaCO_3$ becomes appreciable (that is, the temperature at which the reaction becomes spontaneous). We can make a reliable estimate of that temperature as follows. First we calculate $\Delta H°$ and $\Delta S°$ for the reaction at 25°C, using the data in Appendix 2. To determine $\Delta H°$ we apply Equation (6.16):

$$\begin{aligned}
\Delta H° &= [\Delta H_f°(CaO) + \Delta H_f°(CO_2)] - [\Delta H_f°(CaCO_3)] \\
&= [(1 \text{ mol})(-635.6 \text{ kJ/mol}) + (1 \text{ mol})(-393.5 \text{ kJ/mol})] \\
&\quad - [(1 \text{ mol})(-1206.9 \text{ kJ/mol})] \\
&= 177.8 \text{ kJ}
\end{aligned}$$

Next we apply Equation (18.4) to find $\Delta S°$

$$\begin{aligned}
\Delta S° &= [S°(CaO) + S°(CO_2)] - [S°(CaCO_3)] \\
&= [(1 \text{ mol})(39.8 \text{ J/K} \cdot \text{mol}) + (1 \text{ mol})(213.6 \text{ J/K} \cdot \text{mol})] \\
&\quad - [(1 \text{ mol})(92.9 \text{ J/K} \cdot \text{mol})] \\
&= 160.5 \text{ J/K}
\end{aligned}$$

For reactions carried out under standard-state conditions, Equation (18.7) takes the form

$$\Delta G° = \Delta H° - T\Delta S°$$

so we obtain

$$\Delta G° = 177.8 \text{ kJ} - (298 \text{ K})(160.5 \text{ J/K})\left(\frac{1 \text{ kJ}}{1000 \text{ J}}\right)$$

$$= 130.0 \text{ kJ}$$

Figure 18.6 *The production of CaO in a rotary kiln.*

Figure 18.7 *Equilibrium pressure of CO_2 from the decomposition of $CaCO_3$, as a function of temperature. This curve is calculated by assuming that $\Delta H°$ and $\Delta S°$ of the reaction do not change with temperature.*

Because $\Delta G°$ is a large positive quantity, we conclude that the reaction is not favored at 25°C (or 298 K). In order to make $\Delta G°$ negative, we first have to find the temperature at which $\Delta G°$ is zero; that is,

$$0 = \Delta H° - T\Delta S°$$

or

$$T = \frac{\Delta H°}{\Delta S°}$$

$$T = \frac{(177.8 \text{ kJ})(1000 \text{ J/1 kJ})}{160.5 \text{ J/K}}$$

$$= 1108 \text{ K or } 835°C$$

At a temperature higher than 835°C, $\Delta G°$ becomes negative, indicating that the decomposition is spontaneous. For example, at 840°C, or 1113 K,

$$\Delta G° = \Delta H° - T\Delta S°$$

$$= 177.8 \text{ kJ} - (1113 \text{ K})(160.5 \text{ J/K})\left(\frac{1 \text{ kJ}}{1000 \text{ J}}\right)$$

$$= -0.8 \text{ kJ}$$

Two points are worth making about such a calculation. First, we used the $\Delta H°$ and $\Delta S°$ values at 25°C to calculate changes that occur at a much higher temperature. Because both $\Delta H°$ and $\Delta S°$ change with temperature, this approach will not give us an accurate value of $\Delta G°$, but it is good enough for "ball park" estimates. Second, we should not be misled into thinking that nothing happens below 835°C and that at 835°C $CaCO_3$ suddenly begins to decompose. Far from it. The fact that $\Delta G°$ is a positive value at some temperature below 835°C does not mean that no CO_2 is produced, but rather that the pressure of the CO_2 gas formed at that temperature will be below 1 atm (its standard-state value; see Table 18.2). As Figure 18.7 shows, the pressure of CO_2 at first increases very slowly with temperature; it becomes easily measurable above 700°C. The significance of 835°C is that this is the temperature at which the equilibrium pressure of CO_2 reaches 1 atm. Above 835°C, the equilibrium pressure of CO_2 exceeds 1 atm. (In Section 18.5 we will see how $\Delta G°$ is related to the equilibrium constant of a given reaction.)

If a system is at equilibrium, then there is no tendency for spontaneous change in either direction. The condition $\Delta G = 0$ applies to any phase transition.

Phase Transitions

At the temperature at which a phase transition occurs (the melting point or boiling point) the system is at equilibrium ($\Delta G = 0$), so Equation (18.7) becomes

$$0 = \Delta H - T\Delta S$$

$$\Delta S = \frac{\Delta H}{T}$$

Let us first consider the ice-water equilibrium. For the ice $\longrightarrow$ water transition, ΔH is the molar heat of fusion (see Table 12.7) and T is the melting point. The entropy change is therefore

The equilibrium constant of this reaction is $K_P = P_{CO_2}$.

The melting of ice is an endothermic process (ΔH is positive), and the freezing of water is exothermic (ΔH is negative).

$$\Delta S_{\text{ice} \longrightarrow \text{water}} = \frac{6010 \text{ J/mol}}{273 \text{ K}}$$

$$= 22.0 \text{ J/K} \cdot \text{mol}$$

Thus when 1 mole of ice melts at 0°C, there is an increase in entropy of 22.0 J/K. The increase in entropy is consistent with the increase in disorder from solid to liquid. Conversely, for the water $\longrightarrow$ ice transition, the decrease in entropy is given by

$$\Delta S_{\text{water} \rightarrow \text{ice}} = \frac{-6010 \text{ J/mol}}{273 \text{ K}}$$

$$= -22.0 \text{ J/K} \cdot \text{mol}$$

In the laboratory we normally carry out unidirectional changes, that is, either ice to water or water to ice. We can calculate entropy change in each case using the equation $\Delta S = \Delta H/T$ as long as the temperature remains at 0°C. The same procedure can be applied to the water-steam transition. In this case ΔH is the heat of vaporization and T is the boiling point of water.

Example 18.5 The molar heats of fusion and vaporization of benzene are 10.9 kJ/mol and 31.0 kJ/mol, respectively. Calculate the entropy changes for the solid $\longrightarrow$ liquid and liquid $\longrightarrow$ vapor transitions for benzene. At 1 atm pressure, benzene melts at 5.5°C and boils at 80.1°C.

Reasoning and Solution At the melting point, solid benzene and liquid benzene are at equilibrium. Therefore, $\Delta G = 0$ and the entropy of fusion is given by

$$\Delta G = \Delta H - T\Delta S$$
$$0 = \Delta H_{\text{fus}} - T_{\text{f}}\Delta S_{\text{fus}}$$
$$\Delta S_{\text{fus}} = \frac{\Delta H_{\text{fus}}}{T_{\text{f}}}$$
$$= \frac{(10.9 \text{ kJ/mol})(1000 \text{ J/1 kJ})}{(5.5 + 273) \text{ K}}$$
$$= 39.1 \text{ J/K} \cdot \text{mol}$$

Liquid and solid benzene in equilibrium.

Similarly, at the boiling point, $\Delta G = 0$ and we have

$$\Delta S_{\text{vap}} = \frac{\Delta H_{\text{vap}}}{T_{\text{bp}}}$$
$$= \frac{(31.0 \text{ kJ/mol})(1000 \text{ J/1 kJ})}{(80.1 + 273) \text{ K}}$$
$$= 87.8 \text{ J/K} \cdot \text{mol}$$

Comment Because vaporization creates more disorder than the melting process, $\Delta S_{\text{vap}} > \Delta S_{\text{fus}}$.

Similar problem: **18.59.**

Practice Exercise The molar heats of fusion and vaporization of argon are 1.3 kJ/mol and 6.3 kJ/mol, and argon's melting point and boiling point are −190°C and −186°C, respectively. Calculate the entropy changes for fusion and vaporization.

18.5 Free Energy and Chemical Equilibrium

Suppose we start a reaction in solution with all the reactants in their standard states (that is, all at 1 M concentration). As soon as the reaction starts, the standard-state condition no longer exists for the reactants or the products because their concentrations are different from 1 M. Under conditions that are *not* standard state, we must use ΔG rather than ΔG° to predict the direction of the reaction. The relationship between ΔG and ΔG°, which can be derived from thermodynamics, is

$$\Delta G = \Delta G^\circ + RT \ln Q \tag{18.9}$$

where R is the gas constant (8.314 J/K · mol), T is the absolute temperature of the reaction, and Q is the reaction quotient (see p. 482). We see that ΔG depends on two quantities: ΔG° and $RT \ln Q$. For a given reaction at temperature T the value of ΔG° is fixed but that of $RT \ln Q$ is not, because Q varies according to the composition of the reacting mixture. Let us consider two special cases:

The sign of ΔG, not ΔG°, determines the direction of reaction spontaneity.

Case 1: If ΔG° is a large negative value, the $RT \ln Q$ term will not become positive enough to match the ΔG° term until a significant amount of product has formed.

Case 2: If ΔG° is a large positive value, the $RT \ln Q$ term will be more negative than ΔG° is positive only as long as very little product formation has occurred and the concentration of the reactant is high relative to that of the product.

Sooner or later a reversible reaction will reach equilibrium.

At equilibrium, by definition, $\Delta G = 0$ and $Q = K$, where K is the equilibrium constant. Thus

$$0 = \Delta G^\circ + RT \ln K$$

or

$$\Delta G^\circ = -RT \ln K \tag{18.10}$$

In this equation, K_P is used for gases and K_c for reactions in solution. Note that the larger the K is, the more negative ΔG° is. For chemists, Equation (18.10) is one of the most important equations in thermodynamics because it enables us to find the equilibrium constant of a reaction if we know the change in standard free energy and vice versa.

It is significant that Equation (18.10) relates the equilibrium constant to the *standard* free-energy change ΔG° rather than to the *actual* free-energy change ΔG. The actual free energy of the system changes as the reaction proceeds and becomes zero at equilibrium. On the other hand, both ΔG° and K are constants for a particular reaction at a given temperature. Figure 18.8 shows plots of free energy of a reacting system versus the extent of the reaction for two types of reactions. As you can see, if $\Delta G^\circ < 0$, the products are favored over reactants at equilibrium. Conversely, if $\Delta G^\circ > 0$, there will be more reactants than products at equilibrium. Table 18.4 summarizes the three possible relations between ΔG° and K, as predicted by Equation (18.10).

For reactions having very large or very small equilibrium constants, it is generally very difficult, if not impossible, to measure the K values by monitoring the concentrations of all the reacting species. Consider, for example, the formation of nitric oxide from molecular nitrogen and molecular oxygen:

$$N_2(g) + O_2(g) \rightleftharpoons 2NO(g)$$

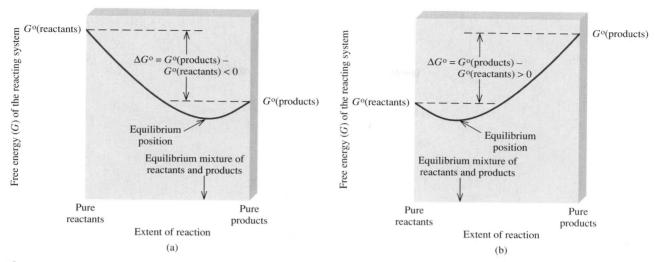

Figure 18.8 (a) $G°_{products} < G°_{reactants}$, so $\Delta G° < 0$. At equilibrium, there is a significant conversion of reactants to products. (b) $G°_{products} > G°_{reactants}$, so $\Delta G° > 0$. At equilibrium, reactants are favored over products.

At 25°C, the equilibrium constant K_c is

$$K_c = \frac{[NO]^2}{[N_2][O_2]} = 4.0 \times 10^{-31}$$

The very small value of K_c means that the concentration of NO at equilibrium will be exceedingly low. In such a case the equilibrium constant is more conveniently obtained from $\Delta G°$. (As we have seen, $\Delta G°$ can be calculated from $\Delta H°$ and $\Delta S°$.) On the other hand, the equilibrium constant for the formation of hydrogen iodide from molecular hydrogen and molecular iodine is near unity at room temperature:

$$H_2(g) + I_2(g) \rightleftharpoons 2HI(g)$$

For this reaction it is easier to measure K and then calculate $\Delta G°$ using Equation (18.10) than to measure $\Delta H°$ and $\Delta S°$ and use Equation (18.7).

A cautionary note in using Equations (18.9) and (18.10). To be consistent with units, we need to express both $\Delta G°$ and ΔG in units of kJ/mol in these equations. Here "per mol" means "per mole of the reaction as it is written." For example, $\Delta G°$ for the reaction

$$H_2(g) + I_2(g) \longrightarrow 2HI(g)$$

is 2.6 kJ/mol at 298 K, meaning that this is the standard free-energy change per mole of H_2, per mole of I_2, and per 2 moles of HI.

K	$\ln K$	$\Delta G°$	Comments
Table 18.4	Relation between $\Delta G°$ and K as Predicted by the Equation $\Delta G° = -RT \ln K$		
K	$\ln K$	$\Delta G°$	Comments
>1	Positive	Negative	Products are favored over reactants at equilibrium.
$=1$	0	0	Products and reactants are equally favored at equilibrium.
<1	Negative	Positive	Reactants are favored over products at equilibrium.

Example 18.6 Using data listed in Appendix 2, calculate the equilibrium constant (K_P) for the following reaction at 25°C:

$$2H_2O(l) \rightleftharpoons 2H_2(g) + O_2(g)$$

Reasoning and Solution According to Equation (18.8)

$$\Delta G°_{rxn} = [2\Delta G°_f(H_2) + \Delta G°_f(O_2)] - [2\Delta G°_f(H_2O)]$$
$$= [(2 \text{ mol})(0 \text{ kJ/mol}) + (1 \text{ mol})(0 \text{ kJ/mol})] - [(2 \text{ mol})(-237.2 \text{ kJ/mol})]$$
$$= 474.4 \text{ kJ}$$

Using Equation (18.10)

$$\Delta G°_{rxn} = -RT \ln K_P$$

$$474.4 \text{ kJ/mol} \times \frac{1000 \text{ J}}{1 \text{ kJ}} = -(8.314 \text{ J/K} \cdot \text{mol})(298 \text{ K}) \ln K_P$$

$$\ln K_P = -191.5$$
$$K_P = e^{-191.5}$$
$$= 7 \times 10^{-84}$$

Comment This extremely small equilibrium constant is consistent with the fact that water does not decompose into hydrogen and oxygen gases at 25°C. Thus a large positive $\Delta G°$ favors reactants over products at equilibrium.

Practice Exercise Calculate the equilibrium constant (K_P) for the reaction

$$2O_3(g) \longrightarrow 3O_2(g)$$

at 25°C.

As stated earlier, it is difficult to measure very large (and very small) values of *K* directly because the concentrations of some of the species are very small.

Similar problems: 18.23 and 18.26.

Example 18.7 In Chapter 17 we discussed the solubility product of slightly soluble substances. Using the solubility product of silver chloride at 25°C (1.6×10^{-10}), calculate $\Delta G°$ for the process

$$AgCl(s) \rightleftharpoons Ag^+(aq) + Cl^-(aq)$$

Reasoning and Solution Because this is a heterogeneous equilibrium, the solubility product is the equilibrium constant.

$$K_{sp} = [Ag^+][Cl^-] = 1.6 \times 10^{-10}$$

Using Equation (18.10) we obtain

$$\Delta G° = -(8.314 \text{ J/K} \cdot \text{mol})(298 \text{ K}) \ln 1.6 \times 10^{-10}$$
$$= 5.6 \times 10^4 \text{ J/mol}$$
$$= 56 \text{ kJ/mol}$$

Comment The large, positive $\Delta G°$ indicates that AgCl is slightly soluble and that the equilibrium lies mostly to the left.

Similar problem: 18.25.

Practice Exercise Calculate $\Delta G°$ for the following process at 25°C:

$$BaF_2(s) \rightleftharpoons Ba^{2+}(aq) + 2F^-(aq)$$

The K_{sp} of BaF_2 is 1.7×10^{-6}.

Example 18.8 The standard free-energy change for the reaction

$$N_2(g) + 3H_2(g) \rightleftharpoons 2NH_3(g)$$

is -33.2 kJ and the equilibrium constant K_P is 6.59×10^5 at 25°C. In a certain experiment, the initial pressures are $P_{H_2} = 0.250$ atm, $P_{N_2} = 0.870$ atm, and $P_{NH_3} = 12.9$ atm. Calculate ΔG for the reaction at these pressures, and predict the direction of reaction.

Reasoning and Solution Equation (18.9) can be written as

$$\Delta G = \Delta G° + RT \ln Q_P$$

$$= \Delta G° + RT \ln \frac{P_{NH_3}^2}{P_{H_2}^3 P_{N_2}}$$

$$= -33.2 \times 1000 \text{ J/mol} + (8.314 \text{ J/K} \cdot \text{mol})(298 \text{ K}) \times \ln \frac{(12.9)^2}{(0.250)^3(0.870)}$$

$$= -33.2 \times 10^3 \text{ J/mol} + 23.3 \times 10^3 \text{ J/mol}$$

$$= -9.9 \times 10^3 \text{ J/mol} = -9.9 \text{ kJ/mol}$$

Because ΔG is negative, the net reaction proceeds from left to right.

Comment As an exercise, confirm this prediction by calculating the reaction quotient Q_P and comparing it with the equilibrium constant K_P.

Similar problems: 18.27 and 18.28.

Practice Exercise The $\Delta G°$ for the reaction

$$H_2(g) + I_2(g) \rightleftharpoons 2HI(g)$$

is 2.60 kJ at 25°C. In one experiment, the initial pressures are $P_{H_2} = 4.26$ atm, $P_{I_2} = 0.024$ atm, and $P_{HI} = 0.23$ atm. Calculate ΔG for the reaction and predict the direction of reaction.

18.6 Thermodynamics in Living Systems

Many biochemical reactions are nonspontaneous (that is, they have a positive $\Delta G°$ value), yet they are essential to the maintenance of life. In living systems these reactions are coupled to an energetically favorable process, one that has a negative $\Delta G°$ value. The principle of *coupled reactions* is based on a simple concept: we can use a thermodynamically favorable reaction to drive an unfavorable one. Consider an industrial process. Suppose we wish to extract zinc from the ore sphalerite (ZnS). The following reaction will not work because it is highly nonspontaneous:

$$ZnS(s) \longrightarrow Zn(s) + S(s) \qquad \Delta G° = 198.3 \text{ kJ}$$

A mechanical analog for coupled reactions. We can make the smaller weight move upward (a nonspontaneous process) by coupling it with the falling of a larger weight.

On the other hand, the combustion of sulfur to form sulfur dioxide is favored because of its large negative $\Delta G°$ value:

$$S(s) + O_2(g) \longrightarrow SO_2(g) \qquad \Delta G° = -300.1 \text{ kJ}$$

By coupling the two processes we can bring about the separation of zinc from zinc sulfide. In practice, this means heating ZnS in air so that the tendency of S to form SO_2 will promote the decomposition of ZnS:

The price we pay for this procedure is acid rain.

$$
\begin{array}{ll}
\text{ZnS}(s) \longrightarrow \text{Zn}(s) + \text{S}(s) & \Delta G° = 198.3 \text{ kJ} \\
\underline{\text{S}(s) + \text{O}_2(g) \longrightarrow \text{SO}_2(g)} & \underline{\Delta G° = -300.1 \text{ kJ}} \\
\text{ZnS}(s) + \text{O}_2(g) \longrightarrow \text{Zn}(s) + \text{SO}_2(g) & \Delta G° = -101.8 \text{ kJ}
\end{array}
$$

Coupled reactions play a crucial role in our survival. In biological systems, enzymes facilitate a wide variety of nonspontaneous reactions. For example, in the human body, food molecules, represented by glucose ($C_6H_{12}O_6$), are converted to carbon dioxide and water during metabolism with a substantial release of free energy:

$$C_6H_{12}O_6(s) + 6O_2(g) \longrightarrow 6CO_2(g) + 6H_2O(l) \quad \Delta G° = -2880 \text{ kJ}$$

In a living cell, this reaction does not take place in a single step (as burning glucose in a flame would); rather, the glucose molecule is broken down with the aid of enzymes in a series of steps. Much of the energy released along the way is used to synthesize adenosine triphosphate (ATP) from adenosine diphosphate (ADP) and phosphoric acid (Figure 18.9):

$$\text{ADP} + H_3PO_4 \longrightarrow \text{ATP} + H_2O \qquad \Delta G° = +31 \text{ kJ}$$

The function of ATP is to store energy until it is needed by cells. Under appropriate conditions, 1 mole of ATP undergoes hydrolysis to give ADP and phosphoric acid, with a release of 31 kJ of free energy, which can be used to drive energetically unfavorable reactions, such as protein synthesis.

Proteins are polymers made of amino acids. The stepwise synthesis of a protein molecule involves the joining of individual amino acids. Consider the formation of the dipeptide (a two-amino-acid unit) alanylglycine from alanine and glycine. This reaction represents the first step in the synthesis of a protein molecule:

$$\text{Alanine} + \text{Glycine} \longrightarrow \text{Alanylglycine} \qquad \Delta G° = +29 \text{ kJ}$$

Adenosine triphosphate
(ATP)

Adenosine diphosphate
(ADP)

Figure 18.9 *Structure of ATP and ADP in ionized forms. The adenine group is in blue, the ribose group in black, and the phosphate group in red. Note that ADP has one fewer phosphate group than ATP.*

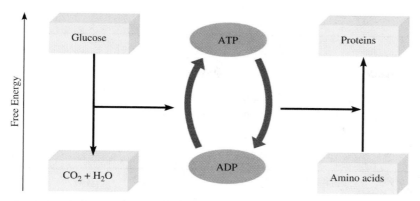

Figure 18.10 *Schematic representation of ATP synthesis and coupled reactions in living systems. The conversion of glucose to carbon dioxide and water during metabolism releases free energy. This release of free energy is used to convert ADP into ATP. The ATP molecules are then used as an energy source to drive nonspontaneous reactions, such as the synthesis of proteins from amino acids.*

As you can see, this reaction is not spontaneous, and so only a little of the dipeptide would be formed at equilibrium. However, with the aid of an enzyme, the reaction is coupled to the hydrolysis of ATP as follows:

$$\text{ATP} + \text{H}_2\text{O} + \text{Alanine} + \text{Glycine} \longrightarrow \text{ADP} + \text{H}_3\text{PO}_4 + \text{Alanylglycine}$$

The overall free-energy change is given by $\Delta G° = -31 \text{ kJ} + 29 \text{ kJ} = -2 \text{ kJ}$, which means that the coupled reaction is spontaneous and an appreciable amount of alanyl-glycine will be formed under this condition. Figure 18.10 shows the ATP-ADP inter-conversions that act as energy storage (from metabolism) and free energy release (from ATP hydrolysis) to drive essential reactions.

Summary

Entropy is a measure of the disorder of a system. Any spontaneous process must lead to a net increase in entropy in the universe (second law of thermodynamics) or, stated mathematically, $\Delta S_{\text{univ}} = \Delta S_{\text{sys}} + \Delta S_{\text{surr}} > 0$. The standard entropy change of a chemical reaction ($\Delta S°_{\text{rxn}}$) can be calculated from the absolute entropies of reactants and products.

The third law of thermodynamics states that the entropy of a perfect crystalline substance is zero at 0 K. This law enables us to measure the absolute entropies of substances.

Under conditions of constant temperature and pressure, the free-energy change ΔG is less than zero for a spontaneous process and greater than zero for a nonspontaneous process. For an equilibrium process, $\Delta G = 0$. For a chemical or physical process at constant temperature and pressure, $\Delta G = \Delta H - T\Delta S$. This equation can be used to predict the spontaneity of a process. The standard free-energy change for a reaction, $\Delta G°$, can be found from the standard free energies of formation of reactants and products. The equilibrium constant of a reaction and the standard free-energy change of the reaction are related by the equation $\Delta G° = -RT \ln K$.

Most biological reactions are nonspontaneous. They are made to occur by coupling them to the hydrolysis of ATP, which has a negative $\Delta G°$ value.

Key Words

Entropy (S) p. 577
Free energy (G) p. 586
Gibbs free energy (G)
    p. 586

Second law of
    thermodynamics
    p. 581
Standard entropy of
    reaction (ΔS°_{rxn}) p. 581

Standard free energy of
    reaction (ΔG°_{rxn})
    p. 587
Standard free energy of
    formation (ΔG°_f) p. 587

Third law of
    thermodynamics
    p. 585

Questions and Problems

Spontaneous Processes and Entropy

Review Questions

18.1 Explain what is meant by a spontaneous process. Give two examples each of spontaneous and nonspontaneous processes.

18.2 Which of these processes are spontaneous and which are nonspontaneous? (a) dissolving table salt (NaCl) in hot soup; (b) climbing Mt. Everest; (c) spreading fragrance in a room by removing the cap from a perfume bottle; (d) separating helium and neon from a mixture of the gases.

18.3 Which of these processes are spontaneous and which are nonspontaneous at a given temperature?

(a) $NaNO_3(s) \xrightarrow{H_2O} NaNO_3(aq)$ saturated soln
(b) $NaNO_3(s) \xrightarrow{H_2O} NaNO_3(aq)$ unsaturated soln
(c) $NaNO_3(s) \xrightarrow{H_2O} NaNO_3(aq)$ supersaturated soln

18.4 Define entropy. What are the units of entropy?

18.5 How does the entropy of a system change for each of these processes?

(a) A solid melts.
(b) A liquid freezes.
(c) A liquid boils.
(d) A vapor is converted to a solid.
(e) A vapor condenses to a liquid.
(f) A solid sublimes.
(g) Urea dissolves in water.

Problems

18.6 Referring to the setup in Figure 18.1(a), calculate the probability of all the molecules ending up in the same flask if the number is (a) 6, (b) 60, (c) 600.

The Second Law of Thermodynamics

Review Questions

18.7 State the second law of thermodynamics in words and express it mathematically.

18.8 State the third law of thermodynamics and explain its usefulness in calculating entropy values.

Problems

18.9 For each pair of substances listed here, choose the one having the larger standard entropy value at 25°C. The same molar amount is used in the comparison. Explain the basis for your choice. (a) Li(s) or Li(l); (b) $C_2H_5OH(l)$ or $CH_3OCH_3(l)$ (*Hint:* Which molecule can hydrogen-bond?); (c) Ar(g) or Xe(g); (d) CO(g) or $CO_2(g)$; (e) $O_2(g)$ or $O_3(g)$; (f) $NO_2(g)$ or $N_2O_4(g)$.

18.10 Arrange these substances (1 mole each) in order of increasing entropy at 25°C: (a) Ne(g), (b) $SO_2(g)$, (c) Na(s), (d) NaCl(s), (e) $H_2(g)$. Give the reasons for your arrangement.

18.11 Using the data in Appendix 2, calculate the standard entropy changes for the following reactions at 25°C:

(a) $S(s) + O_2(g) \longrightarrow SO_2(g)$
(b) $MgCO_3(s) \longrightarrow MgO(s) + CO_2(g)$

18.12 Using the data in Appendix 2, calculate the standard entropy changes for the following reactions at 25°C:

(a) $H_2(g) + CuO(s) \longrightarrow Cu(s) + H_2O(g)$
(b) $2Al(s) + 3ZnO(s) \longrightarrow Al_2O_3(s) + 3Zn(s)$
(c) $CH_4(g) + 2O_2(g) \longrightarrow CO_2(g) + 2H_2O(l)$

The S° values of Zn(s) and ZnO(s) are 41.6 J/K · mol and 43.9 J/K · mol, respectively.

18.13 Without consulting Appendix 2, predict whether the entropy change is positive or negative for each of these reactions. Give reasons for your predictions.

(a) $2KClO_4(s) \longrightarrow 2KClO_3(s) + O_2(g)$
(b) $H_2O(g) \longrightarrow H_2O(l)$
(c) $2Na(s) + 2H_2O(l) \longrightarrow 2NaOH(aq) + H_2(g)$
(d) $N_2(g) \longrightarrow 2N(g)$

18.14 State whether the sign of the entropy change expected for each of these processes will be positive or negative, and explain your predictions.

(a) $PCl_3(l) + Cl_2(g) \longrightarrow PCl_5(s)$
(b) $2HgO(s) \longrightarrow 2Hg(l) + O_2(g)$
(c) $H_2(g) \longrightarrow 2H(g)$
(d) $U(s) + 3F_2(g) \longrightarrow UF_6(s)$

Gibbs Free Energy

Review Questions

18.15 Define free energy. What are its units?

18.16 Why is it more convenient to predict the direction of a reaction in terms of ΔG_{sys} instead of ΔS_{univ}? Under what conditions can ΔG_{sys} be used to predict the spontaneity of a reaction?

Problems

18.17 Calculate $\Delta G°$ for these reactions at 25°C:

(a) $N_2(g) + O_2(g) \longrightarrow 2NO(g)$
(b) $H_2O(l) \longrightarrow H_2O(g)$
(c) $2C_2H_2(g) + 5O_2(g) \longrightarrow 4CO_2(g) + 2H_2O(l)$

(*Hint:* Look up the standard free energies of formation of the reactants and products in Appendix 2.)

18.18 Calculate $\Delta G°$ for these reactions at 25°C:

(a) $2Mg(s) + O_2(g) \longrightarrow 2MgO(s)$
(b) $2SO_2(g) + O_2(g) \longrightarrow 2SO_3(g)$
(c) $2C_2H_6(g) + 7O_2(g) \longrightarrow 4CO_2(g) + 6H_2O(l)$

See Appendix 2 for thermodynamic data.

18.19 From the values of ΔH and ΔS, predict which of these reactions would be spontaneous at 25°C: Reaction A: $\Delta H = 10.5$ kJ, $\Delta S = 30$ J/K; reaction B: $\Delta H = 1.8$ kJ, $\Delta S = -113$ J/K. If either of the reactions is nonspontaneous at 25°C, at what temperature might it become spontaneous?

18.20 Find the temperatures at which reactions with these ΔH and ΔS values would become spontaneous: (a) $\Delta H = -126$ kJ, $\Delta S = 84$ J/K; (b) $\Delta H = -11.7$ kJ, $\Delta S = -105$ J/K.

Free Energy and Chemical Equilibrium

Review Questions

18.21 Explain the difference between ΔG and $\Delta G°$.

18.22 Explain why Equation (18.10) is of great importance in chemistry.

Problems

18.23 Calculate K_P for this reaction at 25°C:

$$H_2(g) + I_2(g) \rightleftharpoons 2HI(g) \qquad \Delta G° = 2.60 \text{ kJ}$$

18.24 For the autoionization of water at 25°C,

$$H_2O(l) \rightleftharpoons H^+(aq) + OH^-(aq)$$

K_w is 1.0×10^{-14}. What is $\Delta G°$ for the process?

18.25 Consider this reaction at 25°C:

$$Fe(OH)_2(s) \rightleftharpoons Fe^{2+}(aq) + 2OH^-(aq)$$

Calculate $\Delta G°$ for the reaction. K_{sp} for $Fe(OH)_2$ is 1.6×10^{-14}.

18.26 Calculate $\Delta G°$ and K_P for this equilibrium reaction at 25°C.

$$2H_2O(g) \rightleftharpoons 2H_2(g) + O_2(g)$$

18.27 (a) Calculate $\Delta G°$ and K_P for this equilibrium reaction at 25°C. The $\Delta G_f°$ values are 0 for $Cl_2(g)$, -286 kJ/mol for $PCl_3(g)$, and -325 kJ/mol for $PCl_5(g)$.

$$PCl_5(g) \rightleftharpoons PCl_3(g) + Cl_2(g)$$

(b) Calculate ΔG for the reaction if the partial pressures of the initial mixture are $P_{PCl_5} = 0.0029$ atm, $P_{PCl_3} = 0.27$ atm, and $P_{Cl_2} = 0.40$ atm.

18.28 The equilibrium constant (K_P) for the reaction

$$H_2(g) + CO_2(g) \rightleftharpoons H_2O(g) + CO(g)$$

is 4.40 at 2000 K. (a) Calculate $\Delta G°$ for the reaction. (b) Calculate ΔG for the reaction when the partial pressures are $P_{H_2} = 0.25$ atm, $P_{CO_2} = 0.78$ atm, $P_{H_2O} = 0.66$ atm, and $P_{CO} = 1.20$ atm.

18.29 Consider the decomposition of calcium carbonate:

$$CaCO_3(s) \rightleftharpoons CaO(s) + CO_2(g)$$

Calculate the pressure in atm of CO_2 in an equilibrium process (a) at 25°C and (b) at 800°C. Assume that $\Delta H° = 177.8$ kJ and $\Delta S° = 160.5$ J/K for the temperature range.

18.30 The equilibrium constant K_P for the reaction

$$CO(g) + Cl_2(g) \rightleftharpoons COCl_2(g)$$

is 5.62×10^{35} at 25°C. Calculate $\Delta G_f°$ for $COCl_2$ at 25°C.

18.31 At 25°C, $\Delta G°$ for the process

$$H_2O(l) \rightleftharpoons H_2O(g)$$

is 8.6 kJ. Calculate the vapor pressure of water at this temperature.

18.32 Calculate $\Delta G°$ for the process

$$C(\text{diamond}) \longrightarrow C(\text{graphite})$$

Is the reaction spontaneous at 25°C? If so, why is it that diamonds do not become graphite on standing?

Thermodynamics in Living Systems

Review Questions

18.33 What is a coupled reaction? What is its importance in biological reactions?

18.34 What is the role of ATP in biological reactions?

Problems

18.35 Referring to the metabolic process involving glucose on p. 596, calculate the maximum number of moles of ATP that can be synthesized from ADP from the breakdown of 1 mole of glucose.

18.36 In the metabolism of glucose, the first step is the conversion of glucose to glucose 6-phosphate:

glucose + $H_3PO_4 \longrightarrow$ glucose 6-phosphate + H_2O
$$\Delta G° = 13.4 \text{ kJ}$$

Because $\Delta G°$ is positive, this reaction is not spontaneous. Show how this reaction can be made to proceed by coupling it with the hydrolysis of ATP. Write an equation for the coupled reaction and estimate the equilibrium constant for the coupled process.

Additional Problems

18.37 Explain this nursery rhyme in terms of the second law of thermodynamics.

> Humpty Dumpty sat on a wall;
> Humpty Dumpty had a great fall.
> All the King's horses and all the King's men
> Couldn't put Humpty together again.

18.38 Calculate ΔG for the reaction

$$H_2O(l) \rightleftharpoons H^+(aq) + OH^-(aq)$$

at 25°C for these conditions:

(a) $[H^+] = 1.0 \times 10^{-7} M$, $[OH^-] = 1.0 \times 10^{-7} M$
(b) $[H^+] = 1.0 \times 10^{-3} M$, $[OH^-] = 1.0 \times 10^{-4} M$
(c) $[H^+] = 1.0 \times 10^{-12} M$, $[OH^-] = 2.0 \times 10^{-8} M$
(d) $[H^+] = 3.5 M$, $[OH^-] = 4.8 \times 10^{-4} M$

18.39 Which of these thermodynamic functions are associated only with the first law of thermodynamics: $S, E, G,$ and H?

18.40 A student placed 1 g of each of three compounds A, B, and C in a container and found that after 1 week no change had occurred. Offer some possible explanations for the fact that no reactions took place. Assume that A, B, and C are totally miscible liquids.

18.41 Give a detailed example of each of these, with an explanation: (a) a thermodynamically spontaneous process; (b) a process that would violate the first law of thermodynamics; (c) a process that would violate the second law of thermodynamics; (d) an irreversible process; (e) an equilibrium process.

18.42 Predict the signs of ΔH, ΔS, and ΔG of the system for these processes at 1 atm: (a) ammonia melts at $-60°C$, (b) ammonia melts at $-77.7°C$, (c) ammonia melts at $-100°C$. (The normal melting point of ammonia is $-77.7°C$.)

18.43 Consider these facts: Water freezes spontaneously at $-5°C$ and 1 atm, and ice has a more ordered structure than liquid water. Explain how a spontaneous process can lead to a decrease in entropy.

18.44 Ammonium nitrate (NH_4NO_3) dissolves spontaneously and endothermically in water. What can you deduce about the sign of ΔS for the solution process?

18.45 Calculate the equilibrium pressure of CO_2 due to the decomposition of barium carbonate ($BaCO_3$) at 25°C. (The $\Delta G_f°$ values of $BaCO_3$ and BaO are -1138.9 kJ/mol and -528.4 kJ/mol, respectively.)

18.46 (a) Trouton's rule states that the ratio of the molar heat of vaporization of a liquid (ΔH_{vap}) to its boiling point in kelvins is approximately 90 J/K · mol. Use these data to show that this is the case and explain why Trouton's rule holds true:

	t_{bp} (°C)	ΔH_{vap} (kJ/mol)
Benzene	80.1	31.0
Hexane	68.7	30.8
Mercury	357	59.0
Toluene	110.6	35.2

(b) Use the values in Table 12.5 to calculate the same ratio for ethanol and water. Explain why Trouton's rule does not apply to these two substances as well as it does to other liquids.

18.47 Referring to Problem 18.46, explain why the ratio is considerably smaller than 90 J/K · mol for liquid HF.

18.48 Carbon monoxide (CO) and nitric oxide (NO) are polluting gases contained in automobile exhaust. Under suitable conditions, these gases can be made to react to form nitrogen (N_2) and the less harmful carbon dioxide (CO_2). (a) Write an equation for this reaction. (b) Identify the oxidizing and reducing agents. (c) Calculate the K_P for the reaction at 25°C. (d) Under normal atmospheric conditions, the partial pressures are $P_{N_2} = 0.80$ atm, $P_{CO_2} = 3.0 \times 10^{-4}$ atm, $P_{CO} = 5.0 \times 10^{-5}$ atm, and $P_{NO} = 5.0 \times 10^{-7}$ atm. Calculate Q_P and predict the direction toward which the reaction will proceed. (e) Will raising the temperature favor the formation of N_2 and CO_2?

18.49 For reactions carried out under standard-state conditions, Equation (18.7) takes the form $\Delta G° = \Delta H° - T\Delta S°$. (a) Assuming $\Delta H°$ and $\Delta S°$ are independent of temperature, derive the equation

$$\ln \frac{K_2}{K_1} = \frac{\Delta H°}{R} \left(\frac{T_2 - T_1}{T_1 T_2} \right)$$

where K_1 and K_2 are the equilibrium constants at T_1 and T_2, respectively. (b) Given that at 25°C K_c is 4.63×10^{-3} for the reaction

$$N_2O_4(g) \rightleftharpoons 2NO_2(g) \qquad \Delta H° = 58.0 \text{ kJ}$$

calculate the equilibrium constant at 65°C.

18.50 The K_{sp} of AgCl is given in Table 17.2. What is its value at 60°C? [*Hint:* You need the result of Problem 18.49(a). The $\Delta H_f°$ values for Ag^+ and AgCl are 105.9 kJ/mol and -127.0 kJ/mol, respectively.]

18.51 Under what conditions does a substance have a standard entropy of zero? Can a substance ever have a negative standard entropy?

18.52 Water gas, a mixture of H_2 and CO, is a fuel made by reacting steam with red-hot coke (a by-product of coal distillation):

$$H_2O(g) + C(s) \rightleftharpoons CO(g) + H_2(g)$$

From the data in Appendix 2, estimate the temperature at which the reaction becomes favorable.

18.53 Consider this Brønsted acid-base reaction at 25°C:

$$HF(aq) + Cl^-(aq) \rightleftharpoons HCl(aq) + F^-(aq)$$

(a) Predict whether K will be greater or smaller than unity.
(b) Does $\Delta S°$ or $\Delta H°$ make a greater contribution to $\Delta G°$?
(c) Is $\Delta H°$ likely to be positive or negative?

18.54 Crystallization of sodium acetate from a supersaturated solution occurs spontaneously (see p. 406). What can you deduce about the signs of ΔS and ΔH?

18.55 Consider the thermal decomposition of $CaCO_3$:

$$CaCO_3(s) \rightleftharpoons CaO(s) + CO_2(g)$$

The equilibrium vapor pressures of CO_2 are 22.6 mmHg at 700°C and 1829 mmHg at 950°C. Calculate the standard enthalpy of the reaction. [*Hint:* See Problem 18.49(a).]

18.56 A certain reaction is spontaneous at 72°C. If the enthalpy change for the reaction is 19 kJ, what is the *minimum* value of ΔS (in joules per kelvin) for the reaction?

18.57 Predict whether the entropy change is positive or negative for each of these reactions:

(a) $Zn(s) + 2HCl(aq) \rightleftharpoons ZnCl_2(aq) + H_2(g)$
(b) $O(g) + O(g) \rightleftharpoons O_2(g)$
(c) $NH_4NO_3(s) \rightleftharpoons N_2O(g) + 2H_2O(g)$
(d) $2H_2O_2(l) \rightleftharpoons 2H_2O(l) + O_2(g)$

18.58 The reaction $NH_3(g) + HCl(g) \longrightarrow NH_4Cl(s)$ proceeds spontaneously at 25°C even though there is a decrease in disorder in the system (gases are converted to a solid). Explain.

18.59 Use these data to determine the normal boiling point, in kelvins, of mercury. What assumptions must you make in order to do the calculation?

$$Hg(l): \quad \Delta H_f° = 0 \text{ (by definition)}$$
$$S° = 77.4 \text{ J/K} \cdot \text{mol}$$
$$Hg(g): \quad \Delta H_f° = 60.78 \text{ kJ/mol}$$
$$S° = 174.7 \text{ J/K} \cdot \text{mol}$$

18.60 The molar heat of vaporization of ethanol is 39.3 kJ/mol and the boiling point of ethanol is 78.3°C. Calculate ΔS for the vaporization of 0.50 mol ethanol.

18.61 A certain reaction is known to have a $\Delta G°$ value of -122 kJ. Will the reaction necessarily occur if the reactants are mixed together?

18.62 In the Mond process for the purification of nickel, carbon monoxide is reacted with heated nickel to produce $Ni(CO)_4$, which is a gas and can therefore be separated from solid impurities:

$$Ni(s) + 4CO(g) \rightleftharpoons Ni(CO)_4(g)$$

Given that the standard free energies of formation of $CO(g)$ and $Ni(CO)_4(g)$ are -137.3 kJ/mol and -587.4 kJ/mol, respectively, calculate the equilibrium constant of the reaction at 80°C. Assume that $\Delta G_f°$ is temperature independent.

18.63 Calculate $\Delta G°$ and K_P for these processes at 25°C:

(a) $H_2(g) + Br_2(l) \rightleftharpoons 2HBr(g)$
(b) $\frac{1}{2}H_2(g) + \frac{1}{2}Br_2(l) \rightleftharpoons HBr(g)$

Account for the differences in $\Delta G°$ and K_P obtained for (a) and (b).

18.64 Calculate the pressure of O_2 (in atm) over a sample of NiO at 25°C if $\Delta G° = 212$ kJ for the reaction

$$NiO(s) \rightleftharpoons Ni(s) + \frac{1}{2}O_2(g)$$

18.65 Comment on the statement: "Just talking about entropy increases its value in the universe."

18.66 For a reaction with a negative $\Delta G°$ value, which of these statements is false? (a) The equilibrium constant K is greater than one, (b) the reaction is spontaneous when all the reactants and products are in their standard states, and (c) the reaction is always exothermic.

18.67 Consider the reaction

$$N_2(g) + O_2(g) \rightleftharpoons 2NO(g)$$

Given that $\Delta G°$ for the reaction at 25°C is 173.4 kJ, (a) calculate the standard free energy of formation of NO, and (b) calculate K_P of the reaction. (c) One of the starting substances in smog formation is NO. Assuming that the temperature in a running automobile engine is 1100°C, estimate K_P for the given reaction. (d) As farmers know, lightning helps to produce a better crop. Why?

18.68 Heating copper(II) oxide at 400°C does not produce any appreciable amount of Cu:

$$CuO(s) \rightleftharpoons Cu(s) + \frac{1}{2}O_2(g) \quad \Delta G° = 127.2 \text{ kJ}$$

However, if this reaction is coupled to the conversion of graphite to carbon monoxide, it becomes spontaneous. Write an equation for the coupled process and calculate the equilibrium constant for the coupled reaction.

18.69 The internal engine of a 1200-kg car is designed to run on octane (C_8H_{18}), whose enthalpy of combustion is 5510 kJ/mol. If the car is moving up a slope, calculate the maximum height (in meters) to which the car can be driven on 1.0 gallon of the fuel. Assume that the engine cylinder temperature is 2200°C and the exit temperature is 760°C, and neglect all forms of friction. The mass of 1 gallon of fuel is 3.1 kg. [*Hint:* See p. 574. The work done in moving the car over a vertical distance is *mgh*, where *m* is the mass of the car in kg, *g* the acceleration due to gravity (9.81 m/s^2), and *h* the height in meters.]

18.70 A carbon monoxide (CO) crystal is found to have entropy greater than zero at absolute zero of temperature. Give two possible explanations for this observation.

18.71 (a) Over the years there have been numerous claims about "perpetual motion machines," machines that will produce useful work with no input of energy. Explain why the first law of thermodynamics prohibits the possibility of such a machine existing. (b) Another kind of machine, sometimes called a "perpetual motion of the second kind," operates as follows. Suppose an ocean liner sails by scooping up water from the ocean and then extracting heat from the water, converting the heat to electric power to run the ship, and dumping the water back into the ocean. This process does not violate the first law of thermodynamics, for no energy is created—energy from the ocean is just converted to electrical energy. Show that the second law of thermodynamics prohibits the existence of such a machine.

18.72 The activity series in Chapter 4 shows that reaction (a) is spontaneous while reaction (b) is nonspontaneous at 25°C:

(a) $Fe(s) + 2H^+ \longrightarrow Fe^{2+}(aq) + H_2(g)$
(b) $Cu(s) + 2H^+ \longrightarrow Cu^{2+}(aq) + H_2(g)$

Use the data in Appendix 2 to calculate the equilibrium constant for these reactions and hence confirm that the activity series is correct.

18.73 The rate constant for the elementary reaction

$$2NO(g) + O_2(g) \longrightarrow 2NO_2(g)$$

is $7.1 \times 10^9/M^2 \cdot s$ at 25°C. What is the rate constant for the reverse reaction at the same temperature?

18.74 This reaction is described as the cause of sulfur deposits formed at volcanic sites:

$$2H_2S(g) + SO_2(g) \rightleftharpoons 3S(s) + 2H_2O(g)$$

It may also be used to remove SO_2 from power-plant stack gases. (a) Identify the type of redox reaction it is. (b) Calculate the equilibrium constant (K_P) at 25°C and comment on whether this method is feasible for removing SO_2. (c) Would this procedure become more or less effective at a higher temperature?

18.75 Describe two ways that you could measure $\Delta G°$ of a reaction.

18.76 This reaction represents the removal of ozone in the stratosphere:

$$2O_3(g) \rightleftharpoons 3O_2(g)$$

Calculate the equilibrium constant (K_P) for the reaction. In view of the magnitude of the equilibrium constant, explain why this reaction is not considered a major cause of ozone depletion in the absence of manmade pollutants such as the nitrogen oxides and chlorofluorohydrocarbons? Assume the temperature of the stratosphere to be −30°C and $\Delta G_f°$ to be temperature independent.

18.77 A 74.6-g ice cube floats in the Arctic Sea. The temperature and pressure of the system and surroundings are at 1 atm and 0°C. Calculate ΔS_{sys}, ΔS_{surr}, and ΔS_{univ} for the melting of the ice cube. What can you conclude about the nature of the process from the value of ΔS_{univ}? (The molar heat of fusion of water is 6.01 kJ/mol.)

18.78 Comment on the feasibility of extracting copper from its ore chalcocite (Cu_2S) by heating:

$$Cu_2S(s) \longrightarrow 2Cu(s) + S(s)$$

Calculate the $\Delta G°$ for the overall reaction if this process is coupled to the conversion of sulfur to sulfur dioxide. Given that $\Delta G_f°(Cu_2S) = -86.1$ kJ/mol.

18.79 Active transport is the process in which a substance is transferred from a region of lower concentration to one of higher concentration. This is a nonspontaneous process and must be coupled to a spontaneous process, such as the hydrolysis of ATP. The concentrations of K^+ ions in the blood plasma and in nerve cells are 15 mM and 400 mM, respectively (1 mM = 1×10^{-3} M). Use Equation (18.9) to calculate ΔG for the process at the physiological temperature of 37°C:

$$K^+(15 \text{ m}M) \longrightarrow K^+(400 \text{ m}M)$$

In this calculation, the $\Delta G°$ term can be set to zero. What is the justification for this step?

18.80 Large quantities of hydrogen are needed for the synthesis of ammonia. One preparation of hydrogen involves the reaction between carbon monoxide and steam at 300°C in the presence of a copper-zinc catalyst:

$$CO(g) + H_2O(g) \rightleftharpoons CO_2(g) + H_2(g)$$

Calculate the equilibrium constant (K_P) for the reaction and the temperature at which the reaction favors the formation of CO and H_2O. Will a larger K_P be attained at the same temperature if a more efficient catalyst is used?

18.81 Consider two carboxylic acids (acids that contain the —COOH group): CH_3COOH (acetic acid, $K_a = 1.8 \times 10^{-5}$) and $CH_2ClCOOH$ (chloroacetic acid, $K_a = 1.4 \times 10^{-3}$). (a) Calculate $\Delta G°$ for the ionization of these acids at 25°C. (b) From the equation $\Delta G° = \Delta H° - T\Delta S°$, we see that the contributions to the $\Delta G°$ term are an enthalpy term ($\Delta H°$) and a temperature times entropy term ($T\Delta S°$). These contributions are listed here for the two acids:

	$\Delta H°$ (kJ)	$T\Delta S°$ (kJ)
CH_3COOH	−0.57	−27.6
$CH_2ClCOOH$	−4.7	−21.1

Which is the dominant term in determining the value of $\Delta G°$ (and hence K_a of the acid)? (c) What processes contribute to $\Delta H°$? (Consider the ionization of the acids as a Brønsted acid-base reaction.) (d) Explain why $T\Delta S°$ is more negative for CH_3COOH.

18.82 One of the steps in the extraction of iron from its ore (FeO) is the reduction of iron(II) oxide by carbon monoxide at 900°C:

$$FeO(s) + CO(g) \rightleftharpoons Fe(s) + CO_2(g)$$

If CO is allowed to react with an excess of FeO, calculate the mole fractions of CO and CO_2 at equilibrium. State any assumptions.

18.83 Derive this equation

$$\Delta G = RT \ln (Q/K)$$

where Q is the reaction quotient and describe how you would use it to predict the spontaneity of a reaction.

Special Problems

18.84 The standard free energies of formation (ΔG_f°) of the three isomers of pentane (see p. 336) in the gas phase are: *n*-pentane: -8.37 kJ/mol; 2-methylbutane: -14.8 kJ/mol; 2,2-dimethylpropane: -15.2 kJ/mol. (a) Determine the mole percent of these molecules in an equilibrium mixture at 25°C. (b) How does the stability of these molecules depend on the extent of branching?

18.85 Carry out the following experiments: Quickly stretch a rubber band (at least 0.5 cm wide) and press it against your lips. You will feel a warming effect. Next, carry out the reverse procedure; that is, first stretch a rubber band and hold it in position for a few seconds. Now quickly release the tension and press the rubber band against your lips. You will feel a cooling effect. (a) Apply Equation (18.7) to these processes to determine the signs of ΔG, ΔH, and hence ΔS in each case. (b) From the signs of ΔS, what can you conclude about the structure of rubber molecules?

Answers to Practice Exercises

18.1 (a) Entropy decreases, (b) entropy decreases, (c) entropy increases, (d) entropy increases. **18.2** (a) -173.6 J/K, (b) -139.8 J/K, (c) 215.3 J/K. **18.3** (a) $\Delta S > 0$, (b) $\Delta S < 0$, (c) $\Delta S \approx 0$. **18.4** (a) -106.4 kJ, (b) -2935.0 kJ. **18.5** $\Delta S_{fus} = 16$ J/K · mol; $\Delta S_{vap} = 72$ J/K · mol. **18.6** 1.93×10^{57}. **18.7** 33 kJ/mol. **18.8** $\Delta G = 0.97$ kJ/mol; direction is from right to left.

"Prometheus brought fire to mankind; for electricity we owe it to Faraday."*

Michael Faraday is regarded as the greatest experimentalist of the nineteenth century and one of the greatest scientists of all time. This is all the more remarkable considering that he had no formal schooling beyond the primary grades. Born in England in 1791, Faraday was one of 10 children of a poor blacksmith. At the age of 14, he was apprenticed to a bookbinder. He became interested in science after reading a book on chemistry. Faraday attended a series of lectures by the renowned chemist Sir Humphry Davy and took neat and copious notes on them. Later, when he applied to Davy for a job at the Royal Institution, he submitted his lecture notes as proof of his earnestness. Davy promptly hired Faraday as his laboratory assistant. On a tour through Europe, Faraday and Davy used a lens to focus the sun's rays on a diamond to convert it to carbon dioxide, thus proving that diamond, like the more common substance graphite, is made of carbon atoms.

During his career, Faraday made an amazing number of important contributions. He discovered benzene and determined its composition. Faraday was the first to liquefy many gases and produced the first stainless steel. He discovered the rotation of polarized light in a magnetic field (now known as the Faraday effect) and demonstrated electromagnetic induction, using a moving magnet to generate electric current in a wire. He also developed the laws of electrolysis and coined the terms "anion," "cation," "electrode," and "electrolyte."

In 1822 Faraday replaced Davy as director of the Royal Institution. When Faraday was nominated for membership in the Royal Society, Davy cast the only dissenting vote, possibly out of spite. Nevertheless, Faraday was elected in 1824.

Faraday could have made a fortune from some of his discoveries, but he deliberately dropped every project when it acquired commercial value. Born in poverty, he also died in poverty, in 1867. His consuming work, he felt, was sufficient reward.

*Quotation attributed to Sir Lawrence Bragg, a British physicist.

Michael Faraday at work in his laboratory.

Redox Reactions and Electrochemistry

Chapter 19

ESSENTIAL CONCEPTS

 Redox Reactions and Electrochemical Cells Equations representing redox reactions can be balanced using the ion-electron method. These reactions involve the transfer of electrons from a reducing agent to an oxidizing agent. Using separate compartments, such a reaction can be used to generate electrons that flow externally in an arrangement called a galvanic cell.

 Thermodynamics of Galvanic Cells The voltage measured in a galvanic cell can be broken down into the electrode potentials of the anode (where oxidation takes place) and cathode (where reduction takes place). This voltage can be related to the Gibbs free energy change and the equilibrium constant of the redox process. The Nernst equation relates the cell voltage to the cell voltage under standard-state conditions and the concentrations of reacting species.

 Batteries Batteries are electrochemical cells that can supply direct electric current at a constant voltage. There are many different types of batteries used in automobiles, flashlights, and pacemakers. Fuel cells are a special type of electrochemical cell that generate electricity by the oxidation of hydrogen or hydrocarbons.

 Corrosion Corrosion is a spontaneous redox reaction that results in the formation of rust from iron, silver sulfide from silver, and patina (copper carbonate) from copper. Corrosion causes enormous damage to buildings, structures, ships, and cars. Many methods have been devised to prevent or minimize the effect of corrosion.

 Electrolysis Electrolysis is the process in which electrical energy is used to cause a nonspontaneous redox reaction to occur. The quantitative relationship between the current supplied and the products formed is provided by Faraday. Electrolysis is the major method for producing active metals and nonmetals and many essential industrial chemicals.

19.1 Redox Reactions

Electrochemistry is the branch of chemistry that deals with the interconversion of electrical energy and chemical energy. Electrochemical processes are redox (oxidation-reduction) reactions in which the energy released by a spontaneous reaction is converted to electricity or in which electrical energy is used to cause a nonspontaneous reaction to occur. Although redox reactions were discussed in Chapter 4, it is helpful to review some of the basic concepts that will come up again in this chapter.

In redox reactions electrons are transferred from one substance to another. The reaction between magnesium metal and hydrochloric acid is an example of a redox reaction:

$$\overset{0}{Mg}(s) + 2\overset{+1}{H}Cl(aq) \longrightarrow \overset{+2}{Mg}Cl_2(aq) + \overset{0}{H_2}(g)$$

Recall that the numbers written above the elements are the oxidation numbers of the elements. The loss of electrons by an element during oxidation is marked by an increase in the element's oxidation number. In reduction, there is a decrease in oxidation number resulting from a gain of electrons by an element. In the reaction shown here Mg metal is oxidized and H^+ ions are reduced; the Cl^- ions are spectator ions.

Balancing Redox Equations

Equations for redox reactions like the one just discussed are relatively easy to balance. However, in the laboratory we often encounter more complex redox reactions involving oxoanions such as chromate (CrO_4^{2-}), dichromate ($Cr_2O_7^{2-}$), permanganate (MnO_4^-), nitrate (NO_3^-), and sulfate (SO_4^{2-}). In principle, we can balance any redox equation using the procedure outlined in Section 3.7, but there are some special techniques for handling redox reactions, techniques that also give us insight into electron transfer processes. Here we will discuss one such procedure, called the *ion-electron method*. In this approach, the overall reaction is divided into two half-reactions, one for oxidation and one for reduction. The equations for the two half-reactions are balanced separately and then added together to give the overall balanced equation.

Suppose we are asked to balance the equation showing the oxidation of Fe^{2+} ions to Fe^{3+} ions by dichromate ions ($Cr_2O_7^{2-}$) in an acidic medium. As a result, the $Cr_2O_7^{2-}$ ions are reduced to Cr^{3+} ions. The following steps will help us balance the equation.

Step 1. Write the unbalanced equation for the reaction in ionic form.

$$Fe^{2+} + Cr_2O_7^{2-} \longrightarrow Fe^{3+} + Cr^{3+}$$

Step 2. Separate the equation into two half-reactions.

$$\text{Oxidation:} \quad \overset{+2}{Fe^{2+}} \longrightarrow \overset{+3}{Fe^{3+}}$$

$$\text{Reduction:} \quad \overset{+6}{Cr_2}O_7^{2-} \longrightarrow \overset{+3}{Cr^{3+}}$$

Step 3. Balance the atoms other than O and H in each half-reaction separately.

The oxidation half-reaction is already balanced for Fe atoms. For the reduction half-reaction we multiply the Cr^{3+} by 2 to balance the Cr atoms.

$$Cr_2O_7^{2-} \longrightarrow 2Cr^{3+}$$

Step 4. For reactions in an acidic medium, add H_2O to balance the O atoms and H^+ to balance the H atoms.

Because the reaction takes place in an acidic medium, we add seven H_2O molecules to the right-hand side of the reduction half-reaction to balance the O atoms:

$$Cr_2O_7^{2-} \longrightarrow 2Cr^{3+} + 7H_2O$$

To balance the H atoms, we add 14 H^+ ions on the left-hand side:

$$14H^+ + Cr_2O_7^{2-} \longrightarrow 2Cr^{3+} + 7H_2O$$

Step 5. Add electrons to one side of each half-reaction to balance the charges. If necessary, equalize the number of electrons in the two half-reactions by multiplying one or both half-reactions by appropriate coefficients.

For the oxidation half-reaction we write

$$Fe^{2+} \longrightarrow Fe^{3+} + e^-$$

We added an electron to the right-hand side so that there is a 2+ charge on each side of the half-reaction.

In the reduction half-reaction there are net 12 positive charges on the left-hand side and only six positive charges on the right-hand side. Therefore, we add six electrons on the left.

$$14H^+ + Cr_2O_7^{2-} + 6e^- \longrightarrow 2Cr^{3+} + 7H_2O$$

To equalize the number of electrons in both half-reactions, we multiply the oxidation half-reaction by 6:

$$6Fe^{2+} \longrightarrow 6Fe^{3+} + 6e^-$$

Step 6. Add the two half-reactions together and balance the final equation by inspection. The electrons on both sides must cancel.

The two half-reactions are added to give

$$14H^+ + Cr_2O_7^{2-} + 6Fe^{2+} + \cancel{6e^-} \longrightarrow 2Cr^{3+} + 6Fe^{3+} + 7H_2O + \cancel{6e^-}$$

The electrons on both sides cancel, and we are left with the balanced net ionic equation:

$$14H^+ + Cr_2O_7^{2-} + 6Fe^{2+} \longrightarrow 2Cr^{3+} + 6Fe^{3+} + 7H_2O$$

This reaction can be carried out by dissolving potassium dichromate and iron(II) sulfate in a dilute sulfuric acid solution.

Step 7. Verify that the equation contains the same types and numbers of atoms and the same charges on both sides of the equation.

A final check shows that the resulting equation is "atomically" and "electrically" balanced.

For reaction in a basic medium, we would balance the atoms as we did in step 4 for an acidic medium. Then, for every H^+ ion we would add an equal number of OH^- ions to *both* sides of the equation. Where H^+ and OH^- appeared on the same side of the equation, we would combine the ions to give H_2O. Example 19.1 illustrates the use of this procedure.

Example 19.1   Write a balanced ionic equation to represent the oxidation of iodide ion (I^-) by permanganate ion (MnO_4^-) in basic solution to yield molecular iodine (I_2) and manganese(IV) oxide (MnO_2).

Reasoning and Solution   We follow the procedure on p. 607 for dealing with redox reactions in basic medium.

Step 1. The unbalanced equation is

$$MnO_4^- + I^- \longrightarrow MnO_2 + I_2$$

Step 2. The two half-reactions are

$$\text{Oxidation:} \quad \overset{-1}{I^-} \longrightarrow \overset{0}{I_2}$$

$$\text{Reduction:} \quad \overset{+7}{MnO_4^-} \longrightarrow \overset{+4}{MnO_2}$$

Step 3. To balance the I atoms in the oxidation half-reaction, we write

$$2I^- \longrightarrow I_2$$

Step 4. In the reduction half-reaction, we add two H_2O molecules on the right to balance the O atoms:

$$MnO_4^- \longrightarrow MnO_2 + 2H_2O$$

To balance the H atoms, we add four H^+ ions on the left:

$$MnO_4^- + 4H^+ \longrightarrow MnO_2 + 2H_2O$$

Because the reaction occurs in a basic medium and there are four H^+ ions, we add four OH^- ions to both sides of the equation:

$$MnO_4^- + 4H^+ + 4OH^- \longrightarrow MnO_2 + 2H_2O + 4OH^-$$

Combining the H^+ and OH^- ions to form H_2O and canceling $2H_2O$ from both sides, we write

$$MnO_4^- + 2H_2O \longrightarrow MnO_2 + 4OH^-$$

Step 5. Next, we balance the charges of the two half-reactions as:

$$2I^- \longrightarrow I_2 + 2e^-$$

$$MnO_4^- + 2H_2O + 3e^- \longrightarrow MnO_2 + 4OH^-$$

To equalize the number of electrons, we multiply the oxidation half-reaction by 3 and the reduction half-reaction by 2:

$$6I^- \longrightarrow 3I_2 + 6e^-$$

$$2MnO_4^- + 4H_2O + 6e^- \longrightarrow 2MnO_2 + 8OH^-$$

Step 6. The two half-reactions are added together to give

$$6I^- + 2MnO_4^- + 4H_2O + 6e^- \longrightarrow 3I_2 + 2MnO_2 + 8OH^- + 6e^-$$

After canceling the electrons on both sides, we obtain

$$6I^- + 2MnO_4^- + 4H_2O \longrightarrow 3I_2 + 2MnO_2 + 8OH^-$$

Step 7. A final check shows that the equation is balanced in terms of both atoms and charges.

To carry out this reaction, mix KI and $KMnO_4$ in a basic medium.

Similar problems: 19.1, 19.2.

Practice Exercise Balance this equation for the reaction in an acidic medium by the ion-electron method:

$$Fe^{2+} + MnO_4^- \longrightarrow Fe^{3+} + Mn^{2+}$$

19.2 Galvanic Cells

In Section 4.4 we saw that when a piece of zinc metal is placed in a $CuSO_4$ solution, Zn is oxidized to Zn^{2+} ions while Cu^{2+} ions are reduced to metallic copper (see Figure 4.11):

$$Zn(s) + Cu^{2+}(aq) \longrightarrow Zn^{2+}(aq) + Cu(s)$$

The electrons are transferred directly from the reducing agent (Zn) to the oxidizing agent (Cu^{2+}) in solution. However, if we physically separate the oxidizing agent from the reducing agent, the transfer of electrons can take place via an external conducting medium (a metal wire). As the reaction progresses, it sets up a constant flow of electrons and hence generates electricity (that is, it produces electrical work such as driving an electric motor).

The experimental apparatus for generating electricity through the use of a spontaneous redox reaction is called a *galvanic cell* or *voltaic cell,* after the Italian scientists Luigi Galvani and Alessandro Volta, who constructed early versions of the device. Figure 19.1 shows the essential components of such a cell. A zinc bar is immersed in a $ZnSO_4$ solution, and a copper bar is immersed in a $CuSO_4$ solution. The cell operates on the principle that the oxidation of Zn to Zn^{2+} and the reduction of Cu^{2+} to Cu can be made to take place simultaneously in separate locations with the transfer of electrons between them occurring through an external wire. The zinc and copper bars are called *electrodes*. This particular arrangement of electrodes (Zn and Cu) and solutions ($ZnSO_4$ and $CuSO_4$) is called the Daniell cell. By definition, the **anode** in a galvanic cell is *the electrode at which oxidation occurs* and the **cathode** is *the electrode at which reduction occurs.*

For the Daniell cell, the **half-cell reactions,** that is, *the oxidation and reduction reactions at the electrodes,* are

Zn electrode (anode): $\qquad Zn(s) \longrightarrow Zn^{2+}(aq) + 2e^-$

Cu electrode (cathode): $Cu^{2+}(aq) + 2e^- \longrightarrow Cu(s)$

Note that unless the two solutions are separated from each other, the Cu^{2+} ions will react directly with the zinc bar:

$$Cu^{2+}(aq) + Zn(s) \longrightarrow Cu(s) + Zn^{2+}(aq)$$

and no useful electrical work will be obtained.

To complete the electrical circuit, the solutions must be connected by a conducting medium through which the cations and anions can move from one electrode compartment to the other. This requirement is satisfied by a *salt bridge,* which, in its simplest form, is an inverted U tube containing an inert electrolyte solution, such as KCl or NH_4NO_3, whose ions will not react with other ions in solution or with the electrodes (see Figure 19.1). During the course of the overall redox reaction, electrons

Galvanic cells.

Alphabetically anode precedes cathode and oxidation precedes reduction. Therefore, anode is where oxidation occurs and cathode is where reduction takes place.

Half-cell reactions are similar to the half-reactions discussed in Section 19.1.

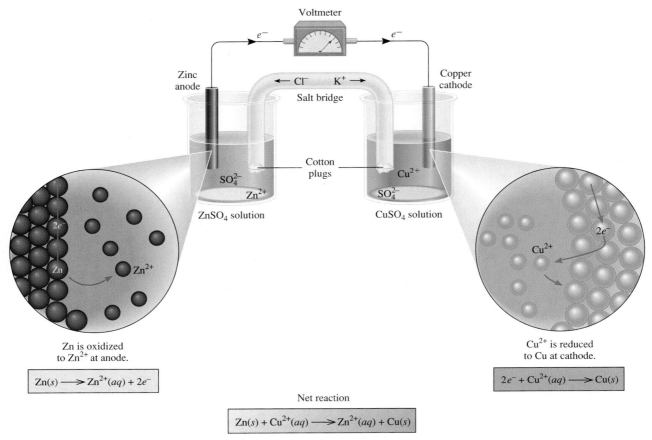

Figure 19.1 *A galvanic cell. The salt bridge (an inverted U tube) containing a KCl solution provides an electrically conducting medium between two solutions. The openings of the U tube are loosely plugged with cotton balls to prevent the KCl solution from flowing into the containers while allowing the anions and cations to move across. Electrons flow externally from the Zn electrode (anode) to the Cu electrode (cathode).*

flow externally from the anode (Zn electrode) through the wire and voltmeter to the cathode (Cu electrode). In the solution, the cations (Zn^{2+}, Cu^{2+}, and K^+) move toward the cathode, while the anions (SO_4^{2-} and Cl^-) move toward the anode. Without the salt bridge connecting the two solutions, the buildup of positive charge in the anode compartment (due to the formation of Zn^{2+} ions) and negative charge in the cathode compartment (created when some of the Cu^{2+} ions are reduced to Cu) would quickly prevent the cell from operating.

An electric current flows from the anode to the cathode because there is a difference in electrical potential energy between the electrodes. This flow of electric current is analogous to that of water down a waterfall, which occurs because there is a difference in gravitational potential energy, or the flow of gas from a high-pressure region to a low-pressure region. Experimentally the *difference in electrical potential between the anode and the cathode* is measured by a voltmeter (Figure 19.2) and the reading (in volts) is called **cell voltage.** However, two other terms, **electromotive force** or **emf (E)** and *cell potential* are also used to denote cell voltage. We will see that the

Figure 19.2 *Practical setup of the Daniell cell described in Figure 19.1. Note the U tube (salt bridge) connecting the two beakers. When the concentrations of ZnSO₄ and CuSO₄ are 1 molar (1 M) at 25°C, the cell voltage is 1.10 V.*

voltage of a cell depends not only on the nature of the electrodes and the ions, but also on the concentrations of the ions and the temperature at which the cell is operated.

The conventional notation for representing galvanic cells is the *cell diagram*. For the Daniell cell shown in Figure 19.1, if we assume that the concentrations of Zn^{2+} and Cu^{2+} ions are 1 M, the cell diagram is

$$Zn(s) \mid Zn^{2+}(1\ M) \parallel Cu^{2+}(1\ M) \mid Cu(s)$$

The single vertical line represents a phase boundary. For example, the zinc electrode is a solid and the Zn^{2+} ions (from $ZnSO_4$) are in solution. Thus we draw a line between Zn and Zn^{2+} to show the phase boundary. The double vertical lines denote the salt bridge. By convention, the anode is written first, to the left of the double lines and the other components appear in the order in which we would encounter them in moving from the anode to the cathode.

19.3 Standard Reduction Potentials

When the concentrations of the Cu^{2+} and Zn^{2+} ions are both 1.0 M, we find that the voltage or emf of the Daniell cell is 1.10 V at 25°C (see Figure 19.2). This voltage must be related directly to the redox reactions, but how? Just as the overall cell reaction can be thought of as the sum of two half-cell reactions, the measured emf of the cell can be treated as the sum of the electrical potentials at the Zn and Cu electrodes. Knowing one of these electrode potentials, we could obtain the other by subtraction (from 1.10 V). It is impossible to measure the potential of just a single electrode, but if we arbitrarily set the potential value of a particular electrode at zero, we can use it to determine the relative potentials of other electrodes. The hydrogen electrode, shown in Figure 19.3, serves as the reference for this purpose. Hydrogen gas is bubbled into a hydrochloric acid solution at 25°C. The platinum electrode has two functions. First, it provides a surface on which the dissociation of hydrogen molecules can take place:

$$H_2 \longrightarrow 2H^+ + 2e^-$$

Second, it serves as an electrical conductor to the external circuit.

H₂ gas at 1 atm

Pt electrode

1 M HCl

Figure 19.3 *A hydrogen electrode operating under standard-state conditions. Hydrogen gas at 1 atm is bubbled through a 1 M HCl solution. The platinum electrode is part of the hydrogen electrode.*

The choice of an arbitrary reference for measuring electrode potential is analogous to choosing the surface of the ocean as the reference for altitude, calling it zero meters, and then referring to any terrestrial altitude as being a certain number of meters above or below sea level.

Standard states are defined in Table 18.2 (p. 587).

Under standard-state conditions (when the pressure of H_2 is 1 atm and the concentration of the HCl solution is 1 M), the potential for the reduction of H^+ at 25°C is taken to be *exactly* zero:

$$2H^+(1\ M) + 2e^- \longrightarrow H_2(1\ atm) \qquad\qquad E° = 0\ V$$

The superscript "°" denotes standard-state conditions, and $E°$ is the **standard reduction potential,** or *the voltage associated with a reduction reaction at an electrode when all solutes are 1 M and all gases are at 1 atm.* Thus, the standard reduction potential of the hydrogen electrode is defined as zero. The hydrogen electrode is called the *standard hydrogen electrode (SHE).*

We can use the SHE to measure the potentials of other kinds of electrodes. For example, Figure 19.4(a) shows a galvanic cell with a zinc electrode and a SHE. In this case the zinc electrode is the anode and the SHE is the cathode. We deduce this fact from the decrease in mass of the zinc electrode during the operation of the cell, which is consistent with the loss of zinc to the solution as Zn^{2+} ions caused by the oxidation reaction:

$$Zn(s) \longrightarrow Zn^{2+}(aq) + 2e^-$$

The cell diagram is

$$Zn(s)\ |\ Zn^{2+}(1\ M)\ \|\ H^+(1\ M)\ |\ H_2(1\ atm)\ |\ Pt(s)$$

As mentioned earlier, the Pt electrode provides the surface on which the reduction takes place. When all the reactants are in their standard states (that is, H_2 at 1 atm, H^+ and Zn^{2+} ions at 1 M), the emf of the cell is 0.76 V at 25°C. We can write the half-cell reactions as:

Anode (oxidation): $Zn(s) \longrightarrow Zn^{2+}(1\ M) + 2e^-$
Cathode (reduction): $\underline{2H^+(1\ M) + 2e^- \longrightarrow H_2(1\ atm)}$
Overall: $Zn(s) + 2H^+(1\ M) \longrightarrow Zn^{2+}(1\ M) + H_2(1\ atm)$

By convention, the **standard emf** of the cell, $E°_{cell}$, which is composed of a contribution from the anode and a contribution from the cathode, is given by

$$E°_{cell} = E°_{cathode} - E°_{anode} \qquad\qquad (19.1)$$

Figure 19.4 *(a) A cell consisting of a zinc electrode and a hydrogen electrode. (b) A cell consisting of a copper electrode and a hydrogen electrode. Both cells are operating under standard-state conditions. Note that in (a) the SHE acts as the cathode, but in (b) it is the anode.*

where *both* E°_{cathode} and E°_{anode} are the standard reduction potentials of the electrodes. For the Zn-SHE cell, we write

$$E^\circ_{\text{cell}} = E^\circ_{\text{H}^+/\text{H}_2} - E^\circ_{\text{Zn}^{2+}/\text{Zn}}$$
$$0.76 \text{ V} = 0 - E^\circ_{\text{Zn}^{2+}/\text{Zn}}$$

where the subscript H^+/H_2 means $2\text{H}^+ + 2e^- \rightarrow \text{H}_2$ and the subscript Zn^{2+}/Zn means $\text{Zn}^{2+} + 2e^- \rightarrow \text{Zn}$. Thus the standard reduction potential of zinc, $E^\circ_{\text{Zn}^{2+}/\text{Zn}}$, is -0.76 V.

The standard electrode potential of copper can be obtained in a similar fashion, by using a cell with a copper electrode and a SHE [Figure 19.4(b)]. In this case, the copper electrode is the cathode because its mass increases during the operation of the cell, as is consistent with the reduction reaction:

$$\text{Cu}^{2+}(aq) + 2e^- \longrightarrow \text{Cu}(s)$$

The cell diagram is

$$\text{Pt}(s) \mid \text{H}_2(1 \text{ atm}) \mid \text{H}^+(1 \text{ } M) \parallel \text{Cu}^{2+}(1 \text{ } M) \mid \text{Cu}(s)$$

and the half-cell reactions are

Anode (oxidation): $\quad\quad\quad\quad \text{H}_2(1 \text{ atm}) \longrightarrow 2\text{H}^+(1 \text{ } M) + 2e^-$
Cathode (reduction): $\quad\quad \underline{\text{Cu}^{2+}(1 \text{ } M) + 2e^- \longrightarrow \text{Cu}(s)}$
Overall: $\text{H}_2(1 \text{ atm}) + \text{Cu}^{2+}(1 \text{ } M) \longrightarrow 2\text{H}^+(1 \text{ } M) + \text{Cu}(s)$

Under standard-state conditions and at 25°C, the emf of the cell is 0.34 V, so we write

$$E^\circ_{\text{cell}} = E^\circ_{\text{cathode}} - E^\circ_{\text{anode}}$$
$$0.34 \text{ V} = E^\circ_{\text{Cu}^{2+}/\text{Cu}} - E^\circ_{\text{H}^+/\text{H}_2}$$
$$= E^\circ_{\text{Cu}^{2+}/\text{Cu}} - 0$$

In this case the standard reduction potential of copper, $E^\circ_{\text{Cu}^{2+}/\text{Cu}}$, is 0.34 V, where the subscript means $\text{Cu}^{2+} + 2e^- \rightarrow \text{Cu}$.

For the Daniell cell shown in Figure 19.1, we can now write

Anode (oxidation): $\quad\quad\quad\quad \text{Zn}(s) \longrightarrow \text{Zn}^{2+}(1 \text{ } M) + 2e^-$
Cathode (reduction): $\underline{\text{Cu}^{2+}(1 \text{ } M) + 2e^- \longrightarrow \text{Cu}(s)}$
Overall: $\text{Zn}(s) + \text{Cu}^{2+}(1 \text{ } M) \longrightarrow \text{Zn}^{2+}(1 \text{ } M) + \text{Cu}(s)$

The emf of the cell is

$$E^\circ_{\text{cell}} = E^\circ_{\text{cathode}} - E^\circ_{\text{anode}}$$
$$= E^\circ_{\text{Cu}^{2+}/\text{Cu}} - E^\circ_{\text{Zn}^{2+}/\text{Zn}}$$
$$= 0.34 \text{ V} - (-0.76 \text{ V})$$
$$= 1.10 \text{ V}$$

This example illustrates how we can use the *sign* of the emf of a cell to predict the spontaneity of a redox reaction. Under standard-state conditions for reactants and products, the redox reaction is spontaneous in the forward direction if the standard emf of the cell is positive. If it is negative, the reaction is spontaneous in the opposite direction. It is important to keep in mind that a negative E°_{cell} does *not* mean that a reaction will not occur if the reactants are mixed at 1 M concentrations. It merely means that the equilibrium of a redox reaction, when reached, will lie to the left. We will examine the relationships among E°_{cell}, ΔG°, and K later in this chapter.

The activity series in Figure 4.12 is based on data given in Table 19.1.

Table 19.1 lists standard reduction potentials for a number of half-cell reactions. By definition, the SHE has an $E°$ value of 0.00 V. Below the SHE the negative standard reduction potentials increase, and above it the positive standard reduction potentials increase. It is important to know these points about the table in calculations:

- The $E°$ values apply to the half-cell reactions as read in the forward (left to right) direction.
- The more positive $E°$ is, the greater the tendency for the substance to be reduced. For example, the half-cell reaction

$$F_2(1 \text{ atm}) + 2e^- \longrightarrow 2F^-(1 \text{ } M) \qquad E° = 2.87 \text{ V}$$

has the highest positive $E°$ value among all the half-cell reactions. Thus F_2 is the *strongest* oxidizing agent because it has the greatest tendency to be reduced. At the other extreme is the reaction

$$Li^+(1 \text{ } M) + e^- \longrightarrow Li(s) \qquad E° = -3.05 \text{ V}$$

which has the most negative $E°$ value. Thus Li^+ is the *weakest* oxidizing agent because it is the most difficult species to reduce. Conversely, we say that F^- is the weakest reducing agent and Li metal is the strongest reducing agent. Under standard-state conditions, the oxidizing agents (the species on the left-hand side of the half-reactions in Table 19.1) increase in strength from bottom to top and the reducing agents (the species on the right-hand side of the half-reactions) increase in strength from top to bottom.

- The half-cell reactions are reversible. Depending on the conditions, any electrode can act either as an anode or as a cathode. Earlier we saw that the SHE is the cathode (H^+ is reduced to H_2) when coupled with zinc in a cell and that it becomes the anode (H_2 is oxidized to H^+) when used in a cell with copper.

- Under standard-state conditions, any species on the left of a given half-cell reaction will react spontaneously with a species that appears on the right of any half-cell reaction located *below* it in Table 19.1. This principle is sometimes called the *diagonal rule*. In the case of the Daniell cell

$$Cu^{2+}(1 \text{ } M) + 2e^- \longrightarrow Cu(s) \qquad E° = 0.34 \text{ V}$$
$$Zn^{2+}(1 \text{ } M) + 2e^- \longrightarrow Zn(s) \qquad E° = -0.76 \text{ V}$$

We see that the substance on the left of the first half-cell reaction is Cu^{2+} and the substance on the right in the second half-cell reaction is Zn. Therefore, as we saw earlier, Zn spontaneously reduces Cu^{2+} to form Zn^{2+} and Cu.

- Changing the stoichiometric coefficients of a half-cell reaction *does not* affect the value of $E°$ because electrode potentials are intensive properties. This means that the value of $E°$ is unaffected by the size of the electrodes or the amount of solutions present. For example,

$$I_2(s) + 2e^- \longrightarrow 2I^-(1 \text{ } M) \qquad E° = 0.53 \text{ V}$$

but $E°$ does not change if we multiply the half-reaction by 2:

$$2I_2(s) + 4e^- \longrightarrow 4I^-(1 \text{ } M) \qquad E° = 0.53 \text{ V}$$

- Like ΔH, ΔG, and ΔS, the sign of $E°$ changes but its magnitude remains the same when we reverse a reaction.

As Examples 19.2 and 19.3 show, Table 19.1 enables us to predict the outcome of redox reactions under standard-state conditions, whether they take place in a galvanic

Table 19.1 | Standard Reduction Potentials at 25°C*

Half-Reaction	$E°(V)$
$F_2(g) + 2e^- \longrightarrow 2F^-(aq)$	+2.87
$O_3(g) + 2H^+(aq) + 2e^- \longrightarrow O_2(g) + H_2O$	+2.07
$Co^{3+}(aq) + e^- \longrightarrow Co^{2+}(aq)$	+1.82
$H_2O_2(aq) + 2H^+(aq) + 2e^- \longrightarrow 2H_2O$	+1.77
$PbO_2(s) + 4H^+(aq) + SO_4^{2-}(aq) + 2e^- \longrightarrow PbSO_4(s) + 2H_2O$	+1.70
$Ce^{4+}(aq) + e^- \longrightarrow Ce^{3+}(aq)$	+1.61
$MnO_4^-(aq) + 8H^+(aq) + 5e^- \longrightarrow Mn^{2+}(aq) + 4H_2O$	+1.51
$Au^{3+}(aq) + 3e^- \longrightarrow Au(s)$	+1.50
$Cl_2(g) + 2e^- \longrightarrow 2Cl^-(aq)$	+1.36
$Cr_2O_7^{2-}(aq) + 14H^+(aq) + 6e^- \longrightarrow 2Cr^{3+}(aq) + 7H_2O$	+1.33
$MnO_2(s) + 4H^+(aq) + 2e^- \longrightarrow Mn^{2+}(aq) + 2H_2O$	+1.23
$O_2(g) + 4H^+(aq) + 4e^- \longrightarrow 2H_2O$	+1.23
$Br_2(l) + 2e^- \longrightarrow 2Br^-(aq)$	+1.07
$NO_3^-(aq) + 4H^+(aq) + 3e^- \longrightarrow NO(g) + 2H_2O$	+0.96
$2Hg^{2+}(aq) + 2e^- \longrightarrow Hg_2^{2+}(aq)$	+0.92
$Hg_2^{2+}(aq) + 2e^- \longrightarrow 2Hg(l)$	+0.85
$Ag^+(aq) + e^- \longrightarrow Ag(s)$	+0.80
$Fe^{3+}(aq) + e^- \longrightarrow Fe^{2+}(aq)$	+0.77
$O_2(g) + 2H^+(aq) + 2e^- \longrightarrow H_2O_2(aq)$	+0.68
$MnO_4^-(aq) + 2H_2O + 3e^- \longrightarrow MnO_2(s) + 4OH^-(aq)$	+0.59
$I_2(s) + 2e^- \longrightarrow 2I^-(aq)$	+0.53
$O_2(g) + 2H_2O + 4e^- \longrightarrow 4OH^-(aq)$	+0.40
$Cu^{2+}(aq) + 2e^- \longrightarrow Cu(s)$	+0.34
$AgCl(s) + e^- \longrightarrow Ag(s) + Cl^-(aq)$	+0.22
$SO_4^{2-}(aq) + 4H^+(aq) + 2e^- \longrightarrow SO_2(g) + 2H_2O$	+0.20
$Cu^{2+}(aq) + e^- \longrightarrow Cu^+(aq)$	+0.15
$Sn^{4+}(aq) + 2e^- \longrightarrow Sn^{2+}(aq)$	+0.13
$2H^+(aq) + 2e^- \longrightarrow H_2(g)$	0.00
$Pb^{2+}(aq) + 2e^- \longrightarrow Pb(s)$	−0.13
$Sn^{2+}(aq) + 2e^- \longrightarrow Sn(s)$	−0.14
$Ni^{2+}(aq) + 2e^- \longrightarrow Ni(s)$	−0.25
$Co^{2+}(aq) + 2e^- \longrightarrow Co(s)$	−0.28
$PbSO_4(s) + 2e^- \longrightarrow Pb(s) + SO_4^{2-}(aq)$	−0.31
$Cd^{2+}(aq) + 2e^- \longrightarrow Cd(s)$	−0.40
$Fe^{2+}(aq) + 2e^- \longrightarrow Fe(s)$	−0.44
$Cr^{3+}(aq) + 3e^- \longrightarrow Cr(s)$	−0.74
$Zn^{2+}(aq) + 2e^- \longrightarrow Zn(s)$	−0.76
$2H_2O + 2e^- \longrightarrow H_2(g) + 2OH^-(aq)$	−0.83
$Mn^{2+}(aq) + 2e^- \longrightarrow Mn(s)$	−1.18
$Al^{3+}(aq) + 3e^- \longrightarrow Al(s)$	−1.66
$Be^{2+}(aq) + 2e^- \longrightarrow Be(s)$	−1.85
$Mg^{2+}(aq) + 2e^- \longrightarrow Mg(s)$	−2.37
$Na^+(aq) + e^- \longrightarrow Na(s)$	−2.71
$Ca^{2+}(aq) + 2e^- \longrightarrow Ca(s)$	−2.87
$Sr^{2+}(aq) + 2e^- \longrightarrow Sr(s)$	−2.89
$Ba^{2+}(aq) + 2e^- \longrightarrow Ba(s)$	−2.90
$K^+(aq) + e^- \longrightarrow K(s)$	−2.93
$Li^+(aq) + e^- \longrightarrow Li(s)$	−3.05

Left margin: Increasing strength as oxidizing agent

Right margin: Increasing strength as reducing agent

*For all half-reactions the concentration is 1 M for dissolved species and the pressure is 1 atm for gases. These are the standard-state values.

cell, where the reducing agent and oxidizing agent are physically separated from each other, or in a beaker, where the reactants are all mixed together.

Example 19.2   Predict what will happen if molecular bromine (Br_2) is added to a solution containing NaCl and NaI at 25°C. Assume all species are in their standard states.

Reasoning and Solution   To predict what redox reaction(s) will take place, we need to compare the standard reduction potentials for the following half-reactions:

$$Cl_2(1 \text{ atm}) + 2e^- \longrightarrow 2Cl^-(1 \text{ } M) \qquad E° = 1.36 \text{ V}$$
$$Br_2(l) + 2e^- \longrightarrow 2Br^-(1 \text{ } M) \qquad E° = 1.07 \text{ V}$$
$$I_2(s) + 2e^- \longrightarrow 2I^-(1 \text{ } M) \qquad E° = 0.53 \text{ V}$$

Applying the diagonal rule we see that Br_2 will oxidize I^- but will not oxidize Cl^-. Therefore, the only redox reaction that will occur appreciably under standard-state conditions is

Oxidation: $\qquad 2I^-(1 \text{ } M) \longrightarrow I_2(s) + 2e^-$
Reduction: $\qquad \underline{Br_2(l) + 2e^- \longrightarrow 2Br^-(1 \text{ } M)}$
$\qquad$ Overall: $2I^-(1 \text{ } M) + Br_2(l) \longrightarrow I_2(s) + 2Br^-(1 \text{ } M)$

Comment   Note that the Na^+ ions are inert and do not enter into the redox reaction.

Practice Exercise   Can Sn reduce $Zn^{2+}(aq)$ under standard-state conditions?

Similar problems: 19.14, 19.17.

Example 19.3   A galvanic cell consists of a Mg electrode in a 1.0 M Mg(NO$_3$)$_2$ solution and a Ag electrode in a 1.0 M AgNO$_3$ solution. Calculate the standard emf of this electrochemical cell at 25°C.

Reasoning and Solution   Table 19.1 gives the standard reduction potentials of the two electrodes:

$$Ag^+(1.0 \text{ } M) + e^- \longrightarrow Ag(s) \qquad E° = 0.80 \text{ V}$$
$$Mg^{2+}(1.0 \text{ } M) + 2e^- \longrightarrow Mg(s) \qquad E° = -2.37 \text{ V}$$

Applying the diagonal rule, we see that Ag^+ will oxidize Mg:

Anode (oxidation): $\qquad Mg(s) \longrightarrow Mg^{2+}(1.0 \text{ } M) + 2e^-$
Cathode (reduction): $\qquad \underline{2Ag^+(1.0 \text{ } M) + 2e^- \longrightarrow 2Ag(s)}$
$\qquad$ Overall: $Mg(s) + 2Ag^+(1.0 \text{ } M) \longrightarrow Mg^{2+}(1.0 \text{ } M) + 2Ag(s)$

Note that in order to balance the overall equation we multiplied the reduction of Ag^+ by 2. We can do so because, as an intensive property, $E°$ is not affected by this procedure. We find the emf of the cell by using Equation (19.1) and Table 19.1:

$$E°_{cell} = E°_{cathode} - E°_{anode}$$
$$= E°_{Ag^+/Ag} - E°_{Mg^{2+}/Mg}$$
$$= 0.80 \text{ V} - (-2.37 \text{ V})$$
$$= 3.17 \text{ V}$$

Comment The positive value of $E°$ shows that the overall reaction is spontaneous.

Practice Exercise What is the standard emf of a galvanic cell made of a Cd electrode in a 1.0 M $Cd(NO_3)_2$ solution and a Cr electrode in a 1.0 M $Cr(NO_3)_3$ solution operating at 25°C?

19.4 Spontaneity of Redox Reactions

Our next step is to see how $E°_{cell}$ is related to thermodynamic quantities such as $\Delta G°$ and K. In a galvanic cell, chemical energy is converted to electrical energy. Electrical energy in this case is the product of the emf of the cell and the total electrical charge (in coulombs) that passes through the cell:

$$\text{electrical energy} = \text{volts} \times \text{coulombs}$$
$$= \text{joules}$$

The total charge is determined by the number of moles of electrons (n) that pass through the circuit. By definition

$$\text{total charge} = nF$$

n is the number of moles of electrons exchanged between the reducing agent and the oxidizing agent in the overall redox equation.

where F, the Faraday constant, is the electrical charge contained in 1 mole of electrons. Experiments have shown that *1 faraday* is *equivalent to 96,485.3 coulombs,* or 96,500 coulombs, rounded off to three significant figures. Thus

$$1 \ F = 96,500 \ \text{C/mol}$$

Because

$$1 \ \text{J} = 1 \ \text{C} \times 1 \ \text{V}$$

we can also express the units of faraday as

$$1 \ F = 96,500 \ \text{J/V} \cdot \text{mol}$$

The measured emf is the *maximum* voltage that the cell can achieve. This value is used to calculate the maximum amount of electrical energy that can be obtained from the chemical reaction. This energy is used to do electrical work (w_{ele}), so

$$w_{max} = w_{ele}$$
$$= -nFE_{cell}$$

where w_{max} is the maximum amount of work that can be done. The negative sign on the right-hand side indicates that the electrical work is done by the system on the surroundings. In Chapter 18 we defined free energy as the energy available to do work. Specifically, the change in free energy (ΔG) represents the maximum amount of useful work that can be obtained from a reaction:

The sign convention for electrical work is the same as that for *P-V* work, discussed in Section 6.3.

$$\Delta G = w_{max}$$

Therefore we can write

$$\Delta G = -nFE_{cell} \qquad (19.2)$$

Both n and F are positive quantities and ΔG is negative for a spontaneous process, so E_{cell} must be positive. For reactions in which reactants and products are in their standard states, Equation (19.2) becomes

As in Equations (18.9) and (18.10), the units of $\Delta G°$ and ΔG are kJ/mol in Equations (19.2) and (19.3).

$$\Delta G° = -nFE°_{\text{cell}} \tag{19.3}$$

Here again, $E°_{\text{cell}}$ is positive for a spontaneous process.

Now we can relate $E°_{\text{cell}}$ to the equilibrium constant (K) of a redox reaction. In Section 18.5 we saw that the standard free-energy change $\Delta G°$ for a reaction is related to its equilibrium constant as [see Equation (18.10)]:

$$\Delta G° = -RT \ln K$$

Therefore, if we combine Equations (18.10) and (19.3) we obtain

$$-nFE°_{\text{cell}} = -RT \ln K$$

Solving for $E°_{\text{cell}}$

$$E°_{\text{cell}} = \frac{RT}{nF} \ln K \tag{19.4}$$

When $T = 298\ K$, Equation (19.4) can be simplified by substituting for R and F:

$$E°_{\text{cell}} = \frac{(8.314\ \text{J/K} \cdot \text{mol})(298\ \text{K})}{n(96{,}500\ \text{J/V} \cdot \text{mol})} \ln K$$

$$= \frac{0.0257\ \text{V}}{n} \ln K \tag{19.5}$$

Alternatively, Equation (19.5) can be written using the base-10 logarithm of K:

$$E°_{\text{cell}} = \frac{0.0592\ \text{V}}{n} \log K \tag{19.6}$$

Thus, if any one of the three quantities $\Delta G°$, K, or $E°_{\text{cell}}$ is known, the other two can be calculated using Equation (18.10), Equation (19.3), or Equation (19.4) (Figure 19.5). We summarize the relationships among $\Delta G°$, K, and $E°_{\text{cell}}$ and characterize the spontaneity of a redox reaction in Table 19.2. For simplicity, we will omit the subscript "cell" in $E°$ and E.

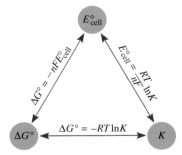

Figure 19.5 *Relationships among $E°_{\text{cell}}$, K, and $\Delta G°$.*

Example 19.4   Calculate the equilibrium constant for this reaction at 25°C:

$$\text{Sn}(s) + 2\text{Cu}^{2+}(aq) \rightleftharpoons \text{Sn}^{2+}(aq) + 2\text{Cu}^{+}(aq)$$

Reasoning and Solution   To calculate the equilibrium constant we need to know the $E°$ for a galvanic cell made up of Sn^{2+}/Sn and Cu^{2+}/Cu couples. The two half-reactions for the overall process are

Oxidation: $\text{Sn}(s) \longrightarrow \text{Sn}^{2+}(aq) + 2e^{-}$
Reduction: $2\text{Cu}^{2+}(aq) + 2e^{-} \longrightarrow 2\text{Cu}^{+}(aq)$

In Table 19.1 we find that $E°_{\text{Sn}^{2+}/\text{Sn}} = -0.14\ \text{V}$ and $E°_{\text{Cu}^{2+}/\text{Cu}^{+}} = 0.15\ \text{V}$. Thus

Table 19.2 | Relationships among $\Delta G°$, K, and $E°_{cell}$

$\Delta G°$	K	$E°_{cell}$	Reaction under Standard-State Conditions
Negative	>1	Positive	Spontaneous
0	=1	0	At equilibrium
Positive	<1	Negative	Nonspontaneous. Reaction is spontaneous in the reverse direction.

$$E° = E°_{Cu^{2+}/Cu^+} - E°_{Sn^{2+}/Sn}$$
$$= 0.15\ V - (-0.14\ V)$$
$$= 0.29\ V$$

Equation (19.5) can be written

$$\ln K = \frac{nE°}{0.0257\ V}$$

In the overall reaction we find $n = 2$. Therefore

$$\ln K = \frac{(2)(0.29\ V)}{0.0257\ V} = 22.6$$
$$K = e^{22.6} = 6.5 \times 10^9$$

Practice Exercise Calculate the equilibrium constant for the following reaction at 25°C:

$$Fe^{2+}(aq) + 2Ag(s) \rightleftharpoons Fe(s) + 2Ag^+(aq)$$

Because the reaction does not occur in a galvanic cell, we omit the $E°_{cathode}$ and $E°_{anode}$ terms.

Similar problems: 19.21, 19.22.

Example 19.5 Calculate the standard free-energy change for the following reaction at 25°C:

$$2Au(s) + 3Ca^{2+}(1\ M) \longrightarrow 2Au^{3+}(1\ M) + 3Ca(s)$$

Reasoning and Solution To calculate $\Delta G°$ for the reaction we need to know the $E°$ value for the redox process. First, we break up the overall reaction into half-reactions:

Oxidation: $\qquad\qquad 2Au(s) \longrightarrow 2Au^{3+}(1\ M) + 6e^-$
Reduction: $3Ca^{2+}(1\ M) + 6e^- \longrightarrow 3Ca(s)$

In Table 19.1 we find that $E°_{Au^{3+}/Au} = 1.50\ V$ and $E°_{Ca^{2+}/Ca} = -2.87\ V$. Therefore

$$E° = E°_{Ca^{2+}/Ca} - E°_{Au^{3+}/Au}$$
$$= -2.87\ V - 1.50\ V$$
$$= -4.37\ V$$

Now we use Equation (19.3):

$$\Delta G° = -nFE°$$

The overall reaction shows that $n = 6$, so

$$\Delta G° = -(6)(96,500 \text{ J/V} \cdot \text{mol})(-4.37 \text{ V})$$
$$= 2.53 \times 10^6 \text{ J/mol}$$
$$= 2.53 \times 10^3 \text{ kJ/mol}$$

Comment   The large positive value of $\Delta G°$ tells us that the reaction is not spontaneous under standard-state conditions at 25°C.

Similar problem: 19.24.

Practice Exercise   Calculate $\Delta G°$ for the following reaction at 25°C:

$$2Al^{3+}(aq) + 3Mg(s) \rightleftharpoons 2Al(s) + 3Mg^{2+}(aq)$$

19.5 The Effect of Concentration on Cell Emf

So far we have focused on redox reactions in which reactants and products are in their standard states, but standard-state conditions are often difficult, and sometimes impossible, to maintain. Nevertheless, there is a mathematical relationship between the emf of a cell and the concentration of reactants and products in a redox reaction under non-standard-state conditions. This equation is derived here.

The Nernst Equation

Consider a redox reaction of the type

$$aA + bB \longrightarrow cC + dD$$

From Equation (18.9)

$$\Delta G = \Delta G° + RT \ln Q$$

Because $\Delta G = -nFE$ and $\Delta G° = -nFE°$, the equation can be expressed as

$$-nFE = -nFE° + RT \ln Q$$

Dividing the equation through by $-nF$, we get

$$E = E° - \frac{RT}{nF} \ln Q \tag{19.7}$$

where Q is the reaction quotient (see Section 15.3). Equation (19.7) is known as the *Nernst equation* (after the German chemist Walter Nernst). At 298 K, Equation (19.7) can be rewritten as

$$E = E° - \frac{0.0257 \text{ V}}{n} \ln Q \tag{19.8}$$

or, expressing Equation (19.8) using the base-10 logarithm of Q:

$$E = E° - \frac{0.0592 \text{ V}}{n} \log Q \tag{19.9}$$

At equilibrium, there is no net transfer of electrons, so $E = 0$ and $Q = K$, where K is the equilibrium constant.

The Nernst equation enables us to calculate E as a function of reactant and product concentrations in a redox reaction. For example, for the Daniell cell in Figure 19.1

$$Zn(s) + Cu^{2+}(aq) \longrightarrow Zn^{2+}(aq) + Cu(s)$$

The Nernst equation for this cell at 25°C can be written as

$$E = 1.10 \text{ V} - \frac{0.0257 \text{ V}}{2} \ln \frac{[Zn^{2+}]}{[Cu^{2+}]}$$

If the ratio $[Zn^{2+}]/[Cu^{2+}]$ is less than 1, $\ln([Zn^{2+}]/[Cu^{2+}])$ is a negative number, so that the second term on the right-hand side of this equation is positive. Under this condition E is greater than the standard emf $E°$. If the ratio is greater than 1, E is smaller than $E°$.

Remember that concentrations of pure solids (and pure liquids) do not appear in the expression for Q.

Example 19.6 Predict whether the following reaction would proceed spontaneously as written at 298 K:

$$Co(s) + Fe^{2+}(aq) \longrightarrow Co^{2+}(aq) + Fe(s)$$

given that $[Co^{2+}] = 0.15 \text{ M}$ and $[Fe^{2+}] = 0.68 \text{ M}$.

Reasoning and Solution We can use the value of E for the redox process to determine the spontaneity of the reaction. To do so, we need to know $E°$ and $\ln Q$ in the Nernst equation. The half-reactions are

$$\text{Oxidation:} \qquad Co(s) \longrightarrow Co^{2+}(aq) + 2e^-$$
$$\text{Reduction:} \quad Fe^{2+}(aq) + 2e^- \longrightarrow Fe(s)$$

In Table 19.1 we find that $E°_{Co^{2+}/Co} = -0.28 \text{ V}$ and $E°_{Fe^{2+}/Fe} = -0.44 \text{ V}$. Therefore, the standard emf is

$$E° = E°_{Fe^{2+}/Fe} - E°_{Co^{2+}/Co}$$

$$= -0.44 \text{ V} - (-0.28 \text{ V})$$

$$= -0.16 \text{ V}$$

From Equation (19.8) we write

$$E = E° - \frac{0.0257 \text{ V}}{n} \ln \frac{[Co^{2+}]}{[Fe^{2+}]}$$

$$= -0.16 \text{ V} - \frac{0.0257 \text{ V}}{2} \ln \frac{0.15}{0.68}$$

$$= -0.16 \text{ V} + 0.019 \text{ V}$$

$$= -0.14 \text{ V}$$

Because E is negative, the reaction is *not* spontaneous in the direction written.

Similar problems: 19.29, 19.30.

Practice Exercise Will the following reaction occur spontaneously at 25°C, given that $[Fe^{2+}] = 0.60 \text{ M}$ and $[Cd^{2+}] = 0.010 \text{ M}$?

$$Cd(s) + Fe^{2+}(aq) \longrightarrow Cd^{2+}(aq) + Fe(s)$$

Now suppose we want to determine at what ratio of $[Co^{2+}]$ to $[Fe^{2+}]$ the reaction in Example 19.6 would become spontaneous. We can use Equation (19.8) as follows:

$$E = E° - \frac{0.0257 \text{ V}}{n} \ln Q$$

When $E = 0$, $Q = K$.

We set E equal to zero, which corresponds to the equilibrium situation.

$$0 = -0.16 \text{ V} - \frac{0.0257 \text{ V}}{2} \ln \frac{[Co^{2+}]}{[Fe^{2+}]}$$

$$\ln \frac{[Co^{2+}]}{[Fe^{2+}]} = -12.5$$

$$\frac{[Co^{2+}]}{[Fe^{2+}]} = e^{-12.5} = K$$

or

$$K = 4 \times 10^{-6}$$

Thus for the reaction to be spontaneous, the ratio $[Co^{2+}]/[Fe^{2+}]$ must be smaller than 4×10^{-6} so that E would be positive.

As Example 19.7 shows, if gases are involved in the cell reaction, their concentrations should be expressed in atm.

Example 19.7   Consider the galvanic cell shown in Figure 19.4(a). In a certain experiment, the emf (E) of the cell is found to be 0.54 V at 25°C. Suppose that $[Zn^{2+}] = 1.0 \ M$ and $P_{H_2} = 1.0$ atm. Calculate the molar concentration of H^+.

Reasoning and Solution   This is an application of the Nernst equation. Note that the concentration of the gaseous component is expressed in atmosphere.

The overall cell reaction is

$$Zn(s) + 2H^+(? \ M) \longrightarrow Zn^{2+}(1.0 \ M) + H_2(1.0 \text{ atm})$$

As we saw earlier (p. 613), the standard emf for the cell is 0.76 V. From Equation (19.8), we write

$$E = E° - \frac{0.0257 \text{ V}}{n} \ln \frac{[Zn^{2+}]P_{H_2}}{[H^+]^2}$$

$$0.54 \text{ V} = 0.76 \text{ V} - \frac{0.0257 \text{ V}}{2} \ln \frac{(1.0)(1.0)}{[H^+]^2}$$

$$-0.22 \text{ V} = -\frac{0.0257 \text{ V}}{2} \ln \frac{1}{[H^+]^2}$$

$$17.1 = \ln \frac{1}{[H^+]^2}$$

$$17.1 = -2 \ln [H^+]$$

$$\ln [H^+] = -8.6$$

$$[H^+] = e^{-8.6} \ M = 1.8 \times 10^{-4} \ M$$

Similar problem: 19.32.

Practice Exercise   What is the emf of a cell consisting of a Cd^{2+}/Cd half-cell and a $Pt/H^+/H_2$ half-cell if $[Cd^{2+}] = 0.20 \ M$, $[H^+] = 0.16 \ M$, and $P_{H_2} = 0.80$ atm?

Example 19.7 shows that an electrochemical cell whose cell reaction involves H^+ ions can be used to measure $[H^+]$ or pH. The pH meter described in Section 16.3 is based on this principle, but for practical reasons the electrodes used in a pH meter are quite different from the SHE and zinc electrode in the electrochemical cell (Figure 19.6).

Concentration Cells

Because electrode potential depends on ion concentrations, it is possible to construct a cell from two half-cells composed of the *same* material but differing in ion concentrations. Such a cell is called a *concentration cell*.

Consider a situation in which zinc electrodes are put into two aqueous solutions of zinc sulfate at 0.10 M and 1.0 M concentrations. The two solutions are connected by a salt bridge, and the electrodes are joined by a piece of wire in an arrangement like that shown in Figure 19.1. According to Le Châtelier's principle, the tendency for the reduction

$$Zn^{2+}(aq) + 2e^- \longrightarrow Zn(s)$$

increases with increasing concentration of Zn^{2+} ions. Therefore, reduction should occur in the more concentrated compartment and oxidation should take place on the more dilute side. The cell diagram is

$$Zn(s) \mid Zn^{2+}(0.10\ M) \parallel Zn^{2+}(1.0\ M) \mid Zn(s)$$

and the half-reactions are

Oxidation: $$Zn(s) \longrightarrow Zn^{2+}(0.10\ M) + 2e^-$$
Reduction: $$Zn^{2+}(1.0\ M) + 2e^- \longrightarrow Zn(s)$$
Overall: $$Zn^{2+}(1.0\ M) \longrightarrow Zn^{2+}(0.10\ M)$$

The emf of the cell is

$$E = E° - \frac{0.0257\ V}{2} \ln \frac{[Zn^{2+}]_{dil}}{[Zn^{2+}]_{conc}}$$

where the subscripts "dil" and "conc" refer to the 0.10 M and 1.0 M concentrations, respectively. The $E°$ for this cell is zero (the same electrode and the same type of ions are involved), so

$$E = 0 - \frac{0.0257\ V}{2} \ln \frac{0.10}{1.0}$$
$$= 0.0296\ V$$

The emf of concentration cells is usually small and decreases continually during the operation of the cell as the concentrations in the two compartments approach each other. When the concentrations of the ions in the two compartments are the same, E becomes zero, and no further change occurs.

A biological cell can be compared to a concentration cell for the purpose of calculating its *membrane potential*. Membrane potential is the electrical potential that exists across the membrane of various kinds of cells, including muscle cells and nerve cells. It is responsible for the propagation of nerve impulses and heart beat. A membrane potential is established whenever there are unequal concentrations of the same type of ion in the interior and exterior of a cell. For example, the concentrations of K^+ ions in the interior and exterior of a nerve cell are 400 mM and 15 mM, respectively.

Figure 19.6 *A glass electrode that is used in conjunction with a reference electrode in a pH meter.*

1 mM = 1 × 10⁻³ M.

Figure 19.7 *Interior section of a dry cell of the kind used in flashlights and transistor radios. Actually, the cell is not completely dry, as it contains a moist electrolyte paste. The anode and cathode are insulated from each other.*

Treating the situation as a concentration cell and applying the Nernst equation for just one kind of ion, we can write

$$E = E° - \frac{0.0257 \text{ V}}{1} \ln \frac{[K^+]_{ex}}{[K^+]_{in}}$$

$$= -(0.0257 \text{ V}) \ln \frac{15}{400}$$

$$= 0.084 \text{ V or } 84 \text{ mV}$$

where "ex" and "in" denote exterior and interior. Note that we have set $E° = 0$ because the same type of ion is involved. Thus an electrical potential of 84 mV exists across the membrane due to the unequal concentrations of K^+ ions.

19.6 Batteries

A **battery** is *a galvanic cell, or a series of combined galvanic cells, that can be used as a source of direct electric current at a constant voltage.* Although the operation of a battery is similar in principle to that of the galvanic cells described in Section 19.2, a battery has the advantage of being completely self-contained and requiring no auxiliary components such as salt bridges. Here we will discuss several types of batteries that are in widespread use.

The Dry Cell Battery

The most common dry cell, that is, a cell without fluid component, is the *Leclanché cell* used in flashlights and transistor radios. The anode of the cell consists of a zinc can or container that is in contact with manganese dioxide (MnO_2) and an electrolyte. The electrolyte consists of ammonium chloride and zinc chloride in water, to which starch is added to thicken the solution to a pastelike consistency so that it is less likely to leak (Figure 19.7). A carbon rod serves as the cathode, which is immersed in the electrolyte in the center of the cell. The cell reactions are

Anode: $\quad\quad\quad\quad\quad\quad\quad\quad\quad Zn(s) \longrightarrow Zn^{2+}(aq) + 2e^-$
Cathode: $2NH_4^+(aq) + 2MnO_2(s) + 2e^- \longrightarrow Mn_2O_3(s) + 2NH_3(aq) + H_2O(l)$

Overall: $Zn(s) + 2NH_4^+(aq) + 2MnO_2(s) \longrightarrow Zn^{2+}(aq) + 2NH_3(aq) + H_2O(l) + Mn_2O_3(s)$

Actually, the equation is an oversimplification of a complex process. The voltage produced by a dry cell is about 1.5 V.

The Mercury Battery

The mercury battery is used extensively in medicine and electronic industries and is more expensive than the common dry cell. Contained in a stainless steel cylinder, the mercury battery consists of a zinc anode (amalgamated with mercury) in contact with a strongly alkaline electrolyte containing zinc oxide and mercury(II) oxide (Figure 19.8). The cell reactions are

Anode: $\quad Zn(Hg) + 2OH^-(aq) \longrightarrow ZnO(s) + H_2O(l) + 2e^-$
Cathode: $HgO(s) + H_2O(l) + 2e^- \longrightarrow Hg(l) + 2OH^-(aq)$

Overall: $\quad\quad Zn(Hg) + HgO(s) \longrightarrow ZnO(s) + Hg(l)$

Figure 19.8 *Interior section of a mercury battery.*

Because there is no change in electrolyte composition during operation—the overall cell reaction involves only solid substances—the mercury battery provides a more constant voltage (1.35 V) than the Leclanché cell. It also has a considerably higher capacity and longer life. These qualities make the mercury battery ideal for use in pacemakers, hearing aids, electric watches, and light meters.

The Lead Storage Battery

The lead storage battery commonly used in automobiles consists of six identical cells joined together in series. Each cell has a lead anode and a cathode made of lead dioxide (PbO_2) packed on a metal plate (Figure 19.9). Both the cathode and the anode are immersed in an aqueous solution of sulfuric acid, which acts as the electrolyte. The cell reactions are

$$\text{Anode:} \quad Pb(s) + SO_4^{2-}(aq) \longrightarrow PbSO_4(s) + 2e^-$$
$$\text{Cathode:} \quad PbO_2(s) + 4H^+(aq) + SO_4^{2-}(aq) + 2e^- \longrightarrow PbSO_4(s) + 2H_2O(l)$$
$$\overline{\text{Overall: } Pb(s) + PbO_2(s) + 4H^+(aq) + 2SO_4^{2-}(aq) \longrightarrow 2PbSO_4(s) + 2H_2O(l)}$$

Under normal operating conditions, each cell produces 2 V; a total of 12 V from the six cells is used to power the ignition circuit of the automobile and its other electrical systems. The lead storage battery can deliver large amounts of current for a short time, such as the time it takes to start up the engine.

Unlike the Leclanché cell and the mercury battery, the lead storage battery is rechargeable. Recharging the battery means reversing the normal electrochemical reaction by applying an external voltage at the cathode and the anode. (This kind of process is called *electrolysis,* see p. 632.) The reactions that replenish the original materials are

$$PbSO_4(s) + 2e^- \longrightarrow Pb(s) + SO_4^{2-}(aq)$$
$$\underline{PbSO_4(s) + 2H_2O(l) \longrightarrow PbO_2(s) + 4H^+(aq) + SO_4^{2-}(aq) + 2e^-}$$
$$\text{Overall: } 2PbSO_4(s) + 2H_2O(l) \longrightarrow Pb(s) + PbO_2(s) + 4H^+(aq) + 2SO_4^{2-}(aq)$$

Removable cap

Anode

Cathode

H_2SO_4 electrolyte

Negative plates (lead grills filled with spongy lead)

Positive plates (lead grills filled with PbO_2)

Figure 19.9 *Interior section of a lead storage battery. Under normal operating conditions, the concentration of the sulfuric acid solution is about 38 percent by mass.*

The overall reaction is exactly the opposite of the normal cell reaction.

Two aspects of the operation of a lead storage battery are worth noting. First, because the electrochemical reaction consumes sulfuric acid, the degree to which the battery has been discharged can be checked by measuring the density of the electrolyte with a hydrometer, as is usually done at gas stations. The density of the fluid in a "healthy," fully charged battery should be equal to or greater than 1.2 g/mL. Second, people living in cold climates sometimes have trouble starting their cars because the battery has "gone dead." Thermodynamic calculations show that the emf of many electrochemical cells decreases with decreasing temperature. However, for a lead storage battery, the temperature coefficient is about 1.5×10^{-4} V/°C; that is, there is a decrease in voltage of 1.5×10^{-4} V for every degree drop in temperature. Thus, even allowing for a 40°C change in temperature, the decrease in voltage amounts to only 6×10^{-3} V, which is about

$$\frac{6 \times 10^{-3}\,\text{V}}{12\,\text{V}} \times 100\% = 0.05\%$$

of the operating voltage, an insignificant change. The real cause of a battery's apparent breakdown is an increase in the viscosity of the electrolyte as the temperature decreases. For the battery to function properly, the electrolyte must be fully conducting. However, the ions move much more slowly in a viscous medium, so the resistance of the fluid increases, leading to a decrease in the power output of the battery. If an apparently "dead battery" is warmed to near room temperature on a frigid day, it recovers its ability to deliver normal power.

The Solid-State Lithium Battery

Unlike the batteries discussed so far, a solid-state battery employs a solid (rather than an aqueous solution or a water-based paste) as the electrolyte connecting the electrodes. Figure 19.10 shows a schematic solid-state lithium battery. The advantage in

Figure 19.10 *A schematic diagram of a solid-state lithium battery. Lithium metal is the anode, and TiS_2 is the cathode. During operation, Li^+ ions migrate through the solid polymer electrolyte from the anode to the cathode while electrons flow externally from the anode to the cathode to complete the circuit.*

choosing lithium as the anode is that it has the most negative $E°$ value, as Table 19.1 shows. Furthermore, lithium is a lightweight metal, so that only 6.941 g of Li (its molar mass) are needed to produce 1 mole of electrons. The electrolyte is a polymer material that permits the passage of ions but not of electrons. The cathode is made of either TiS_2 or V_6O_{13}. The cell voltage of a solid-state lithium battery can be as high as 3 V, and it can be recharged in the same way as a lead storage battery. Although these batteries do not yet have high reliability and long lifetime, they have been hailed as the batteries of the future.

Fuel Cells

Fossil fuels are a major source of energy, but conversion of fossil fuel into electrical energy is a highly inefficient process. Consider the combustion of methane:

$$CH_4(g) + 2O_2(g) \longrightarrow CO_2(g) + 2H_2O(l) + \text{energy}$$

To generate electricity, heat produced by the reaction is first used to convert water to steam, which then drives a turbine that drives a generator. An appreciable fraction of the energy released in the form of heat is lost to the surroundings at each step; even the most efficient power plant converts only about 40 percent of the original chemical energy into electricity. Because combustion reactions are redox reactions, it is more desirable to carry them out directly by electrochemical means, thereby greatly increasing the efficiency of power production. This objective can be accomplished by a device known as *fuel cell, a galvanic cell that requires a continuous supply of reactants to keep functioning.*

In its simplest form, a hydrogen-oxygen fuel cell consists of an electrolyte solution, such as potassium hydroxide solution, and two inert electrodes. Hydrogen and oxygen gases are bubbled through the anode and cathode compartments (Figure 19.11), where the following reactions take place:

Figure 19.11 *A hydrogen-oxygen fuel cell. The Ni and NiO embedded in the porous carbon electrodes are electrocatalysts.*

Oxidation

$$2H_2(g) + 4OH^-(aq) \longrightarrow 4H_2O(l) + 4e^-$$

Reduction

$$O_2(g) + 2H_2O(l) + 4e^- \longrightarrow 4OH^-(aq)$$

$$\text{Anode:} \quad 2H_2(g) + 4OH^-(aq) \longrightarrow 4H_2O(l) + 4e^-$$
$$\underline{\text{Cathode:} \quad O_2(g) + 2H_2O(l) + 4e^- \longrightarrow 4OH^-(aq)}$$
$$\text{Overall:} \quad 2H_2(g) + O_2(g) \longrightarrow 2H_2O(l)$$

The standard emf of the cell can be calculated as follows, with data from Table 19.1:

$$E°_{cell} = E°_{cathode} - E°_{anode}$$
$$= 0.40 \text{ V} - (-0.83 \text{ V})$$
$$= 1.23 \text{ V}$$

Thus the cell reaction is spontaneous under standard-state conditions. Note that the reaction is the same as the hydrogen combustion reaction, but the oxidation and reduction are carried out separately at the anode and the cathode. Like platinum in the standard hydrogen electrode, the electrodes have a twofold function. They serve as electrical conductors, and they provide the necessary surfaces for the initial decomposition of the molecules into atomic species, prior to electron transfer. They are *electrocatalysts*. Metals such as platinum, nickel, and rhodium are good electrocatalysts.

In addition to the H_2-O_2 system, a number of other fuel cells have been developed. Among these is the propane-oxygen fuel cell. The half-cell reactions are

$$\text{Anode:} \quad C_3H_8(g) + 6H_2O(l) \longrightarrow 3CO_2(g) + 20H^+(aq) + 20e^-$$
$$\underline{\text{Cathode:} \quad 5O_2(g) + 20H^+(aq) + 20e^- \longrightarrow 10H_2O(l)}$$
$$\text{Overall:} \quad C_3H_8(g) + 5O_2(g) \longrightarrow 3CO_2(g) + 4H_2O(l)$$

The overall reaction is identical to the burning of propane in oxygen.

Unlike batteries, fuel cells do not store chemical energy. Reactants must be constantly resupplied, and products must be constantly removed from a fuel cell. In this respect, a fuel cell resembles an engine more than it does a battery. However, the fuel cell does not operate like a heat engine and therefore is not subject to the same kind of thermodynamic limitations in energy conversion (see p. 574).

Properly designed fuel cells may be as much as 70 percent efficient, about twice as efficient as an internal combustion engine. In addition, fuel-cell generators are free of the noise, vibration, heat transfer, thermal pollution, and other problems normally associated with conventional power plants. Nevertheless, fuel cells are not yet in

Figure 19.12 *A hydrogen-oxygen fuel cell used in the space program. The pure water produced by the cell is consumed by the astronauts.*

widespread use. A major problem lies in the lack of cheap electrocatalysts able to function efficiently for long periods of time without contamination. The most successful application of fuel cells to date has been in space vehicles (Figure 19.12).

19.7 Corrosion

Corrosion is the term usually applied to *the deterioration of metals by an electrochemical process*. We see many examples of corrosion around us. Rust on iron, tarnish on silver, and the green patina formed on copper and brass are a few of them (Figure 19.13). Corrosion causes enormous damage to buildings, bridges, ships, and cars. The cost of metallic corrosion to the U.S. economy has been estimated to be well over 100 billion dollars a year! This section discusses some of the fundamental processes that occur in corrosion and methods used to protect metals against it.

By far the most familiar example of corrosion is the formation of rust on iron. Oxygen gas and water must be present for iron to rust. Although the reactions involved are quite complex and not completely understood, the main steps are believed to be as follows. A region of the metal's surface serves as the anode, where oxidation occurs:

$$Fe(s) \longrightarrow Fe^{2+}(aq) + 2e^-$$

The electrons given up by iron reduce atmospheric oxygen to water at the cathode, which is another region of the same metal's surface:

$$O_2(g) + 4H^+(aq) + 4e^- \longrightarrow 2H_2O(l)$$

The overall redox reaction is

$$2Fe(s) + O_2(g) + 4H^+(aq) \longrightarrow 2Fe^{2+}(aq) + 2H_2O(l)$$

With data from Table 19.1, we find the standard emf for this process:

$$E^\circ_{cell} = E^\circ_{cathode} - E^\circ_{anode}$$
$$= 1.23 \text{ V} - (-0.44 \text{ V})$$
$$= 1.67 \text{ V}$$

The positive standard emf means that the process will occur spontaneously.

Note that this reaction occurs in an acidic medium; the H^+ ions are supplied in part by the reaction of atmospheric carbon dioxide with water to form H_2CO_3.

The Fe^{2+} ions formed at the anode are further oxidized by oxygen:

$$4Fe^{2+}(aq) + O_2(g) + (4 + 2x)H_2O(l) \longrightarrow 2Fe_2O_3 \cdot xH_2O(s) + 8H^+(aq)$$

This hydrated form of iron(III) oxide is known as rust. The amount of water associated with the iron oxide varies, so we represent the formula as $Fe_2O_3 \cdot xH_2O$.

Figure 19.14 shows the mechanism of rust formation. The electrical circuit is completed by the migration of electrons and ions; this is why rusting occurs so rapidly in salt water. In cold climates, salts (NaCl or $CaCl_2$) spread on roadways to melt ice and snow are a major cause of rust formation on automobiles.

Metallic corrosion is not limited to iron. Consider aluminum, a metal used to make many useful things, including airplanes and beverage cans. Aluminum has a much greater tendency to oxidize than iron does; in Table 19.1 we see that Al has a more negative standard reduction potential than Fe. Based on this fact alone, we might expect to see airplanes slowly corrode away in rainstorms, and soda cans transformed into piles of corroded aluminum. These processes do not occur because the layer of

Figure 19.13 *Examples of corrosion: (a) a rusted ship, (b) a half-tarnished silver dish, and (c) the Statue of Liberty coated with patina before its restoration in 1986.*

(a)

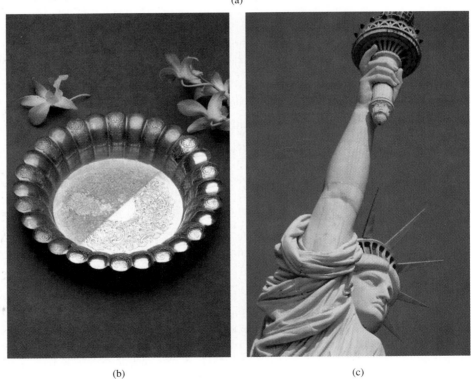

(b)   (c)

insoluble aluminum oxide (Al_2O_3) that forms on its surface when the metal is exposed to air serves to protect the aluminum underneath from further corrosion. The rust that forms on the surface of iron, however, is too porous to protect the underlying metal.

Coinage metals such as copper and silver also corrode, but much more slowly.

$$Cu(s) \longrightarrow Cu^{2+}(aq) + 2e^-$$
$$Ag(s) \longrightarrow Ag^+(aq) + e^-$$

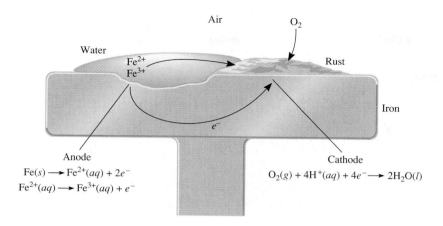

In normal atmospheric exposure, copper forms a layer of copper carbonate ($CuCO_3$), a green substance also called patina, that protects the metal underneath from further corrosion. Likewise, silverware that comes into contact with foodstuffs develops a layer of silver sulfide (Ag_2S).

A number of methods have been devised to protect metals from corrosion. Most of these methods are aimed at preventing rust formation. The most obvious approach is to coat the metal surface with paint. However, if the paint is scratched, pitted, or dented to expose even the smallest area of bare metal, rust will form under the paint layer. The surface of iron metal can be made inactive by a process called *passivation*. A thin oxide layer is formed when the metal is treated with a strong oxidizing agent such as concentrated nitric acid. A solution of sodium chromate is often added to cooling systems and radiators to prevent rust formation by a similar mechanism.

The tendency for iron to oxidize is greatly reduced when it is alloyed with certain other metals. For example, in stainless steel, an alloy of iron and chromium, a layer of chromium oxide forms that protects the iron from corrosion.

An iron container can be covered with a layer of another metal such as tin or zinc. A "tin" can is made by applying a thin layer of tin over iron. Rust formation is prevented as long as the tin layer remains intact. However, once the surface has been scratched, rusting occurs rapidly. If we look up the standard reduction potentials, according to the diagonal rule, we find that iron acts as the anode and tin as the cathode in the corrosion process:

$$Sn^{2+}(aq) + 2e^- \longrightarrow Sn(s) \qquad E° = -0.14 \text{ V}$$
$$Fe^{2+}(aq) + 2e^- \longrightarrow Fe(s) \qquad E° = -0.44 \text{ V}$$

The protective process is different for zinc-plated, or *galvanized,* iron. Zinc is more easily oxidized than iron (see Table 19.1):

$$Zn^{2+}(aq) + 2e^- \longrightarrow Zn(s) \qquad E° = -0.76 \text{ V}$$

So even if a scratch exposes the iron, the zinc is still attacked. In this case, the zinc metal serves as the anode and the iron is the cathode.

Cathodic protection is a process in which the metal that is to be protected from corrosion is made the cathode in what amounts to a galvanic cell. Figure 19.15 shows how an iron nail can be protected from rusting by connecting the nail to a piece of zinc. Without such protection, an iron nail quickly rusts in water. Rusting of

Figure 19.15 *An iron nail that is cathodically protected by a piece of zinc strip does not rust in water, while an iron nail without such protection rusts readily.*

underground iron pipes and iron storage tanks can be prevented or greatly reduced by connecting them to metals such as zinc and magnesium, which oxidize more readily than iron does (Figure 19.16).

19.8 Electrolysis

In contrast to spontaneous redox reactions, which result in the conversion of chemical energy into electrical energy, *electrolysis* is the process in which *electrical energy is used to cause a nonspontaneous chemical reaction to occur*. An ***electrolytic cell*** is *an apparatus for carrying out electrolysis*. The same principles underlie electrolysis and the processes that take place in galvanic cells. Here we will discuss three examples of electrolysis based on those principles. Then we will look at the quantitative aspects of electrolysis.

Electrolysis of Molten Sodium Chloride

In its molten state, sodium chloride, an ionic compound, can be electrolyzed to form sodium metal and chlorine. Figure 19.17(a) is a diagram of a *Downs cell*, which is used for large-scale electrolysis of NaCl. In molten NaCl, the cations and anions are

Figure 19.16 *Cathodic protection of an iron storage tank (cathode) by magnesium, a more electropositive metal (anode). Because only the magnesium is depleted in the electrochemical process, it is sometimes called the sacrificial anode.*

Oxidation: $Mg(s) \longrightarrow Mg^{2+}(aq) + 2e^-$ Reduction: $O_2(g) + 4H^+(aq) + 4e^- \longrightarrow 2H_2O(l)$

Figure 19.17 *(a) A practical arrangement called a Downs cell for the electrolysis of molten NaCl (m.p. = 801°C). The sodium metal formed at the cathodes is in the liquid state. Since liquid sodium metal is lighter than molten NaCl, the sodium floats to the surface, as shown, and is collected. Chlorine gas forms at the anode and is collected at the top. (b) A simplified diagram showing the electrode reactions during the electrolysis of molten NaCl. The battery is needed to drive the nonspontaneous reactions.*

the Na^+ and Cl^- ions, respectively. Figure 19.17(b) is a simplified diagram showing the reactions that occur at the electrodes. The electrolytic cell contains a pair of electrodes connected to the battery. The battery serves as an "electron pump," driving electrons to the cathode, where reduction occurs, and withdrawing electrons from the anode, where oxidation occurs. The reactions at the electrodes are

$$
\begin{array}{ll}
\text{Anode (oxidation):} & 2Cl^-(l) \longrightarrow Cl_2(g) + 2e^- \\
\text{Cathode (reduction):} & \underline{2Na^+(l) + 2e^- \longrightarrow 2Na(l)} \\
\text{Overall:} & 2Na^+(l) + 2Cl^-(l) \longrightarrow 2Na(l) + Cl_2(g)
\end{array}
$$

This process is a major source of pure sodium metal and chlorine gas.

Theoretical estimates show that the $E°$ value for the overall process is about -4 V, which means that this is a nonspontaneous process. Therefore, a *minimum* of 4 V must be supplied by the battery to carry out the reaction. In practice, a higher voltage is necessary because of inefficiencies in the electrolytic process and because of overvoltage, to be discussed shortly.

Electrolysis of Water

Water in a beaker under atmospheric conditions (1 atm and 25°C) will not spontaneously decompose to form hydrogen and oxygen gas because the standard free-energy change for the reaction is a large positive quantity:

$$2H_2O(l) \longrightarrow 2H_2(g) + O_2(g) \qquad \Delta G° = 474.4 \text{ kJ}$$

However, this reaction can be induced in a cell like the one shown in Figure 19.18. This electrolytic cell consists of a pair of electrodes made of a nonreactive metal, such as platinum, immersed in water. When the electrodes are connected to the battery, nothing happens because there are not enough ions in pure water to carry much of an

Figure 19.18 *Apparatus for small-scale electrolysis of water. The volume of hydrogen gas generated (left column) is twice that of oxygen gas (right column).*

electric current. (Remember that at 25°C, pure water has only 1×10^{-7} M H^+ ions and 1×10^{-7} M OH^- ions.)

On the other hand, the reaction occurs readily in a 0.1 M H_2SO_4 solution because there are a sufficient number of ions to conduct electricity (Figure 19.19). Immediately, gas bubbles begin to appear at both electrodes. The process at the anode is

$$2H_2O(l) \longrightarrow O_2(g) + 4H^+(aq) + 4e^-$$

while at the cathode we have

$$H^+(aq) + e^- \longrightarrow \tfrac{1}{2}H_2(g)$$

The overall reaction is given by

Anode (oxidation): $2H_2O(l) \longrightarrow O_2(g) + 4H^+(aq) + 4e^-$
Cathode (reduction): $\underline{4[H^+(aq) + e^- \longrightarrow \tfrac{1}{2}H_2(g)]}$
Overall: $2H_2O(l) \longrightarrow 2H_2(g) + O_2(g)$

> **What is the minimum voltage needed for this electrolytic process?**

Note that no net H_2SO_4 is consumed in this reaction.

Electrolysis of an Aqueous Sodium Chloride Solution

This is the most complicated of the three examples of electrolysis considered here because aqueous sodium chloride solution contains several species that could be oxidized and reduced. The oxidation reactions that might occur at the anode are

(1) $2Cl^-(aq) \longrightarrow Cl_2(g) + 2e^-$
(2) $2H_2O(l) \longrightarrow O_2(g) + 4H^+(aq) + 4e^-$

Referring to Table 19.1, we find

$$Cl_2(g) + 2e^- \longrightarrow 2Cl^-(aq) \qquad E° = 1.36 \text{ V}$$
$$O_2(g) + 4H^+(aq) + 4e^- \longrightarrow 2H_2O(l) \qquad E° = 1.23 \text{ V}$$

> **Because Cl_2 is more easily reduced than O_2, it follows that it would be more difficult to oxidize Cl^- than H_2O at the anode.**

The standard reduction potentials of (1) and (2) are not very different, but the values do suggest that H_2O should be preferentially oxidized at the anode. However, by experiment we find that the gas liberated at the anode is Cl_2, not O_2! In studying electrolytic processes, we sometimes find that the voltage required for a reaction is considerably higher than the electrode potential indicates. The **overvoltage** is *the*

Figure 19.19 *A diagram showing the electrode reactions during the electrolysis of water.*

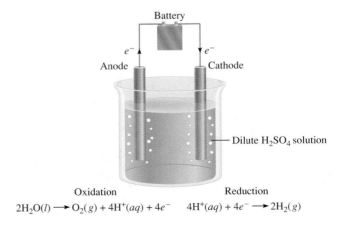

Oxidation   Reduction
$2H_2O(l) \longrightarrow O_2(g) + 4H^+(aq) + 4e^-$ $4H^+(aq) + 4e^- \longrightarrow 2H_2(g)$

difference between the electrode potential and the actual voltage required to cause electrolysis. The overvoltage for O_2 formation is quite high. Therefore, under normal operating conditions Cl_2 gas is actually formed at the anode instead of O_2.

The reductions that might occur at the cathode are

$$(3) \quad 2H^+(aq) + 2e^- \longrightarrow H_2(g) \qquad E° = 0.00 \text{ V}$$
$$(4) \quad 2H_2O(l) + 2e^- \longrightarrow H_2(g) + 2OH^-(aq) \qquad E° = -0.83 \text{ V}$$
$$(5) \quad Na^+(aq) + e^- \longrightarrow Na(s) \qquad E° = -2.71 \text{ V}$$

Reaction (5) is ruled out because it has a very negative standard reduction potential. Reaction (3) is preferred over (4) under standard-state conditions. At a pH of 7 (as is the case for a NaCl solution), however, they are equally probable. We generally use (4) to describe the cathode reaction because the concentration of H^+ ions is too low (about 1×10^{-7} M) to make (3) a reasonable choice.

Thus, the half-cell reactions in the electrolysis of aqueous sodium chloride are

$$\begin{array}{ll} \text{Anode (oxidation):} & 2Cl^-(aq) \longrightarrow Cl_2(g) + 2e^- \\ \text{Cathode (reduction):} & \underline{2H_2O(l) + 2e^- \longrightarrow H_2(g) + 2OH^-(aq)} \\ \text{Overall:} & 2H_2O(l) + 2Cl^-(aq) \longrightarrow H_2(g) + Cl_2(g) + 2OH^-(aq) \end{array}$$

As the overall reaction shows, the concentration of the Cl^- ions decreases during electrolysis and that of the OH^- ions increases. Therefore, in addition to H_2 and Cl_2, the useful by-product NaOH can be obtained by evaporating the aqueous solution at the end of the electrolysis.

Keep in mind the following from our analysis of electrolysis: cations are likely to be reduced at the cathode and anions are likely to be oxidized at the anode, and in aqueous solutions water itself may be oxidized and/or reduced. The outcome depends on the nature of other species present.

Example 19.8 An aqueous Na_2SO_4 solution is electrolyzed, using the apparatus shown in Figure 19.18. If the products formed at the anode and cathode are oxygen gas and hydrogen gas, respectively, describe the electrolysis in terms of the reactions at the electrodes.

Reasoning and Solution Before we look at the electrode reactions, we should consider these facts: (1) Because Na_2SO_4 does not hydrolyze in water, the pH of the solution is close to 7. (2) The Na^+ ions are not reduced at the cathode, and the SO_4^{2-} ions are not oxidized at the anode. These conclusions are drawn from the electrolysis of water in the presence of sulfuric acid and in aqueous sodium chloride solution. Therefore, the electrode reactions are

$$\begin{array}{ll} \text{Anode:} & 2H_2O(l) \longrightarrow O_2(g) + 4H^+(aq) + 4e^- \\ \text{Cathode:} & 2H_2O(l) + 2e^- \longrightarrow H_2(g) + 2OH^- \end{array}$$

The overall reaction, obtained by doubling the cathode reaction coefficients and adding the result to the anode reaction, is

$$6H_2O(l) \longrightarrow 2H_2(g) + O_2(g) + 4H^+(aq) + 4OH^-(aq)$$

If the H^+ and OH^- ions are allowed to mix, then

$$4H^+(aq) + 4OH^-(aq) \longrightarrow 4H_2O(l)$$

The SO_4^{2-} ion is the conjugate base of the weak acid HSO_4^- ($K_a = 1.3 \times 10^{-2}$). However, the extent to which SO_4^{2-} hydrolyzes is negligible. Also, the SO_4^{2-} ion is not oxidized at the anode.

and the overall reaction becomes

$$2H_2O(l) \longrightarrow 2H_2(g) + O_2(g)$$

Practice Exercise   An aqueous solution of $Mg(NO_3)_2$ is electrolyzed. What are the gaseous products at the anode and cathode?

Electrolysis has many important applications in industry, mainly in the extraction and purification of metals. We will discuss some of these applications in Section 19.9.

Quantitative Aspects of Electrolysis

The quantitative treatment of electrolysis was developed primarily by Faraday. He observed that the mass of product formed (or reactant consumed) at an electrode is proportional to both the amount of electricity transferred at the electrode and the molar mass of the substance in question. For example, in the electrolysis of molten NaCl, the cathode reaction tells us that one Na atom is produced when one Na^+ ion accepts an electron from the electrode. To reduce 1 mole of Na^+ ions, we must supply Avogadro's number (6.02×10^{23}) of electrons to the cathode. On the other hand, the stoichiometry of the anode reaction shows that oxidation of two Cl^- ions yields one chlorine molecule. Therefore, the formation of 1 mole of Cl_2 results in the transfer of 2 moles of electrons from the Cl^- ions to the anode. Similarly, it takes 2 moles of electrons to reduce 1 mole of Mg^{2+} ions and 3 moles of electrons to reduce 1 mole of Al^{3+} ions:

$$Mg^{2+} + 2e^- \longrightarrow Mg$$
$$Al^{3+} + 3e^- \longrightarrow Al$$

In an electrolysis experiment, we generally measure the current (in amperes, A) that passes through an electrolytic cell in a given period of time. The relationship between charge (in coulombs, C) and current is

$$1 \text{ C} = 1 \text{ A} \times 1 \text{ s}$$

that is, a coulomb is the quantity of electrical charge passing any point in the circuit in 1 second when the current is 1 ampere.

Figure 19.20 shows the steps involved in calculating the quantities of substances produced in electrolysis. Let us illustrate the approach by considering molten $CaCl_2$ in an electrolytic cell. Suppose a current of 0.452 A is passed through the cell for 1.50 hr. How much product will be formed at the anode and at the cathode? In solving electrolysis problems of this type, the first step is to determine which species will be oxidized at the anode and which species will be reduced at the cathode. Here the choice is straightforward because we only have Ca^{2+} and Cl^- ions in molten $CaCl_2$. Thus we write the half- and overall reactions as

Figure 19.20 *Steps involved in calculating amounts of substances reduced or oxidized in electrolysis.*

Anode (oxidation): $2Cl^-(l) \longrightarrow Cl_2(g) + 2e^-$

Cathode (reduction): $Ca^{2+}(l) + 2e^- \longrightarrow Ca(l)$

Overall: $Ca^{2+}(l) + 2Cl^-(l) \longrightarrow Ca(l) + Cl_2(g)$

The quantities of calcium metal and chlorine gas formed depend on the number of electrons that pass through the electrolytic cell, which in turn depends on current × time, or charge:

$$? \, C = 0.452 \, A \times 1.50 \, hr \times \frac{3600 \, s}{1 \, hr} \times \frac{1 \, C}{1 \, A \cdot s} = 2.44 \times 10^3 \, C$$

Because 1 mol e^- = 96,500 C and 2 mol e^- are required to reduce 1 mole of Ca^{2+} ions, the mass of Ca metal formed at the cathode is calculated as:

$$? \, g \, Ca = 2.44 \times 10^3 \, C \times \frac{1 \, mol \, e^-}{96,500 \, C} \times \frac{1 \, mol \, Ca}{2 \, mol \, e^-} \times \frac{40.08 \, g \, Ca}{1 \, mol \, Ca} = 0.507 \, g \, Ca$$

The anode reaction indicates that 1 mole of chlorine is produced per 2 mol e^- of electricity. Hence the mass of chlorine gas formed is

$$? \, g \, Cl_2 = 2.44 \times 10^3 \, C \times \frac{1 \, mol \, e^-}{96,500 \, C} \times \frac{1 \, mol \, Cl_2}{2 \, mol \, e^-} \times \frac{70.90 \, g \, Cl_2}{1 \, mol \, Cl_2} = 0.896 \, g \, Cl_2$$

Example 19.9 A current of 1.26 A is passed through an electrolytic cell containing a dilute sulfuric acid solution for 7.44 hr. Write the half-cell reactions and calculate the volume of gases generated at STP.

Reasoning and Solution Earlier in this chapter (see p. 634) we saw that the half-cell reactions for this process are

Anode (oxidation): $2H_2O(l) \longrightarrow O_2(g) + 4H^+(aq) + 4e^-$

Cathode (reduction): $4[H^+(aq) + e^- \longrightarrow \frac{1}{2}H_2(g)]$

Overall: $2H_2O(l) \longrightarrow 2H_2(g) + O_2(g)$

First we calculate the number of coulombs of electricity that pass through the cell:

$$? \, C = 1.26 \, A \times 7.44 \, hr \times \frac{3600 \, s}{1 \, hr} \times \frac{1 \, C}{1 \, A \cdot s} = 3.37 \times 10^4 \, C$$

We see that for every mole of O_2 formed at the anode, 4 moles of electrons are generated so that

$$? \, g \, O_2 = 3.37 \times 10^4 \, C \times \frac{1 \, mol \, e^-}{96,500 \, C} \times \frac{1 \, mol \, O_2}{4 \, mol \, e^-} \times \frac{32.00 \, g \, O_2}{1 \, mol \, O_2} = 2.79 \, g \, O_2$$

The volume of 2.79 g O_2 at STP is given by

$$V = \frac{nRT}{P}$$

$$= \frac{(2.79 \, g/32.00 \, g \, mol^{-1})(0.0821 \, L \cdot atm/K \cdot mol)(273 \, K)}{1 \, atm}$$

$$= 1.95 \, L$$

Similarly, for hydrogen we write

$$? \text{ g H}_2 = 3.37 \times 10^4 \text{ } \cancel{C} \times \frac{1 \text{ } \cancel{\text{mol } e^-}}{96,500 \text{ } \cancel{C}} \times \frac{1 \text{ } \cancel{\text{mol H}_2}}{2 \text{ } \cancel{\text{mol } e^-}} \times \frac{2.016 \text{ g H}_2}{1 \text{ } \cancel{\text{mol H}_2}} = 0.352 \text{ g H}_2$$

The volume of 0.352 g H_2 at STP is given by

$$V = \frac{nRT}{P}$$

$$= \frac{(0.352 \text{ g}/2.016 \text{ g mol}^{-1})(0.0821 \text{ L} \cdot \text{atm/K} \cdot \text{mol})(273 \text{ K})}{1 \text{ atm}}$$

$$= 3.91 \text{ L}$$

Comment Note that the volume of H_2 is twice that of O_2, which is what we would expect based on Avogadro's law (at the same temperature and pressure, volume is directly proportional to the number of moles of gases).

Practice Exercise   A constant current is passed through an electrolytic cell containing molten $MgCl_2$ for 18 hr. If 4.8×10^5 g of Cl_2 are obtained, what is the current in amperes?

Similar problem: 19.92.

19.9 Electrometallurgy

Electrolysis methods are useful for obtaining a pure metal from its ores or for refining (purifying) the metal. Collectively, these processes are called *electrometallurgy*. In the last section we saw how an active metal, sodium, can be obtained by electrolytically reducing its cation in the molten NaCl salt (p. 632). Here we will consider two other examples.

Production of Aluminum Metal

Aluminum is usually prepared from bauxite ore ($Al_2O_3 \cdot 2H_2O$). The ore is first treated to remove various impurities and then heated to obtain the anhydrous Al_2O_3. The oxide is dissolved in molten cryolite (Na_3AlF_6) in a Hall electrolytic cell (Figure 19.21). The cell contains a series of carbon anodes; the cathode is also made of carbon and constitutes the lining inside the cell. The solution is electrolyzed to produce aluminum and oxygen gas:

$$
\begin{array}{rl}
\text{Anode:} & 3[2O^{2-} \longrightarrow O_2(g) + 4e^-] \\
\text{Cathode:} & \underline{4[Al^{3+} + 3e^- \longrightarrow Al(l)]} \\
\text{Overall:} & 2Al_2O_3 \longrightarrow 4Al(l) + 3O_2(g)
\end{array}
$$

Oxygen gas reacts with the carbon anodes at 1000°C (the melting point of cryolite) to form carbon monoxide, which escapes as a gas. The liquid aluminum metal (m.p. 660°C) sinks to the bottom of the vessel, from which it can be drained.

Purification of Copper Metal

The copper metal obtained from its ores usually contains a number of impurities such as zinc, iron, silver, and gold. The more electropositive metals are removed by an

Carbon anodes

Carbon cathode

Al_2O_3 in molten cryolite

Molten aluminum

Figure 19.21 *Electrolytic production of aluminum based on the Hall process.*

electrolysis process in which the impure copper acts as the anode and *pure* copper acts as the cathode in a sulfuric acid solution containing Cu^{2+} ions (Figure 19.22). The half-reactions are

Anode: $Cu(s) \longrightarrow Cu^{2+}(aq) + 2e^-$

Cathode: $Cu^{2+}(aq) + 2e^- \longrightarrow Cu(s)$

Reactive metals in the copper anode, such as iron and zinc, are also oxidized at the anode and enter the solution as Fe^{2+} and Zn^{2+} ions. They are not reduced at the cathode, however. The less electropositive metals, such as gold and silver, are not oxidized at the anode. Eventually, as the copper anode dissolves, these metals fall to the bottom of the cell. Thus the net result of this electrolysis process is the transfer of copper from the anode to the cathode. Copper prepared this way has a purity greater than 99.5 percent. It is interesting to note that the metal impurities (mostly silver and gold) from the copper anode are valuable by-products, the sale of which often pays for the electricity used to drive the electrolysis.

Figure 19.22 *Electrolytic purification of copper.*

Summary

Redox reactions involve the transfer of electrons. Equations representing redox processes can be balanced using the ion-electron method.

In a galvanic cell, electricity is produced by a spontaneous chemical reaction. The oxidation at the anode and the reduction at the cathode take place separately, and the electrons flow through an external circuit. The two parts of a galvanic cell are the half-cells, and the reactions at the electrodes are the half-cell reactions. A salt bridge allows ions to flow between the half-cells.

The electromotive force (emf) of a cell is the voltage difference between the two electrodes. In the external circuit, electrons flow from the anode to the cathode in a galvanic cell. In solution, the anions move toward the anode and the cations move toward the cathode. The quantity of electricity carried by 1 mole of electrons is called a faraday, which is equal to 96,500 coulombs.

Standard reduction potentials show the relative likelihood of half-cell reduction reactions and can be used to predict the products, direction, and spontaneity of redox reactions between various substances. The decrease in free energy of the system in a spontaneous redox reaction is equal to the electrical work done by the system on the surroundings, or $\Delta G = -nFE$. The equilibrium constant for a redox reaction can be found from the standard emf of a cell.

The Nernst equation gives the relationship between the cell emf and the concentrations of the reactants and products under conditions other than the standard state.

Batteries, which consist of one or more galvanic cells, are used widely as self-contained power sources. Some of the better-known batteries are the dry cell, such as the Leclanché cell; the mercury battery; and the lead storage battery used in automobiles. Fuel cells produce electrical energy from a continuous supply of reactants.

The corrosion of metals, the most familiar example of which is the rusting of iron, is an electrochemical phenomenon.

In electrolysis, electric current from an external source is used to drive a nonspontaneous chemical reaction. The amount of product formed or reactant consumed in an electrolytic cell depends on the quantity of electricity transferred at the electrode. Electrolysis plays an important role in obtaining pure metals from their ores and in purifying metals.

Key Words

Questions and Problems

Balancing Redox Equations

Problems

19.1   Balance these redox equations by the ion-electron method:

   (a) $H_2O_2 + Fe^{2+} \longrightarrow Fe^{3+} + H_2O$ (in acidic solution)
   (b) $Cu + HNO_3 \longrightarrow Cu^{2+} + NO + H_2O$ (in acidic solution)
   (c) $CN^- + MnO_4^- \longrightarrow CNO^- + MnO_2$ (in basic solution)
   (d) $Br_2 \longrightarrow BrO_3^- + Br^-$ (in basic solution)
   (e) $S_2O_3^{2-} + I_2 \longrightarrow I^- + S_4O_6^{2-}$ (in acidic solution)

19.2   Balance these redox equations by the ion-electron method:

   (a) $Mn^{2+} + H_2O_2 \longrightarrow MnO_2 + H_2O$ (in basic solution)
   (b) $Bi(OH)_3 + SnO_2^{2-} \longrightarrow SnO_3^{2-} + Bi$ (in basic solution)
   (c) $Cr_2O_7^{2-} + C_2O_4^{2-} \longrightarrow Cr^{3+} + CO_2$ (in acidic solution)
   (d) $ClO_3^- + Cl^- \longrightarrow Cl_2 + ClO_2$ (in acidic solution)

Galvanic Cells and Standard Emfs

Review Questions

19.3   Define these terms: anode, cathode, electromotive force, standard oxidation potential, standard reduction potential.

19.4   Describe the basic features of a galvanic cell. Why are the two components of the cell separated from each other?

19.5   What is the function of a salt bridge? What kind of electrolyte should be used in a salt bridge?

19.6   What is a cell diagram? Write the cell diagram for a galvanic cell consisting of an Al electrode placed in a 1 *M* $Al(NO_3)_3$ solution and a Ag electrode placed in a 1 *M* $AgNO_3$ solution.

19.7   What is the difference between the half-reactions discussed in redox processes in Chapter 4 and the half-cell reactions discussed in Section 19.2?

19.8   After operating a Daniell cell (see Figure 19.1) for a few minutes, a student notices that the cell emf begins to drop. Why?

19.9   Use the information in Table 2.1, and calculate the Faraday constant.

19.10   Discuss the spontaneity of an electrochemical reaction in terms of its standard emf (E_{cell}°).

Problems

19.11   Calculate the standard emf of a cell that uses the Mg/Mg^{2+} and Cu/Cu^{2+} half-cell reactions at 25°C. Write the equation for the cell reaction that occurs under standard-state conditions.

19.12   Calculate the standard emf of a cell that uses Ag/Ag^+ and Al/Al^{3+} half-cell reactions? Write the cell reaction that occurs under standard-state conditions.

19.13   Predict whether Fe^{3+} can oxidize I^- to I_2 under standard-state conditions.

19.14   Which of these reagents can oxidize H_2O to $O_2(g)$ under standard-state conditions? $H^+(aq)$, $Cl^-(aq)$, $Cl_2(g)$, $Cu^{2+}(aq)$, $Pb^{2+}(aq)$, $MnO_4^-(aq)$ (in acid).

19.15   Consider these half-reactions:

$$MnO_4^-(aq) + 8H^+(aq) + 5e^- \longrightarrow Mn^{2+}(aq) + 4H_2O(l)$$
$$NO_3^-(aq) + 4H^+(aq) + 3e^- \longrightarrow NO(g) + 2H_2O(l)$$

Predict whether NO_3^- ions will oxidize Mn^{2+} to MnO_4^- under standard-state conditions.

19.16   Predict whether these reactions would occur spontaneously in aqueous solution at 25°C. Assume that the initial concentrations of dissolved species are all 1.0 *M*.

   (a) $Ca(s) + Cd^{2+}(aq) \longrightarrow Ca^{2+}(aq) + Cd(s)$
   (b) $2Br^-(aq) + Sn^{2+}(aq) \longrightarrow Br_2(l) + Sn(s)$
   (c) $2Ag(s) + Ni^{2+}(aq) \longrightarrow 2Ag^+(aq) + Ni(s)$
   (d) $Cu^+(aq) + Fe^{3+}(aq) \longrightarrow Cu^{2+}(aq) + Fe^{2+}(aq)$

19.17   Which species in each pair is a better oxidizing agent under standard-state conditions? (a) Br_2 or Au^{3+}, (b) H_2 or Ag^+, (c) Cd^{2+} or Cr^{3+}, (d) O_2 in acidic media or O_2 in basic media.

19.18 Which species in each pair is a better reducing agent under standard-state conditions? (a) Na or Li, (b) H_2 or I_2, (c) Fe^{2+} or Ag, (d) Br^- or Co^{2+}.

Spontaneity of Redox Reactions

Review Questions

19.19 Write the equations relating $\Delta G°$ and K to the standard emf of a cell. Define all the terms.

19.20 Compare the ease of measuring the equilibrium constant electrochemically with that by chemical means [see Equation (18.10)].

Problems

19.21 What is the equilibrium constant for the following reaction at 25°C?

$$Mg(s) + Zn^{2+}(aq) \rightleftharpoons Mg^{2+}(aq) + Zn(s)$$

19.22 The equilibrium constant for the reaction

$$Sr(s) + Mg^{2+}(aq) \rightleftharpoons Sr^{2+}(aq) + Mg(s)$$

is 2.69×10^{12} at 25°C. Calculate $E°$ for a cell made up of Sr/Sr^{2+} and Mg/Mg^{2+} half-cells.

19.23 Use the standard reduction potentials to find the equilibrium constant for each of these reactions at 25°C:

(a) $Br_2(l) + 2I^-(aq) \rightleftharpoons 2Br^-(aq) + I_2(s)$
(b) $2Ce^{4+}(aq) + 2Cl^-(aq) \rightleftharpoons Cl_2(g) + 2Ce^{3+}(aq)$
(c) $5Fe^{2+}(aq) + MnO_4^-(aq) + 8H^+(aq) \rightleftharpoons$
$\qquad Mn^{2+}(aq) + 4H_2O + 5Fe^{3+}(aq)$

19.24 Calculate $\Delta G°$ and K_c for these reactions at 25°C:

(a) $Mg(s) + Pb^{2+}(aq) \rightleftharpoons Mg^{2+}(aq) + Pb(s)$
(b) $Br_2(l) + 2I^-(aq) \rightleftharpoons 2Br^-(aq) + I_2(s)$
(c) $O_2(g) + 4H^+(aq) + 4Fe^{2+}(aq) \rightleftharpoons$
$\qquad 2H_2O(l) + 4Fe^{3+}(aq)$
(d) $2Al(s) + 3I_2(s) \rightleftharpoons 2Al^{3+}(aq) + 6I^-(aq)$

19.25 Under standard-state conditions, what spontaneous reaction will occur in aqueous solution among the ions Ce^{4+}, Ce^{3+}, Fe^{3+}, and Fe^{2+}? Calculate $\Delta G°$ and K_c for the reaction.

19.26 Given that $E° = 0.52$ V for the reduction $Cu^+(aq) + e^- \rightarrow Cu(s)$, calculate $E°$, $\Delta G°$, and K for this reaction at 25°C:

$$2Cu^+(aq) \longrightarrow Cu^{2+}(aq) + Cu(s)$$

The Effect of Concentration on Cell Emf

Review Questions

19.27 Write the Nernst equation and explain all the terms.

19.28 Write the Nernst equation for these processes at some temperature T:

(a) $Mg(s) + Sn^{2+}(aq) \longrightarrow Mg^{2+}(aq) + Sn(s)$
(b) $2Cr(s) + 3Pb^{2+}(aq) \longrightarrow 2Cr^{3+}(aq) + 3Pb(s)$

Problems

19.29 What is the potential of a cell made up of Zn/Zn^{2+} and Cu/Cu^{2+} half-cells at 25°C if $[Zn^{2+}] = 0.25\ M$ and $[Cu^{2+}] = 0.15\ M$?

19.30 Calculate $E°$, E, and ΔG for these cell reactions:

(a) $Mg(s) + Sn^{2+}(aq) \longrightarrow Mg^{2+}(aq) + Sn(s)$
$\quad [Mg^{2+}] = 0.045\ M,\ [Sn^{2+}] = 0.035\ M$
(b) $3Zn(s) + 2Cr^{3+}(aq) \longrightarrow 3Zn^{2+}(aq) + 2Cr(s)$
$\quad [Cr^{3+}] = 0.010\ M,\ [Zn^{2+}] = 0.0085\ M$

19.31 Calculate the standard potential of the cell consisting of the Zn/Zn^{2+} half-cell and the SHE. What will the emf of the cell be if $[Zn^{2+}] = 0.45\ M$, $P_{H_2} = 2.0$ atm, and $[H^+] = 1.8\ M$?

19.32 What is the emf of a cell consisting of a Pb^{2+}/Pb half-cell and a $Pt/H^+/H_2$ half-cell if $[Pb^{2+}] = 0.10\ M$, $[H^+] = 0.050\ M$, and $P_{H_2} = 1.0$ atm?

19.33 Referring to the arrangement in Figure 19.1, calculate the $[Cu^{2+}]/[Zn^{2+}]$ ratio at which this reaction is spontaneous at 25°C:

$$Cu(s) + Zn^{2+}(aq) \longrightarrow Cu^{2+}(aq) + Zn(s)$$

19.34 Calculate the emf of this concentration cell:

$$Mg(s)\ |\ Mg^{2+}(0.24\ M)\ \|\ Mg^{2+}(0.53\ M)\ |\ Mg(s)$$

Batteries and Fuel Cells

Review Questions

19.35 Explain the differences between a primary galvanic cell—one that is not rechargeable—and a storage cell (for example, the lead storage battery), which is rechargeable.

19.36 Discuss the advantages and disadvantages of fuel cells over conventional power plants in producing electricity.

Problems

19.37 The hydrogen-oxygen fuel cell is described in Section 19.6. (a) What volume of $H_2(g)$, stored at 25°C at a pressure of 155 atm, would be needed to run an electric motor drawing a current of 8.5 A for 3.0 hr? (b) What volume (liters) of air at 25°C and 1.00 atm will have to pass into the cell per minute to run the motor? Assume that air is 20 percent O_2 by volume and that all the O_2 is consumed in the cell. The other components of air do not affect the fuel-cell reactions. Assume ideal gas behavior.

19.38 Calculate the standard emf of the propane fuel cell discussed on p. 628 at 25°C, given that $\Delta G_f°$ for propane is -23.5 kJ/mol.

Corrosion

Review Questions

19.39 Steel hardware, including nuts and bolts, is often coated with a thin plating of cadmium. Explain the function of the cadmium layer.

19.40 "Galvanized iron" is steel sheet that has been coated with zinc; "tin" cans are made of steel sheet coated with tin. Discuss the functions of these coatings and the electrochemistry of the corrosion reactions that occur if an electrolyte contacts the scratched surface of a galvanized iron sheet or a tin can.

19.41 Tarnished silver contains Ag_2S. The tarnish can be removed by placing silverware in an aluminum pan containing an inert electrolyte solution, such as NaCl. Explain the electrochemical principle for this procedure. [The standard reduction potential for the half-cell reaction $Ag_2S(s) + 2e^- \longrightarrow 2Ag(s) + S^{2-}(aq)$ is -0.71 V.]

19.42 How does the tendency of iron to rust depend on the pH of solution?

Electrolysis

Review Questions

19.43 What is the difference between a galvanic cell (such as a Daniell cell) and an electrolytic cell?

19.44 What is Faraday's contribution to quantitative electrolysis?

Problems

19.45 The half-reaction at an electrode is

$$Mg^{2+}(molten) + 2e^- \longrightarrow Mg(s)$$

Calculate the number of grams of magnesium that can be produced by supplying 1.00 F to the electrode.

19.46 Consider the electrolysis of molten barium chloride, $BaCl_2$. (a) Write the half-reactions. (b) How many grams of barium metal can be produced by supplying 0.50 A for 30 min?

19.47 Considering only the cost of electricity, would it be cheaper to produce a ton of sodium or a ton of aluminum by electrolysis?

19.48 If the cost of electricity to produce magnesium by the electrolysis of molten magnesium chloride is $155 per ton of metal, what is the cost (in dollars) of the electricity necessary to produce (a) 10.0 tons of aluminum, (b) 30.0 tons of sodium, (c) 50.0 tons of calcium?

19.49 One of the half-reactions for the electrolysis of water is

$$2H_2O(l) \longrightarrow O_2(g) + 4H^+(aq) + 4e^-$$

If 0.076 L of O_2 is collected at 25°C and 755 mmHg, how many faradays of electricity had to pass through the solution?

19.50 How many faradays of electricity are required to produce (a) 0.84 L of O_2 at exactly 1 atm and 25°C from aqueous H_2SO_4 solution; (b) 1.50 L of Cl_2 at 750 mmHg and 20°C from molten NaCl; (c) 6.0 g of Sn from molten $SnCl_2$?

19.51 Calculate the amounts of Cu and Br_2 produced in 1.0 hr at inert electrodes in a solution of $CuBr_2$ by a current of 4.50 A.

19.52 In the electrolysis of an aqueous $AgNO_3$ solution, 0.67 g of Ag is deposited after a certain period of time. (a) Write the half-reaction for the reduction of Ag^+. (b) What is the probable oxidation half-reaction? (c) Calculate the quantity of electricity used, in coulombs.

19.53 A steady current was passed through molten $CoSO_4$ until 2.35 g of metallic cobalt were produced. Calculate the number of coulombs of electricity used.

19.54 A constant electric current flows for 3.75 hr through two electrolytic cells connected in series. One contains a solution of $AgNO_3$ and the second a solution of $CuCl_2$. During this time 2.00 g of silver are deposited in the first cell. (a) How many grams of copper are deposited in the second cell? (b) What is the current flowing, in amperes?

19.55 What is the hourly production rate of chlorine gas (in kg) from an electrolytic cell using aqueous NaCl electrolyte and carrying a current of 1.500×10^3 A? The anode efficiency for the oxidation of Cl^- is 93.0 percent.

19.56 Chromium plating is applied by electrolysis to objects suspended in a dichromate solution, according to the following (unbalanced) half-reaction:

$$Cr_2O_7^{2-}(aq) + e^- + H^+(aq) \longrightarrow Cr(s) + H_2O(l)$$

How long (in hours) would it take to apply a chromium plating 1.0×10^{-2} mm thick to a car bumper with a surface area 0.25 m^2 in an electrolytic cell carrying a current of 25.0 A? (The density of chromium is 7.19 g/cm^3.)

19.57 The passage of a current of 0.750 A for 25.0 min deposited 0.369 g of copper from a $CuSO_4$ solution. From this information, calculate the molar mass of copper.

19.58 A quantity of 0.300 g of copper was deposited from a $CuSO_4$ solution by passing a current of 3.00 A through the solution for 304 s. Calculate the value of the faraday constant.

19.59 In a certain electrolysis experiment, 1.44 g of Ag were deposited in one cell (containing an aqueous $AgNO_3$ solution), while 0.120 g of an unknown metal X was deposited in another cell (containing an aqueous XCl_3 solution) in series with the $AgNO_3$ cell. Calculate the molar mass of X.

19.60 One of the half-reactions for the electrolysis of water is

$$2H^+(aq) + 2e^- \longrightarrow H_2(g)$$

If 0.845 L of H_2 is collected at 25°C and 782 mmHg, how many faradays of electricity had to pass through the solution?

Additional Problems

19.61 For each of these redox reactions, (i) write the half-reactions; (ii) write a balanced equation for the whole reaction, (iii) determine in which direction the reaction will proceed spontaneously under standard-state conditions:

(a) $H_2(g) + Ni^{2+}(aq) \longrightarrow H^+(aq) + Ni(s)$
(b) $MnO_4^-(aq) + Cl^-(aq) \longrightarrow$
$$Mn^{2+}(aq) + Cl_2(g) \text{ (in acid solution)}$$
(c) $Cr(s) + Zn^{2+}(aq) \longrightarrow Cr^{3+}(aq) + Zn(s)$

19.62 The oxidation of 25.0 mL of a solution containing Fe^{2+} requires 26.0 mL of 0.0250 M $K_2Cr_2O_7$ in acidic solution. Balance this equation and calculate the molar concentration of Fe^{2+}:

$$Cr_2O_7^{2-} + Fe^{2+} + H^+ \longrightarrow Cr^{3+} + Fe^{3+}$$

19.63 The SO_2 present in air is mainly responsible for the phenomenon of acid rain. Its concentration of SO_2 can be determined by titrating against a standard permanganate solution as follows:

$$5SO_2 + 2MnO_4^- + 2H_2O \longrightarrow$$
$$5SO_4^{2-} + 2Mn^{2+} + 4H^+$$

Calculate the number of grams of SO_2 in a sample of air if 7.37 mL of 0.00800 M $KMnO_4$ solution are required for the titration.

19.64 A sample of iron ore weighing 0.2792 g was dissolved in an excess of a dilute acid solution. All the iron was first converted to Fe(II) ions. The solution then required 23.30 mL of 0.0194 M $KMnO_4$ for oxidation to Fe(III) ions. Calculate the percent by mass of iron in the ore.

19.65 The concentration of a hydrogen peroxide solution can be conveniently determined by titration against a standardized potassium permanganate solution in an acidic medium according to the following unbalanced equation:

$$MnO_4^- + H_2O_2 \longrightarrow O_2 + Mn^{2+}$$

(a) Balance the above equation. (b) If 36.44 mL of a 0.01652 M $KMnO_4$ solution are required to completely oxidize 25.00 mL of a H_2O_2 solution, calculate the molarity of the H_2O_2 solution.

19.66 Oxalic acid ($H_2C_2O_4$) is present in many plants and vegetables. (a) Balance the following equation in acid solution:

$$MnO_4^- + C_2O_4^{2-} \longrightarrow Mn^{2+} + CO_2$$

(b) If a 1.00-g impure sample of $H_2C_2O_4$ requires 24.0 mL of 0.0100 M $KMnO_4$ solution for complete oxidation, what is the percent by mass of $H_2C_2O_4$ in the sample?

19.67 Complete the following table. State whether the cell reaction is spontaneous, nonspontaneous, or at equilibrium.

E	ΔG	Cell Reaction
> 0		
	> 0	
$= 0$		

19.68 Calcium oxalate (CaC_2O_4) is insoluble in water. This property has been used to determine the amount of Ca^{2+} ions in blood. The calcium oxalate isolated from blood is dissolved in acid and titrated against a standardized $KMnO_4$ solution as described in Problem 19.66. In one test it is found that the calcium oxalate isolated from a 10.0-mL sample of blood requires 24.2 mL of 9.56×10^{-4} M $KMnO_4$ for titration. Calculate the number of milligrams of calcium per milliliter of blood.

19.69 From the following information, calculate the solubility product of AgBr:

$$Ag^+(aq) + e^- \longrightarrow Ag(s) \qquad E° = 0.80 \text{ V}$$
$$AgBr(s) + e^- \longrightarrow Ag(s) + Br^-(aq) \quad E° = 0.07 \text{ V}$$

19.70 Consider a galvanic cell composed of the SHE and a half-cell using the reaction $Ag^+(aq) + e^- \to Ag(s)$. (a) Calculate the standard emf of the cell. (b) What is the spontaneous cell reaction under standard-state conditions? (c) Calculate the emf of the cell when $[H^+]$ in the hydrogen electrode is changed to (i) 1.0×10^{-2} M and (ii) 1.0×10^{-5} M, all other reagents being held at standard-state conditions. (d) Based on this cell arrangement, suggest a design for a pH meter.

19.71 A galvanic cell consists of a silver electrode in contact with 346 mL of 0.100 M $AgNO_3$ solution and a magnesium electrode in contact with 288 mL of 0.100 M $Mg(NO_3)_2$ solution. (a) Calculate E for the cell at 25°C. (b) A current is drawn from the cell until 1.20 g of silver have been deposited at the silver electrode. Calculate E for the cell at this stage of operation.

19.72 Explain why chlorine gas can be prepared by electrolyzing an aqueous solution of NaCl but fluorine gas cannot be prepared by electrolyzing an aqueous solution of NaF.

19.73 Calculate the emf of the following concentration cell at 25°C:

$$Cu(s) \mid Cu^{2+}(0.080 \text{ } M) \parallel Cu^{2+}(1.2 \text{ } M) \mid Cu(s)$$

19.74 The cathode reaction in the Leclanché cell is given by

$$2MnO_2(s) + Zn^{2+}(aq) + 2e^- \longrightarrow ZnMn_2O_4(s)$$

If a Leclanché cell produces a current of 0.0050 A, calculate how many hours this current supply will last if there are initially 4.0 g of MnO_2 present in the cell. Assume that there is an excess of Zn^{2+} ions.

19.75 Suppose you are asked to verify experimentally the electrode reactions shown in Example 19.8. In addition to the apparatus and the solution, you are also given two pieces of litmus paper, one blue and the other red. Describe what steps you would take in this experiment.

19.76 For a number of years it was not clear whether mercury(I) ions existed in solution as Hg^+ or as Hg_2^{2+}. To distinguish between these two possibilities, we could set up this system:

$$Hg(l) \mid \text{soln A} \parallel \text{soln B} \mid Hg(l)$$

where soln A contained 0.263 g mercury(I) nitrate per liter and soln B contained 2.63 g mercury(I) nitrate per liter. If the measured emf of such a cell is 0.0289 V at 18°C, what can you deduce about the nature of the mercury(I) ions?

19.77 An aqueous KI solution to which a few drops of phenolphthalein have been added is electrolyzed using an apparatus like the one shown below:

Describe what you would observe at the anode and the cathode. (*Hint:* Molecular iodine is only slightly soluble in water, but in the presence of I^- ions, it forms the brown color of I_3^- ions.)

19.78 A piece of magnesium metal weighing 1.56 g is placed in 100.0 mL of 0.100 M $AgNO_3$ at 25°C. Calculate $[Mg^{2+}]$ and $[Ag^+]$ in solution at equilibrium. What is the mass of the magnesium left? The volume remains constant.

19.79 Describe an experiment that would enable you to determine which is the cathode and which is the anode in a galvanic cell using copper and zinc electrodes.

19.80 An acidified solution was electrolyzed using copper electrodes. A constant current of 1.18 A caused the anode to lose 0.584 g after 1.52×10^3 s. (a) What is the gas produced at the cathode and what is its volume at STP? (b) Given that the charge of an electron is 1.6022×10^{-19} C, calculate Avogadro's number. Assume that copper is oxidized to Cu^{2+} ions.

19.81 In a certain electrolysis experiment involving Al^{3+} ions, 60.2 g of Al are recovered when a current of 0.352 A is used. How many minutes did the electrolysis last?

19.82 Consider the oxidation of ammonia:

$$4NH_3(g) + 3O_2(g) \longrightarrow 2N_2(g) + 6H_2O(l)$$

(a) Calculate the $\Delta G°$ for the reaction. (b) If this reaction were used in a fuel cell, what would the standard cell potential be?

19.83 A galvanic cell is constructed by immersing a piece of copper wire in 25.0 mL of a 0.20 M $CuSO_4$ solution and a zinc strip in 25.0 mL of a 0.20 M $ZnSO_4$ solution. (a) Calculate the emf of the cell at 25°C and predict what would happen if a small amount of concentrated NH_3 solution were added to (i) the $CuSO_4$ solution and (ii) the

$ZnSO_4$ solution. Assume that the volume in each compartment remains constant at 25.0 mL. (b) In a separate experiment, 25.0 mL of 3.00 M NH_3 are added to the $CuSO_4$ solution. If the emf of the cell is 0.68 V, calculate the formation constant (K_f) of $Cu(NH_3)_4^{2+}$.

19.84 In an electrolysis experiment, a student passes the same quantity of electricity through two electrolytic cells, one containing a silver salt and the other a gold salt. Over a certain period of time, she finds that 2.64 g of Ag and 1.61 g of Au are deposited at the cathodes. What is the oxidation state of gold in the gold salt?

19.85 People living in cold-climate countries where there is plenty of snow are advised not to heat their garages in the winter. What is the electrochemical basis for this recommendation?

19.86 Given that

$$2Hg^{2+}(aq) + 2e^- \longrightarrow Hg_2^{2+}(aq) \qquad E° = 0.92 \text{ V}$$
$$Hg_2^{2+}(aq) + 2e^- \longrightarrow 2Hg(l) \qquad E° = 0.85 \text{ V}$$

calculate $\Delta G°$ and K for this process at 25°C:

$$Hg_2^{2+}(aq) \longrightarrow Hg^{2+}(aq) + Hg(l)$$

(This reaction is an example of a *disproportionation reaction* in which an element in one oxidation state is both oxidized and reduced.)

19.87 Fluorine (F_2) is obtained by the electrolysis of liquid hydrogen fluoride (HF) containing potassium fluoride (KF). (a) Write the half-cell reactions and the overall reaction for the process. (b) What is the purpose of KF? (c) Calculate the volume of F_2 (in liters) collected at 24.0°C and 1.2 atm after electrolyzing the solution for 15 hr at a current of 502 A.

19.88 A 300-mL solution of NaCl was electrolyzed for 6.00 min. If the pH of the final solution was 12.24, calculate the average current used.

19.89 Industrially, copper is purified by electrolysis. The impure copper acts as the anode, and the cathode is made of pure copper. The electrodes are immersed in a $CuSO_4$ solution. During electrolysis, copper at the anode enters the solution as Cu^{2+} while Cu^{2+} ions are reduced at the cathode. (a) Write half-cell reactions and the overall reaction for the electrolytic process. (b) Suppose the anode was contaminated with Zn and Ag. Explain what happens to these impurities during electrolysis. (c) How many hours will it take to obtain 1.00 kg of Cu at a current of 18.9 A?

19.90 An aqueous solution of a platinum salt is electrolyzed at a current of 2.50 A for 2.00 hr. As a result, 9.09 g of metallic Pt are formed at the cathode. Calculate the charge on the Pt ions in this solution.

19.91 Consider a galvanic cell consisting of a magnesium electrode in contact with 1.0 M $Mg(NO_3)_2$ and a cadmium electrode in contact with 1.0 M $Cd(NO_3)_2$. Calculate $E°$ for the cell, and draw a diagram showing the cathode, anode, and direction of electron flow.

19.92 A current of 6.00 A passes through an electrolytic cell containing dilute sulfuric acid for 3.40 hr. If the volume of O_2 gas generated at the anode is 4.26 L (at STP), calculate the charge (in coulombs) on an electron.

19.93 Gold will not dissolve in either concentrated nitric acid or concentrated hydrochloric acid. However, the metal does dissolve in a mixture of the acids (one part HNO_3 and three parts HCl by volume), called *aqua regia*. (a) Write a balanced equation for this reaction. (*Hint:* Among the products are $HAuCl_4$ and NO_2.) (b) What is the function of HCl?

19.94 Explain why most useful galvanic cells give voltages of no more than 1.5 to 2.5 V. What are the prospects for developing practical galvanic cells with voltages of 5 V or more?

19.95 A silver rod and a SHE are dipped into a saturated aqueous solution of silver oxalate, $Ag_2C_2O_4$, at 25°C. The measured potential difference between the rod and the SHE is 0.589 V, the rod being positive. Calculate the solubility product constant for silver oxalate.

19.96 Zinc is an amphoteric metal; that is, it reacts with both acids and bases. The standard reduction potential is -1.36 V for the reaction

$$Zn(OH)_4^{2-}(aq) + 2e^- \longrightarrow Zn(s) + 4OH^-(aq)$$

Calculate the formation constant (K_f) for the reaction

$$Zn^{2+}(aq) + 4OH^-(aq) \rightleftharpoons Zn(OH)_4^{2-}(aq)$$

19.97 Use the data in Table 19.1 to determine whether or not hydrogen peroxide will undergo disproportionation in an acid medium: $2H_2O_2 \rightarrow 2H_2O + O_2$.

19.98 The magnitudes (but *not* the signs) of the standard reduction potentials of two metals X and Y are

$$\begin{array}{ll} Y^{2+} + 2e^- \longrightarrow Y & |E°| = 0.34 \text{ V} \\ X^{2+} + 2e^- \longrightarrow X & |E°| = 0.25 \text{ V} \end{array}$$

where the $|\ \ |$ notation denotes that only the magnitude (but not the sign) of the $E°$ value is shown. When the half-cells of X and Y are connected, electrons flow from X to Y. When X is connected to a SHE, electrons flow from X to SHE. (a) Are the $E°$ values of the half-reactions positive or negative? (b) What is the standard emf of a cell made up of X and Y?

19.99 A galvanic cell is constructed as follows. One half-cell consists of a platinum wire immersed in a solution containing 1.0 M Sn^{2+} and 1.0 M Sn^{4+}; the other half-cell has a thallium rod immersed in a solution of 1.0 M Tl^+. (a) Write the half-cell reactions and the overall reaction. (b) What is the equilibrium constant at 25°C? (c) What is the cell voltage if the Tl^+ concentration is increased tenfold? ($E°_{Tl^+/Tl} = -0.34$ V.)

19.100 Given the standard reduction potential for Au^{3+} in Table 19.1 and

$$Au^+(aq) + e^- \longrightarrow Au(s) \qquad E° = 1.69 \text{ V}$$

answer these questions. (a) Why does gold not tarnish in air? (b) Will this disproportionation occur spontaneously?

$$3Au^+(aq) \longrightarrow Au^{3+}(aq) + 2Au(s)$$

(c) Predict the reaction between gold and fluorine gas.

19.101 The ingestion of a very small quantity of mercury is not considered too harmful. Would this statement still hold if the gastric juice in your stomach were mostly nitric acid instead of hydrochloric acid?

19.102 When 25.0 mL of a solution containing both Fe^{2+} and Fe^{3+} ions is titrated with 23.0 mL of 0.0200 M $KMnO_4$ (in dilute sulfuric acid), all of the Fe^{2+} ions are oxidized to Fe^{3+} ions. Next, the solution is treated with Zn metal to convert all of the Fe^{3+} ions to Fe^{2+} ions. Finally, 40.0 mL of the same $KMnO_4$ solution are added to the solution in order to oxidize all the Fe^{2+} ions back to Fe^{3+}. Calculate the molar concentrations of Fe^{2+} and Fe^{3+} in the original solution.

19.103 Consider the Daniell cell in Figure 19.1. When viewed externally, the anode appears negative and the cathode positive (electrons are flowing from the anode to the cathode). Yet in solution anions are moving toward the anode, which means that it must appear positive to the anions. Because the anode cannot simultaneously be negative and positive, give an explanation for this apparently contradictory situation.

19.104 Lead storage batteries are rated by ampere hours, that is, the number of amperes they can deliver in an hour. (a) Show that 1 A · hr = 3600 C. (b) The lead anodes of a certain lead-storage battery have a total mass of 406 g. Calculate the maximum theoretical capacity of the battery in ampere hours. Explain why in practice we can never extract this much energy from the battery. (*Hint:* Assume all of the lead will be used up in the electrochemical reaction and refer to the electrode reactions on p. 625.) (c) Calculate $E°_{cell}$ and $\Delta G°$ for the battery.

19.105 The concentration of sulfuric acid in the lead-storage battery of an automobile over a period of time has decreased from 38.0 percent by mass (density = 1.29 g/mL) to 26.0 percent by mass (1.19 g/mL). Assume the volume of the acid remains constant at 724 mL. (a) Calculate the total charge in coulombs supplied by the battery. (b) How long (in hours) will it take to recharge the battery back to the original sulfuric acid concentration using a current of 22.4 amperes.

19.106 Consider a Daniell cell operating under non-standard-state conditions. Suppose that the cell's reaction is multiplied by 2. What effect does this have on each of these quantities in the Nernst equation: (a) E, (b) $E°$, (c) Q, (d) $\ln Q$, and (e) n?

19.107 A spoon was silver-plated electrolytically in a $AgNO_3$ solution. (a) Sketch a diagram for the process. (b) If 0.884 g of Ag was deposited on the spoon at a constant current of 18.5 mA, how long (in minutes) did the electrolysis take?

19.108 Comment on whether F_2 will become a stronger oxidizing agent in solution with increasing H^+ concentration.

19.109 In recent years there has been much interest in electric cars. List some advantages and disadvantages of electric cars compared to automobiles with internal combustion engines.

19.110 Calculate the pressure of H_2 (in atm) required to maintain equilibrium with respect to the following reaction at 25°C:

$$Pb(s) + 2H^+(aq) \rightleftharpoons Pb^{2+}(aq) + H_2(g)$$

Given that $[Pb^{2+}] = 0.035\ M$ and the solution is buffered at pH 1.60.

19.111 Calculate $E°$ for the reactions of mercury with (a) 1 M HCl and (b) 1 M HNO_3. Which acid will oxidize Hg to Hg_2^{2+} under standard-state conditions? Can you identify which test tube below contains HNO_3 and Hg and which contains HCl and Hg?

19.112 Because all alkali metals react with water, it is not possible to measure the standard reduction potentials of these metals directly as in the case of, say, zinc. An indirect method is to consider this hypothetical reaction

$$Li^+(aq) + \tfrac{1}{2}H_2(g) \longrightarrow Li(s) + H^+(aq)$$

Use the appropriate equation presented in this chapter and the thermodynamic data in Appendix 2, calculate $E°$ for $Li^+(aq) + e^- \rightarrow Li(s)$ at 298 K. Compare your result with that listed in Table 19.1. (See back endpaper for the Faraday constant.)

Special Problems

19.113 A piece of magnesium ribbon and a copper wire are partially immersed in a 0.1 M HCl solution in a beaker. The metals are joined externally by another piece of metal wire. Bubbles are seen to evolve at both the Mg and Cu surfaces. (a) Write equations representing the reactions occurring at the metals. (b) What visual evidence would you seek to show that Cu is not oxidized to Cu^{2+}? (c) At some stage, NaOH solution is added to the beaker to neutralize the HCl acid. Upon further addition of NaOH, a white precipitate forms. What is it?

19.114 The zinc-air battery shows much promise for electric cars because it is lightweight and rechargeable:

The net transformation is $Zn(s) + \tfrac{1}{2}O_2(g) \rightarrow ZnO(s)$. (a) Write the half-reactions at the zinc-air electrodes and calculate the standard emf of the battery at 25°C. (b) Calculate the emf under actual operating conditions when the partial pressure of oxygen is 0.21 atm. (c) What is the energy density (measured as the energy in kilojoules that can be obtained from 1 kg of the metal) of the zinc electrode? (d) If a current of 2.1×10^5 A is to be drawn from a zinc-air battery system, what volume of air (in liters) would need to be supplied to the battery every second? Assume that the temperature is 25°C and the partial pressure of oxygen is 0.21 atm.

Answers to Practice Exercises

19.1 $5Fe^{2+} + MnO_4^- + 8H^+ \rightarrow 5Fe^{3+} + Mn^{2+} + 4H_2O$. **19.2** No. **19.3** 0.34 V. **19.4** 1.1×10^{-42}. **19.5** $\Delta G^\circ = -4.1 \times 10^2$ kJ/mol. **19.6** Yes. **19.7** 0.38 V. **19.8** Anode, O_2; cathode, H_2. **19.9** 2.0×10^4 A.

Luck often plays a role in major scientific breakthroughs, but it takes an alert and well-trained person to recognize the significance of an accidental discovery and to take full advantage of it. Such was the case when in 1964 the biophysicist Barnett Rosenberg and his research group at Michigan State University were studying the effect of an electric field on the growth of

Cisplatin disrupts DNA replication and transcription by binding to the double helix. The structure of this major DNA adduct, depicted here, was elucidated by Professor Stephen Lippard's group at M.I.T.

cis-Pt(NH₃)₂Cl₂

bacteria. They suspended a bacterial culture between two platinum electrodes and passed an electric current through it. To their surprise, they found that after an hour or so the bacteria cells ceased dividing. It did not take long for the group to determine that a platinum-containing substance that was extracted from the bacterial culture inhibited cell division.

Rosenberg reasoned that the platinum compound might be useful as an anticancer agent, because cancer involves uncontrolled division of cancerous cells, so he set out to identify the substance. Given the presence of ammonia and chloride ions in solution during electrolysis, Rosenberg synthesized a number of platinum compounds containing ammonia and chlorine. The one that proved most effective at inhibiting cell division was *cis*-diamminedichloroplatinum(II) [Pt(NH$_3$)$_2$Cl$_2$], also called *cisplatin.* Cisplatin is one of a group of transition metal compounds called *coordination compounds,* which have some unique properties.

Interestingly, cisplatin has been known since 1845, but its ability to block cell division was not established until 1967. Today it is one of the most widely used anticancer drugs. Presumably, cisplatin binds to DNA in a specific manner that leads to a mistake (or mutation) in the DNA's replication and the eventual destruction of the cancerous cell.

Cisplatin, a bright yellow compound, is administered intravenously to cancer patients.

The Chemistry of Coordination Compounds

Chapter 20

ESSENTIAL CONCEPTS

Coordination Compounds A coordination compound contains one or more complex ions in which a small number of molecules or ions surround a central metal atom or ion, usually of the transition metal family. Common geometries of coordination compounds are linear, tetrahedral, square planar, and octahedral.

Bonding in Coordination Compounds Crystal field theory explains the bonding in a complex ion in terms of electrostatic forces. The approach of the ligands toward the metal causes a splitting in energy in the five *d* orbitals. The extent of the splitting, called crystal-field splitting, depends on the nature of the ligands. Crystal field theory successfully accounts for the color and magnetic properties of many complex ions.

20.1 Properties of the Transition Metals

Transition metals characteristically have incompletely filled d subshells or readily give rise to ions with incompletely filled d subshells (Figure 20.1). This attribute is responsible for several notable properties, including distinctive coloring, formation of paramagnetic compounds, catalytic activity, and especially a great tendency to form complex ions. In this chapter we focus on the first-row elements from scandium to copper, the most common transition metals (Figure 20.2).

As we read across any period from left to right, atomic numbers increase, electrons are added to the outer shell, and the nuclear charge increases by the addition of protons. In the third-period elements—sodium to argon—the outer electrons weakly shield one another from the extra nuclear charge. Consequently, atomic radii decrease rapidly from sodium to argon and the electronegativities and ionization energies increase steadily (see Figures 8.5, 8.9, and 9.5).

For the transition metals, the trends are different. In Table 20.1 on p. 652, we see that reading across from scandium to copper the nuclear charge, of course, increases, but electrons are being added to the inner $3d$ subshell. These $3d$ electrons shield the $4s$ electrons from the increasing nuclear charge somewhat more effectively than outer-shell electrons can shield one another, so the atomic radii decrease less rapidly. For the same reason, electronegativities and ionization energies increase only slightly from scandium across to copper compared with the increases from sodium to argon in the third period.

Most transition metals are inert toward acids or react slowly because of a protective layer of oxide. A case in point is chromium: it is quite inert chemically because of the formation on its surface of chromium(III) oxide, Cr_2O_3. Consequently, chromium is commonly used as a protective and noncorrosive plating on other metals. On automobile bumpers and trim, chromium plating serves a decorative and a functional purpose.

1 1A																	18 8A
1 H	2 2A											13 3A	14 4A	15 5A	16 6A	17 7A	2 He
3 Li	4 Be											5 B	6 C	7 N	8 O	9 F	10 Ne
11 Na	12 Mg	3 3B	4 4B	5 5B	6 6B	7 7B	8	9 8B	10	11 1B	12 2B	13 Al	14 Si	15 P	16 S	17 Cl	18 Ar
19 K	20 Ca	21 Sc	22 Ti	23 V	24 Cr	25 Mn	26 Fe	27 Co	28 Ni	29 Cu	30 Zn	31 Ga	32 Ge	33 As	34 Se	35 Br	36 Kr
37 Rb	38 Sr	39 Y	40 Zr	41 Nb	42 Mo	43 Tc	44 Ru	45 Rh	46 Pd	47 Ag	48 Cd	49 In	50 Sn	51 Sb	52 Te	53 I	54 Xe
55 Cs	56 Ba	57 La	72 Hf	73 Ta	74 W	75 Re	76 Os	77 Ir	78 Pt	79 Au	80 Hg	81 Tl	82 Pb	83 Bi	84 Po	85 At	86 Rn
87 Fr	88 Ra	89 Ac	104 Rf	105 Db	106 Sg	107 Bh	108 Hs	109 Mt	110	111	112	(113)	114	(115)	(116)	(117)	(118)

Figure 20.1 *The transition metals (blue squares). Note that although the Group 2B elements (Zn, Cd, Hg) are described as transition metals by some chemists, neither the metals nor their ions possess incompletely filled d subshells.*

Scandium (Sc)

Titanium (Ti)

Vanadium (V)

Chromium (Cr)

Manganese (Mn)

Iron (Fe)

Cobalt (Co)

Nickel (Ni)

Copper (Cu)

Figure 20.2 *The first-row transition metals.*

Electron Configurations

The electron configurations of the first-row transition metals were discussed in Section 7.9. Calcium has the electron configuration [Ar]$4s^2$. From scandium across to copper, electrons are added to the $3d$ orbitals. Thus, the outer electron configuration of scandium is $4s^2 3d^1$, that of titanium is $4s^2 3d^2$, and so on. The two exceptions are chromium and copper, whose outer electron configurations are $4s^1 3d^5$ and $4s^1 3d^{10}$, respectively. These irregularities are the result of the extra stability associated with half-filled and completely filled $3d$ subshells.

When the first-row transition metals form cations, electrons are removed first from the $4s$ orbitals and then from the $3d$ orbitals. (This is the opposite of the order in which orbitals are filled in neutral atoms.) For example, the outer electron configuration of Fe^{2+} is $3d^6$, not $4s^2 3d^4$.

Oxidation States

As noted in Chapter 4, the transition metals exhibit variable oxidation states in their compounds. Figure 20.3 shows the oxidation states from scandium to copper. Note

Table 20.1 | Electron Configurations and Other Properties of the First-Row Transition Metals

	Sc	Ti	V	Cr	Mn	Fe	Co	Ni	Cu
Electron configuration									
M	$4s^2 3d^1$	$4s^2 3d^2$	$4s^2 3d^3$	$4s^1 3d^5$	$4s^2 3d^5$	$4s^2 3d^6$	$4s^2 3d^7$	$4s^2 3d^8$	$4s^1 3d^{10}$
M^{2+}	—	$3d^2$	$3d^3$	$3d^4$	$3d^5$	$3d^6$	$3d^7$	$3d^8$	$3d^9$
M^{3+}	[Ar]	$3d^1$	$3d^2$	$3d^3$	$3d^4$	$3d^5$	$3d^6$	$3d^7$	$3d^8$
Electronegativity	1.3	1.5	1.6	1.6	1.5	1.8	1.9	1.9	1.9
Ionization energy (kJ/mol)									
First	631	658	650	652	717	759	760	736	745
Second	1235	1309	1413	1591	1509	1561	1645	1751	1958
Third	2389	2650	2828	2986	3250	2956	3231	3393	3578
Radius (pm)									
M	162	147	134	130	135	126	125	124	128
M^{2+}	—	90	88	85	80	77	75	69	72
M^{3+}	81	77	74	64	66	60	64	—	—
Standard reduction potential (V)*	−2.08	−1.63	−1.2	−0.74	−1.18	−0.44	−0.28	−0.25	0.34

*The half-reaction is $M^{2+}(aq) + 2e^- \rightarrow M(s)$ (except for Sc and Cr, where the ions are Sc^{3+} and Cr^{3+}, respectively).

that the common oxidation states for each element include +2, +3, or both. The +3 oxidation states are more stable at the beginning of the series, whereas toward the end the +2 oxidation states are more stable. The reason is that the ionization energies increase gradually from left to right. However, the third ionization energy (when an electron is removed from the $3d$ orbital) increases more sharply than the first and second ionization energies. Because it takes more energy to remove the third electron

Figure 20.3 *Oxidation states of the first-row transition metals. The most stable oxidation numbers are shown in color. The zero oxidation state is encountered in some compounds, such as $Ni(CO)_4$ and $Fe(CO)_5$.*

from the metals near the end of the row than from those near the beginning, the metals near the end tend to form M^{2+} ions rather than M^{3+} ions.

The highest oxidation state is $+7$, for manganese ($4s^2 3d^5$). Transition metals usually exhibit their highest oxidation states in compounds with very electronegative elements such as oxygen and fluorine—for example, VF_5, CrO_3, and Mn_2O_7.

Recall that oxides in which the metal has a high oxidation number are covalent and acidic, whereas those in which the metal has a low oxidation number are ionic and basic (see Section 16.9).

20.2 Coordination Compounds

Transition metals have a distinct tendency to form complex ions (see p. 561). A *coordination compound* *typically consists of a complex ion and counter ion.* [Note that some coordination compounds such as $Fe(CO)_5$ do not contain complex ions.] Our understanding of the nature of coordination compounds stems from the classic work of the Swiss chemist Alfred Werner, who prepared and characterized many coordination compounds. In 1893, at the age of 26, Werner proposed what is now commonly referred to as *Werner's coordination theory.*

Recall that a complex ion contains a central metal ion bonded to one or more ions or molecules (see Section 17.7).

Nineteenth-century chemists were puzzled by a certain class of reactions that seemed to violate valence theory. For example, the valences of the elements in cobalt(III) chloride and in ammonia seem to be completely satisfied, and yet these two substances react to form a stable compound having the formula $CoCl_3 \cdot 6NH_3$. To explain this behavior, Werner postulated that most elements exhibit two types of valence: *primary valence* and *secondary valence*. In modern terminology, primary valence corresponds to the oxidation number and secondary valence to the coordination number of the element. In $CoCl_3 \cdot 6NH_3$, according to Werner, cobalt has a primary valence of 3 and a secondary valence of 6.

Today we use the formula $[Co(NH_3)_6]Cl_3$ to indicate that the ammonia molecules and the cobalt atom form a complex ion; the chloride ions are not part of the complex but are held to it by ionic forces. Most, but not all, of the metals in coordination compounds are transition metals.

The molecules or ions that surround the metal in a complex ion are called *ligands* (Table 20.2). The interactions between a metal atom and the ligands can be thought of as Lewis acid-base reactions. As we saw in Section 16.10, a Lewis base is a substance capable of donating one or more electron pairs. Every ligand has at least one unshared pair of valence electrons, as these examples show:

$$\overset{\cdot\cdot}{\underset{H \quad\quad H}{O}} \qquad \overset{\cdot\cdot}{\underset{H\ \underset{H}{|}\ H}{N}} \qquad :\overset{\cdot\cdot}{\underset{\cdot\cdot}{Cl}}:^- \qquad :C\equiv O:$$

Therefore, ligands play the role of Lewis bases. On the other hand, a transition metal atom (in either its neutral or positively charged state) acts as a Lewis acid, accepting (and sharing) pairs of electrons from the Lewis bases. Thus the metal-ligand bonds are usually coordinate covalent bonds (see Section 9.7).

The atom in a ligand that is bound directly to the metal atom is known as the *donor atom.* For example, nitrogen is the donor atom in the $[Cu(NH_3)_4]^{2+}$ complex ion. The *coordination number* in coordination compounds is defined as *the number of donor atoms surrounding the central metal atom in a complex ion.* For example, the coordination number of Ag^+ in $[Ag(NH_3)_2]^+$ is 2, that of Cu^{2+} in $[Cu(NH_3)_4]^{2+}$ is 4, and that of Fe^{3+} in $[Fe(CN)_6]^{3-}$ is 6. The most common coordination numbers are 4 and 6, but coordination numbers such as 2 and 5 are also known.

In a crystal lattice the coordination number of an atom (or ion) is defined as the number of atoms (or ions) surrounding the atom (or ion).

Depending on the number of donor atoms present, ligands are classified as *monodentate, bidentate,* or *polydentate* (see Table 20.2). H_2O and NH_3 are monodentate

Table 20.2 | Some Common Ligands

Name	Structure
	Monodentate ligands
Ammonia	H—N̈—H with H below
Carbon monoxide	:C≡O:
Chloride ion	:C̈l:⁻
Cyanide ion	[:C≡N:]⁻
Thiocyanate ion	[:S̈—C≡N:]⁻
	Bidentate ligands
Ethylenediamine	H₂N̈—CH₂—CH₂—N̈H₂
Oxalate ion	oxalate structure, charge 2−
	Polydentate ligand
Ethylenediaminetetraacetate ion (EDTA)	EDTA structure, charge 4−

ligands with only one donor atom each. One bidentate ligand is ethylenediamine (commonly abbreviated "en"):

$$H_2\ddot{N}-CH_2-CH_2-\ddot{N}H_2$$

The two nitrogen atoms can coordinate with a metal atom as shown in Figure 20.4.

Figure 20.4 (a) *Structure of metal-ethylenediamine complex. Each ethylenediamine molecule provides two N donor atoms and is therefore a bidentate ligand. (b) Simplified structure of the same complex.*

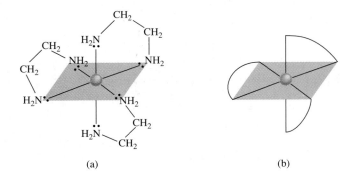

(a)   (b)

Bidentate and polydentate ligands are also called ***chelating agents*** because of *their ability to hold the metal atom like a claw* (from the Greek *chele,* meaning "claw"). One example is ethylenediaminetetraacetate ion (EDTA), a polydentate ligand used to treat metal poisoning (Figure 20.5). Six donor atoms enable EDTA to form a very stable complex ion with lead, which is removed from the blood and excreted from the body. EDTA is also used to clean up spills of radioactive metals.

Figure 20.5 *EDTA complex of lead. The complex bears a net charge of −2, because each O donor atom has one negative charge and the lead ion carries two positive charges. Note the octahedral geometry around the Pb^{2+} ion.*

Oxidation Number of Metals in Coordination Compounds

The net charge of a complex ion is the sum of the charges on the central metal atom and its surrounding ligands. In the $[PtCl_6]^{2-}$ ion, for example, each chloride ion has an oxidation number of −1, so the oxidation number of Pt must be +4. If the ligands do not bear net charges, the oxidation number of the metal is equal to the charge of the complex ion. Thus in $[Cu(NH_3)_4]^{2+}$ each NH_3 is neutral, so the oxidation number of Cu is +2.

Example 20.1   Specify the oxidation number of the central metal atom in each of the following compounds: (a) $[Ru(NH_3)_5(H_2O)]Cl_2$, (b) $[Cr(NH_3)_6](NO_3)_3$, (c) $[Fe(CO)_5]$, and (d) $K_4[Fe(CN)_6]$.

Reasoning and Solution   (a) Both NH_3 and H_2O are neutral species. Because each chloride ion carries a −1 charge, and there are two Cl^- ions, the oxidation number of Ru must be +2.

(b) Each nitrate ion has a charge of −1; therefore, the cation must be $[Cr(NH_3)_6]^{3+}$. NH_3 is neutral, so the oxidation number of Cr is +3.

(c) Because the CO species are neutral, the oxidation number of Fe is zero.

(d) Each potassium ion has a charge of +1; therefore, the anion is $[Fe(CN)_6]^{4-}$. Next, we know that each cyanide group bears a charge of −1, so Fe must have an oxidation number of +2.

Practice Exercise   Write the oxidation numbers of the metals in the compound $K[Au(OH)_4]$.

Similar problems: 20.13, 20.14.

Naming Coordination Compounds

Now that we have discussed the various types of ligands and the oxidation numbers of metals, our next step is to learn what to call these coordination compounds. The rules for naming coordination compounds are:

- The cation is named before the anion, as in other ionic compounds. The rule holds regardless of whether the complex ion bears a net positive or a negative charge. For example, in $K_3[Fe(CN)_6]$ and $[Co(NH_3)_4Cl_2]Cl$ compound, we name the K^+ and $[Co(NH_3)_4Cl_2]^+$ cations first, respectively.
- Within a complex ion the ligands are named first, in alphabetical order, and the metal ion is named last.
- The names of anionic ligands end with the letter *o*, whereas a neutral ligand is usually called by the name of the molecule. The exceptions are H_2O (aquo), CO (carbonyl), and NH_3 (ammine). Table 20.3 lists some common ligands.

Table 20.3 | Names of Common Ligands in Coordination Compounds

Ligand	Name of Ligand in Coordination Compound
Bromide, Br^-	Bromo
Chloride, Cl^-	Chloro
Cyanide, CN^-	Cyano
Hydroxide, OH^-	Hydroxo
Oxide, O^{2-}	Oxo
Carbonate, CO_3^{2-}	Carbonato
Nitrite, NO_2^-	Nitro
Oxalate, $C_2O_4^{2-}$	Oxalato
Ammonia, NH_3	Ammine
Carbon monoxide, CO	Carbonyl
Water, H_2O	Aquo
Ethylenediamine	Ethylenediamine
Ethylenediaminetetraacetate	Ethylenediaminetetraacetato

- When several ligands of a particular kind are present, we use the Greek prefixes *di-, tri-, tetra-, penta-,* and *hexa-* to name them. Thus the ligands in the cation $[Co(NH_3)_4Cl_2]^+$ are "tetraamminedichloro." (Note that prefixes are ignored when alphabetizing ligands.) If the ligand itself contains a Greek prefix, we use the prefixes *bis* (2), *tris* (3), and *tetrakis* (4) to indicate the number of ligands present. For example, the ligand ethylenediamine already contains *di;* therefore, if two such ligands are present the name is *bis(ethylenediamine).*
- The oxidation number of the metal is written in Roman numerals following the name of the metal. For example, the Roman numeral III is used to indicate the +3 oxidation state of chromium in the $[Cr(NH_3)_4Cl_2]^+$ ion, which is called tetraamminedichlorochromium(III) ion.
- If the complex is an anion, its name ends in *-ate.* For example, in $K_4[Fe(CN)_6]$ the anion $[Fe(CN)_6]^{4-}$ is called hexacyanoferrate(II) ion. Note that the Roman numeral II indicates the oxidation state of iron. Table 20.4 gives the names of anions containing metal atoms.

Table 20.4 | Names of Anions Containing Metal Atoms

Metal	Name of Metal in Anionic Complex
Aluminum	Aluminate
Chromium	Chromate
Cobalt	Cobaltate
Copper	Cuprate
Gold	Aurate
Iron	Ferrate
Lead	Plumbate
Manganese	Manganate
Molybdenum	Molybdate
Nickel	Nickelate
Silver	Argentate
Tin	Stannate
Tungsten	Tungstate
Zinc	Zincate

Example 20.2 Write the systematic names of these compounds: (a) $Ni(CO)_4$, (b) $[Co(NH_3)_4Cl_2]Cl$, (c) $K_3[Fe(CN)_6]$, (d) $[Cr(en)_3]Cl_3$.

Reasoning and Solution (a) The CO ligands are neutral species and the nickel atom bears no net charge, so the compound is called *tetracarbonylnickel(0)*, or, more commonly, *nickel tetracarbonyl.*

(b) Starting with the cation, each of the two chloride ligands bears a negative charge and the ammonia molecules are neutral. Thus, the cobalt atom must have an oxidation number of +3 (to balance the chloride anion). The compound is called *tetraamminedichlorocobalt(III) chloride.*

(c) The complex ion is the anion and it bears three negative charges. Thus, the iron atom must have an oxidation number of +3. The compound is *potassium hexacyanoferrate(III)*. This compound is commonly called *potassium ferricyanide.*

(d) As we noted earlier, *en* is the abbreviation for the ligand ethylenediamine. Because there are three *en* groups present and the name of the ligand already contains *di*, the name of the compound is *tris(ethylenediamine)chromium(III) chloride*.

Similar problems: 20.15, 20.16.

Practice Exercise What is the systematic name of $[Cr(H_2O)_4Cl_2]Cl$?

Example 20.3 Write the formulas for these compounds:
(a) pentaamminechlorocobalt(III) chloride, (b) dichlorobis(ethylenediamine)platinum(IV) nitrate, (c) sodium hexanitrocobaltate(III).

Reasoning and Solution (a) The complex cation contains five NH_3 groups, a chloride ion, and a cobalt ion with a +3 oxidation number. The net charge on the cation must be 2+. Therefore, the formula for the compound is $[Co(NH_3)_5Cl]Cl_2$.

(b) There are two chloride ions, two ethylenediamine groups, and a platinum ion with an oxidation number of +4 in the complex cation. Therefore, the formula for the compound is $[Pt(en)_2Cl_2](NO_3)_2$.

(c) The complex anion contains six nitro groups and a cobalt ion with an oxidation number of +3. Therefore, the formula for the compound is $Na_3[Co(NO_2)_6]$.

Similar problems: 20.17, 20.18.

Practice Exercise Write the formula for this compound: tris(ethylenediamine)cobalt(III) sulfate.

20.3 Geometry of Coordination Compounds

Figure 20.6 shows four different geometric arrangements for metal atoms with monodentate ligands. In these diagrams we see that the structure and the coordination number of the metal atom relate to each other as follows:

Coordination Number	Structure
2	Linear
4	Tetrahedral or square planar
6	Octahedral

Coordination Number = 2

The $[Ag(NH_3)_2]^+$ complex ion, formed by the reaction between Ag^+ ions and ammonia (see Table 17.4), has a coordination number of 2 and a linear geometry. Other examples are $[CuCl_2]^-$ and $[Au(CN)_2]^-$.

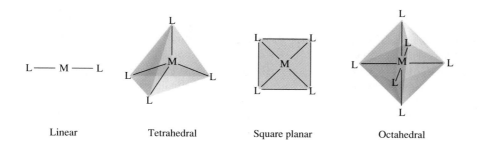

Figure 20.6 *Common geometries of complex ions. In each case M is a metal and L is a monodentate ligand.*

Linear   Tetrahedral   Square planar   Octahedral

Figure 20.7 *The (a) cis and (b) trans isomers of diamminedichloroplatinum(II). Note that the two Cl atoms are adjacent to each other in the cis isomer and diagonally across from each other in the trans isomer.*

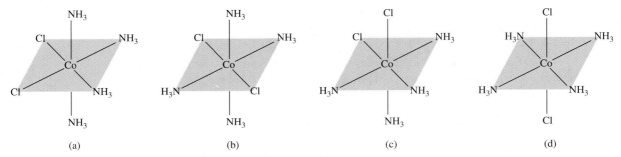

Figure 20.8 *The (a) cis and (b) trans isomers of tetraamminedichlorocobalt(III) ion. The structure shown in (c) can be generated by rotating that in (a), and the structure shown in (d) can be generated by rotating that in (b). The ion has only two geometric isomers, (a) [or (c)] and (b) [or (d)].*

Coordination Number = 4

There are two types of geometries with a coordination number of 4. The $[Zn(NH_3)_4]^{2+}$ and $[CoCl_4]^{2-}$ ions have tetrahedral geometry, whereas the $[Pt(NH_3)_4]^{2+}$ ion has the square planar geometry. In Chapter 11 we discussed geometric isomers of alkenes (see p. 342). Square planar complex ions with two different monodentate ligands can also exhibit geometric isomerism. Figure 20.7 shows the *cis* and *trans* isomers of diamminedichloroplatinum(II). Note that although the types of bonds are the same in both isomers (two Pt—N and two Pt—Cl bonds), the spatial arrangements are different. These two isomers have different properties (melting point, boiling point, color, solubility in water, and dipole moment).

Coordination Number = 6

Complex ions with a coordination number of 6 all have octahedral geometry (see Section 10.1). Geometric isomerism is possible in octahedral complexes when two or more different ligands are present. An example is the tetraamminedichlorocobalt(III) ion shown in Figure 20.8. The two geometric isomers have different colors and other properties even though they have the same ligands and the same number and types of bonds (Figure 20.9).

In addition to geometric isomerism, certain octahedral complex ions can also exhibit optical isomerism (that is, give rise to optical isomers, as discussed in Chapter 11). Figure 20.10 shows the *cis* and *trans* isomers of dichlorobis(ethylenediamine)cobalt(III) ion and their mirror images. Careful examination reveals that the *trans* isomer and its mirror image are superimposable, but the *cis* isomer and its

Figure 20.9 *Left: cis-tetraamminedichlorocobalt(III) chloride. Right: trans-tetraamminedichlorocobalt(III) chloride.*

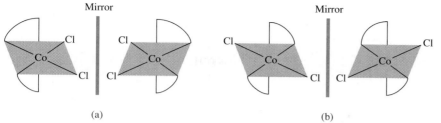

Figure 20.10 *The (a) cis and (b) trans isomers of dichlorobis(ethylenediamine)cobalt(III) ion and their mirror images. If you could rotate the mirror image in (b) 90° clockwise about the vertical position and place it over the trans isomer, you would find that the two are superimposable. No matter how you rotated the cis isomer and its mirror image, however, you could not superimpose one on the other.*

mirror image are not. Therefore, the *cis* isomer and its mirror image are optical isomers, or enantiomers. It is interesting to note that unlike the case in most organic compounds, there are no asymmetric carbon atoms in these compounds.

20.4 Bonding in Coordination Compounds: Crystal Field Theory

A satisfactory theory of bonding in coordination compounds must account for properties such as color and magnetism, as well as stereochemistry and bond strength. No single theory as yet does all this for us. Rather, several different approaches have been applied to transition metal complexes. We will consider only one of them here—crystal field theory—because it accounts for both the color and magnetic properties of many coordination compounds.

We will begin our discussion of crystal field theory with complex ions having octahedral geometry. Then we will see how it is applied to tetrahedral and square-planar complexes.

Crystal Field Splitting in Octahedral Complexes

Crystal field theory explains the bonding in complex ions purely in terms of electrostatic forces. In a complex ion, two types of electrostatic interaction come into play. One is the attraction between the positive metal ion and the negatively charged ligand or the negatively charged end of a polar ligand. This is the force that binds the ligands to the metal. The second type of interaction is electrostatic repulsion between the lone pairs on the ligands and the electrons in the *d* orbitals of the metals.

As we saw in Chapter 7, *d* orbitals have different orientations, but in the absence of external disturbance they all have the same energy. In an octahedral complex, a central metal atom is surrounded by six lone pairs of electrons (on the six ligands), so all five *d* orbitals experience electrostatic repulsion. The magnitude of this repulsion depends on the orientation of the *d* orbital that is involved. Take the $d_{x^2-y^2}$ orbital as an example. In Figure 20.11, we see that the lobes of this orbital point toward corners of the octahedron along the *x*- and *y*-axes, where the lone-pair electrons are positioned. Thus, an electron residing in this orbital would experience a greater repulsion from the ligands than an electron would in, say, the d_{xy} orbital. For this reason, the energy of the $d_{x^2-y^2}$ orbital is increased relative to the d_{xy}, d_{yz}, and d_{xz} orbitals. The d_{z^2} orbital's energy is also greater, because its lobes are pointed at the ligands along the *z*-axis.

As a result of these metal-ligand interactions, the five *d* orbitals in an octahedral complex are split between two sets of energy levels: a higher level with two orbitals ($d_{x^2-y^2}$ and d_{z^2}) having the same energy and a lower level with three equal-energy

The name "crystal field" is associated with the theory used to explain the properties of solid, crystalline materials. The same theory is used to study coordination compounds.

Figure 20.11 *The five d orbitals in an octahedral environment. The metal atom (or ion) is at the center of the octahedron, and the six lone pairs on the donor atoms of the ligands are at the corners.*

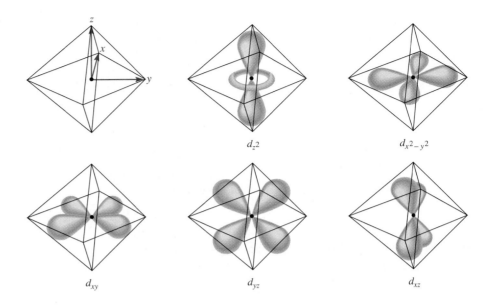

d_{z^2} $d_{x^2-y^2}$

d_{xy} d_{yz} d_{xz}

orbitals (d_{xy}, d_{yz}, and d_{xz}), as shown in Figure 20.12. The ***crystal field splitting (Δ)*** is *the energy difference between two sets of d orbitals in a metal atom when ligands are present.* The magnitude of Δ depends on the metal and the nature of the ligands; it has a direct effect on the color and magnetic properties of complex ions.

Color

In Chapter 7 we learned that white light, such as sunlight, is a combination of all colors. A substance appears black if it absorbs all the visible light that strikes it. If it absorbs no visible light, it is white or colorless. An object appears green if it absorbs all light but reflects the green component. An object also looks green if it reflects all colors except red, the *complementary* color of green. (Figure 20.13).

What has been said of reflected light also applies to transmitted light (that is, the light that passes through the medium, for example, a solution). Consider the hydrated cupric ion, $[Cu(H_2O)_6]^{2+}$, which absorbs light in the orange region of the spectrum so that a solution of $CuSO_4$ appears blue to us. Recall from Chapter 7 that when the energy of a photon is equal to the difference between the ground state and an excited

Figure 20.12 *Crystal field splitting between d orbitals in an octahedral complex.*

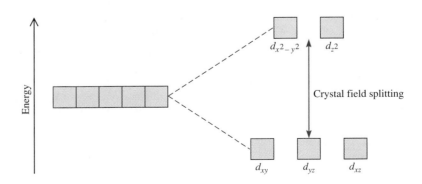

state, absorption occurs as the photon strikes the atom (or ion or compound), and an electron is promoted to a higher level. This knowledge allows us to calculate the energy change involved in the electron transition. The energy of a photon, given by Equation (7.2), is:

$$E = h\nu$$

where h represents Planck's constant (6.63×10^{-34} J s) and ν is the frequency of the radiation, which is 5.00×10^{14}/s for a wavelength of 600 nm. Here $E = \Delta$, so we have

$$\begin{aligned}
\Delta &= h\nu \\
&= (6.63 \times 10^{-34} \text{ J s})(5.00 \times 10^{14}\text{/s}) \\
&= 3.32 \times 10^{-19} \text{ J}
\end{aligned}$$

(Note that this is the energy absorbed by *one* ion.) If the wavelength of the photon absorbed by an ion lies outside the visible region, then the transmitted light looks the same (to us) as the incident light—white—and the ion appears colorless.

The best way to measure crystal field splitting is to use spectroscopy to determine the wavelength at which light is absorbed. The $[Ti(H_2O)_6]^{3+}$ ion provides a straightforward example, because Ti^{3+} has only one $3d$ electron (Figure 20.14). The $[Ti(H_2O)_6]^{3+}$ ion absorbs light in the visible region of the spectrum (Figure 20.15). The wavelength corresponding to maximum absorption is 498 nm [Figure 20.14(b)]. This information enables us to calculate the crystal field splitting as follows. We start by writing

$$\Delta = h\nu \tag{20.1}$$

Also

$$\nu = \frac{c}{\lambda}$$

Figure 20.13 *A color wheel with appropriate wavelengths. Complementary colors, such as red and green, are on opposite sides of the wheel.*

650 nm   580 nm
700 nm / 400 nm   560 nm
430 nm   490 nm

Equation (7.1) shows that $c = \lambda\nu$.

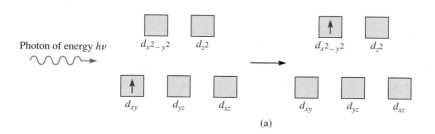

Photon of energy $h\nu$

$d_{x^2-y^2}$ d_{z^2} $d_{x^2-y^2}$ d_{z^2}

d_{xy} d_{yz} d_{xz} d_{xy} d_{yz} d_{xz}

(a)

Absorption

400   500   600   700

Wavelength (nm)

(b)

Figure 20.14 *(a) The process of photon absorption and (b) a graph of the absorption spectrum of $[Ti(H_2O)_6]^{3+}$. The energy of the incoming photon is equal to the crystal field splitting. The maximum absorption peak in the visible region occurs at 498 nm.*

Figure 20.15 *Colors of some of the first-row transition metal ions in solution. From left to right: Ti^{3+}, Cr^{3+}, Mn^{2+}, Fe^{3+}, Co^{2+}, Ni^{2+}, Cu^{2+}. The Sc^{3+} and V^{5+} ions are colorless.*

where c is the speed of light and λ is the wavelength. Therefore

$$\Delta = \frac{hc}{\lambda} = \frac{(6.63 \times 10^{-34}\,\text{J s})(3.00 \times 10^{8}\,\text{m/s})}{(498\,\text{nm})(1 \times 10^{-9}\,\text{m/1 nm})}$$

$$= 3.99 \times 10^{-19}\,\text{J}$$

This is the energy required to excite *one* $[Ti(H_2O)_6]^{3+}$ ion. To express this energy difference in the more convenient units of kilojoules per mole, we write

$$\Delta = (3.99 \times 10^{-19}\,\text{J/ion})(6.02 \times 10^{23}\,\text{ions/mol})$$

$$= 240{,}000\,\text{J/mol}$$

$$= 240\,\text{kJ/mol}$$

Aided by spectroscopic data for a number of complexes, all having the same metal ion but different ligands, chemists calculated the crystal splitting for each ligand and established a ***spectrochemical series,*** which is *a list of ligands arranged in increasing order of their abilities to split the d orbital energy levels:*

$$I^- < Br^- < Cl^- < OH^- < F^- < H_2O < NH_3 < en < CN^- < CO$$

These ligands are arranged in the order of increasing value of Δ. CO and CN^- are called *strong-field ligands,* because they cause a large splitting of the *d* orbital energy levels. The halide ions and hydroxide ion are *weak-field ligands,* because they split the *d* orbitals to a lesser extent.

The order in the spectrochemical series is the same no matter which metal atom (or ion) is present.

Magnetic Properties

The magnitude of the crystal field splitting also determines the magnetic properties of a complex ion. The $[Ti(H_2O)_6]^{3+}$ ion, having only one *d* electron, is always paramagnetic. However, for an ion with several *d* electrons, the situation is less clearcut. Consider, for example, the octahedral complexes $[FeF_6]^{3-}$ and $[Fe(CN)_6]^{3-}$ (Figure 20.16). The electron configuration of Fe^{3+} is $[Ar]3d^5$, and there are two possible ways to distribute the five *d* electrons among the *d* orbitals. According to Hund's rule (see Section 7.8), maximum stability is reached when the electrons are placed in five separate orbitals with parallel spins. But this arrangement can be achieved only at a cost; two of the five electrons must be promoted to the higher-energy $d_{x^2-y^2}$ and d_{z^2} orbitals. No such energy investment is needed if all five electrons enter the d_{xy}, d_{yz}, and d_{xz} orbitals. According to Pauli's exclusion principle (p. 214), there will be only one unpaired electron present in this case.

Figure 20.17 shows the distribution of electrons among *d* orbitals that results in low- and high-spin complexes. The actual arrangement of the electrons is determined

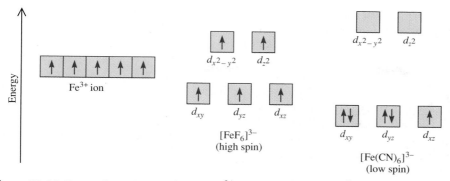

Figure 20.16 *Energy-level diagrams for the* Fe^{3+} *ion and for the* $[FeF_6]^{3-}$ *and* $[Fe(CN)_6]^{3-}$
complex ions.

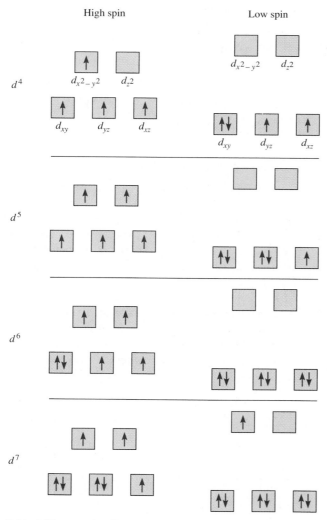

Figure 20.17 *Orbital diagrams for the high-spin and low-spin octahedral complexes corresponding to the electron configurations* d^4, d^5, d^6, *and* d^7. *No such distinctions can be made for* d^1, d^2, d^3, d^8, d^9, *and* d^{10}.

by the amount of stability gained by having maximum parallel spins versus the invest-ment in energy required to promote electrons to higher d orbitals. Because F^- is a weak-field ligand, the five d electrons enter five separate d orbitals with parallel spins to create a high-spin complex (see Figure 20.16). On the other hand, the cyanide ion is a strong-field ligand, so it is energetically preferable for all five electrons to be in the lower orbitals and therefore a low-spin complex is formed. High-spin complexes are more paramagnetic than low-spin complexes.

The magnetic properties of a complex ion depend on the number of unpaired electrons present.

The actual number of unpaired electrons (or spins) in a complex ion can be found by magnetic measurements, and in general, experimental findings support predictions based on crystal field splitting. However, a distinction between low- and high-spin complexes can be made only if the metal ion contains more than three and fewer than eight d electrons, as shown in Figure 20.17.

Example 20.4 Predict the number of unpaired spins in the $[\text{Cr(en)}_3]^{2+}$ ion.

Reasoning and Solution The electron configuration of Cr^{2+} is $[Ar]3d^4$. Because en is a strong-field ligand, we expect $[\text{Cr(en)}_3]^{2+}$ to be a low-spin complex. According to Figure 20.17, all four electrons will be placed in the lower-energy d orbitals (d_{xy}, d_{yz}, and d_{xz}) and there will be a total of two unpaired spins.

Similar problem: 20.30.

Practice Exercise How many unpaired spins are in $[\text{Mn(H}_2\text{O)}_6]^{2+}$? ($H_2O$ is a weak-field ligand.)

Tetrahedral and Square-Planar Complexes

So far we have concentrated on octahedral complexes. The splitting of the d orbital energy levels in two other types of complexes—tetrahedral and square-planar—can also be accounted for satisfactorily by the crystal field theory. In fact, the splitting pattern for a tetrahedral ion is just the reverse of that for octahedral complexes. In this case, the d_{xy}, d_{yz}, and d_{xz} orbitals are more closely directed at the ligands and therefore have more energy than the $d_{x^2-y^2}$ and d_{z^2} orbitals (Figure 20.18). Most tetrahedral complexes are high-spin complexes. Presumably, the tetrahedral arrangement reduces the magni-tude of metal-ligand interactions, resulting in a smaller Δ value. This is a reasonable assumption because the number of ligands is smaller in a tetrahedral complex.

As Figure 20.19 shows, the splitting pattern for square-planar complexes is the most complicated. Clearly, the $d_{x^2-y^2}$ orbital possesses the highest energy (as in the octahedral case), and the d_{xy} orbital the next highest. However, the relative placement

Figure 20.18 *Crystal field split-ting between d orbitals in a tetrahedral complex.*

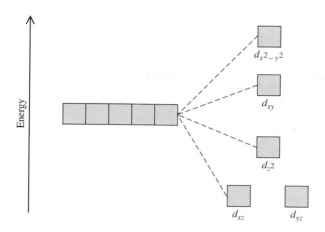

Figure 20.19 *Energy-level diagram for a square-planar complex. Because there are more than two energy levels, we cannot define crystal field splitting as we can for octahedral and tetrahedral complexes.*

of the d_{z^2} and the d_{xz} and d_{yz} orbitals cannot be determined simply by inspection and must be calculated.

20.5 Reactions of Coordination Compounds

Complex ions undergo ligand exchange (or substitution) reactions in solution. The rates of these reactions vary widely, depending on the nature of the metal ion and the ligands.

In studying ligand exchange reactions, it is often useful to distinguish between the stability of a complex ion and its tendency to react, which we call *kinetic lability*. Stability in this context is a thermodynamic property, which is measured in terms of the species' formation constant K_f (see p. 563). For example, we say that the complex ion tetracyanonickelate(II) is stable because it has a large formation constant ($K_f \approx 1 \times 10^{30}$)

$$Ni^{2+} + 4CN^- \rightleftharpoons [Ni(CN)_4]^{2-}$$

By using cyanide ions labeled with the radioactive isotope carbon-14, chemists have shown that $[Ni(CN)_4]^{2-}$ undergoes ligand exchange very rapidly in solution. This equilibrium is established almost as soon as the species are mixed:

$$[Ni(CN)_4]^{2-} + 4^*CN^- \rightleftharpoons [Ni(^*CN)_4]^{2-} + 4CN^-$$

This reaction has a small activation energy.

where the asterisk denotes a ^{14}C atom. Complexes like the tetracyanonickelate(II) ion are termed **labile complexes** because they *undergo rapid ligand exchange reactions.* Thus a thermodynamically stable species (that is, one that has a large formation constant) is not necessarily unreactive. (In Section 14.4 we saw that the smaller the activation energy, the larger the rate constant, and hence the greater the rate.)

A complex that is thermodynamically *unstable* in acidic solution is $[Co(NH_3)_6]^{3+}$. The equilibrium constant for this reaction is about 1×10^{20}.

$$[Co(NH_3)_6]^{3+} + 6H^+ + 6H_2O \rightleftharpoons [Co(H_2O)_6]^{3+} + 6NH_4^+$$

When equilibrium is reached, the concentration of the $[Co(NH_3)_6]^{3+}$ ion is very low. However, this reaction requires several days to complete because of the inertness of the $[Co(NH_3)_6]^{3+}$ ion. This is an example of an **inert complex,** *a complex ion that undergoes very slow exchange reactions* (on the order of hours or even days). It shows that a thermodynamically unstable species is not necessarily chemically reactive. The rate of reaction is determined by the energy of activation, which is high in this case.

Summary

Transition metals usually have incompletely filled d orbitals and have a pronounced tendency to form complexes. The first-row transition metals (scandium to copper) are the most common transition metals; their chemistry is characteristic, in many ways, of the entire group.

A coordination compound is a neutral species in which a small number of molecules or ions surround a central metal atom or ion. The donor atoms in the ligands each contribute an electron pair to the central metal ion in a complex. Coordination compounds may display geometric and/or optical isomerism.

Crystal field theory explains bonding in complexes in terms of electrostatic interactions. According to crystal field theory, the d orbitals are split into two higher-energy and three lower-energy orbitals in an octahedral complex. The energy difference between these two sets of d orbitals is the crystal field splitting. Strong-field ligands cause a large splitting, and weak-field ligands cause a small splitting. Electron spins tend to be parallel with weak-field ligands and paired with strong-field ligands, where a greater investment of energy is required to promote electrons into the high-lying d orbitals. The splitting of d orbitals in a tetrahedral complex is the opposite of that in an octahedral complex, and that in a square planar complex is the most complicated.

Complex ions undergo ligand exchange reactions in solution.

Key Words

Chelating agent p. 655	Coordination number p. 653	Donor atom p. 653	Spectrochemical series p. 662
Coordination compound p. 653	Crystal field splitting (Δ) p. 660	Inert complex p. 665 Labile complex p. 665 Ligand p. 653	

Questions and Problems

Properties of Transition Metals

Review Questions

20.1 What distinguishes a transition metal from a representative metal?

20.2 Why is zinc not considered a transition metal?

20.3 Explain why atomic radii decrease very gradually from scandium to copper.

20.4 Without referring to the text, write the ground-state electron configurations of the first-row transition metals. Explain any irregularities.

20.5 Write the electron configurations of these ions: V^{5+}, Cr^{3+}, Mn^{2+}, Fe^{3+}, Cu^{2+}, Sc^{3+}, Ti^{4+}.

20.6 Why do transition metals have more oxidation states than other elements? Give the highest oxidation states for scandium to copper.

20.7 As we read across the first-row transition metals from left to right, the +2 oxidation state becomes more stable in comparison with the +3 state. Why is this so?

20.8 Chromium exhibits several oxidation states in its compounds, whereas aluminum exhibits only the +3 oxidation state. Explain.

Coordination Compounds: Nomenclature; Oxidation Number

Review Questions

20.9 Define the following terms: coordination compound, ligand, donor atom, coordination number, chelating agent.

20.10 Describe the interaction between a donor atom and a metal atom in terms of a Lewis acid-base reaction.

Problems

20.11 Complete these statements for the complex ion $[Co(en)_2(H_2O)CN]^{2+}$. (a) The term "en" is the abbreviation for _____. (b) The oxidation number of Co is _____. (c) The coordination number of Co is _____. (d) _____ is a bidentate ligand.

20.12 Complete these statements for the complex ion $[Cr(C_2O_4)_2(H_2O)_2]^-$. (a) The oxidation number of Cr is _____. (b) The coordination number of Cr is _____. (c) _____ is a bidentate ligand.

20.13 Give the oxidation numbers of the metals in these species: (a) $K_3[Fe(CN)_6]$, (b) $K_3[Cr(C_2O_4)_3]$, (c) $[Ni(CN)_4]^{2-}$.

20.14 Give the oxidation numbers of the metals in these species: (a) Na_2MoO_4, (b) $MgWO_4$, (c) $K_4[Fe(CN)_6]$.

20.15 What are the systematic names for these ions and compounds: (a) $[Co(NH_3)_4Cl_2]^+$, (b) $Cr(NH_3)_3Cl_3$, (c) $[Co(en)_2Br_2]^+$, (d) $Fe(CO)_5$?

20.16 What are the systematic names for these ions and compounds: (a) $[cis\text{-}Co(en)_2Cl_2]^+$, (b) $[Pt(NH_3)_5Cl]Cl_3$, (c) $[Co(NH_3)_6]Cl_3$, (d) $[Co(NH_3)_5Cl]Cl_2$, (e) $trans$-$Pt(NH_3)_2Cl_2$?

20.17 Write the formulas for each of these ions and compounds: (a) tetrahydroxozincate(II), (b) pentaaquochlorochromium(III) chloride, (c) tetrabromocuprate(II), (d) ethylenediaminetetraacetatoferrate(II).

20.18 Write the formulas for each of these ions and compounds: (a) bis(ethylenediamine) dichlorochromium(III), (b) pentacarbonyliron(0), (c) potassium tetracyanocuprate(II), (d) tetraammineaquochlorocobalt(III) chloride.

Structure of Coordination Compounds

Problems

20.19 Draw structures of all the geometric and optical isomers of each of these cobalt complexes:

    (a) $[Co(NH_3)_6]^{3+}$
    (b) $[Co(NH_3)_5Cl]^{2+}$
    (c) $[Co(NH_3)_4Cl_2]^+$
    (d) $[Co(en)_3]^{3+}$
    (e) $[Co(C_2O_4)_3]^{3-}$

20.20 How many geometric isomers are in these species: (a) $[Co(NH_3)_2Cl_4]^-$, (b) $[Co(NH_3)_3Cl_3]$?

20.21 A student prepared a cobalt complex that has one of the following structures: $[Co(NH_3)_6]Cl_3$, $[Co(NH_3)_5Cl]Cl_2$, or $[Co(NH_3)_4Cl_2]Cl$. Explain how the student would distinguish among these possibilities by an electrical conductance experiment. At the student's disposal are three strong electrolytes: $NaCl$, $MgCl_2$, and $FeCl_3$, which may be used for comparison purposes.

20.22 The complex ion $[Ni(CN)_2Br_2]^{2-}$ has a square planar geometry. Draw the structures of the geometric isomers of this complex.

Bonding, Color, Magnetism

Review Questions

20.23 Briefly describe the crystal field theory. Define the following terms: crystal field splitting, high-spin complex, low-spin complex, spectrochemical series.

20.24 What is the origin of color in a compound?

20.25 Compounds containing the Sc^{3+} ion are colorless, whereas those containing the Ti^{3+} ion are colored. Explain.

20.26 What factors determine whether a given complex will be diamagnetic or paramagnetic?

Problems

20.27 For the same type of ligands, explain why the crystal field splitting for an octahedral complex is always greater than that for a tetrahedral complex.

20.28 Transition metal complexes containing CN^- ligands are often yellow, whereas those containing H_2O ligands are often green or blue. Explain.

20.29 The $[Ni(CN)_4]^{2-}$ ion, which has a square planar geometry, is diamagnetic, whereas the $[NiCl_4]^{2-}$ ion, which has a tetrahedral geometry, is paramagnetic. Show the crystal field splitting diagrams for those two complexes.

20.30 Predict the number of unpaired electrons in these complex ions: (a) $[Cr(CN)_6]^{4-}$, (b) $[Cr(H_2O)_6]^{2+}$.

20.31 The absorption maximum for the complex ion $[Co(NH_3)_6]^{3+}$ occurs at 470 nm. (a) Predict the color of the complex and (b) calculate the crystal field splitting in kilojoules per mole.

20.32 A solution made by dissolving 0.875 g of $Co(NH_3)_4Cl_3$ in 25.0 g of water freezes 0.56°C below the freezing point of pure water. Calculate the number of mole of ions produced when 1 mole of $Co(NH_3)_4Cl_3$ is dissolved in water, and suggest a structure for the complex ion present in this compound.

Reactions of Coordination Compounds

Review Questions

20.33 Define the terms (a) labile complex, (b) inert complex.

20.34 Explain why a thermodynamically stable species may be chemically reactive and a thermodynamically unstable species may be unreactive.

Problems

20.35 Oxalic acid, $H_2C_2O_4$, is sometimes used to clean rust stains from sinks and bathtubs. Explain the chemistry underlying this cleaning action.

20.36 The $[Fe(CN)_6]^{3-}$ complex is more labile than the $[Fe(CN)_6]^{4-}$ complex. Suggest an experiment that would prove that $[Fe(CN)_6]^{3-}$ is a labile complex.

20.37 Aqueous copper(II) sulfate solution is blue in color. When aqueous potassium fluoride is added, a green precipitate is formed. When aqueous potassium chloride is added instead, a bright-green solution is formed. Explain what is happening in these two cases.

20.38 When aqueous potassium cyanide is added to a solution of copper(II) sulfate, a white precipitate, soluble in an excess of potassium cyanide, is formed. No precipitate is formed when hydrogen sulfide is bubbled through the solution at this point. Explain.

20.39 A concentrated aqueous copper(II) chloride solution is bright green in color. When diluted with water, the solution becomes light blue. Explain.

20.40 In a dilute nitric acid solution, Fe^{3+} reacts with thiocyanate ion (SCN^-) to form a dark-red complex:

$$[Fe(H_2O)_6]^{3+} + SCN^- \rightleftharpoons H_2O + [Fe(H_2O)_5NCS]^{2+}$$

The equilibrium concentration of $[Fe(H_2O)_5NCS]^{2+}$ may be determined by how darkly colored the solution is (measured by a spectrometer). In one such experiment, 1.0 mL of 0.20 M $Fe(NO_3)_3$ was mixed with 1.0 mL of 1.0×10^{-3} M KSCN and 8.0 mL of dilute HNO_3. The color of the solution quantitatively indicated that the $[Fe(H_2O)_5NCS]^{2+}$ concentration was 7.3×10^{-5} M. Calculate the formation constant for $[Fe(H_2O)_5NCS]^{2+}$.

Additional Problems

20.41 Explain these facts: (a) Copper and iron have several oxidation states, whereas zinc exists in only one. (b) Copper and iron form colored ions, whereas zinc does not.

20.42 The formation constant for the reaction $Ag^+ + 2NH_3 \rightleftharpoons$ $[Ag(NH_3)_2]^+$ is 1.5×10^7 and that for the reaction $Ag^+ + 2CN^- \rightleftharpoons [Ag(CN)_2]^-$ is 1.0×10^{21} at 25°C (see Table 17.4). Calculate the equilibrium constant at 25°C for the reaction

$$[Ag(NH_3)_2]^+ + 2CN^- \rightleftharpoons [Ag(CN)_2]^- + 2NH_3$$

20.43 Hemoglobin is the oxygen-carrying protein. In each hemoglobin molecule, there are four heme groups. In each heme group an Fe(II) ion is octahedrally bound to five N atoms and to either a water molecule (in deoxyhemoglobin) or an oxygen molecule (in oxyhemoglobin). Oxyhemoglobin is bright red, whereas deoxyhemoglobin is purple. Show that the difference in color can be accounted for qualitatively on the basis of high-spin and low-spin complexes. (*Hint:* O_2 is a strong-field ligand.)

20.44 Hydrated Mn^{2+} ions are practically colorless (see Figure 20.15) even though they possess five $3d$ electrons. Explain. (*Hint:* Electronic transitions in which there is a change in the number of unpaired electrons do not occur readily.)

20.45 Which of these hydrated cations are colorless: $Fe^{2+}(aq)$, $Zn^{2+}(aq)$, $Cu^+(aq)$, $Cu^{2+}(aq)$, $V^{5+}(aq)$, $Ca^{2+}(aq)$, $Co^{2+}(aq)$, $Sc^{3+}(aq)$, $Pb^{2+}(aq)$? Explain your choice.

20.46 In each of these pairs of complexes, choose the one that absorbs light at a longer wavelength: (a) $[Co(NH_3)_6]^{2+}$, $[Co(H_2O)_6]^{2+}$; (b) $[FeF_6]^{3-}$, $[Fe(CN)_6]^{3-}$; (c) $[Cu(NH_3)_4]^{2+}$, $[CuCl_4]^{2-}$.

20.47 A student in 1895 prepared three chromium coordination compounds having the same formulas of $CrCl_3(H_2O)_6$ with these properties:

Color	Cl⁻ Ions in Solution per Formula Unit
Violet	3
Light green	2
Dark green	1

Write modern formulas for these compounds and suggest a method for confirming the number of Cl^- ions present in solution in each case. (*Hint:* Some of the compounds may exist as hydrates, which are compounds that have a specific number of water molecules attached to them.)

20.48 Complex ion formation has been used to extract gold, which exists in nature in the uncombined state. To separate it from other solid impurities, the ore is treated with a sodium cyanide (NaCN) solution in the presence of air to dissolve the gold by forming the soluble complex ion $[Au(CN)_2]^-$. (a) Balance the following equation:

$$Au + CN^- + O_2 + H_2O \longrightarrow [Au(CN)_2]^- + OH^-$$

(b) The gold is obtained by reducing the complex ion with zinc metal. Write a balanced ionic equation for this process. (c) What is the geometry and coordination number of the $[Au(CN)_2]^-$ ion?

20.49 These molecules (AX_4Y_2) have octahedral geometry. Group the molecules that are equivalent to each other.

(a)   (b)

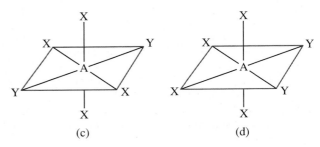

(c)   (d)

20.50 Which is a stronger oxidizing agent, Mn^{3+} or Cr^{3+}? Explain.

20.51 Suggest a method that would enable you to distinguish between *cis*-$Pt(NH_3)_2Cl_2$ and *trans*-$Pt(NH_3)_2Cl_2$.

20.52 You are given two solutions, one containing $FeCl_2$ and the other $FeCl_3$ at the same concentration. One solution is light yellow and the other one is brown. Identify these solutions based only on color.

20.53 The label of a certain brand of mayonnaise lists EDTA as a food preservative. How does EDTA prevent the spoilage of mayonnaise?

Special Problems

20.54 Carbon monoxide binds to the iron atom in hemoglobin some 200 times more strongly than oxygen. This is the reason why CO is a toxic substance. The metal-to-ligand sigma bond is formed by donating a lone pair from the donor atom to an empty sp^3d^2 orbital on Fe. (a) On the basis of electronegativities, would you expect the C or O atom to form the bond to Fe? (b) Draw a diagram illustrating the overlap of the orbitals involved in bonding.

20.55 The compound 1,1,1-trifluoroacetylacetone (tfa) is a bidentate ligand:

$$\overset{\displaystyle O}{\overset{\|}{CF_3C}}\overset{}{CH_2}\overset{\displaystyle O}{\overset{\|}{C}}CH_3$$

It forms a tetrahedral complex with Be^{2+} and a square planar complex with Cu^{2+}. Draw structures of these complex ions and identify the type of isomerism exhibited by these ions.

Answers to Practice Exercises

20.1 K: $+1$; Au: $+3$. **20.2** Tetraaquodichlorochromium(III) chloride. **20.3** $[Co(en)_3]_2(SO_4)_3$. **20.4** 5.

In December 1938, Lise Meitner, an Austrian physicist living in Sweden, received a puzzling letter from her long-time collaborator at the University of Berlin, Otto Hahn. The letter described an experiment in which Hahn and another colleague, Fritz Strassman, had bombarded uranium with neutrons in the hope of creating heavier elements. When they analyzed the products, the two chemists were astonished to find barium, which has roughly half the mass of uranium. On a walk through the Swedish countryside, Meitner and her nephew Otto Frisch, who was also a physicist, concluded that Hahn and Strassman had split apart the uranium nucleus. Borrowing the term "fission" from biology, where it applies to cell division, the scientists called the newly discovered phenomenon *nuclear fission*. They calculated that the process would release enormous amounts of energy.

Lise Meitner was a well-known and respected figure in nuclear physics; Einstein nicknamed her the "German Madam Curie." She was born in 1878 in Vienna, Austria, and showed a marked bent for mathematics and natural science from childhood. In 1906 she

Lise Meitner and Otto Hahn in their laboratory.

received a Ph.D. in physics from the University of Vienna, the second woman to do so in the university's 500-year history. Soon afterward she began a lifelong collaboration with Hahn. Because she was a woman, Meitner was given only a carpenter's workshop as her laboratory, and she worked for five years without pay! Gradually her situation improved, and in 1922 she was named a lecturer at the University of Berlin.

Meitner's research was interrupted in 1938 when Hitler seized Austria, and she lost her citizenship. Being Jewish, she fled to Sweden, where she continued to work with Hahn by mail.

In 1944 Hahn was awarded the Nobel Prize in chemistry for his work on nuclear fission. Meitner received no credit for her contribution. Today, Lise Meitner's statue is displayed among those of other prominent physicists at the Deutsche Museum in Munich, Germany, and element 109 is named meitnerium (Mt) in her honor. The inscription on her gravestone in Cambridge, England, reads: "Lise Meitner, 1878–1968, a great scientist who never forgot her humanity."

Nuclear Chemistry

Chapter 21

ESSENTIAL CONCEPTS

 Nuclear Stability To maintain nuclear stability, the ratio of neutrons-to-protons must fall within a certain range. A quantitative measure of nuclear stability is nuclear binding energy, which is the energy required to break up a nucleus into its component protons and neutrons. Nuclear binding energy can be calculated from the masses of protons and neutrons and that of the nucleus using Einstein's mass-energy equivalence relationship.

 Natural Radioactivity and Nuclear Transmutation Unstable nuclei undergo spontaneous decay with the emission of radiation and particles. All nuclear decays obey first-order kinetics. The half-lives of several radioactive nuclei have been used to date objects. Stable nuclei can also be made radioactive by bombardment with elementary particles or atomic nuclei. Many new elements have been created artificially in particle accelerators where such bombardments occur.

 Nuclear Fission and Nuclear Fusion Certain nuclei, when bombarded with neutrons, undergo fission to produce smaller nuclei, additional neutrons, and a large amount of energy. When enough nuclei are present to reach critical mass, a nuclear chain reaction, a self-sustaining sequence of nuclear fission reactions, takes place. Nuclear fissions find applications in the construction of atomic bombs and nuclear reactors. Nuclear fusion is the process in which nuclei of light elements are made to fuse at very high temperatures to form a heavier nucleus. Such a process releases even a greater amount of energy than nuclear fission and is used in making hydrogen, or thermonuclear, bombs.

 Uses of Isotopes Isotopes, especially radioactive isotopes, are used as tracers to study the mechanisms of chemical and biological reactions and as medical diagnostic tools.

 Biological Effects of Radiation The penetrating and harmful effects of radiation on biological systems have been thoroughly studied and are well understood.

21.1 The Nature of Nuclear Reactions

With the exception of hydrogen ($_1^1H$), all nuclei contain two kinds of fundamental particles, called *protons* and *neutrons*. Some nuclei are unstable; they emit particles and/or electromagnetic radiation spontaneously (see Section 2.2). The name for this phenomenon is *radioactivity*. All elements having an atomic number greater than 83 are radioactive. For example, the isotope of polonium, polonium-210 ($_{84}^{210}Po$), decays spontaneously to $_{82}^{206}Pb$ by emitting an α particle.

Another type of radioactivity, known as ***nuclear transmutation,*** *results from the bombardment of nuclei by neutrons, protons, or other nuclei.* An example of a nuclear transmutation is the conversion of atmospheric $_7^{14}N$ to $_6^{14}C$ and $_1^1H$, which results when the nitrogen isotope captures a neutron (from the sun). In some cases, heavier elements are synthesized from lighter elements. This type of transmutation occurs naturally in outer space, but it can also be achieved artificially, as we will see in Section 21.4.

Radioactive decay and nuclear transmutation are ***nuclear reactions,*** which differ significantly from ordinary chemical reactions. Table 21.1 summarizes the differences.

Balancing Nuclear Equations

To discuss nuclear reactions in any depth, we need to understand how to write and balance the equations. Writing a nuclear equation differs somewhat from writing equations for chemical reactions. In addition to writing the symbols for various chemical elements, we must also explicitly indicate protons, neutrons, and electrons. In fact, we must show the numbers of protons and neutrons present in *every* species in such an equation.

The symbols for elementary particles are as follows:

$_1^1p$ or $_1^1H$	$_0^1n$	$_{-1}^0e$ or $_{-1}^0\beta$	$_{+1}^0e$ or $_{+1}^0\beta$	$_2^4He$ or $_2^4\alpha$
proton	neutron	electron	positron	α particle

In accordance with the notation used in Section 2.3, the superscript in each case denotes the mass number (the total number of neutrons and protons present) and the subscript is the atomic number (the number of protons). Thus, the "atomic number" of a proton is 1, because there is one proton present, and the "mass number" is also 1, because

---

Table 21.1 | Comparison of Chemical Reactions and Nuclear Reactions

Chemical Reactions	Nuclear Reactions
1. Atoms are rearranged by the breaking and forming of chemical bonds.	1. Elements (or isotopes of the same elements) are converted from one to another.
2. Only electrons in atomic or molecular orbitals are involved in the breaking and forming of bonds.	2. Protons, neutrons, electrons, and other elementary particles may be involved.
3. Reactions are accompanied by absorption or release of relatively small amounts of energy.	3. Reactions are accompanied by absorption or release of tremendous amounts of energy.
4. Rates of reaction are influenced by temperature, pressure, concentration, and catalysts.	4. Rates of reaction normally are not affected by temperature, pressure, and catalysts.

there is one proton but no neutrons present. On the other hand, the "mass number" of a neutron is 1, but its "atomic number" is zero, because there are no protons present. For the electron, the "mass number" is zero (there are neither protons nor neutrons present), but the "atomic number" is -1, because the electron possesses a unit negative charge.

The symbol $_{-1}^{0}e$ represents an electron in or from an atomic orbital. The symbol $_{-1}^{0}\beta$ represents an electron that, although physically identical to any other electron, comes from a nucleus (in a decay process in which a neutron is converted to a proton and an electron) and not from an atomic orbital. The **positron** *has the same mass as the electron, but bears a +1 charge.* The α particle has two protons and two neutrons, so its atomic number is 2 and its mass number is 4.

In balancing any nuclear equation, we observe these rules:

- The total number of protons plus neutrons in the products and in the reactants must be the same (conservation of mass number).
- The total number of nuclear charges in the products and in the reactants must be the same (conservation of atomic number).

If we know the atomic numbers and mass numbers of all the species but one in a nuclear equation, we can identify the unknown species by applying these rules, as shown in the following example, which illustrates how to balance nuclear decay equations.

Example 21.1 Balance these nuclear equations (that is, identify the product X):

(a) $_{84}^{212}\text{Po} \longrightarrow \ _{82}^{208}\text{Pb} + \text{X}$

(b) $_{55}^{137}\text{Cs} \longrightarrow \ _{56}^{137}\text{Ba} + \text{X}$

Reasoning and Solution In balancing nuclear equations, note that the sum of atomic numbers and that of mass numbers must match on both sides of the arrow.

(a) The mass number and atomic number are 212 and 84, respectively, on the left-hand side and 208 and 82, respectively, on the right-hand side. Thus, X must have a mass number of 4 and an atomic number of 2, which means that it is an α particle. The balanced equation is

$$_{84}^{212}\text{Po} \longrightarrow \ _{82}^{208}\text{Pb} + \ _{2}^{4}\alpha$$

(b) In this case the mass number is the same on both sides of the equation, but the atomic number of the product is one more than that of the reactant. The only way this change can come about is to have a neutron in the Cs nucleus transformed into a proton and an electron; that is, $_{0}^{1}\text{n} \longrightarrow \ _{1}^{1}\text{p} + \ _{-1}^{0}\beta$ (note that this process does not alter the mass number). Thus, the balanced equation is

$$_{55}^{137}\text{Cs} \longrightarrow \ _{56}^{137}\text{Ba} + \ _{-1}^{0}\beta$$

We use the $_{-1}^{0}\beta$ notation here because the electron came from the nucleus.

Comment Note that the equations in (a) and (b) are balanced for nuclear particles but not for electrical charges. To balance the charges, we would need to add two electrons on the right-hand side of (a) and express barium as a cation (Ba^{+}) in (b).

Similar problems: 21.5, 21.6.

Practice Exercise Identify X in this nuclear equation:

$$_{33}^{78}\text{As} \longrightarrow \ _{-1}^{0}\beta + \text{X}$$

21.2 Nuclear Stability

The nucleus occupies a very small portion of the total volume of an atom, but it contains most of the atom's mass because both the protons and the neutrons reside there. In studying the stability of the atomic nucleus, it is helpful to know something about its density, because it tells us how tightly the particles are packed together. As a sample calculation, let us assume that a nucleus has a radius of 5×10^{-3} pm and a mass of 1×10^{-22} g. These figures correspond roughly to a nucleus containing 30 protons and 30 neutrons. Density is mass/volume, and we can calculate the volume from the known radius (the volume of a sphere is $\frac{4}{3}\pi r^3$, where r is the radius of the sphere). First we convert the pm units to cm. Then we calculate the density in g/cm^3:

$$r = 5 \times 10^{-3} \text{ pm} \times \frac{1 \times 10^{-12} \text{ m}}{1 \text{ pm}} \times \frac{100 \text{ cm}}{1 \text{ m}} = 5 \times 10^{-13} \text{ cm}$$

$$\text{density} = \frac{\text{mass}}{\text{volume}} = \frac{1 \times 10^{-22} \text{ g}}{\frac{4}{3}\pi r^3} = \frac{1 \times 10^{-22} \text{ g}}{\frac{4}{3}\pi (5 \times 10^{-13} \text{ cm})^3}$$

$$= 2 \times 10^{14} \text{ g/cm}^3$$

To dramatize the almost incomprehensibly high density, it has been suggested that it is equivalent to packing the mass of all the world's automobiles into one thimble.

This is an exceedingly high density. The highest density known for an element is 22.6 g/cm^3, for osmium (Os). Thus the average atomic nucleus is roughly 9×10^{12} (or 9 trillion) times more dense than the densest element known!

The enormously high density of the nucleus prompts us to wonder what holds the particles together so tightly. From electrostatic interaction we know that like charges repel and unlike charges attract one another. We would thus expect the protons to repel one another strongly, particularly when we consider how close they must be to each other. This indeed is so. However, in addition to the repulsion, there are also short-range attractions between proton and proton, proton and neutron, and neutron and neutron. The stability of any nucleus is determined by the difference between electrostatic repulsion and the short-range attraction. If repulsion outweighs attraction, the nucleus disintegrates, emitting particles and/or radiation. If attractive forces prevail, the nucleus is stable.

The principal factor that determines whether a nucleus is stable is the *neutron-to-proton ratio (n/p)*. For stable atoms of elements having low atomic number, the n/p value is close to 1. As the atomic number increases, the neutron-to-proton ratios of the stable nuclei become greater than 1. This deviation at higher atomic numbers arises because a larger number of neutrons is needed to counteract the strong repulsion among the protons and stabilize the nucleus. The following rules are useful in predicting nuclear stability:

- Nuclei that contain 2, 8, 20, 50, 82, or 126 protons or neutrons are generally more stable than nuclei that do not possess these numbers. For example, there are 10 stable isotopes of tin (Sn) with the atomic number 50 and only 2 stable isotopes of antimony (Sb) with the atomic number 51. The numbers 2, 8, 20, 50, 82, and 126 are called *magic numbers*. The significance of these numbers for nuclear stability is similar to the numbers of electrons associated with the very stable noble gases (that is, 2, 10, 18, 36, 54, and 86 electrons).
- Nuclei with even numbers of both protons and neutrons are generally more stable than those with odd numbers of these particles (Table 21.2).
- All isotopes of the elements with atomic numbers higher than 83 are radioactive. All isotopes of technetium (Tc, $Z = 43$) and promethium (Pm, $Z = 61$) are radioactive.

Table 21.2	Number of Stable Isotopes with Even and Odd Numbers of Protons and Neutrons	
Protons	**Neutrons**	**Number of Stable Isotopes**
Odd	Odd	4
Odd	Even	50
Even	Odd	53
Even	Even	164

Figure 21.1 shows a plot of the number of neutrons versus the number of protons in various isotopes. The stable nuclei are located in an area of the graph known as the *belt of stability.* Most radioactive nuclei lie outside this belt. Above the stability belt, the nuclei have higher neutron-to-proton ratios than those within the belt (for the same number of protons). To lower this ratio (and hence move down toward the belt of stability), these nuclei undergo the following process, called *β-particle emission:*

$$\,_0^1 n \longrightarrow \,_1^1 p + \,_{-1}^0 \beta$$

Beta-particle emission leads to an increase in the number of protons in the nucleus and a simultaneous decrease in the number of neutrons. Some examples are

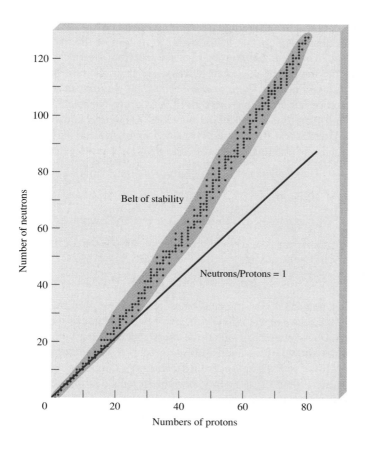

Figure 21.1 *Plot of neutrons versus protons for various stable isotopes, represented by dots. The straight line represents the points at which the neutron-to-proton ratio equals 1. The shaded area represents the belt of stability.*

$$\ce{^{14}_{6}C} \longrightarrow \ce{^{14}_{7}N} + \ce{^{0}_{-1}\beta}$$

$$\ce{^{40}_{19}K} \longrightarrow \ce{^{40}_{20}Ca} + \ce{^{0}_{-1}\beta}$$

$$\ce{^{97}_{40}Zr} \longrightarrow \ce{^{97}_{41}Nb} + \ce{^{0}_{-1}\beta}$$

Below the stability belt the nuclei have lower neutron-to-proton ratios than those in the belt (for the same number of protons). To increase this ratio (and hence move up toward the belt of stability), these nuclei either emit a positron

$$\ce{^{1}_{1}p} \longrightarrow \ce{^{1}_{0}n} + \ce{^{0}_{+1}\beta}$$

or undergo electron capture. An example of positron emission is

$$\ce{^{38}_{19}K} \longrightarrow \ce{^{38}_{18}Ar} + \ce{^{0}_{+1}\beta}$$

Electron capture is the capture of an electron—usually a $1s$ electron—by the nucleus. The captured electron combines with a proton to form a neutron so that the atomic number decreases by one while the mass number remains the same. This process has the same net effect as positron emission:

$$\ce{^{37}_{18}Ar} + \ce{^{0}_{-1}e} \longrightarrow \ce{^{37}_{17}Cl}$$

$$\ce{^{55}_{26}Fe} + \ce{^{0}_{-1}e} \longrightarrow \ce{^{55}_{25}Mn}$$

We use $_{-1}^{0}e$ rather than $_{-1}^{0}\beta$ here because the electron came from an atomic orbital and not from the nucleus.

Nuclear Binding Energy

A quantitative measure of nuclear stability is the **nuclear binding energy**, which is *the energy required to break up a nucleus into its component protons and neutrons.* This quantity represents the conversion of mass to energy that occurs during an exothermic nuclear reaction that leads to the formation of the nucleus.

The concept of nuclear binding energy evolved from studies of nuclear properties showing that the masses of nuclei are always less than the sum of the masses of the **nucleons,** which is *a general term for the protons and neutrons in a nucleus.* For example, the $\ce{^{19}_{9}F}$ isotope has an atomic mass of 18.9984 amu. The nucleus has 9 protons and 10 neutrons and therefore a total of 19 nucleons. Using the known masses of the $\ce{^{1}_{1}H}$ atom (1.007825 amu) and the neutron (1.008665 amu), we can carry out the following analysis. The mass of 9 $\ce{^{1}_{1}H}$ atoms (that is, the mass of 9 protons and 9 electrons) is

$$9 \times 1.007825 \text{ amu} = 9.070425 \text{ amu}$$

and the mass of 10 neutrons is

$$10 \times 1.008665 \text{ amu} = 10.08665 \text{ amu}$$

Therefore, the atomic mass of a $\ce{^{19}_{9}F}$ atom calculated from the known numbers of electrons, protons, and neutrons is

$$9.070425 \text{ amu} + 10.08665 \text{ amu} = 19.15708 \text{ amu}$$

There is no change in the electron's mass because it is not a nucleon.

which is larger than 18.9984 amu (the measured mass of $\ce{^{19}_{9}F}$) by 0.1587 amu.

The difference between the mass of an atom and the sum of the masses of its protons, neutrons, and electrons is called the **mass defect.** Relativity theory tells us that the loss in mass shows up as energy (heat) given off to the surroundings. Thus the formation of $\ce{^{19}_{9}F}$ is exothermic. According to *Einstein's mass-energy equivalence relationship* ($E = mc^2$, where E is energy, m is mass, and c is the velocity of light), we can calculate the amount of energy released. We start by writing

This is the only equation listed in *Bartlett's Quotations.*

$$\Delta E = (\Delta m)c^2 \tag{21.1}$$

where ΔE and Δm are defined as:

$$\Delta E = \text{energy of product} - \text{energy of reactants}$$
$$\Delta m = \text{mass of product} - \text{mass of reactants}$$

Thus we have for the change in mass

$$\Delta m = 18.9984 \text{ amu} - 19.15708 \text{ amu}$$
$$= -0.1587 \text{ amu}$$

Because $^{19}_{9}F$ has a mass that is less than the mass calculated from the number of electrons and nucleons present, Δm is a negative quantity. Consequently, ΔE is also a negative quantity; that is, energy is released to the surroundings as a result of the formation of the fluorine-19 nucleus. So we calculate ΔE as follows:

$$\Delta E = (-0.1587 \text{ amu})(3.00 \times 10^8 \text{ m/s})^2$$
$$= -1.43 \times 10^{16} \text{ amu m}^2/\text{s}^2$$

With the conversion factors

$$1 \text{ kg} = 6.022 \times 10^{26} \text{ amu}$$
$$1 \text{ J} = 1 \text{ kg m}^2/\text{s}^2$$

we obtain

$$\Delta E = \left(-1.43 \times 10^{16} \frac{\text{amu m}^2}{\text{s}^2}\right) \times \left(\frac{1.00 \text{ kg}}{6.022 \times 10^{26} \text{ amu}}\right) \times \left(\frac{1 \text{ J}}{1 \text{ kg m}^2/\text{s}^2}\right)$$
$$= -2.37 \times 10^{-11} \text{ J}$$

This is the amount of energy released when one fluorine-19 nucleus is formed from 9 protons and 10 neutrons. The nuclear binding energy of the nucleus is 2.37×10^{-11} J, which is the amount of energy needed to decompose the nucleus into separate protons and neutrons. In the formation of 1 mole of fluorine nuclei, for instance, the energy released is

$$\Delta E = (-2.37 \times 10^{-11} \text{ J})(6.022 \times 10^{23}/\text{mol})$$
$$= -1.43 \times 10^{13} \text{ J/mol}$$
$$= -1.43 \times 10^{10} \text{ kJ/mol}$$

The nuclear binding energy, therefore, is 1.43×10^{10} kJ for 1 mole of fluorine-19 nuclei, which is a tremendously large quantity when we consider that the enthalpies of ordinary chemical reactions are of the order of only 200 kJ. The procedure we have followed can be used to calculate the nuclear binding energy of any nucleus.

As we have noted, nuclear binding energy is an indication of the stability of a nucleus. However, in comparing the stability of any two nuclei we must account for the fact that they have different numbers of nucleons. For this reason it is more meaningful to use the *nuclear binding energy per nucleon,* defined as

$$\text{nuclear binding energy per nucleon} = \frac{\text{nuclear binding energy}}{\text{number of nucleons}}$$

For the fluorine-19 nucleus,

$$\text{nuclear binding energy per nucleon} = \frac{2.37 \times 10^{-11} \text{ J}}{19 \text{ nucleons}}$$
$$= 1.25 \times 10^{-12} \text{ J/nucleon}$$

Figure 21.2 *Plot of nuclear binding energy per nucleon versus mass number.*

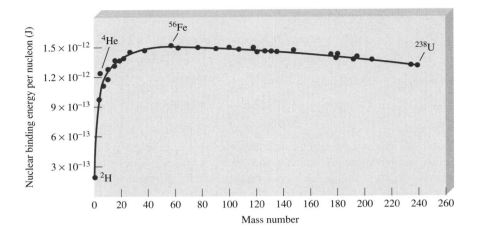

The nuclear binding energy per nucleon enables us to compare the stability of all nuclei on a common basis. Figure 21.2 shows the variation of nuclear binding energy per nucleon plotted against mass number. As you can see, the curve rises rather steeply. The highest binding energies per nucleon belong to elements with intermediate mass numbers—between 40 and 100—and are greatest for elements in the iron, cobalt, and nickel region (the Group 8B elements) of the periodic table. This means that the *net* attractive forces among the particles (protons and neutrons) are greatest for the nuclei of these elements.

Example 21.2   The atomic mass of $^{127}_{53}\text{I}$ is 126.9004 amu. Calculate the nuclear binding energy of this nucleus and the corresponding nuclear binding energy per nucleon.

Reasoning and Solution   To calculate the nuclear binding energy, we first determine the difference between the mass of the nucleus and the total mass of all the protons and neutrons, which gives us the mass defect. Next, we apply Einstein's mass-energy relationship [Equation (21.1)].

There are 53 protons and 74 neutrons in the iodine nucleus. The mass of 53 $^{1}_{1}\text{H}$ atom is

$$53 \times 1.007825 \text{ amu} = 53.41473 \text{ amu}$$

and the mass of 74 neutrons is

$$74 \times 1.008665 \text{ amu} = 74.64121 \text{ amu}$$

Therefore, the predicted mass for $^{127}_{53}\text{I}$ is 53.41473 + 74.64121 = 128.05594 amu, and the mass defect is

$$\Delta m = 126.9004 \text{ amu} - 128.05594 \text{ amu}$$
$$= -1.1555 \text{ amu}$$

The energy released is

$$\Delta E = (\Delta m)c^2$$
$$= (-1.1555 \text{ amu})(3.00 \times 10^8 \text{ m/s})^2$$
$$= -1.04 \times 10^{17} \text{ amu m}^2/\text{s}^2$$

$$\Delta E = \left(-1.04 \times 10^{17} \frac{\text{amu m}^2}{\text{s}^2}\right) \times \left(\frac{1.00 \text{ kg}}{6.022 \times 10^{26} \text{ amu}}\right) \times \left(\frac{1 \text{ J}}{1 \text{ kg m}^2/\text{s}^2}\right)$$

$$= -1.73 \times 10^{-10} \text{ J}$$

Thus the nuclear binding energy is 1.73×10^{-10} J. The nuclear binding energy per nucleon is obtained as:

$$\frac{1.73 \times 10^{-10} \text{ J}}{127 \text{ nucleons}} = 1.36 \times 10^{-12} \text{ J/nucleon}$$

The neutron-to-proton ratio is 1.4, which places iodine-127 in the belt of stability.

Similar problems: 21.19, 21.20.

Practice Exercise Calculate the nuclear binding energy (in J) and the nuclear binding energy per nucleon of $^{209}_{83}\text{Bi}$ (208.9804 amu).

21.3 Natural Radioactivity

Nuclei outside the belt of stability, as well as nuclei with more than 83 protons, tend to be unstable. The spontaneous emission by unstable nuclei of particles or electromagnetic radiation, or both, is known as radioactivity. The main types of radiation are: α particles (or doubly charged helium nuclei, He^{2+}); β particles (or electrons); γ rays, which are very-short-wavelength (0.1 nm to 10^{-4} nm) electromagnetic waves; positron emission; and electron capture.

The disintegration of a radioactive nucleus is often the beginning of a ***radioactive decay series,*** which is *a sequence of nuclear reactions that ultimately result in the formation of a stable isotope*. Table 21.3 shows the decay series of naturally occurring uranium-238, which involves 14 steps. This decay scheme, known as the *uranium decay series,* also shows the half-lives of all the products.

It is important to be able to balance the equation of a nuclear reaction for each of the steps in a radioactive decay series. For example, the first step in the uranium decay series is the decay of uranium-238 to thorium-234, with the emission of an α particle. Hence, the reaction is

$$^{238}_{92}\text{U} \longrightarrow \,^{234}_{90}\text{Th} + \,^{4}_{2}\alpha$$

The next step is represented by

$$^{234}_{90}\text{Th} \longrightarrow \,^{234}_{91}\text{Pa} + \,^{0}_{-1}\beta$$

and so on. In a discussion of radioactive decay steps, the beginning radioactive isotope is called the *parent* and the product, the *daughter.*

Radioactive decay

Kinetics of Radioactive Decay

All radioactive decays obey first-order kinetics. Therefore the rate of radioactive decay at any time t is given by

$$\text{rate of decay at time } t = \lambda N$$

Table 21.3 | The Uranium Decay Series

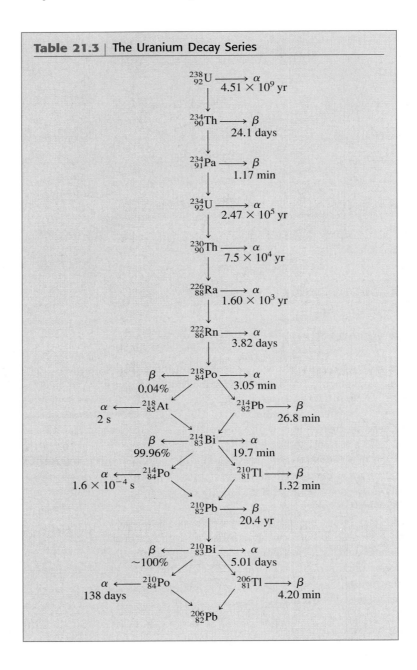

where λ is the first-order rate constant and N is the number of radioactive nuclei present at time t. (We use λ instead of k for rate constant in accord with the notation used by nuclear scientists.) According to Equation (14.3), the number of radioactive nuclei at time zero (N_0) and time t (N_t) are given by

$$\ln \frac{N_t}{N_0} = -\lambda t$$

and the corresponding half-life of the reaction is given by Equation (14.5):

$$t_{\frac{1}{2}} = \frac{0.693}{\lambda}$$

The half-lives (hence the rate constants) of radioactive isotopes vary greatly from nucleus to nucleus. For example, looking at Table 21.3, we find two extreme cases:

$$^{238}_{92}\text{U} \longrightarrow {}^{234}_{90}\text{Th} + {}^{4}_{2}\alpha \qquad t_{\frac{1}{2}} = 4.51 \times 10^9 \text{ yr}$$
$$^{214}_{84}\text{Po} \longrightarrow {}^{210}_{82}\text{Pb} + {}^{4}_{2}\alpha \qquad t_{\frac{1}{2}} = 1.6 \times 10^{-4} \text{ s}$$

The ratio of these two rate constants after conversion to the same time unit is about 1×10^{21}, an enormously large number. Furthermore, the rate constants are unaffected by changes in environmental conditions such as temperature and pressure. These highly unusual features are not seen in ordinary chemical reactions (see Table 21.1).

We do not have to wait 4.51×10^9 yr to make a half-life measurement of uranium-238. Its value can be calculated from the rate constant using Equation (14.5).

Dating Based on Radioactive Decay

The half-lives of radioactive isotopes have been used as "atomic clocks" to determine the ages of certain objects. Some examples of dating by radioactive decay measurements will be described here.

Radiocarbon Dating

The carbon-14 isotope is produced when atmospheric nitrogen is bombarded by cosmic rays:

$$^{14}_{7}\text{N} + {}^{1}_{0}\text{n} \longrightarrow {}^{14}_{6}\text{C} + {}^{1}_{1}\text{H}$$

The radioactive carbon-14 isotope decays according to the equation

$$^{14}_{6}\text{C} \longrightarrow {}^{14}_{7}\text{N} + {}^{0}_{-1}\beta \qquad t_{\frac{1}{2}} = 5730 \text{ yr}$$

The carbon-14 isotopes enter the biosphere as CO_2, which is taken up in plant photosynthesis. Plant-eating animals in turn exhale carbon-14 in CO_2. Eventually, carbon-14 participates in many aspects of the carbon cycle. The ^{14}C lost by radioactive decay is constantly replenished by the production of new isotopes in the atmosphere until a dynamic equilibrium is established whereby the ratio of ^{14}C to ^{12}C remains constant in living matter. But when an individual plant or an animal dies, the carbon-14 isotope in it is no longer replenished, so the ratio decreases as ^{14}C decays. This same change occurs when carbon atoms are trapped in coal, petroleum, or wood preserved underground, and in mummified bodies. After a number of years, there are proportionately fewer ^{14}C nuclei in a mummy than in a living person.

The decreasing ratio of ^{14}C to ^{12}C can be used to estimate the age of a specimen. Using Equation (14.3), we can write

$$\ln \frac{N_0}{N_t} = kt$$

in which N_0 and N_t are the number of ^{14}C nuclei present at $t = 0$ and $t = t$, and k is the first-order rate constant ($1.21 \times 10^{-4} \text{ yr}^{-1}$). Because the decay rate is proportional to the amount of the radioactive isotope present, we have

The age of the Shroud of Turin was shown by carbon-14 dating to be between A.D. 1260 and A.D. 1390, and therefore the shroud cannot be the burial cloth of Jesus Christ.

$$t = \frac{1}{k} \ln \frac{N_0}{N_t}$$

$$= \frac{1}{1.21 \times 10^{-4} \text{ yr}^{-1}} \ln \frac{\text{decay rate of fresh sample}}{\text{decay rate of old sample}}$$

Thus, by measuring the decay rates of the fresh sample and the old sample, we can calculate t, which is the age of the old sample. Radiocarbon dating is a valuable tool for estimating the age of objects (containing C atoms) dating back 1000 to 50,000 years.

Dating Using Uranium-238 Isotopes

Because some of the intermediate products in the uranium decay series have very long half-lives (see Table 21.3), this series is particularly suitable for estimating the age of rocks in the earth and of extraterrestrial objects. The half-life for the first step ($^{238}_{92}$U to $^{234}_{90}$Th) is 4.51×10^9 yr. This is about 20,000 times the second largest value (that is, 2.47×10^5 yr), which is the half-life for $^{234}_{92}$U to $^{230}_{90}$Th. Therefore, as a good approximation, we can assume that the half-life for the overall process (that is, from $^{238}_{92}$U to $^{206}_{82}$Pb) is governed solely by the first step:

$$^{238}_{92}\text{U} \longrightarrow {}^{206}_{82}\text{Pb} + 8\,^4_2\alpha + 6\,^0_{-1}\beta \qquad t_{\frac{1}{2}} = 4.51 \times 10^9 \text{ yr}$$

In naturally occurring uranium minerals we should and do find some lead-206 isotopes formed by radioactive decay. Assuming that no lead was present when the mineral was formed and that the mineral has not undergone chemical changes that would allow the lead-206 isotope to be separated from the parent uranium-238, it is possible to estimate the age of the rocks from the mass ratio of $^{206}_{82}$Pb to $^{238}_{92}$U. The above equation tells us that for every mole, or 238 g, of uranium that undergoes complete decay, 1 mole, or 206 g, of lead is formed. If only half a mole of uranium-238 has undergone decay, the mass ratio ^{206}Pb/^{238}U becomes

$$\frac{206 \text{ g}/2}{238 \text{ g}/2} = 0.866$$

and the process would have taken a half-life of 4.51×10^9 yr to complete (Figure 21.3). Ratios lower than 0.866 mean that the rocks are less than 4.51×10^9 yr old, and higher ratios suggest a greater age. Interestingly, studies based on the uranium series as well as other decay series put the age of the oldest rocks and, therefore, probably the age of Earth itself at 4.5×10^9, or 4.5 billion, years.

Dating Using Potassium-40 Isotopes

This is one of the most important techniques employed in geochemistry. The radioactive potassium-40 isotope decays by several different modes, but the relevant one as far as dating is concerned is that of electron capture:

$$^{40}_{19}\text{K} + {}^0_{-1}e \longrightarrow {}^{40}_{18}\text{Ar} \qquad t_{\frac{1}{2}} = 1.2 \times 10^9 \text{ yr}$$

The accumulation of gaseous argon-40 is used to gauge the age of a specimen. When a potassium-40 atom in a mineral decays, argon-40 is trapped in the lattice of the mineral and can escape only if the material is melted. Melting, therefore, is the procedure for analyzing a mineral sample in the laboratory. The amount of argon-40 present can be conveniently measured with a mass spectrometer (see p. 64).

We can think of the first step as the rate-determining step in the overall process.

Figure 21.3 *After one half-life, half of the original uranium-238 is converted to lead-206.*

Knowing the ratio of argon-40 to potassium-40 in the mineral and the half-life of decay makes it possible to establish the ages of rocks ranging from millions to billions of years old.

21.4 Nuclear Transmutation

The scope of nuclear chemistry would be rather narrow if study were limited to natural radioactive elements. An experiment performed by Rutherford in 1919, however, suggested the possibility of producing radioactivity artificially. When he bombarded a sample of nitrogen with α particles, the following reaction took place:

$$^{14}_{7}N + ^{4}_{2}\alpha \longrightarrow ^{17}_{8}O + ^{1}_{1}p$$

An oxygen-17 isotope was produced with the emission of a proton. This reaction demonstrated for the first time the feasibility of converting one element into another, by the process of nuclear transmutation. Nuclear transmutation differs from radioactive decay in that the former is brought about by the collision of two particles.

This reaction can be abbreviated as $^{14}_{7}N(\alpha,p)^{17}_{8}O$. Note that in the parentheses the bombarding particle is written first, followed by the ejected particle. Example 21.3 illustrates the use of this notation to represent nuclear transmutations.

Example 21.3   Write the balanced equation for the nuclear reaction $^{56}_{26}Fe(d,\alpha)^{54}_{25}Mn$, where d represents the deuterium nucleus (that is, $^{2}_{1}H$).

Reasoning and Solution   The abbreviation tells us that when iron-56 is bombarded with a deuterium nucleus, it produces the manganese-54 nucleus plus an α particle, $^{4}_{2}He$. Thus, the equation for this reaction is

$$^{56}_{26}Fe + ^{2}_{1}H \longrightarrow ^{4}_{2}\alpha + ^{54}_{25}Mn$$

Practice Exercise   Write a balanced equation for $^{106}_{46}Pd(\alpha,p)^{109}_{47}Ag$.

Similar problems: 21.33, 21.34.

Although light elements are generally not radioactive, they can be made so by bombarding their nuclei with appropriate particles. As we saw earlier, the radioactive carbon-14 isotope can be prepared by bombarding nitrogen-14 with neutrons. Tritium, $^{3}_{1}H$, is prepared according to the following bombardment:

$$^{6}_{3}Li + ^{1}_{0}n \longrightarrow ^{3}_{1}H + ^{4}_{2}\alpha$$

Tritium decays with the emission of β particles:

$$^{3}_{1}H \longrightarrow ^{3}_{2}He + ^{0}_{-1}\beta \qquad\qquad t_{\frac{1}{2}} = 12.5 \text{ yr}$$

Many synthetic isotopes are prepared by using neutrons as projectiles. This approach is particularly convenient because neutrons carry no charges and therefore are not repelled by the targets—the nuclei. In contrast, when the projectiles are positively charged particles (for example, protons or α particles), they must have considerable kinetic energy to overcome the electrostatic repulsion between themselves and the target nuclei. The synthesis of phosphorus from aluminum is one example:

$$^{27}_{13}Al + ^{4}_{2}\alpha \longrightarrow ^{30}_{15}P + ^{1}_{0}n$$

Figure 21.4 *Schematic diagram of a cyclotron particle accelerator. The particle (an ion) to be accelerated starts at the center and is forced to move in a spiral path through the influence of electric and magnetic fields until it emerges at a high velocity. The magnetic fields are perpendicular to the plane of the dees (so-called because of their shape), which are hollow and serve as electrodes.*

A *particle accelerator* uses electric and magnetic fields to increase the kinetic energy of charged species so that a reaction will occur (Figure 21.4). Alternating the polarity (that is, + and −) on specially constructed plates causes the particles to accelerate along a spiral path. When they have the energy necessary to initiate the desired nuclear reaction, they are guided out of the accelerator into a collision with a target substance.

Various designs have been developed for particle accelerators, one of which accelerates particles along a linear path of about 3 km (Figure 21.5). It is now possible to accelerate particles to a speed well above 90 percent of the speed of light. (According to Einstein's theory of relativity, it is impossible for a particle to move *at* the speed of light. The only exception is the photon, which has a zero rest mass.) The extremely energetic particles produced in accelerators are employed by physicists to smash atomic nuclei to fragments. Studying the debris from such disintegrations provides valuable information about nuclear structure and binding forces.

The Transuranium Elements

Particle accelerators made it possible to synthesize the so-called ***transuranium elements,*** *elements with atomic numbers greater than 92.* Neptunium (Z = 93) was first prepared in 1940. Since then, 21 other transuranium elements have been synthesized. All isotopes of these elements are radioactive. Table 21.4 lists the transuranium elements up to Z = 109 and the reactions through which they are formed.

Figure 21.5 *A section of a linear particle accelerator.*

Table 21.4 | **The Transuranium Elements***

Atomic Number	Name	Symbol	Preparation
93	Neptunium	Np	$^{238}_{92}U + ^{1}_{0}n \longrightarrow ^{239}_{93}Np + ^{0}_{-1}\beta$
94	Plutonium	Pu	$^{239}_{93}Np \longrightarrow ^{239}_{94}Pu + ^{0}_{-1}\beta$
95	Americium	Am	$^{239}_{94}Pu + ^{1}_{0}n \longrightarrow ^{240}_{95}Am + ^{0}_{-1}\beta$
96	Curium	Cm	$^{239}_{94}Pu + ^{4}_{2}\alpha \longrightarrow ^{242}_{96}Cm + ^{1}_{0}n$
97	Berkelium	Bk	$^{241}_{95}Am + ^{4}_{2}\alpha \longrightarrow ^{243}_{97}Bk + 2^{1}_{0}n$
98	Californium	Cf	$^{242}_{96}Cm + ^{4}_{2}\alpha \longrightarrow ^{245}_{98}Cf + ^{1}_{0}n$
99	Einsteinium	Es	$^{238}_{92}U + 15^{1}_{0}n \longrightarrow ^{253}_{99}Es + 7^{0}_{-1}\beta$
100	Fermium	Fm	$^{238}_{92}U + 17^{1}_{0}n \longrightarrow ^{255}_{100}Fm + 8^{0}_{-1}\beta$
101	Mendelevium	Md	$^{253}_{99}Es + ^{4}_{2}\alpha \longrightarrow ^{256}_{101}Md + ^{1}_{0}n$
102	Nobelium	No	$^{246}_{96}Cm + ^{12}_{6}C \longrightarrow ^{254}_{102}No + 4^{1}_{0}n$
103	Lawrencium	Lr	$^{252}_{98}Cf + ^{10}_{5}B \longrightarrow ^{257}_{103}Lr + 5^{1}_{0}n$
104	Rutherfordium	Rf	$^{249}_{98}Cf + ^{12}_{6}C \longrightarrow ^{257}_{104}Rf + 4^{1}_{0}n$
105	Dubnium	Db	$^{249}_{98}Cf + ^{15}_{7}N \longrightarrow ^{260}_{105}Db + 4^{1}_{0}n$
106	Seaborgium	Sg	$^{249}_{98}Cf + ^{18}_{8}O \longrightarrow ^{263}_{106}Sg + 4^{1}_{0}n$
107	Bohrium	Bh	$^{209}_{83}Bi + ^{54}_{24}Cr \longrightarrow ^{262}_{107}Bh + ^{1}_{0}n$
108	Hassium	Hs	$^{208}_{82}Pb + ^{58}_{26}Fe \longrightarrow ^{265}_{108}Hs + ^{1}_{0}n$
109	Meitnerium	Mt	$^{209}_{83}Bi + ^{58}_{26}Fe \longrightarrow ^{266}_{109}Mt + ^{1}_{0}n$

*Elements 110–112 and 114 have been synthesized but have not been named (see Problem 21.75).

21.5 Nuclear Fission

Nuclear fission is the process in which *a heavy nucleus (mass number > 200) divides to form smaller nuclei of intermediate mass and one or more neutrons*. Because the heavy nucleus is less stable than its products (see Figure 21.2), this process releases a large amount of energy.

The first nuclear fission reaction to be studied was that of uranium-235 bombarded with slow neutrons, whose speed is comparable to that of air molecules at room temperature. Under these conditions, uranium-235 undergoes fission, as shown in Figure 21.6. Actually, this reaction is very complex: More than 30 different elements have been found among the fission products (Figure 21.7). A representative reaction is

$$^{235}_{92}U + ^{1}_{0}n \longrightarrow ^{90}_{38}Sr + ^{143}_{54}Xe + 3^{1}_{0}n$$

Figure 21.6 *Nuclear fission of U-235. When a U-235 nucleus captures a neutron (red dot), it undergoes fission to yield two smaller nuclei. On the average, 2.4 neutrons are emitted for every U-235 nucleus that divides.*

Figure 21.7 *Relative yields of the products resulting from the fission of U-235, as a function of mass number.*

Although many heavy nuclei can be made to undergo fission, only the fission of naturally occurring uranium-235 and of the artificial isotope plutonium-239 has any practical importance. Table 21.5 shows the nuclear binding energies of uranium-235 and its fission products. As the table shows, the binding energy per nucleon for uranium-235 is less than the sum of the binding energies for strontium-90 and xenon-143. Therefore, when a uranium-235 nucleus is split into two smaller nuclei, a certain amount of energy is released. Let us estimate the magnitude of this energy. The difference between the binding energies of the reactants and products is $(1.23 \times 10^{-10} + 1.92 \times 10^{-10})$ J $- (2.82 \times 10^{-10})$ J, or 3.3×10^{-11} J per uranium-235 nucleus. For 1 mole of uranium-235, the energy released would be $(3.3 \times 10^{-11})(6.02 \times 10^{23})$, or 2.0×10^{13} J. This is an extremely exothermic reaction, considering that the heat of combustion of 1 ton of coal is only about 5×10^7 J.

The significant feature of uranium-235 fission is not just the enormous amount of energy released, but the fact that more neutrons are produced than are originally captured in the process. This property makes possible a ***nuclear chain reaction,*** which is *a self-sustaining sequence of nuclear fission reactions.* The neutrons generated during the initial stages of fission can induce fission in other uranium-235 nuclei, which in turn produce more neutrons, and so on. In less than a second, the reaction can become uncontrollable, liberating a tremendous amount of heat to the surroundings.

Figure 21.8 shows two types of fission reactions. For a chain reaction to occur, enough uranium-235 must be present in the sample to capture the neutrons. Otherwise, many of the neutrons will escape from the sample and the chain reaction will not occur, as depicted in Figure 21.8(a). In this situation the mass of the sample is said to be *subcritical.* Figure 21.8(b) shows what happens when the amount of the fissionable material is equal to or greater than the ***critical mass,*** *the minimum mass of fissionable material required to generate a self-sustaining nuclear chain reaction.* In this case, most of the neutrons will be captured by uranium-235 nuclei, and a chain reaction will occur.

Table 21.5 Nuclear Binding Energies of ^{235}U and Its Fission Products	
	Nuclear Binding Energy
^{235}U	2.82×10^{-10} J
^{90}Sr	1.23×10^{-10} J
^{143}Xe	1.92×10^{-10} J

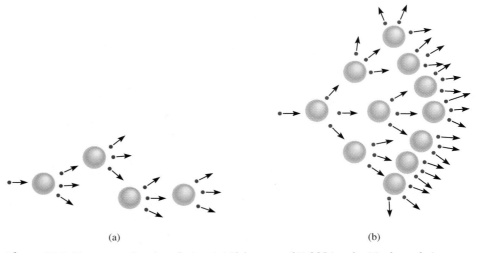

(a)   (b)

Figure 21.8 *Two types of nuclear fission. (a) If the mass of U-235 is subcritical, no chain reaction will result. Many of the neutrons produced will escape to the surroundings. (b) If a critical mass is present, many of the neutrons emitted during the fission process will be captured by other U-235 nuclei. Above the critical mass a chain reaction will occur.*

The Atomic Bomb

The first application of nuclear fission was in the development of the atomic bomb. How is such a bomb made and detonated? The crucial factor in the bomb's design is the determination of the critical mass for the bomb. A small atomic bomb is equivalent to 20,000 tons of TNT (trinitrotoluene). Because 1 ton of TNT releases about 4×10^9 J of energy, 20,000 tons would produce 8×10^{13} J. Earlier we saw that 1 mole, or 235 g, of uranium-235 liberates 2.0×10^{13} J of energy when it undergoes fission. Thus the mass of the isotope present in a small bomb must be at least

$$235 \text{ g} \times \frac{8 \times 10^{13} \text{ J}}{2.0 \times 10^{13} \text{ J}} \approx 1 \text{ kg}$$

For obvious reasons, an atomic bomb is never assembled with the critical mass already present. Instead, the critical mass is formed by using a conventional explosive, such as TNT, to force the fissionable sections together, as shown in Figure 21.9. Neutrons from a source at the center of the device trigger the nuclear chain reaction. Uranium-235 was the fissionable material in the bomb dropped on Hiroshima, Japan, on August 6, 1945. Plutonium-239 was used in the bomb exploded over Nagasaki three days later. The fission reactions generated were similar in these two cases, as was the extent of the destruction.

TNT explosive

Subcritical U-235 wedge

Figure 21.9 *Schematic cross section of an atomic bomb. The TNT explosives are set off first. The explosion forces the sections of fissionable material together to form an amount considerably larger than the critical mass.*

Nuclear Reactors

A peaceful but controversial application of nuclear fission is the generation of electricity using heat from a controlled chain reaction in a nuclear reactor. Currently, nuclear reactors provide about 20 percent of the electrical energy in the United States. This is a small but by no means negligible contribution to the nation's energy production. Several different types of nuclear reactors are in operation; we will briefly discuss the main features of three of them, along with their advantages and disadvantages.

In Europe, nuclear reactors provide about 40 percent of the electrical energy consumed.

Light Water Reactors

Most of the nuclear reactors in the United States are *light water reactors*. Figure 21.10 is a schematic diagram of such a reactor, and Figure 21.11 shows the refueling process in the core of a nuclear reactor.

An important aspect of the fission process is the speed of the neutrons. Slow neutrons split uranium-235 nuclei more efficiently than do fast ones. Because fission reactions are highly exothermic, the neutrons produced usually move at high velocities. For greater efficiency they must be slowed down before they can be used to induce nuclear disintegration. To accomplish this goal, scientists use **moderators,** which are *substances that can reduce the kinetic energy of neutrons.* A good moderator must satisfy several requirements: It should be nontoxic and inexpensive (as very large quantities of it are necessary); and it should resist conversion into a radioactive substance by neutron bombardment. Furthermore, it is advantageous for the moderator to be a fluid so that it can also be used as a coolant. No substance fulfills all these requirements, although water comes closer than many others that have been considered. Nuclear reactors that use light water (H_2O) as a moderator are called light water reactors because ^1_1H is the lightest isotope of the element hydrogen.

The nuclear fuel consists of uranium, usually in the form of its oxide, U_3O_8 (Figure 21.12). Naturally occurring uranium contains about 0.7 percent of the

Figure 21.11 *Refueling the core of a nuclear reactor.*

Shield

Steam

To steam turbine

Water

Shield

Pump

Control rod

Uranium fuel

Figure 21.10 *Schematic diagram of a nuclear fission reactor. The fission process is controlled by cadmium or boron rods. The heat generated by the process is used to produce steam for the generation of electricity via a heat exchange system.*

Figure 21.12 *Uranium oxide,* U_3O_8.

uranium-235 isotope, which is too low a concentration to sustain a small-scale chain reaction. For effective operation of a light water reactor, uranium-235 must be enriched to a concentration of 3 or 4 percent. In principle, the main difference between an atomic bomb and a nuclear reactor is that the chain reaction that takes place in a nuclear reactor is kept under control at all times. The factor limiting the rate of the reaction is the number of neutrons present. This can be controlled by lowering cadmium or boron rods between the fuel elements. These rods capture neutrons according to the equations

$$^{113}_{48}\text{Cd} + ^{1}_{0}\text{n} \longrightarrow ^{114}_{48}\text{Cd} + \gamma$$
$$^{10}_{5}\text{B} + ^{1}_{0}\text{n} \longrightarrow ^{7}_{3}\text{Li} + ^{4}_{2}\alpha$$

where γ denotes gamma rays. Without the control rods the reactor core would melt from the heat generated and release radioactive materials into the environment.

Nuclear reactors have rather elaborate cooling systems that absorb the heat given off by the nuclear reaction and transfer it outside the reactor core, where it is used to produce enough steam to drive an electric generator. In this respect, a nuclear power plant is similar to a conventional power plant that burns fossil fuel. In both cases, large quantities of cooling water are needed to condense steam for reuse. Thus, most nuclear power plants are built near a river or a lake. Unfortunately this method of cooling causes thermal pollution (see Section 13.4).

Heavy Water Reactors

Another type of nuclear reactor uses D_2O, or heavy water, as the moderator, rather than H_2O. Deuterium absorbs neutrons much less efficiently than does ordinary hydrogen. Because fewer neutrons are absorbed, the reactor is more efficient and does not require enriched uranium. The fact that deuterium is a less efficient moderator has a negative impact on the operation of the reactor, because more neutrons leak out of the reactor. However, this is not a serious disadvantage.

The main advantage of a heavy water reactor is that it eliminates the need for building expensive uranium enrichment facilities. However, D_2O must be prepared by either fractional distillation or electrolysis of ordinary water, which can be very expensive considering the amount of water used in a nuclear reactor. In countries where hydroelectric power is abundant, the cost of producing D_2O by electrolysis can be reasonably low. At present, Canada is the only nation successfully using heavy water nuclear reactors. The fact that no enriched uranium is required in a heavy water reactor allows a country to enjoy the benefits of nuclear power without undertaking work that is closely associated with weapons technology.

Breeder Reactors

A **breeder reactor** uses uranium fuel, but unlike a conventional nuclear reactor, it *produces more fissionable materials than it uses.*

We know that when uranium-238 is bombarded with fast neutrons, these reactions take place:

$$^{238}_{92}U + {}^{1}_{0}n \longrightarrow {}^{239}_{92}U$$
$$^{239}_{92}U \longrightarrow {}^{239}_{93}Np + {}^{0}_{-1}\beta \qquad t_{\frac{1}{2}} = 23.4 \text{ min}$$
$$^{239}_{93}Np \longrightarrow {}^{239}_{94}Pu + {}^{0}_{-1}\beta \qquad t_{\frac{1}{2}} = 2.35 \text{ days}$$

Plutonium-239 forms plutonium oxide, which can be readily separated from uranium.

In this manner the nonfissionable uranium-238 is transmuted into the fissionable isotope plutonium-239 (Figure 21.13).

In a typical breeder reactor, nuclear fuel containing uranium-235 or plutonium-239 is mixed with uranium-238 so that breeding takes place within the core. For every uranium-235 (or plutonium-239) nucleus undergoing fission, more than one neutron is captured by uranium-238 to generate plutonium-239. Thus, the stockpile of fissionable material can be steadily increased as the starting nuclear fuels are consumed. It takes about 7 to 10 yr to regenerate the sizable amount of material needed to refuel the original reactor and to fuel another reactor of comparable size. This interval is called the *doubling time.*

Another fertile isotope is $^{232}_{90}Th$. Upon capturing slow neutrons, thorium is transmuted to uranium-233, which, like uranium-235, is a fissionable isotope:

$$^{232}_{90}Th + {}^{1}_{0}n \longrightarrow {}^{233}_{90}Th$$
$$^{233}_{90}Th \longrightarrow {}^{233}_{91}Pa + {}^{0}_{-1}\beta \qquad t_{\frac{1}{2}} = 22 \text{ min}$$
$$^{233}_{91}Pa \longrightarrow {}^{233}_{92}U + {}^{0}_{-1}\beta \qquad t_{\frac{1}{2}} = 27.4 \text{ days}$$

Uranium-233 is stable enough for long-term storage.

Although the amounts of uranium-238 and thorium-232 in Earth's crust are relatively plentiful (4 ppm and 12 ppm by mass, respectively), the development of breeder reactors has been very slow. To date, the United States does not have a single operating breeder reactor, and only a few have been built in other countries, such as France and Russia. One problem is economics; breeder reactors are more expensive to build

Figure 21.13 *The red glow of the radioactive plutonium-239 isotope. The orange color is due to the presence of its oxide.*

than conventional reactors. There are also more technical difficulties associated with the construction of such reactors. As a result, the future of breeder reactors, in the United States at least, is rather uncertain.

Hazards of Nuclear Energy

Many people, including environmentalists, regard nuclear fission as a highly undesirable method of energy production. Fission products such as strontium-90 are dangerous radioactive isotopes with long half-lives. Plutonium-239, used as a nuclear fuel and produced in breeder reactors, is one of the most toxic substances known. It is an alpha emitter with a half-life of 24,400 yr.

Accidents, too, present many dangers. An accident at the Three Mile Island reactor in Pennsylvania in 1979 first brought the potential hazards of nuclear plants to public attention. In this instance very little radiation escaped the reactor, but the plant remained closed for more than a decade while repairs were made and safety issues addressed. Only a few years later, on April 26, 1986, a reactor at the Chernobyl nuclear plant in Ukraine surged out of control. The fire and explosion that followed released much radioactive material into the environment. People working near the plant died within weeks as a result of the exposure to the intense radiation. The long-term effect of the radioactive fallout from this incident have not yet been clearly assessed, although agriculture and dairy farming were affected by the fallout. The number of potential cancer deaths attributable to the radiation contamination is estimated to be between a few thousand and more than 100,000.

In addition to the risk of accidents, the problem of radioactive waste disposal has not been satisfactorily resolved even for safely operated nuclear plants. Many suggestions have been made as to where to store or dispose of nuclear waste, including burial underground, burial beneath the ocean floor, and storage in deep geologic formations. But none of these sites has proved absolutely safe in the long run. Leakage of radioactive wastes into underground water, for example, can endanger nearby communities. The ideal disposal site would seem to be the sun, where a bit more radiation would make little difference, but this kind of operation requires 100 percent reliability in space technology.

Because of the hazards, the future of nuclear reactors is clouded. What was once hailed as the ultimate solution to our energy needs in the twenty-first century is now being debated and questioned by both the scientific community and laypeople. It seems likely that the controversy will continue for some time.

Molten glass is poured over nuclear waste before burial.

21.6 Nuclear Fusion

In contrast to the nuclear fission process, ***nuclear fusion,*** *the combining of small nuclei into larger ones,* is largely exempt from the waste disposal problem.

Figure 21.2 showed that for the lightest elements, nuclear stability increases with increasing mass number. This behavior suggests that if two light nuclei combine or fuse together to form a larger, more stable nucleus, an appreciable amount of energy will be released in the process. This is the basis for ongoing research into the harnessing of nuclear fusion for the production of energy.

Nuclear fusion occurs constantly in the sun (Figure 21.14). The sun is made up mostly of hydrogen and helium. In its interior, where temperatures reach about 15 million degrees Celsius, the following fusion reactions are believed to take place:

Figure 21.14 *Nuclear fusion keeps the temperature in the interior of the sun at about 15 million °C.*

$$\begin{aligned}
{}^1_1\text{H} + {}^2_1\text{H} &\longrightarrow {}^3_2\text{He} \\
{}^3_2\text{He} + {}^3_2\text{He} &\longrightarrow {}^4_2\text{He} + 2\,{}^1_1\text{H} \\
{}^1_1\text{H} + {}^1_1\text{H} &\longrightarrow {}^2_1\text{H} + {}^{\ \ 0}_{+1}\beta
\end{aligned}$$

Because *fusion reactions take place only at very high temperatures,* they are often called ***thermonuclear reactions.***

Fusion Reactors

A major concern in choosing the proper nuclear fusion process for energy production is the temperature necessary to carry out the process. Some promising reactions are

Reaction	Energy Released
${}^2_1\text{H} + {}^2_1\text{H} \longrightarrow {}^3_1\text{H} + {}^1_1\text{H}$	6.3×10^{-13} J
${}^2_1\text{H} + {}^3_1\text{H} \longrightarrow {}^4_2\text{He} + {}^1_0\text{n}$	2.8×10^{-12} J
${}^6_3\text{Li} + {}^2_1\text{H} \longrightarrow 2\,{}^4_2\text{He}$	3.6×10^{-12} J

These reactions take place at extremely high temperatures, on the order of 100 million degrees Celsius, to overcome the repulsive forces between the nuclei. The first reaction is particularly attractive because the world's supply of deuterium is virtually inexhaustible. The total volume of water on Earth is about 1.5×10^{21} L. Because the natural abundance of deuterium is 1.5×10^{-2} percent, the total amount of deuterium present is roughly 4.5×10^{21} g, or 5.0×10^{15} tons. The cost of preparing deuterium is minimal compared with the value of the energy released by the reaction.

In contrast to the fission process, nuclear fusion looks like a very promising energy source, at least "on paper." Although thermal pollution would be a problem, fusion has the following advantages: (1) The fuels are cheap and almost inexhaustible and (2) the process produces little radioactive waste. If a fusion machine were turned off, it would shut down completely and instantly, without any danger of a meltdown.

If nuclear fusion is so great, why isn't there even one fusion reactor producing energy? Although we command the scientific knowledge to design such a reactor, the technical difficulties have not yet been solved. The basic problem is finding a way to hold the nuclei together long enough, and at the appropriate temperature, for fusion to occur. At temperatures of about 100 million degrees Celsius, molecules cannot exist, and most or all of the atoms are stripped of their electrons. This *state of matter, a gaseous mixture of positive ions and electrons,* is called ***plasma.*** The problem of containing this plasma is a formidable one. What solid container can exist at such temperatures? None, unless the amount of plasma is small; but then the solid surface would immediately cool the sample and quench the fusion reaction. One approach to solving this problem is to use *magnetic confinement.* Because a plasma consists of charged particles moving at high speeds, a magnetic field will exert force on it. As Figure 21.15 shows, the plasma moves through a doughnut-shaped tunnel, confined by a complex magnetic field. Thus the plasma never comes in contact with the walls of the container.

Another promising design employs high-power lasers to initiate the fusion reaction. In test runs a number of laser beams transfer energy to a small fuel pellet, heating it and causing it to *implode,* that is, to collapse inward from all sides and compress into a small volume (Figure 21.16). Consequently, fusion occurs. Like the magnetic confinement approach, laser fusion presents a number of technical difficulties that still need to be overcome before it can be put to practical use on a large scale.

Figure 21.15 *A magnetic plasma confinement design called tokamak.*

Plasma   Magnet

The Hydrogen Bomb

The technical problems inherent in the design of a nuclear fusion reactor do not affect the production of a hydrogen bomb, also called a thermonuclear bomb. In this case the objective is all power and no control. Hydrogen bombs do not contain gaseous hydrogen or gaseous deuterium; they contain solid lithium deuteride (LiD), which can be packed very tightly. The detonation of a hydrogen bomb occurs in two stages—first a fission reaction and then a fusion reaction. The required temperature for fusion is achieved with an atomic bomb. Immediately after the atomic bomb explodes, the following fusion reactions occur, releasing vast amounts of energy (Figure 21.17):

$$^6_3\text{Li} + {}^2_1\text{H} \longrightarrow 2\,{}^4_2\alpha$$
$$^2_1\text{H} + {}^2_1\text{H} \longrightarrow {}^3_1\text{H} + {}^1_1\text{H}$$

Figure 21.16 *This small-scale fusion reaction was created at the Lawrence Livermore National Laboratory using the world's most powerful laser, Nova.*

Figure 21.17 *Explosion of a thermonuclear bomb.*

There is no critical mass in a fusion bomb, and the force of the explosion is limited only by the quantity of reactants present. Thermonuclear bombs are described as being "cleaner" than atomic bombs because the only radioactive isotopes they produce are tritium, which is a weak β-particle emitter ($t_{\frac{1}{2}} = 12.5$ yr), and the products of the fission starter. Their damaging effects on the environment can be aggravated, however, by incorporating in the construction some nonfissionable material such as cobalt. Upon bombardment by neutrons, cobalt-59 is converted to cobalt-60, which is a very strong γ-ray emitter with a half-life of 5.2 yr. The presence of radioactive cobalt isotopes in the debris or fallout from a thermonuclear explosion would be fatal to those who survived the initial blast.

21.7 Uses of Isotopes

Radioactive and stable isotopes alike have many applications in science and medicine. We have previously described the use of isotopes in dating artifacts (Section 21.3). In this section we will discuss a few more examples.

Structural Determination

The formula of the thiosulfate ion is $S_2O_3^{2-}$. For some years chemists were uncertain as to whether the two sulfur atoms occupied equivalent positions in the ion. The thiosulfate ion is prepared by treatment of the sulfite ion with elemental sulfur:

$$SO_3^{2-}(aq) + S(s) \longrightarrow S_2O_3^{2-}(aq)$$

When thiosulfate is treated with dilute acid, the reaction is reversed. The sulfite ion is reformed and elemental sulfur precipitates:

$$S_2O_3^{2-}(aq) \xrightarrow{\text{H}^+} SO_3^{2-}(aq) + S(s) \tag{21.2}$$

If this sequence is started with elemental sulfur enriched with the radioactive sulfur-35 isotope, the isotope acts as a "label" for S atoms. All the labels are found in the sulfur precipitate in Equation (21.2); none of them appears in the final sulfite ions. Clearly, then, the two atoms of sulfur in $S_2O_3^{2-}$ are not structurally equivalent, as would be the case if the structure were

$$\left[\overset{\cdot\cdot}{\underset{\cdot\cdot}{:}O} - \overset{\cdot\cdot}{\underset{\cdot\cdot}{S}} - \overset{\cdot\cdot}{\underset{\cdot\cdot}{O}} - \overset{\cdot\cdot}{\underset{\cdot\cdot}{S}} - \overset{\cdot\cdot}{\underset{\cdot\cdot}{O}:} \right]^{2-}$$

Otherwise, the radioactive isotope would be present in both the elemental sulfur precipitate and the sulfite ion. Based on spectroscopic studies, we now know that the structure of the thiosulfate ion is

$$\left[\begin{array}{c} : \overset{\cdot\cdot}{S} : \\ \| \\ : \overset{\cdot\cdot}{O} - S - \overset{\cdot\cdot}{O} : \\ \| \\ : \overset{\cdot\cdot}{O} : \end{array} \right]^{2-}$$

$S_2O_3^{2-}$

Study of Photosynthesis

The study of photosynthesis is also rich with isotope applications. The overall photosynthesis reaction can be represented as

$$6CO_2 + 6H_2O \longrightarrow C_6H_{12}O_6 + 6O_2$$

The radioactive ^{14}C isotope has helped to determine the path of carbon in photosynthesis. Starting with $^{14}CO_2$, it was possible to isolate the intermediate products during photosynthesis and measure the amount of radioactivity of each carbon-containing compound. In this manner the path from CO_2 through various intermediate compounds to carbohydrate could be clearly charted. *Isotopes, especially radioactive isotopes that are used to trace the path of the atoms of an element in a chemical or biological process,* are called **tracers.**

Isotopes in Medicine

Tracers are used also for diagnosis in medicine. Sodium-24 (a β emitter with a half-life of 14.8 hr) injected into the bloodstream as a salt solution can be monitored to trace the flow of blood and detect possible constrictions or obstructions in the circulatory system. Iodine-131 (a β emitter with a half-life of 8 days) has been used

Figure 21.18 *A compound labeled with ^{123}I is used to image the brain. Left: A normal brain. Right: The brain of a patient with Alzheimer's disease.*

Figure 21.19 *Schematic diagram of a Geiger counter. Radiation (α, β, or γ rays) entering through the window ionized the argon gas to generate a small current flow between the electrodes. This current is amplified and is used to flash a light or operate a counter with a clicking sound.*

to test the activity of the thyroid gland. A malfunctioning thyroid can be detected by giving the patient a drink of a solution containing a known amount of $Na^{131}I$ and measuring the radioactivity just above the thyroid to see if the iodine is absorbed at the normal rate. Of course, the amounts of radioisotope used in the human body must always be kept small; otherwise, the patient might suffer permanent damage from the high-energy radiation. Another radioactive isotope of iodine, iodine-123 (a γ-ray emitter), is used to image the brain (Figure 21.18).

Technetium is one of the most useful elements in nuclear medicine. Although technetium is a transition metal, all its isotopes are radioactive. In the laboratory it is prepared by the nuclear reactions

$$^{98}_{42}Mo + ^{1}_{0}n \longrightarrow ^{99}_{42}Mo$$
$$^{99}_{42}Mo \longrightarrow ^{99m}_{43}Tc + ^{0}_{-1}\beta$$

where the superscript m denotes that the technetium-99 isotope is produced in its excited nuclear state. This isotope has a half-life of about 6 hours, decaying by γ radiation to technetium-99 in its nuclear ground state. Thus it is a valuable diagnostic tool. The patient either drinks or is injected with a solution containing ^{99m}Tc. By detecting the γ rays emitted by ^{99m}Tc, doctors can obtain images of organs such as the heart, liver, and lungs.

A major advantage of using radioactive isotopes as tracers is that they are easy to detect. Their presence even in very small amounts can be detected by photographic techniques or by devices known as counters. Figure 21.19 is a diagram of a Geiger counter, an instrument widely used in scientific work and medical laboratories to detect radiation.

Technetium was the first artificially prepared element.

21.8 Biological Effects of Radiation

In this section we will examine briefly the effects of radiation on biological systems. But first let us define quantitative measures of radiation. The fundamental unit of radioactivity is the *curie* (Ci); 1 Ci corresponds to exactly 3.70×10^{10} nuclear disintegrations per second. This decay rate is equivalent to that of 1 g of radium. A *millicurie* (mCi) is one-thousandth of a curie. Thus, 10 mCi of a carbon-14 sample is the quantity that undergoes

$$(10 \times 10^{-3})(3.70 \times 10^{10}) = 3.70 \times 10^{8}$$

disintegrations per second. The intensity of radiation depends on the number of disintegrations as well as on the energy and type of radiation emitted. One common unit for the absorbed dose of radiation is the *rad* (*r*adiation *a*bsorbed *d*ose), which is the amount of radiation that results in the absorption of 1×10^{-5} J per gram of irradiated material. The biological effect of radiation depends on the part of the body irradiated and the type of radiation. For this reason the rad is often multiplied by a factor called *RBE* (*r*elative *b*iological *e*ffectiveness). The product is called a *rem* (*r*oentgen *e*quivalent for *m*an):

$$1 \text{ rem} = 1 \text{ rad} \times 1 \text{ RBE}$$

Of the three types of nuclear radiation, α particles usually have the least penetrating power. Beta particles are more penetrating than α particles, but less so than γ rays. Gamma rays have very short wavelengths and high energies. Furthermore, because they carry no charge, they cannot be stopped by shielding materials as easily as α and β particles. However, if α or β emitters are ingested, their damaging effects are greatly aggravated because the organs will be constantly subject to damaging radiation at close range. For example, strontium-90, a β emitter, can replace calcium in bones, where it does the greatest damage.

Table 21.6 lists the average amounts of radiation an American receives every year. It should be pointed out that for short-term exposures to radiation, a dosage of 50–200 rem will cause a decrease in white blood cell counts and other complications, while a dosage of 500 rem or greater may result in death within weeks. Current safety standards permit nuclear workers to be exposed to no more than 5 rem per year and specify a maximum of 0.5 rem of human-made radiation per year for the general public.

The chemical basis of radiation damage is that of ionizing radiation. Radiation of either particles or γ rays can remove electrons from atoms and molecules in its path, leading to the formation of ions and radicals. For example, when water is irradiated with γ rays, the following reactions take place:

$$H_2O \xrightarrow{\text{radiation}} H_2O^+ + e^-$$

$$H_2O^+ + H_2O \longrightarrow H_3O^+ + \underset{\text{hydroxyl radical}}{\cdot OH}$$

The electron (in the hydrated form) can subsequently react with water or with a hydrogen ion to form atomic hydrogen, and with oxygen to produce the superoxide ion, O_2^- (a radical):

$$e^- + O_2 \longrightarrow \cdot O_2^-$$

Table 21.6 | Average Yearly Radiation Doses for Americans

Source	Dose (mrem/yr)*
Cosmic rays	20–50
Ground and surroundings	25
Human body†	26
Medical and dental X rays	50–75
Air travel	5
Fallout from weapons tests	5
Nuclear waste	2
Total	133–188

*1 mrem = 1 millirem = 1×10^{-3} rem.
†The radioactivity in the body comes from food and air.

In the tissues, the superoxide ions and other radicals attack cell membranes and a host of organic compounds, such as enzymes and DNA molecules. Organic compounds can themselves be directly ionized and destroyed by high-energy radiation.

It has long been known that exposure to high-energy radiation can induce cancer in humans and other animals. Cancer is characterized by uncontrolled cellular growth. On the other hand, it is also well established that cancer cells can be destroyed by proper radiation treatment. In radiation therapy, a compromise is sought. The radiation to which the patient is exposed must be sufficient to destroy cancer cells without killing too many normal cells and, it is hoped, without inducing another form of cancer.

Radiation damage to living systems is generally classified as *somatic* or *genetic*. Somatic injuries are those that affect the organism during its own lifetime. Sunburn, skin rash, cancer, and cataracts are examples of somatic damage. Genetic damage means inheritable changes or gene mutations. For example, a person whose chromosomes have been damaged or altered by radiation may have deformed offspring.

Chromosomes are parts of the cell structure that contain the genetic material (DNA).

Summary

Nuclear chemistry is the study of changes in atomic nuclei. Such changes are termed nuclear reactions. Radioactive decay and nuclear transmutation are nuclear reactions.

For stable nuclei of low atomic number, the neutron-to-proton ratio is close to 1. For heavier stable nuclei, the ratio becomes greater than 1. All nuclei with 84 or more protons are unstable and radioactive. Nuclei with even atomic numbers are more stable than those with odd atomic numbers. A quantitative measure of nuclear stability is the nuclear binding energy, which can be calculated from a knowledge of the mass defect of the nucleus.

Radioactive nuclei emit α particles, β particles, positrons, or γ rays. The equation for a nuclear reaction includes the particles emitted, and both the mass numbers and the atomic numbers must balance. Uranium-238 is the parent of a natural radioactive decay series. A number of radioactive isotopes, such as ^{238}U and ^{14}C, can be used to date objects. Artificially radioactive elements are created by the bombardment of other elements by accelerated neutrons, protons, or α particles.

Nuclear fission is the splitting of a large nucleus into smaller nuclei plus neutrons. When these neutrons are captured efficiently by other nuclei, an uncontrollable chain reaction can occur. Nuclear reactors use the heat from a controlled nuclear fission reaction to produce power. The three important types of reactors are light water reactors, heavy water reactors, and breeder reactors.

Nuclear fusion, the type of reaction that occurs in the sun, is the combination of two light nuclei to form one heavy nucleus. Fusion takes place only at very high temperatures—so high that controlled large-scale nuclear fusion has so far not been achieved.

Radioactive isotopes are easy to detect and thus make excellent tracers in chemical reactions and in medical practice. High-energy radiation damages living systems by causing ionization and the formation of reactive radicals.

Key Words

Breeder reactor p. 689
Critical mass p. 686
Mass defect p. 676
Moderators p. 687

Nuclear binding energy
   p. 676
Nuclear chain reaction
   p. 686

Nuclear fission p. 685
Nuclear fusion p. 690
Nuclear reaction
   p. 672

Nuclear transmutation
   p. 672
Nucleon p. 676
Plasma p. 691

Positron p. 673
Radioactive decay series
    p. 679

Thermonuclear reaction
    p. 691
Tracer p. 694

Transuranium elements
    p. 684

Questions and Problems

Nuclear Reactions

Review Questions

21.1 How do nuclear reactions differ from ordinary chemical reactions?

21.2 What are the steps in balancing nuclear equations?

21.3 What is the difference between $_{-1}^{0}e$ and $_{-1}^{0}\beta$?

21.4 What is the difference between an electron and a positron?

Problems

21.5 Complete these nuclear equations and identify X in each case:

(a) $_{12}^{26}\text{Mg} + _{1}^{1}\text{p} \longrightarrow _{2}^{4}\alpha + \text{X}$
(b) $_{27}^{59}\text{Co} + _{1}^{2}\text{H} \longrightarrow _{27}^{60}\text{Co} + \text{X}$
(c) $_{92}^{235}\text{U} + _{0}^{1}\text{n} \longrightarrow _{36}^{94}\text{Kr} + _{56}^{139}\text{Ba} + 3\text{X}$
(d) $_{24}^{53}\text{Cr} + _{2}^{4}\alpha \longrightarrow _{0}^{1}\text{n} + \text{X}$
(e) $_{8}^{20}\text{O} \longrightarrow _{9}^{20}\text{F} + \text{X}$

21.6 Complete these nuclear equations and identify X in each case:

(a) $_{53}^{135}\text{I} \longrightarrow _{54}^{135}\text{Xe} + \text{X}$
(b) $_{19}^{40}\text{K} \longrightarrow _{-1}^{0}\beta + \text{X}$
(c) $_{27}^{59}\text{Co} + _{0}^{1}\text{n} \longrightarrow _{25}^{56}\text{Mn} + \text{X}$
(d) $_{92}^{235}\text{U} + _{0}^{1}\text{n} \longrightarrow _{40}^{99}\text{Sr} + _{52}^{135}\text{Te} + 2\text{X}$

Nuclear Stability

Review Questions

21.7 State the general rules for predicting nuclear stability.

21.8 What is the belt of stability?

21.9 Why is it impossible for the isotope $_{2}^{2}\text{He}$ to exist?

21.10 Define nuclear binding energy, mass defect, and nucleon.

21.11 How does Einstein's equation, $E = mc^2$, enable us to calculate nuclear binding energy?

21.12 Why is it preferable to use nuclear binding energy per nucleon for a comparison of the stabilities of different nuclei?

Problems

21.13 The radius of a uranium-235 nucleus is about 7.0×10^{-3} pm. Calculate the density of the nucleus in g/cm^3. (Assume the atomic mass is 235 amu.)

21.14 For each pair of isotopes listed, predict which one is less stable: (a) $_{3}^{6}\text{Li}$ or $_{3}^{9}\text{Li}$, (b) $_{11}^{23}\text{Na}$ or $_{11}^{25}\text{Na}$, (c) $_{20}^{48}\text{Ca}$ or $_{21}^{48}\text{Sc}$.

21.15 For each pair of elements listed, predict which one has more stable isotopes: (a) Co or Ni, (b) F or Se, (c) Ag or Cd.

21.16 In each pair of isotopes shown, indicate which one you would expect to be radioactive: (a) $_{10}^{20}\text{Ne}$ and $_{10}^{17}\text{Ne}$, (b) $_{20}^{40}\text{Ca}$

and $_{20}^{45}\text{Ca}$, (c) $_{42}^{95}\text{Mo}$ and $_{43}^{92}\text{Tc}$, (d) $_{80}^{195}\text{Hg}$ and $_{80}^{196}\text{Hg}$, (e) $_{83}^{209}\text{Bi}$ and $_{96}^{242}\text{Cm}$.

21.17 Given that

$$\text{H}(g) + \text{H}(g) \longrightarrow \text{H}_2(g) \quad \Delta H^\circ = -436.4 \text{ kJ}$$

calculate the change in mass (in kg) per mole of H_2 formed.

21.18 Estimates show that the total energy output of the sun is 5×10^{26} J/s. What is the corresponding mass loss in kg/s of the sun?

21.19 Calculate the nuclear binding energy (in J) and the nuclear binding energy per nucleon of the following isotopes: (a) $_{3}^{7}\text{Li}$ (7.01600 amu) and (b) $_{17}^{35}\text{Cl}$ (34.95952 amu).

21.20 Calculate the nuclear binding energy (in J) and the nuclear binding energy per nucleon of the following isotopes: (a) $_{2}^{4}\text{He}$ (4.0026 amu) and (b) $_{74}^{184}\text{W}$ (183.9510 amu).

Natural Radioactivity

Review Questions

21.21 Discuss factors that lead to nuclear decay.

21.22 Outline the principle for dating materials using radioactive isotopes.

Problems

21.23 Fill in the blanks in these radioactive decay series:

(a) $^{232}\text{Th} \xrightarrow{\alpha} \underline{\quad} \xrightarrow{\beta} \underline{\quad} \xrightarrow{\beta} {}^{228}\text{Th}$
(b) $^{235}\text{U} \xrightarrow{\alpha} \underline{\quad} \xrightarrow{\beta} \underline{\quad} \xrightarrow{\alpha} {}^{227}\text{Ac}$
(c) $\underline{\quad} \xrightarrow{\alpha} {}^{233}\text{Pa} \xrightarrow{\beta} \underline{\quad} \xrightarrow{\alpha} \underline{\quad}$

21.24 A radioactive substance undergoes decay as:

Time (days)	Mass (g)
0	500
1	389
2	303
3	236
4	184
5	143
6	112

Calculate the first-order decay constant and the half-life of the reaction.

21.25 The radioactive decay of Tl-206 to Pb-206 has a half-life of 4.20 min. Starting with 5.00×10^{22} atoms of Tl-206, calculate the number of such atoms left after 42.0 min.

21.26 A freshly isolated sample of ^{90}Y was found to have an activity of 9.8×10^5 disintegrations per minute at 1:00 P.M. on December 3, 2000. At 2:15 P.M. on December 17, 2000, its activity was redetermined and found to be 2.6×10^4 disintegrations per minute. Calculate the half-life of ^{90}Y.

21.27 Why do radioactive decay series obey first-order kinetics?

21.28 In the thorium decay series, thorium-232 loses a total of 6 α particles and 4 β particles in a 10-stage process. What is the final isotope produced?

21.29 Strontium-90 is one of the products of the fission of uranium-235. This strontium isotope is radioactive, with a half-life of 28.1 yr. Calculate how long (in yr) it will take for 1.00 g of the isotope to be reduced to 0.200 g by decay.

21.30 Consider the decay series

$$A \longrightarrow B \longrightarrow C \longrightarrow D$$

where A, B, and C are radioactive isotopes with half-lives of 4.50 s, 15.0 days, and 1.00 s, respectively, and D is non-radioactive. Starting with 1.00 mole of A, and none of B, C, or D, calculate the number of moles of A, B, C, and D left after 30 days.

Nuclear Transmutation

Review Questions

21.31 What is the difference between radioactive decay and nuclear transmutation?

21.32 How is nuclear transmutation achieved in practice?

Problems

21.33 Write balanced nuclear equations for these reactions and identify X: (a) $X(p,\alpha)^{12}_{6}C$, (b) $^{27}_{13}Al(d,\alpha)X$, (c) $^{55}_{25}Mn(n,\gamma)X$

21.34 Write balanced nuclear equations for these reactions and identify X: (a) $^{80}_{34}Se(d,p)X$, (b) $X(d,2p)^{9}_{3}Li$, (c) $^{10}_{5}B(n,\alpha)X$

21.35 Describe how you would prepare astatine-211, starting with bismuth-209.

21.36 A long-cherished dream of alchemists was to produce gold from cheaper and more abundant elements. This dream was finally realized when $^{198}_{80}Hg$ was converted into gold by neutron bombardment. Write a balanced equation for this reaction.

Nuclear Fission

Review Questions

21.37 Define nuclear fission, nuclear chain reaction, and critical mass.

21.38 Which isotopes can undergo nuclear fission?

21.39 Explain how an atomic bomb works.

21.40 Explain the functions of a moderator and a control rod in a nuclear reactor.

21.41 Discuss the differences between a light water and a heavy water nuclear fission reactor. What are the advantages of a breeder reactor over a conventional nuclear fission reactor?

21.42 What makes water particularly suitable for use as moderator in a nuclear reactor?

Nuclear Fusion

Review Questions

21.43 Define nuclear fusion, thermonuclear reaction, and plasma.

21.44 Why do heavy elements such as uranium undergo fission while light elements such as hydrogen and lithium undergo fusion?

21.45 How does a hydrogen bomb work?

21.46 What are the advantages of a fusion reactor over a fission reactor? What are the practical difficulties in operating a large-scale fusion reactor?

Uses of Isotopes

Problems

21.47 Describe how you would use a radioactive iodine isotope to demonstrate that the following process is in dynamic equilibrium:

$$PbI_2(s) \rightleftharpoons Pb^{2+}(aq) + 2I^-(aq)$$

21.48 Consider this redox reaction:

$$IO_4^-(aq) + 2I^-(aq) + H_2O(l) \longrightarrow$$
$$I_2(s) + IO_3^-(aq) + 2OH^-(aq)$$

When KIO_4 is added to a solution containing iodide ions labeled with radioactive iodine-128, all the radioactivity appears in I_2 and none in the IO_3^- ion. What can you deduce about the mechanism for the redox process?

21.49 Explain how you might use a radioactive tracer to show that ions are not completely motionless in crystals.

21.50 Each molecule of hemoglobin, the oxygen carrier in blood, contains four Fe atoms. Explain how you would use the radioactive $^{59}_{26}Fe$ ($t_{\frac{1}{2}} = 46$ days) to show that the iron in a certain food is converted into hemoglobin.

Additional Problems

21.51 How does a Geiger counter work?

21.52 Nuclei with an even number of protons and an even number of neutrons are more stable than those with an odd number of protons and/or an odd number of neutrons. What is the significance of the even numbers of protons and neutrons in this case?

21.53 Tritium, ^3H, is radioactive and decays by electron emission. Its half-life is 12.5 yr. In ordinary water the ratio of ^1H to ^3H atoms is 1.0×10^{17} to 1. (a) Write a balanced nuclear equation for tritium decay. (b) How many disintegrations will be observed per minute in a 1.00-kg sample of water?

21.54 (a) What is the activity, in millicuries, of a 0.500-g sample of $^{237}_{93}Np$? (This isotope decays by α-particle emission and has a half-life of 2.20×10^6 yr.) (b) Write a balanced nuclear equation for the decay of $^{237}_{93}Np$.

21.55 These equations are for nuclear reactions that are known to occur in the explosion of an atomic bomb. Identify X.

(a) $^{235}_{92}U + ^{1}_{0}n \longrightarrow ^{140}_{56}Ba + 3^{1}_{0}n + X$
(b) $^{235}_{92}U + ^{1}_{0}n \longrightarrow ^{144}_{55}Cs + ^{90}_{37}Rb + 2X$
(c) $^{235}_{92}U + ^{1}_{0}n \longrightarrow ^{87}_{35}Br + 3^{1}_{0}n + X$
(d) $^{235}_{92}U + ^{1}_{0}n \longrightarrow ^{160}_{62}Sm + ^{72}_{30}Zn + 4X$

21.56 Calculate the nuclear binding energies, in J/nucleon, for these species: (a) ^{10}B (10.0129 amu), (b) ^{11}B (11.00931 amu), (c) ^{14}N (14.00307 amu), (d) ^{56}Fe (55.9349 amu).

21.57 Write complete nuclear equations for these processes: (a) tritium, ^3H, undergoes β decay; (b) ^{242}Pu undergoes α-particle emission; (c) ^{131}I undergoes β decay; (d) ^{251}Cf emits an α particle.

21.58 The nucleus of nitrogen-18 lies above the stability belt. Write an equation for a nuclear reaction by which nitrogen-18 can achieve stability.

21.59 Why is strontium-90 a particularly dangerous isotope for humans?

21.60 How are scientists able to tell the age of a fossil?

21.61 After the Chernobyl accident, people living close to the nuclear reactor site were urged to take large amounts of potassium iodide as a safety precaution. What is the chemical basis for this action?

21.62 Astatine, the last member of Group 7A, can be prepared by bombarding bismuth-209 with α particles. (a) Write an equation for the reaction. (b) Represent the equation in the abbreviated form as discussed in Section 21.4.

21.63 To detect bombs that may be smuggled onto airplanes, the Federal Aviation Administration (FAA) will soon require all major airports in the United States to install thermal neutron analyzers. The thermal neutron analyzer will bombard baggage with low-energy neutrons, converting some of the nitrogen-14 nuclei to nitrogen-15, with simultaneous emission of γ rays. Because nitrogen content is usually high in explosives, detection of a high dosage of γ rays will suggest that a bomb may be present. (a) Write an equation for the nuclear process. (b) Compare this technique with the conventional X-ray detection method.

21.64 Explain why achievement of nuclear fusion in the laboratory requires a temperature of about 100 million degrees Celsius, which is much higher than that in the interior of the sun (15 million degrees Celsius).

21.65 Tritium contains one proton and two neutrons. There is no proton-proton repulsion present in the nucleus. Why, then, is tritium radioactive?

21.66 The carbon-14 decay rate of a sample obtained from a young tree is 0.260 disintegration per second per gram of

the sample. Another wood sample prepared from an object recovered at an archaeological excavation gives a decay rate of 0.186 disintegration per second per gram of the sample. What is the age of the object?

21.67 The usefulness of radiocarbon dating is limited to objects no older than 50,000 years. What percent of the carbon-14, originally present in the sample, remains after this period of time?

21.68 The radioactive potassium-40 isotope decays to argon-40 with a half-life of 1.2×10^9 yr. (a) Write a balanced equation for the reaction. (b) A sample of moon rock is found to contain 18 percent potassium-40 and 82 percent argon by mass. Calculate the age of the rock in years.

21.69 Both barium (Ba) and radium (Ra) are members of Group 2A and are expected to exhibit similar chemical properties. However, Ra is not found in barium ores. Instead, it is found in uranium ores. Explain.

21.70 Nuclear waste disposal is one of the major concerns of the nuclear industry. In choosing a safe and stable environment to store nuclear wastes, consideration must be given to the heat released during nuclear decay. As an example, consider the β decay of ^{90}Sr (89.907738 amu):

$$^{90}_{38}Sr \longrightarrow ^{90}_{39}Y + ^{0}_{-1}\beta \qquad t_{\frac{1}{2}} = 28.1 \text{ yr}$$

The ^{90}Y (89.907152 amu) further decays as follows:

$$^{90}_{39}Y \longrightarrow ^{90}_{40}Zr + ^{0}_{-1}\beta \qquad t_{\frac{1}{2}} = 64 \text{ hr}$$

Zirconium-90 (89.904703 amu) is a stable isotope. (a) Use the mass defect to calculate the energy released (in joules) in each of the above two decays. (The mass of the electron is 5.4857×10^{-4} amu.) (b) Starting with 1 mole of ^{90}Sr, calculate the number of moles of ^{90}Sr that will decay in a year. (c) Calculate the amount of heat released (in kilojoules) corresponding to the number of moles of ^{90}Sr decayed to ^{90}Zr in (b).

21.71 Which of the following poses a greater health hazard: a radioactive isotope with a short half-life or a radioactive isotope with a long half-life? Explain. [Assume same type of radiation (α or β) and comparable energetics per particle emitted.]

21.72 As a result of being exposed to the radiation released during the Chernobyl nuclear accident, the dose of iodine-131 in a person's body is 7.4 mC ($1 \text{ mC} = 1 \times 10^{-3}$ Ci). Use the relationship rate $= \lambda N$ to calculate the number of atoms of iodine-131 to which this radioactivity corresponds. (The half-life of ^{131}I is 8.1 days.)

21.73 Bismuth-214 is an α-emitter with a half-life of 19.7 min. A 5.26-mg sample of the isotope is placed in a sealed, evacuated flask of volume 20.0 mL at 40°C. Assuming that all the α particles generated are converted to helium gas and that the other decay product is nonradioactive, calculate the pressure (in mmHg) inside the flask after 78.8 min. Use 214 amu for the atomic mass of bismuth.

21.74 From the definition of curie, calculate Avogadro's number. Given that the molar mass of ^{226}Ra is 226.03 g/mol and that it decays with a half life of 1.6×10^3 yr.

21.75 (a) Since 1994, elements 110, 111, and 112 have been synthesized. Element 110 was created by bombarding ^{208}Pb with ^{62}Ni; element 111 was created by bombarding ^{209}Bi with ^{64}Ni; and element 112 was created by bombarding ^{208}Pb with ^{66}Zn. Write an equation for each synthesis. Predict the chemical properties of these elements. Use X for element 110, Y for element 111, and Z for element 112. (b) In 1999 element 114 was also synthesized by bombarding ^{244}Pu with ^{48}Ca. Write an equation for its synthesis and predict its chemical property.

21.76 Sources of energy on Earth include fossil fuels, geothermal, gravitational, hydroelectric, nuclear fission, nuclear fusion, solar, and wind. Which of these have a "nuclear origin," either directly or indirectly?

21.77 A person received an anonymous gift of a decorative cube which he placed on his desk. A few months later he became ill and died shortly afterward. After investigation, the cause of his death was linked to the box. The box was air-tight and had no toxic chemicals on it. What might have killed the man?

21.78 Identify two of the most abundant radioactive elements that exist on Earth. Explain why they are still present. (You may need to consult a handbook of chemistry.)

21.79 (a) Calculate the energy released when an ^{238}U isotope decays to ^{234}Th. The atomic masses are given by: ^{238}U: 238.0508 amu; ^{234}Th: 234.0436 amu; ^{4}He: 4.0026 amu. (b) The energy released in (a) is transformed into the kinetic energy of the recoiling ^{234}Th nucleus and the α particle. Which of the two will move away faster? Explain.

21.80 Cobalt-60 is an isotope used in diagnostic medicine and cancer treatment. It decays with γ ray emission. Calculate the wavelength of the radiation in nanometers if the energy of the γ ray is 2.4×10^{-13} J/photon.

21.81 Americium-241 is used in smoke detectors because it has a long half-life (458 yr) and its emitted α particles are energetic enough to ionize air molecules. Given the schematic diagram of a smoke detector below, explain how it works.

Current

^{241}Am   Battery

21.82 The constituents of wine contain, among others, carbon, hydrogen, and oxygen atoms. A bottle of wine was sealed about 6 yr ago. To confirm its age, which of the isotopes would you choose in a radioactive dating study? The half-lives of the isotopes are: ^{13}C: 5730 yr; ^{15}O: 124 s; ^{3}H: 12.5 yr. Assume that the activities of the isotopes were known at the time the bottle was sealed.

21.83 Name two advantages of a nuclear-powered submarine over a conventional submarine.

21.84 In 1997 a scientist at a nuclear research center in Russia placed a thin shell of copper on a sphere of highly enriched uranium-235. Suddenly, there was a huge burst of radiation, which turned the air blue. Three days later, the scientist died of radiation damage. Explain what caused the accident. (*Hint:* Copper is an effective metal for reflecting neutrons.)

Special Problems

21.85 Describe, with appropriate equations, nuclear processes that lead to the formation of the noble gases He, Ne, Ar, Kr, Xe, and Rn. (*Hint:* Helium is formed from radioactive decay, neon is formed from the positron emission of ^{22}Na, the formation of Ar, Xe, and Rn are discussed in the chapter, and Kr is produced from the fission of ^{235}U.)

21.86 Write an essay on the pros and cons of nuclear power (based on nuclear fission), paying particular attention to its effect on global warming, nuclear reactor safety and weapon risks, and nuclear waste disposal.

Answers to Practice Exercises

21.1 $^{78}_{34}$Se. **21.2** 2.63×10^{-10} J; 1.26×10^{-12} J/nucleon.
21.3 $^{106}_{46}$Pd $+ \, ^4_2\alpha \longrightarrow \, ^{109}_{47}$Ag $+ \, ^1_1$p.

Our planet probably condensed from interstellar dust about 4.5 billion years ago. The land then was barren; while water was present, the sterile seas harbored nothing but minerals. The primitive atmosphere was reducing, containing NH_3, CH_4, H_2O, and CO_2. It was proposed in the 1920s by some scientists that organic compounds could be synthesized nonbiologically under primitive Earth conditions. The energy required for these processes may have

Electrical discharge

Ammonia, methane, and hydrogen gases plus water vapor

Condenser

Water

Water

Boiling water

Solution containing amino acids and other simple organic compounds

been supplied by the sun, whose ultraviolet radiation would have penetrated Earth's atmosphere in the absence of the ozone layer. Without molecular oxygen to oxidize them, these organic molecules would be stable and would accumulate to form a "primordial soup" out of which the first living organism evolved. Photosynthesis probably arose as CO_2 became plentiful and the oxygen released from the process then could slowly change the atmosphere from reducing to oxidizing.

In 1953 the chemist Stanley Miller carried out an experiment that gave much credibility to the chemical origin of life theory. Using an apparatus such as that shown, Miller subjected a mixture of H_2, CH_4, and NH_3 to an electric discharge, in the presence of steam. After running the process for a week, the products were found to include several amino acids, and a number of other metabolic compounds such as urea and lactic acid. Since then, many other biologically important compounds such as sugars, purines, and pyrimidines have been synthesized under

simulated primitive Earth conditions, using UV radiation as well as heating and radiation from radioactive sources.

Miller's work provided strong support for the chemical basis of evolution. In addition to the results of simulation experiments, in 1996 scientists made another startling discovery. Analysis of a meteorite from Mars that year revealed polyaromatic hydrocarbons (PAHs), which are large complex organic molecules that may be remnants of biological activity. (It is believed that an impact between a meteorite and a rock on Mars' surface blasted fragments of the rock into space, some of which eventually landed on Earth.) Assuming the PAHs are not contaminants, this finding lends support for extraterrestrial life. Eventually scientists hope to find traces of amino acids in the rock, which are better markers for life.

Another question that must be addressed is: Because many biological compounds are chiral and the reactions they undergo are specific only for one of the enantiomers, how was the choice made in the first place? No clear answer is available, although in recent years astronomers have detected polarized light from distance stars, which may be the "template" required for photochemically synthesizing one particular isomer of an enantiomeric pair.

Considering the number of galaxies in the universe and the multitudes of stars and planets in each galaxy, the existence of life on other planets should be almost a foregone conclusion. The only question that remains is, What form of life may have evolved under conditions characteristic of these planets?

Organic Polymers— Synthetic and Natural

Chapter 22

ESSENTIAL CONCEPTS

 Synthetic Organic Polymers Many organic polymers have been synthesized by various chemical processes. They mimic and sometimes surpass the properties of those that occur naturally. Nylon is the best known of all synthetic organic polymers.

 Proteins Proteins are natural polymers made of amino acids. They perform a host of functions ranging from catalysis, transport and storage of vital substances, coordinated motion, and protection against diseases. The complex structures of proteins have been analyzed in terms of their primary, secondary, tertiary, and quaternary structures. The three-dimensional integrity of protein molecules is maintained by various intermolecular forces and hydrogen bonding.

 Nucleic Acids Deoxyribonucleic acid (DNA) carries all the genetic information and ribonucleic acid (RNA) controls the synthesis of proteins. The elucidation of the double-helical structure of DNA is one of the major accomplishments in science in the twentieth century.

22.1 Properties of Polymers

A *polymer* is *a molecular compound distinguished by a high molar mass, ranging into thousands and millions of grams, and made up of many repeating units.* The physical properties of these so-called macromolecules differ greatly from those of small, ordinary molecules, and special techniques are required to study them.

Synthetic polymers have many useful properties. The strength of one kind of polymer called Lexan is so great that it is used to make bullet-proof windows.

Naturally occurring polymers include proteins, nucleic acids, cellulose (polysaccharides), and rubber (polyisoprene). Most synthetic polymers are organic compounds. Familiar examples are nylon, poly(hexamethylene adipamide); Dacron, poly(ethylene terephthalate); and Lucite or Plexiglas, poly(methyl methacrylate).

The development of polymer chemistry began in the 1920s with the investigation into a puzzling behavior of certain materials, including wood, gelatin, cotton, and rubber. For example, when rubber, with the known empirical formula of C_5H_8, was dissolved in an organic solvent, the solution displayed several unusual properties—high viscosity, low osmotic pressure, and negligible freezing-point depression. These observations strongly suggested the presence of solutes of very high molar mass, but chemists were not ready at that time to accept the idea that such giant molecules could exist. Instead, they postulated that materials such as rubber consist of aggregates of small molecular units, like C_5H_8 or $C_{10}H_{16}$, held together by intermolecular forces. This misconception persisted for a number of years, until the German chemist Hermann Staudinger clearly showed that these so-called aggregates are, in fact, enormously large molecules, each of which contains many thousands of atoms held together by covalent bonds.

Once the structures of these macromolecules were understood, the way was open for manufacturing polymers, which now pervade almost every aspect of our daily lives. About 90 percent of today's chemists, including biochemists, work with polymers.

22.2 Synthetic Organic Polymers

Because of their size, we might expect molecules containing thousands of carbon and hydrogen atoms to form an enormous number of structural and geometric isomers (if C=C bonds are present). However, these molecules are made up of *monomers, simple repeating units,* and this type of composition severely restricts the number of possible isomers. Synthetic polymers are created by joining monomers together, one at a time, by means of addition reactions and condensation reactions.

Addition Reactions

Addition reactions were described on p. 345.

Addition reactions involve unsaturated compounds containing double or triple bonds, particularly C=C and C≡C. Hydrogenation and reactions of hydrogen halides and halogens with alkenes and alkynes are examples of addition reactions.

In Chapter 11 we saw that polyethylene is formed by addition reaction. Polyethylene is an example of a *homopolymer,* which is *a polymer made up of only one type of monomer.* Other homopolymers that are synthesized by the radical mechanism are Teflon, polytetrafluoroethylene and poly(vinyl chloride) (PVC):

$$\left(\!\!-CF_2\!\!-\!CF_2\!\!-\!\right)_{\!\overline{n}} \qquad \left(\!\!-CH_2\!\!-\!\underset{\underset{\displaystyle Cl}{|}}{CH}\!\!-\!\right)_{\!\overline{n}}$$

  Teflon   PVC

The chemistry of polymers is more complex if the starting units are asymmetric:

propylene   polypropylene

Several stereoisomers can result from an addition reaction of propylenes (Figure 22.1). If the additions occur randomly, we obtain *atactic* polypropylenes, which do not pack together well. These polymers are rubbery, amorphous, and relatively weak. Two other possibilities are an *isotactic* structure, in which the R groups are all on the same side of the asymmetric carbon atoms, and a *syndiotactic* form, in which the R groups alternate on the left and right of the asymmetric carbons. Of these, the isotactic isomer has the highest melting point and greatest crystallinity and is endowed with superior mechanical properties.

Catalysts are available for designing polymers of different shapes.

Rubber is probably the best known organic polymer and the only true hydrocarbon polymer found in nature. It is formed by the radical addition of the monomer isoprene. Actually, polymerization can result in either poly-*cis*-isoprene or poly-*trans*-isoprene—or a mixture of both, depending on reaction conditions:

isoprene   poly-*cis*-isoprene   and/or   poly-*trans*-isoprene

Note that in the *cis* isomer the two CH_2 groups are on the same side of the $C=C$ bond, whereas the same groups are across from each other in the *trans* isomer. Natural rubber is poly-*cis*-isoprene, which is extracted from the tree *Hevea brasiliensis* (Figure 22.2).

An unusual and very useful property of rubber is its elasticity. Rubber will stretch up to 10 times its length and, if released, will return to its original size. In contrast, a piece of copper wire can be stretched only a small percentage of its length and still

(a)

(b)

(c)

Figure 22.1 *Stereoisomers of polymers. When the R group (green sphere) is CH_3, the polymer is polypropylene. (a) When the R groups are all on one side of the chain, the polymer is said to be isotactic. (b) When the R groups alternate from side to side, the polymer is said to be syndiotactic. (c) When the R groups are disposed at random, the polymer is atactic.*

Figure 22.2 *Latex (aqueous suspension of rubber particles) being collected from a rubber tree.*

(a)

(b)

(c)

Figure 22.3 *Rubber molecules ordinarily are bent and convoluted. Parts (a) and (b) represent the long chains before and after vulcanization, respectively; (c) shows the alignment of molecules when stretched. Without vulcanization these molecules would slip past one another, and rubber's elastic properties would be gone.*

Condensation reactions were described on p. 353.

return to its original size. Unstretched rubber has no regular X-ray diffraction pattern and is therefore amorphous. Stretched rubber, however, possesses a fair amount of crystallinity and order.

The elastic property of rubber is due to the flexibility of the long-chain molecules. In the bulk state, however, rubber is a tangle of polymeric chains, and if the external force is strong enough, individual chains slip past one another, thereby causing the rubber to lose most of its elasticity. In 1839, the American chemist Charles Goodyear discovered that natural rubber could be cross-linked with sulfur (using zinc oxide as the catalyst) to prevent chain slippage (Figure 22.3). His process, known as *vulcanization,* paved the way for many practical and commercial uses of rubber, such as in automobile tires and dentures.

During World War II a shortage of natural rubber in the United States prompted an intensive program to produce synthetic rubber. Most synthetic rubbers (called *elastomers*) are made from petroleum products such as ethylene, propylene, and butadiene. For example, chloroprene molecules polymerize readily to form polychloroprene, commonly known as *neoprene,* which has properties that are comparable or even superior to those of natural rubber:

$$H_2C=CH-CCl=CH_2 \qquad \left(\begin{array}{c} CH_2 \qquad H \\ C=C \\ Cl \qquad CH_2 \end{array} \right)_n$$

chloroprene   polychloroprene

Another important synthetic rubber is formed by the addition of butadiene to styrene in a 3:1 ratio to give styrene-butadiene rubber (SBR). Because styrene and butadiene are different monomers, SBR is called a ***copolymer,*** which is *a polymer containing two or more different monomers.* Table 22.1 shows a number of common and familiar homopolymers and one copolymer produced by addition reactions.

Condensation Reactions

One of the best-known polymer condensation processes is the reaction between hexamethylenediamine and adipic acid, shown in Figure 22.4. The final product, called nylon 66 (because there are six carbon atoms each in hexamethylenediamine and adipic acid), was first made by the American chemist Wallace Carothers at DuPont

Table 22.1 | Some Monomers and Their Common Synthetic Polymers

Monomer		Polymer	
Formula	**Name**	**Name and Formula**	**Uses**
$H_2C{=}CH_2$	Ethylene	Polyethylene $-(CH_2{-}CH_2)_n$	Plastic piping, bottles, electrical insulation, toys
$H_2C{=}\underset{\underset{CH_3}{\vert}}{\overset{\overset{H}{\vert}}{C}}$	Propylene	Polypropylene $\left(\underset{\underset{CH_3}{\vert}}{CH}{-}CH_2{-}\underset{\underset{CH_3}{\vert}}{CH}{-}CH_2\right)_n$	Packaging film, carpets, crates for soft-drink bottles, lab wares, toys
$H_2C{=}\underset{\underset{Cl}{\vert}}{\overset{\overset{H}{\vert}}{C}}$	Vinyl chloride	Poly(vinyl chloride) (PVC) $-(CH_2{-}\underset{\underset{Cl}{\vert}}{CH})_n$	Piping, siding, gutters, floor tile, clothing, toys
$H_2C{=}\underset{\underset{CN}{\vert}}{\overset{\overset{H}{\vert}}{C}}$	Acrylonitrile	Polyacrylonitrile (PAN) $\left(CH_2{-}\underset{\underset{CN}{\vert}}{CH}\right)_n$	Carpets, knitwear
$F_2C{=}CF_2$	Tetrafluoro-ethylene	Polytetrafluoroethylene (Teflon) $-(CF_2{-}CF_2)_n$	Coating on cooking utensils, electrical insulation, bearings
$H_2C{=}\underset{\underset{CH_3}{\vert}}{\overset{\overset{COOCH_3}{\vert}}{C}}$	Methyl methacrylate	Poly(methyl methacrylate) (Plexiglas) $-(CH_2{-}\underset{\underset{CH_3}{\vert}}{\overset{\overset{COOCH_3}{\vert}}{C}})_n$	Optical equipment, home furnishing
$H_2C{=}\overset{\overset{H}{\vert}}{C}$ ⌬	Styrene	Polystyrene $-(CH_2{-}CH)_n$ ⌬	Containers, thermal insulation (ice buckets, water coolers), toys
$H_2C{=}\overset{\overset{H}{\vert}}{C}{-}\overset{\overset{H}{\vert}}{C}{=}CH_2$	Butadiene	Polybutadiene $-(CH_2CH{=}CHCH_2)_n$	Tire tread, coating resin
See above structures	Butadiene and styrene	Styrene-butadiene rubber (SBR) $-(CH{-}CH_2{-}CH_2{-}CH{=}CH{-}CH_2)_n$ ⌬	Synthetic rubber

Bubble gums contain synthetic styrene-butadiene rubber.

Figure 22.4 *The formation of nylon by the condensation reaction between hexamethylenediamine and adipic acid.*

$$H_2N-(CH_2)_6-NH_2 \quad + \quad HOOC-(CH_2)_4-COOH$$

Hexamethylenediamine   Adipic acid

Condensation ↓

$$H_2N-(CH_2)_6-\overset{\overset{\displaystyle O}{\|}}{\underset{\underset{\displaystyle H}{|}}{N}}-C-(CH_2)_4-COOH + H_2O$$

Further condensation reactions ↓

$$-(CH_2)_4-\overset{\overset{\displaystyle O}{\|}}{C}-\overset{\underset{\displaystyle H}{|}}{N}-(CH_2)_6-\overset{\underset{\displaystyle H}{|}}{N}-\overset{\overset{\displaystyle O}{\|}}{C}-(CH_2)_4-\overset{\overset{\displaystyle O}{\|}}{C}-\overset{\underset{\displaystyle H}{|}}{N}-(CH_2)_6-$$

in 1931. The versatility of nylons is so great that the annual production of nylons and related substances now amounts to several billion pounds. Figure 22.5 shows how nylon 66 is prepared in the laboratory.

Condensation reactions are also used in the manufacture of Dacron (polyester)

$$n HO-\overset{\overset{\displaystyle O}{\|}}{C}-\langle \bigcirc \rangle-\overset{\overset{\displaystyle O}{\|}}{C}-OH + n HO-(CH_2)_2-OH \longrightarrow \left(\overset{\overset{\displaystyle O}{\|}}{C}-\langle \bigcirc \rangle-\overset{\overset{\displaystyle O}{\|}}{C}-O-CH_2CH_2-O\right)_n + n H_2O$$

terephthalic acid   1,2-ethylene glycol   Dacron

Polyesters are used in fibers, films, and plastic bottles.

22.3 Proteins

Proteins *are polymers of amino acids.* They play a key role in nearly all biological processes. Enzymes, the catalysts of biochemical reactions, are mostly proteins. Proteins also facilitate a wide range of other functions, such as transport and storage of vital sub-

Figure 22.5 *The nylon rope trick. Adding a solution of adipoyl chloride (an adipic acid derivative in which the OH groups have been replaced by Cl groups) in cyclohexane to an aqueous solution of hexamethylenediamine causes nylon to form at the interface of the two solutions, which do not mix. It can then be drawn off.*

stances, coordinated motion, mechanical support, and protection against diseases. The human body contains an estimated 100,000 different kinds of proteins, each of which has a specific physiological function. As we will see in this section, the chemical composition and structure of these complex natural polymers are the basis of their specificity.

Amino Acids

Proteins have high molar masses, ranging from about 5000 g to 1×10^7 g, and yet the percent composition by mass of the elements in proteins is remarkably constant: carbon, 50 to 55 percent; hydrogen, 7 percent; oxygen, 23 percent; nitrogen, 16 percent; and sulfur, 1 percent.

Elements in proteins.

The basic structural units of proteins are *amino acids*. An **amino acid** is *a compound that contains at least one amino group* (—NH$_2$) *and at least one carboxyl group* (—COOH):

amino group   carboxyl group

Twenty different amino acids are the building blocks of all the proteins in the human body. Table 22.2 shows the structures of these vital compounds, along with their three-letter abbreviations.

Amino acids in solution at neutral pH exist as *dipolar ions,* meaning that the proton on the carboxyl group has migrated to the amino group. Consider glycine, the simplest amino acid. The un-ionized form and the dipolar ion of glycine are shown below:

Un-ionized form   Dipolar ion

The first step in the synthesis of a protein molecule is a condensation reaction between an amino group on one amino acid and a carboxyl group on another amino acid. The molecule formed from the two amino acids is called a *dipeptide,* and the bond joining them together is a *peptide bond:*

peptide bond

It is interesting to compare this reaction with the one shown in Figure 22.4.

where R$_1$ and R$_2$ represent a H atom or some other group; —CO—NH— is called the *amide group.* Because the equilibrium of the reaction joining two amino acids lies to the left, the process is coupled to the hydrolysis of ATP (see p. 596).

Either end of a dipeptide can engage in a condensation reaction with another amino acid to form a *tripeptide,* a *tetrapeptide,* and so on. The final product, the protein molecule, is a *polypeptide;* it can also be thought of as a polymer of amino acids.

Table 22.2 | The 20 Amino Acids Essential to Living Organisms*

Name	Abbreviation	Structure
Alanine	Ala	$H_3C-\underset{\underset{NH_3^+}{|}}{\overset{\overset{H}{|}}{C}}-COO^-$
Arginine	Arg	$H_2N-\underset{\underset{NH}{\|}}{C}-N-CH_2-CH_2-CH_2-\underset{\underset{NH_3^+}{|}}{\overset{\overset{H}{|}}{C}}-COO^-$
Asparagine	Asn	$H_2N-\underset{\overset{\|}{O}}{C}-CH_2-\underset{\underset{NH_3^+}{|}}{\overset{\overset{H}{|}}{C}}-COO^-$
Aspartic acid	Asp	$HOOC-CH_2-\underset{\underset{NH_3^+}{|}}{\overset{\overset{H}{|}}{C}}-COO^-$
Cysteine	Cys	$HS-CH_2-\underset{\underset{NH_3^+}{|}}{\overset{\overset{H}{|}}{C}}-COO^-$
Glutamic acid	Glu	$HOOC-CH_2-CH_2-\underset{\underset{NH_3^+}{|}}{\overset{\overset{H}{|}}{C}}-COO^-$
Glutamine	Gln	$H_2N-\underset{\overset{\|}{O}}{C}-CH_2-CH_2-\underset{\underset{NH_3^+}{|}}{\overset{\overset{H}{|}}{C}}-COO^-$
Glycine	Gly	$H-\underset{\underset{NH_3^+}{|}}{\overset{\overset{H}{|}}{C}}-COO^-$
Histidine	His	$HC=C-CH_2-\underset{\underset{NH_3^+}{|}}{\overset{\overset{H}{|}}{C}}-COO^-$ with ring $N\underset{\underset{H}{|}}{C}NH$
Isoleucine	Ile	$H_3C-CH_2-\underset{\underset{H}{|}}{\overset{\overset{CH_3}{|}}{C}}-\underset{\underset{NH_3^+}{|}}{\overset{\overset{H}{|}}{C}}-COO^-$

*The shaded portion is the R group of the amino acid.

(Cont.)

Table 22.2 | Continued

Name	Abbreviation	Structure
Leucine	Leu	H_3C, H_3C—CH—CH$_2$—$\overset{\overset{\text{H}}{\mid}}{\underset{\underset{NH_3^+}{\mid}}{C}}$—COO$^-$
Lysine	Lys	H_2N—CH$_2$—CH$_2$—CH$_2$—CH$_2$—$\overset{\overset{\text{H}}{\mid}}{\underset{\underset{NH_3^+}{\mid}}{C}}$—COO$^-$
Methionine	Met	H_3C—S—CH$_2$—CH$_2$—$\overset{\overset{\text{H}}{\mid}}{\underset{\underset{NH_3^+}{\mid}}{C}}$—COO$^-$
Phenylalanine	Phe	—CH$_2$—$\overset{\overset{\text{H}}{\mid}}{\underset{\underset{NH_3^+}{\mid}}{C}}$—COO$^-$
Proline	Pro	$H_2\overset{+}{N}$—$\overset{\overset{\text{H}}{\mid}}{C}$—COO$^-$ (ring: H_2C—CH$_2$—CH$_2$)
Serine	Ser	HO—CH$_2$—$\overset{\overset{\text{H}}{\mid}}{\underset{\underset{NH_3^+}{\mid}}{C}}$—COO$^-$
Threonine	Thr	H_3C—$\overset{\overset{OH}{\mid}}{\underset{\underset{H}{\mid}}{C}}$—$\overset{\overset{\text{H}}{\mid}}{\underset{\underset{NH_3^+}{\mid}}{C}}$—COO$^-$
Tryptophan	Trp	(indole ring)—C—CH$_2$—$\overset{\overset{\text{H}}{\mid}}{\underset{\underset{NH_3^+}{\mid}}{C}}$—COO$^-$
Tyrosine	Tyr	HO—(ring)—CH$_2$—$\overset{\overset{\text{H}}{\mid}}{\underset{\underset{NH_3^+}{\mid}}{C}}$—COO$^-$
Valine	Val	H_3C, H_3C—CH—$\overset{\overset{\text{H}}{\mid}}{\underset{\underset{NH_3^+}{\mid}}{C}}$—COO$^-$

Figure 22.6 *The formation of two dipeptides from two different amino acids. Alanylglycine is different from glycylalanine in that in alanylglycine the amino and methyl groups are bonded to the same carbon atom.*

Alanine

Glycine

Alanylglycine

Glycylalanine

An amino acid unit in a polypeptide chain is called a *residue*. Typically, a polypeptide chain contains 100 or more amino acid residues. The sequence of amino acids in a polypeptide chain is written conventionally from left to right, starting with the amino-terminal residue and ending with the carboxyl-terminal residue. Let us consider a dipeptide formed from glycine and alanine. Figure 22.6 shows that alanylglycine and glycylalanine are different molecules. With 20 different amino acids to choose from, 20^2, or 400, different dipeptides can be generated. Even for a very small protein such as insulin, which contains only 50 amino acid residues, the number of chemically different structures that is possible is of the order of 20^{50} or 10^{65}! This is an incredibly large number when you consider that the total number of atoms in our galaxy is about 10^{68}. With so many possibilities for protein synthesis, it is remarkable that generation after generation of cells can produce identical proteins for specific physiological functions.

Protein Structure

The type and number of amino acids in a given protein along with the sequence or order in which these amino acids are joined together determine the protein's structure. In the 1930s Linus Pauling and his coworkers conducted a systematic investigation of protein structure. First they studied the geometry of the basic repeating group, that is, the amide group, which is represented by the following resonance structures:

Because it is more difficult (that is, it would take more energy) to twist a double bond than a single bond, the four atoms in the amide group become locked in the same plane (Figure 22.7). Figure 22.8 depicts the repeating amide group in a polypeptide chain.

On the basis of models and X-ray diffraction data, Pauling deduced that there are two common structures for protein molecules, called the *α helix* and the *β-pleated* sheet. The α-helical structure of a polypeptide chain is shown in Figure 22.9. The helix is stabilized by *intramolecular* hydrogen bonds between the NH and CO groups of the main chain, giving rise to an overall rodlike shape. The CO group of each amino acid is hydrogen-bonded to the NH group of the amino acid that is four residues

Figure 22.7 *The planar amide group in protein. Rotation about the peptide bond in the amide group is hindered by its double-bond character. The black atoms represent carbon; blue, nitrogen; red, oxygen; green, R group; and gray, hydrogen.*

Figure 22.8 *A polypeptide chain. Note the repeating units of the amide group. The symbol R represents part of the structure characteristic of the individual amino acids. For glycine, R is simply an H atom.*

away in the sequence. In this manner all the main-chain CO and NH groups take part in hydrogen bonding. X-ray studies have shown that the structure of a number of proteins, including myoglobin and hemoglobin, is to a great extent α-helical in nature.

The β-pleated structure is markedly different from the α helix in that it is like a sheet rather than a rod. The polypeptide chain is almost fully extended, and each chain forms many *intermolecular* hydrogen bonds with adjacent chains. Figure 22.10 shows the two different types of β-pleated structures, called *parallel* and *antiparallel*. Silk molecules possess the β structure. Because its polypeptide chains are already in extended form, silk lacks elasticity and extensibility, but it is quite strong due to the many intermolecular hydrogen bonds.

It is customary to divide protein structure into four levels of organization. The *primary structure* refers to the unique amino acid sequence of the polypeptide chain. The *secondary structure* includes those parts of the polypeptide chain that are stabilized by a regular pattern of hydrogen bonds between the CO and NH groups of the backbone, for example, the α helix. The term *tertiary structure* applies to the three-dimensional structure stabilized by dispersion forces, hydrogen bonding, and other intermolecular forces. It differs from secondary structure in that the amino acids taking part in these interactions may be far apart in the polypeptide chain. A protein molecule may be made up of more than one polypeptide chain. Thus, in addition to the various interactions *within* a chain that give rise to the secondary and tertiary structures, we must also consider the interaction *between* chains. The overall arrangement of the polypeptide chains is called the *quaternary structure*. For example, the hemoglobin molecule consists of four separate polypeptide chains, or *subunits*. These subunits are held together by van der Waals forces and ionic forces (Figure 22.11).

Pauling's work was a great triumph in protein chemistry. It showed for the first time how to predict a protein structure purely from a knowledge of the geometry of its fundamental building blocks—amino acids. However, there are many proteins whose structures do not correspond to the α-helical or β structure. Chemists now know that the three-dimensional structures of these biopolymers are maintained by several types of intermolecular forces in addition to hydrogen bonding. The delicate balance of the various interactions can be appreciated by considering an example: When glutamic acid, one of the amino acid residues in two of the four polypeptide chains in hemoglobin, is replaced by valine, another amino acid, the protein molecules aggregate to form insoluble polymers, causing the disease known as sickle-cell anemia.

In spite of all the forces that give proteins their structural stability, most proteins have a certain amount of flexibility. Enzymes, for example, are flexible enough to change their geometry to fit substrates of various sizes and shapes. Another interesting example of protein flexibility is found in the binding of hemoglobin to oxygen. Each of the four polypeptide chains in hemoglobin contains a heme group that can bind to an oxygen molecule. In deoxyhemoglobin, the affinity of each of the heme groups for oxygen is about the same. However, as soon as one of the heme groups becomes oxygenated, the affinity of the other three hemes for oxygen is greatly enhanced. This phenomenon, called *cooperativity,* makes hemoglobin a particularly

Figure 22.9 *The α-helical structure of a polypeptide chain. The structure is held in position by intramolecular hydrogen bonds, shown as dotted lines. For color key, see Figure 22.7.*

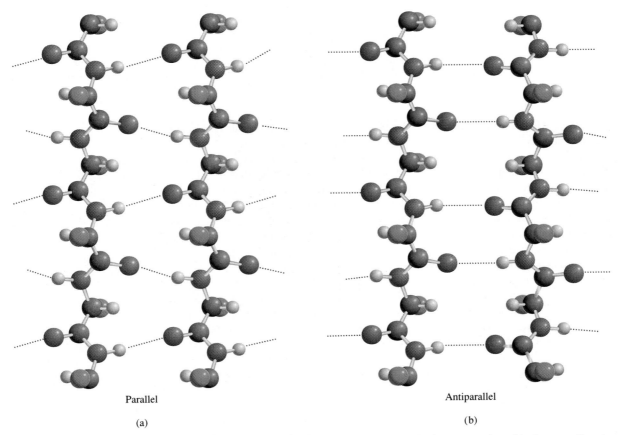

Parallel

(a)

Antiparallel

(b)

Figure 22.10 *Hydrogen bonds (a) in a parallel β-pleated sheet, in which all the polypeptide chains are oriented in the same direction, and (b) in an antiparallel β-pleated sheet, in which adjacent polypeptide chains run in opposite directions. For color key, see Figure 22.7.*

suitable substance for the uptake of oxygen in the lungs. By the same token, once a fully oxygenated hemoglobin molecule releases an oxygen molecule (to myoglobin in the tissues), the other three oxygen molecules will depart with increasing ease. The cooperative nature of the bonding is such that information about the presence (or absence) of oxygen molecules is transmitted from one subunit to another along the polypeptide chains, a process made possible by the flexibility of the three-dimensional structure (Figure 22.12). It is believed that the Fe^{2+} ion has too large a radius to fit into the porphyrin ring of deoxyhemoglobin. When O_2 binds to Fe^{2+}, however, the ion shrinks somewhat so that it can fit into the plane of the ring. As the ion slips into the ring, it pulls the histidine residue toward the ring and thereby sets off a sequence of structural changes from one subunit to another. Although the details of the changes are not clear, biochemists believe that this is how the binding of an oxygen molecule to one heme group affects another heme group. The structural changes drastically affect the affinity of the remaining heme groups for oxygen molecules.

When proteins are heated above body temperature or when they are subjected to unusual acid or base conditions or treated with special reagents called *denaturants,* they lose some or all of their tertiary and secondary structure. Called ***denatured proteins,*** proteins in this state *no longer exhibit normal biological activities.* Figure 22.13 shows the variation of rate with temperature for a typical enzyme-catalyzed reaction.

Hard-boiling an egg denatures the proteins in the egg white.

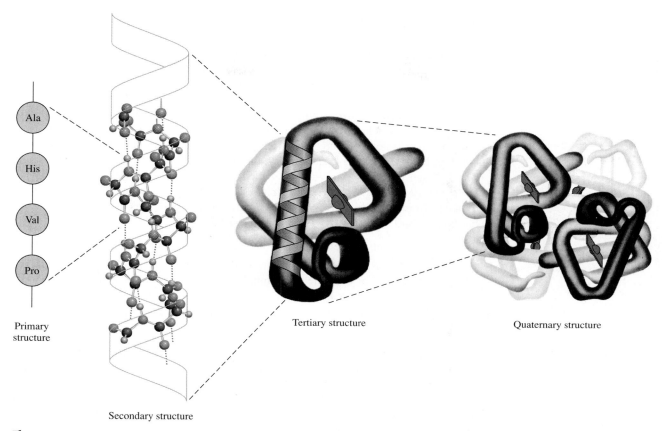

Figure 22.11 *The primary, secondary, tertiary, and quaternary structure of the hemoglobin molecule.*

Figure 22.12 *The structural changes that occur when the heme group in hemoglobin binds to an oxygen molecule. (a) The heme group in deoxyhemoglobin. (b) Oxyhemoglobin.*

Figure 22.13 *Dependence of the rate of an enzyme-catalyzed reaction on temperature. Above the optimum temperature at which an enzyme is most effective, its activity drops off as a consequence of denaturation.*

Initially, the rate increases with increasing temperature, as we would expect. Beyond the optimum temperature, however, the enzyme begins to denature and the rate falls rapidly. If a protein is denatured under mild conditions, its original structure can often be regenerated by removing the denaturant or by restoring the temperature to normal conditions. This process is called *reversible denaturation*.

22.4 Nucleic Acids

Nucleic acids are *high molar mass polymers that play an essential role in protein synthesis.* **Deoxyribonucleic acid (DNA)** and **ribonucleic acid (RNA)** are *the two types of nucleic acid.* DNA molecules are among the largest molecules known; they have molar masses of up to tens of billions of grams. On the other hand, RNA molecules vary greatly in size, some having a molar mass of about 25,000 g. Compared with proteins, which are made of up to 20 different amino acids, nucleic acids are fairly simple in composition. A DNA or RNA molecule contains only four types of building blocks: purines, pyrimidines, furanose sugars, and phosphate groups (Figure 22.14). Each purine or pyrimidine is called a *base.*

In the 1940s, the American biochemist Erwin Chargaff studied DNA molecules obtained from various sources and observed certain regularities. *Chargaff's rules,* as his findings are now known, describe these patterns:

1. The amount of adenine (a purine) is equal to that of thymine (a pyrimidine); that is, A = T, or A/T = 1.

2. The amount of cytosine (a pyrimidine) is equal to that of guanine (a purine); that is, C = G, or C/G = 1.

3. The total number of purine bases is equal to the total number of pyrimidine bases; that is, A + G = C + T.

Based on chemical analyses and information obtained from X-ray diffraction measurements, the American biologist James Watson and the British biologist Francis Crick formulated the double-helical structure for the DNA molecule in 1953. Watson and Crick determined that the DNA molecule has two helical strands. Each strand is made up of **nucleotides,** which *consist of a base, a deoxyribose, and a phosphate group linked together* (Figure 22.15).

The key to the double-helical structure of DNA is the formation of hydrogen bonds between bases in the two strands of a molecule. Although hydrogen bonds can form between any two bases, called *base pairs,* Watson and Crick found that the most favorable couplings are between adenine and thymine and between cytosine and guanine (Figure 22.16). Note that this scheme is consistent with Chargaff's rules, because every purine base is hydrogen-bonded to a pyrimidine base, and vice versa (A + G = C + T). Other attractive forces such as dipole-dipole interactions and van der Waals forces between the base pairs also help to stabilize the double helix.

The structure of RNA differs from that of DNA in several respects. First, as shown in Figure 22.14, the four bases found in RNA molecules are adenine, cytosine, guanine, and uracil. Second, RNA contains the sugar ribose rather than the 2-deoxyribose of DNA. Third, chemical analysis shows that the composition of RNA does not obey Chargaff's rules. In other words, the purine-to-pyrimidine ratio is not equal to 1 as in the case of DNA. This and other evidence rule out a double-helical structure. In fact, the RNA molecule exists as a single-strand polynucleotide. There are actually three types of RNA molecules—messenger RNA (*m*RNA), ribosomal RNA (*r*RNA), and transfer RNA (*t*RNA). These RNAs have similar nucleotides but differ from one another in molar mass, overall structure, and biological functions.

DNA and RNA molecules direct the synthesis of proteins in the cell, a subject that is beyond the scope of this book. Introductory texts in biochemistry and molecular biology explain this process.

If the DNA molecules from all the cells in a human were stretched and joined end to end, the length would be many times the distance from Earth to the sun!

An electron micrograph of a DNA molecule. The double-helical structure is evident.

In the 1980s chemists discovered that certain RNAs can function as enzymes.

Figure 22.14 *The components of the nucleic acids DNA and RNA.*

Figure 22.15 *Structure of a nucleotide, one of the repeating units in DNA.*

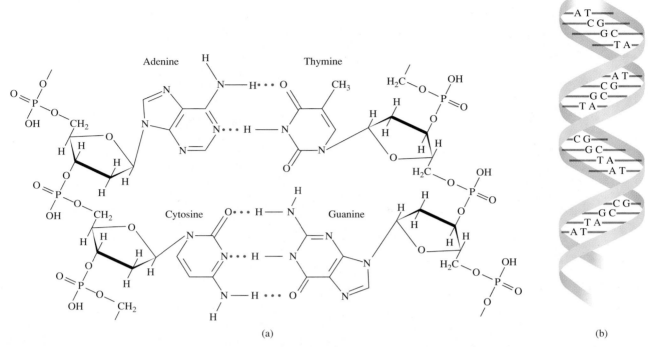

Figure 22.16 *(a) Base-pair formation by adenine and thymine and by cytosine and guanine. (b) The two strands of a DNA molecule are held together by hydrogen bonds (and other intermolecular forces) between base pairs A-T and C-G.*

Summary

Polymers are large molecules made up of small, repeating units called monomers. Proteins, nucleic acids, cellulose, and rubber are natural polymers. Nylon, Dacron, and Lucite are examples of synthetic polymers. The properties of a polymer vary, depending on whether its structure is atactic, isotactic, or syndiotactic.

Organic polymers can be synthesized via addition reactions or condensation reactions. Synthetic rubbers include polychloroprene and styrene-butadiene rubber, which is made by the copolymerization of styrene and butadiene.

Proteins are polymers of amino acids. Twenty different amino acids make up all the proteins found in the human body; all except glycine are chiral. Most enzymes are proteins. Structure is very closely related to the function and properties of proteins. Hydrogen bonding and other intermolecular forces largely determine the three-dimensional structure of proteins.

Nucleotides are the building blocks of DNA and RNA. The DNA nucleotides each contain a purine or pyrimidine base, a deoxyribose molecule, and a phosphate group. The nucleotides of RNA are similar but contain different bases and ribose instead of deoxyribose.

Key Words

Amino acid p. 709
Copolymer p. 706
Denatured protein p. 714

Deoxyribonucleic acid
  (DNA) p. 716
DNA p. 716
Homopolymer p. 704

Monomer p. 704
Nucleic acids p. 716
Nucleotide p. 716
Polymer p. 704

Protein p. 708
Ribonucleic acid (RNA)
  p. 716
RNA p. 716

Questions and Problems

Synthetic Organic Polymers

Review Questions

22.1 Define the following terms: monomer, polymer, homopolymer, copolymer.

22.2 Name 10 objects that contain synthetic organic polymers.

22.3 Calculate the molar mass of a particular polyethylene sample, $-(CH_2-CH_2)_n$, where $n = 4600$.

22.4 Describe the two major mechanisms of organic polymer synthesis.

22.5 Discuss the three isomers of polypropylene.

22.6 In Chapter 13 you learned about the colligative properties of solutions. Which of the colligative properties is suitable for determining the molar mass of a polymer? Why?

Problems

22.7 Teflon is formed by a radical addition reaction involving the monomer tetrafluoroethylene. Show the mechanism for this reaction.

22.8 Vinyl chloride, $H_2C=CHCl$, undergoes copolymerization with 1,1-dichloroethylene, $H_2C=CCl_2$, to form a polymer commercially known as Saran. Draw the structure of the polymer, showing the repeating monomer units.

22.9 Kevlar is a copolymer used in bullet-proof vests. It is formed in a condensation reaction between the following two monomers:

Sketch a portion of the polymer chain showing several monomer units. Write the overall equation for the condensation reaction.

22.10 Describe the formation of polystyrene.

22.11 Deduce plausible monomers for polymers with the following repeating units:

(a) $-(CH_2-CF_2)_n$

(b)

22.12 Deduce plausible monomers for polymers with the following repeating units:

(a) $-(CH_2-CH=CH-CH_2)_n$
(b) $-(CO-(CH_2)_6 NH)_n$

Proteins

Review Questions

22.13 Discuss the characteristics of an amide group and its importance in protein structure.

22.14 What is the α-helical structure in proteins?

22.15 Describe the β-pleated structure present in some proteins.

22.16 Discuss the main functions of proteins in living systems.

22.17 Briefly explain the phenomenon of cooperativity exhibited by the hemoglobin molecule in binding oxygen.

22.18 Why is sickle-cell anemia called a molecular disease?

Problems

22.19 Draw the structures of the dipeptides that can be formed from the reaction between the amino acids glycine and alanine.

22.20 Draw the structures of the dipeptides that can be formed from the reaction between the amino acids glycine and lysine.

22.21 The amino acid glycine can be condensed to form a polymer called polyglycine. Draw the repeating monomer unit.

22.22 The following are data obtained on the rate of product formation of an enzyme-catalyzed reaction:

Temperature (°C)	Rate of Product Formation (M/s)
10	0.0025
20	0.0048
30	0.0090
35	0.0086
45	0.0012

Comment on the dependence of rate on temperature. (No calculations are required.)

Nucleic Acids

Review Questions

22.23 Describe the structure of a nucleotide.

22.24 What is the difference between ribose and deoxyribose?

22.25 What are Chargaff's rules?

22.26 Describe the role of hydrogen bonding in maintaining the double-helical structure of DNA.

Additional Problems

22.27 Discuss the importance of hydrogen bonding in biological systems. Use proteins and nucleic acids as examples.

22.28 Proteins vary widely in structure, whereas nucleic acids have rather uniform structures. How do you account for this major difference?

22.29 If untreated, fevers of 104°F or higher may lead to brain damage. Why?

22.30 The "melting point" of a DNA molecule is the temperature at which the double-helical strand breaks apart. Suppose you are given two DNA samples. One sample contains 45 percent C-G base pairs while the other contains 64 percent C-G base pairs. The total number of bases is the same in each sample. Which of the two samples has a higher melting point? Why?

22.31 When fruits such as apples and pears are cut, the exposed parts begin to turn brown. This is the result of an oxidation reaction catalyzed by enzymes present in the fruit. Often the browning action can be prevented or slowed by adding a few drops of lemon juice to the exposed areas. What is the chemical basis for this treatment?

22.32 "Dark meat" and "white meat" are one's choices when eating a turkey. Explain what causes the meat to assume different colors. (*Hint:* The more active muscles in a turkey have a higher rate of metabolism and need more oxygen.)

22.33 Nylon can be destroyed easily by strong acids. Explain the chemical basis for the destruction. (*Hint:* The products are the starting materials of the polymerization reaction.)

22.34 Despite what you may have read in science fiction novels or seen in horror movies, it is extremely unlikely that insects can ever grow to human size. Why? (*Hint:* Insects do not have hemoglobin molecules in their blood.)

22.35 How many different tripeptides can be formed by lysine and alanine?

22.36 Chemical analysis shows that hemoglobin contains 0.34 percent Fe by mass. What is the minimum possible molar mass of hemoglobin? The actual molar mass of hemoglobin is four times this minimum value. What conclusion can you draw from these data?

22.37 The folding of a polypeptide chain depends not only on its amino acid sequence but also on the nature of the solvent. Discuss the types of interactions that might occur between water molecules and the amino acid residues of the polypeptide chain. Which groups would be exposed on the exterior of the protein in contact with water and which groups would be buried in the interior of the protein?

22.38 What kind of intermolecular forces are responsible for the aggregation of hemoglobin molecules that leads to sickle-cell anemia?

22.39 Draw structures of the nucleotides containing the following components: (a) deoxyribose and cytosine, (b) ribose and uracil.

22.40 When a nonapeptide (containing nine amino acid residues) isolated from rat brains was hydrolyzed, it gave the following smaller peptides as identifiable products: Gly-Ala-Phe, Ala-Leu-Val, Gly-Ala-Leu, Phe-Glu-His, and His-Gly-Ala. Reconstruct the amino acid sequence in the nonapeptide, giving your reasons. (Remember the convention for writing peptides.)

22.41 At neutral pH amino acids exist as dipolar ions. Using glycine as an example, and given that the pK_a of the carboxyl group is 2.3 and that of the ammonium group is 9.6, predict the predominant form of the molecule at pH 1, 7, and 12. Justify your answers using Equation (17.3).

22.42 In Lewis Carroll's tale "Through the Looking Glass," Alice wonders whether "looking-glass milk" on the other side of the mirror would be fit to drink. Based on your knowledge of chirality and enzyme action, comment on the validity of Alice's concern.

Special Problem

22.43 Nylon was designed to be a synthetic silk. (a) The average molar mass of a batch of nylon 66 is 12,000 g/mol. How many monomer units are there in this sample? (b) Which part of nylon's structure is similar to a polypeptide's structure? (c) How many different tripeptides (made up of three amino acids) can be formed from the amino acids alanine (Ala), glycine (Gly), and serine (Ser), which account for most of the amino acids in silk?

Units for the Gas Constant

In this appendix we will see how the gas constant R can be expressed in units J/K $\cdot$ mol. Our first step is to derive a relationship between atm and pascal. We start with

$$\text{pressure} = \frac{\text{force}}{\text{area}}$$

$$= \frac{\text{mass} \times \text{acceleration}}{\text{area}}$$

$$= \frac{\text{volume} \times \text{density} \times \text{acceleration}}{\text{area}}$$

$$= \text{length} \times \text{density} \times \text{acceleration}$$

By definition, the standard atmosphere is the pressure exerted by a column of mercury exactly 76 cm high of density 13.5951 g/cm^3, in a place where acceleration due to gravity is 980.665 cm/s^2. However, to express pressure in N/m^2 it is necessary to write

$$\text{density of mercury} = 1.35951 \times 10^4 \, \text{kg/m}^3$$
$$\text{acceleration due to gravity} = 9.80665 \, \text{m/s}^2$$

The standard atmosphere is given by

$$1 \, \text{atm} = (0.76 \, \text{m Hg})(1.35951 \times 10^4 \, \text{kg/m}^3)(9.80665 \, \text{m/s}^2)$$
$$= 101{,}325 \, \text{kg m/m}^2 \cdot \text{s}^2$$
$$= 101{,}325 \, \text{N/m}^2$$
$$= 101{,}325 \, \text{Pa}$$

From Section 5.4 we see that the gas constant R is given by 0.082057 L $\cdot$ atm/K $\cdot$ mol. Using the conversion factors

$$1 \, \text{L} = 1 \times 10^{-3} \, \text{m}^3$$
$$1 \, \text{atm} = 101{,}325 \, \text{N/m}^2$$

we write

$$R = \left(0.082057 \, \frac{\text{L atm}}{\text{K mol}}\right)\left(\frac{1 \times 10^{-3} \, \text{m}^3}{1 \, \text{L}}\right)\left(\frac{101{,}325 \, \text{N/m}^2}{1 \, \text{atm}}\right)$$

$$= 8.314 \, \frac{\text{N m}}{\text{K mol}}$$

$$= 8.314 \, \frac{\text{J}}{\text{K mol}}$$

$$1 \, \text{L} \cdot \text{atm} = (1 \times 10^{-3} \, \text{m}^3)(101{,}325 \, \text{N/m}^2)$$
$$= 101.3 \, \text{N m}$$

and
$$= 101.3 \, \text{J}$$

Appendix 2

Selected Thermodynamic Data at 1 atm and 25°C*

Inorganic Substances			
Substance	ΔH_f^o (kJ/mol)	ΔG_f^o (kJ/mol)	S^o (J/K · mol)
Al(s)	0	0	28.3
Al^{3+}(aq)	−524.7	−481.2	−313.38
Al$_2$O$_3$(s)	−1669.8	−1576.4	50.99
Br$_2$(l)	0	0	152.3
Br$^-$(aq)	−120.9	−102.8	80.7
HBr(g)	−36.2	−53.2	198.48
C(graphite)	0	0	5.69
C(diamond)	1.90	2.87	2.4
CO(g)	−110.5	−137.3	197.9
CO$_2$(g)	−393.5	−394.4	213.6
CO$_2$(aq)	−412.9	−386.2	121.3
CO$_3^{2-}$(aq)	−676.3	−528.1	−53.1
HCO$_3^-$(aq)	−691.1	−587.1	94.98
H$_2$CO$_3$(aq)	−699.7	−623.2	187.4
CS$_2$(g)	115.3	65.1	237.8
CS$_2$(l)	87.3	63.6	151.0
HCN(aq)	105.4	112.1	128.9
CN$^-$(aq)	151.0	165.69	117.99
(NH$_2$)$_2$CO(s)	−333.19	−197.15	104.6
(NH$_2$)$_2$CO(aq)	−319.2	−203.84	173.85
Ca(s)	0	0	41.6
Ca^{2+}(aq)	−542.96	−553.0	−55.2
CaO(s)	−635.6	−604.2	39.8
Ca(OH)$_2$(s)	−986.6	−896.8	76.2
CaF$_2$(s)	−1214.6	−1161.9	68.87
CaCl$_2$(s)	−794.96	−750.19	113.8
CaSO$_4$(s)	−1432.69	−1320.3	106.69
CaCO$_3$(s)	−1206.9	−1128.8	92.9
Cl$_2$(g)	0	0	223.0
Cl$^-$(aq)	−167.2	−131.2	56.5
HCl(g)	−92.3	−95.27	187.0

*The ΔH_f^o, ΔG_f^o, and S^o values for ions are based on the reference states ΔH_f^o(H$^+$) = 0, ΔG_f^o(H$^+$) = 0, and S^o(H$^+$) = 0.

(Cont.)

Substance	ΔH°_f (kJ/mol)	ΔG°_f (kJ/mol)	S° (J/K · mol)
Cu(s)	0	0	33.3
Cu$^+$(aq)	51.88	50.2	40.6
Cu^{2+}(aq)	64.39	64.98	−99.6
CuO(s)	−155.2	−127.2	43.5
Cu$_2$O(s)	−166.69	−146.36	100.8
CuCl(s)	−134.7	−118.8	91.6
CuCl$_2$(s)	−205.85	?	?
CuS(s)	−48.5	−49.0	66.5
CuSO$_4$(s)	−769.86	−661.9	113.39
F$_2$(g)	0	0	203.34
F$^-$(aq)	−329.1	−276.48	−9.6
HF(g)	−271.6	−270.7	173.5
Fe(s)	0	0	27.2
Fe^{2+}(aq)	−87.86	−84.9	−113.39
Fe^{3+}(aq)	−47.7	−10.5	−293.3
Fe$_2$O$_3$(s)	−822.2	−741.0	90.0
Fe(OH)$_2$(s)	−568.19	−483.55	79.5
Fe(OH)$_3$(s)	−824.25	?	?
H(g)	218.2	203.2	114.6
H$_2$(g)	0	0	131.0
H$^+$(aq)	0	0	0
OH$^-$(aq)	−229.94	−157.30	−10.5
H$_2$O(g)	−241.8	−228.6	188.7
H$_2$O(l)	−285.8	−237.2	69.9
H$_2$O$_2$(l)	−187.6	−118.1	?
I$_2$(s)	0	0	116.7
I$^-$(aq)	55.9	51.67	109.37
HI(g)	25.9	1.30	206.3
K(s)	0	0	63.6
K$^+$(aq)	−251.2	−282.28	102.5
KOH(s)	−425.85	?	?
KCl(s)	−435.87	−408.3	82.68
KClO$_3$(s)	−391.20	−289.9	142.97
KClO$_4$(s)	−433.46	−304.18	151.0
KBr(s)	−392.17	−379.2	96.4
KI(s)	−327.65	−322.29	104.35
KNO$_3$(s)	−492.7	−393.1	132.9
Li(s)	0	0	28.0
Li$^+$(aq)	−278.46	−293.8	14.2
Li$_2$O(s)	−595.8	?	?
LiOH(s)	−487.2	−443.9	50.2

(Cont.)

Substance	ΔH_f^o (kJ/mol)	ΔG_f^o (kJ/mol)	S^o (J/K · mol)
$Mg(s)$	0	0	32.5
$Mg^{2+}(aq)$	−461.96	−456.0	−117.99
$MgO(s)$	−601.8	−569.6	26.78
$Mg(OH)_2(s)$	−924.66	−833.75	63.1
$MgCl_2(s)$	−641.8	−592.3	89.5
$MgSO_4(s)$	−1278.2	−1173.6	91.6
$MgCO_3(s)$	−1112.9	−1029.3	65.69
$N_2(g)$	0	0	191.5
$N_3^-(aq)$	245.18	?	?
$NH_3(g)$	−46.3	−16.6	193.0
$NH_4^+(aq)$	−132.80	−79.5	112.8
$NH_4Cl(s)$	−315.39	−203.89	94.56
$NH_3(aq)$	−80.3	−263.76	111.3
$N_2H_4(l)$	50.4	?	?
$NO(g)$	90.4	86.7	210.6
$NO_2(g)$	33.85	51.8	240.46
$N_2O_4(g)$	9.66	98.29	304.3
$N_2O(g)$	81.56	103.6	219.99
$HNO_3(aq)$	−207.4	−111.3	146.4
$Na(s)$	0	0	51.05
$Na^+(aq)$	−239.66	−261.87	60.25
$Na_2O(s)$	−415.89	−376.56	72.8
$NaCl(s)$	−411.0	−384.0	72.38
$NaI(s)$	−288.0	?	?
$Na_2SO_4(s)$	−1384.49	−1266.8	149.49
$NaNO_3(s)$	−466.68	−365.89	116.3
$Na_2CO_3(s)$	−1130.9	−1047.67	135.98
$NaHCO_3(s)$	−947.68	−851.86	102.09
$O(g)$	249.4	230.1	160.95
$O_2(g)$	0	0	205.0
$O_3(aq)$	−12.09	16.3	110.88
$O_3(g)$	142.2	163.4	237.6
$P(white)$	0	0	44.0
$P(red)$	−18.4	13.8	29.3
$PO_4^{3-}(aq)$	−1284.07	−1025.59	−217.57
$P_4O_{10}(s)$	−3012.48	?	?
$PH_3(g)$	9.25	18.2	210.0
$HPO_4^{2-}(aq)$	−1298.7	−1094.1	−35.98
$H_2PO_4^-(aq)$	−1302.48	−1135.1	89.1
$S(rhombic)$	0	0	31.88
$S(monoclinic)$	0.30	0.10	32.55
$SO_2(g)$	−296.1	−300.4	248.5
$SO_3(g)$	−395.2	−370.4	256.2
$SO_3^{2-}(aq)$	−624.25	−497.06	43.5
$SO_4^{2-}(aq)$	−907.5	−741.99	17.15
$H_2S(g)$	−20.15	−33.0	205.64

(Cont.)

Substance	ΔH_f^o (kJ/mol)	ΔG_f^o (kJ/mol)	S^o (J/K · mol)
$HSO_3^-(aq)$	−627.98	−527.3	132.38
$HSO_4^-(aq)$	−885.75	−752.87	126.86
$H_2SO_4(l)$	−811.3	−690.0	156.9
$H_2SO_4(aq)$	−909.3	−744.5	20.1
$SF_6(g)$	−1096.2	−1105.3	291.8
$Zn(s)$	0	0	41.6
$Zn^{2+}(aq)$	−152.4	−147.2	−112.1
$ZnO(s)$	−348.0	−318.2	43.9
$ZnCl_2(s)$	−415.89	−369.26	108.37
$ZnS(s)$	−202.9	−198.3	57.7
$ZnSO_4(s)$	−978.6	−871.6	124.7

Organic Substances				
Substance	Formula	ΔH_f^o (kJ/mol)	ΔG_f^o (kJ/mol)	S^o (J/K · mol)
Acetic acid(l)	CH_3COOH	−484.2	−389.45	159.83
Acetaldehyde(g)	CH_3CHO	−166.35	−139.08	264.2
Acetone(l)	CH_3COCH_3	−246.8	−153.55	198.74
Acetylene(g)	C_2H_2	226.6	209.2	200.8
Benzene(l)	C_6H_6	49.04	124.5	172.8
Ethanol(l)	C_2H_5OH	−276.98	−174.18	161.04
Ethane(g)	C_2H_6	−84.7	−32.89	229.49
Ethylene(g)	C_2H_4	52.3	68.1	219.45
Formic acid(l)	$HCOOH$	−409.2	−346.0	128.95
Glucose(s)	$C_6H_{12}O_6$	−1274.5	−910.56	212.1
Methane(g)	CH_4	−74.85	−50.8	186.19
Methanol(l)	CH_3OH	−238.7	−166.3	126.78
Sucrose(s)	$C_{12}H_{22}O_{11}$	−2221.7	−1544.3	360.24

Appendix 3   Mathematical Operations

Logarithms

Common Logarithms

The concept of the logarithm is an extension of the concept of exponents, which is discussed in Chapter 1. The *common*, or base-10, logarithm of any number is the power to which 10 must be raised to equal the number. These examples illustrate this relationship:

Logarithm	Exponent
$\log 1 = 0$	$10^0 = 1$
$\log 10 = 1$	$10^1 = 10$
$\log 100 = 2$	$10^2 = 100$
$\log 10^{-1} = -1$	$10^{-1} = 0.1$
$\log 10^{-2} = -2$	$10^{-2} = 0.01$

In each case the logarithm of the number can be obtained by inspection.

Because the logarithms of numbers are exponents, they have the same properties as exponents. Thus, we have

Logarithm	Exponent
$\log AB = \log A + \log B$	$10^A \times 10^B = 10^{A+B}$
$\log \dfrac{A}{B} = \log A - \log B$	$\dfrac{10^A}{10^B} = 10^{A-B}$

Furthermore, $\log A^n = n \log A$.

Now suppose we want to find the common logarithm of 6.7×10^{-4}. On most electronic calculators, the number is entered first and then the log key is punched. This operation gives us

$$\log 6.7 \times 10^{-4} = -3.17$$

Note that there are as many digits *after* the decimal point as there are significant figures in the original number. The original number has two significant figures and the "17" in -3.17 tells us that the log has two significant figures. The 3 in 3.17 serves only to locate the decimal point in the number 6.7×10^{-4}. Other examples are

Number	Common Logarithm
62	1.79
0.872	−0.0595
1.0×10^{-7}	−7.00

Sometimes (as in the case of pH calculations) it is necessary to obtain the number whose logarithm is known. This procedure is known as taking the antilogarithm; it is simply the reverse of taking the logarithm of a number. Suppose in a certain calculation we have pH = 1.46 and are asked to calculate $[H^+]$. From the definition of pH (pH $= -\log [H^+]$) we can write

$$[H^+] = 10^{-1.46}$$

Many calculators have a key labeled $\log^{-1}$ or INV log to obtain antilogs. Other calculators have a 10^x or y^x key (in which x corresponds to -1.46 in our example and y is 10 for base-10 logarithm). Therefore, we find that $[H^+] = 0.035\ M$.

Natural Logarithms

Logarithms taken to the base e instead of 10 are known as natural logarithms (denoted by ln or $\log_e$); e is equal to 2.7183. The relationship between common logarithms and natural logarithms is:

$$\log 10 = 1 \qquad 10^1 = 10$$
$$\ln 10 = 2.303 \qquad e^{2.303} = 10$$

Thus

$$\ln x = 2.303 \log x$$

To find the natural logarithm of 2.27, say, we first enter the number on the electronic calculator and then punch the ln key to get

$$\ln 2.27 = 0.820$$

If no ln key is provided, we can proceed as:

$$2.303 \log 2.27 = 2.303 \times 0.356$$
$$= 0.820$$

Sometimes we may be given the natural logarithm and asked to find the number it represents. For example,

$$\ln x = 59.7$$

On many calculators, we simply enter the number and punch the e key:

$$e^{59.7} = 8.5 \times 10^{25}$$

The Quadratic Equation

A quadratic equation takes the form

$$ax^2 + bx + c = 0$$

If coefficients a, b, and c are known, then x is given by

$$x = \frac{-b \pm \sqrt{b^2 - 4ac}}{2a}$$

Suppose we have this quadratic equation:

$$2x^2 + 5x - 12 = 0$$

Solving for x, we write

$$x = \frac{-5 \pm \sqrt{(5)^2 - 4(2)(-12)}}{2(2)}$$

$$= \frac{-5 \pm \sqrt{25 + 96}}{4}$$

Therefore

$$x = \frac{-5 + 11}{4} = \frac{3}{2}$$

and

$$x = \frac{-5 - 11}{4} = -4$$

Appendix 4

The Elements and the Derivation of Their Names and Symbols[*]

Element	Symbol	Atomic No.	Atomic Mass[†]	Date of Discovery	Discoverer and Nationality[‡]	Derivation
Actinium	Ac	89	(227)	1899	A. Debierne (Fr.)	Gr. *aktis*, beam or ray
Aluminum	Al	13	26.98	1827	F. Wochler (Ge.)	Alum, the aluminum compound in which it was discovered; derived from L. *alumen*, astringent taste
Americium	Am	95	(243)	1944	A. Ghiorso (USA) R. A. James (USA) G. T. Seaborg (USA) S. G. Thompson (USA)	The Americas
Antimony	Sb	51	121.8	Ancient		L. *antimonium* (*anti*, opposite of; *monium*, isolated condition), so named because it is a tangible (metallic) substance which combines readily; symbol, L. *stibium*, mark
Argon	Ar	18	39.95	1894	Lord Raleigh (GB) Sir William Ramsay (GB)	Gr. *argos*, inactive
Arsenic	As	33	74.92	1250	Albertus Magnus (Ge.)	Gr. *aksenikon*, yellow pigment; L. *arsenicum*, orpiment; the Greeks once used arsenic trisulfide as a pigment
Astatine	At	85	(210)	1940	D. R. Corson (USA) K. R. MacKenzie (USA) E. Segre (USA)	Gr. *astatos*, unstable

Source: From "The Elements and Derivation of Their Names and Symbols," G. P. Dinga, *Chemistry* **41** (2), 20–22 (1968). Copyright by the American Chemical Society.

[*]At the time this table was drawn up, only 103 elements were known to exist.

[†]The atomic masses given here correspond to the 1961 values of the Commission on Atomic Weights. Masses in parentheses are those of the most stable or most common isotopes.

[‡]The abbreviations are (Ar.) Arabic; (Au.) Austrian; (Du.) Dutch; (Fr.) French; (Ge.) German; (GB) British; (Gr.) Greek; (H.) Hungarian; (I.) Italian; (L.) Latin; (P.) Polish; (R.) Russian; (Sp.) Spanish; (Swe.) Swedish; (USA) American.

(Cont.)

Element	Symbol	Atomic No.	Atomic Mass	Date of Discovery	Discoverer and Nationality	Derivation
Barium	Ba	56	137.3	1808	Sir Humphry Davy (GB)	barite, a heavy spar, derived from Gr. *barys*, heavy
Berkelium	Bk	97	(247)	1950	G. T. Seaborg (USA) S. G. Thompson (USA) A. Ghiorso (USA)	Berkeley, Calif.
Beryllium	Be	4	9.012	1828	F. Woehler (Ge.) A. A. B. Bussy (Fr.)	Fr. L. *beryl*, sweet
Bismuth	Bi	83	209.0	1753	Claude Geoffroy (Fr.)	Ge. *bismuth*, probably a distortion of *weisse masse* (white mass) in which it was found
Boron	B	5	10.81	1808	Sir Humphry Davy (GB) J. L. Gay-Lussac (Fr.) L. J. Thenard (Fr.)	The compound borax, derived from Ar. *buraq*, white
Bromine	Br	35	79.90	1826	A. J. Balard (Fr.)	Gr. *bromos*, stench
Cadmium	Cd	48	112.4	1817	Fr. Stromeyer (Ge.)	Gr. *kadmia*, earth; L. *cadmia*, calamine (because it is found along with calamine)
Calcium	Ca	20	40.08	1808	Sir Humphry Davy (GB)	L. *calx*, lime
Californium	Cf	98	(249)	1950	G. T. Seaborg (USA) S. G. Thompson (USA) A. Ghiorso (USA) K. Street, Jr. (USA)	California
Carbon	C	6	12.01	Ancient		L. *carbo*, charcoal
Cerium	Ce	58	140.1	1803	J. J. Berzelius (Swe.) William Hisinger (Swe.) M. H. Klaproth (Ge.)	Asteroid Ceres
Cesium	Cs	55	132.9	1860	R. Bunsen (Ge.) G. R. Kirchhoff (Ge.)	L. *caesium*, blue (cesium was discovered by its spectral lines, which are blue)
Chlorine	Cl	17	35.45	1774	K. W. Scheele (Swe.)	Gr. *chloros*, light green
Chromium	Cr	24	52.00	1797	L. N. Vauquelin (Fr.)	Gr. *chroma*, color (because it is used in pigments)
Cobalt	Co	27	58.93	1735	G. Brandt (Ge.)	Ge. *Kobold*, goblin (because the ore yielded cobalt instead of the expected metal, copper, it was attributed to goblins)
Copper	Cu	29	63.55	Ancient		L. *cuprum*, copper, derived from *cyprium*, Island of Cyprus, the main source of ancient copper
Curium	Cm	96	(247)	1944	G. T. Seaborg (USA) R. A. James (USA) A. Ghiorso (USA)	Pierre and Marie Curie
Dysprosium	Dy	66	162.5	1886	Lecoq de Boisbaudran (Fr.)	Gr. *dysprositos*, hard to get at

(Cont.)

Element	Symbol	Atomic No.	Atomic Mass	Date of Discovery	Discoverer and Nationality	Derivation
Einsteinium	Es	99	(254)	1952	A. Ghiorso (USA)	Albert Einstein
Erbium	Er	68	167.3	1843	C. G. Mosander (Swe.)	Ytterby, Sweden, where many rare earths were discovered
Europium	Eu	63	152.0	1896	E. Demarcay (Fr.)	Europe
Fermium	Fm	100	(253)	1953	A. Ghiorso (USA)	Enrico Fermi
Fluorine	F	9	19.00	1886	H. Moissan (Fr.)	Mineral fluorspar, from L. *fluere,* flow (because fluorspar was used as a flux)
Francium	Fr	87	(223)	1939	Marguerite Perey (Fr.)	France
Gadolinium	Gd	64	157.3	1880	J. C. Marignac (Fr.)	Johan Gadolin, Finnish rare earth chemist
Gallium	Ga	31	69.72	1875	Lecoq de Boisbaudran (Fr.)	L. *Gallia,* France
Germanium	Ge	32	72.59	1886	Clemens Winkler (Ge.)	L. *Germania,* Germany
Gold	Au	79	197.0	Ancient		L. *aurum,* shining dawn
Hafnium	Hf	72	178.5	1923	D. Coster (Du.) G. von Hevesey (H.)	L. *Hafnia,* Copenhagen
Helium	He	2	4.003	1868	P. Janssen (spectr) (Fr.) Sir William Ramsay (isolated) (GB)	Gr. *helios,* sun (because it was first discovered in the sun's spectrum)
Holmium	Ho	67	164.9	1879	P. T. Cleve (Swe.)	L. *Holmia,* Stockholm
Hydrogen	H	1	1.008	1766	Sir Henry Cavendish (GB)	Gr. *hydro,* water; *genes,* forming (because it produces water when burned with oxygen)
Indium	In	49	114.8	1863	F. Reich (Ge.) T. Richter (Ge.)	Indigo, because of its indigo blue lines in the spectrum
Iodine	I	53	126.9	1811	B. Courtois (Fr.)	Gr. *iodes,* violet
Iridium	Ir	77	192.2	1803	S. Tennant (GB)	L. *iris,* rainbow
Iron	Fe	26	55.85	Ancient		L. *ferrum,* iron
Krypton	Kr	36	83.80	1898	Sir William Ramsay (GB) M. W. Travers (GB)	Gr. *kryptos,* hidden
Lanthanum	La	57	138.9	1839	C. G. Mosander (Swe.)	Gr. *lanthanein,* concealed
Lawrencium	Lr	103	(257)	1961	A. Ghiorso (USA) T. Sikkeland (USA) A. E. Larsh (USA) R. M. Latimer (USA)	E. O. Lawrence (USA), inventor of the cyclotron
Lead	Pb	82	207.2	Ancient		Symbol, L. *plumbum,* lead, meaning heavy
Lithium	Li	3	6.941	1817	A. Arfvedson (Swe.)	Gr. *lithos,* rock (because it occurs in rocks)
Lutetium	Lu	71	175.0	1907	G. Urbain (Fr.) C. A. von Welsbach (Au.)	*Lutetia,* ancient name for Paris
Magnesium	Mg	12	24.31	1808	Sir Humphry Davy (GB)	*Magnesia,* a district in Thessaly; possibly derived from L. *magnesia*

(Cont.)

Element	Symbol	Atomic No.	Atomic Mass	Date of Discovery	Discoverer and Nationality	Derivation
Manganese	Mn	25	54.94	1774	J. G. Gahn (Swe.)	L. *magnes*, magnet
Mendelevium	Md	101	(256)	1955	A. Ghiorso (USA) G. R. Choppin (USA) G. T. Seaborg (USA) B. G. Harvey (USA) S. G. Thompson (USA)	Mendeleev, Russian chemist who prepared the periodic chart and predicted properties of undiscovered elements
Mercury	Hg	80	200.6	Ancient		Symbol, L. *hydrargyrum*, liquid silver
Molybdenum	Mo	42	95.94	1778	G. W. Scheele (Swe.)	Gr. *molybdos*, lead
Neodymium	Nd	60	144.2	1885	C. A. von Welsbach (Au.)	Gr. *neos*, new; *didymos*, twin
Neon	Ne	10	20.18	1898	Sir William Ramsay (GB) M. W. Travers (GB)	Gr. *neos*, new
Neptunium	Np	93	(237)	1940	E. M. McMillan (USA) P. H. Abelson (USA)	Planet Neptune
Nickel	Ni	28	58.69	1751	A. F. Cronstedt (Swe.)	Swe. *kopparnickel*, false copper; also Ge. *nickel*, referring to the devil that prevented copper from being extracted from nickel ores
Niobium	Nb	41	92.91	1801	Charles Hatchett (GB)	Gr. *Niobe*, daughter of Tantalus (niobium was considered identical to tantalum, named after *Tantalus*, until 1884; originally called columbium, with symbol Cb)
Nitrogen	N	7	14.01	1772	Daniel Rutherford (GB)	Fr. *nitrogene*, derived from L. *nitrum*, native soda, or Gr. *nitron*, native soda, and Gr. *genes*, forming
Nobelium	No	102	(253)	1958	A. Ghiorso (USA) T. Sikkeland (USA) J. R. Walton (USA) G. T. Seaborg (USA)	Alfred Nobel
Osmium	Os	76	190.2	1803	S. Tennant (GB)	Gr. *osme*, odor
Oxygen	O	8	16.00	1774	Joseph Priestley (GB) C. W. Scheele (Swe.)	Fr. *oxygene*, generator of acid, derived from Gr. *oxys*, acid, and L. *genes*, forming (because it was once thought to be a part of all acids)
Palladium	Pd	46	106.4	1803	W. H. Wollaston (GB)	Asteroid Pallas
Phosphorus	P	15	30.97	1669	H. Brandt (Ge.)	Gr. *phosphoros*, light bearing
Platinum	Pt	78	195.1	1735 1741	A. de Ulloa (Sp.) Charles Wood (GB)	Sp. *platina*, silver

(Cont.)

Element	Symbol	Atomic No.	Atomic Mass	Date of Discovery	Discoverer and Nationality	Derivation
Plutonium	Pu	94	(242)	1940	G. T. Seaborg (USA) E. M. McMillan (USA) J. W. Kennedy (USA) A. C. Wahl (USA)	Planet Pluto
Polonium	Po	84	(210)	1898	Marie Curie (P.)	Poland
Potassium	K	19	39.10	1807	Sir Humphry Davy (GB)	Symbol, L. *kalium,* potash
Praseodymium	Pr	59	140.9	1885	C. A. von Welsbach (Au.)	Gr. *prasios,* green; *didymos,* twin
Promethium	Pm	61	(147)	1945	J. A. Marinsky (USA) L. E. Glendenin (USA) C. D. Coryell (USA)	Gr. mythology, *Prometheus,* the Greek Titan who stole fire from heaven
Protactinium	Pa	91	(231)	1917	O. Hahn (Ge.) L. Meitner (Au.)	Gr. *protos,* first; *actinium* (because it disintegrates into actinium)
Radium	Ra	88	(226)	1898	Pierre and Marie Curie (Fr.; P.)	L. *radius,* ray
Radon	Rn	86	(222)	1900	F. E. Dorn (Ge.)	Derived from radium with suffix "on" common to inert gases (once called nitron, meaning shining, with symbol Nt)
Rhenium	Re	75	186.2	1925	W. Noddack (Ge.) I. Tacke (Ge.) Otto Berg (Ge.)	L. *Rhenus,* Rhine
Rhodium	Rh	45	102.9	1804	W. H. Wollaston (GB)	Gr. *rhodon,* rose (because some of its salts are rose-colored)
Rubidium	Rb	37	85.47	1861	R. W. Bunsen (Ge.) G. Kirchoff (Ge.)	L. *rubidius,* dark red (discovered with the spectroscope, its spectrum shows red lines)
Ruthenium	Ru	44	101.1	1844	K. K. Klaus (R.)	L. *Ruthenia,* Russia
Samarium	Sm	62	150.4	1879	Lecoq de Boisbaudran (Fr.)	Samarskite, after Samarski, a Russian engineer
Scandium	Sc	21	44.96	1879	L. F. Nilson (Swe.)	Scandinavia
Selenium	Se	34	78.96	1817	J. J. Berzelius (Swe.)	Gr. *selene,* moon (because it resembles tellurium, named for the earth)
Silicon	Si	14	28.09	1824	J. J. Berzelius (Swe.)	L. *silex, silicis,* flint
Silver	Ag	47	107.9	Ancient		Symbol, L. *argentum,* silver
Sodium	Na	11	22.99	1807	Sir Humphry Davy (GB)	L. *sodanum,* headache remedy; symbol, L. *natrium,* soda
Strontium	Sr	38	87.62	1808	Sir Humphry Davy (GB)	Strontian, Scotland, derived from mineral strontionite
Sulfur	S	16	32.07	Ancient		L. *sulphurium* (Sanskrit, *sulvere*)

(Cont.)

Element	Symbol	Atomic No.	Atomic Mass	Date of Discovery	Discoverer and Nationality	Derivation
Tantalum	Ta	73	180.9	1802	A. G. Ekeberg (Swe.)	Gr. mythology, *Tantalus*, because of difficulty in isolating it (Tantalus, son of Zeus, was punished by being forced to stand up to his chin in water that receded whenever he tried to drink)
Technetium	Tc	43	(99)	1937	C. Perrier (I.)	Gr. *technetos*, artificial (because it was the first artificial element)
Tellurium	Te	52	127.6	1782	F. J. Müller (Au.)	L. *tellus*, earth
Terbium	Tb	65	158.9	1843	C. G. Mosander (Swe.)	Ytterby, Sweden
Thallium	Tl	81	204.4	1861	Sir William Crookes (GB)	Gr. *thallos*, a budding twig (because its spectrum shows a bright green line)
Thorium	Th	90	232.0	1828	J. J. Berzelius (Swe.)	Mineral thorite, derived from *Thor*, Norse god of war
Thulium	Tm	69	168.9	1879	P. T. Cleve (Swe.)	*Thule*, early name for Scandinavia
Tin	Sn	50	118.7	Ancient		Symbol, L. *stannum*, tin
Titanium	Ti	22	47.88	1791	W. Gregor (GB)	Gr. giants, the Titans, and L. *titans*, giant deities
Tungsten	W	74	183.9	1783	J. J. and F. de Elhuyar (Sp.)	Swe. *tung sten*, heavy stone; symbol, wolframite, a mineral
Uranium	U	92	238.0	1789 1841	M. H. Klaproth (Ge.) E. M. Peligot (Fr.)	Planet Uranus
Vanadium	V	23	50.94	1801 1830	A. M. del Rio (Sp.) N. G. Sefstrom (Swe.)	*Vanadis*, Norse goddess of love and beauty
Xenon	Xe	54	131.3	1898	Sir William Ramsay (GB) M. W. Travers (GB)	Gr. *xenos*, stranger
Ytterbium	Yb	70	173.0	1907	G. Urbain (Fr.)	Ytterby, Sweden
Yttrium	Y	39	88.91	1843	C. G. Mosander (Swe.)	Ytterby, Sweden
Zinc	Zn	30	65.39	1746	A. S. Marggraf (Ge.)	Ge. *zink*, of obscure origin
Zirconium	Zr	40	91.22	1789	M. H. Klaproth (Ge.)	Zircon, in which it was found, derived from Ar. *zargum*, gold color

Glossary*

A

absolute temperature scale. A temperature scale on which absolute zero (0 K) is the lowest temperature (also called the Kelvin temperature scale). (5.3)

absolute zero. Theoretically the lowest attainable temperature. (5.3)

accuracy. The closeness of a measurement to the true value of the quantity that is being measured. (1.6)

acid. A substance that yields hydrogen ions (H^+) when dissolved in water. (2.7)

acid ionization constant (K_a). The equilibrium constant for acid ionization. (16.5)

actinide series. Elements that have incompletely filled $5f$ subshells or readily give rise to cations that have incompletely filled $5f$ subshells. (7.9)

activated complex. The species temporarily formed by reactant molecules as a result of a collision before they form the product. (14.4)

activation energy. The minimum amount of energy required to initiate a chemical reaction. (14.4)

activity series. A summary of the results of many possible displacement reactions. (4.4)

actual yield. The amount of product actually obtained in a reaction. (3.9)

addition reaction. A reaction in which one molecule is added to another. (11.2)

adhesion. Attraction between unlike molecules. (12.3)

alcohol. An organic compound containing the hydroxyl group (—OH). (11.4)

aldehydes. Compounds with a carbonyl functional group and the general formula RCHO, in which R is an H atom, an alkyl, or an aromatic hydrocarbon group. (11.4)

aliphatic hydrocarbons. Hydrocarbons that do not contain the benzene group or the benzene ring. (11.1)

alkali metals. The Group 1A elements (Li, Na, K, Rb, Cs, and Fr). (2.4)

alkaline earth metals. The Group 2A elements (Be, Mg, Ca, Sr, Ba, and Ra). (2.4)

alkanes. Hydrocarbons having the general formula C_nH_{2n+2}, in which $n = 1, 2, \ldots$. (11.2)

alkenes. Hydrocarbons that contain one or more carbon-carbon double bonds. They have the general formula C_nH_{2n}, in which $n = 2, 3, \ldots$. (11.2)

alkynes. Hydrocarbons that contain one or more carbon-carbon triple bonds. They have the general formula C_nH_{2n-2}, in which $n = 2, 3, \ldots$. (11.2)

allotropes. Two or more forms of the same element that differ significantly in chemical and physical properties. (2.6)

alpha particles. See alpha rays.

alpha (α) rays. Helium ions with a charge of +2. (2.2)

amines. Organic bases that have the functional group —NR_2, in which R may be H, an alkyl group, or an aromatic hydrocarbon group. (11.4)

amino acid. A special kind of carboxylic acid that contains at least one carboxyl group (—COOH) and at least one amino group (—NH_2). (22.3)

amphoteric oxide. An oxide that exhibits both acidic and basic properties. (8.6)

amplitude. The vertical distance from the middle of a wave to the peak or trough. (7.1)

anion. An ion with a net negative charge. (2.5)

anode. The electrode at which oxidation occurs. (19.2)

antibonding molecular orbital. A molecular orbital that is of higher energy and lower stability than the atomic orbitals from which it was formed. (10.6)

aqueous solution. A solution in which the solvent is water. (4.1)

aromatic hydrocarbon. A hydrocarbon that contains one or more benzene rings. (11.1)

atmospheric pressure. The pressure exerted by Earth's atmosphere. (5.2)

atom. The basic unit of an element that can enter into chemical combination. (2.2)

atomic mass. The mass of an atom in atomic mass units. (3.1)

atomic mass unit (amu). A mass exactly equal to one-twelfth the mass of one carbon-12 atom. (3.1)

atomic number (Z). The number of protons in the nucleus of an atom. (2.3)

atomic orbital. The wave function of an electron in an atom. (7.5)

atomic radius. One-half the distance between the nuclei in two adjacent atoms of the same element in a metal. For elements that exist as diatomic units, the atomic radius is one-half the distance between the nuclei of two atoms in a particular molecule. (8.3)

Aufbau principle. As protons are added one by one to the nucleus to build up the elements, electrons similarly are added to the atomic orbitals. (7.9)

Avogadro's law. At constant pressure and temperature, the volume of a gas is directly proportional to the number of moles of the gas present. (5.3)

Avogadro's number (N_A). 6.022×10^{23}; the number of particles in a mole. (3.2)

B

barometer. An instrument that measures atmospheric pressure. (5.2)

base. A substance that yields hydroxide ions (OH^-) when dissolved in water. (2.7)

base ionization constant (K_b). The equilibrium constant for the ionization of a base. (16.6)

*The number in parentheses is the number of the section in which the term first appears.

battery. A galvanic cell or a series of several connected galvanic cells that can be used as a source of direct electric current at a constant voltage. (19.6)

beta particles. See beta rays.

beta (β) rays. Streams of electrons emitted during the decay of certain radioactive substances. (2.2)

bimolecular reaction. An elementary step involving two molecules that is part of a reaction mechanism. (14.5)

binary compounds. Compounds containing just two elements. (2.7)

boiling point. The temperature at which the vapor pressure of a liquid is equal to the external atmospheric pressure. (12.6)

bond energy. The enthalpy change required to break a bond in a mole of gaseous molecules. (9.8)

bond length. The distance between the centers of two bonded atoms in a molecule. (9.2)

bond order. The difference between the number of electrons in bonding molecular orbitals and antibonding molecular orbitals, divided by two. (10.6)

bonding molecular orbital. A molecular orbital that is of lower energy and greater stability than the atomic orbitals from which it was formed. (10.6)

boundary surface diagram. Diagram of the region containing about 90 percent of the electron density in an atomic orbital. (7.7)

Boyle's law. The volume of a fixed amount of gas is inversely proportional to the gas pressure at constant temperature. (5.3)

breeder reactor. A nuclear reactor that produces more fissionable material than it uses. (21.5)

Brønsted acid. A substance capable of donating a proton in a reaction. (4.3)

Brønsted base. A substance capable of accepting a proton in a reaction. (4.3)

buffer solution. A solution of (a) a weak acid or base and (b) its salt; both components must be present. A buffer solution has the ability to resist changes in pH when small amounts of either acid or base are added to it. (17.2)

C

calorimetry. The measurement of heat changes. (6.5)

carboxylic acids. Acids that contain the carboxyl group (—COOH). (11.4)

catalyst. A substance that increases the rate of a chemical reaction without being consumed during the reaction. (14.6)

cathode. The electrode at which reduction occurs. (19.2)

cation. An ion with a net positive charge. (2.5)

cell voltage. Difference in electrical potential between the anode and the cathode of a galvanic cell. (19.2)

Charles's and Gay-Lussac's law. See Charles's law.

Charles's law. The volume of a fixed amount of gas is directly proportional to the absolute temperature of the gas when the pressure is held constant. (5.3)

chelating agent. A substance that forms complex ions with metal ions in solution. (20.2)

chemical energy. Energy stored within the structural units of chemical substances. (6.1)

chemical equation. An equation that uses chemical symbols to show what happens during a chemical reaction. (3.7)

chemical equilibrium. A state in which the rates of the forward and reverse reactions are equal and no net changes can be observed. (4.1)

chemical formula. An expression showing the chemical composition of a compound in terms of the symbols for the atoms of the elements involved. (2.6)

chemical kinetics. The area of chemistry concerned with the speeds, or rates, at which chemical reactions occur. (14.1)

chemical property. Any property of a substance that cannot be studied without converting the substance into some other substance. (1.4)

chemical reaction. Chemical change. (3.7)

chemistry. The science that studies the properties of substances and how substances react with one another. (1.3)

chiral. Compounds or ions that are not superimposable with their mirror images. (11.5)

closed system. A system that allows the exchange of energy (usually in the form of heat) but not mass with its surroundings. (6.2)

cohesion. The intermolecular attraction between like molecules. (12.3)

colligative properties. Properties of solutions that depend on the number of solute particles in solution and not on the nature of the solute. (13.6)

complex ion. An ion containing a central metal cation bonded to one or more molecules or ions. (17.7)

compound. A substance composed of atoms of two or more elements chemically united in fixed proportions. (1.3)

concentration of a solution. The amount of solute present in a given quantity of solution. (4.5)

condensation. The phenomenon of going from the gaseous state to the liquid state. (12.6)

conformations. Different spatial arrangements of a molecule that are generated by rotation about single bonds. (11.2)

conjugate acid-base pair. An acid and its conjugate base or a base and its conjugate acid. (16.1)

coordinate covalent bond. A bond in which the pair of electrons is supplied by one of the two bonded atoms. (9.7)

coordination compound. A neutral species containing a complex ion. (20.2)

coordination number. In a crystal lattice it is defined as the number of atoms (or ions) surrounding an atom (or ion) (12.4). In coordination compounds it is defined as the number of donor atoms surrounding the central metal atom in a complex. (20.2)

corrosion. The deterioration of metals by an electrochemical process. (19.7)

covalent bond. A bond in which two electrons are shared by two atoms. (9.2)

covalent compounds. Compounds containing only covalent bonds. (9.2)

critical mass. The minimum mass of fissionable material required to generate a self-sustaining nuclear chain reaction. (21.5)

critical pressure (P_c). The minimum pressure necessary to bring about liquefaction at the critical temperature. (12.6)

critical temperature (T_c). The temperature above which a gas will not liquefy. (12.6)

crystal-field splitting. The energy difference between two sets of d orbitals of a metal atom in the presence of ligands. (20.4)

crystalline solid. A solid that possesses rigid and long-range structural order; its atoms, molecules, or ions occupy specific positions. (12.4)

crystallization. The process in which dissolved solute comes out of solution and forms crystals. (13.1)

cycloalkanes. Hydrocarbons having the general formula C_nH_{2n} in which $n = 3, 4, \ldots$. (11.2)

D

Dalton's law of partial pressures. The total pressure of a mixture of gases is just the sum of the pressures that each gas would exert if it were present alone. (5.5)

delocalized molecular orbital. A molecular orbital that is not confined between two adjacent bonding atoms but actually extends over three or more atoms. (11.3)

density. The mass of a substance divided by its volume. (1.5)

deposition. The process in which vapor molecules are converted directly to the solid phase. (12.6)

diagonal relationship. Similarities between pairs of elements in different groups and periods of the periodic table. (8.6)

diamagnetic. Repelled by a magnet; a diamagnetic substance contains only paired electrons. (7.8)

diatomic molecule. A molecule that consists of two atoms. (2.5)

diffusion. The gradual mixing of molecules of one gas with the molecules of another by virtue of their kinetic properties. (5.6)

dilution. A procedure for preparing a less concentrated solution from a more concentrated solution. (4.5)

dipole moment. The product of charge and the distance between the charges in a molecule. (10.2)

dipole-dipole forces. Forces that act between polar molecules. (12.2)

diprotic acid. Each unit of the acid yields two hydrogen ions. (4.3)

dispersion forces. The attractive forces that arise as a result of temporary dipoles induced in the atoms or molecules. (12.2)

displacement reaction. A reaction in which an atom or an ion in a compound is replaced by an atom of another element. (4.4)

DNA. A type of nucleic acid. (22.4)

donor atom. The atom in a ligand that is bonded directly to the metal atom. (20.2)

double bond. A covalent bond in which two atoms share two pairs of electrons. (9.2)

dynamic equilibrium. The condition in which the rate of a forward process is exactly balanced by the rate of the reverse process. (12.6)

E

electrochemistry. The branch of chemistry that deals with the interconversion of electrical energy and chemical energy. (19.1)

electrolysis. A process in which electrical energy is used to bring about a nonspontaneous chemical reaction. (19.8)

electrolyte. A substance that, when dissolved in water, results in a solution that can conduct electricity. (4.1)

electrolytic cell. An apparatus for carrying out electrolysis. (19.8)

electromagnetic radiation. A wave that has an electric field component and a magnetic field component. (7.1)

electromotive force (emf). The voltage difference between electrodes. (19.3)

electron. A subatomic particle that has a very low mass and carries a single negative electric charge. (2.2)

electron affinity. The negative of the energy change that takes place when an electron is

accepted by an atom (or an ion) in the gaseous state. (8.5)

electron configuration. The distribution of electrons among the various orbitals in an atom or molecule. (7.8)

electron density. The probability that an electron will be found at a particular region in an atomic orbital. (7.5)

electronegativity. The ability of an atom to attract electrons toward itself in a chemical bond. (9.3)

element. A substance that cannot be separated into simpler substances by chemical means. (1.3)

elementary steps. A series of simple reactions that represent the overall progress of a reaction at the molecular level. (14.5)

emission spectrum. The continuous or line spectrum of electromagnetic radiation emitted by a substance. (7.3)

empirical formula. An expression using chemical symbols to show the types of elements in a substance and the ratios of the different kinds of atoms. (2.6)

enantiomers. Optical isomers, that is, compounds and their nonsuperimposable mirror images. (11.5)

end point. Occurs in a titration when the indicator changes color. (17.4)

endothermic processes. Processes that absorb heat from the surroundings. (6.2)

energy. The capacity to do work or to produce change. (6.1)

enthalpy (H). A thermodynamic quantity used to describe heat changes taking place at constant pressure. (6.4)

enthalpy of reaction (ΔH). The difference between the enthalpies of the products and the enthalpies of the reactants. (6.4)

entropy (S). A direct measure of the randomness or disorder of a system. (18.2)

enzyme. A biological catalyst. (14.6)

equilibrium constant. A number equal to the ratio of the equilibrium concentrations of products to the equilibrium concentrations of reactants, each raised to the power of its stoichiometric coefficient. (15.1)

equilibrium vapor pressure. The vapor pressure measured for a dynamic equilibrium of condensation and evaporation. (12.6)

equivalence point. The point at which an acid is completely reacted with or neutralized by a base. (4.6)

esters. Compounds that have the general formula RCOOR′, in which R can be H or an alkyl group or an aromatic hydrocarbon group and R′ is an alkyl group or an aromatic hydrocarbon group. (11.4)

ether. An organic compound containing the R—O—R′ linkage, in which R and R′ are

alkyl and/or aromatic hydrocarbon groups. (11.4)

evaporation. The escape of molecules from the surface of a liquid; also called vaporization. (12.6)

excess reagent. A reactant present in a quantity greater than necessary to react with the amount of the limiting reagent present. (3.9)

excited level (or state). A state that has higher energy than the ground state of the system. (7.3)

exothermic processes. Processes that give off heat to the surroundings. (6.2)

extensive property. A property that depends on how much matter is being considered. (1.4)

F

family. The elements in a vertical column of the periodic table. (2.4)

Faraday. Charge contained in 1 mole of electrons, equivalent to 96,485 coulombs. (19.4)

first law of thermodynamics. Energy can be converted from one form to another, but cannot be created or destroyed. (6.3)

first-order reaction. A reaction whose rate depends on reactant concentration raised to the first power. (14.3)

formal charge. The electrical charge difference between the number of valence electrons in an isolated atom and the number of electrons assigned to that atom in a Lewis structure. (9.5)

formation constant (K_f). The equilibrium constant for the complex ion formation. (17.7)

free energy (G). The energy available to do useful work. (18.4)

frequency (ν). The number of waves that pass through a particular point per unit time. (7.1)

fuel cell. A galvanic cell that requires a continuous supply of reactants to keep functioning. (19.6)

functional group. That part of a molecule characterized by a special arrangement of atoms that is largely responsible for the chemical behavior of the parent molecule. (11.1)

G

galvanic cell. An electrochemical cell that generates electricity by means of a spontaneous redox reaction. (19.2)

gamma (γ) rays. High-energy radiation. (2.2)

gas constant (R). The constant that appears in the ideal gas equation ($PV = nRT$). It is expressed as 0.08206 L · atm/K · mol, or 8.314 J/K · mol. (5.4)

geometric isomers. Compounds with the same type and number of atoms and the same chemical bonds but different spatial arrangements; such isomers cannot be interconverted without breaking a chemical bond. (11.2)

Gibbs free energy. See free energy.

gravimetric analysis. An experimental procedure that involves the measurement of mass to identify an unknown component of a substance. (4.6)

ground level (or state). The lowest energy state of a system. (7.3)

group. The elements in a vertical column of the periodic table. (2.4)

H

half-cell reactions. Oxidation and reduction reactions that occur at the electrodes. (19.2)

half-life. The time required for the concentration of a reactant to decrease to half its initial concentration. (14.3)

half-reaction. A reaction that explicitly shows electrons involved in either oxidation or reduction. (4.4)

halogens. The nonmetallic elements in Group 7A (F, Cl, Br, I, and At). (2.4)

heat. Transfer of energy between two bodies that are at different temperatures. (6.2)

heat capacity (*C*). The amount of heat required to raise the temperature of a given quantity of a substance by one degree Celsius. (6.5)

heat content. See enthalpy.

Heisenberg uncertainty principle. It is impossible to know simultaneously both the momentum and the position of a particle with certainty. (7.5)

Henry's law. The solubility of a gas in a liquid is proportional to the pressure of the gas over the solution. (13.5)

Hess's law. When reactants are converted to products, the change in enthalpy is the same whether the reaction takes place in one step or in a series of steps. (6.6)

heterogeneous equilibrium. An equilibrium state in which the reacting species are not all in the same phase. (15.2)

heterogeneous mixture. The individual components of such a mixture remain physically separate and can be seen as separate components. (1.3)

homogeneous equilibrium. An equilibrium condition in which all reacting species are in the same phase. (15.2)

homogeneous mixture. The composition of the mixture is the same throughout the solution. (1.3)

homonuclear diatomic molecule. A diatomic molecule containing atoms of the same element. (10.6)

Hund's rule. The most stable arrangement of electrons in atomic subshells is the one with the greatest number of parallel spins. (7.8)

hybrid orbitals. Atomic orbitals obtained when two or more nonequivalent orbitals of the same atom combine before covalent bond formation. (10.4)

hybridization. The process of mixing the atomic orbitals in an atom (usually the central atom) to generate a set of new atomic orbitals before covalent bond formation. (10.4)

hydration. A process in which an ion or a molecule is surrounded by water molecules arranged in a specific manner. (4.1)

hydrocarbons. Compounds made up of only carbon and hydrogen. (11.1)

hydrogen bond. A special type of dipole-dipole interaction between the hydrogen atom bonded to an atom of a very electronegative element (F, N, O) and another atom of one of the three electronegative elements. (12.2)

hydrogenation. The addition of hydrogen, especially to compounds with double and triple carbon-carbon bonds. (11.2)

hydronium ion. H_3O^+. (4.3)

hypothesis. A tentative explanation for a set of observations. (1.2)

I

ideal gas. A hypothetical gas whose pressure-volume-temperature behavior can be completely accounted for by the ideal gas equation. (5.4)

ideal gas equation. An equation expressing the relationships among pressure, volume, temperature, and amount of gas ($PV = nRT$, in which R is the gas constant). (5.4)

ideal solution. Any solution that obeys Raoult's law. (13.6)

indicators. Substances that have distinctly different colors in acidic and basic media. (4.6)

induced dipole. The separation of positive and negative charges in an atom (or a nonpolar molecule) caused by the proximity of an ion or a polar molecule. (12.2)

inert complex. A complex ion that undergoes very slow ligand exchange reactions. (20.5)

intensive property. A property that does not depend on how much matter is being considered. (1.4)

intermediate. A species that appears in the mechanism of the reaction (that is, in the elementary steps) but not in the overall balanced equation. (14.5)

intermolecular forces. Attractive forces that exist among molecules. (12.2)

International System of Units. A revised metric system (abbreviated SI) that is widely used in scientific research. (1.5)

intramolecular forces. Forces that hold atoms together in a molecule. (12.2)

ion. An atom or group of atoms that has a net positive or negative charge. (2.5)

ion pair. A species made up of at least one cation and at least one anion held together by electrostatic forces. (13.6)

ion-dipole forces. Forces that operate between an ion and a dipole. (12.2)

ionic bond. The electrostatic force that holds ions together in an ionic compound. (9.3)

ionic compound. Any neutral compound containing cations and anions. (2.5)

ionic equation. An equation that shows dissolved ionic compounds in terms of their free ions. (4.2)

ionic radius. The radius of a cation or an anion as measured in an ionic compound. (8.3)

ionization energy. The minimum energy required to remove an electron from an isolated atom (or an ion) in its ground state. (8.4)

ion-product constant. Product of hydrogen ion concentration and hydroxide ion concentration (both in molarity) at a particular temperature. (16.2)

isoelectronic. Ions, or atoms and ions, that possess the same number of electrons, and hence the same ground-state electron configuration, are said to be isoelectronic. (8.2)

isolated system. A system that does not allow the transfer of either mass or energy to or from its surroundings. (6.2)

isotopes. Atoms having the same atomic number but different mass numbers. (2.3)

J

Joule. Unit of energy given by newtons × meters. (5.6)

K

Kelvin temperature scale. See absolute temperature scale.

ketones. Compounds with a carbonyl functional group and the general formula RR′CO, in which R and R′ are alkyl and/or aromatic hydrocarbon groups. (11.4)

kinetic energy (KE). Energy available because of the motion of an object. (5.6)

kinetic molecular theory of gases. A theory that describes the physical behavior of gases at the molecular level. (5.6)

L

labile complex. Complexes that undergo rapid ligand exchange reactions. (20.5)

lanthanide series. Elements that have incompletely filled $4f$ subshells or readily give rise to cations that have incompletely filled $4f$ subshells. (7.9)

lattice points. The positions occupied by atoms, molecules, or ions that define the geometry of a unit cell. (12.4)

law. A concise verbal or mathematical statement of a relationship between phenomena that is always the same under the same conditions. (1.2)

law of conservation of energy. The total quantity of energy in the universe is constant. (6.1)

law of conservation of mass. Matter can be neither created nor destroyed. (2.1)

law of definite proportions. Different samples of the same compound always contain its constituent elements in the same proportions by mass. (2.1)

law of multiple proportions. If two elements can combine to form more than one type of compound, the masses of one element that combine with a fixed mass of the other element are in ratios of small whole numbers. (2.1)

Le Châtelier's principle. If an external stress is applied to a system at equilibrium, the system will adjust itself in such a way as to partially offset the stress. (15.4)

Lewis acid. A substance that can accept a pair of electrons. (16.10)

Lewis base. A substance that can donate a pair of electrons. (16.10)

Lewis dot symbol. The symbol of an element with one or more dots that represent the number of valence electrons in an atom of the element. (9.1)

Lewis structure. A representation of covalent bonding using Lewis symbols. Shared electron pairs are shown either as lines or as pairs of dots between two atoms, and lone pairs are shown as pairs of dots on individual atoms. (9.2)

ligand. A molecule or an ion that is bonded to the metal ion in a complex ion. (20.2)

limiting reagent. The reactant used up first in a reaction. (3.9)

line spectrum. Spectrum produced when radiation is absorbed or emitted by a substance only at some wavelengths. (7.3)

liter. The volume occupied by 1 cubic decimeter. (1.5)

lone pairs. Valence electrons that are not involved in covalent bond formation. (9.2)

M

macroscopic properties. Properties that can be measured directly. (1.5)

manometer. A device used to measure the pressure of gases. (5.2)

many-electron atoms. Atoms that contain two or more electrons. (7.5)

mass. A measure of the quantity of matter contained in an object. (1.5)

mass defect. The difference between the mass of an atom and the sum of the masses of its protons, neutrons, and electrons. (21.2)

mass number. The total number of neutrons and protons present in the nucleus of an atom. (2.3)

matter. Anything that occupies space and possesses mass. (1.3)

melting point. The temperature at which solid and liquid phases coexist in equilibrium. (12.6)

metalloid. An element with properties intermediate between those of metals and nonmetals. (2.4)

metals. Elements that are good conductors of heat and electricity and have the tendency to form positive ions in ionic compounds. (2.4)

microscopic properties. Properties that must be measured indirectly with the aid of a microscope or other special instrument. (1.5)

miscible. Two liquids that are completely soluble in each other in all proportions are said to be miscible. (13.2)

mixture. A combination of two or more substances in which the substances retain their identities. (1.3)

moderator. A substance that can reduce the kinetic energy of neutrons. (21.5)

molality. The number of moles of solute dissolved in 1 kilogram of solvent. (13.3)

molar concentration. See molarity.

molar heat of fusion (ΔH_{fus}). The energy (in kilojoules) required to melt 1 mole of a solid. (12.6)

molar heat of sublimation (ΔH_{sub}). The energy (in kilojoules) required to sublime 1 mole of a solid. (12.6)

molar heat of vaporization (ΔH_{vap}). The energy (in kilojoules) required to vaporize 1 mole of a liquid. (12.6)

molar mass ($\mathcal{M}$). The mass (in grams or kilograms) of 1 mole of atoms, molecules, or other particles. (3.2)

molar solubility. The number of moles of solute in 1 liter of a saturated solution (mol/L). (17.5)

molarity (M). The number of moles of solute in 1 liter of solution. (4.5)

mole (mol). The amount of substance that contains as many elementary entities (atoms, molecules, or other particles) as there are atoms in exactly 12 grams (or 0.012 kilograms) of the carbon-12 isotope. (3.2)

mole fraction. Ratio of the number of moles of one component of a mixture to the total number of moles of all components in the mixture. (5.5)

mole method. An approach for determining the amount of product formed in a reaction. (3.8)

molecular equations. Equations in which the formulas of the compounds are written as though all species existed as molecules or whole units. (4.2)

molecular formula. An expression showing the exact numbers of atoms of each element in a molecule. (2.6)

molecular mass. The sum of the atomic masses (in amu) present in a given molecule. (3.3)

molecular orbital. An orbital that results from the interaction of atomic orbitals of the bonding atoms. (10.6)

molecularity of a reaction. The number of molecules reacting in an elementary step. (14.5)

molecule. An aggregate of at least two atoms in a definite arrangement held together by special forces. (2.5)

monatomic ion. An ion that contains only one atom. (2.5)

monomer. Simple repeating units in a polymer. (22.2)

monoprotic acid. Each unit of the acid yields one hydrogen ion. (4.3)

multiple bonds. Bonds formed when two atoms share two or more pairs of electrons. (9.2)

N

Nernst equation. The relation between the emf of a galvanic cell and the standard emf and the concentrations of the oxidizing and reducing agents. (19.5)

net ionic equation. An equation that includes only the ionic species that actually take part in the reaction. (4.2)

neutralization reaction. A reaction between an acid and a base. (4.3)

neutron. A subatomic particle that bears no net electric charge. Its mass is slightly greater than a proton's. (2.2)

Newton (N). The SI unit for force. (5.2)

noble gas core. The noble gas that most nearly precedes the element being considered; used in writing electron configurations. (7.9)

noble gases. Nonmetallic elements in Group 8A (He, Ne, Ar, Kr, Xe, and Rn). (2.4)

node. A point at which the amplitude of a wave is zero. (7.4)

nonelectrolyte. A substance that, when dissolved in water, gives a solution that is not electrically conducting. (4.1)

nonmetals. Elements that are usually poor conductors of heat and electricity. (2.4)

nonpolar molecule. A molecule that does not possess a dipole moment. (10.2)

nonvolatile. Does not have a measurable vapor pressure. (13.6)

nuclear binding energy. The energy required to break up a nucleus into protons and neutrons. (21.2)

nuclear chain reaction. A self-sustaining sequence of nuclear fission reactions. (21.5)

nuclear fission. The process in which a heavy nucleus (mass number > 200) divides to form small nuclei of intermediate mass and one or more neutrons. (21.5)

nuclear fusion. The combining of small nuclei into larger ones. (21.6)

nuclear reaction. A reaction involving change in an atomic nucleus. (21.1)

nuclear transmutation. The change undergone by a nucleus as a result of bombardment by neutrons or other particles. (21.1)

nucleic acid. High molar mass polymers that play an essential role in protein synthesis. (22.4)

nucleons. Protons and neutrons in a nucleus. (21.2)

nucleotide. The repeating unit in each strand of a DNA molecule which consists of a base-deoxyribose-phosphate group. (22.4)

nucleus. The central core of an atom. (2.2)

O

octet rule. An atom other than hydrogen tends to form bonds until it is surrounded by eight valence electrons. (9.2)

open system. A system that can exchange mass and energy (usually in the form of heat) with its surroundings. (6.2)

optical isomers. Compounds that are non-superimposable mirror images. (11.5)

orbital. See atomic orbital.

organic chemistry. The branch of chemistry that deals with carbon compounds. (11.1)

osmosis. The net movement of solvent molecules through a semipermeable membrane from a pure solvent or from a dilute solution to a more concentrated solution. (13.6)

osmotic pressure (π). The pressure required to stop osmosis. (13.6)

overvoltage. The additional voltage required to cause electrolysis. (19.8)

oxidation number. The number of charges an atom would have in a molecule if electrons were transferred completely in the direction indicated by the difference in electronegativity. (4.4)

oxidation reaction. The half-reaction that represents the loss of electrons in a redox process. (4.4)

oxidation state. See oxidation number.

oxidation-reduction reaction. See redox reaction.

oxidizing agent. A substance that can accept electrons from another substance or increase the oxidation numbers of another substance. (4.4)

oxoacid. An acid containing hydrogen, oxygen, and another element (the central element). (2.7)

oxoanion. An anion derived from an oxoacid. (2.7)

P

paramagnetic. Attracted to a magnet. A paramagnetic substance contains one or more unpaired electrons. (7.8)

partial pressure. The pressure of one component in a mixture of gases. (5.5)

Pascal (Pa). A pressure of one newton per square meter (1 N/m^2). (5.2)

Pauli exclusion principle. No two electrons in an atom can have the same four quantum numbers. (7.8)

percent composition. The percent by mass of each element in a compound. (3.5)

percent ionization. The ratio of ionized acid concentration at equilibrium to the initial concentration of acid. (16.5)

percent by mass. The ratio of the mass of a solute to the mass of the solution, multiplied by 100%. (13.3)

percent yield. The ratio of the actual yield of a reaction to the theoretical yield, multiplied by 100%. (3.9)

period. A horizontal row of the periodic table. (2.4)

periodic table. A tabular arrangement of the elements by similarities in properties and by increasing atomic number. (2.4)

pH. The negative logarithm of the hydrogen ion concentration in an aqueous solution. (16.3)

phase. A homogeneous part of a system that is in contact with other parts of the system but separated from them by a well-defined boundary. (12.6)

phase change. Transformation from one phase to another. (12.6)

phase diagram. A diagram showing the conditions at which a substance exists as a solid, liquid, and vapor. (12.7)

photon. A particle of light. (7.2)

physical equilibrium. An equilibrium in which only physical properties change. (15.1)

physical property. Any property of a substance that can be observed without transforming the substance into some other substance. (1.4)

pi bond (π bond). A covalent bond formed by sideways overlapping orbitals; its electron density is concentrated above and below the plane of the nuclei of the bonding atoms. (10.5)

pi molecular orbital. A molecular orbital in which the electron density is concentrated above and below the line joining the two nuclei of the bonding atoms. (10.6)

plasma. A gaseous state of matter consisting of positive ions and electrons. (21.6)

polar covalent bond. In such a bond, the electrons spend more time in the vicinity of one atom than the other. (9.3)

polar molecule. A molecule that possesses a dipole moment. (10.2)

polarimeter. The instrument for studying interaction between plane-polarized light and chiral molecules. (11.5)

polarizability. The ease with which the electron density in a neutral atom (or molecule) can be distorted. (12.2)

polyatomic ion. An ion that contains more than one atom. (2.5)

polyatomic molecule. A molecule that consists of more than two atoms. (2.5)

polymer. A molecular compound distinguished by a high molar mass and made up of many repeating units. (22.1)

positron. A particle that has the same mass as the electron but bears a +1 charge. (21.1)

potential energy. Energy available by virtue of an object's position. (6.1)

precipitate. An insoluble solid that separates from a supersaturated solution. (4.2)

precipitation reaction. A reaction characterized by the formation of a precipitate. (4.2)

precision. The closeness of agreement of two or more measurements of the same quantity. (1.6)

pressure. Force applied per unit area. (5.2)

product. The substance formed as a result of a chemical reaction. (3.7)

protein. A polymer of amino acids. (22.3)

proton. A subatomic particle having a single positive electric charge. The mass of a proton is about 1840 times that of an electron. (2.2)

Q

qualitative. Consisting of general observations about the system. (1.2)

qualitative analysis. The determination of the types of ions present in a solution. (17.8)

quantitative. Comprising numbers obtained by various measurements of the system. (1.2)

quantum. The smallest quantity of energy that can be emitted or absorbed in the form of electromagnetic radiation. (7.1)

quantum numbers. Numbers that describe the distribution of electrons in atoms. (7.6)

R

racemic mixture. An equimolar mixture of two enantiomers. (11.5)

radiant energy. Energy transmitted in the form of waves. (6.1)

radiation. The emission and transmission of energy through space in the form of particles and/or waves. (2.2)

radical. Any species containing one or more unpaired electrons. (11.2)

radioactive decay series. A sequence of nuclear reactions that ultimately result in the formation of a stable isotope. (21.3)

radioactivity. The spontaneous breakdown of a nucleus by the emission of particles and/or radiation. (2.2)

Raoult's law. The partial pressure of the solvent over a solution is given by the product of the vapor pressure of the pure solvent and the mole fraction of the solvent in the solution. (13.6)

rare earth series. See lanthanide series.

rare gases. See noble gases.

rate constant. Proportionality constant relating reaction rate to the concentrations of reactants. (14.2)

rate law. An expression relating the rate of a reaction to the rate constant and the concentrations of the reactants. (14.2)

rate-determining step. The slowest step in the sequence of steps leading to the formation of products. (14.5)

reactants. The starting substances in a chemical reaction. (3.7)

reaction mechanism. The sequence of elementary steps that leads to product formation. (14.5)

reaction order. The sum of the powers to which all reactant concentrations appearing in the rate law are raised. (14.2)

reaction quotient. A number equal to the ratio of product concentrations to reactant concentrations, each raised to the power of its stoichiometric coefficient at some point other than equilibrium. (15.3)

reaction rate. The change in concentration of reactant or product with time. (14.1)

redox reaction. A reaction in which there is either a transfer of electrons or a change in the oxidation numbers of the substances taking part in the reaction. Also called oxidation-reduction reaction. (4.4)

reducing agent. A substance that can donate electrons to another substance or decrease the oxidation numbers in another substance. (4.4)

reduction reaction. The half-reaction that represents the gain of electrons in a redox process. (4.4)

representative elements. Elements in Groups 1A through 7A, all of which have at least an incompletely filled s or p subshell of the highest principal quantum number. (8.2)

resonance. The use of two or more Lewis structures to represent a particular molecule. (9.6)

resonance structure. One of two or more alternative Lewis structures for a single molecule that cannot be described fully with a single Lewis structure. (9.6)

reversible reaction. A reaction that can occur in both directions. (4.1)

RNA. A type of nucleic acid. (22.4)

root-mean-square (rms) speed (u_{rms}). A measure of the average molecular speed at a given temperature. (5.6)

S

salt. An ionic compound made up of a cation other than H^+ and an anion other than OH^- or O^{2-}. (4.3)

salt hydrolysis. The reaction of the anion or cation, or both, of a salt with water. (16.8)

saponification. Soapmaking. (11.4)

saturated hydrocarbons. Hydrocarbons that contain only single covalent bonds. (11.2)

saturated solution. At a given temperature, the solution that results when the maximum amount of a substance has dissolved in a solvent. (13.1)

scientific method. A systematic approach to research. (1.2)

second law of thermodynamics. The entropy of the universe increases in a spontaneous process and remains unchanged in an equilibrium process. (18.3)

second-order reaction. A reaction whose rate depends on the concentration of one reactant raised to the second power or on the concentrations of two different reactants, each raised to the first power. (14.3)

semipermeable membrane. A membrane that allows solvent molecules to pass through, but blocks the movement of solute molecules. (13.6)

sigma bond (σ bond). A covalent bond formed by orbitals overlapping end-to-end; its electron density is concentrated between the nuclei of the bonding atoms. (10.5)

sigma molecular orbital. A molecular orbital in which the electron density is concentrated around a line between the two nuclei of the bonding atoms. (10.6)

significant figures. The number of meaningful digits in a measured or calculated quantity. (1.6)

single bond. Two atoms held together by one electron pair are joined by a single bond. (9.2)

solubility. The maximum amount of solute that can be dissolved in a given quantity of solvent at a specific temperature. (4.2)

solubility product (K_{sp}). The product of the molar concentrations of constituent ions, each raised to the power of its stoichiometric coefficient in the equilibrium equation. (17.5)

solute. The substance present in the smaller amount in a solution. (4.1)

solution. A homogeneous mixture of two or more substances. (4.1)

solvation. The process in which an ion or molecule is surrounded by solvent molecules arranged in an ordered manner. (13.2)

solvent. The substance present in the larger amount in a solution. (4.1)

specific heat (s). The amount of heat energy required to raise the temperature of one gram of a substance by one degree Celsius. (6.5)

spectator ions. Ions that are not involved in the overall reaction. (4.2)

spectrochemical series. A list of ligands arranged in order of their abilities to split the d-orbital energies. (20.4)

standard atmospheric pressure (1 atm). The pressure that supports a column of mercury exactly 76-cm high at 0°C at sea level. (5.2)

standard emf ($E°$). The difference of the standard reduction potentials of the substance that undergoes reduction and the substance that undergoes oxidation in a redox process. (19.3)

standard enthalpy of formation (ΔH_f^o). The heat change that results when 1 mole of a compound is formed from its elements in their standard states. (6.6)

standard enthalpy of reaction (ΔH_{rxn}^o). The enthalpy change that occurs when a reaction is carried out under standard-state conditions. (6.6)

standard entropy of reaction (ΔS_{rxn}^o). The entropy change when the reaction is carried out under standard-state conditions. (18.3)

standard free energy of formation (ΔG_f°). The free-energy change when 1 mole of a compound is synthesized from its elements in their standard states. (18.4)

standard free energy of reaction (ΔG_{rxn}°). The free energy change when the reaction is carried out under standard-state conditions. (18.4)

standard reduction potential. The voltage measured as a reduction reaction occurs at the electrode when all solutes are 1 M and all gases are at 1 atm. (19.3)

standard solution. A solution of accurately known concentration. (4.6)

standard state. The condition of 1 atm of pressure. (6.6)

standard temperature and pressure (STP). 0°C and 1 atm. (5.4)

state function. A property that is determined by the state of the system. (6.3)

state of a system. The values of all pertinent macroscopic variables (for example, composition, volume, pressure, and temperature) of a system. (6.3)

stoichiometric amounts. The exact molar amounts of reactants and products that appear in a balanced chemical equation. (3.9)

stoichiometry. The mass relationships among reactants and products in chemical reactions. (3.8)

strong acid. An acid that is a strong electrolyte. (16.4)

strong base. A base that is a strong electrolyte. (16.4)

structural formula. A representation that shows how atoms are bonded to one another in a molecule. (2.6)

structural isomers. Molecules that have the same molecular formula but different structures. (11.2)

sublimation. The process in which molecules go directly from the solid phase into the vapor phase. (12.6)

substance. A form of matter that has a definite or constant composition (the number and type of basic units present) and distinct properties. (1.2)

supersaturated solution. A solution that contains more of the solute than is present in a saturated solution. (13.1)

surface tension. The amount of energy required to stretch or increase the surface of a liquid by a unit area. (12.3)

surroundings. The rest of the universe outside a system. (6.2)

system. Any specific part of the universe that is of interest to us. (6.2)

T

termolecular reaction. An elementary step involving three molecules. (14.5)

ternary compounds. Compounds consisting of three elements. (2.7)

theoretical yield. The amount of product predicted by the balanced equation when all of the limiting reagent has reacted. (3.9)

theory. A unifying principle that explains a body of facts and/or those laws that are based on them. (1.2)

thermal energy. Energy associated with the random motion of atoms and molecules. (6.1)

thermal pollution. The heating of the environment to temperatures that are harmful to its living inhabitants. (13.4)

thermochemical equation. An equation that shows both the mass and enthalpy relations. (6.4)

thermochemistry. The study of heat changes in chemical reactions. (6.2)

thermodynamics. The scientific study of the interconversion of heat and other forms of energy. (6.3)

thermonuclear reactions. Nuclear fusion reactions that occur at very high temperatures. (21.6)

third law of thermodynamics. The entropy of a perfect crystalline substance is zero at the absolute zero of temperature. (18.3)

titration. The gradual addition of a solution of accurately known concentration to another solution of unknown concentration until the chemical reaction between the two solutions is complete. (4.6)

tracers. Isotopes, especially radioactive isotopes, that are used to trace the path of the atoms of an element in a chemical or biological process. (21.7)

transition metals. Elements that have incompletely filled d subshells or readily give rise to cations that have incompletely filled d subshells. (7.9)

transuranium elements. Elements with atomic numbers greater than 92. (21.4)

triple bond. A covalent bond in which two atoms share three pairs of electrons. (9.2)

triple point. The point at which the vapor, liquid, and solid states of a substance are in equilibrium. (12.7)

triprotic acid. Each unit of the acid yields three hydrogen ions. (4.3)

U

unimolecular reaction. An elementary step involving one molecule. (14.5)

unit cell. The basic repeating unit of the arrangement of atoms, molecules, or ions in a crystalline solid. (12.4)

unsaturated hydrocarbons. Hydrocarbons that contain carbon-carbon double bonds or carbon-carbon triple bonds. (11.2)

unsaturated solution. A solution that contains less solute than it has the capacity to dissolve. (13.1)

V

valence electrons. The outer electrons of an atom, which are the ones involved in chemical bonding. (8.2)

valence shell. The outermost electron-occupied shell of an atom, which holds the electrons that are usually involved in bonding. (10.1)

valence-shell electron-pair repulsion (VSEPR) model. A model that accounts for the geometrical arrangements of shared and unshared electron pairs around a central atom in terms of the repulsive forces between electron pairs. (10.1)

van der Waals equation. An equation that describes the relationships among P, V, n, and T for a nonideal gas. (5.7)

van der Waals forces. The collective name for certain attractive forces between atoms and molecules, namely, dipole-dipole, dipole-induced dipole, and dispersion forces. (12.2)

vaporization. The escape of molecules from the surface of a liquid; also called evaporation. (12.6)

viscosity. A measure of a fluid's resistance to flow. (12.3)

volatile. Having a measurable vapor pressure. (13.6)

volume. Length cubed. (1.5)

W

wave. A vibrating disturbance by which energy is transmitted. (7.1)

wavelength (λ). The distance between identical points on successive waves. (7.1)

weak acid. An acid that is a weak electrolyte. (16.4)

weak base. A base that is a weak electrolyte. (16.4)

weight. The force that gravity exerts on an object. (1.5)

work. Directed energy change resulting from a process. (6.1)

Answers to Even-Numbered Problems

Chapter 1

1.8 (a) Physical change, (b) chemical change, (c) physical change, (d) chemical change, (e) physical change. **1.10** (a) Extensive, (b) intensive, (c) intensive. **1.12** (a) Compound, (b) element, (c) compound, (d) element. **1.18** 1.30×10^3 g. **1.20** (a) 41°C, (b) 11.3°F, (c) 1.1×10^4°F. **1.22** (a) 7.49×10^{-1}, (b) 8.026×10^2, (c) 6.21×10^{-7}. **1.24** (a) 0.00003256, (b) 6030000. **1.26** (a) 1.8×10^{-2}, (b) 1.14×10^{10}, (c) -5×10^4, (d) 1.3×10^3. **1.28** (a) Three, (b) one, (c) one or two, (d) two. **1.30** (a) 1.28, (b) 3.18×10^{-3} mg, (c) 8.14×10^7 dm. **1.32** (a) 1.10×10^8 mg, (b) 6.83×10^{-5} m^3. **1.34** 3.1557×10^7 s. **1.36** (a) 81 in/s, (b) 1.2×10^2 m/min, (c) 7.4 km/h. **1.38** (a) 8.35×10^{12} mi, (b) 2.96×10^3 cm, (c) 9.8×10^8 ft/s, (d) 8.6°C, (e) -459.67°F, (f) 7.12×10^{-5} m^3, (g) 7.2×10^3 L. **1.40** 6.25×10^{-4} g/cm^3. **1.42** 4.33×10^7 ton. **1.44** 2.6 g/cm^3. **1.46** 0.882 cm. **1.48** 10.5 g/cm^3. **1.50** 767 mph. **1.52** 75.6°F $\pm$ 0.2°F. **1.54** 500 mL. **1.56** 5.4×10^{10} L. **1.58** $-40°$. **1.60** (a) 0.5%, (b) 3.1%. **1.62** 6.0×10^{12} g Au; 7.4×10^{13}. **1.64** 30 times. **1.66** 1.450 $\times 10^{-2}$ mm. **1.68** 2.3×10^4 kg NaF; 99%. **1.70** 4.2×10^{-19} g/L.

Chapter 2

2.8 0.62 mi. **2.12** 145. **2.14** ^{15}N: 7 protons, 7 electrons, and 8 neutrons; ^{33}S: 16 protons, 16 electrons and 17 neutrons; ^{63}Cu: 29 protons, 29 electrons, and 34 neutrons; ^{84}Sr: 38 protons, 38 electrons, and 46 neutrons; ^{130}Ba: 56 protons, 56 electrons and 74 neutrons; ^{186}W: 74 protons, 74 electrons, and 112 neutrons; ^{202}Hg: 80 protons, 80 electrons, and 122 neutrons. **2.16** (a) $^{186}_{74}$W, (b) $^{201}_{80}$Hg. **2.34** C_2H_6O. **2.36** (a) $AlBr_3$, (b) $NaSO_2$, (c) N_2O_5, (d) $K_2Cr_2O_7$. **2.40** K$^+$(19,18), Mg^{2+}(12,10), Fe^{3+}(26,23), Br$^-$(35,36), Mn^{2+}(25,23), C^{4-}(6,10), Cu^{2+}(29,27). **2.42** Ionic: NaBr, BaF$_2$, CsCl; molecular: CH$_4$, CCl$_4$, ICl, NF$_3$. **2.44** (a) Potassium hypochlorite, (b) silver carbonate, (c) iron(II) chloride, (d) potassium permanganate, (e) cesium chlorate, (f) hypoiodous acid, (g) iron(II) oxide, (h) iron(III) oxide, (i) titanium(IV) chloride, (j) sodium hydride, (k) lithium nitride, (l) sodium oxide, (m) sodium peroxide. **2.46** (a) CuCN, (b) Sr(ClO$_2$)$_2$, (c) HClO$_4$, (d) HI, (e) Na$_2$NH$_4$PO$_4$, (f) PbCO$_3$, (g) SnF$_2$, (h) P$_4$S$_{10}$, (i) HgO, (j) Hg$_2$I$_2$. **2.48** (c). **2.50** (a) H or H$_2$?, (b) NaCl is an ionic compound. **2.52** (a) Molecule and compound, (b) element and molecule, (c) element, (d) molecule and compound, (e) element, (f) element and molecule, (g) element and molecule, (h) molecule and compound, (i) compound, (j) element, (k) element and molecule, (l) compound. **2.54** (a) CO$_2$(solid), (b) NaCl, (c) N$_2$O, (d) CaCO$_3$, (e) CaO, (f) Ca(OH)$_2$, (g) NaHCO$_3$, (h) Mg(OH)$_2$. **2.56** (a) Metals in groups 1A, 2A, and aluminum and nonmetals such as nitrogen, oxygen, and the halogens, (b) the transition metals. **2.58** ^{23}Na. **2.60** Mercury (Hg) and bromine (Br$_2$). **2.62** H$_2$, N$_2$, O$_2$, O$_3$, F$_2$, Cl$_2$, He, Ne, Ar, Kr, Xe, Rn. **2.64** He, Ne, and Ar are chemically inert; Kr, Xe, and Rn only form a few compounds. **2.66** All isotopes of radium are radioactive. It is a radioactive decay product of ^{238}U. **2.68** (a) NaH (sodium hydride), (b) B$_2$O$_3$ (diboron trioxide), (c) Na$_2$S (sodium sulfide), (d) AlF$_3$ (aluminum fluoride), (e) OF$_2$ (oxygen difluoride), (f) SrCl$_2$ (strontium chloride).

Chapter 3

3.10 ^{6}Li: 7.5%; ^{7}Li: 92.5%. **3.12** 5.82×10^3 light years. **3.14** 5.1×10^{24} amu. **3.16** 9.96×10^{-15} mol. **3.18** 3.01×10^3 g Au. **3.20** (a) 1.244×10^{-22} g/atom, (b) 9.746×10^{-23} g/atom. **3.22** 2.98×10^{22} Cu atoms. **3.24** Lead. **3.26** (a) 256.6 g, (b) 76.15 g, (c) 119.37 g, (d) 176.12 g. **3.28** 1.11×10^{-2} mol. **3.30** 1.68×10^{26} C atoms; 6.72×10^{26} H atoms; 3.36 $\times 10^{26}$ N atoms; 1.68×10^{26} O atoms. **3.32** 8.56×10^{22} molecules. **3.36** Seven peaks. **3.42** (a) Na: 54.75%; F: 45.25%, (b) Na: 39.34%; Cl: 60.66%, (c) Na: 22.34%; Br: 77.66%, (d) Na: 15.34%; I: 84.66%. **3.44** Both are C$_6$H$_{10}$S$_2$O. **3.46** Ammonia. **3.48** 39.3 g. **3.50** 5.97 g F. **3.52** O: 66.1%; C$_2$H$_3$NO$_5$. **3.54** C$_5$H$_8$O$_4$NNa. **3.60** (a) 2N$_2$O$_5$ $\longrightarrow$ 2N$_2$O$_4$ + O$_2$, (b) 2KNO$_3$ $\longrightarrow$ 2KNO$_2$ + O$_2$, (c) NH$_4$NO$_3$ $\longrightarrow$ N$_2$O + 2H$_2$O, (d) NH$_4$NO$_2$ $\longrightarrow$ N$_2$ + 2H$_2$O, (e) 2NaHCO$_3$ $\longrightarrow$ Na$_2$CO$_3$ + H$_2$O + CO$_2$, (f) P$_4$O$_{10}$ + 6H$_2$O $\longrightarrow$ 4H$_3$PO$_4$, (g) 2HCl + CaCO$_3$ $\longrightarrow$ CaCl$_2$ + H$_2$O + CO$_2$, (h) 2Al + 3H$_2$SO$_4$ $\longrightarrow$ Al$_2$(SO$_4$)$_3$ + 3H$_2$, (i) CO$_2$ + 2KOH $\longrightarrow$ K$_2$CO$_3$ + H$_2$O, (j) CH$_4$ + 2O$_2$ $\longrightarrow$ CO$_2$ + 2H$_2$O. **3.66** 1.01 mol. **3.68** (a) 2NaHCO$_3$ $\longrightarrow$ Na$_2$CO$_3$ + H$_2$O + CO$_2$, (b) 78.3 g. **3.70** 255.9 g, 0.324 L. **3.72** 0.294 mole. **3.74** (a) NH$_4$NO$_3$ $\longrightarrow$ N$_2$O + 2H$_2$O, (b) 2.0×10^1 g. **3.76** 18.0 g. **3.80** 0.709 g NO$_2$; O$_3$; 6.9×10^{-3} mole NO. **3.82** HCl; 23.4 g. **3.86** (a) 7.05 g, (b) 92.9%. **3.88** 3.48×10^3 g. **3.90** 18. **3.92** 65.4 amu, Zn. **3.94** 89.6%. **3.96** CH$_2$O, C$_6$H$_{12}$O$_6$. **3.98** (a) 4.3×10^{22} Mg atoms, (b) 1.6×10^2 pm. **3.100** 3.1×10^{23} molecules/mol. **3.102** X$_2$O$_3$ + 3CO $\longrightarrow$ 2X + 3CO$_2$. **3.104** Mg$_3$N$_2$, magnesium nitride.

Chapter 4

4.6 (a) Strong electrolyte, (b) nonelectrolyte, (c) weak electrolyte, (d) strong electrolyte. **4.8** (a) Solid NaCl does not conduct electricity, (b) molten NaCl conducts, (c) aqueous solution of NaCl conducts. **4.10** In water HCl breaks up as H$^+$ and Cl$^-$ ions; in benzene it remains as a molecule. **4.14** (a) Insoluble, (b) soluble, (c) soluble, (d) insoluble, (e) soluble. **4.16** (a) Ionic: 2Na$^+$(aq) + S^{2-}(aq) + Zn^{2+}(aq) + 2Cl$^-$(aq) $\longrightarrow$ ZnS(s) + 2Na$^+$(aq) + 2Cl$^-$(aq); net ionic: Zn^{2+}(aq) + S^{2-}(aq) $\longrightarrow$ ZnS(s). (b) Ionic: 6K$^+$(aq) + 2PO$_4^{3-}$(aq) + 3Sr^{2+}(aq) + 6NO$_3^-$(aq) $\longrightarrow$ Sr$_3$(PO$_4$)$_2$(s) + 6K$^+$(aq) + 6NO$_3^-$(aq); net ionic: 3Sr^{2+}(aq) + 2PO$_4^{3-}$(aq) $\longrightarrow$ Sr$_3$(PO$_4$)$_2$(s). (c) Ionic: Mg^{2+}(aq) + 2NO$_3^-$(aq) + 2Na$^+$(aq) + 2OH$^-$(aq) $\longrightarrow$ Mg(OH)$_2$(s) + 2Na$^+$(aq) + 2NO$_3^-$(aq); net ionic: Mg^{2+}(aq) + 2OH$^-$(aq) $\longrightarrow$ Mg(OH)$_2$(s). **4.18** (a) Add chloride ions. KCl is soluble; AgCl is not. (b) Add hydroxide ions. Ba(OH)$_2$ is soluble but Pb(OH)$_2$ is not. (c) Add carbonate ions. (NH$_4$)$_2$CO$_3$ is soluble, but CaCO$_3$ is insoluble. (d) Add sulfate ions. CuSO$_4$ is soluble, but BaSO$_4$ is insoluble. **4.26** (a) Brønsted base, (b) Brønsted base, (c) Brønsted acid, (d) Brønsted acid and Brønsted base. **4.28** (a) Ionic: CH$_3$COOH(aq) + K$^+$(aq) + OH$^-$(aq) $\longrightarrow$ CH$_3$COO$^-$(aq) + K$^+$(aq) + H$_2$O(l); net ionic: CH$_3$COOH(aq) + OH$^-$(aq) $\longrightarrow$ CH$_3$COO$^-$(aq) + H$_2$O(l). (b) Ionic: H$_2$CO$_3$(aq) + 2Na$^+$(aq) + 2OH$^-$(aq) $\longrightarrow$ 2Na$^+$(aq) + CO$_3^{2-}$(aq) + 2H$_2$O(l); net ionic: H$_2$CO$_3$(aq) + 2OH$^-$(aq) $\longrightarrow$ CO$_3^{2-}$(aq) + 2H$_2$O(l). (c) Ionic: 2H$^+$(aq) + 2NO$_3^-$(aq) + Ba^{2+}(aq) + 2OH$^-$(aq) $\longrightarrow$ Ba^{2+}(aq) +

$2NO_3^-(aq) + 2H_2O(l)$; net ionic: $2H^+(aq) + 2OH^-(aq) \longrightarrow 2H_2O(l)$.
4.32 (a) Reducing agent: $Fe \longrightarrow Fe^{3+} + 3e^-$; oxidizing agent: $O_2 + 4e^- \longrightarrow 2O^{2-}$. (b) Reducing agent: $2Br^- \longrightarrow Br_2 + 2e^-$; oxidizing agent: $Cl_2 + 2e^- \longrightarrow 2Cl^-$. (c) Reducing agent: $Si \longrightarrow Si^{4+} + 4e^-$; oxidizing agent: $F_2 + 2e^- \longrightarrow 2F^-$. (d) Reducing agent: $H_2 \longrightarrow 2H^+ + 2e^-$; oxidizing agent: $Cl_2 + 2e^- \longrightarrow 2Cl^-$. **4.36** (a) $+5$, (b) $+1$, (c) $+3$, (d) $+5$, (e) $+5$, (f) $+5$. **4.38** (a) $+1$, (b) -1, (c) $+3$, (d) $+3$, (e) $+4$, (f) $+6$, (g) $+2$, (h) $+4$, (i) $+2$, (j) $+3$, (k) $+5$. **4.40** (a) -3, (b) $-\frac{1}{2}$, (c) -1, (d) $+4$, (e) $+3$, (f) -2, (g) $+3$, (h) $+6$. **4.42** SO_3 (S is in its maximum oxidation state of $+6$). **4.44** Li and Ca. **4.48** $0.131\ M$. **4.50** 10.8 g. **4.52** (a) $1.37\ M$, (b) $0.426\ M$, (c) $0.716\ M$. **4.54** (a) 6.50 g, (b) 2.45 g, (c) 2.65 g, (d) 7.36 g, (e) 3.95 g. **4.58** $0.0433\ M$. **4.60** 126 mL. **4.62** $1.09\ M$. **4.66** $S^{2-}(aq) + Cu^{2+}(aq) \longrightarrow CuS(s)$, $2.31 \times 10^{-4}\ M$. **4.72** (a) 6.0 mL, (b) 8.0 mL, (c) 23 mL. **4.74** The Ba^{2+} ions combine with the SO_4^{2-} ions to form $BaSO_4$ precipitate. **4.76** Physical test: Only the NaCl solution would conduct electricity; chemical test: Add $AgNO_3$ solution. Only the NaCl solution would give AgCl precipitate. **4.78** Mg, Na, Ca, Ba, K, or Li. **4.80** Oxidation number of C is $+2$ in CO and $+4$ (maximum) in CO_2. **4.82** $1.3\ M$. **4.84** $0.171\ M$. **4.86** $0.115\ M$. **4.88** 0.80 L. **4.90** $1.73\ M$. **4.92** $NaHCO_3$. NaOH is caustic and expensive. **4.94** 43.97% NaCl; 56.03% KCl. **4.96** (a) The precipitate $CaSO_4$ formed over Ca prevents the Ca from reacting with the sulfuric acid, (b) aluminum is protected by a tenacious oxide layer (Al_2O_3), (c) these metals react more readily with water, (d) the metal should be placed below Fe and above H. **4.98** (a) $8.320 \times 10^{-7}M$, (b) 3.286×10^{-5} g. **4.100** 4.99 grains. **4.102** (a) $NH_4^+ + OH^- \longrightarrow NH_3 + H_2O$, (b) 97.99%. **4.104** Because the volume of the solution changes (increases or decreases) when the solid dissolves.

Chapter 5

5.14 0.797 atm. **5.18** 53 atm. **5.20** (a) 0.69 L, (b) 61 atm. **5.22** $-196°C$; $-269°C$; $5.7 \times 10^3°C$. **5.24** 1.3×10^2 K. **5.26** ClF_3. **5.32** 6.2 atm. **5.34** $472°C$. **5.36** 1.9 atm. **5.38** 0.82 L. **5.40** 33.6 mL. **5.42** 6.1×10^{-3} atm. **5.44** 35.1 g/mol. **5.46** N_2: 2.1×10^{22} molecules; O_2: 5.7×10^{21} molecules; Ar: 2.7×10^{20} atoms. **5.48** 2.98 g/L. **5.50** SF_4. **5.52** 18 g. **5.54** P_2F_4. **5.58** (a) 0.89 atm, (b) 1.4 L. **5.60** 349 mmHg. **5.62** 19.8 g. **5.64** N_2: 217 mmHg; H_2: 650 mmHg. **5.72** N_2: 471.7 m/s; O_2: 441.4 m/s; O_3: 360.4 m/s. **5.74** Average speed: 2.7 m/s; rms speed: 2.8 m/s. Squaring favors the larger values compared to just taking the average value. **5.80** Not ideal. Ideal gas pressure is 164 atm. **5.82** Ideal. **5.84** 1.7×10^2 L; CO_2: 0.49 atm; N_2: 0.25 atm; H_2O: 0.41 atm; O_2: 0.041 atm. **5.86** (a) $NH_4NO_2(s) \longrightarrow N_2(g) + 2H_2O(l)$, (b) 0.273 g. **5.88** No. An ideal gas would not condense. **5.90** Higher in the winter because of reduced photosynthesis. **5.92** (a) 0.86 L, (b) $NH_4HCO_3(s) \longrightarrow NH_3(g) + CO_2(g) + H_2O(l)$. Advantage: more gases (CO_2 and NH_3) generated; disadvantage: odor of ammonia! **5.94** 3.88 L. **5.96** (a) $C_3H_8(g) + 5O_2(g) \longrightarrow 3CO_2(g) + 4H_2O(g)$, (b) 11.4 L. **5.98** O_2: 0.166 atm; NO_2: 0.333 atm. **5.100** (a) $CO_2(g) + CaO(s) \longrightarrow CaCO_3(s)$; $CO_2(g) + BaO(s) \longrightarrow BaCO_3(s)$, (b) 10.5% CaO; 89.5% BaO. **5.102** 1.7×10^{12} O_2 molecules. **5.104** 3.7 nm; 0.31 nm. The water molecules are touching each other in the liquid phase.

Chapter 6

6.16 48 J. **6.18** -3.1×10^3 J. **6.24** -3.43×10^4 kJ. **6.26** No work done. $\Delta E = -553.8$ kJ. **6.32** 728 kJ. **6.34** $50.7°C$. **6.36** $26.26°C$. **6.44** O_2. **6.46** (a) $Br_2(l) = 0$, $Br_2(g) > 0$; (b) $I_2(s) = 0$, $I_2(g) > 0$. **6.48** -1411 kJ, (b) -1124 kJ. **6.50** 218.2 kJ/mol. **6.52** 2.70×10^2 kJ. **6.54** -84.6 kJ. **6.56** -780 kJ. **6.58** $\Delta H_2 - \Delta H_1$. **6.60** (a) -336.5 kJ. (b) (1) $N_2H_4 + O_2 \longrightarrow N_2 + 2H_2O$, -622.0 kJ; (2) $4NH_3 + 3O_2 \longrightarrow 2N_2 + 6H_2O$, -1529.6 kJ; NH_3 is a better fuel. **6.62** 0. **6.64** -366.1 kJ. **6.66** 0.492 J/g $\cdot$ °C. **6.68** The first (exothermic) reaction promotes the second (endothermic) reaction. **6.70** 1.09×10^4 kJ. **6.72** 4.10 L. **6.74** (a). **6.76** (a) 0, (b) -9.1 J, (c) -48 J; 2.4 L. **6.78** (a) Greater mass and hence greater heat capacity, (b) tea and coffee have more water content. **6.80** -1.84×10^3 kJ. **6.82** 3.0×10^9. **6.84** 13.70 kJ/°C. **6.86** -5.2×10^6 kJ. **6.88** (a)

1.4×10^2 kJ, (b) 3.9×10^2 kJ. **6.90** -110.5 kJ. The reaction will also produce some CO_2. **6.92** -277.0 kJ/mol. **6.94** $304°C$. **6.96** (a) $CaC_2 + 2H_2O \longrightarrow Ca(OH)_2 + C_2H_2$, (b) 1.51×10^3 kJ.

Chapter 7

7.8 (a) 6.58×10^{14}/s, (b) 1.22×10^8 nm. **7.10** 2.5 min. **7.12** 4.95×10^{14}/s. **7.16** (a) 4.0×10^2 nm, (b) 5.0×10^{-19} J. **7.18** 1.2×10^2 nm (UV). **7.20** (a) 3.70×10^2 nm; 3.70×10^{-7} m, (b) UV, (c) 5.38×10^{-19} J. **7.26** Use a prism. **7.28** Excited atoms emit the same characteristic frequencies or lines on Earth as they do from a distant star. **7.30** 3.027×10^{-19} J. **7.32** 6.17×10^{14}/s; 4.86×10^2 nm. **7.34** 5. **7.40** 1.37×10^{-6} nm. **7.42** 1.7×10^{-23} nm. **7.54** $\ell = 2$: $m_\ell = -2, -1, 0, 1, 2$; $\ell = 1$: $m_\ell = -1, 0, 1$; $\ell = 0$: $m_\ell = 0$. **7.56** (a) $n = 4$, $\ell = 1$, $m_\ell = -1, 0, 1$, (b) $n = 3$, $\ell = 2$, $m_\ell = -2, -1, 0, 1, 2$, (c) $n = 3$, $\ell = 0$, $m_\ell = 0$, (d) $n = 5$, $\ell = 3$, $m_\ell = -3, -2, -1, 0, 1, 2, 3$. **7.58** Different orientations in space. **7.60** Values of ℓ: 0 ($6s$, one orbital), 1 ($6p$, three orbitals), 2 ($6d$, five orbitals), 3 ($6f$, seven orbitals), 4 ($6g$, nine orbitals), 5 ($6h$, eleven orbitals). **7.62** $2n^2$. **7.64** (a) 3, (b) 6, (c) 0. **7.68** (a) $2s$, (b) $3p$, (c) $3s$, (d) $4d$. **7.78** Al: $1s^22s^22p^63s^23p^1$; B: $1s^22s^22p^1$; F: $1s^22s^22p^5$. **7.80** B(1), Ne(0), P(3), Sc(1), Mn(5), Se(2), Zr(2), Ru(4), Cd(0), I(1), W(4), Pb(2), Ce(2), Ho(3). **7.82** Ge: $[Ar]4s^23d^{10}4p^2$; Fe: $[Ar]4s^23d^6$; Zn: $[Ar]4s^23d^{10}$; Ru: $[Kr]5s^14d^7$; W: $[Xe]6s^24f^{14}5d^4$; Tl: $[Xe]6s^24f^{14}5d^{10}6p^1$. **7.84** S^+. Hund's rule. **7.86** (a) Incorrect, (b) correct, (c) incorrect. **7.88** (a) 4 (one in $2s$ and three in $2p$), (b) 6 (2 in $4p$, 2 in $4d$, and 2 in $4f$), (c) 10 (2 in each of $3d$), (d) 1 (in $2s$), (e) 2 (in $4f$). **7.90** Wave properties. **7.92** (a) 1.05×10^{-25} nm, (b) 8.86 nm. **7.94** 419 nm; no. **7.96** He^+: $n = 3 \longrightarrow 2$, $\lambda = 164$ nm; $n = 4 \longrightarrow 2$, $\lambda = 121$ nm; $n = 5 \longrightarrow 2$, $\lambda = 109$ nm; $n = 6 \longrightarrow 2$, $\lambda = 103$ nm. H: $n = 3 \longrightarrow 2$, $\lambda = 657$ nm; $n = 4 \longrightarrow 2$, $\lambda = 487$ nm; $n = 5 \longrightarrow 2$, $\lambda = 434$ nm; $n = 6 \longrightarrow 2$, $\lambda = 411$ nm. **7.98** $1/\lambda_1 = 1/\lambda_2 + 1/\lambda_3$. **7.100** 2.0×10^{-5} m/s. **7.102** 2.8×10^6 K. **7.104** Yellow light will generate more electrons; blue light will generate electrons with greater kinetic energy.

Chapter 8

8.18 (a) and (d); (b) and (e); (c) and (f). **8.20** (a) Group 1A, (b) Group 5A, (c) Group 8A, (d) Group 8B. **8.22** Fe. **8.28** (a) [Ne], (b) [Ne], (c) [Ar], (d) [Ar], (e) [Ar], (f) $[Ar]3d^6$, (g) $[Ar]3d^9$, (h) $[Ar]3d^{10}$. **8.30** (a) Cr^{3+}, (b) Sc^{3+}, (c) Rh^{3+}, (d) Ir^{3+}. **8.32** Be^{2+} and He; F^- and N^{3-}; Fe^{2+} and Co^{3+}; S^{2-} and Ar. **8.38** Na > Mg > Al > P > Cl. **8.40** F. **8.42** The effective nuclear charge on the outermost electrons increases from left to right as a result of incomplete shielding by the inner electrons. **8.44** $Mg^{2+} < Na^+ < F^- < O^{2-} < N^{3-}$. **8.46** Te^{2-}. **8.48** $-199.4°C$. **8.52** The $3p^1$ electron of Al is effectively shielded by the inner electrons and $3s^2$ electrons. **8.54** 2080 kJ/mol is for $1s^22s^22p^6$. **8.56** 8.43×10^6 kJ/mol. **8.60** Cl. **8.62** Alkali metals have ns^1 and so can accept another electron. **8.66** Low ionization energy, reacts with water to form FrOH, and with oxygen to form oxide and superoxide. **8.68** The ns^1 electron of Group 1B metals is incompletely shielded by the inner d electrons and therefore they have much higher first ionization energies. **8.70** (a) $Li_2O(s) + H_2O(l) \longrightarrow 2LiOH(aq)$, (b) $CaO(s) + H_2O(l) \longrightarrow Ca(OH)_2(aq)$, (c) $CO_2(g) + H_2O(l) \longrightarrow H_2CO_3(aq)$. **8.72** BaO. **8.74** (a) Bromine, (b) nitrogen, (c) rubidium, (d) magnesium. **8.76** (a) $Mg^{2+} < Na^+ < F^- < O^{2-}$, (b) $O^{2-} < F^- < Na^+ < Mg^{2+}$. **8.78** M is potassium (K) and X is bromine (Br_2). **8.80** O^+ and N; Ar and S^{2-}; Ne and N^{3-}; Zn and As^{3+}; Cs^+ and Xe. **8.82** (a) and (d). **8.84** First: bromine; second: iodine; third: chlorine; fourth: fluorine. **8.86** Fluorine. **8.88** H^-. **8.90** Li_2O (basic), BeO (amphoteric), B_2O_3 (acidic), CO_2 (acidic), N_2O_5 (acidic). **8.92** It can form H^+ and H^- ions. **8.94** 418 kJ/mol. **8.96** About $23°C$. **8.98** (a) Mg [in $Mg(OH)_2$], (b) liquid Na, (c) Mg (in $MgSO_4 \cdot 7H_2O$), (d) Na (in $NaHCO_3$), (e) K (in KNO_3), (f) Mg, (g) Ca (in CaO), (h) Ca, (i) Na (in NaCl); Ca (in $CaCl_2$). **8.100** Physical properties: Solid, high m.p; $2NaAt + 2H_2SO_4 \longrightarrow At_2 + SO_2 + Na_2SO_4 + 2H_2O$. **8.102** 242 nm. **8.104** (a) F_2, (b) Na, (c) B, (d) N_2, (e) Al.

Chapter 9

9.14 C—H < Br—H < F—H < Li—Cl < Na—Cl < K—F.
9.16 Cl—Cl < Br—Cl < Si—C < Cs—F. **9.18** (a) Covalent, (b) polar covalent, (c) ionic, (d) polar covalent.

9.22 (a) $\left[:\ddot{O}{-}\ddot{O}:\right]^{2-}$

(b) $\left[:C{\equiv}C:\right]^{2-}$

(c) $\left[:N{\equiv}O:\right]^{+}$

(d) $\left[\begin{array}{c} H \\ | \\ H{-}N{-}H \\ | \\ H \end{array}\right]^{+}$

9.24 (a) Neither O atom has a complete octet; one H atom forms a double bond.

(b) $H{-}\overset{\displaystyle H}{\underset{\displaystyle H}{C}}{-}\overset{\displaystyle :\!O\!:}{C}{-}\ddot{O}{-}H$

9.30 $\ddot{O}{=}\overset{+}{\underset{}{Cl}}{-}\ddot{O}:^{-} \longleftrightarrow {}^{-}:\ddot{O}{-}\overset{+}{Cl}{=}\ddot{O}: \longleftrightarrow {}^{-}:\ddot{O}{-}\overset{+}{Cl}{=}\ddot{O}$

9.32 $H{-}\overset{\displaystyle H}{C}{=}\overset{+}{N}{=}\ddot{N}:^{-} \longleftrightarrow H{-}\overset{\displaystyle H}{\underset{}{C}}{-}\overset{+}{N}{\equiv}N:$

9.34 ${}^{-}:\ddot{N}{=}\overset{+}{N}{=}\ddot{O}: \longleftrightarrow :N{\equiv}\overset{+}{N}{-}\ddot{O}:^{-} \longleftrightarrow {}^{2-}:\ddot{N}{-}\overset{+}{N}{\equiv}O:^{+}$

9.40 $^{+}\ddot{Cl}{=}\overset{2-}{Be}{=}\ddot{Cl}^{+}$

9.42 $\underset{\displaystyle Cl}{\overset{\displaystyle Cl}{Cl{-}Sb{-}Cl}}$ No.

9.44 $:\ddot{Cl}{-}\underset{\displaystyle :\ddot{Cl}:}{\overset{}{Al}}{-}\ddot{Cl}: + :\ddot{Cl}:^{-} \longrightarrow \left[:\ddot{Cl}{-}\underset{\displaystyle :\ddot{Cl}:}{\overset{\displaystyle :\ddot{Cl}:}{Al}}{-}\ddot{Cl}:\right]^{-}$

9.48 303.0 kJ/mol. **9.50** (a) −2759 kJ, (b) −2855 kJ. **9.52** Covalent: SiF_4, PF_5, SF_6, ClF_3; Ionic: NaF, MgF_2, AlF_3. **9.54** KF is a solid; has a high melting point; is an electrolyte. CO_2 is a gas; is a molecular compound.

9.56 $:\ddot{N}{=}\overset{+}{N}{=}\ddot{N}: \longleftrightarrow {}^{2-}:\ddot{N}{-}\overset{+}{N}{\equiv}N: \longleftrightarrow :N{\equiv}\overset{+}{N}{-}\ddot{N}:^{2-}$

9.58 (a) $AlCl_4^-$, (b) AlF_6^{3-}, (c) $AlCl_3$. **9.60** C has incomplete octet in CF_2; C has an expanded octet in CH_5; F and H can only form a single bond; the I atoms are too large to surround the P atom. **9.62** (a) False, (b) true, (c) false, (d) false. **9.64** −67 kJ/mol. **9.66** N_2. **9.68** NH_4^+ and CH_4; CO and N_2; $B_3N_3H_6$ and C_6H_6.

9.70 $H{-}\ddot{N}:^{-}\underset{\displaystyle H}{} + H{-}\ddot{O}:\underset{\displaystyle H}{} \longrightarrow H{-}\overset{\displaystyle H}{\underset{\displaystyle H}{N}}{-}H + {}^{-}:\ddot{O}{-}H$

9.72 F cannot form an expanded octet.

9.74 $H{-}\overset{\displaystyle H}{\underset{\displaystyle H}{C}}{-}\ddot{N}{=}C{=}\ddot{O} \longleftrightarrow H{-}\overset{\displaystyle H}{\underset{\displaystyle H}{C}}{-}\overset{+}{N}{\equiv}C{-}\ddot{O}:^{-}$

9.76 (a) Important, (b) important, (c) not important, (d) not important.

9.78 $\underset{\displaystyle Cl}{\overset{\displaystyle Cl}{F{-}C{-}Cl}}$ $\underset{\displaystyle Cl}{\overset{\displaystyle Cl}{F{-}C{-}F}}$ $\underset{\displaystyle Cl}{\overset{\displaystyle F}{H{-}C{-}F}}$ $\underset{\displaystyle F \; F}{\overset{\displaystyle F \; H}{F{-}C{-}C{-}F}}$

9.80 (a) −9.2 kJ, (b) −9.2 kJ. **9.82** (a) $^{-}:C{\equiv}O:^{+}$ (b) $:N{\equiv}O:^{+}$ (c) $^{-}:C{\equiv}N:$ (d) $:N{\equiv}N:$ **9.84** True.

9.86 $:C{\equiv}\overset{+}{N}{-}\ddot{O}:^{-} \longleftrightarrow {}^{2-}:\ddot{C}{=}\overset{+}{N}{=}\ddot{O} \longleftrightarrow {}^{3-}:\ddot{C}{-}\overset{+}{N}{\equiv}O:^{+}$

The importance decreases from left to right.

Chapter 10

10.8 (a) Trigonal planar, (b) linear, (c) tetrahedral. **10.10** (a) Tetrahedral, (b) bent, (c) trigonal planar, (d) linear, (e) square planar, (f) tetrahedral, (g) trigonal bipyramid, (h) trigonal pyramid, (i) tetrahedral. **10.12** $SiCl_4$, CI_4, $CdCl_4^{2-}$. **10.18** Electronegativity decreases from F to I. **10.20** Higher. **10.22** (b) = (d) < (c) < (a). **10.32** sp^3. **10.34** B: sp^2 to sp^3; N: remains sp^3. **10.36** (a) sp^3; (b) sp^3, sp^2, sp^2; (c) sp^3, sp, sp, sp^3; (d) sp^3, sp^2; (e) sp^3, sp^2. **10.38** sp. **10.40** sp^3d. **10.42** Nine pi bonds and nine sigma bonds. **10.48** Electron spins must be paired in H_2. **10.50** $Li_2^- = Li_2^+ < Li_2$. **10.52** B_2^+. **10.54** Only MO theory predicts O_2 to be paramagnetic. **10.56** $O_2^{2-} < O_2^- < O_2 < O_2^+$. **10.58** B_2 contains a pi bond; C_2 contains two pi bonds. **10.60** Linear. Dipole moment measurement. **10.62** The large size of Si results in poor sideways overlap of p orbitals to form pi bonds. **10.64** XeF_3^+: T-shaped; XeF_5^+: square pyramid; SbF_6^-: octahedral. **10.66** (a) 180°, (b) 120°, (c) 109.5°, (d) about 109.5°, (e) 180°, (f) about 120°, (g) about 109.5°, (h) 109.5°. **10.68** sp^3d. **10.70** ICl_2^- and $CdBr_2$. **10.72** (a) sp^2, (b) molecule on the right. **10.74** The pi bond in *cis*-dichloroethylene prevents rotation. **10.76** Si has $3d$ orbitals so water can add to Si (valence shell expansion). **10.78** C: sp^2; N:N atom that forms a double bond is sp^2, the others are sp^3. **10.80** 43.6%.

Chapter 11

11.12 $CH_3CH_2CH_2CH_2CH_2Cl$. $CH_3CH_2CH_2CH_2CHClCH_3$. $CH_3CH_2CHClCH_2CH_3$.

11.14
$\underset{\displaystyle Br}{\overset{\displaystyle H}{}}C{=}C\underset{\displaystyle H}{\overset{\displaystyle CH_3}{}}$ $\underset{\displaystyle H}{\overset{\displaystyle Br}{}}C{=}C\underset{\displaystyle H}{\overset{\displaystyle CH_3}{}}$

$\underset{\displaystyle H}{\overset{\displaystyle H}{}}C{=}C\underset{\displaystyle H}{\overset{\displaystyle CH_2Br}{}}$ $\underset{\displaystyle H}{\overset{\displaystyle H}{C}}\underset{H \quad Br}{\overset{}{C{-}C}}$

11.16 (a) Alkene or cycloalkane, (b) alkyne, (c) alkane, (d) like (a), (e) alkyne.

11.18

A   B

C   D

Stability decreases from A to D. **11.20** No. **11.22** (a) is alkane and (b) is alkene. Only an alkene reacts with a hydrogen halide. **11.24** −630.8 kJ. **11.26** (a) *cis*-1,2-dichlorocyclopropane, (b) *trans*-1,2-dichlorocyclopropane. **11.28** (a) 2-methylpentane, (b) 2,3,4-trimethylhexane, (c) 3-ethylhexane, (d) 3-methyl-1,4-pentadiene, (e) 2-pentyne. **11.32** (a) 1,3-dichloro-4-methylbenzene, (b) 2-ethyl-1,4-dinitrobenzene, (c) 1,2,4,5-tetramethylbenzene, (d) 3-phenyl-1-butene. **11.36** (a) Ether, (b) amine, (c) aldehyde, (d) ketone, (e) carboxylic acid, (f) alcohol, (g) amino acid. **11.38** $HCOOH + CH_3OH \longrightarrow HCOOCH_3 + H_2O$. **11.40** $(CH_3)_2CH—O—CH_3$. **11.42** (a) Ketone, (b) ester, (c) ether. **11.46** (a) The C atoms bonded to the methyl group and the amino group and the H atom. (b) The C atoms bonded to Br. **11.48** −174 kJ. **11.50** $+CF_2—CF_2\frac{}{}_n$. **11.52** (a) Rubbing alcohol, (b) vinegar, (c) moth balls, (d) organic synthesis, (e) solvent, (f) antifreeze, (g) natural gas, (h) synthetic polymer. **11.54** (a) 3, (b) 16, (c) 6. **11.56** (a) C: 15.81 mg; H: 1.33 mg; O: 3.49 mg, (b) C_6H_6O, (c) phenol. **11.58** Empirical and molecular formula: $C_5H_{10}O$.

11.60 (a)

11.62 CH_3CH_2CHO. **11.64** (a) Alcohol, (b) ether, (c) aldehyde, (d) carboxylic acid, (e) amine. **11.66** The acids in lemon juice convert the amines to the ammonium salts, which have very low vapor pressures. **11.68** Methane (CH_4); ethanol (C_2H_5OH); methanol (CH_3OH); isopropanol (C_3H_7OH); ethylene glycol (CH_2OHCH_2OH); naphthalene ($C_{10}H_8$); acetic acid (CH_3COOH). **11.70** (a) 1, (b) 2, (c) 5. **11.72** Br_2 dissociates into Br atoms, which react with CH_4 to form CH_3Br and HBr.

Chapter 12

12.8 Methane; it has the lowest boiling point. **12.10** (a) Dispersion forces, (b) dispersion and dipole-dipole forces, (c) same as (b), (d) ionic and dispersion forces, (e) dispersion forces. **12.12** (e). **12.14** Only 1-butanol can form hydrogen bonds so it has a higher boiling point. **12.16** (a) Xe (greater dispersion forces), (b) CS_2 (greater dispersion forces), (c) Cl_2 (greater dispersion forces), (d) LiF (ionic compound), (e) NH_3 (hydrogen bond). **12.18** (a) Hydrogen bonding, dipole-dipole forces, dispersion forces, (b) dispersion forces, (c) dispersion forces, (d) the attractive forces of the covalent bond. **12.20** The compound on the left can form hydrogen bonds with itself (intramolecular hydrogen bonding). **12.32** Its viscosity is between that of ethanol and glycerol. **12.44** Simple cubic: 1 sphere; body-centered cubic: 2 spheres; face-centered cubic: 4 spheres. **12.46** 6.17×10^{23} atom/mol. **12.48** 458 pm. **12.50** XY_3. **12.52** Each C atom in diamond is covalently bonded to four other C atoms. **12.72** 2670 kJ. **12.74** 47.03 kJ/mol. **12.76** First step: freezing; second step: sublimation. **12.78** Additional heat is liberated when steam condenses at 100°C. **12.80** 331 mmHg. **12.84** The pressure exerted by the blades on the ice lowers the m.p. of the ice. A film of liquid water between the blades and the ice provides lubrication for the movement of the skater. **12.86** (a) Water will boil and turn into steam, (b) water vapor will turn into ice, (c) liquid water will vaporize. **12.88** (a) Dipole-dipole, dispersion and hydrogen bonding, (b) dispersion, (c) ionic, hydrogen bonding, dispersion, (d) metallic bond. **12.90** HF molecules form hydrogen bonds; HI molecules do not. **12.92** (a) Solid, (b) vapor. **12.94** The critical temperature of CO_2 is 31°C. The liquid CO_2 in the fire extinguisher is converted to critical fluid on a hot summer day. **12.96** Initially water boils under reduced pressure. During boiling, water cools and it freezes. Eventually it sublimes. **12.98** (a) Two, (b) diamond, (c) apply high pressure and high temperature. **12.100** W: Au; X: PbS; Y: I_2; Z: SiO_2. **12.102** 8.3×10^{-3} atm. **12.104** Smaller ions have high charge density and are more effective in ion-dipole interaction. **12.106** $3Hg + O_3 \longrightarrow 3HgO$. Conversion to solid HgO changes its surface tension. **12.108** 66.8%.

Chapter 13

13.8 Cyclohexane cannot form hydrogen bonds with ethanol. **13.10** The longer the chain gets, the more nonpolar the molecule becomes. The —OH group can form hydrogen bonds with water, but the rest of the molecule cannot. **13.14** (a) 25.9 g, (b) 1.72×10^3 g. **13.16** (a) 2.68 m, (b) 7.82 m. **13.20** 5.0×10^2 m; 18.3 M. **13.22** (a) 2.41 m, (b) 2.13 M, (c) 0.0587 L. **13.26** 45.9 g. **13.34** At the bottom of the mine the carbon dioxide pressure is greater and the dissolved gas is not released from the solution. Coming up, the pressure decreases and carbon dioxide is released from the solution. **13.36** 0.28 L. **13.50** 1.3×10^3 g. **13.52** Ethanol: 30.0 mmHg; propanol: 26.3 mmHg. **13.54** 128 g. **13.56** 0.59 m. **13.58** 120 g/mol; $C_4H_8O_4$. **13.60** −8.6°C. **13.62** 4.3×10^2 g/mol; $C_{24}H_{20}P_4$. **13.64** 1.75×10^4 g/mol. **13.66** 343 g/mol. **13.72** Boiling-point, vapor-pressure, osmotic pressure. **13.74** (b) > (c) > (a). **13.76** 0.9420 m. **13.78** 7.6 atm. **13.80** 1.6 atm. **13.82** 3.5 atm. **13.84** (a) 104 mmHg, (b) 116 mmHg. **13.86** 2.95×10^3 g/mol. **13.88** No. Compound is assumed to be pure, monomeric, and a nonelectrolyte. **13.90** 12.3 M. **13.92** Soln A: 124 g/mol; Soln B: 248 g/mol. Dimerization in benzene. **13.94** (a) Boiling under reduced pressure, (b) CO_2 boils off, expands and cools, condensing water vapor to form fog. **13.96** (a) Seawater contains many solutes, (b) solubility of CO_2 in water decreases at reduced pressure, (c) density of solution is close to that of water, (d) molality is independent of temperature, (e) methanol is volatile. **13.98** C_6H_{12}: 36%; $C_{10}H_8$: 65%. **13.100** First row: positive, positive; second row: A $\longleftrightarrow$ A, B $\longleftrightarrow$ B < A $\longleftrightarrow$ B, negative; third row: A $\longleftrightarrow$ A, B $\longleftrightarrow$ B = A $\longleftrightarrow$ B, no deviation. **13.102** Ethanol and water have an intermolecular attraction that results in an overall smaller volume.

Chapter 14

14.6 (a) 0.049 M/s, (b) 0.025 M/s. **14.16** (a) rate $= k[F_2][ClO_2]$, (b) $1.2/M \cdot s$, (c) 2.4×10^{-4} M/s. **14.18** (a) rate $= k[X]^2[Y]^3$; fifth order, (b) 0.68 M/s. **14.20** (a) 0.046 s^{-1}, (b) 0.13 M^{-1} s^{-1}. **14.22** three half-lives or $2.08/k$. **14.24** (a) 0.0198 s^{-1}, (b) 151 s. **14.26** (a) 3.6 s, (b) 3.0 s, 6.6 s. **14.34** 135 kJ/mol. **14.36** 368°C. **14.38** 51.0 kJ/mol. **14.48** (a) rate $= k[X_2][Y]$, (b) Z does not appear in the rate-determining step, (c) $X_2 + Y \longrightarrow XY + X$ (slow); $X + Z \longrightarrow XZ$ (fast). **14.56** rate $= (k_1k_2/k_{-1})[E][S]$. **14.58** (a) Temperature, (b) activation energy, (c) concentration of reactant, (d) catalyst. **14.60** Area of large sphere: 22.6 cm^2; area of 8 small spheres: 44.9 cm^2. The sphere with the larger surface area is the more effective catalyst. **14.62** $[H_2O]$ can be treated as a constant. **14.64** rate $= k[H^+][CH_3COCH_3]$, (b) $3.8 \times 10^{-3}/M \cdot s$. **14.66** 2.63 atm. **14.68** M^{-2} s^{-1}. **14.70** 56.4 min. **14.72** (b), (d), and (e). **14.74** 0.098%. **14.76** (a) Increased, (b) decreased, (c) decreased, (d) increased. **14.78** 0.0896 min^{-1}. **14.80** 1.12×10^3 min. **14.82** (a) rate $= k[X][Y]^2$, (b) 1.9×10^{-2} M^{-2} s^{-1}. **14.84** 3.1. **14.86** During the first 10 min or so the engine is relatively cold so the reaction will be slow. **14.90** At very high H_2 concentration, rate $= (k_1/k_2)[NO]^2$, at very low H_2 concentration, rate $= k_1[NO]^2[H_2]$. **14.92** $NO_2 + NO_2 \longrightarrow NO_3 + NO$ (slow); $NO_3 + CO \longrightarrow NO_2 + CO_2$ (fast). **14.94** At high pressures, all the sites on W are occupied so the reaction is zero order in $[PH_3]$. **14.96** (a) $\Delta[B]/\Delta t = k_1[A] - k_2[B]$, (b) $[B] = (k_1/k_2)[A]$. **14.98** Use H_2O enriched with the ^{18}O isotope. Only mechanism (a) would give acetic acid with that isotope and only mechanism (b) would give methanol with that isotope.

Chapter 15

15.12 1.08×10^7. **15.14** 1.1×10^{-5}. **15.16** (a) 0.082, (b) 0.29. **15.18** 0.105; 2.05×10^{-3}. **15.20** 7.09×10^{-3}. **15.22** 3.3. **15.24** 0.0356. **15.28** $[N_2]$ and $[H_2]$ will decrease and $[NH_3]$ will increase. **15.30** 0.50 atm and 0.020 atm. **15.32** $[I] = 8.58 \times 10^{-4}$ M; $[I_2] = 0.0194$ M. **15.34** (a) 0.52, (b) $[CO_2] = 0.48$ M; $[H_2] = 0.020$ M; $[CO] = 0.075$ M; $[H_2O] = 0.065$ M. **15.36** $[H_2] = [CO_2] = 0.05$ M; $[H_2O] = [CO] = 0.11$ M. **15.42** (a) Shift to the right, (b) no effect, (c) no effect. **15.44** (a) No effect, (b) no effect, (c) shift to the left, (d) no effect, (e) shift to the left. **15.46** (a) Shift to the right, (b) shift to the left, (c) shift to the right, (d) shift to the left, (e) no effect. **15.48** No change. **15.50** (a) Shift to the right, (b) no effect, (c) no effect, (d) shift to the left, (e) shift to the right, (f) shift to the left, (g) shift to the right. **15.52** (a) NO: 0.24 atm; Cl_2: 0.12 atm, (b) 0.017. **15.54** (a) No, (b) yes. **15.56** (a) 8×10^{-44}, (b) the reaction needs an input of energy to get started. **15.58** (a) 1.7, (b) $P_A = 0.69$ atm, $P_B = 0.81$ atm. **15.60** 1.5×10^5. **15.62** H_2: 0.28 atm, Cl_2: 0.049 atm, HCl: 1.67 atm. **15.64** 50 atm. **15.66** 3.86×10^{-2}. **15.68** 3.13. **15.70** N_2: 0.860 atm; H_2: 0.366 atm; NH_3: 4.40×10^{-3} atm. **15.72** (a) 1.16, (b) 53.7%. **15.74** (a) 0.49, (b) 0.23, (c) 0.037, (d) 0.0368 mol. **15.76** $[H_2] = 0.070$ M, $[I_2] = 0.182$ M, $[HI] = 0.825$ M. **15.78** (c). **15.80** (a) Color deepens, (b) increases, (c) decreases, (d) increases, (e) unchanged. **15.82** K is more volatile than Na. Therefore, its removal shifts the equilibrium to the right. **15.84** (a) High temperature and low pressure, (b) (i) 1.4×10^5, (ii) CH_4; H_2O: 2 atm; CO: 13 atm; H_2: 39 atm. **15.86** 1.2. **15.88** (a) React Ni with CO above 50°C, (b) the decomposition is endothermic. Heat $Ni(CO)_4$ above 200°C. **15.90** (a) $K_P = 2.6 \times 10^{-6}$, $K_c = 1.1 \times 10^{-7}$, (b) 2.2 g; 22 mg/m^3; yes.

Chapter 16

16.4 (a) Nitrite ion: NO_2^-, (b) hydrogen sulfate ion: HSO_4^-, (c) hydrogen sulfide ion: HS^-, (d) cyanide ion: CN^-, (e) formate ion: $HCOO^-$. **16.6** (a) H_2S, (b) H_2CO_3, (c) HCO_3^-, (d) H_3PO_4, (e) $H_2PO_4^-$, (f) HPO_4^{2-}, (g) H_2SO_4, (h) HSO_4^-, (i) HNO_2, (j) HSO_3^-.

16.8 (a)

$$^-O-\overset{\overset{O}{\|}}{C}-\overset{\overset{O}{\|}}{C}-OH \qquad ^-O-\overset{\overset{O}{\|}}{C}-\overset{\overset{O}{\|}}{C}-O^-$$

(b) Acid: H^+ and $C_2H_2O_4$; base: $C_2O_4^{2-}$; both acid and base: $C_2HO_4^-$. **16.16** (a) 6.3×10^{-6} M, (b) 1.0×10^{-16} M, (c) 2.7×10^{-6} M. **16.18** 6.72. **16.20** (a) Acidic, (b) neutral, (c) basic. **16.22** 1.98×10^{-3} mol; 0.444. **16.24** 2.2×10^{-3} g. **16.30** (a) Strong base, (b) weak base, (c) weak base, (d) weak base, (e) strong base. **16.32** (a) False, (b) true, (c) true, (d) false. **16.34** To the left. **16.36** Only the anion of phenol can be stabilized by resonance. **16.42** $[H^+] = [CH_3COO^-] = 5.8 \times 10^{-4}$ M; $[CH_3COOH] = 0.0181$ M. **16.44** 2.3×10^{-3} M. **16.46** (a) 3.5%, (b) 9.0%, (c) 33%, (d) 79%. Extent of ionization increases with dilution. **16.48** (a) 3.9%, (b) 0.30%. **16.52** $[H^+] = [HCO_3^-] = 1.0 \times 10^{-4}$ M; $[CO_3^{2-}] = 4.8 \times 10^{-11}$ M. **16.56** 7.1×10^{-7}. **16.58** 1.5%. **16.66** HZ < HY < HX. **16.68** 4.82. **16.70** >7. **16.74** $AlCl_3$ is a Lewis acid, Cl^- is a Lewis base. **16.76** CO_2 and BF_3. **16.78** 0.106 L. **16.80** No. **16.82** No. **16.84** CrO is ionic and basic; CrO_3 is covalent and acidic. **16.86** 4.0×10^{-2}. **16.88** 0.028. **16.90** $[Na^+] = 0.200$ M; $[HCO_3^-] = 4.6 \times 10^{-3}$ M; $[H_2CO_3] = 2.4 \times 10^{-8}$ M; $[OH^-] = 4.6 \times 10^{-3}$ M; $[H^+] = 2.2 \times 10^{-12}$ M. **16.92** $NaCN + HCl \longrightarrow NaCl + HCN$. HCN is a very weak acid so it escapes into the vapor phase. **16.94** 1.000. **16.96** (a) Increases, (b) decreases, (c) no effect, (d) increases. **16.98** 1.6×10^{-4}. **16.100** 4.40.

Chapter 17

17.6 (a) and (c). **17.8** 4.74; (a) Higher concentration. **17.10** 7.03. **17.12** 10; more effective against an added acid. **17.14** (a) 4.82, (b) 4.64. **17.16** HC. pK_a is closest to pH. **17.18** 90.1 g/mol. **17.20** (a) 110 g/mol, (b) 1.6×10^{-6}. **17.26** Red. **17.34** (a) 7.8×10^{-10}, (b) 1.8×10^{-18}. **17.36** 2.6×10^{-6} M. **17.38** 1.7×10^{-2} g/L. **17.40** 2.3×10^{-9}. **17.42** $[NO_3^-] = 0.076$ M; $[Na^+] = 0.045$ M; $[Sr^{2+}] = 1.6 \times 10^{-2}$ M; $[F^-] = 1.1 \times 10^{-4}$ M. **17.46** (a) 0.013 M, (b) 2.2×10^{-4} M, (c) 3.3×10^{-3} M. **17.48** (a) 1.0×10^{-5} M, (b) 1.1×10^{-10} M. **17.52** Complex ion formations. (a) $[Cu(NH_3)_4]^{2+}$, (b) $[Ag(CN)_2]^-$, (c) $[HgCl_4]^{2-}$. For equations for the formation of the complex ions see Table 17.2. **17.54** $[Cd^{2+}] = 1.1 \times 10^{-18}$ M; $[Cd(CN)_4^{2-}] = 4.2 \times 10^{-3}$ M; $[CN^-] = 0.48$ M. **17.56** 3.5×10^{-5} M. **17.60** 0.011 M. **17.62** Chloride ions will only precipitate Ag^+ ions, or a flame test for Cu^{2+} ions. **17.64** (a) 1.37, (b) 5.28, (c) 8.85. **17.66** 2.51–4.41. **17.68** 1.28 M. **17.70** $[H^+] = 2.99 \times 10^{-13}$ M; $[CH_3COO^-] = 0.0500$ M; $[CH_3COOH] = 8.3 \times 10^{-10}$ M; $[OH^-] = 0.0335$ M; $[Na^+] = 0.0835$ M. **17.72** 9.25; 9.18. **17.74** 9.97 g; 13.04. **17.76** $[Ag^+] = 2.0 \times 10^{-9}$ M; $[Cl^-] = 0.080$ M; $[Zn^{2+}] = 0.070$ M; $[NO_3^-] = 0.060$ M. **17.78** pH is greater than 2.68 but less than 8.11. **17.80** All except (a). **17.82** 2.4×10^{-13}. **17.84** A precipitate of $Fe(OH)_2$ will form. **17.86** The original precipitate is HgI_2. At higher concentration of KI, the complex ion HgI_4^{2-} forms so mass of HgI_2 decreases. **17.88** 7.82–10.38. **17.90** (a) AgI, (b) 1.6×10^{-7} M, (c) 0.0016%. **17.92** (a) $MCO_3 + 2HCl \longrightarrow MCl_2 + H_2O + CO_2$, $HCl + NaOH \longrightarrow NaCl + H_2O$, (b) 197 g/mol; Ba. **17.94** 2. **17.96** (a) Sulfate, (b) sulfide, (c) iodide. **17.98** 13 mL. **17.100** (c). **17.102** (a) 1.7×10^{-7} M, (b) because $MgCO_3$ is fairly soluble, (c) 12.40, (d) 1.9×10^{-8} M, (e) Ca^{2+} because it is present in larger amounts.

Chapter 18

18.6 (a) 0.02, (b) 9×10^{-19}, (c) 2×10^{-181}. **18.10** (c) < (d) < (e) < (a) < (b). **18.12** (a) 47.5 J/K, (b) -12.5 J/K, (c) -242.8 J/K. **18.14** (a) $\Delta S < 0$, (b) $\Delta S > 0$, (c) $\Delta S > 0$, (d) $\Delta S < 0$. **18.18** (a) -1139 kJ, (b) -140.0 kJ, (c) -2935.0 kJ. **18.20** (a) At all temperatures, (b) below 111 K. **18.24** 8.0×10^4 kJ. **18.26** 4.572×10^2 kJ; 4.5×10^{-81}. **18.28** (a) -24.6 kJ, (b) -1.10 kJ. **18.30** -341 kJ/mol. **18.32** -2.87 kJ. **18.36** 1×10^3. **18.38** (a) 8.0×10^4 J, (b) 4.0×10^4 J, (c) -3.2×10^4 J, (d) 6.4×10^4 J. **18.40** (i) No reaction is possible because $\Delta G > 0$. (ii) The reaction has a very large activation energy. (iii) Reactants already at their equilibrium concentrations. **18.42** In all cases $\Delta H > 0$ and $\Delta S > 0$. $\Delta G < 0$ for (a), $= 0$ for (b), and > 0 for (c). **18.44** $\Delta S > 0$. **18.46** (b) ΔS_{vap} are larger for ethanol and water because of hydrogen bonding. **18.48** (a) $2CO + 2NO \longrightarrow 2CO_2 + N_2$, (b) oxidizing agent: NO; reducing agent: CO, (c) 1×10^{121}, (d) 1.2×10^{18}; from left to right, (e) no. **18.50** $2.6 \times$

10^{-9}. **18.52** 976 K. **18.54** $\Delta S < 0$; $\Delta H < 0$. **18.56** 55 J/K. **18.58** Increase in entropy of the surroundings offsets the decrease in entropy of the system. **18.60** 56 J/K. **18.62** 4.5×10^5. **18.64** 4.5×10^{-75} atm. **18.66** (a) True, (b) true, (c) false. **18.68** $C + CuO \rightleftharpoons CO + Cu$. 6.1 **18.70** Crystal structure has disorder or has impurity. **18.72** (a) 7.6×10^{14}, (b) 4.1×10^{-12}. **18.74** (a) A reverse disproportionation reaction, (b) 8.2×10^{15}, (c) less effective. **18.76** 1.8×10^{70}. Reaction has a large activation energy. **18.78** -214.3 kJ. **18.80** $K_P = 36$. 981 K. No. **18.82** $X_{CO} = 0.45$; $X_{CO_2} = 0.55$.

Chapter 19

19.2 (a) $Mn^{2+} + H_2O_2 + 2OH^- \longrightarrow MnO_2 + 2H_2O$, (b) $2Bi(OH)_3 + 3SnO_2^{2-} \longrightarrow 2Bi + 3H_2O + 3SnO_3^{2-}$, (c) $Cr_2O_7^{2-} + 14H^+ + 3C_2O_4^{2-} \longrightarrow 2Cr^{3+} + 6CO_2 + 7H_2O$, (d) $2Cl^- + 2ClO_3^- + 4H^+ \longrightarrow Cl_2 + 2ClO_2 + 2H_2O$. **19.12** 2.46 V. **19.14** $Cl_2(g)$ and $MnO_4^-(aq)$. **19.16** Only (a) and (d) are spontaneous. **19.18** (a) Li, (b) H_2, (c) Fe^{2+}, (d) Br^-. **19.22** 0.368 V. **19.24** (a) -432 kJ, 5×10^{75}, (b) -104 kJ; 2×10^{18}, (c) -178 kJ; 1×10^{31}, (d) -1.27×10^3 kJ; 8×10^{211}. **19.26** 0.37 V; -36 kJ; 2×10^6. **19.30** (a) 2.23 V; 2.23 V; -430 kJ, (b) 0.02 V; 0.04 V; -23 kJ. **19.32** 0.083 V. **19.34** 0.010 V. **19.38** 1.09 V. **19.46** (b) 0.64 g. **19.48** (a) $\$2.09 \times 10^3$, (b) $\$2.46 \times 10^3$, (c) $\$4.70 \times 10^3$. **19.50** (a) 0.14 F, (b) 0.121 F, (c) 0.10 F. **19.52** (a) $Ag^+ + e^- \longrightarrow Ag$, (b) $2H_2O \longrightarrow O_2 + 4H^+ + 4e^-$, (c) 6.0×10^2 C. **19.54** (a) 0.589 g Cu, (b) 0.133 A. **19.56** 2.3 hr. **19.58** 9.66 $\times 10^4$ C. **19.60** 0.0710 F. **19.62** 0.156 M. **19.64** 45.1%. **19.66** (b) 5.40%. **19.68** 0.231 mg Ca^{2+}/mL blood. **19.70** (a) 0.80 V, (b) $2Ag^+ + H_2 \longrightarrow 2Ag + 2H^+$, (c) (i) 0.92 V; (ii) 1.10 V, (d) the cell operates as a pH meter. **19.72** Fluorine gas reacts with water. **19.74** 2.5×10^2 hr. **19.76** Hg_2^{2+}. **19.78** $[Mg^{2+}] = 0.0500 M$; $[Ag^+] = 7 \times 10^{-55}M$; 1.44 g. **19.80** (a) 0.206 L H_2, (b) 6.09 $\times 10^{23}$/mol. **19.82** (a) -1356.8 kJ, (b) 1.17 V. **19.84** +3. **19.86** 6.8 kJ; 0.064. **19.88** 1.4 A. **19.90** +4. **19.92** 1.60×10^{-19} C/e^-. **19.94** A cell made of Li^+/Li and F_2/F^- gives the maximum voltage of 5.92 V. **19.96** 2×10^{20}. **19.98** (b) 0.59 V. **19.100** (a) The reduction potential of O_2 is insufficient to oxidize gold, (b) yes, (c) $2Au + 3F_2 \longrightarrow 2AuF_3$. **19.102** $[Fe^{2+}] = 0.0920 M$; $[Fe^{3+}] = 0.0680 M$. **19.104** (b) 104 A · hr, (c) 2.01 V; -3.88×10^5 J. **19.106** (a) Unchanged, (b) unchanged, (c) squared, (d) doubled, (e) doubled. **19.108** Stronger. **19.110** 4.4×10^2 atm. **19.112** 3.05 V.

Chapter 20

20.12 (a) +3, (b) 6, (c) oxalate. **20.14** (a) Na: +1; Mo: +6, (b) Mg: +2; W: +6, (c) K: +1; Fe: +2. **20.16** (a) *cis*-dichlorobis(ethylenediamine)-cobalt(III), (b) pentaamminechloroplatinum(IV) chloride, (c) hexaamminecobalt(III) chloride, (d) pentaamminechlorocobalt(III) chloride, (e) *trans*-diamminedichloroplatinum(II). **20.18** (a) $[Cr(en)_2Cl_2]^+$, (b) $Fe(CO)_5$, (c) $K_2[Cu(CN)_4]$, (d) $[Co(NH_3)_4(H_2O)Cl]Cl_2$. **20.20** (a) 2, (b) 2.

20.22

cis isomer   *trans* isomer

20.28 CN^- is a strong-field ligand. Absorption occurs near the high-energy end of the spectrum (blue) and the complex ion looks yellow. H_2O is a weaker field ligand and absorption occurs in the orange or red part of the spectrum. Consequently, the complex ion appears green or blue. **20.30** (a) 2 unpaired spins, (b) 4 unpaired spins. **20.32** 2; $[Co(NH_3)_4Cl_2]Cl$. **20.36** Use $^{14}CN^-$ label (in NaCN). **20.38** First $Cu(CN)_2$ (white) is formed. It redissolves as $Cu(CN)_4^{2-}$. **20.40** 1.4×10^2. **20.42** 6.7×10^{13}. **20.44** The $Mn(H_2O)_6^{2+}$ ion is a high-spin complex, containing five unpaired spins. **20.46** These will absorb at longer wavelength: (a) $[Co(H_2O)_6]^{2+}$, (b) $[FeF_6]^{3-}$, (c) $[CuCl_4]^{2-}$. **20.48** (a) $4Au(s) + 8CN^-(aq) + O_2(g) + 2H_2O(l) \longrightarrow 4[Au(CN)_2]^-(aq) + 4OH^-(aq)$, (b) $Zn(s) + 2[Au(CN)_2]^-(aq) \longrightarrow [Zn(CN)_4]^{2-}(aq) + 2Au(s)$, (c) linear (Au is sp-hybridized). **20.50** Mn^{3+} ($3d^4$) because it is less stable than Cr^{3+} ($3d^5$). **20.52** Fe^{3+} ($3d^5$) has a lighter color.

Chapter 21

21.6 (a) $_{-1}^0\beta$, (b) $_{20}^{40}Ca$, (c) $_2^4\alpha$, (d) $_0^1n$. **21.14** (a) $_3^9Li$, (b) $_{11}^{25}Na$, (c) $_{21}^{48}Sc$. **21.16** (a) $_{10}^{17}Ne$, (b) $_{20}^{45}Ca$, (c) $_{43}^{92}Tc$, (d) $_{80}^{195}Hg$, (e) $_{96}^{242}Cm$. **21.18** 6×10^9 kg/s. **21.20** (a) 4.55×10^{-12} J; 1.14×10^{-12} J/nucleon, (b) 2.36×10^{-10} J; 1.28×10^{-12} J/nucleon. **21.24** 2.77 day. **21.26** 2.7 day. **21.28** $_{82}^{208}Pb$. **21.30** A: 0; B: 0.25 mole; C: 0; D: 0.75 mole. **21.34** (a) $_{34}^{80}Se + _1^2H \longrightarrow _1^1p + _{34}^{81}Se$, (b) $_4^9Be + _1^2H \longrightarrow 2_1^1p + _3^9Li$, (c) $_5^{10}B + _0^1n \longrightarrow _2^4\alpha + _3^7Li$. **21.36** $_{80}^{198}Hg + _0^1n \longrightarrow _{79}^{198}Au + _1^1p$. **21.48** IO_3^- is only formed from IO_4^-. **21.50** Incorporate ^{59}Fe into a person's body. After a few days isolate red blood cells and monitor radioactivity from the hemoglobin molecules. **21.52** An analogous Pauli exclusion principle for nucleons. **21.54** (a) 0.343 mCi, (b) $_{93}^{237}Np \longrightarrow _2^4\alpha + _{91}^{233}Pa$. **21.56** (a) 1.040×10^{-12} J/nucleon, (b) 1.111×10^{-12} J/nucleon, (c) 1.199×10^{-12} J/nucleon, (d) 1.410×10^{-12} J/nucleon. **21.58** $_7^{18}N \longrightarrow _8^{18}O + _{-1}^0\beta$. **21.60** Radioactive dating. **21.62** $_{83}^{209}Bi + _2^4\alpha \longrightarrow _{85}^{211}At + 2_0^1n$, (b) $_{83}^{209}Bi(\alpha, 2n)_{85}^{211}At$. **21.64** The sun exerts a much greater gravity on the particles. **21.66** 2.77×10^3 yr. **21.68** (a) $_{19}^{40}K \longrightarrow _{18}^{40}Ar + _{+1}^0\beta$, (b) 3.0×10^9 yr. **21.70** (a) 5.59×10^{-15} J; 2.84×10^{-13} J, (b) 0.024 mol, (c) 4.06×10^6 kJ. **21.72** 2.7×10^{14} I-131 atoms. **21.74** 5.9×10^{23}/mol. **21.76** All except gravitational. **21.78** U-238 and Th-232. **21.80** 8.3×10^{-4} nm. **21.82** $_1^3H$. **21.84** The reflected neutrons induced a nuclear chain reaction.

Chapter 22

22.8 $-(CH_2-CHCl-CH_2-CCl_2)-$. **22.10** By an addition reaction involving styrene monomers. **22.12** (a) $CH_2=CH-CH=CH_2$. (b) $HO_2C(CH_2)_6NH_2$. **22.22** At 35°C the enzyme begins to denature. **22.28** Proteins are made of 20 amino acids. Nucleic acids are made of four building blocks (purines, pyrimidines, sugar, phosphate group) only. **22.30** C-G base pairs have three hydrogen bonds; A-T base pairs have two hydrogen bonds. **22.32** Leg muscles are active, have a high metabolic rate and hence a high concentration of myoglobin. The iron content in Mb make the meat look dark. **22.34** Insects have blood that contains no hemoglobin. It is unlikely that a human-sized insect could obtain sufficient oxygen for metabolism by diffusion. **22.36** There are four Fe atoms per hemoglobin molecule. **22.38** Mostly dispersion forces. **22.40** Gly-Ala-Phe-Glu-His-Gly-Ala-Leu-Val. **22.42** No. Enzymes only act on one of the two optical isomers of a compound.

Photograph Credits

About the Author

Michele Migdal

Table of Contents

Chapter 1

Opener: © Bettmann/Corbis Images; Fig. 1.2: © Fritz Goro/Life Magazine@Timepix; Fig. 1.3a-b, p. 9: © The McGraw-Hill Companies, Inc./Photo by Ken Karp; p. 11: © NASA; p. 13: © Comstock; p. 14: © The McGraw-Hill Companies, Inc./Photo by Ken Karp; Fig. 1.9: Courtesy of Mettler; p. 22: © Ed Wheeler/Stock Market; p. 27: NASA/JPL.

Chapter 2

Opener (top): © Bettmann/Corbis Images; Opener (bottom): © The Curie and Joliot-Curie Association; Fig. 2.4a-b: © The McGraw-Hill Companies, Inc./Photo by Ken Karp; Fig. 2.12c: © Dr. E.R. Degginger/Color-Pic; p. 45: © The McGraw-Hill Companies, Inc./Photo by Ken Karp.

Chapter 3

Opener: © The Granger Collection; p. 59: © Andrew Popper; Fig. 3.1: © The McGraw-Hill Companies, Inc./Photo by Ken Karp; pp. 61–62: © Dr. E.R. Degginger/Color-Pic; p. 63: © American Gas Association; p. 64: © The McGraw-Hill Companies, Inc./Photo by Ken Karp; p. 68: Ward's Natural Science Establishment; pp. 72–73, p. 76: © The McGraw-Hill Compa-nies, Inc./Photo by Ken Karp; p. 81: Courtesy, Merlin Metalworks, Inc.

Chapter 4

Opener: © Dirck Halstead/Getty Images; Fig. 4.1a-c, Fig. 4.3, Fig. 4.4, Fig. 4.5, Fig. 4.6, p. 97 (both), p. 100, Fig. 4.9a-c, p. 103 (bottom), Fig. 4.13a-b: © The McGraw-Hill Companies, Inc./Photo by Ken Karp; p. 108 (bottom): © Mula & Haramaty/Photo-take; Fig. 4.17, p. 110 (bottom), Fig. 4.16a-c, Fig. 4.17a-b, p. 114 (top): © The McGraw-Hill Companies, Inc./Photo by Ken Karp; Problem 4.105: Cour-tesy Dow Chemical USA.

Chapter 5

Opener: © The Granger Collection; p. 126, Fig. 5.10: © The McGraw-Hill Companies, Inc./Photo by Ken Karp; p. 140: Courtesy of General Motors Corpora-tion; p. 149 (top): © NASA.

Chapter 6

Opener (top): © UPI/Corbis Images; Opener (bot-tom): © NASA; p. 162: © Farrell Grehan/Photo Re-searchers; Fig. 6.2: © E.R. Degginger/Color-Pic; p. 164 (bottom), p. 166, p. 171, p. 173: © The McGraw-Hill Companies, Inc./Photo by Ken Karp; p. 174: © Raymond Chang; p. 181: © Dr. E.R. Degginger/Color-Pic; p. 179 (top): © The McGraw-Hill Companies, Inc./Photo by Ken Karp; p. 179 (bot-tom): Courtesy of the Diamond Information Center.

Chapter 7

Opener (top): © Deutsches Museum; Opener (bot-tom): © Royal Observatory, Edinburgh/AATB/SPL/Photo Researchers; Fig. 7.2: © B.S.I.P./Custom Medical Stock Photo; Fig. 7.5: © Joel Gordon; Fig. 7.11(both): © Educational Development Center.

Chapter 8

Opener: © Special Collections Van Pelt Library, Uni-versity of Pennsylvania; Fig. 8.12(Li), (Na): © The McGraw-Hill Companies, Inc./Photo by Ken Karp; Fig. 8.12(K): Life Science Library/MATTER, photo Albert Fenn @ Time-Life Books, Inc.; Fig. 8.12(Rb), (Cs): © L.V. Bergman/The Bergman Collection; Fig. 8.13(Be), (Mg), (Ca), (Sr), (Ba): © L.V. Bergman/The Bergman Collection; Fig. 8.13(Ra): © The Curie and Joliot-Curie Association; Fig. 8.14(B), (Al), (Ga), (In): © L.V. Bergman/The Bergman Collection; Fig. 8.15(graphite): © The McGraw-Hill Companies, Inc./Photo by Ken Karp; Fig. 8.15(diamond): © Dia-mond Information Center; Fig. 8.15(Si): © Frank Wing/Stock Boston; Fig. 8.15(Ge): © The McGraw-Hill Companies, Inc./Photo by Ken Karp; Fig. 8.15(Sn): © L.V. Bergman/The Bergman Collection; Fig. 8.15(Pb): © The McGraw-Hill Companies, Inc./Photo by Ken Karp; Fig. 8.16(N2): © Joe McNally; Fig. 8.16(P): Life Science Library/MAT-TER, photo Albert Fenn @ Time-Life Books, Inc.; Fig. 8.16(As), (Sb), (Bi): © L.V. Bergman/The Bergman Collection; Fig. 8.17(S8), (Se8): © L.V. Bergman/The Bergman Collection; Fig. 8.17(Te): © The McGraw-Hill Companies, Inc./Photo by Ken Karp; Fig. 8.18: © Joel Gordon; Fig. 8.19(He), (Ne), (Ar), (Kr), (Xe): © The McGraw-Hill Companies, Inc./Photo by Ken Karp; Fig. 8.20a-b: © Neil Bartlett; Fig. 8.21: Courtesy of Argonne National Laboratory.

Chapter 9

Opener (left): Bancroft Library, University of Califor-nia; Opener (right): From G.N. Lewis, Valence, Dover Publications, Inc., New York 1966; p. 274: © The McGraw-Hill Companies, Inc./Photo by Ken Karp; p. 277 (top): © Courtesy of James O. Schreck, Professor of Chemistry, University of Northern Colorado; p. 277 (bottom): © IBM's Almaden Research Center.

Chapter 10

Opener: © Raymond Chang; Fig. 10.20: © Donald Clegg.

Chapter 11

Opener: Courtesy, Laura Stern, U.S. Geological Sur-vey—Menlo Park, California, photo by John Pinkston & Laura Stern; p. 335: © J.H. Robinson/Photo Re-searchers; p. 337: Reprinted courtesy of the American Petroleum Institute; Fig. 11.11, p. 347 (top): © The McGraw-Hill Companies, Inc./Photo by Ken Karp; p. 347 (bottom): © E.R. Degginger/Color-Pic; p. 353 (both), p. 356: © The McGraw-Hill Companies, Inc./Photo by Ken Karp; p. 358: © BioPhoto/Photo Re-searchers; Fig. 11.23: © Joel Gordon.

Index

A

Absolute entropy, 585
Absolute temperature scale, 133
Absolute zero, 133
Absorption spectrum, 661
Accuracy, 19
Acetaldehyde (CH_3CHO), 354
Acetate ion, 524
Acetic acid (CH_3COOH), 94, 99, 502, 509, 512
   ionization constants of, 512
   titrations of, 547
Acetic acid-sodium acetate system, 546
Acetone (CH_3COCH_3), 354
Acetylene (C_2H_2), 347
   bonding in, 268, 316
   combustion of, 163, 347
   properties and reactions of, 347
Acetylsalicylic acid (aspirin), 512
Achiral molecules, 359
Acid(s), 48, 99, 502, 540. *See also* Strong
     acids; Weak acids
   Arrhenius, 98
   Brønsted, 98, 502
   diprotic, 99, 517
   general properties of, 98
   ionization constants of. *See* Ionization
     constants
   Lewis, 530
   monoprotic, 99, 511
   nomenclature, 48
   polyprotic, 99, 517
   properties, 98
   strength of, 99, 507
   strong and weak, defined, 507
   triprotic, 99
Acid ionization constants (K_a), 511, 519
   of diprotic and polyprotic acids, 519
   of monoprotic acids, 511
   relation between base ionization constants
     and, 522
Acid rain, 529

Acid strength, 99, 507
Acid-base indicators, 115, 551, 553 (*table*)
Acid-base properties, 98
   of hydroxides, 564
   of oxides, 257, 528
   of salt solutions, 523
   of water, 503
Acid-base reactions, 101, 114, 544
Acid-base theory
   Arrhenius, 98
   Brønsted, 98, 502
   Lewis, 530
Acid-base titrations, 114, 544
   indicators for. *See* Acid-base indicators
Acidic oxides, 257, 528
Actinide series, 222
Activated complex, 449
Activation energy (E_a), 449
Active site, 461
Activity series, 106
Actual yield, 80
Addition reactions, 346, 704
Adenine, 716
Adenosine diphosphate (ADP), 596
Adenosine triphosphate (ATP), 596
Adhesion, 377
Adipic acid, 706
Air, composition of, 126
Alcohol(s), 352
   condensation reactions of, 353
   denatured, 353
   oxidation of, 354
Alcohol dehydrogenase, 352
Aldehydes, 354
Aliphatic hydrocarbons, 334
Alkali metal(s), 39, 248
   common properties of, 248
   electronegativity of, 270
   group trends of, 248
   ionization energy of, 242
   reactions of, with oxygen, 248
Alkali metal hydroxides, 508

Alkaline earth metal(s), 39, 249
   electronegativity of, 270
   ionization energy of, 242
   properties of, 249
Alkaline earth metal hydroxides, 508
Alkanes, 334
   nomenclature of, 338
   optical isomerism of substituted, 359
   reactions of, 340
Alkenes, 342
   nomenclature of, 343
   properties and reactions of, 344
Alkyl group, 339
Alkyl halides, 340
Alkynes, 347
Allotropes, 41
   carbon, 41, 179, 252
   oxygen, 41, 179
   phosphorus, 181
   sulfur, 181, 252
Alpha helix, 712
Alpha (α) particles, 33
Alpha (α) rays. *See* Alpha particles
Aluminum, 250, 638
   electrometallurgy of, 638
Aluminum chloride ($AlCl_3$), 525
Aluminum hydroxide [$Al(OH)_3$], 564
Aluminum oxide (Al_2O_3), 256, 529, 638
Alzheimer's disease, 694
Amide group, 709
Amide ion, 509
Amines, 357
Amino acids, 355, 709
Ammonia (NH_3)
   as base, 100, 502, 509
   geometry of, 296, 308
   ionization constant of, 522
   as Lewis base, 530
   preparation of, 470
Ammonium chloride (NH_4Cl), 524
Ammonium ion, 44, 524
Ampere (A), 636

Fundamental Constants

Avogadro's number	6.0221367×10^{23}
Electron charge (e)	1.6022×10^{-19} C
Electron mass	9.109387×10^{-28} g
Faraday constant (F)	96,485.3 C/mol e$^-$
Gas constant (R)	8.314 J/K · mol (0.08206 L · atm/K · mol)
Planck's constant (h)	6.6256×10^{-34} J s
Proton mass	1.672623×10^{-24} g
Neutron mass	1.674928×10^{-24} g
Speed of light in vacuum	2.99792458×10^8 m/s

Useful Conversion Factors and Relationships

1 lb = 453.6 g
1 in = 2.54 cm (exactly)
1 mi = 1.609 km
1 km = 0.6215 mi
1 pm = 1×10^{-12} m = 1×10^{-10} cm
1 atm = 760 mmHg = 760 torr = 101,325 N/m^2 = 101,325 Pa
1 cal = 4.184 J (exactly)
1 L atm = 101.325 J
1 J = 1 C × 1 V

$$?°C = (°F - 32°F) \times \frac{5°C}{9°F}$$

$$?°F = \frac{9°F}{5°C} \times (°C) + 32°F$$

$$? K = (°C + 273.15°C)\left(\frac{1 K}{1°C}\right)$$

Color Codes for Molecular Models

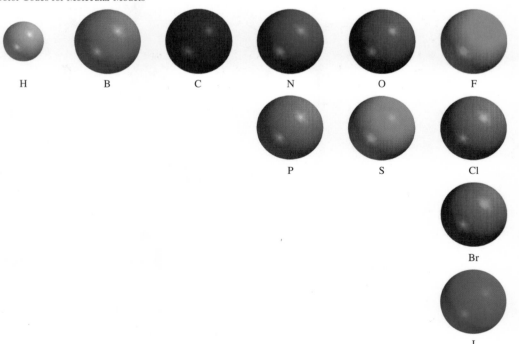